MECHANICS OF SOLIDS I

EDITOR

C. TRUESDELL

MECHANICS OF SOLIDS I – IV

EDITOR

C. TRUESDELL

VOLUME I

The Experimental Foundations of Solid Mechanics

VOLUME II

Linear Theories of Elasticity and Thermoelasticity
Linear and Nonlinear Theories of Rods, Plates, and Shells

VOLUME III

Theory of Viscoelasticity, Plasticity, Elastic Waves, and Elastic Stability

VOLUME IV

Waves in Elastic and Viscoelastic Solids
(Theory and Experiment)

Originally published as ENCYCLOPEDIA OF PHYSICS VIa/1–4

MECHANICS OF SOLIDS

VOLUME I

James F. Bell

The Experimental Foundations of Solid Mechanics

With 481 Figures

Springer-Verlag
Berlin Heidelberg New York Tokyo 1984

Professor Dr. James F. Bell

The Johns Hopkins University, Baltimore, MD 21218, USA

Volume Editor

Professor Dr. C. Truesdell

The Johns Hopkins University, Baltimore, MD 21218, USA

Chief Editor of Encyclopedia of Physics

Professor Dr. S. Flügge

Fakultät für Physik, Universität Freiburg
D-7800 Freiburg, Fed. Rep. of Germany

This book originally appeared in hardcover as Volume VIa/1 of
Encyclopedia of Physics
© by Springer-Verlag Berlin, Heidelberg 1973
ISBN 3-540-05873-7 Springer-Verlag Berlin Heidelberg New York
ISBN 0-387-05873-7 Springer-Verlag New York Heidelberg Berlin

ISBN 3-540-13160-4 Springer-Verlag Berlin Heidelberg New York Tokyo
ISBN 0-387-13160-4 Springer-Verlag New York Heidelberg Berlin Tokyo

Offset printing: Beltz Offsetdruck, 6944 Hemsbach
Bookbinding: J. Schäffer OHG, 6718 Grünstadt
2153/3130-543210

Preface.

Reissue of
Encyclopedia of Physics / Handbuch der Physik, Volume VIa

The mechanical response of solids was first reduced to an organized science of fairly general scope in the nineteenth century. The theory of small elastic deformations is in the main the creation of CAUCHY, who, correcting and simplifying the work of NAVIER and POISSON, through an astounding application of conjoined scholarship, originality, and labor greatly extended in breadth the shallowest aspects of the treatments of particular kinds of bodies by GALILEO, LEIBNIZ, JAMES BERNOULLI, PARENT, DANIEL BERNOULLI, EULER, and COULOMB. Linear elasticity became a branch of mathematics, cultivated wherever there were mathematicians. The magisterial treatise of LOVE in its second edition, 1906 – clear, compact, exhaustive, and learned – stands as the summary of the classical theory. It is one of the great "gaslight works" that in BOCHNER's words[1] "either do not have any adequate successor[s] ... or, at least, refuse to be superseded ...; and so they have to be reprinted, in ever increasing numbers, for active research and reference", as long as State and Society shall permit men to learn mathematics by, for, and of men's minds.

Abundant experimentation on solids was done during the same century. Usually the materials arising in nature, with which experiment most justly concerns itself, do not stoop easily to the limitations classical elasticity posits. It is no wonder that the investigations LOVE's treatise collects, condenses, and reduces to symmetry and system were in the main ill at ease with experiment and unconcerned with practical applications. In LOVE's words, they belong to "an abstract conceptual scheme of Rational Mechanics". He concluded thus his famous Historical Introduction:

> The history of the mathematical theory of Elasticity shows clearly that the development of the theory has not been guided exclusively by considerations of its utility for technical Mechanics. Most of the men by whose researches it has been founded and shaped have been more interested in Natural Philosophy than in material progress, in trying to understand the world than in trying to make it more comfortable. From this attitude of mind it may possibly have resulted that the theory has contributed less to the material advance of mankind than it might otherwise have done. Be this as it may, the intellectual gain which has accrued from the work of these men must be estimated very highly. The discussions that have taken place concerning the number and meaning of the elastic constants have thrown light on most recondite questions concerning the nature of molecules and the mode of their interaction. The efforts that have been made to explain optical phenomena by means of the hypothesis of a medium having the same physical character as an elastic solid body led, in the first instance, to

[1] SALOMON BOCHNER: "Einstein between centuries", Rice Univ. Stud. 65 (3), 54 (1979).

the understanding of a concrete example of a medium which can transmit transverse vibrations, and, at a later stage, to the definite conclusion that the luminiferous medium has not the physical character assumed in the hypothesis. They have thus issued in an essential widening of our ideas concerning the nature of the aether and the nature of luminous vibrations. The methods that have been devised for solving the equations of equilibrium of an isotropic solid body form part of an analytical theory which is of great importance in pure mathematics. The application of these methods to the problem of the internal constitution of the Earth has led to results which must influence profoundly the course of speculative thought both in Geology and in cosmical Physics. Even in the more technical problems, such as the transmission of force and the resistance of bars and plates, attention has been directed, for the most part, rather to theoretical than to practical aspects of the questions. To get insight into what goes on in impact, to bring the theory of the behaviour of thin bars and plates into accord with the general equations - these and such-like aims have been more attractive to most of the men to whom we owe the theory than endeavours to devise means for effecting economies in engineering constructions or to ascertain the conditions in which structures become unsafe. The fact that much material progress is the indirect outcome of work done in this spirit is not without significance. The equally significant fact that most great advances in Natural Philosophy have been made by men who had a first-hand acquaintance with practical needs and experimental methods has often been emphasized; and, although the names of Green, Poisson, Cauchy show that the rule is not without important exceptions, yet it is exemplified well in the history of our science.

Love's treatise mentions experiment rarely and scantly. Its one passage concerning experiment in general, § 63, in effect warns its reader to have a care of experimental data because of their indirectness.

In an irony of history the ever-increasing use of mathematical notation in physical science, to the point that now often works on experiment are dominated by their authors' seemingly compulsive recourse to mathematical formulae interconnected by copied or adapted bits of old mathematical manipulation, Love's treatise is sometimes in reproaches upon modern "pure" or "abstract" researchers held up as a model of practical, applied theory.

Experiment on the mechanical properties of solids became in the later nineteenth century a science nearly divorced from theory. Nevertheless, no great treatise on experiment fit to be set beside Love's on theory ever appeared. Even such books of experiment as were published seem to have in the main taken positions either dominated by theory, usually crude, verbose, and ill presented, or flatly opposed to theory.

The modern reader will cite as objections against the foregoing coarse summary many individual masterpieces that do not support it: brilliant comparisons of theory with experiment by St. Venant, independent experiments of fundamental importance by Wertheim, Cauchy's marvellously clear mathematical apparatus for conceiving stress and arbitrarily large strains and rotations, theories of internal friction and plasticity proposed by Boltzmann, St. Venant, and others. If he is searching for antecedents of what has happened in the second half of the twentieth century, he is abundantly right in citing these and other achievements of the nineteenth while passing over the work of the ruck, but in that century's gross product of solid mechanics they are exceptions that prove rules.

In planning this volume on the mechanics of solids for the *Encyclopedia of Physics* I designed

1) To provide a treatise on experimental mechanics of solids that, not dominated by mathematical theory and not neglecting the work of the eighteenth and nineteenth centuries in favor of recent, more popular, and more costly forays, should be comparable in authority, breadth, and scholarship with LOVE's.
2) To provide treatises on basic, mathematical theory that would stand at the level of LOVE's while in their own, narrower scope supplanting it by compact and efficient development of fundamentals, making use of modern, incisive, yet elementary mathematics to weave together old and recent insights and achievements.
3) To illustrate the power of modern mathematical theory and modern experiment by articles on selected topics recently developed for their intellectual and practical importance, these two qualities being closer to each other than to some they may seem.

I encouraged the authors to meet the standard established by LOVE in just citation and temperate respect for the discoverers.

The reader will be able to form his own judgment of such success and failure as did accrue.

On the first head, experiment in general, the reader will find the treatise by Mr. BELL, filling all of Part 1. While it is not primarily a historical work, the historian S. G. BRUSH pronounced it in 1975 "the most important new publication by a single author" on the history of physics.

On the second head, the reader should not expect to find the basic ideas of solids treated *ab novo* or in isolation. The general and unified mechanics of EULER and CAUCHY, in which fluids, solids, and materials of other kinds are but instances, has come into its own in our day. No wise scientist now can afford to shut out solids when studying fluids or to forget the nature and peculiarities of fluids when studying solids. The two are but extreme examples in the class of systems comprised by mechanics. Articles in Parts 1 and 3 of Volume III of the *Encyclopedia: The Classical Field Theories* and *The Non-Linear Field Theories of Mechanics,* are cited so often by the authors writing in Volume VIa as to make it fatuous to deny that they provide the basic concepts, structures, and mathematical apparatus for the articles on theoretical mechanics of solids. In particular *The Non-Linear Field Theories* goes into such detail regarding large mechanical deformation as to allow most of the text in Volume VIa to concentrate upon small strain.

This much understood, we see that while Mr. BELL's volume provides, at last, a monument of exposition and scholarship on experiment, the articles by Messrs. GURTIN, CARLSON, FICHERA, NAGHDI, ANTMAN, and FISHER & LEITMAN, by Mrs. GEIRINGER, and by Mr. TING together provide a modern treatise on mathematical theories of the classical kinds. The survey of theories of elastic stability by Messrs. KNOPS & WILKES, now justly regarded as the standard reference for its field, necessarily considers deformations that need not be small.

Coming finally to application, in which theory and experiment complement one another, the reader will find major examples in the articles by Messrs. CHEN; NUNZIATO, WALSH, SCHULER and BARKER; and THURSTON. Many more topics of application might have been included. I regret that I could not secure articles about them. The most serious want is a survey of applications of linear elasticity to problems of intrinsic or applied interest that have arisen in this century and that illustrate the power of new mathematical analysis in dealing with special problems. A long article of that kind, a veritable treatise, was twice contracted and twice defaulted. Fortunately the gap thus left has been abundantly and expertly filled by Mr. VILLAGGIO, *Qualitative Methods in Elasticity,* Leyden, Noordhoff, 1977.

Baltimore, December, 1983 C. TRUESDELL

Contents.

The Experimental Foundations of Solid Mechanics.

By

JAMES F. BELL.

With 481 Figures.

This treatise was finished as a memorial to a gentle, intelligent, 21 year old son, CHRISTOPHER J. BELL, killed in the Vietnam war, May 27, 1969. The depths of his affection, his pungent humor, his warm concern for other people, his independent intellect, his love for the mountains, and the haunting beauty of his oboe, are his legacy.

I. Introduction.

As with the unique contributions of the major theorist, there is a timeless quality to the discoveries of the gifted experimentist. No treatise on the experimental foundations of a field of physics as old and as important as mechanics can confine itself to the past one or two decades of meaningful research and escape distortion. Even a casual tracing of the 300 years of significant experimentation in solid mechanics would reveal that each decade chooses its problems, its methods of solution, and its judgments of excellence, from its own peculiar historical point of view. Any global approach to explanation which does not include the sum of the trustworthy observations of the past is inviting obsolescence. The remarkable experiments of COULOMB and CHLADNI in the 1780's, of DUPIN and DULEAU at the end of the first decade of the 19th century, of WEBER and VICAT in the 1830's, of WERTHEIM, TRESCA, and KOHLRAUSCH in the mid-19th century, of STRAUBEL and GRÜNEISEN at the turn of the 20th century, are as definitive in the 1970's as the best experimental research of today.

The history of solid mechanics to the present moment is replete with examples of incompetence in experiment which appeared at the time to "verify" some then-popular theory, thereby producing a confusion which in a few notable instances persisted as long as a half century. Thus we see CAGNIARD DE LATOUR in 1827 claiming to have measured the volume change in a wire under tension by a method which later easily was shown to be inadequate for reaching *any* conclusion, and POISSON, also in 1827, stating that CAGNIARD DE LATOUR's work was an experimental correlation with his own newly developed molecular theory of elasticity. Long after the experiments of WERTHEIM, in the 1840's and 1850's, which to CAUCHY's satisfaction had disposed of this presumed correlation, SAINT VENANT and many others in the 1860's and 1870's were still quoting the results of CAGNIARD DE LATOUR as support for the applicability of the POISSON-CAUCHY theory of uni-constant linear elasticity, despite what was by then a mountain of opposing experimental evidence. Thus, too, the experiments of KUPFFER in the 1850's pertaining to the determination of elastic constants, their thermo-viscous dependence, and the relations among them, were known to be erroneous by the 1860's yet were still dominating the thought of PEARSON while he was writing his portion of the famous history of elasticity in the 1880's and 1890's.

In many instances an experimentist who has made an interesting observation in one specific area is interpreted as having done something quite different; thence for tens of years the unread original work is cited in support of developing explanations which in fact have no basis in experiment at all. Interesting examples of this are certain aspects of the classic work of TRESCA in the 1860's in quasi-static plasticity, or of JOHN HOPKINSON in 1872 and LUDWIK in 1909 in dynamic plasticity and viscoplasticity.

It is often said that whereas all theories whose objective is the interpretation of physical observation sooner or later face the prospect of being supplanted, the results of the outstanding experiment must be coped with forever. Although there are some areas in which increasing precision with time necessitates reinterpretation, it is nonetheless true that in solid mechanics in the past 300 years there are many experiments which remain crucial, classic for their qualitative revelation, independent of the continual movement toward the next decimal place.

As an experimentist interested in dynamic plasticity, I became convinced several years ago that it was essential to an understanding of experiment that I, or my students and I, should repeat in my laboratory the vital experiments in this field during the past century in which it has been in existence. The experience has deeply influenced the writing of the present treatise in which it has been my wish to provide a critical evaluation of experiments by an experimentist,[1] as I probed the experimental foundations of solid mechanics so that both the theorist and the experimentist may come to some understanding of what these foundations actually are.

Since the work is written with the theorist always in mind, I have been aware of the only real gift the experimentist has for the theorist: a set of interrelated, reliable physical facts devoid of misleading preconceptions and hidden auxiliary empirical assumptions. The hypotheses required to simplify a mathematical statement so that an acceptable analytical solution may be obtained, unfortunately all too often do not correlate with the physical situation which must be utilized to simplify the experimental statement so that a meaningful laboratory solution may be obtained. The experimentist's role would be a very limited one if he only gathered bits and pieces of data to correlate with a theory devised because an analytical solution was available. It was well understood in 19th century physics that a function of the experimentist was to establish sets of reliable observations in previously unexplored and unexplained areas, to stimulate theoretical study. I believe this was in no small measure a factor in the scholarly lives of the major theorists of that time. However, many a 19th century experimentist, notably VICAT in the 1830's and WERTHEIM in the 1850's, pleaded with the geometers of the time to consider in theoretical terms this or that unexpected experimental observation. Some of these matters, such as WERTHEIM's discovery in 1850 of the rudiments of what is now known as the POYNTING effect, had to await nearly a century before competent geometers paid heed to the call, long after the original experiments had been forgotten and were replaced by others' observations of the same event.

[1] The term "experimenter", or one who performs experiments, is preferred by many over the term "experimentalist," which has fallen upon evil days. The latter term is loosely employed by persons ranging from those who measure anything for any purpose to some in the mathematically oriented sciences who, like the proverbial squaw, dutifully trot in the footsteps of their theoretical husbands. I have chosen to return to the term "experimentist," in respectable use as early as the 1670's as my learned editor has pointed out to me. In addition to being one who performs experiments, the "experimentist," as I interpret him, is a person who is engaged in the independent and systematic use of experiment to disclose patterns in nature which may or may not have been successfully considered by the theorist.

It is essential to view the role of the experimentist as somewhat different from the currently accepted image. He may well enter into areas which are either analytically fallow because they present great difficulty, or which have lost interest because all the problems are claimed solved. Since within some degree of precision several theories based upon different assumptions, may square with the same experiment; and, since in any given situation only one such theory may be currently available, with adjacent theory or theories yet to be produced; it is obvious that an experimentist does not "verify" theories. Moreover, inasmuch as adjacent theories are based upon different sets of initial assumptions, it is fallacious to presume that a correlation between data and prediction implies the validity of any one set of such assumptions. Arbitrarily choosing that set which leads to further analytical success, as is so often the practice, ultimately may, and almost inevitably does obscure rather than reveal the physical relations. It makes little difference whether a few people for a brief time or most people for a whole generation concur in choosing a particular direction by ignoring some observations in favor of others; the final result is the same, namely, the need to retrace ideas from some point in the process where absurdity had been ignored. In view of the fact that sooner or later some sort of experimental data usually can be produced to coincide with a plausible analytically successful hypothesis, independent experimental study is essential to retain contact with Mother Nature. One major experimentist after another, including such 18th and 19th century figures as Coulomb, Biot, Dupin, and Wertheim, has clearly stated his objective of developing an experimental study uninfluenced by the bias or the bounds of the then current theory, though that is a point of view which history has shown is ill-designed to win easy acceptance of new discoveries.

The true experimentist sticks to his last, masters the matter of the theory-sensitivity of experiment, recognizes that at a given time, as with proper theory, there are subjects as yet incapable of being explored in terms of proper experiment, and seeks to measure accurately from different points of view every pertinent variable, and cross-check all germane empirical assumptions before publishing his results. In this treatise I have endeavored to make these criteria the basis for the choice and subsequent criticism of the various experiments which have been included.

The experimentist in solid mechanics has two main challenges. The first and by far the most important is to establish the form and content of constitutive equations in the wide variety of solids extant in nature. This is a function which he and he alone can perform. The second, which may be placed under the general heading of dimensionality, is to determine experimentally whether or not restricted specified forces and/or displacements applied to finite bodies of defined geometry actually do produce the stress, strain, displacement, particle velocity, rotational distributions, and magnetic, electrical, and thermal fields, which the assumptions of a given set of constitutive equations would suggest.

In fact, these two aspects of experimental study are almost always inseparable. Experiments are performed on bounded specimens to which loads and displacements are applied. The measurement of the resulting stress and strain distributions provides a constitutive equation. Probably no other single factor in experimental mechanics has led to more error in the determination of constitutive equations in some necessarily highly restricted domain than the misinterpretation of stress-strain functions, arising out of the fact that presumed distributions actually were not obtained. Since one is dealing with six components of stress and strain when only the most elegant experiments can measure even two or three of these at once, it has been necessary to seek ways to apply loads and select geometries which would provide zero values of several components of stress, strain, displacement, and rotation throughout the solid.

For well over a century, even among some savants, it has been the custom to dismiss unwelcome experimental results on the basis of an unsubstantiated argument that the specimens must have been inhomogeneous or anisotropic, if homogeneity and isotropy were assumed, or improperly oriented, if some specific form of anisotropy were imputed. Such gross rejections may be suspect, but in fact, specimen preparation with due regard for the total prior thermal and mechanical histories does present to the experimentist one of his most fundamental difficulties. With a few notable exceptions, it has been possible to make measurements only on the surface of a solid, and those exceptions have required some process of integration through the solid in order to interpret the data. In an individual experiment a pre-eminent question always is, whether the specimen is indeed the one the experimentist thinks it is.

Hence, excluding those persons who perform multitudes of load vs deformation measurements without any personal contact with or possibly even cognizance of the method of production, the chemical content, the substructure, let alone the prior thermal and mechanical histories, the serious experimentist is faced with the gigantic laboratory task of establishing to his own satisfaction that he is dealing with known quantities in his experiment. As will be shown in considerable detail below, historically the first experimentist to understand this problem fully was one of the greatest the field has produced, GUILLAUME WERTHEIM, who in the 1840's prepared his own specimens, determined their density and purity, and described in considerable detail the manner in which those procedures had been carried out. It is no coincidence that he is in fact the founder of the modern science of experimental solid mechanics.

In 1787 when CHLADNI discovered the remarkable sand patterns which could be produced by the lateral vibration of thin plates of various shapes, dimensionality itself became a specific subject for experimental study. CHLADNI's experiments fascinated two generations and surely inspired the extensive theoretical and experimental studies of the vibrations of plates and shells, results of which have occupied a large proportion of the papers on experiment from the late 18th century to the present. Much of the work of SAVART in the 1820's and 1830's, of KIRCHHOFF in the 1850's, of Lord RAYLEIGH at the end of the 19th century, and of MINDLIN in the 20th century, for example, was designed to explore the variety of boundary value problems which arise when the linear theory of elasticity is presumed to model the behavior of the material considered.

Beginning from the early 18th century experimental studies of MUSSCHENBROEK on the buckling of compressed struts and the classic theoretical studies of EULER on the same subject, an enormous literature on experiment has grown which has described the complex buckling of all manner of geometrical shapes. However, unlike the boundary value problems in the field of vibration, for which many precise experiments in the 19th and 20th centuries led to truly striking correlations between prediction and observation, the experimental data in the field of elastic stability has been beset from the first small deformation measurements of ALPHONSE DULEAU in 1812, to the present, with basic difficulties. The widely variable experimental data have arisen from the fact that buckling behavior is keenly sensitive to small details in matters of load application, alignment, and local peculiarities in the specimens.

A third area of dimensionality which first was extensively studied experimentally by EATON HODGKINSON in the 1830's was that of the impact of objects of different or similar shapes. Apart from the experimentists who have used impact so as to study the waves generated in the solid, there has been a procession of persons after HODGKINSON who have attempted to reduce the problem of impact to simple terms, chiefly toward technological applications. Although little worthwhile for science has come out of such approximations, they have demonstrated the importance of the shape of the specimen in the study of the dynamics of the solid continuum.

There are many instances in which dimensional studies performed for purely technological reasons have contributed directly to, or have stimulated the growth of fundamental thought in the field of solid mechanics. And, tests upon bounded specimens, whatever the objective, have provided by the sheer bulk of their results considerable insight into the limitations which must be anticipated in the design of experimental apparatus. It is fortunate for technology that reliable numbers alone suffice for design purposes; the "factor of safety" or its reciprocal, the "factor of ignorance" bridges the gap between industrial success and shaky scientific foundations.

For the most part, however, and in particular for the enormous numbers of papers describing boundary value problems in the linear theory of elasticity, few of the results from this type of experimental study have been of intrinsic scientific interest. In this treatise I have not undertaken the impossible task of tracing the details of the growth and present status, successes and failures, of the study of boundary value problems, let alone the evaluation of their technological importance. I also decided early in this work to exclude most of the vast experimental literature on the subject of rupture, primarily because centuries of breaking specimens in every manner of material, from whalebone to steel, under nearly every conceivable set of test conditions, have not yet revealed any generalized patterns of behavior. The main body of this treatise, therefore, is concerned with the prime problem of experimental solid mechanics: the determination of constitutive equations.

In conception this experimental problem is simple. One produces forces and displacements on the surface or through the interior of the solid with precisely known space-time histories, and observes the distribution of stress, strain, displacement, and rotation throughout the solid for a sufficiently long time. The arbitrary variation of the applied loading and displacement histories, spatially among the numerous components and temporally in intervals varying from fractions of a microsecond to years, is accompanied by known histories in ambient temperature and magnetic and electric fields. The chemistry of the surrounding atmosphere, the prior thermal, mechanical, and chemical specimen histories, applied radiation, and the like provide an infinitude of situations in which one might examine the countless variety of solids in nature.

To no small extent, as with the theorist, experimental excellence is inseparable from insight and taste in the choice of issues to be studied. A proper problem thoroughly explored in the hands of a great experimentist is capable of revealing general order and patterns of understanding of new phenomena among large classes of solids, as well as demonstrating the fascinating peculiarities present in special situations. In a large proportion of the research in the continuum mechanics of solids since the beginning of the 19th century, choice of problem has been overly dictated and influenced by the practical needs of technology. This attempt to serve two masters has led to compromise, in that esoteric steel specimens, intricate non-ferrous metallic alloys, or very preparation-sensitive non-metallic substances have been chosen for extensive study, restudy, and voluminous report in the general literature. The profusion of varied results deeply has influenced the general attitude toward the subject by giving rise to the widespread suspicion that to a large extent the outcome of an experiment on a solid depends strongly on the specimen itself. Time and again the free experimentist who seeks simplicity and is aware of the prior history of his solids as well as their present condition has demonstrated that precision and order in solid mechanics experimentation does indeed exist.

The quantities to be measured, including stress, strain, displacement, particle velocity, crystallographic orientation, rigid body rotations, ambient and deformation generated thermal, electrical, and magnetic fields, may be measured as is well known, by a wide variety of techniques which are applicable to this or that particular situation. Many experimentists with deep interests in some particular type of measurement capable of determining a specific quantity, choose problems to be studied solely on this basis and thus spend all of their time exploring some specific, limited set of questions. No laboratory has succeeded in mastering all of the many techniques to obtain the flexibility which many theorists have achieved in applying the tools of their trade. It goes without saying that the mastery of some varied set of techniques is implicit, although most of the great experimentists have had very little interest in that aspect of the subject for its own sake. It is curious that, nevertheless, they have been responsible for most of the innovations in experimental method.

An experimentist can do a certain number of things in a finite time. I believe that it is essential that a sufficient number of methods of measurement of the prime variables should be readily available in the laboratory in order that the experimentist may be as free as possible of dependence upon technique in choosing the directions in which to go. Every decade, certainly since the mid-19th century, has seen the over-use of some contemporary method of measurement whose limitations are many times unconsciously suppressed in the excitement of the new possibilities. A recent example among many is ultrasonics, where tens of thousands of measurements of wave speed have been made in literally hundreds of compounds and elements over a good part of the temperature scale under various ambient pressures, etc., providing a literature that is so vast during the past 15 years that it is a major undertaking merely to tabulate it, let alone to review it critically. Relatively few studies have been undertaken with ultrasonics to explore various aspects of the general continuum mechanics of solids, and even fewer have questioned the use of linear elasticity in interpreting measurements.

Another example is the technique of photoelasticity, most of whose practitioners have concentrated intensively on model analysis of structural configurations. A relatively small proportion of a very large number of papers in the area exhibit any interest in exploring the properties and generalizations which may exist in birefringent substances as prime specimens.

In this treatise, when choosing the experiments which seemed to me in fact to provide the best indication of the sum and substance of what is known, or samples of those which exemplify persistent weaknesses in experimental concept and practice, I have given critical attention to this matter of technique and the individual experimentist's attitude toward what he has accomplished, as expressed both in the content of his published papers and through my own detailed study of the results he obtained. In writing the work I have sought to summarize my findings and refer the interested reader to the original paper. On the other hand, I have deemed it essential to give explicitly the results of every experiment within the scope of this treatise, those that are of major importance and those which are illustrative of experimental confusion, while I am discussing the work of that particular experimentist. I consider it a waste of time to present a stream of value judgments about one successive experimentist after another without the detailed, simultaneous presentation of the gist of his argument, which lies in his data. To omit such details would be similar to writing a history of mathematics, or rather, a history of the excellence of mathematicians, devoid of the mathematics itself.

Following the prescriptions and deletions which I have outlined above, I found early in planning this treatise that there was still a very large literature requiring consideration. It also became obvious to me that much of the important research, even after 300 years, was still in the process of treating the simplest geometric situations and solid substances. The problem which has been most frequently considered experimentally has been that first introduced as a recorded observation by ROBERT HOOKE[1] in 1678, namely, the uniaxial loading of a cylindrical specimen which was presumed to be a homogeneous and isotropic solid. By 1820, and particularly by 1830, this simplest of all experiments in solid mechanics was disclosing a wealth of patterns in nature which still are being explored in the present time.

Sufficient accuracy was obtained early in the 1830's and 1840's to provide definitive load vs deformation histories in small and in finite deformation. The discovery of such phenomena as creep, the SAVART-MASSON (PORTEVIN-LE CHATELIER) effect, the temperature dependence of elastic constants, the dependence of certain solids' response upon the electric and magnetic fields present, the elastic after-effect, thermal-elastic behavior, etc., all appeared in a period of intense growth of the subject before the mid-19th century. It became obvious then, as it is still obvious now, that it will take a very long time to explore even the response of solids in uniaxial stress.

There has been much experimental research in torsion, beginning with COULOMB in the 1780's, followed by DULEAU in 1813, which by the mid-19th century included emphasis upon the torsion of hollow specimens of various cross-sections. For uniaxial loading and for torsion, quasi-static and vibration experiments were compared extensively throughout the 19th century. For uniaxial loading there were numerous efforts to consider simultaneously the problem of one dimensional wave propagation in linear elasticity.

A third stress configuration for solids, study of which began in the mid-19th century, is that of hydrostatic pressure. Measurement of the deformation of compressed solids, initiated by REGNAULT and WERTHEIM led ultimately, as is well known, to the comprehensive studies of BRIDGMAN, spanning more than 50 years of the present century.

Experiments involving the more complex situation of flexure in which the deformation as well as the loads were determined, began very early in the 19th century in quasi-static measurement. This situation was already well studied in dynamic measurement by the latter part of the 18th century.

It is enlightening in our time to trace through the 19th and 20th centuries the discussions among experimentists with respect to persistent differences which were found between the predictions of elementary theory, and observation. As early as 1811 it was known from definitive experiments that the deflection of wooden beams was not linear and that the curvature of the elastic line could be approximated better to a hyperbola than to the predictions of the elementary linear beam theory. One experimentist after another throughout all the remaining decades of the 19th century demonstrated in one solid after another, for torsion,

[1] The idea of elongating a specimen by a weight preceded HOOKE, as one of LEONARDO DA VINCI's drawings illustrates. (*See* TRUESDELL [1960, *1*], pp. 19–20.) LEONARDO DA VINCI's proposal, however, was to measure the load required to cause rupture. As TRUESDELL observed, there is no indication as to whether or not the actual measurement ever was made.

flexure, and axial loading, in tension and in compression, that careful measurement revealed an essential and, by the end of the 19th century, a reproducible nonlinearity, capable of being generalized analytically, which governed the small deformation of many solids, common metals included. The simultaneous measurement of flexure and torsion in a specimen was considered by KIRCHHOFF in the 1850's, and the uniaxial deformation in the presence of hydrostatic pressure by VON KÁRMÁN in 1911. Concern with the deformation properties of human tissue, i.e., bones, muscles, nerves, etc., began in the 1840's and generated for the next three decades a wide and provocative study of the tensile deformation properties of living and post-mortem organic substances. In the 1860's TRESCA's classic study on the flow of solids inaugurated an experimental subject which was followed by a century of discussion and explanation. The original experiments of TRESCA to this day remain singularly significant.

The 20th century, which is firmly based upon foundations from the 19th, has to some extent merely reexamined with increased precision much of what had been studied earlier. Improvement of technique which in the 19th century had enabled experimental physicists in each successive decade to study new problems, has of course greatly influenced the 20th century, particularly in the field of dynamic measurement. High frequency wave studies have permitted the consideration of more generalized stress states. The exploration of the finite deformation of single crystals and of polycrystals under loading conditions which produced strain rates over a range of 13 orders of magnitude, including quasi-static stress histories extending over days, months, or years, or stress histories producing measurable large deformation in fractions of a microsecond, indeed have extended the experimental horizons and are to a large extent directly the consequence of the development of new optical and electronic techniques of measurement.

Following BRIDGMAN's lead in studying the response of solids to very slow compression at high pressure, other experimentists more recently have used explosive loading to produce even higher stress fields in microsecond intervals. A new area may open for studies not limited to single, double, or even triple stress states if it becomes possible to determine what such space-time stress distributions actually are.

The experiments of RIVLIN in the 1950's on the finite elasticity of rubber stand as a classic model; they emphasize what may be achieved in solid mechanics when rare insight is simultaneously focused upon both experiment and theory.

During the past decade and a half, my own experimental studies have revealed that there exists an heretofore undiscovered order in the large deformation properties of annealed crystalline solids, which is expressible in terms of generalized, linearly temperature dependent constitutive equations for radial and for nonradial loading paths. Those same experimental studies revealed the existence of a material stability structure in crystalline solids in the form of a discrete distribution of deformation modes and second order transitions which occur at fixed, predictable strains, and the existence of a related quantum structure for the elastic constants of the elements.

We thus find at this point in the 20th century that the experimental foundations of solid mechanics consist of a large body of knowledge relevant to the constitutive equations of classes of solids, including many beyond those few briefly mentioned above, in which success invariably was achieved through the consideration of the simplest geometries and the simplest materials.

In many areas, within the framework of what is known, current theory would seem satisfactory for explaining the trustworthy experimental observations. In many instances, restricted theories have been found to correlate with equally restricted portions of what is known from experiments. For a surprisingly large amount of the available, reliable knowledge in this field there are as yet no parallel, plausible theories which can account for the experimental observations that already have been made.

That this branch of physics remains a vital and provocative subject for fundamental study nearly three fourths of the way through the 20th century, is one of the lessons to be learned from perusing the 300 year history of the growth of the experimental foundations of solid mechanics since the inaugural measurements of ROBERT HOOKE in the 17th century.

Comment on Sources. In writing this treatise, for the work of the 17th and 18th centuries to the time of COULOMB, I found invaluable the classic volume of C. TRUESDELL on *The Rational Mechanics of Flexible or Elastic Bodies, 1638–1788*. For the period from 1780 to the present, it early became obvious that I would have to examine systematically the successive volumes of the journals of the pertinent scientific and technological literature.

For the 19th century, I found that my effort to develop an independent bibliography of sources on experimental solid mechanics was aided considerably by the *Royal Society of London Catalogue of Scientific Papers (1800–1900)*, as that volume enabled me to check for possible omissions, and it provided new leads. KARL PEARSON, who contributed the sections on experiment for TODHUNTER and PEARSON's *A History of the Theory of Elasticity*, included a great many experiments of minor significance. Often his judgments were influenced strongly by factors not pertinent to the evaluation of experiment. A number of major studies were dismissed by PEARSON on theoretical grounds, and many works of lesser importance were emphasized on the same grounds. However valuable they may be as a historical source on the theory of elasticity, on experiment the volumes by TODHUNTER and PEARSON must be consulted with caution.

O. D. CHWOLSON in his *Traité de Physique* in 1908 described and listed details of many of the experimental studies of the 19th century which have been considered in the present treatise. CHWOLSON's treatment of the subject was more descriptive than critical, and I found many of his references only loosely related to the subject he was discussing. Nevertheless, his volume provided an additional check on 19th century sources. Similarly, volume 3 of JAMES R. PARTINGTON's *An Advanced Treatise on Physical Chemistry*, published in 1952, which aimed to give a broad view of experimental results rather than to examine their sources, provided a few leads to earlier work. S. P. TIMOSHENKO's *History of Strength of Materials* in 1953, while providing a non-experimental background in applied mechanics for some of the subjects considered here, contains little on actual experiment.

Of the vast literature on experimental solid mechanics which has accumulated during the past three centuries, I could include here only a small fraction. I have selected almost exclusively those studies which seemed to me to be the major sources for describing the foundations of this branch of the physics of solids.

Acknowledgments. Rarely has a writer been so fortunate as to editor. My appreciation for CLIFFORD TRUESDELL's perceptive criticism and for his laborious pruning of nouns and awkward phrases is exceeded only by my admiration for his scholarship. In this treatise, whatever lapses there be from the impeccable prose sought by my editor, are my own insuppressible solecisms.

FAITH FARCHIONE PACQUET MOECKEL has been of singular assistance, tracking down volumes of obscure references and even more obscure units of measure; being concerned with the translations of the French and, occasionally, the German papers; drawing with artistry as well as with accuracy, re-drawing, and keeping track of figures; sometimes typing; doing a host of other tedious labors; and with it all, always loyally presenting an indefatigible air.

To ANNEMARIE ALBAUGH go my sincere thanks for her concern with the translations of papers written in German, and for the numerous other chores she cheerfully and readily assumed. My thanks, too, to CATHERINE KELLER, PEGGY BROUGHAM, and RUTH STINEHART of the Johns Hopkins University for typing many of these pages. The librarians who aided me, preserve their reputation for professional competence; my thanks especially go to LUCIE GECKLER, ADELAIDE EISENHART, and MARTHA HUBBARD of the Johns Hopkins University.

I want to thank JERALD ERICKSEN, who, with his usual generosity, took the time to read these chapters; and WILLIAM HARTMAN and AKHTAR KHAN, who kindly read the manuscript.

To my daughter, JANE, whose vivacity often has revived this tired writer, I reserve a special thanks. And, I can but lamely, in wholly inadequate measure here refer to my wife, PERRA, who was more than figuratively beside me, helping me in innumerable ways, during thousands of hours, as I wrote and rewrote this treatise.

II. Small deformation nonlinearity.

2.1. Introduction.

The explicit form and functional dependence of constitutive equations in the continuum mechanics of solids, as GOTTFRIED WILHELM LEIBNIZ[1] declared nearly three centuries ago, are solely a matter for the experimentist to determine. To make such a determination in some particular solid in some range of deformation, it is necessary to know from experiment that the pertinent deformation variables actually are distributed in space as assumed, at all times during the measurement. Ideally, this would require that the entire simultaneous histories of the stress, strain, thermal, electrical, and magnetic distributions throughout the solid, including all components of stress and all components of strain at every point, be specified precisely for some arbitrary generalized stress state produced by clearly defined surface tractions, surface displacements, and body forces. In the laboratory this ideal is approached by delimiting the situations to be considered so that many of the parameters remain constant during the interval of the experiment.

With respect to any given solid one must consider reversibility, reproducibility, and type of measurement, and one must examine all auxiliary empirical assumptions by separate, independent experiments. Of equal importance is the detailed knowledge of the *prior* mechanical, thermal, and physical, chemical, or metallurgical histories of the particular specimen. For either small or large deformation, the experimentist must consider systematically the functional dependence of the constitutive equation upon anisotropic, viscous, thermal, electrical, and magnetic parameters, being aware of the possibility of critical points associated with material stability, geometric stability, and the transition from one region of deformation to another.

In lieu of determining constitutive equations in the ideal sense, the experimentist has resorted to the use of a far simpler, although clearly more limited, situation. A solid, whose prior thermal-mechanical history and degree of anisotropy

---

[1] LEIBNIZ [1690, *1*]. Letter written by LEIBNIZ to JAMES BERNOULLI, 24 September, 1690. *See* C. TRUESDELL [1960, *1*; p. 63] who states, "... he (LEIBNIZ) replies, in effect, that *the relation between extension and stretching force should be determined by experiment* ..."

and homogeneity it is hoped are adequately known,[1] is subjected to a distribution of surface tractions, surface displacements, or body forces such as may be expected to provide a known simple distribution of stress and deformation. By measuring such applied surface tractions or displacements and simultaneously measuring strain or displacement over the surface of the solid, one then can compare the observed stress history and the observed strain history to provide a stress-strain function.

Both in the past and in the present the all-too-common failure to establish that expected simple distributions actually do occur, has led to confusion and controversy. This has been particularly true in experiments on dynamic behavior.

The 17th and 18th century experimentists in solid mechanics confined their studies largely to the problem of rupture, which from the 17th century until now has refused to submit to any generally plausible, let alone satisfactory, theory. Included among those early scientists concerned with solid mechanics were some of the most imaginative, such as MERSENNE,[2] MARIOTTE,[3] PARENT,[4] and MUSSCHENBROEK.[5]

The stress and strain at which rupture occurs in a well-defined situation is of course a bound of some interest for constitutive equations, provided that all pertinent variables are included and that in terms of them, the experiment may be reproduced. The literature on ultimate loads is vast, comprising a sizeable portion of all the papers published on solid mechanics during the past 300 years. In the 18th century one finds long lists of breaking loads for various kinds of rocks, wood, glass, cardboard, paper, whale bone, ivory, walrus teeth, leather, metals, etc., usually without referring to the dimensions of the specimen, and very often without stating the manner in which the test was performed.[6] The impetus for those measurements was the hope of unravelling the mysteries of solid cohesion.

During the first half of the 19th century, one again finds long lists of ultimate loads for a wide variety of solids. By then, however, the dimensions of the specimens usually were given, so that those data could be, or were, expressed in terms of ultimate stress. The practical engineer concerned with the design and failure of structures had become interested in such data. Thousands of ultimate-strength experiments have been reported in the engineering literature during the past 150 years which, while contributing very little to science, obviously have been of inestimable significance in practical engineering. From 300 years ago to the present time, numerous persons making ultimate load measurements have attempted to establish empirical relations, grouping portions of such data. Except under very restricted conditions, and even then with something less than modest success, no plausible general theory of rupture has evolved.

During the 19th century, when experimentists began to be interested in pre-rupture phenomena in the deformation of solids, it became generally accepted

---

[1] JOHN T. RICHARDS [1952, *1*], pp. 99–100, while contributing an important paper at a Symposium on the Determination of Elastic Constants, provided a relevant comment on this matter: "Unfortunately, much of the published data on elastic constants is incomplete with respect to composition, temper, thermal treatment, test equipment, accuracy, straining rate, etc. Physicists are frequently guilty of composition and temper omissions, while metallurgists are prone to neglect accuracy and equipment details. For example, in surveying published E-modulus data for copper from 1828 through 1949, fully 40% of the 45 listed authorities failed to give the temper of the copper tested."

[2] MARIN MERSENNE (1588–1648).

[3] EDMÉ MARIOTTE (1620–1684).

[4] ANTOINE PARENT (1666–1716).

[5] PIETER VAN MUSSCHENBROEK (1693–1761).

[6] Exceptions are the elegant diagrams of testing machines for tension and compression of MUSSCHENBROEK [1739, *1*] and his data on the cross-sections of specimens.

that for metals, let alone for stone, plaster, leather, rubber, wood, glass, silk, cat gut, frogs' tongue muscles, bone, and human tissue, all of which were studied, the linear law of infinitesimal elasticity of ROBERT HOOKE[1] was only an approximation. Rather than being concerned whether or not solids, including metals, had nonlinear stress-strain functions in small deformation, experimentists decided that the major question, which was still a matter of controversy at the beginning of the 20th century, was the form which a generalized nonlinear law of *infinitesimal* elasticity in metals and other solids should take.[2]

It has become fashionable in the second half of the 20th century to state that the field of theoretical solid mechanics was constructed with minimal reference to experiment and that the great development of linear elasticity in the first half of the 19th century with its dynamic analog in electromagnetism, and the major nonlinear developments of the present century, included many individuals who paid small heed to experiment, particularly where constitutive equations were concerned. One might well point out, however, that the state of the field usually is characterized by the degree of analytical success which has been achieved for general or specific issues which have been found to be *amenable* to analytical description, i.e., characterized by the logic of mathematical tractability, rather than in terms of rational experimental observations which, in many instances, have proceeded far beyond what to the experimentist are often limits imposed by the present competence of theorists.

It is equally true that important experiments were sparse during the late 18th and early 19th centuries. The major papers published in the 1780's by CHARLES AUGUSTIN COULOMB[3] and ERNST FLORENS FRIEDRICH CHLADNI,[4] were written with insight and rigor equal to that of the best of today's experimentists in solid mechanics. Only two experiments of similar stature were published within the next 40 years. These were the works of PIERRE CHARLES FRANÇOIS DUPIN[5] in 1811 on wood, and of ALPHONSE JEAN CLAUDE BOURGUIGNON DULEAU[6] in 1813 on iron, both of whom conducted extensive series of experiments on small deformation far below the rupture load. Between the late 1820's and the mid-19th century, experimental studies relating to the deformation of solids laid the founda-

[1] HOOKE [1678, *1*; reprinted 1931].

[2] Two interesting examples of this are, first, the casual statement of a reviewer for the British Institute of Metals with the initials, "C. H. D." [CECIL H. DESCH], who, in an abstract of a paper by GUSTAVO COLONNETTI on the elasticity of copper wire which was subjected to repeated tensile loading, stated:

> "It is concluded that the behaviour of copper which is usually considered to diverge very widely from the accepted theory of elasticity conforms to the latter in a satisfactory manner when the cycle is of very small amplitude between the limits of stress. The rate at which the alterations are carried out is almost without influence on the general character of the cycle, although it alters the individual numerical values" (DESCH [1914, *1*], p. 273);

and, second, the comment in 1900 of GEORGE FREDERICK CHARLES SEARLE [1900, *1*], whose reputation for experimental precision is almost legendary, and who, upon perfecting a level to allow the determination of tensile strain of 10^{-6} in long wires and at the same time eliminate thermal effects, stated: "With this apparatus the students at the Cavendish Laboratory find it easy to investigate the deviations from Hooke's Law for copper wire." In EGON LARSEN's *The Cavendish Laboratory* [1962, *1*], p. 48, SEARLE is described as "... a lecturer with great enthusiasm for teaching and a genius for devising experiments ..."

[3] COULOMB [1784, *1*].

[4] CHLADNI [1787, *1*].

[5] DUPIN did not publish these experiments until 1815. See [1815, *1*] DUPIN was born in 1784, the year of COULOMB's famous, pioneering memoir on the torsion of metal wires.

[6] DULEAU [1820, *1*]. The memoir of 1813 rapidly became unavailable; in fact, by the late 1830's it was unobtainable. The many scientists who referred to DULEAU's work during the next 50 years depended, as I have, upon reading the republication of the work in 1820.

tion for much of what dominates the field of solid mechanics, both theoretically and experimentally, more than 100 years later, in the second half of the 20th century.

In addition to the discovery of the essential nonlinearity of the small deformation of wood, cement, plaster, gut, human tissue, frog muscles, bones, stones, rubber, leather, silk, cork, and clay, nonlinearity also was found for the infinitesimal deformation of every metal considered. The phenomenon of the recoverable elastic after-effect in silk, human muscle, and metals; the thermal after-effect in metals; the occurrence of micro-permanent deformation in metals for extremely small strains; the phenomenon of short and long time creep in metals; the variation of elastic moduli for different amounts of permanent deformation; the relation between magnetization, permanent deformation, electric resistance, temperature, and elastic constants; the effect upon deformation behavior of anisotropy, inhomogeneity, and prior thermal histories; the parameters affecting the internal friction or attenuation properties of solids; the deformation instability phenomenon known today, from a study of 1923, as the PORTEVIN-LE CHATELIER effect; and the essential experimental features of the plasticity of metals, including short-time loading phenomena, were all properties of constitutive equations which were the subject of extensive, often definitive experimentation before 1850.

2.2. Nonlinear vs linear elasticity in the 18th century.

The first nonlinear law of elasticity related to experiment was introduced in a letter written by LEIBNIZ[1] in 1690. LEIBNIZ stated that the experimental data which JAMES BERNOULLI[2] had sent him in December, 1687 from a tensile test in a gut string 3 ft. long seemed to fit a hyperbolic curve, in contrast to the experiments of others such as HOOKE[3] and MARIOTTE,[4] which had supported a linear law. In 1695 BERNOULLI[5] himself proposed a parabolic law, where t is the elongation, and x the longitudinal force.

$$t = k \cdot x^m. \tag{2.1}$$

As CLIFFORD AMBROSE TRUESDELL[6] has pointed out, since x is proportional to the stretching force, BERNOULLI is, in effect, assuming that strain $\propto$ (force)^m. GEORG BERNHARD BÜLFFINGER,[7] who has been credited wrongly by a number of subsequent authors with the BERNOULLI parabolic law,[8] compared BERNOULLI's gut string data with Eq. (2.1) for a value of $m = {}^3/_2$. These data provided for the sequence of forces of 2, 4, 6, and 8, the observed corresponding sequence of extensions, 9, 17, 23, and 27.

From the point of view of what was considered to be acceptable empirical curve fitting in physics at the beginning of the 18th century, the comparison shown in Fig. 2.1 is very interesting. BERNOULLI's experimental points[9] are com-

[1] LEIBNIZ [1690, *1*]. See TRUESDELL [1960, *1*].
[2] BERNOULLI [1687, *1*].
[3] HOOKE [1678, *1*].
[4] MARIOTTE [1700, *1*].
[5] BERNOULLI [1694, *1*].
[6] TRUESDELL [1960, *1*].
[7] BÜLFFINGER [1729, *1*].
[8] For example, RUDOLF MEHMKE [1897, *2*] and STEPHEN PROKOFIEVITCH TIMOSHENKO [1953, *1*].
[9] BERNOULLI's data became known through a much later publication (BERNOULLI [1705, *1*], Schol. after Lemme III), in which he revealed that the forces were measured in pounds and the extensions in lines.

pared with the hyperbola proposed by LEIBNIZ[1] and with Eq. (2.1) for $m=3/2$ proposed by BÜLFFINGER.

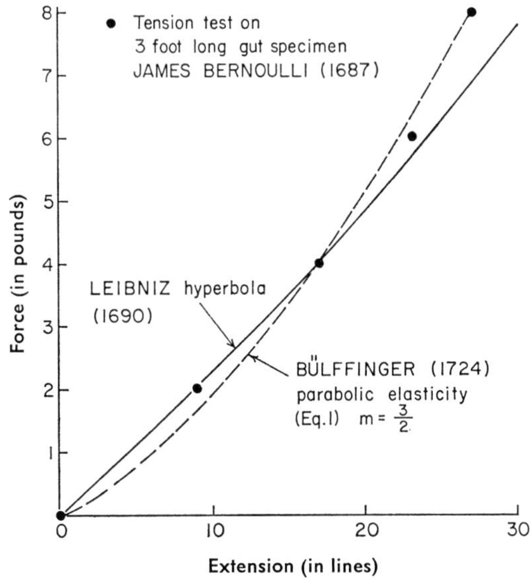

Fig. 2.1. JAMES BERNOULLI's gut string measurement.

The failure of the linear law of HOOKE and MARIOTTE to fit these data caused LEIBNIZ some concern.[2] It may be recalled that HOOKE's classic experiments not only included the well-known tests on helical and spiral springs and the lesser known measurements of the flexure of dry wood, but also the measurements which I think were the most important, those on the tension of vertical metal wires 30 ft. to 40 ft. long. In the manner of many experimentists or theorists prematurely generalizing an important discovery, HOOKE stated somewhat extravagantly with respect to his linear relation between the force and elongation: "Nor is it observable in these bodies only, but in all other springy bodies whatsoever, whether Metal, Wood, Stones, baked Earths, Hair, Silk, Bones, Sinews, Glass, and the like."[3]

It was left to the 19th century experimentists to demonstrate that for every one of those solids, HOOKE's law was only an approximation. Although by the 1890's there was much discussion as to the response function of the nonlinear stress-strain law for different solids, including metals, there was ample experimental evidence that the nonlinearity of small deformation was a reproducible fact.

[1] LEIBNIZ [1690, *1*].

[2] These measurements of JAMES BERNOULLI on a gut string were repeated in 1721 by JAMES RICCATI [1721, *1*], who is quoted by TRUESDELL [1960, *1*] as stating: "I have repeated JAMES BERNOULLI's experiment various times in strings of different material; often I have found true what that famous author says, but often experiment showed me just the opposite ... But when further equal weights were added until the string broke, the extensions which went on increasing up to a certain point then began to decrease again in inverse order."

[3] HOOKE [1678, *1*].

2.3. The small deformation nonlinearity of wood: Dupin (1815).

The first 19th-century experimentist who attempted to correlate small deformation measurements with a nonlinear law of elasticity was Dupin.[1] Following his graduation from the École Polytechnique in 1803, he went to the Ionian island of Corcyre (Corfu or Kerhyra), where he became interested in the change of shape which occurred in wooden ships after launching. In 1811, under the most primitive circumstances,[2] Dupin performed a series of experiments in which he measured the central deflection and elastic curves of simply supported, 2 m long, prismatic beams of cypress, beech, oak, and pine. For each beam which had a square section 3 cm on a side the specific weight was determined. Two types of applied loads were compared: a central load and a uniformly distributed load. The central load occurred in even multiples of 4 kg to the maximum value of 28 kg. The central deflection measurements were made using a graduated square,[3] allowing an accuracy of $^2/_{10}$ mm. In explaining his reasons for carrying out such a series of unusual measurements, which had certainly not been the custom of his predecessors, Dupin stated:

> Up to now one has attempted mainly to determine the resistance of which woods are capable at the instant of rupture, either by rupturing them perpendicularly to their fibers, or by collapsing them under weights which act in the direction of the fibers themselves.
>
> Doubtless it is necessary to know this extreme point, this limit strength of wood, in order to use always materials endowed with a strength greater than that to which they are subjected in constructions and in machines of which they are a part; but one must always remain at some distance from this limit, and when one wishes to make durable works, one must remain even farther, since time incessantly decreases the strength of wood, and a thousand things contribute to the deterioration of its original qualities.
>
> There is another type of study no less useful, more useful perhaps, which nonetheless seems to me to have been the least pursued. It is to determine the comparative resistances of woods when they are subjected to the forces able to very slightly alter their shape, and to find, if I may express it so, their *virtual resistance*.[4]

Dupin found that for very small loads the deflection was proportional to the load. However, as the load increased, for equal increments of load the corresponding increments of displacements were not constant. From the tabulated numbers of the first series of experiments it is difficult to see how Dupin reached this conclusion, since the deflections varied considerably, with only a trend toward an increase with increasing load.

---

[1] Dupin [1815, *1*].

[2] Dupin [1815, *1*], p. 148. Said Dupin: "Let us always remember that I was in a port where I lacked everything needed for my working with the ultimate precision, even perfect balances, and one will see that none of these small differences of observation and of calculation is beyond the total limit assigned to the accuracy of these operations."

[3] The care with which Dupin prepared his beams and conducted his experiments is reflected in the following footnote from his *Memoir*, which will be appreciated by any experimentist who takes his data seriously: "In the naval arsenal at Corcyre where I conducted my experiments, I selected from my company of military workers two extremely adroit men, one a joiner, the other a machinist. Why not name them here, since they were so very useful to me? Their names were Chorlet and Raymond. Every day they prepared what was needed for the experiment of the following day. I then spent from six to seven hours with them. These young artisans, abounding in natural talent, attempted to understand and follow my work. I watched with pleasure as they reflected on their own and struggled, so to speak, with me. I had them do geometry and mechanics on the actual objects, while they taught me a thousand practical results, by which I was able to see what a large degree of finesse and correct observation is brought to the arts by men otherwise devoid of instruction. I believe young engineers could not be too close to their good workers, since they would improve the latter, and they themselves would become more adept in profiting from such an experience, the only one which they cannot acquire of themselves, that which is born of manual work." [*Ibid.*, pp. 140 to 141.]

[4] Dupin [1815, *1*], pp. 138–139.

After having discussed the source of the several anomalies in the data and examined the total data from many experiments, DUPIN from his study of the second differences provided a parabolic law:

$$\delta = bF + cF^2 \tag{2.2}$$

where δ was the central deflection; F, the central load; and b and c, constants for a given type of wood determined from the analysis of the first and second deflection differences.

DUPIN reflected at considerable length on the importance of the observed nonlinearity, in terms of the second difference, for the construction of wooden ships. He found experimentally that the resistance to deflection increased with the density of the wood for simply supported beams of exactly the same dimensions and loading histories. What makes DUPIN's experiments the outstanding contribution to solid mechanics that they are,[1] and what caused ADHÉMARD JEAN CLAUDE BARRÉ DE SAINT-VENANT,[2] forty-five years later, to remark upon them with such favor as "des belles expériences," is that DUPIN was the first to determine precisely the magnitude of the central deflection of a simply supported prismatic beam in the initial region of small strain as a function of the length, width, and height and, for a given load, the precise shape of the loaded beam.

2.4. Dupin's 18th century predecessors: Buffon (1741), Duhamel (1742), and Gauthey (1774).

Before briefly discussing those data, it is interesting to look at the state of slow loading experiment during the century which preceded DUPIN's work. In 1742 after noting that MARIOTTE and LEIBNIZ had observed that "every material, glass included, extends slightly before rupturing," and further, that this was "a principle adopted by both PIERRE VARIGNON and PARENT," HENRI LOUIS DUHAMEL DU MONCEAU[3] became interested in studying in very general terms the position of the invariable line and the associated regions of longitudinal contraction and elongation in prismatic beams just before rupture. Like EATON HODGKINSON[4] 80 years later, DUHAMEL sought to investigate this by experiment by mechanically changing the position of the neutral line and observing its effect upon the rupture load. In the lateral section at the center he sawed to different depths 24 simply supported wooden beams of willow, 3 ft. long and $1^1/_2$ in. square, after which he introduced an oak wedge at the saw cut. The details of those tests are of somewhat minor interest except for what he considered a paradox. The simply supported willow beams of equal dimensions which had been sawed three-quarters of the way through, with an oak wedge inserted, were stronger with respect to rupture for a central force than were similar willow beams upon which no cuts had been made.[5]

---

[1] In TODHUNTER and PEARSON's *History of the Theory of Elasticity and of the Strength of Materials*, Vol. I, p. 162, KARL PEARSON, whose appraisal of experimental papers in solid mechanics is in many instances less than profound, says of this classic memoir of DUPIN in a brief dismissal without discussion, "There is nothing of mathematical value to note in the paper" [1886, *1*].

[2] SAINT-VENANT [1856, *1*].

[3] DUHAMEL [1742, *1*], p. 456.

[4] HODGKINSON [1824, *1*].

[5] DUHAMEL [1742, *1*], p. 468 stated, "It would be advancing a highly paradoxical proposition to say that a piece of wood would be greatly strengthened, that it would render it capable of supporting much greater weight if it were sawed to half or three-quarters of its thickness; yet this is what my experiments proclaim."

In a related series of experiments in oak, reported upon to the Académie Royale des Sciences in 1741, one year before[1] those of DUHAMEL, GEORGES LOUIS LECLERC Comte de BUFFON described not only the force necessary to rupture simply supported wooden beams, but also, in an important innovation, the magnitude of the central deflection just prior to rupture. He carefully recorded at great length the conditions, probable age, and date of cutting of nearly 90 oak trees from which he had selected his cylindrical and square beams. The beams had come from the heartwood section, from the periphery, and from the sapwood at a variety of heights in both living and dead trees estimated to be between 33 and 110 years old. Then BUFFON proceeded to a protracted series of experiments which are described in what seem like endless detail. At one point he stated:

> Despite all the precautions and all the care I gave to my work, I often was scarcely satisfied. I sometimes perceived irregularities and variations which disturbed the conclusions I wished to draw from my experiments, and my register contains over a thousand reports made at various times from which I was unable to draw anything conclusive, and which left me in manifest uncertainty from many points of view. Since all the experiments were made with pieces of wood 1, $1^1/_2$, and 2 in. square section, a very scrupulous attention was needed in the selection of wood, almost perfect equality in weight, equal number of layers [tree rings] and, in addition, the almost unavoidable problem of the angle of the fibers, which often considerably reduced the strength of the wood; I do not mention the knots, faults, and extreme oblique angles of the layers, since such specimens were automatically discarded without bothering to test them. Thus, from this enormous number of tests on small specimens, I was able to draw with assurance only those results given above, and they were insufficient for establishing tables showing in a generalized manner the strength of wood.
>
> These considerations, and a deep regret for lost pains and wasted time, inspired me to undertake experiments on a much larger scale. I clearly saw the difficulties, but I could not reconcile myself to abandon the project, and, happily, the results were much more satisfactory than I had expected.[2]

The more satisfactory results to which he refers were his comparisons of the rupture loads with GALILEO's rule that, "The resistance is in inverse ratio to the length, in direct ratio to the width, and in double ratio to the height." With reference to comparing rupture data and calculation by means of GALILEO's rule, BUFFON observed that for 5 in. square beams ranging from 7 to 12 Paris ft.[3] in length, the rule applied only approximately; the necessary modifications which had to be made to utilize the Galilean rule increased with increasing length.[4]

Perhaps of most importance with respect to the much later experiments of DUPIN were, first, BUFFON's long tabulations of the central deflection of simply supported beams of wood just prior to cracking and just prior to rupture; and, second, his experimental studies of the effect upon the rupture load and maximum central deflection of the number of layers (tree rings), the density of the wood, and whether the layers were vertical or horizontal.[5]

---

[1] BUFFON [1741, *1*] cut down his first oak tree for science on March 31, 1734. BUFFON noted that DUHAMEL had conducted experiments on this subject prior to the publication of the *Mémoires* in 1741.

[2] BUFFON [1741, *1*], pp. 303–304, stated that the total number of tests he performed exceeded a thousand. Since each test was destructive, we may assume that BUFFON could command a considerable forest of trees. It is interesting that 96 years later, PETER BARLOW [1837, *1*], pp. 100–103, who himself made no real contribution to the problem, and whose mathematical limitations are a matter of public record, ascribed BUFFON's failure to solve the problem of rupture to his lack of preconceived theories and his minimal knowledge of mathematics.

[3] One Paris foot equals 30.4794416 cm.

[4] BUFFON [1741, *1*]. Here was possibly the first "fudge factor" in engineering.

[5] This seems to be the first reference in the scientific literature to what later came to be called "anisotropy"; of course the fact to which BUFFON referred had been known to carpenters since time immemorial. Theorists in the 18th century did not consider anisotropy, at least in mechanics. It remained for CAUCHY and POISSON in the 19th century to represent this concept in a theory.

In 1774 Emiland Marie Gauthey[1] indicated that he was aware that previous experimentists such as Mariotte had asserted that all solids were slightly deformed before rupture, but he himself could find no evidence of such behavior when he compared the breaking strengths of simply supported beams of "tender" and "tough" stones. He found that Galileo's rule was vaguely satisfied, but not with sufficient correlation between prediction and experiment to carry conviction. Besides mentioning Gauthey's work as representative of 18th century slow loading experiment prior to Dupin's, I particularly point out that Gauthey did appreciate the complexity of the problem with which he was concerned and that as an experimentist he was unwilling to rest content with simple assumptions without questioning them.

Perhaps the major contribution of Gauthey to the field of solid mechanics was indirect. As the uncle of Claude Louis Marie Henri Navier he was instrumental in influencing both the education and career of a person who played a most important role many years later in the initial development of the linear theory of elasticity. Upon the death of his father, Navier, then only 14 years old, moved into the home of his uncle, and following Gauthey's death in 1807, Navier edited his uncle's three-volume treatise on bridges and channels. In 1804 Navier had been a student at the École des Ponts et Chaussées; he later joined its faculty, thus also paralleling the career of his uncle who had studied and taught mathematics at that institution.[2] The influences of the uncle upon the nephew cannot be dismissed when considering the experimental antecedents of the initial development of the linear theory of elasticity[3] in 1821–1822.

2.5. Details of Dupin's experiments on wooden beams (1815).

Returning to the experiments of Dupin, we see that his study of the central deflection of simply supported wooden beams as a function of the length, height, and thickness in the initial region of small deflection provided a measure of the accuracy of his experiments with respect to the nonlinear relation between force and central deflection. Dupin, who had been trained in mathematics at the École Polytechnique, indicated in his memoir his familiarity with earlier theoretical proposals regarding the magnitude of the central deflection of a simply supported beam as a function of its length, height, and width. However, he proposed "to see the elastic properties of wood deduced in a rigorous manner from the simpler experimental results."[4] In Fig. 2.2 are shown his measurements of the central deflection for a 2 cm × 3 cm oak beam, and for a 2 cm × 5 cm Northern fir beam, for nine different lengths between 1 and 2 m long. The applied central weight had the constant value of 10 kg in every instance. The central deflection was plotted against the cube of the length. From simply supported beams of 2 m in length with different cross-sections, he established experimentally that the central deflection was inversely proportional to the width and inversely proportional to the cube of the height.

[1] Gauthey [1774, *1*].

[2] See Timoshenko [1953, *1*], p. 71.

[3] It is interesting to note that besides Navier, two other participants in the 1820's in the development of the linear theory of elasticity were Cauchy and Poisson, who with Girard in 1819 wrote a summary report for the Académie on the experimental works of Duleau of 1813 [1819, *1*]. Like the experimental study of wood by Dupin at approximately the same time, Duleau's work is notable for providing the first definitive experimental results on the small compression, tension, flexure, and torsion of iron. These data of Duleau became the mainstay of discussion and argument with respect to the small deformation of metal for the next third of a century.

[4] Dupin [1815, *1*], p. 167.

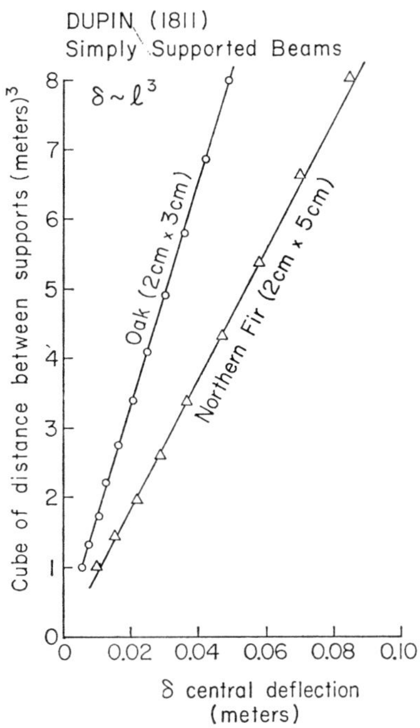

Fig. 2.2. DUPIN's experimental results on wooden beams compared with elementary theory.

To check this point further on a single beam, DUPIN noted that for a beam of rectangular section the ratio of the central deflection of the beam in the flat position to the cant position should be as the ratio of the lateral dimensions squared. A simple calculation reveals that this is the inverse ratio of the moments of inertia for the two positions. In Table 1 are shown his experimental results for four Northern fir wood beams of 2 m length, at the designated loads.

The straightness of the experimental data in Fig. 2.2 and the correlation between prediction and measurement in Table 1 not only establish the relations

Table 1. DUPIN (1811). *Simply supported fir wood beams, 2 m between supports*

Test	Cross-section of beam (cm)	Central load (kg)	Ratio of flat to cant deflection	Calculated ratio of lateral dimension squared
1	2×1	0.5	4.19	4.0
2	3×1	1.0	11.44	9.0
3	3×2	2.0	2.36	2.25
		4.0	2.28	2.25
		6.0	2.25	2.25
		8.0	2.25	2.25
		10.0	2.13	2.25
		Average	2.25	2.25
4	5×2	10.0	6 to 6.2	6.25

between central deflection and length, height, and width for simply supported prismatic beams, and to this extent show the BERNOULLI-EULER theory was in accord with DUPIN's measurements, but also bear witness to the care with which DUPIN performed his experiments.

DUPIN stated his results as

$$y = H + (m-1)\,\delta' + \left(\frac{m-1}{1}\right)\left(\frac{m-2}{2}\right)\delta'' \tag{2.3}$$

where y is the central deflection. For all his experiments the first load was 4 kg, and subsequent loads were 4 kg intervals to the maximum value of 28 kg. Thus m in Eq. (2.3) refers to the integral multipliers of the initial 4 kg which had produced the initial central deflection, H; δ' designates the first difference, i.e., $\delta' = H$; and δ'' is the second difference.

Table 2 shows the experimental results for the four types of wooden beams considered: oak, cypress, beech, and Northern fir. The central deflections at the maximum of 28 kg ranged from 0.04 m to 0.06 m. DUPIN's Eq. (2.3) is consistent with his observation of linearity up to $m=2$ for the 8 kg load, after which the second differences occur.

Table 2. DUPIN (1811). *Comparative table of the elements of the flexure of wood.*

	H (m)	δ' (m)	δ'' (m)
Oak	0.0056	0.0056	0.0001
Cypress	0.0071	0.0071	0.00008666...
Beech	0.0084	0.0084	0.00031
Northern fir	0.0130	0.0132	0.0002

Sect. V of DUPIN's memoir "On the Curve presented by the Flexure of Woods resting on two supports" not only provides the first precise experimental study of this important question but also demonstrates the high quality of his general approach to the entire matter of the flexure of wood, including the observed nonlinearity. DUPIN remarked:

"After having carefully examined the shape of this curve in the first experiments, and in thought having related it to the shapes which are most familiar to me, I judged that it could be a hyperbola: so I supposed it, and here is how I verified this hypothesis."[1]

Choosing from the most flexible of the woods a piece 10 cm wide, 1 cm thick, with 2 m between the simple supports, he produced a deflection of 13 cm at the center by means of a central load. To a stated accuracy of $^2/_{10}$ mm he determined the deflection for 21 equally spaced points along the beam. Letting y represent the distance along the beam from the center, x the deflection, a and b the real and imaginary axes of a hyperbola with its origin at the rather awkwardly chosen maximum central point of deflection, he found the empirical relation,

$$\frac{a^2}{b^2}\,y^2 = x^2 + 2ax. \tag{2.4}$$

DUPIN concluded[2] for his curve fitting "... Whatever the elastic curve produced by the flexion of prismatic specimens between two support points, it is permissible

[1] DUPIN [1815, *1*], p. 175.

[2] In this connection it is interesting to examine the Doctoral dissertation of MAX BORN in 1906. BORN's analysis of various shapes in the large deflection of the elastic line was accompanied by pictorial illustrations showing that a metal band in flexure indeed did adopt shapes consistent with predicted curves. BORN [1906, *1*].

to confuse it with the hyperbola without fear of appreciable errors in practice and even in calculation, where the approximation would be pushed quite far."[1]

Table 3.

Distance from center of beam (m)	Deflection, left side (m)	Deflection, right side (m)	Average experimental deflection (m)	DUPIN's calculated hyperbola (m)	Deflection calculated from elementary linear beam theory (m)
0	0	0	0	0	0
0.2	0.007	0.007	0.007	0.00699	0.0074
0.3	0.0152	0.0156	0.0154	0.0154	0.0155
0.4	0.0267	0.0265	0.0266	0.0265	0.0270
0.5	0.040	0.040	0.040	0.040	0.0406
0.6	0.055	0.055	0.055	0.0554	0.0562
0.7	0.0725	0.0725	0.0725	0.07238	0.0731
0.8	0.0910	0.0912	0.0911	0.0906	0.0910
1.0	0.1300	0.1300	0.1300	0.1300	0.1300

All deflections are measured from the central deflected position.

In Table 3, DUPIN's experimental data and hyperbola are compared with the BERNOULLI-EULER theory for a Northern fir beam of the above dimensions. The second and third columns compare the symmetry of the data on the two sides of the center; the fourth, their averages. Distances were determined from the center in centimeters, with the origin at the deflected position. Using the central deflection, the point of support, and quarter distance at 0.05 m, DUPIN calculated the empirical hyperbolic elastic curve of column 5. I have added column 6 of calculation from the BERNOULLI-EULER theory, from which it is indeed possible if perhaps not permissible to "confuse" the data with a hyperbola.

In view of the observed nonlinearity of the force-deflection functions, the fact that the elastic curve calculated from linear beam theory fails to provide as close a fit as some other curve such as DUPIN's hyperbola, is not unexpected.

DUPIN described an equally close correlation for other wood and other values of central deflections, thus providing a modicum of experimental generality for his hyperbolic elastic line. From the latter, he studied the maximum curvature as a function of the load. Other topics in this long memoir include rupture, the maximum curvature at rupture, and the forced bending of beams over surfaces with defined curvatures.

The obvious care and accuracy of DUPIN in his experiments, revealed in the flexure studies described above, leave no doubt as to the validity of his conclusion that a nonlinear stress-strain function described the small deformation of wood.[2] The discovery was important especially because experimentists, theorists, and the new specialists in technology were coming more and more to accept the linear response function of HOOKE as the only basis for representing the small deformation of solids.

[1] DUPIN [1815, *1*], p. 179.

[2] This *Mémoire* of 1815 on the deflection of wood, and a paper on geometry written while he was a student at the École Polytechnique, a paper which introduced what became known as "cycles" of DUPIN and DUPIN's "indicatrix" constitute the entire contribution of original work by DUPIN to science, on the basis of which he was elected to the French Academy in 1817 at the age of 33. DUPIN became a dominant scientific politician, which he remained until his death at the age of 88. During his career, he wrote a six-volume history of England in 1825, published his notes from lectures on geometry given at Metz, and was active in government politics in his later years. See JOSEPH LOUIS FRANÇOIS BERTRAND's "Éloge historique de Pierre-Charles-François Dupin", *Éloges Académiques* (Paris, 1890), pp. 221–246.

From the perspective of 160 years, one might ask whether the observed nonlinearity was associated with the increasing curvature of the beam. However, a calculation for deformation where the nonlinearity was observed reveals that the curvature is not excessive. The important point is that in 1811 precise experiments revealed that a measured stress-strain function in one supposedly linear solid was *not* in accord with the expected linear relation. It remained for later studies to establish that the slightly nonlinear small deformation elasticity observed in 1811 is indeed present in every solid.

2.6. Experiments on the nonlinear response of wood, iron, and stone, and the introduction of the concept of microplasticity: Hodgkinson (1824–1844).

The early 19th century English engineers, PETER BARLOW, WILLIAM FAIRBAIRN, EATON HODGKINSON, W. J. MACQUORN RANKINE, GEORGE RENNIE, and THOMAS TREDGOLD are justly cited with favor in the history of engineering for their practical ingenuity in the design and completion of structures with the newly developed fabricated iron members which were then coming into increasing use. One finds a large literature containing descriptions of the design of iron bridges and other structures, all too often followed by later memoirs attempting to explain the reason for their collapse.[1] The success of these men and of others in resolving many of the practical difficulties was to a large extent based upon the vast number of tests which measured the ultimate loads in beams as a function of their shapes and measured the creep and failure properties of iron wires beginning to be used in suspension bridges. When these 19th century engineers are

[1] An example of this may be found in Vol. V of the *Memoirs of the Literary and Philosophical Society of Manchester* for 1831. The paper of HODGKINSON on the deformation of iron, [1831, *1*] to be discussed immediately below, is preceded (on p. 384) by an article by the same author on the "Chain Bridge at Broughton" over the river Irwell near Manchester [1831, *2*]. He described in interesting detail the design of the bridge, including in the latter section a professional discussion of the possible modes of failure. In the Appendix in the same volume, by the same author, the subsequent collapse of this Broughton bridge is described:

> "The accident happened through the vibration caused in the structure, by the marching of a troop of Soldiers over it. They were four abreast, and about 60 of them on the bridge, the foremost being half way across when they heard a tremendous crash like a continuous discharge of musketry, and in a moment one side of the bridge sunk down, sloping into the river, dragging after it the main pillar, which they had passed on their right, with the stone to which it was attached, and throwing every one upon the bridge into the river or among the chains. Some of the men escaped unhurt, several were injured, but fortunately no lives were lost."

The failure was due to the improper design of a single bolt. HODGKINSON went on to say:

> "We here see an absurd consequence of the theory of Galileo;—for if the strength of the bolt be estimated by that theory, the bolt would appear to be just as strong as the other parts of the chain; and therefore the failure of the structure may perhaps have arisen from the application of an erroneous theory."

That a rule rejected by the major theorists from the day it was published and demonstrated experimentally to be incorrect 100 years before, could produce disaster two centuries after it was propounded, should provide reflection for casual rule-makers in our own time who have the imagination or interest to project to possible 21st or 22nd century technological fiascos.

Lest the reader think I have been too severe in judging the importance of the misconceptions of some individuals in 1830, I suggest he peruse the work of JOHN LESLIE [1823, *1*], Professor of Natural Philosophy at the University of Edinburgh, in which he accepted GALILEO's rule as a proven fact. In support, LESLIE cited the experiments of BUFFON, who, as we have just seen, had had to introduce a "modifying factor" to obtain a presumed correlation; BARLOW, whose experiments on cohesion were less than nothing; and DUPIN, who we know was interested in small deformation studies rather than rupture, as DUPIN specifically pointed out. Thus LESLIE, like the designer of the Broughton Bridge, was propounding GALILEO's rule solely on the basis of the prestige of the 17th century author.

cited in histories of the theory of elasticity, it is usual to emphasize either their limited knowledge of the subject or their perpetuation of errors in the linear theory, such as the location of the neutral line of the beam.

Certainly an exception to this characterization (aside from the fact that RANKINE is recognized as an important theorist in linear elasticity, thermodynamics, and hydrodynamics), and the outstanding person of the aforementioned group from the point of view of experimental solid mechanics was HODGKINSON, who had been taught mathematics by JOHN DALTON,[1] who introduced him to the work of the BERNOULLIS, EULER, and LAGRANGE. HODGKINSON, who was elected a Fellow of the Royal Society in 1841 and became a Professor of the Mechanical Principles of Engineering in University College, London in 1847, performed an interesting and important series of studies, particularly in the areas of flexure, impact, and stability during the 30 years following his first experimental treatise on wood in 1822.[2]

In English histories of the strength of materials, HODGKINSON is credited with having corrected in 1824 the misconceptions regarding the neutral line in the flexure of rectangular beams, a misconception which had been introduced into the English literature by JOHN ROBISON at the end of the 18th century in an article on the strength of materials in the 4th edition of the *Encyclopaedia Britannica*[3]. PETER BARLOW, one of ROBISON's inexplicably numerous early 19th century admirers, included ROBISON's misconceptions of beam deformation in his treatise in 1817 entitled *Essay on the Strength and Stress of Timber.*[4] This standard engineering work unfortunately went into several editions and was one of the most widely read technical works among early and mid-19th century English engineers.[5] Although in 1822 HODGKINSON pointed out and emphasized both

[1] *See* TIMOSHENKO [1953, *1*], p. 127.

[2] HODGKINSON published in [1824, *1*].

[3] Sir DAVID BREWSTER edited the republication of ROBISON's Encyclopaedia articles. See ROBISON [1822, *1*].

[4] BARLOW [1817, *1*]. By the time of the edition in 1837, this title had become: *A Treatise on the Strength of Timber, Malleable Iron and Other Materials with Rules for Application in Architecture, Construction of Suspension Bridges, Railways, Etc.*

[5] The popularity of BARLOW's book was so great among the English engineers that his two sons continued with revisions and brought out a sixth and new edition in 1867, five years after their father's death in 1862, and fifty years after the work was first published. The dominance of this engineering manual is seen in the complaints of DAVID KIRKALDY who pioneered in founding and managing a large commercial testing company in England from 1860–1890, in which various iron manufacturers publicly tested the quality of their products. KIRKALDY's initially successful venture almost ended in disaster when he was attacked by BARLOW's son because some of KIRKALDY's numbers failed to agree with those published in an earlier edition of the elder BARLOW's book. In 1891 writing a history of DAVID KIRKALDY's life in commercial testing, WILLIAM G. KIRKALDY referred to this attack, together with the comment that a later edition of the BARLOW book incorporated his controversial numbers without referring to the nearly disastrous controversy. WILLIAM KIRKALDY stated: ([1891, *1*], pp. 272–273).

"When the material was undergoing preparation for testing, Mr. Barlow called at the grove, producing some printed results of experiments upon extremely soft STEEL, which had been made shortly after the opening of the works, for the Barrow Haematite Steel Company; he proceeded to state that they must have been wrong, because the extensions were greater than in some IRON bars which his (Mr. Barlow's) father had tested many years before. This was his sole argument, and he did not give his father's figures, or mention which of the figures he objected to. Such action was not kind or gentlemanly towards the man who had just laid out all his means upon the machine, which, moreover, bore ample evidence of the anxiety of its inventor and owner to make it accurate. David Kirkaldy took this criticism in quite a friendly spirit, and cordially invited Mr. Barlow to bring any measuring instrument that he liked at whatever time would suit him; that then they would go quietly into the matter, as he was only anxious to arrive at the truth, and consequently was prepared to use any instrument when such was found to be more accurate and reliable than his own. Strange to record, Mr. Barlow never availed himself

the nature and seriousness of the ROBISON-BARLOW error, it was not until the 4th edition in 1837 that BARLOW finally grudgingly conceded the point, and corrected the treatise.[1]

HODGKINSON was strongly of the opinion that the actual behavior of beams in flexure in any solid was a matter for experiment to disclose and that in considering such a matter, the experimentist should minimize or eliminate prior conceptions. Therefore, in his first extensive series of experiments in three kinds of wood in 1822 he did not assume that a linear stress-strain function necessarily was applicable or that the stress-strain function, whether linear or nonlinear, was necessarily the same for tension and for compression; consequently he reasoned that the precise location of the neutral line in a given solid could be determined only through measurement. HODGKINSON seems not to have been familiar with the work of DUPIN, but he departed from the work of his predecessors of the previous century, such as BUFFON, and of his contemporaries, THOMAS TREDGOLD and BARLOW. HODGKINSON considered it necessary to place considerable emphasis not only on the deformation before rupture but also on the very small strain region. Indeed, this was one of the most important aspects of HODGKINSON's work. His experiments on wood and his experiments on iron in the 1830's were performed on large beams so that he could achieve accuracy in the measurement of deflection and also because, as an engineer whose experiments were financed[2] by the industralist FAIRBAIRN, he was motivated to obtain data which would be useful in engineering practice for full-scale structural members.

To determine the position of the "neutral line" as he referred to it[3] for deformation limited to the small strain region, HODGKINSON subjected simply supported

of this straightforward and courteous offer, but still persisted in the absurd assertion, and what was worse, deliberately circulated the statement that Kirkaldy's measuring dial was wrong, yet without ever mentioning the invitation extended to him, to bring any instruments to check or verify the same (see the gist of Mr. John Fowler's remarks in reference to Mr. Barlow, near the top of p. xviii, of Appendix)."*

"Before leaving this matter of Mr. Barlow's charge of inaccuracy—which he NEVER SUBSTANTIATED, and yet did not withdraw, but circulated it, thus prejudicing manufacturers and engineers from sending work—it is as well to mention that Mr. Barlow allowed the reports of all those experiments made for the Barrow Haematite Company, which he called in question (also some on Swedish iron), to be printed, occupying ten pages, in a fresh edition of his father's book: which appeared in August, 1867. Comment is hardly needed; he had evidently judged them to be accurate, and that they would help to bring that book up to date, for it is not possible that he would have allowed such to be given forth to the profession if he had really believed those results to be inaccurate."

*[This appendix contains a greatly detailed, highly personal account of Mr. BARLOW's visit. It is worth reading as an intimate picture of the extent to which personalities can influence the progress of events in the development of technology.]

[1] SAINT-VENANT [1856, *1*] referred to this HODGKINSON-BARLOW controversy and the 15 years which had elapsed between HODGKINSON's comments and BARLOW's concession, pointing out that DULEAU in 1812 also had erred in his placement of the neutral axis, because he had introduced a false principle. However, DULEAU's error was confined to his introduction. The experimental results were presented as measured, uninfluenced by the erroneous conjecture.

[2] It is relevant in this day when enormous sums are allocated to the detailed dissection of minutia to consider briefly the early 19th century state of affairs in financing fundamental experimental research. The *Report of the Thirteenth Meeting of the British Association for the Advancement of Science* held in 1843 records the recommendation "That Mr. Eaton Hodgkinson be requested to continue his experiments on the Strength of Materials and the changes which take place in their internal constitution, with 100 *l* at his disposal for the purpose" (p. xxiii). One year later one finds the grant suffering what in today's jargon would be called a "no-cost extension," inasmuch as the Report states "That E. Hodgkinson, Esq., be requested to continue his Experiments on the Strength of Materials." [*Report of the Fourteenth Meeting of the British Association for the Advancement of Science*, 1844], p. xxii.

[3] This was referred to by various authors as the "neutral line," the "invariable line," or the "line of passage."

9 ft. long, 1 in. square beams of pine, Danzig fir, and Quebec oak to central loads. By means of a graduated 9 ft. long tin scale which was sufficiently flexible that it could be made to conform to the curvature of the concave or convex side of the deflected beam, he measured the change in length of the outer fibers, from which he determined that the depth ratio of tension to compression for very small deformation was 169 to 190 for pine, 17 to 20 for Danzig fir, and 3 to 4 for Quebec oak, having a mean of approximately 4 to 5. He contrasted this with BARLOW's widely quoted measurements of 3 to 5 for tension to compression, critically pointing out that BARLOW's measurement had been made for very large deflection just prior to the rupture of the beam.

Directly influenced by the ideas of DUHAMEL, which he discussed in his memoir of 1824 in the form of quotations from the above-mentioned encyclopedia article of the highly over-esteemed ROBISON, HODGKINSON wished to find a means of placing the entire beam in distributed compression. In separate experiments in distributed tension, he then sought to determine the force vs deflection function in each situation. To accomplish this, he nailed two thin strips of iron to a flat face of the beam which contained a tongue and groove at the precise center of the beam where the load was applied. By placing the iron strips on the top or on the bottom of the beam, and carefully measuring the central deflection for the application of known central forces to simply supported 9 ft. long beams, he discovered that the behavior in both instances was parabolic:

$$y = aF^m, \tag{2.5}$$

where m for tension varied from 0.91 to 1.03, with an average value of 0.97; and for compression, from 0.798 to 0.901, with an average value of 0.895. Thus the nonlinearity in tension and compression were unequal in magnitude.

In a later memoir also published by the Manchester Literary and Philosophical Society, HODGKINSON[1] presented a long and detailed discussion of 35 tests on commercially produced iron beams for a wide variety of cross-sections. He not only performed the experiment to determine the effect of shape upon the breaking loads, as was the custom of the time, but also, which was unusual, he sought to determine the effect of shape upon the stress-strain function. These experiments led to his most important discovery in the field, a discovery which followed from his practice of removing each load before adding the next increase in load, in order to observe the amount of permanent set, if any, which had occurred.

The paper contained a long discussion of the relative merits of one cross-sectional shape or of one type of iron vs another, which I have omitted. Anteceding today's experimentists on microplasticity, HODGKINSON observed that some measurable permanent set occurred at even the smallest strains. With this permanent set the experimentally determined stress-strain function had invariably the parabolic form of Eq. (2.5), with $m=2$, which included both the recoverable elastic and the non-recoverable permanent deformation. The fiber stress for the beam was:

$$\sigma = aF - bF^2 \tag{2.6}$$

where σ is the axial fiber stress, F is the applied central force, and a and b are to be determined experimentally for a given iron and a given beam cross-section and length, depending upon whether the beam was cantilevered or simply supported. The observed parabolas are shown in Fig. 2.3, where the circles represent the averages of 40 tests performed by HODGKINSON[2] himself on 18 different kinds of cast iron; and the triangles, the additional data supplied by WILLIAM FAIR-

[1] HODGKINSON [1831, *1*].
[2] HODGKINSON [1843, *1*].

BAIRN,[1] the industrialist for whom HODGKINSON had worked. FAIRBAIRN's data provided the average of 100 tests for 44 kinds of cast iron.

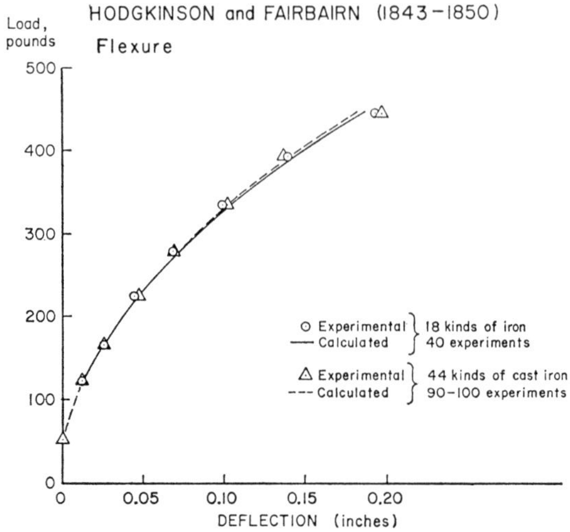

Fig. 2.3. The average results of experiments on simply supported beams.

The "elastic defect," as HODGKINSON called it, and the concomitant discovery of a functional regularity of the permanent set even at small strain, considerably influenced the thinking of experimentists during the remainder of the 19th century with regard to the determination of elastic constants, etc. Although HODGKINSON thought that stress was indeed a linear function of strain in extremely small strains, he stated: "This element is a defect of elasticity, or set, to which all bodies made to undergo a change of form, however small, seem to be liable." The apparatus used in obtaining all these data of HODGKINSON and FAIRBAIRN, which is known as "Fairbairn's Lever," is shown in Fig. 2.4.

HODGKINSON performed tension and compression experiments on iron,[2] for which he obtained the relation

$$\sigma = a\,\varepsilon - b\,\varepsilon^2, \tag{2.7}$$

where a and b are again experimental constants. He attributed the difference between Eqs. (2.5) and (2.7) to the combined effects of tension and compression in the flexure, since the constants in tension and compression in Eq. (2.7) were not necessarily the same, as he had shown in his earlier research on wood.[3]

[1] FAIRBAIRN [1850, *1*].

[2] HODGKINSON [1843, *1*]. To achieve accuracy in these experiments, HODGKINSON chose some specimens as long as 50 ft. As stated in the Iron Commission Report [1851, *1*], pp. 209–210:

> "Experiments were made to determine precisely the longitudinal compressions and extensions of cast iron and forged iron bars. The elongations were measured by attaching a bar 50 feet long and one inch square to the roof of a high building, and suspending weights from its lower end."

[3] Because of British ton conversion, the values for a and b in the Iron Commission's Report are given to seven significant figures: $a = 13\,934\,040$ psi and $b = 2\,907\,432\,000$ psi for tension, and the different values of $a = 12\,931\,560$ psi and $b = 522\,979\,200$ psi for compression. These values are the averages for four kinds of cast iron. The compression data should be viewed with caution since:

> "The compressions were measured by enclosing a bar 10 feet long and one inch square in a groove in a cast iron framework, which allowed the bar to slide freely and without friction, and yet did not allow lateral flexion. The bar was compressed by means of a lever loaded with a weight which could be varied. All possible precautions were taken to obtain precise results." [*Ibid.*, p. 209].

HODGKINSON suspended his 50 ft. tensile specimens vertically from the top of a building. The specimens were composed of a series of sections joined by screws threaded in opposite directions. The use of this type of composite specimen was to be criticized severely by ARTHUR JULES MORIN[1] twenty years later. Since

Fig. 2.4. FAIRBAIRN's lever. Early 19th century testing machine.

HODGKINSON's micrometer screw was capable of measuring elongations of 0.0025 mm, his potential strain resolution for the 600 in. gage length was 1.6×10^{-7}. As in all of his other experiments, his test procedure was to load the specimen to a prescribed value, record the total strain, unload to zero stress, and record permanent set. He then re-loaded to a new value and repeated the unloading, etc. The averaged experimental results for nine bars of cast iron are shown in Fig. 2.5, together with the corresponding permanent deformation.

From these data it is clear that Eq. (2.7) refers to the stress as a function of the total strain. Further, the moduli determined from comparing the elastic components after the subtraction of the permanent deformation provide a decreasing modulus. This experimental fact received much emphasis in the second half of the 19th century. HODGKINSON's initial values for this modulus are in very close accord with the very accurate measurements made later in cast iron. Unlike FRANZ JOSEPH VON GERSTNER, whose experiments will be described shortly,[2] HODGKINSON unfortunately did not measure the strain during the unloading cycle; hence, during that interval he did not know whether or not the unloading response function was linear.

HODGKINSON contended that no matter how little a metal was deformed, some permanent deformation ensued, a prediction in the 1830's which in a sense was a forerunner of some of the research today in microplasticity. A combination of HODGKINSON's experiments on long bars in tension with his experiments on long

[1] MORIN [1862, *1*].
[2] See Sect. 2.7.

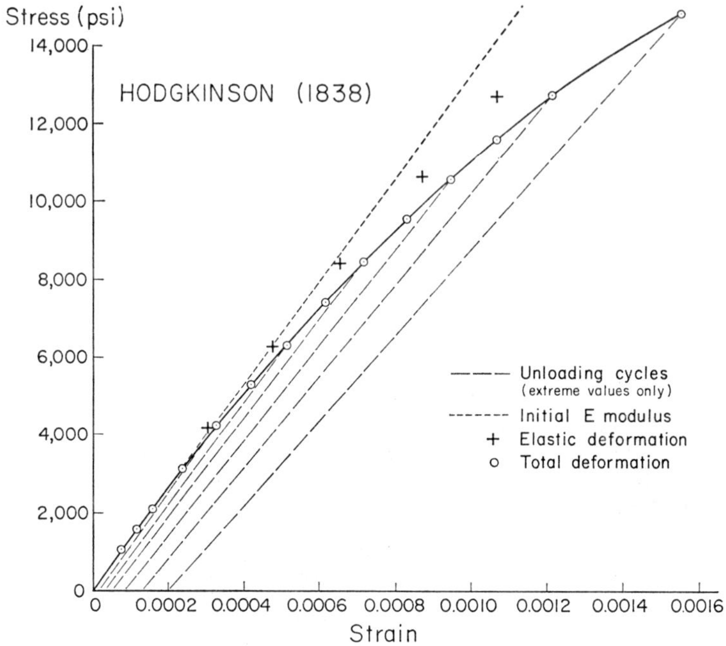

Fig. 2.5. The average of 9 tensile tests on long bars of cast iron.

bars in compression provides the reproducible nonlinearity on either side of zero stress for cast iron shown in Fig. 2.6.

In 1824 Hodgkinson[1] had observed nonlinear elasticity in wooden beams. He noted in 1839 that the stress in cast iron[2] also was a nonlinear function of the strain. An elaboration of these observations was included in his Report[3] for the British Association for the Advancement of Science in 1843. His data subsequently were subjected to the criticism[4] that the permanent deformation and nonlinear effect he was reporting arose from friction at the supports. In answer to this objection, Hodgkinson in 1844 described a new series of experiments he had performed on vertical beams simply supported by rollers and loaded by horizontal forces. He went to considerable lengths to improve on the graduated scale used in his earlier experiments, by introducing a micrometer screw capable of measuring distances as small as $^1/_{10\,000}$ in. The length of the bars was 6 ft. 6 in., and the depth of different bars ranged from $^3/_{10}$ in. to 1 in. in the direction in which the bar was bent. Each test lasted from three hours to a whole day, so that small effects due to the vibrations "in the neighborhood of a large manufactory"[5] could be nullified. In a series of experiments with this apparatus, he again obtained the same parabolic form with permanent set at the smallest strains, thus nullifying the objections of his critics.

In addition to the tests on cast iron, using this new apparatus Hodgkinson performed tests on soft stone, "each sawn 7 ft. long, 4 in. broad, and about

[1] Hodgkinson [1824, *1*].
[2] Hodgkinson [1839, *1*].
[3] Hodgkinson [1843, *1*].
[4] One of his chief critics was his former Professor, John Dalton, at the University of Manchester.
[5] Hodgkinson [1844, *1*].

1 in. thick."[1] He bent them in the direction of their least dimension. The experimental procedure was to lay on the same weight gently about four times in three minutes, each time with the bar unloaded for five minutes between each series of four loadings. After each loading series, he observed the permanent set. He found that stone, like cast iron, also followed a parabolic rule and experienced permanent set, the parabolic rule having a power coefficient which ranged from 1.650 to 1.957 in different experiments, which provided an average value power law index of 1.786 for all the tests.

Similar experiments were performed on wrought iron and steel. Although nonlinearity of the load deflection function was observed for those materials, it differed somewhat in form and magnitude from those in cast iron and stone. HODGKINSON stated with respect to all of these experiments in wood, cast iron, wrought iron, and stone: "In these, as in materials of every description tried,

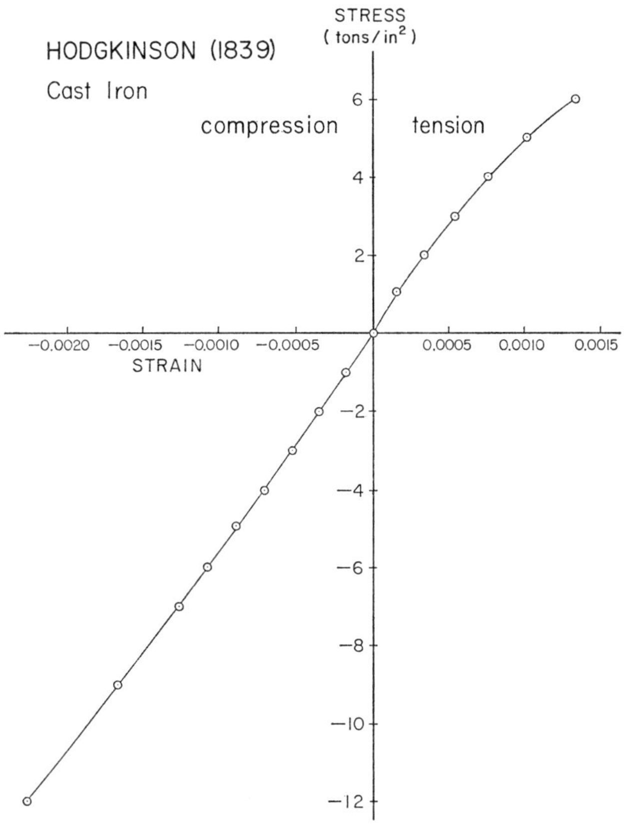

Fig. 2.6. An early comparison of tension and compression in long bars.

the weights, however small, seemed to produce a permanent set; no body recovering of itself its original form after a change of figure had been produced in it."[2] He noticed that if a small weight were carefully laid on the beam a number of times without acceleration, it ultimately produced an observable permanent set.

[1] *Ibid.*, p. 26.
[2] *Ibid.*

In view of the experiments of WILHELM WEBER in 1835 on silk thread, to be described below,[1] it is interesting that HODGKINSON observed elastic recovery to his permanent set in time intervals ranging from a few minutes to some hours after unloading, although, as he was careful to point out, in no instance was the elastic recovery complete. HODGKINSON noted that elastic recovery in those solids was rapid at first and reached stability after five minutes. He therefore took readings one minute, five minutes, and a half-hour after unloading. The importance of these experiments was not merely the nonlinearity observed but also the fact that the small permanent deformations present with increasing load were stable and expressible with fair reproducibility in terms of an empirical function.

In any experiment designed to discover a particular functional relation between two variables, it is fairly obvious that only the experimentist of considerable caliber excercising great care will establish within some prescribed limits a correlation, or lack of it, between prediction and experiment. The less competent large majority may be expected to design their experiments to provide random "verifications" of whatever empirical guess prompted their study unless the guiding light of a currently fashionable mathematically tractable theory such as linear elasticity should prejudge the issue.

One might dismiss the nonlinearity of the response functions observed in the experiments of HODGKINSON and DUPIN as merely an interesting historical development in the field of solid mechanics, were it not for the fact that by the end of the 19th century the increasing accuracy of measurement and improved experiments demonstrated that that was indeed the precise manner in which such solids deformed.

2.7. "Gerstner's law" for steel piano wire (1824).

Before HODGKINSON performed his experiments in iron and stone, the elder VON GERSTNER, FRANZ JOSEPH,[2] in 1824 had conducted a series of tension measurements on steel piano wire, which were published in 1831 in Vol. I of his *Handbuch der Mechanik.* The experiments also were reported by his son, FRANZ ANTON, Ritter VON GERSTNER,[3] in POGGENDORFF'S *Annalen der Physik und Chemie* in 1832, after the death of the elder GERSTNER. FRANZ JOSEPH VON GERSTNER used a lever having a ratio of 54. He loaded his wires with a series of weights differing by four Lower Austrian pounds. For each weight added, he recorded the deflection. Upon reaching a certain value he successively removed the same weights, again recording the deflection and finally the permanent set. This loading cycle was then repeated with a higher maximum with each repetition. After imposing the weight, he let ten or twelve minutes pass before he measured the elongation. The result for a piano wire having a diameter 0.063 cm and a length of 147 cm is shown in Fig. 2.7. This curve is plotted from GERSTNER's table in which the units are given as Lower Austrian pounds vs Lower Austrian lines.[4] Since the dimensions of the wire were provided, the data have been converted to kg/mm² vs strain.

In a study in which "... those experiments in which the wire strings revealed sudden yielding or some other irregularity were left out, and only those were

[1] WILHELM WEBER [1835, *1*]. See Sect. 2.12 below.

[2] FRANZ JOSEPH GERSTNER [1831, *1*]. He was the author of a volume entitled *Theorie der Wellen*, published in 1804 and highly regarded in the early 19th century.

[3] FRANZ ANTON GERSTNER [1832, *1*].

[4] One Lower Austrian pound = 373.3 g; one Lower Austrian line is one-twelfth of one Lower Austrian inch, which equals 2.54 cm.

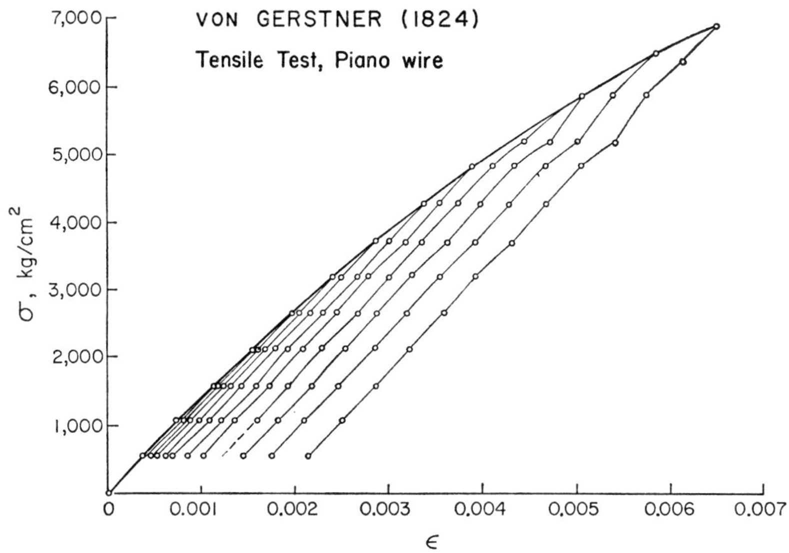

Fig. 2.7. Repeated loading experiments showing increasing permanent deformation.

included in which the properties of the wire had shown themselves completely uniform up to fracture,"[1] GERSTNER discovered that the data were fitted by the parabola

$$p = A e - B e^2 \tag{2.8}$$

where p is the tensile stress and e the corresponding strain. He observed that when he could measure deflections during the unloading and reloading up to the point of the previous loading, he obtained linearity. Differentiating (2.8) provided a maximum load P, with a corresponding strain E, in terms of which GERSTNER re-wrote his nonlinear load elongation formula of (2.8) as

$$e = E\left(1 - \sqrt{1 - \frac{p}{P}}\right). \tag{2.9}$$

This became widely known throughout the remainder of the 19th century as "GERSTNER's law" for the determination of maximum stress.

These experiments of GERSTNER of course preceded by at least a decade and a half HODGKINSON's similar observation in large iron bars. They provide the same nonlinear stress-strain function as HODGKINSON's tension and compression experiments.

GERSTNER's observation of the increasing elastic domain with increasing permanent deformation for linear or nearly linear elasticity had been studied in great detail by COULOMB much earlier, in 1784. COULOMB had observed, too, that the torsion modulus in such a region varied with increasing permanent deformation; such a reduction in modulus also was present in GERSTNER's data, but GERSTNER failed to comment upon it. The experiments of GERSTNER, along with those of DUPIN and HODGKINSON, are of historical importance: by 1835 the stress-strain functions of all of the commonly studied solids even in the region of small deforma-

[1] GERSTNER [1832, *1*], p. 270.

tion in quasi-static testing were considered by most experimentists to be essentially nonlinear.[1]

2.8. The discovery of the creep of metals: Coriolis and Vicat (1830–1834).

In 1834 Louis Joseph Vicat[2] made an important discovery in a series of experiments which obviously must have been performed considerably earlier than their publication. Vicat discovered, and described in detail over 135 years ago, the long time aspects of the now widely studied phenomenon known as "creep."

Gustave Gaspard Coriolis[3] in 1830 who while studying the effect of oxide on the deformation of lead noted that an increase in strain could occur at constant stress. Although a number of Continental and English engineers had been conversationally concerned during the previous decade with the long time stability of iron wires and chains used in suspension bridges, there was no actual experimental study of the phenomenon of creep. Navier,[4] four years before Vicat, in a series of twenty-seven experiments on metal sheets, cylindrical tubes, and spheres under internal pressure, had observed that lead, copper, and iron in tension continued to deform to rupture if the constant load were a sufficiently high fraction of the load required to produce immediate fracture. However, Navier provided no measurements of this behavior since he was almost totally concerned with tabulating the usual rupture data.

Vicat observed in the tensile testing of wire that if the axial loading was large enough, wires under a fixed load would continue to elongate beyond the first instantaneous elongation.[5] He subjected four unannealed iron wires to the constant load indicated in Fig. 2.8. These loads were approximately $^1/_4$, $^1/_3$, $^1/_2$, and $^3/_4$ of the ultimate load of 48.5 kg which the wires could support. He suspended the four wires from an "extemely secure stabilized oak beam extending from wall to wall of a vaulted chamber. The beam was supported above and below by a total of six supports."[6] The wires were 1 m long and oiled to preserve them from corrosion. Vicat made measurements by means of a comparative lever device so that the elongation data of Fig. 2.8 for each load were presented in the form of the "sines of the travelled arcs."

The data of Fig. 2.9 show the elongation of each wire at the end of 33 months. Since for these data the results were converted to millimeter elongation, they are expressible in the form of strain. These experimental data on creep obtained 14 decades ago show at once that if the load in iron is high, stability of the permanent deformation, although observable at small loads, no longer occurs for large ones. As is seen in Fig. 2.9 the magnitude of the creep deformation is a function of the size of the load.

[1] The controversy between 1830 and 1834 introduced by Louis Joseph Vicat, who pleaded as an experimentist for something more realistic than linear elasticity, and his Academy opponents including Dupin, who failed to understand the point Vicat was making, is fairly reported by Saint-Venant some years later. Saint-Venant emphasized the self-imposed limitations of the linear theory.

[2] Vicat [1834, *1*].

[3] Four years earlier than Vicat, Coriolis, [1830, *1*] had observed that 24 mm diameter, 19 mm high lead cylinders loaded in axial compression for intervals from five seconds to several hours never achieved stability, but continued to deform for as long as the load was applied. (See Sect. 4.2 for a discussion of these experiments of Coriolis.)

[4] Navier [1826, *1*].

[5] For some reason Vicat failed to record the details of this first instantaneous elongation.

[6] Vicat [1834, *1*].

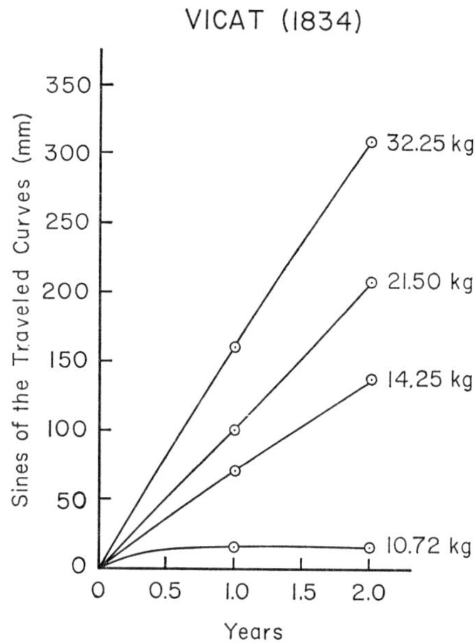

Fig. 2.8. The first observations of creep in the tensile loading of iron.

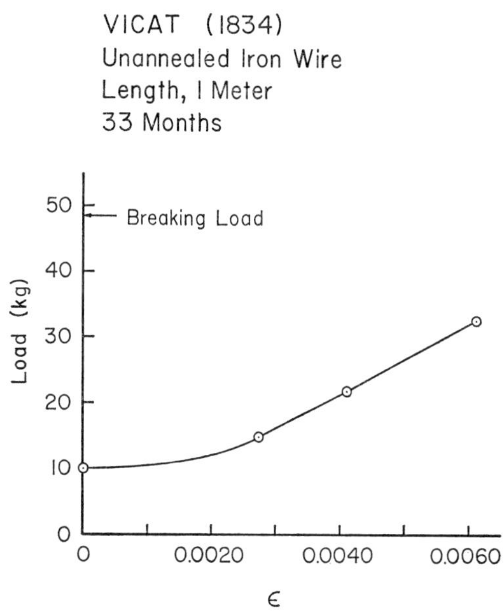

Fig. 2.9. The first quantitative determination of the relation between the magnitude of the creep and that of the constant applied stress.

These discoveries of both stable and unstable permanent deformation, even for relatively small strains, during the ensuing 50 years provided the impetus for many experiments concerning phenomena excluded at the outset by the mathematical theory of elasticity, which was developing at the same time.

2.9. The first resolution of microstrain: Vicat (1831).

For experimentists of today it is surprising to discover that 140 years ago, in the 1830's, experimental physicists were accurately determining contact times of 100 μsec and with amazing precision were able to measure strains or obtain a potential strain resolution as low as 1×10^{-7}. The short contact times were achieved in 1837 by CLAUDE SERVAIS MATHAIS ROLAND POUILLET,[1] and the low strain resolution by LOUIS JOSEPH VICAT in 1831.[2] POUILLET's time of contact experiments will be described in Sect. 3.35 on experiments related to wave propagation induced by impact.

In 1829 VICAT was assigned to inspect a series of eight suspension bridges over the Rhône River. That this was a very complete survey of all aspects of the bridges from fundamentals to façade may be learned from the fact that he became curious as to whether or not the methods of manufacturing cables from iron wires provided a product for which any reliable design calculations could be made. When he saw how such cables were manufactured, he began to suspect that poor workmanship and crude standards of production might nullify calculations made in the design of bridges. That expedition modified VICAT's experimentation, which resulted not only in his discovery of creep as described above, but also in his obtaining more reliable stress-strain functions in a metal than any ever before achieved.

VICAT achieved his results in two ways: first, he used wire specimens with a length of 63.82 m; and second, he developed a mechanical extensometer with a ratio 80:1. In this latter series, the specimen was a vertical wire 2 m long. He performed the experiments with long wires, horizontally, utilizing the frame structure of the cable-making apparatus. To eliminate the initial friction, he did not begin his measurements of elongation until the wire was under a considerable load. Thus for the two experiments shown in Fig. 2.10, the stress-strain functions began at a finite point on the stress ordinate.

In these experiments, with his measuring apparatus VICAT could determine the elongation of 0.1 mm; the potential strain resolution was 1.4×10^{-7}. To ascertain whether such a resolution actually had been achieved, I examined VICAT's tabulated data for increments of stress and strain, and noted that experimental equivalence indeed was obtained in the initial linear region, consistent with the number of significant figures of that resolution. Recorded variations for equal increments of stress varied by 1 mm, which suggested a measured resolution of 1×10^{-5} for the long wires.

VICAT made a second check by performing experiments in his vertical wires 2 m long, with the 80:1 mechanical extensometer he had developed. The calculated resolution in this situation for a 0.10 mm division was 5×10^{-7}. An example of such an experiment also is shown in Fig. 2.10 for an annealed iron wire. Once again, he obtained the expected accuracy when comparing increments of strain for equivalent increments of stress. Equal increments of stress furnished measured variations of only 5×10^{-6} strain.

I might add that VICAT, as indicated by his creep studies, knew well that he needed to support his wires securely at the point of vertical attachment, a feat which he accomplished by means of the huge beams in the vaulted chambers described above.

[1] The original experiments of POUILLET in 1837 were improved and extended during the succeeding years, these results being contained in a French publication in 1844 [1844, *1*] and in a German publication in 1845.

[2] VICAT [1831, *1*].

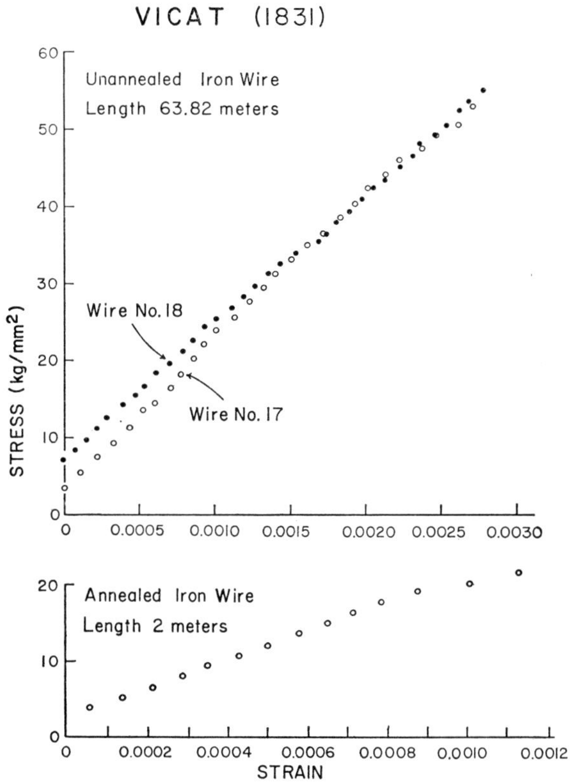

Fig. 2.10. The first precise measurements in the tensile deformation of metals.

That only an experimentist of more than ordinary talents could achieve a strain resolution of 10^{-6} in fact and 10^{-7} potentially, may be seen from comparing the data of VICAT with those of PETER BARLOW,[1] published in the edition of his book in 1837. BARLOW ascertained the stress-strain function in tension in seven long iron bars, six of which presumably were identical, being referred to as "remanufactured malleable iron." The seventh was a newly manufactured bar of the same material. BARLOW's extensometer, gage clamps, and a typical tensile specimen, are shown in Fig. 2.11.

His mechanical extensometer permitted a ratio of only 10:1. Hence, with the stated scale division of 0.1 English inch and a gage length for his iron specimens of 100 in., BARLOW's strain resolution was only 1×10^{-4}. His seven tests are shown in Fig. 2.12. From them it may be seen how much scatter could be present in measurements which some experimentists deemed it worthwhile to publish.

Despite the fact that these data were not reproducible, and rather obviously not linear, BARLOW averaged all increments of elongation between all load points for all of the tests so as to obtain an elongation per ton. This provided an overall modulus for iron which not surprisingly was considerably different from that obtained by other experimentists, including VICAT. The moduli obtained from VICAT's data, which were visibly linear in the small strain region, were 18080 kg/mm² for the unannealed long wire #18 in direct measurement, and 18120 kg/mm²

[1] BARLOW [1837, *1*].

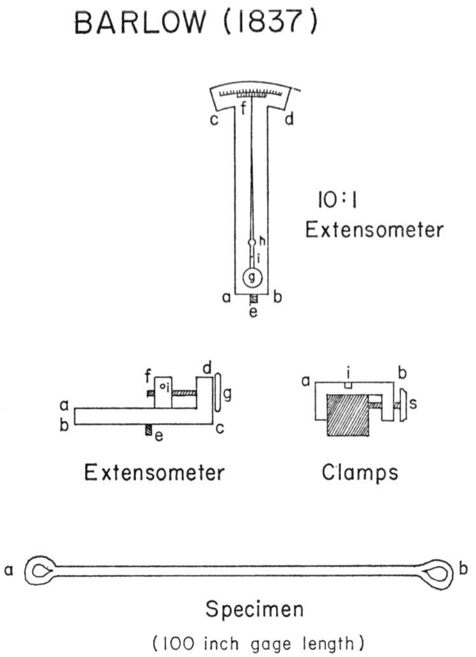

Fig. 2.11. An example of extensometers and specimens in the early 19th century.

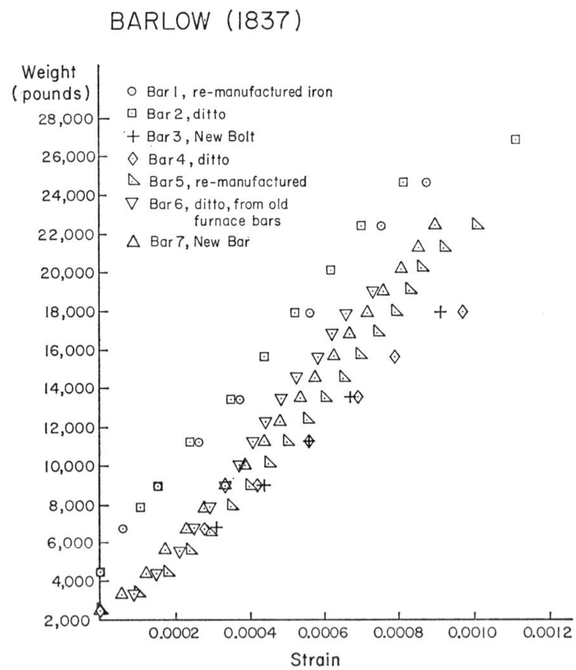

Fig. 2.12. An example of the nonlinearity and the lack of reproducibility in some early 19th century tensile measurements which were widely used in technology. The original data were tabulated.

for the annealed short wires using the 80:1 extensometer, a difference of only 0.2%. Indeed, VICAT's value was very close to that of 18045 obtained in accurate elongation experiments by WERTHEIM[1] 25 years later in as-received and subsequently annealed iron wire.

In HODGKINSON's tensile data described above, with measured increments of strain of 75×10^{-6} the calculated modulus was 10000 kg/mm² which is almost precisely that predicted for cast iron from the data which EDUARD AUGUST GRÜNEISEN[2] obtained under conditions of extreme accuracy in 1906 in the same strain range. HODGKINSON with his strain resolution of 10^{-6}, like VICAT, was capable of providing data in the 1830's comparable to that in modern studies of deformation. It is difficult to evaluate the precise strain resolution which actually was achieved by a given experimentist. Dividing the known elongation resolution by the gage length, of course, does not indicate all of the factors pertinent in determining strain resolution. As I have noted, the comparison of linear increments in VICAT's data revealed an actual strain resolution of 10^{-6} with a potential of 10^{-7}. Nevertheless, it is interesting to compare this potential strain resolution over the years with the improvement of the elongation resolution over the same period of time. Such a comparison in Fig. 2.13 reveals that for an improvement of five orders of magnitude in elongation resolution over a 150 year interval, the major accomplishment was to achieve a reduction in specimen gage lengths for a nearly constant potential strain resolution.

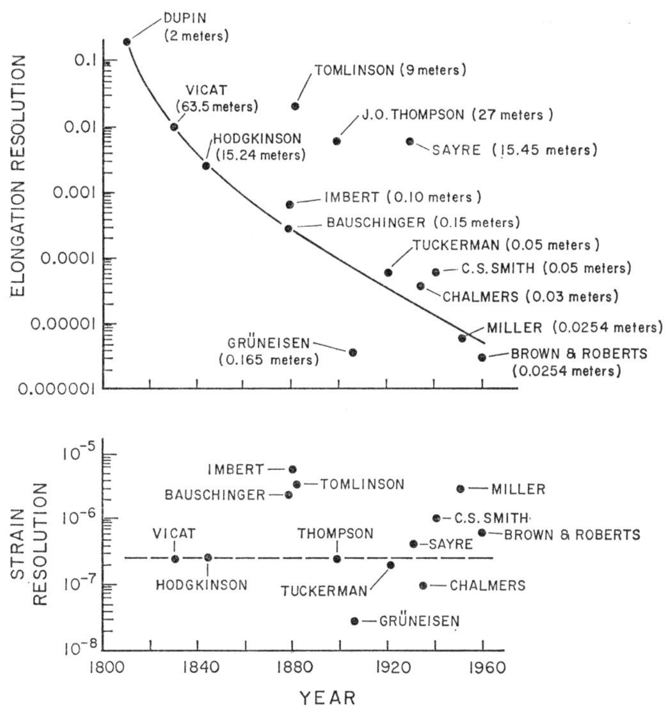

Fig. 2.13. 160 years of increasing precision in the measurement of elongation. The corresponding decrease in length of the specimen provided the nearly equal strain resolution shown.

[1] WERTHEIM [1844, *1a*].

[2] HODGKINSON [1843, *1*]; GRÜNEISEN [1906, *1*].

Two other major experimental discoveries of the 1830's were made by WILHELM WEBER[1] in his classical experimental paper on the specific heats of metals in which he observed a thermal after-effect, and in his equally important experiments on the elasticity of silk threads[2] in which he discovered what he termed the "elastic after-effect."[3]

The newly found fact of creep deformation in iron wires under load had serious implications upon the durability of suspension bridges using cables composed of twisted strands of the same solid. Among the several persons who investigated the phenomenon after 1834 was FÉLIX LEBLANC[4] who in 1839 published one of the most interesting papers on the subject.

2.10. Experiments on the stability of the permanent deformation of iron wire: Leblanc (1839).

The first question LEBLANC examined is of considerable importance in evaluating the quality of the materials being investigated at that time. He designed his first experimental series to determine whether a long wire, with its greater number of flaws, offered less average resistance than a very short wire. The answer LEBLANC provided was no. The question obviously had reference to the homogeneity of the solid. Therefore, LEBLANC had cut 150 m lengths of drawn iron wire with diameters ranging from 0.003 to 0.0035 m, into sections 2 m long and 26 m long. His 24 rupture tests, 12 for each length, are compared in Table 4. The observed distribution of breaking loads surely indicates that the iron wire of the 1830's was sufficiently homogeneous for serious experimentation.

Table 4. LEBLANC (1839).

Test no.	Resistance of wires	
	2 m long (kg)	26 m long (kg)
1	700	670
2	710	720
3	650	660
4	640	640
5	690	750
6	620	650
7	620	640
8	580	620
9	640	670
10	690	650
11	690	670
12	690	680
Average	660	668

LEBLANC's second experimental series[4] had to do with the matter of where failure occurred. He noted that 14 times out of 17 the wire broke near the point of attachment. Further reference to this will be made in Sect. 4.25 below, when

[1] WEBER [1830, *1*].

[2] WEBER [1835, *1*].

[3] WEBER, of course, is better known for his contribution to magnetism and electricity, or as the physicist collaborating with his mathematical friend and colleague at Göttingen, KARL FRIEDRICH GAUSS, on a number of projects including that of a 9000 ft. long crude telegraph in 1833, which preceded by four years SAMUEL MORSE's development.

[4] LEBLANC [1839, *1*].

I discuss the experiments of JOHN HOPKINSON[1] of 1872 on dynamic plasticity. Omitting a series of experiments relating to the effects of curvature and of prior straightening, we may note that LEBLANC raised the most important question in the present context, namely, whether a wire could be subjected to near rupture tension for a long time as well as a short time, without any loss of strength. He demonstrated that such was indeed the situation.

The first experiments in that series consisted of nine tests which established that 690 kg was the average load at rupture of an iron wire described as of "good appearance." The gage length over which he measured subsequent strains[2] was 1.30 m, determined after a small initial load of 20 kg was placed on the wire to straighten it. In the first experiment the total load was 640 kg, or 93% of the rupture load; it produced an elongation of 5.60 mm. He obtained that result at noon on January 6, 1839. He measured elongations, shown in Table 5, every day for 13 days until January 19, when he added 30 kg and obtained elongations of 0.0069 m. Next, he added 10 kg, which provided an elongation of 0.0070 m; 10 kg more produced rupture. The total load at rupture at the end of the 13-day test was 690 kg just as in the previous 9 tests.

Table 5. LEBLANC (1839).

Date	Temperature in degrees Reaumur	Elongation (mm)
January		
6	+5	5.60
7	+4	5.70
8	+4	5.8
9	+4	5.8
10	+2	5.8
11	+3	5.9
12	+7	6.2
13	+8	6.3
14	+8	6.3
15	+7	6.3
16	+5.5	6.4
17	+4.5	6.4
18	+3	6.4
19	+3	6.4

The second experiment in similar wire, with an initial load of 20 kg and a subsequently measured gage length of 1.30 m, LEBLANC loaded with 300 kg at noon on January 22, 1839; this caused an elongation of 0.00130 m. An additional load, making a total of 600 kg, produced an elongation of 0.00630 m; another 20 kg, 620 kg in all, produced an elongation of 0.00680 m. He then left the wire loaded from January 22 through March 22, recording measurements of deformation each day. By the sixteenth day, on February 7, he achieved stability of deformation, with an elongation of 7.40 mm, corresponding to an incremental increase in strain of 0.00046. This remained unchanged under load until the 60th day March 22, when he unloaded and reloaded the specimen and still did not observe any change in the deformation. Twenty-one days later, on April 12, he again unloaded the wire and left it for twelve days, then reloaded it with the

[1] JOHN HOPKINSON [1872, *1*, *2*].
[2] Strain resolution was approximately 1×10^{-5}.

same weight of 620 kg on April 24. The elongation of 7.3 mm he measured was slightly lower than the 7.50 mm which he had recorded on April 12. Within 24 h, the elongation had returned to 7.50 mm, where it remained for seven additional days until April 30 when he successively subjected the wire to additional 20 kg weights: 640 kg produced an elongation of 0.00760 m; 660 kg brought an elongation of 0.00775 m; and finally, after ninety-nine days the wire broke at a load of 680 kg, i.e., under a final load nearly identical with that of the original nine short-time tests and that of the thirteen-day experiment. The load of 620 kg which had

Table 6. Leblanc (1839).

Dates	Temperature (°R)	Elongations (mm)	Dates	Temperature (°R)	Elongations (mm)
January			March		
22	+ 5	6.80	6	+ 5	7.40
23	+ 4	7.10	7	+ 4	7.40
24	+ 3	7.10	8	+ 4.25	7.40
25	+ 4	7.20	9	+ 4.75	7.40
26	+ 3.50	7.20	10	+ 6.25	7.40
27	+ 1	7.20	11	+ 6	7.40
28	0	7.20	12	+ 8.50	7.40
29	+ 1.50	7.20	13	+ 10	7.40
30	+ 1.70	7.20	14	+ 10.25	7.40
31	0	7.20	15	+ 9.75	7.40
			16	+ 8.50	7.40
February			17	+ 6	7.40
1	0	7.20	18	+ 4.75	7.40
2	+ 1.75	7.20	19	+ 5	7.40
3	+ 4	7.30	20	+ 6	7.40
4	+ 6	7.30	21	+ 7.75	7.40
5	+ 7.25	7.30	22	+ 7.75	7.40
6	+ 8	7.30	23	+ 8	7.40
7	+ 8.75	7.40	24	+ 10	7.40
8	+ 8.50	7.40	25	+ 9.75	7.40
9	+ 8	7.40	26	+ 9.50	7.40
10	+ 8	7.40	27	+ 10.50	7.40
11	+ 8	7.40	28	+ 9.50	7.40
12	+ 7.75	7.40	29	+ 9.50	7.40
13	+ 8.25	7.40	30	+ 10	7.40
14	+ 8	7.40	31	+ 11	7.40
15	+ 6.25	7.40			
16	+ 7	7.40	April		
17	+ 5.25	7.40	1	+ 10.50	7.40
18	+ 4.25	7.40	2	+ 9	7.40
19	+ 4.50	7.40	3	+ 8.50	7.40
20	+ 6.50	7.40	4	+ 9	7.40
21	+ 4.25	7.40	5	+ 9	7.40
22	+ 7	7.40	6	+ 8	7.40
23	+ 8.25	7.40	7	+ 9	7.50
24	+ 8.25	7.40	8	+ 7.50	7.50
25	+ 6	7.40	9	+ 6.50	7.50
26	+ 6	7.40	10	+ 7	7.50
27	+ 6	7.40	11	+ 9	7.50
28	+ 6.75	7.40	12	+ 8	7.50
			24	+ 12	7.30
March			25	+ 10	7.50
1	+ 7.50	7.40	26	+ 11	7.50
2	+ 8	7.40	27	+ 11	7.50
3	+ 8.50	7.40	28	+ 11.50	7.50
4	+ 10	7.40	29	+ 12	7.50
5	+ 8.50	7.40	30	+ 12.50	7.50

been maintained hence was 91% of the rupture load. This experiment and the temperature in degrees Réaumur[1] at the time at which the elongations were recorded are shown in Table 6. LEBLANC, of course, had discovered primary creep as a prelude to VICAT's long time, or secondary creep.

2.11. On the phenomenon discovered by Savart (1837) and Masson (1841), now known as the Portevin-Le Chatelier (1923) effect.

It is perhaps unfortunate for GERSTNER that he decided to eliminate experiments "which revealed a sudden yielding or other irregularity,"[2] for otherwise he might have been the first to discover what is today referred to as the "PORTEVIN-LE CHATELIER" effect.[3] In historical truth the discovery of this effect should be credited to FÉLIX SAVART[4] in 1837 and to ANTOINE PHILIBERT MASSON[5] in 1841. MASSON described the steep vertical rise of the stress accompanied by extremely small strain up to a value at which suddenly a large increase in strain occurred at constant stress. For dead weight tests of the type performed in the 19th century, this phenomenon had an appearance which later gave rise to the use of the term "staircase" effect.[6]

This phenomenon in large deformation has been studied during the past two decades by a number of experimentists, including ANDREW WETHERBEE MCREYNOLDS,[7] OSCAR W. DILLON,[8] WILLIAM SHARPE,[9] and myself.[10] An example of this

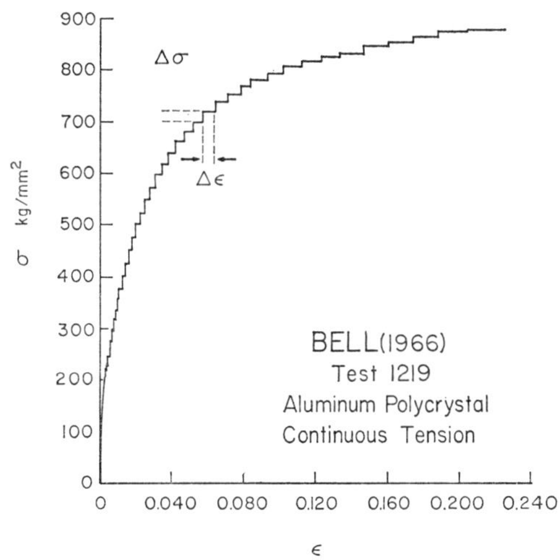

Fig. 2.14. An example of discontinuous deformation in a tensile test at constant load rate.

---

[1] Temperature in degrees Réaumur: 1° Réaumur = $^5/_4$° Centigrade.
[2] GERSTNER [1832, *1*].
[3] PORTEVIN and LE CHATELIER [1923, *1*].
[4] SAVART [1837, *1*].
[5] MASSON [1841, *1*].
[6] Since most tests in the 20th century are performed on "hard" testing machines, this phenomenon appears as an oscillation in the response function, and the "staircase" appearance no longer is observed. See PORTEVIN and LE CHATELIER [1923, *1*].
[7] MCREYNOLDS [1949, *1*].
[8] DILLON [1963, *1*].
[9] SHARPE [1966, *1*].
[10] BELL [1968, *1*]; BELL and STEIN [1962, *7*].

behavior for large deformation in polycrystalline aluminum is shown in Fig. 2.14 from a dead weight tensile experiment of mine in 1966 on fully annealed aluminum in which the loading was at a constant stress rate, the entire experiment lasting one hour.

Another experiment, in 1962, also by me, in the small deformation region of present interest took five days to perform and was an incremental dead-weight test in compression; the results for the first 36 hours are shown in Fig. 2.15. The

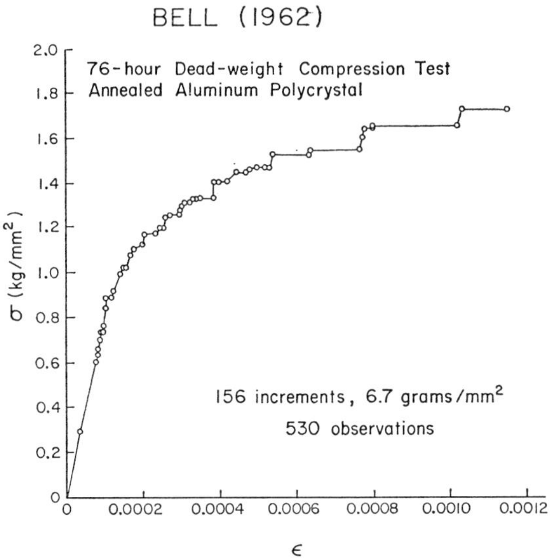

Fig. 2.15. An example of discontinuous deformation in an incremental test in compression.

deformation proceeded slowly up the vertical portion of the step until an unstable stress increment was reached, after which there was a sudden increase in strain to an equilibrium value. Until large stress was reached the equilibrium value could be maintained for a matter of days. The longest time during which such a permanent deformation actually was observed was 153 days in the large deformation dead weight tensile experiment performed in 1931 by DANIEL HANSON and M.A. WHEELER.[1] Fig. 2.16 shows their results for stress in kg/mm² (not psi).

This SAVART-MASSON phenomenon described in considerable detail in Sect. 4.31 is superimposed upon the response function which is obtained when the phenomenon is absent. On connecting the points of permanent deformation, after the strain steps in Fig. 2.14, a reproducible stress-strain function is obtained. One notes in the data of GERSTNER in 1824 that the maximum strain of 0.006 is such that the deformation still may be categorized as relatively small, as may be seen from the example I have included from one of my own experiments, shown in Fig. 2.15. The essential feature of the SAVART-MASSON phenomenon described above, whether or not the incremental loading was sufficiently small to see the step behavior, was the fact that the strain remained unchanged at the foot of the step for many hours, or even for days.

In 1840 (and indeed, even today), it was difficult to reconcile the various experimental facts in specifying strain: WEBER's thermal after-effect, GERSTNER's

[1] HANSON and WHEELER [1931, *1*].

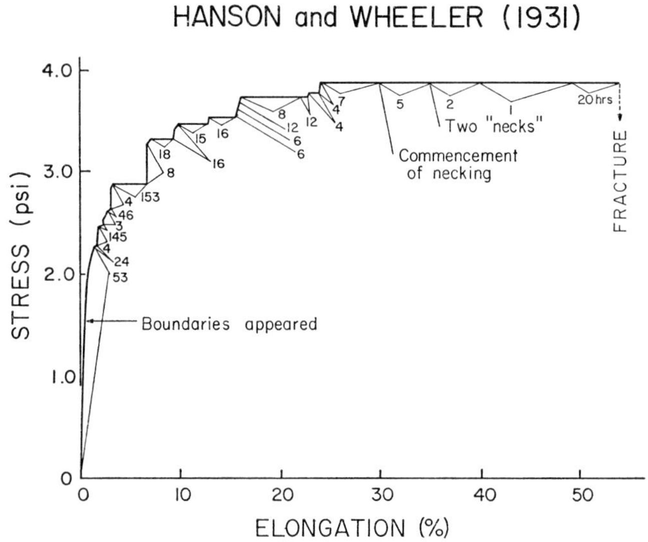

Fig. 2.16. An example of discontinuous deformation in an incremental tensile test in polycrystalline aluminium. This test lasted 18 months.

and HODGKINSON's permanent elongation phenomenon, VICAT's creep, and LEBLANC's observations of stability above 90% of the fracture load following initial short-time instability.

MASSON[1] conducted a series of experiments on brass, iron, steel, copper, and zinc in the laboratory of SAVART where he had done research during SAVART's serious illness. The first experiments MASSON performed in that laboratory were a repetition, on the very same specimens SAVART had used, of a series of dead weight tensile tests which SAVART had made four years earlier. Understandably, MASSON became interested in the differences in the results of the two series. With respect to the comparison of SAVART's data[2] with his own, he commented:

> This table shows that solid bodies do not elongate in a continuous manner, but by sudden jumps, which phenomenon has also been observed in dilatation by heat. Large differences which exist between Savart's results and mine for brass and iron rods at first made me suspect experimental difficulties which I could not overcome, despite the care I took to repeat in succession several times, the same measurement, which remained invariable, but I was reassured by the measures of iron and steel rods, and by Savart, who had often observed this phenomenon of discontinuous elongation, which he regarded as inherent to the nature of the body.[3]

MASSON stated further:

> Savart thought that there was no stable equilibrium at all for the molecules of solid bodies exposed to the forces of traction; they had to displace themselves from one another indefinitely, so that the loaded rods had to stretch. It seems to me however that the old construction proves that there is for solid bodies a limit for elongation corresponding to a given tension, and I wish to place in evidence through experiments which unfortunately did not last long enough but which, imperfect as they are, seem to confirm the idea of an elastic limit.[4]

[1] MASSON [1841, *1*].

[2] These data will be discussed in some detail in Chap. IV. What is meant here is an "elastic limit" in strain, not stress, a concept familiar only to those who study dead weight loading.

[3] MASSON [1841, *1*], p. 454.

[4] *Ibid.*

As one finds in present-day dead weight experiments, LEBLANC, too, had been unable to obtain stability in zinc[1] although he achieved it for copper. In dead weight experiments, the stress is prescribed and the strain, with all of the interesting information which it provides, is the measured variable. When the strain is prescribed and the stress is the quantity to be measured, as in modern testing machines, many of these effects either are not observed or are markedly altered so that when seen, they customarily were attributed to the testing machine itself. The recent large-scale revival of dead weight testing in polycrystalline solids in my laboratory[2] since 1950, in the laboratory of DILLON[3] in the 1960's, and in a few isolated measurements in other laboratories, has reintroduced those questions of 1840 regarding the stability of strain in deformation.

2.12. The experiments of Gough (1805) and Wilhelm Weber (1830), introducing thermoelasticity; Weber's discovery of the elastic after-effect (1835).

In 1806 JOHN GOUGH commented upon what he referred to as a "singular property" of caoutchouc, or India rubber, which he stated "has never been taken notice of in print, at least by any English writer."[4] Referring to a slip of rubber which previously had been steeped for a few minutes in warm water or held in the fist somewhat longer, GOUGH stated under the somewhat dubious label of "Experiment I":

> Hold one end of the slip, thus prepared, between the thumb and fore-finger of each hand, bring the middle of the piece into slight contact with the edges of the lips, taking care to keep it straight at the time, but not to stretch it much beyond its natural length: after taking these preparatory steps, extend the slip suddenly, and you will immediately perceive a sensation of warmth in that part of the mouth which touches it, arising from an augmentation of temperature in the caoutchouc; for this resin evidently grows warmer the further it is extended, and the edges of the lips possess a high degree of sensibility, which enables them to discover these changes with greater facility than other parts of the body. The increase of temperature, which is perceived upon extending a piece of caoutchouc, may be destroyed in an instant, by permitting the slip to contract again; which it will do quickly by virtue of its own spring, as oft as the stretching force ceases to act as soon as it has been fully exerted. Perhaps it will be said, that the preceding experiment is conducted in a negligent manner; that a person who wishes for accuracy will not trust his own sense of feeling in inquiries of this description, but will contrive to employ a thermometer in the business. Should the objection be stated, the answer to it is obvious; for the experiment in its present state demonstrates the reality of a singular fact, by convincing that sense, which is the only direct judge in the case, that the temperature of a piece of caoutchouc may be changed by compelling it to change its dimensions. The use of a thermometer determines the relative magnitudes of these variations, by referring the question of temperature to the eye: experiments of this sort are therefore of a mathematical nature, and afford a kind of knowledge with which we have nothing to do at present; for we are not inquiring after proportions, but endeavoring to establish the certainty of a fact, which may assist in discovering the reason of the uncommon elasticity observable in caoutchouc.[5]

In the experiments of CORIOLIS referred to in Sect. 2.8, which consisted of allowing a heavy carriage wheel to compress a lead cylinder held between two steel plates, the top plate having a steel button over the specimen which was in

[1] Modern handbooks, for example, always provide a wide range of values, rather than a single modulus for zinc.

[2] BELL [1956, *1*], BELL and STEIN [1962, *7*], BELL [1968, *1*].

[3] DILLON [1963, *1*], [1966, *1*].

[4] GOUGH [1806, *1*], p. 39.

[5] GOUGH [1806, *1*], p. 40. See also [1805, *1*], pp. 290–291.

contact with the wheel for a determined time, CORIOLIS had observed while commenting on the insensitivity of the plastic deformation to prior thermal treatment between 0 and 50° C: "... one notes that merely the crushing develops so much heat that one can barely touch the lead at the moment it has just been compressed ..."[1] These observations of thermal effects associated with deformation called for further experiments to be undertaken so as to establish the bounds of perfect linear elasticity.

In contrast to the somewhat vague suggestions of GOUGH and CORIOLIS, WILHELM WEBER in 1830 provided an elegant experimental study which launched the subject of thermal elasticity.[2] WEBER's first research in this area was an experimental classic, equalling in importance and originality the experimental contributions of HOOKE, MARIOTTE, COULOMB, CHLADNI, and DUPIN. Following the discussion of a number of abortive attempts to effect sudden and partial loading or unloading of a metal string with a known load differential, including, for example, the use of a device similar to the friction pin of the piano, WEBER finally evolved the beautifully simple experimental arrangement shown in Fig. 2.17.

WILHELM WEBER (1830)

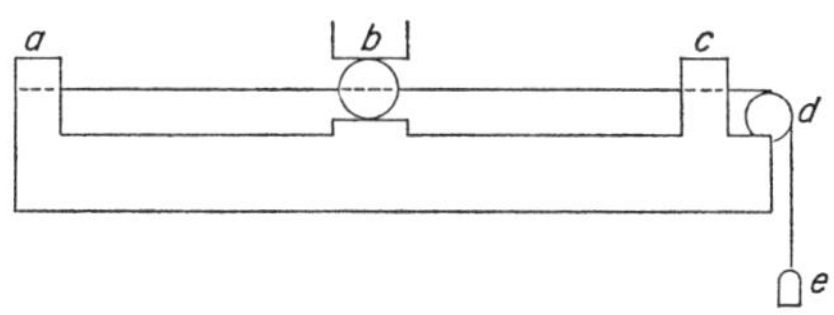

Fig. 2.17. WILHELM WEBER's apparatus for the first experiments in thermo-elasticity.

Applying an initial weight at e, WEBER could provide a tension P in both $a-b$ and $b-c$, which were metal strings of equal initial length. After he had clamped the sphere b, he increased the load at e to produce a tension Q in $b-c$, while $a-b$ remained under tension P. He then tightened the clamp at c. The clamps at a and c were grooved split cylinders in which the wire was placed; the clamp at b was a grooved split sphere. WEBER provided these attachments in order to insure that there was no injury to the wire during clamping, since he repeated the experiment many times on each wire. When he opened clamp b, with the clamps a and c remaining closed, the tension in the wire was $\frac{1}{2}(P+Q)$. When he reclosed the clamp after an interval of time and re-subjected the strings $a-b$ or $b-c$ to lateral vibration, by means of Eq. (2.10) he could determine the frequency n from the tension, providing the linear density π were known.

$$n=\frac{1}{l}\sqrt{\frac{2g}{\pi}}\sqrt{\frac{P+Q}{2}}\,. \tag{2.10}$$

Conversely, if the frequency were accurately measured, he could determine the tensile force in a metal string. To make an accurate determination of the frequency, WEBER chose P and Q to be such that $\frac{1}{2}(P+Q)$ would provide a frequency slightly below that of a tuning fork with the tone F. The experiment proceeded as follows. After loading P and Q to the desired values, he opened the clamp at b for approximately $^1/_4$ sec and then closed it. He then set into vibration wire $a-b$ or $b-c$ and recorded the number of beats with respect to the tuning fork. After an interval of time, he again opened and immediately closed the clamp

[1] CORIOLIS [1830, *1*], pp. 105–106.
[2] WILHELM WEBER [1830, *1*].

at b, and again recorded the number of beats. The observed data are contained in Table 7 for tests on iron, copper, silver, and platinum wires of the specified lengths and values of P and Q.

Table 7. WEBER (1830).

No.	Wire	Its vibration coinciding with F of the tuning fork for 5 sec after momentary communication	As in preceding columns, after a longer communication	Increase of the coinciding vibrations	Average of all experiments
			Iron wire		
1	ab	$7^1/_4$	$3^3/_4$	$-3^1/_2$	$-3^3/^8$
	bc	$9^3/_4$	$12^3/_4$	$+3$	$+3^1/_4$
2	ab	$11^1/_{16}$	$8^5/_8$	$-2^7/_{16}$	
	bc	$15^{11}/_{16}$	$18^7/_8$	$+3^3/_{16}$	
3	ab	$8^3/_8$	$4^7/_{32}$	$-4^5/_{32}$	
	bc	$10^1/_4$	$13^{25}/_{32}$	$+3^{17}/_{32}$	
			Copper wire		
1	ab	19	$14^{19}/_{32}$	$-4^{13}/_{32}$	$-3^1/_3$
	bc	17	$20^{25}/_{32}$	$+3^{25}/_{32}$	$+3^1/_{32}$
2	ab	$14^1/_4$	$10^{13}/_{16}$	$-3^7/_{16}$	
	bc	$13^3/_4$	$16^7/_{16}$	$+2^{11}/_{16}$	
3	ab	$18^3/_{16}$	$15^{31}/_{32}$	$-2^7/_{32}$	
	bc	$20^3/_4$	$23^{13}/_{32}$	$+2^{21}/_{32}$	
			Silver wire		
1	ab	$18^3/_4$	$16^1/_4$	$-2^1/_2$	$-2^5/_{16}$
	bc	$14^3/_4$	$17^1/_2$	$+2^3/_4$	$+3$
2	ab	$20^5/_{16}$	$17^1/_2$	$-2^{13}/_{16}$	
	bc	$14^{15}/_{16}$	$16^3/_4$	$+1^{13}/_{16}$	
3	ab	$19^1/_{32}$	$15^3/_8$	$-3^{21}/_{32}$	
	bc	$16^7/_{32}$	$18^5/_8$	$+2^{13}/_{32}$	
			Platinum wire		
1	ab	21	12	-9	-8
	bc	$^1/_2$	8	$+7^1/_2$	$+7^1/_3$
2	ab	$46^3/_4$	39	$-7^3/_4$	
	bc	$23^1/_4$	31	$+7^3/_4$	
3	ab	$45^1/_2$	$38^1/_4$	$-7^1/_4$	
	bc	$23^1/_4$	30	$+6^3/_4$	

Iron wire: $ab=257$ lines; $bc=257$ lines; a portion of 1300 lines length weighed 2.878 g; $P=1006$ g; $Q=7000$ g; at 4000 g tension, the tone of the iron wire was a bit lower than the tone F of a tuning fork.

Copper wire: $ab=240$ lines; $bc=240$ lines; a portion of 1300 lines length weighed 3.332 g; $P=1700$ g; $Q=5700$ g; at 3700 g tension, the tone of the copper wire was a bit lower than the F of the tuning fork.

Silver wire: $ab=184$ lines; $bc=184$ lines; a portion of 1300 lines length weighed 4.4815 g; $P=1000$ g; $Q=5000$ g; at 3000 g tension, the tone of the silver wire was a bit lower than the tone F of a tuning fork.

Platinum wire: $ab=113$ lines; $bc=113$ lines; a portion of 800 lines length weighed 6.253 g; $P=800$ g; $Q=4000$ g; at 2400 g tension, the tone of the platinum wire was a bit lower than the tone F of a tuning fork.

On the initial opening of the clamp, there was a sudden increase in the tensile force in $a-b$ and an equivalent sudden decrease in the wire $b-c$ in the amount of $\frac{1}{2}(Q-P)$. The increased tension cooled the wire $a-b$, which subsequently heated to equilibrium, producing a decrease in the tension after the clamp was closed. The opposite effect of heating and subsequent cooling to equilibrium, and hence an increase in tension, occurred in the wire $b-c$. On the second opening and closing of the clamps WEBER could determine the amount of change in tension in either situation after the wires had reached equilibrium with the surroundings. Thus we see in the data of Table 7 that the number of beats decreased in $a-b$ as the temperature decreased, and increased by almost the same amount in $b-c$ as the temperature increased. It should be noted that WEBER always chose a tone of the tuning fork to be slightly higher than that of the wire under the mean load.

By separate experiments WEBER established for a fixed end wire the relation between tension and temperature. Thus, from those experimental data he was able to determine the temperature changes which had occurred in the four wires. He obtained a value of 1.092° for iron, 0.883° for copper, 0.960° for silver, and 2.073° for platinum for the experimental data shown in Table 7. Since the decrease in volume was but a small fraction of the change in strain produced by $\frac{1}{2}(P+Q)$, the strain itself being an extremely small quantity, WEBER used these data to determine the specific heat at constant volume which he compared with the specific heats at constant pressure in the experiments of PIERRE LOUIS DULONG.[1] These comparisons, shown in Table 8, demonstrated that the ratio of specific heats for these four metals was near unity.[2]

Table 8.

	Specific heats		
	Constant pressure DULONG (1819)	Constant volume WEBER (1830)	Ratio C_p/C_v
Iron	0.1100	0.1026	1.072
Copper	0.0949	0.0872	1.088
Silver	0.0557	0.0525	1.061
Platinum	0.0314	0.0259	1.212

WEBER's other series of experiments of interest here was performed in 1835; it contained[3] a second major discovery regarding nonlinear elasticity in the small deformation of silk threads.[4] In that paper in 1835 WEBER does not seem to be aware of the fact that in 1781 COULOMB had performed an important series of experiments on the torsion of silk threads and human hair. [The memoir of

[1] DULONG [1826, *1*].

[2] I am indebted to Dr. JOHN S. THOMSEN of The Johns Hopkins University for examining in detail the assumptions and arguments WEBER introduced into his analysis of these measurements.

WEBER's primitive thermodynamics emphasizes the necessity of viewing his theoretical results with caution but does not detract from the elegance, importance, and originality of the experiment in concept and performance. With these measurements, however interpreted, WEBER discovered that thermal influences in the small deformation of crystalline solids indeed were present.

[3] WEBER [1835, *1*].

[4] They were, in fact, cocoon threads.

Coulomb will be described in Sect. 3.3.] After Weber, experimentists who considered the small deformation of silk (among other organic solids) generally found marked nonlinearity in the stress-strain function. Although Weber emphasized an observed deviation from Hooke's law, he was not specifically interested in determining the stress-strain relation. Perhaps this is understandable in view of the fact that while performing the tests he had made the major discovery of what he termed the "elastic after-effect" (elastische Nachwirkung).

Unable to specify a diameter for his threads, Weber resorted to Thomas Young's[1] "height of the modulus" of elasticity,[2] which differed from the standard modulus by the factor of the weight density (see Sect. 3.7). Weber found that the length of a silk thread whose weight would be neccessary to double the original length (i.e., the height of the modulus), was 864400 m, a number which he used in estimating the elongations associated with applied loads in his studies of the elastic after-effect. This elastic after-effect following the initial application of a load and the instantaneous strain it produces, is the subsequent slow increase in the deformation to an asymptotic value. Upon removal of the load, a similar but opposite behavior occurs; with sufficient time, this strain decreases to the value before the load cycle.

In silk Weber determined that these two opposite after-effects were the same for equal stress differences. He took great care to emphasize that this observed phenomenon was different from that of Hodgkinson's or Gerstner's permanent deformation by subjecting his strings to a large deformation three separate times before testing. He noted that no further permanent deformation occurred in the last loading. He then performed his elastic after-effect experiments at a much lower level of stress. Gauss provided the following simple empirical law to describe Weber's observations:

$$l = 3900 + \frac{23.7}{7.4 + T} \tag{2.11}$$

where l represents length including the after-strain (3900 is the assumed equilibrium state), and T represents the time. Weber's comparison of experimental results and calculations are shown in Table 9.

Table 9. Wilhelm Weber (1835).

No.	Time	Stress (g)	Measured length (mm)	Calculated length (mm)	Difference
1	0.0	9.341	3921.90		
2	2.1	4.215	3902.55	3902.50	−0.05
3	3.6		3902.08	3902.15	+0.07
4	4.6		3901.84	3901.90	+0.14
5	18.5		3901.61	3901.49	−0.12
6	11.0		3901.38	3901.29	−0.09
7	12.7		3901.23	3901.18	−0.05
8	26.2		3900.99	3901.00	+0.01
9	25.7		3900.75	3900.72	−0.03
10	36.0		3900.51	3900.55	+0.04
11	68.0		3900.14	3900.31	+0.17
12	250.0		3900.14	3900.09	−0.05

[1] Young [1807, *1*].

[2] This was a somewhat questionable procedure in view of the known nonlinearity.

This elastic after-effect and thermal after-effect of WEBER, the permanent deformation measurements of GERSTNER and HODGKINSON, the creep observations of VICAT, and the stability phenomenon of MASSON, became the major topics of discussion in this field for over 70 years. These phenomena were used to support numerous arguments both for and against the correlation, or lack of it, of ensuing experiments with one theory or another. RUDOLPH JULIUS EMMANUEL CLAUSIUS[1] in 1849, for example, attacked WEBER's hypothesis that the thermal after-effect was produced by a temperature change, and at the same time implied that the elastic after-effect in silk also was present to a similar degree in metals. Both these erroneous arguments were introduced to attempt to prove that on the basis of a reinterpretation of WEBER's experiments one could set aside GUILLAUME WERTHEIM's experimental demonstration that POISSON's ratio in many solids differed from $^1/_4$. WERTHEIM's discovery had disallowed the then attractive atomistic theories of elasticity of SIMÉON DENIS POISSON and LOUIS AUGUSTIN CAUCHY, which CLAUSIUS implicitly accepted.[2] By the 1890's, precise experiments of JOSEPH OSGOOD THOMPSON[3] and others in very long specimens had separated the various after-effects in metal wires.

In 1841, six years after finding the elastic after-effect, WEBER[4] discovered that the phenomenon had a perplexing aspect. He established the apparently paradoxical fact that a specimen could be made to shorten after an increase in the tensile load and could be made to increase in length after a decrease in the tensile load. WEBER accomplished this in the manner shown in Fig. 2.18 which not only is self-explanatory but also provides a further example of the beautiful simplicity of his experimental conception.

WILHELM WEBER (1841)

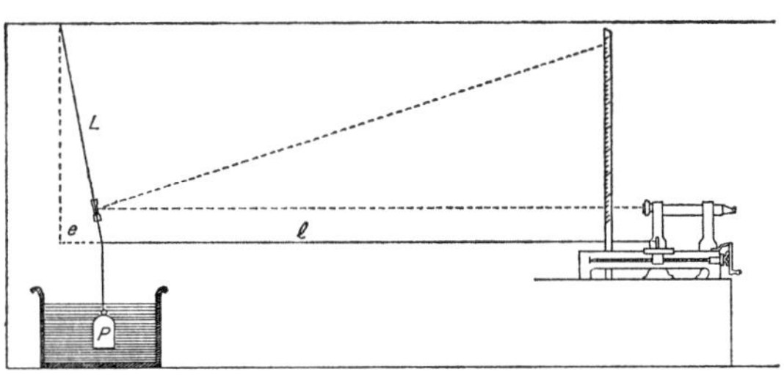

Fig. 2.18. WILHELM WEBER's apparatus for studying elastic after-effect when specimen shortened under increasing tensile load, and lengthened under a decreasing tensile load.

A mirror was supported by two wires, L, at an angle in a bifilar suspension; the threads were spaced 15 cm apart at the ceiling and lay in the plane of the mirror. The horizontal untwisted silk specimen was somewhat over 5 m in length. His calculated constants for the apparatus were given as Eq. (2.12) for the elongation, e, and Eq. (2.13) for the horizontal component T. The applied weight of 1782 g was suspended in water to limit the effects of vibration.

$$e = 0.2105\, n, \tag{2.12}$$

$$T = 0.179\, n \tag{2.13}$$

[1] CLAUSIUS [1849, *1*].

[2] See the reference to WERTHEIM in Sect. 2.14 immediately below, and Sects. 3.16 and 3.19 for a detailed discussion of the controversy.

[3] J. O. THOMPSON [1891, *1*].

[4] WILHELM WEBER [1841, *1*].

where n is the number of partitions on the vertical scale viewed in reflection from the mirror through the telescope at the micrometer screw.

One day before the first experiment Weber tightened the silk specimen. Just before the experiment began he determined the elongation $e = 117.06$ mm and the sum of the elongation and the unloaded length of the specimen, $e+l = 5457.88$ mm. Next, he suddenly turned the screw to produce a release of 201.15 mm. After the release, $e+l = 5256.73$ mm. The behavior which followed is recorded in Table 10. A gradual *increase* in the load from 13.05 to 29.76 g was accompanied by a gradual *shortening* of the specimen from 5241.39 mm to 5221.76 mm "in plain opposition to the above law, according to which a lengthening was to be expected."[1]

Table 10. Wilhelm Weber (1841).

Time (min)	Displacement n	Time (min)	Displacement n	Time (min)	Displacement n
0.00	72.93	11.47	117.58	57.47	138.41
0.47	78.88	11.80	117.87	60.47	138.95
0.80	83.84	12.14	118.37	63.47	139.55
1.14	86.91	12.47	118.47	66.47	139.89
1.47	89.59	13.47	119.81	69.47	140.38
1.80	91.98	14.47	120.75	72.47	140.88
2.14	94.46	15.47	121.74	81.47	142.22
2.47	95.75	16.47	122.49	90.47	143.46
2.80	97.43	17.47	123.28	99.47	144.51
3.14	99.22	18.47	124.12	108.47	145.40
3.47	100.01	19.47	124.77	117.47	146.20
3.80	101.70	20.47	125.46	126.47	146.79
4.14	102.79	21.47	126.06	135.47	147.63
4.47	104.08	22.47	126.60	144.47	148.08
4.80	104.97	23.47	127.20	153.47	148.62
5.14	105.76	24.47	127.70	162.47	149.06
5.47	106.65	25.47	128.35	171.44	149.61
5.80	107.65	26.47	128.84	180.47	150.00
6.14	108.54	27.47	129.24	207.47	151.50
6.47	109.24	28.47	129.79	234.47	152.84
6.80	110.03	29.47	130.28	261.47	153.94
7.14	110.63	30.47	130.68	288.47	154.69
7.47	111.52	31.47	131.02	315.47	155.38
7.80	112.02	32.47	131.52	342.47	156.18
8.14	112.61	33.47	131.71	369.47	157.37
8.47	113.31	34.47	132.06	396.47	158.01
8.80	113.80	35.47	132.50	423.47	158.61
9.14	114.20	36.47	132.84	450.47	158.85
9.47	114.69	39.47	133.79	477.47	159.00
9.80	115.29	42.47	134.78	504.47	159.20
10.14	115.69	45.47	135.68	585.47	160.09
10.47	116.19	48.47	136.57	666.47	161.04
10.80	116.58	51.47	137.22	747.47	162.19
11.14	116.78	54.47	137.76	1233.47	166.08

For the second experiment the thread which had been released for some time was tightened by a 116.27 mm displacement of the screw. Before tightening, he measured the deflection $e = 35.72$ mm and original length l summed to $e+l = 5256.73$ mm. Following tightening, $e+l = 5373.00$ mm. During the subsequent interval a gradual *decrease* of the load from 84.37 to 71.29 g was accompanied by a gradual *elongation* from 5274.06 to 5289.41 mm, "likewise in plain opposition

[1] Weber [1841, *1*], p. 1.

to the above law according to which a shortening was to be expected"[1] (see Table 11).

Table 11. WILHELM WEBER (1841).

Time (min)	Displacement n	Time (min)	Displacement n	Time (min)	Displacement n
0.00	469.80	8.78	444.10	25.78	432.89
0.45	465.52	9.12	443.81	26.78	432.35
0.78	463.34	9.45	443.31	27.78	431.95
1.12	461.56	9.78	443.01	28.78	431.70
1.45	459.80	10.12	442.62	29.78	431.30
1.78	458.00	10.45	442.32	30.78	431.01
2.12	456.91	10.78	442.02	31.78	430.71
2.45	455.91	11.12	441.82	32.78	430.26
2.78	454.92	11.45	441.53	33.78	429.91
3.12	453.92	11.78	441.23	34.78	429.51
3.45	453.03	12.12	440.93	35.78	429.27
3.78	452.23	12.45	440.63	36.78	429.02
4.12	451.64	12.78	440.23	39.78	428.18
4.45	450.84	13.78	439.43	42.78	427.53
4.78	450.05	14.78	438.69	45.78	426.89
5.12	449.46	15.78	438.10	48.78	426.29
5.45	448.86	16.78	437.45	51.78	425.65
5.78	448.27	17.78	436.76	54.78	425.05
6.12	447.77	18.78	436.21	57.78	424.61
6.45	447.08	19.78	435.72	60.78	424.16
6.78	446.78	20.78	435.02	63.78	423.72
7.12	446.28	21.78	434.73	66.78	423.32
7.45	445.89	22.78	434.28	69.78	422.98
7.78	445.39	23.78	433.63	72.78	422.58
8.12	444.99	24.78	433.24	75.78	422.23
8.45	444.50				

WEBER emphasized that the observed apparently paradoxical behavior was reproducible. The experimental curves for the two types of loading could be adjusted to coincide. This may be seen in Fig. 2.19, which is a plot of the data of Tables 10 and 11.

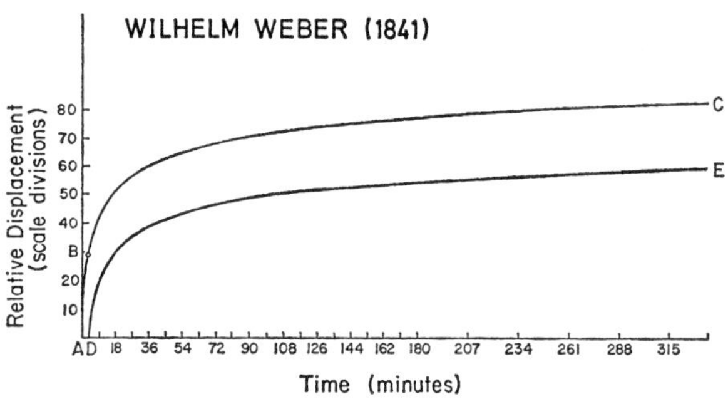

Fig. 2.19. Experimental results showing the shortening of the specimen during increasing load (C), and elongation during decreasing load (E). WILHELM WEBER's apparent paradox in the elastic after-effect.

WEBER stated an observed fact which has been ignored by those who later referred to his work, namely:

[1] *Ibid.*, p. 2.

It may be regarded as a fact that the thread after a sudden change of its tension does not immediately attain a state of complete equilibrium. Even aside from the vibrations it makes afterward, and if one pays heed only to the mean site from either side of which the vibrations deviate equally far, which is commonly considered to be the new site of the equilibrium, the result is that this mean site also, which was always observed in the above experiments, cannot be regarded as a site of complete equilibrium; rather, it changes gradually yet considerably. At best, in relation to those vibrations it may be considered a site of complete equilibrium. In fact the thread approaches the site of complete equilibrium only asymptotically with time, probably without ever reaching it completely.[1]

WEBER observed further that Eq. (2.11), which was the solution of Eq. (2.14) for $M=2$ was inadequate. However, he found that when the value of M was determined from experiment the measured results could be represented adequately:

$$\frac{dx}{dt} = B x^M. \tag{2.14}$$

WEBER considered at length the relation between the instantaneous strain and the subsequent elastic after-effect as a function of the previous loadings, which by their nature and timing influenced strongly both the primary and the secondary deformation. These observations mark the beginning of the study of memory as a parameter in the mechanics of solids. WEBER emphasized that Eq. (2.14) was empirical, having been obtained initially for constant load and then used for a variable load. He justified this on the basis of the slowness of the secondary deformation.

WEBER's study provided dramatic experimental evidence that any general theory of the deformation of solids must include behavior such as he described, particularly since he had shown the elastic after-effect to be reproducible and to be present in the response functions for very dissimilar histories of deformation.

Predicting the controversy which was to follow, WEBER was very much concerned about the possibility that some form of permanent deformation was responsible for his thermal measurements. In addition to subjecting all of his wires to large deformation and demonstrating that no further permanent set occurred under much smaller loads, he also made measurements to settle the question. He left the central clamp open for a number of seconds before closing it, and found no change in the tension at subsequent intervals of time during which permanent deformation would have been expected to occur. Of major importance for later studies, including many being performed today on automatic testing machines, is the fact that by these means WEBER was able to establish that the length of time for iron, copper, silver, and platinum to come to thermal equilibrium was of the order of 6 sec. Measurements from far more accurate experiments performed 60 years later on very long wires[2] were found to be in close agreement with those of WEBER.

2.13. The large deformation of gut strings: Karmarsch (1841).

With regard to the studies of the deformation of silk thread by WEBER in 1835 and 1841, it is interesting that the experiments of JAMES BERNOULLI in the 17th century, which had provided the first experimental evidence that the response function of a solid could be nonlinear, also had been performed on an organic substance, namely, gut.[3] In his article in the fourth edition of the *Encyclopaedia*

[1] WILHELM WEBER [1841, *1*], p. 9.

[2] J. O. THOMPSON [1891, *1*].

[3] "Darmseiten" usually is translated into English as "cat gut",* giving rise to quips such as that of HUGH REGINALD HAWEIS [1898, *1*],** who said:

"'To scrape the inside of a cat with the outside of a horse' is far from an accurate or exhaustive description of violin playing, nor can I understand why violin strings are called

Britannica, written at the end of the 18th century, JOHN ROBISON with an implicit belief in HOOKE's law, had dismissed the nonlinear results BERNOULLI had obtained from his own experiments on gut, on the basis that the latter's specimens had been composed of twisted strands and therefore the observed nonlinearity could be ignored. To dismiss the study of aggregates because they are aggregates, would provide a severe restriction indeed on experimental mechanics.[1] An example of ROBISON's carefree attitude toward science is his claim, in the same article, that he had pulled a rubber band to ten times its original length, attaining a perfectly linear relation between load and elongation. Although he

---

> cat-gut at all, since they are made from the intestines of the sheep, goat, or lamb ..." (p. 153).

It might be added that pig-gut also has been used. Somewhat later in his chapter on violin strings, HAWEIS, after referring to the fact that there was no reason to suppose that any advance in the manufacture of gut strings had been made since the 17th century and that works on the subject as early as 1570 and 1647 could be cited on this matter, proceeded to describe the manner of preparing specimens. (One might note that the experimentist studying metals would be castigated for omitting reference to such prior procedures.)

> "Putting aside mature sheep and goats, we kill our young Italian lamb in September. We open him at once, and take the intestine whilst still warm; stretch it on an inclined plane; scrape it and clean it thoroughly without delay. We then steep it for about fifteen hours in cold water, with a little carbonate of soda, and then substitute tepid water for a few hours more.
>
> "Now we are ready to remove the fibrous or muscular membrane from between the peritoneal and mucous membrane. This is done by women, who scrape it with a cane. The precious selected membranes are then rubbed through the fingers three times a day, treated with permanganate of potash, cleaned, sorted, cut, and split; and, finally, the threads are spun—three or four thin threads for the first violin strings, three or four thicknesses for the second, six or seven for the 'D' string. Double-bass strings take up to eighty-five threads. Further twistings, soakings, and polishings take place, into which we need not enter, and the strings are finally dressed with olive oil and then coiled" (pp. 155 to 156).

* See for example the English translation of the title of the paper by KARMARSCH [1841, *1*], discussed immediately below, in the *Royal Society Catalogue of Scientific Papers* for the 19th century [1909, *1*], p. 317.

** This date of publication is assumed from the context. See note in the references of the present treatise.

[1] In ROBISON's article entitled "Strength of Materials" reprinted in *A System of Mechanical Philosophy*, Vol. 1, pp. 369–495 [1822, *1*], we find:

> "In the same dissertation, Hooke mentions all the facts which John Bernoulli afterwards adduced in support of Leibnitz's whimsical doctrine of the force of bodies in motion, or the doctrine of the *vires vivae*; a doctrine which Hooke might justly have claimed as his own, had he not seen its futility.
>
> "Experiments made since the time of Hooke show that this law is strictly true in the extent to which we have limited it, viz. in all the changes of form which will be completely undone by the elasticity of the body. It is nearly true to a much greater extent. James Bernoulli, in his dissertation on the elastic curve, relates some experiments of his own, which seem to deviate considerably from it; but on close examination they do not" (p. 384).

Then, going back to the same subject fifty pages later where he recorded BERNOULLI's data, ROBISON inadvertently demonstrated the absurdity of his previous statement:

> "James Bernoulli, in his second dissertation in the elastic curve, calls in question this law, and accommodates his investigation to any hypothesis concerning the relation of the forces and extensions. He relates some experiments of lute strings where the relation was considerably different. Strings of three feet long,

Stretched by	2,	4,	6,	8,	10 pds.
Were lengthened	9,	17,	23,	27,	30 lines.

> But this is a most exceptional form of the experiment. The strings were twisted, and the mechanism of the extensions is here exceedingly complicated, combined with compressions and with transverse twists, &c" (p. 434).

had provided no experimental details, this statement by ROBISON was to be widely quoted during the first part of the 19th century, until definitive experiments on rubber revealed marked nonlinearity at a mere fraction of this extension. BERNOULLI had not revealed how he had constructed his specimens of gut, i.e., we do not know whether or not he had studied a single strand, but we may assume that he had used the strings of musical instruments, which usually are composed of many strands.

In 1841 KARL KARMARSCH, completely ignorant of the fact that JAMES BERNOULLI had considered the deformation of gut 154 years before, or that the tests had been repeated and extended by JAMES RICCATI in 1721, 120 years before his own, began his paper with the statement: "As far as I know, no studies on this subject have been published."[1] KARMARSCH's experiments were on "good" Italian gut strings, which included that from the contrabass, the D-string of the violoncello, the D and A strings of the violin, the E-string of the guitar, and the fifth string of the harp. The measured diameters, number of strands, and the twists per Hannoverian in. for each of these strings are shown in Table 12.

Table 12. KARMARSCH (1841).

No.	Designation of the strings	Diameter, Hannoverian in.	No. of strands of which string is composed	Twists per Hannoverian in.
1	Contrebass	0.166	48	1–3/5
2	Contrebass	0.146	45	1–2/3
3	Cello-D	0.089	24	2–1/3
4	Cello-D	0.070	12	4–2/3
5	Violin-D	0.048	9	5–2/3
6	Violin-A	0.040	4	6
7	Guitar-e	0.027	3	7–1/2
8	Harp-quinte	0.029	4	6

KARMARSCH showed that the absolute tensile strength of these strings was roughly in accord with the variations of diameter from the largest number of strands, i.e., 48, to the smallest, which was 3. An order thus was exhibited in the deformation of twisted, many-stranded gut strings, which, of course disposes of ROBISON's objections to JAMES BERNOULLI's data and inferences.

Two additional facts are noteworthy from this otherwise scarcely important experimental series. First, the stress-strain function was nonlinear, being concave to the loading axis. If KARMARSCH's data were expressed in arbitrary units, as were JAMES BERNOULLI's data of Fig. 2.1, a close correlation between the two sets of data would be seen. Second, KARMARSCH observed no permanent deformation until he reached strains of the order of 9 to 10%, which corresponded to a load of approximately 70% of the load at rupture.

To remove the change of curvature, KARMARSCH first placed a small weight on his strings; he then let the strings rest for an unspecified period of time before he performed the test. He applied the load by carefully placing weights on a scale, using an iron lever to increase the load for the thicker strings. Although he did not specifically mention the WEBER elastic after-effect, the fact that KARMARSCH specified that he always waited approximately five minutes after placing the load before he made the measurement of elongation indicates that he recognized that the phenomenon existed.

[1] KARMARSCH [1841, *1*], p. 427.

His measurements of elongation were made between two marks in the central region of the gut string, with a gage length that differed from one test to another, varying from 4 to 6 Hannoverian in. One of the problems for the present-day experimentist who wishes to study the details of early 19th century experiments is that of traversing the morass of purely local units. KARMARSCH's experiments were plotted as Cologne lbs. vs Hannoverian in.[1] Since KARMARSCH used a measuring rod and a series of specific weights, it seems obvious that the weights must have been purchased in one place and the scale in another. In Fig. 2.20 is shown the load-strain function for the contrabass gut string for a gage length of 6 Hannoverian in. It is possible to convert the tabulated data to stress in kg/mm² and to strain, since KARMARSCH had provided all of the dimensions for such a calculation.

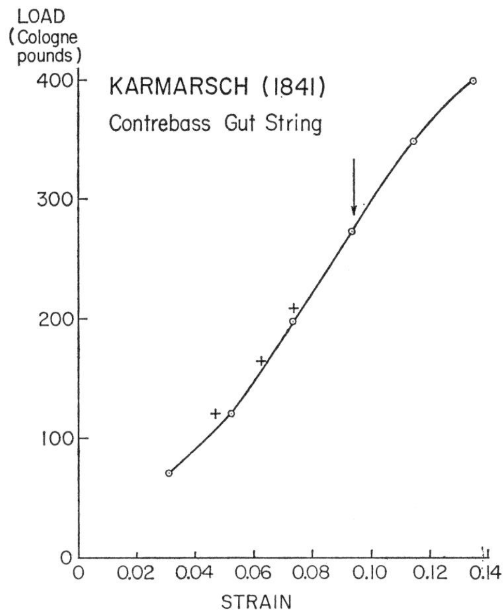

Fig. 2.20. The tensile loading of gut (two tests) showing the point of inflection at the onset of permanent deformation.

JAMES RICCATI[2] had noted in 1721 that at higher loads, changes in curvature could occur in his data. A similar inflection point is seen in KARMARSCH's data of Fig. 2.20 at a strain of 9.3%. There was no permanent deformation until this inflection point, at which he observed a permanent deformation of approximately 1% on unloading. Curiously enough, there was no further increase in the permanent deformation as the load increased. In some of his smaller strings, KARMARSCH had found no permanent deformation and no inflection point before rupture. In the violoncello string, which contained half as many strands as that of the contrabass, he observed no permanent deformation until 10% strain, just before rupture when he noted the permanent strain of 1.8%. Because the point of inflection was so close to the point of rupture he could not make measurements

[1] 1 Cologne lb. = 467.8 g; 1 Hannoverian in. = 2.43 cm.

[2] RICCATI [1721, *1*]. (See TRUESDELL [1960, *1*].)

between the two. KARMARSCH conjectured that the permanent deformation associated with the point of inflection was due to a rearrangement of the twisted fibers.

The remainder of KARMARSCH's paper[1] was concerned primarily with rupture; he compared the ultimate stresses in gut with those from wire-wound silk, guitar strings, hemp rope, commercial twine, and brass piano strings. The data for two contrabass strings included in Fig. 2.20 indicate that he obtained a modicum of reproducibility.

These data of KARMARSCH provide a nonlinear response function but do not strictly belong in a chapter devoted to small deformation. They are included here because they reveal that the 17th and early 18th century confusion regarding the linear response function of HOOKE and the nonlinear function of BERNOULLI and RICCATI might not have occurred[2] if the distinction between small and large deformation had been appreciated.

2.14. Experiments on the elasticity and cohesion of the principal tissues of the human body: Wertheim (1846–1847).

GUILLAUME WERTHEIM certainly was one of the finest, perhaps the most important 19th century experimentist in the continuum physics of solids. He also was the most controversial figure which this field has produced. His belief that a subject should be examined exhaustively by various experiments from different perspectives in order to be able to scrutinize closely extant theory, placed him and his work squarely in the center of 60 years of intense debate. His measurements, which were admired by all his contemporaries, have withstood the test of time and constitute the majority of values cited at the turn of this century; they still appear, usually unacknowledged, in current handbooks. Today, from the perspective of over 125 years, it is evident that practically every conclusion of the now-forgotten WERTHEIM is accepted as obvious. This is not surprising since he formed his judgments on the basis of observations from his own experiments, often ingenious and always careful, the superiority and integrity of which almost never were questioned by any of the numerous persons who quoted and used his data during the three-quarters of a century in which they were widely discussed to support one or another now also largely forgotten interpretation.

Although WERTHEIM received a medical degree in Vienna in 1839, he gave up medicine. After studying mathematics and physics at the University of Berlin for two years, he went to Paris, and spent the remaining 22 years of his brief life of 46 years in experimental research in solid mechanics. He received a doctor of science degree in physics in 1853. His experimental output during this interval was prodigious. He made the first definitive study of dynamic and quasi-static elastic constants, an original, major comparative study based on extensive measurement in nearly all of the then available metals.[3] His research included the first serious examination of the effects of ambient temperature, magnetic fields, electric currents, and prior permanent deformation upon the elastic coefficients

[1] KARMARSCH [1841, *1*].

[2] TRUESDELL [1960, *1*], p. 62, has shown that CHRISTIAAN HUYGENS in a letter on 5 May 1691 to LEIBNIZ, did emphasize the necessity for making this distinction. HUYGENS' letter was in reply to one from LEIBNIZ dated 20 April 1691 indicating that LEIBNIZ questioned HOOKE's results and asking whether HUYGENS ever had performed any similar experiments.

[3] WERTHEIM [1844, *1*(*a*)]. Included in this publication in 1844 were papers presented in 1842 and 1843. The paper referred to above had been presented on 18 July, 1842.

for isotropic solids in extension, torsion, and flexural modes of deformation.[1] From an experimentally elegant study to be described in Sect. 3.16 to 3.19 below, he made the then very original and upsetting discovery that the philosophically attractive atomistic theories of elasticity based upon POISSON's and CAUCHY's considerations of central forces between adjacent atoms were experimentally unacceptable for isotropic solids. WERTHEIM's discovery in 1848[2] that the CAUCHY relations did not describe the deformation of crystalline solids was still reverberating among some theorists over fifty years later, at the turn of the century, but is questioned by none in the present generation.

This century's recognition that the CAUCHY relations are incompatible with experimental data has come too late to provide recompense for a bitter man[3] who, after a dozen years of being the target of scholarly violence, committed suicide by leaping from the Cathedral at Tours in 1861. Twenty-five to fifty years after WERTHEIM's death, we still find in memoirs and monographs caustic references to his interpretations and praise for his experiments. There is additional irony in the fact that the major portion of the experimental numbers in such publications are drawn from and referenced to WERTHEIM.[4] It became a standard

---

[1] WERTHEIM [1844, *1* (a), (b), (c)].

[2] WERTHEIM [1848, *1*].

[3] Although WERTHEIM is described by his contemporaries as having been decidedly disenchanted with medicine, that was the only advanced field of study in the 1830's open in Austria to an Austrian of Jewish descent. WERTHEIM's father was a banker of great wealth and was described as the leader of the Viennese Hebrew community. WERTHEIM was born in 1815; his father had died already, and his mother died a few days afterward.

Two years in Berlin attending lectures in physics and studying mathematics with JACOBI, STEINER, and DIRICHLET preceded WERTHEIM's years of research in Paris in the laboratories of the École Polytechnique and the Collège de France; his inherited wealth made formal appointment unnecessary as a source of support.

A few years later, WERTHEIM briefly returned to Vienna, having been misinformed with respect to a major University appointment in his native land. He would have had to foreswear his religion in order to receive the appointment; otherwise only a minor position was available for him, by then a famous experimental physicist.

WERTHEIM returned to France, sought and obtained French citizenship (1848), and thereafter gave allegiance to his adopted country. For several years he was an "Examinateur d'Entrée à L'École Polytechnique" and was a Visiting Professor for a short time on the Faculty of Montpellier (1854). As a member of the jury of the Paris Exhibition of 1855, he was granted the Cross of the French Legion of Honor.

WERTHEIM never was elected to the French Academy, but on two occasions, in 1851 and in 1859, he was high on the list of candidates. He was elected Corresponding Member of both the Vienna and Berlin Academies, in 1848 and 1853 respectively.

WERTHEIM, who remained a bachelor, is described in the *Almanach der Kaiserlichen Akademie der Wissenschaften* (probably by the General Secretary, Dr. Anton Schötter) as a person with a "free and gay spirit" who lived a withdrawn life almost exclusively concerned with scientific research. One only can surmise from the increasing bitterness visible in his later papers that what were described as "moods of depression" preceding his death had their origin in the scientific controversy. The *Journal d'Indre-et-Loire* of 22 January, 1861, tells us that on the day of his death WERTHEIM demanded with impatience that he be led to the Tower of the Cathedral of St. Gathien in Tours, a city to which he had travelled on the advice of his doctor. After having climbed in great haste to the platform of the Tower, WERTHEIM, visibly sick and very agitated, leaped to the parapet and plunged down to the square below, before the accompanying guard could stop him.

See: *Almanach der Kaiserlichen Akademie der Wissenschaften* (Wertheim Eulogy), pp. 176–188 (1861 and 1862); J. C. POGGENDORFF: *Biographisch-Literarisches Handwörterbuch zur Geschichte der Exacten Wissenschaften*, vol. 2, pp. 1302–1303. Leipzig (1863); O. TERQUEM: "Wertheim (Guillaume)." *Archives Israelites 22*, 142–147 (1861); MARCEL EMILE VERDET: Notices sur les Travaux Scientifiques de M. Guillaume Wertheim, Membre de la Société Philomathique de Paris, Redigée sur la demande de la Société. *L'Institut*, 29, Nos. 1432 and 1433, and 1434. Footnotes pp. 198–201, 205–209, and 213–216 respectively (1861).

[4] See, for example, Lord KELVIN [1880, *1*] and also TODHUNTER and PEARSON [1886, *1*], [1893, *1*], [1960, *1*].

practice when describing the excellence of his experiments to refer regretfully to his mathematical limitations, as if the fact that some of his data indicated the physical inapplicability of the then popular theories could be explained away on the basis of his presumed or actual limited understanding of those theories. WERTHEIM himself did not always exhibit intuitive insight regarding theoretical elasticity, as may be seen in such instances as that in 1848 when he used infinitesimal linear elasticity in calculations which he compared with data for the very large deformation of rubber in compression. This error, incidentally, was one which none of his critics observed.[1] His real problem was that in the mid-19th century he was asking the right experimental questions in the context of the wrong theories.

But WERTHEIM's experimental achievements were many. WERTHEIM must be credited with having made the most important early experimental contribution to photo-elasticity;[2] the first large study of the torsion of prisms with a wide variety of cross-sections;[3] the first definitive study of the compressive effects accompanying the axial deformation of hollow prisms;[4] the first study of one-dimensional tensile wave propagation in wire over 3 km long;[5] and, of particular interest today, he must be classified as an early biomedical physicist since in 1846 he studied the elasticity of bones, tissues, muscles, arteries, and nerves of the human body as a function of sex, age, and the effects of the loss of moisture which occurred after a lapse of time following death.[6] For what we may anachronistically call his "biomedical engineering" study, WERTHEIM dissected both female and male cadavers, eight in all, and he found every one of his specimens to be governed by a nonlinear stress-strain function. The degree of nonlinearity decreased with the time during which the specimens dried, a new and unusual constitutive variable.

In 1846 when the discovery of micro-permanent deformation at small strain, i.e. the "elastic defect," the elastic after-effect, the thermal after-effect, the stability or instability associated with the SAVART-MASSON effect, and VICAT's creep, were important and exciting new influences on the thought of the period and thus were dominating discussion and interpretation, it was difficult to decide which measured strain to record.[7]

WERTHEIM, in his studies both of organic tissues and of metals, carefully stated his procedure; he noted that he had loaded the specimen to a specified value, had immediately unloaded it, and then had recorded the strain or elongation at both extremes. WERTHEIM recorded the unloaded elongation within minutes after the removal of the load. The data provided a division of the strain into permanent deformation[8] and the immediately recoverable elastic deforma-

[1] Some months after writing the above comment I found that WERTHEIM's error did not escape the attentive eyes of JAMES CLERK MAXWELL.

[2] WERTHEIM [1854, *1*]. The linear stress-optics law in photoelasticity is referred to as "WERTHEIM's law."

[3] WERTHEIM [1857, *1*, *2*]. Earlier torsional studies by DULEAU [1813, *1*], [1820, *1*] and SAVART [1830, *1*] were important preliminaries of this far more definitive study of WERTHEIM.

[4] WERTHEIM [1848, *1*].

[5] WERTHEIM [1851, *1*].

[6] WERTHEIM [1847, *1*].

[7] Again we must recall that this was the day of dead weight loading, in which many of the phenomena associated with strain were revealed and easily studied, contrary to the current practice of masking such behavior on standardized hard testing machines which prescribe that the strain be an input variable.

[8] In the present connection it should be remarked that although WERTHEIM's main interest was in linear phenomena, he noted that GERSTNER's nonlinear law applied to all the metals for which he had observed permanent deformation. Like HODGKINSON earlier, and BAUSCHINGER later, WERTHEIM stated his belief that improvement of accuracy in measurement probably would reveal that some permanent deformation always accompanied any deformation, however small; i.e., microplasticity was suggested in 1831, 1844, and 1879.

tion. He repeated this process with increasing loads, usually until rupture occurred. Contrary to the statements of his critics, WERTHEIM was aware of and specifically considered the problems associated with WEBER's after-strain in organic solids. He made the point that since the strain could vary with time, it was important to state precisely and unequivocally what was done.

Table 13. WERTHEIM (1847).

Substance	Sex	Age (years)	Specific weight	*A*	*B*	Elasticity coefficient	Cohesion
Bones:							
Femur	F	21	1.968	0.4585		2181	6.87
Fibula	F	21	1.940	0.3690		2710	10.26
Femur	M	30	1.984	0.5498		1819	10.50
Fibula	M	30	1.997	0.4857		2059	15.03
Femur	F	60	1.849	0.4130		2421	6.40
Fibula	F	60	1.799	—		—	3.30
Femur	M	74	1.987	0.3791		2638	7.30
Fibula	M	74	1.947	—		—	4.335
Tendons:							
Small plantar	F	21	1.115	48.21	80.86	164.71	10.38
Small plantar	M	35	1.125	51.04	55.85	139.42	4.91
Big toe flexor	M	35	1.132	60.58	9.91	128.39	—
Big toe flexor after slight air-drying	M	35	1.132	29.72	5.36	183.44	—
Big toe flexor completely dried	M	35	1.132	28.64	0.867	186.85	4.11
Small plantar	M	40	1.124	54.69	48.22	134.78	7.10
Small plantar	F	70	1.114	34.53	67.20	169.21	5.61
Small plantar	M	74	1.105	24.35	105.38	200.50	5.39
Muscles:							
Sartorial	M	1	1.071	607700	13832	1.271	0.070
Sartorial	F	21	1.049	1351875	8219	0.857	0.040
Sartorial	M	30	1.058	7960000	38860	0.352	0.026
Sartorial	F	60	1.040	—	—	—	—
Sartorial	M	74	1.045	14549333	23863	0.261	0.017
Nerves:							
Internal popliteal	F	21	1.038	—	—	—	0.769
Sciatic	F	21	1.030	9890.0	36.56	10.053	0.900
Sciatic	M	35	1.071	1720.4	573.00	23.943	0.963
Posterior tibial	M	35	1.040	—	—	—	1.959
Posterior tibial	M	40	1.041	1426.2	149.28	26.427	1.300
Sciatic	F	60	1.028	5417.5	755.4	13.517	0.800
Cutaneous fibular	F	70	1.052	1708.8	1078.1	23.878	3.530
Sciatic	M	74	1.014	5032.0	936.8	14.004	0.590
Posterior tibial	M	74	1.041	905.0	960.2	32.417	—
External saphena	M	74	1.050	—	—	—	—
External saphena, dried	M	74	1.129	36.79	49.18	164.198	9.46
Arteries:							
Femoral	F	21	1.056	—	—	—	0.1403
Femoral	M	30	1.014	257747000	5784200	0.052	0.1660
Femoral	F	70	1.085	—	—	—	0.1070
Veins:							
Femoral	F	21	1.055	1174780	193970	0.844	0.0969
Internal saphena	F	21	1.048	—	—	—	0.3108
Femoral	F	70	1.019	1091550	169699	0.883	0.1490

Wertheim observed that human bone stress-strain functions departed only slightly from linearity, with a tendency for the tangent modulus to increase with stress, but that all of the other organic human specimens obeyed the following nonlinear stress-strain rule:

$$\varepsilon^2 = A\sigma^2 + B\sigma \tag{2.15}$$

where ε was the longitudinal strain, σ the corresponding stress in kg/mm², and A and B experimental constants found to be positive in all instances. In Table 13 are the experimental values of A and B for the designated bone or tissue. Also included are Wertheim's measurements of the specific weight and an elastic modulus in the sense of Young's height of the modulus[1] or the calculated load necessary to double the length, and finally, the rupture strength.

Wertheim had considerable difficulty with gripping whole bones in order to perform tension tests. Therefore, those measurements were made from small uniform samples. All other experiments were performed on the longest length of

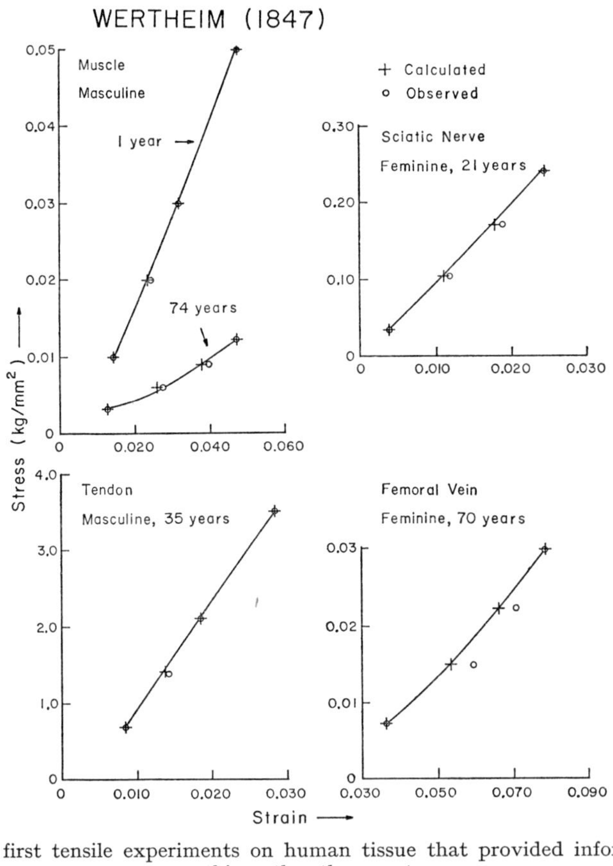

Fig. 2.21. The first tensile experiments on human tissue that provided information about something other than rupture.

[1] Wertheim [1847, *1*]. This was another somewhat questionable piece of analysis on the part of Wertheim, although he was not alone in this, since a similar proposal had been made earlier by Weber for silk threads. Wertheim obtained the values of the elastic coefficients listed by differentiating the nonlinear Eq. (2.15) with respect to the strain, to determine the value at a strain corresponding to double the length. See Chap. III, Sect. 3.7 for the discussion of Young's "height of the modulus."

tissue which he was able to extract. Instead of resorting to the method of least squares in order to determine the coefficients A and B, an omission for which he apologized, WERTHEIM, except in certain instances for very large deformation, determined his values from the smallest and largest deformation, and then plotted his data to see how they conformed in the central region of his nonlinear curve. Fig. 2.21 shows diagrammatically the comparisons of the tabulated experimental data with the nonlinear stress-strain function of Eq. (2.15) determined in that questionable mode of calculation.[1] From these data WERTHEIM was able to draw a number of conclusions besides the general agreement of a single nonlinear law with the data.

Since WERTHEIM's were the first modern measurements of the elasticity of specimens from the human body, it is interesting to mention some details. He noted that the elasticity of bones, tendons, and nerves seemed to increase[2] with

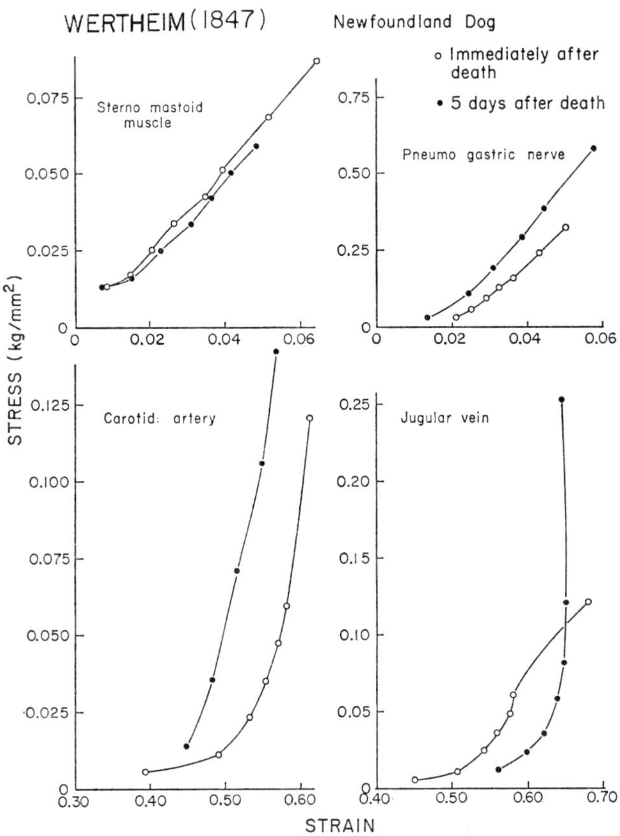

Fig. 2.22. Tensile tests on the tissues of a dog, providing the first study of the effects of *rigor mortis*.

[1] These calculations and the comparisons with experiment are complicated further by the fact that the strain at each point is referred to the current length including the small but measurable permanent deformation determined after the unloading which followed every increase in load. However, the tabulated data in WERTHEIM's paper [1847, *1*] are given in complete detail.

[2] A biochemist associate, FRED HENEMAN, has suggested that since an increase in elasticity to a physiologist means more flexibility of the tissue, it should be pointed out that the opposite, i.e. increase in stiffness, is understood by the physicist.

age whereas that of muscles decreased considerably. The cohesion or load at rupture decreased with age. If the various tissues were arranged according to the magnitude of their elasticity, or their cohesion, the following order was obtained: bones, tendons, nerves, muscles, veins, and arteries. The elasticity and cohesion generally increased with drying.

To study this last phenomenon, WERTHEIM compared data determined from specimens extracted from the right side of a large Newfoundland dog. The specimens consisted of a muscle, a tendon, a nerve, a vein, and an artery. Five days later, he studied similar specimens extracted from the left side. Comparison of the data for the sterno mastoid muscle, pneumo gastric nerve, carotid artery, and jugular vein shown in Fig. 2.22 indicates that while for the muscle tissue, the deformation was altered only slightly by the five day period of drying, for the other tissues there was considerable shift in one strain direction or the other, with some change in shape.

Thus WERTHEIM's conclusion that his post-mortem studies on human tissue could be extrapolated to live tissue was valid only for some types of organic solids. His deep concern with this aspect of the problem perhaps was as important as his unprecedented studies of the human tissue itself. WERTHEIM's elastic coefficients for these experiments, calculated in the odd manner described above, are tabulated in Table 14.

Table 14. WERTHEIM (1847).

Examined part	Newfoundland dog					
	Immediately after death			Five days after death		
	Density	Elasticity coefficient	Cohesion	Density	Elasticity coefficient	Cohesion
Sterno-mastoid muscle	1.060	1.425	0.124	1.059	1.234	0.086
External tibial tendon	1.136	—	5.061	1.132	166.969	6.001
Pneumo-gastric nerve	1.016	17.768	0.732	1.024	26.453	1.461
Primary carotid artery	1.077	—	0.364	1.039	—	0.512
External jugular vein	1.045	—	0.363	1.042	—	0.505

2.15. Further experiments on the elasticity of organic tissue: The comparison of response functions for live and dead specimens. Wundt (1858), Volkmann (1859).

Following WERTHEIM's pioneering study in 1846, a number of physiologists became interested in tissue elasticity, including WILHELM MAX WUNDT,[1] who was critical of WERTHEIM's choice of strain, implying that WERTHEIM had ignored the earlier experiments of WEBER in silk, and that if WERTHEIM's strains had been properly chosen, either immediately after loading or after a long lapse of time,[2] he probably would have found linearity. WUNDT's own data were so

[1] WUNDT [1858, *1*]. He was Privatdozent in physiology at Heidelberg at the time.
[2] This was a measurement which WUNDT conceded was not possible in studies of tissue.

lacking in precision, as ALFRED WILHELM VOLKMANN pointed out[1] in 1859, that experimental points fell on either side of linearity in such a manner that no real conclusion could be drawn.

VOLKMANN himself carefully approached the problem of how strains should be measured by attaching a pen with a hair point to the moving end of the specimen to record the strain on a rotating smoked-glass cylinder. Thus, one could select for study any strain whatever: either the very first strain as proposed by WUNDT or the strain after very long times. An example of such a measurement by VOLKMANN is shown in Fig. 2.23.

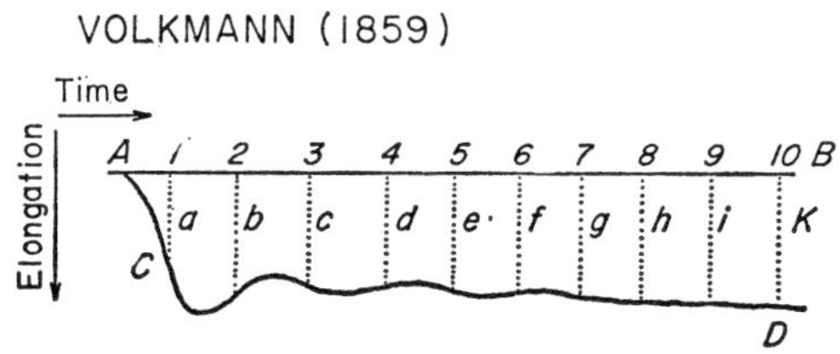

Fig. 2.23. The first automatic recording of elongation vs time results for the study of elastic after-effect in organic solids.

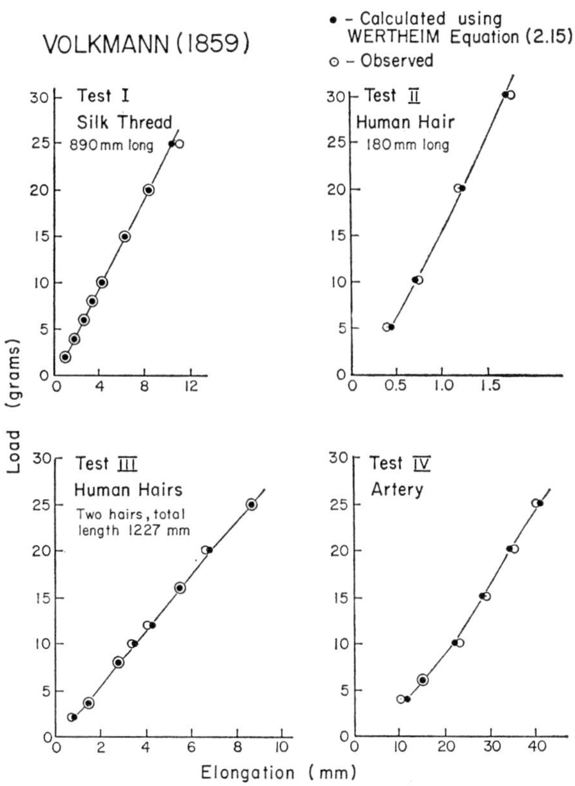

Fig. 2.24. Tensile results compared with prediction from WERTHEIM's equation.

[1] VOLKMANN [1859, *1*].

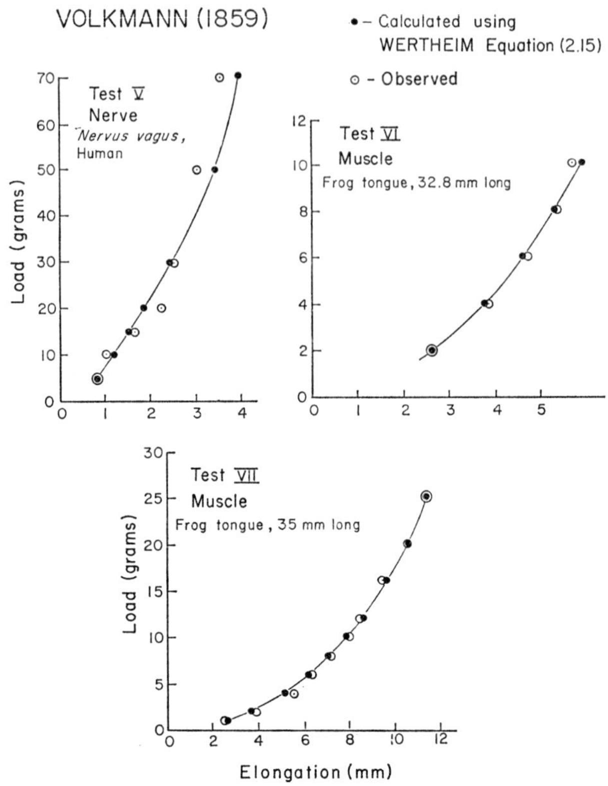

Fig. 2.25. Tensile results compared with prediction from WERTHEIM's equation.

So as to avoid impact on the specimen when the weights were being attached, VOLKMANN placed weights on the specimen at zero load, on a platform which subsequently fell away, producing the elongation diagram shown. His data were from experiments on a silk thread 890 mm long, a woman's hair 180 mm long, two human hairs bound together 1.227 m long, the head artery of a dog 72 mm long, a human vagus nerve 136 mm long, and three experiments on the tongue muscle of a frog, the muscles being from 32.8 mm long to 35 mm long. The data are shown in Figs. 2.24 and 2.25, and they are compared with WERTHEIM's nonlinear Eq. (2.15), the constants A and B having been determined by VOLKMANN from the total data by a study of the differences of all of the experimental points.

VOLKMANN's paper[1] concluded with a long discussion of WUNDT's errors in conception and experiment, apropos of the very small permanent elongations which had been observed by VOLKMANN for large times in his experiment on the tongue muscle of a frog. These data of VOLKMANN firmly establish the nonlinearity of such solids and indicate that an empirical formula such as WERTHEIM's Eq. (2.15) indeed is applicable. During the next half century this equation was generally assumed to describe adequately the deformation of organic substances.[2]

[1] *Ibid.*

[2] An exception was rubber, which by the 1880's had been shown to have a somewhat different form of stress-strain function.

After carefully reading WUNDT's monograph of 1858, *Die Lehre von der Muskelbewegung*, and making a detailed analysis of his data which included tensile tests on human hair, on tendons and veins of an ox, and on thigh muscles of living and fresh-killed frogs, one must conclude that the criticisms made by VOLKMANN are valid. WUNDT used a microscope to read a scale of $^1/_{50}$ mm divisions, attached to the specimen; he claimed that he thus could ascertain elongations to $^1/_{500}$ mm. VOLKMANN estimated that the time required for WUNDT to make such a measurement was of the order of 30 sec, at which time, based upon VOLKMANN's total elongations vs time records described above, the maximum variation of the elastic after-effect was occurring and was partly responsible for the scatter of WUNDT's elongation data, a scatter of as much as 10% to 20% for some tests. Despite the fact that his own data belied the hypothesis, WUNDT was carried away with a desire to demonstrate that at relatively small strain HOOKE's law applied to organic tissue.

There is some question whether anything is accomplished in recreating the bitterness of past scientific controversy, particularly of one that is over 110 years old and led nowhere. WUNDT's sustained attack upon WERTHEIM in his monograph was not only gratuitous but also incorrect. It is curious that WUNDT's own measurements were such that he himself was guilty of some of the errors which he had unjustly ascribed to WERTHEIM. In his eagerness to establish that for organic tissues, "ut tensio sic vis," and to disallow the hyperbolic relation of Eq. (2.15) of WERTHEIM, WUNDT committed the error all too common among the present century's theory-dominated experimentists of failing to consult critically his own experimental data, tabulated in his monograph.

WUNDT's confusion regarding the distinction between elongation and strain is discernible in his comparison of the experimental data of WERTHEIM on a sterno-mastoid muscle of a freshly killed Newfoundland dog with the data of EDUARD WEBER from experiments[1] on the muscle of a frog.[2] WUNDT stated:

> But neither Weber's experiments nor those reported above by us can be squared with the law of the hyperbola, and in fact the entire agreement between Weber and Wertheim is merely a seeming one, since the decrease of extensibility, as has been remarked already, is far more rapid according to Weber than according to Wertheim. The curves which may be sketched by use of the numbers reported by the two investigators thus have very different shapes.[3]

WUNDT failed to take into account the fact that while WERTHEIM's[4] elongations ranged from 8.913 to 64.555 mm and WEBER's ranged from 7.92 to 12.49 mm, nevertheless the length of WERTHEIM's gage was 1 m long while that of WEBER was only 0.02495 m. Thus the range of strain, which is the important parameter, for WERTHEIM was only from 0.0085 to 0.06455, whereas WEBER's was in the enormously higher region of from 0.317 to 0.500. Therefore, a comparison of the differences in these data in order to expunge the hyperbolic law from the physics of organic tissue is somewhat ludicrous.[5] These data of WERTHEIM and WEBER are shown in Fig. 2.26.

I have plotted WUNDT's data, as did VOLKMANN over 110 years ago, and I have found that despite WUNDT's constant emphasis upon his accuracy to within $^1/_{500}$ mm, the scatter ranged as high as 22%. In one summary table containing

[1] EDUARD WEBER [1846, *1*].
[2] Musculus hyglossus.
[3] WUNDT [1858, *1*], p. 33.
[4] WERTHEIM [1847, *1*].
[5] It should be noted, too, that WUNDT's comparison of these data also failed to take cognizance of WERTHEIM's having tabulated stress vs strain, whereas WEBER had tabulated load vs elongation.

sixteen measurements, five are in error when compared with the tabulated data. One of the errors[1] was of considerable importance in WUNDT's argument concerning linearity; he failed to realize that he, too, had permanent deformation occurring in his experiments. Despite his claim that HOOKE's law was basic to small deformation of organic tissue, the curious thing about WUNDT's data when plotted is that not only are the data not linear, as VOLKMANN pointed out, but also they turn upward in the same sense as do those of EDUARD WEBER, WERTHEIM, or VOLKMANN for organic tissues.

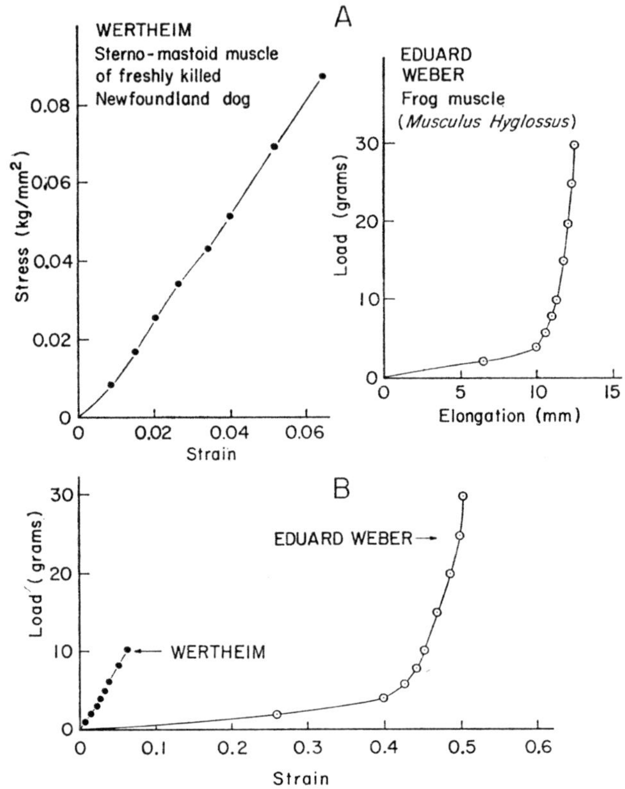

Fig. 2.26. WUNDT's comparison of WERTHEIM's tensile stress-strain results and EDUARD WEBER's tensile load vs elongation results (*A*), shown with a proper comparison of load and strain for both, (*B*).

The reason for including WUNDT in this discussion beyond merely referring to the comments of VOLKMANN with respect to WUNDT's faulty attack upon WERTHEIM, is that two qualitative aspects of his experiments are certainly worth noting. One of these is his comparison of the stress-strain functions of living thigh muscles of the frog to those within 30 min after death, which he compared in turn with the effects of both fatigue and constant load on the living muscle. The second interesting aspect of WUNDT's experiments is his study of the relation

[1] VOLKMANN [1859, *1*], p. 312, comments on this error as follows:
"Wundt has found in his calculation in the Table (p. 29) totally different numbers, in his opinion better fitting numbers which have remained unclear to me."

between primary and secondary deformation regarding the elastic after-effect following many increments of loading and many increments of unloading specimens of human hair and the vena cava of an ox.

Following what to someone who is not a physiologist is a rather grim, detailed depiction of the manner of preparing specimens, including reference to the necessity for being concerned with how the "misery"[1] of the specimen affected its elasticity, are a number of tables containing the experimental results of WUNDT's living frog muscle tests. For brevity, the tabulated data are summarized in Fig. 2.27.

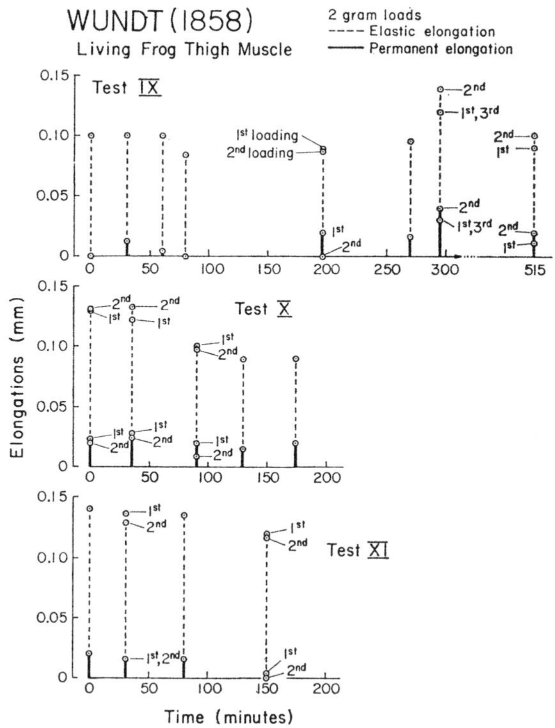

Fig. 2.27. Successive loading of living muscles, showing the variation in elastic elongation and permanent elongation.

At the designated times, he applied a 2 g load and recorded elastic elongation and permanent deformation on unloading. In some instances, he made many loadings in immediate succession. In test X, unlike tests IX and XI in which the frogs remained vigorous, WUNDT described the specimen as "miserable." He noted that under these conditions the elasticity fell off and all measurements included a considerable amount of permanent deformation.

In two other experimental series on living frog muscles, WUNDT compared the load vs elongation for a specimen which just previously he had loaded by 20 g for two hours, with that for a specimen which he had loaded repeatedly to 20 g during the five minutes before the recorded test. After several cycles of loading, as may be seen in Fig. 2.28, the slopes of the two types of tests were approximately the same.

[1] "Elend."

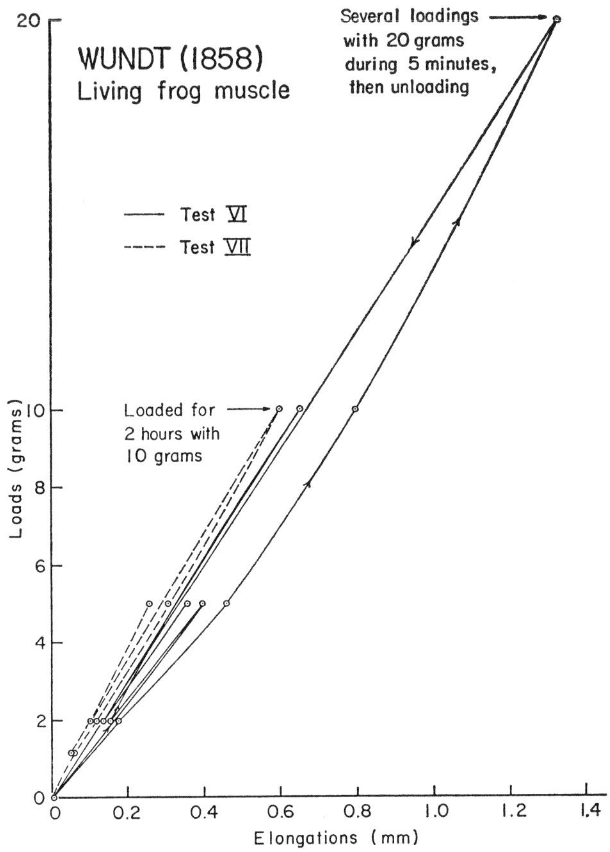

Fig. 2.28. Experiments performed to study the effects of prior loading histories upon the subsequent tensile behavior of living muscles.

Referring to untabulated elastic moduli, WUNDT observed a major change in the slope of the load vs elongation curves in the presence of *rigor mortis*. In his discussion he referred to untabulated elastic moduli. The data which he did provide are compared with the averages of seven living muscle tests in Fig. 2.29. For the 70 min experiment with the living muscles, for those just after death, and for those 20 min after death, all slopes, or moduli, were similar, but there was a shift in the origin of the strain.

Between 70 min and $18^1/_2$ h, WUNDT removed the specimens from the apparatus and kept them in "water steam" for the nearly $17^1/_2$ h, after which he again tested them and observed the obvious increase in slope which he attributed to *rigor mortis*. A biochemist colleague[1] has pointed out to me that more likely, the specimens were pretty thoroughly cooked during the $17^1/_2$ h, thus introducing another odd constitutive variable into mechanics. In any event, one notes that further testing at intervals up to $43^1/_4$ h, including a second 18-h steaming, provided an essentially constant slope, again with a shift of strain origin toward the load ordinate with time.

[1] Professor CHARLES TESAR of The Johns Hopkins University.

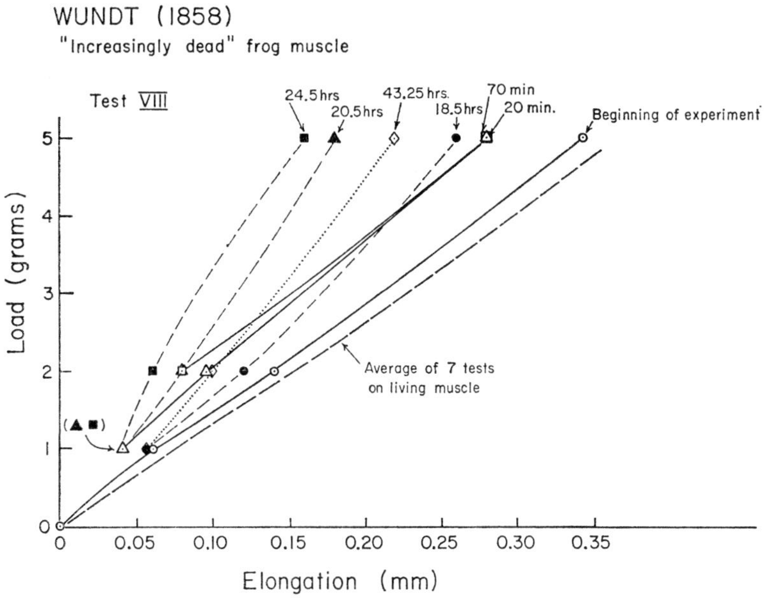

Fig. 2.29. Tensile measurements of muscle tissue at successive intervals after death, compared with the average of 7 tests on living muscle. Note that during the $17\frac{1}{2}$ hour and 18 hour intervals the muscles were removed and kept in "water steam."

The variations of elasticity for given loads for these specimens as a function of time are compared in Fig. 2.30 with the average of the living muscle. One notes that the elongation changes notably with the load under the specified condition.

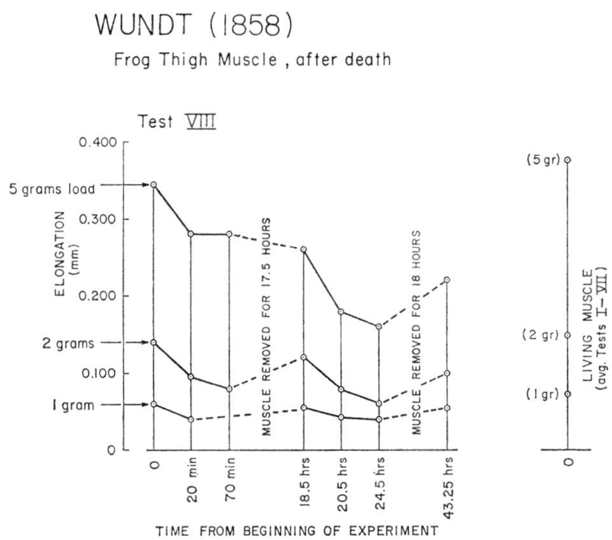

Fig. 2.30. The variation with time after death, of the elongation of a muscle for the same 3 tensile loads compared with the living muscle.

Following this addition of "misery", *rigor mortis*, and possibly cooking, to the list of constitutive variables, we come to the second set of experiments on the tensile deformation of human hair and the vena cava of the ox, shown in Fig. 2.31. One may see the elastic recovery, although as VOLKMANN pointed out,[1] the major portion of the recovery was missed by WUNDT since a large part had occurred in the first half-minute, which presumably he was incapable of observing.

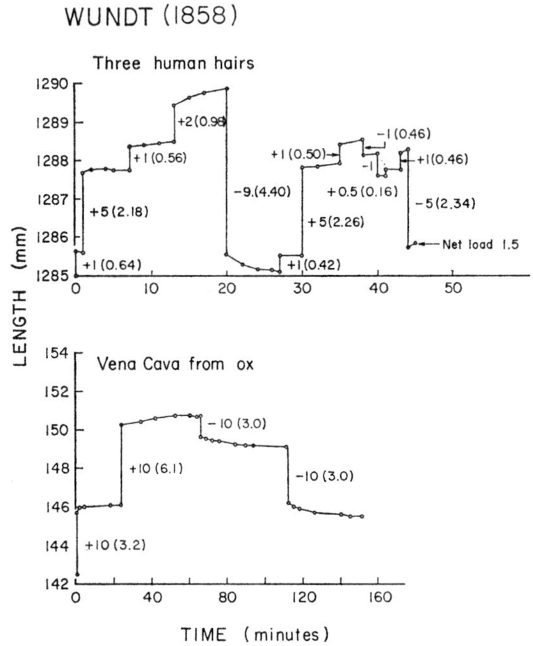

Fig. 2.31. The elastic after-effect in organic tissue. The accompanying numbers designate the load applied in each instance.

These first studies of the phenomenon of relaxation for multiple incremental loading of organic tissues were the source of a series of interesting comments by WUNDT, extending the earlier, original ideas of WILHELM WEBER with respect to the influence upon a primary incremental load and the subsequent elastic after-effect of the entire past history of the deformation.

2.16. The repeal of Hooke's law by the British Royal Iron Commission in 1849.

In 1849 the British Royal Commission "appointed to inquire into the application of iron to railway structures"[2] recommended to the engineering profession

[1] VOLKMANN [1859, *1*].

[2] Iron Commission Report [1849, *1*]. The sole copy in English of which I am aware is the two-volume edition in the library of the Iron and Steel Institute, London. I have been able to consult only the French translation of this report [1851, *1*]. The edition of 1851 had omitted the Appendix containing the additional experimental data.

that henceforth HOOKE's linear law of elasticity for iron in tension, compression, and flexure, should be replaced[1] by the parabolic law

$$\sigma = A\,\varepsilon - B\,\varepsilon^2. \tag{2.16}$$

Their conclusion was based on detailed study of available experimental data in iron, mainly that obtained by HODGKINSON[2] in nearly twenty years of experimentation.

One year later Professor GEORGE GABRIEL STOKES read a paper on behalf of HOMERSHAM COX, who proposed that those same data in the Iron Commission Report were better fitted by what he termed the hyperbolic law of elasticity:

$$\sigma = \frac{\alpha\,\varepsilon}{1+\beta\,\varepsilon}. \tag{2.17}$$

This paper by COX,[3] which was published in its entirety in the *Transactions of the Cambridge Philosophical Society* in 1856, contained an extensive analysis of the errors in HODGKINSON's data on tension, compression, and flexure included in the Iron Commission report, with the objective of developing a nonlinear law of elasticity for iron which would reduce the mean error. COX's success in that empirical enterprise is shown in Table 15 for the averaged data of four different

Table 15. Cox (1856). *Extension.*

Extension in parts of an inch (e)	Corresponding weight by experiment (w)	Weight computed by hyperbolic formula	Difference; weight by hyperbolic formula minus real weight	Weight computed by parabolic formula	Difference; weight by parabolic formula minus real weight
0.0090	1053.77	1040.84	−12.83	1028.70	−25.07
0.0137	1580.65	1567.06	−13.59	1552.91	−27.74
0.0186	2107.54	2103.43	− 4.11	2089.92	−17.62
0.0287	3161.31	3171.73	+10.42	3166.25	+ 4.94
0.0391	4215.08	4222.23	+ 7.15	4231.46	+16.38
0.0500	5268.85	5272.51	+ 3.66	5301.09	+32.24
0.0613	6322.62	6310.84	−11.78	6358.27	+35.55
0.0734	7376.39	7369.11	− 7.28	7435.22	+58.83
0.0859	8430.16	8409.10	−21.06	8484.63	+54.47
0.0995	9483.94	9482.86	− 1.08	9554.78	+70.74
0.1136	10537.71	10538.28	+ 0.57	10585.31	+47.60
0.1283	11591.48	11579.18	−12.30	11574.28	−17.20
0.1448	12645.25	12683.16	+37.91	12580.39	−64.86
		13	143.74	13	473.24
		Mean Error	11.05	Mean Error	36.40

Parabolic formula $w = 116117e - 201905e^2$. Hyperbolic formula $w = 118156.424 - \left(\frac{1}{e} + 2.41\right)$.

[1] There is no intent here to imply that the British engineering profession complied with this suggestion, which it obviously did not, but merely to note the interesting fact that a Royal Commission in 1849 felt it necessary to make such a factual proposal.

[2] These data included a large number of HODGKINSON's measurements beyond those described above, the latter having been performed in the 1840's. Among them were a number of measurements in direct tension and compression of 10 ft. long, 1 in. square iron bars of many types. The experiments in compression were performed inside of tubes to avoid buckling, and therefore must be judged with caution.

[3] COX [1856, *1*].

sorts of cast iron in tension, and for the central deflection of a 13.5 ft. long, 1.552 in. deep, 3.066 in. broad beam of Blaenavon iron. The coefficients of the parabolic formula recommended by the Iron Commission for these experiments and for the hyperbolic formula proposed by COX are included in the table.

The difference in the mean errors is consistent with COX's conclusion that his hyperbolic law of elasticity provided a better fit than the proposed parabolic law. It is interesting that 47 years later, in 1897, RUDOLF MEHMKE[1] reexamined this matter of curve fitting with various analytical forms, including parabolic and hyperbolic functions and other nonlinear stress-strain laws which had been proposed by that time. MEHMKE concluded that JAMES BERNOULLI's parabolic law of infinitesimal elasticity as rediscovered by CARL BACH[2] that same year, 1897, provided the best fit to the data. The differences were exceedingly small. The analysis of MEHMKE and the data of BACH will be described below in Sect. 2.24.

In the introduction to his paper COX felt it necessary to inform his readers that nonlinear infinitesimal elasticity had a long history, beginning with the work of JAMES BERNOULLI[3] and LEIBNIZ[4] referred to above. With respect to the nature of the iron castings used in the experiments COX discussed, it is relevant to point out his observation that the experimental constants α and β varied with the dimensions of the specimens.[5] In this connection, the smaller the casting the higher the elasticity. This effect of size and its associated permanent deformation for very small strain he described in HODGKINSON's terminology; it was, he said, a "defect of elasticity." COX ended his paper with the statement:

> "The great desideratum for perfecting the Hyperbolic or any other hypothetical law of elasticity, is the want of knowledge of these variations of the strength and elasticity of the material, which depend on the magnitude of the castings. It is greatly to be desired that this defect of experimental data may not long continue unsupplied."[6]

At any time, most of those who attempt to eliminate such a "defect of experimental data" succeed merely in contributing further to the general confusion; they report as definitive, conclusions drawn from widely scattered data, from ill conceived experiments which are in fact replete with auxiliary empirical assumptions designed to "verify" the fashionable consensus explanation. An example of this in the present context is the contribution to the French Academy by ARTHUR JULES MORIN[7] in 1862. He attempted to demonstrate experimentally that the extensive uniaxial data of HODGKINSON of 20 to 30 years earlier, with its attendant nonlinearity, could be dismissed on the basis of three crude tests upon each of three wires 22 to 24 m long. MORIN confidently concluded that permanent microdeformation did not occur in very small strain and that "*ut tensio sic vis* was finally confirmed by observation." Concerning his ill-described tests, two in copper and one in iron, he mentioned only the persistent difficulty of uncoiling the wires from 0.6 to 0.7 m diameter packets; he included no experimental detail, neither how he measured elongation, nor how the wires were supported in the experimental gallery of the Conservatoire des Arts et Métiers.[8]

---

[1] MEHMKE [1897, *1*].
[2] BACH [1897, *1*].
[3] JAMES BERNOULLI [1694, *1*].
[4] LEIBNIZ [1690, *1*].
[5] Similar observations were made by KARMARSCH [1859, *1*] who provided an analytical relation between the diameter of the specimen and its strength at rupture for the iron of the time, thus emphasizing the non-homogeneity of the solid.
[6] COX [1856, *1*], p. 190.
[7] MORIN [1862, *1*].
[8] As will be shown below, even ROBERT HOOKE in 1678 referred to the nail on which he supported his 40-ft. wire.

Permanent deformation,[1] which varied from somewhat negative to expected positive values of 400×10^{-6}, was given without reference to the magnitude of the total strain at which it had occurred. The elastic after-effect and time of data recording were not mentioned. MORIN should have perceived that his data were so inconclusive as to be worthless. Instead, he focused upon the small magnitude of his permanent strains, vaguely assumed the variation in his data was caused by thermal effects, and concluded that he had finally disposed of prior observations of nonlinearity in small deformation. HOOKE's law, said he, obviously was universally applicable.

Ignoring the many ten foot long specimens studied by HODGKINSON in tension or compression to provide a check on data from long bars, and completely unaware of the data of GERSTNER on iron wires, MORIN attributed all measured permanent deformation and nonlinearity of metals to the composite structure of HODGKINSON's fifty foot specimens. Again, without specifying the range of strain or accuracy of measurement, MORIN gave moduli data in the form of EULER's or YOUNG's "height of the modulus."[2] As to reproducibility, he obtained numbers which differed from his own average from -11% to $+20\%$, and which were from 10 to 50% less than E-moduli data in copper and iron obtained by others 50 years on either side of 1862.[3] Indeed, MORIN's tests on long wire were made within three years of the careful temperature-compensated experiments on double wires in the Glasgow University tower by KELVIN in 1865.

I cite MORIN's mid-19th century experiments as an example not only of shoddy data and superficially drawn conclusions, but also of the hasty and persistent acceptance of such experiments, made "respectable" by their having appeared in the *Comptes Rendus*. They were referenced in the literature of the next 30 years as if they were of scientific importance, instead of being seen as an erroneous, unfounded attack upon the serious work of an earlier, competent scholar who had carefully described all of his experimental methods and assumptions.[4] It was a mere coincidence that subsequent generations ignored the recommendations of the British Royal Iron Commission which were based on sound experimental results, and adopted the point of view of MORIN whose experiments were practically worthless. Mathematical simplicity, rather than attention to "minor" experimental details, however well defined, gave impetus to the assumption of linearity.

[1] It is curious that MORIN used these data to dismiss the results of HODGKINSON. The latter had measured accurately permanent strains from 2×10^{-6} to 260×10^{-6} in the first 13 loadings for an overall total strain of 0.00120; for subsequent loading he found that relatively large permanent deformation occurred. The magnitude of MORIN's permanent elongations were the same as HODGKINSON's. Hence MORIN's entire argument rested on the one questionable negative value he had obtained.

[2] See Chap. III, Sect. 3.7.

[3] MORIN averaged earlier researchers' data for the height of the modulus, 12000000000 km for drawn red copper and 10500000000 km for annealed red copper, obtaining 11250000000. He compared this average with his average of three tests, each of which was made on two identical wires of 7338740405 km, with unexplained variations from as low as 6521770186 km to a high of 8777809696 km. He dismissed the previous data on moduli and reported higher density values, as if his own obviously inaccurate data were a definitive first proof of HOOKE's law.

[4] On the positive side of the ledger, MORIN must be cited for having encouraged and made possible the research of TRESCA on plastic flow. In 1860, just before TRESCA began his classic series of experiments, he and MORIN shared a historical moment, which from a 20th century perspective is of no little interest. The first specimen of aluminum bronze was cast in MORIN's foundry at Nantorre in March, 1860 (10% aluminum, 90% copper).

A "Mr. Burg" had compared its rupture load not only to that of iron, but also to that of a small bar of the element, aluminum. Although the element aluminum had been reduced

2.17. Experiments on stress relaxation in glass and brass: The origins of nonlinear viscoelasticity. Kohlrausch (1863).

A mid-19th century contrast to that superficial study by MORIN is the penetrating experimental research of FRIEDRICH KOHLRAUSCH[1] in 1863. KOHLRAUSCH, too, was concerned with reevaluating and extending a discovery of the 1830's. He noted that except for a few comments on the subject by his father, R. KOHLRAUSCH, no study of WEBER's elastic after-effect had been made in the intervening 30 years.[2] KOHLRAUSCH then described his own discovery of the important related phenomenon of stress relaxation. With the thoroughness characteristic of those who have had the relatively rare pleasure of finding a genuinely new, and obviously important, experimental pattern in nature, KOHLRAUSCH performed four separate types of experiments which to this day are essential to an understanding of stress relaxation.[3]

In many of the studies of small deformation in solids performed at the time of the experiments of KOHLRAUSCH,[4] the phenomena of microplasticity, the elastic and thermal after-effects, and creep, were simultaneously present in different degrees. To confine the study of deformation solely to a consideration of the elastic after-effect, KOHLRAUSCH performed his experiments in torsion on fine, 35 mm long glass threads for which he could discern no measurable permanent deformation in the region of deformation[5] he was examining. The interval of time for making the initial measurements was sufficiently long that thermal equilibrium was reached.

KOHLRAUSCH's objective was to study the variation with time of the torsion moment required to maintain the thread at a fixed angle of twist, for comparison with data giving the variation with time of the torsion angle for a fixed torsional moment. He also wanted to study the effect upon these histories of maintaining the specimen in one fixed position for different amounts of time before introducing

in 1827, to my knowledge there had been no measurement of an E-modulus* either by WERTHEIM or by anyone else. Not merely the price of aluminum, which in 1856 was $90 a pound, had discouraged study of it, but also its apparent lack of prospects as a practical material. The discovery of the high strength of the aluminum bronze alloy awakened the interest of MORIN and TRESCA to determine the modulus of aluminum itself.

Two measurements were made: one on a cast aluminum bar, 25.6 cm long by 1.125 cm in a square cross-section; and the other on a straight and polished meter bar, one meter long, 0.42 cm high, and 3.6 cm wide. Measurement of the central deflection of the simply supported beam provided for the cast bar a modulus of 6603 kg/mm², and for the meter bar, a modulus of 6911 kg/mm². These were average values for increments of loading and deflection which provided individual moduli varying from as low as 5900 to 7500 kg/mm².

For this first E-modulus determination for aluminum of unknown purity, MORIN and TRESCA proposed that the average of the two bars be chosen, i.e., 6757 kg/mm². This number may be compared with the modern room temperature modulus for aluminum, of 7200 kg/mm² (MORIN and TRESCA [1860, *1*]).

* The E-modulus is a material constant given by the slope of the linear response function obtained in an uniaxial tension or compression experiment.

[1] KOHLRAUSCH [1863, *1*].

[2] He was apparently unaware of the studies of organic tissue of VOLKMANN and WUNDT five years before. Since WUNDT's experiments also were performed at Göttingen, it is especially curious that KOHLRAUSCH did not know of them. Apparently by the 1860's physiology and physics, represented at Göttingen by HELMHOLTZ and WILHELM WEBER, already were ceasing to be in communication.

[3] These experiments of KOHLRAUSCH may be compared below with 20th century experimental studies. See Sect. 3.43.

[4] KOHLRAUSCH [1863, *1*].

[5] Unlike WEBER, KOHLRAUSCH did not have to pre-stress his specimens to large deformation before testing. He was of the perhaps justifiable opinion that he thereby avoided introducing unknown influences into his experimental results.

the fixed torque or fixed angle conditions which preceded the study of the elastic after-effect with respect to time. To accomplish this he modified a device known as a sine-electrometer. One end of the vertical thread could be rotated so that its relative motion with respect to a magnet suspended at the opposite end could be determined precisely. With respect to the earth's magnetic field, three rotations, or 1080°, caused the magnetic element to rotate to somewhat less than the 90° maximum position. Perpendicularity was achieved at approximately 1200° rotation. In separate experiments, using a damped galvanometer, he observed, over several hours, small variations in the declination of the earth's magnetism, and, from the duration of the vibration of a suspended magnet, he determined small variations in intensity.

In the first series of tests with an initial twist of three rotations KOHLRAUSCH maintained the fixed angle by rotating the system so that the angle of the magnet with respect to the earth's magnetic field was such as to provide the required torsional moment. The results were given as the fraction of this required torsional moment to the maximum at 90°. In two such tests, shown in Fig. 2.32 a and b, KOHLRAUSCH made the first observation of stress relaxation.

KOHLRAUSCH's attempt to establish reproducibility by performing experiments on presumably identical glass threads revealed that the shape of the curve representing the torsional moment as a function of time was the same, but the magnitude of the initial value was different when he began his measurements approximately one minute after applying the first twist of three rotations. Three-quarters of a minute were required to damp the oscillation of the magnetic needle.

To ensure reproducibility and to study the behavior under a different fixed angle, KOHLRAUSCH performed the following series of fascinating experiments. The specimen was twisted to a fixed angle for a prescribed amount of time, the duration of which was varied as one of the experimental parameters. It was then returned to its initial position of repose, and the time required for the torsion moment to maintain that position was measured at short intervals over a period of several hours.

In the first two of these experiments KOHLRAUSCH compared the variation of the torsional moment of a specimen which had been twisted three rotations to 1080° and held for 2790 min before returning to the original position of repose, with that of a specimen at a fixed angle of two rotations or 720° for approximately

Table 16. KOHLRAUSCH (1863).

	Torsion angle		
	1080°	720°	
t	x	x_1	x/x_1
4.33	0.0259	0.0184	1.41
5.77	0.0249	0.0177	1.41
6.97	0.0247	0.0175	1.41
7.97	0.0246	0.0173	1.42
14	0.0221	0.0154	1.44
20	0.0210	0.0149	1.41
30	0.0199	0.0141	1.41
54	0.0178	0.0127	1.40
70	0.0168	0.0121	1.39
140	0.0147	0.0095	1.55
177	0.0141	0.0094	1.50
1570	0.0067	0.0045	1.49

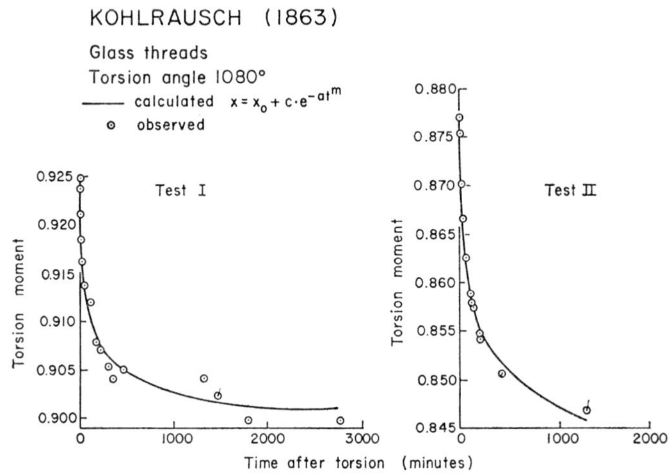

a

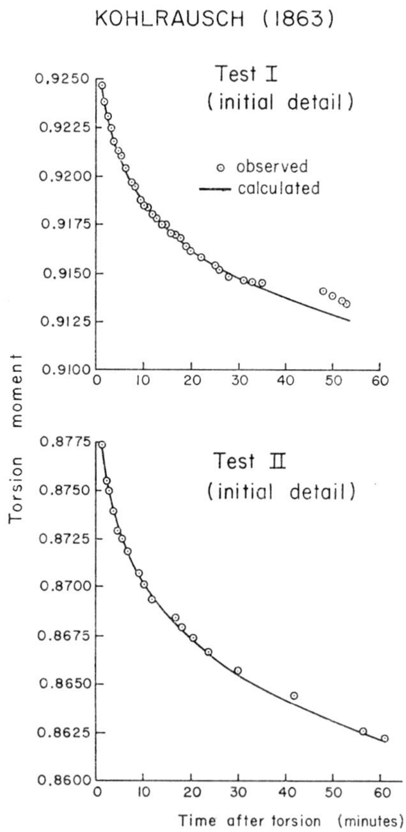

b

Fig. 2.32a and b. Stress relaxation experimental results compared with calculation (a); the initial details of these first measurements of this phenomenon (b). The initial torsion was 1080°.

the same interval of 2866 min before being returned to its initial position. These results were tabulated (see Table 16) as the ratios of the maximum moment. Also included in the table were the ratios of the two moments at equal times for the two experiments whose initial angle of twist was in the ratio of $^3/_2$. The average value of the aforementioned ratio was 1.347; there had been an increase from the initial 1.41 to the expected value of 1.5 at long times. As KOHLRAUSCH pointed out, if one corrects for the small difference in time in the ratio 2790 to 2866, the expected value would be 1.46, rather than 1.50. Thus one sees that in this experiment fairly predictable magnitudes were obtained.

To study the effect of maintaining the initial fixed angle of twist for different lengths of time, KOHLRAUSCH provided data for the 1080° initial angle held for 10, 20, 40, and 1380 min, before returning in each instance to the fixed angle of the repose position. The results of those four tests are shown in Fig. 2.33a and b.

One sees that the magnitude of the torsional moment for a given time after a return to the fixed zero-angle position increased markedly with the duration of the initial twist, everything else being equal. KOHLRAUSCH did not comment on the fact that the data shown in Table 16 had contained an identical test in which 1080° was held for the much longer time of 2790 min. The result had been that a small torsional moment was attained at less than half the total time of the test, suggesting that there was a maximum somewhere between 2790 and 1380 min.

KOHLRAUSCH's method of procedure shows something of the power of empiricism in the hands of one who, without leaning toward the predicted or expected results, waits to analyze the data. He sought a function which would approximate the data in at least one range of times. By varying the numerical values for each empirical fit, and then studying the values for two different functions, he concluded that to attain generality a constitutive equation would have to be of the kind now used in nonlinear infinitesimal viscoelasticity.

In his original experiments in 1835 WILHELM WEBER[1] had studied the elastic after-effect following the application of a fixed load maintained in tension. As I pointed out above, WEBER, with the aid of GAUSS, had obtained Eq. (2.11), and the results of his continued study[2] in 1841 had led to the necessary modification in Eq. (2.14). Using a modified version of an empirical equation proposed by his father, who had found it applicable in studies of similar electrical effects, KOHLRAUSCH introduced the empirical formula

$$t^{m-1}\frac{dM}{dT} = m\,a\,M, \tag{2.18}$$

which leads to the following expression for the torsional moment:

$$M = c\,e^{-a t^m}. \tag{2.19}$$

KOHLRAUSCH showed that it was not possible to fit either of WEBER's equations to the data of Fig. 2.33. On the other hand, KOHLRAUSCH's second empirical relation provided a fair fit to the data up to relatively large times, as may be seen from the comparison of experiment and calculation in Fig. 2.33. The values of the constants for each of these four experimental situations are given in Table 17, to which I have added a calculation of the values for the experiment in Fig. 2.33 at 1080° held for 2790 min.

The differences in the constants obtained in the different situations when applying Eq. (2.19) led KOHLRAUSCH to test the applicability of his formula

[1] WILHELM WEBER [1835, *1*].
[2] WILHELM WEBER [1841, *1*].

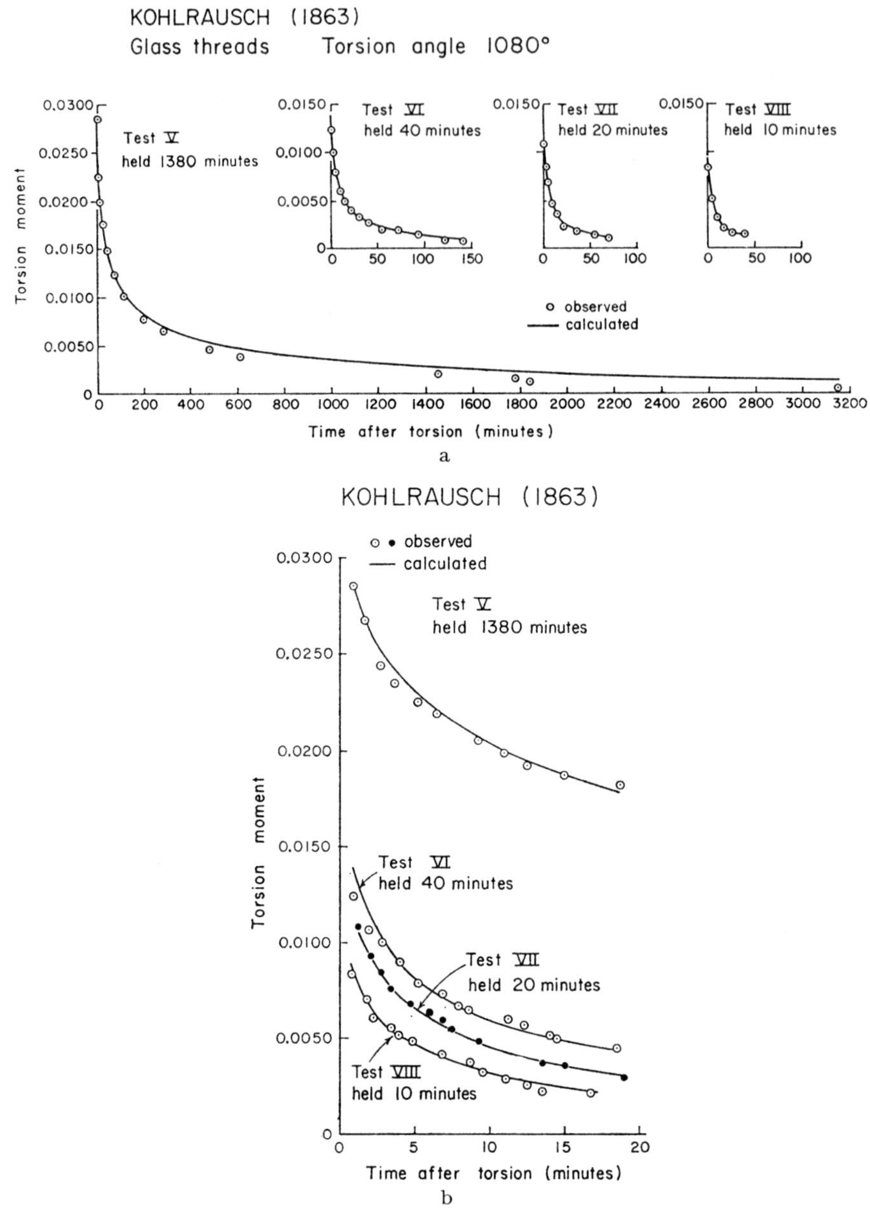

Fig. 2.33a and b. Stress relaxation experimental results compared with calculation; the specimens initially were held for the different times indicated (a). Initial detail of these same measurements is shown in (b).

against a new series of experiments in which the torsional moment was held constant and the angle was made to vary. To achieve a situation similar to that of the previous experiments, he held the angle of twist fixed during two different short intervals of time and then released the specimen so that the torsional moment during the observation of the angle would be zero. He provided a pool of oil so as to damp out immediately the vibrations following this release. The glass threads he had used previously were not of sufficient strength to be damped in this manner without affecting the result. Accordingly, KOHLRAUSCH employed 131 mm long, 0.011 mm diameter hard drawn brass wire specimens.[1]

Table 17. KOHLRAUSCH (1863).

Duration of initial torsion 1080° angle (min)	c	a
10	0.03240	1.3086
20	0.04178	1.2497
40	0.04225	1.0886
1380	0.04551	0.45204
2790	0.0385	0.290

KOHLRAUSCH reasoned that when both fixed values were chosen as zero, the same function should apply for the histories of the torsional moment at a fixed angle as for histories of the angle of torsion at a fixed moment. A comparison of WEBER's and KOHLRAUSCH's formulae with the latter's experimental data, is tabulated in Table 18. The durations of torsion were one minute in the first experiment and two minutes in the second. Also in the table are the values of the constants for the two calculations for each experiment.

To provide a better comparison with the previous plots of experimental results for glass threads, these data also are plotted in Fig. 2.34a and b.

From an examination of these results, including the change in the constants, a, c, and m, among the different experiments, one notes that $m={}^1/_4$ in all of the fixed angle experiments on glass thread, but m is much smaller, i.e., $m=0.0170$, for the fixed-moment experiments on brass.

In order to generalize his experimental results KOHLRAUSCH concluded that it was necessary to introduce a differential equation of the form:

$$\frac{d^2 x}{d t^2} = -a x - \gamma(x). \tag{2.20}$$

In Eq. (2.20) x is the elongation at time t; $a x$ is the rotational moment when the equilibrium condition has been reached; and $\gamma(x)$ is the moment due to the elastic after-effect which operates before this equilibrium condition can be reached. Thus KOHLRAUSCH obtained for the angular velocity,

$$v^2 = v_0^2 - a x^2 - 2 \int \gamma \, d x. \tag{2.21}$$

The time as a function of displacement was given by

$$t = \int \frac{d x}{[a x^2 + 2 \int \gamma(x) \, d x]^{\frac{1}{2}}} + \text{const.} \tag{2.22}$$

KOHLRAUSCH further divided γ, which was an unknown function of x, into two terms designated as γ_1 when moving away from the position of repose, and γ_2,

[1] With respect to a phenomenon such as elastic after-effect, of course, one can raise a valid objection against comparing glass with brass since this presumes no dependence on crystal structure. However, this is no worse than many of today's experiments in which atomistic theoretical models are "verified" for all crystalline structures on the basis of tests in some relatively obscure, specifically prepared alloy or compound of remarkable purity, or tests on a single solid particularly tractable experimentally for special purposes, such as lithium fluoride.

Table 18. Kohlrausch (1863).

Duration of torsion, 1 min				Duration of torsion, 2 min			
t	x			t	x		
	Observed	Calculation according to equation			Observed	Calculation according to equation	
		$x=bt^{-m}$	$x=ce^{-at^m}$			$x=bt^{-m}$	$x=ce^{-at^m}$
0.77	28.0	28.00	27.92	0.73	45.3	45.26	45.16
0.95	27.0	27.05	27.00	0.97	43.3	43.20	43.17
1.21	26.0	25.99	25.98	1.11	42.3	42.25	42.24
1.53	25.0	25.01	25.01	1.28	41.3	41.27	41.29
1.73	24.5	24.51	24.52	1.49	40.3	40.25	40.30
1.97	24.0	24.00	24.02	1.73	39.3	39.28	39.33
2.23	23.5	23.53	23.54	2.02	38.3	38.29	38.37
2.55	23.0	23.00	23.04	2.35	37.3	37.35	37.43
2.91	22.5	22.51	22.54	2.76	35.3	36.38	36.46
3.35	22.0	21.99	22.03	3.27	35.3	35.38	35.48
3.91	21.5	21.44	21.48	3.92	34.3	34.34	34.44
4.57	21.0	20.90	20.95	4.68	33.3	33.36	33.46
5.26	20.5	20.42	20.46	5.67	32.3	32.32	32.42
6.10	20.0	19.93	19.98	6.92	31.3	31.28	31.37
7.07	19.5	19.46	19.49	6.45	30.3	30.27	30.35
8.25	19.0	18.97	19.00	10.37	29.3	29.27	29.33
9.62	18.5	18.50	18.53	11.50	28.8	28.78	28.84
11.38	18.0	17.99	18.02	12.75	28.3	28.29	28.35
13.58	17.5	17.48	17.49	14.07	27.8	27.84	27.89
16.47	17.0	16.94	16.94	15.70	27.3	27.34	27.38
19.92	16.5	16.41	16.40	17.75	26.8	26.80	26.81
24.23	16.0	15.89	15.87	20.13	26.3	26.25	26.26
28.83	15.5	15.45	15.41	23.12	25.8	25.66	25.65
35.13	15.0	14.95	14.90	26.25	25.3	25.13	25.11
40.92	14.5	14.59	14.52	29.62	24.8	24.64	24.61
46.30	14.0	14.29	14.22	33.47	24.3	24.14	24.10
53.28	13.5	13.97	13.88	37.62	23.8	23.69	23.63
68.17	13.0	13.41	13.31	42.50	23.3	23.22	23.14
80.0	12.5	13.07	12.95	46.45	22.8	22.88	22.80
89.5	12.0	12.83	12.70	50.17	22.3	22.59	22.50
99.8	11.5	12.60	12.46	54.00	21.8	22.32	22.22
118.7	11.0	12.25	12.10	60.58	21.3	21.90	21.78
143	9.8	11.88	11.74	67.67	20.3	21.51	21.37
190	8.7	11.34	11.15	82.00	19.3	20.84	20.68
300	7.4	10.52	10.29	104.25	18.3	20.03	19.84
500	6.2	9.67	9.40	125.6	17.3	19.43	19.21
620	5.4	9.34	9.05	137	16.8	19.16	18.93
1440	2.1	8.13	7.78	183	14.3	18.23	17.96
		$b=26.80$	$c=322\,540$	217	13.3	17.76	17.46
		$m=0.16426$	$m=0.0170$	307	10.7	16.78	16.43
			$a=9.3964$	425	8.3	15.91	15.52
				492	5.3	15.53	15.12
				569	4.9	15.16	14.73
				1440	2.7	13.02	12.47
						$b=42.98$	$c=517\,310$

when moving toward it, where, from the experimental data, $\gamma_1(x)>\gamma_2(x)$. Thus he chose the function γ in Eq. (2.22) to be the difference of the two: $\gamma=\gamma_1-\gamma_2$. Lacking 20th century perspective in such matters, Kohlrausch did not consider that his experiments were sufficiently definitive to determine the unknown function γ.

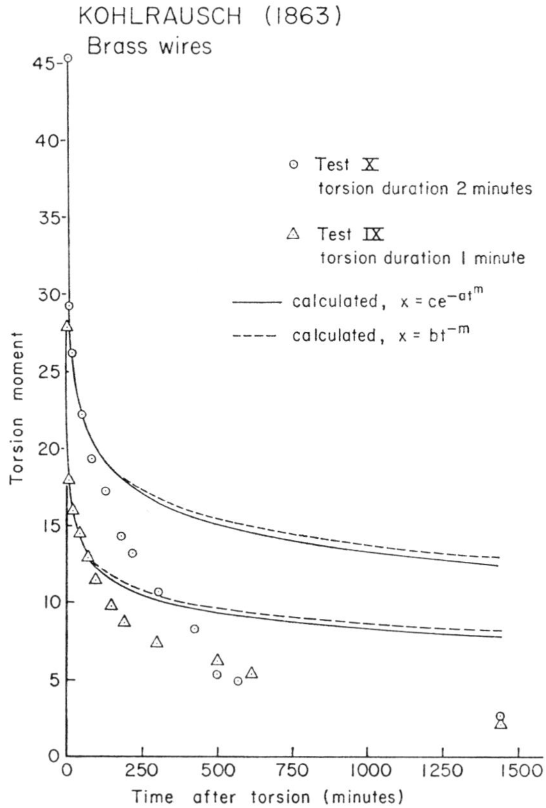

a

Fig. 2.34a and b. Observations of stress relaxation, compared with different calculations (a). The initial detail of these experimental results (b). (See p. 82).

In the first studies of small deformation during stress relaxation, besides his suggestion that nonlinearity must be introduced, KOHLRAUSCH reflected upon a number of other problems which he believed were influenced by his observations. He proffered what he thought was fresh insight: due to the after-effects, the damping of vibrations and related phenomena must possess an associated heat release. Apparently he was familiar with no experimental data on the subject.[1]

In a second examination of the possible influences of the elastic after-effect and stress relaxation on deformation, KOHLRAUSCH proposed that the elastic modulus should be modified when a specimen was set into a strong rotational vibration over a period of time. He then determined the modulus by means of the torsional pendulum experiment. As COULOMB[2] had observed in 1784 and WERTHEIM[3] in 1842, when permanent deformation was applied the modulus was reduced. KOHLRAUSCH first reported a similar reduction following a dynamic prior

[1] This is curious in view of the work of KELVIN on metals during the previous 5 or 6 years, and the several references in the literature to the thermal aspects of rubber undergoing deformation.

[2] COULOMB [1784, *1*].

[3] WERTHEIM [1842, *1*], [1844, *1* (a)].

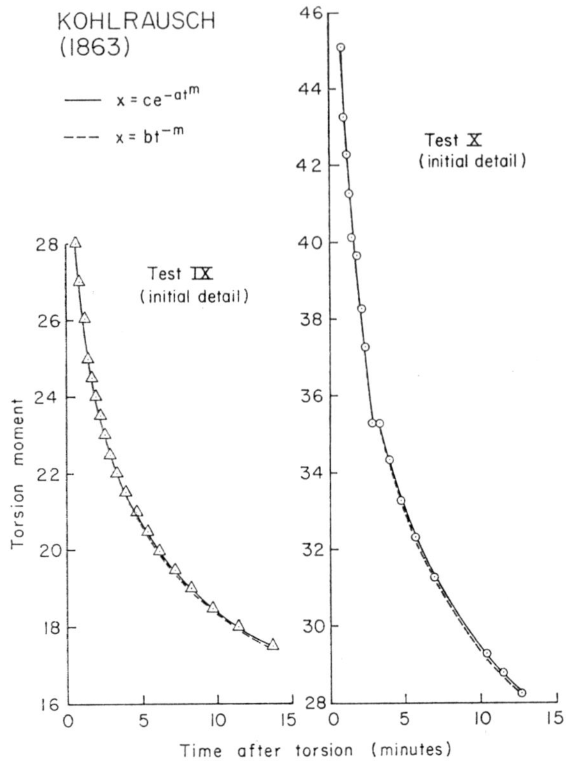

Fig. 2.34b

history.[1] Presumably he had introduced no permanent deformation. His experiments in annealed iron wire show that the period of oscillation gradually increased from 6.256 to 6.411 sec after the period of strong vibration, i.e., the torsion modulus itself was affected by the stress relaxation phenomenon.

Moreover, Kohlrausch raised a serious question regarding the efforts to calculate specific heats by means of the comparison of Wertheim's[2] quasi-static and dynamic measurements of moduli in tension and compression. Reacting within the framework of the rather limited thermodynamic concepts of the time, he pointed out that heat could be generated in torsion experiments in which the volume did not change. In tension or in compression experiments in which a small change of volume did occur, he noted that the variation of moduli also could arise from the generation of heat due to elastic after-effect.[3]

Of these experiments of Kohlrausch, which provided both the discovery and the first definitive study of a phenomenon as important as stress relaxation, not one is cited in any modern paper on anelasticity I have seen.

[1] It seems to me highly unlikely that Kelvin's intense interest in this matter, as evinced in his paper in 1865 on the viscosity of metals, was unrelated to these observations of Kohlrausch. Kohlrausch had made these observations while working in Weber's laboratory.

[2] Wertheim [1842, *1*], [1844, *1*(a)].

[3] Kohlrausch [1863, *1*].

2.18. On the change of volume during plastic deformation: The experiments of Bauschinger (1879).

The dependence of the magnitude of the experimentally determined shear modulus or E-modulus upon the amount of the prior deformation as well as upon the thermal treatment the specimen had received,[1] was another phenomenon related to nonlinearity extensively studied in 1844 by WERTHEIM[2] in tension experiments on numerous metals. COULOMB[3] in 1784 had discovered that the shear modulus[4] decreased with increasing permanent strain in the torsion of iron and brass wires, and, after WERTHEIM, KELVIN[5] finally definitively examined the phenomenon. Indeed, KELVIN thought the reduction in modulus with increasing plastic strain was so important that in 1865 he referred to his study of it as his chief interest in experimental solid mechanics.

I recently showed that this dependence of the elasticity upon prior mechanical and prior thermal histories has a major influence upon the degree of apparent linearity in dead weight experiments for infinitesimal deformation.[6] My experimental data and the comparison of them with the data discussed above will be examined at length in the next chapter on the approximate linearity of infinitesimal stress-strain functions (see Sect. 3.44).

The data of KELVIN had been obtained from very carefully performed experiments with long-wire torsional pendulums, the amplitude of the oscillations being in the range of small deformation. The effect of plastic deformation upon moduli, whether in the form of micro-strain or large deformation, has been one of the main sources of small deformation nonlinear elasticity, certainly in the metals which have been described up to this point.

September 24, 1877, marks the inauguration of a new era in experimental solid mechanics, for on that date JOHANN BAUSCHINGER performed the first accurate experiment on compression of a cast iron bar of rectangular cross-section, the bar being sufficiently short that strains at large stress could be measured before buckling occurred. Of even more importance in that initial experiment and those that followed was the fact that for the first time sufficient accuracy was achieved to measure simultaneously both the lateral and longitudinal deformation. Thus POISSON's ratio and volume changes could be determined with precision in the range of small deformation below the elastic limit and also in the range of large strain and large plastic deformation.

The experimental series in cast iron, wrought iron, Bessemer steel, and sandstone, which BAUSCHINGER performed between 24 September, 1877 and 5 February, 1878, was described in a large memoir[7] published in January 1879. That study and the voluminous outpouring of experimental results in the years that followed were made possible by BAUSCHINGER's development of a remarkable double-mirror extensometer.[8] He devoted many pages to a detailed description of the construc-

[1] WERTHEIM recorded the thermal aspects of the problem in later studies.

[2] WERTHEIM [1844, *1*(a), *3*].

[3] COULOMB [1784, *1*].

[4] The μ modulus is a material constant given by the slope of the linear response function in a torsion experiment.

[5] KELVIN [1865, *1*].

[6] BELL [1968, *1*].

[7] BAUSCHINGER [1879, *1*].

[8] Very few experimentists in the 300-year history of solid mechanics have found it possible to publish the amount of experimental detail which the meticulous BAUSCHINGER succeeded in getting into print. One has the impression that one can follow the daily details of his scientific life, the progress of his thought, his lecture days and his holidays, inasmuch as his laboratory notes, published in successive volumes over the years (see, for example [1886, *1*]), included the advances and retreats which preceded each scientific publication.

tion and operation of his mirror extensometer, which permitted him to determine elongations to within 0.0002 mm.

Whether the elongation was lateral or axial, he invariably used two mirror levers and two telescopes. He made a measurement on opposite sides of the specimen so that he could determine slight deviations from axiality. Hence, in most of his experiments BAUSCHINGER found it necessary to align four mirror systems. During the test he had to read four telescopes since he was interested in determining both lateral and longitudinal strains for compression as well as for tension. The reader is referred to the original papers for the detailed description of this device, which is shown here in Fig. 2.35.

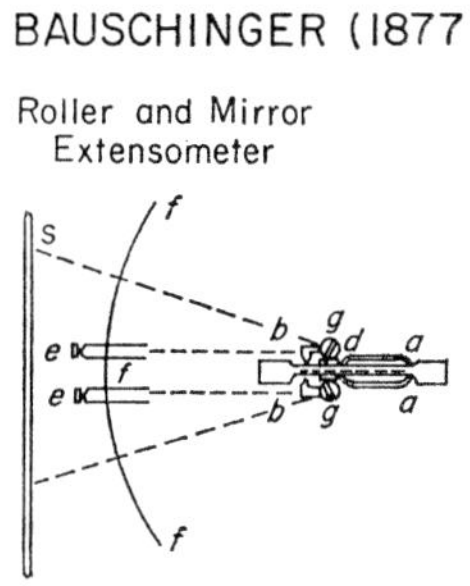

Fig. 2.35. The first optical extensometer for precise measurements.

In much of the experimental research on stress-strain functions described above, such as the results of HODGKINSON[1] or FRANZ ANTON GERSTNER,[2] the permanent deformation was a small fraction of the total strain in the region of interest, and the total strain was confined to the region of relatively small deformation. The study by BAUSCHINGER in 1879 for the first time provided a reversal of this situation: he was concerned with the small change in volume while large permanent deformation was in progress. In many studies of plasticity today it is assumed, perhaps incorrectly, that the volume does not undergo a permanent change when the load is removed.

BAUSCHINGER began his paper[3] with a brief review of the earlier work of CHARLES, Baron CAGNIARD DE LATOUR[4] and WERTHEIM,[5] both of whom had measured the relative increase of volume which a prismatic specimen undergoes, the former by observing the lowering of water contained in a tube when a brass wire in the tube was subjected to elongation, and the latter by observing by means of a capillary tube the interior volume change of a hollow specimen undergoing axial deformation.[6] BAUSCHINGER noted that except for those volume experiments, the usual manner of attempting to determine POISSON's ratio for isotropic solids was through a comparison of the measurements of E-moduli and μ-moduli in small deformation, i.e.

$$\nu = \frac{E}{2\mu} - 1 \tag{2.23}$$

where ν is POISSON's ratio.

[1] HODGKINSON [see 1824–1844] see *supra*, Sect. 2.6.

[2] GERSTNER [1832, *1*].

[3] BAUSCHINGER [1879, *1*].

[4] CAGNIARD DE LATOUR [1828, *1*] and see POISSON [1827, *1*].

[5] WERTHEIM [1848, *1*].

[6] These experiments will be described in some detail in Sect. 3.16 since they provide the major experimental basis of the rari-constant vs multi-constant controversy which extended over the entire second half of the 19th century. As will be seen, WERTHEIM severely criticized the limitations of the experiments of CAGNIARD DE LATOUR.

BAUSCHINGER proceeded to demonstrate for the first time the previously suspected and now well known experimental fact that the problem of determining a ratio of large numbers, required for ascertaining POISSON's ratio in this manner, leads to unreliable values.[1] It probably was this variation that was the main source of the 19th century controversy on elastic constants; the rari-constant theory of POISSON and CAUCHY required that POISSON's ratio should have the fixed value of $\nu = {}^1/_4$.

BAUSCHINGER determined E from two tension tests and two compression tests for nine Bessemer steels of different carbon contents from 0.19 to 0.96%. In the same solids he determined the shear moduli μ. He then showed that Eq.(2.23) led to values of POISSON's ratio for steel which varied from 0.25 to 0.36, with an average of 0.305.

In a second experimental series,[2] described as "Siemens-Martinstahl von Neuberg-Mariazell," two tensile measurements of E, two flexural measurements of E, and two torsional measurements were made on five samples of presumably the same solid, steel. Comparing the 30 measurements, BAUSCHINGER found that in terms of Eq. (2.23), POISSON's ratio varied from 0.24 to 0.30. For all these tests and an additional series consisting of 30 separate specimens of steel, described as "Bessemerstahl von Teschen in Oesterr.-Schlesien", the average was 0.290, a number which will be recognized as agreeing with the now generally accepted value of POISSON's ratio for this solid. The variation from 0.24 to 0.36 in individual measurements, and the fact that very few measurements gave precisely the same number, may account for the insistence of those who, on the basis of one or two tests, chose sides in the rari-constant vs multi-constant controversy.

With his mirror extensometer BAUSCHINGER was able to measure POISSON's ratio directly, during loading and during unloading, for both small and large deformation. His study of the variation of this quantity will be described in Sect. 2.19. Of major interest here is the behavior of the measured small dilatation in the presence of large permanent deformation.

As to nonlinearity, BAUSCHINGER observed two main features. They are shown in Fig. 2.36 in a plot of his data on cast iron, wrought iron, Bessemer steel, and sandstone. There were separate tests in compression and in tension, although in his diagram they appear to proceed smoothly through the point of zero stress. One observes that in every instance, whether tension or compression, BAUSCHINGER found that the relation between dilatation and axial stress was nonlinear when measured during the plastic deformation of the solid. (Since these were uniaxial tests, the hydrostatic pressure may be chosen as ${}^1/_3$ the axial stress.)

A second pattern of behavior BAUSCHINGER observed in his study of nonlinearity recently was re-discovered in dynamic plasticity experiments by WILLIAM FRANCIS HARTMAN,[3] namely, the presence of unexpectedly large dilatation during the plastic deformation of homogeneous solids accompanied by small permanent dilatational deformation when the load was removed. For the past 90 years this discovery of BAUSCHINGER's generally has been ignored by theorists and experimentists in their development of both quasi-static and dynamic plasticity. As may be seen in Fig. 2.36, BAUSCHINGER found that the sudden changes could occur at a specific strain and could be relatively large in comparison with the total dilatation. The axial plastic strains, of course, were an order of magnitude larger than those of the measured dilatation. A comparison of these total axial strains with the

---

[1] For many contemporary textbook writers and not a few modern experimentists BAUSCHINGER's 19th century pronouncements on this matter might as well never have been made.

[2] BAUSCHINGER [1879, *1*].

[3] HARTMAN [1967, *1*], [1969, *1*].

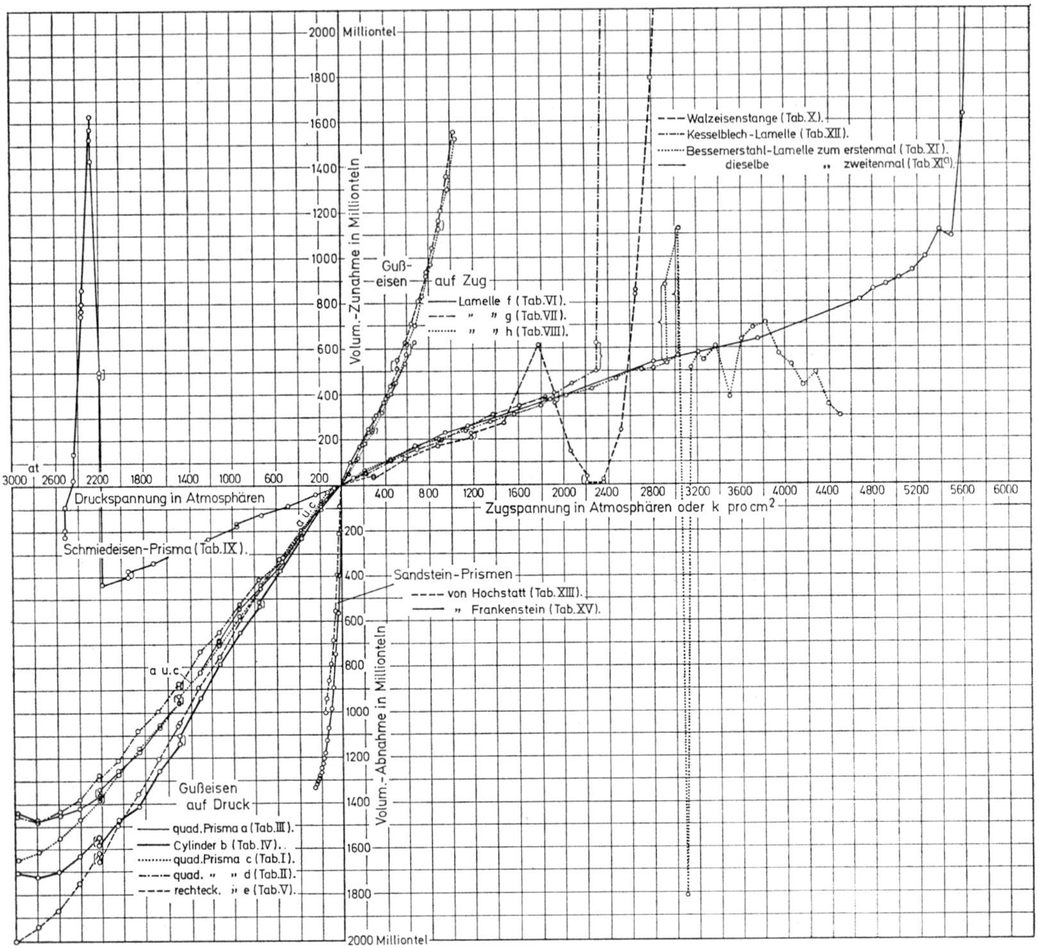

Fig. 2.36. The dilatation accompanying the loading in tension and compression of cast iron, wrought iron, Bessemer steel, and sandstone. These are the first direct measurements of volume changes associated with the axial deformation of metals.

dilatations of interest here will be made below in Chap. IV, on finite strain. The presence of dilatation during plastic deformation has been found to be of importance for a modern theory of plasticity. (See Sect. 4.35.).

Bauschinger first found the jump in dilatation in his preliminary experiment on Huntsman steel. In his subsequent experiments he became very much interested in establishing without question that it was indeed a fact of nature and not merely an experimental anomaly. This he succeeded in doing by comparing the behavior in many tests and by observing the changes in temperature which accompanied the deformation. Having encased a Bessemer steel specimen (tests XI and XIa) in a pool of mercury with a thermometer capable of being read easily to one-tenth of a degree, he followed the thermal history qualitatively as he subjected the specimen to a tensile test. Up to the elastic limit the temperature gradually decreased, as would be expected in a tensile test. During the initial phases of plasticity, to approximately 0.5% longitudinal strain, no change of temperature occurred. Then, as the plastic strain proceeded to 3% (the additional loads being applied in three-minute intervals), the temperature gradually rose. Between 3 and 4% plastic strain a large increase in dilatation occurred; this was accompanied by a rapid drop in temperature.

In 1829 CORIOLIS[1] had shown that lead specimens undergoing relatively rapid plastic deformation became exceedingly hot. Although only qualitative, BAUSCHINGER's tests provided the first demonstration of the coupling of the increase in temperature due to plasticity with the decrease in temperature due to an increase of volume of metals in tension. It was certainly a milestone in a subject which received surprisingly little attention in the decades that followed.

The second aspect of BAUSCHINGER's tests, namely the permanent deformation in the dilatation, also is shown in Fig. 2.36.

The name, BAUSCHINGER, of course, is throughly familiar to every student or practitioner of either continuum mechanics or metallurgy. He is credited with having noted the changes in the elastic limit of metals undergoing prescribed stress cycles. This change is known as the "BAUSCHINGER effect." BAUSCHINGER's discovery of it was made possible by his mirror extensometer, which permitted a test to be conducted in compression followed by tension, or vice-versa, a procedure essential for such a study.[2]

Since the elastic limit and the BAUSCHINGER effect are associated with the inauguration of plastic deformation, they also will be discussed below, in Sect. 4.7 of the chapter on finite strain. However, the nature of the experiments is relevant here too because BAUSCHINGER, like nearly every experimentist who preceded him, was wont to load and unload his specimen continuously as a tensile test proceeded. Thus in Fig. 2.36 we see an increase in the magnitude of the permanent dilatation deformation when the experiment was carried out for increasing values of the total plastic strain. In order to avoid wrecking the apparatus, BAUSCHINGER removed his mirror extensometer from the specimen a considerable time before rupture became imminent. Thus we can be certain that the dilatational behavior described is not associated with near-rupture phenomena such as "necking," etc.

The importance of the "BAUSCHINGER effect" certainly justifies the 90 years of acclaim for the first definitive study of it.[3] It is deserving of such attention particularly in applied technology where elastic limits are related to practical failure criteria, and also in crystal plasticity, where the phenomenon has stimulated or constrained numerous fundamental studies. However, in the light of recent experimentation on dynamic plasticity I am of the opinion that BAUSCHINGER's studies of compressibility ultimately will be viewed as his major contribution to science. It is remarkable that these discoveries have lain in limbo along with much of the rest of 19th century experimental mechanics.

2.19. Nonlinear torsion including influence on magnetization, 1857 to 1881.

One problem in elasticity which had fascinated successive experimentists from DULEAU[4] in 1813 to BAUSCHINGER[5] in 1881, was the torsion of rods of non-circular

[1] CORIOLIS, see [1830, *1*].

[2] BAUSCHINGER [1879, *1*].

[3] The *discovery* of the "BAUSCHINGER effect" actually was not BAUSCHINGER's, but must be credited to GUSTAV HEINRICH WIEDEMANN in 1859; the time-lapse aspect was discovered by ROBERT HENRY THURSTON in 1874 (WIEDEMANN [1859, *1*], THURSTON [1874, *1*]). As will be described in Sect. 4.5 below, BAUSCHINGER's efforts to establish some form of undeserved precedence for himself were eminently successful.

[4] DULEAU [1813, *1*].

[5] BAUSCHINGER [1881, *2*].

cross-section. The torsion of square or rectangular prisms received most attention because of the measurements of DULEAU and the early interest of CAUCHY in the theory for that cross-section. In 1853, SAINT-VENANT[1] presented in an author's summary to the French Academy his classic study of torsion within the theory of infinitesimal elasticity. The full paper[2] did not appear until 1856, after an Academy commision composed of CAUCHY, PONCELET, PIOBERT, and LAMÉ[3] had glowingly recommended it for publication.

The year of its publication is significant in the present context, because the equally important memoir on torsional experiments of WERTHEIM[4] had been presented in 1855 and published in 1857. WERTHEIM was much angered by what he considered to be SAINT-VENANT's unethical use of a few preliminary experimental numbers conveyed in private correspondence. SAINT-VENANT had published them with adverse comments which obviously infuriated WERTHEIM. Thus, the outstanding experimental and theoretical contributions on this subject, published within a year of each other, were not cooperative. WERTHEIM confined his comparison with theory solely to providing an (unexplained) empirical correction factor to the theory of CAUCHY, which he considered SAINT-VENANT had modified only slightly, but in fact WERTHEIM was the first to establish the need for a nonlinear theory of torsion.

WERTHEIM's experiments on torsion in the judgment of time may exceed in importance the epoch-making theory of SAINT-VENANT. WERTHEIM discovered that in the small quasi-static deformation of solid and hollow brass, iron, and steel cylinders of circular and non-circular cross-section, the torsional response function was nonlinear. He therefore declined to represent the data as shear moduli. He was not at all surprised, hence, to find changes in volume proportional to the square of the twist, and changes in axial dimensions which also were not proportional to the angle of twist. Such anomalies in the context of a linear response function were explicable once he had found that the problem to be resolved was nonlinear.

That he observed nonlinear response functions in torsion for solids in which he had observed linearity in tension, WERTHEIM ascribed to the greater accuracy which could be attained when angles rather than the elongation were magnified. This was a fact BAUSCHINGER[5] demonstrated 24 years later when he found a nonlinear response function in iron while making a similar comparison of experiments in torsion and in tension.[6] WERTHEIM had noted that nonlinearity in quasi-static measurement was not inconsistent with dynamic measurement since in vibration experiments the frequency increases as the sound fades away.[7]

WERTHEIM conducted experiments on 65 specimens: 6 full cylinders of steel, brass, iron, and glass; 10 hollow cylinders, 6 brass and 4 iron; 4 solid elliptical

[1] SAINT-VENANT [1853, *1*].

[2] SAINT-VENANT [1856, *2*].

[3] It should be noted that SAINT-VENANT, born in 1797, was 71 years old in 1868 when finally elected to the French Academy. His anti-war student activity in opposition to NAPOLEON in 1814 had adversely influenced his entire career. His final election presumably was due not only to his outstanding eminence but to strong attacks on the French Academy by non-French scholars at the continued lack of recognition at home, of such a world-famous elastician.

[4] WERTHEIM [1857, *2*, *1*].

[5] BAUSCHINGER [1881, *2*].

[6] In the 1890's VOIGT refused to accept the nonlinearity he observed in some of his own torsional measurements on at least one single crystal and one polycrystal. He attributed such deviation from linearity to the apparatus and to faulty specimens.

[7] Viewed from today's knowledge of nonlinear mechanics, of course, WERTHEIM's argument is not conclusive. The important point is that before it became commonplace experimentally to study nonlinear phenomena, he appreciated that a nonlinear response function could lead to the unexpected, from the perspective of a linear theory, even for quasi-linear problems.

specimens, 2 steel and 2 brass; 12 iron prisms, 3 square and 10 with one side 24 mm and the other ranging from 1 mm to 24 mm; 5 prisms of cast steel of rectangular bases, with ratios varying from 1 to 36; 21 rectangular prisms in steel, iron, sheet iron, brass, and different kinds of glass; 3 hollow rectangular prisms of brass; and 4 prisms of oak and fir wood. The changes of volume in the hollow tubes, the measurement of which was a unique precursor to such measurement in the 20th century, WERTHEIM determined by means of capillary tubes[1] attached to the water-filled specimens. Because he decided that he would not present his results in terms of moduli or tangent moduli but, due to their being nonlinear, would give them in the form of numerous tables showing dimensions and measured angles, it is difficult to summarize specific experimental results from his 172 page memoir.[2]

Before making measurements in torsion, WERTHEIM determined E for each specimen in tension, and ascertained the accompanying changes of volume of the hollow bars. He expected changes in volume and found that his measurements on brass roughly agreed with his expectations. That his measurements on iron and steel did not so accord, he ascribed to conditions the specimen had encountered prior to the test. His torsional experiments on glass were accompanied by photoelastic observations. Despite the experimental complications which made quantitative determination difficult, and despite the fact that torsional loading precluded comparison with NEUMANN's analysis, WERTHEIM's description of the photoelastic phenomena during loading is of interest.

For all the cylinders, WERTHEIM noticed that in addition to the nonlinearity of the relation between torque and angle, the amount of the departure from the linear approximation depended upon the length of the specimen. Only when very long specimens were compared could he achieve the independence of length assumed in the elementary theory. WERTHEIM included the sheet iron and the wooden specimens to study the effects of anisotropy for three elastic axes. He found in these pioneering torsional experiments that he had to let elastic constants vary with direction if he were to achieve some correlation between measurement and his estimate of how anisotropy would influence the results, but the agreement still was far from satisfactory.

To WERTHEIM, the necessity of supplying empirical correction factors to CAUCHY's theory[3] to correlate with his own results probably seemed of prime importance. In fact, his observation that the changes of volume were proportional to the square of the radius and to the square of the angle of twist was of greater import. No theory had been proposed to account for change of volume of this magnitude, occurring in this manner. He had observed it for every shape of cross-section, and he tells of his unsuccessful labors to account for it. He hoped in vain that his experimental observations would stimulate others to tackle the subject of torsion with sufficient generality to include the behavior he had observed.

The final section of WERTHEIM's remarkable memoir dealt with the connection between torsion and magnetization in iron, a subject of great interest during the 19th century. In his customary historical introduction, WERTHEIM referred to BADEN POWELL's observations in 1829 concerning the loss of magnetization with impact; to GAY-LUSSAC, who had reported that twisting did not affect permanent magnetism, but that unloading did; and to BECQUEREL who had suggested that torsional loading in either direction induced an electric current of the same sign, while unloading from either direction produced a current of opposite sign.

[1] These studies led WERTHEIM in 1854 to engage in an important experimental investigation of the subject of capillarity in general [1854, *2*], [1861, *1*].

[2] WERTHEIM [1857, *1*, *2*].

[3] CAUCHY [1830, *1*].

WERTHEIM then recounted at length the controversy MATTEUCCI had generated over WERTHEIM's observations in 1844 and 1852. MATTEUCCI had claimed that torsion had no influence upon the direction of an induced electric current. WERTHEIM showed in detail that twisting a magnetized iron wire in either direction caused a loss of magnetization which was restored upon unloading. If the specimen were twisted in one direction to permanent deformation, then subsequent loading in that direction produced magnetization, and loading away from that direction, demagnetization. WERTHEIM's care in these difficult measurements is indicated by his having built an alternative apparatus of wood to make sure his metal device for torsion measurements did not interfere with demagnetization of the specimen. Finally, he pointed out that MATTEUCCI's theory that the magnetic strength was proportional to the dilatation was not in accord with WERTHEIM's own observations that the volume varied with the square of the angle of twist, while the magnetic strength varied linearly. WERTHEIM, too, discussed the less than satisfactory agreement between the theoretical explanations of AMPÈRE and his own observations in 1857 of the effects of deformation upon magnetization in iron and steel.

BAUSCHINGER's experiments[1] in which he also studied the torsion of circular, elliptical, square, and rectangular bars, had the advantage of a quarter of a century of perspective with respect to SAINT-VENANT's theory. However, BAUSCHINGER likewise found that torsional measurement was sufficiently sensitive that the essential nonlinearity was readily visible, but he did not object to tabulating his experimental results as tangent moduli. In Fig. 2.37 are shown BAUSCHINGER's tangent moduli for the various cross-sections for cast iron.

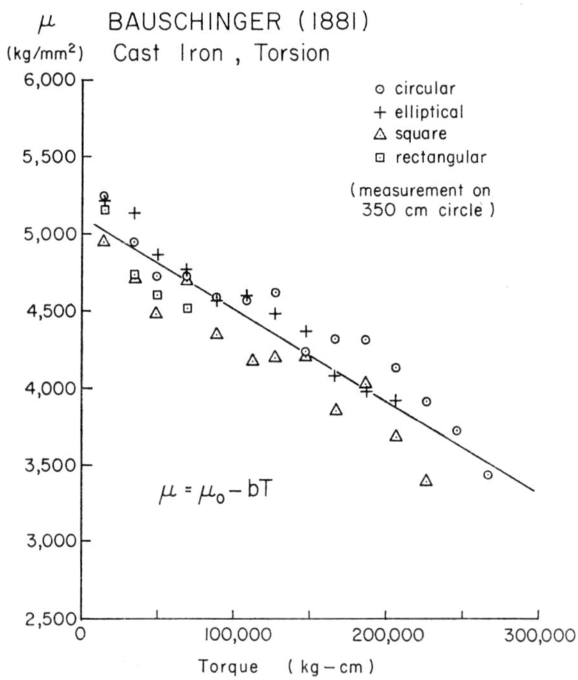

Fig. 2.37. The decrease of the tangent modulus, μ, with increasing torque for the torsion of cast iron specimens of various cross-sections.

[1] BAUSCHINGER [1881, *2*].

As will be shown below in Sect. 2.23, the empirical description in 1881 of the linear decrease in the tangential shear modulus with increasing torque, $dT/d\gamma = \mu_0 - bT$, preceded HARTIG's[1] generalization of the nonlinearity of E-moduli by more than a decade. BAUSCHINGER merely tabulated the data he observed; he was more interested in comparing the moduli with values of POISSON's ratio measured from the axial loading of identical specimens.

Using his mirror extensometer, BAUSCHINGER measured the lateral deformation during tension and compression tests. He noted that as the load increased, not only did POISSON's ratio increase, but at the same time both E and μ decreased, just as they had when he had made direct torsional measurements. One may compare these results in Fig. 2.38. With increasing strain in compression, E and μ decreased as shown, the measured increase in POISSON's ratio being present at the same time. The observed relation among tangent moduli, $E = 1.9\,\mu + 2600$ kg/mm², gave, assuming isotropy, $\nu = 1300/\mu - 0.05$ for the variation of POISSON's ratio with the tangent to the nonlinear response function in torsion.

BAUSCHINGER's subsequent observations on the torsional stability of the various cross-sections, like the final series of experiments in the first part of WERTHEIM's great memoir on stability and rupture, essentially were studies in plastic deformation.

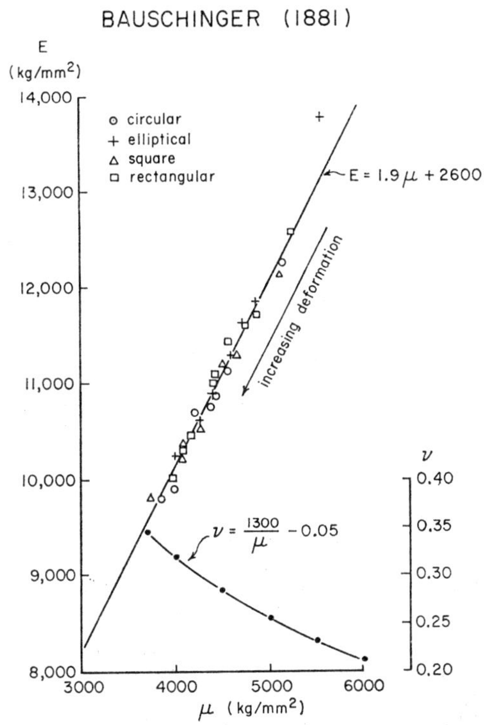

Fig. 2.38. The dependence of the measured E modulus upon the measured μ modulus as a function of the magnitude of the deformation. Also shown are the corresponding measured values of POISSON's ratio. The correlation of experimental results from four different cross-sections may be noted.

---

[1] HARTIG [1893, *1*].

2.20. The decrease of moduli with permanent deformation: The experiments on metals of Wertheim (1844–1848), Kelvin (1865), Tomlinson (1881), and Fischer (1882).

Percy Williams Bridgman once stated that original discovery in experimental physics seldom is made by going to the next decimal point through the refinement of the existing apparatus, but rather, by new and imaginative experimental approaches to the problem, which achieved this objective as a by-product.[1] It would be difficult to argue with this point of view. On the other hand, over the years one infrequently finds an individual experimentist who has developed a known experiment to a higher precision, and who, while perhaps making no discovery of his own, may promote a better understanding of the relative importance or unimportance of the previously observed phenomenon. Such a man was Herbert Tomlinson[2] who, between 1881 and 1886, published four mammoth papers describing over 70 experiments with numerous repetitions of each and achieved an accuracy of measurement more than an order of magnitude greater than had Wertheim's study of 1844.

Tomlinson's research, which was made possible by a £4000 government grant, aimed to determine the influence of stress and strain on all of the then known thermal, electrical, magnetic, and mechanical properties not only of the commonly used metals, but also of silver, platinum, aluminum, tin, and lead. He reexamined and discussed nearly every aspect of deformation which had been discovered through experiment in the previous fifty years.

The apparatus Tomlinson[3] used was based upon an experiment developed in 1865 by Kelvin[4] in which two long wires were suspended from a common support and initially weighted by equal, small loads to straighten the wires. Kelvin's wires had been 24 m long; Tomlinson's were 30 ft. (9.196 m). The latter's experiment was performed inside an enclosed tower to minimize thermal effects. One wire was used to check extensions and contractions produced by thermal changes, while the second wire, adjacent, was the specimen which was loaded to study the primary deformation. The measurements of extension were made by means of a microscope operating as an optical cathetometer. On his wires, Tom-

[1] Bridgman [1943, *1*], pp. 14–15; also, paper No. 140 in [1964, *1*], pp. 3518–3519. "It has always been the experience in physics that new phenomena lie concealed beyond the next decimal place, and these phenomena may be of great and even revolutionary importance, as shown by the whole domain of quantum phenomena. The situation is no different in the high-pressure field; in fact I had early found that for liquids there are small-scale phenomena characteristic of each individual liquid, and later discovered a great wealth of small-scale phenomena in the behavior of solids with complicated structures, such as alloys with order-disorder transitions. Nevertheless, in spite of the undoubted possibilities, I was not personally of a temperament to look with enthusiasm on the exploitation of the phenomena of the next decimal place, particularly if it was to be done simply by a refinement of the techniques I had already practiced, though there was doubtless room for that. In this view I think I am not greatly different from many of my fellow physicists. One recalls the consternation with which many physicists in the eighteen-nineties contemplated the drab prospect of a future dedicated to the exploration of the next decimal. As a matter of fact, the next decimal place has seldom been exploited except by the development of a radically new technique."

[2] Tomlinson [1883, *1*].

[3] Tomlinson [*Ibid.*] achieved greater accuracy by repeating a large number of measurements using very long wires. His deflection sensitivity of 0.01 mm, which was the same as Wertheim's, had been far exceeded a few years earlier, in 1877, by Johann Bauschinger [1879, *1*]. As was seen in Sect. 2.18, by means of an optical lever Bauschinger had achieved a sensitivity of 0.0002 mm which enabled him to study for the first time the deformation of short specimens in both compression and tension.

[4] Kelvin (Sir William Thomson) [1865, *1*].

LINSON could read an extension of $^1/_{100}$ mm and thus could observe a strain of the order of 1×10^{-6}. He isolated thermal effects by comparing the elongation of the two wires with each reading. After the experiment he separated the wire into a number of pieces which he then tested individually as a check of homogeneity.

In every metal studied, TOMLINSON observed: "A departure, as far as temporary elongation is concerned, from 'HOOKE's law,' more or less decided, always ensues after recent permanent extension, even when the weights employed to produce the temporary elongation do not exceed one-tenth of the breaking-load of the wire."[1] After a series of experiments in which he alternately loaded and unloaded the wires, observing the phenomena of recovery and permanent deformation, TOMLINSON went on to investigate the departure from HOOKE's law by comparing the observed increments of elongation for equal increments of load. Some of his results for soft copper and annealed iron are shown in Fig. 2.39. He stated that he obtained similar results for annealed platinum-silver, aluminum, German-silver, and zinc. He stated further that in all three experiments, the wires returned to zero deformation after the removal of the loads, i.e., there was perfect elasticity.

In TOMLINSON's first experiment on copper, designated as experiment VIII, the wire studied had been heavily loaded and frequently tested during a period of three weeks before the measurement was made. In the second experiment, IX, the specimen was treated in the same manner for a period of only six days because

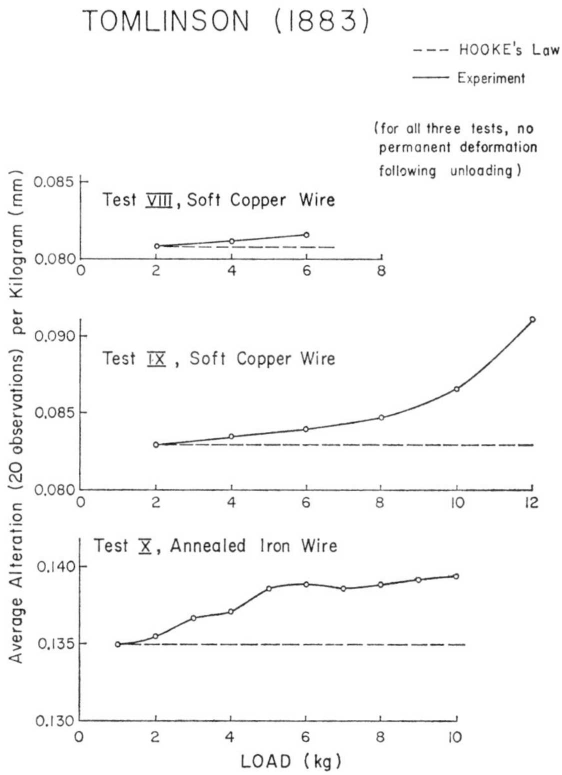

Fig. 2.39. Precise measurement in the tensile loading of 30 m long metal wires exhibiting the departure from linearity shown.

[1] TOMLINSON [1883, *1*], p. 36.

TOMLINSON then was studying KELVIN's earlier observation that the elasticity of metals, even in the range of small deformation, definitely was affected by prior mechanical histories, including vibrations of small amplitude over a period of days. Also, according to KELVIN elasticity was influenced particularly by long periods of rest prior to the experiment. Each of TOMLINSON's measurements shown in Fig. 2.39 consisted in twenty observations. It is interesting that whenever he referred to HOOKE's law or YOUNG's modulus, TOMLINSON always enclosed the terms in quotation marks.

These few experiments of TOMLINSON's series[1] have been described here merely to indicate that as experimental precision increased, consistent nonlinearity of the stress-strain function, whether or not accompanied by permanent deformation, continued to be observed in all metals studied.

In the present context TOMLINSON's major contribution was, first, his recognition of the importance of distinguishing between a "defect of elasticity" and a "defect of HOOKE's law"; and second, his recognition of the fact that the widely used quasi-static moduli obtained by WERTHEIM were averages of measurements made following unloading with different amounts of permanent deformation. Since in all of the solids studied by WERTHEIM those moduli decreased with increasing permanent deformation, the average had no particular significance, as TOMLINSON pointed out. For example, for the determination of specific heats, WERTHEIM's averaged value was represented as the isothermal component to compare with WERTHEIM's vibration moduli for the adiabatic component. (This failure to appreciate the complexities added by having different ranges of strain when comparing dynamic and static data, was shared by both WERTHEIM and his opponent CLAUSIUS, although CLAUSIUS's criticism of WERTHEIM's faulty use of a three-dimensional analysis in a one-dimensional situation is the only remnant of the controversy of nearly 120 years ago which has survived for contemporary reference.)

WERTHEIM's data of 1844 on moduli in metals became the experimental basis for much of the discussion and research of the next sixty years. WERTHEIM had performed experiments on the longitudinal vibration of rods for all of the many metals with different prior thermal treatments which he had investigated; he had separately studied the lateral vibrations of rods, and thus he had determined two dynamic moduli. These were in addition to the quasi-static determinations of E he had obtained from unloading the given specimen from a large number of prior permanent deformations. Before TOMLINSON emphasized the implications of describing the quasi-static E-modulus as the average of all of these data, there had been disagreement as to which set of moduli was the more accurate, the dynamic set or the quasi-static set.

TOMLINSON[2] quoted the following excerpt from KELVIN's article on Elasticity in the *Encyclopaedia Britannica* of 1880: "It is probable that his [WERTHEIM's] moduluses determined by static elongation are minutely accurate; the discrepancies of those found by vibrations are probably due to imperfections of the arrangements for carrying out the vibrational method."[3] Twenty-five years later, EDUARD AUGUST GRÜNEISEN[4] showed experimentally that the dynamic moduli were surprisingly accurate and that when quasi-static measurements were made at the smallest strains, E-moduli were found which were in close agreement with the dynamic values. This same agreement between the dynamic and the initial

[1] TOMLINSON [1883, *1*].
[2] *Ibid.*, p. 14.
[3] KELVIN [1880, *1*].
[4] GRÜNEISEN [1907, *1*].

quasi-static moduli had been found by WERTHEIM. However, WERTHEIM preferred to use the averaged values which included the variation with permanent deformation.

The variation of the elastic moduli of metals with permanent strain or perhaps more correctly, the variation of the tangent modulus for small deformation, is of great importance for current studies of the plastic deformation of crystalline solids where atomistic hypotheses introduce the erroneous assumption that the shear modulus μ for the elastic field around a dislocation is constant, despite the increasing amount of permanent deformation. This dependence of the modulus upon prior deformation had been discovered by COULOMB in 1784,[1] as will be shown in Sect. 3.3 where his data on torsion will be described. WERTHEIM's variations of moduli with permanent deformation are shown in Fig. 2.40 for some of the metals for which he[2] obtained E. The data were from quasi-static tests.[3] Dynamic data determined from KELVIN's[4] studies of the torsional pendulum, which were, in effect, a repetition in 1865 of COULOMB's experiments of ninety years before, are shown in Fig. 2.41 for copper, soft iron, and brass.

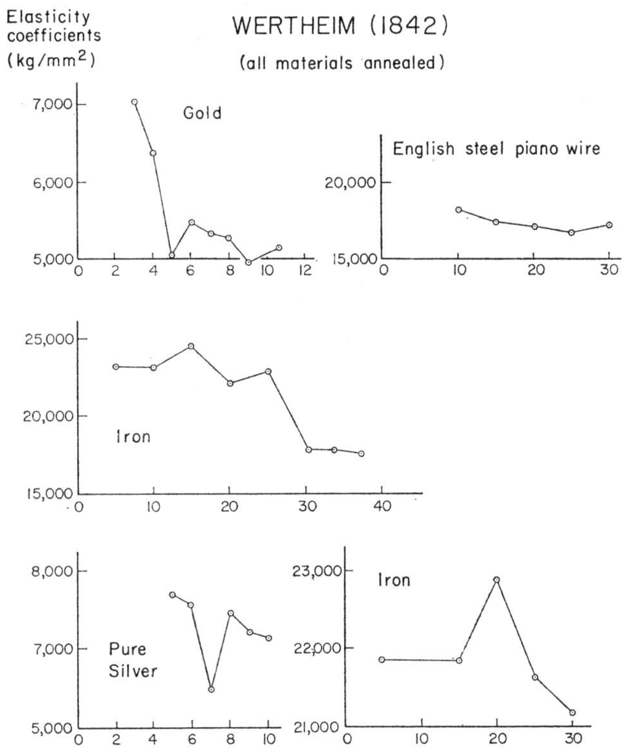

Fig. 2.40. The decreasing modulus with increasing permanent deformation for repeated quasi-static testing.

[1] COULOMB [1784, *1*].

[2] WERTHEIM [1844, *1* (a)].

[3] Both WERTHEIM and KELVIN were aware that they had to consider the change in cross-sectional area with permanent deformation in uniaxial tests. They referred to the experiments of CAGNIARD DE LATOUR [1828, *1*] in describing their concern that a change of volume during plastic deformation also might be a factor contributing to the observed reduction of the modulus.

[4] KELVIN [1865, *1*].

The numerical values of these moduli as a function of the permanent deformation could be altered, as KELVIN discovered, by changing the prior mechanical history of the specimen. Thus the modulus of a specimen which had been maintained for several days under the load producing the permanent deformation was found to be markedly lower than one which had been loaded to the same stress and unloaded to zero stress a number of days before the dynamic moduli were determined; the modulus appeared to recover with time.

The dependence of the moduli upon permanent deformation also was affected by the length of time during which the specimen had or had not been subjected to continuous oscillations of small amplitude. In 1882, with tensile tests on phosphor-bronze wires FRIEDRICH HUGO ROBERT FISCHER[1] studied these phenomena in terms of the dependence of the unloading and reloading for various amounts of permanent deformation. With wires $5^1/_2$ m in length in order to obtain accuracy, he used a self-recording tensile apparatus so that his experiment could proceed without interruption to unload, reload, and continue to the next permanent deformation, bringing a single specimen through a very large number of loading cycles to rupture.[2] FISCHER observed that the response functions during unloading and reloading were nonlinear, as may be seen in Fig. 2.42 taken from his paper.

The cycles shown in Fig. 2.42 were merely the first few of a large series in which the stress P_1 ultimately reached the value of P. FISCHER noted the gradual increase

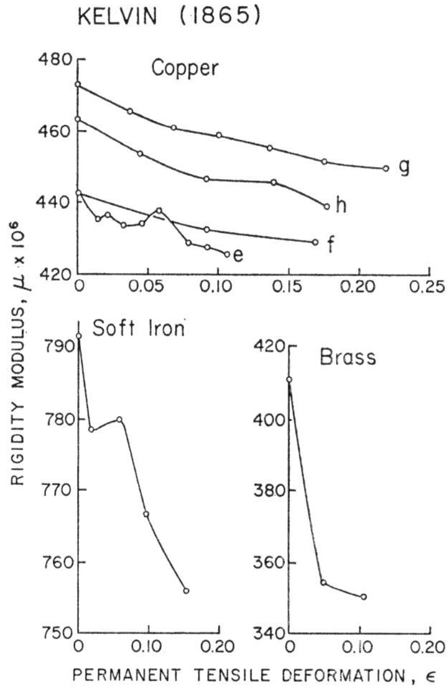

Fig. 2.41. The decreasing modulus with increasing permanent deformation for repeated dynamic testing.

[1] FISCHER [1882, *1*].

[2] Following the development in 1872 by ROBERT HENRY THURSTON at the Stevens Institute of Technology of the first testing machine to record automatically the load and deformation, there was a tremendous amount of interest in automatic recording testing machines of all types during the next twenty years, for all sorts of mechanical testing. THURSTON's machine was a torsion apparatus. That used by FISCHER was the self-registering device developed at the Technisches Institut at Dresden by REUSCH in 1880. The velocity of the recording pencil for the continuous testing data of Fig. 2.42 was 0.3 mm/sec.

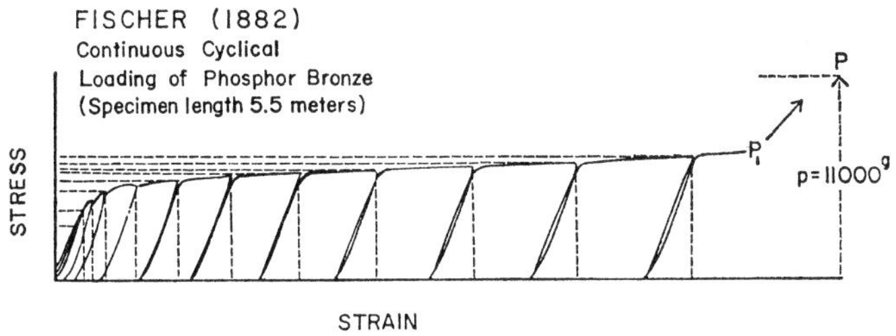

Fig. 2.42. The automatically recorded cyclical loading of a metal, showing the gradual change in the reloading response.

in the difference between the loading and unloading portions of the cycle as the permanent deformation increased. He was convinced that no particular significance should be attached to what is normally called the elastic limit of the material. Rather he viewed as important the "elasticity" of the solid just before rupture, in contrast to the "non-elasticity" associated with the permanent deformation. FISCHER plotted the ratio of the initial elastic recovery and the non-elastic strain, obtaining the curves shown in Fig. 2.43, from which he observed, as had COULOMB[1] for torsion in 1784 and as JOHANN BAUSCHINGER[2] was to find for tension and compression four years after FISCHER, that the magnitude of the "elastic" component remained constant.

Shown in Fig. 2.44 are FISCHER's comparisons of the "elastic" stress-strain curves at different permanent strains, including the region near rupture. He regarded these data as providing a demonstration of the fundamental "elastic" properties of the solid which in effect should be studied as functions of the various prior deformation and thermal histories to which the solid had been submitted. The zero point for such studies would be established by beginning with the annealed solid. As a matter of fact, FISCHER re-annealed his wires for this purpose.

2.21. The cyclical loading of raw silk: Müller (1882).

The self-registering continuous testing machine of REUSCH at Dresden, which was the basis of FISCHER's study, was used by ERNST MÜLLER[3] in 1882 to delineate the different ways the elastic after-effect and permanent deformation contributed to the nonlinearity in the tensile deformation of raw silk. Without any reference to the elegant experiments on silk threads in which WEBER, nearly fifty years before, had discovered the elastic after-effect,[4] MÜLLER in the introduction to his paper justified his experiments as originating in the desire to expand to organic tissues the study of these interesting phenomena in metals. His specimens, which were from 250 to 300 mm in length, varied in cross-section from circular and elliptical to "bean"-shaped. He thus used the weight density and YOUNG's height of the modulus to express the tensile load in comparable units of length. As in some of the earlier studies of organic tissues described above, the strain was not limited to small deformation, which is our major interest here. However, since permanent deformation did not appear until 2.5% strain on the initial loading, and as much as 8% strain on subsequent loadings, MÜLLER's experiments provide the opportunity to study the elastic after-effect and energy distribution under conditions not

[1] COULOMB [1784, *1*].
[2] BAUSCHINGER [1886, *1*].
[3] MÜLLER [1882, *1*].
[4] WEBER [1835, *1*].

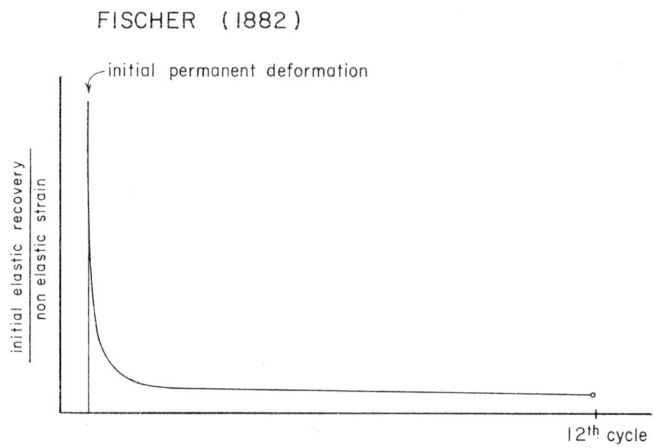

Fig. 2.43. The ratio of the initial elastic recovery to the non-elastic strain with increasing numbers of cycles.

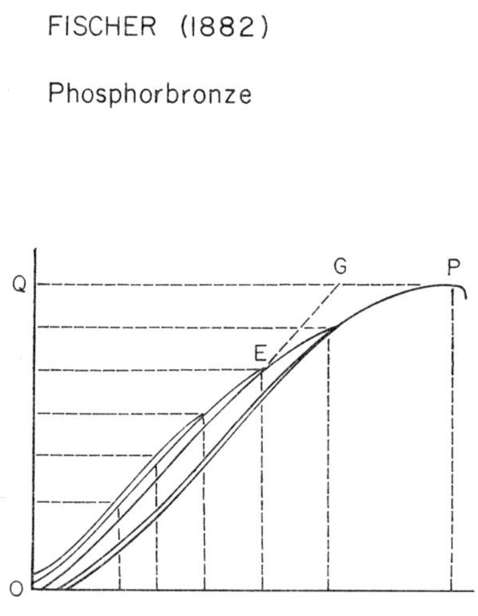

Fig. 2.44. The variation of the reloading behavior with increasing numbers of cycles.

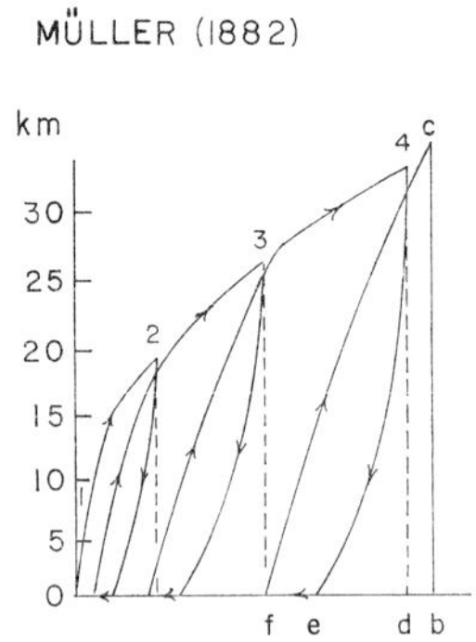

Fig. 2.45. The cyclical loading of silk thread. The automatic recording shows the magnitude of the permanent deformation.

demanding extreme accuracy. The response of raw silk to cyclical loading, as described by MÜLLER is shown in Fig. 2.45. (Note: 33 km provides 44.8 kg/mm²).

The unloading in this continuous test occurred at the positions numbered 1, 2, 3, 4. For the first unloading, perfect elasticity was obtained within the limits of observation. At all subsequent unloadings there was a final permanent set in recovery, not only during unloading, but along the strain abscissa at zero-stress. (I have added the arrows showing the progress of the deformation.) The rate of this recovery at zero stress for two different initial loads is shown in Fig. 2.46 where the times of elastic recovery are noted.

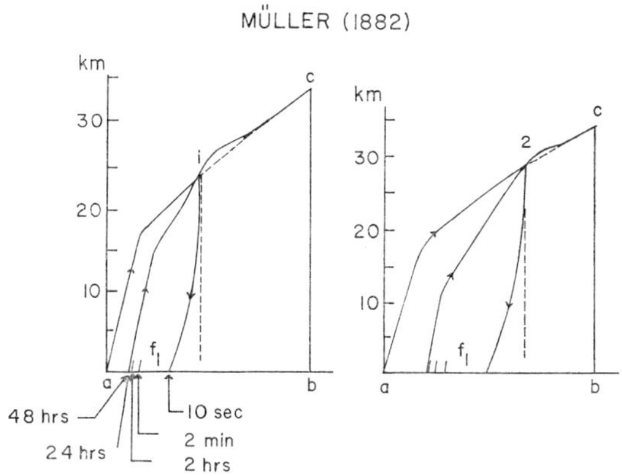

Fig. 2.46. The cyclical loading of silk thread, showing the elastic after-effect.

Not only did the amount of permanent deformation and the magnitude of the elastic after-effect 10 sec after unloading depend upon the load at the onset of unloading, but the unloading and reloading curves were altered. In each instance the reloading curve passed through the previous point at which unloading occurred, then continued on as if no such cycle had been effected.

MÜLLER[1] provided the study of the reloading behavior shown in Fig. 2.47. He obtained each unloading curve by the use of different specimens, having taken proper account of the variations in cross-section and length. The average rupture load was 33 km so that the maximum load in Fig. 2.47 was near the ultimate.

Also shown in Fig. 2.47 is the amount of permanent deformation and recoverable elastic deformation for each unloading. The initial nonlinearity of stress-strain function had occurred before the first permanent deformation was observed at 2.5% strain. Some of the observed nonlinearity must be attributed to the unrecorded temporal details of the elastic after-effect. As was shown in Fig. 2.46, the history of such recovery varied with the pre-load. Each re-load in Fig. 2.47 returned to the previous unloading point. Like FISCHER[2] in his studies on metals, MÜLLER was interested in determining the maximum elasticity for a final reloading from a point near rupture. From Fig. 2.47 one notes that the final loading curves possessed an inflection point above which they turned toward the load axis. In all of WEBER's[3] experiments on silk 47 years earlier, the specimens had been pre-loaded to a maximum stress far beyond that reached in the later measure-

[1] MÜLLER [1882, *1*].
[2] FISCHER [1884, *1*].
[3] WEBER [1835, *1*], [1841, *1*].

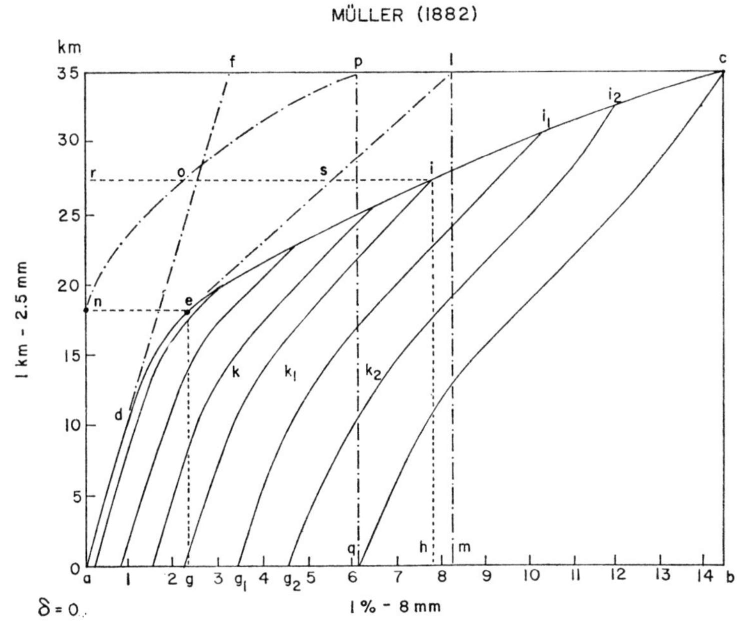

Fig. 2.47. The reloading behavior of pre-loaded silk. Each loading curve represents a different specimen.

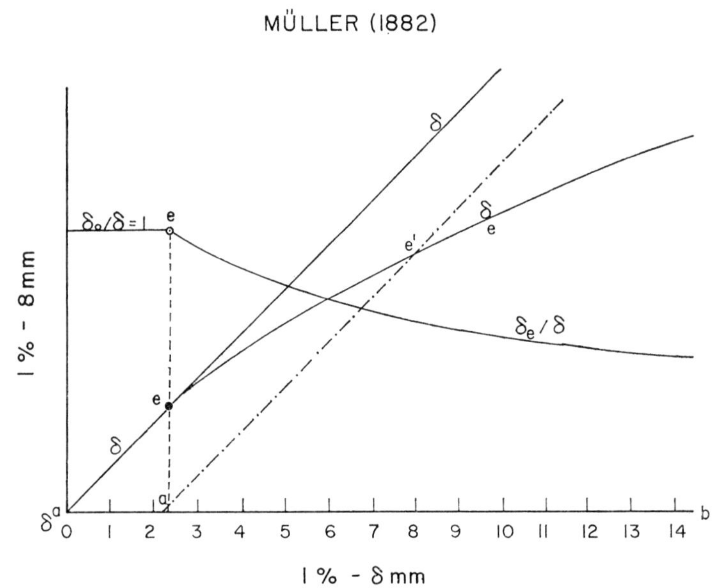

Fig. 2.48. The ratio of the elastic strain δ_e to the total strain, δ, as a function of the total strain for the tensile loading of silk threads.

ments. The ratio of the elastic strain δ_ε to the total strain δ for MÜLLER's experiments is shown in Fig. 2.48 as a function of the total strain.

In addition to this ratio, MÜLLER included the measured values for which, in terms of FISCHER's concept of metals, the final maximum value was considered the fundamental elastic behavior of the solid. Moreover, MÜLLER considered the energy relations as expressed in terms of the ratio of areas, as well as the problems which might arise through cross-sectional variations during deformation, and the manner in which a reduced stress for constant volume might be determined if so desired.[1]

That analysis in 1882 of total strain, permanent deformation, and the elastic after-effect including observations from 10 sec to 48 h after unloading in a solid where such phenomena are visible on an enlarged scale, presented close qualitative parallels with the related work of FISCHER.[2] Combined with the papers of WEBER in 1835 and 1841 described above, the paper of MÜLLER in 1882 provided a fairly thorough experimental introduction to what in modern terminology is called the anelasticity of solids.

Two years later FISCHER[3] extended his studies in phosphor-bronze to a number of other metals. That more detailed paper, together with his earlier work, appeared at the beginning of a twenty-five year period of experimental study of the non-linear response of metals in small deformation which culminated in two directions: the extremely precise optical experiments of GRÜNEISEN[4] 1906–1908 and the nonlinear torsion experiments of J. HENRY POYNTING[5] between 1909 and 1912.

2.22. The first precise measurement of the nonlinearity of metals for infinitesimal deformation: Joseph Thompson (1891).

The major experimentists who studied nonlinear elasticity for small deformation were CARL BACH, ERNST KARL HARTIG, JOSEPH OSGOOD THOMPSON, RUDOLF MEHMKE, and EDUARD AUGUST GRÜNEISEN, whose work was done between 1887 and 1906.[6] In a series of experiments by GEORG STRADLING,[7] which previously had interested his professor, FRIEDRICH KOHLRAUSCH,[8] the question of nonlinearity for small deformation of metals arose. KOHLRAUSCH induced another of his former students, JOSEPH OSGOOD THOMPSON[9], to perform a second series of

[1] MÜLLER [1882, *1*].
[2] FISCHER [1882, *1*].
[3] FISCHER [1884, *1*].
[4] GRÜNEISEN [1906, *1*], [1907, *1*], [1908, *1*].
[5] POYNTING [1909, *1*], [1912, *1*].
[6] Reminiscent of the Royal Commision pronouncement of 1849 concerning HOOKE's law and the engineering profession, was the widely used text on strength of materials entitled, *Elasticität und Festigkeit* by CARL BACH, which had gone into six editions by 1911. BACH's book was unique in that it introduced a parabolic law, generalized to include many solids in the region of infinitesimal strain or "stress range considered in industry," as the basis of discussion. AUGUSTUS EDWARD HOUGH LOVE in his classic, *A Treatise on the Mathematical Theory of Elasticity* [1892 ... 1927] retained the following statement on the matter of experiment and theory, even in his fourth edition in 1927:

> "It is known that many materials used in engineering structures, e.g. cast iron, building stone, cement, do not obey Hooke's Law for any strains that are large enough to be observed ... Although there exists much experimental knowledge in regard to the behavior of bodies which are not in the conditions to which the mathematical theory is applicable, yet it appears that the appropriate extensions of the theory which would be needed in order to incorporate such knowledge within it cannot be made until much fuller experimental knowledge has been obtained."

[7] STRADLING [1890, *1*]. These experiments were designed to re-study WILHELM JACOB STORM s'GRAVESANDE's method for determining moduli [1720, *1*] in terms of a formula derived by KOHLRAUSCH. The data on elongation for wires 23 m long in the control experiment were not proportional to the load, and the moduli decreased markedly with increasing load. (See Sect. 3.2).
[8] I discussed above, in Sect. 2.17, KOHLRAUSCH's experiments in WEBER's laboratory nearly 30 years earlier, in which he discovered stress relaxation in solids.
[9] J. O. THOMPSON [1891, *1*].

experiments on specially prepared wires, 27 m long, of several metals, with sufficient accuracy to settle experimentally the question whether HOOKE's law were applicable to *any* crystalline solid.

THOMPSON himself did not seem to have been aware that the problem which his former professor, KOHLRAUSCH, had suggested to him had ever before been seriously considered. "As far as I know, up to date everybody regarded the old law [HOOKE's law] as valid, and it has never been subjected to a strict criticism."[1] THOMPSON apparently was equally unaware that the experiment he used to study the question had been introduced by KELVIN twenty-five years earlier and extensively used by TOMLINSON in the previous decade. Since THOMPSON in his version of the experiment succeeded in achieving what must be regarded as the most accurate determination of a stress-strain function in the 19th century, one which still compares favorably with the best obtainable even today, it is of interest to provide some experimental detail.

THOMPSON's optical cathetometers were capable of reading accurately to 0.005 mm. To measure strains of less than 10^{-6} he used a gage length of 27 m. To minimize thermal variation he performed the experiments in the tower of the Physical Institute at Strasbourg, where the initial measurements of STRADLING had been made. Six thermometers located at various levels in the tower demonstrated that the variation in temperature over a morning's testing differed by no more than 0.5° C. For a wire 27 m long, however, even this small change in ambient temperature produces a considerable deflection. Therefore, as both KELVIN and TOMLINSON had done before him, THOMPSON introduced a thermometer wire which was hung from the same suspension system as the specimen and remained throughout the test under constant load.[2] It was sufficiently close to the specimen wire of the same length that its elongation due to ambient temperature fluctuations at the time in which the measurement was made could be read on the same optical cathetometer. Both wires were attached to a large beam at the top of the tower. This supporting beam was submitted to a load four times greater than the maximum in the experiments. An optical cathetometer located at the point of attachment under the beam demonstrated that the measurable deflection of the support was of the order of 0.013 mm; thus for the maximum load in the test, only one-fourth as great, the deflection of the support would be negligible. Nonetheless, throughout the experiment this optical cathetometer at the top of the wires was checked for constant length by means of a comparison with a fixed optical cathetometer at the zero point, 27 m below. A third optical cathetometer at the bottom of the specimen was used to measure the changing deformation. THOMPSON's thermometer wire was sufficiently long that he could measure a change in temperature of $^1/_{40}$° C and estimate a change of $^1/_{80}$° C.

THOMPSON, like all his predecessors and followers in dead-weight testing, was perplexed with respect to the strain which should be recorded. WERTHEIM had waited from 5 to 10 min before recording strains in metals so that the thermal after-effect and the elastic after-effect would have reached an approximate stability. ANDREAS MILLER[3] was criticized by THOMPSON for having chosen to measure the elongations two seconds after applying the load since, as THOMPSON[4] noted, thermal conditions certainly were not stationary at that time. THOMPSON himself was criticized a decade and a half later by GRÜNEISEN[5] for having chosen to meas-

---

[1] *Ibid.*, p. 555.
[2] However, THOMPSON attributed this technique to a suggestion of Professor KOHLRAUSCH.
[3] MILLER [1882, *1*].
[4] THOMPSON [1891, *1*].
[5] GRÜNEISEN [1906, *1*].

ure his strain precisely 13 sec after applying the load, since, as GRÜNEISEN noted, the elastic after-effect was not stationary and its magnitude varied with the load. THOMPSON, however, provided an example in which a change in length of 0.07 mm occurred in the first 10 or 12 sec; in the subsequent minute the further change was only 0.01 mm. In any event, following the prescription of WERTHEIM in this complicated matter, THOMPSON tells us precisely what he did.

To improve his accuracy in the measurements of strains whose threshold determination was 2×10^{-7}, THOMPSON made 20 precise repetitions of every experiment, always correcting for temperature from his thermometer wire and always checking that after the removal of the load the wire returned exactly to the initial zero-point as determined from the fixed cathetometer and as corrected, if any slight temperature shift occurred, by means of the thermometer wire.

In addition to these precautions, THOMPSON used an iron, nickel-silver thermocouple to observe by these means the transient temperature changes which occurred during load, obtaining a time interval of 6.5 sec for a return to equilibrium comparable to the 6 sec obtained by WEBER[1] sixty years before.

THOMPSON criticized the values obtained experimentally by ERIK EDLUND[2] and JAMES PRESCOTT JOULE.[3] JOULE had provided an approximate experimental correlation of KELVIN's theoretical relation between temperature increment Δt produced by a load increment Δp. EDLUND had found values which were smaller by about 40% than KELVIN's predicted values. THOMPSON was able to study the same phenomenon using the results of EDLUND which would have predicted thermal effects producing a motion of up to 0.10 mm in THOMPSON's measurements. The values which THOMPSON obtained in the instance cited above were much smaller, i.e. 0.07 mm.

THOMPSON performed his experiments[4] on brass, copper, steel, and silver wires. He concluded that the tangent modulus varied with the load in every one of these metals. His measured data were two orders of magnitude more accurate than those WERTHEIM had obtained nearly fifty years earlier. Nevertheless, as THOMPSON pointed out, the variation of the moduli with stress in the two sets of data could be correlated directly in many cases. He provided two instances of this comparison in polycrystalline silver and gold. In Table 19 is shown the reproducibility in the compilation of the individual data for a series of ten measurements on brass wire.

Table 19. J. O. THOMPSON (1891).

kg	mm										Avg. (mm)
	1	2	3	4	5	6	7	8	9	10	
0.2	7.10	7.13	7.12	7.11	7.105	7.12	7.11	7.11	7.12	7.09	7.111
0.4	14.25	14.265	14.26	14.28	14.285	14.27	14.265	14.27	14.28	14.27	14.269
0.6	21.47	21.50	21.50	21.50	21.485	21.495	21.49	21.475	21.49	21.49	21.489
0.8	28.76	28.775	28.78	28.77	28.75	28.785	28.77	28.77	28.79	28.77	28.772
1.0	36.11	36.125	36.14	36.145	36.115	36.13	36.13	36.11	36.12	36.12	36.124
1.2	43.53	43.57	43.56	43.55	43.555	—	43.57	43.55	43.545	43.56	43.554
1.4	51.05	51.07	51.065	51.07	51.065	51.06	51.065	51.09	51.055	51.08	51.067
1.6	58.66	58.71	58.685	58.71	58.66	58.69	58.675	58.68	58.69	58.675	58.683
1.8	66.33	66.355	66.335	66.36	66.32	66.35	66.31	66.32	66.34	66.32	66.334

[1] WEBER [1830, *1*].
[2] EDLUND [1865, *1*].
[3] JOULE [1859, *1*].
[4] J. O. THOMPSON [1891, *1*].

Thompson proposed a simple cubic stress-strain function for fitting the data in all four metals:

$$\varepsilon = a\sigma + b\sigma^2 + c\sigma^3. \tag{2.24}$$

In Table 20 is shown the average of 18 measurements in brass. Using three data points to determine a, b, and c, Thompson compared observation and calculation from Eq. (2.24).

Table 20. J. O. Thompson (1891).

p (kg)	x Observed (mm)	x Calculated (mm)	Observed-calculated (mm)
0.2	7.111	7.110	+0.001
0.4	14.272	14.271	+1
0.6	21.488	21.488	±0
0.8	28.770	28.770	±0
1.0	36.119	36.122	−3
1.2	43.554	43.554	±0
1.4	51.076	51.071	+5
1.6	58.679	58.681	−2
1.8	66.341	—	—
Average temperature		9°	
Length of the wire		22700 mm	
Cross-section of the wire		0.0627 mm²	
Density of the wire		8.42	
Initial loading		0.665 kg	

p is the additional loading; x is the strain.

Similar comparisons for the copper, steel, and silver averaged data are shown in Table 21:

Table 21. J. O. Thompson (1891).

p (kg)	x Observed (mm)	x Calculated (mm)	Observed-calculated (mm)
		Copper	
0.2	5.531	5.529	+0.002
0.4	11.084	11.086	−2
0.6	16.671	16.673	−2
0.8	22.298	22.294	+4
1.0	27.949	27.951	−2
1.2	33.646	33.646	±0
Average temperature		13.5°	
Density of the wire		8.99	
Cross-section of the wire		0.0641 mm²	
Length of the wire		22.690 mm	
Initial loading		0.192 kg	
		Steel	
0.2	7.078	7.077	+0.001
0.4	14.196	14.197	−1
0.6	21.358	21.358	±0
0.8	28.558	28.558	±0
1.0	35.792	35.793	−1
Average temperature		13°	
Density of the wire		7.74	
Cross-section of the wire		0.03263 mm²	
Length of the wire		22.70 m	
Initial loading		0.491 kg	

Table 21. (Continued)

p (kg)	x Observed (mm)	x Calculated (mm)	Observed-calculated (mm)
		Silver	
0.2	7.898	7.896	+0.002
0.4	15.820	15.822	−2
0.6	23.775	23.776	−1
0.8	31.758	31.756	+2
1.0	39.762	39.762	±0
Average temperature		14°	
Density of the wire		10.00	
Cross-section of the wire		0.0687 mm²	
Length of the wire		22.69 m	
Initial loading		0.593 kg	

Rewriting Eq. (2.24) as load P vs elongation X, THOMPSON obtained the empirical values of the coefficients a, b, c, shown for the four metals in Eq. (2.25):

$$\begin{aligned}
&\text{Steel:} && X=34.672\,P+0.6498\,P^2-0.0525\,P^3\\
&\text{Brass:} && X=34.924\,P+0.2386\,P^2+0.1487\,P^3\\
&\text{Silver:} && X=38.907\,P+0.4462\,P^2-0.313\,P^3\\
&\text{Copper:} && X=27.461\,P+0.2883\,P^2+0.0538\,P^3
\end{aligned} \tag{2.25}$$

The modulus E, or preferably the tangent modulus,[1] of course could be obtained by differentiating Eq. (2.24) at any value of stress. To emphasize the amount of variation that had taken place when no permanent deformation was occurring (i.e., in every instance, unloading produced a return to zero-strain), THOMPSON compared E at the smallest load with the secant modulus[2] E_{I} and the tangent modulus E_{II} for the last increment of load. In Table 22 these comparison are shown:

Table 22. J. O. THOMPSON (1891).

	d	p (mm²)	E	E_{I}	E_{II}
Steel	7.74	0.03263	20050	19430	19230
Brass	8.42	0.0627	10370	9820	9450
Silver	10.00	0.0687	8490	8300	8250
Copper	8.99	0.0641	12890	12620	12420

2.23. Hartig's nonlinear law: A general response function for the small deformation of solids (1893).

A contemporary contrast to the precise experiments of JOSEPH THOMPSON was the research of ERNST KARL HARTIG in 1893.[3] Doing his work at the end of what is now sometimes slightingly referred to as the century of phenomenological

[1] The tangent modulus is the slope of the response function at a specified stress or strain in an uniaxial tension or compression experiment, or in a simple torsion experiment.

[2] The secant modulus is the ratio of stress to strain irrespective of the form of the response function obtained in an uniaxial tension or compression experiment, or in a simple torsion experiment.

[3] HARTIG [1893, *1*].

physics, he had a notion of what constituted a reasonable fit of empirical relations to experimental data reminiscent of that of LEIBNIZ[1] and of BÜLFFINGER[2] two centuries earlier. However, HARTIG's emphasis upon the study of tangent moduli in tension and in compression in order to ascertain a value of E at zero stress and the manner in which E varied for small changes of stress at either side of this zero value, led to a new approach in the effort to understand nonlinearity in small deformation. This proposal and HARTIG's formula for the tangent modulus of metals contained in it, thirteen to fifteen years later led to GRÜNEISEN's definitive, precise experimental statement describing the nonlinearity of small deformation.

HARTIG, after giving a handwaving reference to the importance of the work of HODGKINSON fifty years before, rested content as to other possible predecessors, and analyzed only the nonlinear small deformation experimental studies of the decade and a half preceding his own contribution. After viewing the experiments of the next quarter of a century, one can see that the main point revealed by HARTIG is that for the tension data then available in metals, the tangent modulus decreased with increasing stress, and in compression it increased with increasing stress. In CARL BACH's[3] data of 1897 in cast iron, listed in Table 23, the decrease of the modulus with increasing tensile stress is obvious. JACOB JOHANN WEYRAUCH[4] in 1888 observed a similar decrease from $E=956400$ to $E=672300$ kg/cm² when the load increased from 148 to 1040 kg/cm².

Table 23. BACH (1897).

Boundaries of the specific stress σ (kg/cm²)	Elasticity modulus E (kg/cm²)
0–100	1220000
100–200	1115000
200–300	1085000
300–400	1020000

ARMAND GABRIEL CONSIDÈRE[5] described experiments in grey cast iron by a Lieutenant BUBBÉ, in which the modulus changed from the initial value of $E=950000$ to $E=180000$ kg/cm², near rupture. The data from CONSIDÈRE's book and the data of ROBERT HENRY THURSTON[6] obtained on an autographic testing machine in 1878 are reproduced in Fig. 2.49, from HARTIG's paper of 1893. In Fig. 2.50, from HARTIG[7] also, are the tension-compression data in cast iron of ALEXANDER BLACKIE WILLIAM KENNEDY[8] in 1887 (solid line) and the cast iron data published in 1892 by HUDSON BEARE[9] (dashed line). In both sets of data the permanent deformation was excluded; KENNEDY had used a testing machine described by WILLIAM CAWTHORNE UNWIN,[10] which was designed to permit the elimi-

[1] LEIBNIZ [1690, *1*].
[2] BÜLFFINGER [1729, *1*].
[3] BACH [1897, *1*].
[4] WEYRAUCH [1888, *1*], p. 94.
[5] CONSIDÈRE [1888, *1*], p. 78.
[6] THURSTON [1878, *1*]. THURSTON was a Professor of Engineering at the Stevens Institute of Technology, U.S.A.
[7] HARTIG [1893, *1*].
[8] KENNEDY [1887, *1*]. KENNEDY was a widely known British engineer, once President of the British Institute of Mechanical Engineers.
[9] BEARE [1892, *1*].
[10] UNWIN [1910, *1*]. The machine actually was based on the autographic torsion tester invented by THURSTON in 1872. UNWIN was a Professor of Civil Engineering at Central Technical College, England, and was A. E. H. LOVE's chief authority on "technical" elasticity.

nation of the permanent deformation at each stress; and BEARE had pre-stressed his specimen.

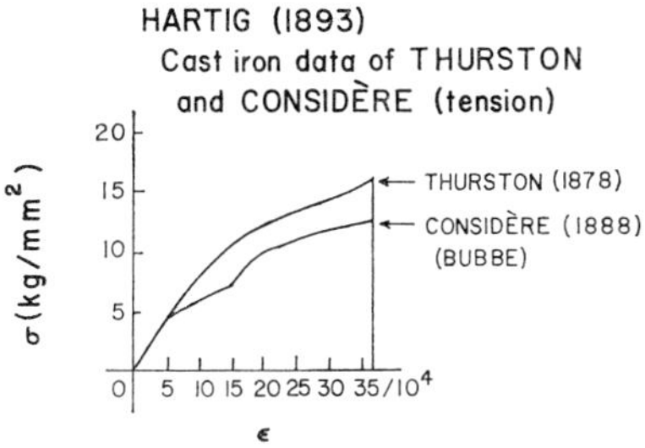

Fig. 2.49. The small deformation tensile testing of cast iron used by HARTIG in his analysis.

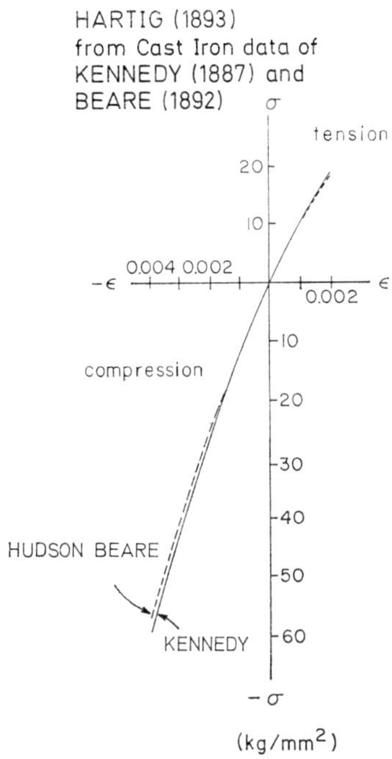

Fig. 2.50. The tension and compression data in cast iron used by HARTIG in his analysis.

From all the data HARTIG proposed for both tension and compression stresses the relation

$$E = \frac{d\sigma}{d\varepsilon} = E_0 - b\sigma, \tag{2.26}$$

where E_0 is the zero stress modulus. Integration yielded the following stress-strain relation of Eq. (2.27):

$$\sigma = \frac{E_0}{b}\left(1 - e^{-b\varepsilon}\right) \tag{2.27}$$

which may be rewritten as Eq. (2.28):

$$\sigma = E_0\,\varepsilon - \frac{b E_0}{2}\,\varepsilon^2 + \frac{b^2 E_0}{6}\,\varepsilon^3 + \cdots. \tag{2.28}$$

The first two terms of Eq. (2.28) are GERSTNER's stress-strain function[1,2] for piano wire and HODGKINSON's empirical function for cast iron, given more than half a century earlier, in 1824 and 1839, respectively.[3]

HARTIG noted that the tensile date of BACH in copper provided a decrease of the tangent modulus from 1.1×10^6 to 0.704×10^4 kg/cm² as the load increased from 100 to 600 kg/cm². Although THOMPSON's Eq. (2.24) used a different empirical form from HARTIG's Eq. (2.27), the tensile tangent moduli of course decreased for both copper and steel.

The phosphor-bronze data of FISCHER,[4] who was a colleague of HARTIG's at Dresden, have been reexamined in the present context. One observes a decrease of E from 1.165×10^6 kg/cm² for the stress interval between 0 and 1270 atmospheres,[5] and the much lower value of 90000 kg/cm² between 3130 and 3300 atmospheres, with a smooth transition between the two. HARTIG saw the necessity for distinguishing between data such as those of FISCHER which contained permanent deformation and those of THOMPSON, for example, which did not, but he used both their data for the purpose of interpolating E_0 at zero stress from empirical functions. FISCHER's data in phosphor-bronze are reproduced from HARTIG's paper in Fig. 2.51.

The extensive experimental studies of BACH on the stress-strain functions of leather in tension revealed an increase of the tangent modulus with increasing stress. As was shown above,[6] this behavior is characteristic of organic tissues in

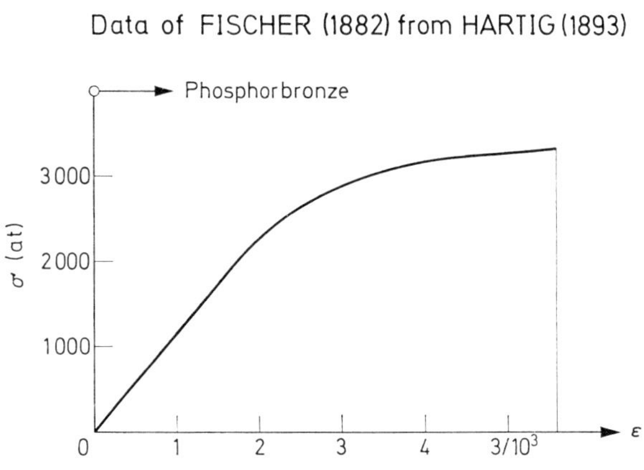

Fig. 2.51. The phosphorbronze tensile data of FISCHER, used by HARTIG in his analysis.

[1] However, "Gerstner's Law" of 1824 was concerned primarily with small deformation permanent set.

[2] The load-deflection function for wooden beams of DUPIN in 1811, Eq. (2.2) also may be rewritten in the same form as Eq. (2.28), i.e.

$$F = \frac{1}{a}\,\delta - \frac{b}{a^3}\,\delta^2 + \frac{2b^2}{a^5}\,\delta^3 \ldots$$

and thus is the first experimental observation of this form.

[3] See Sects. 2.6 and 2.7 above. The third term of Eq. (2.28) does not constitute an adjustable correction but was determined uniquely by the same data that determined the GERSTNER-HODGKINSON function.

[4] FISCHER [1882 *1*].

[5] 1 atm ≅ 14.7 psi ≅ 1.03 kg/cm².

[6] See for example Sects. 2.13, 2.14 and 2.15.

general. HARTIG proposed that BACH's data, shown in Fig. 2.52 could be fitted to Eq. (2.26) with a change of sign for extrapolation to E_0:

$$E = \frac{d\sigma}{d\varepsilon} = E_0 + b\sigma, \tag{2.29}$$

leading to Eqs. (2.30) and (2.31) for the stress-strain relation for leather:

$$\sigma = \frac{E_0}{b}\left(e^{b\varepsilon} - 1\right), \tag{2.30}$$

$$\sigma = E_0\,\varepsilon + \frac{b E_0}{2}\,\varepsilon^2 + \frac{b^2 E_0}{6}\,\varepsilon^3 + \cdots. \tag{2.31}$$

For the compression vs tension data on vulcanized rubber given by EMIL WINKLER[1] in 1878, shown in Fig. 2.53 in both nominal form as measured and in reduced form assuming incompressibility, HARTIG conceded that Eq. (2.26) for the upward-turning stress vs strain function for leather was not applicable. He noted that the empirical equation for rubber introduced by A. IMBERT[2] in 1880, Eq. (2.32), provided the tangent moduli relation of Eq. (2.33):

$$\varepsilon = e^n - 1, \tag{2.32}$$

$$E = \frac{d\sigma}{d\varepsilon} = \frac{1/n}{1+\varepsilon}. \tag{2.33}$$

The latter equation approximated the data in the small strain region for $E_0 \cong 10\ \mathrm{kg/cm^2}$.

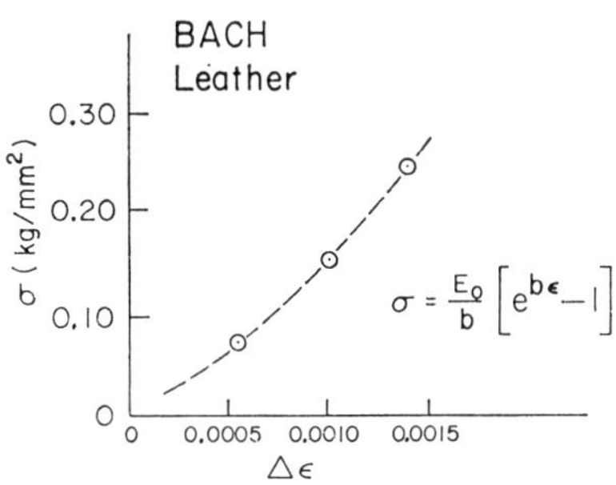

Fig. 2.52. The tensile data of BACH compared with HARTIG's equation.

HARTIG thought it to be of considerable importance to determine whether or not the nonlinear stress-strain function proceeded smoothly through the zero stress position. If the function were smooth, and if no point of inflection were observable then the point of zero stress in itself would be of no particular significance.[3]

As his own experimental contribution to this effort to demonstrate nonlinearity for small deformation through the point of zero stress, HARTIG[4] presented his data

[1] WINKLER [1878, *1*]. The strain was measured at long times to include the after-effect.

[2] IMBERT [1880, *1*].

[3] It is interesting to contrast these ideas of HARTIG with those introduced by TRUESDELL six decades later in his theory of hypo-elasticity. See TRUESDELL [1955, *1*], p. 85. With respect to the generalization of Eq. (2.29), it also is of interest to examine the work of W. PRAGER carried out between 1938 and 1942. See PRAGER [1938, *1*], [1941, *1*], [1942, *1*].

[4] HARTIG [1893, *1*].

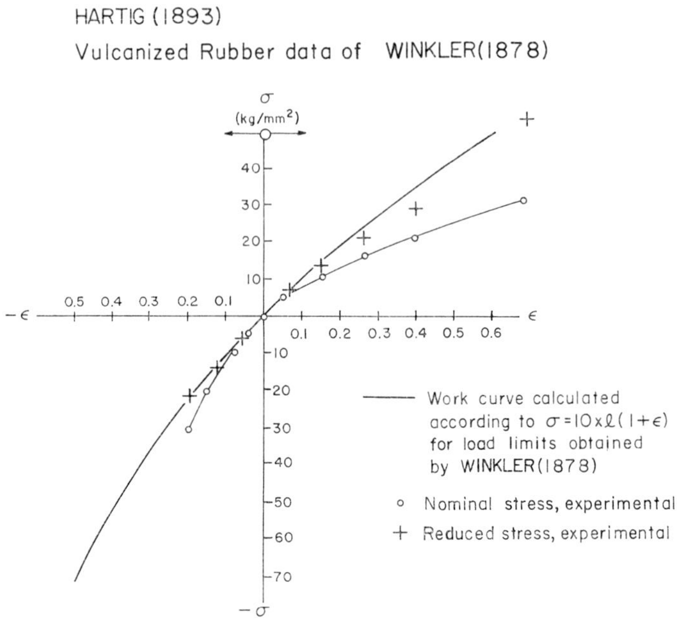

Fig. 2.53. HARTIG's analysis of the tensile data of WINKLER in vulcanized rubber.

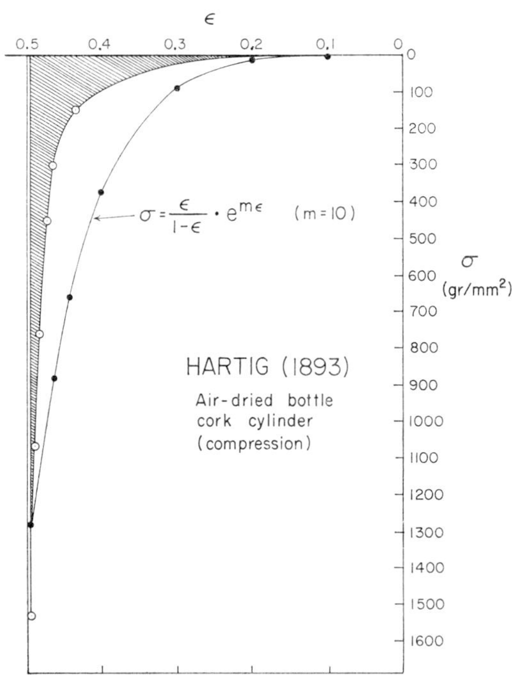

Fig. 2.54. Measurements on cork in compression (open circles), compared with HARTIG's prediction from small strain results. The lack of agreement is obvious.

on compression of a 2.52 cm high, 6.895 cm^2 cross-section specimen of a "practically flawless" air-dried specimen of bottle cork. He twice compressed the specimen to large deformation to eliminate further permanent deformation and to minimize the elastic after-effect. Since the latter still was present 36 h later, he made his measurements of deformation as soon as possible after the application of the load. These data, plotted in Fig. 2.54, he considered to be approximately representable by means of Eq. (2.34), in terms of which he provided a questionable estimate of E_0 at the origin.

$$\sigma = \frac{\varepsilon}{1-\varepsilon} e^{m\varepsilon}. \tag{2.34}$$

From Eq. (2.34) one obtains

$$E = \frac{d\sigma}{d\varepsilon} = \frac{(m+1)}{(1+\varepsilon)^2} e^{m\varepsilon} \tag{2.35}$$

which gave a value of $E_0 = 11$ kg/cm². He noted that when compressed to 50% of its unloaded height, the tangent modulus of elasticity for cork had increased to $E' = 6595$ kg/cm².

HARTIG's[1] study of small deformation nonlinearity, together with the precise data of JOSEPH OSGOOD THOMPSON[2] two years earlier, set a pattern for intensive study during the next fifteen years. In terms of sheer bulk of data, the major contributor was CARL BACH,[3] who between 1884 and 1897 conducted experiments and made thousands of measurements in the region of small strain, in an effort to find a generalized stress-strain function to replace HOOKE's law for small deformation. BACH had in mind the needs of engineering technology.

2.24. The Bach-Schüle law (1897): A rediscovery of the parabolic response function of James Bernoulli (1695) and of Hodgkinson (1824).

In all of the solids BACH considered, he observed deviations from linearity. He stated that HOOKE's law, which had formed the basis for the linear theory of elasticity, could be regarded as valid only for a minority of materials, and then only within certain limits. In 1897, as the result of his own experiments and his review of the very careful experiments of J. O. THOMPSON, BACH[4] concluded that it was no longer realistic to regard the linear relation as a general law. This observation then became the basis for his more general study of elastic behavior. He pointed out that when submitted to very careful tests, the important engineering materials such as cast iron and steel, for which the applicablity of HOOKE's law was generally assumed, did not behave as predicted by that law.

BACH's extensive experiments on tension, compression, torsion, and flexure included measurements in cast iron, copper, granite, pure cement, cement plaster, concrete, leather, and sandstone. At the end of ten years of experimentation, he asked a former student, Dr. WILHELM SCHÜLE, to use the experimental results to try to ascertain a single generalized nonlinear stress-strain function applicable to all of these solids, since he believed the search for a nonlinear law between strains and stress was more meaningful than the establishment of an approximate elastic

[1] *Ibid.*
[2] THOMPSON [1891, *1*].
[3] BACH [1884, *1*], [1887, *1*], [1897, *1*].
[4] BACH [1897, *1*].

modulus E. In a short time, SCHÜLE, like others before him, rediscovered the BERNOULLI parabolic law[1] of 1694:

$$\varepsilon = \alpha \sigma^m, \tag{2.36}$$

which provided various values of α and m for the different solids. In Table 24 may be seen the values of α and m which SCHÜLE obtained.

Table 24. BACH-SCHÜLE (1897).

	$1/\alpha$	m
Cast iron (compression)	1381700	1.0663
Cast iron (tension)	1132700	1.395
Copper (tension)	2084000	1.093
Granite I (compression)	249540	1.13207
Granite II (compression)	339750	1.1089
Granite III (tension)	234600	1.374
Pure cement I (compression)	254841	1.0903
Pure cement II (compression)	259134	1.0950
Pure cement III (compression)	231416	1.09282
Cement-plaster (compression)		
Ratio: 1 part cement; $1^1/_2$ parts sand[a]	355942	1.10984
1 cement; 3 sand[a]	315239	1.14732
1 cement; $4^1/_2$ sand[a]	229026	1.16871
Croncete		
Ratio: 1 cement; 5 sand[a]; 5 gravel[b]	297820	1.14478
1 cement; $2^1/_2$ sand[c]; 5 limestone gravel	456910	1.15749
1 cement; 5 sand[a]; 6 gravel[b]	279981	1.13713
1 cement; 3 sand[a]; 6 limestone gravel	380283	1.16075
1 cement; 5 sand[a]; 10 gravel[b]	217260	1.15662
1 cement; 5 sand[c]; 10 limestone gravel	367018	1.20677

[a] Danube sand.
[b] Danube gravel.
c Egginger sand.

At the end of his paper describing this discovery, BACH[2] with considerable enthusiasm stated that he had found the sought-for law of elasticity in the small stress range used in industry and thus had established a sufficiently firm basis for more general studies. BACH's experiments had been performed not only in tension and compression, which SCHÜLE had considered, but also in torsion and flexure, and he expressed concern that the supposedly general nonlinear function had not been extended to torsion and flexure.

BACH's enthusiasm must have been short-lived. In the latter part of the same year, 1897, RUDOLF MEHMKE,[3] like SCHÜLE, used the small strain data of BACH but considered a number of different stress-strain functions. MEHMKE's demonstration that different types of functions also gave a close fit to the data, further indicates the inherent dangers in fitting data for single experimental situations to curves. In his paper MEHMKE provided a tabulation of the infinitesimal deformation laws of uniaxial stress-strain elasticity which had been proposed until 1897. In Fig. 2.55 are shown MEHMKE's comparisons of the indicated different

[1] By the end of the 19th century it had become customary to make no reference to JAMES BERNOULLI's parabolic law of 1694, let alone to HODGKINSON's observations in 1824.

[2] BACH [1897, *1*].

[3] MEHMKE [1897, *1*].

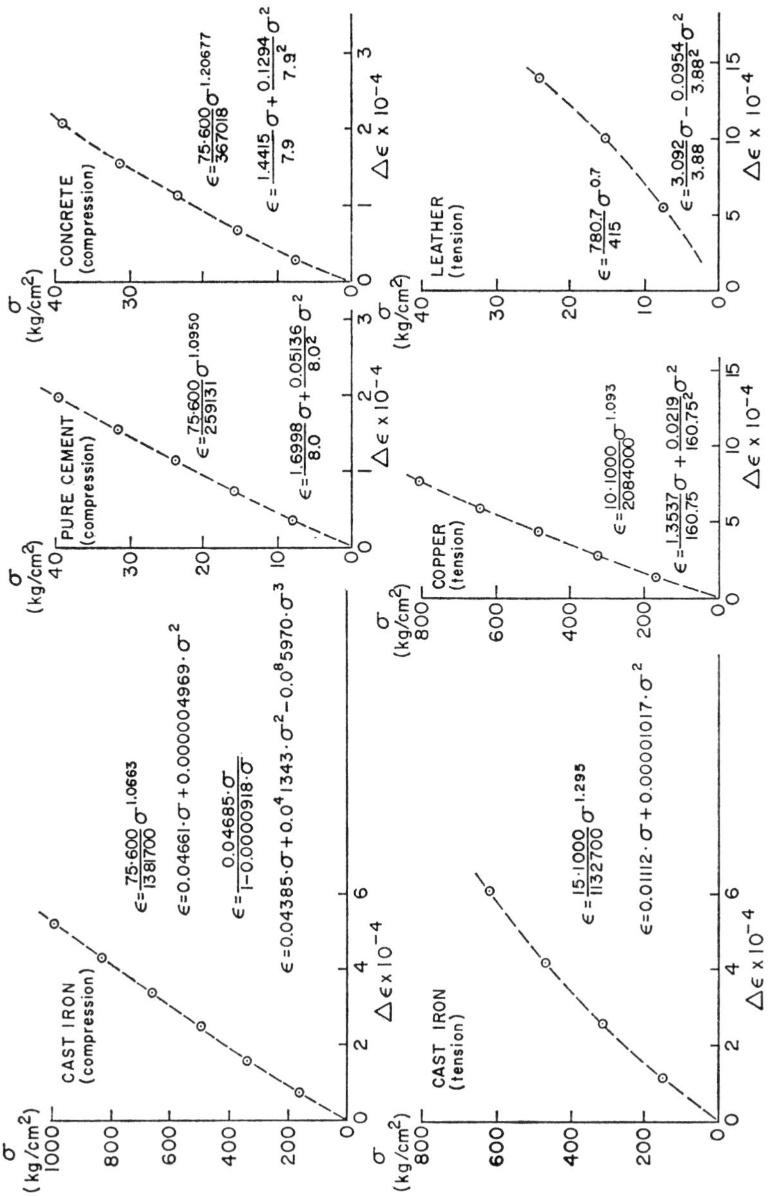

Fig. 2.55. The data of BACH from different solids fitted by MEHMKE to various empirical equations. The differences are too small to be delineated in the scale of the figures.

stress-strain functions with BACH's[1] experimental data. It should be emphasized that BACH's figures were the averages of data from hundreds of tests beginning with his efforts to determine an elastic modulus for leather in 1884.

RUDOLF MEHMKE was motivated to study the BACH-SCHÜLE data because of his concern that the exponential formula Eq. (2.36), despite its being "charming

[1] Overlooking BÜLFFINGER's own clear acknowledgment to VARIGNON (1702) (see TRUESDELL [1960, *1*], p. 102) and BERNOULLI [1695, *1*], MEHMKE incorrectly credited the parabolic law to BÜLFFINGER [1729, *1*].

by its elegance,"[1] presented difficulty with respect to the modulus or slope at zero stress. MEHMKE made one of the strongest attacks of that time upon the physicists' unquestioning acceptance of the linear law of HOOKE. This effort to interest the scientific community in the essential nonlinearity of metals, etc., is reminiscent of the similar efforts of VICAT sixty-five years before. MEHMKE pointed out that HOOKE's law never had been totally uncontested and had been confirmed only speciously by the experiments of several physicists[2] beginning with WERTHEIM in 1848; that nevertheless it had "threatened to become a dogma" and had been represented as a "self-proven law."[3]

> "We see here the peculiar case in which the physicists have hampered for a while the progress of perception of the truth and indirectly have held back the development of an important branch of engineering".[4]

His review of the accumulated nonlinear stress-strain functions of his predecessors, listed in Table 25, was published for the purpose of sparing engineers and scientists a repetition of older and forgotten studies, and with the expressed hope that he would stimulate the study of elasticity without the assumption of HOOKE's law.

In the scale of the drawings shown in Fig. 2.55 the calculated numerical values for the different empirical functions indicated are too close to be delineated. From the tabulated data in MEHMKE's paper one observes that in every situation shown, the BERNOULLI-BÜLFFINGER-HODGKINSON-BACH-SCHÜLE exponential stress-strain function provided the closest fit. MEHMKE commented that even the cubic relation with three constants did not fit the data better.[5] Only in the case of granite in compression (not shown) did the exponential law diverge more from the data.

The important point made by both MEHMKE and BACH was that the exponential nonlinear stress-strain relation described the small deformation of such diverse materials as copper, pure cement, or leather. It seemed to them entirely reasonable to think that this discovery was as fundamental as that of HOOKE himself, over two centuries earlier.

---

[1] MEHMKE [1897, *1*], p. 329.

[2] The experimentists of course include VOIGT and KELVIN. WERTHEIM attributed his variations of modulus with permanent deformation solely to density changes, an assumption demonstrated by GRÜNEISEN to be fallacious, as will be shown in Sect. 2.25 below. For mathematical simplicity KELVIN, like VOIGT, preferred to consider linearity. KELVIN would say of his own torsion data:

> "The results proved a deviation from Hooke's law by showing a diminution of the torsional rigidity, of about 1.6 per cent produced by hanging a weight of 112 lb. on the wire. Of this 1.2 per cent is accounted for by elongation and by shrinkage of the diameter, leaving 0.4 per cent of the diminution of the rigidity-modulus." (See KELVIN [1878, *1*], p. 23.)

Nevertheless, in the same article KELVIN merely had asserted without demonstration that linearity must hold for infinitesimally small deformation. KELVIN stated:

> "The answer happily for mathematicians and engineers is *that Hooke's law is fulfilled, as accurately as any experiments hitherto made can tell,* for all metals and hard solids each through the whole range within its limits of elasticity; and for woods, cork, India-rubber, jellies, when the elongation is not more than two or three per cent, or the angular distortion not more than a few hundredths of a radian (or not more than about two or three degrees)." (*Ibid.*, p. 9.)

KELVIN then added in parentheses that the experiments done in his laboratory by his assistant, MACFARLANE, were performed *after* the previous section had been set in type. [Certainly the extensive, nonlinear data from 1811 to 1878 had been available to him.]

[3] MEHMKE [1897, *1*], p. 327.

[4] *Ibid.*, p. 328.

[5] MEHMKE [1897, *1*] stated that after making these calculations he learned that HOMERSHAM COX had shown that his hyperbolic relation gave a better fit to HODGKINSON's data for iron than did the parabolic relation. For BACH's data for the same solid, MEHMKE commented that the hyperbolic and parabolic laws agreed almost precisely. MEHMKE felt that this difference between the data of HODGKINSON and BACH needed clarification.

Table 25. MEHMKE (1897).

1. *Linear law:*	$\varepsilon = a\sigma$	HOOKE (1678)
2. *Exponential law:*	$\varepsilon = a\sigma^m$	JAMES BERNOULLI (1694)[a]; BÜLFFINGER (1729), tension; HODGKINSON (1822); BACH-SCHÜLE (1897)
3. *Parabolic law:*	$\sigma = a\varepsilon - b\varepsilon^2$	HODGKINSON (1849), cast iron; HARTIG (1893), cast iron, cement, cement plaster; GERSTNER (1824)[b], iron piano wire
4. *Hyperbolic law:*		
A.	$\varepsilon = \dfrac{\sigma}{a - b\sigma}$	COX (1850), cast iron; LANG (1896), cast iron, stones, plaster
B.	$\varepsilon^2 = a\sigma^2 + b\sigma$	WERTHEIM (1847), organic tissues
5. *Cubic and biquadric-parabolic law:*		
A.	$\sigma = a\varepsilon + b\varepsilon^2 + c\varepsilon^3$	COX (1850), cast iron
	$\varepsilon = a\sigma + \beta\sigma^2 + \gamma\sigma^3$	J. O. THOMPSON (1891), metals, tension
B.	$\sigma = a\varepsilon + b\varepsilon^2 + c\varepsilon^3 + d\varepsilon^4$	HODGKINSON (1849), cast iron
6. *Exponential law:*		
A.	$\sigma = c e^{-1/\varepsilon}$	RICCATI (1731)
B.	$\varepsilon = e^{m\sigma} - 1$	IMBERT (1880), India rubber
C.	$\sigma = c(e^{m\varepsilon} - 1)$	HARTIG (1893), leather, tension; burned red clay, compression
D.	$\varepsilon = \sigma(a + b e^{m\sigma})$	PONCELET (1839), brass, tension
E.	$\sigma = \dfrac{\varepsilon}{1-\varepsilon} \cdot e^{m\varepsilon}$	HARTIG (1893), cork, compression

[a] Recall that MEHMKE incorrectly credited the exponential relation (2.1) to BÜLFFINGER (1729).

[b] Omitted by MEHMKE; MEHMKE omitted also DUPIN's Eq. (2.3) of 1811.

At the end of the 19th century the dire predictions by BACH, MEHMKE, and others, with respect to the continued use of linear elasticity in technology, failed to prevent those who were participants in the fantastic growth of the vast 20th century industrial complex from using the linear approximation in their engineering calculations for small deformation. From the point of view of the experimental science of continuum physics, however, as well as in the efforts to correlate microscopic and macroscopic concepts in terms of atomistic physics and perhaps of even the technology of the 21st century, the persistence of non-linearity to the zero stress point is of no small moment. BACH wrote what must be the only strength of materials text for engineers based on a nonlinear stress-strain function. His *Elasticität und Festigkeit*,[1] which went into six editions between 1889 and 1911, contained a long development[2] based upon his exponential law.

None of the participants in this controversy at the last turn of a century appear to have been sufficiently well informed in the subtleties of continuum mechanics to explore the far-reaching implications of a nonlinear stress-strain function for infinitesimal deformation which extended from tension to compression through zero stress. At about the same time one finds WILHELM SCHÜLE[3] attempting to generalize the exponential law to the study of flexure, and MARCEL BRILLOUIN,[4]

[1] BACH [1902, *1*].
[2] This development is in the 4th edition, published in 1902.
[3] SCHÜLE [1902, *1*].
[4] BRILLOUIN [1898, *1*].

by using HERTZ's theory of contact, attempting to show that the effects of the grips and of the corresponding displacement at the point of application of the load could account for the nonlinearity observed, i.e., HOOKE's law only *seemed* to be violated. That argument of BRILLOUIN[1] apparently interested no one, and the experiments of GRÜNEISEN[2] eight years later, in effect, if not in intent, disposed of it.

SCHÜLE assumed that in flexure, plane sections remained plane and that the constants α and m of Eq. (2.36) had different values for compression and for tension as the experiments of BACH indicated. He attempted to derive an expression for the central deflection of a simply supported cast iron beam. SCHÜLE's comparisons of this central deflection as a function of the applied load were in close agreement with his data, which led him to believe he had made an important first step towards the development of an experimentally satisfactory general theory of flexure based upon what he should have known was JAMES BERNOULLI's nonlinear stress-strain function of 1695.

One physicist who certainly does not fall in MEHMKE's category is KOHLRAUSCH, who, the reader may recall, had stimulated the research of his former student, JOSEPH OSGOOD THOMPSON in 1891. Noting that the tangent modulus of the BACH-SCHÜLE law at zero stress was either zero or infinity unless $m=1$, KOHLRAUSCH in 1901 carried out with GRÜNEISEN a series of experiments on cast iron rods 20 mm wide, 2 mm thick, and 922 mm long, loaded at the center by flexural forces of only 0.1 to 50 g, which produced maximum stresses in the range of 0.173 to 86.5 kg/cm². They concluded that for these data on exceedingly small deformations of cast iron, the closest fit which avoided the BACH-SCHÜLE zero stress paradox was given by:

$$\varepsilon=\alpha\sigma+\beta\sigma^{1.5}. \tag{2.37}$$

This paper[3] in 1901 established historically that GRÜNEISEN, who was to perform the definitive experiments on this subject five or six years later, from the first favored nonlinear stress-strain functions in small deformation.[4]

SCHÜLE[5] was to reject all but his own nonlinear stress-strain function. For example, he rejected the functions of THOMPSON, KOHLRAUSCH and GRÜNEISEN, and HARTIG, solely on the basis of the rather curious argument that such stress-strain laws could not be incorporated in a conventional manner into a flexural calculation. With not too much success, SCHÜLE attempted to fit the data of KOHLRAUSCH and GRÜNEISEN at very small strains to the BACH-SCHÜLE exponential formula, assuming linearity up to the stress of 0.1 kg/mm². The KOHLRAUSCH-GRÜNEISEN relation, of course, would provide for a finite modulus at zero stress.

2.25. Grüneisen's experiments (1906) using an interferometer, which established Hartig's law for the infinitesimal deformation of metals.

In his experiments of 1906 and 1907 EDUARD AUGUST GRÜNEISEN[6] provided the definitive and culminating study of the nonlinearity of metals for small defor-

---

[1] *Ibid.*

[2] GRÜNEISEN [1906, *1*]. These experiments will be described in Sect. 2.25 below.

[3] KOHLRAUSCH and GRÜNEISEN [1901, *1*].

[4] In the later literature, GRÜNEISEN erroneously is credited with having disposed of the 19th century concern with small deformation nonlinearity. (See, for example, TIMOSHENKO [1953, *1*].)

[5] SCHÜLE [1902, *1*].

[6] GRÜNEISEN [1906, *1*], [1907, *1*].

mation by successfully measuring stresses and strains for a range of the latter between 1.7×10^{-6} to 7×10^{-6} with an accuracy of 2×10^{-8}. This he accomplished with interference fringes and half-silvered, parallel plane glass plates attached to a rod in the manner shown in Fig. 2.56. Using the 5461 λ green line of the mercury arc and a gage length of approximately 16.5 cm, he was able to measure a strain of 1×10^{-6} with the stated accuracy of 1%, i.e., to $\varepsilon = 10^{-8}$.

GRÜNEISEN

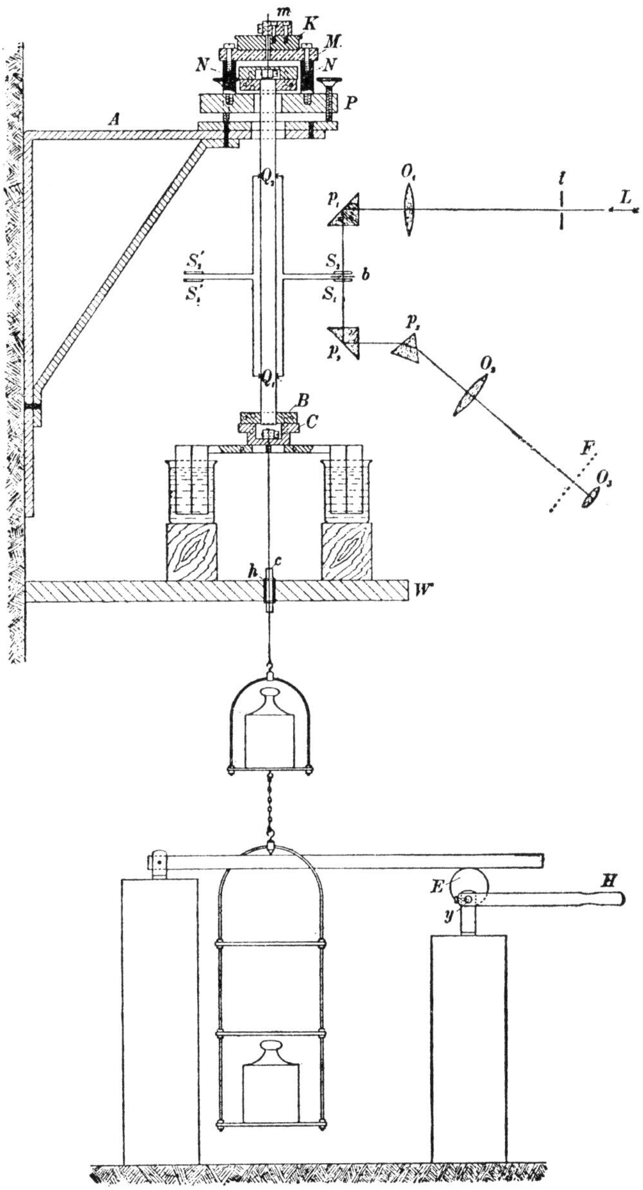

Fig. 2.56. GRÜNEISEN's experiment employing interference optics for the precise determination of axial strain.

In GRÜNEISEN's first series of experiments on iron, in 1906, the specimens had been provided by BACH and were the very ones[1] on which BACH had determined the stress-strain function at strains in a range approximately 200 times greater than that studied by GRÜNEISEN. GRÜNEISEN's specific objective was to compare the predictions of BACH's exponential formula at small strain in the vicinity of zero stress, Eq. (2.36), with the equation of HARTIG, (2.26) which certainly was preferable since it provided for a finite slope at zero stress.

Omitting the fascinating detail of GRÜNEISEN's interference technique and his discussion of his studies of the elastic after-effect, the thermal after-effect, and problems arising in meeting the requirement of precise axiality, we see in the first portion of columns 1 and 2 in Table 26 the data he obtained for cast iron GK and cast iron A. This table also contains BACH's observations of stress and strain in the high strain region on the same specimens. In column 3, GRÜNEISEN compared the small strains predicted from the DUPIN—GERSTNER—HODGKINSON—HARTIG stress-strain function, Eq. (2.27), and in column 4 he showed in the same small strain region the same prediction from the BERNOULLI—BACH—SCHÜLE exponential function, Eq. (2.36).

On the basis of these comparisons, GRÜNEISEN rejected BACH's exponential formula[2] of column 4. As to columns 1 and 2 however, the aspect of remarkable experimental significance is the almost precise agreement obtained in small deformation for HARTIG's nonlinear stress-strain function. In columns 5 and 6, GRÜNEISEN compared the observed tangent modulus E with values calculated from HARTIG's tangent modulus relation, Eq. (2.26), which again provided an extremely close correlation.

In Table 27 are the numerical constants which GRÜNEISEN used.

It is curious that the later literature often refers to GRÜNEISEN as if he had done nothing but demonstrate HOOKE's law valid in sufficiently small strains.[3] In fact, he himself thought he had established the truth of HARTIG's nonlinear stress-strain function right down to the zero stress.[4] For strains between 1.7×10^{-6} and 7×10^{-6}, the departure of HARTIG's relation from linearity is less than the limit, 1%, or 10^{-8} strain, of the accuracy of GRÜNEISEN's interference technique.[5]

[1] This cooperation which led to probably the most important small deformation experiment in crystalline solids since HOOKE's, and which, in the final analysis, rejected the prior constitutive hypotheses of both participants, should be noted by those who proliferate proposed experimental constitutive relations on an institutional basis.

[2] Also rejected without comment was the GRÜNEISEN—KOHLRAUSCH relation, Eq. (2.37) for cast iron.

[3] See, for example, TIMOSHENKO's *History of the Strength of Materials* [1953, *1*], pp. 355–356, where he confined his discussion of this major controversy to the comment:

"Better accuracy in determining the modulus of elasticity was introduced by E. Grüneisen [*Verhandl. physik. Ges.*, vol. 4, p. 469, 1906]. Using the interference of light in measuring small elongations, he showed that such a material as cast iron accurately follows Hooke's law for small stresses (below 140 lb. per in.²) and that the exponential formula $\varepsilon = a\sigma^m$ proposed by G. B. Bülffinger [*Comm. Acad. Petrop.*, vol. 4, p. 164, 1729] and E. Hodgkinson [*Mem. Proc. Manchester Lit. Phil. Soc.*, vol. 4, p. 225, 1824] and widely used by C. Bach [*Abhandl. u. Ber.*, Stuttgart, 1897] and W. Schüle [*Dinglers Polytech., J.*, vol. 317, p. 149, 1902] is completely unsatisfactory for very small deformations."

[4] GRÜNEISEN presented this conclusion in 1906 in his paper on cast iron; his paper in 1907 reported his having extended the studies of small deformation to over 20 metals, for which he had found the same results.

[5] CORNU in 1869 was the first to use optical interference to study deformation parameters related to stress. CORNU was followed in 1888 by CANTONE and in 1899 by both SHAKESPEAR and STRAUBEL. FIZEAU in 1864, and a number of other persons after him, had used interference optics to determine the coefficient of thermal expansion of solids. CORNU [1869, *1*], CANTONE [1888, *1*], SHAKESPEAR [1899, *1*], STRAUBEL [1899, *1*].

Table 26. GRÜNEISEN (1906).

1 σ (kg/cm²)	2 $\Delta\varepsilon \times 10^6$ observed	3 $\Delta\varepsilon \times 10^6$ calc. 1	4 $\Delta\varepsilon \times 10^6$ calc. 2	5 $\Delta E \times 10^{-2}$ observed	6 $\Delta E \times 10^{-2}$ calc. 1 A
		Cast iron GK			
GRÜNEISEN					
0	—	—	—	—	10603
1.576	Initial loading	—	—	—	588
3.457	1.79	1.78	0.09	—	571
5.332	3.56	3.55	0.22	10540	554
7.215	5.33	5.34	0.39	(17.3° C)	536
9.042	7.09	7.08	0.58	—	520
BACH					
159.2	Initial loading	—	—	—	9133
—	—	—	—	7500	—
477.7	425	421	425	—	6193
—	—	—	—	4610	—
796.2	1115	1118	1115	—	3253
1114.6	(2077)	3651	2030	—	—
		Cast iron A			
GRÜNEISEN					
0	—	—	—	—	14019
1.587	Initial loading	—	—	—	011
4.049	1.84	1.76	0.32	13290	13999
6.467	3.68	3.49	0.72	(16.4° C)	986
8.893	5.50	5.22	1.18		974
BACH					
159.2	Initial loading	—	—	—	13215
—	—	—	—	12740	—
477.7	250	257	239	—	11607
—	—	—	—	10520	—
796.2	553	553	551	—	9998
—	—	—	—	9150	—
1114.6	901	899	914	—	8390
—	—	—	—	7560	—
1433.1	1322	1321	1318	—	6782
1751.6	1862	1857	1757	—	5173

Table 27. GRÜNEISEN (1906).

Iron	$E_0 \times 10^{-2}$	b	$1/\alpha \times 10^{-2}$	m
GK	10603	923	674170	1.691
A	14019	505	153250	1.371

Two facts are of great importance here. The first is that a single predictable nonlinear stress-strain function was shown to describe qualitatively and quantitatively the response from the largest to the smallest deformations in a range considered small, and, second, as will be shown below,[1] the value of E_0 extrapolated to zero stress, closely agreed with dynamic moduli determined from longitudinal vibrations in the same region of 10^{-6} strain.

[1] See Sect. 3.27.

GRÜNEISEN found that as he increased the deformation above his imposed upper limit of $\varepsilon = 7 \times 10^{-6}$ the interference fringes began to wander due to an inseparable combination of elastic and thermal after-effects. Hence, being the excellent experimentist he was, he confined his study to the region in which both were negligible. It is interesting that in the upper part of the range of strain he considered, he found that the magnitudes of the combined after-effects were either marked or negligible, depending upon the specimen studied, but he could not explain this fact. To control ambient temperature, he had chosen metallic cylinders as holders for his glass plates so that the coefficient of thermal expansion would be the same as that of the equally long specimen under study.

The interference technique which GRÜNEISEN used for making the measurements of the coefficients of thermal expansion presented difficulties at zero load. To overcome this difficulty he had introduced small initial loadings of 1.576 kg/cm² and 1.587 kg/cm² (see Table 26) which produced an initial strain of the order of 1×10^{-6}. Shown in Fig. 2.57a and b are the comparisons of predicted and observed behavior for large and small deformation which began at the designated initial stress. In his calculations GRÜNEISEN took into account the shift of origin.

In a paper published the following year, GRÜNEISEN[1] extended these studies of small deformation to a large number of metals. He had to depend upon measurements on different specimens, rather than being able to compare the small and large deformations on the same specimens as he had done, in cooperation with BACH, on cast iron. Specifically using HARTIG's expression for the tangent modulus, Eq. (2.26), GRÜNEISEN compared his data with those of J. O. THOMPSON[2] described above. His results were:

$$\left.\begin{array}{lll} E = 10603 - 923\,\sigma & \text{GK3 cast iron} \\ E = 14019 - 506\,\sigma & \text{A15 cast iron} \\ E = 11732 - 184.7\,\sigma & \text{soft copper} \end{array}\right\} \text{C. BACH}$$
$$\left.\begin{array}{ll} E = 10576 - 29.829\,\sigma & \text{brass} \\ E = 12920 - 23.665\,\sigma & \text{copper} \\ E = 8462 - 9.786\,\sigma & \text{silver} \\ E = 19975 - 17.730\,\sigma & \text{steel} \end{array}\right\} \text{J. O. THOMPSON} \qquad (2.38)$$

GRÜNEISEN obtained the static values of E_0 by applying a load of 1 kg which produced measured strain increments from less than 1×10^{-6} to approximately 3×10^{-6}. The values may be compared with the E_0 or constant term in Eq. (2.38): cast iron GK, 10540 kg/mm²; cast iron A, 13200 kg/mm²; copper, 12500 kg/mm²; silver, 8056 kg/mm²; and steel, 21320 kg/mm². Thus one sees that although the predicted values of the modulus at zero stress for the data of THOMPSON were for thin wire specimens 27 m long, while the specimens of GRÜNEISEN were approximately 16 cm long (between 1 cm and 2 cm in diameter); nevertheless, both sets of data are consistent with HARTIG's nonlinear response function.[3]

GRÜNEISEN also became interested in comparing his data with those obtained earlier by VOIGT[4] and KOHLRAUSCH.[5] He observed that, in general, at extremely small strain the E of those experimentists fell below his own, more accurate, determinations. The differences were greater than would be expected from com-

[1] GRÜNEISEN [1907, *1*].
[2] J. O. THOMPSON [1891, *1*].
[3] Obviously it would be of interest to repeat such experiments on the same specimens, as was done for cast iron.
[4] VOIGT [1893, *1*], see also VOIGT [1910, *1*].
[5] KOHLRAUSCH [1905, *1*], see also KOHLRAUSCH [1872, *1*].

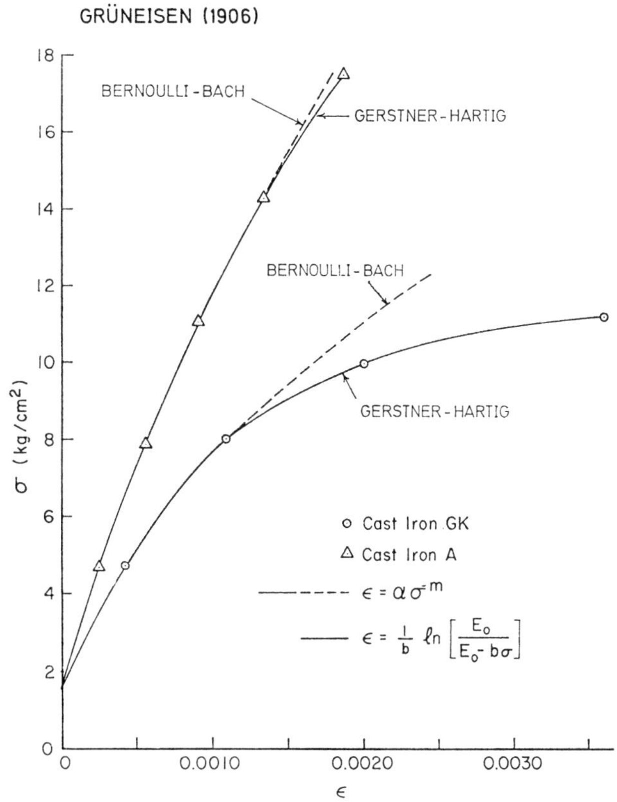

a

Fig. 2.57a and b. In (a) is shown the close comparison with experiment of BACH's and HARTIG's formulae at relatively large strain. In (b) (see p. 122) is shown the comparison of GRÜNEISEN's infinitesimal strain measurements with the extrapolated prediction (in microstrain) from these two formulae, demonstrating the close agreement obtained from HARTIG's equation.

paring what amounts to a secant modulus at the larger strains, with the slope at zero stress. In fact, as GRÜNEISEN pointed out, the data of VOIGT differed by amounts greater than might be expected from the variations in the preparation and the purity of the specimen. He emphasized that in very accurate measurements of moduli, a small variation for the individual specimen did reflect the influence of the prior history of the specimen, both thermal and mechanical.

Since his threshold of measurement was for a strain of 10^{-8}, GRÜNEISEN was able to consider the problems associated with determining strain down to the point at which phenomena such as microscopic porosity of the surface, oxides, minor inhomogeneity, and small differences in the mechanical forming of the specimen began to dominate the data, making the results not reproducible. These measurements will be discussed in more detail in the next chapter. Here we note only that in addition to the quasi-static data for some 20 metals, GRÜNEISEN obtained E for longitudinal vibration and for lateral vibration including rotary inertia, in which the magnitude of the dynamic strain also was of the order of 1 to 3 times 10^{-6}. He found that the dynamic value of E from the longitudinal vibrations coincided exactly with the quasi-static value, while the transverse moduli almost always were slightly higher. This difference he attributed to the maximum stress for the

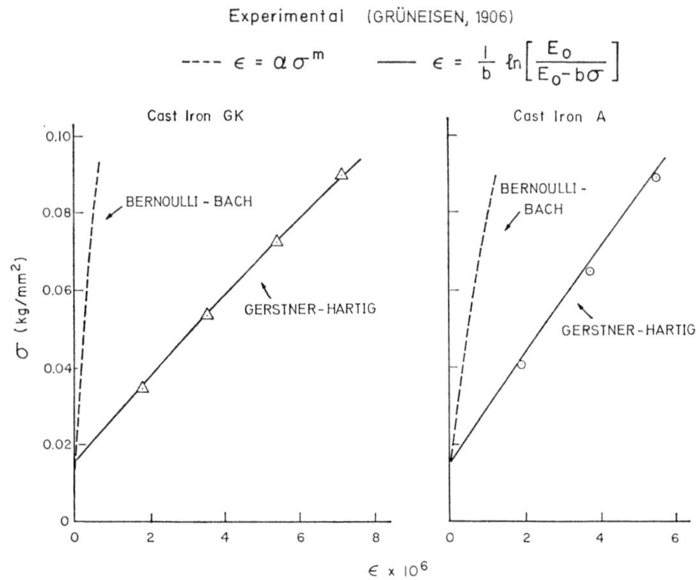

Fig. 2.57b.

transverse vibration occurring at the surface; he thought the material properties at the surface might be different from those in the interior of the solid.

That even a very accurate measurement failed to distinguish between adiabatic and isothermal effects, as in the comparison of the longitudinal dynamic and longitudinal quasi-static data, is not as surprising as it may seem. As GRÜNEISEN pointed out, with the ratio of specific heats being as close to 1 as it is for most of the metals, the experimental distinction of an adiabatic modulus from an isothermal one was at the limit of his own extremely accurate technique. Zinc and cadmium which are different from the other metals in that such a distinction should be possible, were unsuitable for such exact measurement for other reasons.[1]

Despite the fact that nonlinearity of the stress-strain function in crystalline solids for nearly zero stress has far-reaching implications, both with respect to the internal structure of the solid and with respect to phenomena in continuum mechanics such as stability and wave propagation, the experimental study of this nonlinear behavior went into virtual eclipse following GRÜNEISEN's definitive work. A careful examination of the literature on experimental continuum physics, as well as that in the fields of metal physics and metallurgy, reveals almost no reference to his studies even when apparently isolated re-discoveries of the nonlinear small deformation have occurred. A history of work on this subject in the 20th century would have to include an effort to understand why the nonlinear behavior ceased to receive the attention which its importance would seem to have demanded.

Considerable confusion about the implications of nonlinear stress-strain functions in small deformation is reflected by comments of SEARLE[2] in the introduction to his monograph of 1908, entitled *Experimental Elasticity*.

[1] At the present time, zinc and cadmium are studied as examples of unstable solids subject to transitions in moduli which are extremely sensitive to minor variations in prior thermal and mechanical histories and in the method of testing. See BELL [1968, *1*].

[2] SEARLE [1908, *1*].

Within the range where Hooke's law holds, we may speak of the body as being *perfectly elastic*. If the forces on the body be increased, a more or less definite point is reached where Hooke's law begins to fail. When Hooke's law fails, we may say that the *elastic limit* of the body has been passed.[1]

TOMLINSON's admonition 27 years earlier, that one had to distinguish between a "defect of HOOKE's law" and a "defect of elasticity" still was pertinent to a monograph which devoted a number of pages to discussing the nonlinearity of the infinitesimal deformation of copper.

SEARLE's experiments were based on KELVIN's bifilar measurement. The wire was 285.7 cm long, of "about 0.0119 cm² in cross-section." The relative elongation of the two equally long wires for a suspended level was 0.001 mm, permitting a strain resolution of 3×10^{-7}. SEARLE was able to measure hysteresis loops in copper for extension. So as to present the measured nonlinearity of the stress-strain function on a diagram, SEARLE considered the parameter $d = z - w\frac{Z}{W}$, where Z was the deflection due to a relatively large load W, and z was the deflection produced by a small load w. Linearity would require that the factor d be zero for all w. Two of SEARLE's loading and unloading cycles are shown in Fig. 2.58, where $W = 6$ kg, $Z = 0.1159$ cm, and $Z/W = 0.01932$, for one loading and unloading measurement; and $W = 4$ kg, $Z = 0.0763$ cm, and $Z/W = 0.01908$ for the second cycle.

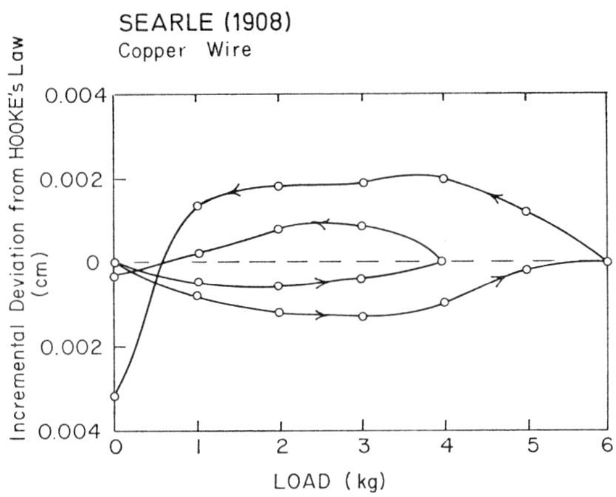

Fig. 2.58. The deviations from linearity in the tensile loading and unloading of long copper wires.

The cyclical nonlinearity of the stress-strain function in the small deformation of copper demonstrates once again the departure from HOOKE's law in these otherwise nearly perfectly elastic, highly precise measurements.

The experimental work of GRÜNEISEN[2] marks a point of division between the 19th and the 20th centuries: a shift of interest and a shift of emphasis occurred in experimental solid mechanics. From then until now, nearly all of the studies of moduli or elastic constants, whether accurate or not, have been based upon dynamic determinations of constants, either from longitudinal, flexural, and torsional vibrations, or, in more recent years, by means of ultrasonic wave propagation. In vibration experiments the magnitude of the strain usually has been

[1] *Ibid.*, p. 2.

[2] GRÜNEISEN [1906, *1*], [1907, *1*], [1908, *1*], [1910, *1*, *2*, *3*].

of the order of 10^{-5}, whereas in ultrasonic experiments the amplitude of the pulse has involved strains of the order of 10^{-7}.

As GRÜNEISEN showed, experimental accuracy[1] for strains of the order of 10^{-6} to 10^{-5} must be better than 10^{-8} to detect nonlinearity in a well-prepared

Table 28.

Date	Name	Elongation (mm)	L (mm)	ε	Comment
1811	DUPIN	0.2	2000	1×10^{-4}	
1830	VICAT	0.01	63500	1.5×10^{-7}	
1844	HODGKINSON	0.0025	15240	1.6×10^{-7}	Micrometer
1844	WERTHEIM	0.020	800	2.5×10^{-5}	Microscope cathetometer
1879	BAUSCHINGER	0.001	150	6.6×10^{-7}	Optical mirror
1881	TOMLINSON	0.020	9000	2.2×10^{-6}	Micrometer cathetometer
1891	J. O. THOMPSON	0.005	27000	1.5×10^{-7}	Microscope cathetometer
1906	GRÜNEISEN	0.00000273	165	1.7×10^{-8}	Interference fringes
1923	TUCKERMAN	0.000051	50	1×10^{-7}	Optical mirror
1930	SAYRE	0.0050	15450	3.2×10^{-7}	Microscope cathetometer
1935	CHALMERS	0.00000273	30	9×10^{-8}	Interference fringes
1940	C. S. SMITH	0.000050	50	1×10^{-6}	Tuckerman optical gages
1952	MILLER	0.000005	25.4	2×10^{-7}	Tuckerman optical gages
1956	LAURIENTE and POND	—	3.2	2.5×10^{-8}	Electric resistance self-calibrating potentiometer
1970	Electric resistance	—	3–25	10^{-7} to 10^{-8}	Commercial equipment

polycrystalline metal specimen, while in the range from 10^{-5} to 10^{-4}, the accuracy must be of the order of 10^{-7}. We have seen that GRÜNEISEN's conclusion that the small deformation of metals was nonlinear of course arose from his having been able to predict, quantitatively, the modulus at near zero stress from tangent moduli determined in the range of 10^{-4} to 10^{-3}.

Because of the importance of strain resolution in reliably detecting departures from the linear approximation, I have included Table 28 which gives estimates of the elongation resolution, strain resolution, and gage length for a number of experimentists between 1811 and 1970 (all have been referred to in this treatise). These were the experimentists whose objectives were dependent upon obtaining a known accuracy of strain.

[1] If the uniaxial nonlinear stress-strain function has a tangent modulus which decreases linearly with stress in the manner proposed by HARTIG in Eq. (2.26), then it is a simple matter, if one refers to Eq. (2.28), to determine the difference in strain $\Delta\varepsilon$ for a fixed stress given by the nonlinear and linear stress-strain function with the same tangent modulus at zero stress. From this relation $\Delta\varepsilon = \frac{b}{2}\varepsilon^2$, one can determine the accuracy of measurement necessary to detect nonlinearity in a given range of strain ε for a prescribed value of b. The experimental values of b range from approximately 10 for silver to approximately 1000 for cast iron.

Thus we see that for $b=20$, an accuracy of 10^{-5} is required to determine nonlinearity when ε is in the range of 10^{-3}, whereas an accuracy of 10^{-9} would be required for this value of b to determine nonlinearity in the range of 10^{-5}. On the other hand, for cast iron or concrete, for b of the order of 1000, an accuracy of 10^{-3} is all that is required to see nonlinearity in the range of 10^{-3}, or an accuracy of only 10^{-7} is needed to detect nonlinearity in the range of 10^{-5}. Even for this relatively large value of b, however, it would require an accuracy of 10^{-9} to detect nonlinearity in the range of 10^{-6}. The table below lists these ranges for the designated values of b.

2.26. On some examples of unwitting rediscovery in the 20th century of nonlinear phenomena first observed in the 19th century.

The overwhelming emphasis in the 20th century upon dynamically determined moduli for which strain amplitudes were extremely small, was not solely responsible for the fact that after World War I the experimental research of the 19th and early 20th centuries on nonlinear mechanics was forgotten.[1] Among testing engineers, international agreement as to the shape and size of specimens, the type of machine, and the manner of performing a test led to standardized specifications for materials, defined by the needs of technology and hence confined to a few stable solids. Much became known about the properties of the few stable solids, but relatively little was learned about the properties of the unstable solids which were, scientifically, decidedly more interesting.

For quasi-static measurement, the single factor most responsible for the great shift in perspective after 1920 was the abandonment of the dead-weight apparatus made in the individual laboratory, in favor of commercially-made, standardized "hard" testing machines. As I have pointed out elsewhere in this treatise, "hard" testing machines, whatever their accuracy, are designed to force upon the specimen a prescribed strain history. The measured quantity is the stress history experienced by the loading elements in meeting the input recipe of the strain history. With dead-weight experiments, the stress is prescribed and the resulting measured strain history describes the accommodation of the material to the applied loads. Anyone who has performed and compared the results from both types of measurements is aware that for many solids the differences among the data are far from trivial, and represent fundamental changes in the response of the material.

Obviously, therefore, the justice of an experimentist's conclusion that a stress-strain function is linear in some range, depends upon the accuracy he was able to achieve.

	Range of strain ε	Accuracy required for $\Delta\varepsilon$
$b=20$	10^{-3}	10^{-5}
	10^{-4}	10^{-7}
	10^{-5}	10^{-9}
$b=200$	10^{-3}	10^{-4}
	10^{-4}	10^{-6}
	10^{-5}	10^{-8}
	10^{-6}	10^{-10}
$b=2000$	10^{-3}	10^{-3}
	10^{-4}	10^{-5}
	10^{-5}	10^{-7}
	10^{-6}	10^{-9}

The above, of course, represents the limits for the competent experimentist who successfully has minimized all other difficulties. BAUSCHINGER [1886, *1*] was the first to note that what is called the "proportional limit" in technology, decreases with increasing accuracy of determination of strain and thus is not a material parameter but a measure of resolution for the experimental technique.

[1] The experiments of GIORDANO RICCATI, COULOMB, and CHLADNI in the 18th century, and those of YOUNG and BIOT in the first decade of the 19th century were limited to dynamic measurements. Beginning with DULEAU's in the second decade of the 19th century, a hundred years of dead weight, quasi-static measurement gave rise to different problems and attitudes toward problems. The abrupt and almost complete shift back to dynamic measurement of moduli in the 20th century demonstrates the connection between concept and fashion in science.

Early in the 20th century the discovery of x-ray diffraction and its importance with respect to the deformation of single crystals gave rise to many interesting new problems. The ability to analyze the crystallographic orientation and structure eventually was responsible for crystal plasticity being viewed in terms of crystal imperfections or dislocations. Since 1925, much of the literature on the large deformation of crystalline solids has presented the macroscopic deformation as merely incidental to investigating or supporting this or that atomistic model, for a wide variety of material parameters including purity, grain size, orientation, prior thermal and mechanical histories, diffusion, etc., etc.

For whatever reason, whether it be new techniques, new attitudes, or new interests, it is a fact, easily discerned from an examination of the literature following the first World War, that the 19th century experimental discoveries in nonlinear elasticity were as forgotten as was the general agreement among the technologists and scientists at the end of the 19th century that linear response functions were inadequate to describe the small deformation of solids. A systematic reading of the technical and scientific literature of the past half century in the field of solids reveals many experimentists who, surrounded by overwhelming emphasis upon linearity, yet observed nonlinear phenomena, unaware that such nonlinearity previously had been studied extensively. As I describe the work of a few of these later experimentists, we should note the interesting fact that their observations generally were expressed in a form which was either that of BACH's parabola[1] or HARTIG's law[2] from the 1890's. They cited neither BACH's nor HARTIG's response functions, nor earlier versions of the same response functions, nor GRÜNEISEN's contrast of the two formulae by experiment. The illustrative examples I give below in chronological order are but a few from a fair sized list; they demonstrate the wasted motion in science which is the price of the nearly universal custom of concentrating almost exclusively on the research and fashionable interpretations of one's own decade.

2.26a. A law for paints and varnish: Nelson (1921).

JAMES BERNOULLI's parabolic response function of 1694, as I have shown, has been re-discovered again and again during the past 275 years and claimed to describe the response of a wide variety of solids, from gut to cast iron and copper. A 20th century, independent re-discovery of this relation was made by HARLEY A. NELSON[3] in 1921 in a series of simple tensile tests on films of drying oils, paints, and varnishes. On flat samples 40 mm long, 10 mm wide, and from 0.075 to 0.180 mm thick, the response functions for various paints were described empirically by Eq. (2.39). The results on which this adoption of BERNOULLI's law were based, are shown in the log-log plot of load and elongation of Fig. 2.59.

$$y = a x^n. \tag{2.39}$$

These data, like those from the experiments on tissue by WERTHEIM, WUNDT, and VOLKMANN, perhaps are more pertinent to the chapter on large deformation below. However, we should heed here NELSON's conclusion that not only is there no evidence of a measurable elastic limit, but it is highly improbable that any small linear region obeying HOOKE's law is present. The values of n and a for these paints and enamels are given in Table 29.

[1] BACH [1897, *1*].
[2] HARTIG [1893, *1*].
[3] NELSON [1921, *1*], [1923, *1*, *2*].

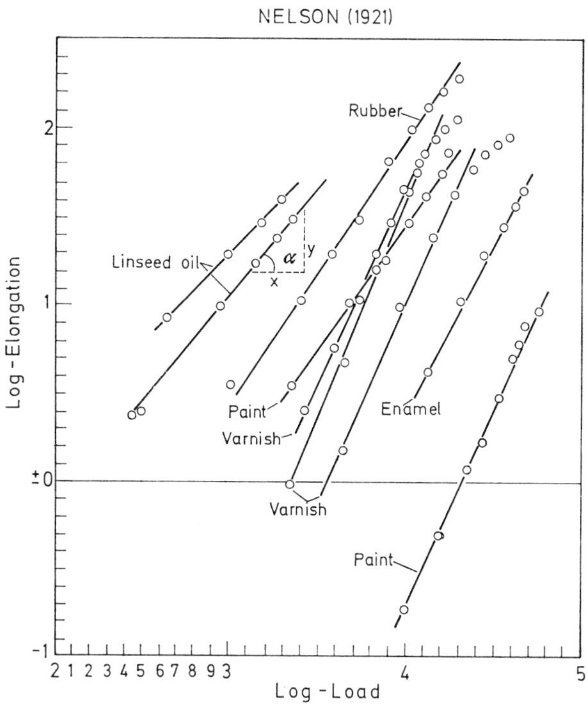

Fig. 2.59. Logarithmic plots of experiments on rubber and paint, showing agreement with exponential stress-strain formula of HODGKINSON and BACH for different integral indices.

Table 29. NELSON (1921).

Experiment no.	Material	n	a
1	Rubber	1.6	0.000040
2	Varnish	2.5	0.0000000048
3	Varnish	2.4	0.0000000031
4	Interior varnish	2.2	0.000000083
5	Paint	1.5	0.000032
6	Paint (highly oxidized)	2.2	0.00000000038
7	Enamel	1.8	0.00000023
8	Kettle bodied oil	1.0	0.0218
9	Kettle bodied oil	1.2	0.0032

2.26b. Sayre's nonlinear law for the small deformation of steel (1930).

A rediscovery of HARTIG's law[1] was made by MORTIMER FREEMAN SAYRE,[2] which added his name to a long list of co-discoverers, including DUPIN,[3] GERSTNER,[4] HODGKINSON.[5] Also, SAYRE used the same experiment as had been intro-

[1] HARTIG [1893, *1*].
[2] SAYRE [1930, *1*].
[3] DUPIN [1811; published in 1815, *1*].
[4] GERSTNER (1824) [1831, *1*].
[5] HODGKINSON [1824, *1*], [1839, 1], [1843, *1*], [1844, 1].

duced by KELVIN[1] in 1865 and subsequently employed during the next three decades by TOMLINSON,[2] STRADLING,[3] J. O. THOMPSON,[4] and SEARLE,[5] among others.

SAYRE, a Professor at Union College, Buffalo, New York, performed two types of tensile tests. One, with a mirror extensometer which he had developed, permitted a measurement of elongation to one part in 500000; he used this apparatus on relatively short specimens. In a second series of experiments on long wires, he could resolve elongations of only 0.005 mm with his micrometer microscope, but since he used a gage length of 15.75 m, his strain resolution, which is the important quantity, was 10^{-7}. This resolution is less than that from GRÜNEISEN's[6] interference measurements with a 16 cm gage length. SAYRE's experiments, which he referred to in the oral discussion of a paper of CYRIL STANLEY SMITH in 1940 as "a subject that has been a hobby of mine for a number of years,"[7] were performed on steel and aluminum wires.

Table 30. SAYRE (1930).

Specimen no.	Material	Temperature °C	Temperature °F	Modulus of elasticity (lb. per sq. in)[a]
481	0.67% carbon heat-treated steel wire, 0.0281 in. in diameter	16.8	62.6	30040000 — 6.6 S[b]
479	0.67% carbon heat-treated steel wire, 0.0465 in. in diameter	20.7	69.8	30160000 — 6.5 S[b]
111	0.68% carbon heat-treated steel wire, $^1/_8$ by $^1/_2$ in. in diameter	24.4	76	29700000 — 7.7 S
112	0.68% carbon heat-treated steel wire, $^1/_8$ by $^1/_2$ in. in diameter	24.4	76	30100000 — 6.8 S
113	0.68% carbon heat-treated steel wire, $^1/_8$ by $^1/_2$ in. in diameter	24.4	76	29500000 — 8.3 S
530, 538	$3^1/_2$% nickel steel, $^1/_2$ in. in diameter	25.6	78	29140000 — 4.3 S
476	$3^1/_2$% nickel steel, $^1/_2$ in. in diameter	22.8	73	28930000 — 10.9 S
306	Spring tempered phosphor-bronze, $^1/_8$ by $^1/_2$ in. in diameter	22.2	72	14680000 — 10.8 S
483	17 SRT aluminum alloy wire, 0.065 in. in diameter	22.5	72.5	10280000 — 10.9 S[b]

[a] S = unit stress in tension, pounds per square inch.
[b] These figures were obtained using dead weight loading and long gage length, and are somewhat more reliable than the other figures.

SAYRE, without reference to KELVIN's work,[8] repeated KELVIN's experiment of 1865 on two long wires in an enclosed tower. One wire was the specimen, and the other was used for temperature compensation in the manner which KELVIN had introduced. SAYRE's wires of aluminum and heat treated 0.76% carbon steel were not quite 16 m long. He placed both the specimen wire and the thermal control

[1] KELVIN [1865, *1*].
[2] TOMLINSON [1881; published in 1883, *1*].
[3] STRADLING [1890, *1*].
[4] J. O. THOMPSON [1891, *1*].
[5] SEARLE [1900, *1*], [1908, *1*].
[6] GRÜNEISEN [1906, *1*].
[7] See SMITH [1940, *1*], p. 874. Perhaps by 1940 this had become typical of the general attitude toward the matter of the nonlinearity of metals in small deformation.
[8] SAYRE referred to none of his 19th century predecessors. One has the impression that he viewed as original his use of this experiment.

wire inside a college laboratory standpipe 5 ft. in diameter and over 50 ft. high. Using a micrometer microscope cathetometer with scale divisions of 0.005 mm, he could determine a strain of the same magnitude as J. O. THOMPSON's, i.e., $\varepsilon = 10^{-7}$.

The main result of SAYRE's research was that when strain was measured with sufficient resolution, the tangent modulus of carbon steel, nickel steel, aluminum alloy, and cold rolled phosphor bronze decreased linearly with increasing stress: HARTIG's law. The results from the first series of experiments on short specimens are given in Table 30 for the designated solids.

For the far more accurate measurements using KELVIN's two wire experiment, SAYRE obtained the results shown in Fig. 2.60 for the designated carbon steel and aluminum alloy. It may be seen that the tangent modulus indeed does decrease linearly with increasing stress, in a simple tensile test.

SAYRE made one observation which had not been considered in earlier research, namely that this linear decrease of the tangent modulus in a tensile test was independent of the magnitude of the pre-stress. Results which he obtained for phosphor bronze are shown in Fig. 2.61.

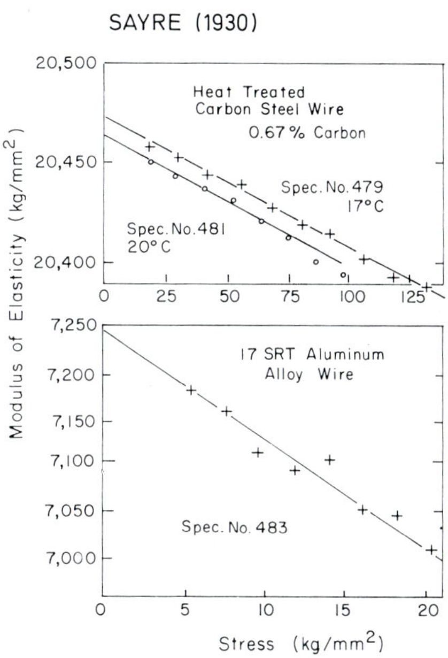

Fig. 2.60. SAYRE's measurements on steel and aluminum compared with prediction from HARTIG's equation. These were precise data.

Using his apparatus, SAYRE established that there was negligible permanent deformation during these tests. He also described in some detail the presence of hysteresis loops, the observation of which his precise measurements made possible. In reporting these observations, including also the negligible permanent deformation for cyclical loading, he seems to have been unaware of the long 19th century history of the elastic after-effect[1] discovered by WEBER in 1831.

[1] Another index of the relatively little interest in the subject of nonlinearity in small deformation might be the fact that in the large volume of the *Proceedings ... of the American Society for Testing and Materials* (see SAYRE [1930, *1*]), SAYRE's is one of a very few papers which was not followed by any discussion, either written or oral.

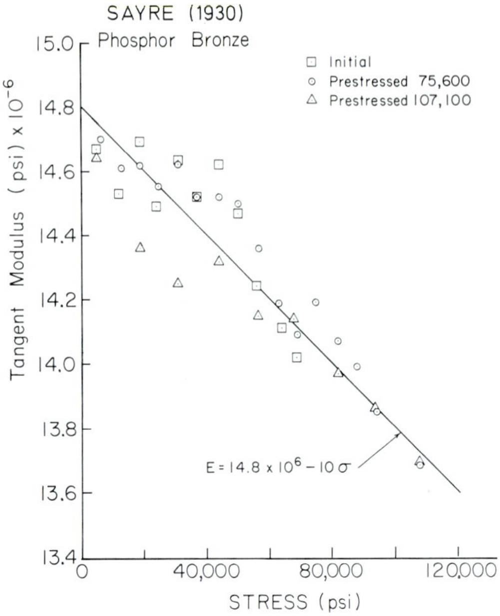

Fig. 2.61. The observed decrease of modulus for phosphorbronze as a function of increasing stress, compared with HARTIG's equation. The results for no pre-stress and for two different values of prestress are essentially in agreement.

One point of great importance is that SAYRE not only considered nonlinear response functions for PIOLA-KIRCHHOFF stress (nominal, or engineering stress), but also for CAUCHY stress ("true" stress). The nonlinearity of both belies the contention often casually made that nonlinearity apparent in small deformation could be eliminated by including the small correction due to the lateral contraction of the specimen.

2.26c. Nonlinearity in tensile experiments on copper alloys: Smith (1940–1948).

In 1940, using an optical lever known as the TUCKERMAN optical extensometer which had been developed a few years earlier, CYRIL STANLEY SMITH[1] on a 2 in. gage length in polycrystalline copper alloys which allowed a strain resolution of 1×10^{-6} again examined nonlinearity in small deformation. SMITH approached the problem by measuring the deviations from the linear modulus as a function of stress.

SMITH used a standard testing machine, the loads being calibrated with proving rings good to 0.1%, by means of which he estimated that errors of load were unimportant until a stress of 35000 psi had been reached. Fitting a straight line through the initial data in a stress-strain plot, SMITH then recorded the deviations after various amounts of prior deformation for a series of experiments in 70-30 α brass. These deviations from the assumed initial modulus are shown in Fig. 2.62 for a specimen annealed for two hours at 400° C, which produced a grain size of 0.008 mm.

[1] SMITH [1940, *1*].

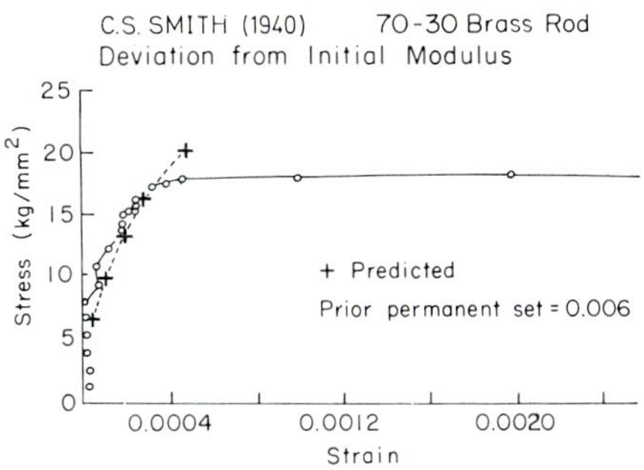

Fig. 2.62. Observations by SMITH in 1940 of the deviation from the initial modulus in tensile testing of 70–30 α brass. I have added the crosses to correlate with HARTIG's equation.

One notes that within the prescribed limit of resolution of 4×10^{-6} shown, the deviation from nonlinearity is not detectable until a stress level is reached which varies with the different amounts of permanent deformation. If the behavior is governed solely by HARTIG's tangent modulus, then by combining the response function of Eq. (2.28) with $\Delta\varepsilon=\frac{b}{2}\varepsilon^2$ it is a simple matter to determine not only the value of b for the condition from the observed point of departure from nonlinearity, but also the form of the deviation curve beyond this point. As may be seen for the initial curve shown in Fig. 2.62, the magnitude of the permanent deformation is such that only the first calculation may be considered. I have added crosses to the first deviation curve of Fig. 2.62, calculated from introducing the experimentally determined b into HARTIG's formula, Eq. (2.28).

Among those who made comments during the oral discussion of SMITH's paper was SAYRE,[1] who pointed out that in his own experiments (described above), it was only when he had gone to a strain resolution one order of magnitude beyond that observed by SMITH that the problems of nonlinearity had become readily visible. LOUIS BRYANT TUCKERMAN[2] also offered some interesting suggestions regarding the limitations and advantages of his optical extensometer, and, finally, ARTHUR ROLAND ANDERSON[3] made lengthy remarks which, curiously enough, were widely quoted in the later literature as indicating that ANDERSON was the first to recognize that PIOLA-KIRCHHOFF and CAUCHY stresses would provide slightly different nonlinear response functions.

The next year, SMITH and VAN WAGNER[4] extended these experiments to include a large variety of copper alloys. As far as the nonlinear deviation from the linear modulus for the response function is concerned, they obtained essentially the same results as in SMITH's previous experiments. In those two papers, SMITH had provided experimental evidence of a nonlinear response function in copper alloys, consistent with that of HARTIG over four decades earlier, although SMITH did not compare his results with those of HARTIG. He considered this aspect of the research reported in those two papers[5,6] to be less important than other metallurgical considerations, as is made evident by a quotation from the last paragraph of SMITH and VAN WAGNER's paper:

[1] SAYRE. See SMITH [1940, *1*], p. 874.
[2] TUCKERMAN. See *ibid.*, p. 875.
[3] ANDERSON. See *ibid.*, p. 877.
[4] SMITH and VAN WAGNER [1941, *1*].
[5] SMITH [1940, *1*].
[6] SMITH and VAN WAGNER [1941, *1*].

The engineer might be cautioned regarding the significance of the precise tests reported herewith. Slight departure from Hooke's law is without practical significance and without relation to possible service stresses. Small offsets are rarely accompanied by equivalent permanent sets.[1]

From this quotation we see, as also is made manifest in detail in the other papers cited here, that the observed nonlinearity did not arise through the presence of significant plastic deformation, i.e., these experiments fall within the realm of nonlinear elasticity.

2.26d. An exhaustive study of a single solid in simple loading: The analysis of the small deformation of beryllium copper by Richards (1952).

Despite the fairly obvious fact that broad generalization in physical theory ultimately must be paired with an equally broad generalization of the known physical facts, most experiments in physics are performed as isolated investigations with the aim of checking upon some particular detail of a currently popular hypothesis. However, a few persons in the past century and a half have provided critical historical surveys of measured moduli, with a global objective in mind.[2] On experiments on nonlinearity in small deformation, JOHN T. RICHARDS[3] wrote one of the most important papers of either the 19th or 20th centuries. He described no experiment of his own but was unique in making his survey by analyzing the results of a wide variety of dynamic and quasi-static measurements of E, μ, and ν made in laboratories he selected from different countries, on specimens of the same solid, a solid which he had prepared and supplied.

All of the tests were conducted on rods of beryllium copper containing 1.85% beryllium. The specimens of one group, designated as $^1/_2$H, were cold drawn; and the other, designated as $^1/_2$HT, were hardened by precipitation. Each group was divided into three lots based upon diameter: the first, identified as H (0.091 in. diameter); the second, J (0.219 in. diameter); the third, K (0.560 in. diameter). RICHARDS' paper listed in detail the composition and properties of the three lots in each group. He sent specimens of each type to cooperating laboratories in the United States and Europe, requesting that each laboratory determine moduli by its most accurate technique. RICHARDS then collected, tabulated, and analyzed this large amount of varied data, all obtained on specimens in presumably identical material.

Apart from the wide scatter in the data from the different laboratories, some of which, as RICHARDS pointed out, may be attributed to the necessity for squaring and cubing geometric dimensions to interpret the data as elastic constants, there are a number of interesting trends both with respect to comparison of ex-

[1] *Ibid.*, p. 845.

[2] Large tabular collections of data do exist, such as those of LANDOLT-BÖRNSTEIN (see, for example, HEARMON [1966, *1*]), PARTINGTON [1952, *1*], etc. Critical surveys of large cross sections of data are more infrequent, although they, too, do exist. Examples are found in the writings of PONCELET [1841, *1*] for moduli; in the Appendix SAINT-VENANT added to the new edition of NAVIER's treatise of 1833 [1851, *1*], [1864, *1*]; in the reviews of earlier data in nearly all of the papers by WERTHEIM (1842 ... 1860); in the much less critical collections of numbers by UNWIN and by EVERETT between 1867 and 1890; and, more recently, in my own survey of moduli in isotropic solids and response functions for finite strain in single crystals and polycrystals (BELL [1968, *1*]). RICHARDS tells us of a survey he made of measurements of E in copper obtained between 1829 and 1849, which I have not been able to locate. We shall see evidence of RICHARDS' interest in 19th century experiment in mechanics when I discuss the work of KIRCHHOFF [1850, *1*], [1859, *1*]. (See Sect. 3.20 below.)

[3] RICHARDS [1952, *1*].

perimental techniques and behavior of material. For the focus of the present chapter, the most important trend in the behavior of the material is the nonlinearity of the metallic solid, beryllium copper, in small deformation. The present discussion will be confined to this aspect of the analysis in RICHARDS' paper.

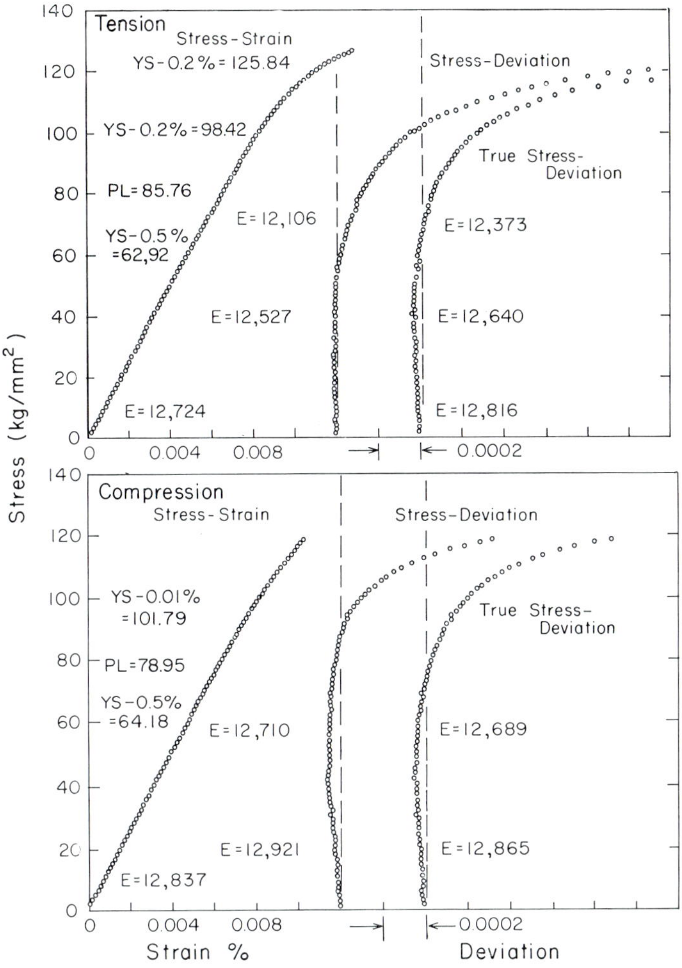

Fig. 2.63. Tension and compression stress-strain observations of MILLER, using the TUCKERMAN optical extensometer. Also shown are the deviations in both nominal and true stress provided by RICHARDS, demonstrating that the lateral contraction was not the source of the observed nonlinearity.

Two of the experiments which RICHARDS considered were those performed in tension and in compression on lot K, $^1/_2$HT beryllium copper rod, by J. A. MILLER[1] of the United States National Bureau of Standards, using a TUCKERMAN optical extensometer on a gage length of 2.54 cm, allowing a strain resolution of 2×10^{-6}. These data, shown in Fig. 2.63, are accompanied by diagrams showing deviation from linearity for both nominal stress and for the CAUCHY or so called

[1] MILLER (see RICHARDS [1952, *1*], pp. 74–75).

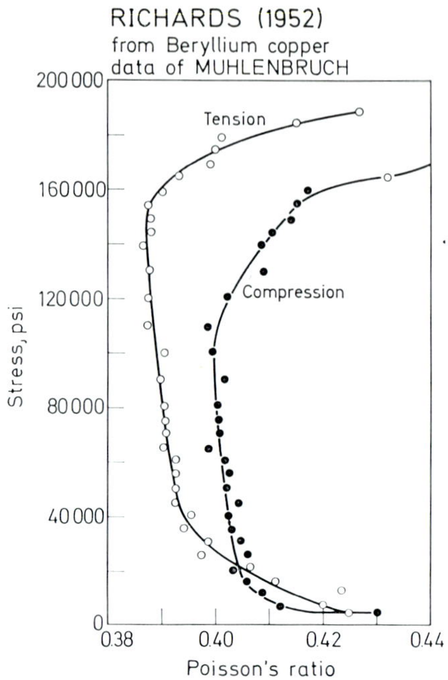

Fig. 2.64. The tension and compression variation in Poisson's ratio with increasing stress, determined by Richards from Muhlenbruch's observations on beryllium copper.

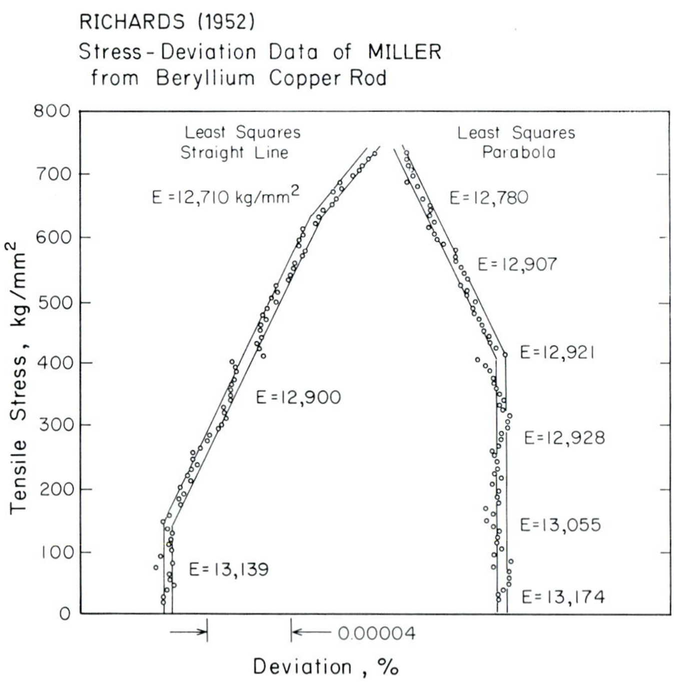

Fig. 2.65. Richards' analysis of the data of Miller, showing the decreasing modulus with increasing stress.

"true" stress, on the basis of the calculation of ANDERSON[1] in his discussion of the paper of SMITH in 1940.

It may be seen that neither the nominal stress-strain function nor the "true" stress-strain function was linear for either tension or compression. The calculation of CAUCHY stress for uniaxial tests requires a knowledge of POISSON's ratio. RICHARDS described the measurements by CARL W. MUHLENBRUCH[2] of Northwestern University, Evanston, Illinois, also obtained on lot K, $^1/_2$HT beryllium copper, who used paired electric resistance gages. These data for tension and compression (shown in Fig. 2.64) exhibited some variation in POISSON's ratio with increasing stress.

MILLER[3] also conducted experiments in tension on lot J, $^1/_2$HT beryllium copper for which RICHARDS provided deviations from HOOKE's law calculated on the basis of the least squares straight line, and also on the basis of the least squares parabola. From these calculations (see Fig. 2.65), one may observe that to a stress of 60000 psi, the stress-strain data fit the nonlinear parabola much better than the linear HOOKE's law.

Tension and compression tests, with electric resistance strain gages, on the lot K, $^1/_2$HT beryllium copper rods similar to those in Fig. 2.63 were provided by MUHLENBRUCH.[4] These data are shown in Fig. 2.66, where *a* designates compression; *b*, tension; *c*, compressive stress deviation; and *d*, tensile stress deviation.

MUHLENBRUCH also obtained data on flexural load deflection and torsional torque-twist. The latter exhibited little departure from linearity, which is to be expected since his resolution of angle did not exceed 0.0001 radians. Of course if the response function for small deformation is nonlinear, then with insufficient resolution of strain it becomes difficult to interpret measurements either in flexure or in torsion on solid specimens. This also was a problem in interpreting the helical spring data contained in RICHARDS' paper.

MUHLENBRUCH determined E by means of experiments interpreted in terms of EULER's column buckling formula for a given lot, the K, $^1/_2$HT beryllium copper rod. With each determination of the critical stress, the length of the rod was decreased for a new measurement. As RICHARDS pointed out, this method had been used over a decade earlier by OTTO FRANZ MEISSER[5] for observing thermal elastic characteristics and had been described in MEISSER's paper entitled "The Deflection of Thin Straight Rods under Longitudinal Loading and the Application of this Test Method for Static Elastic Modulus." MUHLENBRUCH's variation of moduli with increasing stress is shown in Fig. 2.67. Generally higher values were obtained, which suggest some difficulties were associated with this experimental method. However, the trend of the data certainly is that of a decrease of modulus with an increase of stress, as prescribed by HARTIG.

An interesting question is raised regarding the behavior as the response function crosses the point of zero stress from tension to compression. As was noted above, HARTIG had considered this to be a matter of great importance although the data which he considered did not permit him to discern the phenomenon in the immediate vicinity of the point of zero stress. Two possibilities are shown in the schematic diagram of Fig. 2.68. For one of these possibilities, that consistent with HARTIG's formulation, I have included the departure from linearity of the tangent modulus with stress, at the beginning of significant plastic deformation.

---

[1] ANDERSON (see SMITH [1940, *1*]).
[2] MUHLENBRUCH (see RICHARDS [1952, *1*], p. 73).
[3] MILLER (see RICHARDS [1952, *1*], p. 76).
[4] MUHLENBRUCH (see *supra*, p. 73).
[5] MEISSER [1939, *1*].

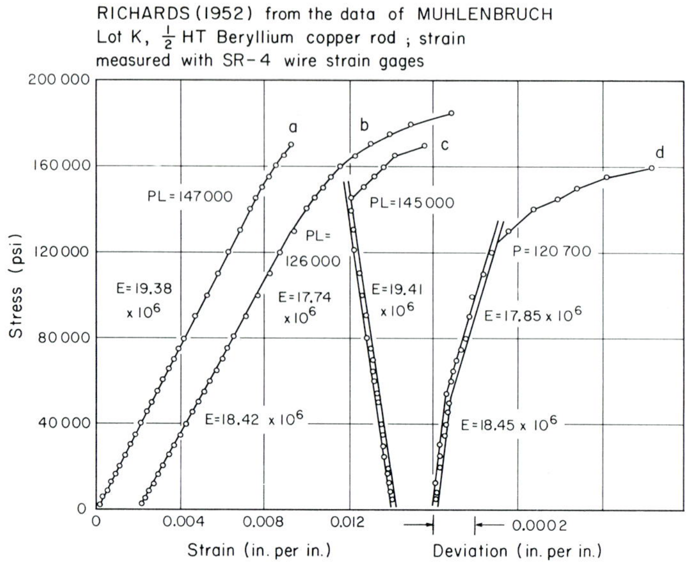

Fig. 2.66. Tension and compression stress-strain observations by Muhlenbruch and their deviations from linearity determined by Richards.

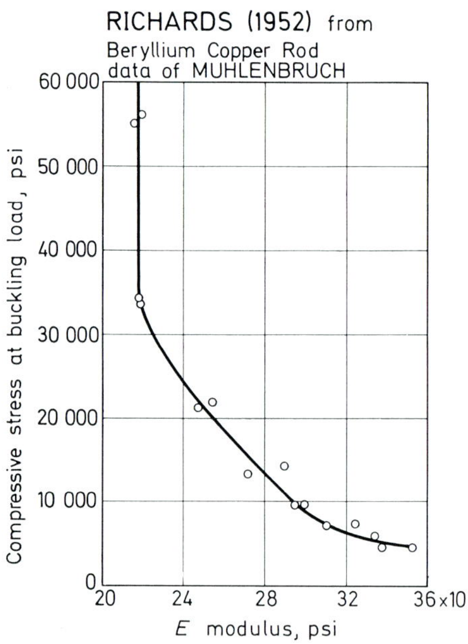

Fig. 2.67. Muhlenbruch's determination of the decrease in modulus with increasing stress, from the compression buckling of beryllium copper.

The tangent modulus decreases linearly with increasing stress for tension. Therefore, the change in sign for compression in Hartig's equation leads to the expectation that the tangent modulus should increase with increasing compressive stress. Starting with Hodgkinson in the 1840's, nearly every experimentist who has measured E in tension and in compression on the same solid has noted the higher values for compression. Among those who have thought that this was of particular importance are RICHARDS and BAUSCHINGER, the latter's measurements in 1879 with a mirror extensometer having been sufficiently accurate to establish that the behavior was beyond experimental error.

A schematic drawing of the response anticipated from HARTIG's formula is shown in Fig. 2.69. Usually, if sufficient accuracy is not obtained, a measured linear E is a secant modulus, as is shown in the dashed lines of that figure. Any summary of tension and compression measurement, therefore, will provide higher values of E for compression.

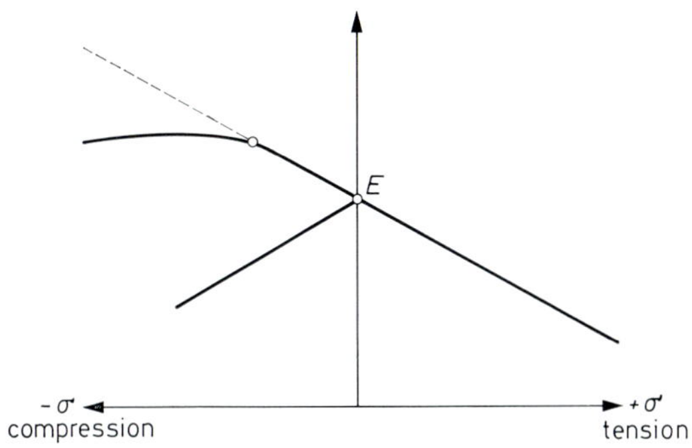

Fig. 2.68. A schematic diagram of the expected variation in the modulus through the zero stress point based upon HARTIG's equation, and upon the assumption that the decrease is present in both tension and compression.

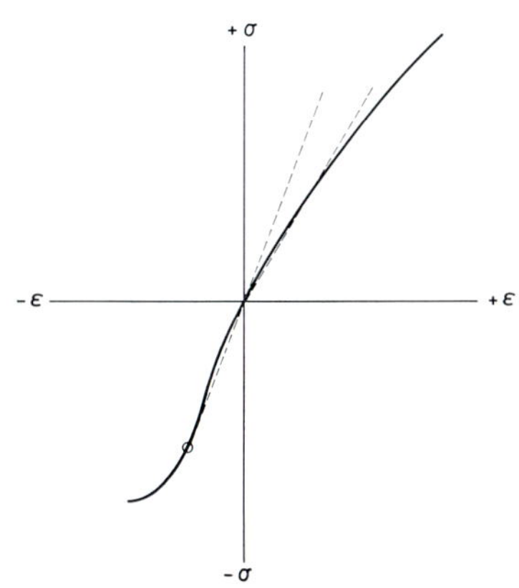

Fig. 2.69. Schematic diagram of expected stress-strain behavior from HARTIG's equation (solid line). Note that with increasing stress in compression a point is reached at which permanent deformation assumes major importance. The dashed lines show the difference in moduli in tension and in compression if one approximates linearity.

In Fig. 2.70 are shown the tangent moduli in tension and compression which RICHARDS obtained from the slopes of MILLER's data. In Fig. 2.71, I have made a similar plot of HODGKINSON's data on cast iron obtained in 1839.

Neither of these results is for a continuous test through zero stress on a single specimen. HODGKINSON's compression tests were performed with lateral rollers to prevent buckling. These two examples, separated by over a century, serve to

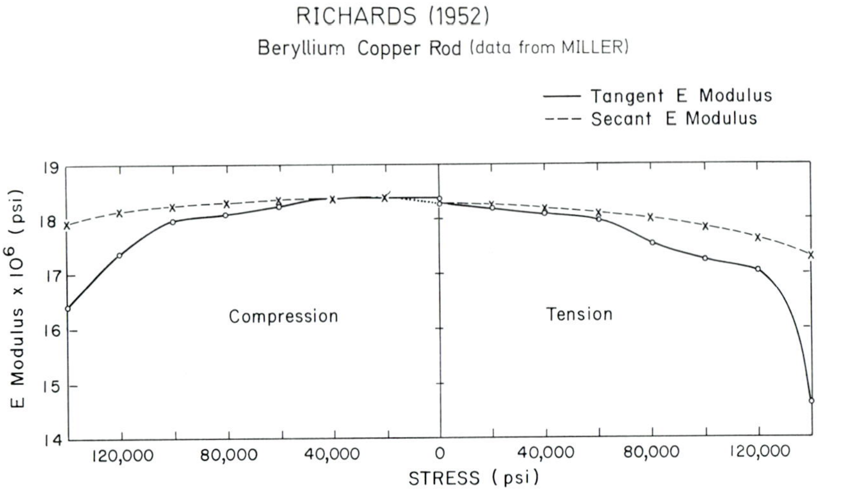

Fig. 2.70. RICHARDS' analysis of the moduli in tension and compression for the data of MILLER on beryllium copper. Note the generally higher values in compression as expected from HARTIG's formulation.

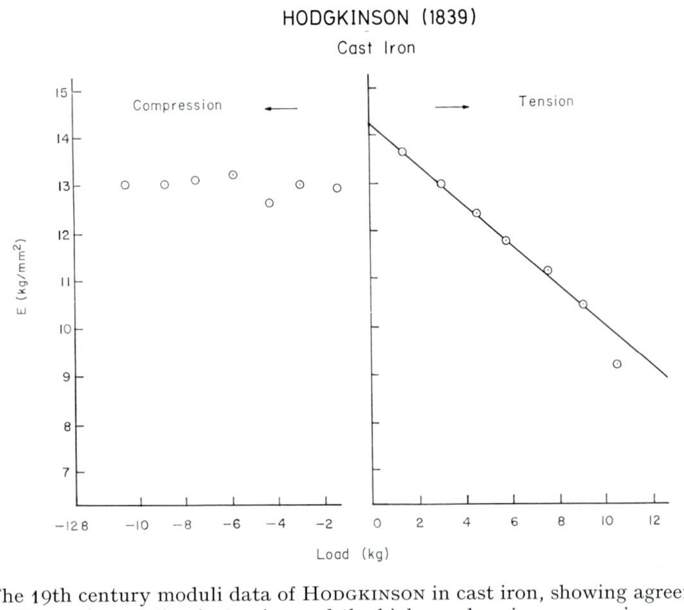

Fig. 2.71. The 19th century moduli data of HODGKINSON in cast iron, showing agreement with HARTIG's equation in tension and the higher values in compression.

support the conclusions in the 19th century which had been based on results obtained from dead weight testing. In modern studies, oscillations for very small amplitudes around zero stress can exhibit nonlinearity of the response function solely from the presence of anharmonic phenomena.

The only experiment I have found for a continuous axial test from tension through zero stress to compression is that of S. F. GROVER, W. MUNRO, and

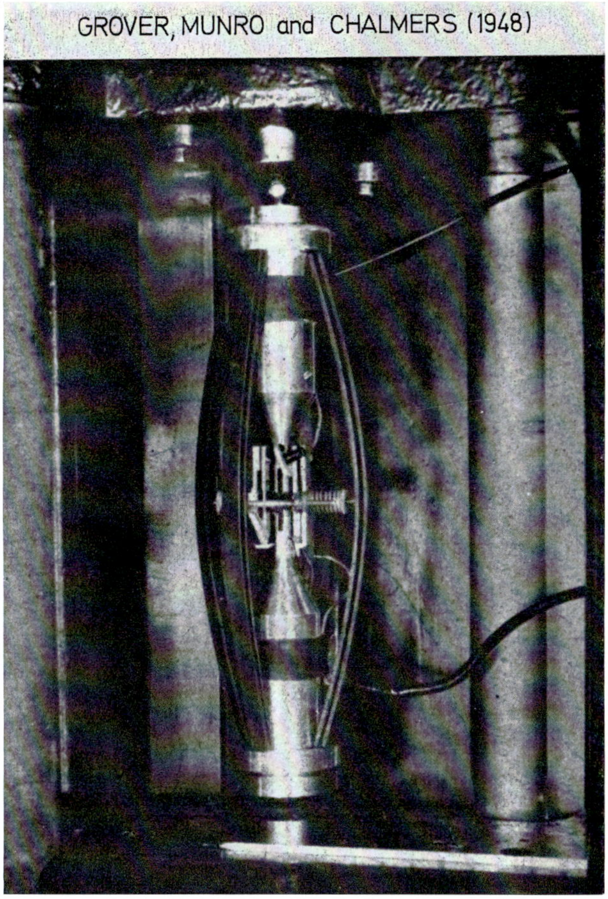

Fig. 2.72. Experimental apparatus for a continuous tension-compression experiment through the zero stress point.

BRUCE CHALMERS[1] in 1948. They first compressed, to a fixed value, a cylindrical specimen of a polycrystalline aluminum alloy having a diameter of 0.4 in. They determined the compressive strain by means of three HUGGENBERGER mechanical extensometers located at 120° intervals around the specimen. They then glued three wire resistance strain gages on the test piece, also at 120° intervals. This attachment under load was a precaution taken so that the electric resistance gages would be under tension at all times during load. Next, the tensile load was applied by means of eight buckled struts. The way the struts applied the initial tensile load to the specimen may be seen in the photograph shown as Fig. 2.72, reproduced from their paper.

GROVER, MUNRO, and CHALMERS then placed the whole assembly in a compression machine so that they could obtain stress-strain data, starting from

[1] This experiment is described in the Appendix of a paper, "The Young's Modulus of Some Aluminum Alloys," by N. DUDZINSKI, J. R. MURRAY, B. N. MOTT, and BRUCE CHALMERS [1948, *1*]. The Appendix is entitled "The Moduli of Aluminum Alloys in Tension and Compression." GROVER, MUNRO, and CHALMERS [1948, *1*].

initial tension and proceeding smoothly through the zero stress into compression. The results of such an experiment are in Fig. 2.73, which shows the smooth transition from tension to compression.

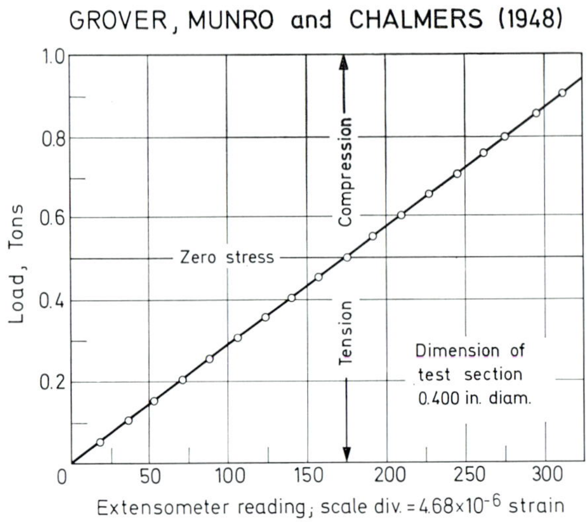

Fig. 2.73. The results of a test in tension and compression proceeding smoothly through the zero stress point.

Since the resolution of strain did not exceed 10^{-6} in this experiment, it was not possible to detect nonlinearity definitively. However, GROVER, MUNRO, and CHALMERS, after stating that they had performed similar experiments on sheets of the aluminum alloy, and that neither for the round bars nor for the flat sheets could a difference be seen between the moduli for tension and compression, nevertheless wrote:

Statistical curve fitting with the test results indicated that there was some significant deviation from the straight line, and that the deviation was of the right form to be accounted for by the increase in the gauge length and decrease in cross-section. However, this effect was of an order of less than 1%, and it was thought that greater care would have to be taken in the extensometry and load application before such results could be accepted as significant. In consequence, no details of the curve fitting are given in this report.[1]

2.26e. Hodgkinson's parabola and "elastic defect": The microplasticity experiments of Thomas and Averbach (1959) and of Bilello and Metzger (1969).

In 1843 HODGKINSON,[2] while describing the tensile experiments on iron and stone which led to his parabolic response function, stated:

It appears from the above-stated experiments, and others that were made, that the sets produced in bodies are as the squares of the weights applied. Hence there is no weight, however small, that will not produce a set and permanent change in a body.[3]

In ascribing his parabolic response function to what he called the "elastic defect," and in his insistence well before 1850 that with every load, however small, there was associated some permanent deformation, HODGKINSON, like

[1] GROVER, MUNRO, and CHALMERS [1948, *1*], pp. 313–314.
[2] HODGKINSON [1843, *1*].
[3] *Ibid.*, p. 24.

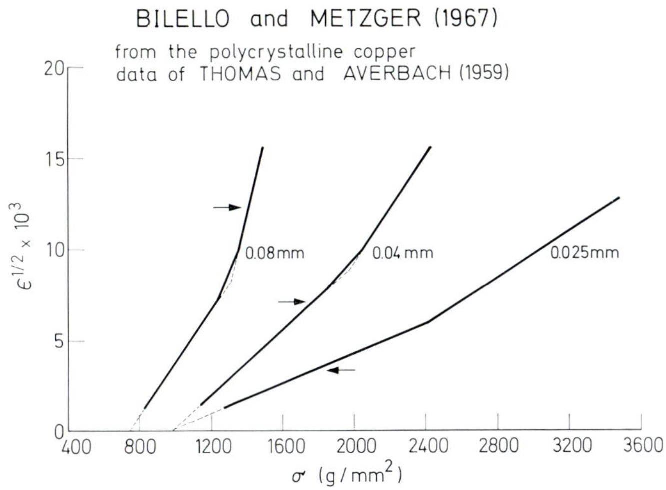

Fig. 2.74. The parabolicity in the tensile stress-strain behavior in the earlier experiments of THOMAS and AVERBACH, plotted by BILELLO and METZGER. Note transition in the linear slopes.

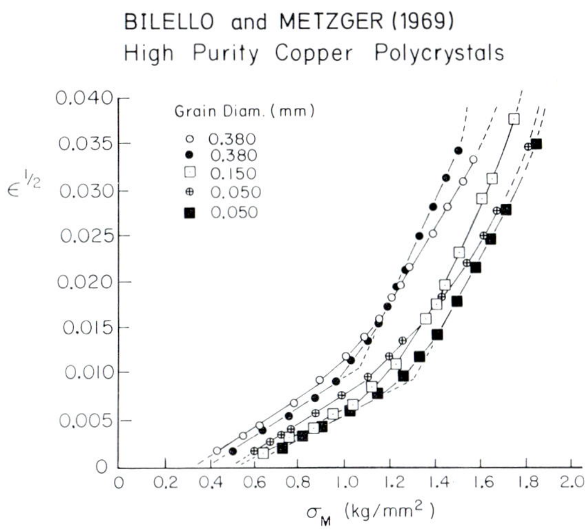

Fig. 2.75. The parabolicity of high purity copper polycrystals of various grain diameters. Note transition in the linear slopes.

WERTHEIM and BAUSCHINGER who later came to the same conclusion, had introduced what is generally considered a modern innovation: microplasticity.

In 1959 THOMAS and AVERBACH,[1] expressly stimulated by SMITH's[2] and SMITH and VAN WAGNER's[3] observation of nonlinearity in small deformation nearly two decades earlier, performed a series of tensile tests on 99.999% purity polycrystalline copper. Using electric resistance strain gages for a strain resolution of 1×10^{-6}, they examined the nonlinearity of the observed response function for small deformation as a function of grain size (and, therefore, also as a function of prior thermal history).

[1] THOMAS and AVERBACH [1959, *1*].

[2] SMITH [1940, *1*].

[3] SMITH and VAN WAGNER [1941, *1*].

Ten years later, in 1969, BILELLO and METZGER,[1] now in the context of microplasticity, likewise studied the effects of tensile tests upon 99.999% purity polycrystalline copper specimens of 3 mm² cross-section, also having different grain sizes. Without discussing the role which microplasticity may play in atomistic model analysis, we may note that the form of their observed response function for small deformation is like that found earlier.

Using two matched linear variable differential transformer transducers to measure the displacement over a 19 mm gage length, BILELLO and METZGER achieved a strain resolution of 5×10^{-7}, nearly one order of magnitude beyond that of SMITH. As is the modern custom, they performed all their experiments upon commercial testing machines. They plotted their results and the earlier measurements of THOMAS and AVERBACH as σ vs $\varepsilon^{\frac{1}{2}}$. The straight lines in Figs. 2.74 and 2.75 once again give evidence of the parabolic response which for 145 years has been associated with the presence of very small permanent deformation.

2.26f. A comparison of the response of fibre and whole muscle: The experiments of Sichel (1935).

In 1935, FERDINAND J. M. SICHEL[2] again raised the question debated by WUNDT and VOLKMANN over 75 years before, namely, whether HOOKE's law was applicable as a response function for the leg muscle of the frog. SICHEL cited the research of EDUARD WEBER and WERTHEIM in the 1840's as primary references on the matter. With some confusion regarding the distinction between infinitesimal and finite strain, SICHEL developed an apparatus in which the extension of the muscle with application of the load caused the flexure of a calibrated microneedle, the motion of whose tip was observed against a micrometer scale divided into 0.010 mm. He used a microscope to make the reading. The main objective of the experiments was to consider the response function of an individual muscle fibre for a comparison with the response function for the aggregate or the whole frog muscle. SICHEL observed and conceded that the response function of the whole muscle was essentially nonlinear, curving toward the load axis in the same manner as the data of WERTHEIM,[3] EDUARD WEBER,[4] WUNDT,[5] and VOLKMANN.[6]

In his first series of experiments on individual fibres, SICHEL's gage lengths were extremely short, between 0.1 and 1.0 mm. In every one of the several experiments tabulated, the measured E of the fibre decreased with increasing strain. When he compared the numerical values of the moduli from one fibre to another, he saw a considerable difference. The range of values for the tangent modulus were of the same order of magnitude as those obtained by WEBER in 1846 for the whole frog muscle (hyglossus).

In a second series of experiments designed to improve upon the accuracy of measurement for the fibre studied, SICHEL used longer pieces of fibre (0.7 to 30.0 mm), higher magnification, "and above all, ... taking measurements only on a length of fibre that is in sound physiological condition throughout the duration of the experiment."[7] Three tension versus percentage elongation experiments

[1] BILELLO and METZGER [1969, *1*].
[2] SICHEL [1935, *1*].
[3] WERTHEIM [1847, *1*].
[4] EDUARD WEBER [1846, *1*], and see WUNDT [1858, *1*].
[5] WUNDT [1858, *1*].
[6] VOLKMANN [1859, *1*].
[7] SICHEL [1935, *1*], p. 35.

on fibre are shown in Fig. 2.76 where the tension was given in arbitrary units and thus could not be compared quantitatively with the elongation curve for the whole muscle in tension, which was included in the figure.

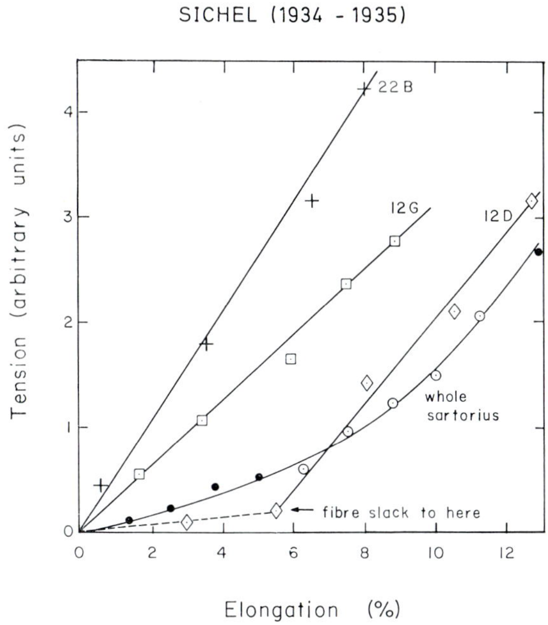

Fig. 2.76. Tension-elongation measurements of individual muscle fibers compared with the behavior of the whole muscle (circles) showing linearity for the former and nonlinearity for the aggregate.

The response of two of the individual fibres was approximately linear. Certain of the fibres, one of which is in Fig. 2.76, required an initial stage of stretching to provide for "kinks due to local injury."[1] SICHEL's demonstrating that the deformation of the individual fibre was nearly linear while that of the whole muscle was nonlinear is of obvious significance for the aggregate problem in organic tissues.

VOLKMANN's experiments[2] were an order of magnitude more accurate than SICHEL's, and as I pointed out above, VOLKMANN could determine also the elongation vs time history so that he could take into account elastic and thermal after-effects. SICHEL noted neither the thermal nor the elastic after-effect in his paper. Since he did not include the time after the application of the load, it is difficult to compare his data with data from mid-19th century experiments in which those effects were matters of major moment.

2.26g. The nonlinear response of artificial stone: The experiments of Powers (1938).

In 1938, TREVAL CLIFFORD POWERS[2] compared dynamic and quasi-static values of E in concrete or mortar specimens. He made the dynamic measurements

[1] *Ibid.*, p. 38.
[2] VOLKMANN [1859, *1*].
[3] POWERS [1938, *1*].

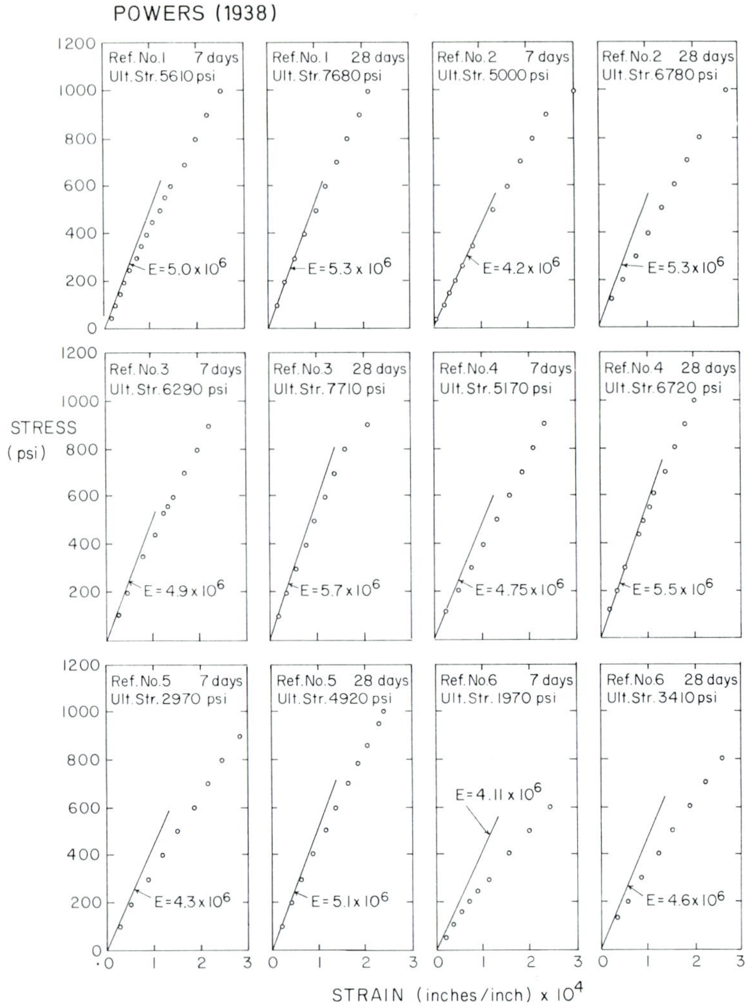

Fig. 2.77. Quasi-static compression observations in concrete aged either 7 or 28 days (circles) compared with the initial slopes dynamically determined in flexure by means of piezo electric quartz crystals. It is readily apparent that the behavior is consistent with that of Hartig's formulation.

in flexure by means of piezoelectric quartz crystals attached to the specimens. He kept the strains as small as possible, and his calculation of E from the measured frequency included corrections for both rotary and lateral inertia. Powers' data were obtained on $5''\times 5''\times 18''$ bars. The static data were from $6''\times 12''$ cylinders which were made and cured from the same batch of concrete as that used for the dynamic tests. He determined strain by rotating mirror extensometers, one on each side of the specimen. He observed some nonaxiality. The data from the flexural vibration are given in Table 31 for six batches where d is the density of the specimen, and N is the measured vibration frequency.

In Fig. 2.77 the average static results (circles) are compared with the results from the vibration test from Table 31 (solid line). The maximum strain for the vibration is far less than that for the quasi-static tests. In his measurements in this solid, POWERS achieved more accuracy than the 19th century predecessors who had investigated that particular material, but his conclusions merely re-affirmed the 19th century appraisal of this solid. One notes that HARTIG's formula, Eq. (2.29) has a positive sign for concrete, as it had for leather and a number of other solids.

Table 31. POWERS (1938). YOUNG's *modulus of 5 by 5 by 18-in. bars as determined by the sonic method.*

Series	Reference no.	Age (days)	d	N	$N^2 \times 10^{-6}$	$N^2 d \times 10^{-6}$	$E = 0.525$ $N^2 d \times 10^{-6}$
J 294	1	7	2.43	1975	3.90	9.48	5.0
	1	28	2.44	2040	4.16	10.15	5.3
	2	7	2.42	1825	3.32	8.03	4.2
	2	28	2.43	2035	4.14	10.06	5.3
	3	7	2.47	1950	3.80	9.39	4.9
	3	28	2.48	2100	4.40	10.91	5.7
	4	7	2.48	1910	3.65	9.05	4.75
	4	28	2.49	2050	4.20	10.46	5.5
	5	7	2.44	1830	3.35	8.17	4.3
	5	28	2.45	1990	3.96	9.70	5.1
	6	7	2.44	1800	3.24	7.90	4.1
	6	28	2.45	1900	3.61	8.84	4.6

2.26h. The "after-effect" in lead single crystals: Chalmers (1935).

In 1935, CHALMERS[1] re-introduced GRÜNEISEN's[2] interference technique to obtain accurate data on elongation at small strain in lead and tin. GRÜNEISEN, thirty years earlier, had used two interference systems, one on each side of the specimen; CHALMERS limited his measurement to a single side. CHALMER's resolution of strain was confined to a range below 7×10^{-6} to avoid the elastic and thermal after-effects which GRÜNEISEN had found were negligible in this region of his solids. Both investigators could measure displacement to within $^1/_{100}$ of the half fringe width of the green line of the mercury arc, namely 2.73×10^{-6} mm. Since GRÜNEISEN had used specimens $16^1/_2$ cm long whereas CHALMERS used specimens of 3 cm, the difference in overall experimental accuracy was one order of magnitude. Thus, the detection of nonlinearity was difficult in the region of 10^{-5} strains which CHALMERS studied. The elastic after-effect which WILHELM WEBER[3] had found 100 years before, in silk, was called "recoverable creep" by CHALMERS. From GRÜNEISEN's and J. O. THOMPSON's[4] results, of course, one also should expect the thermal after-effect to be present in data in the range of 10^{-5} strain.

Hysteresis, or recoverable creep, an important aspect of the nonlinearity in small deformation, had been of interest to experimentists in the previous century; CHALMERS, among other results, provided the two loading and unloading cycles shown in Fig. 2.78. These were observed only when the maximum stress did not exceed the elastic limit.

[1] CHALMERS [1935, *1*].
[2] GRÜNEISEN [1906, *1*]. CHALMERS did not refer to this earlier work.
[3] WILHELM WEBER [1835, *1*], [1841, *1*].
[4] J. O. THOMPSON [1891, *1*].

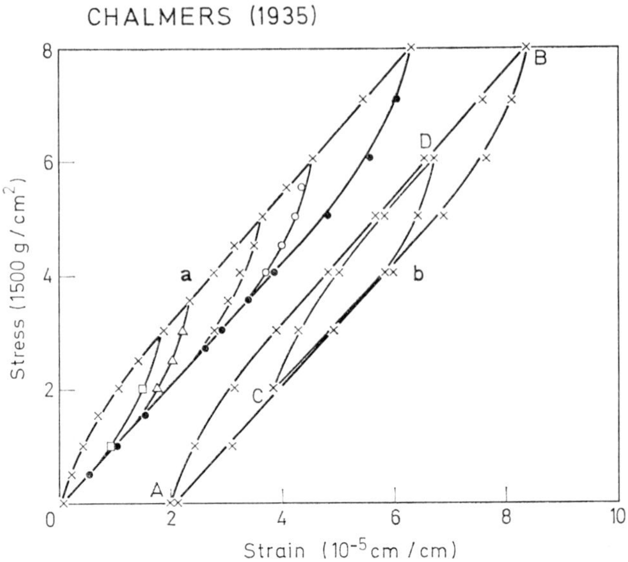

Fig. 2.78. An interference optics extensometer measurement of the elasticity of lead at infinitesimal strain, exhibiting recoverable deformation in the loading and unloading cycle.

Here we should note CHALMER's observation of initial deviation from HOOKE's law:

> The deviation from Hooke's law (see Fig. 5) is small, and may be a property either of the true lattice or of the block structure.[1]

Referring to the creep behavior of tin crystallites, CHALMERS[2] observed that the change of length with time may be related to the formula

$$l = \frac{l_0}{1 + bt} \tag{2.40}$$

where l is length at the time t, and l_0 is the initial length at the time $t=0$. While silk and tin may not necessarily be substances for fruitful, simultaneous study, it is interesting that an empirical formula for the elastic after-effect for silk reappears 100 years later to describe the same effect in tin. CHALMERS found that his data was represented better by Eq. (2.40) than by the proposed $t^{\frac{1}{3}}$ law of EDWARD NEVILLE DA COSTA ANDRADE.[3] One finds that formula (2.40) is similar to that which KARL FRIEDRICH GAUSS[4] fitted to WEBER's data for silk thread. We may derive it from the relation $dl/dt = c\,l^2$, which WEBER[5] in 1841 had generalized to

$$\frac{dl}{dt} = c\,l^m \tag{2.41}$$

in the hope of including the effects of different kinds of loading.

---

[1] CHALMERS [1936, *1*], p. 442.
[2] CHALMERS [1937, *1*].
[3] ANDRADE [1910, *1*].
[4] GAUSS (see WILHELM WEBER [1835, *1*]).
[5] WILHELM WEBER [1841, *1*].

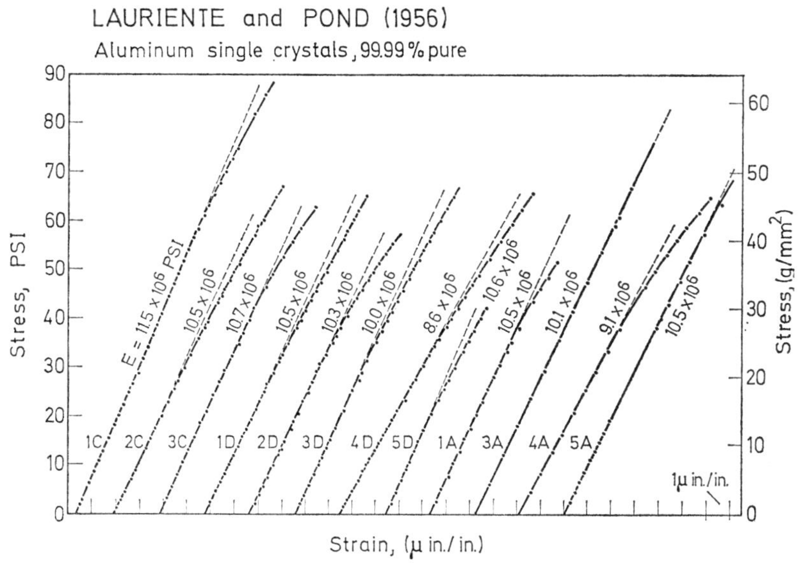

Fig. 2.79. Precisely determined tensile stress-strain curves for 99.99% pure aluminum single crystals. Note the nonlinearity at very small strain.

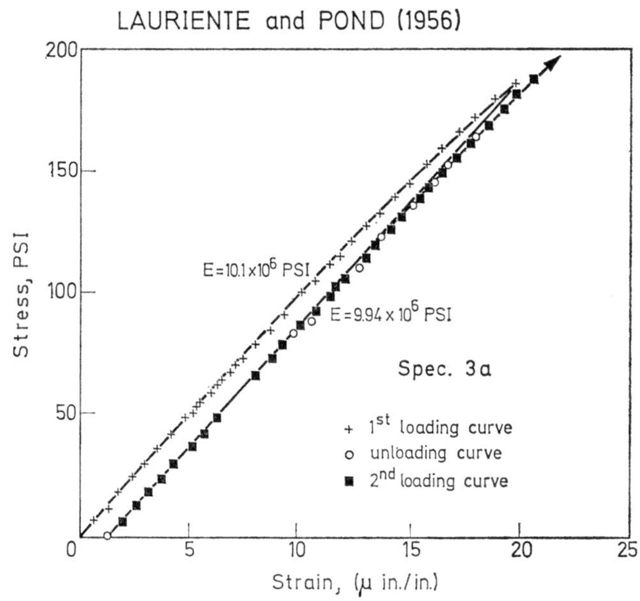

Fig. 2.80. The cyclical tensile loading of 99.99% pure aluminum single crystals showing decrease in modulus on second loading.

2.26i. The decrease in *E* with micro-permanent deformation: Lauriente and Pond's experiments on aluminum crystals (1956).

In 1956 Michael Lauriente and Robert Barrett Pond[1] studied the region of small deformation of 99.99+% aluminum single crystals for a wide variety of

[1] Lauriente and Pond [1956, *1*].

initial crystallographic orientations. Using a bridge network of four electric resistance strain gages, two on the strained specimen and two on an identical unstrained specimen, with strains determined by means of a self-calibrating potentiometer, they succeeded in obtaining strain resolutions which varied, depending upon thermal and electronic shielding conditions, from the lowest extreme of 2.5×10^{-8} to 1×10^{-6}. The dead-weight tensile loading apparatus consisted in a bucket suspended from the crystal by a chain and counter-weighted yoke. LAURIENTE and POND checked the measured loads against a load cell in the system, with a second flexible chain between cell and specimen. Stresses were measured to a resolution of less than 1 psi. For the experiments discussed here, the strain resolution was 1×10^{-7}. Fig. 2.79 shows 12 single crystal tensile stress-strain curves in which observable departures from linearity at strains as low as 2×10^{-6} may be seen.

In Fig. 2.80, a loading, unloading, and second loading curve for one of these single crystals shows that permanent deformation was visible in such a single crystal not only for strains of the order of 10^{-5} but even for a permanent deformation as small as approximately 1×10^{-6}. For the latter, E dropped from 10.1×10^6 psi to 9.94×10^6 psi, the lower value being reproducible in both the first unloading and the second loading.

These experiments are notable for the care with which they were performed and the precision which was achieved. The manner of loading insured axiality. The range of microstrain was the same as that which GRÜNEISEN[1] had considered in 1906. However, LAURIENTE and POND's strain resolution was 10^{-7} rather than GRÜNEISEN's 10^{-8}. This loss of sensitivity was offset by the reduction in the length of the gage, from GRÜNEISEN's 16.5 cm to LAURIENTE and POND's 0.32 cm for their electric resistance elements. For the same resolution as that of J. O. THOMPSON,[2] 65 years later it had become possible to reduce the length of specimen by a factor of over 8000, thus permitting the study of very small single crystals.

Reminiscent of the "elastic defect" which HODGKINSON,[3] WERTHEIM,[4] and BAUSCHINGER,[5] believed would be observed if sufficient precision were attained, LAURIENTE and POND[6] found permanent deformation of 1×10^{-6} when they examined magnitudes of strain as small as 10^{-5}. Of even greater interest was their finding that for the high purity single crystal even so small an amount of permanent deformation lowered the value of E. Thus, COULOMB's[7] discovery of the decrease of moduli on reloading after introducing a permanent strain, which WERTHEIM, KELVIN, and others in the 19th century had explored in detail, once again was observed. This time, however, the range of strain was far below that seen by any previous experimentist.[8]

2.27. Some recent experiments on the nonlinearity of infinitesimal deformation in crystalline solids.

During the past 15 years three new avenues of research on the subject of nonlinearity in small deformation have appeared which are neither a repetition,

[1] GRÜNEISEN [1906, *1*].
[2] J. O. THOMPSON [1891, *1*].
[3] HODGKINSON [1843, *1*].
[4] WERTHEIM [1844, *1*].
[5] BAUSCHINGER [1886, *1*].
[6] LAURIENTE and POND [1956, *1*].
[7] COULOMB [1784, *1*] also [1884, *1*].
[8] LAURIENTE and POND made no reference to the extensive 19th century study of these phenomena.

nor a readaptation, nor merely an improvement of experiments introduced in the 19th or early 20th centuries. The determination of elastic constants from wave speeds by experiments employing ultrasonics will be described in Chap. III, Sect. 3.39. In general, the amplitudes of the waves were exceedingly small. Recent studies have employed somewhat larger amplitudes, often called finite amplitude waves, which in fact are finite only in relation to the very small amplitudes more commonly used. A nonlinear response function for infinitesimal deformation gives rise to anharmonic phenomena, the experimental detection of which provides a measure of the departure from the usually assumed linear law of ROBERT HOOKE. Such studies, together with the determination, in a second type of experiment,[1] of pressure coefficients by ascertaining the speeds of ultrasonic waves for various ambient pressures from which third order elastic constants were determined, point to a decidedly new and interesting direction for exploration.

Another new perspective in recent experimental research is based upon the systematic exploration of prior thermal and mechanical histories which, for many annealed crystalline solids, accentuates certain aspects of nonlinearity in the small deformation of solids. Such studies have been the objective of a sizeable fraction of my own research since 1955. Measurements on specimens which had been held at temperatures within 90% of the melting point for periods of two to twenty hours, furnace cooled, and examined for grain size, were made on specially built dead weight loading apparatuses. Three machines were constructed which allowed for the direct axial tension of the specimen either by increments of load or by continuous loading in a prescribed manner. One such machine was designed for single crystals; it provided for X-ray exposures to be made during the test. A second machine allowed the study of very long specimens, and a third, besides being useful for simple tensile tests, was designed so an independent uncoupled torsion component could be added if desired.

Two dead weight machines also were built, one for tension and one for compression. They included a counterbalanced lever arm, carefully located knife edges, and provision for obtaining a constant stress rate axial loading. The measurement of strain was made by means of an optical cathetometer (or traveling microscope), electric resistance gages which were applied directly and which eliminated flexure, and the use of a clip gage in the form of an arc of a thin beryllium copper strip on which had been placed highly sensitive electric resistance foil gages. The tests lasted from a few minutes to many hours and consisted in several cycles of loading, unloading, and reloading of specimens to observe second order transitions or discrete changes in the slopes or moduli which provided a discontinuity in the derivative.

The first series of experiments were performed[2] to establish whether or not nonlinearity could be detected from simple loading on these apparatuses, and whether, if found, discrete slopes would occur in the expected[3] quantized sequence. This quantized sequence had been found in my earlier studies comparing the elastic constants of 59 elements (see Chap. III, Sect. 3.44, below). I had predicted second order transitions in elastic moduli on the basis of results from experiments in large deformation from which constitutive equations obtained from one-dimensional finite amplitude waves were compared with quasi-static axial experiments on the same solid.

[1] See Chap. III, Sect. 3.45.

[2] BELL [1968, *1*]. These experiments, first described in 1968, had been performed during the years 1955 to 1968. I delayed publication until I had accumulated sufficient data to be able to affirm, precisely, not only that second order transitions occur in elastic moduli, but also that there was order in their distribution.

[3] *Ibid.*

Some of the results of my experiments and those of my students, on second order transitions were given in a monograph[1] in 1968, in which the term "multiple elasticities" was introduced. A series of second order transitions provided a

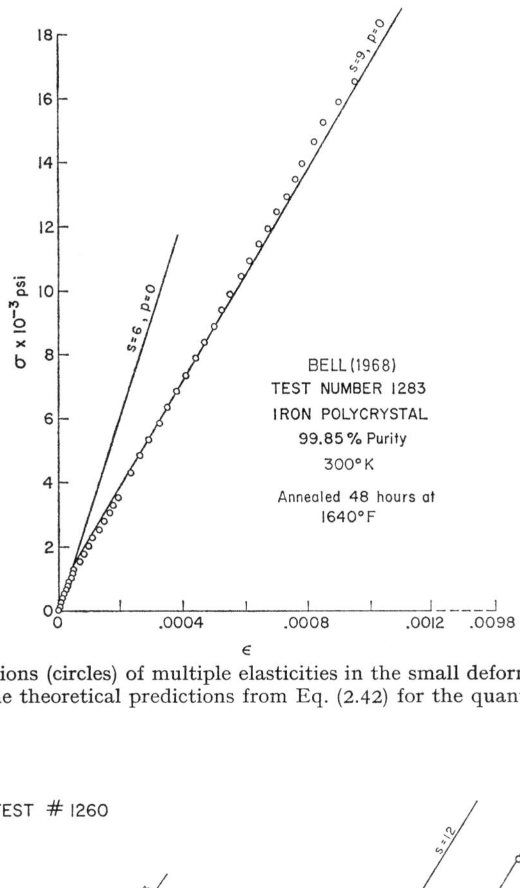

Fig. 2.81. Observations (circles) of multiple elasticities in the small deformation of iron. Solid lines are the theoretical predictions from Eq. (2.42) for the quantized moduli.

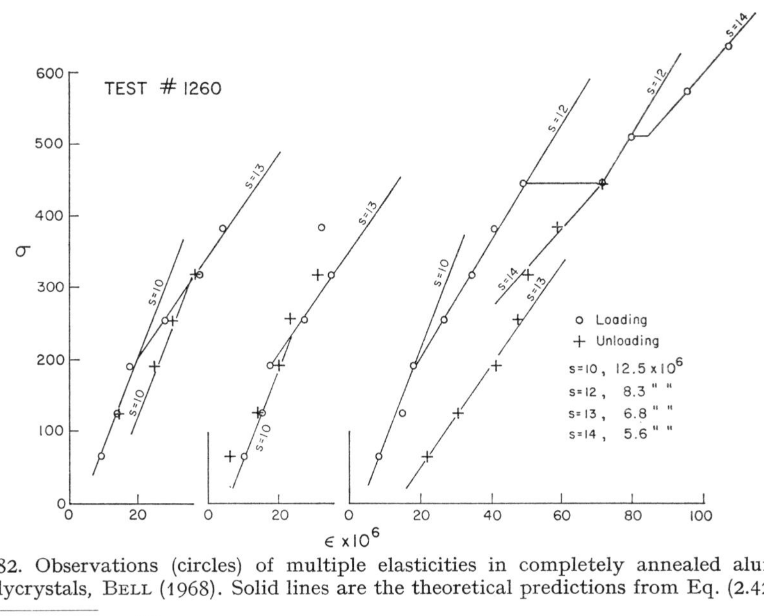

Fig. 2.82. Observations (circles) of multiple elasticities in completely annealed aluminum polycrystals, BELL (1968). Solid lines are the theoretical predictions from Eq. (2.42).

[1] *Ibid.*, see Chap. VI.

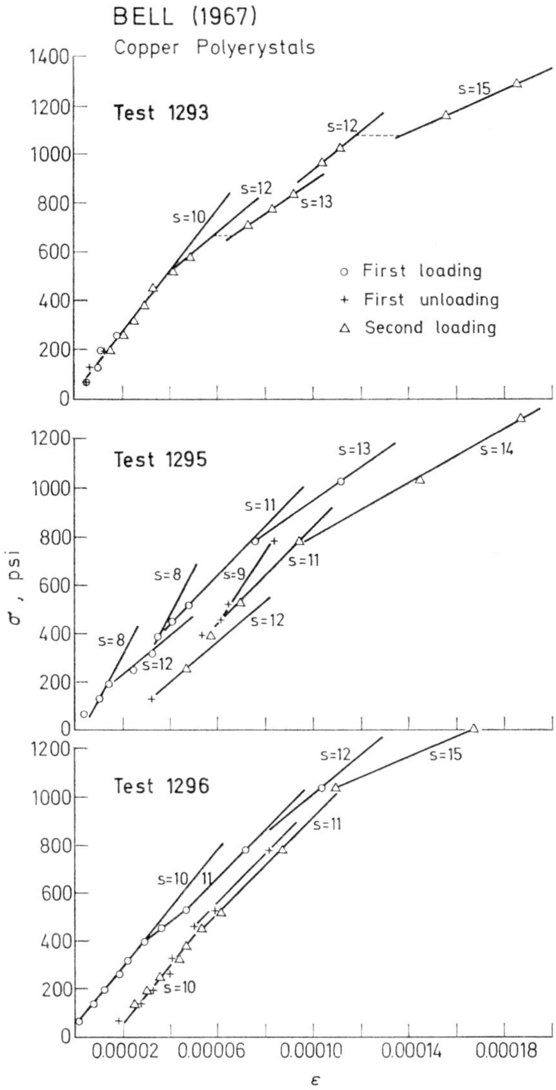

Fig. 2.83. Observations (circles) of multiple elasticities in copper polycrystals. Solid lines are the theoretical predictions from Eq. (2.42).

response function consisting of a sequence of linear segments, each having a slope corresponding to a value of the integer $s=1, 2, 3, \ldots$ for a factor $(\frac{2}{3})^{s/2}$ times a universal constant. (See Sect. 3.44 below, for details concerning this quantized sequence for stable elastic constants.) Given ν, T, and T_m (POISSON's ratio, the temperature of the test in K°, and the melting point of the specimen), the value of E for any temperature, I found was given by

$$E = 2.06\,(\tfrac{2}{3})^{\,s/2+p/4} A(1+\nu)\,(1-T/2T_m) \tag{2.42}$$

where $s=1, 2, 3, 4, \ldots$, $p=0$ or 1 (fixed value for a given solid), and $A=2.89\times 10^4$ kg/mm², a universal constant among 59 elements.

An example of a second order transition in an axial compression test on an iron specimen of 99.85% purity, annealed at 1640° F for 48 h and furnace cooled, is shown in Fig. 2.81. One sees two distinct moduli with a single second order transition. The first, having an integral index, $s=6$, corresponds to the well known standard value at room temperature of $E=20900$ kg/mm². The second slope, for $s=9$, corresponds to the value of $E=11350$ kg/mm². The experimental results (circles) and the predicted slope (solid lines) were for stresses well below the measured yield limit of the solid of $Y=16000$ psi (11.25 kg/mm²).

In Fig. 2.82 are shown three loading cycles in a 99.2% purity, fine grained polycrystal of aluminum which was annealed for ten minutes at 20° F under the melting point, followed by an annealing of two hours at 120° F under the melting point, the specimen then having been furnace cooled.

A similar behavior is shown in Fig. 2.83 for a dead annealed, 99.9% purity, fine grained polycrystal of copper.

And, in Fig. 2.84 are shown three loading cycles in a fully annealed specimen of 70-30 α brass from a dead weight measurement described in a doctoral dissertation by William Francis Hartman.[1] Hartman's was a tensile test made on the same laboratory apparatus which I had used in performing the similar experiments on other solids. All of the straight line segments shown, were predicted from the discrete distribution of Eq. (2.42).

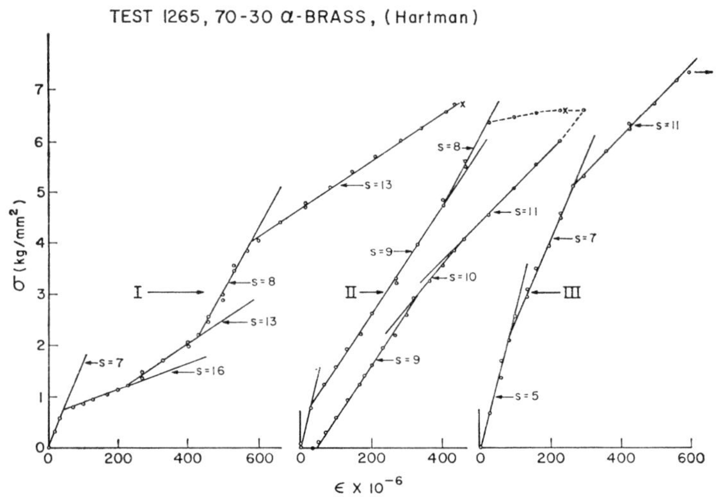

Fig. 2.84. Observations (circles) of multiple elasticities in 70–30 α brass. Solid lines are the theoretical predictions from Eq. (2.42).

These three response functions illustrate those found from the many simple, dead weight loading experiments performed in my laboratory during the past few years. The experiments provided evidence of nonlinearity in small deformation which, as will be seen below, not only is consistent with 19th century observation but also indicates the necessity for reexamining the material stability properties of crystalline solids in terms of such second order transitions, or their absence among a restricted group of stable solids, namely, the solids produced for the needs of practical technology from metallurgical recipes for chemical composition and prior thermal and mechanical histories. The great emphasis

[1] Hartman [1967, *1*].

upon the study of this restricted group of solids doubtless has been the main cause of the long delay in identifying second order transitions in the infinitesimal deformation of crystalline solids.[1]

2.28. New problems for the critic in reviewing experiments described in the literature during the past decade.

To ferret out the major contributions to the subject of nonlinearity in small deformation from the 17th century to a decade or so ago requires no more than long labor, understanding, and criticism. To carry out the same program for the past decade is more difficult. Obviously most of the recent experiments, like past ones, will fail to outlive the change of perspective and fashion which time always brings. But there is an additional hazard for the survival of most of the recent experiments. From the experimental point of view, time well may judge this to have been the decade of the "black box." We have fallen victims to those who unknowingly describe deformation obtained from amplifiers incapable of responding to relevant frequencies; electronic apparatus, the checking of whose daily calibration is in the hands of the gods; automatic testing machines linked to automatic data plotters after briefly consulting associated computers, which leave all important experimental matters to be believed, on trust, from the manufacturers' brochures; vast arrays of data obtained from the observation of new effects in specimens whose purity, crystallinity, and prior histories shall remain forever obscure; and a large assortment of packaged transducers the use of which supposedly eliminates the necessity for considering elementary assumptions.

That this big-business experimental science is a passing phase is, of course, obvious, even in its present heyday. The final phase most certainly will be a return to simple statement with a prudent, probably limited use of some of the recent developments, where applicable, without distorting the intellectual content of the idea in question. This writer, who himself has invented, developed, and used "sophisticated" experimental methods not available earlier, is not impressed with apparatus *per se*; I believe that a simple acquaintance with the past will restore a balance of emphasis between apparatus and fundamental mechanics. Parallels easily are found in other areas, in, for example, what one eminent mathematical friend (now deceased) used to refer to as some of the "lace curtains" of modern mathematics, and in some of the misuses of what has become known as "computer science."

In short, from published reports of far too many of the experimental studies in our time, the reader cannot determine the details of the experiment; moreover, the current fashion of using atomistic interpretation as a sole guide has obfuscated the data by presenting them in a way which sometimes makes it impossible to extract the measured quantities from the surrounding assumptions made, often tacitly, by the experimentist.

How different was the scene in 1879 when BAUSCHINGER, with his newly developed mirror extensometer, placed a strain resolution of 10^{-6} on a firm footing. His description of the apparatus, his methods of ascertaining the various errors, and his calibration procedures occupied several pages, which give the reader not only a clear picture of the experiment (and admiration for its author), but also sufficient detail to build the apparatus and repeat the experiment.

[1] An historical account of nearly two centuries of apparently divergent observations which now are clearly understood in terms of the quantized distribution of elastic constants, "multiple elasticities," and second order transitions, is given in Sect. 3.44 below.

Similarly, J. O. THOMPSON in 1891, while establishing an accurate resolution of strain of 10^{-7} on specimens 27 m long, described his apparatus and testing procedures completely, as did SEARLE in 1900, who succeeded in obtaining the same resolution on specimens an order of magnitude shorter. SEARLE, in fact, provided pages of precise detail describing all of the difficulties and pitfalls. The same can be said of GRÜNEISEN in 1906 when he wrote of his measurements with his optical interference extensometer which provided a resolution of strain of 1.7×10^{-8}. This remained the finest resolution of strain until the present decade, although, as has been shown above, it has become possible to decrease the size of the specimen. The development of the electric resistance gage in the 1940's and 1950's was accompanied by a large and detailed literature on its possibilities and limitations for the measurement of strain. With ordinary care, one could achieve a resolution of strain of 10^{-6}. Some experimentists, such as LAURIENTE and POND[1] in 1956, provided enough detail to demonstrate that with sufficient care a resolution of strain of 10^{-7} could be achieved with electric resistance gages even for very short gage lengths, under fairly complex experimental conditions.

In decided contrast to the manner in which experimentists for over a century have described their experiments, thereby making it possible for us to have confidence in the results they obtained, we, during the past decade, have entered upon a period of mass production of data, the manner of obtaining it having ceased to be viewed as of prime importance. There always have been isolated examples of poorly described experiments, but today these abound. In the mid-1960's we find the history-making claim of strain resolution of 10^{-9}, yet the method of determining the strain is completely unrevealed; a single sentence vaguely refers to "a strain sensing system." The omission of all experimental detail in describing results, and, instead, the providing of imprecise guides to partial theories and proposed models, demonstrate the regrettable fact that for many persons the interest in experiment in this field has come to a low estate indeed. For the critic or for the scholar in research who wishes to review that literature objectively, today or a century hence, there is nothing to write about.

Having traced the long history of nonlinearity in small deformation, one sees that nonlinearity has been observed in apparently perfectly elastic solids; in solids in which the recoverable elastic after-effect was measurably present; in solids for which the SAVART-MASSON (PORTEVIN-LE CHATELIER) effect was observable at extremely small strains; and, in solids in which small permanent deformation occurred when the specimens were unloaded. A further study of the subject would require (a) the consideration of solids whose nonlinear response function exhibited strain differences sufficiently large to be determinable from careful observations within the framework of present measuring techniques; or (b) the improvement of the strain resolution[2] from GRÜNEISEN's 10^{-8}, obtained in 1906, to 10^{-9} or 10^{-10}.

---

[1] LAURIENTE and POND [1956, *1*]. See Sect. 2.26i above.

[2] In this connection the very recent research of C. HARVEY PALMER is of great interest. PALMER has shown that the use of moiré-type fringes produced by the superposition of the image of one Ronchi grid on another allows ultrasensitive angle measurements to be made. PALMER [1969, *1*, *2*], PALMER and HOLLMANN [1972, *1*]. With a relatively simple system he could measure, routinely, angular changes of 10^{-9} radians or even less. The instrument can be constructed to indicate either direct angular changes or differential changes between two reflecting surfaces. Frequencies from zero to megahertz can be studied. Preliminary measurements were made of strain and attenuation of cylindrical annealed copper, brass, and aluminum specimens at strains in the range 10^{-10} to 10^{-4}. With a moderate increase in the specimen length, PALMER's method makes it possible to consider the measurement of strains of 10^{-11}. PALMER described his development in detail.

2.29. Summary.

The dilemma of LEIBNIZ in the 17th century over the apparently conflicting experiments of HOOKE and JAMES BERNOULLI has been resolved in favor of the latter. The experiments of 280 years have demonstrated amply for every solid substance examined with sufficient care, that the strain resulting from small applied stress is not a linear function thereof.

Such failure of linearity might have been expected because of experimental error and the common imperfections of individual specimens. It may not be surprising to find departures from a straight line, but it is surprising, and it is of fundamental physical significance to find in wholly independent, individual experiments from 1811 to the present the rediscovery of the same, intermittently forgotten, nonlinear response function for one solid after another, including all of the metals. DUPIN's flexural studies in wood in 1811, "GERSTNER's law" in 1824, HODGKINSON's "elastic defect" of 1839, BAUSCHINGER's torsion data of 1881, "HARTIG's law" of 1893, GRÜNEISEN's interferometric experiments in 1906, "SAYRE's law" of 1930, and RICHARDS' conclusions in 1952, were all experimental expressions of the same behavior. The relation between load and small deformation is nonlinear, but its tangent modulus is a continuous linear function of the stress.

The 19th century often is characterized by an emphasis on linearity, yet every one of its ten decades, and each of the seven which followed, has witnessed an effort by one or more experimentist to call attention to the fact that for all solids seriously studied the response for small deformation was essentially nonlinear. The overwhelming insistence, by technologists and atomistically oriented physicists, that the quasi-static and dynamic infinitesimal elasticity of solids are basically linear, time and again has isolated successive intervals of fundamental research on nonlinear small deformation in the continuum mechanics of solids. One wonders whether the experimental study of third order elastic constants in the present decade heralds the beginning of a new, lasting, broad understanding of the importance of nonlinearity in small deformation, or whether it is to be just another isolated instance of an interval of experimentation to be forgotten in future decades.

Related to the fact that HOOKE's law did not apply were the experimental discoveries of the relation between approximated moduli and prior permanent deformation, of microplasticity, of creep, of thermoelasticity, of the elastic after-effect, of discontinuous deformation, and of the conditions for stability of deformation in crystalline solids, all during the decade of the 1830's. These important discoveries not only emphasized the experimental difficulties of defining the measured strain but also delineated what came to be known as the fundamental problems characterizing the continuum mechanics of solids. In this chapter of the present treatise I have shown how experimentists after 1840 have pursued the consequences of these observations.

For the many natural phenomena, such as the critical points in elastic stability, the ultra- and sub-harmonics and other common peculiarities of anharmonic small oscillations, or the dispersion of waves in non-viscous media, a nonlinear response function must be used since a linear one rules out these effects altogether. When closely examined, all solids are seen to possess the requisite property. The apparent dichotomy between the observed facts and the proposed explanations arose from a century of overemphasis upon linear theory.

In the continuum mechanics of solids and in other branches of physics, one of the most significant aspects of the present time is that rational mechanics, not impoverished by oversimplification and allowing nonlinearity as a matter of course, is providing a basis for an understanding of physical behavior compatible with observation from over two centuries of serious experimental study.

III. Small deformation: The linear approximation.

3.1. The 17th century origins: Hooke and Mariotte.

Certainly ROBERT HOOKE'S discovery in 1678[1] that force was a linear function of elongation in his experiments with springs and long wires must be regarded as unique for having no rival in the extent to which it has dominated three centuries of scientific thought.[2] The discovery was somewhat dependent upon the choice of solids actually studied and the limited strain resolution. As is well known, HOOKE first stated the law in an anagram at the end of a paper on helioscopes in 1676. The anagram was preceded by this sentence: "To fill the vacancy of the ensuing page, I have here added a *decimate* of the *centesme* of the inventions I intend to publish ...."[3] In his list of nine items, the third was:

3. *The true Theory of* Elasticity *or* Springiness, *and a particular Explication thereof in several Subjects in which it is to be found*: *And the way of computing the velocity of Bodies moved by them.* ceiiinosssttuu.[4]

HOOKE noted that in 1660 he had first discovered the law which now bears his name. With something less than scholarly objectivity he had refrained from publishing it in order to protect his invention of a spiral watch spring; he had used his principle to design the spring.

"From this appears the reason, as I shall shew by and by, why a Spring applied to the balance of a Watch does make the Vibrations thereof equal, whether they be greater or smaller, one of which kind I shewed to the right Honourable the Lord Viscount *Brounker*, the Honourable *Robert Boyle*, Esq; and Sir *Robert Morey* in the year 1660 in order to have gotten Letters Patents for the use and benefit thereof."[5]

Lest one place too much emphasis upon the secrecy implied by the use of an anagram, entries in the diary of HOOKE during the months of September and October 1675, as CLIFFORD AMBROSE TRUESDELL[6] has emphasized, not only contain the statement of the linear relation, but also refer to discussions about springs with Sir CHRISTOPHER WREN and to his demonstration of his spring experiments on 6 October, 1675, for the King of England, CHARLES II. Three years later, in the now famous paper entitled *De Potentia Restitutiva*, the anagram was deciphered as "*Ut tensio sic vis.*"[7]

The same law was discovered independently in 1680 by EDMÉ MARIOTTE, two years after HOOKE'S explanation of his anagram in the open literature. During the following century many continental writers attributed the linear relation solely to MARIOTTE. In 1690 LEIBNIZ, aware of the contributions of both HOOKE and MARIOTTE, questioned the former's generalization, thereby revealing that two decades were insufficient for HOOKE'S observations to have assumed the form of a "self proven law" which so irked MEHMKE 20 decades later. In this context TRUESDELL provides an interesting quotation from a letter written by LEIBNIZ on 20 April, 1691 to CHRISTIAAN HUYGENS in which he stated: "In England they have published a little book on springs, I believe by Mr. HOOK[E], but it seems to me I have found something wrong in it. I beg you to tell me the experiments

[1] HOOKE [1678, *1*].
[2] ROBERT HOOKE in his remarkable originality as an experimentist, unaware of and hence unlimited by the logic of mathematical tractability, can in no way be held responsible for the succeeding centuries of overemphasis upon the physics of inaudible sounds, invisible vibrations, and linear oscillations in potential wells.
[3] HOOKE [1676, *1*], p. 151.
[4] *Ibid.*
[5] HOOKE [1678, *1*], p. 337.
[6] TRUESDELL [1960, *1*], § 8.
[7] HOOKE [1678, *1*], p. 333.

you say you have made on this subject."[1] TRUESDELL indicates that in subsequent correspondence HUYGENS revealed that he agreed with HOOKE's results, but only when springs were slightly extended.

In his *De Potentia Restitutiva* HOOKE referred to four specific types of experiments he had performed in making his discovery: the deflection of a metal wire in the form of a cylindrical helix or spring; the twist of a flat metal spiral spring; the tensile deformation of 20, 30, or 40 ft. long metal wires; and the end deflection of a cantilevered wooden beam. From the experimental point of view the first two and the last presented relatively complex stress distributions. Since HOOKE did not provide numerical values, it is not possible to ascertain whether he made gross observations of relatively large deformation or relatively refined measurements of small deformation. The small nonlinearity of the response functions of iron and wood would not have been observed in either circumstance.

The third experiment is of particular significance in relation to the studies described in Chap. II above. With a 40 ft. long wire as a gage length, and with HOOKE's statement that his measurements were made to the unit of one line, i.e., one-twelfth the English in., we know that he would have been able to detect a strain of approximately 2×10^{-4} provided that all of the relevant experimental conditions were equally well taken into account. We may doubt HOOKE's rigor in this respect when we read his description of the manner in which the wire was supported, a matter of major concern in the experiments of VICAT,[2] TOMLINSON,[3] and JOSEPH OSGOOD THOMPSON,[4] for example, as we have seen. HOOKE stated:

> Or take a Wire string of twenty, or thirty, or forty foot long, and fasten the upper part thereof to a nail, and to the other end fasten a Scale to receive the weights: Then with a pair of Compasses take the distance of the bottom of the scale from the ground or floor underneath, and set down the said distance, then put in weights into the said scale in the same manner as in the former trials, and measure the several stretchings of the said string, and set them down. Then compare the several stretchings of the said string, and you will find that they will always bear the same proportions one to the other that the weights do that made them.[5]

Obviously a support such as a nail could have contributed considerably to the elongation measured below that point. Perhaps some readers may view these comments[6] as quibbling since HOOKE's interest was not in providing a quantitative statement of the important phenomenon he had discovered, but simply to show that equal ratios of weights and elongations could be observed. In HOOKE's diagram, Fig. 3.1, which depicted the three types of experiments he had made on iron wire,[7] the nail may be seen.

---

[1] TRUESDELL, *op. cit.*, p. 62 (footnote).

[2] VICAT [1834, *1*].

[3] TOMLINSON [1883, *1*].

[4] J. O. THOMPSON [1891, *1*].

[5] HOOKE [1678, *1*], p. 335.

[6] In histories of mathematical subjects rigor, esthetics, relevance, performance, and conception are judged not only in the context of their times but against a broader background of universal understanding. The experimentist, on the other hand, when he does not exhibit the same timeless qualities, is excused historically on the basis of presumed or actual primitiveness of technical apparatus and expertise which, in fact, are irrelevancies, since the understanding of an experimental issue at any time is only loosely related to the limitations of technique. Such limitations are merely a matter of frustration, not of logic. Judged by the same rigorous standards as are applied to mathematicians, one notes that the fundamental contributions of experimentists in any period, past or present, reveal a remarkable gap between the excellent and the ordinary. Bumbling precedence is no more of a criterion of excellence for the experimental physicist than for the mathematician.

[7] This important nail is omitted in some redrawn accounts today. (See, for example, TIMOSHENKO, *History of the Strength of Materials* [1953, *1*], p. 19.) Famous experiments have a way of becoming legend, devoid of all assumption, details, and limitations, cited whenever convenient in support of all manner of argument.

HOOKE (1678)

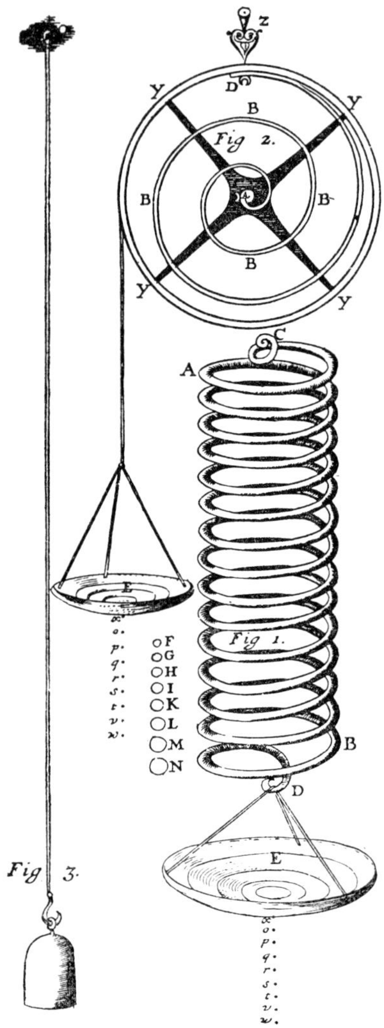

Fig. 3.1. The three experimental situations considered by HOOKE.

HOOKE had made no assertions regarding the magnitude of the ratio of the increment of force to the increment of elongation. In terms of the proper geometric considerations, long wire experiments ultimately led to the discovery of the E-modulus of a solid; the extension of the helical spring would be found proportional to the shear modulus μ. But such considerations were left for experimentists in another century.

The fact that for sufficiently small deformation, a linear relation between stress and strain was observed in metals and other solids and ultimately was expressed in what is now referred to as the generalized HOOKE's law (whether or not it was merely an approximation in terms of increased precision of strain resolution) has provided a powerful tool for the experimental exploration of the nature of the solid continuum undergoing deformation. If in the 17th century, solely nonlinear response functions had been observed for solids, most of the developments of the past 200 years in physics, and particularly in technology, would have been delayed for many centuries. Even in the 17th century there were ample data regarding the rupture loads of solids, and hence it is a simple matter to calculate that if Hooke actually had achieved his potential strain resolution of 10^{-4}, the maximum strains which could have been observed would have been of the order of 10^{-3}. Even had Hooke exercised the greatest experimental care, we can see now from later knowledge of the nature and magnitude of small deformation nonlinearity of iron described in Chap. II above, that he still would have found a linear relation in iron wire.

In the more than 290 years since Hooke first unravelled his anagram, vast numbers of papers have been written on the constants relating stress and strain in nearly every known solid substance. Experimentally, the overwhelming portion of those studies has consisted of the determination either of E (which in the past 150 years has been called, with no historical justice, "Young's modulus"), or of μ, which before 1850 sometimes was referred to as the "slide" modulus, for what we may assume were isotropic solids. Such numbers were determined from

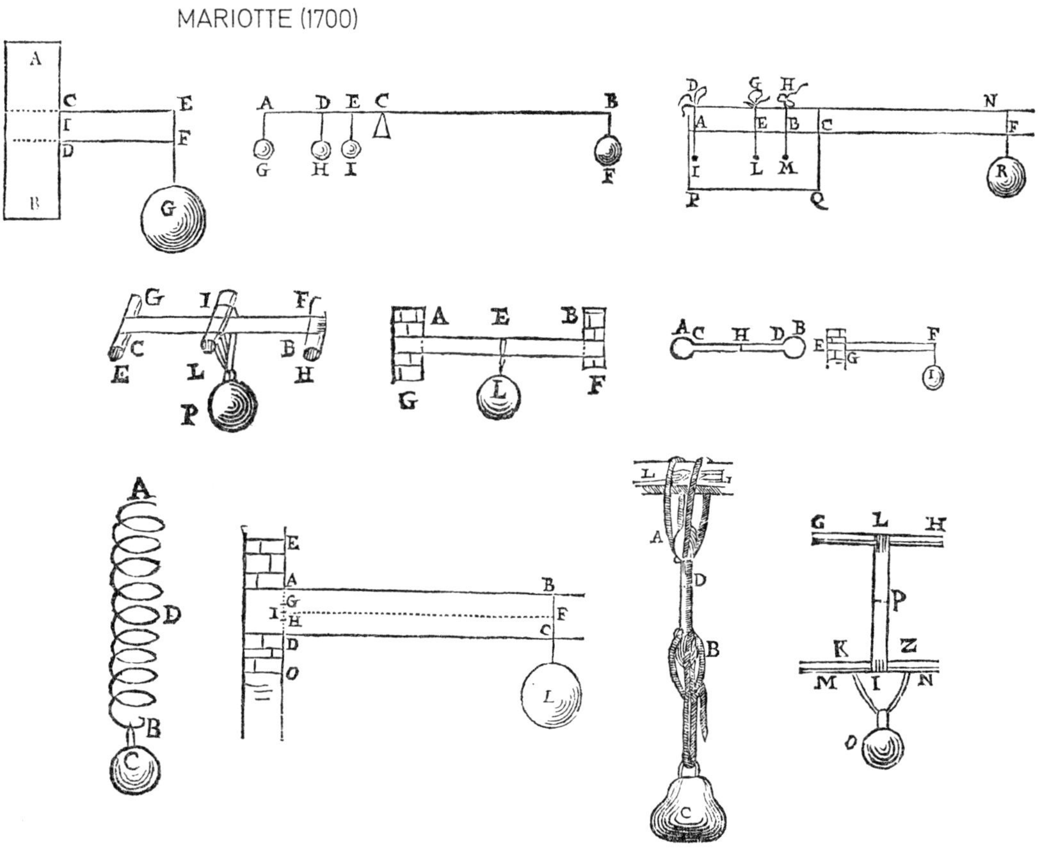

Fig. 3.2. Specimens and experiments of Mariotte.

direct quasi-static load deformation measurements; from the time elapsed in one-dimensional wave propagation experiments in relatively large specimens; from the frequency of longitudinal, lateral, or torsional vibration of bars; or, more recently, by the use of ultrasonic wave propagation techniques.

There is a more modest but still substantial literature beginning in the 19th century which is concerned with the legitimacy or accuracy of those various methods of determining moduli. More correctly, the primary concern was whether or not the presumed simple distributions of stress and strain throughout the solid actually were maintained during the entire time in which the measurements were made. Since we are discussing here the magnitude of these moduli in both isotropic and anisotropic solids and the effect of a wide variety of influences upon their numerical values in a given solid, we shall trace the growth of the idea of material moduli, which ultimately provided a quantitative expression of HOOKE's law.

Some of MARIOTTE's specimens, shown in Fig. 3.2, indicate the accoutrements of an outstanding 17th century experimentist.

In discussing the deformation of the helical spring, MARIOTTE was content to regard the deformation as uniform since all coils separated by the same amount. This uniformity of the elongation also was observed in a long cord supported by a weight. MARIOTTE stated:

> ... for all of its parts suffer the same elongation, so that the upper ones no more lessen the elongations of the lower ones than these latter do theirs, and a long and a short one support always the same weight, unless it happen that within a long cord there occur some defect where it will break sooner than a lesser one.[1]

He observed this same uniformity in bands of white cast iron.

In Part 5 of that same volume, discussing the general properties of wood, iron, and other solids whose fibers are joined together, interlacing each other, MARIOTTE may be said to share with HOOKE the discovery of linear elasticity, since he independently pointed out that when specimens are pulled in the direction of their length:

> ... that these parts can elongate more or less by different weights, and finally that there is an elongation they cannot suffer without breaking, so that (for example) if a block of wood has to be elongated by two lines to be broken, and that a weight of 500 pounds can effect this elongation, a weight of about 125 pounds will make it elongate but about half of a line, one of 250 pounds only about a line, etc. and thus each elongation will strike a balance with a certain weight.[2]

MARIOTTE, of course, was referring to a quasi-static experiment.

3.2. Experiments before 1780: Riccati, Musschenbroek, s'Gravesande, Coulomb; Euler's introduction of the concept of an elastic modulus.

TRUESDELL records that in the 1690's LEIBNIZ had seen that the elastic and acoustic properties of a solid were related, for example, in that the frequency of oscillation increased with the stiffness or hardness of the body.[3] In 1748 in a paper which TRUESDELL[4] indicates was based upon work extending back to

[1] MARIOTTE [1700, *1*], pp. 372–373.
[2] *Ibid.*, p. 352.
[3] TRUESDELL [1960, *1*].
[4] *Ibid.*, p. 115 (footnote).

1720, JAMES RICCATI,[1] while curiously disdaining experiment as a basis for decision, proposed that the elastic properties of a body could be inferred from the frequency of vibration.

From the experimental point of view, JORDAN RICCATI was most fortunate in the parental lottery. Enlarging upon his father's comment of 60 years earlier[2] he determined, from the experimental frequency of flexural vibrations in steel and brass cylinders, that the ratio of their E-moduli[3] was $E_{\text{steel}}/E_{\text{brass}} \approx 2.06$. In so doing, JORDAN RICCATI provided the initial experimental study of material moduli, years before THOMAS YOUNG in 1807. Also, the number JORDAN RICCATI obtained is comparable to that which results from present-day experiments.

As to the source of what is commonly referred to as "Young's modulus" TRUESDELL has traced the idea to a manuscript of LEONHARD EULER written in 1727, eighty years before THOMAS YOUNG[4] introduced into the literature of solid mechanics his version of the "height of the modulus" and the "weight of the modulus." While this manuscript of EULER's was not published until 1862, EULER had explained the modulus and used it in a paper published by 1766, and JORDAN RICCATI had used it in a paper published in 1767.[5] To be a material constant of a given solid, the numerical value, of course, must be independent of the size and shape of the specimen. TRUESDELL has shown that EULER, after he had introduced E in "precisely its modern sense", chose to use instead a material modulus which included a weight density factor ϱg, namely, $h = E/\varrho g$. Of course this modulus, which has the dimensions of length, is also independent of size and shape. As we shall see below, this h is exactly what YOUNG 32 years later called "the height of the modulus." In modern terminology E is the bar modulus, which is usually referred to as "Young's modulus."[6] YOUNG himself never introduced such a concept. His "height of the modulus" depended upon the material density and his "weight of the modulus," upon specimen dimensions. The latter is not a proper elastic constant and the former is not a desirable one; although it does not depend upon size and shape, it does depend upon and require the measurement of the density of each individual specimen.

Although without experimental foundation, the concept of a modulus was in the theoretical literature for over 20 years, so it is not surprising that in the experimental explorations of the 1780's both CHARLES AUGUSTIN COULOMB[7] and JORDAN RICCATI[8] should have sought experimentally for numerical values. COULOMB, with complete understanding, was the first to measure a material elastic constant in the modern continuum sense, i.e., the shear modulus μ; JORDAN RICCATI had measured a ratio of E-moduli.

Continuing to follow TRUESDELL's incisive recounting of the growth of thought in solid mechanics in the 18th century, we find DANIEL BERNOULLI[9] in the 1740's

[1] JAMES RICCATI: Verae, et germanae virium elasticarum leges ex phaenomenis demonstratae. De Bonononiensi sci. art. inst. acad. comm. **1**, 523–544 = Opera **3**, 239–257. (See TRUESDELL [1960, *1*], p. 115.)

[2] *See* TRUESDELL, *op. cit.*, p. 115.

[3] JORDAN RICCATI: Delle vibrazioni sonore dei cilindri. Mem. mat. fis. soc. Italiana **1**, 444–525 (1782) (or *see* TRUESDELL [1960, *1*], pp. 328–329).

[4] YOUNG [1807, *1*].

[5] TRUESDELL [1960, *1*], p. 402.

[6] Obviously for historical accuracy E should be referred to as the "EULER modulus" although the prolific EULER would doubtless be less than moderately interested. In the modern sense, probably ALPHONSE DULEAU in 1812 was the first experimentist to define properly this material constant, independent of the density of the material and the shape and size of the specimen.

[7] COULOMB [1780, *1*].

[8] JORDAN RICCATI [1782, *1*].

[9] DANIEL BERNOULLI [1751, *1*].

referring to experiments on the vibration of rods for a variety of end conditions, for which he found nodal ratios for different vibrations in agreement with his theoretical prediction. Since BERNOULLI supplied few details and no data other than some drawings depicting modal shapes and nodal points, it is difficult for an experimentist to comment on those measurements or upon their correlation with the theory. As TRUESDELL pointed out, however, BERNOULLI was the first to attempt such correlation. The situation parallels that in which a theorist, without describing either his initial assumptions or his proof, merely asserts his final results.

In examining other antecedents to the sudden appearance of modern experimental solid mechanics in that singular decade of the 1780's, one also must refer to the extensive and systematic investigations of PIETER VAN MUSSCHENBROEK,[1] in whose *Physicae Experimentales et Geometricae* in 1729 we find the first testing machines, namely, for tension, compression, and flexure. These are shown in Figs. 3.3, 3.4, and 3.5.

These testing machines, with minor modifications, still could be of some use in the laboratory of an experimentist who has retained an interest in deadweight testing. Unfortunately, nearly all of MUSSCHENBROEK's investigations were related to the problem of rupture, but, like the Comte de BUFFON about the same time, he did record the magnitude of the deformation just prior to failure. If a man of his excellence had been interested in pre-rupture constitutive equations, the impact of experimental solid mechanics upon theory and interpretation might have been considerable even in the mid-18th century. MUSSCHENBROEK did do some important experiments on buckling.

As we have seen in the previous Chapter (Sect 2.18), until the time of BAUSCHINGER in the 1880's nearly all experiment was performed in tension. Exceptions of course were the rather crude measurements of HODGKINSON in the 1840's, which have been described in Chap. II, Sect. 2.6.

We have referred to the experiments of GEORGES LOUIS LECLERC, Comte de BUFFON, of DUHAMEL, and of ÉMILAND MARIE GAUTHEY in the mid-18th century as a prelude to the experiments of PIERRE DUPIN in the field of small deformation nonlinearity. It must be said that their experiments, together with those of MUSSCHENBROEK and MARIOTTE, with their emphasis almost solely on the problem of rupture, did not provide a major contribution to the ideas of COULOMB, CHLADNI, and JORDAN RICCATI in the 1780's, who obviously were far more influenced by the earlier theoretical works of EULER and the BERNOULLIS.

To provide a somewhat more complete picture of the nature of experimental apparatus prior to the 1780's, the drawings of tensile tests performed by MARIOTTE in 1680 and the sketches by LEONARDO DA VINCI[2] are shown in Fig. 3.6. They were interested in using the apparatus to determine rupture. TRUESDELL provides us with an amusing comment by MUSSCHENBROEK with respect to the apparatus of MARIOTTE for the tensile test: "In this method I noticed the inconvenience that the feet of him who performs the experiment are always exposed to danger of injury when the weight falls."[3]

No description of experiment and apparatus in the 17th century would be complete without GALILEO's figure in 1638 for the breaking of a beam with an end load.[4]

[1] MUSSCHENBROEK [1729, *1*], [1739, *1*].

[2] LEONARDO DA VINCI (1452–1519).

[3] *See* TRUESDELL [1960, *1*], p. 151.

[4] GALILEO's drawing is Fig. 3.7, p. 166 below. Apparatus diagrams complete with such details as weeds and bricks were still being drawn on occasion, up until the mid-19th century; witness, for example, the figures contained in PETER BARLOW's *Treatise on the Strength of Timber, Cast Iron, Malleable Iron, and Other Materials* published in 1837 [1837, *1*].

MUSSCHENBROEK (1729)

Fig. 3.3. 18th century apparatus for the tensile test.

In 1720 s'GRAVESANDE[1] proposed that HOOKE's relation between load and elongation might be tested by means of the central loading of a wire in the manner shown in Fig. 3.8. That experiment was widely quoted until nearly the middle of the 19th century as an important milestone in experimental solid mechanics.

[1] s'GRAVESANDE [1720, *1*].

MUSSCHENBROEK (1729)

Fig. 3.4. Musschenbroek's apparatus for compression tests.

MUSSCHENBROEK (1729)

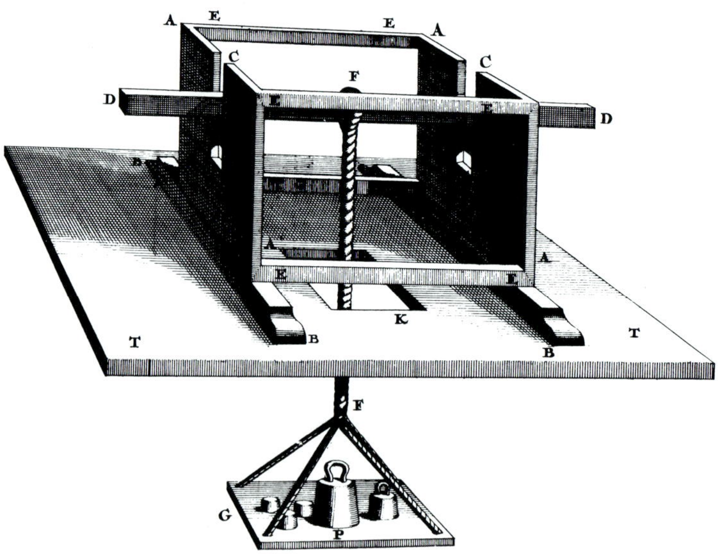

Fig. 3.5. Musschenbroek's apparatus for flexural tests.

Truesdell[1] has shown in considerable detail that if the wire in the s'Gravesande experiment were initially taut, having been tightened by applied forces after having been laid over fixed wedges, then the load P would be found to be proportional to the elongation δ, no matter what the stress-strain relation of the

[1] Truesdell [1960, *1*], p. 117.

LEONARDO DA VINCI

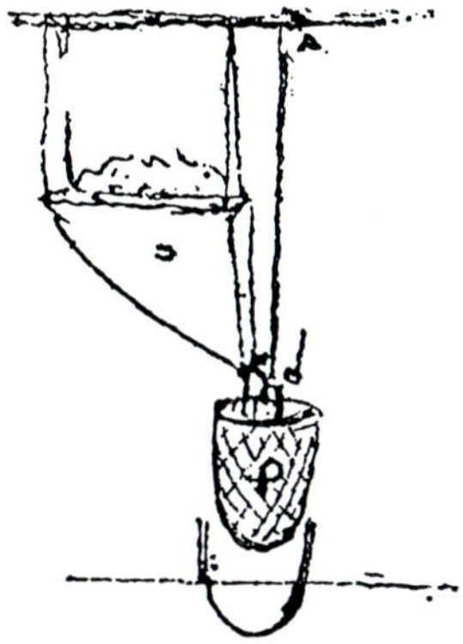

a

MARIOTTE (1680)

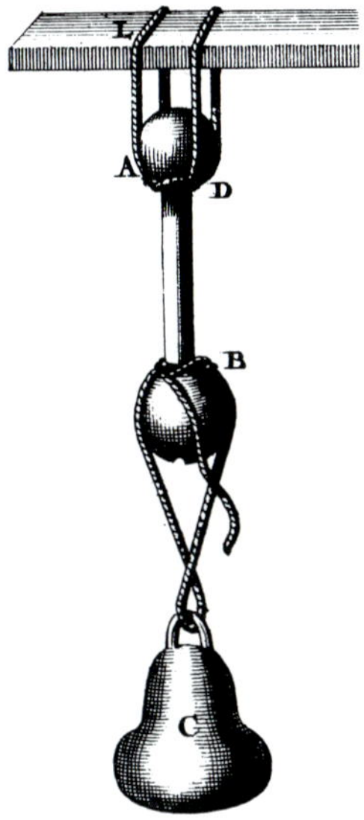

b

Fig. 3.6a and b. Drawings of (a) an experiment described by DA VINCI, and (b) an experiment described by MARIOTTE.

wire. If, on the other hand, the ends of the wire were fixed, proportionality between deflection and load would be limited to small deformations and would be obtained only if the wire initially were not taut. If the fixed wire were taut, then, as TRUESDELL has shown, the load would have been proportional to the cube of the deflection.

In 1889, 169 years later, EDUARD TACKE[1] became interested in performing the s'GRAVESANDE experiment. On the basis of his study, he dismissed the

[1] TACKE [1889, *1*].

Fig. 3.7. Galileo's beam experiment in 1638.

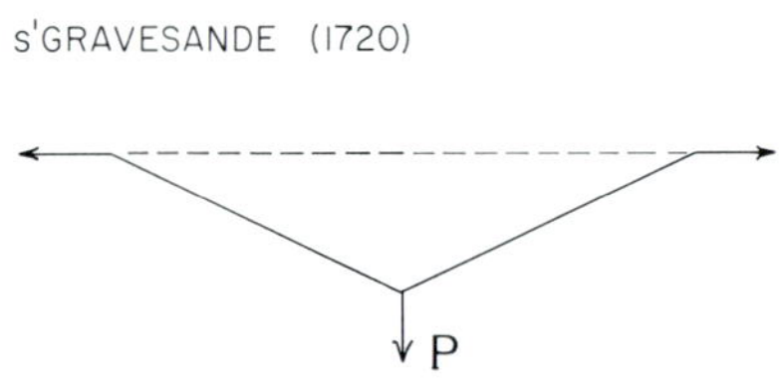

Fig. 3.8. The experiment of s'Gravesande in 1720.

experiment as incapable of providing any meaningful information. Friedrich Kohlrausch[1] subsequently re-evaluated the validity of this conclusion. He suggested that one start with an initially deflected wire under a primary load, add a second load to produce the corresponding additional deflection, and hence obtain a relation from which a modulus of elasticity might be determined. This, of course, accords with Truesdell's[2] independent, later analysis of the problem, as the only condition under which linearity of the stress-strain relation might be tested by an experiment of this kind.

Kohlrausch asked a student, Georg Stradling[3] to determine experimentally whether the moduli obtained from the s'Gravesande test would compare with accurate values from a longitudinal tension experiment in the same wire. Stradling's measured values are shown in Table 32. Despite Stradling's optimism,

[1] Kohlrausch (see Stradling [1890, *1*]).
[2] Truesdell *op. cit.*
[3] Stradling [1890, *1*].

s'GRAVESANDE's experiment obviously does not provide a method for accurately determining an elastic constant.[1]

Table 32. STRADLING (1890).

	Iron		Brass		Nickel-Silver
	I	II	III	IV	V
s'GRAVESANDE method	20560	20450	9610	9470	15380
Longitudinal method	20300	19740	10140	9130	15680

Of the tens of thousands of rupture experiments which have been performed in the past three or four centuries, COULOMB's comparisons in 1773 of rupture in tension with rupture in shear certainly are among the most important. A "block of white stone, fine-grained and homogeneous"[2] was suspended as shown in Fig. 3.9. The dimensions of the block were 1 sq. Paris ft. and 1 Paris in. thick.[3] A footnote description was given as follows: "This stone is found around Bordeaux, and is used for the façades of the large buildings of this town."[4] A load of 430 lbs. was needed to rupture the block at *ef*, giving a cohesion stress of 215 psi.[5]

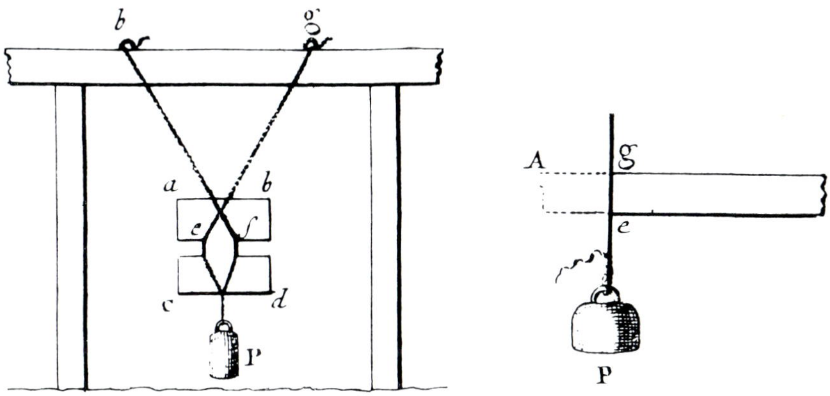

Fig. 3.9. COULOMB's drawing for rupture in tension and rupture in shear, (1773).

COULOMB stated: "I wished to see whether to break a piece of stone with a force directed along the rupture plane, the same weight had to be used as to break it in the preceding experiment, by a stress perpendicular to that plane."[6] He found

[1] It is perhaps ironic that this much-quoted experiment of s'GRAVESANDE which did not, and perhaps as conceived in the early 18th century could not have made the contribution claimed for it, did in fact interest Professor KOHLRAUSCH in nonlinear small deformation and thus led directly to the definitive work of JOSEPH OSGOOD THOMPSON and GRÜNEISEN, as was pointed out above. This did not come about because of s'GRAVESANDE's experiment itself, but from the nonlinearity which STRADLING observed in his tensile tests on long wires, performed in the tower at Dresden for comparison with s'GRAVESANDE's experimental data.

[2] COULOMB [1773, *1*], p. 348.

[3] 1 Paris ft. = 0.3248 m; 1 Paris in. = 2.707 cm.

[4] COULOMB, *op. cit.*, p. 348.

[5] As TRUESDELL pointed out, this is the first time an experimentist employed the concept of stress, rather than force [1960, *1*], p. 397.

[6] COULOMB [1773, *1*], p. 348.

that the two were equivalent: 215 psi for the tension and 220 psi for the shear. He further showed an experiment for the rupture and compression through failure in shear (see Fig. 3.10). He then made the following comment with respect to the ideas which MUSSCHENBROEK had proposed concerning the failure of masonry pillars.

Finally, I must give notice that the way in which Mr. Musschenbroek determines the force of a masonry pillar has no relation with that which I have just used. A pillar pressed by a force directed along its length breaks, says this celebrated physicist, only because it begins to bend; otherwise it would support a weight of any amount. On the basis of this principle he determines the force of square pillars as being in the inverse ratio to the squares of their lengths and triple ratio to their sides; so that if the pillar the force of which we have just calculated had had only the half of its length, it would have supported four times the weight, that is, 832 pounds, while I believe I have shown that it would scarcely have supported the same weight, 208 pounds.[1]

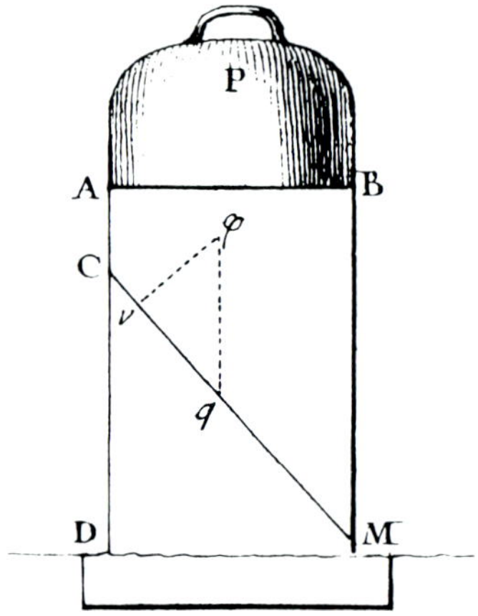

Fig. 3.10. COULOMB's drawing for rupture in compression through failure in shear (1773).

3.3. The origins of an experimental science of solid mechanics: The torsion studies of Coulomb in the 1780's.

It is interesting to examine this memoir, written by COULOMB at the age of 37, as a precursor to the great experimental classic on the torsion of wires which he was to write eleven years later,[2] in 1784, a classic which to the experimentist, if not necessarily to the theorist, may deserve to rank as the most important paper on experiment in solid mechanics up to that time.

In 1777 COULOMB had presented a long essay on the theoretical and experimental determination of magnetic forces.[3] Included in it were studies on the best method for making magnetized needles and, of great interest here, a series of six experiments on the torsion strength of hairs and silk threads which were used in the suspension system. This experimental paper related to continuum mechanics probably is the one best known today among persons outside the field. We may well look at some of its details since these measurements began the series which culminated in his torsion experiments on metal wires seven years later.

From the perspective of the 20th century it seems remarkable that COULOMB could begin the torsion section of his memoir with the certainty that "We cannot

[1] *Ibid.*, pp. 355–356.
[2] COULOMB [1784, *1*].
[3] COULOMB [1780, *1*]. The final paper was published at this later date.

cite here the experiments of any author ..." This unusually competent experimentist then gave a hint of his attitude toward science: "... but those we are about to report are so simple and so easy to repeat, that I hope they will deserve some confidence."[1]

Other parallels among the features that distinguish excellence for the experimentist may be noted. Apart from the obvious prerequisite of originality, there is perceptivity and taste in the choice of problem, an intuitive feeling for that which may be expected to broaden the understanding rather than terminate the branches; aesthetic simplicity; completeness within the framework of well defined assumptions; and a logical development utilizing different experimental perspectives which lead to new patterns of understanding and either definitively separate the physical applicability of prior plausible explanations, or define precisely the main structure which must be included in any new proposed explanation. In the process, the region investigated must be quantitatively and qualitatively established, and associated behavior beyond such boundaries be given at least rudimentary consideration. COULOMB was the first to understand such an approach to experiment in solid mechanics. His work has remained to the present a paragon of the method he inaugurated, and he must be regarded as one of the few outstanding experimentists in this branch of physics.

In his study of 1777 COULOMB performed six experiments in the following logical order:

In the first, a round copper disk, $^3/_4$ Paris in. in diameter and 250 grains in weight was suspended at its center from a vertical human hair 6 Paris in. long. The disk was in a horizontal plane. Twisting the disk about the axis of the hair, he noted that the period of oscillation remained constant and equal to 8 sec, "whether this plane made one, two, or up to six or seven revolutions in its oscillation ...," from which he concluded that "the forces of torsion which bring a body back to its natural situation are necessarily proportional to the angle of torsion."[2]

In the second experiment on the same 6 in. long hair, COULOMB successively added identical disks one at a time until he had a total of 7; he used a bit of wax to prevent slipping. He noted that since the radii of the disks were all equal and had been aligned in a vertical stack, his prior analysis of the problem would indicate that the square of the period of oscillation should be proportional to the number of disks. His results may be seen in Fig. 3.11; they are indeed consistent with prediction.

To provide a more precise comparison of theory and experiment we may examine COULOMB's calculation of the predicted oscillation of each successive test determined from the period of the first. From this comparison, shown in Table 33, COULOMB concluded that there was a close agreement and furthermore: "the mass of the bodies held up by hairs, or, what amounts to the same thing, the tension of these hairs, has no influence at all on the force of torsion."[3] As COULOMB indicated, his latter conclusion was established only when the total weight was relatively small.

The focus of interest in solid mechanics evident from the mid-19th century onward makes equally important the rest of COULOMB's conclusions from this second experiment. Not only do they delineate the region of the observed behavior, but also they introduce a number of important questions, some of which remain unanswered today. COULOMB stated:

---

[1] *Ibid.*, p. 201.
[2] COULOMB [1780, *1*], p. 202.
[3] *Ibid.*, p. 203.

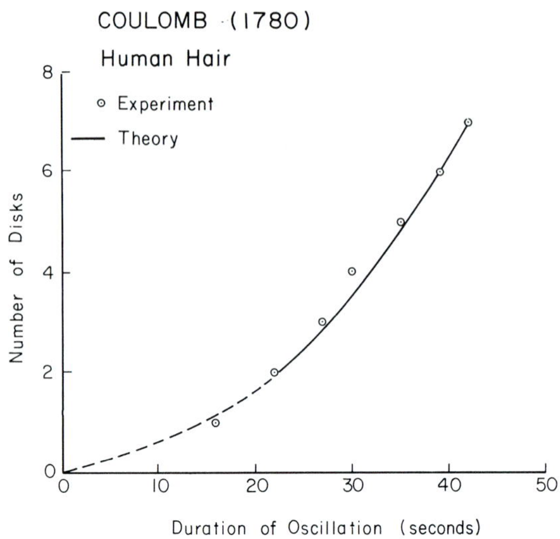

Fig. 3.11. A comparison of the observation and prediction by Coulomb of the relation between the period of oscillation and the number of disks.

Table 33. Coulomb (1780).

Test no.	No. of disks	Duration of oscillation	
		Theory (sec)	Experiment (sec)
I	1		16
II	2	22.6	22
III	3	27.7	27
IV	4	32.0	30
V	5	35.8	35
VI	6	39	39
VII	7	42.3	42

It must be noted nevertheless that when the weight of the bodies is much increased, and when the hairs or silken threads are ready to break, the same law does not hold precisely; but the force of torsion seems much lessened; the oscillations are no longer isochronic, and the times of the large ones are considerably greater than those of the small ones: in this case what happens is that the thread from too great tension loses its elasticity, much like a band which retains its spring only so long as it is bent only up to a certain point.[1]

In a third experiment using a single copper weight Coulomb examined the influence of the length of the hair on the duration of the oscillation. Table 34

Table 34. Coulomb (1780).

Test no.	Length of hair (in.)	Duration of oscillation (experiment) (sec)	Compared with test I (theory) (sec)
I	3	11/2	—
II	6	16/2	15/2 + 1/4
III	12	22/2	22/2

[1] *Ibid.*

contains both the durations and the prediction based upon the calculated differences for the 6 in. and 12 in. lengths as they increased from the initial length of 3 in.

The fourth experiment does not provide agreement with his prior prediction. Using homogeneous silk threads as well as hairs, he attempted to determine the influence of the diameter on the force of torsion. Because of the difficulty of measuring the diameters of such specimens, a problem which, as we have seen, was still limiting the experimental analyses of WEBER in 1830 and of WERTHEIM, WUNDT, and VOLKMANN in the 1840's and 1850's, COULOMB compared the results of a great number of tests, most frequently those on homogeneous silk of identical lengths. He obtained results incompatible with prediction. The force of torsion was in proportion to the "cube" of the diameter. One wonders whether this disagreement provided the impetus for his later study on the torsion of wires. In 1784 he saw that the measurement of mean diameter could be accomplished with relative ease by weighing specimens of known length.

In the fifth experiment, COULOMB turned to a single silk thread "such as comes from cocoons." It was only 1 in. in length and to it was attached a brass wire, more closely representing a magnetic needle as far as the torsional pendulum aspect of the problem is concerned. Having replaced the copper disk with the brass needle, and the hair with the silk, COULOMB still found that the oscillations were isochronic. The period was 40 sec.

In the sixth experiment COULOMB examined a 20 in. long silk thread composed of twelve strands which were stuck together and twisted. Again the oscillations were isochronic, but with a period of only 29 sec. From the two periods he determined that the force of torsion which previously had been shown to be in inverse ratio to the square of the duration was in the ratio of 1.90, or approximately double the force of torsion of the previous experiment. He then introduced this ratio into his earlier calculations. Thus he demonstrated the practical consequence that "the torsion of silk [threads] can have only an insignificant influence on the position of magnetic needles suspended from them."[1] COULOMB's final comment referred to the possible importance of air friction as a source of error. He explored this matter in his memoir on the oscillation of metal wires.[2]

In 1784, with the publication of the classical *Mémoire* of COULOMB entitled, "Recherches théoriques et expérimentales sur la Force de Torsion et sur l'Élasticité des Fils de Métal,"[3] a new era in solid mechanics was inaugurated. COULOMB's understanding of experiment and its relation to theory and his logical style of presenting his data make his memoir a model strongly recommended to the student in experimental mechanics.

In that research COULOMB stated that he had two objectives: first, he sought to determine the law of torsional forces, assuming such forces were proportional to the angle of torsion. In particular he wished to determine the elastic torsional strength of iron and brass wires relative to their length, their size, and their degree of tension. He noted that for his first objective it was necessary to confine the measurements to small amplitudes in order to obtain consistency. His second objective was to study the defects of the elasticity of metal wires in an effort to determine the laws which govern the alteration of elasticity in relatively large oscillations. His ultimate purpose was to provide an understanding of the coherence and elasticity of metals.

---

[1] *Ibid.*, p. 208.
[2] COULOMB [1784, *1*].
[3] *Ibid.*

COULOMB described a total of 21 experiments in terms of 67 individual measurements. He stated that he had included only a portion of the total number of measurements since the others had given identical results. The study of torsional forces was the objective of his first 13 experiments. Three commercially distributed, iron harpsichord wires, designated as #12, #7, and #1, each with a length of 9 Paris in.[1] (24.36 cm), were suspended vertically with a circular cylinder of lead attached to the bottom. In the first, third, and fifth iron wire experiments in each of the three sizes, the weight of the cylinder was $^1/_2$ lb., with a diameter of 19 Paris lines (4.29 cm) and a height of $6^1/_2$ Paris lines (1.47 cm). In the second, fourth, and sixth experiments for identical wires, the weight of the lead cylinder was increased to 2 lbs., i.e., increased by a factor of 4, with the diameter of the attached lead cylinder remaining the same. COULOMB did not provide the diameters of the wires, but, as he indicated, each could be calculated from the given data with respect to the given weight of a 6 Paris ft. length. The calculation provides a diameter $D = 0.015$ cm for the #12 wire; $D = 0.025$ cm for the #7 wire; and $D = 0.051$ cm for the #1 wire. (In the latter part of the treatise, COULOMB calculated the diameter of the #12 wire as almost exactly $^1/_{15}$ of a Paris line.)

Those six experiments were repeated for #12, #7, and #1 brass wires having the same diameters and lengths, to which were attached the same lead cylinders.

It would be interesting to trace the systematic experimental logic of COULOMB; since there is not space here to do so, and since his *Mémoire* is easy to obtain in libraries, it suffices to tabulate all those data in Table 35. Included in the table also are the results of experiment No. 13 in which the length of a #7 brass wire upon which a 2 lb. lead cylinder was attached, was increased by a factor of 4, from 9 Paris in. to 36 Paris in. (97.45 cm) in order to study the effect of changing the length upon the torsional oscillations. In his discussion of experiment No. 13, COULOMB again indicated that he had conducted many similar experiments which exactly confirmed the observed results, so that he felt it unnecessary to enlarge his treatise with them.

Table 35. COULOMB (1784).

Experiment no.	Specimen	Cylinder weight (lb.)	Length of wire (in.)	Trial no.	Time for 20 oscill. (sec)	Torsion angle (degrees) ≦
1	Iron wire #12	0.5	9	1	120	180
				2	—	1080
2	Iron wire #12	2	9	1	242	180
3	Iron wire #7	0.5	—	1	42	180
4	Iron wire #7	2	—	1	85	—
5	Iron wire #1	0.5	9	1	—	—
6	Iron wire #1	2	9	1	23	45
7	Brass wire #12	0.5	9	1	220	360
				2	225	1080
8	Brass wire #12	2	9	1	442	360
				2	444	1080
9	Brass wire #7	0.5	9	1	57	360
10	Brass wire #7	2	9	1	110	360
				2	111	720
11	Brass wire #1	0.5	9	1	Uncertain	—
12	Brass wire #1	2	9	1	32	50
				2	33.5	450
13	Brass wire #7	2	36	1	222	1080

[1] 6 Paris ft. = 1.94904 m; 1 Paris ft. = 0.3248 m; 1 Paris in. = 2.707 cm; 1 Paris line = $^1/_{12}$ Paris in. = 0.2256 cm.

COULOMB'S portrayal of the apparatus is shown in Fig. 3.12. From an analysis of the data, he discovered that the torsional force M was proportional to the fourth power of the diameter, as well as proportional to the angle of torsion [see Eq. (3.1)].

$$M = \mu \frac{d^4}{l} \Theta. \tag{3.1}$$

COULOMB (1784)

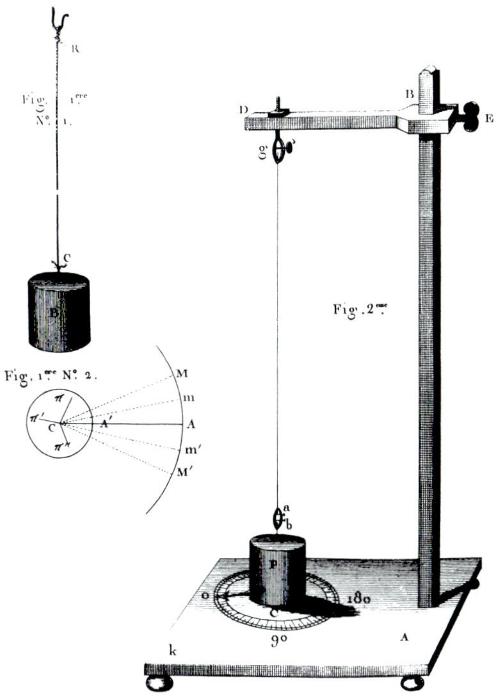

Fig. 3.12. COULOMB'S drawing of his torsion vibration apparatus of 1784.

3.4. Coulomb's first measurement of an elastic modulus and his experiments on viscosity and plasticity (1784).

The most remarkable experimental discovery in that portion of his treatise is the constant material coefficient μ which, as COULOMB stated, was dependent upon the material rigidity of each metal for all wires of the same metal, and, as he indicated, is easily determined experimentally. This, the first *experimental* statement[1] that a given solid has a material elastic constant, independent of the dimensions of the specimen and the density of the solid, is of course a major landmark in the history of continuum physics. COULOMB compared the values of this shear modulus for iron and brass and obtained[2] the ratio of 3.34. His

[1] COULOMB [1784, *1*].

[2] In the 1884 Centennial reprinting of the famous treatise of COULOMB by the Société Française de Physique, the editor, A. POTIER, calculated the values of COULOMB'S shear moduli. Allowing a specific weight of 7.8 for COULOMB'S iron wire, he obtained a value of 7480 kg/mm²; for the brass, assuming a specific weight of 8.6, he obtained a value of 2730 kg/mm². The ratio of these two moduli is considerably higher than JORDAN RICCATI'S ratio of 2.06 for the values of E for these two metals. This difference, too, will be commented upon later in the present treatise (*see* Sect. 3.44).

value for iron is fairly close to the average of 8200 kg/mm² I obtained[1] in the 1960's. However, for brass, as Alfred Potier[2] pointed out, Coulomb's value of 2730 kg/mm² is notably smaller than the 19th century values of Lord Kelvin, Guillaume Wertheim, and Joseph David Everett, who obtained values of 4100 kg/mm² and 3470 kg/mm². The differences among those numbers in terms of experimental multiple elasticity also will be discussed in Sect. 3.44.

To the experimentist in the second half of the 20th century, the remaining portion of Coulomb's paper, in which he makes an effort to develop experimental generalizations with respect to the imperfections of linear elasticity, assumes an importance every bit as great as that of his studies of infinitesimal linear elasticity in the first part. One does see constant reference in the literature after 1784 to the experiments on the torsional pendulum, but because of the subsequent emphasis upon mathematical development in infinitesimal linear elasticity, Coulomb's experiments on plasticity and viscosity have been largely ignored in areas where the theory to this day still is inadequate.

In this connection it is interesting to note that the experiments on linear elastic vibration by Ernst Florens Friedrich Chladni,[3] one of the finest experimentists in the field of solid mechanics, also were completed before any theoretical explanation was available. Chladni between 1776 and 1787 performed his famous experiments on rods and upon the nodal figures on plates. One wonders to what extent the acclaim for Chladni's figures was related to the fact that in a short time success was achieved in extending the theory of linear elasticity to include the lateral oscillations of thin plates, although it took a very long time to explain some of the detail of even the simplest of Chladni's figures.[4]

In Coulomb's first series of five experiments on the imperfections of elasticity, each experiment was divided into two parts.[5] First, keeping the maximum twist below what would now be referred to as the "elastic limit", a value, incidentally, which Coulomb supplied in his paper and which he was the first to discover, he set his torsional pendulum in oscillation and in the ensuing diminution of the maximum amplitude he noted how many oscillations were required to produce a loss of 10°. In those experiments, all of which were for 2 lb. lead cylinders, the length of the wire was considerably shortened from 9 Paris in. (24.36 cm) to 6 Paris in., 6 Paris lines (17.60 cm). On reducing the initial angle of torsion, Coulomb observed that the number of oscillations required to produce a 10° loss markedly increased. The first experiment was for the #1 iron wire with the largest diameter. He noted that the diminution of the amplitude was very uncertain if the original angle were more than 90°, which was the highest value he recorded for the experiment. These data are shown in Table 36. It was in this experiment that Coulomb discovered that an angle of torsion was reached between 90° and 180° rotation, at which point the cylinder did not return to its original position.

---

[1] Bell [1968, *1*].

[2] Potier (*see* Coulomb [1884, *1*], centennial edition, p. 85).

[3] Chladni [1787, *1*], [1802, *1*], [1817, *1*].

[4] The first extensive experimental efforts to explain some of the complications of the Chladni figures were made by Savart in the 1820's and 1830's. These were followed by those of Lord Rayleigh toward the end of the 19th century, with occasional studies by several persons in the years that followed. Having myself a few years ago performed some of Chladni's experiments, I can personally attest to the fact that there are many aspects of the observed phenomena which still remain unexplained. Nevertheless, the development of the linear theory of the lateral oscillations of the plate led to the general belief that the experimental complications were merely details to which the theory would apply in due time. In this context, I leave to later discussion the "deep tone" in longitudinal oscillations (see Sect. 3.18). Savart [1820, *1*], [1824, *1*], [1829, *1*], [1830, *1*], [1837, *1*]. Rayleigh [1894, *1*].

[5] Coulomb [1784, *1*].

Table 36. COULOMB (1784).

	Torsion angle (degrees)	Loss of 10° in
First trial	90	3.5 oscillations
Second trial	45	10.5 oscillations
Third trial	22.5	23.0 oscillations
Fourth trial	11.25	46.0 oscillations

Following those interesting discoveries, COULOMB in a series of nine trials continued the testing of the suspended wire, each time using a sufficiently high angle to produce a further permanent set until on the ninth trial at a total angle of twist of 13258°, the wire broke. He noted that after the experiment the wire was straight and very rigid, and when examined with a magnifying glass the fracture resembled the shape of a cord made up of two strands.

As a prelude to modern studies of the differences between stress-strain functions for loading and unloading in the large plastic deformation of metals, COULOMB noted further that although recovery was rather irregular in the first two trials, in all of the six remaining trials the amount of recovery was approximately the same. There was only a 10% difference between the recovery for the fourth and seventh trials. The data for that experiment are provided in Table 37 where I have added additional columns to include the total angle of twist from the initial position, the total permanent set, and the amount of recovery for each trial.

Table 37. COULOMB (1784).

Trial	Torsion (circumferences)	Total angle of twist (degrees)	Displacement of the indicator from the center of torsion (degrees)		Total permanent set (degrees)	Recovery (degrees)
			First experiment: Iron wire # 1			
1	1/2	180	8	8	8	172
2	1	368	50	50	58	310
3	2	778	310	310	368	410
4	3	1448	1 circumference + 300 =	660	1028	420
5	4	2468	2 circumferences + 290 =	1010	2038	430
6	5	3838	3 circumferences + 280 =	1360	3398	440
7	6	5558	4 circumferences + 260 =	1700	5098	460
8	10	8698	8 circumferences + 240 =	3120	8218	480
9	14	13258	(wire broke)		13258	
			Second experiment: Iron wire # 7			
1	3	1080	300		300	780
2	4	1740	1 circumference + 180 =	540	840	900
3	6	3000	3 circumferences + 90 =	1170	2010	990
4	8	4890	5 circumferences + 90 =	1890	3900	990
5	12	8220	9 circumferences + 40 =	3280	7180	1040
6	20	14380	16 circumferences + 310 =	6070	13250	1130
7	30	23850	26 circumferences + 180 =	9540	22790	1260
8	50	40790	46 circumferences + 20 =	16580	49370	1420
9	7	(wire broke)				
			Third experiment: Iron wire # 12			
1	4	1440	300	300	300	300
2	6	2460	2 circumferences + 40 =	760	760	1090
3	6	(wire broke)			2160	3250

The second and third experiments were for successively smaller diameter #7 and #12 iron wires. The damping data for those experiments are shown in Fig. 3.13. A close inspection of the iron wire data reveals why COULOMB was unable to formulate a rule governing attenuation.

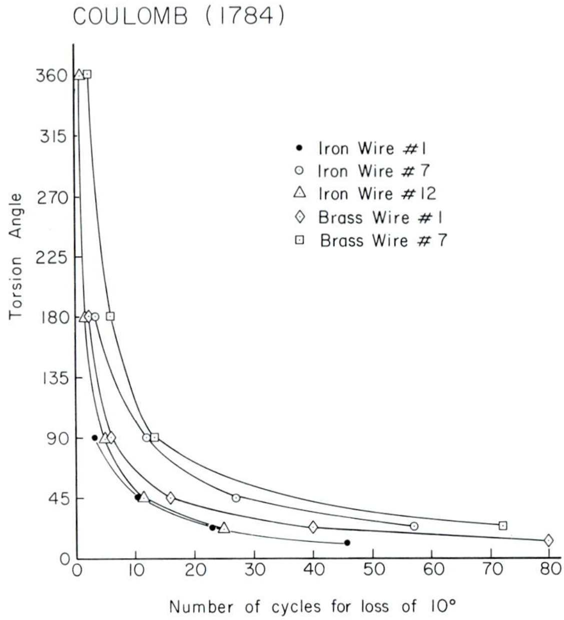

Fig. 3.13. The results of COULOMB's viscous damping experiments on iron and brass wires.

In addition, Fig. 3.13 includes attenuation for two brass wires, also in the form of initial angle vs number of cycles for a loss of 10 degrees.

Data, too, were included for similar elastic imperfection experiments in #1 and #7 brass wires having the same shortened length of 6 Paris in., 6 Paris lines (17.60 cm) as the iron wires. In Table 38 are shown the data on brass for the two wires.

Table 38. COULOMB (1784).

Trial	Torsion (circumferences)	Total Angle of twist (degrees)	Displacement of the indicator from the center of torsion (degrees)	Total permanent set (degrees)	Recovery (degrees)
			Fourth experiment: Brass wire #1		
1	2	720			
2	4	1600	2 circumferences + 0 = 720	160	560
3	6	3040	3 circumferences + 300 = 1380	880	720
4	10	5860	7 circumferences + 300 = 2820	2260	780
5	20	12280	17 circumferences + 340 = 6460	11540	740
6	Wire broke after 28th circ.				
			Fifth experiment: Brass wire #7		
1	4	1440	220	220	1220
2	Wire broke after 6th circ.				

COULOMB noted that in comparing iron and brass wires of the same diameter, the latter sustained amplitudes much larger than the former. He suggested that the irregularity of the data probably could be attributed to the manner of manufacture, including the temperature to which the wires were subjected in order to reduce their diameters.

COULOMB also was the first to investigate the possibility that air damping could be a factor contributing to the variability of his results.[1] He wrapped lightweight paper around the lead cylinder so as to produce a total height three times that in which his initial oscillatory motion had been studied, thinking that this would triple the influence of air resistance. He noted that the amplitude never increased by more than $^1/_{10}$ with this change, and more often than not, there was no detectable difference in frequencies. He concluded, therefore, that air damping had a negligible influence upon the data.

An inspection of the attenuated oscillation data of Fig. 3.13 reveals inconsistencies with respect to the wire diameter. Being such an excellent experimentist, COULOMB, by inaugurating an additional series of experiments, went beyond merely conjecturing that the reduction of diameter was responsible for the variation in the times of oscillation. He annealed a $\#1$ diameter brass wire by heating it by fire until it was red hot in order to make it lose the greater part of its elasticity. He found that the wire could be twisted by 140 complete turns before failure. For the first turn, the elastic recovery was only 50 degrees, but after 90 complete turns it had acquired an elastic domain close to 500 degrees. He observed that from the second to the third complete turn, the elastic reaction was increased by 12 degrees; from the 40th to 41st, by 6 degrees; and from the 90th to the 91st complete turn, it was increased by only about 1 degree. Hence, after the center of torsion had been displaced by a certain angle, the increase of the elastic region was approximately in inverse ratio to the angle of displacement.

Comparing the results of that experiment with the permanent deformation data from the unannealed specimens, COULOMB concluded that with a single turn it was possible to give an annealed wire all of the linear elasticity to which it was susceptible; hard hammering or drawing could add nothing. He later noted that before annealing, the brass wire could support up to 22 lbs. at the moment of breaking, while after annealing it bore barely 12 to 14 lbs. He also remarked that although this major change had taken place in the plastic deformation of the solid, in the reduction of the ultimate torque, and in the same magnitude of reduction in what now would be called the elastic limit, nevertheless the speed of torsional oscillation for small angles was identical under the same conditions of wire diameter, length, and lead weight. From this he concluded that the elasticity was unchanged.

Again, in the spirit of a proper experimental study of a phenomenon for which in 1784 no theory was available, COULOMB proceeded to conduct quite a different set of experiments to explore the relation between what he referred to as "coherence" and elasticity. A cantilevered beam such as that shown in Fig. 3.14 was subjected to a series of three weights at point B, and the deflection of the end was measured. The first beam so tested was a steel blade 2.48 cm wide, 0.113 cm thick, and 18.95 cm long from point A to B. The blade was heated until white hot and was given a very rigid temper. The same blade then was heated until it had acquired a violet color and had returned to the consistency of an excellent spring. Finally, the blade was raised to white heat and allowed to cool very slowly to produce an annealed state. Load and elongation then were compared

[1] This was a concern he had expressed several years earlier in his paper describing his experiments on human hair and silk. COULOMB [1780, *1*].

for each situation. The results are shown in Table 39. The prestige of COULOMB was such that after he had stated that load vs elongation data were identical for annealed and hardened steel, no one seriously questioned the matter for half a century, until WERTHEIM demonstrated otherwise.

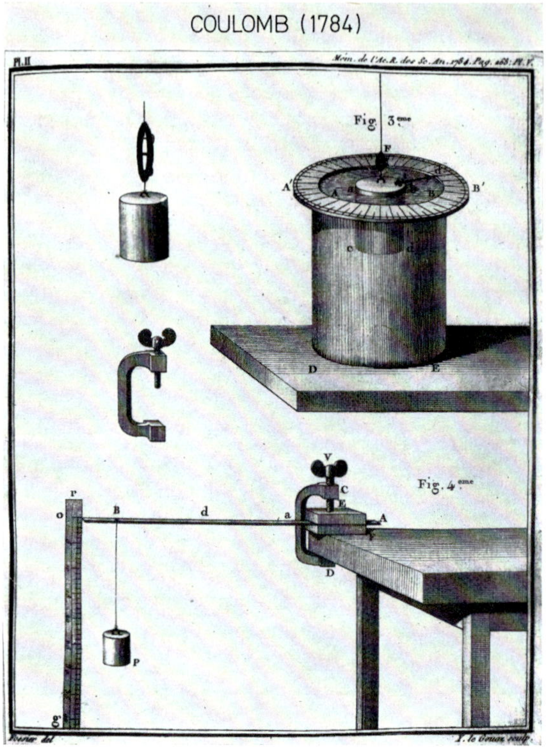

Fig. 3.14. COULOMB's drawings of his method of measurement of the angle of torsion, and of his flexure experiments.

Table 39. COULOMB (1784).

Load no.	Weight in pounds	Grams	Lines	Cm.
1	$^1/_2$	245	8	1.80
2	1	489	15.5	3.49
3	1.5	734	23	5.19

In a second series of supplementary experiments, each of three different steel blades of the same dimensions was treated to the three thermal histories described above. A balance was attached at point D, $2^1/_2$ Paris in. from point a, and the direction of the load was maintained always perpendicular to the direction of the blade. The tempered blade broke under a 6 lb. load. For any angle under that which produced rupture, it resumed its original position exactly. The "excellent spring" blade ruptured under a load of 18 lbs., and it, too, up to rupture resumed its original position when released. The annealed, slowly cooled blade bent as a linear elastic solid, up to a load of 5 to 6 lbs. Having increased the load to 7 lbs.,

COULOMB observed that after the weight was removed, the blade had acquired a permanent set. Recovery approximated the extent of the elastic deformation at the 6 lb. load, i.e., at the elastic limit of the solid.

From all these experiments on large deformation COULOMB drew a distinction between elasticity and what now would be called plasticity. In the latter state, as he suggested, the parts of the metal might slide on each other and become displaced. That insight, together with the recognition of the existence of the elastic limit and the fact that it could be altered by changes in the prior thermal history, constitute an introduction to the major features of quasi-static plasticity, albeit in highly restricted form, which theories of the present day still are inadequate to explain.

3.5. On the measurement of elastic constants.

In view of the overwhelming volume of literature on elastic constants in the 19th and 20th centuries, it is indeed odd that over one hundred years elapsed before it occurred to any experimental physicist that there existed a measurable ratio of possible importance, discovered by HOOKE and MARIOTTE[1] in their linear load-elongation experiments. The intense interest in elastic constants was inaugurated by the sudden appearance in 1842 of the great systematic experimental study of the quasi-static and dynamic moduli of a very large number of metals by GUILLAUME WERTHEIM, followed by the work of ADOLF THEODORE KUPFFER,[2] WOLDEMAR VOIGT, FRIEDRICH KOHLRAUSCH, HERBERT TOMLINSON, and KELVIN (WILLIAM THOMSON), to mention a few of the more prolific 19th century experimentists, and EDUARD GRÜNEISEN, KARL ANTON ERICH GOENS, and WERNER KÖSTER, as representative of the even greater bulk of data published since 1900.

The interest in the subject went far beyond the needs of an expanding technology. In fact, beginning with the work of WERTHEIM nearly every scientific study on the subject began with lamentations over the experimental inadequacy of previous investigators who failed to provide definitive numbers. It may well be that the variety of published moduli led to the general technological use of somewhat arbitrary fixed numbers. The latter were then given "validity" by being included in most of the major engineering manuals and textbooks.

Careful measurement of elastic constants by a few experimentists made possible experimental study in many areas: the similarities among different solids; anisotropy in crystalline solids; the effects of various thermal influences, such as annealing or variation of ambient temperatures or the difference between isothermal and adiabatic behavior; and the influence of electrical and magnetic phenomena upon metallic solids. With such a quantitative base it was possible to explore the effects upon continuum elasticity and atomistic models of various metal states, prior histories, compositions, and structures. Beginning with the values provided by COULOMB, the great majority of the reported elastic constants have been dynamic. The 19th and early 20th century emphasis upon experiments on vibration now has shifted to ultrasonic measurements in small samples.

In the historical past as well as in the present, elastic constants in presumably isotropic solids have been determined chiefly from measurements of E. Of course measured values of μ also were in the literature. Since the second half of the 19th

[1] HOOKE [1676, *1*], [1678, *1*], MARIOTTE [1700, *1*].

[2] In Russian publications of his works, KUPFFER's middle name was given as FEODOR which, of course, is the Russian equivalent of the German name Theodore. In English publications such as TODHUNTER and PEARSON's *Elasticity and Strength of Materials*, he is referred to both as A. *T*. KUPFFER and as A. *F*. KUPFFER.

century there have been a few efforts to provide direct measurements of Poisson's ratio ν and of the bulk modulus K. It was early recognized that the calculation of the two latter values from measured E and μ led not only to widely different determinations, but also in many instances, to obviously incorrect values.

Inasmuch as an experimentist who believed his measurements were inaccurate would be akin to a mathematician who doubted the rigor of his logic, variations in experimental data nearly always were attributed to unseen anisotropy of the solid. For example, when George Frederick Charles Searle in 1908 in his laboratory manual, *Experimental Elasticity*,[1] determined Poisson's ratio for nine different metals from the ratio of E and μ, he found five values in excess of one-half, with numbers ranging from 0.598 to 1.207. Since the range of strain was infinitesimal, Searle concluded that those metals must be exceedingly anisotropic in the polycrystalline state.

Values of Poisson's ratio of 0.608 for annealed copper and 0.614 for prestressed nickel, suggest to me, as similar numbers suggested to Bauschinger and Grüneisen decades ago, that experimental accuracy in the measurement of the frequencies of flexural vibration, in the observation of the period of the torsional pendulum, and in the determination of specimen dimensions was insufficient for E and μ to be introduced into a ratio of large numbers for the calculation of Poisson's ratio.

As was shown above,[2] from an examination of thirty different specimens of steel, Bauschinger in 1879 cast serious doubts on any calculation of either Poisson's ratio or the bulk modulus which used the ratio of the E and μ moduli. The dynamic measurement of E was obtained from both flexural and longitudinal vibration. Flexural values almost invariably were lower than longitudinal, even in the latter half of the 19th century when corrections began to be made for rotary inertia, and in the 20th century when deflections due to shear and those due to lateral contraction also were considered in the computations.

Throughout the entire history of longitudinal measurement of moduli in metals, whether dynamic or quasi-static, compression values were slightly higher than tension values. The implications of this observation with respect to small deformation nonlinearity were described in Chap. II above.The common interpretation of the data from both the torsion of solid cylinders and the flexure of rectangular or circular bars presumes stress distributions in accordance with linear elasticity and thus requires that the experimentist make an *a priori* assumption.

Dynamic measurements in the longitudinal situation also suffer from the practical necessity for regarding as one-dimensional what may be in fact a relatively complex, three-dimensional situation. This is true particularly in the recent ultrasonic analyses, and in 20th century attempts to determine stress-strain functions in short cylindrical specimens subject to impact, where, for the initial strain as well as for the large deformation region, to interpret the data it seems to be necessary to assume in advance that deformation and stress are both one-dimensional, though there is no way to check such an assumption. In quasi-static determinations of E in prisms, at least it is possible to check over the entire surface of the specimen to ascertain whether in fact one does have the assumed one-dimensional distribution of strain so as to determine a material coefficient accurately.

Ultrasonic measurements of moduli use strain amplitudes in the range of 10^{-7} to 10^{-8}. The amplitudes of vibratory measurements are in the range of 10^{-5} to 10^{-6}. Quasi-static strains generally are in the range of 10^{-3} to 10^{-6}, depending

[1] Searle [1908, *1*], p. 115.

[2] See Chap. II, Sect. 2.18.

upon the accuracy of the apparatus used. It is of utmost importance to realize that the differences in values thus determined depend as much upon the variation of material properties in these three different ranges of deformation as upon the thermal and viscous phenomena to which they are often ascribed. Nonlinearity, or third-order elastic constants, will appear in the dynamic measurements as alterations in the wave profile or as ultraharmonics or superharmonics. For driven systems, such nonlinearity appears in the jump phenomenon or in bifurcation.

In quasi-static measurement, as was described above in the work of GRÜNEISEN, it is possible to compare different regions of deformation and thus separate the various attributes of solids, including such matters as the discrete distribution of parameters defining material stability.

Regarding the deformation of metals, two more points must be remembered in evaluating the data of the past 90 years. First is the fact that shortly before the turn of the century technologists standardized the shapes and prior thermal and mechanical histories of specimens, and universally agreed upon testing procedures. The standardization imposed by commercial testing machines which varied only slightly from one laboratory to another, thereby presented the possibility for systematic anomalies, or errors, in what appeared to be independent investigations.

Second, the demands of technology imposed upon the applied metallurgist the need to provide stable solids with at least roughly reproducible moduli to a relatively high elastic limit. Because the composition, prior annealing, purity, grain size, etc., were so prescribed, many of the most interesting deformation properties of the solid were minimized or apparently eliminated.

By returning to 19th century dead weight loading on metallic specimens whose prior thermal and mechanical histories were very different from the prescribed recipes, I discovered[1] during the past two decades that the stable elastic constants of 60 elements have a quantized distribution. Furthermore, I have shown that for any given solid, transitions occur from one discrete state to another, in the same integral distribution. The stable, technologically useful material thus has been shown to represent but a single one of many possible elastic states of a solid. Hence, with respect to the history of experiment in small deformation from 1784 to the present, one must keep in mind constantly that after 1820, engineering tests on iron and later on steel equalled in number the tests on all other solids combined, and, since the latter part of the 19th century, the standardization procedures just described have obscured fundamental physical behavior, by characterizing and limiting nearly all studies for which the stated objective was to provide an understanding of the small deformation behavior of solids.[2]

If an experimentist on examining his data assumes isotropy, while suspecting anisotropy, we must wonder why he publishes his results. If he assumes an attractive, but otherwise arbitrary theoretical hypothesis in calculating the results he presented, we must suspect his objectivity. And if, as in many instances, the reverse calculation to the raw data cannot be performed, the conclusions must be ignored when later hypotheses are being developed. Experiments designed

[1] See BELL [1968, *1*].

[2] For example, earlier in this century as the operating temperatures and pressures of turbines increased, it became necessary to find steel alloys which would retain high strength and high damping properties under the new conditions. Consequently, many experiments at high temperatures were performed on esoteric solids which possessed unique characteristics in that atmosphere. When the preponderance of available information was derived from such special circumstances, there was danger of misunderstanding, or missing, the general properties of solids at high temperatures.

under highly specialized circumstances solely to give physical support to some generally applied theoretical model, make it imperative that we compare the results with the order established from the total domain of experimental data. When for one reason or another such a comparison is not possible, the conclusions deduced from the hypothesis must be limited.

Clearly, measured experimental parameters must be considered suspect when theories of structural behavior are based upon assumed distributions of stress, strain, displacements, velocities, etc., which it was not possible to determine accurately. Similarly suspect is the imposed constitutive equation which has been assumed, but not demonstrated to be applicable.

3.6. The experiments of Chladni on the longitudinal vibration of bars (1787).

The studies of CHLADNI[1] in 1787 provided one of the major 18th century experimental stimuli to 19th century continuum mechanics. During the many years he traveled throughout Europe giving lectures demonstrating various feats of scientific "magic," CHLADNI fascinated audiences with his famous sand figures produced from the lateral vibration of plates. Obviously the experiments on plates for which no good theory was yet known, gave a distribution of stress and strain too complex to provide a basis for inferring elastic constants. On the other hand, the experiments on longitudinal bars were sufficiently simple and sufficiently understood that a material constant could be determined from them.

Starting with JEAN BAPTISTE BIOT[2] in 1816, numerous experimentists in the 19th century calculated E-moduli of isotropic materials from the experiments of CHLADNI on longitudinal oscillations of bars. CHLADNI himself was content to provide a material number in the form of the dimensionless ratio of the velocity of sound in the solid to that in air. This ratio was determined by comparing the frequencies for the longitudinal vibration of 2 ft. long, free-free bars, with those of the air in an open-ended tube of the same length. The comparisons were made in terms of the musical scale. Such data, of course, were subject to the atmospheric conditions in CHLADNI's laboratory at the time of the measurement. BIOT, in his *Traité de Physique Expérimentale et Mathématique*,[3] extracted those velocity ratios for whale bone, glass, and those woods and metals which CHLADNI had considered. The data are shown in Table 40.

In Table 41, I have compared CHLADNI's velocities of sound in metals determined from the ratios in Table 40 (assuming 334 m/sec as the velocity of sound in air at room temperature), with values given in a recent source: those from the 43rd edition of the *Handbook of Chemistry and Physics*.[4]

To determine the modulus, such velocity measurements must be squared, and one must know the density of the solids which CHLADNI studied. Assuming the densities from averages listed in his time, CHLADNI's moduli[5] for the metals are shown in Table 42, where again they are compared with present-day values.

The high value of the modulus in iron was the source of considerable discussion from the time of BIOT's[6] experiments on wave propagation in 1808 to those of

[1] CHLADNI [1787, *1*].
[2] BIOT [1816, *1*].
[3] See BIOT [1816, *1*], Vol. II, Book 2, Chap. 5.
[4] Handbook of Chemistry and Physics [1961, *1*].
[5] CHLADNI [1802, *1*].
[6] BIOT published in [1809, *1*].

Table 40. Biot (1816).

Type of rod	Value of N'	Relative velocity[a] N'/ut_3
Whalebone	la_5	6.66
Tin	si_5	7.50
Silver	re_6	9.00
Walnut	fa_6	10.66
Yew[b]	fa_6	10.66
Brass, Oak, Plum	fa_6	10.66
Pipe-stem	mi_6	10.00
	sol_6	12.00
Copper	about sol_6	12.00
Pear, Copper beach, Maple	$sol^{\sharp}_6$	12.50
	la_6	13.33
Mahogany, Ebony, Hornbeam, Elm, Alder, Birch	approx. $si^{\flat}_6$	14.40
Linden	almost si_6	15.00
Cherry	si_6	15.00
Willow, Pine[c]	ut_7	16.00
Glass, Iron or steel	ut_7	16.66
Fir	$ut^{\sharp}_7$ or almost re_7	18.00

[a] ut_3 is the tone of the air column.

[b] If the fibers of this wood had been exactly straight, the sound would have been a bit higher.

[c] If the fibers of this wood had been less straight, the sound would have been one-third lower.

Table 41.

Substance	Chladni[a] (m/sec)	Handbook of Chem. and Phys.[b] (m/sec)
Tin	2505	2500
Silver	3006	2610
Copper	4008	3560
Iron	5564	5000
Brass	3560	3500
Glass	5564	5000–6000

[a] See Biot [1816, 1].

[b] *Handbook of Chemistry and Physics*, 43rd edition, p. 2537 (1961). It is interesting to note that these values for Ag, Cu, and Fe, as well as the majority of those for other metals and woods, are those determined by Wertheim 120 years earlier.

Wertheim and Breguet[1] in 1851. Biot, and Wertheim and Breguet also obtained numbers considerably below Chladni's for the ratio of the velocity of sound in iron to that in air. The details of the controversy over these values,

[1] Wertheim and Breguet [1851, *1*].

Table 42.

Substance	Chladni (kg/mm²)	Current tabulated values[a] (kg/mm²)
Tin	4909	4002
Silver	9478	7913
Brass	10696	9209
Copper	14361	10404–12252
Iron	24364	18684–20828

[a] *Handbook of Chemistry and Physics*, 43rd edition, p. 2169.

including the contributions of Félix Savart,[1] will be considered immediately below. Of major interest at this point is the fact that the difference in the numbers was large for iron and was reproducible in the dynamic data of different experimentists.

Chladni's ratio of $E_{\text{steel}}/E_{\text{brass}}=2.26$, was considerably higher than the value of 2.06 obtained five years earlier by Jordan Riccati. Since the moduli for brass and copper calculated from Chladni's velocities of sound are in close accord with 19th and 20th century values, the higher number in iron obtained by Chladni may not be attributed to a lack of experimental expertise, as was implied in the attacks upon him thirty or more years later by Savart and others, when his wave speed for iron was found to be markedly higher than that obtained by Biot in 1808.[2] It will be of some interest to return to this subject later in the present treatise[3] when the accumulated data of many decades has been discussed. It will be shown that the differences in moduli, such as those of Chladni and Biot in iron, not only are too large to be ascribed to experimental error, but reproducibly occur over and over among the data of many experimentists. Such multiple elasticities are now known to have a more fundamental origin than poor specimen preparation, inclusions, inhomogeneity, or extreme anisotropy, as is often asserted.

3.7. An assessment of fact and myth for the modulus in Young's Lectures on Natural Philosophy (1807).

Much has been written on Thomas Young's *A Course of Lectures on Natural Philosophy and the Mechanical Arts*. That it is a vague and rambling account on various scientific and technological matters of interest at the end of the 18th century, is immediately obvious to any person who evaluates the two-volume work by close reading, rather than by reputation. Isaac Todhunter, in Todhunter and Pearson's *A History of the Theory of Elasticity and of the Strength of Materials*, says of the section on elasticity:

> The whole section seems to me very obscure like most of the writings of its distinguished author; among his vast attainments in sciences and languages that of expressing himself clearly in the ordinary dialect of mathematicians was unfortunately not included. The formulae of the section were probably mainly new at the time of their appearance, but they were little likely to gain attention in consequence of the unattractive form in which they were presented.[4]

[1] Savart [1820, *1*]. See also later articles by Savart in the general area: [1829, *1*], [1830, *1*], [1837, *1*].

[2] Biot's paper was read at the Institute in 1808, and published in the *Mémoires de Physique et de Chimie de la Société d'Arcueil* in 1809 [1809, *1*].

[3] See Sect. 3.44, and Bell [1968, *1*], [1965, *2*].

[4] Todhunter and Pearson [1886, *1*], vol. 1, pp. 82–83.

On the other hand, YOUNG's work apparently impressed Lord RAYLEIGH,[1] whose extensive marginal notes suggest he regarded the work as well worth intensive study.

With the provocative contrast of the profound and the superficial often present in the intellect of the truly great eclectic, such a man as YOUNG is misunderstood when viewed against the isolated context of specialization whether it be as a mathematician, an experimental physicist, a philologist, or a medical practitioner. The violence of the personal vendetta of Lord BROUGHTON in YOUNG's lifetime with reference to the undulatory theory of light is counterbalanced by the unstinting eulogy by JOHN TYNDALL.[2] Of course it is possible to read *post hoc* discovery into a massive imprecise work such as YOUNG's Lectures, and therefore it is signally unfair to YOUNG to place undue emphasis upon matters of precedence or lack of precedence for much of what he wrote in those two volumes. The substance of his extensive bibliography clearly implies that he was engaged in producing a scientific and technological cosmos rather than a development of his own original work. It is somewhat fractious to note that he, in awarding asterisks for what he judged were especially significant works, drew far too heavily upon the trivial Encyclopaedia writings of JOHN ROBISON.

Thus, we may re-examine the first volume of Lecture XIII on passive strength and friction, where YOUNG defined his height of the modulus in terms which, as TRUESDELL[3] has shown, were a repetition of the statement of LEONHARD EULER of many years before, with whose works YOUNG obviously was familiar. In the light of TRUESDELL's more precise probing of the historical record, the widely quoted, but clearly inaccurate statements such as that of AUGUSTUS EDWARD HOUGH LOVE in the introduction to his great classic treatise on the *Mathematical Theory of Elasticity*,[4] can be seen as having perpetuated the traditional historical appraisal of YOUNG, rather than representing any claim of discovery made by YOUNG himself regarding "YOUNG's modulus."

Said LOVE:

> This introduction of a definite physical concept, associated with the coefficient of elasticity which descends, as it were from a clear sky, on the reader of mathematical memoirs, marks an epoch in the history of science.[5]

In assessing THOMAS YOUNG as an experimentist in solid mechanics, it must be noted that his references to actual experiment were minimal, and when they do occur, experimental details of any kind are entirely absent, except for one or two vague suggestions. It should be remembered, too, that YOUNG was writing in the time frame of COULOMB, CHLADNI, BIOT, DUPIN, and DULEAU, who were thoroughly aware of the logic of experimental science, both in performance and presentation. YOUNG's initial statement with reference to an elastic modulus occurred on the third page of Lecture XIII[6] following a brief discussion, three pages earlier at the end of Lecture XII, of COULOMB's torsion studies in which, as we have seen, a modulus of elasticity was stated. After noting that extension and compression "follow so nearly the same laws, that they may be best understood by comparison with each other," and after making a clear statement of HOOKE's "analagous" linear force-deflection relation, YOUNG stated:

> According to this analogy, we may express the elasticity of any substance which may be denominated the modulus of its elasticity, and of which the weight is such, that any addition

---

[1] RAYLEIGH [1894, *1*].
[2] TYNDALL [1877, *1*], p. 51.
[3] TRUESDELL [1960, *1*].
[4] LOVE [1892, *1*].
[5] This quotation appears on p. 4 of the fourth edition of LOVE's treatise [1944, 1].
[6] YOUNG [1807, *1*], Vol. I, p. 137.

to it would increase it in the same proportion, as the weight added would shorten, by its pressure, a portion of the substance of equal diameter. Thus if a rod of any kind, 100 in. long, were compressed 1 in. by a weight of 1000 pounds, the weight of the modulus of its elasticity would be 100 thousand pounds, or more accurately 99000, which is to 100000 in the same proportion as 99 to 100. In the same manner, we must suppose that the subtraction of any weight from that of the modulus will also diminish it, in the same ratio that the equivalent force would extend any portion of the substance. The height of the modulus is the same, for the same substance, whatever its breadth and thickness may be: for atmospheric air, it is about 5 miles, and for steel nearly 1500. This supposition is sufficiently confirmed by experiments, to be considered at least as a good approximation: it follows that the weight of the modulus must always exceed the utmost cohesive strength of the substance, and that the compression produced by such a weight must reduce its dimensions to one half: ...[1]

Neither YOUNG's weight of the modulus nor height of the modulus is a material parameter, in the modern sense of elastic constants. The weight of the modulus w, in terms of the E-modulus, should be written as $w = EA$, where A is the area of the specimen; and the height of the modulus h as $h = E/\gamma$, where γ is the weight density of the solid. On the only two occasions in which YOUNG provided numbers for the weight modulus, he specified that he referred to a cross-section of one square inch; thus in the two instances when he used this modulus, he gave it in the modern form. In vol. I, p. 151, he stated:

The weight of the modulus of the elasticity of a square inch of steel, or that weight which would be capable of compressing it to half its dimensions, is about 3 million pounds ...[2]

In vol. II, p. 509, YOUNG provided a comparative table of the physical properties of various substances. Those of interest in the mechanics of solids, for which a height of the modulus is provided, are given in Table 43.

Table 43. YOUNG (1807). *A comparative table of the physical properties of various substances*

Substance	Specific gravity	Height[a]	Cohesive strength[b]	Melting point	Extensibility
Water	1.000	750	—	32°	—
Ice	0.93	850	—	—	—
Steam at 212°	0.0004	90	—	—	—
Hydrogen gas	0.0001	350	—	—	—
Nitrogen gas	0.0012	30	—	—	—
Atmospheric air	0.00128	{28, 40}	—	—	—
Oxygen gas	0.00137	26	—	—	—
Carbonic acid gas	0.00176	19	—	—	—
Fir wood	0.56	10000	8	—	1/304
Elm	0.80	8000	13	—	1/214
Beech	0.85	8000	17	—	1/173
Oak	0.99	5060	17	—	1/128
Box	1.10	5050	—	—	—
Crown glass	2.5	9800	—	—	—
Tin	7.3	2250	6	415°	1/714
Iron and steel	7.8	10000	40–150	1600°	1/846 to 1/226
Brass	8.4	5000	—	—	—
Copper	8.8	5760	36	1450°	1/610
Silver	10.5	3240	42	1000°	1/352
Mercury	13.6	750	—	−39°	—

[a] Height of the modulus of elasticity, in thousands of feet.
[b] Cohesive strength of a square inch, in thousands of pounds.

[1] *Ibid.*

[2] Three million pounds per square inch for steel is only one-tenth the value for steel and one-eleventh the value determined for this solid by YOUNG in terms of the height of the modulus.

That YOUNG was not completely in command of EULER's concept is evidenced by his not realizing that for liquids and gases his height of the modulus would be zero. Failing to understand this distinction, he gave such numbers, relying upon dilatation measurements when he sought to represent the load required to double the original length of a column. This representation indicates YOUNG's purely theoretical approach. If, as he claimed, his interest had been primarily in nature, he would not have included substances to which such a concept obviously did not apply.

Table 43 definitely constitutes the first tabular listing of elasticity-related deformation parameters in solid mechanics.[1] It contains a number of curious discrepancies and errors which, because of the historical uniqueness of the table, are worth commenting upon. For example, in vol. I, p. 137, the height of the modulus of steel was given as 1 500 miles, or 7920000 ft., whereas that given in the table for steel was 10000000 ft. On p. 151, the weight of the modulus for 1 sq. in. of steel was given as 3000000 lbs., which,[2] when converted to the height of the modulus, knowing the weight density of steel, provides the excessively low value of 874000 ft. On p. 86 of vol. II, in describing the height of the modulus for a tuning fork of steel, YOUNG gave the value for the height of the modulus of 8530000 ft. For atmospheric air (on p. 137), he proposed a value of 5 miles, or 26400 ft., whereas in the table, he suggested that the value lay between 28000 and 40000 ft. On p. 372, vol. I, and again on p. 64 of vol. II, he referred to the experiments in water of Mr. CANTON which provided the value of 750000 for the height of the modulus, which coincides with that in the table. That this number is seriously in error provides an interesting commentary on YOUNG's general attitude toward numbers, at least in the field of solid mechanics.

On 18 January, 1826, BENJAMIN BEVAN, in a letter[3] to Dr. THOMAS YOUNG, as Foreign Secretary of the Royal Society, delicately pointed out that in his interpretation of the experiments of CANTON, described twenty years before, YOUNG had confused cubical dilatation with linear elongation[4] and thus was in error by a factor of three, so that instead of 750000 ft. for water, he should have had 2250000 ft. BEVAN's actual purpose in pointing this out, however, was far more serious, because in a series of experiments on ice on a pond near Leighton Bussard in the freezing winter of 1826, he had shown from several experiments on the flexural deformation of 100 in. long, 10 in. wide, cantilevered ice beams with an average thickness of 3.97 in., that the height of the modulus was 2100000 ft., rather than the 850000 ft. given by YOUNG in his table. BEVAN suggested that the same confusion regarding cubic and linear dimensions with respect to the modulus of elasticity had been present in considering the data on ice.[5] YOUNG, at the end of the letter BEVAN had presented in the *Transactions*, appended the following note which strongly suggests that he did not completely understand the distinction between the bulk behavior and the E-modulus.

---

[1] CHLADNI's tables had not included any data on quasi-static deformation.

[2] If this number were 30000000 lbs., the corresponding height of the modulus would be 8740000 ft., similar to YOUNG's other numbers for steel. The corresponding value of E would be 21 100 kg/mm² which closely agrees with that now accepted, i.e., 20900 kg/mm².

[3] BEVAN [1826, *1*], pp. 304–306.

[4] At the conclusion (p. 306) of this contribution, BEVAN expressed the too often forlorn hope of the experimentist in serious conflict with major "authority": "I should not have troubled you with this letter, but as the frost may yet continue for a few days, an opportunity may be found of verifying, or correcting, the result of my experiments."

[5] That moduli experiments on ice offered special hazards to the experimentist in the early 19th century may be seen from the amusing introductory paragraphs to BEVAN's letter dated Norwich, 18 January 1826: "Dear Sir, I have been long desirous of repeating my experiment on the elasticity of ice, but until the present frost have not had an opportunity

It does not appear quite clear from reasoning, that the modulus ought to come out different in experiments on solids and on fluids: for though the linear compression in a fluid may be only $^1/_3$ as much as in a solid, yet the number of particles acting in any given section must be greater in the duplicate ratio of this compression, and ought apparently to make up the same resistance. And in a single experiment made hastily [the same word was used in 1807 in referring to the ice test providing 850000 ft.] some years ago on the sound yielded by a piece of ice, the modulus did appear to be about 800000 ft. only: but the presumption of accuracy is the greater in this case the higher the modulus appears.[1]

Young apparently took a somewhat dim view of physical measurement in suggesting that the haste with which he performed his experiment was an adequate explanation for the discrepancy of a factor of nearly three, when comparing his own measurement of the speed of sound in ice with the larger value obtained by Bevan.[2]

In the English literature on solid mechanics, from Tredgold, Rennie, and Bevan in the 1820's, to Kelvin in the 1880's, one finds linear elasticity data given in the form of Young's awkward height of the modulus. (The weight of the modulus seems to have attracted no interest.) That the height of the modulus is an inconvenient term becomes immediately obvious when one compares data between one experimentist and another where the weight density is not measured on the specimens whose behavior is being observed. Except for Morin who, in 1862 provided height of the modulus data for copper and iron in terms of kilometers, the continental physicists followed the early lead of Coulomb in 1784

---

for two years, when I had but a single specimen, and rather too small, and which broke soon after the commencement of the experiment. From that experiment I had calculated a modulus of 6.000.000 ft. [Except for such direct quotations, throughout this treatise, to avoid confusion I have used neither commas according to the American practice, nor periods, as given here, in writing large numbers.]; but finding the result of my experiments in the present season much less, and pretty uniform amongst themselves, I re-examined the calculations made in 1824, and have discovered an error, which had before escaped me, in the reduction, and which brings the result of that experiment to agree with those made in the present season, or about 2.100.000 ft. I will not answer for the absolute correctness of the second figure, because a variation of the $^1/_{100}$ of an inch in the thickness of the specimen will change this figure.

The present severe frost has enabled me to try experiments on a much larger scale, and upon ice from 1–$^1/_4$ to 4 in. in thickness: to explain the mode I adopted with ice of near 4 in. in thickness may be necessary, and was as follows:

Upon a large pond within less than a mile of my residence at Leighton Bussard, the depth of which I found to be about 4 ft., I had a channel sawed on three sides of a parallelogram, in the following form, *a b c*, separating the specimen to be operated upon, except at the end lettered *a*, where its union was left undisturbed. The dimensions of this prism was 100 inches long, 10 in. wide, and thickness at $a = 3.62$, at $b = 4.00$, and at $c = 3.75$ in. To save some trouble in the calculation, I considered the thickness uniformly 3.97 in. or the mean of the whole: although I am aware that it would not be strictly accurate; but as the experiment was upon a large scale, and with weights up to 25 lbs. and with deflections proportional to the weights applied, I considered the experiment a fair one. In this experiment the weights were place at 98 in. from the line of union with the main body, and the deflection by 25 lbs. was 0.206 in.; from which I estimate the modulus to be 2.100.000 ft.

After this, I repeated the experiment on ice of various dimensions and of different thickness, and in all, the result agreed quite as near as the admeasurement of the thickness could be ascertained, as well when the deflection was tried upon the water, as when the ice was taken out and tried in the manner used with wood and metal."

[1] Young [1826, *1*], p. 306.

[2] After writing the above I found a similar reference to the ice experiments of Bevan and to Young's appended footnote in Lord Kelvin's *Encyclopaedia Britannica* article in 1878 on elasticity, in which he devoted some number of lines to Young's "faulty logic" in this matter as an example of the power of effective experimentation to maintain scientific understanding on a sane course. The illuminating exposition of Pierre Simon, Marquis de Laplace in 1816 which included a discussion of the difference between one-dimensional and three-dimensional propagation of sound in air, as well as a history of the subject starting from the 17th century observations of Isaac Newton, belie the thought that the matter was ill understood in 1826. Laplace [1816, *1*].

and DULEAU in 1812 by expressing measured elastic constants in what is now the accepted form.[1] In 1878, when KELVIN compiled a table of E-moduli for his widely quoted article on elasticity for the Ninth Edition of the *Encyclopaedia Britannica,* his admiration for THOMAS YOUNG was such that he converted the data of WERTHEIM for ten different metals from the continental units kg/mm² into the units of English feet for the height of the modulus, or, as he described it, the "length of the modulus". It is interesting that in comparing these data and those of RANKINE[2] and BEVAN[3] with the original data of YOUNG,[4] KELVIN[5] cited YOUNG as having given a value of "about 9000000 ft." and, under the column describing the method of determination of the modulus, wrote "probably flexure". If one introduces the nominal density for iron or steel, a height of the modulus of about 9000000 ft. provides an E of about 22000 kg/mm² which, of course, is fairly close to the average value commonly obtained for steel but not to the value given in YOUNG's own table.

In YOUNG's treatise, the only description of experimental results related to determination of the height of the modulus was that contained in the Scholium following the theorem on lateral vibration of prismatic and cylindrical rods.[6] Here he introduced a series into the BERNOULLI-EULER beam equation. YOUNG calculated a relation between the height of the modulus and the frequency of vibration for the cantilevered and simply supported beams. The description is as follows:

> SCHOLIUM. All these results are amply confirmed by experiment, and they afford an easy method of comparing the elasticity of various substances. In a tuning fork of steel, l was 2.8 inches, $d=.125$; and $n=512$, hence h is about 8530000 feet. In a plate of brass, held loosely about one fifth of its length from one end, l was 6.2 inches, $d=.072$, and $n=273$, whence $h=4940000$; in a wire of inferior brass, l being 20 inches, $d=.225$, and $n=74$, h appears to be 4700000. A plate of crown glass, 6.2 inches long and .05 thick, produced a sound consisting of 284 vibrations in a second, whence $h=9610000$ feet.
>
> A box scale .012 f. thick, and 1.01 f. long, gave 154 vibrations, hence $h=5050000$ feet. When these substances were held in the middle, the note became higher by an octave and somewhat more than a fourth. Riccati found the difference between the elasticities of steel and brass somewhat greater than this. For ice, h appeared to be about 850000.
>
> Two small rods of deal, one foot in length, produced sounds, consisting of 270 and 384 vibrations in a second; their weights were 153 and 127 grains respectively: hence the formula $.0242\ n^2\ l^2$ gives nearly 35 and 65 pounds for the force under which they would bend; the experiment, which was made somewhat hastily, gave 36 and 50.[7]

As was noted above, this value of 8530000 ft. for steel is considerably below that which was given in YOUNG's table. YOUNG specifically noted that the ratio

---

[1] An exception to this practice has been noted above in the one instance in which the height of the modulus possesses a distinct advantage, namely, for solids such as the silk thread of WEBER or the organic tissue of WERTHEIM, etc., for which the determination of diameters is difficult or impossible, but for which the density is known. In this instance, the height of the modulus provided a relative measure of behavior. DULEAU introduced the use of the elongation of a unit cube under a unit actual load also used during the next half century as a definition of elasticity by SAVART and MASSON in the 1830's and 1840's and in a more awkward form by KUPFFER in the 1850's and 1860's. This formulation does permit of the calculation of E directly from the presented data but certainly does not lend itself to the generalization of linear elastic constitutive equations to include multiple stress states or anisotropy. DULEAU can be excused from this criticism in 1812 since the 1820's were still to be, but MASSON, SAVART, KUPFFER, and others from 1837 to 1870 most certainly had a limited understanding of the great theoretical developments which were occurring in linear elasticity during this time.

[2] RANKINE [1858, *1*].

[3] BEVAN [1826, *1*].

[4] YOUNG [1807, *1*].

[5] KELVIN [1878, *1*].

[6] See YOUNG [1807, *1*], Article 398, vol. II, p. 84.

[7] YOUNG [1807, *1*], vol. II, p. 86.

found by JORDAN RICCATI between the E-moduli of steel and brass was somewhat greater than his own, thus indicating without question that he knew E had appeared in the literature, twenty-five years before the publication of his course of lectures.[1]

Of course, on the basis of this discussion, I am not attempting to evaluate the total scientific contribution of THOMAS YOUNG. Certainly his impact upon his time in the English community, and to a somewhat lesser extent across the Channel, was considerable. Since legend erroneously has bestowed upon him initial preeminence in the specific matters presented here, it has been of no little importance to describe critically, and in some detail, precisely the sum total of his contribution concerning elastic constants in experimental mechanics.

YOUNG was thoroughly familiar with the fact that the elasticity of a substance, including solids, could be determined from the speed of sound. Based upon the fact that Professor ROBISON had informed him that "he heard the sound of a bell transmitted by water at the distance of 1200 ft."[2] (i.e., that sound did propagate through liquids), YOUNG, for water, used the height of the modulus he had calculated from Mr. CANTON's data, to obtain a velocity of 4900 ft./sec. There was no measurement of this velocity available for comparison.[3] As YOUNG pointed out, a Mr. WUNSCH in 1788 united a number of deal rods (wooden boards) in an effort to determine the velocity of sounds in wood and concluded from this measurement that it was infinite.[4] YOUNG himself described his investigation of this subject as follows:

> I have also found that the blow of a hammer on a wall, at the upper part of a high house, is heard as if double by a person standing near it on the ground, the first sound descending through the wall, the second through the air. It appears from experiments on the flexure of solid bodies of all kinds, that their elasticity, compared with their density, is much greater than that of the air: thus, the height of the modulus of elasticity of fir wood, is found, by means of such experiments, to be about 9500000 ft., whence the velocity of an impulse conveyed through it must be 17400 ft., or more than three miles, in a second. It is obvious, therefore, that in all common experiments such a transmission must appear perfectly instantaneous. There are various methods of ascertaining this velocity from the sounds produced under different circumstances by the substances to be examined, and Professor Chladni has in this manner compared the properties of a variety of natural and artificial productions.[5]

[1] We have seen that RICCATI's ratio was $E_{\text{steel}}/E_{\text{brass}} \approx 2.06$, whereas CHLADNI obtained the value of 2.25. Converting YOUNG's moduli in the Scholium to E-moduli by means of the densities provided by YOUNG in Table 43, one obtains for steel the value of 20200 kg/mm², and for brass, 12650 and 12000 kg/mm², providing the low $E_{\text{steel}}/E_{\text{brass}}$ ratios of from 1.59 to 1.68.

[2] YOUNG [1807, *1*], vol. I, p. 372. A similar observation was made earlier by BENJAMIN FRANKLIN over a distance of half a mile, by means of the noise of two stones struck against each other. Having noted the transmission of sounds in mines and quarries from one tunnel to another, numerous physicists, including GAY-LUSSAC and CHLADNI, had curiously commented upon the high velocity of sound which must exist in solids.

[3] This velocity of 4900 ft./sec is not too different from the first experimental value of 4708 ft./sec obtained in 1826 by JEAN DANIEL COLLADON [1838, *1*], in experiments at 8.1° C in the Lake of Geneva, or the value of 5013 obtained by WERTHEIM in the Seine River at a temperature of 30° C. In the light of the comments on the cubical vs linear difficulty of YOUNG's interpretation of the linear height of the modulus, he had used the value of 750000 ft. to determine the relatively accurate velocity of sound in the three-dimensional medium and thus, perhaps, should not have been confused by the comments made by BEVAN. To relate the elastic compressibility with sound velocity was a suggestion often attributed independently to both YOUNG and LAPLACE. The spirit of this suggestion in the analysis of small vibration, however, obviously antedates both by a century. See TRUESDELL [1960, *1*].

[4] WUNSCH [*Berlin Memoirs*, 1788]. Although I have not been able to locate the reference given by YOUNG, it has been brought to my attention that WUNSCH wrote a treatise on physics which is remembered historically for its high proportion of error.

[5] YOUNG [1807, *1*], vol. I, p. 373. JEAN BAPTISTE BIOT [1809, *1*] described an experiment somewhat similar to YOUNG's which had been performed by unspecified physicists in Denmark. A horizontally stretched, 600 ft. long metal wire had at one end a suspended piece of sonorous metal which was struck with small blows. At the other end, a person held the

This reference to Professor CHLADNI incidentally provides further evidence of YOUNG's familiarity (before he wrote his course of lectures) with parameters specifying the elastic properties of a material since, as was indicated above, the publication by CHLADNI[1] in 1787 contains an expressed understanding of these matters.

3.8. Biot's use of the new Paris water pipes to obtain the first direct measurement of the velocity of sound in a solid (1809).

The opportunity for measuring the velocity of sound over long distances in a metal occurred to BIOT[2] when, in 1808, he observed the construction of an aqueduct in the form of cast iron conduit pipes in the city of Paris. These pipes were cylindrical; their average length, obtained by measuring the total length of 12 of them placed end to end, was 2.515 m. The pipes were separated by lead disks covered by tarred cloth, which were tightened with strong screws under extreme compression so as to permit no leakage of water. The average thickness of each disk, based upon the measurement of 12 different ones, was 0.14256 m. The total assembly unfortunately formed a curved line which had two inflections toward the middle of the length. Since they were not all joined at the beginning, BIOT and his assistant were able to work successively on different lengths as the assembly took place. Since BIOT's experiments, and the differences between them and the earlier ones on longitudinal oscillation perfomed by CHLADNI, are cited over and over throughout the 19th century, they are worth describing in some detail.

The first experiment was conducted by BIOT and a Mr. BOUVARD to examine the velocity of sound in an assembly of 78 pipes, having a total length of 196.17 m. The 77 lead disks between them added 1.10 m; the total length thus was 197.27 m. A length of iron with a bell at its center was placed in the last pipe, with a hammer which struck the bell and simultaneously struck the pipe. Therefore, at the same instant a sound was inaugurated for transmission through the air inside the pipe as well as through the metal of the pipe itself. Time was recorded from a half-second chronometer. On different measurements, both sexigesimal and decimal watches were used to vary the observed numbers. To perform the entire measurement on the same chronometer in the first series of experiments, the time was noted at the opposite end of the conduit between the arrival of the sound in the pipe and that in the air.[3] From 53 observations with a measured air temperature of 11° C and a barometric pressure of 0.76, this difference was found to be 0.542. Referring to experiments of the French Academy which had given 334.02 m/sec for the velocity of sound in air at 0° C and at 0.76 barometric pressure, BIOT determined a sound velocity of 340.84 m/sec in air at 11° C, from which he

wire "between his teeth or touched it to some solid part of the hearing organ," and was able to detect two distinct sound velocities. It was concluded that the sounds arrived through the wire almost instantaneously. HASSENFRATZ, with GAY-LUSSAC, conducted essentially the same experiments with essentially the same results in the quarry of Paris. As BIOT emphasized, the velocity of sound in a solid had been shown to be not only finite but also measurable by CHLADNI in 1787 by means of longitudinal oscillations in relatively short rods. BIOT also referred to experiments of the British Royal Society which undoubtedly were those of YOUNG, in which some results were reported for which he had been unable to obtain details.

[1] CHLADNI [1787, *1*].

[2] BIOT [1809, *1*].

[3] BIOT made these measurements both with the sound in the air, having the bell as the source, and also with simply the hammer blow as the source, to be certain that the difference in velocities was in no way affected by the nature of the initial signal.

could calculate the length of time for the transmission of sound in air. Thence, with the observed difference of time between the two sounds, he obtained a time of propagation of 0.037 sec through 197.27 m of metal.

BIOT's interest in obtaining a ratio of the velocity of sound in the metal to that in air presumably was for a comparison with CHLADNI's earlier ratios. If one calculates the velocity of sound in cast iron from the elapsed time, one obtains 5200 m/sec, or, with BIOT's calculated air velocity, the ratio of 15.25, which is considerably below CHLADNI's value of 16.6 for iron.

BIOT considered his measurements on 78 pipes as too crude to do other than establish for the first time experimentally that the velocity of sound in a solid is finite. He was particularly influenced in this decision by a second series of experiments which were conducted by Mr. BOUVARD with the help of a Mr. MALUS when the construction had reached exactly twice the first number of pipes for a total length, with disks, of 394.55 m. For twice the number of pipes those two gentlemen obtained from 64 observations an interval of 0.81 sec of time between the two sounds. As BIOT pointed out from his calculation the time for propagation in the air was 1.158 sec, leaving a time of 0.348 sec for the propagation through the solid body. BIOT dismissed those measurements and questioned his own as being susceptible to error. To overcome this difficulty he waited until 376 pipes had been assembled, which, with a length of 5.61 m for the 375 disks, had a total length of 951.25 m.

With his assistant, Mr. MARTIN,[1] BIOT performed more than 200 separate tests with either the hammer or the bell, obtaining an interval of time of 2.50 sec between the sound transmitted in the metal and that in the air. He stated that he had found no variation in that value of 2.50 sec. "I had Mr. Martin observe it as well without informing him of my result, and he found the same thing." For the distance of 951.25 m at a temperature of 11° C, the calculated time of sound through air was 2.79 sec. Subtracting the 2.50 sec, the time for propagation in the metal was 0.29 sec. Although BIOT regarded this result as very accurate because it involved beats of the half-second chronometer, he desired to modify the method of measurement so as to increase the accuracy still further.

He struck the pipe, and with two chronometers he recorded the time of transit of the sound.

> I stationed Mr. Martin at one end of the canal with a half-second watch. I took my place at the other end with a similar watch, carefully compared with the first one both at the beginning and at the end of the experiments; a comparison which, as will shortly be seen, has no influence on the results. When the watch of Mr. Martin pointed to 0″ or 30″, he struck a blow with a hammer, and when my watch pointed to 15″ or 45″, I replied to him with a similar blow. We each noted the arrivals of the sounds which were sent to us, and we recorded the times. Besides that we took great pains to strike just at the second agreed upon, and with a little practice success was easily achieved, as the series of our observations itself confirms. But, whatever the difference of the watches, even if variable, so long as it does not change appreciably in the interval of 30 sec, it is cancelled out exactly when the mean of two consecutive observations is taken, and the result becomes entirely independent of it.[2]

This method of measurement, which now allowed the velocity of sound in the metal itself to be measured with reference to the same chronometer, provided the data shown in Table 44 which gave a time of propagation of 0.26 sec through the cast iron conduit, 951.25 m long.

The time of propagation through the air, which also could be measured, was 2.76 sec, giving a measured ratio of 10.75. Since the calculated value for the velocity of sound in air was 2.79 sec, a difference of only 0.03 sec, BIOT

[1] Mr. MARTIN was an expert in the construction of chronometers and watches.

[2] BIOT [1809, *1*], pp. 415–416.

Table 44. BIOT (1809).

$p-r$ (sec)	$p+r$ (sec)	Sum or value of $2p$ (sec)
	First series from 52 to 59 min	
−2.0	+2.5	0.5
2.0	2.5	0.5
2.0	2.5	0.5
2.0	2.5	0.5
2.0	2.5	0.5
2.0	2.5	0.5
2.0	2.5	0.5
2.0	2.5	0.5
2.0	2.5	0.5
	Second series, 1 h 27 min to 1 h 32 min	
−2.8	+3.5	0.7
2.9	3.5	0.6
3.0	3.5	0.5
2.9	3.5	0.6
3.0	3.5	0.5
3.0	3.5	0.5
3.0	3.5	0.5
2.9	3.5	0.6
3.0	3.5	0.5
3.0	3.5	0.5
3.1	3.5	0.4
	Average value of $2p=$	0.52
	Which gives for the value of $p=$	0.26

concluded that the closeness "seems of a nature to inspire a certain confidence in the results."

The difference between CHLADNI's velocity ratio of 16.6 for iron to air in longitudinal oscillations and BIOT's ratio of approximately 10.5 was far too great to be dismissed casually as observational error. The stage was set for a controversy lasting over forty years, in which it was ultimately assumed that one or the other set of data must be fundamentally incorrect. BIOT, for whom CHLADNI expressed considerable respect as an experimentist, was never a vitriolic participant in this controversy.[1]

This discrepancy between BIOT's measured sound velocities in iron over large distances and the results obtained from the longitudinal oscillations of a short rod, which came to be known as the "CHLADNI" process, was still unexplained in 1851. In that year WERTHEIM and BREGUET[2] conducted a series of experiments over a much longer distance than that of BIOT. For a length of 4067.2 m of iron telegraph wires, WERTHEIM and BREGUET obtained a velocity of 3485 m/sec, which is quite close to the value of 3658 m/sec obtained by BIOT in his most accurate measurement. WERTHEIM and BREGUET were more puzzled by the fact that when performing the CHLADNI experiment, they found that two meters of

[1] BIOT [1816, *1*]. Nevertheless, BIOT had aided SAVART in performing the experiments which gave rise to the latter's criticism of CHLADNI. The collaboration of BIOT and SAVART is far better known for the law which bears their names. One often finds BIOT's name mentioned as having participated in his colleagues' experiments. Certainly the most daring of such participations must have been his being the passenger in GAY-LUSSAC's first balloon ascension in 1804.

[2] WERTHEIM and BREGUET [1851, *1*].

the same iron wire provided the much higher sound velocity of 4634 m/sec. These mid-19th century experiments, which will be discussed below in Sect. 3.44, lend some credence to BIOT's suggestion that for CHLADNI's experiment the longitudinal vibrations were not audibly exciting the fundamental vibration mode.[1] When, in 1816, BIOT first published his widely read two-volume work,[2] he discussed in great detail his own experiments on the velocity of sound and, separately, the experiments of CHLADNI, without dwelling in either section on the major differences between the velocities each of them had measured for sound in iron.

Meanwhile, in 1811, LEHOT, DÉSORMES, and CLÉMENT[3] in a repetition of BIOT's experiments found the extremely low velocity of sound of 593 m/sec in iron pipes. CHLADNI[4] dismissed those later experiments; he pointed out that their pipes were very poorly connected, unlike those of BIOT which had been securely attached. BIOT, too, was challenged in 1811. BENZENBERG[5] attacked the experiments of BIOT on the basis that he had not used good instruments for his observations, i.e., BENZENBERG was critical of the use of chronometers in general. When CHLADNI in 1817 referred to the experiments of LEHOT *et al.*, he was reviewing the discrepancies between his own and BIOT's results, and suggested that any form of attachment of multiple sections of pipes would slow the sound velocity and, moreover, that perhaps even nearly a kilometer was too short a distance to measure the velocity of sound. (Actually, BIOT's demonstration that he could measure the velocity of sound in air through the pipes with extremely close accuracy to a known value effectively eliminated the substance of BENZENBERG's comment, and, as was pointed out above, we know in retrospect from the measurements of WERTHEIM and BREGUET on continuous specimens that the criticism with regard to the slowing of the velocity because of the method of attachment was not necessarily valid. Of course this had not been demonstrated in 1817).

CHLADNI, who had entered the discussion to respond to a paper by GEORG WILHELM MUNKE[6] on the discrepancy, added the curious statement that although BIOT's velocity ratio was considerably below his own, the results agreed entirely with his findings. He attributed the difference to the fact that there were multiple pipe sections and connecting parts, and that wood was used in the joints. He was incorrect regarding the joints[7] because all connections were made by means of lead disks. CHLADNI somewhat incongruously then proceeded to criticize MUNKE for introducing the same arguments, and took strong issue with MUNKE's additional objection that BIOT's pipes had probably not been iron[8] because that is "not a durable material for water pipes."

[1] It should be noted in this connection, however, that CHLADNI's wave speeds provide moduli which were not too different from those obtained by others, such as YOUNG, from lateral oscillations, i.e. the problem arises from the low wave speed values in long bars when compared to any type of test, dynamic or static, in a short specimen. A few years ago I found similar differences in comparing wave speed data from a 500 ft. long aluminum wire with vibratory measurements in 12 ft. sections of the same wire. Because of interest in this very matter, I performed these experiments a decade ago, but I have not yet found time to publish the results.

[2] *Traité de Physique Expérimentale et Mathématique* [1816, *1*].

[3] LEHOT, DÉSORMES, and CLÉMENT [1811, *1*].

[4] CHLADNI [1817, *1*].

[5] BENZENBERG [1811, *1*].

[6] MUNKE [1814, *1*].

[7] BIOT did use wooden clamps externally, to hold the joints in place, but they were not in the path of the propagating sound.

[8] BIOT was not the only experimental scholar intrigued by the possibilities of the new Paris pipe lines. P. S. GIRARD in an 1831 paper entitled, Mémoire sur la pose des conduites d'eau dans la ville de Paris, Mém. Acad. Sci. (Paris), **10**, 405–456 (1831) reminisces on his earlier thermal dilatation experiments in 1805–1815 on the same Paris iron water pipes, i.e., cast iron pipes. [1831, *1*].

This controversy took on more serious proportions when by 1819 it was claimed that CHLADNI had not been able actually to excite longitudinal oscillations in rods, with one end fixed and the other free. This attack was made in a memoir by FÉLIX SAVART,[1] to which CHLADNI replied bitterly in a note published two years later.[2] To the suggestion of SAVART that CHLADNI must have used an analogy based upon observations of the behavior of air vibrations in an organ pipe plugged at one end, CHLADNI pointed out that he had performed his experiment on rods in public on numerous occasions. Replying to another similar attack, CHLADNI noted that as a matter of fact he had performed many of those experiments in precisely the manner which SAVART had claimed was impossible, on metal, wood, and glass, in front of Professor GILBERT. CHLADNI stated in 1822:

> Thus I do not deserve the reproof of having given as the result of experiments something that was only a play of imagination or a false conjecture drawn from analogy; a reproof which, if founded, would suffice to deprive a physicist of all confidence he might claim.[3]

Observations of discrete differences in moduli during the 19th century are not limited to these early data. At the end of the present chapter, in terms of recently discovered multiple elasticities in isotropic solids, it will be shown that it is at least conceptually possible to reconcile the observed wide differences, and also to conclude that the experiments of CHLADNI, BIOT, WERTHEIM, etc., did indeed possess the experimental accuracy claimed for them.

At the end of BIOT's paper[4] he described a series of acoustical experiments in the long pipe which, in our day of telephone and radio, are singularly amusing. He found it amazing to be able to talk at a whisper with his assistant, who was nearly a kilometer away.

> It was not necessary to speak into the pipe in order to be heard; ordinary conversation, at 2 m from the opening, was transmitted perfectly, and in writing my observations I asked Mr. Martin the time on his chronometer as if I were asking it of someone standing two feet away from me. This manner of conversing with an invisible neighbor is so strange that, even when one knows the cause of it, one cannot help being astonished by it.[5]

In the 951 m pipe, the noises of Paris made BIOT at first conclude that there was a limit to the length at which such a transmission could occur, but when he went to the location with "Mr. Martin and two intelligent workers" between one and four o'clock in the morning, he could hear even the lowest whisper. He also found that when he discharged a pistol at one end, he was able to blow out a candle half a meter from the other end of the pipe, nearly a kilometer away. To examine the attenuation as a possible function of frequency, he had a flutist play airs at one end, from which he observed that the deeper tones were much better heard than the higher ones. The conclusions drawn from this observation were completely incorrect, as BIOT himself pointed out in an amusing footnote shortly thereafter:

> After this memoir was read, I ascertained that the person who played the flute, since he had a very weak chest, has much trouble in blowing the high notes and sometimes must skip them entirely. Thus it is clear why I did not hear them. But I have wished that this peculiarity of my first version be let stand, so that it would be seen that I faithfully reported the least details of these phenomena, and so that my truthfulness in this circumstance should corroborate the other results I have noted.[6]

---

[1] SAVART [1820, *1*].
[2] CHLADNI [1822, *1*].
[3] CHLADNI [1822, *1*], p. 75.
[4] BIOT [1809, *1*].
[5] *Ibid.*, p. 419. One hopes that BIOT remembered that in obtaining the time from Mr. MARTIN at the other end of the pipe, he had to have a knowledge of the results of the experiment he was attempting to perform.
[6] BIOT [1809, *1*], p. 422.

One last point of interest at the end of BIOT's paper is his comment that one could distinctly note as many as six separate echoes when a sound was made into one end of the pipe, the last being approximately the time necessary for the sound to propagate at the other end of the pipe. These echoes occurred at perfectly equal time intervals of "very nearly 0.5 sec" when generated from either end of the pipe, but could be heard only from the end where the sound had been inaugurated.

3.9. Duleau's introduction of quasi-static measurements into the study of linear elasticity (1813).

In the second half of the 20th century, when papers on experiment in print more than five years are almost never referenced in any detail and, if referenced, are cited only in the introduction, extend back but thirty years and are viewed as part of the misty, but supposedly accurate historical origins of the subject, we see as a decided contrast the over fifty years of detailed discussion following the important experiments of ALPHONSE JEAN CLAUDE BOURGUIGNON DULEAU[1] in 1811. In that year, DULEAU had been commissioned to plan a bridge of forged iron over the Dordogne River at Cubzac. In preparing for this task, he performed at Bordeaux a series of experiments on malleable iron. For that period his experiments were unique in that they were confined to small deformation which would correspond to the loading requirements of the proposed bridge. The specimens in every instance were full-sized iron members, the total number of tests being in excess of one hundred.

After reviewing at considerable length the existing theories regarding the behavior of prismatic bars subject to flexure, axial tension, axial compression, and torsion, for rectangular, square, and circular cross-sections, DULEAU proceeded to perform numerous experiments, checking against the various computations, including EULER's buckling formula for columns, by varying the dimensions of the specimens from one test to another. He also performed experiments on arches formed from these rods, as well as upon composite structures made from assemblies of prismatic bars, checking such matters as the friction between two adjacent bars in flexure, etc. He was concerned with the "line of passage from tension to compression" in malleable iron beams[2] (i.e., the neutral line), and with linearity of the applicable stress-strain function.

The success of this first broad quasi-static experimental study of infinitesimal elasticity is manifested in the fact that the experiments of DULEAU became the primary basis for discussion and criticism, both with respect to matters of further experimentation, and in the later development of the linear theory of elasticity, throughout the entire first half of the 19th century. Few experimentists have made such an impact upon their time in this field. In 1842, when WERTHEIM[3] presented the first great experimental study of elastic constants in the modern sense, the data of DULEAU were still considered the definitive experimental work on the subject. His results were first published in a memoir in 1813. In 1819 CAUCHY, POISSON, and GIRARD presented an analysis of DULEAU's memoir in the process

---

[1] DULEAU published in [1813, *1*].

[2] His misconceptions on the matter of the location of the neutral line in flexure were referred to by several persons in later years, including SAINT-VENANT in 1857. This criticism applies only to his introductory discussions of the subject and is in no way of importance for the experiments which presented the observed dimensions, forces, and deflections. It is well to contrast this separating of data and hypothesis with the common modern practice of inextricably entwining measurement and interpretation.

[3] WERTHEIM [1844, *1*] (presented in 1842, published in 1844).

of recommending its acceptance by the French Academy. In the following year the memoir was republished in the form of a brochure[1] which was the main reference for all subsequent study of DULEAU's data.[2]

DULEAU indeed was dealing with the origins of quasi-static elasticity. He himself had commented early in the paper that he had been unable to find any experiments in the previous literature except two tests of a Monsieur AUBREY in 1790 which had recorded quasi-static results for small deformation. All other quasi-static data available to DULEAU when he began his work, as he stated, had the sole objective of determining the force necessary to rupture a bar pulled in the direction of its length. AUBREY, who had been an Inspector General of Bridges and Highways,[3] had loaded centrally what was apparently a simply supported beam and had observed a proportionality between the deflections and the loads. DULEAU noted that this was consistent with the theory of an elastic strip and set himself the task of what is certainly the first experimental study in quasi-static linear elasticity: to find how general was the correlation between theory and experiment in 1811. He referred to the linear property of materials as "perfect elasticity." The evaluation which follows is based upon my reading of DULEAU's brochure of 1820, my comparison of the discussions and re-tabulations of DULEAU's tabulated data by the numerous persons who considered them at great length, as well as upon my reading of the review in 1819 of the paper for the French Academy by CAUCHY, POISSON, and GIRARD.

DULEAU performed a total of 105 experiments, many of which consisted of different types of tests upon the same specimen; these he categorized as a single experiment. The measurements, which were all made on iron specimens, included: (a) the flexure of simply supported cantilevered beams of various cross-sections with loads applied at various positions; (b) the same bars axially loaded in compression with various end conditions for the purpose of determining the EULER buckling load (DULEAU was unable to measure the elongation of these specimens with sufficient accuracy to determine their linearity, let alone an E-modulus); (c) the torsion of both solid and hollow bars of different cross-sections; (d) the deformation of systems of iron pieces joined in a variety of ways which DULEAU considered of practical interest in the design of bridges; (e) the deformation of arches with fixed supports.

---

[1] DULEAU [1820, *1*]. The brochure in 1820 also contained an Appendix evaluating the experiments of others who had been stimulated by DULEAU's original (1813) memoir during the years between 1813 and 1820. DULEAU noted that all of the ten experimentists who supposedly had been stimulated by his publication of 1813, whether in England, Sweden, France, or Russia, simply had continued to perform rupture load experiments, ignoring in experiment if not in comment the original, fundamental significance of DULEAU's contribution.

[2] PEARSON, in TODHUNTER and PEARSON's *History of the Theory of Elasticity and of the Strength of Materials*, also was limited to a discussion of this brochure, since the memoir of 1813 apparently had ceased to be available, even by 1886. PEARSON's comments on this important experimental paper are confined to a criticism of DULEAU's unimportant review of existing theories and a comparison of some of his conclusions regarding these calculations in the light of pronouncements of JOHN ROBISON, etc. When it came to the important experimental part of DULEAU's paper, the study is dismissed with the sole comment: "The rest of the book is occupied with the discussion of experiments on iron bars" (vol. I, p. 121). It is interesting to list a few of those who have examined this work by DULEAU in some detail in the years which followed. They include TREDGOLD (1824, *1*), SAVART [1829, *1*], LAGERHJELM [1829, *1*], BARLOW [1837, *1*], HODGKINSON [1831, *1*], NAVIER [1833, *1*], PONCELET [1829, *1* and 1841, *1*] (two editions), WERTHEIM [1844, *1*], CLAPEYRON [1858, *1*], and SAINT-VENANT [1856, *1*].

[3] AUBREY: *Mémoire sur Différentes Questions de la Science des Constructions Publiques et Économiques* (Lyon, 1790). This is the reference given by DULEAU; I have not seen the publication.

As was noted above, the year in which the experiments were performed was 1811, which was only four years after YOUNG had introduced his limited concepts of the height of the modulus and the weight of the modulus, so it is of considerable interest to find not only a proper statement of E in modern terms in DULEAU's treatise, but also a detailed quasi-static measurement of this elastic modulus in iron, based upon the detailed study of 25 carefully performed experiments. From these data DULEAU obtained the also surprisingly modern average value of $E = 20430$ kg/mm^2.

In examining his data DULEAU found fluctuations from as high above the average value as 24922 kg/mm^2 and as low as 16121 kg/mm^2. For convenience he selected the value of $E = 20000$ kg/mm^2, which became the accepted value among physicists and technologists for the next four decades, until the more precise experiments of WERTHEIM became the standard for the second half of the 19th century. Like all experimentists of first calibre, DULEAU performed many cross-checks and gave a detailed, critical discussion of all the many experiments described.

In a footnote to the summary section on "The Laws of the Resistance of Wrought Iron based upon the results of the Preceding Experiments," DULEAU wrote that according to a well-known formula his quasi-statically determined modulus provided in iron a velocity of sound of 5018 m/sec.[1] The determination of this value preceded the controversy surrounding the differences between the dynamically determined moduli of CHLADNI and BIOT, referred to above. There might have been no issue for controversy if this aspect of DULEAU's experimental results had not been completely ignored; too often one finds similar historical lapses.[2]

In presenting his data DULEAU in a set of notes described the prior history of each specimen as well as particular observations about the test. He tabulated the measured distance between the supports, the measured cross-sectional dimensions which were checked by weighing the bar and determining the total specimen length, the deflection under a load of 10 kg as measured for a central load and as calculated from elementary beam theory, employing the modulus which, as indicated above, had been determined from the analysis of 25 of the experiments. He provided the strongest load for which the specimen could remain "perfectly elastic," i.e., the stress at the elastic limit.

In addition, for each of the very large number of bars he tabulated the measured vertical load, which he compared with calculations from EULER's buckling formula. based on the fixed or free end conditions which were used in the experiment.

DULEAU's results, even though they are historically of major importance, are far too extensive to be presented here in detail. The reader is referred to the memoir of 1820 which, from the experimental point of view, is a fascinating document. Because perhaps it would be of more contemporary interest, I have chosen to present a table of the selections from those experiments of DULEAU in 1811 which are contained in the 1851 edition of NAVIER's book[3] edited by

---

[1] Note that in the review article of CAUCHY, POISSON, and GIRARD, the value of the load for a $^1/_{10}$ mm elongation for a specimen 1 m long is erroneously quoted as 4 kg/mm^2 leading to an E-modulus of 40000 kg/mm^2 for which those three reviewers stated: "From this result one may deduce the velocity of sound in forged iron by the known formula; this velocity is 7087.82 m per second." DULEAU [1819, *1*], p. 145.

[2] For that matter ignored, too, was YOUNG's velocity which, like DULEAU's, would have supported CHLADNI's side of the argument.

[3] NAVIER [1851, *1*]. This is the third edition of NAVIER's book published in 1833. The 1851 edition contained the Appendix by SAINT-VENANT. In the Appendix SAINT-VENANT had referred to a series of 16 unsuccessful experiments performed at the École des Ponts et Chaus-

SAINT-VENANT (NAVIER had died in 1836), to which I have added more calculations for E. (See Table 45.)

The tabulation had omitted the identifying numbers with which DULEAU invariably specified every experiment. However, I found it possible to ascertain them and check the accuracy of the table.

Many of the 105 tests of DULEAU became the basis of individual discussion during the next half-century not only with respect to the presumed or actual national superiority of one country's iron versus another's, but also as to details of individual tests. (See, for example, HODGKINSON or BARLOW.)[1] Thus DULEAU's experiment on the flexure of a simply supported, centrally loaded beam having a cross-section in the form of an equilateral triangle, for which he found no difference in the load vs deflection curve when the beam was supported on a vertex or on a flat side, provoked a very large amount of discussion in the literature of the 1820's and 1830's. In Fig. 3.15 I have plotted DULEAU's tabulated data for the two situations; he drew his conclusions on the basis of the similarity of the results.

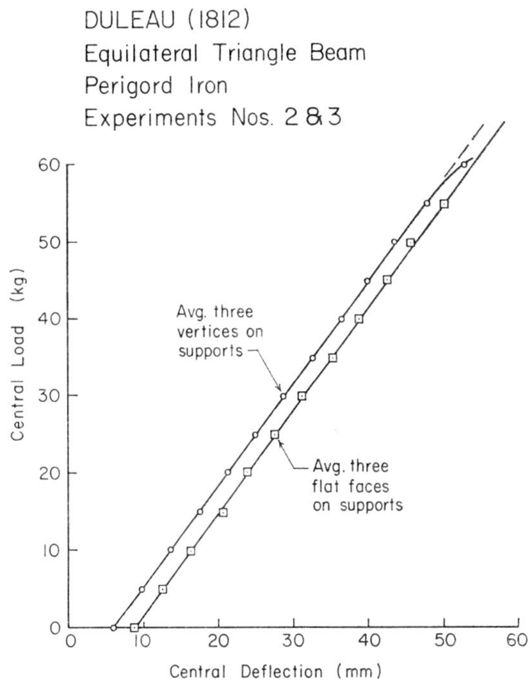

Fig. 3.15. Simply supported beam experiments of DULEAU for a triangular cross-section showing a comparison of force vs deflection behavior when vertices and flat faces are in contact with the simple supports.

Since DULEAU had varied the length, width, and height of his specimens, he was able to show that the deflection was as the length cubed, and inversely,

sées by NAVIER and a Monsieur BRUYERE. NAVIER had attributed the difficulties to the machine employed, from which probably arose his subsequent distrust of all testing machines. Perhaps, as DULEAU had pointed out, the main source of NAVIER's difficulty had been more in choosing specimens in which the length was too short compared to the thickness, to obtain any meaningful results in flexure for a comparison with theory.

[1] HODGKINSON [1831, *1*], BARLOW [1837, *1*].

Table 45. Duleau (1813). [See Navier (1851)].

Specimens subjected to experiment	Interval between supports (m)	Specimen width (mm)	Specimen height (mm)	Central deflection (mm)	E (kg/mm²)
Perigord iron. The transverse section is an equilateral triangle, 0.038 m/side. (The flex angle is the same when setting the specimen on edge or on its side)	3	—	—	7.6	—
Perigord iron	1	61	5.5	12.57	19596.89
Same specimen	0.5	61	5.5	1.71	18006.86
English iron, as it comes from large forges	3.035	34	8.56	136	24977.72
Same specimen	3.075	8.56	34	13.5	16004.07
Perigord iron	2	30	11	24	20869.86
Same specimen	1	30	11	3	20869.86
Perigord iron, soft (intended for horseshoes)	2	70	11.2	9.5	21390.60
Perigord iron	1	68	11	1.5	18414.58
Idem, as found in the forge	2	45	12	12	21433.47
Perigord iron	2	40	11.5	21	15653.86
Same specimen	1	40	11.5	2.5	16436.55
Same specimen	2	11.5	40	1.67	16271.80
Perigord iron (as found in the forge)	3	77	14	14.4	22185.36
English iron, marked B (as found in the forge)	1.5	67.8	14.7	2	19585.60
Perigord iron	3	25	15	37	21621.62
Same specimen	3	15	25	14	20571.43
Perigord iron	1	58	16.3	0.57	17460.20
Idem	3	39	19.6	10.8	21282.39
Same specimen	3	19.6	39	2.8	20734.61
Perigord iron	2	60	20	2	20833.33
Idem	3	60	20	6.6	21306.82
Same specimen	3	20	60	0.75	20833.33
Perigord iron	5	120	20	15	21701.39
Landes iron	2	120	21	1	17996.62
Perigord iron	3	39	24.5	6	14615.23
Same specimen	3	24.5	39	2.33	19933.70
Perigord iron (as found in the forge)	3	67	26	2.3	24921.90
Perigord iron	5	108	30	4.75	22561.55
Same specimen	5	30	108	0.4	20672.71
Perigord iron	2.92	31	31	3	22465.65
Same specimen placed on edge	—	—	—	3.35	—
		Diameter (mm)			
Round Arriege iron, as it comes from large forges	3.69	21.49	—	48.25	—
Idem	2.99	21.51	—	27.5	—
Round English iron, *indem*	2.93	23.52	—	18	—
Round Arriege iron, *indem*	2.92	26.82	—	10	—
Round Bilbao iron, very soft	2.92	31.8	—	5	—
		Specimen width (mm)			
Cast steel of England, marked Huntsman	0.98	13.3	5.9	32.05	26921.33
Same specimen	0.98	5.9	13.3	8.4	20173.26

Table 45. (Continued).

Specimens subjected to experiment	Interval between supports (m)	Specimen width (mm)	Specimen height (mm)	Central deflection (mm)	$E\,(\text{kg/mm}^2)$
Cement steel, German, marked Fortsman, for razors	0.68	14.5	7.8	8	14246.82
Same specimen	0.68	7.8	14.5	2.1	15718.00
Steel of the same type	1.845	25.7	21.6	2.8	21648.81
Same specimen	1.845	21.6	25.7	2.2	19463.16
Same type of steel	1.845	28.5	21.9	2.6	20172.89
Same specimen	1.845	21.9	28.5	1.8	17204.85
Steel of the same type	1.35	54.8	25.5	0.55	12306.12
Same specimen	1.35	25.5	54.8	0.27	5427.86
Steel of the same type	1.35	52	26.6	0.5	12567.79
Same specimen	1.35	26.6	52	0.3	5481.03

as the width and the height cubed, which was consistent with the expectations of the elementary linear beam theory he wished to use. Every 19th century author who discussed DULEAU's work pointed out that in his introduction he had been in error in determining the neutral axis for a beam; nevertheless, he had successfully compared central deflections for equivalent loads of square and round beams.

In the Appendix or fourth section of DULEAU's treatise in 1820 in which, as I have indicated above, he considered the work of the various experimentists in the field between 1813 and 1820, he included what I believe is the first compression experiment not primarily concerned with buckling. It had been performed by a Monsieur PICTET and was described in the Bibliothèque Universelle for March 1816. In that experiment PICTET determined the axial load necessary to shorten an iron bar by $^1/_{10\,000}$ of its original length, obtaining the value of 1.3 kg/mm². This provided a modulus of $E=13\,000$ kg/mm² in contrast to DULEAU's flexural value of $E=20\,000$ kg/mm², and is closer to the low modulus value calculated from BIOT's wave propagation experiments in iron in 1807 which, as we shall see later, was also the value obtained by WERTHEIM in 1851, likewise from wave propagation experiments in very long specimens of iron.

There thus existed by 1816, quasi-static experimental results supporting the dynamic results of both BIOT and CHLADNI. Although it is now well established that such multiple elasticities may occur,[1] and, as will be shown later, those early 19th century numbers are themselves in accord with the differences obtained in contemporary studies, no experimentist before WERTHEIM in the 1840's paid serious attention to the influence of prior thermal and mechanical histories upon the deformation properties of solids. Indeed, as was indicated above, COULOMB in the 1780's in observing no difference in one instance for tempered and hardened steel, seemed to prohibit any other experimentist for the next 50 years from seriously questioning the matter.

MUSSCHENBROEK early in the 18th century had used his ingenious testing machines to study the phenomenon of compression buckling. With proper credits to that predecessor DULEAU investigated the same subject for a very large number of specimens. For a variety of length to width ratios ranging from 200 to 24, he obtained an average ratio of 1.16, of experimental measurement of the buckling load to the calculation from the EULER formula. DULEAU was not certain that

[1] BELL [1968, *1*], Chap. VI.

the experimental results necessarily cast doubt upon the applicability of EULER's theory. DULEAU pointed out in those first definitive measurements what every modern experimentist studying the buckling problem knows too well, namely that friction and the manner of holding the specimen make such experiments extremely difficult to perform.[1]

Probably the most provocative of DULEAU's experiments with respect to their impact upon the development of linear elastic theory was his large series on the torsion of round and square, long iron bars. (He also considered the torsion of tubular specimens, in which he was much interested.) From the time of COULOMB's experiments on torsion in 1784 to CAUCHY's theory[2] in 1829, experimentists generally considered that the torsion of a rectangular bar could be calculated in the same manner as a circular cross-section. With respect to the relation between theory and experiment, BIOT once wrote:

> But in order that this association be useful, two indispensable conditions should be observed with the greatest care: that the analysis relied upon be rigorous, and that the experiments compared with it or entrusted to it be very exact. I tend to think this latter point is the more important to recommend. For, after all, if the analysis is false, observation will soon reveal it; while if the data furnished by experiment are false, the analysis has scarcely any means of recognizing that, since it merely puts them together and strictly deduces some false consequences from them.[3]

The experiments of DULEAU in 1812 are certainly a case in point, for when they were repeated in later years by SAVART in the 1830's and, more particularly, by WERTHEIM as described in his memoir on torsion in the 1850's, there seemed to be an agreement between experiment and CAUCHY's predictions. If one merely calculated the modulus using the COULOMB analysis, assuming that a rectangular prism, like a circular one, does not warp, one obtained lower values from the rectangular cross-section. It was not until 1857, after SAINT-VENANT had reconsidered the whole problem of torsion and at the same time reexamined in detail the torsion data of DULEAU, SAVART, and WERTHEIM, that a proper correlation between square and round cross-sections was obtained in which the averaged shear modulus thus determined was identical in both instances. DULEAU was the first to perform torsion experiments on beams of non-circular cross-section. That a correlation was not achieved between a proper experiment and a proper theory, was not due to any lack of interest in the subject during the intervening years.[4] SAINT-VENANT's tabulation[5] of DULEAU's experiments in iron for round and square cross-sections are given in Table 46.

The shear modulus μ for square bars calculated by the extensions of COULOMB's analysis and by the SAINT-VENANT theory which introduced a factor in the denominator of 0.843462, are compared in the table. The close agreement between the averaged values of the shear modulus for the round and square bars when viewed in the light of the SAINT-VENANT theory, were, as SAINT-VENANT stated, supported by these experiments of DULEAU.

---

[1] The Commission appointed by the French Academy in 1819 to consider DULEAU's memoir of 1813, which consisted of CAUCHY, POISSON, and GIRARD, were not completely in accord with DULEAU's accounting for the discrepancy. The two famous theorists thought that the difference also could be attributed to the fact that EULER's formula applied to the situation was in some sense an approximation, and suggested that the experiments indicated that it did indeed give too weak a result. DULEAU [1819, *1*].

[2] CAUCHY; published in [1830, *1*].

[3] BIOT [1816, *1*], p. xviii.

[4] As is immediately the situation when theory and experiment fail to coincide for isotropic solids, whether it be 1812, 1938 or 1972, unmeasured anisotropy in the experimental data is called upon to relieve the conflict.

[5] SAINT-VENANT's list excluded many of the repetitions of a given experiment which DULEAU had introduced to ascertain the amount of variation in his experimental results.

Table 46. DULEAU (1812). [From SAINT-VENANT (1857).]

Test no.	Designation of the iron	Length of the twisted part	Diameter of round specimens and side of square specs. calculated by weight	Angle by which one end is twisted while the other is maintained fixed (degrees sexagesimal)	$\frac{M_x}{J\theta}=G$ for round bars only	$\frac{M_x}{0.843462 J\theta}=G$ for square bars	Partial average values of coefficient G
86	Round Perigord iron, as it comes from the forge	2.81	0.0152	13.4	7336700000		
87	Round Perigord iron	3.17	0.0196	6	6685800000		
91	Round Perigord iron	3.19	0.02205	3.32	7590900000		6577070000
92	Round Perigord iron	2.89	0.02303	3	6395500000		
94	Round Perigord iron	2.94	0.0265	1.82	6117400000		
95	Round Perigord iron	3.35	0.02673	1.87	6553600000		
97	Round Perigord iron	2.92	0.03572	0.625	5359600000		
89	Round Ariege iron	3.57	0.02149	4.8	6512600000		
90	Round Ariege iron	2.89	0.02151	4.5	5602700000		6058270000
96	Round Ariege iron	2.77	0.02682	1.65	6059500000		
88	Round English iron, Dawlays	2.40	0.01983	4	7246600000		7848250000
93	Round English iron	3.24	0.02352	2.34	8449900000		
			General average for round irons		6659230000		
100	Square Perigord iron	2.52	0.02035	3.08	5248300000	6222400000	6268150000
101	Square Perigord iron	3.39	0.03260	0.62	5325500000	6313900000	
98	English iron 02	4.12	0.01846	6.5	6004600000	7119000000	7097350000
99	Same specimen	2.52	0.01846	4	5968100000	7075700000	
			General average for square irons		5636625000	6682750000	

Remarks: The measured diameters were, for test # 86 = 0.0142; for test # 87 = 0.0197.
The measured sides were, for tests # 98 and # 99 = 0.020.

Duleau had recognized the existence in quasi-static torsion testing of a relevant material constant, or shear modulus. He seemed unfamiliar with the work of Coulomb, as is evinced by his confusion in the way he chose to represent such a value for the twisting of solid specimens. Many others, including Savart, Wertheim, and Saint-Venant, whose calculations I have given in Table 46, had calculated proper values from the experimental results of 1811.[1]

Like Wertheim and Bauschinger much later in the 19th century, Duleau actually did observe a nonlinear relation between torque and twist, as my diagram of his tabulated results for his experiment No. 87 demonstrates in Fig. 3.16. Both Wertheim and Bauschinger recognized the phenomenon and the fact that it would be more readily observable in torsion than in tension or compression. Unlike his followers, and for that matter unlike his contemporaries[2] Dupin and Hodgkinson, Duleau insisted that twist must be a linear function of torque since the measurements were below the elastic limit. He proposed to ignore the deviation from linearity by providing the calculated values (dashed line) for a comparison with his experimental values (circles). The calculation was based upon extrapolating the initial measurement.

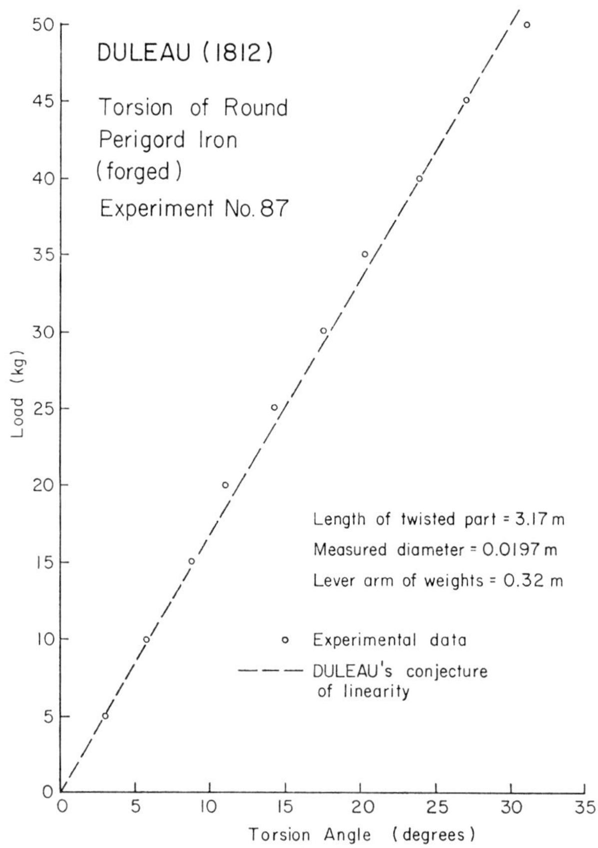

Fig. 3.16. The experimental torsion results from which Duleau in 1813 concluded that a linear response function was applicable.

[1] Duleau [1813, *1*], [1820, *1*].

[2] In 1832 Duleau died of cholera, at the age of 42.

Before the work of DULEAU in 1812 and DUPIN in 1811, all experimental determinations of the E-modulus by JORDAN RICCATI, CHLADNI, YOUNG, and BIOT, and of the shear modulus μ by COULOMB, had been dynamic, based upon measurement of the frequency of oscillation, or, in the single case of BIOT, of a determined wave speed. DUPIN's and DULEAU's were the first quasi-static experimental studies in the region of genuinely small deformation. DULEAU's comprehensive analysis of prismatic bars of various cross-sections subject to the wide variety of loading conditions constitute a major milestone not only in the historical development of experimental solid mechanics, but also in the theoretical foundations of linear elasticity which were to be rapidly developed in the years which followed.

Because of the importance of these results in the stimulation of theoretical studies in linear elasticity, I have included three of the four figures from DULEAU's publication in 1820 as Figs. 3.17, 3.18, and 3.19. The experiments obviously are well defined, including the patent difficulties which were encountered in the compression measurements. The experiments in second mode buckling which are referred to in the paper should be noted particularly, and also the different kinds of loading considered in the studies of flexure. I have not described the work on arches and frames since it is primarily of only technological interest in the special situations considered, but for their historical importance I have included in the fourth figure (3.20) DULEAU's drawings referring to these experimental problems.

The first wide publication of DULEAU's memoir was in 1820, just at the time when NAVIER, CAUCHY, and POISSON were publishing their classic memoirs on linear elasticity. The timing of DULEAU's experiments and the emphasis placed upon them in succeeding years belies the oft-expressed belief that the growth of the linear theory in the 1820's and 1830's was essentially independent of contemporary experiment.

3.10. Research on elastic moduli in the three decades (1811–1841) before Wertheim.

In 1841 GUILLAUME WERTHEIM presented a sealed packet to the French Academy containing the results of experiments which were the basis of a memoir read before the Academy the following year. This work, which was published[1] in French in 1844 and was translated into German in 1848, provided the first great definitive study of dynamic and quasi-static elastic constants as a function of temperature, prior permanent deformation, prior thermal history, alloy content, and susceptibility to the presence of electrical and magnetic fields during the deformation. WERTHEIM's enormous array of experiments in most of the then known metals for which it was possible to prepare a specimen for study, dominated both the theoretical and experimental study of solid mechanics from the mid-19th century until well into the present century.

To obtain a view of the state of affairs in the year preceding WERTHEIM's report, one may consult the remarkably comprehensive work by JEAN VICTOR PONCELET[2] which surveys the field of experimental solid mechanics. PONCELET had been taken prisoner during the French retreat from Moscow in November 1812, and had spent two years at Saratov on the Volga River.[3] Probably as a conse-

[1] WERTHEIM [1844, *1*].
[2] PONCELET [1841, *1*].
[3] *See* BERTRAND [1879, *1*].

DULEAU (1820)

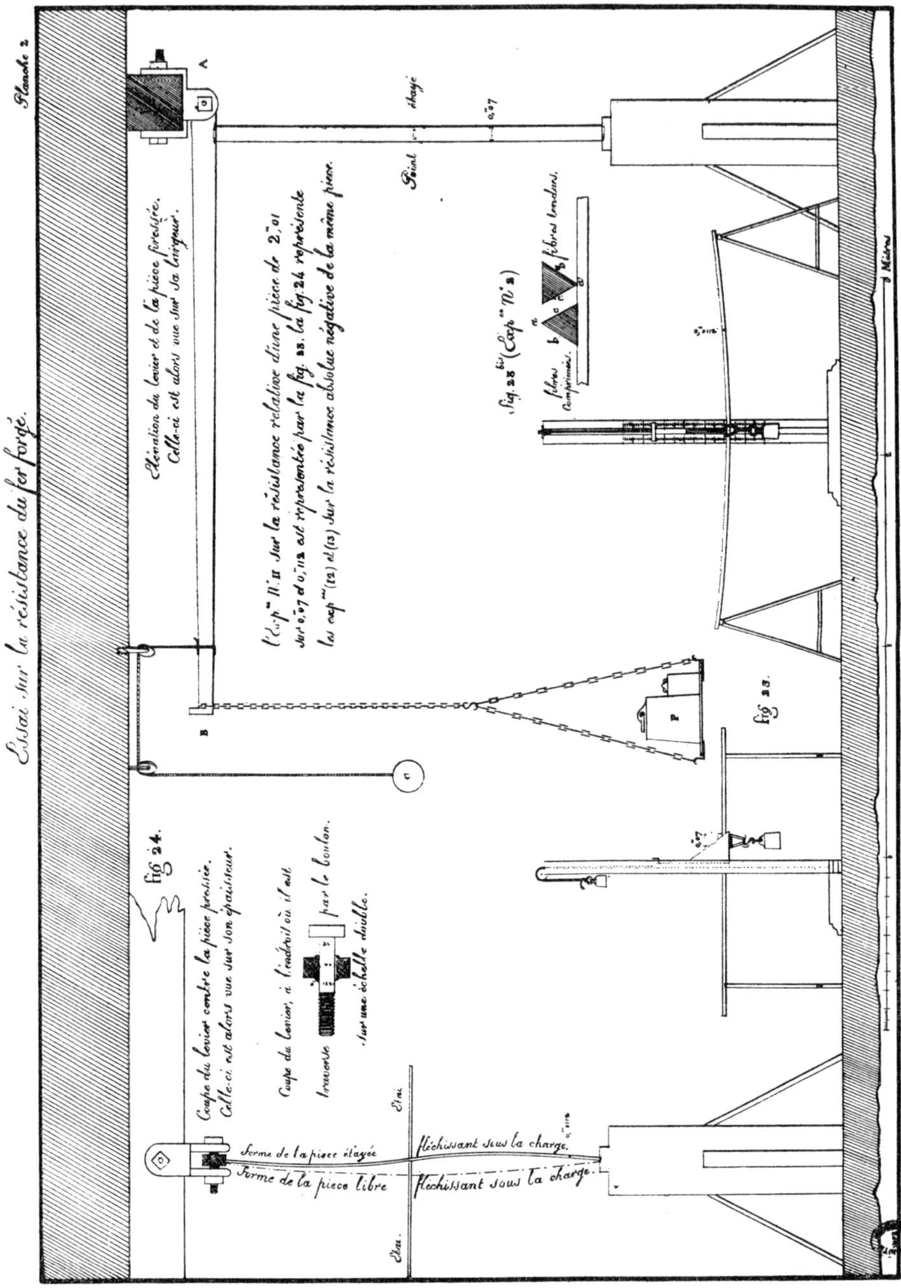

Fig. 3.17. A drawing by DULEAU of experimental situations he consierded.

quence of this confinement, much of his scholarly life was constrained by the vicissitudes of ill health. After the war he returned to his birthplace at Metz to take a position in the arsenal; in 1825 he became a Professor in engineering mechanics at the military school. In 1827 he gave a series of talks for the workers of the town of Metz; a member of the audience, GOSSELIN, asked for permission

DULEAU (1820)

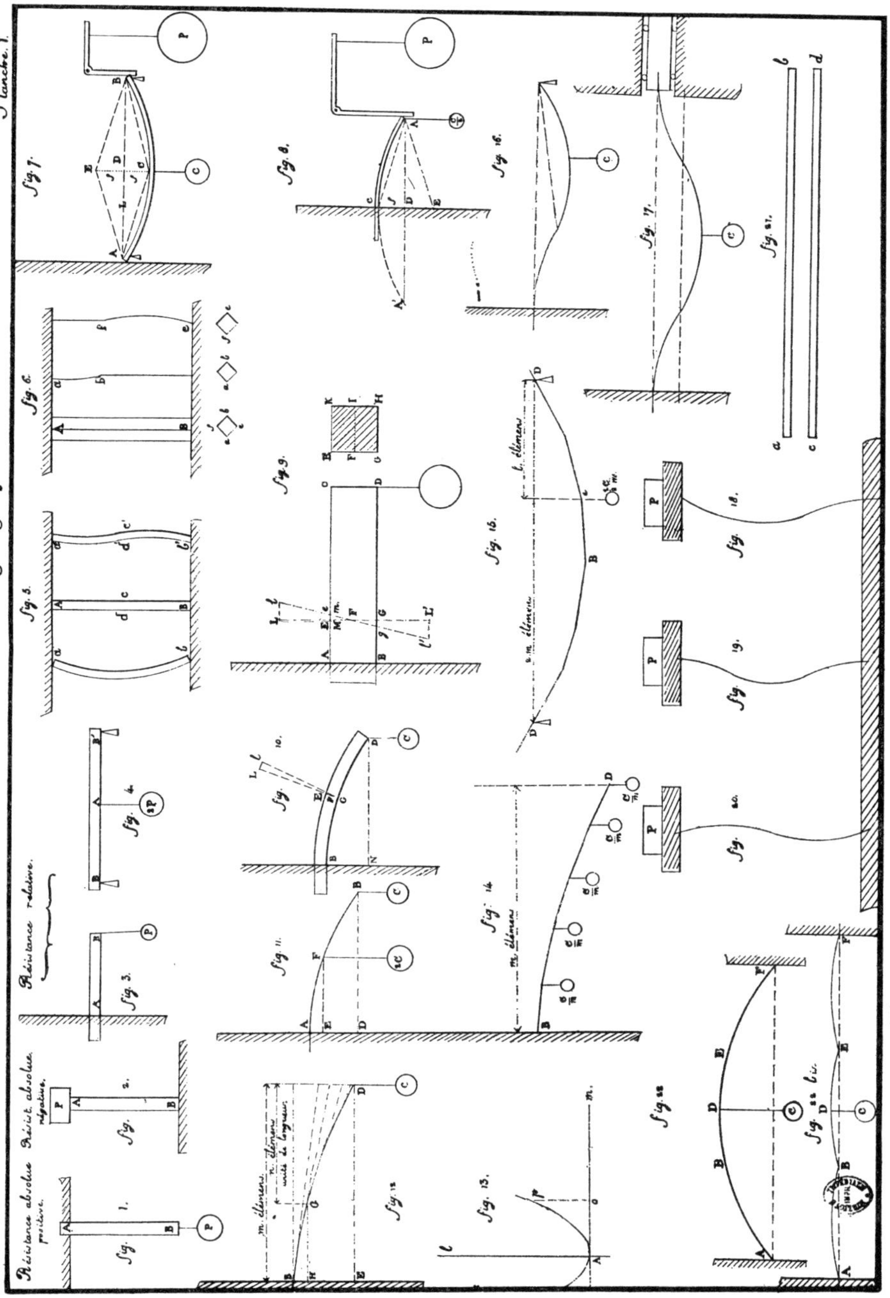

Fig. 3.18. A drawing by DULEAU of experimental situations he considered.

to publish the notes he had written on PONCELET's lectures. PONCELET, ill and tired, gave his permission but did not himself participate in the work. The notes were published[1] in three parts between 1827 and 1829. They were an immediate success, and a reprinting in 1830 also immediately sold out.

[1] The first edition was entitled: *Cours de Mécanique Industrielle*.

DULEAU (1820)

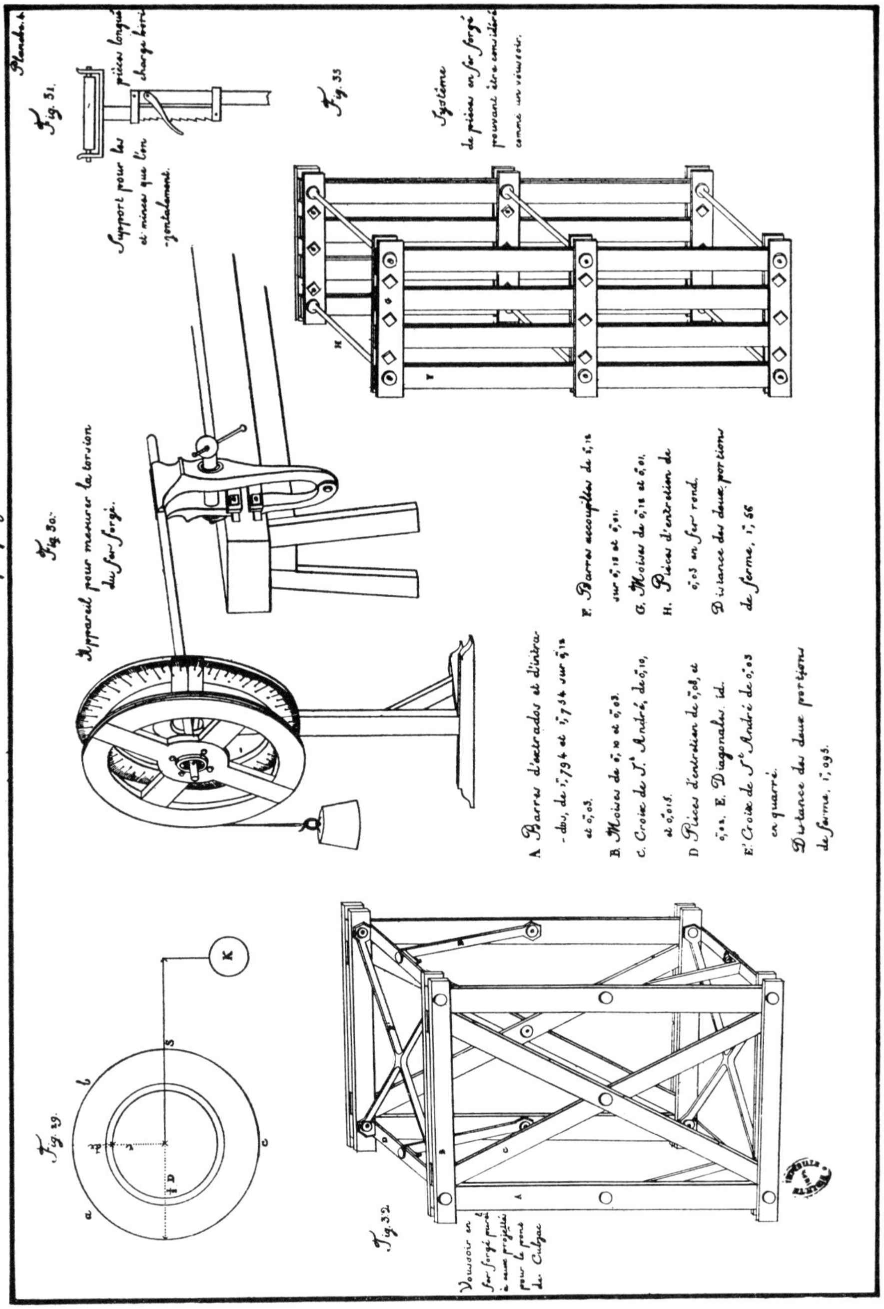

Fig. 3.19. A drawing by DULEAU of experimental situations he considered.

PONCELET was elected to the Académie in 1834; in 1839, having been appointed Professor of physical and experimental mechanics at the Sorbonne, he moved to Paris, where he began to rewrite the volume on his Metz lectures. A second edition was published in 1841, with the new title: *Introduction à la Mécanique Industrielle, Physique ou Expérimentale.* Since I have been able to consult only the second

DULEAU (1820)

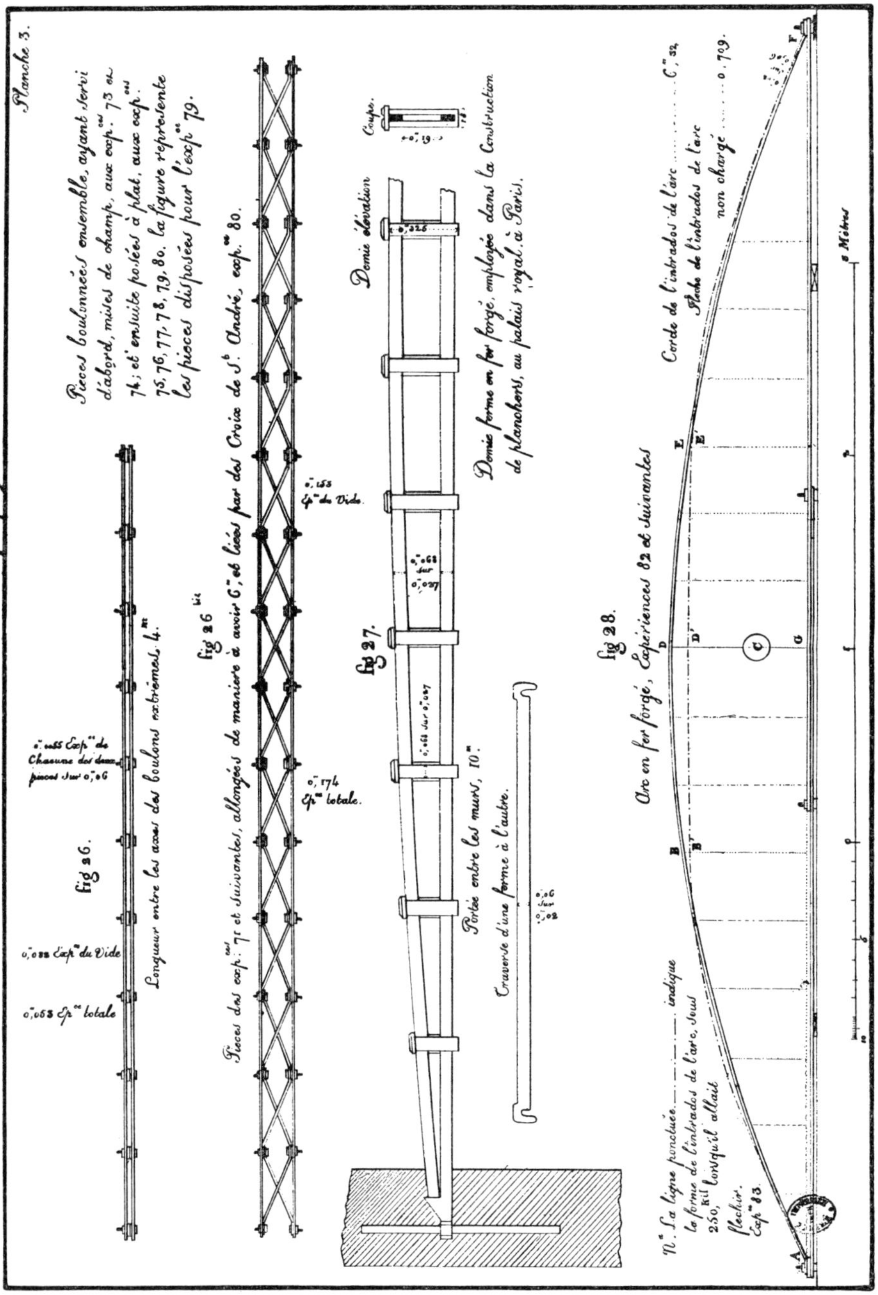

Fig. 3.20. A drawing by DULEAU of experimental situations he considered.

edition, I cannot say which portions of the work were completed between 1827 and 1829 from the notes on his lecture series, and which were written in 1839 when PONCELET began to prepare the second edition. In the later edition, PONCELET included some new material, such as results ARDANT had supplied him of some experiments in 1835.

In any event, with minor additions, it is sufficient to consult this later work[1] of Poncelet to obtain an adequate view of the experimental research on elastic constants from the year 1813 in which Duleau's memoir appeared, to the year 1842, in which the definitive work of Wertheim was read before the Académie. Besides its general excellence, this monograph of Poncelet is valuable for having inaugurated a new practice in the presentation of experimental data. Only a few previous experimentists, such as James Riccati in 1721, had published data in such a way, namely, displaying stress-strain functions diagrammatically. Poncelet, for the first time, presented data on plots which contained quantitative information for both ordinate and abscissa.

As we have seen, one of the difficulties for the present-day experimentist interested in early studies is that of grappling with the limitations of results presented solely in tabulated form. By presenting data in both forms, Poncelet introduced the advantages of accuracy in the tabulation, and a comprehensive view of the over-all behavior in the graphs.[2] In Table 47 are listed the results obtained by Seguin (1826), Bornet (1834), and Ardant (1835) from tensile tests on iron, brass, steel, and lead wires whose stress-strain diagrams from Poncelet's monograph are shown[3] in Fig. 3.21. One notices that in these experiments stresses were carried beyond the elastic limit.

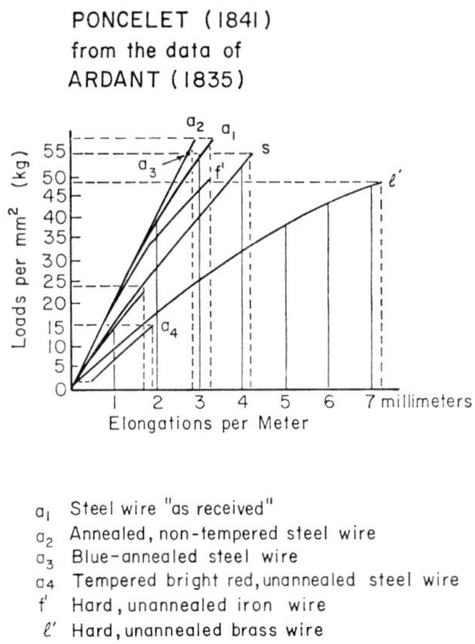

Fig. 3.21. Poncelet's (1841) graphical presentation of tensile tests by Seguin (1826), Bornet (1834), and Ardant (1835) on iron, brass, steel, and lead wires. These were the first quantitative diagrams of stress-strain behavior in solid mechanics.

---

[1] Poncelet [1841, *1*].

[2] The point made here is not trivial. Data published solely in the form of graphs provide additional sources of error for those who think it is important to re-examine, in close numerical detail, the work of others. On the other hand, data considered and published only in tabulated form, on many occasions have led authors and their readers to miss important aspects of the phenomena studied, and obviously not every reader can be asked to plot curves of tabulated data in every paper which he thinks might interest him.

[3] Apropos of the discussion in the previous section, it is interesting to note that as Poncelet pointed out, only the data of Morin is found to deviate widely from the elastic constants for iron for which a modern value of 20800 kg/mm² (29.7×10^6 psi) would be given.

Table 47. Poncelet (1841). *Table of the elongations of various metals under successively increasing loads from zero up to that which produces rupture.*

(Seguin) Iron wire precisely annealed diam. 1.06 mm length 1.50 m		(Bornet) Cable iron ductile diam. 49.50 mm length 6.42 m		Experimental results of Mr. Ardant on metallic wires from 1 to 1.5 m long; 4 mm diameter for lead; 0.40 to 1.6 mm for the other metals											
					Elongations per meter length (mm)							Tempered at cherry red, unannealed Steel wire		Melted cupel (cup) cold-drawn Lead wire	
					Iron wires		Brass wires		Steel wires						
Load (kg/mm²) (f'')	Elongation (mm/m) (f'')	Load (kg/mm²) (F)	Elongation (mm/m) (F)	Load (kg/mm²)	Soft, or annealed (f)	Hard, unannealed (f')	Soft, or annealed (l)	Hard, unannealed (l')	From the factory (a_1)	Annealed, non temp. (a_2)	Blue-anneald (a_3)	Load (kg/mm²) (a_4)	Elongation (mm/m) (a_4)	Load (kg/mm²) (p)	Elongation (mm/m) (p)
25.90	2	2	0.08	5.0	0.294	0.26	0.45	0.55	0.25	0.24	0.23	2.49	0.59	0.10	0.17
27.07	3	4	0.16	10.0	0.588	0.52	0.90	1.11	0.56	0.48	0.48	4.97	0.83	0.30	0.41
28.20	4	6	0.31	15.0	0.882	0.78	1.35	1.70	0.81	0.72	0.72	7.46	1.08	0.43	0.62
29.33	5	8	0.36	20.0	1.176	1.04	1.80	2.28	1.02	0.96	0.96	9.95	1.39	0.50	0.81
30.45	6	10	0.47	25.0	1.470	1.30	2.25	2.98	1.25	1.20	1.20	12.44	1.58	0.70	31.60
32.60	30	12	0.55	30.0	2.500	1.56	7.30	3.70	1.50	1.44	1.44	14.92	1.87	0.90	70.20
33.78	58	14	0.69	32.5	13.000	—	—	—	—	—	—	15.57	rupture	1.10	127.20
34.91	72	16	0.86	35.0	14.100	2.22	10.80	4.43	1.80	1.68	1.68			1.30	324.60
36.04	86	18	2.20	40.0	18.000	2.40	49.90	5.20	2.10	1.92	1.92			1.36	rupture
36.71	110	20	15.76	42.5	20.500	—	—	—	—	—	—				
37.16	118	22	24.34	45.0	rupture	2.82	115.00	6.15	2.36	2.16	2.16				
37.84	120	24	34.79	49.0		3.10	rupture	7.19	—	—	—				
rupture		26	46.96	50.0		rupture		rupture	2.65	2.40	2.40				
		28	67.70	52.5					—	—	2.52				
		30	89.39	55.0					3.00	2.66	rupture				
		32	132.48	57.5					3.15	2.76					
		33	rupture						rupture	rupture					
11400		21200			17000	19200	11111	9091	19400	20800	20800	10000			

E (kg/mm²)

Table 48. Poncelet (1841).

Indication of the nature of the material	Elongations relative to the natural elastic limit	Load per mm corre-sponding to the limit	Ratio of this load to that of rupture	Value of E (kg/mm²)
Forged iron bars, flexure experiments (Duleau)				
Strongest result	0.00167	—	—	24000
Weakest result	0.00044	—	—	16000
Average	0.00062	12.4	—	20000
Forged iron, flexure (Tredgold)				
Average result	0.00071	12.1	0.30	20000
The same, in bars, hammer welded, or cylinders (traction) (Lagerhjelm)				
Welded Swedish iron	0.00093	17.2	0.44	20680
Welded English cable	0.00052	13.3	0.37	20750
Average	0.00072	15.0	0.40	20700
Large bars of strong iron (traction) (Navier)	0.00093	18.0	0.45	19400
Thick iron wire, unannealed (traction) (Vicat)	-	-	-	18000
1.20 mm diameter iron wire (traction) (Ardant)				
Strong, unannealed	0.00084	15.0	0.33	18300
Soft, annealed	0.00088	15.0	0.50	17000
Steel and cast iron				
Bars of English steel, melted, Huntsman, nontempered (flexure) (Duleau)				
Average	—	—	—	24000
Bars of forged steel, soft, annealed or unannealed (flexure) (tredgold)	0.00140	29.0	—	20400
Blades, English steel, melted, annealed and blue-tempered (flexure) (dyanometric springs) (Morin)				
Average	0.00222	66.0	0.67	30000
Steel wire, melted, drawn, unannealed, commercial (traction) (Ardant)				
First elongations	—	—	—	20800
Subsequent elongations	—	—	—	19000
The same, red-annealed, nontempered, pliant				
First elongations	—	—	—	23600
Subsequent elongations	—	—	—	20800
The same, red-tempered, blue-annealed, of spring (traction) (Ardant)				
The same, unannealed, cherry-red tempered, fragile, (traction) (Ardant)				
First elongations	—	—	—	11000
Subsequent elongations	—	—	—	10000
Cast iron (flexure) (Rondelet)				
Average	—	—	—	9840
The same (flexure) (Tredgold)				
Average	0.00083	10.0	—	12000

On p. 352 of his monograph, PONCELET listed moduli measured from either flexure or axial tension data, as indicated, for eight experimentists from DULEAU to ARDANT. PONCELET's tabulation is reproduced here as Table 48. I have added my calculations of E determined from the small deformation region.

WERTHEIM in his classic memoir on elastic constants, presented to the French Academy on July 18, 1842, summarized the work which preceded his on this subject:

> The simple metals, which are the subject of this treatise, have already been studied by many. It would be too involved to mention all the numbers discovered and their errors. One finds them all in *Mécanique Industrielle* by Poncelet.
>
> Most of these studies were undertaken with the aim of practical application. One dealt mainly with metals used for construction, without being able to take into account their purity, their density, and their changes which might result or have resulted from mechanical use or exposure to heat. The discrepancies in the results and the shortcoming of each general law can probably be attributed to the variability of the conditions.[1]

A similar castigation of previous investigations has appeared in the introductory passages of nearly every large study of elastic constants which has been published in the nearly 130 years after that of WERTHEIM, but his was the first, and one of the few specific statements of the sources of conflict and error in the pursuit of constitutive equations for solids. Characteristically, WERTHEIM also gave proper emphasis to the outstanding achievements of his predecessors such as DULEAU, GERSTNER, and VICAT, for whose work he expressed considerable admiration.

A major topic for investigation for experimentists interested in the elasticity of metals in the thirty years between DULEAU and WERTHEIM, was whether or not the elastic constants and cohesion measurably were influenced by the prior heat treatment of the solid. COULOMB, as we have seen, from his study of a steel blade subjected to different heat treatments, had found that a solid was produced which varied from being an "excellent spring" to being a very poor one when in an highly annealed state. Nevertheless, he had concluded that for small deformation no difference in the elastic modulus was observed.[2] In a letter written in 1823 to THOMAS YOUNG, THOMAS TREDGOLD[3] reported similar results from a series of experiments for simply supported beams. TREDGOLD's apparatus is shown in Fig. 3.22. His first experiment was made with a bar of blistered steel of "very good quality."

TREDGOLD's letter further stated:

> It [the bar] was drawn out by the hammer to the width and thickness I had fixed upon, and then filed true and regular. It was then hardened, and tempered to the same degree of hardness as common files.[4]

A bar 14 in. long, with a breadth of 0.95 in. and depth of 0.375 in., with 13 in. between the simple supports, gave the load deflection characteristic shown in Fig. 3.23.

The bar tempered to a "rather deep straw yellow," and then again until the color was "uniform blue, or spring temper," had the same deflections for the same loads. This bar was then hardened and loaded beyond the previous range, as shown in Fig. 3.23 (triangles), providing the observed continued linearity.

A second experiment was on a bar of much greater length, with a breadth of 0.92 in., a depth of 0.36 in., and a distance between the supports of 24 in. The bar first was tested in a state so soft as to "yield easily to the file," producing the linear load-deflection behavior of the second test shown in Fig. 3.23 (dots). The

---

[1] WERTHEIM [1844, *1* (a)], p. 388.
[2] COULOMB did note that the elastic limit was reduced.
[3] TREDGOLD [1824, *1*].
[4] *Ibid.*, p. 355.

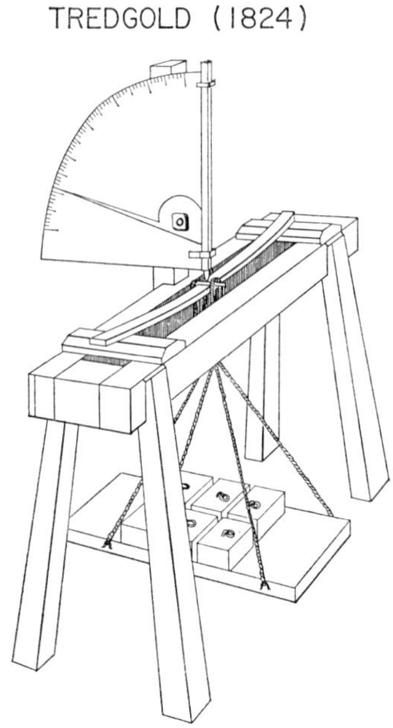

Fig. 3.22. The flexural apparatus of TREDGOLD (1824).

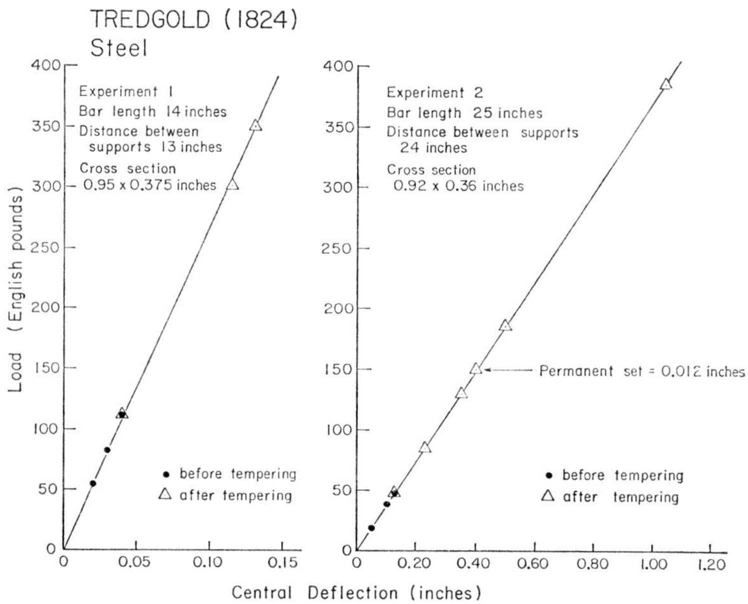

Fig. 3.23. Load-central deflection results of TREDGOLD in steel, using the apparatus shown in Fig. 3.22. I have constructed the diagram from his tabulated data.

bar then was hardened until a file could make no impression on any part of it, "and the same loads did not produce flexures that were sensibly different from those in the soft state." The temper of the bar was then lowered to "an uniform straw color," and again tested; the same linearity was observed to an even higher load. (*See* Fig. 3.23, triangles.)

TREDGOLD summarized these experiments, comparing them with those by YOUNG, whose data on vibration as we have seen, provided the value of 8530000 ft. for the height of the modulus. TREDGOLD obtained a value of 8827300 ft. from his first experiment and 8810000 ft. from his second, i.e., for a weight density of 0.286 lbs./in.^{3}: 21500 kg/mm^2 and 21650 kg/mm^2, respectively. He further compared these with a curious selection of values from the data of DULEAU, described above: 9400000 ft. for cast steel, and the extremely low value of 6600000 ft. for German steel, i.e., 23100 kg/mm^2 and 16250 kg/mm^2, respectively.

LAGERHJELM,[1] a Scandinavian, implied that the Englishman TREDGOLD had revealed something less than scholarly objectivity in assessing the Frenchman DULEAU's comparison of the relative merits of English and German steel. As may be seen in the tabulation of my calculation of DULEAU's moduli (Table 45), there are two extremely low values for the German steel. LAGERHJELM stated that DULEAU had regarded both of these as highly questionable, in that the measurements of elongation had been extremely irregular. TREDGOLD selected one of these tests (experiment #61), in his average of four, which resulted in the international imbalance reported. Incidentally, in examining the same tabulation, I have failed to find where TREDGOLD obtained the very high value of 23100 kg/mm^2 for English steel, except for that denoted as "Huntsman steel", which has an abnormally high value in comparison with the remainder of the data, the average of which is 20000 kg/mm^2. However, LAGERHJELM agreed with TREDGOLD and with COULOMB with regard to what appeared to be variations in the prior thermal history. He, too, thought that the annealing and tempering recipes did not produce a marked effect upon the elastic constants.

In his memoir of 1841 on the elasticity of solid bodies, MASSON reported that SAVART at the end of a paper several years before[2] on "the mechanical properties of fluids" had expressed a desire to submit solids to comparative studies, with the ultimate aim of establishing the groundwork for a general theory of deformation.[3] MASSON continued:

> He had entrusted me with a part of this work. Helped by his advice, working under his direction and in his rooms, I had begun the researches which have been interrupted by the misfortune which has submerged all his friends and me especially in a grief which nothing could assuage, unless it were the recollection of his kindness and friendship.[4]

The reference, of course, was to SAVART's death. Neither MASSON nor SAVART carried out such a program, and one may raise serious question as to whether they could have succeeded, had they tried. As was stated earlier, MASSON continued work in SAVART's laboratory during the latter's illness and after SAVART's death March 1, 1841. On specimens of brass, iron, steel, and copper, one meter long, MASSON repeated experiments which had been performed by SAVART in 1837. The specimens employed in the two series were identical. A comparison of the results of the two experiments four years apart, on the same specimens, is shown in Fig. 3.24 for three of these four metals. The large amount of variation evident precludes any meaningful discourse on elastic constants from these tests.[5]

---

[1] LAGERHJELM [1829, *1*].

[2] SAVART [1837, *1*]. This paper contained SAVART's data, which MASSON described. I have been unable to locate among SAVART's voluminous early publications the passage referred to here.

[3] MASSON [1841, *1*].

[4] MASSON [1841, *1*], p. 451.

[5] As I have shown in Chap. II, part of the difficulty in trying to compare these tests arose from the presence of what is now known as the "PORTEVIN-LE CHATELIER" effect, which should

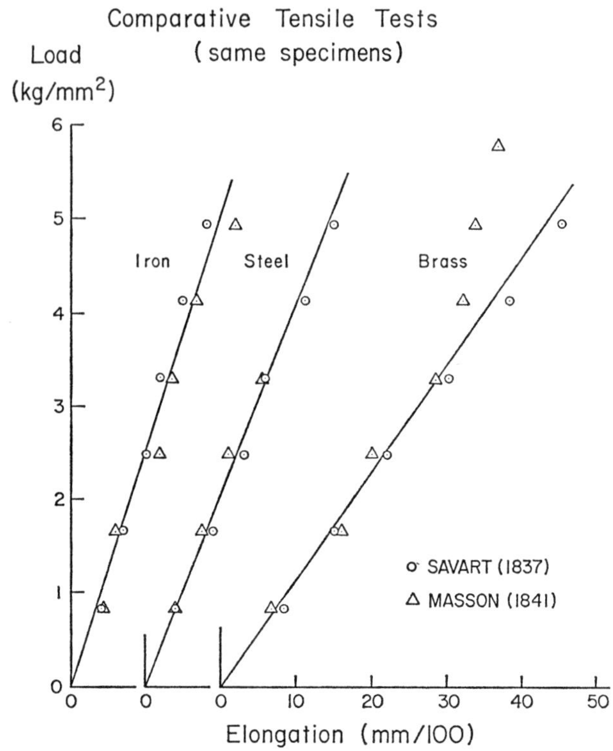

Fig. 3.24. A comparison of the experimental results of SAVART and MASSON on exactly the same specimens. I have constructed the diagram from the data tabulated by MASSON.

MASSON, following the practice of DULEAU, gave an elastic constant in the form of the elongation a unit cube of the solid would undergo when subjected to an axial unit load in one perpendicular direction. Converting these to the E-moduli, one obtains the abnormal values shown in Table 49.

Table 49. MASSON (1841).

	E-moduli		Sound velocity (ratio to that in air)	
	(SAVART) (kg/mm²)	(MASSON) (kg/mm²)	(SAVART)	(MASSON)
Rods:				
Copper	14150	—	12.21	—
Steel	24000	24800	15.22	—
Brass	10750	11200	10.56	—
Iron	29500	24500	15.17	—
Zinc	-	12450	11.85	—
Tin	—	—	—	7.89
Lead	—	—	—	4.2782

be attributed to MASSON. Nevertheless, the fact that in these tests there is no reference to matters of purity, prior thermal or mechanical histories, nor to experimental details, suggests a somewhat limited understanding of the experimental requirements for a definitive study of the subject.

MASSON described in some detail the comparison of experiments in lead and tin, the latter being described as "weeping tin" which had been passed through a rolling mill until it became as soft as lead, and "gave forth no cry when bent." Both the tin and lead rods were drawn to 3 mm diameter. The frequencies were determined as usual by comparison with the notes of the scale, giving for tin a sound velocity ratio to that of air of 7.89, and for lead, 4.2782. For some inexplicable reason MASSON was considerably struck with this latter value when he noted that it provided a velocity of sound of 1443.48 m/sec in lead, which was almost exactly the 1435 m/sec he ascribed to water.[1]

We have already referred to the great scatter in the data of BARLOW (see Fig. 2.12) and have contrasted his tension tests on iron with those of VICAT. All these experimentists were concerned with the cohesion or ultimate strength of these solids, but no sensible order had yet appeared for such a parameter. As we shall see, many of the reasons for the observed lack of order were first discovered in WERTHEIM's far more definitive studies. MASSON, too, was curious about COULOMB's statement that tempering had no influence on the elasticity in either torsion or flexure. He took a watch spring, measured its elongation, annealed it, and tested it again, obtaining for successive loads of 4 kg, the values shown in Table 50.

Table 50. MASSON (1841).

Total load (kg)	Additional elongation for each Successive 4 kg load	
	Tempered spring	Annealed spring
0	0	0
4	1.4	0
8	12.2	0
12	8.4	12.0
16	4.8	13.6
20	0	9
24	12.4	5
28	6.6	11
32	(no add. load)	8
36	(no add. load)	4
Average elongation for 1 kg	1.63 c.mm	1.739 c.mm

From these highly questionable results, MASSON concluded that there was no tempering effect in the elasticity. However, without describing the experiments, he reported that in similar tests in zinc he had observed marked differences. He remarked also, as had many others before him, including COULOMB, that annealing did have a pronounced effect upon the magnitude of the elastic limit.[2] His paper is described here primarily so as to provide the background of the state of experiment in elastic constants in the year in which WERTHEIM submitted his sealed

[1] Presumably MASSON was referring to the famous experiments in 1838 in Lake Geneva of COLLADON and STURM [1838, *1*], who obtained this value in their experiments.

[2] One must await the monumental experimental study of the elastic limit by BAUSCHINGER in 1886 before the full implication of this complicated behavior is appreciated.

packet to the Academy,[1] 1841. MASSON initiated an atomistic approach to the consideration of elastic constants which was to be developed in great detail by WERTHEIM and, still later, by VOIGT. MASSON noted that for iron, copper, zinc, tin, and silver, the product of the elasticity and atomic weight was nearly constant. The numbers he obtained for the product for these five metals were 2.63, 2.39, 2.48, 2.31, and 2.45, respectively. He attributed this coincidence to chance, but added:

I thought I should point this out, since in studies so difficult as those concerning molecular physics, studies so little advanced, no datum should be neglected, whatever its value might be.[2]

3.11. Guillaume Wertheim: A Faraday without a Maxwell.

Few experimentists in solid mechanics during the past 300 years have had the great enthusiasm or ability necessary to engage in major programs which provide the fundamental base for future theoretical or experimental research in a specific area. Nearly all of those in the 19th century who might lay claim in some measure to such an endeavor were academicians or university professors in command of one or more research assistants and of more or less adequate laboratory facilities. Such men as KELVIN, KUPFFER, BAUSCHINGER, VOIGT, BACH, and GRÜNEISEN were established and respected authorities when they entered upon the surveys which are associated with their names. It is not enough merely to have run a thousand tests, as had GEORGES LOUIS LECLERC Comte de BUFFON[3] and DAVID KIRKALDY.[4] A proper, systematic experimental study which lays the foundation of a particular subject must be clearly stated, rigorously performed, and conceived with sufficient imagination to eliminate all of the auxiliary empirical assumptions which otherwise would diminish the whole to nearly meaningless conjecture when more perceptive studies follow.

It is indeed remarkable that GUILLAUME WERTHEIM at the age of twenty-five, a newly arrived foreigner in Paris with a year-old medical degree, should, as a private citizen, in the space of three short years lay the experimental foundations for the comparative study of elastic constants in fifteen metals and sixty-four alloys, all prepared by himself, all tested for density, composition, purity, elastic limit, ultimate strength, maximum elongation, with elastic constants determined dynamically from both lateral and longitudinal vibrations, and quasi-statically by unloading from every region of strain to rupture for both drawn and annealed metals. When one adds to this that his experiments were performed for the pure metals at room temperature, at 100° C, at 200° C, and at −20° C, so that his was the first study of the effects of temperature upon elastic constants, and that in addition, the effects of electric and magnetic fields were included where relevant, we can see indeed that his was one of the most extraordinary personal feats in the history of experimental physics.

WERTHEIM immediately followed this work with similar studies in elasticity and cohesion of different kinds of glass, different kinds of wood, and on the principal tissues of the human body, all before the end of 1846. It was not until 1854,

[1] In 1859, MASSON and WERTHEIM were on the ballot under identical ratings as "fourth line" candidates for a place in the French Academy vacated by the death of CAGNIARD-LATOUR. This constituted a demotion for WERTHEIM, who in 1851 had been considered a third rank candidate for the Academy opening created by the death of GAY-LUSSAC. If WERTHEIM had been given recognition commensurate with his achievements, one wonders whether experimental solid mechanics would have flourished, rather than declined, in France during the years that followed.

[2] MASSON [1841, *1*], p. 460.

[3] BUFFON [1741, *1*].

[4] KIRKALDY [1891, *1*].

fourteen years after coming to Paris and one year after receiving the Doctor of Science degree, that WERTHEIM held for a year his only academic appointment, the position of Substitute Professor at the Faculté de Montpellier. From 1855 until his premature death at the age of 46, when, "in a fit of depression", he leaped from the tower of the Cathedral of St. Gathien at Tours, he was an entrance examiner at the École Polytechnique in Paris.[1]

I have already commented on the controversy which almost invariably surrounded the experimental studies of WERTHEIM. Challenges or attacks, however, did not characterize the reception of his first memoirs on elastic constants. On these, presented at the French Academy in 1842 and 1843 and subsequently published in the *Annales de Chimie et de Physique* in 1844, the report by PONCELET, DUHAMEL, PELOUZE, and BABINET (the Commission appointed by the Academy), which was read in 1844, prophetically stated:

> Consideration of the whole of Mr. Wertheim's work shows that contrary to custom, he has not allowed himself to be drawn into premature theoretical deductions; that the relations he has pointed out between the results of his experiments and the molecular constitution of bodies are of great importance; but, more than this, that his work, regarded from the experimental side, contains such a number of new determinations, of numerical constants, of exact measurements, that the Academy cannot do otherwise than give complete approval of this work, conscientious and rich in facts.
>
> Your Commission thus proposes that you approve these two Memoirs of Mr. Wertheim, and order them to be printed in the *Recueil des Savants Étrangers*.[2]

The 1841 sealed packet to the French Academy reveals that WERTHEIM's interest in the physics of solids had extended back to 1837 when he was a medical student in Vienna. At that time he stated that he had communicated to a Mr. d'ESTINGSHAUSEN in Vienna a comparative analysis of the deformation experiments of GUYTON-MORVEAU, RENNIE, TREDGOLD, and CHLADNI, from which he had concluded that although general trends were discernible, the results that had been obtained were in fact insufficiently accurate, not only in experimental method but also in specification of the chemical purity of the specimens, to serve as a basis for the definitive study of the behavior of solids. His own objective, he stated, was to make a systematic study of chemically pure solids.

WERTHEIM repeated in one memoir after another from 1842 to 1860 that it was his purpose to provide a broad base from which to test the physical applicability of the various theories which had been proposed, without accepting in advance any particular hypothesis as the initial basis for an experimental study. It was his belief that such an approach would lead inevitably to a better understanding of the subject, not only by the rejection of hypotheses and theories seen to describe the physical situation improperly but also by increasing the plausibility of new theoretical approaches which would replace them. It is strange that while he was in pursuit of this clearly stated and logical approach to physics WERTHEIM was constantly under attack for failing to substantiate popular hypotheses. He was accused of being either mathematically incapable of understanding the points of view he experimentally dismissed, or of not assuming the role of a major theorist in developing new theories to replace or extend the earlier ones. This is the all too common fate of experimentists whose new discoveries are not in accord with fashionable science. It is interesting that several major theorists of WERTHEIM's time, such as CAUCHY, DUHAMEL, PONCELET, and on some occasions,

[1] This association with the entrance exams at the École Polytechnique, an appointment of some significance at the time, was stated in POGGENDORFF's eulogy [1863, *1*]. I examined the École Polytechnique list of examiners for the period and found no reference to WERTHEIM as having had senior responsibility in this capacity. For further details concerning his life, see Chap. II, Sect. 2.14. See also Sects. 2.19 and 2.20.

[2] BABINET *et al.* [1844, *1*].

SAINT-VENANT, appeared to understand and be in sympathy with his objectives, even when earlier theories of their own had been revealed by WERTHEIM's experiments to have limited physical application. Thus we see in an Academy report, with CAUCHY as "rapporteur", the committee's summation of WERTHEIM's study in 1848 on the compressibility of solids, a study which revealed that CAUCHY's and POISSON's uniconstant and molecular theory of elasticity was not applicable:

> In summary, the Commissioners think that Mr. Wertheim in the new memoirs submitted for their examination, after having given an experimental solution of an important question that concerns both physicists and geometers, has discussed this question with the shrewdness he has shown in former studies. Hence, the Commission is of the opinion that these memoirs are worthy of the Academy's approval and of being printed in the *Recueil des Savants Étrangers*.[1]

3.12. Wertheim's memoir of 1842: Values of *E* for 15 elements and the first study of the effect of ambient temperature, prior history of the specimen, rate of loading, and atomic spacing.

Like the memoirs of COULOMB, WERTHEIM's writings provide a model for the clear exposition of experimental research. They are a refreshing contrast to the rambling half-complete writings common in the solid mechanics of his time. WERTHEIM invariably began with a long, complete, and extremely useful historical survey containg a well documented bibliography. This was followed by an explicit statement of purpose, a detailed and careful description of the apparatus and its limitations, a carefully tabulated section containing all of the results in a form easily understood, and an enumerated, definitive summary of his conclusions.

The initial memoir was divided into three parts, the first being presented to the Academy on 18 July, 1842; the second on 8 May, 1843; and the third on 22 July, 1844. In this work,[2] his first research in physics, WERTHEIM, at the age of twenty-five, laid the experimental foundations for our knowledge of the elasticity of crystalline solids under small strain; he immediately acquired an international reputation as the leading authority on this subject. In the first section of this memoir, he reviewed the earlier studies on elastic coefficients and on the speed of sound in solids including iron, copper, lead, zinc, silver, and platinum. He observed that not only had there been a wide variety of numbers reported for such measurements, but also the composition and purity of the solids considered almost invariably were unknown. He stated that no studies had been made of elastic parameters as a function of temperature, or as related to careful determinations of densities, or as influenced by the presence of electric and magnetic fields in terms of a defined and observed elastic limit, or as a function of well-defined prior thermal histories, or in terms of the distinctions between dynamic and quasi-static measurements with sufficient precision to effect a meaningful distinction.

WERTHEIM's first program, carried out in lead, tin, cadmium, gold, silver, zinc, palladium, copper, platinum, cast steel, three kinds of iron wire, and English steel wire, determined E for cast, drawn, annealed, and previously ruptured specimens. The experiments at room temperature consisted in a measurement of density both before and after rupture, a determination of a modulus from longitudinal vibration, a modulus from a lateral vibration, and quasi-static moduli from unloading a tensile specimen from a whole series of plastic loading states, for

[1] *See* "Rapport sur divers Mémoires de M. Wertheim", WERTHEIM [1851, *9*], pp. 329–330. While all three of the Academy commissions appointed to consider various memoirs of WERTHEIM provided strong recommendations that they be included in the *Recueil des Savants Étrangers*, none was ever published therein.

[2] WERTHEIM [1844, *1* (a), (b), (c)].

each of which the permanent deformation was specified. For a given specimen he often made as many as 40 separate determinations of the quasi-static modulus. Whether the specimen was cast, annealed, or drawn, he estimated the visible elastic limit, the rupture load, and the maximum permanent elongation. In addition, in each instance for each condition, he computed a velocity of sound. A few examples of WERTHEIM's quasi-static results are shown in Figs. 3.25 and 3.26 which are taken from his tabulated results. The final E was given as the average of all slopes.

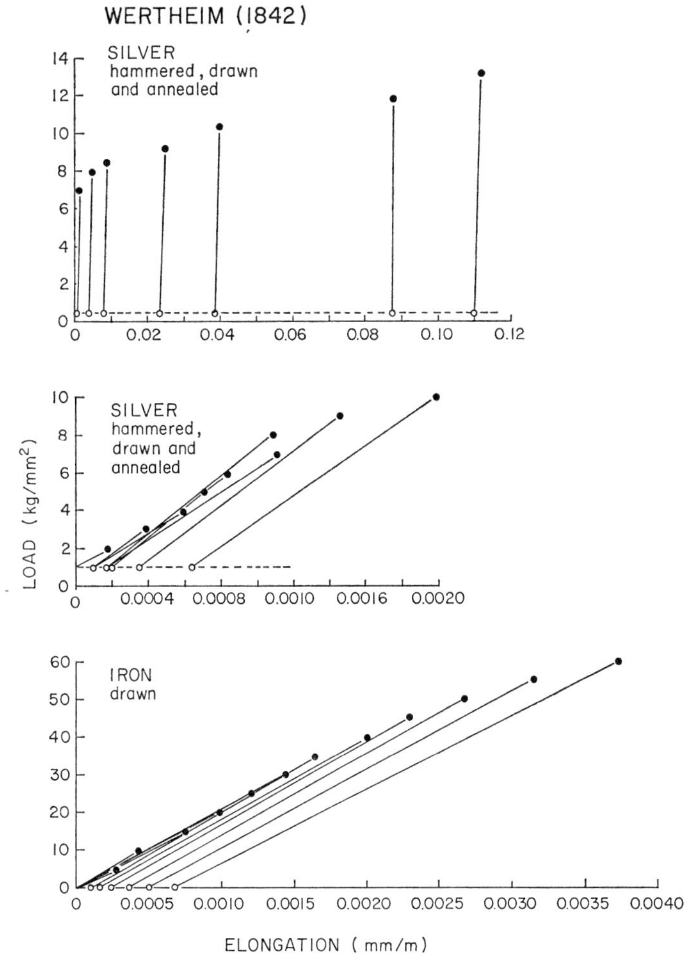

Fig. 3.25. Examples of WERTHEIM's repeated loadings quasi-static results in silver and iron.

The moduli determined from longitudinal vibrations were measured by CHLADNI's technique, in which the centrally held rods were rubbed at the end[1] and wires were clamped at both ends and rubbed at the center. The frequency of vibration was matched to a sonometer. The frequencies of transverse vibration were determined through the use of a rather elaborate apparatus in which a small

[1] A woolen cloth either impregnated with resin for metal rods or dampened with water for glass rods was the most common vibration generator.

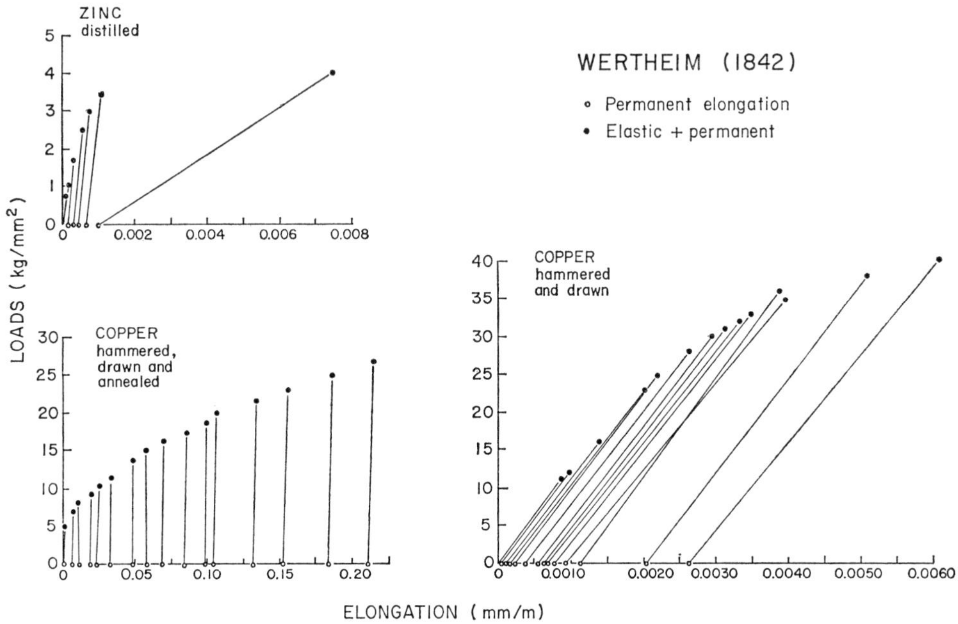

Fig. 3.26. Examples of Wertheim's repeated loading quasi-static results in zinc and copper.

wire at the end of the rod was in contact with a rotating disk covered with graphite. The disk was driven by a series of weights and the timing was calibrated by comparison with the vibratory markings produced on it by a carefully calibrated tuning fork, a type of calibration which Wertheim attributed to Duhamel, who had suggested it in a lecture. The whole apparatus was constructed on a heavy oak table containing a foot pedal by means of which the entire system could be started.

To check the longitudinal measurements made with the sonometer against the calibrated transverse measurements, Wertheim noted that in the generation of the latter, small longitudinal vibrations could be made visible, even at the nodes of the lateral vibration. Counting the number of such longitudinal oscillations per cycle of the transverse vibration made it possible to compare the results of the two experimental approaches. In Table 51 such a comparison for cast steel and brass reveals that the number of longitudinal vibrations per second given by the product of the number of transverse vibrations times the number of longitudinal vibrations per transverse cycle, is very similar to the number of directly measured longitudinal vibrations per second according to the sonometer. Wertheim showed that for the boundary conditions considered for rods of different lengths the ratio of the lateral frequency to the longitudinal frequency should be 0.55958. The value for cast steel was 0.56184, and for brass, 0.55974. Thus, as was characteristic of all of Wertheim's studies and unlike those of most of his contemporaries in this field, cross-checks of experimental results were provided which led to confidence in the data presented.

Wertheim proposed that since the stress distributions differed in the two situations described above, such a correlation as he had obtained showed that the solids considered were indeed isotropic and homogeneous. For bismuth and antimony specimens, which he regarded as non-homogeneous metals, the longitu-

Table 51. WERTHEIM (1842).

	Specific weight	Radius of rods (mm)	Number of transverse vibrations	Number of longitudinal vibrations over each transverse vibration	Products of these two numbers	Number of longitudinal vibrations according to the sonometer
Cast steel	7.841	5.060	7.204	175.4	1263.6	1267.0
Brass	8.452	5.029	4.867	179.0	871.2	864.5

dinal and transverse experiments, as may be seen in Table 52, provided widely different moduli and ratios of the velocity of sound in the solid to that in air. The final tabulation for the measurements at room temperature, excluding the uninterpretable ones from bismuth and antimony, is shown in Table 53.

Table 52. WERTHEIM (1842).

	Elasticity coefficient from		Sound velocity from (ratio to that in air)	
	Longitudinal vibrations (kg/mm²)	Transverse vibrations (kg/mm²)	Longitudinal vibrations	Transverse vibrations
Bismuth	3290	2473	5.455	4.731
Antimony	4817	3144	8.028	6.486

The elasticity coefficient and calculated velocity of sound are based on the average of moduli measurements made for different amounts of permanent set in tensile loading. Following the experimental practices of GERSTNER and HODGKINSON, WERTHEIM loaded his specimen to a prescribed value of stress and strain and then removed the load; he then measured the amount of permanent deformation and recovered elastic deformation by means of an optical cathetometer which provided him with a strain resolution of $\varepsilon = 10^{-5}$ for the meter-long specimen. A higher load was then placed on the specimen, and this process was repeated. One example of the very large amount of data he tabulated is given in Table 54. When calculating moduli from WERTHEIM's data, one must note that he subtracted initial loads which did not produce an observed elongation in the third decimal place.

The various parameters which WERTHEIM used in describing his data are:

P = load (kg/mm²)
L' = distance between reference points under tension of load P (mm)
L = distance between reference points without load (mm)
a = elastic elongation per meter (mm)
a' = permanent elongation per meter (mm)
E = elasticity coefficient (kg/mm²)

One notes that WERTHEIM measured an elastic limit which he defined as the strain at which a permanent deformation of $\varepsilon = 10^{-5}$ was observed. This of course is far below the 0.2% often used in present-day technology. It was WERTHEIM's opinion, as it had been HODGKINSON's, and as it was to be BAUSCHINGER's, that in reality there was no elastic limit below which no permanent deformation was present. WERTHEIM thought it was merely a matter of the refinement of measure-

Table 53. Wertheim (1842)

	Elasticity coefficients determined from			Sound velocities determined from (ratio to those in air)			Specific weight
	Longitudinal vibrations (kg/mm²)	Transverse vibrations (kg/mm²)	Elongations (kg/mm²)	Longitudinal vibrations	Transverse vibrations	Elongations	
Lead							
cast	1993.4	1985.2	1775.0	3.974	3.966	3.561	11.215
drawn	2278.0	1781.2	1803.0	4.257	3.764	3.787	11.169
annealed	2146.0	1854.2	1727.5	4.120	3.841	3.697	11.232
ruptured		1788.6			3.749		11.308
Tin							
drawn	4006.0	3839.7		7.480	6.829		7.313
annealed	4418.0	3703.4		7.338	6.719		7.290
as received,							
cast	4643.0	4172.0		7.465	7.076		7.404
drawn	4564.0	4148.0		7.401	7.086		7.342
ruptured		3918.0			6.909		7.293
Cadmium							
drawn	6090.3	5424.0		7.903	7.456		8.665
annealed	4241.0	5313.0		6.651	7.444		8.520
ruptured		4084.0			6.518		8.541
Gold							
drawn	8599.0	8644.6	8131.5	6.424	6.441	6.247	18.514
annealed	6372.0	5989.0	5584.6	5.603	5.432	5.245	18.035
ruptured		5833.0			5.212		19.077
Silver							
drawn	7576.0	7820.4	7357.7	8.057	8.186	7.940	10.369
annealed	7242.0	7533.0	7140.5	7.903	8.060	7.847	10.304
ruptured		7648.0			8.115		10.320
Zinc, distilled							
sand cast	7536.0	6778.0		9.683	9.188		7.134
cast in mold	9338.0	9423.0	9021.0	10.774	10.823	10.591	7.146

Zinc, as received							
drawn	9555.0	8793.6	8734.5	11.007	10.560	10.524	7.008
annealed	9292.0	9641.0		10.814	11.016		7.060
ruptured		9324.0			10.822		6.997
Palladium							
drawn		12395	11759		10.066	9.804	11.350
annealed		11281	9789.0		9.450	8.803	11.225
Copper							
drawn	12536	12513	12449	11.167	11.157	11.128	8.933
annealed	12540	11833	10519	11.167	10.847	10.703	8.936
ruptured		12040			10.970		8.890
Platinum wire							
thin	16176	15928		8.241	8.177		21.166
annealed	14292	14373		7.823	7.845		20.753
medium	17165	17153	17044	8.467	8.456	8.437	21.275
annealed	15611	15355	15518	8.111	8.045	8.087	21.083
thick	16159	15814		8.218	6.130		21.259
annealed	15560	15683		8.074	8.106		21.207
ruptured		16748			8.421		20.987
Iron (Berry)							
drawn	19903	18547	20869	15.108	14.584	15.472	7.748
annealed	19925	19410	20794	15.108	14.913	15.433	7.757
ruptured		17538			14.179		7.751
Iron wire, as received			18613			14.798	7.553
Cast steel							
drawn	19823	18247	19549	15.108	14.495	15.003	7.717
annealed	19828	18811	19561	15.108	14.716	15.006	7.919
ruptured		16728			13.965		7.710
Steel wire, English							
drawn	19445	20714	18809	14.961	15.442	14.716	7.718
annealed	19200	21070	17278	14.961	14.673	14.193	7.622
ruptured		16728			13.965		7.710
Iron wire, as received							
blue annealed			18045			14.700	

Table 54. WERTHEIM (1842). *Copper: Hammered and drawn; Density = 8.933, Rod length = 450 mm, Radius = 1.4986 mm.*

P Load (kg/mm²)	L' Distance between reference points under tension of P (mm)	L Distance between reference points without load (mm)	a Elastic elongation per meter (mm)	a' Permanent elongation per meter (mm)	E Elasticity coefficient (kg/mm²)
0	—	787.22	—	—	—
3	787.39	787.22	0.216	—	—
4	787.45	787.22	0.292	—	—
5	787.53	787.22	0.393	—	12700
6	787.64	787.22	0.533	—	11249
7	787.65	787.22	0.546	—	12820
8	787.74	787.22	0.660	—	12121
9	787.80	787.22	0.736	—	12218
10	787.86	787.22	0.813	—	12300
11	787.90	787.22	0.863	—	12737
12	787.99	787.26	0.927	0.050	12954
13	788.09	787.28	1.028	0.076	12637
14	788.17	787.27	1.143	0.063	12250
15	788.20	787.27	1.181	0.063	12700
16	788.31	787.30	1.282	0.101	12474
17	788.36	787.30	1.346	0.101	12628
18	788.44	787.31	1.435	0.114	12543
19	788.49	787.31	1.498	0.114	12679
20	788.55	787.32	1.562	0.127	12804
21	788.62	787.32	1.651	0.127	12720
22	788.69	787.32	1.740	0.127	12644
23	788.80	787.34	1.858	0.152	12384
24	788.90	787.40	1.905	0.228	12599
25	788.95	787.40	1.968	0.228	12700
26	789.06	787.43	2.070	0.267	12560
27	789.15	787.43	2.148	0.267	12360
28	789.28	787.50	2.260	0.355	12386
29	789.43	787.62	2.298	0.508	12616
30	789.53	787.64	2.400	0.533	12499
31	789.68	787.71	2.502	0.622	12390
32	789.84	787.77	2.627	0.698	12182
33	789.97	787.83	2.715	0.774	12151
34	790.13	787.94	2.779	0.914	12234
35	790.33	787.96	3.007	0.939	11637
36	790.50	788.20	2.619	1.244	12334
38	791.22	788.83	3.033	2.043	12529
40	792.00	789.28	3.452	2.614	11588

Chemical composition: 98.82% Cu; 0.38% Ag; 0.80% Fe.
$E = 12513$ kg/mm² (7.675 transverse vibrations per second).
$E = 12536$ kg/mm² (1855 longitudinal vibrations per second).

ment which, if made more precise, would reveal microplastic strain at whatever strain level one was considering.[1]

WERTHEIM's quasi-static apparatus contained a device which allowed him to place the loads on the specimen in a very gradual manner "without the slightest jolt". The apparatus was such that it could be surrounded by three united casings,

[1] This is a point of view adopted by the most careful experimentists but very strongly objected to by a number of minor theorists and lesser experimentists of the 19th century. Modern microplasticity indeed confirms the conjecture of HODGKINSON, WERTHEIM, and BAUSCHINGER.

two inner ones of copper and an outer one of tin. Sand was poured between the two copper casings. An oven produced heat to bring the temperature of the apparatus in the inner section to a specified value determined by thermometers located along the specimen. Elongation moduli were determined for the metals considered at 100 and 200 °C. The apparatus then was modified to contain a mixture of crushed ice and sulphuric acid, so that similar measurements could be made from −15 to −20 °C. Since he did not believe that reliable measurements of dynamic moduli could be made under these conditions, his comparison of elastic coefficients and velocity ratios of the velocities of sound in the metal to those in air at the four temperatures, shown in Table 55, was based upon the quasi-static elongation measurements. This was the first study of the dependence of elastic constants upon temperature.

Table 55. WERTHEIM (1842).

Annealed metals	Density	Elasticity coefficients (kg/mm²)	Sound velocities (ratios to those in air)	Density (calculated)	Elasticity coefficients (kg/mm²)	Sound velocities (ratio to those in air)	Density (calculated)	Elasticity coefficients (kg/mm²)	Sound velocities (ratios to those in air)
	at 10 to 15° C			at 100° C			at 200° C		
Lead	11.232	1727	3.697	11.075	1630	3.616			
Gold	18.035	5584	5.245	17.953	5408	5.174	17.873	5482	5.221
Silver	10.304	7140	7.847	10.245	7274	7.943	10.187	6374	7.456
Copper	8.936	10519	10.703	8.891	9827	9.910	8.840	7862	8.890
Platinum	21.083	15518	8.087	21.027	14178	7.740	20.969	12964	7.412
Iron	7.757	20794	15.433	7.729	21877	15.859	7.696	17700	14.295
Iron wire, as received	7.553	18613	14.798	7.543	19995	15.347			
Cast steel	7.719	19561	15.006	7.694	19014	14.819	7.669	17926	14.412
Steel wire, English	7.622	17278	14.193	7.597	21292	15.781	7.573	19278	15.040
Steel wire, blue annealed	7.420	18045	14.700	7.410	18977	15.085			
Unannealed metals	at + 10° C			at − 15° C					
Gold	18.889	8603	6.362	18.896	9351	6.631			
Silver	10.458	7411	7.935	16.463	7800	8.139			
Palladium	10.661	10289	9.261	10.664	10659	9.424			
Platinum	20.513	15647	8.233	20.518	16224	8.382			
Copper	8.906	12200	11.042	8.910	13052	11.399			
Iron wire, as received	7.553	18613	14.798	7.555	17743	14.446			
Steel, blue annealed	7.420	18045	14.700	7.422	17690	14.553			
Brass	8.247	9005	9.744	8.431	9782	10.151			

In his analysis of this large body of results WERTHEIM was the first to study systematically the small deformation of metals.[2] No one before had considered, and certainly no one previously had compared, metals in a defined, relatively pure state. WERTHEIM showed that the elasticity coefficient decreased when

[2] WERTHEIM [1844, *1*], [1845, *1*], [1850, *2*]; and see BAUDRIMONT [1850, *1*].

temperatures rose from -15 to 200 °C for all metals except steel and iron. For iron, in going from -15 to 200 °C, the modulus of elasticity rose to a maximum somewhere between 100 and 200 °C, at which latter temperature it fell below the value of the modulus at 100 °C. He observed that dynamically determined moduli were systematically larger than the averages obtained from the quasi-static determinations of elongation. WERTHEIM attributed this to the differences between what we should term today the adiabatic and isothermal situations. In attempting to calculate a ratio of specific heats from these data, he used a relation proposed by DUHAMEL:

$$K = 1.8\frac{v'^2}{v^2} - 0.8 \tag{3.2}$$

where K is the ratio of specific heats; v', the velocity determined from longitudinal vibrations; and v, that calculated from the quasi-static measurements.

As RUDOLPH JULIUS EMMANUEL CLAUSIUS[1] pointed out in a forceful if partially incorrect attack on both WERTHEIM and WEBER, the dynamic velocity in DUHAMEL's development was the dilatational wave speed in an infinite medium, a speed which would be considerably higher than that determined from the longitudinal vibrations in a rod. CLAUSIUS attempted to dismiss WEBER's thermal measurements (discussed in a previous chapter)[2] and his determinations of specific heats from them, on the basis not of the theoretical limitations and approximations of the thermodynamic analysis involved, but on the assumption that WEBER had ignored the elastic after-effect, which CLAUSIUS assumed must be present in metals as it was in silk. Recalculating WERTHEIM's ratios on the basis of the bar velocity, CLAUSIUS obtained ratios of specific heats which, as he observed, were untenable. He concluded hence that WERTHEIM, too, must have ignored the elastic after-effect in metals. In a strongly worded reply to this proposal that the elastic after-effect could account for the differences between dynamic and quasi-static measurements by WEBER and WERTHEIM, WERTHEIM in his last memoir in 1860 disposed of CLAUSIUS's conjecture that WEBER's elastic after-effect was the explanation of the difference.[3]

Neither the dilatation nor the shear wave speeds predicted by the linear theory of elasticity for the infinite solid could be measured in any solid in 1842, or for that matter, for many decades thereafter. Not only WERTHEIM but also many other persons[4] were confused as to how longitudinal and transverse wave speeds in bars were related to dilatational and shear wave speeds from the linear theory. Efforts were made to resort to analogy by comparing measured wave speeds of fluids in tubes with those in large bodies of water, and experimental studies were made of longitudinal wave speeds and transverse wave speeds in large plates in which LAMB waves (the extensional wave in the plane of the thin plate) mistakenly were assumed to propagate with the dilatational wave speed of the linear theory.

Thus WERTHEIM's use of a longitudinal bar wave speed in DUHAMEL's Eq. (3.2), while erroneous, is best viewed in the historical light of the problem of the 1840's. For obvious reasons, neither WERTHEIM's data nor CLAUSIUS's correction is tabulated here. CLAUSIUS's attack on WEBER's experiments is simply incorrect. The experimental source of the defect in WERTHEIM's comparison of dynamic and quasi-static moduli arises from the fact, first observed by COULOMB in 1784,

[1] CLAUSIUS [1849, *1*].
[2] See Chap. II, Sect. 2.12.
[3] WERTHEIM [1860, *1*]. See also [1852, *3*].
[4] Included, for example, were CAUCHY, DUHAMEL, and the major experimentist, REGNAULT.

that the magnitude of the modulus decreases with increasing permanent deformation; hence, an average of quasi-static measurements at many different permanent deformation states for relatively large strain is lower than the dynamic modulus which for both the longitudinal and flexural situations are measured only at exceedingly small strain. WERTHEIM's dynamic measurements were for strain amplitudes which were always below the minimum quasi-static strain observed. GRÜNEISEN, in the first decade of the 20th century, examined this question of the adiabatic vs isothermal moduli in the same range of strain of $\varepsilon = 10^{-6}$ for both dynamic and quasi-static situations and showed that in the metals WERTHEIM had studied, the difference was exceedingly small, i. e., in the fourth decimal of the E-moduli.[1]

With respect to the measurement of the elastic limit, WERTHEIM noted that by introducing the arbitrary assumption that it occurs at the weight that causes a permanent strain of 0.00005, "one could find as many small permanent elongations as the instrument is capable of measuring." He stated that he had performed very slow experiments in silver, copper, gold, and platinum, never changing the weight abruptly by a large amount. From this he concluded that the permanent elongations were an unknown function of the time during which the weight operated. A weight which after a time caused no measurable permanent elongation, did so when its effect was continued for a longer time.

> It is not improbable that the same thing occurs for elongations smaller than 0.00001 of the unit length, although our instruments cannot measure them. It can be said therefore that the numbers expressing the limits of elasticity must diminish as the measurement instruments are perfected and as the weights are allowed to act for longer times.[2]

WERTHEIM's results revealed that although longitudinal and transverse vibrations led to nearly the same elasticity coefficient, the value obtained from the former was almost invariably slightly higher than that from the latter. A small systematic variation of that sort can be expected for a transverse modulus computed without including rotary inertia or the deflection due to the effects of shear and lateral contraction. None of these three aspects of response to dynamic flexure was seen to require some correction to the elementary theory[3] until long after 1842. The errors, of course, are contained in all values of E calculated from dynamic flexure, from those of THOMAS YOUNG in 1807 to EDUARD GRÜNEISEN's in 1907; E. GOENS[4] in 1931 was the first to take account of both rotary inertia and deflection due to shear in such experiments.

Another aspect of elastic constants which had been of some interest before 1840 was the possible dependence of the elasticity of the solid upon small variations in density. Except in wood, in which such a variation had been observed, experiments had not been performed with sufficient accuracy to settle the question. WERTHEIM demonstrated that in metals, too, the elasticity did vary a little with the density. He showed that GERSTNER's law, described above, was applicable

---

[1] Many writers of contemporary textbooks on ultrasonics generally have assumed, without experimental demonstration, that their considerably larger differences between dynamic and quasi-static measurements may be attributed solely to expected differences between adiabatic and isothermal deformation; as we have seen in Chap. II, amplitudes would have to be specified carefully in order to make a proper comparison.

[2] WERTHEIM [1844, *1* (c)], p. 439. (In the German translation of this article, in 1848, the quotation is on p. 57; see 1844, *1* (c), note.)

[3] See recent papers by HARDIE and PARKINS [1968, *1*] and by HART [1968, *1*], where a detailed comparison of E-moduli, including all known corrections, demonstrates that the slight experimental difference observed by WERTHEIM, using his elementary calculations, is indeed to be expected.

[4] GOENS [1931, *1*].

in all of the metals in which permanent deformation was occurring. He also noted in one of his conclusions that:

> The permanent elongations do not occur by jumps or by jerks, but in a continuous manner; by appropriately modifying the load and the duration of its action any desired permanent elongation could be produced.[1]

In other words, WERTHEIM had concluded that the serrated stress-strain curve commented upon by MASSON[2] and known today (inaccurately, with reference to precedence) as the "PORTEVIN-LE CHATELIER effect", can be eliminated in dead-weight loading if the load increments are sufficiently large. This fact is well known to those currently studying this phenomenon by means of dead-weight loading.

One further observation of WERTHEIM'S, which was the origin of a considerable amount of study in later years by TOMLINSON, VOIGT, and others,[3] was that E for metals decreased as the atomic volume increased. WERTHEIM noted that the product of the modulus and the seventh power of the atomic distance was nearly constant. Table 56 lists the specific weights, S; the weight of a single atom, A; the log of the atomic distance, α; the elasticity modulus, E; the log of the product, $E\alpha^7$; and finally, the log of the average value at room temperature for each of the metals he considered. From the experimental point of view, the finding of a relation between an elastic constant and a parameter for a crystal lattice is a historical landmark in solid state physics.[4]

WERTHEIM made similar comparisons of the relation log $E\alpha^7$ for the different metals at 100° C, noting that since the elasticity decreased with increased temperature at a higher rate than would occur with elongation itself, there must be some effect. He realized that for these data or for those at the lower temperature, the temperature difference was insufficient for a definitive study. It is interesting to note that in his memoir of 1842 WERTHEIM emphasized the desirability of the experimental study of elastic constants in anisotropic solids; he noted the difficulties which would be entailed in obtaining specimens. This program, which was carried out by VOIGT approximately half a century later in numerous crystals, was pursued by WERTHEIM only in wood, as will be seen.

3.13. Wertheim's memoir of 1843: The first experiments on binary and tertiary alloys including, for 64 combinations, the influence upon *E* of composition and rate of loading.

The second section of the memoir presented to the Academy in 1843 provided, as the Academy Commission emphasized, the first study of the deformation of binary and tertiary alloys. As WERTHEIM indicated in his introduction to this section of his paper, brass and bell metal were the only alloys whose behavior in deformation had ever been studied, beyond, of course, the usual determination of cohesion or rupture strength. For each of the 64 alloys, which were in every instance prepared by WERTHEIM himself, a check was made of the percentage composition by an analysis of samples from both ends of the wire specimens. WERTHEIM wrote that he had thrown away a large number of rods which exhibited a difference in composition at the two ends. The molten mixture was poured

[1] WERTHEIM [1844, *1*(a)], p. 452.

[2] MASSON [1841, *1*].

[3] See, for example, JAMES REDDICK PARTINGTON, who cited WERTHEIM'S study as the first of its kind [1952, *1*], p. 196.

[4] MASSON [1841, *1*] a year earlier had suggested that for iron, copper, zinc, tin, and silver, the product of E and the atomic weight was a constant.

Table 56. WERTHEIM (1842).

	S	A	log α	E (kg/mm²)	log $E\alpha^7$	Log of the average value of $E\alpha^7$
Lead						
cast	11.215	1294.498	0.684733	1775	8.04233	
drawn	11.169		0.688030	1803	8.07221	8.05469
annealed	11.232		0.687213	1727	8.04778	
ruptured	11.038		0.686240	1788	8.05605	
Tin						
pure	7.313	735.294	0.667453	3839	8.25637	
annealed	7.290		0.667910	3703	8.24397	
Tin, as received						
cast	7.404		0.665663	4172	8.27998	
drawn	7.342		0.666883	4148	8.28602	8.26715
ruptured	7.293		0.667850	3918	8.26801	
Cadmium						
drawn	8.665	696.770	0.635063	5424	8.17976	
annealed	8.520		0.637507	5313	8.18789	8.14944
ruptured	8.541		0.637150	4084	8.07114	
Gold						
drawn	18.514	1243.013	0.608990	8131	8.17307	
annealed	18.035		0.612783	5584	8.03643	8.07533
ruptured	19.077		0.604320	5833	7.99613	
Silver						
drawn	10.369	675.803	0.600783	7357	8.07218	
annealed	10.304		0.601693	7140	8.06555	8.07737
ruptured	10.320		0.601470	7648	8.09384	
Zinc						
pure, cast	7.146	403.226	0.583830	9021	8.04206	
ordinary, drawn	7.008		0.586653	8734	8.04778	8.06002
annealed	7.060		0.586880	9641	8.09228	
Palladium						
drawn	11.539	665.900	0.589360	11759	8.19589	8.16337
annealed	11.225		0.591070	9789	8.12823	
Platinum						
drawn	21.259	1233.499	0.587867	15814	8.31411	
annealed	21.207		0.588220	15683	8.31297	
drawn	21.275		0.587757	17044	8.34587	
annealed	21.083		0.589070	15518	8.31433	8.31789
drawn	21.166		0.588500	15928	8.32167	
annealed	20.753		0.591353	14373	8.29702	
ruptured	20.987		0.589730	16748	8.35207	
Copper						
drawn	8.933	395.695	0.548790	12450	7.93670	
annealed	8.936		0.548743	10519	7.86318	7.91020
ruptured	8.890		0.549490	12040	7.92706	
Iron						
drawn	7.748	339.205	0.547090	20869	8.14913	
annealed	7.757		0.546923	20794	8.14639	
ruptured	7.751		0.547035	17538	8.07322	
Iron wire						
ordinary	7.553		0.550780	18613	8.12528	
Cast steel						
drawn	7.717		0.547670	19549	8.12482	8.11330
annealed	7.719		0.547633	19561	8.12482	

Table 56 (Continued).

	S	A	$\log\alpha$	E (kg/mm²)	$\log E\alpha^7$	Log of the average value of $E\alpha^7$
Steel wire						
drawn	7.718		0.547653	18809	8.10794	
annealed	7.622		0.549463	17278	8.08373	
ruptured	7.710		0.547805	16728	8.05808	
Steel wire						
blue annealed	7.420		0.553353	18045	8.12981	

into a 50 cm long cast iron mold, and the specimens were then drawn, before the analysis was made. Dynamic moduli were determined for each alloy, in 20 instances by means of the longitudinal vibration method, and in 45, by the flexural oscillation method. In 8 of the alloys the moduli were determined in quasi-static tensile elongation in the way described above for the pure metals. These data are shown in Table 57.

In analyzing the data WERTHEIM noted that, in general, if the atoms were evenly distributed, the moduli increased with decreasing atomic distance. He observed that as had been the situation for the pure metals (described above), elastic limits and ultimate strengths did not lend themselves to a representation in a generalized pattern. In particular, for the alloys he concluded that neither the elastic limit, nor maximum elongation, nor cohesion could be calculated for a binary alloy *a priori* from the values for the two pure solids and a knowledge of the chemical composition. On the other hand, he found that E varied linearly

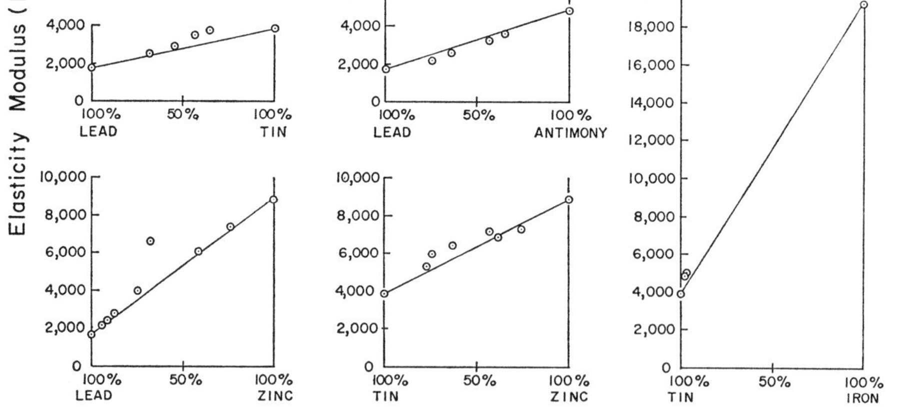

Fig. 3.27a and b. WERTHEIM's E moduli measurements with various percentage composition for the indicated alloys. These were the first measurements of the elasticity of metal alloys.

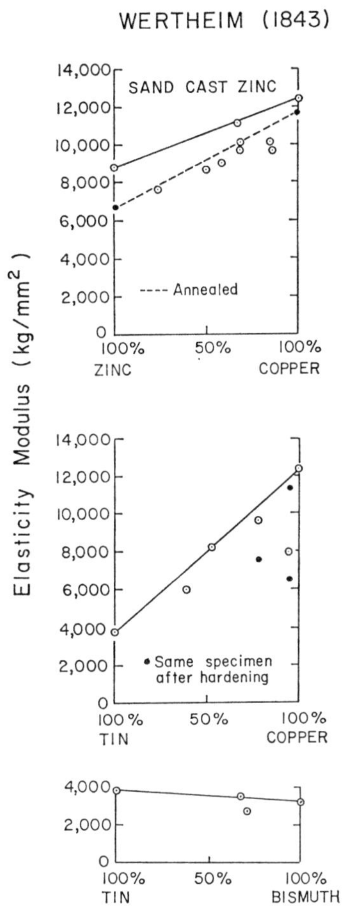

Fig. 3.27b

with the percentage composition. The end points were determined from the measured values of E for the pure metals.[1] The linear relation may be seen in Fig. 3.27a and b for those alloys for which a sufficient number of composition percentages were tested.

WERTHEIM had calculated the expected moduli on this linear basis,[2] to compare with the experimental values. If one averages all of the 59 values for which such a calculation was made for both binary and tertiary alloys, one obtains a number which differs by less than 2% from that obtained by a similar average of all of the corresponding experimental observations. All of this, of course, is without reference to phase diagrams, variations in melting points, etc.

In this first comparative study of the behavior of 64 alloys and 16 pure metals, many of WERTHEIM's contemporaries thought that his observation of such a linear relation between elasticity and composition was of enormous importance in understanding the structure of solids. While WERTHEIM had shown for pure

[1] WERNER KÖSTER, one hundred and five years later, provided an interesting extended study of this matter which will be considered below. (*See*, KÖSTER and RAUSCHER [1948, *1*].)

[2] See Table 57.

Table 57. Wertheim (1843)

Metal mixture	Composition	Density: Calculated	Density: Found	Measured percentage composition		Elasticity Coefficients (kg/mm²): Calculated	Found by vibrations: Longitudinal	Found by vibrations: Transverse	Elongations	Sound velocity according to (ratio to vel. in air): Vibrations	Sound velocity: Elongations	Elastic limit	Elongation maximum (mm)	Cohesion (kg/mm²)
Lead, tin	$Pb_{16}Sn_{13}$	9.919	10.073	68.50	31.50	2430	2596			4.786		<0.2	0.552	0.93
	PbSn[a]	9.801	9.408	64.39	35.61	2514	2969			5.295		<0.2	2.077	2.46
	Pb_3Sn_7	8.996	8.750	43.42	56.58	2946	3512			5.973		<0.1	1.591	2.07
	Pb_2Sn_7	8.584	8.378	32.98	67.02	3162	3700			6.265		<0.2	0.340	1.07
Lead, bismuth	$Pb_{12}Bi_7$	10.692	11.037	62.44	37.56	2322	2021			4.034		<0.2	0.262	1.52
	PbBi	10.498	10.790	48.60	51.40	2446	2367			4.415		0.1	0.440	1.79
	PbBi	10.316	10.403	35.45	64.55	2564		2838		4.923			0.025	5.22
Lead, antimony	Pb_2Sb[b]	10.059	10.101	74.27	25.73	2489		2183		4.382				1.87
	PbSb	9.592	10.064	64.51	35.49	2693		2592		4.784				5.59
	$PbSb_2$	8.636	8.946	43.63	56.37	3109		3242		5.674				
	$PbSb_3$	8.241	8.499	34.99	65.01	3282		3536		6.080				
Lead, gold	$Pb_{20}Au$	11.540	11.301	95.56	4.44	2281		2227		4.185			0.055	
Lead, silver	Pb_3Ag_2	10.958	10.743	73.64	26.36	3373		3095		5.060				
Lead, platinum	$Pb_{85}Pt$	11.325	11.473	98.88	1.12	2134	2684			4.560		0.4; 0.6	0.026	1.65
	Pb_8Pt	11.930	12.207	87.08	12.92	3713		3107		4.756				
Lead, zinc	Pb_6Zc	11.013	11.199	95.15	4.85	2356		2144		4.125			0.069	2.75
	Pb_4Zc	10.890	11.172	91.91	8.09	2548	2493			4.454		0.1	0.060	2.02
	Pb_2Zc	10.705	11.130	87.90	12.16	2902	2833			4.756		0.2	0.060	2.02
	PbZc	10.165	9.430	74.72	25.28	3921		4007		6.145				3.47
	Pb_2Zc_3	9.880	9.043	67.88	32.12	4444		6647		8.081				3.40
	$PbZc_5$	8.758	8.397	40.88	59.12	6511		6108		8.039				
	$PbZc_{10}$	7.970	7.910	24.31	75.69	7880		7352		9.088			0.004	4.40
Lead, copper	Pb_5Cu	11.083	11.165	94.21	5.79	2555		2113		4.100			0.043	2.13
Tin, bismuth	Sn_9Bi_{10}	8.980	8.685	33.23	66.77	3155		3610		6.078			0.028	8.19
	Sn_3Bi_4	9.064	8.890	29.91	70.09	3127		2874		5.360			0.015	6.63
Tin, antimony	Sn_4Sb	7.155	7.211	20.75	79.25	3761		4033		7.050				8.86
	Sn_8Sb_5	7.027	7.052	59.52	40.48	3815		4695		7.692			0.010	7.82
	Sn_3Sb_2	7.008	7.007	56.60	43.40	3823		5168		8.096				
Tin, zinc	Sn_2Zc	7.243	7.366	77.11	22.89	4974	5336			8.023		0.2	0.246	5.78
	Sn_3Zc_2	7.234	7.255	74.31	25.69	5113	5982			8.560		<0.2	0.252	5.00

	SnZc	7.201	7.143	63.40	36.60		5653	6453			8.959		0.2; 0.5	0.036	4.68
	Sn_3Zc_7	7.141	7.193	43.51	56.49		6639		7113		9.376		0.3; 0.5	0.124	2.44
	$SnZc_3$	7.125	6.746	38.47	61.53		6888		6976		9.525		0.2	0.082	4.32
	$SnZc_5$	7.087	6.957	26.08	73.92		7502	7314			9.665		0.4	0.023	7.52
Tin, platinum	$Sn_{50}Pt$	7.724	7.578	96.77	3.23		4090		5309		7.890				4.75
Tin, copper	Sn_5Cu_6	7.931	8.332	61.07	38.93		6868		6113		8.074				
	$SnCu_2$	8.148	8.531	47.90	52.10		7939		8280		9.287				
tamtam-metal	$SnCu_7$	8.580	8.813	21.62	78.38		10082		9784		9.932				
ditto, hardened	ditto		8.686	21.62	78.38				7716		8.884				
Bronze	$SnCu_{22}$	8.806	8.738	7.87	92.13		11192		8115?		9.085?				
ditto, hardened	ditto		8.537	7.87	92.13				6734?		8.372?				
Tin, iron	$Sn_{20}Fe$	7.300	7.266	97.78	2.22		4052		4881		7.726				
	$Sn_{20}Fe$	7.309	7.418	96.03	3.97		4327	5001			7.740				2.40
Silver, palladium[c]	Ag_2Pd_3	10.957	10.903	38.10	61.90		10542		10003		9.029	9.222	20.23	0.004	50.46
Silver, copper	Ag_5Cu[d]	10.225	10.121	90.00	10.00		8290		8913		8.846				44.05
	AgCu	9.819	9.603	61.72	38.28		9617		8590		8.915	8.831	21.25	0.002	51.97
Gold, platinum	Au_7Pt_2	19.122	19.650	77.87	22.13		10231		9844		6.848				7.12
Gold, iron	$Au_{19}Fe_2$	18.215	18.842	97.22	2.78		8920		9024		6.524			0.016	20.41
Zinc, copper	$Zc_{13}Cu_4$	7.495	7.301	76.80	23.20		10149		7678		9.667				4.10
	ZcCu	7.997	8.265	50.07	49.93		10749		8774		9.713				18.68
	Zc_3Cu_4	8.129	8.310	42.99	57.01		10906		9105		9.867				36.80
Pure brass	$ZcCu_2$	8.313	8.606	33.20	66.80		11123		10163	8543	10.244	9.392	15		60.22
commercial tomback	ditto	8.311	8.655	33.35	66.65		11120		11218	9277	10.732		16.20	0.140	32.5
Berlin brass	Zc_8Cu_{17}	8.329	8.427	32.35	67.65		11142		9823	9394	10.177	9.890			
	Zc_2Cu_{11}	8.640	8.670	15.79	84.21		11509	10290		8644	10.270	9.812	30–40	0.003	48.3
Type of similor	$ZcCu_6$	8.663	8.636	14.96	85.04		11538	9778			10.031	9.431	40	0.001	51.9
Lead, bismuth, tin	$Pb_2Bi_3Sn_2$	9.870	9.795	35.80	46.92	17.28	2653	2626			4.881	4.881	<0.2	0.695	1.74
Lead, antimony, tin	$Pb_4Sb_2Sn_5$	8.977	9.196	49.36	15.05	35.59	2873		2735		5.141			0.032	7.80
	$PbSbSn_3$	8.433	8.317	32.06	17.82	50.12	3202		3232		5.876				5.62
Lead, tin, zinc	$Pb_{14}Sn_4Zc_9$	10.109	10.212	73.63	11.96	14.41	3037	2486			4.651		<0.2	0.162	1.44
Tin, antimony, copper	$Sn_4Sb_2Cu_3$	7.458	7.751	50.57	28.16	21.27	4992		5770		8.133				4.17
Zinc, copper, nickel	$Zc_9Cu_{15}Ni_2$	8.256	8.403	33.96	57.24	8.80		9617		9261	10.032	9.896	20–30	0.001	
art nickel-silver	$Zc_4Cu_{18}Ni_5$	8.525	8.541	18.24	60.54	21.22		10227		10788	10.315	11.594	19–24	0.001	61.88
very pliable packfong	$Zc_5Cu_6Ni_3$	8.170	8.436	36.51	43.00	20.49		11722		11500	11.112	11.006	30–40	0.001	55.0
commercial packfong	$Zc_2Cu_5Ni_3$	8.463	8.615	20.39	50.58	29.03			12250	10333	11.240	10.324	45	0.002	68.1

[a] Soft solder. [b] Type metal. [c] Dental material. [d] French coined silver.

Table 58. WERTHEIM (1844)

	s	F	P	L_1	p	l_1	α	t	E_1	E_2	E_3	$\frac{E_2 - E_3}{E_2}$	$\frac{R}{s}$
Silver	0.154	8.03	13	707.57	3	706.38	1.685	14.8		6539	5936	0.0922	6773
		6.69		707.36		706.19	1.657				6036	0.0769	
		4.55		707.35		706.26	1.543	4.2			6479	0.0092	
		3.96		707.34		706.26	1.529	4.5			6539	0.0000	
		0.00		707.28		706.20	1.529	0.0	6539	7248			
Silver	0.119	0.00	13	776.28	3	775.23	1.354	0.0	7383				
		10.17		777.12		775.95	1.508	38.9			6632	0.0850	8765
		8.50		776.78		775.63	1.483	17.7			6744	0.0738	
		7.26		776.54		775.48	1.367	9.4			7316	−0.0047	
		0.00		776.44		775.35	1.406	0.0	7113				
Copper	0.887	0.00	26	851.45	5.3	850.04	1.659	0.0	12479				1750
		10.30		851.51		850.11	1.647	5.4		12423	12570		
Copper	0.364	10.04	27.5	867.03	5.5	865.54	1.721	2.7		13628	12780	0.0622	4264
		8.50		866.97		865.49	1.710	0.0		13698	12865	0.0608	
		8.03		866.98		865.52	1.687	1.9		13677	13042	0.0464	
		0.00		866.86		865.46	1.606	0.0	13698				
		0.00		866.87		865.49							
Copper	0.241	13.78	21	812.82	5	811.35	1.565	15.7		11225	10222	0.0894	6426
		9.79		812.45		811.29	1.430	13.8		11255	11190	0.0058	
		8.50		812.41		811.22	1.467	8.2		11307	10907	0.0354	
		7.54		812.25		811.12	1.393	2.4		11361	11485	0.0109	
		0.00		812.23		811.09	1.405	0.0	11384				
Gold	0.0276	3.96	14	538.85	4	538.00	1.580	142.0		6985	6329	0.0939	63260
		3.64		538.81		537.95	1.599	135.5		7391	7255	0.1537	
		2.45		538.71		537.96	1.394	137.9		7383	7173	0.0284	
		1.60		538.67		537.90	1.431	130.1		7408	6986	0.0570	
		0.00		538.56		537.87	1.280	0.0	7810				
Brass	0.143	13.78	30	712.48	10	710.64	2.589	179.5		9193	7724	0.1598	39175
		10.30		712.33		710.61	2.420	181.8		9155	8263	0.0974	
		10.04		712.02		710.38	2.309	167.4		9284	8663	0.0669	
		8.03		711.04		709.48	2.199	102.3		9861	9096	0.0776	
		5.00		710.42		708.95	2.074	27.7		10522	9645	0.0834	
		3.40		710.10		708.68	2.004	9.2		10686	9981	0.0660	
		2.00		709.98		708.58	1.976	2.3		10123	10747	0.0581	
		0.00		709.81		708.54	1.857	0.0	10767				
		0.00		709.79		708.45							

Iron	0.169	6.69	50	817.09	12	815.33	2.159	91.4		20800	17604	0.1537	52006
		6.31		816.94		815.24	2.085	84.0		20677	18222	0.1187	
		4.70		816.54		814.88	2.037	47.9		20107	18654	0.0723	
		2.50		816.22		814.62	1.964	12.4		19361	19347	0.0007	
		0.00		816.00		814.40	1.965	0.0	19342				
Iron, Swedish	0.0216	2.05	60.5	783.31	14.3	781.31	2.560				18048		
		1.95		790.92		789.03	2.395	137.6		23024	19287	0.1623	406900
		1.90		790.81		788.91	2.408	124.3		22798	19183	0.1585	
		1.75		790.47		788.59	2.384	90.6		22223	19380	0.1279	
		1.25		789.97		788.14	2.322	43.9		21426	19897	0.0714	
		0.60		789.60		787.80	2.285	8.3		20819	20220	0.0288	
		0.00		789.45		787.71	2.234	0.0	20677				
		0.00		789.49		787.71							
Steel	0.1293	5.10		778.32		776.22	2.705	137.3		20355	17187	0.1556	67973
		4.55		777.93		775.85	2.681	93.8		19883	17345	0.1277	
		3.85		777.52		775.57	2.514	65.4		19577	18494	0.0553	
		2.50		777.13		775.21	2.476	23.7		19125	18775	0.0183	
		0.00		776.92		775.01	2.464	0.0	18868				
Steel	0.0502	4.32	60.0	752.96	20.0	751.06	2.530	183.2		21480 ?	15812		171080
		3.96		752.87		751.10	2.356	196.2		21641 ?	16974		
		2.50		751.54		749.81	2.307	39.0		19646	17336	0.1176	
		1.50		751.13		749.53	2.135	12.4		19352	18738	0.0317	
		0.00		750.98		749.41	2.082	0.0	19216				
		0.00		750.96		749.41							
Platinum	0.329	9.60	30	669.32	15	668.51	1.212	105.1		15178	12380	0.1843	30395
		6.97		668.79		668.06	1.093	42.4		16336	13728	0.1596	
		5.69		668.55		667.92	0.943	35.6		16461	15903	0.0339	
		5.00		668.47		667.88	0.883	35.6		16461	16980	—0.0315	
		3.96		668.31		667.74	0.854	15.2		16839	17572	—0.0435	
		0.00		668.26		667.68	0.876	0.0	17120				
		0.00		668.24		667.65							
Platinum	0.0940	6.31	26	748.36	12	747.49	1.164	177.0		13854	12028	0.1318	106383
		5.00		748.01		747.15	1.51	126.8		14781	12163	0.1771	
		4.18		747.39		746.68	0.951	75.0		15738	14723	0.0645	
		3.40		746.99		746.38	0.817	42.5		16339	17113	—0.0474	
		2.30		746.86		746.25		22.8		16702	17127	—0.0255	
		0.00		746.71		746.10	0.818	0.0	17124				

s = the cross-section of the wire in mm²; F = current intensity; P = load/mm²; L_1 and l_1 = the distances of marked positions under load and current; p = load/mm² needed to tighten the wire; α = the thermal coefficient of linear elongation; t = temperature in °C with current; R = electric resistance per unit area.

metals that the modulus varied as the inverse seventh power of the atomic distance, he found that no such relation holds for alloys.

As a result of all this research, WERTHEIM had established a fundamental basis for nearly every succeeding experimental or theoretical study in the mechanics of solids for the next 60 years. With respect to the question raised by COULOMB and incorrectly answered by TREDGOLD, LAGERHJELM, and others, as to the influence of annealing on elastic constants, WERTHEIM had shown that a carefully defined prior thermal history indeed may effect a major change in the value of E. That thermal-mechanical histories, small differences in chemical composition, ambient temperature, and manner of experiment needed to be clearly defined in any rational effort to compare experiment and explanation was first understood by WERTHEIM and provided the main structure for his approach to this branch of physics.

3.14. Wertheim's memoir of 1844: The first study of the dependence of E upon the strength of electric and magnetic fields.

To complete his study WERTHEIM in the third section of this memoir investigated the effect of electric current and magnetic fields upon the elastic modulus and cohesion. The specimens, which were protected from air drafts, were connected with an electric source. He measured the elongation for a fixed load before applying the current, then again while the current was flowing through the specimen, and, finally, after he had shut off the current. In each instance he measured the temperature. He observed for a given specimen as many as 4 or 5 intensities of the current as measured on a galvanometer. For all the many pure metals, WERTHEIM discovered that the elastic modulus always decreased, the magnitude of the reduction increasing with the intensity of the electric current. He heated specimens to various recorded temperatures depending upon their electric resistance and the current intensity applied, both of which he determined. Since he already had ascertained the temperature dependence of the elastic constants in the range of temperatures under consideration, he was able to make a correction for the effects of temperature. In Table 58 are shown WERTHEIM's results for all of the metals for which temperature data were available.

In addition to tabulating E as E_1, the measured modulus at room temperature without an electric current, he tabulated E_2, the calculated modulus based upon the measured temperature change without the electric current, and E_3, the elastic coefficient, with the current flowing at the measured temperature. A difference between E_2 and E_3 thus represents an effect of the electric current upon the modulus, independent of the temperature. WERTHEIM compared the ratios $\frac{E_2 - E_3}{E_2}$, demonstrating that the value of the modulus did depend upon the intensity of the electric current for all of the metals considered. He further noted that the size of the decrease in modulus depended also to some extent on the electrical resistance of the metal. He observed, too, that the cohesion or ultimate strength of the wires definitely decreased with the current, but he was unable to separate the possible influence of the thermal effect upon this change.

As cross-checks upon this conclusion that the elastic modulus was a function of electric current intensity, WERTHEIM performed experiments in vibration. He found that when the current flow was interrupted there was a distinct, noticeable change in the frequency, corresponding to the change in modulus. He performed these experiments on approximately $3^1/_2$ m long wires, tightened on both ends and connected to an electrical source. Longitudinal tones were produced at the center by rubbing, giving the measured change in vibration frequency shown in Table 59.

Table 59. WERTHEIM (1844)

	Diameter (in mm)	Current intensity	No. of longitudinal vibrations
Copper	0.59	0.00	1058
		7.80	1041
Steel	0.31	0.00	1358
		1.50	1326
		2.10	1313
Steel	0.14	0.00	1403
		1.50	1391

For gold and steel, Table 60 shows the variations in the rupture load with an electric current.

Table 60. WERTHEIM (1844)

	Cross-section of the wire (mm^2)	Load required to rupture the wire (kg/mm^2)		Current intensity
		Without current	With current	
Gold	0.0276	12.2	7.6	3.64
Iron wire, Swedish	0.0216	114–118	110.0	1.20
	—	—	99.0	2.00
Iron wire, ordinary	0.169	59.2	60.5	5.00
Steel	0.0508	102.0	102.0	2.50
	—	—	99.5	4.18
Steel	0.1293	89.1	80.6	4.55

WERTHEIM considered it to be of considerable importance that this change in modulus abruptly disappeared when the electric current was removed. No such abrupt change had occurred for magnetized iron. In an experiment on annealed iron in which an intensive current flowed uninterruptedly for six hours, he did not notice the slightest amount of permanent elongation.

Winding a cotton-covered copper wire of more than 1 mm in diameter around two glass tubes 80 cm in length to obtain a double layer of wire with 900 winding convolutions, WERTHEIM produced a magnetic field. His purpose was to study the influence of a magnetic field upon elastic constants. Only in iron and steel was such an effect discernible. Whether the magnetization involved the North pole or the South pole, it was possible to observe in both soft iron and steel a small decrease in the elasticity coefficient. When he removed the magnetic field, WERTHEIM found that the elastic constant did not return completely to the previous value, even after a time as long as twenty-four hours. This change in the elastic constant in the presence of a magnetic field did not occur the moment the field was imposed but appeared gradually over a period of time. WERTHEIM then concluded: "It seems, therefore, that the magnetization does not directly affect the elasticity, but that under its influence a new arrangement of the molecules is produced."[1]

[1] WERTHEIM [1844, *1* (c)], p. 623. (In German translation of 1848, p. 114.) Also [1844, *2*].

3.15. Wertheim's memoirs in 1845–1846 on the elasticity of glass, wood, and human tissue.

WERTHEIM's novel experiments on the effects of electrical currents and magnetic fields upon the deformation properties of metals stimulated a number of experimentists, both on the continent and in England, to perform a variety of dynamic and quasi-static measurements which culminated in 1911 in the investigations on the subject by WALKER.[1] WERTHEIM, continuing his investigation of the response of solids to deformation, during the next two years turned to a study of the dynamic and quasi-static properties of glass and of wood, and in 1846 he was the first to explore definitively the mechanical properties of human tissue.

In the first of these studies, which he carried out with JEAN PIERRE EUGÈNE NAPOLÉON CHEVANDIER,[2] in accordance with his custom WERTHEIM reviewed the history of previous measurement. He noted that JEAN DANIEL COLLADON and JACQUES CHARLES FRANÇOIS STURM[3] had given an elastic coefficient of 10000 kg/mm² for a glass wand of unknown origin; that SAVART's experiments on glass had provided ratios of the velocity of sound in glass to that in air from 15.39 to 16.28; while CHLADNI had obtained a value as high as 16.6. He might have added that YOUNG had provided a height of the modulus from which one might calculate an E of 7390 kg/mm². WERTHEIM also commented that none of the authors he cited had specified the nature of the glass used, or details concerning its preparation (whether the rods had been drawn or poured; whether the specimen had been annealed; etc.); nor had any of the authors indicated the degree of precision with which they had done the experiment. Hence, concluded WERTHEIM, no definite meaning could be found in their work.

To investigate the influence of annealing on E, WERTHEIM and CHEVANDIER had rods of window glass and mirror glass drawn to lengths of 10 to 12 m, from which they selected the most regular pieces of 1 to 2 m in length. In ten unannealed specimens of window glass they found velocities of sound whose ratios to that in air varied from 14.75 to 17.19. Six similar rods of mirror glass gave ratios from 14.04 to 17.41. By means of "machines of great precision which were at our disposal," WERTHEIM and CHEVANDIER produced a perfectly regular set of square glass specimens which then were carefully annealed. From these they obtained ratios in the now much-narrowed range of 16.48 to 16.76 for window glass, and from 15.70 to 16.02 for mirror glass. It thus became possible to perform a systematic, reproducible study of the dynamic and quasi-static moduli for glass. WERTHEIM and CHEVANDIER also measured the density of the glass before and after annealing, and its cohesion (rupture strength). They obtained average coefficients for mirror glass, common goblet glass, and crystal. They determined dynamic moduli by means of a longitudinal vibration, and the quasi-static moduli by means of a tensile test. They extended their study to the consideration of the effects upon the moduli and cohesion of various metallic additives, which produced colored glass. Table 61 compiles these results.

WERTHEIM and CHEVANDIER found that a very slight increase in density resulted from annealing, as well as a slight increase in E. As there had been some question regarding the vibrational characteristics of glass as a function of the length of the specimen, they noted that in careful experiments the frequency of longitudinal vibration was exactly in inverse ratio to the lengths. They found

[1] WALKER [1907, *1*], [1908, *1*], [1911, *1*].

[2] WERTHEIM and CHEVANDIER [1845, *2*], [1847, *3*] (German translation, 1848).

[3] COLLADON and STURM [1838, *1*]; presented to the Academy in 1827.

Table 61. WERTHEIM and CHEVANDIER (1845).

	Rod length (m)	Cross-section (mm^2)	Density		Number of double vibrations per sec	Velocity of sound, that in air being taken as 1	Elasticity coefficient deduced from vibrations (kg/mm^2)	Resistance to rupture in tension (kg/mm^2)
			Before anneal	After anneal				
Window glass	1.668	78.31			1651.6	16.584	7810	1.847
	1.566	27.07	2.517	2.523	1777.7	16.759	7975	1.680
	1.000	50.00			2782.6	16750	7967	—
Mirror glass								
drawn	0.775	95.67			3422.5	15.967	7081	—
poured	2.991	165.07			882.8	15.894	7008	1.424
poured	2.495	123.00	2.454	2.467	1066.7	16.020	7120	1.327
poured	1.662	165.07			1590.0	15.908	7020	1.448
poured	1.325	165.07			1969.2	15.707	6844	—
Goblet glass, sodium base								
common	2.261	123.00	2.447	2.448	1158.4	15.766	6848	0.831
common	2.437	122.95	2.455	2.455	1097.6	16.104	7165	0.935
fine	1.408	79.83	2.430	2.435	1868.6	15.837	6873	1.026
fine	1.327	85.54	2.432	2.437	1984.5	15.852	6892	1.110
Goblet glass, potassium base								
common	1.481	93.72	2.453	2.459	1753.4	15.632	6763	0.774
common	1.422	96.61	2.453	2.460	1855.1	15.879	6980	0.983
common	1.762	90.08	2.452	2.454	1488.4	15.787	6883	1.261
fine	1.001	96.61	2.453	2.460	2585.8	15.581	6720	—
Goblet glass								
opaline	1.087	134.49	2.513	2.525	2285.7	14.956	6356	too brittle to test in extension
opaline	0.922	133.73	2.513	2.525	2694.7	14.956	6356	
violet	1.283	102.00	2.556	2.559	2031.8	15.692	7080	0.848
violet	1.072	—	2.556	2.561	2461.6	15.884	7275	—
Crystal								
white	1.870	182.35	3.326	3.330	1084.8	12.211	5588	0.741
violet	1.383	176.70	3.241	3.250	1467.9	12.220	5462	0.652
blue	2.135	204.06	3.372	3.374	934.3	12.008	5475	0.652
blue	1.074	—	3.361	3.357	1839.1	11.890	5341	—
green	1.284	199.92	3.311	3.315	1580.2	12.214	5565	0.617
green	0.951	—	3.311	3.321	2105.3	12.052	5429	—

that for any individual glass, the elastic constants determined from longitudinal vibrations were invariably greater than those determined from elongation. This had been true to a lesser degree in the metals and was later found to be the case for wood. That this effect is considerable in glass may be seen from their data in Table 62.

WERTHEIM and CHEVANDIER found that density made no difference in the deformation of poured or drawn glass, if, in each instance the material had been annealed. Although the addition of lead markedly decreased both the moduli and the cohesion, addition of manganese which tinted ordinary glass to violet, provided an increase. The addition of cobalt, copper, and manganese to crystal had no discernible effect upon the elastic properties.

The work we have just described, which was presented to the Academy on June 2, 1845, was immediately followed by what must be listed among the most

Table 62. WERTHEIM and CHEVANDIER (1845)

Kind of glass	Elasticity coefficient (kg/mm²) from vibrations	Elasticity coefficient (kg/mm²) from elongation
Mirror glass	6844	6183
Common goblet glass, sodium base	7165	6722
Fine goblet glass	6892	6040
Violet goblet glass	7080	5000

exhaustive programs of experimentation in the field of solid mechanics. In an 136-page manuscript[1] containing 35 large tables and 6 sheets of drawings, which was published as an 135-page brochure,[2] WERTHEIM and CHEVANDIER provided the first definitive study of the elastic anisotropy of wood.[3]

The program began with the felling of 94 trees from the west slope of the Vosges, grown on Vosges sandstone, colored sandstone, and shell limestone. The woods considered were oak, beech, fir, pine, hornbeam, birch, acacia, elm, ash, sycamore, maple, aspen, poplar, and alder. The memoir began with a summary of the results of 23 investigators of the 18th and 19th centuries who had experimented upon wood, and with the observation that despite the large literature on the subject, the response of wood to deformation was still uncertain. The details of this study, even as presented in the *Comptes Rendus* abstract,[4] are far too numerous to be considered here. WERTHEIM and CHEVANDIER set out to examine the differences in elasticity and cohesion of specimens selected from various parts of the same tree under different conditions of humidity in comparison with the characteristics of entire trees. They systematically studied moduli and velocities of sound determined from longitudinal as well as lateral vibrations, and they made quasi-static experiments both in tension and in flexure. Of greatest interest here is their detailed study of the anisotropy of trees,[5] selecting specimens in the direction of the radius and the tangent and perpendicular to the cross-section, for various heights and for various mean radii. They compared the results with moduli determined in the vertical direction along the fibers. These comparisons are shown in Tables 63 and 64.

Among the 16 tabulated conclusions were the following: that, as in metals and glass, moduli determined from longitudinal vibrations invariably were greater than those determined from quasi-static tensile tests; that moduli determined from small specimens were, in general, in close agreement with those determined from large planks taken from the same tree; that small permanent deformation invariably was measurable, even at relatively small total strain; that wood did exhibit marked anisotropy; that in order to get reproducible results one would have to determine the water content (WERTHEIM and CHEVANDIER determined this from slivers of each specimen, and also they compared the results of natural

[1] The size of the original manuscript is referred to in the introduction to the report of the Academy Commission [1847, *2*].

[2] CHEVANDIER and WERTHEIM [1848, *1*].

[3] WERTHEIM and CHEVANDIER [1846, *1*] (abstract); and see: WERTHEIM [1847, *2*] and, CHEVANDIER and WERTHEIM [1848, *1*].

[4] WERTHEIM and CHEVANDIER [1846, *1*].

[5] This study was much appreciated by SAINT-VENANT; his comments appear in the Appendix of the volume containing lectures by NAVIER [1864, *1*] which volume he edited and annotated.

Table 63. WERTHEIM and CHEVANDIER (1846)

Wood	Density	Sound velocity (ratio to that in air)	Elasticity coefficient[a] in the direction of the fibers (kg/mm^2)	Relation of the elasticity coefficients found through vibration, to the ones found through elongation	Elastic limit (kg/mm^2)	Cohesion (kg/mm^2)
Acacia	0.717	14.19	1261.9	1.193	3.188	7.93
Fir	0.493	13.96	1113.2	1.056	2.153	4.18
Hornbeam	0.756	11.80	1085.7	1.105	1.282	2.99
Birch	0.812	13.32	997.2	1.212	1.617	4.30
Beech	0.823	10.06	980.4	1.087	2.317	3.57
Oak	0.808	—	977.8	—	—	6.49
Oak	0.872	11.58	921.3	1.117	2.349	5.66
Pine	0.559	10.00	564.1	1.086	1.633	2.48

[a] Average results of the measurements in the direction of the fibers of several trees of the same specimens, reduced to 20% humidity.

Table 64. WERTHEIM and CHEVANDIER (1846). *Measurements in the directions of the radius and the tangent*

Wood	In direction of the radius			In direction of the tangent		
	Elasticity coefficient (kg/mm^2)	Sound velocity (ratio to that in air)	Cohesion (kg/mm^2)	Elasticity coefficient (kg/mm^2)	Sound velocity (ratio to that in air)	Cohesion (kg/mm^2)
Hornbeam	208.4	10.28	1.007	103.4	7.20	0.608
Aspen	107.6	9.72	0.171	43.4	5.48	0.414
Alder	98.3	8.25	0.329	59.4	6.28	0.175
Sycamore	134.9	9.02	0.522	80.5	6.85	0.610
Maple	157.1	9.26	0.716	72.7	6.23	0.371
Oak	188.7	9.24	0.582	129.8	7.76	0.406
Birch	81.1	6.46	0.823	155.2	9.14	1.063
Beech	269.7	11.06	0.885	159.3	8.53	0.752
Ash	111.3	8.39	0.218	102.0	7.60	0.408
Elm	122.6	8.56	0.345	63.4	6.11	0.366
Poplar	73.3	8.44	0.146	38.9	6.32	0.214
Fir	94.5	8.05	0.220	34.1	4.72	0.297
Pine	97.7	8.53	0.256	28.6	4.78	0.196

and forced drying); that elasticity coefficients and cohesion decreased at the same rate as the age of the tree increased; that variations in the soil had a pronounced effect upon the elastic properties of trees of the same species; that, provided the water content were taken into account, the time of the felling of the trees did not influence their mechanical characteristics; and that the location of the specimen, both along the tree and out from the center, made a measurable difference in its elastic constants and its cohesion, both of these being found to decrease with height and radius.

The final work in this great, systematic study of elasticity and cohesion by WERTHEIM was his memoir[1] on the principal tissues of the human body, presented

[1] WERTHEIM [1847, *1*].

to the French Academy on December 28, 1846. This work, which I have described in some detail in the previous chapter,[1] led to an empirical equation relating to stress and strain in organic solids, an equation which later was found to apply to one solid after another. The sudden appearance of this first comprehensive study of the elastic properties of human tissue stimulated years of experimental research by Continental physiologists, some of which has been referred to in Chap. II.

I do not think it an exaggeration to state that these six major memoirs of WERTHEIM between 1842 and 1846 converted experimental solid mechanics from a collection of a few isolated, poorly defined, and generally not reproducible facts, into an experimental science. Having built this fundamental structure—an understanding of the requirements for a consistent, coherent program of experiment in the mechanics of solids, a delineation of the variables in terms of which the response of solids to applied forces may be specified, and the provision of a representative and reproducible array of accurately measured parameters (over 3000) in a broad selection of well-defined material states—WERTHEIM proceeded to his main interest as stated by him in several places, notably in a paper reviewing his objectives which was presented to the Academy of Vienna[2] in 1850. He wished to test the applicability of the then known theories without assuming in advance that they were valid. In his time, this meant interrelating with physical observation the rapidly growing linear theory of elasticity, the developing ideas with respect to the thermodynamics of solids, and the contemporary hypotheses connecting atomistic and continuum behavior. In all of the remaining memoirs during the 14 years after 1846, except perhaps for his Doctor of Science dissertation[3] in 1853, he systematically developed these objectives.

For over 20 years after the theoretical contributions of POISSON and CAUCHY in the late 1820's, theorist and experimentist alike accepted the idea that on the basis of the consideration of central forces, a successful union had been made relating the atomistic structure of solids with the linear continuum theory of elasticity. This molecular theory, while directed toward the more general description of anisotropic solids for which the number of constants was reduced from 21 to 15, could be considered by the mid-19th century experimentist only in terms of the special case of the isotropic solid. The consideration by experimentists of anisotropic solids in terms of what are now known as the "CAUCHY relations," began in the 1880's after abortive efforts in the preceding decade. For every isotropic solid, according to this theory, POISSON's ratio had to be $^1/_4$, so that in linear elasticity of isotropic bodies only a single constant was left to be determined by experiment.

---

[1] Chap. II, Sect. 2.14.

[2] This was at the time of his election as a Corresponding Member of the Academy. WERTHEIM [1850, *1*].

[3] The obituary for WERTHEIM by J. C. POGGENDORFF [1863, *1*] erroneously gave 1848 as the year WERTHEIM received his doctorate, a date which was copied by several later writers.

The substance of his doctoral work was published in the *Annales de Chimie et de Physique* in 1854. WERTHEIM [1854, *1*], and see [1852, *2*]. This was the first quantitative study of photoelasticity. It led to WERTHEIM's discovery of the linear stress-optics relation for doubly refracting solids, which became known as "WERTHEIM's law." COKER and FILON in their well-known book on photoelasticity [1931, *1*] referred to WERTHEIM's law and described in detail some of his experiments. Subsequent authors have cited only DAVID BREWSTER's observations of the phenomenon in 1816 [1816, *1*] and the theoretical work of FRANZ E. NEUMANN in the 1840's. BREWSTER's observations were insufficient; they merely indicated that the phenomenon existed. Before WERTHEIM's first study in 1852, others had performed some unsuccessful experiments for flexure. Part of WERTHEIM's success was due to his solving the problem of making such studies in simple compression and tension.

3.16. Wertheim's first experiments on Poisson's ratio, which revealed that the Poisson-Cauchy molecular theory failed to describe crystalline solids (1848).

In a memoir in 1848, WERTHEIM[1] first questioned this uniconstant theory. He began by noting that CAGNIARD DE LATOUR[2] in 1828 had provided the only experimental measurement supporting the requirement that POISSON's ratio should have the fixed value of $^1/_4$. SIMÉON DENIS POISSON[3] at once had cited that measurement as indicating the physical applicability of his theory. At the time the usual way to state CAUCHY's relations was to claim that a stretched rod should increase in specific volume by half its longitudinal strain. It was not until 1879, with BAUSCHINGER's mirror extensometer, that it became possible to measure directly the lateral contraction or expansion of a solid rod undergoing small longitudinal extension or compression.

CAGNIARD DE LATOUR had elongated brass wires in a 2 m long glass tube filled with water, measuring the water level before and after he applied a strain of approximately 0.0015. In both cases he extracted a fixed small length of the wire from the water and noted the difference in the change of the water level. CAGNIARD DE LATOUR stated that the difference in specific volume was indeed $^1/_2$ the longitudinal strain. WERTHEIM critically and correctly evaluated this conclusion on the basis of the error introduced by the relatively large diameter of the tube, the withdrawal of water on removal of the specimen, and the fact that at this strain in brass there would have to have been marked permanent deformation. He correctly dismissed the experiment as a test for a theory which applied only to small, linear, perfectly elastic deformations.

HENRI VICTOR REGNAULT,[4] in examining the behavior of the containers in his study of the compressibility of water had noted that his results seemed to disagree with the POISSON-CAUCHY theory. He suggested that WERTHEIM consider this problem in more detail. WERTHEIM's first experiment in this connection[5] was performed on a rubber rod of square cross-section, large enough so that he could make measurements with a caliper. His strains went up to 200%, a magnitude for which, of course, he should not have expected the elementary theory of elasticity to apply, as JAMES CLERK MAXWELL later pointed out. Noting that permanent deformation was minimal, particularly in the lower strain regions, WERTHEIM compared his simultaneously measured values of elongation and lateral contraction with POISSON's ratio, $\nu={}^1/_4$, $\nu={}^1/_3$, and $\nu={}^1/_2$, noting, as may be seen in Fig. 3.28 (which is a plot of these data) that in the lowest strain region the data certainly did not follow the predicted value of $^1/_4$ for isotropic solids.

I have calculated the ratio $\varepsilon_r/\varepsilon$, assuming incompressibility. From this calculation, shown as a solid line in Fig. 3.28, it may be seen that WERTHEIM's measurements were done with sufficient care to make the distinction he sought. Although WERTHEIM provided values over the large range of strain, his primary interest

[1] WERTHEIM [1848, *1*].

[2] CAGNIARD DE LATOUR [1828, *1*].

[3] POISSON [1827, *1*]. In this note POISSON provided a few calculations to emphasize that the recently presented experiment of CAGNIARD DE LATOUR on brass wire was in agreement with his theory. He calculated what the theory would predict if the experiment were to be extended to the consideration of a plate or membrane of constant thickness, though he conceded the difficulties which would be encountered if it were to be performed. As far as I know, no experimentist has ever attempted to follow up this suggestion of POISSON from 1827.

[4] REGNAULT [1842, *1*], [1847, *1*].

[5] WERTHEIM [1848, *1*].

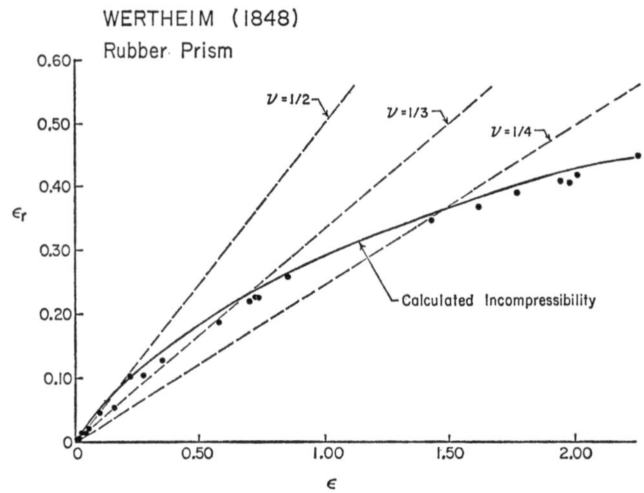

Fig. 3.28. Wertheim's measurements of lateral and longitudinal strain (circles), shown with his comparisons for three values of Poisson's ratio. The material was rubber. I have added the calculated curve for an incompressible material (solid line).

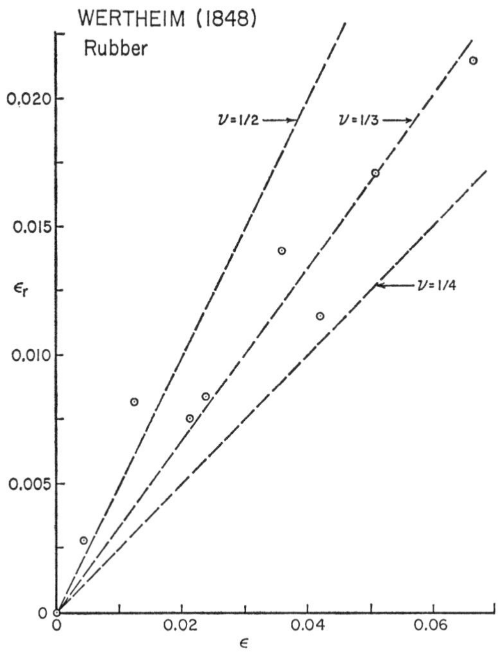

Fig. 3.29. The initial portion of the lateral-longitudinal strain measurements of Wertheim in the rubber prism of Fig. 3.28, from which he showed that Poisson's ratio in this solid was larger than $\frac{1}{4}$.

was in the initial range shown in Fig. 3.29, which in this one solid certainly supported his contention.

This experiment, which Wertheim repeatedly stated he regarded as merely preliminary to the studies which followed, was the source of considerable controversy in later years, as a sequel to the hornet's nest which WERTHEIM had entered in demonstrating that in all the known solids the linear continuum theory was disassociated from the completely accepted, attractive, atomistic theory of POISSON and CAUCHY.[1]

The thoroughness of WERTHEIM's experimental approach to this issue is evinced by the fact that he considered it from seven different points of view, including the preliminary direct measurement referred to above. He used capillary tubes to measure the volume change of hollow tubes in tension; he repeated the compressibility studies REGNAULT had made with a spherical piezometer; he compared longitudinal and torsional frequencies in cylindrical rods; he contrasted the measured velocity of sound in water whether in large bodies or confined in cylindrical tubes, to establish an analogy for comparing the velocity of sound in cylindrical rods and in the infinite solid; he investigated the nodal dimensions of vibrating circular plates; and he made the now perhaps more questionable measurements of the "deep tone" lateral vibrations accompanying longitudinal vibrations in rods.[2] In every instance he obtained POISSON's ratios in excess of the value of $^1/_4$ required by the uniconstant molecular theory of POISSON and CAUCHY. Since for each solid whose small deformation he studied he obtained values approximately $\nu = ^1/_3$, he suggested that the molecular theory ought to be modified in some way so as to accord with what he believed to be the experimental facts for isotropic solids.[3]

The hollow tube experiments were performed on three brass and five crystal glass specimens. The brass specimens were approximately 1 m long and had 5 mm interior diameters with carefully measured wall thicknesses of 1, 2, and 3 mm. The five crystal glass tubes were of various lengths and diameters. The

[1] POISSON (1781–1840) died before WERTHEIM's experiments opened the controversy, and to CAUCHY (1789–1857) the atomistic theory was of small moment. Perhaps nowhere is the perspective of the theorist and the experimentist more divergent than in judging the importance of numerical constants. In the matter of aesthetics, explanation and nature are not necessarily in accord.

[2] Although WERTHEIM had had the foresight to propose measurements of wave speeds of induced vibrations in the crust of the earth, in 1847 it was impossible to measure dilatational and shear wave speeds in a solid without encountering insurmountable problems at the boundaries. Unopposed was the conjecture that wave speeds in the vibrating bar consisted of separate dilatational and shear components of smaller magnitude but in the same ratio as in the unbounded solid. As we shall see, even CAUCHY accepted the concept in his review of WERTHEIM's memoir on these experiments.

[3] In many commentaries during the next half-century, notably in PEARSON's of 1886 and 1893 and in that of WERTHEIM's contemporary, VERDET, WERTHEIM was constantly attacked for having claimed that $\nu = ^1/_3$ was a fundamental property of all isotropic bodies. Although it is fairly apparent that WERTHEIM considered this a highly probable possibility, in both his first memoir on the subject in 1848 and in his last, in 1860, he emphasized that he laid claim only to the solids on which he had experimented and for these, ν was not $^1/_4$ but approximately $^1/_3$. Thus in 1860 in response to VERDET, WERTHEIM stated: [1860, *1*]. (I have taken this quotation from the English translation which appeared in the *Philosophical Magazine* in 1861, p. 449. (See note following WERTHEIM [1860, *1*].))

> "MM. Lamé and Maxwell admit that the ratio above defined, or what comes to the same thing, the ratio between the cubic and linear compressibilities, may vary in different substances. Experiment alone can determine whether this is the case, as I have not failed to remark both in my original memoir, and in several of those I have since published. Mr. Verdet is therefore wrong in asserting, as he does in the extract of a memoir which we shall have to discuss hereafter, and of which Mr. Kirchhoff is the author, that I have 'endeavored to show by numerous experiments that this ratio has in all bodies the same constant value $^1/_3$.' On the contrary, while affirming and maintaining the accuracy of this value for those bodies which were the subject of my researches, I excluded those not yet submitted to experiment."

details of the experiment are contained in a memoir[1] presented to the Academy in February, 1848. The use of capillary tubes to measure volume led Wertheim, with his characteristic thoroughness, to a separate, extensive study of the subject of capillarity. It is contained in a large memoir[2] several years later, in 1857. The change in the water level was determined with an optical cathetometer.

The measurements of elongation of the brass and glass specimens were made between two reference points by utilizing two telescopes which followed the movement by means of micrometer screws with graduated heads reading 0.001 mm. Wertheim's strain resolution for these experiments was thus 10^{-6}, or one order of magnitude more than in his previous experiments. The major problems were described as due to temperature change; the hermetically sealed apparatus wherein the water had been purged of air, became a "veritable thermometer". The memoir in 1848 contained a large number of tables describing in complete detail the elongation and changes of volume of each tube for a large number of axial loads. The data in his summary include the moduli obtained from the elongation measurements for an axial load of 1 kg/mm², which are shown in Table 65, and the changes of interior volume for the specified elongations which are shown in Table 66.

If Poisson's ratio $\nu = {}^1/_4$, the calculated change of volume in infinitesimal strain is $\Delta V = {}^1/_2(\Delta L/L)\,V$; if $\nu = {}^1/_3$, then $\Delta V = {}^1/_3(\Delta L/L)\,V$, where $V = LS$, where S is the initial area of the tube. From the comparisons shown in Table 66 for

Table 65. Wertheim (1848)

Substance	No. of tube	$\Delta L/L$	E (kg/mm²)
Brass	I	0.0000939	10645.2
	II	0.0001015	9855.5
	III	0.0001035	9664.9
Crystal	I	0.0002596	3852.5
	II	0.0002324	4302.6
	III	0.0002873	3481.1
	IV	0.0002258	4429.0
	V	0.0002284	4379.1

Table 66. Wertheim (1848). *Elongations and volume increases*

No. of tube	S (mm²)	s (mm²)	ΔL (mm)	w (mm)	$\Delta L\cdot S$	$\frac{1}{2}\Delta L\cdot S$	ws	$\frac{1}{3}\Delta L\cdot S$
				Substance: Brass				
I	19.182	0.14248	0.08450	3.6508	1.62095	0.81047	0.52017	0.54032
II	19.292	0.14413	0.09109	3.7718	1.75733	0.87866	0.54363	0.58578
III	19.113	0.18358	0.09308	3.0561	1.77898	0.88949	0.56104	0.59299
				Substance: Crystal				
I	48.978	0.18358	0.21908	21.033	10.7301	5.3650	3.8613	3.5767
III	32.710	0.14248	0.24848	16.997	8.1279	4.0639	2.4217	2.7093
IV	15.668	0.14248	0.19508	8.051	3.0565	1.5282	1.1472	1.0188
V	12.032	0.14248	0.19844	5.465	2.3877	1.1938	0.7786	0.7959

[1] Wertheim [1848, *1*].
[2] Published in [1861, *2*].

the various wall thicknesses in brass, and lengths and diameters in glass, the change of volume measured in the capillary tube where s was the area and w the change in height and water level, in every instance was widely different from that expected according to the POISSON-CAUCHY theory for isotropic solids. The average change in volume in seven experiments was 1.40481. If $\nu = 1/3$ the predicted average would be 1.40283, while if $\nu = 1/4$ the theory would predict the very different value of 2.07593. As WERTHEIM pointed out, the moduli of the brass tubes decreased as the thickness of the wall increased. The thinner-walled tubes had been drawn through the die more times. By a comparison with his measurements on wires, WERTHEIM showed that part of this difference was due to changes in density. The variation of his moduli for glass, however, is more puzzling since WERTHEIM had provided the dynamic moduli determined from longitudinal vibrations (shown in Table 67) which not only were greater, but also were more consistent.

Table 67. WERTHEIM (1848)

No. of tube (glass crystal)	D	L_1	n_1	v_1	E (kg/mm²)
I	3.202	0.878	2306.3	12.190	5354.0
II	3.206	0.689	3011.8	12.491	5629.7
III	3.202	0.912	2245.5	12.328	5476.7
IV	3.198	0.908	2281.6	12.471	5597.3
V	3.195	0.874	2348.6	12.356	5489.8

The agreement between the moduli given in Table 67 for the hollow crystal tubes, and those for the solid rods of square cross-section, given in Table 61 above, convinced WERTHEIM the materials on which he was experimenting were isotropic.

In this same memoir WERTHEIM published sufficiently detailed specifications of his specimens, including the dimensions of his capillary tubes at several places, calibrations of volume, and sufficiently accurate measurements of elongation that his results could be assessed critically in 1848 and can be assessed critically in 1972.

The magnitudes of the simultaneously measured moduli and the comparison of data from specimens WERTHEIM had selected for their wide differences in dimensions, should have convinced serious contemporary scientists that for glass and brass WERTHEIM clearly had demonstrated that REGNAULT's suspicions had a basis in fact, i.e., the POISSON-CAUCHY molecular theory for which $\lambda = \mu$ and $\nu = 1/4$ was not applicable.[1]

[1] Twentieth century writers habitually attribute to VOIGT this experimental demonstration that the CAUCHY relations did not describe crystalline solids. In fact, VOIGT's studies of single crystals, made many years after WERTHEIM's, merely confirmed the original discovery. Such writers are unknowingly influenced by the late 19th century and early 20th century prejudices of die-hard advocates of the attractive POISSON-CAUCHY molecular theory, who would yield to experimental results only when overwhelmed with diverse data. The point is perhaps most strongly made by PIERRE MAURICE MARIE DUHEM in 1903, near the end of the uniconstant molecular controversy; in his *L'Évolution de la Mécanique*, he stated.

"The study of isotropic solids leads Poisson to some remarkably simple consequences; thus, when a prism formed of such a body is stretched, the ratio of transverse contraction to the longitudinal extension is fixed and equal to $1/4$; even more, in every isotropic body the ratio of the coefficients of cubical dilatation to the coefficient of tensile elasticity is $2/3$.

"Does experiment verify these conclusions? Cornu and Kirchhoff have found them good in certain special cases; but, according to Wertheim, they are not so for metals. Consequently 'a solid body, even an isotropic one,* cannot be regarded as formed by a system of

REGNAULT's experiments with a piezometer, which at his suggestion WERTHEIM repeated with particular emphasis upon the compressibility of the solid container rather than upon that of the fluid, used a hollow sphere to which both external and internal pressure could be applied. When both pressures were equal, the shell would be under hydrostatic pressure as if it were an interior section of a solid sphere under uniform normal external pressure. When a known internal pressure was applied, a capillary determination of the externally produced volume change could be made. With the same known external pressure, an internal capillary change of volume could be made. From the combined behavior a crude estimate of compressibility of the spherical solid could be calculated from linear elastic theory. PERCY WILLIAMS BRIDGMAN in the historical introduction to his *Physics of High Pressure* in 1949[1] emphasized that the determination of compressibility by means of piezometers had limitations. They are indeed fairly obvious when one examines either REGNAULT's or WERTHEIM's results.

WERTHEIM's calculations of ν for brass and glass using the spherical piezometer, contained an error. E was too high for the POISSON-CAUCHY theory, and also the measured ν was considerably in excess of $^1/_3$. For the cylindrical glass piezometers with hemispherical bases, the calculations were correct; the correlation with $\nu = {}^1/_3$ was somewhat more convincing. Since when $\nu = {}^1/_3$ the bulk modulus is equal to the E-modulus, while if $\nu = {}^1/_4$ the bulk modulus is only $^2/_3$ of the E-modulus, the difference is not negligible. Despite the quantitative limitations of these data,[2] they are sufficient to provide additional, independent evidence that the POISSON-CAUCHY molecular theory of elasticity did not apply to glass and brass. The demonstration, in contrasting experiments, that a supposedly general theory

molecules attracting or repelling one another mutually according to a function of the distance, ... without being subject to certain constraints such as those regarded in analytical mechanics.'

"The partisans of the theory of Poisson, it is true, could always plead a demurrer to the contradictions with experiment by contending that those bodies whose properties do not agree with their formulae are not truly isotropic, but consist of jumbles of crystals; and they have not failed to use this loophole; but they may be opposed by an argument that seems unanswerable.

"Everything that the theory of Poisson asserts of isotropic elastic bodies should in good logic hold equally for liquids. If then, for truly isotropic bodies, the coefficient of cubical dilatation is gotten by multiplying the coefficients of tensile elasticity by $^3/_2$, this proposition should remain true for liquids. However, that cannot be, since for liquids the coefficient of cubical dilatation is not zero, while the coefficient of tensile elasticity is.

"Thus it is impossible to retain the principles on which Poisson wished mechanical physics to rest, except by recourse to subleties and shifts. Besides, even Poisson found himself reduced to desperate means of defense; to be convinced of this it is enough to read the *Notions préliminaires* with which he opens his *Mémoire sur l'équilibre des fluides*. Not only does Poisson there no longer regard the elements of bodies as being points without extension, not only does he treat them as having a shape [particules figurées], but he even evokes under the name of *secondary action* a force which depends upon the form of the molecules, which hinders or eases their mobility, and to which he attributes all the effects that analytical mechanics attributes to forces of constraint.

"When, so as to defend itself, a theory thus multiplies ruses and chicanery, it is futile to pursue it, for it becomes impossible to grasp; but it would be idle to grasp it, since, for every just mind, it is a defeated doctrine. Such is mechanical physics."

[Lorsqu'une théorie, pour se défendre, multiplie ainsi les ruses et les chicanes, il est inutile de la poursuivre, car elle devient insaisissable; mais il serait oiseux de la saisir, car, pour tout esprit juste, c'est une doctrine vaincue. Telle est la Mécanique physique.]

DUHEM [1903, *1*], pp. 86–88.

* É. MATHIEU: Théorie de l'élasticité des corps solides, t. I, p. 6 and 39, Paris, 1890.

[1] BRIDGMAN [1949, *1*].

[2] WERTHEIM [1848, *1*].

for solids did not apply to two such diverse materials, clearly limited to special cases any possible physical foundation for the theory.

With his curious but understandable persistence in this matter, WERTHEIM attacked the issue from still another point of view in his note on torsion[1] in 1848. The ratio of the frequencies of longitudinal and torsional vibrations should equal that of the square root of the E-modulus to the shear modulus μ. Thus we have $n/n' = \sqrt{2(1+\nu)}$, which if $\nu = 1/4$ is 1.5811, while if $\nu = 1/3$ it is 1.6330. Setting 2 m long rods of cast steel, iron, and brass in longitudinal vibration, WERTHEIM obtained the values shown in Table 68.

Table 68. WERTHEIM (1849)

	Length 2 l (m)	n No. of longitudinal vibrations	n'	n/n'	Average
Cast steel	2.00	2585.8	1580.2	1.6364	
Iron	2.06	2560.0	1565.8	1.6350	1.6309
Brass	2.00	1747.4	1077.9	1.6212	

By making both his measurements on the same sample, WERTHEIM minimized the errors which arise when considering the ratio of large numbers, such as those for E and μ, from quasi-static data in different specimens. In this note on torsion he reviewed COULOMB's data[2] on the torsion of iron and brass rods, DULEAU's[3] data on iron rods, SAVART's[4] on rods of cast steel, copper, and brass, and GIULIO's[5] data on iron. WERTHEIM showed that when one assumed $\nu = 1/3$ one calculated values of E comparable to those obtained from direct longitudinal measurement. He emphasized that for brass SAVART several years earlier had determined the ratio of frequencies n/n', obtaining $n/n' = 1.6668$. This ratio not only provided a value of $\nu > 1/4$ and closer to $\nu = 1/3$, but in fact gave $\nu = 0.385$. This latter value, which is the one accepted today, is greater than $\nu = 1/3$ by two thirds as much as $\nu = 1/4$ is less. $\nu = 0.385$ is too high to be regarded as scatter in the data around the value of $\nu = 1/3$. It is interesting that when WERTHEIM in his final memoir[6] of 1860 came to consider this subject again, he attempted to dismiss the value $\nu = 0.387$ which GUSTAV ROBERT KIRCHHOFF[7] had measured directly in 1859, and failed to note the weight of experimental evidence that at least the value for brass differed from $\nu = 1/3$.

3.17. Wertheim succeeds in making the first measurement of the frequency of standing waves in a liquid column (1848).

WERTHEIM's next pursuit of the problem of whether or not the POISSON-CAUCHY molecular theory had any physical counterpart, provided a discovery in liquid acoustics which VERDET[8] in his note of 1861 for *L'Institut* on the scientific

[1] WERTHEIM [1848, *3*], [1849, *1*].
[2] COULOMB [1784, *1*].
[3] DULEAU [1813, *1*], [1820, *1*].
[4] SAVART [1829, *1*].
[5] GIULIO [1842, *1*].
[6] WERTHEIM [1860, *1*].
[7] KIRCHHOFF [1859, *1*].
[8] VERDET [1861, *1*].

life of "the late G. WERTHEIM" was certain would be sufficient alone to assure WERTHEIM a durable place in physics. WERTHEIM described his experimental success with this problem in a note read to the French Academy in 1847, and in a detailed memoir one year later.

In the garden of the Collège de France, using compressed air and water tanks from REGNAULT's laboratory, WERTHEIM succeeded in controlling a liquid jet so as to produce standing vibrations in cylindrical liquid columns.[1]

With his sonometer he compared audible frequencies for the fundamental tone and several overtones. At five different ambient temperatures he made the first determination of the velocity of sound in a column of water. He also determined the velocity of sound for cylindrical columns of eleven other liquids from salt water to alcohol and ether. These results, together with a free-field prediction, are given in Table 69.

Table 69. WERTHEIM (1848)

Name of the liquid	Temperature (° C)	Density	Velocity of sound		Compressibility
			In a column (m/sec)	In an unbounded mass (m/sec)	
Water of the Seine	15.0	0.9996	1173.4	1437.1	0.0000491
idem	30.0	0.9963	1250.9	1528.5	0.0000433
idem	40.0	0.9931	1324.8	1622.5	0.0000388
idem	50.0	0.9893	1349.0	1652.2	0.0000375
idem	60.0	0.9841	1408.2	1724.7	0.0000346
Sea water (artificial)	20.0	1.0264	1187.0	1453.8	0.0000467
Sodium chloride solution	18.0	1.1920	1275.0	1561.6	0.0000349
Sodium sulphate solution	20.0	1.1089	1245.2	1525.1	0.0000393
Sodium sulphate solution	18.8	1.1602	1292.9	1583.5	0.0000348
Sodium carbonate solution	22.2	1.1828	1301.8	1594.4	0.0000337
Sodium nitrate solution	20.9	1.2066	1363.5	1669.9	0.0000301
Calcium chloride solution	22.5	1.4322	1616.3	1979.6	0.0000181
Ordinary alcohol at 36° C	20.0	0.8362	1049.9	1285.9	0.0000733
Absolute alcohol	23.0	0.7960	947.0	1159.8	0.0000947
Essence of turpentine	24.0	0.8622	989.8	1212.3	0.0000800
Sulphuric ether	0.0	0.7529	946.3	1159.0	0.0001002

For WERTHEIM the major importance of these data was that the ratio of the wave speed for water in the free field to that determined in his liquid column was $\sqrt{3/2}$. For his comparison he drew upon the data of COLLADON and STURM,[2] who had measured the velocity of sound in the water of Lake Geneva. Sounding a bell under water, COLLADON and STURM had determined travel times for both still and rough[3] water at various distances from the source. They concluded that the velocity of sound in water at 8.1° C, was 1435 m/sec. WERTHEIM's measurement of the velocity of sound in a water column was 1173.4 m/sec.

[1] WERTHEIM [1848, *2*]. WERTHEIM noted that water organs had been in existence in earlier centuries.

[2] COLLADON and STURM [1838, *1*]. This memoir was presented at the Academy on June 11, 1827.

[3] The description of the near disaster for both the swimmer under water and the boatsman during a heavy storm reveals a physical dedication to science. Mr. VERDET, later a Paris colleague of WERTHEIM and the author of the aforementioned scientific evaluation written in 1861 after WERTHEIM's death, was the oarsman in the boat on that occasion. See COLLADON and STURM [1838, *1*], p. 343.

Multiplying his value by $\sqrt{3/2}$, WERTHEIM obtained a value of 1437.1 for comparison with that of COLLADON and STURM. It was an almost precise agreement. When comparing the predicted dilatational velocity of a solid $c_1=\sqrt{\frac{\lambda+2\mu}{\varrho}}$ with the bar velocity $c_0=\sqrt{E/\varrho}$ one obtains the ratio $\sqrt{3/2}$ if $\nu=1/3$ and $\sqrt{6/5}$ if $\nu=1/4$.

That for $\nu=1/3$ in solids, the ratio of the dilatational wave speed to the longitudinal wave speed in a bar provided exactly the same numerical value as the ratio of wave speeds in water for what were presumed to be analogous situations, convinced WERTHEIM this was a matter of fundamental significance.[1] Had he been more conversant with the earlier writings of POISSON on the subject of the equilibrium of fluids or had he heeded the warnings of MAXWELL, to whom he referred in his memoir of 1860, WERTHEIM might not have placed so much emphasis upon this analogy between the solids and liquids he had considered. At a time when the relation between theory and experiment was such that, among many untested concepts, the existence of either dilatational or shear wave speeds in any solid had not yet been demonstrated, it did not seem unreasonable for an experimentist to expect to find in a bar two wave speeds having lower values but occurring in the same ratio as that predicted for dilatational and shear waves in the free field. Also, at that time it was not unreasonable for an experimentist to conjecture that a shear velocity in water must exist. WERTHEIM wondered whether the double echoes which VERDET had heard[2] when rowing the boat in the vicinity of the bell might not be attributed to a second shear wave in water.[3]

Like YOUNG 35 years earlier, VERDET also was confused as to the significance of volume vs linear compressibility of fluids, as is attested by the comment in his review of WERTHEIM's work in 1861 that the analogy needed further proof. No definitive criticism of WERTHEIM's widely read papers on the analogy appeared for over a decade after the first one was published, which indicates that in the mid-19th century the subject generally was not understood. Indeed, in the 1851 Academy Commission consisting of REGNAULT, DUHAMEL, and CAUCHY[4], CAUCHY as "rapporteur" remarked that the close correlation of WERTHEIM's wave speed prediction from the liquid column, with the data of COLLADON and STURM, was contributing evidence that the molecular theory[5] did not apply to the physical situation.

With the style characteristic of the great experimentist which he was, WERTHEIM surrounded his experiments with a thorough study of auxiliary empirical assumptions. He noted that for over two centuries experiments on columns of air invariably had revealed velocities of sound lower than those in unlimited space, because of embouchure, tube opening, and diametral effects; hence he wished to learn whether or not such effects were present in the velocity of sound in liquid columns. Commenting upon the fact that daily experience in the physics

---

[1] WERTHEIM was further influenced by the fact that if the ratio of $\sqrt{6/5}$ were to apply in water, COLLADON and STURM would have observed 1285 m/sec instead of their 1435 m/sec. Such a large numerical difference exceeded the probability of experimental error.

Sooner or later every serious experimentist comes across such a numerical coincidence but seldom in a situation as dramatic as WERTHEIM's. Although WERTHEIM indeed was strongly influenced by this observation, as were many of his contemporaries, his checks by means of independent experiment of course were the proper approach. What makes his case somewhat unusual is that his independent checks supported his use of the fluid-solid analogy.

[2] This was reported by COLLADON [1842, *1*].

[3] WERTHEIM [1851, *4*].

[4] WERTHEIM [1851, *9*].

[5] See DUHEM's [1903, *1*] comment on this subject, in Sect. 3.16, *supra* (footnote).

of sound production was gleaned from rooms, closets, glasses, and bottles, i.e., from limited space filled with air, WERTHEIM expressed the experimentist's lament with respect to the restrictions and oversimplifications of the available theories of his time. After a long historical review of the study of the effects of embouchure, end conditions, and diameter by DANIEL BERNOULLI, EULER, WILLIAM HOPKINS, PELLISOSANS, BIOT, HAMEL, SAVART, POISSON, LISKOVIOUS, SONDHAUS, and PINARD, WERTHEIM[1] provided a detailed critical experimental study of the subject including the precise data used for the comparison of his results with the work of others. For a wide variety of pipe dimensions and orifices he showed an agreement with a "law of like volumes" MARIN MERSENNE[2] had presented two centuries earlier. Of interest here is the care which WERTHEIM exercised before he assured himself that his previous choice of embouchure in cylinders filled with air and submerged in liquid, had been appropriately designed to provide accurate sound velocities.

3.18. Wertheim on vibration of plates, and the "deep tone" of vibrating rods.

A year after his dynamic measurements on liquids in 1848, WERTHEIM[3] looked at the POISSON-CAUCHY molecular theory from still another point of view. Following and extending a series of experiments on plates by GUSTAV ROBERT KIRCHHOFF[4] a year before, he performed a series of experiments on the lateral vibration of thin circular plates of iron, glass, and brass. His objective was to show that E determined from the fundamental, from the first two octaves, and from measurement of corresponding nodal patterns in CHLADNI's experiment, would coincide with E determined from the then currently published torsion moduli of THEODORE KUPFFER if KUPFFER's data were calculated on the assumption that $\nu = {}^1/_3$, and not ${}^1/_4$, as KUPFFER had assumed.

WERTHEIM examined the isotropy of the sheet material by comparing the longitudinal dynamic moduli of bands cut in perpendicular directions. For brass he obtained $E = 10626.8$ kg/mm² in the direction of the laminate and a similar value of 10797.8 kg/mm² perpendicular to the laminate.[5] He further checked the isotropy by noting that for the averages of several measurements of nodal radii for soft iron, 3 brass plates, and 3 glass plates, the magnitude and direction of the nodal ellipses for the fundamental and first two symmetrical overtones were as expected. The experiments as well as the computations were inadequate for their objectives as WERTHEIM seemed to suspect.

WERTHEIM published one additional paper in his experimental study of the POISSON-CAUCHY theory. It provides an interesting commentary on how numerical coincidence in a non-understood observed behavior, coupled with theoretically expected but as yet unobserved similar behavior, can cause fundamental error to be accepted broadly. Infinitesimal linear elastic theory predicted that isotropic solids would sustain dilatational and shear waves, the difference in whose wave speeds

---

[1] WERTHEIM [1851, *2*, *6*, *7*], [1860, *2*].
[2] MERSENNE [1636–1637, *1*], p. 335.
[3] WERTHEIM [1851, *3*]. This paper was presented to the Academy in 1849.
[4] These experiments were published by KIRCHHOFF three years later: KIRCHHOFF [1850, 1].
[5] Not only was this the first separate experimental demonstration that the solid considered was isotropic, a fact which should have been noted by more of WERTHEIM's later critics who invariably attacked him on this point, but also this check on isotropy preceded by approximately 40 years the experiments of VOIGT, who customarily is credited with having been the first to consider this issue seriously.

was dependent upon POISSON's ratio. Many experimentists had noted that the longitudinal vibration of a rod was accompanied by what was referred to as a deep tone whose audibility varied as the longitudinal oscillation continued.[1] SAVART[2] had identified the source of the deep tone as a transverse vibration with a frequency which was always exactly one octave below the longitudinal tone, whether the longitudinal tone was the fundamental or the first or second overtone. The sound of the deep tone was referred to as harsh and as perceived only intermittently. When the deep tone occurred, the longitudinal tone was noticeably diminished. The phenomenon which WERTHEIM in 1851 described as known to everyone working in this type of experiment, was a common source of fracture during the longitudinal testing of glass or crystal.

WERTHEIM, in the memoir[3] in 1851 presented to the Academy on December 10, 1849, described in detail how the deep tone might be accentuated, and he reported a series of experiments, from his use of CHLADNI figures to his study of the trace made by a needle attached to the end of the rod, upon a carbon coated glass plate. Such data are shown in Fig. 3.30 where (1), (2) and (3) are traces when the needle is perpendicular and the plate is parallel to the axis of the rod. For (1) the needle is fixed on the side of the rectangular rod, and for (2) and (3) it is on the narrow side.

In Fig. 3.30, (4) and (5) show the behavior when the needle is perpendicular and the plate is parallel to the axis of the rod. In WERTHEIM's long discussion he, like SAVART, failed to see the possibility of coupled modes. After describing several plausible explanations for the deep tone, WERTHEIM proposed that it arose from a shear wave propagating in the bar at the free field shear wave speed divided by $\sqrt{3/2}$. It was thus in the same ratio as that of the dilatational wave speed in the free field to the speed of longitudinal waves in a bar. For $\nu = 1/4$ the ratio of dilatational velocity to shear velocity is $\sqrt{3}$ for isotropic solids while for $\nu = 1/3$ the ratio is 2. Since the deep tone invariably was one-half the frequency of the mode sounding, whether it was the fundamental or the first or second overtone, WERTHEIM concluded that this provided further evidence that $\nu = 1/3$ and therefore the POISSON-CAUCHY theory did not apply.[4]

WERTHEIM, among others, saw seismological measurement as the only hope for determining whether the theoretical wave speeds existed in the unbounded solid. Far ahead of his time, he proposed the use of a surface explosion to measure the arrival time of the two waves at a fixed place. He certainly was not alone in thinking that the "deep tone" vibration provided a bar wave speed for the shear wave, as may be seen for example in the independent assertion to that effect by WILLIAM HOPKINS in the British Association Report[5] in that same year, 1847; moreover, I have not been able to find any adverse criticism of the idea by those who commented upon WERTHEIM's work during the next two decades. In the 1851 Academy Commission report,[6] REGNAULT, DUHAMEL, DESPRETS, and CAUCHY concurred

[1] In the experiments in which this phenomenon was observed the longitudinal oscillations were obtained by rubbing the rod along its length with an impregnated cloth. The technique introduced first by CHLADNI was the method used by SAVART and by WERTHEIM for obtaining dynamic moduli. It was a commonly observed phenomenon up to 1880 when, for studies on longitudinal bars, the waves were induced by impact. After that, reports on the deep tone disappeared from the literature. It would seem worthwhile to re-examine the phenomenon, which apparently required some experimental skill to produce.

[2] SAVART [1837, *1*].

[3] WERTHEIM [1851, *4*].

[4] See Sect. 3.17 above.

[5] HOPKINS [1847, *1*], p. 78. HOPKINS, however, referred to the general lateral vibration of the bar.

[6] The report was written by CAUCHY; see WERTHEIM [1851, *9*].

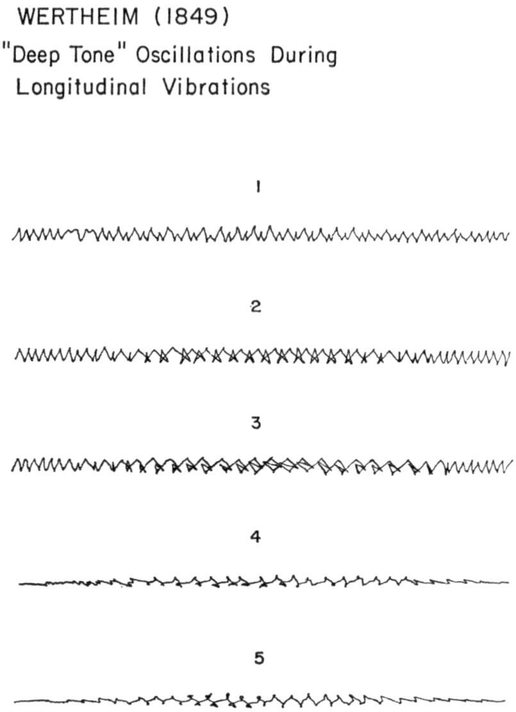

Fig. 3.30. The experimental evidence of WERTHEIM for the existence of "deep tone" transverse oscillation during longitudinal vibration whose frequency was always one octave lower. The numbers refer to the different location of the recording needle which produced the results shown on the rotating disk.

with WERTHEIM's interpretation of the deep tone transverse vibration. They regarded these experiments as supporting WERTHEIM's conclusions that the POISSON-CAUCHY theory, which the authors of the report wryly admitted had been generalized "by one of us," ought to be rejected.

WERTHEIM in 1860 evinced some bitterness relative to the twelve year controversy which had followed his original memoir on the equilibrium of solids. I quote his remarks from the English translation in the *Philosophical Magazine*:[1]

> Several distinguished geometricians, without repeating my experiments, and without disputing their accuracy, have endeavored to bring them in accordance with the ancient theory by various, and unfortunately also, very arbitrary hypotheses. I shall shortly mention and discuss these hypotheses before describing my new experiments on this subject.

The new experiments were presumably those on flexure, which were described in a memoir left unfinished at the time of his death.[2]

---

[1] See p. 448 of the English translation (1861) referred to in the WERTHEIM reference [1860, *1*]. This memoir was first published one month before his suicide at Tours.

[2] VERDET in his article on WERTHEIM in *L'Institut* [1861, *1*] described the state of those manuscripts as being such that he would not attempt to publish them posthumously. He stated that WERTHEIM had measured the change of volume of specimens in flexure and had shown that the change was not the same for rectangular rods bent parallel to the largest or smallest side of the section. VERDET stated further that these experiments appeared to have led WERTHEIM to still another demonstration that the molecular theory did not apply. [My efforts to locate the manuscripts have been unavailing.]

From the many who, like SAINT-VENANT, were loth to abandon the earlier simple molecular basis for linear elasticity, the most common attacks asserted that there "must be" anisotropy, or some form of geometrical inhomogeneity or some effects of tempering.[1] WERTHEIM was particularly disturbed by such assertion from CLAUSIUS and SAINT-VENANT. He pointed out how curious it was that in spite of the various types of homogeneity based upon geometrical shape, proposed by the latter, experiments performed on spheres under pressure, and on cylinders both hollow and solid, in tension, torsion, and flexure all gave essentially the same result, i.e., $\nu \neq {}^1/_4$.

3.19. The Wertheim controversy viewed from the 20th century.

For well over half a century after WERTHEIM's observations, even after overwhelming experimental corroboration had been presented by others, he remained the object of attack in this area for having stated that for metals and glass his experimental results had provided the different constant value of $\nu = {}^1/_3$, i.e., he had found an uniconstant elasticity on some unexplained basis. In his last memoir he made it clear that such attacks, including that of VERDET, who was shortly to be his scientific biographer, were simply wrong. He agreed with MAXWELL and LAMÉ that various ratios of cubic and linear compressibilities might be found in different substances and further, that these differences for rubber in contrast to metals already had been established. That his data provided a value of $\nu = {}^1/_3$ for certain solids for which he hoped some new molecular theory might be developed, was probably WERTHEIM's point of view. He had resembled his critics to some extent when he attempted to dispose of KIRCHHOFF's data in iron and brass; while $\nu \neq {}^1/_4$, also $\nu \neq {}^1/_3$. An experimentist of WERTHEIM's excellence has every right to indulge in the too often forlorn hope that his work will stimulate a new theory.[2]

In suggesting that theorists could well take $\nu = {}^1/_3$, WERTHEIM once again was far ahead of his time. Exactly 100 years after WERTHEIM published his first memoir on elastic constants, WERNER KÖSTER[3] in 1943 summarized the data available on isotropic elastic constants in the mid-20th century in terms of the two linear-elastic formulae relating E, K, μ, and ν, and concluded that although variations in POISSON's ratio of course were observed, an average value of $\nu = {}^1/_3$ was satisfactory for theoretical purposes. An inspection of independently measured

[1] To these might be added a fourth common type of attack, perhaps best illustrated by a statement of PEARSON in the *History* (TODHUNTER and PEARSON [1960, *1*], vol. I, p. 714) in discussing the memoir of 1848 on the equilibrium of solids. He said of the man who had studied mathematics with JACOBI, DIRICHLET, and STEINER: "The memoir is very instructive as shewing the dangers into which a physicist may fall who has not thoroughly grasped the steps of a mathematical process." Since the matter under consideration was purely experimental, this statement also represents one more example of PEARSON's limited understanding of experimental mechanics. It was PEARSON's persistent and usually negative commentary on WERTHEIM, while calling attention again and again to WERTHEIM's data, that over a decade ago first aroused my interest and curiosity to undertake a close look at WERTHEIM's experiments.

[2] WERTHEIM's implied or direct appeals to the geometers were not understood as such. More often, one finds a statement such as that of LOVE, nearly seven decades after WERTHEIM's plea, who, while making no brief for the CAUCHY-POISSON theory, yet seems to imply that WERTHEIM's experimental results were diminished because there was no theory available to explain them:

> "This ratio is often called 'Poisson's ratio'. Poisson deduced from his theory the result that this ratio must be $^1/_4$. The experiments of Wertheim on glass and brass did not support this result, and Wertheim proposed to take the ratio to be $^1/_3$—a value which has no theoretical foundation." LOVE [1927, *1*], p. 13.

[3] KÖSTER [1943, *1*].

values of K, μ, and E convinced Köster that certainly a number of the polycrystalline metals were isotropic or nearly so. Using the ratios of K/E and E/μ, he found comparable values of ν. Accordingly, with one or the other formula, he calculated values of Poisson's ratio for 46 elements; they are shown in Fig. 3.31.

KÖSTER (1948)

□ body centered cubic
⊡ face centered cubic
⬡ hexagonal
▯ tetragonal
▱ rhombohedric
⊠ diamond

Li □ 0.36											Be ⬡ 0.08	B —	C ⊠ —	N —
Na □ 0.32											Mg ⬡ 0.28	Al ⊡ 0.34	Si ⊠ 0.27	P rh
K □ 0.35	Ca ⊡ 0.31	Sc – —	Ti ⬡ 0.36	V □ 0.35	Cr □ 0.31	Mn 0.24	Fe □ 0.29	Co ⊡ 0.31	Ni ⊡ 0.32	Cu ⊡ 0.35	Zn ⬡ 0.27	Ga rh 0.47	Ge ⊠ 0.32	As ▱ —
Rb □ —	Sr ⊡ 0.28	Y ⬡ —	Zn ⬡ 0.37	Nb □ 0.35	Mo □ 0.31	Ma ⬡	Ru ⬡ 0.23	Rh ⊡ 0.26	Pd ⊡ 0.39	Ag ⊡ 0.37	Cd ⬡ 0.29	Jn ▯ 0.45	Sn ▯ 0.33	Sb ▱ 0.33
Cs □ —	Ba □ 0.28	La ⊡ 0.28	Hf ⬡ 0.37	Ta □ 0.35	W □ 0.30	Re ⬡ 0.26	Os ⬡ 0.25	Ir ⊡ 0.26	Pt ⊡ 0.39	Au ⊡ 0.42	Hg ⬡ —	Tl ⬡ 0.45	Pb ⊡ 0.44	Bi ▱ 0.33
87	Ra —	Ac —	Th ⊡ 0.26	Pa —	U 0.30									

Fig. 3.31. Köster's (1948) summary of experimental measurements of Poisson's ratio for the elements.

As will be shown below, where I shall describe an extension of Köster's work in 1948, the average value of Poisson's ratio for the 46 elements is 0.337; with average values of 0.325 for 11 body-centered cubic elements, 0.330 for 16 face-centered cubic elements, 0.303 for 10 hexagonal elements, 0.390 for 2 tetragonal elements, 0.330 for 2 rhombohedral elements, 0.345 for 2 diamond type elements, and 0.337 for 3 elements whose crystallographic structure was unspecified. With respect to "theoretical purposes", perhaps it would be more appropriate to say that for most metals experiment indicated that ν is not much different from $^1/_3$. I shall discuss later Köster's description of how these values of Poisson's ratio vary with position in the periodic system, and the efforts which have been made to study the more general case of anisotropic solids by means of measurement on single crystals. The important point here merely is to note that although Köster referred neither to Wertheim nor to the bitter 19th century controversy which lasted nearly 50 years, Wertheim's conclusion from his own experiments had become simply a matter of fact by the mid-20th century.

In 1954, Cornelis Zwikker[1] in a monograph on the physical properties of solid materials, also without reference either to Wertheim or to the difficulties which had dominated the second half of the 19th century in this matter, presented two figures, reproduced as Figs. 3.32 and 3.33 below, in which he showed that in a plot of the shear modulus vs Young's modulus, the data fit the solid line best, with the slope $\mu = {}^3/_8\, E$. He also provided a second comparison of λ with E, the solid line representing $K = E$, and in this instance found the expected wide variation due to the greater sensitivity of this ratio to variations in Poisson's ratio. These constants of proportionality of course correspond to the value $\nu = {}^1/_3$ found by Wertheim. Zwikker noted that the value of $\nu = {}^1/_3$ was a reasonably good average for metals, but he thought that for stone and glass a value of $^1/_4$ was preferable, while for elastomers the value was $\nu = {}^1/_2$.

[1] Zwikker [1954, *1*].

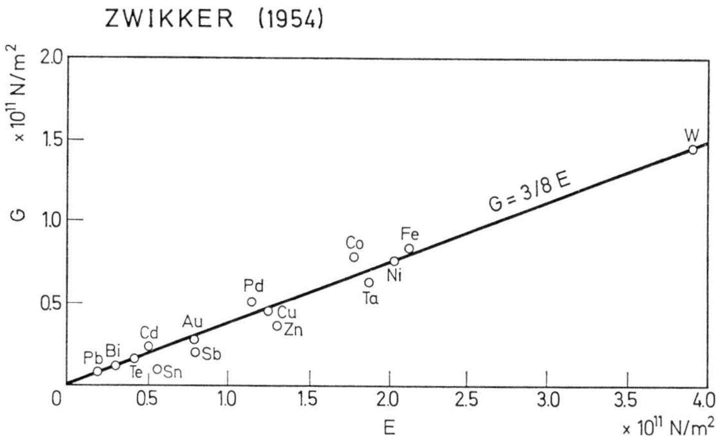

Fig. 3.32. The relation between the shear and E moduli compared with prediction for POISSON's ratio equals $\frac{1}{3}$.

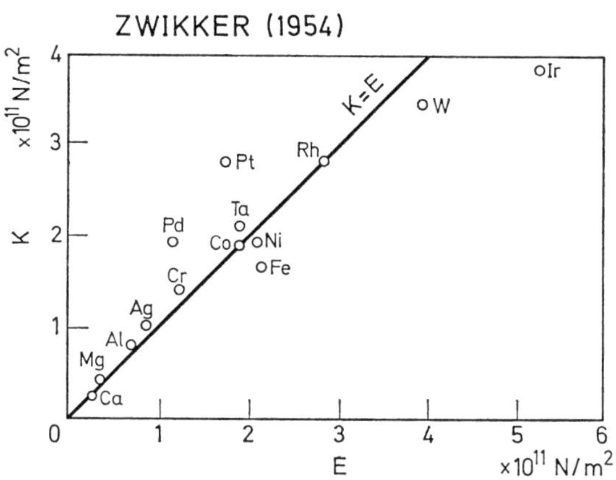

Fig. 3.33. The relation between the bulk and E moduli compared with prediction for POISSON's ratio equals $\frac{1}{3}$.

The weight of experimental evidence during the nearly 125 years after REGNAULT's suspicions had been borne out definitively by WERTHEIM's experiments forces one to conclude that in nearly every crystalline solid the CAUCHY relations fail to hold; therefore, it does not seem in order in the present work to trace all of the voluminous details of the uniconstant vs multi-constant controversy. Some aspects of that controversy will be described, but all too often it revealed the lesser moments of great men, as well as the unfounded assertions of many lesser men. However, a few additional experiments on this subject are so unusual in conception that they must be included at this point.

3.20. Kirchhoff's experiment for the direct measurement of Poisson's ratio (1859).

GUSTAV ROBERT KIRCHHOFF[1] in 1859 proposed an experiment from which POISSON's ratio could be measured directly on a single specimen subjected to

[1] KIRCHHOFF [1859, *1*].

combined torsion and flexure, without requiring that the diameter of the rod be measured (thus eliminating the errors from raising this result to the third and fourth power), and without requiring that either E or μ be determined.[1] The nature of the specimens upon which these experiments were performed is shown in Fig. 3.34.

KIRCHHOFF (1859)

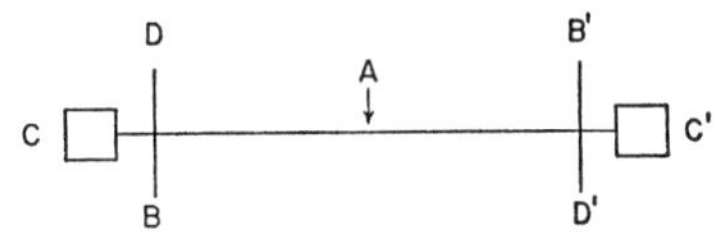

Fig. 3.34. A schematic diagram of the specimens used by KIRCHHOFF in his direct measurement of POISSON's ratio.

The overall apparatus used by KIRCHHOFF is shown in Fig. 3.35. The bar was supported at its center, designated as A. Loads were applied simultaneously on the two oppsite struts, B and B', for the first measurement, then removed and located at D and D' for the second measurement. By means of mirrors located at C, displacements were determined.

KIRCHHOFF (1859)

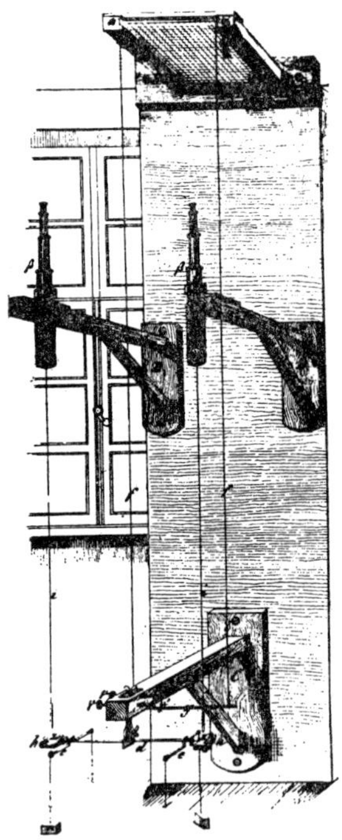

Fig. 3.35. The apparatus used by KIRCHHOFF in his direct measurement of POISSON's ratio.

[1] WERTHEIM in his memoir in 1860 attributed this idea to GAUSS, although there is no mention of this in KIRCHHOFF's paper.

Analysis of the combined problem shows that these measurements provide a direct determination of POISSON's ratio, independent of the diameter of the bar. A comparison between the two sets of measurements provides a cross-check upon the accuracy of each. KIRCHHOFF performed four experiments, three on separate specimens of steel and one on brass. His memoir described the reproducibility of the data for a given specimen when the same loads were placed at different locations, and the constancy when the loads were doubled. Averaging these results for the first steel specimen, KIRCHHOFF obtained a value for POISSON's ratio of $\nu = 0.293$. For the second steel rod, he obtained an average of $\nu = 0.295$; for the third steel rod, $\nu = 0.294$; for a total average value for steel of $\nu = 0.294$, i.e., precisely the modern accepted value of POISSON's ratio for this solid.

The experiment of KIRCHHOFF on brass for which the magnitude of the weights had to be changed to maintain sufficiently small deflections, again provided the modern accepted value of $\nu = 0.387$. This latter value is in close agreement with $\nu = 0.385$ for brass, which WERTHEIM had computed from the ratio of SAVART's longitudinal and torsional frequencies. It is my opinion that WERTHEIM's efforts to dismiss KIRCHHOFF's values on the basis of the flimsiness of the apparatus were not well grounded, nor was KIRCHHOFF's assertion that whereas he was certain the steel rods were isotropic he had some doubts for the brass, and hence thought that the number for that solid was less meaningful than that for the steel.

The significance of KIRCHHOFF's ingenious experiment, in which he had measured POISSON's ratio directly without introducing the experimental difficulty implicit in determining the ratio of measured numbers, was not generally appreciated, as is illustrated by the work of JOSEPH DAVID EVERETT.[1] Seven years later, in 1866, EVERETT repeated KIRCHHOFF's experiment, but claimed to improve upon it by returning to the measurement of the two moduli and computing the ratio. Obviously missing the vital point of KIRCHHOFF's experiment, EVERETT first fixed the end and measured the torsion angle, then removed the couple and the fixture to measure the flexural deflection due to an end load. He made the measurements with the same type of mirror and telescope system as KIRCHHOFF's, but he had to re-set the apparatus and adjust the alignment between the two loadings, thereby introducing what was referred to as a "mechanical correction." The numbers for POISSON's ratio differed widely not only from those of KIRCHHOFF, but also from those of WERTHEIM and MAXWELL.[2] EVERETT's averaged value of POISSON's ratio for glass was 0.239, instead of WERTHEIM's 0.330 and MAXWELL's 0.332. For brass EVERETT calculated the abnormally high average value of 0.469 instead of KIRCHHOFF's 0.387, SAVART's 0.385, or WERTHEIM's $^1/_3$.

These experiments of EVERETT illustrate the general failure of many experimentists of the time to appreciate the difficulties of the problem. But EVERETT's role is of interest here primarily because in 1875 he wrote a manual[3] which became a major source of numerical values in physics for many years. The book, which tabulated EVERETT's faulty values of E and μ along with his calculated compressibilities and POISSON's ratios, went into several English editions and was translated into French, German, Polish, Dutch, and Italian. The uncritical comparison of these data with those of KIRCHHOFF, MAXWELL, AMAGAT, WERTHEIM, MALLOCK, CORNU, PULFRICH, WIEDEMANN, SAVART, KUPFFER, and KELVIN was no small factor in creating the general impression for the remainder of the 19th century that the measurement of elastic constants was at best an imprecise science.

---

[1] EVERETT [1866, *1*], [1867, *1*].

[2] Two years later, CORNU [1869, *1*] obtained a value of 0.237 for Saint Gobain's glass. (See Sect. 3.21 below.)

[3] The manual was on the *C. G. S. System of Units*. EVERETT [1875, *1*].

During the winter of 1862–63 at Heidelberg, MICHAIL OKATOW was encouraged by KIRCHHOFF to repeat with minor modifications KIRCHHOFF's experiments so as to study the effect of prior thermal history and the modifications of anisotropy, if any, produced by the manufacture of different cross-sections. Among the "many well meant as well as informative hints of KIRCHHOFF"[1] was quite probably his suggestion that OKATOW repeat the measurement on the thin specimens (steel knitting needles), the use of which had been criticized by WERTHEIM, in order to compare them with identical measurement on sturdy specimens. The experiments of OKATOW also differed from the original ones in that the central section of the apparatus was free instead of being clamped, and it was used for the optical measurements, while the ends were used for clamping and for applying the eccentric loads. By planning to analyze the simultaneous torsion and flexure for elliptical, square, and rectangular cross-sections in addition to the circular section studied earlier, OKATOW obviously had in mind a far more extensive experimental program than was possible for a single winter. All but two of his reported tests had been for circular cross-sections, and those two were for a steel. OKATOW expressed regret that no circular cross-section studies of this steel had been made for purposes of comparison.

OKATOW was delighted with the reproducibility of his experimental results from one bar to another, and particularly with his agreement with the measurements of KIRCHHOFF since the conditions of the specimens had been the same. For example, in Table 70 for specimens numbered 3, 4, and 5, the two specimens in the condition they were received gave $\nu = 0.2984$ and 0.2994; two specimens in the oil-hardened condition gave $\nu = 0.3199$ and 0.3181; and the two specimens in the completely annealed condition, gradually cooled off gave $\nu = 0.3278$ and 0.3284.

KIRCHHOFF's average for three specimens of steel knitting needles had been $\nu = 0.294$ (i.e., 0.293, 0.294, and 0.295, respectively, for three tests). For a specimen of the same solid in the same condition with his slight alteration of the experiment, OKATOW obtained $\nu = 0.2968$. One also may note the similarity of the measured POISSON's ratio for the three specimens in the condition as received, Nos. 1, 4, and 5, with an average of $\nu = 0.291$. The higher values of ν for the completely annealed specimens and the general reproducibility obtained is consistent with later observations of BAUSCHINGER, for example, and with the earlier measurements by WERTHEIM which had led him to accept a value of slightly less than $^1/_3$ for steel. The diameters of the round specimens, 4 and 5, are in the ratio of 10:18.

OKATOW concluded that, independent of diameter, when proper cognizance was taken of the prior thermal history, the experimental results were reproducible. He regretted not having time to explore further whether the results for the rectangular Huntsman steel specimens, Nos. 6 and 7, also were obtainable in round specimens. Since he performed no such experiments, he failed to achieve his initial major objective. The obvious highly improved reproducibility obtained in this experiment was due partly to the simultaneous measurement of the flexure and torsion on the same specimen. Of perhaps even greater importance was the independence of the radius of the specimen in OKATOW's direct determination of ν. Obviously the dimensional errors in the measurement of the radius were not cubed, thus eliminating an error present when determining the moduli separately. The length of the specimen, of course, still did appear in the analysis of the data.[2]

[1] OKATOW [1863, *1*], p. 12.

[2] CHWOLSON in the *Traité de Physique* [1908, *1*] summarized OKATOW's data for tempered steel (above) as $\nu = 0.294$, or identically the same value as KIRCHHOFF's, and nearly the same as the 0.296 of SCHNEEBELI. For OKATOW's data for annealed metals, CHWOLSON somehow provided the value $\nu = 0.304$.

Table 70. OKATOW (1863)

Type of rod:	Knitting needle rods				Smooth English round wire			Prismatic rods, milled in the hot condition	
Number of rod:	1	2	8	9	3	4	5	6	7
Condition:									
a) as from the factory	$\nu_1 = 0.2750$	—	—	—	—	$\nu_4 = 0.2984$	$\nu_5 = 0.2994$	$\nu_6 = 0.40$	$\nu_7 = 0.3982$
b) hardened in oil	—	$\nu_2' = 0.2968$	—	—	—	$\nu_4' = 0.3199$	$\nu_5' = 0.3181$	—	—
	or, after corrections were made:								
	—	$\nu_2' = 0.293k$[a]	$\nu_2' = 0.294$[a]	$\nu_2' = 0.295$[a]	Average value:		$\nu_{4,5}' = 0.3190$		
c) partially annealed and cooled off	—	$\nu_2'' = 0.2988$	—	—	—	—	$\nu_5'' = 0.3234$	—	—
d) annealed and gradually cooled off	—	$\nu_2''' = 0.3037$	—	—	$\nu_3''' = 0.3278$	—	$\nu_5''' = 0.3284$	—	—
					Average value: $\nu_{3,5}''' = 0.3281$				
Differences:									
between d) and b)	$\delta = \nu_2''' - \nu_2' = 0.0227\, \nu_2'''$ [See below*]				$\delta = \nu_{3,5}''' - \nu_{4,5}' = 0.0277\, \nu_{3,5}'''$				
between d) and a)	$\Delta = \nu_2''' - \nu = 0.0945\, \nu_2'''$				$\Delta = \nu_{3,5}''' - \nu_{4,5} = 0.0890\, \nu_{3,5}'''$				

[a] KIRCHHOFF, Pogg. Annal., 108, 390–391 (1859). *[I have corrected an obvious error]

3.21. Cornu's optical interference experiment for determining Poisson's ratio (1869).

The first truly direct measurement of POISSON's ratio, independent of all dimensions and moduli, also was the first determination of elastic constants in solids by means of interference optics.[1] The remarkable paper of MARIE ALFRED CORNU in 1869 on his direct determination of POISSON's ratio unfortunately included an illogical argument which he had introduced in an attempt to correlate his data with the POISSON-CAUCHY molecular hypothesis. Moreover, CORNU uncritically accepted the dubious volumetric data obtained by CAGNIARD DE LATOUR in 1829, which CORNU described as "differing only slightly from that of KIRCHHOFF." In short, CORNU serves as a sad example of the theory-dominated experimentist.

CORNU's ingenious experiment, however, permitted the direct determination of POISSON's ratio, entirely independent of the specific dimensions of the specimens or their squares and cubes. The accuracy of CORNU's measurements in 1869 was comparable to that obtainable in a modern laboratory. Thereafter, many years elapsed before measuring techniques in solid mechanics again achieved such precision.[2] In his theoretical studies of the flexure of beams, SAINT-VENANT had shown that for a beam with a constant moment the ratio of the principal radii of curvature was equal to POISSON's ratio. CORNU designed an experiment to utilize this result. He optically determined the radii of curvature on the top section, perpendicular to the axis of the beam and in the direction of the axis of the beam, and used end loads and overlapped simple supports to obtain the constant moment. His beams, consisting of strips of Saint Gobain glass, were from 12 to 20 cm long.

A rather obvious typographical error, as well as unfortunate omissions, prevent us from knowing from CORNU's paper[3] the remaining dimensions of the deflected glass prisms or whether the total load, which varied up to 500 g, was the sum of the two end loads or the individual maxima. SAINT-VENANT on various later occasions expressed great admiration for this experiment by CORNU, which he considered as settling definitively the molecular question in favor of the POISSON hypothesis.[4] Apart from the limitations of CORNU's presentation and his failure in some instances to provide experimental conditions prescribed by SAINT-VENANT's theory, the introduction of his experimental technique certainly was a milestone for measurement in solid mechanics. Indeed, CORNU had the honor, as he stated, of presenting at the French Academy the first results from measurements of deformation by means of interference optics, and the relation of those results to the mathematical theory of elasticity.

To determine the two principal radii of curvature in the top central section of a glass beam of constant moment, CORNU used an optical flat maintained at a small distance from the deflected upper surface of the transparent beam. A monochromatic light obtained from an induction spark flashing between two poles of magnesium augmented by the use of a Leyden jar, generated a dominant wave length of 3830 λ. The radii of curvature and consequently POISSON's ratio were determined by two methods. CORNU's optical examination of the deformed surface showed a system of conjugate hyperbolas having the same asymptotes. The tangent of the

[1] FIZEAU's experiments in [1864, *1*], which obviously stimulated CORNU's studies, were concerned with the coefficient of thermal expansion.

[2] BAUSCHINGER's mirror extensometer, which also used optical measurement, was the only other 19th century apparatus in solid mechanics for specimens of reasonable length which even approached the precision obtained by CORNU and by CANTONE [1888, *1*].

[3] CORNU [1869, *1*]. See also SAINT-VENANT [1856, *1*], [1878, *1*].

[4] JESSOP in 1921 observed that most of these experiments of CORNU violated the conditions under which SAINT-VENANT's principle applied, as is noted below. JESSOP [1921, *1*].

angle formed by each asymptote with the axial direction of the prism was equal to the inverse of the square root of the desired POISSON's ratio. By means of a very exact quadrilateral engraved on the optical flat, he found that the tangent of the angle was very close to the numerical value of two. He concluded that POISSON's ratio was close to the theoretically predicted value of $^1/_4$.

For a closer check, CORNU, again pioneering in deformation measurement, succeeded in photographing the interference ring pattern. Hence his micrometric measurements made with a telescope which enlarged the photograph 25 times, corresponding to a real enlargement of the phenomenon by about 6 times, gave directly and very accurately the displacement distribution from which the principal curvatures, and, consequently, POISSON's ratio could be determined.[1] Table 71 provides the results of 8 measurements on 7 different specimens of Saint Gobain glass, which gave an average value of POISSON's ratio of 0.237.

Table 71. CORNU (1869)

Designation of the strips	Thickness (mm)	Ratio of the transverse dimensions	Value of ν
# 2[a]	1.380	18.4	0.225
# 3	2.037	12.3	0.226
# 4	1.370	7.3	0.224
# 5	2.040	6.4	0.257
# 7[b]	1.554	8.7	0.236
# 1[c]	8.50	3.76	{ 0.243 0.250

[a] Specimens 2, 3, 4, and 5 had a distance of 16 mm between supports. As JESSOP first noted, 52 years later, in 1921, the respective widths from columns two and three are 25.4, 25.0, 10.0 and 13.0 mm, so that a comparison of widths for the distance of 8 mm from the center of beam to knife edges violated Saint-Venant's principle; i.e., for these measurements CORNU had failed to provide a constant moment beam. Specimens 1 and 7, however, did meet the required conditions. (JESSOP [1921, 1]).

[b] The distance between the supports was 120 mm.

[c] This strip had much larger dimensions; 60 cm long, 12 cm distance between the supports; Load, 1 kg. The two values of ν which it furnished were obtained by two entirely different methods: first by the direct measurement of the rings, then by another optical method based on the change of focus from parallel rays reflected on the surface before and after deformation.

CORNU then proceeded with the argument. Choosing the single measurement of CAGNIARD DE LATOUR[2] (which by then WERTHEIM had shown to be inaccurate) because it gave a value close to CORNU's own; curiously asserting that it was significant that KIRCHHOFF's few measurements lay between CORNU's and WERTHEIM's, although in fact KIRCHHOFF's results were closer to those of WERTHEIM; and dismissing WERTHEIM's hundreds of measurements from various experiments because they disagreed with CORNU's own results from a single type of test in a single solid; CORNU stated: "I am thus led to conclude that elastic isotropy is characterized by the property that the coefficient of transverse contraction is one-quarter the coefficient of longitudinal elasticity."[3] For this comparison, CORNU

[1] The full significance of this photography in 1869 can be appreciated in CORNU's contrasting the tedious difficulties of his attempts at determining the pattern from visual observations during the time of the spark with the first perusal at leisure of a permanent photographic record.

[2] CAGNIARD DE LATOUR [1828, *1*].

[3] CORNU [1869, *1*], p. 336.

showed KIRCHHOFF's value of $\nu = 0.294$ for steel rather than that of 0.387 for brass. He condemned WERTHEIM's results by gratuitously implying that WERTHEIM had chosen his specimens so carelessly he could not be certain that they were isotropic, and, without any documentation, CORNU equally gratuitously charged that WERTHEIM had unspecified "sources of error serious enough to render suspect the numbers he gave."[1] The gist of CORNU's argument was that transparent glass was the only solid which one could examine optically so as to be certain that one had perfect isotropy. Just how he ascertained this, he did not state.[2]

It is interesting to move ahead 75 years and compare the results of an entirely different direct optical determination of POISSON's ratio in glass, with the measurements by CORNU in 1869. In 1944 W. T. SZYMANOWSKI[3] applied the ultrasonic diffraction method developed by CLEMENS SCHAEFER and LUDWIG BERGMANN[4] in 1934 and 1935 for the determination of elastic constants. The theory for ultrasonic diffraction, from which they interpreted their results, had been developed in 1935 by FUES and LUDLOFF.[5] SZYMANOWSKI stated:

> A strong monochromatic source of light illuminates with polarized light an ordinary adjustable slit. After passing the slit the beam of light is made slightly convergent by a lens, and goes through the investigated transparent sample. The sample is set on a piezoquartz, the electrodes of which are connected to an oscillator of variable frequency. After passing through the sample and a crossed Nicol the light forms a magnified image of the slit which is viewed by a microscope with a filar micrometer eyepiece.
>
> The piezoquartz, which is excited in one of its overtones, sets the sample into strong longitudinal resonance vibrations, which are diffracting the beam of light. The diffraction pattern, as viewed in the microscope, forms then in the first order a line on each side of the central image of the slit. The sample is, however, set at the same time into transverse vibrations giving rise to another diffraction pattern superimposed over the longitudinal one. This is formed also by two lines equidistant from the central image. The mutual distance, however, of these lines is less than two times the mutual distance of the longitudinal diffraction lines.
>
> The measurement of the distance between the two parallel lines, as well in the longitudinal as in the transverse diffraction pattern can be easily and rapidly performed and repeated by means of the traveling wires of the micrometer eyepiece. These distances are the only experimental data, besides the density ϱ of the sample, and the resonant frequency ν, necessary for the determination of the Poisson ratio σ, the shear modulus μ, and the Young's modulus E, because:
>
> $$\sigma = \frac{1 - 2(d_l/d_t)^2}{2 - 2(d_l/d_t)^2}, \quad \mu = \frac{\nu^2 k^2}{(d_t/2)^2}$$
>
> and
>
> $$E = 2\mu(1 + \sigma),$$
>
> where d_l = distance of the 2 parallel lines in the longitudinal diffraction pattern, d_t = distance of the 2 parallel lines in the transverse diffraction pattern, k is an apparatus constant obtained by calibration with a grating having a known grating constant.[6]

In soft glass bars the measurement of d_l and d_t provided a value of POISSON's ratio of 0.2315, which is very close to that of CORNU's average of 0.237. SZYMANOWSKI's determination of POISSON's ratio by these ultrasonic means was, of course, similar to CORNU's, being also a direct measurement independent of specimen dimensions, and not requiring the measurement of either E or μ. As may be seen from this comparison, CORNU had based his very general conclusion upon

---

[1] *Ibid.*

[2] CORNU seemed to presume that optical isotropy implies elastic isotropy. There is no theoretical or experimental basis to this day, for this common assumption. As we have seen, WERTHEIM did properly check directly, if somewhat crudely, for elastic isotropy. The logic and thoroughness were all on the side of WERTHEIM.

[3] SZYMANOWSKI [1944, *1*].

[4] SCHAEFER and BERGMANN [1934, *1*], [1935, *1*].

[5] FUES and LUDLOFF [1935, *1*].

[6] SZYMANOWSKI [1944, *1*], p. 627.

the measurement on a solid which had values of POISSON's ratio in the vicinity of $^1/_4$, but he had not paid heed to the actual value obtained from his very precise measurements. His technique was sufficiently accurate to demonstrate that the actual value lay approximately 6% below the theoretical value. Thus, his results might best have been offered as at least preliminary evidence that POISSON's ratio for glass under close scrutiny differed from the theoretical value, rather than as a conclusive experimental demonstration that the universal constant of $\nu = ^1/_4$ should be adopted.

This particular experiment has been included here because it was used to determine POISSON's ratio. As we shall see in Sect. 3.39, the use of ultrasonics[1] as an experimental tool has been dominant in the exploration of the elastic properties of solids during the past two decades, just as KELVIN's double wire experiment and THURSTON's automatic testing machines were dominant during the last two decades of the 19th century.

The significance of KIRCHHOFF's direct measurement of POISSON's ratio was not completely lost in the years that followed. ARNULPH MALLOCK[2] in 1879, after recognizing the ingenuity of KIRCHHOFF's method for measuring the value in brass and steel rods, devised an experiment which, while perhaps lacking in accuracy, is arresting in its conception. Without mentioning either SAINT-VENANT or CORNU, he noted that if a rectangular bar whose depth was sufficiently small compared with its width (and both depth and width were relatively small, compared to the length), were "bent by opposing couples whose planes are parallel to one pair of sides, the other pair become surfaces of uniform anticlastic curvature, with principal radii of curvature R and $-\mu/R$."[3] MALLOCK then proceeded to describe a variation on CORNU's original experiment. For any kind of accuracy, it is of course necessary that R be sufficiently large. The dimensions of the bars studied were $8'' \times 1'' \times 0.25''$. The manner in which MALLOCK[4] applied the couples is shown in Fig. 3.36.

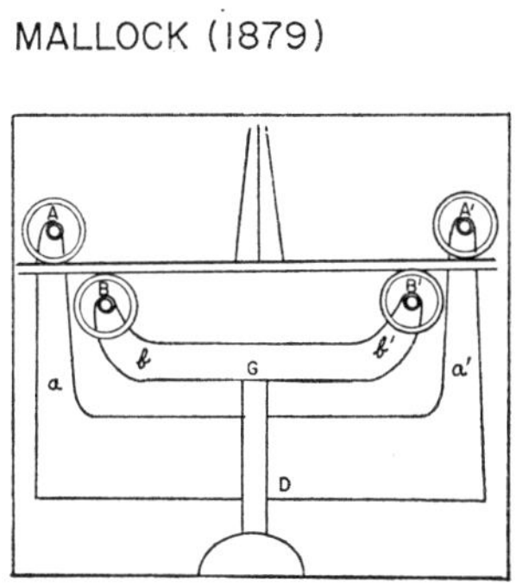

Fig. 3.36. The apparatus for MALLOCK's non-optical version of CORNU's experiment. Note the method for obtaining the constant moment beam.

To determine the two radii of curvature across and along the beam, four thin wires were inserted into very small holes drilled in the plane of the beam, but

[1] SZYMANOWSKI referred to his own experiment as a "supersonic" diffraction method, rather than using the word "ultrasonic" now current.

[2] MALLOCK [1879, *1*].

[3] *Ibid.*, p. 158. MALLOCK used μ for POISSON's ratio.

[4] ARNULPH MALLOCK was long an intimate friend of Lord RAYLEIGH, in whose laboratory he had briefly worked as an assistant (STRUTT [1968, *1*]).

arranged in the manner which is shown in Fig. 3.37a. The ends of the wire were brought under a microscope in the manner shown in Fig. 3.37b, from which the displacements were measured, allowing a calculation of the radii of curvature.

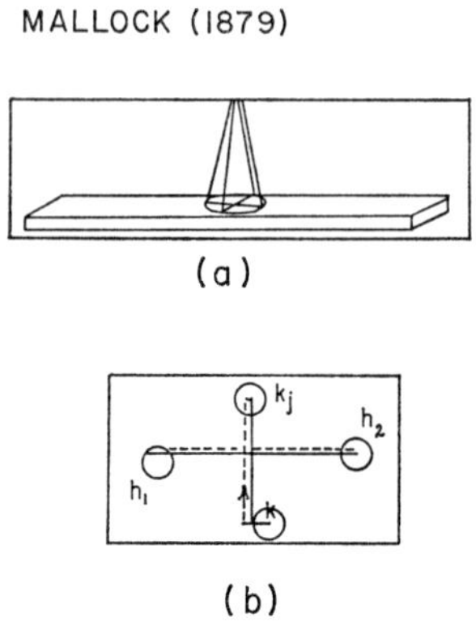

Fig. 3.37a and b. The method of attaching the wires (a) in MALLOCK's experiment, and the movement of the wires as seen in a microscope (b). The latter measurements provided the desired values of the principal radii.

From these measurements and the fact that $\nu = -\frac{R_1}{R_2}$, MALLOCK was able to measure POISSON's ratio directly without use of the large numbers required to determine it from the separate E and μ. He thought it was of particular importance in his experimental method that he could determine a POISSON's ratio for both isotropic and anisotropic solids. His values are included in Table 72 for several metals, woods, and other solids.

Table 72. MALLOCK (1879). *Values of ν for various substances*

Substance	ν	ν_σ	ν_ϱ	ν_τ
Steel	0.253			
Brass	0.325			
Copper	0.348			
Lead	0.375			
Zinc (rolled)	0.180			
Ditto (cast)	0.230			
Ebonite	0.389			
Ivory	0.50			
India-rubber	0.50			
Paraffin	0.50			
Plaster of Paris	0.181			
Cardboard	0.2			
Cork	0.00			
Box-wood	—	0.42	0.406	
Beech-wood	—	0.53	0.408	
Deal (white pine)	—	0.486	0.372	0.227

MALLOCK's conclusion in 1879 that POISSON's theory not only was not applicable to solids but that it had reached the status of a "kind of historical interest," suggests that the uniconstant vs multiconstant controversy had ended; this was certainly far from true, despite the fact that EVERETT[1] in his *C. G. S. System of Units* in 1891 featured MALLOCK's data of 1879 as a "final" experimental demonstration in favor of multiconstant elasticity.

3.22. The experiments of Voigt on the isotropy and moduli of glass (1882).

BAUSCHINGER[2] in 1879 and TOMLINSON[3] in 1883 examined the question of the accuracy of POISSON's ratio as calculated from measurements of E and μ. BAUSCHINGER's measurements, which were described above, included the direct determination of lateral contraction and longitudinal compression or elongation which he could compare with torsion, compression, and tension moduli in the same specimens. He raised serious doubts that any importance should be attached to ν when determined from the ratio of measured E and μ.

TOMLINSON compared the measurements of moduli and the corresponding calculation of POISSON's ratio for annealed and hard drawn specimens of iron, piano steel, platinum, German silver, copper, platinum-silver, brass, zinc, silver, aluminum, and lead. Great variation was found among the POISSON's ratios so calculated, both among the annealed specimens for some solids such as German silver, and particularly for the hard drawn metals for which the calculated ratio varied from extremely small to far in excess of $^1/_2$. Nine of TOMLINSON's calculated values of POISSON's ratio were for metals in the annealed state. As WERTHEIM had done before him, TOMLINSON averaged these values; however, instead of $^1/_3$, he obtained for the average of the metals considered, the numerical value 0.2515, "a number closely according with that assigned by POISSON as the value of σ for *each.*"[4]

In 1878 SAINT-VENANT summarized the experimental argument of those who remained strong proponents of the POISSON-CAUCHY molecular theory with its requirement that POISSON's ratio for isotropic solids be $^1/_4$. After examining the consequences of anisotropy in providing directionally dependent moduli in one-dimensional experiments in anisotropic solids, he emphasized that calculated or directly measured values of POISSON's ratio could vary widely. Referring to a collection of data on copper rods in the January 1878 issue of the London-Edinburgh-Dublin *Philosophical Magazine,*[5] which varied from $\nu = 0.40$ to $\nu = 0.23$, with values of $\nu = 0.26$ and $\nu = 0.23$ for flint glass and of $\nu = 0.27$ for iron, he concluded that the copper used, as well as the metal specimens in WERTHEIM's experiments for which ν was found to be $^1/_3$, were anisotropic, and that only for iron and particularly for glass had isotropic specimens been found. This premise

---

[1] EVERETT [1875, *1*], [1891, *1*]. I was able to obtain, and refer here to the fourth edition of this volume, published in 1891.

[2] BAUSCHINGER [1879, *1*].

[3] TOMLINSON [1883, *1*].

[4] *Ibid.*, p. 29. TOMLINSON used σ for POISSON's ratio.

[5] This is in the form of a table in a paper by Sir WILLIAM THOMSON (KELVIN) [1857, *1*], [1878, *1*], p. 18, in which he listed the data of WERTHEIM, EVERETT, KIRCHHOFF, and MAXWELL, together with a reference to general experience as a justification for the assumption that POISSON's ratio is zero in cork. For vulcanized rubber JOULE's value of less than 0.5 also was included, together with THOMSON's unreferenced summation of values of POISSON's ratio from 0.23 to 0.40 in copper. Most of these numbers, which have been reported upon above, were obtained from flint glass, brass, drawn brass, copper, iron, steel, cast steel, and vulcanized rubber.

was based primarily upon the work of CORNU described above,[1] who had stated without demonstration that his glass specimens were indeed isotropic.

VOIGT[2] in 1882 contested CORNU's conjecture, pointing out that the mere statement of transparency without other demonstration did not provide any evidence from which such a conclusion could be drawn with respect to the isotropy of elasticity. However, he affirmed, and demonstrated, that one could settle the question by choosing specimens from a plate of glass with different orientations and cut from different depths in the glass and then subjecting these specimens to tests in torsion and flexure. For flexure, the neutral plane was chosen to be parallel to both the long and short dimensions of the rectangular cross-section of the specimen. In this manner, by comparing the measured values of E and μ and the calculated values of POISSON's ratio, he was able to establish that he did indeed have an isotropic solid. Although the separate flexure and torsion tests were performed on the same specimens, they were not considered simultaneously as in the experiments by KIRCHHOFF. The details of VOIGT's apparatus had been developed and described in his doctoral dissertation in 1876 on the determination of elastic constants in rock salt.

To achieve even greater accuracy in his subsequent experiments, so that he could check isotropy from the numerical values of the moduli for the different orientations and locations in the master glass plate, VOIGT had to be even more precise in determining the dimensions of his specimens, particularly those for the cross-sections. By means of a scaled micrometer and a slide carriage, he was able to achieve an accuracy of 0.001 mm for each individual observation for measurements which were repeated at equal intervals along the rod. He estimated that his measurement of dimensions was so precise that the accuracy of the determination of individual moduli was within $^1/_{300}$ of its value.[3]

I have summarized the pertinent results of VOIGT's extensive tables for the measurements of the flexure and torsion on 9 specimens cut from 50 mm thick plates of "greenish glass" whose specific weight was 2.540 (refractive index, 1.55). He noted that despite its significant thickness, the glass did not show color in polarized light.[4] From a depth of about 6 mm down, the glass proved to be quite isotropic on the basis of his comparisons of the torsional moduli of 9 measurements of 6 different configurations, as may be seen in Table 73. Specimens noted as I and II in the table were from sections near the surface and had moduli which differed from those at greater depth. In this latter group, the average value of POISSON's ratio was 0.213, with the lowest measured value being 0.211 and the highest in the group being 0.218.

A similar study in "white mirror glass from the Rhine region" (specific weight, 2.56; refractive index, 1.53) provided the results, also shown in Table 73, for which the average POISSON's ratio was 0.2085. Incidentally, VOIGT compared the torsion calculations of a specimen of rectangular cross-section with the theories of SAINT-

[1] In view of the importance which SAINT-VENANT attached to "CORNU's ingenious experiment", while he remained an immovable adherent of uniconstant elasticity, there is some lesson to be learned from his failure to observe that four of the six beams CORNU studied had dimensions which violated his own principle of equipollent loads, i.e., SAINT-VENANT's principle (*see* JESSOP [1921, *1*], p. 552).

[2] VOIGT [1882, *1*].

[3] It is curious that after having sought for such extremes of accuracy, VOIGT was willing to compare measurements made at different temperatures which varied within 20° C, on the argument that in this range differences were negligible. As a matter of fact, however, I have calculated the possible error from this source and found it to be greater than that associated with his $^1/_{300}$ error from the measurement of dimension.

[4] VOIGT thanked NEUMANN for having recommended that GUINAUD in Paris produce the glass.

Table 73. VOIGT (1882)

Rod	Flexure			Torsion			Average E/μ	ν	Remarks
	Gage length (mm)	θ (° C)	E (g/mm²)	Gage length (mm)	θ (° C)	μ (g/mm²)			
				First series: Greenish glass					
I	64.1	14	6430000	48.4	13	2635000	2.438	0.220	Small sides parallel to the plane of the plate immediately below original surface (I, II)
		13	6420000					0.218	
II	64.1	13	6445000	48.85	13	2645000	2.426	0.219	
			6440000					0.218	
III	64.1	13	6480000	47.82	14	2670000	2.426	0.214	Same orientation as I and II, but cut from center of plate (III, IV)
		13.7	6480000	48.5		2672000		0.212	
				24.65		2675000		0.211	
IV	64.1	13.7	6460000	64.75	13	2665000	2.421	0.212	
				25.75		2677000		0.206	
V	62.0	14	6490000	44.9	13	2680000	2.422	0.211	Broad side parallel to plate plane and about 8 mm from free surface
VII	66.1	12	6460000	48.1	13	2665000	2.432	0.212	Square cross-section
			6480000					0.216	
			6505000					0.221	
VIII	72.0	13	6470000	48.3	12	2665000	2.428	0.214	Broad side parallel to the plane of the plate (VIII, IX)
IX	72.0	12.3	6490000	47.2	15	2660000	2.436	0.219	
	68.0	13	6475000					0.217	
							Average	0.215	
				Second series: White mirror glass					
A I	62	13	7375000	44.75	12.5	3040000	2.426	0.213	Small sides parallel to original surface of the plate (A I, A II)
A II	62	13	7340000	43.5	13	3040000	2.414	0.207	
B I	62	11	7350000	45.9	12	3040000	2.418	0.209	Broad sides parallel to original surface of the plate (B I, B II)
B II	62	13	7365000	47.2	12	3055000	2.411	0.206	
						Average	2.417	0.2085	

VENANT and CAUCHY, the respective merits of which had been an issue of some importance 30 years earlier, and found only SAINT-VENANT's to be in close agreement with experiment.

VOIGT concluded that for a solid whose state had been shown experimentally to be nearly isotropic, he had established that POISSON's ratio did not have the theoretical value of $^1/_4$. Instead, for the two glasses considered the smaller values of 0.2130 and 0.2085, respectively, were obtained. VOIGT thus established experimentally that the argument of SAINT-VENANT regarding the data of CORNU was inadmissible. We see once again that one may not simply dismiss data which do not agree with one's conjecture, on the *ad hoc* argument, without specific experimental demonstration, that some unchecked condition such as unmeasured anisotropy explains the lack of correlation between measurement and theory.

3.23 Mercadier's determination of the ratio of elastic constants from the first and second mode frequencies of a vibrating plate (1888).

A quite different approach to the problem of determining POISSON's ratio and hence of the applicability of the POISSON-CAUCHY molecular theory, was introduced in 1887 and developed in a memoir of 1888 by E. MERCADIER.[1] He first measured the frequencies of vibration of the first and second modes of circular steel plates and then used KIRCHHOFF's analytical results[2] to obtain the ratio of LAMÉ's constants: λ/μ. With n as a resonant frequency, n_0 the frequency of the fundamental mode and n_1 that of the first harmonic; $\theta=\lambda/2\mu$; e, the thickness of the disk; l, the diameter; E, the elasticity coefficient; δ, the density; d, the number of diametrical nodal lines; and c, that of the circular nodal lines; KIRCHHOFF's theory yielded

$$n=f(\theta, d, c)\sqrt{\frac{E}{\delta}\,\frac{(1+2\theta)^2}{(1+\theta)(1+3\theta)}}\,\frac{e}{l^2}\,, \tag{3.3}$$

$$\frac{n_1}{n_0}=\frac{f(\theta,0,1)}{f(\theta,2,0)}\,. \tag{3.4}$$

For the values[3] of θ designated in Table 74, the ratio of n_1/n_0 may be calculated as shown.

Table 74.

$\theta=\dfrac{\lambda}{2\mu}$	0.50	0.60	0.70	0.80	0.838	0.90	1.00
$\dfrac{n_1}{n_0}$ or $\dfrac{f(\theta,0,1)}{f(\theta,2,0)}$	1.615	1.641	1.665	1.687	1.695	1.708	1.729

If POISSON's ratio were $^1/_4$, θ would equal 0.50, and the ratio of the frequency of the first mode to that of the fundamental would be 1.615. On the other hand, if WERTHEIM's value of POISSON's ratio of $^1/_3$ were right, θ would be equal to 1, and the frequency ratio would be 1.729. MERCADIER performed a preliminary

[1] MERCADIER [1888, *1*]. Similar experiments determining fundamental and first overtone frequencies had, in fact, been performed on circular plates by WERTHEIM thirty-six years earlier in 1851 [1851, *1*] also with respect to the matter of POISSON's ratio. See Sect. 3.18 above.

[2] KIRCHHOFF [1850, *1*].

[3] See Eq. (3.5), Sect. 3.24.

series of experiments on disks of cast steel of which he knew neither the origin, nor composition, nor physical properties, to examine the possibilities of such an experiment. These preliminary investigations provided for steel an elasticity coefficient E of 20608 kg/mm², which is remarkably close to the accepted modern value. MERCADIER found also that the ratio $\lambda/\mu=2$ was in agreement with WERTHEIM's conjecture. These frequency measurements were followed by a series of careful experiments on six disks of steel of well-defined composition, physical properties, and fabrication, which had been especially prepared by a designated source. In Table 75 is shown MERCADIER's description of these steels, which varied from very mild to hard, along with their chemical composition, the quasi-statically determined elastic limit, rupture load, elastic elongation, and E-modulus.

Table 75. MERCADIER (1888).

Steels	Elastic limit[a]	Rupture load[a]	Elasticity elongation[b]	Elasticity coefficient[b]	Carbon	Silicon	Sulphur	Phosphorus	Manganese
1. Very mild	23.5	36.5	0.0520	19250	0.12	0.022	0.066	0.058	0.26
2. Very mild	23.2	35.5	0.0500	20000	0.12	0.022	0.066	0.058	0.26
3. Semi-hard	31.8	52.0	0.0500	20000	0.43	0.125	0.026	0.028	0.51
4. Semi-hard	26.5	50.5	0.0512	19550	0.43	0.125	0.026	0.028	0.51
5. Hard	39.8	70.1	0.0472	21170	0.64	0.160	0.032	0.033	0.71
6. Hard	40.2	69.4	0.0495	20200	0.64	0.160	0.032	0.033	0.71

[a] The numbers of columns 2 and 3 represent kilograms per millimeter squared.

[b] The numbers of column 4 correspond to 1 kg per millimeter squared and per meter length; they are determined on a length of only 200 mm, so that the *static* coefficients of column 5 do not present great guarantees of precision.

As MERCADIER pointed out in his footnote to his table, the lengths of the specimens cut from the disks were necessarily so short that the quasi-static E-modulus had to be viewed as merely providing an indication of the general range of this elastic constant. The ratio of n_1/n_0 was determined by experiment for each disk by the method of beats. The fundamental note and the first harmonic of each were compared to well-calibrated, graduated tuning forks. The results for the six steels are shown in Table 76.

Table 76. MERCADIER (1888)

Steels	n_1	n_0	n_1/n_0	Average n_1/n_0
1. Very mild	903.0	528	1.710	1.710
2. Very mild	901.0	526.8	1.710	
3. Semi-mild	899.4	530	1.694	1.6966
4. Semi-mild	898.8	530.4	1.697	
5. Hard	906.0	524.65	1.727	1.7275
6. Hard	906.2	524.4	1.728	

Upon comparing these results with the predictions from KIRCHHOFF's analysis listed in Table 74, one notes that for disks one and two, $\lambda=1.8\,\mu$; for disks three and four, $\lambda=1.7\,\mu$; and for disks five and six, $\lambda=2\,\mu$, corresponding to values of POISSON's ratio of $\nu=0.321$, $\nu=0.315$, and $\nu=0.333$, respectively. MERCADIER thus concluded that if the ratio $\lambda/\mu=1$ were, as indicated by SAINT-VENANT, characteristic of the isotropy of an elastic body, then the steels defined above were in no way isotropic. However, he remarked that for the different steels varying from very mild to hard, with different chemical compositions, rupture loads, and elastic limits, he had measured remarkably close values of λ/μ. This of course provided strong evidence that (as is now known) POISSON's ratio was not $^1/_4$ for all isotropic solids. The values obtained by MERCADIER were somewhat higher than KIRCHHOFF's $\nu = 0.294$ for steel, which is now the accepted average value, but were certainly within the range of numbers for various steels given by BAUSCHINGER in 1879.

3.24. The piezometer experiments of Amagat (1884–1889).

Despite the growing experimental evidence that POISSON's ratio did indeed vary from one solid to another, numerous theorists and experimentists remained adamant, loth to abandon the thought that the theoretical correlation between atomistic and continuum concepts which led to the CAUCHY relations did not apply to the infinitesimal deformation of any of the common solids.

One experimentist in that category was ÉMILE HILAIRE AMAGAT,[1] who, as late as 1884, in the face of the growing evidence, including some of his own, clung to the experiments of CAGNIARD DE LATOUR (1828) and of CORNU (1859) and assumed that a perfect solid must indeed have a POISSON's ratio of $^1/_4$. In addition to the usual assumption that unknown anisotropy was partly at fault, an argument that by the late 1880's we have seen VOIGT dismissing with reference to two kinds of glass, AMAGAT proposed that imperfect elasticity, i.e., unmeasured small permanent deformation, was an equally important explanation why the behavior of real solids deviated from "perfect" solids. It is curious that most of those who resisted the implications of the mounting experimental evidence, almost invariably turned to unfounded attacks upon WERTHEIM, as if somehow his having found values of a POISSON's ratio approximating $^1/_3$ for the few solids which he considered, in some vague way provided a valid objection to any experiment which gave values of POISSON's ratio significantly different from $^1/_4$. I remind the reader (see Sect. 3.19 above) that WERTHEIM had performed the widest variety of experiments related to this matter, and had himself stated, twenty to thirty years before many of these arguments were brought forth, that he claimed only to have found a value approximately $^1/_3$ for the solids he had examined, not a single value common to all metals.[2]

AMAGAT also determined a value of POISSON's ratio for India rubber in 1884. As we have seen, WERTHEIM had considered the same solid in an experiment in which he had measured both the lateral contraction and the axial elongation in rubber bars. He had emphasized that he viewed this preliminary experiment as a response to REGNAULT who the previous year had first raised the question with respect to the physical applicability of the POISSON-CAUCHY theory; initially WERTHEIM wished merely to demonstrate that POISSON's ratio for some solids

[1] AMAGAT [1884, *1*].

[2] It is interesting that in the many posthumous adverse criticisms of WERTHEIM, as well as those before his death in 1861, there was never any genuine discussion of his experimental details or of what was the basis of the often caustic statements of attack and dismissal which his experiments in 1849 so often had generated. It is also interesting to note that in modern studies when averaging POISSON's ratio for the elements, one obtains a value of approximately $^1/_3$, with large variations, of course, for individual solids. Perhaps the major criticism which should have been made of WERTHEIM was that he was overly influenced by his own averages. Such a criticism would be difficult to level at most of his critics, who generally confined themselves to a very small number of measurements from some one particular point of view.

in small strain was indeed not $^1/_4$. WERTHEIM then had turned to his piezometer experiments for the study of glass and brass. The latter experiment was drawn upon by AMAGAT[1] in his study of India rubber in 1884. His piezometer was constructed like REGNAULT's forty years earlier, in that pressure could be applied to the inside or the outside of a cylinder or sphere, either separately or at the same time. In AMAGAT's experiments, however, the apparatus was sufficiently large that two spheres of different solids could be placed side by side and thus delicate comparative measurements could be made. The two spheres in those experiments were of rubber and bronze.

Denoting by χ the reciprocal of the bulk coefficient of cubic compressibility, and introducing α as the reciprocal of the E-modulus with λ and μ the LAMÉ constants, AMAGAT[2] noted that for an isotropic solid, POISSON's ratio ν is given by

$$\nu = \frac{\lambda}{2(\lambda+\mu)}\,. \tag{3.5}$$

Also, for the relation between cubic compressibility and the reciprocal of the E-modulus, one has

$$\chi = 3\alpha(1-2\nu) \tag{3.6}$$

and, correspondingly, between the reciprocal of the E-modulus and the shear modulus,

$$\frac{1}{\alpha} = \mu\,\frac{3\lambda+2\mu}{\lambda+\mu}\,. \tag{3.7}$$

If two different bodies are being compared, as in AMAGAT's piezometer experiments, one has for the ratio of the cubic compressibility for two geometrically identical spheres

$$\frac{\chi}{\chi'} = \frac{\alpha(1-2\nu)}{\alpha'(1-2\nu')} \tag{3.8}$$

where primes are introduced to distinguish between two solids.

In the first experiment AMAGAT noted that a very small amount of external pressure raised by 300 divisions the level of the water in the capillary tube attached to the rubber sphere, whereas in the capillary tube of the bronze sphere the same amount of pressure scarely moved the meniscus at all. From this he concluded that whatever the POISSON's ratio might be for bronze, the ratio of α to α' was very large, i.e., the E for rubber was extremely small with respect to that for bronze. He performed direct tensile tests on specimens of the two solids and found what already was known from the experiments of WERTHEIM and others, that the ratio of E for brass to rubber was 60000. Compressing both spheres internally and externally at the same time, he then noted that the variations of the interior volume should be proportional to χ and χ' and thus χ was extremely large with respect to χ'. The effect of the compressibility of the water was negligible; hence he expected that the liquid would have to re-rise in the capillary stem of the rubber sphere, but nothing of that sort occurred; in fact, whatever the pressure, the water descended. Thus he concluded that the rubber must indeed be incompressible, with a POISSON's ratio of $\nu = ^1/_2$.

Erroneously concluding that WERTHEIM had stated that POISSON's ratio was $^1/_3$ for rubber, although in fact he had stated merely that it was not $^1/_4$ (see Fig. 3.28), AMAGAT emphasized that assuming POISSON's ratio to be $^1/_3$ for the two solids was entirely inconsistent with his piezometer observations in rubber. Thus,

[1] AMAGAT [1884, *1*].
[2] *Ibid.*

instead of viewing his experimental achievement as confirmation that in small deformation rubber is essentially incompressible, he regarded it merely as proof that in at least one solid, POISSON's ratio was not equal to $^1/_3$. As to its also not being equal to $^1/_4$ as required by the POISSON-CAUCHY theory, this he ascribed to the possibility that, in fact, Eqs. (3.5) to (3.8) themselves do not apply.

In a paper in 1888, AMAGAT[1] described piezometer experiments on steel and bronze for which the REGNAULT cylinders replaced the spheres, and the two specimens studied were of the same solid with identical interior radii but with different exterior radii. They were terminated by "extremely resistant" plane bases. Using water of maximum density in order to minimize error, he kept the temperature of the bath at 4° C. Following the same program as for the experiments on rubber and bronze described above, he obtained for the two steel cylinders values of POISSON's ratio of $\nu = 0.2609$ and $\nu = 0.2620$. For the two bronze cylinders he found $\nu = 0.3190$ for one, and $\nu = 0.3204$ for the other. He noted that these values were accurate only if the analysis were correct. His curious conclusion was that since the value for steel was so close to $^1/_4$ and thus in agreement with the experiments of CORNU on glass, the contention was supported that the steel, like the glass, was a nearly perfectly isotropic body.

To explain the fact that the bronze cylinders provided a different value he stated that they had been cast vertically, and because they were cooled in that position they had taken on a structure which would provide a difference in the elasticity in the vertical and transverse directions, and thus would have taken on an anisotropy he presumed common to WERTHEIM's cylinders which had been drawn through a die. Therefore, having assumed the steel to be isotropic, AMAGAT concluded that when his results were compared with those of CORNU the POISSON-CAUCHY theory indeed could be seen as having been established experimentally.[2] For metals, at this point AMAGAT in fact had done nothing more than raise some questions regarding the isotropy of cast bronze.

In 1889, in connection with the study of the compressibility of liquid mercury AMAGAT performed the dual piezometer experiments for glass and crystal.[3] He compared his data with those of MICHELE CANTONE, obtained from the same experiment.[4] CANTONE had found for four glass cylinders the following values of POISSON's ratio, ν: 0.246, 0.261, 0.264, and 0.256, with an average value of 0.257. In another series of experiments with both crystal and glass cylinders, he obtained the elongation α and the cubic compressibility χ for the solid, as well as the coefficients of apparent compressibility and absolute compressibility for mercury.[5] These data are shown in Table 77.

Also in 1889, AMAGAT, using piezometers, extended his studies of compressibility to include steel, copper, brass, delta metal, and lead, in addition to the glass, bronze, and crystal described above. For these solids he also applied what he referred to as "the method of WERTHEIM", using the same apparatus as for the glass and crystal, with an arrangement which allowed the measurement of the

[1] AMAGAT [1888, *1*]. ÉMILE HILAIRE AMAGAT is remembered primarily for his research on the compressibility of gases and liquids over a pressure range up to 3000 kg/cm² and a temperature range of 200° C. The major reference on these matters is what P. W. BRIDGMAN described as "the classical paper of AMAGAT" [1893, *1*], in which he had summarized his work over a period of nearly 25 years.

[2] AMAGAT also had submitted his cylinders to longitudinal traction to determine E.

[3] AMAGAT [1889, *1*, *3*]. See also [1888, *1*, *2*], [1889, *2*], and [1890, *1*].

[4] CANTONE [1888, *1*]. AMAGAT referred to him as CAUTONE, and claimed that his was "un travail très intéressant, de très peu postérieur à ma première Note sur ce sujet." The French and the Italian papers in fact were published at the same time.

[5] AMAGAT [1889, *1*].

Table 77. AMAGAT (1889)

Cylinder no.	Crystal and glass			Mercury	
	POISSON's ratio, ν	Coefficients of elongation	Coefficients of cubic compressibility	Coefficients of apparent compressibility	Coefficients of absolute compress. ibility
Glass					
1	0.2476	0.000001434	0.000002202	0.000001696	0.000003898
2	0.2450	0.000001437	0.000002200	0.000001680	0.000003880
3	0.2428	0.000001419	0.000002190	0.000001744	0.000003934
Average	0.2451	0.000001430	0.000002181	0.000001707	0.000003904
Crystal					
1	0.2538	0.000001604	0.000002369	0.000001547	0.000003916
2	0.2481	0.000001603	0.000002423	0.000001502	0.000003925
3	0.2534	0.000001624	0.000002403	0.000001470	0.000003937
4	0.2443	0.000001580	0.000002424	0.000001530	0.000003954
Average	0.2499	0.000001602	0.000002405	0.000001512	0.000003933
			General average for mercury		0.000003918

elongation of the cylinders directly by means of micrometer screws in a manner completely independent of the movements of the apparatus. These two methods provided separate values of POISSON's ratio, ν, the coefficient of cubic compressibility, χ, and the reciprocal E-modulus α. These experimental data are tabulated in Table 78 for the 7 solids.

Table 78, AMAGAT (1889).

	POISSON's coefficient			Coefficient of cubic compressibility	Elasticity coefficient ($1/\alpha$) (kg/mm^a)		
	1st method	2nd method	Average		1st method	2nd method	Average
Glass	0.2451	—	0.2451	0.00002197	6775	—	6775
Crystal	0.2499	—	0.2499	0.00002405	6242	—	6242
Steel	0.2694	0.2679	0.2686	0.00000680	20333	20457	20395
Copper	0.3288	0.3252	0.3270	0.00000857	11979	12312	12145
Brass	0.3305	0.3236	0.3275	0.00000953	10680	11022	10851
Delta metal	0.3333	0.3468	0.3399	0.00001021	12054	11331	11697
Lead	0.4252	0.4313	0.4282	0.00002761	1626	1493	1556

The remainder of this paper of AMAGAT consisted in an attempt to account for the fact that only the values for glass, the crystal, and perhaps the steel, were sufficiently close to $\nu = {}^1/_4$ to lend some credence to the POISSON-CAUCHY theory, of which AMAGAT, even as late as 1889, was still a devoted disciple. With the usual reference to the undefined errors, AMAGAT still believed WERTHEIM's conclusion was "absolutely unacceptable," but he found it difficult to attribute solely to the "flaws of isotropy" WERTHEIM's values of POISSON's ratio. He questioned whether lead were really a solid, and he speculated whether or not POISSON's ratio might not tend toward the theoretical limit of ${}^1/_4$ as the stress went to zero, a hypothesis which he fully realized was not confirmed by the facts.

He suggested that if the higher values of POISSON's ratio were measured, rather than the lower theoretical value, one had to presume that one had imperfect solids. In other words, if it were possible to achieve a state of complete isotropy and perfect elasticity in any solid, one would expect to find experimentally the theoretical value. Ignoring the definitive experiments by VOIGT[1] on glass in 1882, AMAGAT considered that material to be the most perfect, with lead and rubber at the other end of the scale. He stated:

The perfect solid, according to the point of view before us, and in which ν should be rigorously 0.25, would thus be the one that realizes the double condition of being at one and the same time perfectly elastic and perfectly isotropic.[2]

3.25. The experiments of Bock on the dependence of Poisson's ratio upon temperature (1894).

In 1894 ADALBERT MICHAEL BOCK[3] proposed that POISSON's ratio should increase with temperature for small deformation, reaching $^1/_2$ at the melting point. This same suggestion was attributed both to BOCK and to GEORGE GABRIEL STOKES by CLEMENS SCHAEFER[4] in 1902; he did not cite any work of STOKES.[5] BOCK thought that the demonstration of this fact experimentally would shed considerable light on the molecular constitution of solids. Regretting that he was unable to determine POISSON's ratio from the low temperature of "steaming hydrogen" to the melting point, he confined his experiments to the range of temperatures from 0 to 150° C to determine whether any trends could be established. SCHAEFER, who also subscribed to this melting point hypothesis, as we shall see, determined POISSON's ratio at room temperature for such extremely low melting point temperatures as those for selenium, WOOD's alloy, and LIPOWITZ's alloy, obtaining values between 0.45 and 0.49, which presumably lent some credence to the idea. Only a few years later, in 1910, GRÜNEISEN[6] effectively disposed of the melting point hypothesis.

However, BOCK's work on the dependence of POISSON's ratio on temperature stands by itself as an interesting initial study of an important phenomenon. He had copied, with increased precision, the experiments by KIRCHHOFF 35 years earlier, determining the POISSON's ratio directly from the combined torsion-flexure behavior in a manner independent of the dimensions of the cross-section of the specimen. Because the mirror system and all of the other details of the KIRCHHOFF experiment were faithfully reproduced, one need refer only to KIRCHHOFF's paper of 1859 described earlier. To perform tests at various temperatures, BOCK enclosed the apparatus in a rectangular iron plate box, which was set inside a larger one so that the space between the double walls could be heated. Referring to the fact that KIRCHHOFF had faced the problem of considering the opposing opinions of POISSON and WERTHEIM, a matter which had been settled definitively in favor of the latter, but with a different POISSON's ratio for each material, BOCK reexamined the problem of whether in fact KIRCHHOFF's experiment could provide an absolute value of POISSON's ratio. He noted that the corrected and uncorrected results differed at most by 1%, while fluctuations due to the peculiarity of individual specimens exceeded this value, so that it was necessary

[1] VOIGT [1882, *1*]. And see VOIGT's experiments on rock salt [1876, *1*], [1884, *1*].
[2] AMAGAT [1889, *2*], p. 1202.
[3] BOCK [1894, *1*].
[4] SCHAEFER [1902, *2*].
[5] CHWOLSON [1908, *1*] also attributed the suggestion to both BOCK and STOKES, also without citation for the latter
[6] GRÜNEISEN [1910, *2*].

to pay even more particular attention to the prior thermal history and to such matters as the thermal elastic after-effect, which might certainly influence the experimental results.

Letting T be the torsion scale divisions from the mirror reflection as seen by the telescope; and B, the flexure contribution in the same scale divisions; with $2s$ the total length of the specimen under study; and $4l$ the combined length of the two opposing lever arms at the end of which the loads were applied, BOCK gave the following simple linear expression for POISSON's ratio for an isotropic HOOKEAN solid:

$$\nu=\frac{T}{B}\cdot\frac{s}{2l}-1. \tag{3.9}$$

BOCK performed his first experiments upon a 3.38 mm diameter round rod of hard English steel whose $^1/_2$ length s was 149.3 mm with a lever arm length $2l$ of 140 mm, which provided the values of B, T, and POISSON's ratio ν shown in Table 79 for measurements at or near 22 and 120° C, repeated at the times indicated, for an applied load of 100 g. He used two telescopes in determining the values of T and B. The interested reader is referred to the original paper for details.[1]

Table 79. BOCK (1894).

$t°$	B	T	ν	Date and hour
22	23.34	28.05	0.282	4 Dec., 11 h
120	24.19	28.98	0.278	4 Dec., 12 h
22	23.31	27.96	0.279	5 Dec., 9 h
20	23.33	27.92	0.276	7 Dec., 4 h
120	24.13	28.87	0.276	10 Dec., 5 h

Those experiments on 4 and 5 December were repeated a few days later with the applied load increased to 200 g. BOCK expressed considerable surprise that the measured POISSON's ratio was independent of temperature; i.e., it did not increase as he had expected. A few months later he repeated the experiments on a steel rod of precisely the same quality which for each day of one week had been held for many hours in a "very weak red heat," after which he measured its POISSON's ratio. Incidentally, it should be noted that BOCK knew he needed to measure the lengths s and l separately at each ambient temperature. In Table 80 are shown the POISSON's ratios determined by him at the different temperatures.

After each temperature increase, BOCK returned the ambient temperature to 20° C before reheating to the new value. The average value for all of the room temperature measurements was $\nu=0.256$ for this annealed iron rod, which of course was considerably below the values obtained previously for the same type of rod in the hardened state. Even though the measured POISSON's ratios at 20° C were only from 0.254 to 0.259 at different positions in the heating cycle, whereas the values at all temperatures up to 100° C were in the same range (only the value of $\nu=0.261$ at 120° C was slightly out of the range) BOCK concluded that for the annealed iron rods there was a slight increase in POISSON's ratio with increasing temperature.

Also in Table 80 are BOCK's values for copper, silver, and nickel in which in every instance the slight increase of POISSON's ratio with temperature is countered by fluctuations of almost the same amount at room temperature. Comparing

[1] BOCK [1894, *1*].

Table 80. Bock (1894).

$t°$	B	T	ν	Date and hour
	Annealed iron: $s = 149.5$ mm, $2l = 140$ mm, $P = 100$ g			
20	20.02	23.55	0.256	23 July, 11 h
40	20.16	23.71	0.257	23 July, 11 h
20	20.05	23.56	0.255	24 July, 9 h
60	20.30	23.86	0.255	24 July, 11 h
20	20.09	23.58	0.254	25 July, 3 h
80	20.38	23.99	0.258	25 July, 5 h
20	20.02	23.59	0.259	26 July, 9 h
100	20.48	24.11	0.257	26 July, 11 h
20	20.03	23.58	0.258	26 July, 3 h
120	20.58	24.29	0.261	26 July, 6 h
20	20.04	23.57	0.256	27 July, 9 h
150	20.85	24.59	0.259	27 July, 11 h
20	20.08	23.60	0.255	27 July, 4 h
	Copper: $s = 149.3$ mm, $2l = 140$ mm, $P = 50$ g			
20	12.02	15.10	0.340	1 August, 10 h
40	12.02	15.17	0.346	1 August, 12 h
20	12.03	15.05	0.331	1 August, 4 h
60	12.06	15.32	0.355	1 August, 6 h
20	11.86	15.03	0.351	2 August, 9 h
80	12.14	15.42	0.355	2 August, 10 h
20	11.92	15.05	0.346	2 August, 12 h
100	12.29	15.57	0.352	2 August, 4 h
20	11.92	14.99	0.341	2 August, 6 h
120	12.41	15.67	0.347	3 August, 9 h
20	11.80	14.96	0.352	3 August, 10 h
150	12.53	16.08	0.370	3 August, 12 h
20	11.64	14.86	0.361	3 August, 4 h
	Silver: $s = 150.0$ mm, $2l = 140.0$ mm, $P = 50$ g			
20	10.66	13.34	0.337	27 Jan., 3 h
40	10.85	13.63	0.346	27 Jan., 5 h
20	10.68	13.20	0.324	28 Jan., 5 h
60	10.93	13.79	0.352	29 Jan., 11 h
20	10.52	13.16	0.340	30 Jan., 9 h
80	11.26	14.35	0.366	30 Jan., 10 h
20	10.65	13.62	0.370	30 Jan., 11 h
100	11.45	14.65	0.372	30 Jan., 5 h
20	10.42	13.15	0.352	2 Feb., 4 h
120	11.89	15.31	0.381	2 Feb., 6 h
20	10.39	13.15	0.356	5 Feb., 11 h
	Nickel: $s = 156.4$ mm, $2l = 140$ mm, $P = 200$ g			
20	10.17	12.13	0.332	17 March, 2 h
40	10.25	12.21	0.331	17 March, 3 h
20	10.22	12.16	0.329	17 March, 5 h
60	10.34	12.36	0.336	17 March, 6 h
20	10.32	12.09	0.309	18 March, 11 h
80	10.39	12.43	0.337	19 March, 12 h
20	10.20	12.11	0.326	20 March, 10 h
100	10.51	12.51	0.330	20 March, 12 h
20	10.13	12.18	0.343	21 March, 12 h
120	10.56	12.64	0.337	21 March, 4 h
20	10.19	12.16	0.333	21 March, 6 h

the ratio of the POISSON'S ratio at the highest temperature with the average at room temperature, BOCK noted that the percentage increase was in the order of the melting points of the metals. The order from the largest increase to the smallest was: silver, copper, nickel, and iron, which is, of course, the order of the increasing melting point for these metals. There was such scatter in the experimental data that BOCK'S conclusion obviously indicates more of belief than of fact. Nevertheless these measurements of the temperature-dependence of POISSON'S ratio in metals are significant as being the first to attack the question of how the relation between elastic constants in infinitesimal elasticity depends upon temperature.

In 1902, CLEMENS SCHAEFER also was of the strong opinion that POISSON'S ratio should proceed to $^1/_2$ with increase in temperature. This conclusion was based upon his having extrapolated his empirical equation for the temperature dependence of POISSON'S ratio:

$$1+\nu_t=(1+\nu_0)\frac{1-\alpha_t}{1-\beta_t} \tag{3.10}$$

where $t°$ C was the temperature of the test, and α and β were the temperature coefficients of E and μ, respectively. This empirical relation was developed in a series of experiments described in two earlier papers.[1,2] Of interest here[3] is the paper in which SCHAEFER, using the optical interference method of CORNU,[4] based upon the extensive re-study by CONSTANTIN RUDOLPH STRAUBEL[5] in 1899 of CORNU'S experiment on glass, attempted to extend the experiment to other solids with low melting points. The metals he considered were selenium, WOOD'S alloy, and LIPOWITZ'S alloy.

Mirror flat beams of these solids were made by pouring the metal between flat glass plates and then cooling it. SCHAEFER pointed out that it was necessary to prepare many specimens in this manner so as to obtain a few which could be used in the CORNU experiment. The usual end loads and intermediate simple supports were employed to obtain a constant moment. An optical glass flat was located at the center of the beam in a plane parallel to the tangent plane of the beam at that point. Vertical monochromatic light provided a fringe pattern from the anticlastic curvature when the beam was under load. Stimulated by BOCK'S idea, SCHAEFER conjectured that these solids, for which room temperature was very close to their melting points, should have POISSON'S ratios approaching $^1/_2$. For selenium with a melting point of 217° C he obtained $\nu=0.447$; for WOOD'S alloy with a melting point of 65° C, $\nu=0.489$; and for LIPOWITZ'S alloy with a melting point of 75° C, $\nu=0.452$, which SCHAEFER thought supported his conjecture.

SCHAEFER made a more detailed study of selenium by determining POISSON'S ratio at a temperature of 80° C; i.e., he compared the measurements which were made while the plate was still heated, with those at 20° C. The results of this study are interesting. The specimen was heated to 80° C, at which he obtained a value of POISSON'S ratio of 0.490. It then was allowed to cool off to room temperature, with the following values for POISSON'S ratio being determined at ten-minute intervals during the cooling: at 10 min, $\nu=0.480$; at 20 min, $\nu=0.480$; at 30 min, $\nu=0.448$; and at 40 min at which time the specimen had reached room

[1] SCHAEFER [1901, *1*].
[2] SCHAEFER [1902, *1*].
[3] SCHAEFER [1902, *2*].
[4] CORNU [1869, *1*].
[5] STRAUBEL [1899, *1*].

temperature of 20° C, $\nu=0.445$. SCHAEFFER thought all of these measurements strongly supported the hypothesis that POISSON'S ratio increased to $^1/_2$ at the melting point.[1] However, he did note that in considering the elastic behavior of such materials, it was difficult to perform an elastic experiment without at least a modicum of permanent deformation.

In the next chapter on finite deformation where experimental studies by CARL PULFRICH[2] in 1886 on the determination of a geometrical POISSON'S ratio for the large deformation of rubber tubes are described, we shall see the importance of the elastic after-effect in such measurements. A number of the experimentists whose work has been discussed above made minor reference to or were concerned about the possibility of error arising from this phenomenon but proceeded nevertheless as if it had no influence. In general, all of the defects of dead-weight loading strain measurement which had been discovered in the 1830's could contribute to the general variation of the measured data.

3.26. Straubel's definitive study of the Cornu experiment for the direct measurement of Poisson's ratio (1899).

CONSTANTIN RUDOLPH STRAUBEL[3] is among the very few experimentists in this field who have succeeded in actually obtaining maximum precision from an exhaustive independent analysis, by means of experiment, of the influences of all manner of spurious effects. In a paper of 1899 on the measurement of POISSON'S ratio by means of an enormous number of interference optics experiments based upon the original experiment by CORNU, STRAUBEL began by stating that it no longer was necessary to decide by experiment for or against the applicability of the molecular theory of elasticity of NAVIER and POISSON: the theory did not apply. Reviewing the measurements on glass of EVERETT, CORNU, AMAGAT, VOIGT, CANTONE,[4] and J. VON KOWALSKI[5] (all but the experiments of CANTONE and KOWALSKI have been described above), he noted the variation of the data which had been obtained. The seven measurements by CORNU had provided an average value of POISSON'S ratio of $\nu=0.237$. EVERETT had obtained values of $\nu=0.224$ and $\nu=0.258$ for two different specimens of the same glass. For different types of glass, VOIGT had obtained for a known and tested isotropic material very accurate values of $\nu=0.213$ and $\nu=0.208$. CANTONE, who had worked with tubes of Thuringer glass, had determined POISSON'S ratio by measuring the change

[1] In the discussion to the paper of GAROFALO, MALENOCK, and SMITH [1952, *1*]. M. J. MANJOINE referred to a conversation with A. NADAI who commented upon the early speculation that POISSON'S ratio should approach the value of $^1/_2$ as the material approached the melting point. NADAI gave as an illustration the fact that a person can be supported by, and skate on, ice that is practically at its melting point as evidence that normal elastic properties are observed to the vicinity of the melting point. This is a rather curious argument.

[2] PULFRICH [1886, *1*]. See Chap. IV, Sect. 4.39, below.

[3] STRAUBEL [1899, *1*].

[4] CANTONE [1888, *1*]. CANTONE and AMAGAT [1888, *1*] independently repeated, with some improvements in technique, WERTHEIM'S cylindrical piezometer experiments of forty years before. WERTHEIM [1848, *1*]. The use of the words "Nuovo metodo" in the title of CANTONE'S two "Notes" was a bit odd. He mentioned WERTHEIM and REGNAULT, and suggested that the latter's careful studies of compressibility had been diminished by REGNAULT'S use of WERTHEIM'S data. Casually dismissing the results of the "illustre sperimentatore francese," WERTHEIM, in favor of those of CORNU [1869, *1*], CANTONE considered that his own measurements finally had settled the controversy concerning the molecular theory of elasticity by establishing its applicability.

[5] KOWALSKI [1890, *1*].

of the inner volume by exterior pressure, and the change of length by inner pressure,[1] obtaining for four different tubular specimens of the same glass values of POISSON's ratio of $\nu=0.246$, 0.261, 0.264, and 0.256. KOWALSKI had studied the flexure and torsion of drawn glass rods, obtaining in 1889 a value of $\nu=0.226$, and in a later series of experiments on the same specimens, a value of $\nu=0.212$.

STRAUBEL proposed to measure carefully the POISSON's ratio of thirty different glasses which had been specially prepared as to composition and manufacture for his experiments by the Jena Glass Technical Laboratorium. Noting, probably in retrospect after his major study had been completed, that the original experiment by CORNU thirty years before had contained a number of small errors and that, as a matter of fact, CORNU himself, in stating that his value of $\nu=0.237$ was proof of a predicted value of 0.250 had thus recognized the possible magnitude of the errors in his measurement, STRAUBEL set out to dissect the CORNU experiment, utilizing all of the optical and photographic improvements of the previous three decades. Using the interference apparatus constructed by C. PULFRICH,[2] with carefully tested filters to enable him to work with the 6560λ red hydrogen line for monochromatic light, and a specially constructed rigid metal camera and newly designed rotating table, STRAUBEL measured the angle between the hyperbolic asymptotes for constant moment glass beams to determine the ratio of the two principal curvatures and, consequently, POISSON's ratio.

In addition to an exhaustive reexamination of the method of calculation, STRAUBEL investigated the errors arising from the method of applying the load; he found it preferable to use screws applied to the two overhung sections of the simply supported, constant moment beam. He paid considerable attention to the nature of the simple supports and their effect upon the result and made an exhaustive study, running into a very large number of individual measurements, of the effects upon the measured value of changing over a fairly large range the thickness and width of the glass, as well as the points of location of the simple supports and point of application of the screw loads. He found that one of the major sources of error lay in the impossibility of obtaining truly flat plates, devoid of initial small curvature.

STRAUBEL then proceeded to an extensive program of grinding and polishing specimens and of determining initial curvatures and their effect upon the radii of curvature obtained under deflection. He made many optical tests of the method itself, independent of the anticlastic curvature measurements which were the purpose of the study. It would be interesting, if space allowed, to describe these details.[3] The amount of data contained in the paper is enormous, and yet STRAUBEL lamented the fact that he was able to include only a very small fraction of the total number of measurements made. He chose one glass, labelled #1991, with the composition: SiO_2, 65.22; B_2O_3, 2.7; ZnO, 1.5; As_2O_5, 0.5; BaO, 10.0; Na_2O, 5.0; K_2O, 15.0; Mn_2O_3, 0.08; as representative of the experiments made for the various dimensions over a range of deflections from minimum to a maximum, from which the average value of POISSON's ratio, and the calculated error, were included.

[1] Implicit in this method is the need to return to the consideration of the ratio of two moduli, K and E.

[2] PULFRICH [1898, *1*].

[3] For example, STRAUBEL found that the closeness of the knife edges for wider beams made it difficult to obtain consistent results due to a decrease in the amount of cross-bending, i.e., length to width ratios of less than 3. In these instances he inserted thin rubber or cardboard pads between the knife edge and glass, after which he obtained consistency similar to that present with knife edges alone for widths of 2 cm and below with knife edges 5 to 7 cm apart [1899, *1*].

Because of the accuracy of this study and its importance, not for its originality but for the insight it gives into the basic experimental difficulties, difficulties which indeed underlay the controversy of the previous fifty years, I am presenting the data for this glass in Table 81.

Table 81. STRAUBEL (1899).

d	b	m	s_a	s_i	n_1 to n_m	a_1 to a_m	ν
0.3	3.0	6	10	7	3.5 – 6.1	27.01–26.12	0.2126 ± 0.0031
0.3	3.0	7	10	7	3.5 – 6.6	26.89–26.00	0.2173 ± 0.0015
0.3	2.5	4	10	7	4.4 – 6.4	27.96–27.16	0.2213 ± 0.0041
0.3	2.5	6	10	5	3.8 – 7.2	28.21–26.79	0.2195 ± 0.0022
0.3	2.5	5	10	3.4	4.1 – 7.2	27.76–26.60	0.2152 ± 0.0018
0.3	2.0	5	10	7	4.55– 8.4	27.39–26.57	0.2285 ± 0.0007
0.3	2.0	5	10	7	4.6 – 8.7	26.46–26.02	0.2281 ± 0.0009
0.3	2.0	6	10	5	3.9 – 8.7	27.73–26.46	0.2229 ± 0.0010
0.3	2.0	3	10	5	4.6 – 9.0	26.50–25.91	0.2228 ± 0.0000
0.3	2.0	6	10	3.4	3.7 – 8.7	27.70–26.27	0.2193 ± 0.0009
0.3	2.0	3	10	3.4	4.5 – 8.8	26.29–25.82	0.2241 ± 0.0004
0.2	2.0	6	10	7	6.3 –13.6	28.35–26.81	0.2244 ± 0.0007
0.2	2.0	6	10	7	4.2 – 7.65	29.65–27.95	0.2266 ± 0.0021
0.2	2.0	6	10	5	6.2 –13.5	28.26–26.77	0.2248 ± 0.0017
0.2	2.0	6	10	5	4.2 – 8.0	29.58–27.60	0.2205 ± 0.0017
0.2	2.0	6	10	3.4	6.2 –13.6	28.05–26.51	0.2208 ± 0.0013
0.2	2.0	6	10	3.4	4.1 – 7.9	29.61–27.54	0.2152 ± 0.0014
0.25	1.75	6	10	7	4.75– 9.95	28.16–26.75	0.2246 ± 0.0006
0.25	1.75	6	10	7	4.65– 9.65	28.34–26.85	0.2236 ± 0.0010
0.25	1.75	6	10	5	4.8 –10.0	28.19–26.71	0.2227 ± 0.0020
0.25	1.75	6	10	5	4.65– 9.85	28.22–26.85	0.2266 ± 0.0010
0.25	1.75	6	10	3.4	4.85–10.0	27.99–26.55	0.2186 ± 0.0012
0.25	1.75	6	10	3.4	4.9 – 9.9	27.95–26.72	0.2249 ± 0.0014
0.2	1.0	7	10	7	6.1 –12.0	29.46–27.39	0.2236 ± 0.0044
0.2	1.0	7	10	7	5.95–12.1	29.37–27.48	0.2295 ± 0.0028
0.2	1.0	5	10	5	6.05–14.0	29.25–27.12	0.2239 ± 0.0017
0.2	1.0	5	10	5	6.0 –14.0	29.27–27.22	0.2298 ± 0.0013
0.2	1.0	5	10	3.4	6.0 –14.0	29.03–27.17	0.2328 ± 0.0026
0.2	1.0	5	10	3.4	6.0 –14.0	29.50–27.15	0.2217 ± 0.0046

STRAUBEL provided the averaged values for different specimen dimensions for this glass. As reported in his paper, he followed this study with the examination of an additional 29 different glasses of various compositions; he also examined boric acid. These data are summarized in Table 82 which not only includes the total average value of POISSON's ratio for each of the glasses, but also their percentage chemical composition, E as measured by A. WINKELMANN and O. SCHOTT,[1] K, and μ. STRAUBEL calculated the latter two values of K and μ from the isotropic relations among the constants.

If 40 years earlier it had been known that among isotropic glasses, with an experimental accuracy of better than 1%, the value of POISSON's ratio varies from 0.197 to 0.319, perhaps, but not necessarily, it would have minimized the controversy between atomistic theory and continuum experiment.

STRAUBEL continued with a long analysis of the effects of chemical composition upon POISSON's ratio and the effects of hardness, in terms of the penetration measurements of E' modulus by F. AUERBACH.[2] STRAUBEL found individual

[1] WINKELMANN and SCHOTT [1894, *1*].

[2] AUERBACH [1894, *1*].

Table 82. STRAUBEL (1899).

Fabrication no.	SiO_2	B_2O_3	ZnO	PbO	Al_2O_3	As_2O_3	BaO	Na_2O	K_2O	CaO	P_2O_3	Mn_2O_3	POISSON's ratio, ν	Tension modulus E (kg/mm²) (WINKELMANN)	Bulk modulus K (kg/mm²)	Torsion modulus μ (kg/mm²)
1450	71	14	—	—	5	—	—	10	—	—	—	—	0.197	7300	4020	3050
278iii													0.208	6640	3790	2750
2175	68.7	8	—	—	—	1.5	—	5.3	14.5	2	—	0.03	0.210	7460 ~	4290	3080
627	68.2	10	2	—	—	0.2	—	10	9.5	—	—	0.1	0.213	7970	4630	3290
1893[a]	53.5	20	—	—	—	—	—	—	6.5	—	—	—	0.219	5170	3070	2120
714	74.6	—	—	—	—	0.3	—	9	11	5	—	0.1	0.221	6570	3920	2690
20	69.5	2	—	2.5	2.5	0.4	—	7	16	—	—	—	0.221	6340	3790	2600
2154	54.2	1.5	—	33	—	0.2	—	3	8	—	—	0.1	0.222	6100	3660	2500
2106	44.6	—	—	46.6	—	0.3	—	0.5	8	—	—	—	0.222	5390 ~	3230	2210
1571	41	—	—	51.7	—	0.2	—	—	7	—	—	0.1	0.224	5460	3300	2230
709	70.6	—	12	—	—	0.4	—	17	—	—	—	—	0.226	6630	4030	2700
Normal glass	67.3	2	7	—	2.5	—	—	14	—	7	—	0.2	0.228	7400	4530	3010
2158	64.6	2.7	2	—	—	0.4	10.2	5	15	—	—	0.1	0.231	6610	4100	2690
S219[a]	—	3	—	—	10	1.5	—	—	12	—	69.5	—	0.235	6780	4260	2750
500	29.3	—	—	67.5	—	0.2	—	—	3	—	—	—	0.239	5490	3510	2220
658	32.7	31	—	25	7	0.3	—	1	3	—	—	—	0.250	5470	3650	2190
1973	48.1	4.5	10.1	—	—	0.4	28.3	1	7.5	—	—	0.1	0.252	7420	4990	2960
290	58.7	—	—	—	—	0.3	—	—	33	8	—	—	0.253	6010	4060	2400
270	—	3	—	—	8	1.5	28	—	—	—	59.5	—	0.253	6330	4270	2530
2122	37.5	15	—	—	5	1.5	41	—	—	—	—	—	0.256	—	—	—
370	54.8	—	17	—	—	0.2	—	—	28	—	—	—	0.261	5850	4080	2320
S208	20	—	—	80	—	—	—	—	—	—	—	—	0.261	5090	3550	2020
1933	39.64	6	9.2	—	2.5	0.5	42.1	—	—	—	—	0.06	0.266	—	—	—
1299	34.5	10.1	7.8	—	5	0.5	42	—	—	—	—	—	0.271	7970	5800	3140
S95	—	3	—	—	1.5	1.5	38	—	—	—	56	—	0.272	—	—	—
S185[a]	—	71.8	—	—	22.4	—	—	—	—	—	—	—	0.273	—	—	—
S196	—	69.1	—	—	18	0.2	4.7	8	—	—	—	—	0.274	4700	3470	1840
S120	—	42.8	—	52.0	5.0	0.2	—	—	—	—	—	—	0.279	—	—	—
Boric acid	—	100	—	—	—	—	—	—	—	—	—	—	0.283	—	—	—
665	—	41	59	—	—	—	—	—	—	—	—	—	0.319	8170	7520	3100

[a] They contain, furthermore: 1893 20 portions Sb_2O_3; S219 4 MgO; S185 5.8 Li_2O.

differences up to 37.3% occurred, with an average difference of 16.2% in POISSON's ratio between his own direct measurements and those of AUERBACH and WINKELMANN as calculated from the equation, $E = E'(1-\nu^2)$. In the comparison of uniaxial stress and strain in the linear theory, STRAUBEL subscribed to the STOKES-BOCK hypothesis that POISSON's ratio should increase to $^1/_2$ at the melting point, and he suggested that this might be related to the hardness index from AUERBACH's experiments.

In STRAUBEL's study the range of values of POISSON's ratio in glass includes the entire half-century spectrum of experimental results starting from WERTHEIM's initial glass piezometer measurements in 1849. Even for isotropic solids POISSON's ratio is a function of chemical composition, as well as of prior thermal and mechanical histories. A precise, culminating study such as STRAUBEL's may or may not merit lower esteem than the ingenuity and originality of the often cruder first study such as, in this instance, CORNU's,[1] but precision and refinement of the final study placed severe restrictions upon the serious experimentists who followed. Thus we see H. T. JESSOP,[2] approximately two decades after STRAUBEL, again performing the CORNU experiment in plate glass to study the influence of the elastic after-effect upon the measured POISSON's ratio, and we note that JESSOP's work, although imprecise, would have been of some interest if it had been done before STRAUBEL's, but now is worth citing chiefly as an example of the fact that in experimental science there are regresses as well as advances.

Since JESSOP was unable to achieve angular resolution better than 0.5°, rather than the less than 6 min of STRAUBEL, he was obliged to determine fringe spacing, emphasizing behavior at a considerable distance from the point whose radii of curvature were of interest. This limitation, plus his use of unpolished specimens, with unknown initial curvature, and much thicker specimens, i.e., optical flats, resulted in values of POISSON's ratio from 0.138 to 0.229 for the same glass. Thus the observation of changes with time in the fringe pattern which JESSOP attributed to the elastic after-effect provided inconclusive results which might have been thought to have some minor significance had they been published before STRAUBEL's study. Citing STRAUBEL's least square analysis of literally hundreds of tests as too "cumbersome", JESSOP for one pair of a total of eight measurements in six specimens suggested that initial curvature errors could be eliminated by bending the same specimen in both directions so that the sign of the curvatures of the saddle and of the anticlastic behavior would be reversed. The measured difference between the two extremes was 10%, which really showed that it was necessary to resort to STRAUBEL's precise analysis[3] for any definitive study based upon such optical interference experiments as those by CORNU.

In 1908, GEORGE FREDERICK CHARLES SEARLE[4] in his *Experimental Elasticity* compared dynamic moduli which supposedly were determined with great accuracy.

---

[1] That some question existed is revealed in examining successive editions of FRIEDRICH KOHLRAUSCH's great classic, *Lehrbuch der Praktischen Physik* [1905, *1*]. In the first (1870) edition there is no mention of POISSON's ratio or its measurement. The tenth edition in 1905 refers to the experiments of CORNU in 1869 and of STRAUBEL in 1899. By 1940, in the re-edited posthumus publication of this work, CORNU has disappeared, and STRAUBEL (1899) and JESSOP (1921) are given as equivalent sole references, followed by the statement that inhomogeneity or anisotropy for metals introduce errors which preclude the determination of the ratio of moduli to ascertain POISSON's ratio for these solids.

[2] JESSOP [1921, *1*].

[3] P. RICKERT [1928, *1*] in the introductory remarks to a series of tables on deformation parameters in Vol. VI of the *Handbuch der Physik* in 1928 stated that he regarded the measurements in Jena glass made three decades before as among the few reliable moduli data in existence.

[4] SEARLE [1908, *1*].

For the eight different metals, most of which were in the hard drawn state, he obtained nine values of POISSON's ratio, of which five were in excess of one-half for small deformation, ranging from 0.598 for hardened copper, 0.608 for annealed copper, to 1.207 for hard drawn German silver. TOMLINSON's[1] value for hard drawn German silver had been 0.500. Although I have not tabulated the large collection of moduli data in the works of BAUSCHINGER,[2] TOMLINSON, and SEARLE, it is easy to see when examining the data why one may draw the general conclusion, as did BAUSCHINGER, that apart from the matter of anisotropy (arguments about which were introduced always without attempting to see by experiment whether the observed deviations were reasonable), the calculation of POISSON's ratio solely from the directly measured E and μ requires accuracy not achieved by most experimentists even today.

In Fig. 3.38 is shown the variation on the experiment of MALLOCK[3] suggested by SEARLE in 1906 for the direct determination of POISSON's ratio. SEARLE included a discussion of the limitations of this experiment with respect to the magnitude of dimensions and radii of curvature. He obtained for a steel specimen a value of POISSON's ratio of $\nu = 0.285$, which may be compared with KIRCHHOFF's[4] directly measured 0.294, the generally accepted average modern value of POISSON's ratio for steel.

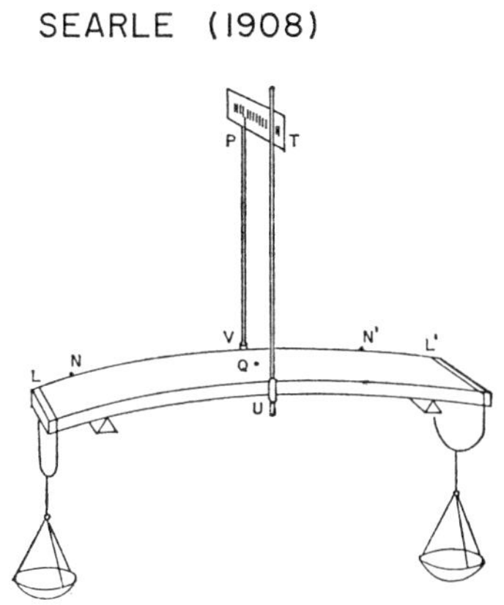

Fig. 3.38. The apparatus used by SEARLE in his version of the CORNU experiment.

3.27. Grüneisen's experiments checking isotropic formulae by the independent measurement of E, μ, and ν

Elementary elasticity textbooks state that POISSON's ratio for the isotropic solid is to be determined from the ratio of the lateral and longitudinal strains during an uniaxial test at infinitesimal strain. For the 19th century this "experiment" remained largely conceptual. The small lateral gage length precluded the use of the main 19th century method of obtaining high strain resolution through the use of long specimens. WERTHEIM[5] in 1848 had measured simultaneously longitudinal and lateral strain in axial tension. As we have seen,[6] however, this first such measurement was made on rubber, for which only the values at large

[1] TOMLINSON [1883, *1*].
[2] BAUSCHINGER [1877, *1*], [1881, *1*], [1886, *1*].
[3] MALLOCK [1879, *1*]. See Sect. 3.21 above.
[4] KIRCHHOFF [1859, *1*].
[5] WERTHEIM [1849, *1*].
[6] See Sect. 3.16.

strain were really precise. The first successful measurement, and the only definitive studies until the turn of the century, were those obtained with the mirror extensometer in iron and steel by BAUSCHINGER[1] in 1879. C.E. STROMEYER[2] in 1894 and J.R. BENTON[3] in 1900, used optical interferometry for the study of lateral contraction of relatively large deformations. J. MORROW[4] in 1903, according to GRÜNEISEN, essentially repeated BAUSCHINGER's experiment with the mirror apparatus but in fact had developed a mirror system to observe the lateral strain during bending. MORROW's concern was the nonlinear dependence of stress on strain and the shifting of the neutral axis toward the compression side. He referred to POISSON's ratio only to emphasize that it could be determined only if the elementary beam theory were assumed to apply.

C.E. STROMEYER's[5] interference optics apparatus for studies of lateral strain in tension, compression, and, curiously enough, torsion, is shown in Fig. 3.39. From nine wrought and cast irons and steels he obtained, as had BAUSCHINGER, a variety of values from $\nu = 0.148$ to $\nu = 0.301$. For three copper specimens including rolled and cast bars ν ranged from 0.319 to 0.380. For four bronze specimens including cold-rolled manganese bronze, values ranged from 0.305 to 0.354. Similar variations were found for Muntz metal in tension and compression. Two tests in delta metal provided values of ν in excess of $^1/_2$. In the three tests in torsion, which included one in rolled copper (the "best selected rolled bar"), STROMEYER reported seeing lateral changes. In view of this, I think his experiments should be repeated.

STROMEYER (1894)

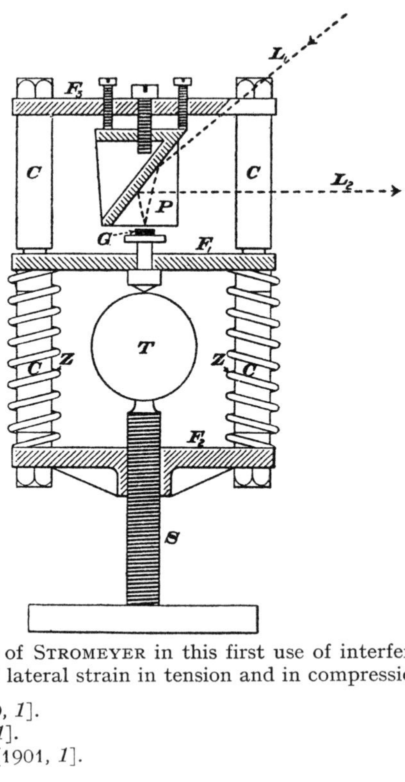

Fig. 3.39. The apparatus of STROMEYER in this first use of interference optics for the study of lateral strain in tension and in compression.

[1] BAUSCHINGER [1879, *1*].
[2] STROMEYER [1894, *1*].
[3] BENTON [1900, *1*], [1901, *1*].
[4] MORROW [1903, *1*].
[5] STROMEYER [1894, *1*].

Rather than viewing STROMEYER's experiments in 1894 as the precursor to the observations of POYNTING in 1912, or as the sequel to the discovery by WERTHEIM in 1854 that the change of volume was proportional to the square of the angle of twist, we may judge STROMEYER as more deserving of distinction for his having used, however faultily, interference optics in experimental elasticity.

The simple textbook experiment referred to above lost its merely conceptual status with the work of GRÜNEISEN[1] in 1908. Using a precise optical interferometry technique, GRÜNEISEN measured lateral and longitudinal strain. His determination of POISSON's ratio was part of a major experimental program to obtain, independently, precise values for E, μ, K, and ν in polycrystalline isotropic metals. Among other objectives, GRÜNEISEN thereby hoped at least to establish the applicability of the formulae for linear, isotropic elasticity by interrelating the four constants, only two of which should be independent. Both dynamic and quasi-static values of E were obtained so that adiabatic-isothermal ratios could be observed.[2] For the torsion modulus μ GRÜNEISEN assumed the difference between μ (isothermal) and μ (adiabatic) would be so minor that he was content only to measure a dynamic value.[3] The experiment was a motorized version of that done by CHLADNI over a century earlier.

A metal ring attached to the center of the rod was clamped in a lathe. At one free end a split cork ring dusted with "kolophonium" was pushed over the rod which, at the application of finger pressure during rotation, produced the basic tone of the torsional vibration. For some solids, such as Al, Ir, Rh, Zn, Pb, Bi, Cu, and Sn, for which either no such tone could be produced, or it died out too fast for measurement, longer rods were used with oscillations being observed in the 19th century manner by means of pine dust nodal traces.[4] The audible tone measurements were calibrated by means of longitudinally vibrating comparison rods. Dynamic E-moduli also were measured by means of the vibration tone to compare directly with the values of CHLADNI, SAVART, and WERTHEIM.

GRÜNEISEN's careful determination of longitudinal strain by means of interference optics[5] has been referred to in the previous chapter. His use of the method for the simultaneous determination of lateral contraction during an uniaxial loading experiment in a rod was no simple adaptation of the previous technique. To obtain equivalent accuracy, he introduced the 11 cm long lever arms shown in Fig. 3.40 carrying one of the parallel glass plates at one end, and at the other, connected to spring loaded steel points which were in lateral contact with the specimen.

Using a mercury arc light source and a movable telescope to view both systems, and surrounding the apparatus with a constant temperature box, GRÜNEISEN

---

[1] GRÜNEISEN [1908, *1*].

[2] There is no reference to the initial experimental studies and analysis in 1844 by WERTHEIM [1844, *1*], which led to the CLAUSIUS-WERTHEIM controversy on the ratio of specific heats, 1849 to 1860. (CLAUSIUS [1849, *1*], WERTHEIM [1860, *1*].)

[3] It is curious that GRÜNEISEN, who already had exhibited an apparent familiarity with WERTHEIM's research, and who, as we have seen, in 1906 made a major contribution to small deformation nonlinearity in extension, would have ignored the nonlinear shear behavior WERTHEIM had discovered in torsion. As I have shown above, WERTHEIM refused to represent his quasi-static data in the form of a shear modulus μ. Those experiments in 1858 were still the most important study of quasi-static torsion in 1910, or even perhaps in 1972.

[4] This is referred to as KUNDT's dust figures, a specialized adaptation of SAVART's version in the 1820's of CHLADNI's experiment of the 18th century. KUNDT's method is described by KOHLRAUSCH in the second edition of his famous treatise (KOHLRAUSCH [1872, *1*]). An English translation with the deliberately altered title, *An Introduction to Physical Measurements*, described the method on pp. 90–91 (KOHLRAUSCH [1886, *1*]).

[5] The first use of interference optics for this particular measurement was that of CANTONE in 1888. CANTONE [1888, *1*].

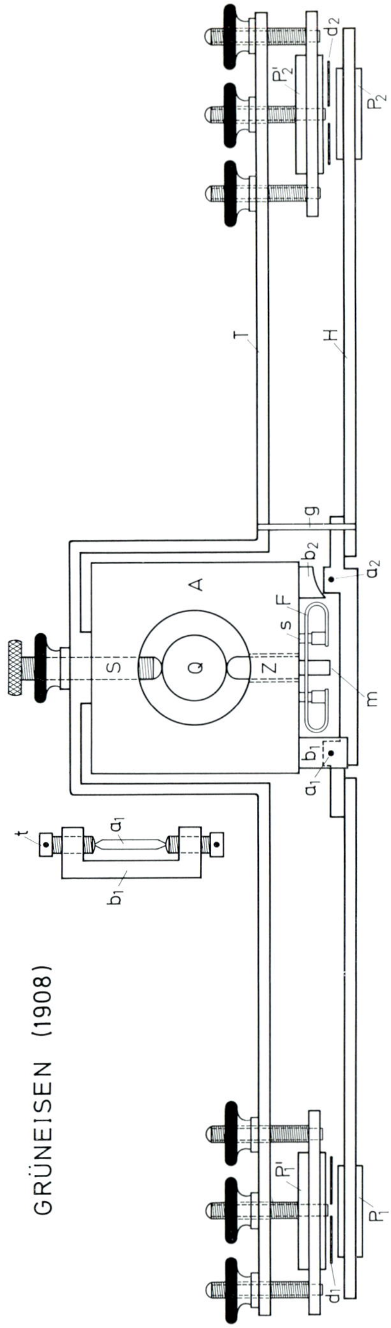

Fig. 3.40. Grüneisen's experimental apparatus for the precise determination of lateral strain by means of interference optics.

obtained the measurements given in Table 83. This paper, like others of Grüneisen, contains a long description of experimental detail and difficulties.

In this table, d is the specimen diameter, ζ the transverse contraction for a 1 kg load, and ε the corresponding longitudinal strain for 1 kg load, with $\nu_{(\text{static})}$

Table 83. GRÜNEISEN (1908)

	d (cm)	$\zeta \times 10^6$ (l/kg)	$\varepsilon \times 10^6$ (l/kg)	ν(static)
Aluminum	1.207	0.417	1.216	0.343
Aluminum I	1.299	0.359	1.059	0.339
Aluminum II	1.300	0.344	1.029	0.334
Copper IVa	1.196	0.248	0.712	0.348
Copper VI	1.300	0.195	0.578	0.337
Silver	1.109	0.488	1.286	0.379
Gold I	1.208	0.475	1.121	0.423
Gold II	1.154	0.494	1.176	0.420
Nickel	1.605	0.0743	0.2407	0.309
Iron I	1.301	0.0974	0.347	0.280
Steel	1.595	0.0674	0.2347	0.287
Constantan	1.997	0.0625	0.1924	0.325
Manganin	1.806	0.1016	0.309	0.329
Palladium	1.610	0.1685	0.428	0.393
Platinum II	1.617	0.1116	0.286	0.387
Red brass	1.468	0.256	0.714	0.358

the desired ratio. The Roman numerals refer to differences in chemical composition. GRÜNEISEN thanked a Mr. GROSCHUFF for providing the chemical composition and electrical conductivity and the temperature coefficient at 15° C for the specimens he studied. GRÜNEISEN's results are given in Table 84.

Table 84. GRÜNEISEN (1908)

				Electrical conductivity $\times 10^4 =$
Aluminum I	1.1% Si	0.5% Fe	0.05% Cu	35 (4.1‰)
Aluminum II	1.6% Si	0.5% Fe	5.7% Cu	32 (3.8‰)
Copper V	0.2% As	0.04% Fe	1.0% Ni	32 (2.2‰)
Copper VI	0.15% As	0.03% Fe	Trace Ni	53 (3.5‰)

The data of Table 85 involved longitudinal changes of up to 2×10^{-5} mm, which GRÜNEISEN estimated could be determined to an accuracy of $\pm 1\%$ with the optical technique. The final summary of all of these experiments, including the direct, independent measurement of E, μ, and ν, is presented in Table 85.

In addition to obtaining the directly measured values of E in column 2, $\nu_{\text{(static)}}$ in column 3, and $\mu_{\text{(torsion) (dynamic)}}$ in column 5, GRÜNEISEN considered the isotropic relation, $E=2(1+\nu)\,\mu$, by comparing a calculated $\nu_{\text{(torsion)}}$ in column 4 from measured E and μ and a calculated $\mu_{\text{(static)}}$ in column 6 from measured E and ν.

The differences between calculated and measured values in accurate experiments finally confirmed BAUSCHINGER's statement of thirty years previous, a statement which had been universally ignored in the interim. Calculations of the ratio of large numbers having small errors, as in the formulae relating elastic constants in isotropic solids, result in inevitable errors in the values obtained. GRÜNEISEN's conclusions that anisotropy was the main reason for the difference in the results, while possibly true, is logically disappointing, since he himself had observed elsewhere that a plus and minus experimental error of 1% in E and μ would lead to a 10% error in the calculated ν. Inasmuch as that was the order of error in his direct measurement of the two moduli, no just conclusion could be

Table 85. Grüneisen (1908)

1 Drawn rods	2 E (kg/mm²)	3 ν(static)	4 ν(torsion) $=\frac{E}{2\mu}-1$	5 μ(torsion) (dynamic) measured (kg/mm²)	6 μ calculated (kg/mm²)
Aluminum	7190	0.343	—	2680	—
Aluminum I	7120	0.339	0.310	2660	2717
Aluminum II	7320	0.334	0.337	2740	2737
Copper IVa	12500	0.348	0.356	4640	4640
Copper V	13110	—	0.391	—	4710
Copper VI	13040	0.337	0.399	4880	4660
Silver	8050	0.379	0.369	2920	2940
Gold I	7780	0.423	0.495	2730	2602
Gold II	8120	0.420	0.435	2860	2822
Nickel	20540	0.309	—	7850	—
Iron I	21680	0.280	—	8470	—
Steel	21320	0.287	0.287	8280	8280
Constantan	16590	0.325	0.329	6260	6230
Manganin	12640	0.329	0.329	4760	4740
Cast rods					
Cadmium	5090	0.30	—	1960	—
Tin	5540	0.33	—	2080	—
Lead	1656	0.446	—	573	—
Bismuth	3250	0.33	—	1220	—
Palladium	11480	0.393	0.101	4120	5210
Platinum II	17080	0.387	0.368	6160	6220
Red brass (bronze)	8240	0.358	0.177	3030	3500

drawn. Grüneisen did note, however, that for bronze and gold, comparative measurements of lateral contraction at the same cross-section had revealed some anisotropy.

The paper also contained compressibilities, χ, calculated from $\chi E = 3(1-2\nu)$, which Grüneisen critically compared with the data of Amagat,[1] Buchanan,[2] and Richards.[3] In view of the errors in such calculations, I shall defer discussing the experiment until I describe Grüneisen's study two years later in which he made a direct measurement of χ, using Mallock's method.

As a final summary of measured values of Poisson's ratio, Grüneisen compiled "all the data known to me," and concluded that the general correlation of directly determined data was satisfactory. To those I have added his average. His failure to cite the direct measurements of Poisson's ratio by Kirchhoff, Okatow, Mallock, and others, set the pattern for the 20th century in regard to citation: eight of his nine references were to papers of the previous two decades.

There is no question but that the experimental research by Grüneisen in this particular subject, and the other work he described, represent milestones in the development of experimental mechanics. Like Duleau a century earlier, and Wertheim a half-century earlier, he posed fundamental questions within the framework of definitive experiments, devoid of auxiliary empirical assumptions which would have reduced the importance of his conclusions.

[1] Amagat [1889, *3*].
[2] Buchanan [1880, *1*], [1904, *1*].
[3] Richards [1907, *1*].

3.28. The mid-20th century repetition of Kirchhoff's experiment for determining Poisson's ratio.

The significance of KIRCHHOFF's[1] insistence upon measuring POISSON's ratio directly was fully appreciated by MALLOCK in 1879, twenty years after KIRCHHOFF's experiments were performed, and by BOCK as late as 1899. Ignoring the fact that KIRCHHOFF had eliminated the errors due to cross-sectional dimensions, a series of experimentists, both in England and on the Continent, tried in vain to measure E and μ with sufficient accuracy to obtain a meaningful POISSON's ratio from their ratios. It was GRÜNEISEN in the 20th century, as we have seen, who finally reached the extremes of accuracy required in such an approach. [I repeat GRÜNEISEN's observation that 1% errors of opposite sign in the measurement of E and μ lead to a 10% error in the value of POISSON's ratio.]

Eighty-five years after KIRCHHOFF had published all the details of his ingenious experiment, not only the significance of his proposal for a direct measurement independent of measured cross-section dimensions, but the experiment itself and all its details had been almost completely forgotten.[2] Thus, in 1944, FRANKLIN L. EVERETT and JULIUS MIKLOWITZ[3], unaware of so doing, developed an experiment which was almost identical with that of KIRCHHOFF of 1859. Their experiment was the end result of an investigation whose purpose had been to find and design a simple test on one specimen for simultaneous measurement of E and μ so that POISSON's ratio ν could be calculated from $\nu = (E/2\mu) - 1$. The manner in which they performed the experiment, however, resembled the study by JOSEPH D. EVERETT in 1867, and contained some, but not all, of the limitations of that approach. JOSEPH EVERETT had pinned the free end of the cantilevered beam while he applied the torsional couple, thus introducing additional error. In their analysis, FRANKLIN EVERETT and MIKLOWITZ considered independent torsion and flexure measurements.

The prime purpose of EVERETT and MIKLOWITZ's study was to determine the temperature dependence[4] of POISSON's ratio for various types of steel. These included hot and cold rolled steel, a medium carbon steel, and two types of steel which were described as "high temperature steel," meaning that they retained a relatively high modulus at high temperatures. The great variation of the calculated POISSON's ratio recalls the calculations in a paper in 1879 by BAUSCHINGER, in which he provided the first major criticism of the use of the moduli ratio formula for isotropic solids in the determination of POISSON's ratio. In general, when proceeding from room temperature to 1000° F in the five steels at six different temperatures, EVERETT and MIKLOWITZ observed that the value of POISSON's ratio increased with increasing temperature. They calculated numerical values in excess of $^1/_2$ for one of the high-temperature steels. The measured E and μ are shown in Fig. 3.41, together with the calculated variation of ν with temperature.

In the paper of EVERETT and MIKLOWITZ the measured values of the E and μ for hot and cold rolled mild steel over the temperature range were compared with those of a number of other experimentists of the preceding two decades. The

---

[1] KIRCHHOFF [1859, *1*].

[2] The experiment was not forgotten by JOHN T. RICHARDS [1952, *1*] whose interest in the history of elastic constants I have noted in Chap. II.

[3] F. EVERETT and J. MIKLOWITZ [1944, *1*].

[4] Once again fifty years was sufficient for an experimental achievement in solid mechanics to disappear. As we have seen, A. BOCK [1894, *1*] not only precisely repeated this experiment by KIRCHHOFF but also used it to study the temperature dependence of POISSON's ratio in iron, steel, copper, silver, and nickel.

data of BACH[1] and BAUMANN[2] and others had been obtained from individual experiments in different specimens for the determination of the two moduli. The general trend of the numbers from all the experimentists was the same, but examination of the scatter of their measurements reemphasizes the importance of KIRCHHOFF's suggestion in the middle of the 19th century that POISSON's ratio should be determined directly from simultaneous flexure and torsion.

This rediscovery of the KIRCHHOFF experiment stimulated F. GAROFALO, P.R. MALENOCK, and G.V. SMITH[3] to carry out a study similar to that of EVERETT and MIKLOWITZ in 42 types of steel at a total of 8 temperatures over the range from room temperature to 1500° F. Their apparatus, shown in Fig. 3.42 duplicates that of EVERETT and MIKLOWITZ, to whom they gave credit for the idea.[4] Their experiment, hence, also resembles the original one by KIRCHHOFF in 1859 (see Fig. 3.35).

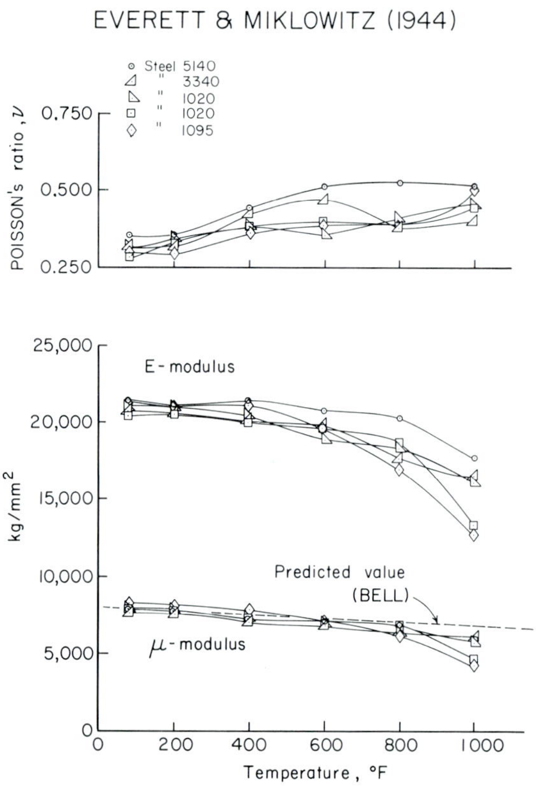

Fig. 3.41. POISSON's ratio as a function of temperature for steel, obtained by EVERETT and MIKLOWITZ in their version of the original KIRCHHOFF experiment.

---

[1] BACH and BAUMANN [1921, *1*].

[2] BAUMANN [1923, *1*].

[3] GAROFALO, MALENOCK, and SMITH [1952, *1*].

[4] JOHN T. RICHARDS [1952, *1*], who was a participant in the ASTM Symposium on the Determination of Elastic Constants at which the paper by GAROFALO, MALENOCK, and SMITH was presented [1952, *1*], pp. 10–27, noted in the discussion which followed (p. 29) that:

> "It should be pointed out that this test method was not developed by Everett and Miklowitz in 1944 but was apparently first reported by Kirchhoff in 1859 and subsequently by Okatow in 1863."

Totally forgotten was the interesting fact, pointed out by WERTHEIM, that KARL FRIEDRICH GAUSS had suggested this experiment before 1859.

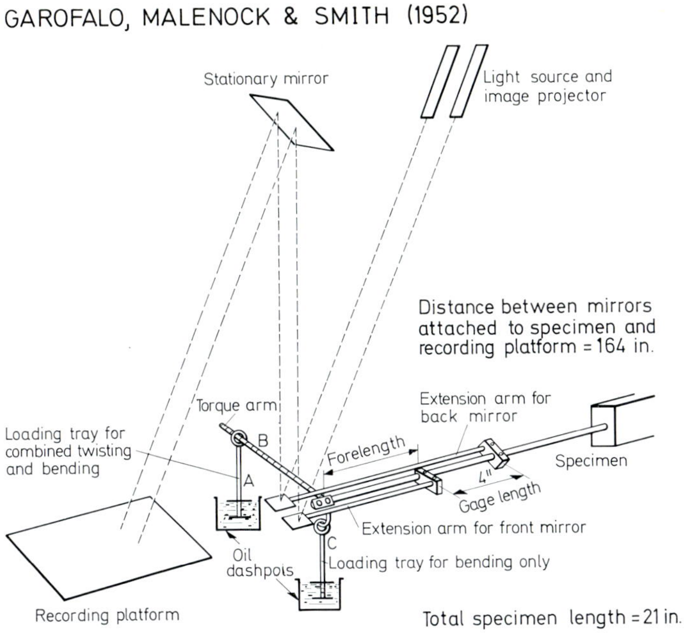

Fig. 3.42. The experimental apparatus of GAROFALO, MALENOCK, and SMITH (1952) in their use of the original KIRCHHOFF experiment.

The apparatus of GAROFALO, MALENOCK, and SMITH was placed in a furnace in order to make the high temperature measurements. To minimize high-temperature creep, they made all measurements while unloading the specimen after having placed the initial load.[1] The test results at room temperature are shown in Fig. 3.43, since the earlier experiments of EVERETT and MIKLOWITZ were thought by some to have been performed with the presence of small amounts of permanent

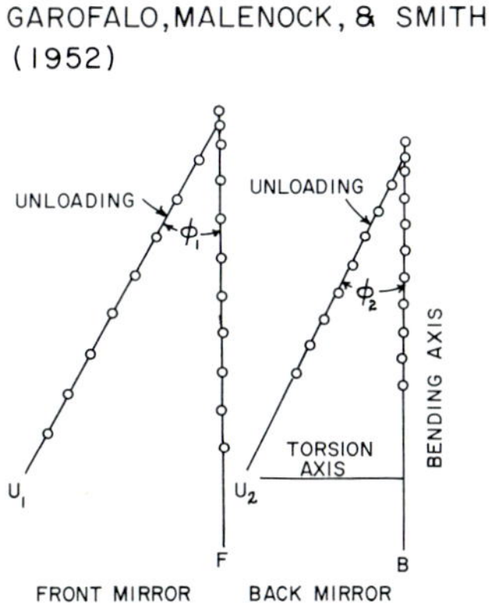

Fig. 3.43. Examples of the recorded data from the two mirrors in the apparatus of Fig. 3.42.

[1] Half a century earlier GILBERT ARDEN SHAKESPEAR, in 1899, in his studies of the temperature dependence of the E modulus had emphasized the influence of the initial portion of the thermal cycle upon subsequent elastic constant determinations (SHAKESPEAR [1899, *1*]).

deformation at high temperatures, an argument reminiscent of AMAGAT's conjecture of 1889.

These measurements during unloading, and the confining of the total load to relatively small values, were part of the experimental plan of GAROFALO, MALENOCK, and SMITH. The very large amount of data presented in their study allows one to make comparisons of the variations in individual measurements, and, with respect to the averages, the temperature dependence of the moduli, and POISSON's ratio for steel. They calculated their POISSON's ratios in the same manner as had EVERETT and MIKLOWITZ in 1944 and JOSEPH EVERETT in 1867. An analysis of their data reveals for the second time[1] a result of very great importance for the small deformation of solids, in that, over a range of fractional melting points, T/T_m, from 0.15 to 0.60, i.e. 297° K to 1100° K, POISSON's ratio is constant for at least one metal. If POISSON's ratio is independent of temperature, the moduli E, λ, and μ for the isotropic solid must depend upon temperature in the same way. In other words, the constitutive equations are the product of a function of strain and a function of temperature. One expects, although separate experiments must be performed to demonstrate the fact, that similar common temperature dependence should characterize the elastic constants of the anisotropic solid or the single crystal.

Recently in a study of elastic constants and their temperature dependence for nearly 60 crystalline solids, I have found[2] a discrete quantized distribution of zero-point shear moduli, as well as a linear dependence upon the fractional melting point temperature for $T/T_m < 0.4$. This generalization concerning crystalline solids will be described more fully below,[3] as well as the fact that POISSON's ratio is found to be a constant. The moduli, E, μ, and ν as functions of temperature, and the averaged values for all of the forty-two types of steel in the experiments by GAROFALO, MALENOCK, and SMITH are shown in Fig. 3.44 (circles), together with the prediction according to my analysis (solid line). This prediction of values for the moduli as a function of the temperature, also is included in Fig. 3.41 which shows the EVERETT and MIKLOWITZ data for μ.

3.29. The confusion generated by the experiments of Kupffer (1848–1863).

One of the unfortunate by-products of an attractive, universally accepted theory which later experiment clearly demonstrates not to apply generally or perhaps not to apply at all, is that the uncritical experimentist may fill the literature over a period of years with erroneous numerical values. Such an uncritical experimentist was ADOLF THEODORE KUPFFER. In a series of annual reports between 1850 and 1861 on experiments in elasticity performed at the Russian Central Laboratory of Weights and Measures, of which KUPFFER had become the Director in 1849, and in five memoirs which he summarized in a 430-page monograph in 1860, every numerical value from literally hundreds of experiments is fundamentally wrong, having been calculated on the basis of inapplicable theories and erroneous hypotheses. Yet PEARSON characterized KUPFFER's work, thus: "probably no more careful and exhaustive experiments than those of KUPFFER

---

[1] The reader may recall that BOCK, in 1894, in the same experiment, had shown that POISSON's ratio for hardened English steel was independent of temperature (BOCK [1894, *1*]).

[2] BELL [1968, *1*].

[3] See Sects. 3.43 and 3.44.

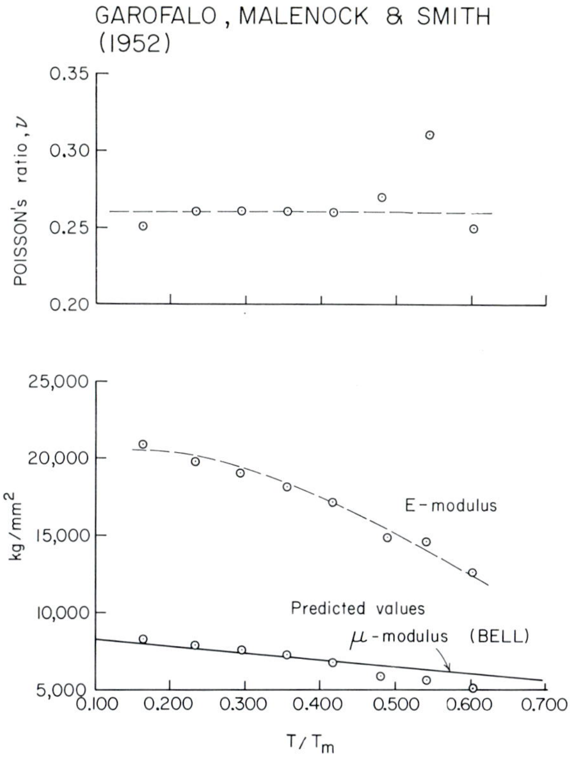

Fig. 3.44. POISSON's ratio and E and μ moduli as a function of the fractional melting point temperature. The μ moduli are compared with the linear prediction by BELL for $T/T_m < 0.4$ [see Eq. (3.27)].

have ever been made on the vibrational constants of elasticity and the temperature-effect."[1]

Throughout the decade KUPFFER and his assistants performed essentially two types of experiments, namely, torsional and flexural, on brass, iron, steel, platinum, silver, copper, and gold. In both series, large added weights and/or long specimens were used, so that the frequencies of flexural or torsional vibration would lie in the extremely low range of two or three cycles per minute. Thus, unlike his predecessors, CHLADNI,[2] SAVART,[3] and WERTHEIM,[4] he did not have to count the number of cycles per second of a rapidly vibrating specimen but could observe the amplitude as a function of the time in individual cycles. However, KUPFFER's modifications made it impossible for him to determine from the measurements he made the moduli he sought. He also insisted upon using the POISSON-CAUCHY theory, dismissing the work of WERTHEIM with the casual and unsubstantiated statement that it was "based on assertions on experiments not completely satisfactory, in truth."[5] Quite incredibly, KUPFFER further argued

[1] TODHUNTER and PEARSON [1886, *1*]. (Excerpt from vol. I, p. 750 of the 1960 Dover edition.) The same view of KUPFFER's excellence as an experimentist is suggested by TIMOSHENKO [1953, *1*], who quoted the above statement from TODHUNTER and PEARSON as evidence.

[2] CHLADNI [1787, *1*], [1802, *1*], [1817, *1*], [1822, *1*].

[3] SAVART [1820, *1*], [1829, *1*], [1830, *1*], [1837, *1*].

[4] WERTHEIM [1844, *1*].

[5] KUPFFER [1852, *1*], p. 193.

that since in an isotropic solid POISSON's ratio could vary from 0 to 0.5, the mean value of 0.250 acquired an added significance independent of the fact that $\nu = 1/4$ was the "required" numerical value for POISSON's ratio.

KUPFFER's papers are extremely difficult to read, not only because they contain numerous errors, some of which have been noted by others, and a considerable amount of fuzzy reasoning,[1] but also because he chose to represent the elasticity of a solid in terms of a single constant, and that constant was defined in a singularly awkward manner. Use of the constant, denoted by δ, indicated a reversion to the state of knowledge at the beginning of the 19th century, since its value depended upon the shape of the cross-section as well as (in hidden form) the units of the applied force. For the axial loading of a cylindrical bar, δ was defined as the elongation produced by the application of a unit force to a specimen of unit length, with a unit radius. Thus δ became $\delta = 1/(\pi E)$. For a bar of square cross-section δ was defined as the elongation produced by a unit force on a bar of unit length with unit sides for the cross-section. For the bar of rectangular cross-section, $\delta = 1/E$. For the latter type bar, in some instances, but unfortunately not always, KUPFFER used δ'. He presented some of his data in Russian pounds and Russian inches.[2] In other instances, he gave δ in centimeters with the applied load in grams, and in one instance he employed English units. As even KUPFFER himself admitted indirectly in a footnote, the reader of his papers often must perform independent calculations to ascertain what KUPFFER had done.

For torsion, KUPFFER defined the elastic constant for a cylindrical rod as the angle of twist for a rod of unit length, with a unit force applied at the unit radius of the bar. Unfortunately, he designated this quantity by the symbol μ, which by that time was being used by many elasticians to represent the shear modulus of an isotropic material. If we designate this quantity of KUPFFER's as μ_K, we see that it was related to the shear modulus by $\mu = \pi/(2\mu_K)$. Assuming, in all instances, the uniconstant theory, so that POISSON's ratio ν was $1/4$, KUPFFER's value of δ obtained from the torsion of a cylindrical rod becomes $\delta = 1/(5\mu_K)$.

Declaring that results at frequencies corresponding to audible sounds were obtained indirectly and therefore were unsatisfactory, KUPFFER attached very large weights to 10 ft. long specimens undergoing torsional vibration. By this means he achieved periods of oscillation of the order of half a minute, in which time intervals he could measure the amplitude as a function of time by means of a finely graduated ring and a chronometer. He recognized that the introduction of the large weight extending a considerable distance from the axis of the rod introduced air friction which would lead not only to incorrect results with respect to the attenuation of the torsional vibration of the solid but also to errors in the determination of the elastic constant. By averaging the results obtained in experiments in the range of 70 to 100 cycles, the total number of observed cycles being as many as 300, he inferred the following relation between the amplitude and the duration of the oscillation:

$$P_0 = P_S - \alpha \sqrt{S} \tag{3.11}$$

where P_S is the period of oscillation at the amplitude S, α is an experimentally determined constant, and P_0 was assumed to be the period of oscillation at 0

[1] Perhaps the outstanding example of this is in his curious effort to determine the mechanical equivalent of heat by comparing the force required to produce an elongation equal to that of the measured thermal expansion between 0 and 100° C. This was too much even for PEARSON, who noted the error of equating force and work in KUPFFER's faulty analysis (KUPFFER [1852, *1*]).

[2] One Russian pound equals 409.4512 g, or 0.9028 English pounds. One Russian inch equals one English inch, or 25.3995 mm.

amplitude. KUPFFER curiously concluded that this value, P_0, would be the same as that which would be obtained if the experiment had been run in a vacuum. He argued equally strangely that although the observed attenuation included air damping, it must contain also material damping for the solid since different values of α were obtained for different metals.

In the introduction to his paper on torsion[1] read before the St. Petersburg Academy on December 1, 1848, KUPFFER referred to the work of COULOMB 70 years earlier as the basis of his study. COULOMB, in providing the first quantitative study of the attenuation of torsional oscillations in metals, had shown that the air damping in his measurements was negligible by surrounding the suspended weights with paper extenders, as has been described above.[2] KUPFFER, who admittedly was unable to eliminate the air damping, did study its variation by surrounding the suspended weights with cardboard forms. Introducing the parameter $\psi=\alpha/P_S$, KUPFFER rewrote Eq. (3.11) in the form

$$P_0 = P_S\,(1-\psi\sqrt{S}). \tag{3.12}$$

Rewriting Eq. (3.12) as

$$P_S = P_0\,(1+\gamma r\sqrt{S/l}), \tag{3.13}$$

where r is the radius of the wire and l is its length, KUPFFER introduced the coefficient γ, which he designated as the "true coefficient of fluidity or ductility."

As I have noted in Sect. 3.4, COULOMB must be credited with the discovery that the attenuation of oscillations in a metallic solid depended upon the amplitude, as did also the period of oscillation. The questionable results obtained by KUPFFER do *not* constitute the discovery of the elastic after-effect in metals, despite PEARSON's having so stated in quite glowing terms at several places in his history of the theory of elasticity.[3] With regard to metals, the honor belongs either to COULOMB (1784) or to KOHLRAUSCH (1863), depending upon whether one is emphasizing the discovery of the existence of the phenomenon, or the proper experimental study of it.[4]

As I noted above, in addition to the yearly reports on the work of the Central Laboratory at St. Petersburg, KUPFFER published five memoirs on the subject of elasticity between 1852–1863. To follow his work, however, we need to consult only the large volume entitled *Recherches Expérimentales sur l'Élasticité des Métaux*,[5] published in 1860. KUPFFER was deeply concerned about the fact that the "fundamental law of elasticity of ROBERT HOOKE" did not strictly characterize the small deformation of the metals he considered. He noted that others, such as HODGKINSON, had had similar concern regarding HOOKE's law. KUPFFER found the response of cast iron to torsion and flexure essentially nonlinear. He planned a work of more than one large volume, one of the later volumes of which was to be devoted to this matter of the evidence that the deformation of metals was essentially

[1] KUPFFER [1853, *1*].

[2] See Sect. 3.4. See also CORNU and BAILLE [1878, *1*].

[3] PEARSON unquestionably regarded KUPFFER as the greatest 19th century experimental physicist in elasticity. Curiously enough, the many pages which he devoted to the discussion of the *Comptes Rendus Annuels* of the St. Petersburg Academy, the memoirs, and the monograph fail to mention one after another of the weaknesses and limitations in KUPFFER's work. Yet PEARSON [1892, *1*] (Sect. 757), after commenting that "we cannot feel thoroughly satisfied with his use of the experimental method of transverse vibrations ..." and that there was some difficulty with his temperature determinations, stated: "still to have demonstrated the existence of after-strain in metals and indicated its changes with temperature is no small service ..."

[4] As a non-oscillatory phenomenon, the elastic after-effect was discovered by WEBER in 1835. (see Chap. II, Sect. 2.12.)

[5] KUPFFER [1860, *1*].

nonlinear, but only the first volume, devoted to the approximately linear properties, ever was published.[1]

Considering what has been written here concerning KUPFFER, one might wonder why his work is being discussed at all in this treatise[2]. KARL ZÖPPRITZ,[3] who joined with nearly every other theorist or experimentist in the second half of the 19th century in expressing admiration for the sheer bulk of KUPFFER's data, lamented the fact that for the flexure of a thin beam on which a heavy weight had been placed, the theory which KUPFFER had used to calculate his elasticity coefficient δ was completely incorrect. KUPFFER's experimental apparatus is shown in Fig. 3.45a and b. Through the telescope shown in his drawing he observed the motion of the slowly moving mass.

KUPFFER (1860)

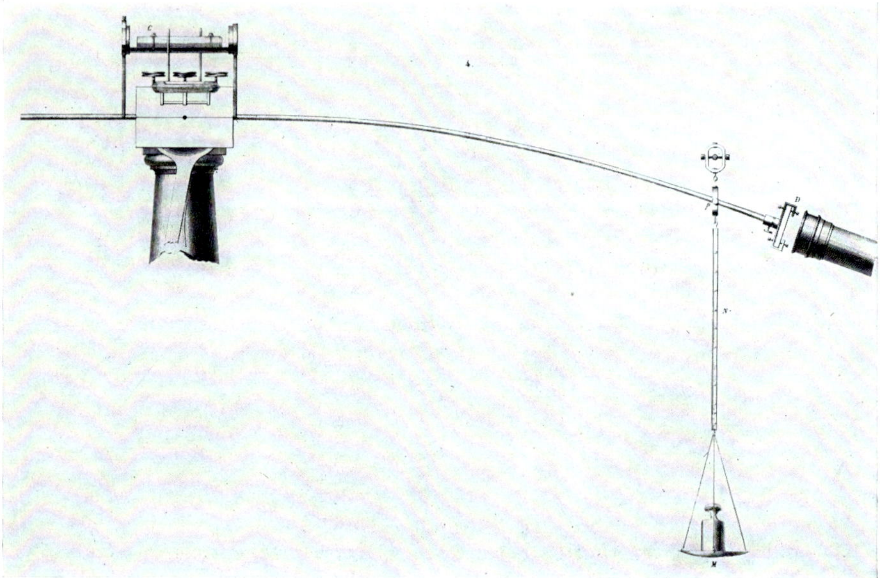

Fig. 3.45a.

Fig. 3.45a and b. (a) KUPFFER's apparatus, including telescope, for the observation of long period vibrations. (b) KUPFFER's apparatus for long period flexural vibrations, showing the manner in which the specimen could be rotated to be above and then below the point of support.

Thinking that he would eliminate the effect of the weight, he had designed an apparatus so that the fixed end of the rod could be placed vertically at either the top or bottom of the vibrating, mass-loaded, cantilevered beam, by rotation of the base. Shown in Fig. 3.45b is a variation of this idea; with this equipment he recorded vibrations, the beam being the same amount above as below the horizontal position. KUPFFER's calculations for torsion contain the assumption that $\nu = {}^1/_4$ for all metals. Although this error arose from his dogged devotion to the unicon-

[1] KUPFFER died in 1865 before the rest of his ambitious plan could be carried out.

[2] Certainly one reason is to document the fact that PEARSON's contributions to the widely quoted *History* with respect to experimental matters must be read with caution. The main reason, however, is the dominant influence the data itself had on numerous persons during the half-century after KUPFFER's death, in contributing to exaggerated notions of imprecision of 19th century experiment.

[3] ZÖPPRITZ [1866, *1*].

KUPFFER (1860)

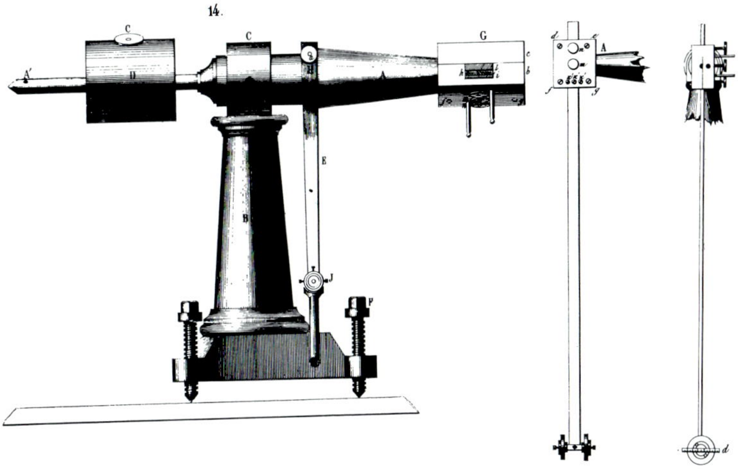

Fig. 3.45 b

stant hypothesis despite the experimental evidence provided by WERTHEIM, we find these results of KUPFFER being cited nearly one century later, in 1953, as providing evidence that the uniconstant hypothesis was inapplicable.[1] Actually, KUPFFER in his insistence upon the physical validity of the uniconstant conjecture paid little heed either to his own or to WERTHEIM's data.

ZÖPPRITZ in his doctoral dissertation had pointed out errors in KUPFFER's analysis and essentially had shown it was impossible to recover the data for the heavy mass loading, particularly for the case when the rod was in the two vertical positions. Subsequently, in a paper in 1866 ZÖPPRITZ[2] supplied corrections for brass, iron, silver, gold, zinc, steel, and copper, for the few experiments of KUPFFER in which only the weight of the rod itself had to be considered, although KUPFFER had used the two-position vertical configuration. Recalculating KUPFFER's δ as the bar modulus E, ZÖPPRITZ compared the recalculated results of KUPFFER's data with the data of ARDANT, DULEAU, LAGERHJELM, RONDOLET, SAVART, and TREDGOLD, as compiled by PONCELET in his *Introduction à la Mécanique Industrielle*,[3] and with WERTHEIM's results. These comparisons, in Table 86, where the capital letter designated the experimentist, show that the corrected values obtained from KUPFFER's data after proper analysis for the flexure of unweighted thin beams, are consistent with previous experimental results of his time.

PEARSON in republishing these comparisons suggested the desirability of an effort more extensive than that of ZÖPPRITZ to recapture by correction, if possible, KUPFFER's voluminous data. Having read the works by KUPFFER, and, in particular,

[1] We see, for example, TIMOSHENKO's *History of the Strength of Materials* in which, after quoting PEARSON's laudatory statement noted above, TIMOSHENKO wrote:

"Kupffer begins with torsion tests from which he determines the shear modulus G. Multiplying G by $^5/_2$ he should, in accordance with the uniconstant hypothesis, obtain the modulus in tension E. But he finds that the values obtained in this way differ considerably from those obtained by tensile or bending tests. Thus his results do not uphold the uniconstant hypothesis." (TIMOSHENKO [1953, *1*], pp. 220–221.)

[2] ZÖPPRITZ [1866, *1*].

[3] PONCELET [1841, *1*].

Table 86. ZÖPPRITZ (1866).

	KUPFFER'S corrected results	From WERTHEIM	Older observation
Brass			
Cast (2, 4)	8439	8543	6450 ? T
ditto, von Hesse (7)	9775		
Milled (5, 6, 9)	10687		
Hammered (1, 8)	11327		
Wire	10375	9005–10348	10000 A, 9600 S
Steel			
Cast steel (6)	20874	19881–20698	20400 T (forged, soft)
von Remscheid (17, 19)	21100		
Wire (10)	19276	19549	19000 A, 20000 S
Iron			
Sheet parallel (1)	21112		
ditto, vertical (2)	19249		
English forge (8)	20061	20869–18547 (Fer du Berry)	20680 L 20000 D, T
Swedish forge (11)	21029		20750 L
Cast (3)	11267		9840 R, 12000 T
Copper			
Annealed, soft (2)	12827		
Strongly milled (4)	12646	12513 (wire)	13100 S (wire)
Zinc, milled	9666	8800–9600	9600 T
Silver, barre	7773	7358–7820	
Gold, barre	7412	8132–8645	

the large volume[1] of 1860, I find it easy to understand why no one has ever followed PEARSON's suggestion. Even KUPFFER conceded that his expertise had increased enormously during the decade but that nevertheless he had included the data from the early experiments "for whatever value they might have."

The serious experimental physicist, while beset with all the intellectual and socio-scientific difficulties facing his theorist colleague, must in addition acquire a laboratory. Few, if any, mid-19th century experimentists in mechanics commanded the facilities, the financial support, and the freedom of choice which KUPFFER acquired in 1849 on accepting the directorship of the new Central Laboratory at St. Petersburg. Following his election in 1828, at the age of 29, to the first rank of the St. Petersburg Academy of Science,[2] he published over sixty notes, reports, and memoirs on such matters as the measurements of magnetic declination and barometric pressure in the various parts of Russia and at various elevations during mountain climbing expeditions, as well as a few papers reporting observations of the growth of organic crystals. His decision to direct the attention of the new Central Laboratory to the pursuit of a grandly conceived plan to provide the experimental foundation for the physics of deformation of solids so as to counterbalance the rapidly expanding developments in the theory of elasticity, surely represented a major and potentially fruitful change of course for a fifty year old

[1] KUPFFER [1860, *1*].

[2] There were 50 Academy members when he was elected. In the 1860 listing of the members of the Academy, 40 of whom had achieved the first rank membership of "ordinaire", KUPFFER was almost alone in not having progressed step by step from the third to the first rank.

experimentist. That more than a decade's intensive experimentation produced an overall contribution far below that of the mere side studies of even the technologically oriented VICAT or HODGKINSON, or of the physicist, WEBER, whose tenure at Göttingen had been interrupted for a period of years, or of WERTHEIM who even with his private fortune never had the command of a personal laboratory, provides an interesting historical antecedent for the meager results derived from the widespread modern practice of approaching fundamental scientific problems by means of team research in specially appointed laboratories.

3.30. The Mallock method for the quasi-static determination of the bulk modulus.

Most studies of elastic constants in the 19th century, as we have seen, were devoted to the direct experimental determination of μ, E, and, finally, ν. A far from satisfactory solution to the problem of obtaining the quasi-static bulk modulus of solids was the work of REGNAULT and WERTHEIM in the 1840's and the studies on the elongation of tubes, by AMAGAT in the 1880's. They all had used piezometers. Only WERTHEIM had been concerned primarily with solids; REGNAULT and AMAGAT had sought to ascertain the absolute compressibility of liquids by using the linear theory of elasticity to determine, indirectly, the compressibility of the container.

The first experimentist to try to determine compressibility directly without resort to the linear theory of elasticity in a relatively complex situation, was JOHN YOUNG BUCHANAN[1] in 1880. Through the walls of a heavy glass tube containing a rod under hydrostatic pressure, he measured by means of a microscope the change in the length of the rod.

PERCY WILLIAMS BRIDGMAN in his Historical Introduction to *The Physics of High Pressure*[2] suggested that the effects of optical refraction of the thick glass probably were what made BUCHANAN's results so poor. In a later chapter on the compressibility of solids, BRIDGMAN pointed out that BUCHANAN's value 2.92×10^{-6} per atmosphere for the cubic compressibility of glass was reasonable enough, but the method of obtaining it could not be checked since the compressibility of glass may vary through wide limits. (Recall, for example, the widely different POISSON's ratios found in thirty glasses by STRAUBEL.)

Several years later, in 1904, BUCHANAN published results[3] determined in the same way for a number of solids, including metals, up to 300 kg/cm² atmospheric pressure. He obtained values which, as BRIDGMAN noted, were "particularly bad for the metals, leading in the worst case, that of platinum, to the value 5.5×10^{-7} for cubic compressibility against 3.6×10^{-7}, the correct value."[4]

ARNULPH MALLOCK[5] in 1904 reintroduced an experimental method for determination of the quasi-static bulk modulus which, although based upon the linear theory of elasticity, provided a simple approach to the problem by requiring the accurate measurement of pressure and a single elongation; i.e., there was no necessity to measure the volume changes as in the piezometer studies. This experi-

[1] BUCHANAN [1880, *1*].
[2] BRIDGMAN [1931, *1*].
[3] BUCHANAN [1904, *1*].
[4] BRIDGMAN [1931, *1*], p. 151.
[5] MALLOCK [1904, *1*]. "MALLOCK's method" in fact was based on an experiment first introduced by WERTHEIM in 1848, and employed independently by AMAGAT and by CANTONE in 1888, both of whom did refer to WERTHEIM. CANTONE had used interference optics to measure elongation. WERTHEIM [1848, *1*]; CANTONE [1888, *1*]. MALLOCK did not mention his predecessors.

ment, which became known as the "MALLOCK method", consisted of observing the change in the length or elongation of a cylindrical tube with closed ends under a known hydrostatic pressure. MALLOCK provided an oversimplified analysis, based upon the assumption that the wall of the tube was extremely thin.

In a paper further on in the volume which contained MALLOCK's paper, CHARLES CHREE[1] reanalyzed the problem, showing that the bulk modulus K would be given by the simple direct expression

$$K = \frac{p_i}{3\varepsilon t} \frac{r_1^2}{r_2 + r_1}, \tag{3.14}$$

where K is the bulk modulus to be determined; p_i is the known hydrostatic internal pressure; ε is the measured strain in the length of the tube due to the application of that pressure; t is the measured wall thickness; r_2 is the radius of the outer wall; and r_1 is the radius of the inner wall of the tube, i.e., $t = r_2 - r_1$. CHREE's corrections for MALLOCK's experiments in steel, brass, and copper in both the hardened and annealed states are shown in Table 87.

Table 87. CHREE (1905).

Material	l (in.)	Outside diameter (in.)	t (in.)	State of material	$\delta l \times 10^3$ (in.)	Bulk modulus K (dynes/cm² $\times 10^{-11}$)
Steel	60	0.75	0.0190	Hard	2.8	18.0
				Annealed	2.75	18.3
Brass	50	0.415	0.0185	Hard	2.12	10.6
				Annealed	2.09	10.7
Copper	50	0.4485	0.0382	Hard	0.562	18.1
				Annealed	0.71	14.3

3.31. Grüneisen's use of Mallock's method to compare elastic constants in isotropic solids (1910).

GRÜNEISEN, who as we have seen above had made direct measurements of E, μ, and ν in isotropic solids in order to introduce all measured quantities into $E = 2(1+\nu)\mu$ for a comparison with prediction, also was interested in making a similar comparison[2] with prediction for $\chi E = 3(1-2\nu)$. This required that he measure directly the compressibilities χ (the reciprocal of K), E, and ν. He measured E and ν by means of interferometry and μ through the study of torsional vibration, reminiscent of CHLADNI's. All of these measurements were on bars. When he used MALLOCK's method to determine the compressibility of the same solids, GRÜNEISEN realized that there would be difficulties in comparing elastic constants determined from the rods and from entirely different specimens in the form of tubes, i.e. E and μ also should have been determined on the same tubes.

[1] CHREE [1905, *1*]. Both AMAGAT and CANTONE gave precisely the same formula in analyzing their data obtained from the same experiment, nearly two decades before either MALLOCK or CHREE considered the problem.

[2] GRÜNEISEN [1910, *3*].

In these studies, GRÜNEISEN was equally interested in another aspect of the compression of isotropic solids, namely, the temperature dependence. MALLOCK's method accordingly was used to determine the compressibility or bulk modulus at -195° C, 17° C, and 100° C. The data will be discussed in Sect. 3.41 below. This paper by GRÜNEISEN, which is still widely referred to in the literature over one-half century later, contains a detailed description of the limitations of the MALLOCK method for accurate measurement. GRÜNEISEN's diagram shows the tube which was suspended at two points, with the ends of the tube plugged and attached to a pressure apparatus (see Fig. 3.46).

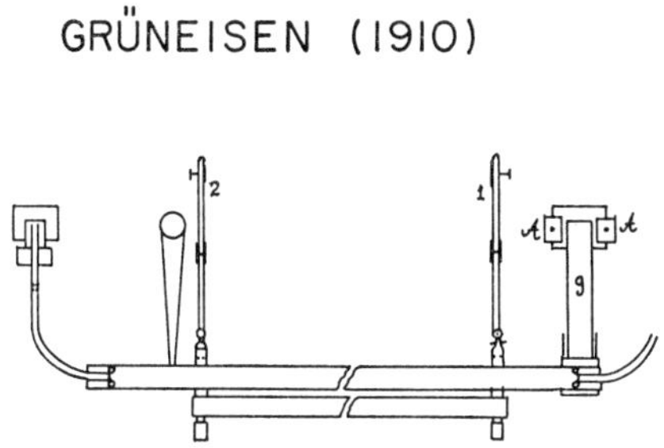

Fig. 3.46. The apparatus used by GRÜNEISEN in his adaptation of MALLOCK's experiment at different ambient temperatures.

GRÜNEISEN found it necessary to consider in detail the effects of thermal expansion on the apparatus, the effects of compression heating as the internal pressure was applied, and the effects associated with the ends of the tube and wall thickness in producing a flexure of the tubular specimen, which could result in error. Problems of interpretation arose, especially for the softer metals, when the elastic limit was slightly exceeded. Serious errors could be introduced by the phenomenon of the elastic after-effect, so that the time of the measurements was of major importance, particularly when this latter phenomenon was considered in conjunction with the opposing effect of compression heating. The accuracy which GRÜNEISEN wished to achieve required care in measurement of dimensions, including both inside and outside radii and initial length, as well as analysis of the precision with which the applied pressure could be determined.

Referring the reader to GRÜNEISEN's paper[1] for the long description of those interesting details together with the discussion of the nature of the metal hangers on which the specimen was suspended, the study of the influences upon the result of their positions along the tube, and the problems associated with the construction of an external box in which boiling liquid nitrogen supplied the low temperature and heated oil gave the higher temperatures, I shall discuss only the data. The experimental results at 18° C for the compressibility χ are given in Table 88, to which I have added its reciprocal the bulk modulus K as the final column.

In his table GRÜNEISEN also provided his values of E and the directly measured ν. The measured data are in standard type and the calculated values are in italic type. In addition, he gave the measured density s for the rods and tubes which served as specimens of all seven of the polycrystalline metals examined. As GRÜNEISEN pointed out, except for silver, for which major differences were observed, there is some agreement among the calculated and measured numbers.

[1] GRÜNEISEN [1910, *3*].

Table 88. GRÜNEISEN (1910).

Material		s	E (kg/mm²)	ν	$\chi \cdot 10^4$ (cm²/kg)	K (kg/mm²)
Aluminum	Rod	2.71	7200	0.34	*1.31*	7630
	Tube	2.705	7330	*0.32*	1.46	6850
Iron	2 Rods	7.83	21500	0.28	*0.60*	16530
	Tube 1	7.82	20900	*0.280*	0.633	15780
	Tube 2	7.83	21250	*0.293*	0.583	17100
Copper	Rod	8.96	12500	0.35	*0.73*	13800
	Tube	8.89	12750	*0.34*	0.77	13000
Silver	Rod	10.53	8050	0.38	*0.90*	11100
	Tube	10.49	7350	*0.407*	0.76	13150
Tin	Rod	7.28	5540	0.33	*1.9*	5260
	Tube	7.27	—	—	(3.1)[a]	3225
Platinum	Rod	21.39	17080	0.39	*0.40*	2500
	Tube	21.44	16770	*0.391*	0.39	2560
Lead	Rod	11.32	1660	0.45	*2.0*	5000
	Tube	—	—	—	(3.2)[a]	3125

[a] GRÜNEISEN mistrusted the absolute values of the numbers in parentheses for tin and lead.

It is not possible to be certain to what extent the observed differences may be attributed to the use of two different shapes of specimen and different methods of determination of moduli. We have seen from GRÜNEISEN's discussion[1] in 1908 of his E, μ, and ν, obtained from directly measured data, that calculations which require the consideration of ratios of measured numbers are extremely sensitive to small errors.

3.32. The linear approximation and one-dimensional wave propagation: Wertheim and Breguet (1851).

All studies of elastic moduli whether in tension, compression, torsion, or flexure, and whether quasi-static or vibratory, have presumed a knowledge of stress and strain distributions throughout the solid as a function of space and time. Unlike MARIOTTE in the 17th century, who saw that it was necessary to ascertain that the deformation along the specimen in quasi-static experiments was really uniform, nearly all of his successors, whether considering 30 m long wire or 30 mm long single crystals, have been content to *assume* uniformity. In vibratory studies the use of characteristic functions and their eigenvalues not only presumes that the statistical space-time sum of the microseismic longitudinal and shear wave propagation, complex reflection, and interaction conforms to approximate theories, but also presumes that in attempting to ascertain the physical applicability of various proposed constitutive equations, it is possible to distinguish geometrical, viscous, and thermal departures from simple assumptions. The refinement of approximate theories of the vibration of rods during the past 90 years has demonstrated time and again the dangers of complacency in these matters.

The use of wave propagation experiments *per se* for the study of the physical applicability of the linear theory or any other theory for the behavior of solids in

[1] GRÜNEISEN [1908, *1*].

small deformation, logically requires that the premises and assumptions of the proposed analysis be demonstrated as exactly coinciding with the experimental conditions before one makes too hasty inferences concerning the significance of the numerical agreement obtained by experimentists who performed the same type of test and made the same auxiliary empirical assumptions. The elementary linear elastic theory requires that certain simple wave profiles shall propagate unchanged with constant wave speeds. The observation of dispersion and the analysis of its detailed wave speed distribution as a function of the amplitude of strain or particle velocity raises the far more serious difficulty of distinguishing between nonlinear stress-strain contributions, three-dimensional effects in a situation presumed to be one-dimensional, linear or nonlinear viscous contributions, and the contribution of unanticipated influences. Radar-like measurement of average arrival times in bounded specimens, particularly for very low strain amplitudes and for measurements over relatively short distances, may provide average wave speeds, but by themselves clearly do not provide the opportunity to assess the applicability of the theory upon which their interpretation is based.

The first wave propagation experiments in 951 m of cast iron water pipes of Paris by BIOT[1] in 1808, and the later experiments of WERTHEIM and BREGUET[2] in 1851 on 4050 m of the Versailles iron electric telegraph wire, of course were solely measurements of arrival time. BIOT, as we have seen, obtained in iron a sound velocity of 10.5 times that in air, which was considerably below the value of 12.2 calculated from the quasi-static modulus and far below that of CHLADNI[3] who had obtained a ratio of 16.66. WERTHEIM and BREGUET made a number of attempts at different locations before finding the most favorable telegraph cable, the Versailles-Paris line between Asnières and the entrance to the tunnel at Puteaux.[4] The wire was tightly stretched on slightly elevated posts. The measured length was 4067.2 m, including curvature of the wire, unequal elevation of posts, and the thickness of the wood through which the initial shock must travel. They tried many ways of making the sound, but only the blow of a hammer on one of the supporting posts produced an effect they could measure at great distances.

WERTHEIM and BREGUET's method of measurement duplicated BIOT's. An assistant at one end struck a blow at a pre-arranged moment. An observer at the same position with his back turned, on hearing the blow recorded the initial time on a chronometer measuring tenths of a second with precision. The observer at the other end of the wire on hearing the sound waited 30 sec as recorded on a second chronometer before sending a shock back along the wire so that the total time of a double traverse could be noted on the same chronometer as that used to record the initial blow. The average of many experiments gave a velocity of 3485 m/sec for iron. This value of 10.5 times the velocity of sound in air, (with a velocity of sound of 332 m/sec in air, obtained by shooting a pistol) is nearly the same as that obtained 43 years earlier by BIOT who actually had obtained 10.6 but had recorded it as being 10.5 times the velocity of sound in air. BIOT's velocity for cast iron was 3658 m/sec; his measured velocity of sound for air in the pipe was 344.6 m/sec. WERTHEIM and BREGUET noted that from the measured quasi-static E-modulus in iron, BIOT's velocity ratio would have been 12.2, i.e. 4075 m/sec.

[1] BIOT [1809, *1*].

[2] WERTHEIM and BREGUET [1851, *1*].

[3] CHLADNI [1817, *1*].

[4] WERTHEIM and BREGUET [1851, *1*] noted the curious obstacle which the tunnel offered to the passage of sound along the wire. They could not detect sound propagated through the tunnel even though they finally tried a relatively thick iron cable which did not touch the walls anywhere.

For a two meter length of precisely the same wire, CHLADNI's experiment with a longitudinal sonometer provided from vibrations the much higher wave velocity of 4634 m/sec in the iron telegraph wire. The earlier controversy raised by the differences between the measurements of CHLADNI and of BIOT thus was reactivated, this time based on the comparison of results from identical specimens of 2 m and 4067.2 m lengths. For a density of 7.9 g/cm³ for iron, WERTHEIM's E for iron determined from wave propagation was 9620 kg/mm². (BIOT gave no density for his cast iron pipe. A value of 7.9 g/cm³ provides $E=10600$ kg/mm².) WERTHEIM's calculated value from his measured quasi-static modulus was $E=13150$ kg/mm²; his value from longitudinal vibration in 2 m long wires was $E=16950$ kg/mm²; and finally, CHLADNI's early value from the same type of longitudinal vibration experiment was $E=24700$ kg/mm². The standard modern value is 20800 kg/mm². This value was obtained first by DULEAU in 1811 and was obtained also by WERTHEIM for some specimens other than telegraph wires.

WERTHEIM's data from four kilometers of continuous wire may have raised new experimental questions,[1] but they certainly eliminated the basis for the criticism of BIOT's experiments on one kilometer of pipe, that the existence of many sections had been the source of the lower wave speed. Wave speeds in segmented iron pipe and unsegmented iron wire had the same low value. The array of numbers above is consistent with a recently discovered distribution of discrete multiple elasticity in metals,[2] which will be considered later in the present chapter when we return to these data in Sect. 3.44.

3.33. Exner's experiments on wave propagation in rubber (1874).

The next important attempt to study wave propagation in solids was based upon JOHN GOUGH's[3] observation in 1802 of the abnormal thermal properties of rubber under stress. In his second "experiment" GOUGH had observed that a loaded rubber specimen contracts when heated. The second half of the 19th century attributed this discovery to a suggestion of KELVIN in 1855. It was the suggestion of KELVIN which led to the experiments of JAMES PRESCOTT JOULE[4] in 1859, demonstrating such behavior. However, JOULE also observed that there was an initial region of deformation in which the specimen of rubber slightly cooled

[1] CHWOLSON in his extensive historical review of experimental mechanics, *Traité de Physique*, in 1908, by way of explanation of the low wave speed, made the far-fetched suggestion that the sound traveled the four kilometers through the ground [1908, *1*].

[2] BELL [1968, *1*]. This quantized distribution of elastic constants for the elements described in Sect. 3.44 gives for the integral indices $s = 5, 6, 7, 8, 9$, and 10, values for the E of iron of:

25500 kg/mm²
20800 kg/mm²
16980 kg/mm²
13850 kg/mm²
11300 kg/mm²
9250 kg/mm²

which, as may be seen, closely agree with the array of 19th century measured values. As is indicated in Sect. 3.44, this distribution was discovered from comparing the isotropic elastic moduli among the elements, and from subsequently finding that a similar distribution could occur in a given element after it had had different prior histories of temperature and deformation. That the distribution of 19th century measurements of E in iron closely agrees with a quantized classification discovered from the study of modern data, while of course to be expected in retrospect, played no role in the discovery itself. It is now clear that the CHLADNI-BIOT controversy, which remained unresolved for a century and a half, did not have its origin in poor experimentation on the part of one or another of the participants, as was so often suggested, but rather arose from the assumption that for every solid there must exist a unique set of elastic constants.

[3] GOUGH [1806, *1*].

[4] JOULE [1859, *1*].

before it became heated at large deformation. This was in contrast to nearly all other solids, for which only cooling was observed in tensile loading.

JACOB SCHMULEWITSCH,[1] who had obtained experimental evidence of similar abnormal behavior in muscle fibers, claimed that in rubber the transition from the region of cooling, to that of heating, occurred at an elongation of double the original length. (He did not refer to the fact that earlier JOULE had observed the transition at 20% strain.) The existence of this "neutral point," as it was called, and the fact that SCHMULEWITSCH's experiments had indicated that it occurred at approximately a unit elongation, which was thought to be especially significant, not only gave rise to a variety of explanations but also aroused interest in the temperature dependence of elastic constants in the different ranges of strain and at the neutral point itself. SCHMULEWITSCH tried to explain the phenomenon by suggesting that the elastic constant of rubber must increase not only with stress but also with ambient temperature. This suggestion led FRANZ EXNER[2] to attempt to measure the velocity of sound in rubber strings, as a function of stress, initial elongation, and ambient temperature.

These results, which demonstrated that SCHMULEWITSCH's assumption about the influence of temperature did not conform with the facts, were based upon an experiment which was suggested by "Director STEFAN," brought to fruition by EXNER. EXNER stretched by prescribed amounts rubber strings of known initial lengths and then attached them to a silk thread under tension. Burning the thread gave rise to a deformation wave at one end. This wave proceeded to the opposite end in a finite time, measurement of which provided the wave speed for the ambient elongation and temperature; this speed, in turn, and a known density, provided a corresponding dynamic elastic modulus. He determined the time interval by means of an instrument referred to as a "Hipp chronoscope," the clockwork of which functioned only when a current which influenced an electromagnetic restricting-bar was interrupted. EXNER described his Hipp chronoscope as new and as capable of responding to 0.012 to 0.016 sec intervals of interruption of current.

EXNER elaborated in some detail the manner in which he checked the proper operation of the *d*, *f*, and *e* switches in the diagram he provided to show the apparatus (see Fig. 3.47). *A* and *B* are Daniell cells; *c*, the chronoscope; *C*, the rubber string; and *a* and *b*, strong springs. The wave front was initiated at *a*.

In the first series of measurements, a load of 200 g was applied to a rubber string which was cut into a series of initial lengths, producing the stretched lengths of Table 89. In each instance the constant load of 200 g was sufficient to produce almost exactly the doubling of length to the interesting neutral point.[3] The chrono-

[1] SCHMULEWITSCH [1866, *1*]. SCHMULEWITSCH's interest in obtaining an experiment analogous to the deformation of the leg muscle of the frog led him to consider the relatively small deformation of rubber. His results disagreed with those of JOULE until sufficient load was applied to double the original length. These data will be referred to in the chapter on finite deformation (Sect. 4.39), but it is interesting to note here that at just beyond 10% deformation in rubber, SCHMULEWITSCH found that ambient temperature changes produced no increase or decrease of elongation. He also noted that in this initial small finite strain region, the stress-strain function was concave toward the strain abscissa, unlike that for frog muscles which, obeying WERTHEIM's law, was concave to the stress axis. It was only after the intial deformation region that an inflection point occurred and the response function of rubber, too, became convex to the stress axis for large finite strain.

[2] EXNER [1874, *1*].

[3] Both SCHMULEWITSCH and EXNER described the "neutral point" as occurring at $\varepsilon = 1.00$ instead of at $\varepsilon \cong 0.20$ of JOULE or $\varepsilon \cong 0.12$ of more recent measurement. They either were committing the same error in very different experiments or were examining rubber specimens prepared in a way markedly unlike that of JOULE.

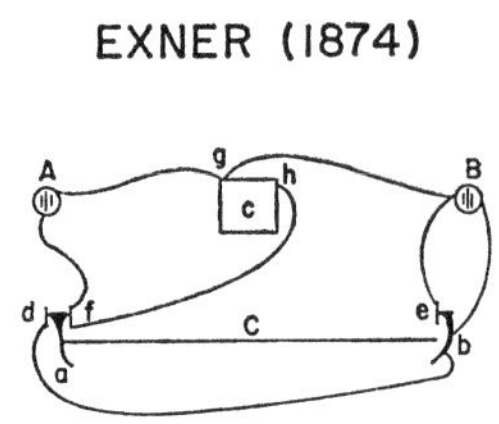

Fig. 3.47. Exner's drawing of the electrical circuitry in his experiments of 1874 on wave propagation

Table 89. Exner (1874).

Length of thread (m)	Chronoscope data for individual experiments (sec)	Sound velocity (m/sec)
I. 0.91	0.020, 0.020, 0.019 0.020, 0.021, 0.020	45.5
II. 2.07	0.044, 0.042, 0.046, 0.046 0.046, 0.045, 0.046	46.0
III. 3.83	0.085, 0.084, 0.084, 0.085	45.3
IV. 6.77	0.144, 0.142, 0.145, 0.145 0.142, 0.144, 0.146, 0.143 0.142	47.1

Table 90. Exner (1874).

Length of thread (m)	Chronoscope data for individual experiments (sec)	Sound velocity (m/sec)
I. 2.0	0.041, 0.044, 0.043, 0.042 0.041, 0.043, 0.042	47.2
II. 3.0	0.054, 0.051, 0.054, 0.052 0.053, 0.054, 0.050, 0.053 0.056, 0.053	56.6
III. 4.0	0.063, 0.064, 0.064, 0.066 0.063, 0.063, 0.063, 0.063 0.065, 0.062	62.9
IV. 5.0	0.077, 0.074, 0.077, 0.077 0.074, 0.075, 0.078, 0.074 0.075, 0.077	65.9

scope data of the individual measurements for the lengths, which varied by a factor of seven, provided nearly the same room temperature velocity, approximately 46 m/sec for this stress.

After having thus checked the general applicability of the method, Exner stretched a rubber string, whose initial zero stress length was one meter, to two, three, four, and five meters. As may be seen in Table 90, he was able to show the increase in velocity, and therefore tangent modulus of the stress-strain function, with increasing stress.

After those preliminary experiments, EXNER examined the dependence of the velocity of sound on temperature by performing his experiments in a chamber in which the temperature could be maintained uniform at any temperature from 0 to 74° C. For the 200 g load which produced a stretched length double that of the unstressed length, he obtained the decrease in the sound velocity and, hence, tangent modulus with increasing temperature shown in Table 91.

Table 91. EXNER (1874).

A. Black vulcanized rubber		B. Red vulcanized rubber	
Temperature (° C)	Sound velocity (m/sec)	Temperature (° C)	Sound velocity (m/sec)
0	54.0	0	69.3
12	47.6	11	57.1
15	47.0	22	44.4
33	37.5	42	39.1
40	33.5	57	36.6
50	30.7	70	33.9
60	30.2		
74	29.0		

The data in Table 91 A were for black vulcanized rubber. EXNER obtained similar results for red vulcanized rubber, as may be seen in Table 91B.

Modifying the apparatus so that an impact could be applied to one end of a 9 cm² cross-section, 1.60 m long, unstressed rod, with metal end caps, but using the same chronoscope technique, EXNER studied the sound velocity vs temperature for zero stress in the initial region. These results, shown in Table 92, demonstrate that in the initial region, as well as at the neutral point separating the first and the second region, the sound velocity decreased with temperature. In thus disposing of SCHMULEWITSCH's conjecture, EXNER neglected to compare tangent moduli determined from quasi-static and vibratory measurements with his results for wave propagation.

Unable to obtain further samples of the original strings, and aware that the variation from one lot of rubber to another precluded a meaningful comparison with the data of the other experiments, EXNER made wave measurements and quasi-static measurements on a newly procured rubber string of the same type. He obtained a value of 34.6 m/sec at room temperature for a rod of 4 cm² cross-section in the stretched condition. At the same stress, the sound velocity determined from the quasi-static elongation was 37.5 m/sec, which EXNER considered a sufficient check in view of the fact that the latter experiment was only a rough approximation.

Finally, for the experiments with zero initial stress, for a 9 cm² cross-section and a 4 cm² cross-section specimen, EXNER provided the demonstration shown in Table 93 that the velocity of sound was not a function of the amplitude of the wave. He showed this by comparing wave speeds from an impact of low intensity with one of high intensity.

Despite the lack of precision in these experiments of EXNER, especially when compared to the work of BIOT or WERTHEIM, EXNER's fresh experimental approach

Table 92. EXNER (1874).

Temperature (° C)	Chronoscope readings for individual experiments (sec)	Sound velocity (m/sec)
0	0.036, 0.036, 0.037 0.038, 0.037, 0.038 0.038, 0.038, 0.038 0.035, 0.037, 0.038 0.035, 0.038	43.2
10	0.040, 0.039, 0.039 0.040, 0.040, 0.039 0.038, 0.039, 0.039 0.040, 0.040, 0.040	40.8
22	0.041, 0.043, 0.043 0.045, 0.042, 0.042 0.045, 0.041, 0.044 0.041, 0.043	37.4
45	0.050, 0.047, 0.047, 0.050 0.050, 0.050, 0.051, 0.049 0.052, 0.047, 0.049, 0.050 0.051, 0.051, 0.050, 0.051 0.050	32.3

Table 93. EXNER (1874).

Impulse	Chronoscope data for individual experiments (sec)	Average
	Rod: 9 cm² cross-section	
Very weak	0.043, 0.042, 0.042, 0.042 0.043, 0.043, 0.043, 0.043	42.6
Very strong	0.043, 0.043, 0.044, 0.039 0.042, 0.042, 0.043, 0.043	42.4
	Rod: 4 cm² cross-section	
Very weak	0.044, 0.044, 0.042, 0.044 0.042, 0.044, 0.045	43.6
Very strong	0.045, 0.043, 0.044, 0.043 0.045, 0.042, 0.045	43.8

to wave propagation constituted an advance in the field. He also had raised further questions for the 1870's in trying to explain the thermal anomaly of stretched rubber. That was the first study of wave propagation in a prestressed solid. Seventy-five years were to elapse before further experiments of this type were reintroduced, by LAZARUS for an utrasonic wave in a pre-stressed high pressure field and by me for an incremental wave in a solid pre-stressed to produce large plastic deformation.[1]

[1] LAZARUS [1949, *1*], BELL [1951, *1*].

3.34. The axial collision of rods with an assumed linear response function: The Boltzmann experiment (1881 *et seq.*) vs Saint-Venant's theory (1867).

To study the behavior of a wave profile in relatively small bounded solids, loads must be applied in an extremely short time interval. The free-flight collision of solids offered the first feasible method for meeting this requirement.

CAUCHY in 1826[1] and POISSON in 1833[2] proposed theories for the elastic impact of cylindrical bars of equal and of unequal lengths. Errors later found in both proposals demonstrated the theoretical difficulties in such a problem, particularly with respect to the time of contact or the amount of time between impact and that at which bars of unequal length would separate due to different particle velocities at the impact face. In 1867 SAINT-VENANT[3] presented a plausible, detailed prediction according to linear elasticity of what could be expected to occur for the one-dimensional impact of rods. LUDWIG BOLTZMANN, who is better known as a theorist than as a person who contributed to experimental solid mechanics, in 1881 became interested in the physical applicability[4] of SAINT-VENANT's theory. In a paper notable for its modest tone in addressing an experimental question, BOLTZMANN described a series of axial impact experiments in smooth grey India rubber rods with brass end caps. He discovered that for rods of equal or of unequal lengths, the separation velocities fell far below SAINT-VENANT's prediction, more, in fact, as BOLTZMANN emphasized, than could be accounted for by such a phenomenon as the elastic after-effect.

The rods were held by silk threads in a bifilar suspension to minimize losses due to external friction. The length of the suspension threads was 153 cm. In the repose position the rods barely touched. One rod was held in a displaced position by means of an additional silk thread. The burning of the thread allowed the rod to swing from its elevated position, resulting in the desired axial collision. Each test was repeated three times to demonstrate reproducibility. The rods were designated as A, A', B, and B'; they had lengths of 100, 104, 230, and 228 mm, respectively. The rods of 100 and 104 mm had a 17 mm diameter, the longer rods 11 mm; the weights of the rods and their suspension hooks were 23.816, 23.7, 23.904, and 23.802 g, respectively for the four lengths cited. H denoted the carefully measured initial height of the impact bar, and S was the divergence from the repose position for the impacted bar. Summarizing BOLTZMANN's data, measured for him by a local Gymnasium teacher named HAMMER, as percentage differences P, i.e. $P=\frac{H-S}{H}$, we have the results in Table 94, in which the order of the designat-

Table 94. BOLTZMANN (1882).

	A′A	BB′	B′B	AB	BA	BA′	A′B′	B′A	B′A′
$H=100$	$P=17.0$	16.5	16.3	20.7	20.5	21.0	21.0	21.0	—
$H=\ 50$	$P=$ —	16.0	16.0	19.4	20.0	—	20.0	—	20.0
$H=\ 30$	$P=13.9$	—	—	—	—	19.4	—	—	—

[1] CAUCHY [1826, *1*].
[2] POISSON [1833, *1*].
[3] SAINT-VENANT [1867, *1*].
[4] BOLTZMANN [1882, *1*]. The paper was dated 5 December, 1881.

ing letters specifies the impacting and the impacted bars. BOLTZMANN noted that the results generally failed to agree with SAINT-VENANT's theory. He remarked also that with increasing velocity there was a curious slight increase in P.

VOIGT admittedly was motivated by BOLTZMANN's experiments and also by recollections of FRANZ E. NEUMANN's lectures on the subject in the 1850's. NEUMANN had discussed the limitations of CAUCHY's and POISSON's theoretical contributions. In VOIGT's opinion, NEUMANN's lectures had contained a theoretical development, never published, which preceded that of SAINT-VENANT by several years. Be that as it may, VOIGT's interest in the subject was aroused sufficiently for him to perform[1] the BOLTZMANN experiment on metals rather than on rubber. He considered steel rods of 8 and 11 mm in diameter and 20 and 40 cm in length; he also used a bifilar suspension from the ceiling of the laboratory. He concluded, as had BOLTZMANN, that SAINT-VENANT's theory was not in accord with the experimental facts.[2] VOIGT carefully aligned the rods before impact. The drawn rod was set in motion by the burning of a paper attachment. The amplitudes following impact were read with great care to 0.1 mm to avoid parallax in the readings. Characteristically, VOIGT included a long discussion of all manner of minute possible sources of experimental error and a lengthy but arbitrary disposal of his auxiliary empirical assumptions.[3] He provided numerous tabulated results, each involving 5 to 12 observations of rods of equal and unequal length, different diameters, lengths, and impact velocities. In every instance the experimental velocity following impact differed from that predicted by the theory.

VOIGT attempted to modify SAINT-VENANT's theory by introducing an effect supposedly due to an elastic layer at the impact face. Noting the effects of curvature and unevenness of the impact surfaces, gas layers condensed upon these surfaces, and differences in the elasticity characteristics of an "in between layer" from the surface to the interior, he introduced an empirical factor to correlate his data. He debated whether or not the empirical coefficient $C = EA/\delta$ (where E is the unknown modulus of the "in between layer", δ its unknown thickness, and A is the contact area) was independent of impact velocity, etc. His "modified theory" was short-lived; it was to be disposed of by HAUSMANINGER and by HAMBURGER during the following three or four years. But there were lengthy discussions before that denouement. At the end of his paper VOIGT noted that after completing his work he had read the paper of HEINRICH RUDOLPH HERTZ[4] on the theory of the contact of elastic bodies; curiously VOIGT then added that HERTZ's theory was somehow similar to the formula he himself had inferred by introducing empirical parameters. On the basis of the large magnitude of his diameter-to-length ratio, i.e., 17:100, VOIGT dismissed the earlier data of BOLTZMANN, which disagreed with his own "modified" theory. He stated, however, that those data had placed restrictions upon the extent of applicability of his theory under "unusual circumstances".

---

[1] VOIGT [1883, *1*].

[2] Twentieth century references on this matter from as early as 1912 omit BOLTZMANN's original contribution, crediting historical precedence to VOIGT. Perhaps the fact that BOLTZMANN used India rubber somewhat mitigates this comment.

[3] Along with some of VOIGT's own contemporaries, such as HAUSMANINGER [1884, *1*] in 1883, I find VOIGT's use of sometimes arbitrary empirical arguments to "correct" his measured data before comparing it with theory, disturbing to say the least. In this instance he modified his numerical results to allow for the effects of an assumed air pressure followed by a vacuum at the impact face, and further to allow for a divergence in the velocity of the impacting rod following impact, and finally, to allow for air resistance.

[4] HERTZ [1882, *1*].

3.35. Hausmaninger's use (1884) of the time of contact technique of Pouillet (1844) in the Boltzmann experiment, and the half century of similar experiments (1884–1936).

In 1884 BOLTZMANN induced VICTOR HAUSMANINGER[1] to enter the controversy. HAUSMANINGER responded by repeating BOLTZMANN's measurements on the very same specimens. He obtained essentially the same results. Further, to answer VOIGT's criticism, at BOLTZMANN's urging he performed a new series of experiments on long rods. The diameter-length ratio which was 1:10 for thick rods and 1:28.5 for the thinner, corresponded to the ratios of 1:25 to 1:36.4 of VOIGT, and thus eliminated the earlier ratio of 17:100 to which VOIGT had objected. HAUSMANINGER noted that the impacting rod always followed by a small amount the impacted rod, as VOIGT had indicated. He found that in the new series of experiments neither SAINT-VENANT's theory nor VOIGT's modification applied. That more questions were raised by these data on rubber than had been raised before, becomes apparent in the second and very important part of HAUSMANINGER's paper where he reported the first measurements of time of contact (duration of the impact) for the impact of two symmetrical specimens.

Before discussing these results it is important to review the work of two persons, POUILLET and SCHNEEBELI: CLAUDE SERVAIS MATHIAS MARIE ROLAND POUILLET, whose remarkable earlier studies had provided the method which HEINRICH SCHNEEBELI, in an unusual experiment,[2] applied to solid mechanics.

In 1822 POUILLET[3] had been interested in determining the threshold electric current for movement of a magnetized needle. By 1844[4] he had extended these ideas to the study of the minimum time an electric current had to be operable to provide an observable deviation of a galvanometer in the circuit. He demonstrated that a current would run through a wire of several thousand meters length in a time of less than $^1/_{7000}$ sec. POUILLET had become interested in the timing of ballistic events to 0.0001 sec accuracy.[5]

He glued a 1 mm width radial strip of tin foil onto an 84 cm diameter glass plate which rotated at a known speed on a central axis. This strip was in continuous electrical contact with a spring loaded brush at the axle. A Daniell cell of six elements connected to 40 m of 1 mm copper wire was used, and the circuit was completed when a second brush near the glass disk periphery made contact with the foil strip. Rotating the disk at various speeds, he was able to calibrate the galvanometer deviation as a function of the time of application of the electric current without having to determine directly the mechanical inertia, etc. A time interval of $^1/_{5000}$ sec caused a 12° displacement of the galvanometer needle which required about 10 sec to travel this arc. Using a more sensitive Melloni galvanometer which gave 15° for $^1/_{5000}$ sec with 20 m of 1 mm copper wire, he found it was "easy to observe $^1/_{10000}$ sec." POUILLET used his technique to determine the travel time of a bullet in a gun barrel. The firing of the trigger closed a circuit which was opened again when the bullet on reaching the muzzle broke a wire in the circuit.

Twenty-seven years later, in 1871, it occurred to HEINRICH SCHNEEBELI[6] that POUILLET's method could be employed to study the impact of elastic bodies.

---

[1] HAUSMANINGER [1884, *1*].
[2] SCHNEEBELI [1871, *1*].
[3] POUILLET [1822, *1*].
[4] POUILLET [1844, *1*].
[5] POUILLET [1844, *1*], [1845, *1*].
[6] SCHNEEBELI [1871, *1*].

Apparently unaware that any significant theoretical advances in the laws of impact had been made since the contributions of HUYGENS in the 17th century, SCHNEEBELI described an empirical situation in which cylinders and spheres of various metals with different dimensions struck the end face of a steel cylinder tightly screwed down on a stationary plane. From four 7 cm long steel cylinders of the same length and various cross-sections, SCHNEEBELI observed that the duration of impact increased with the mass. He neglected to consider the possibility that such an increase in the time of contact equally well could have been a result of the increase in diameter.

From the use of spheres as impacting bodies he discovered that the duration of impact decreased with increasing velocity or height of fall. By screwing half spheres of various radii on the same cylinders, he found that the time of contact decreased as the radius of curvature was increased, but there was only a small influence. By varying the lengths of four steel cylinders he observed that the time of contact increased with increasing length. Apparently realizing the difficulties of analyzing the problem for a specimen fixed to a stationary plane, at the end of his paper he compared the impact of freely suspended spheres of equal and unequal radii, noting for the latter that it made no difference which one of the spheres was the impacting body. For the three situations shown in Fig. 3.48, designated as a, b, and c, SCHNEEBELI stated that the duration should be as $b=3a=4c$. In two series of experiments on 110 g and 155 g steel spheres he obtained values of duration in scale partitions of $a=21.3$, $b=66.0$, $c=15.6$ and $a=20.5$, $b=61.6$, and $c=15.0$, which are in surprising agreement with the empirical equation. Although the data were given in scale partitions, a calibration graph was provided for conversions to time. One example cited by him provided a time for cylinder impact of 0.000190 sec for a height of fall of 33 mm.

SCHNEEBELI (1871)

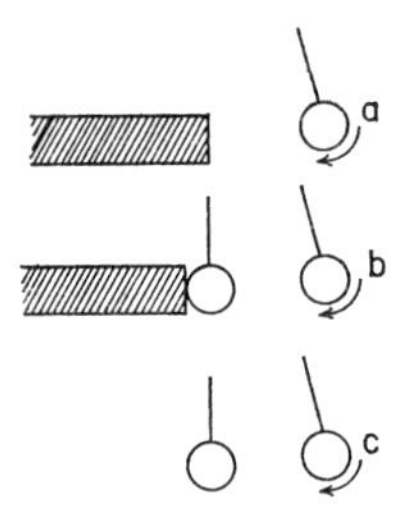

Fig. 3.48. The three experimental situations considered by SCHNEEBELI in his measurements of the duration of impacts.

In a second paper[1] SCHNEEBELI determined durations for metal spheres of equal mass of steel, copper, zinc, brass, silver, tin, and lead, and showed that the contact times were inversely proportional to the square root of E. These data are given in Table 95 for two experimental series. The letter a denotes scale partitions proportional to impact duration. SCHNEEBELI was concerned that he had to alter radii to achieve constant weight for the spheres, and that lead and tin showed some slight permanent deformation.

Thus, in comparing experiments, many of which were poorly conceived from an analytical point of view, SCHNEEBELI discovered the general nature of elastic

[1] SCHNEEBELI [1872, *1*].

Table 95. SCHNEEBELI (1872).

Sphere	Series I				Series II		
	$\sqrt{E}$	a	$a\sqrt{E}$	Corr.	a	$a\sqrt{E}$	Corr.
Steel	140.0	72.5	101.5[a]	+ 0.5	84.2	117.9	+ 2.1
Copper	102.0	94.2	96.1	+ 5.9	115.0	117.3	+ 2.7
Zinc	93.3	111.0	103.6	− 1.6	130.0	121.3	− 1.3
Brass	92.4	110.5	102.1	− 0.1	127.0	117.4	+ 2.6
Silver	84.5	112.0	94.6	+ 7.4	130.0	110.0	+ 10
Tin	63.0	164.0	103.0	− 1.0	194.0	122.0	− 2
Lead	42.0	270.0	113.0	− 11	320.0	134.0	− 14

[a] a is the divergence of the galvanometer in arbitrary units. Obviously SCHNEEBELI has made an error in the decimal points for this column.

HAUSMANINGER (1884)

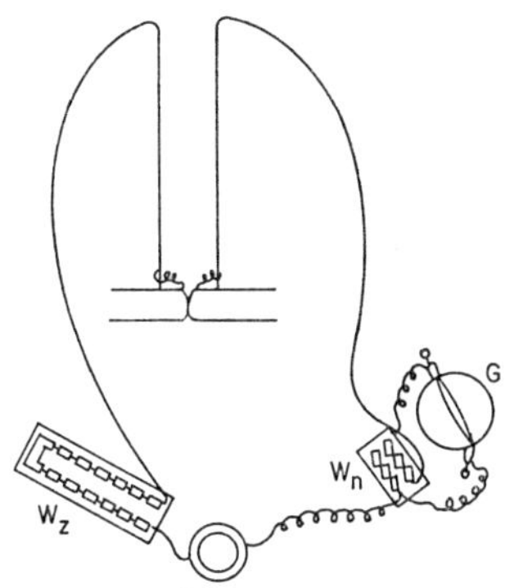

Fig. 3.49. The electrical circuitry used by HAUSMANINGER in his determination of the time of contact in experiments on the impact of bars.

impact. If he had been as imaginative in the design of experiments as in adapting the POUILLET method, his work would have provided a physical foundation for the theory of impact HERTZ was to conceive a decade later.

Returning to the study on impact of bars by HAUSMANINGER in 1883,[1] we find a properly conceived experiment with respect to analysis performed upon a poorly chosen solid, i.e., India rubber with brass end caps. In 1883 one should have had every reason to believe that the linear theory itself did not apply to such material. HAUSMANINGER's schematic diagram of the apparatus is shown in Fig. 3.49.

The electrical contact was made by means of glued fine platinum strips, and a check was made with the earlier tests to be certain that only a slight difference in final velocities was introduced. The fact that no difference in impact durations was observed for an 80% increase in impact velocity for bars of equal length and of unequal length should have raised questions, although reproducibility admittedly was obtained for measurements at each impact velocity.

In a better choice of solid for comparing a linear elastic theory with experiment, HAUSMANINGER described measurements on glass. For three rods of glass, one of

[1] HAUSMANINGER [1884, *1*].

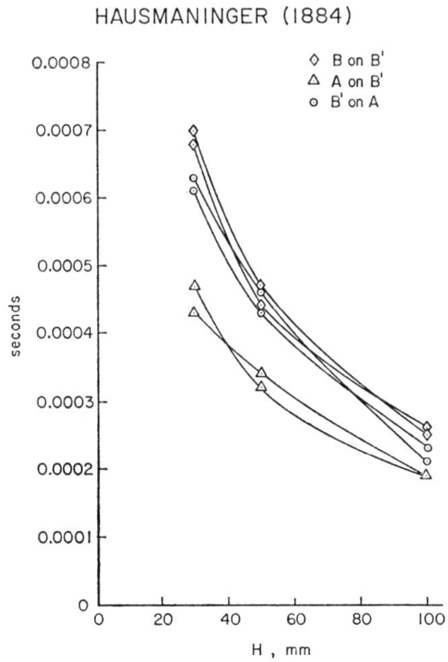

Fig. 3.50. The variation of the time of contact with the height of drop for glass rods of different lengths and diameters.

135 mm length, 13 mm diameter designated A'; and two of 272 mm length, 9 mm diameter designated B and B'; variations with impact velocity were observed, as may be seen in Fig. 3.50 for two tests of each combination of bars at the three initial heights indicated.

Hausmaninger noted that the durations for the glass bars of unequal lengths were curiously short, the behavior being opposite to that of the rubber bar. In 1885, in a second paper[1] he concluded that not only did Saint-Venant's theory fail to apply but also Voigt's modification could not be analyzed experimentally due to its undefinable parameters.

In the 19th century, the definitive experimental study of the elastic impact of two bars was that of Max Hamburger[2] in 1886. Also using the technique of Pouillet for measuring the time of contact, Hamburger studied the impact of hard brass rods. He calibrated the galvanometer by Schneebeli's pendulum method, which after extensive experimentation gave close agreement among the measurements. After first trying and then abandoning as unworkable Schneebeli's experiment with the struck bar fixed, Hamburger turned to the Boltzmann experiment with its double bifilar suspension system. The now freely suspended impacted rod was the same bar in all experiments, having a length of 300 mm and a diameter of 10.3 mm. Hamburger examined the end faces optically after each impact for any evidence of permanent deformation. He filed the impact end faces carefully to permit as close an axial alignment as possible. Five lengths of hard brass impacting rods of four different diameters struck the 300 mm impacted bar with seven different velocities. The tabulated data for the many combinations are shown graphically in Fig. 3.51.

[1] Hausmaninger [1885, *1*].
[2] Hamburger [1886, *1*].

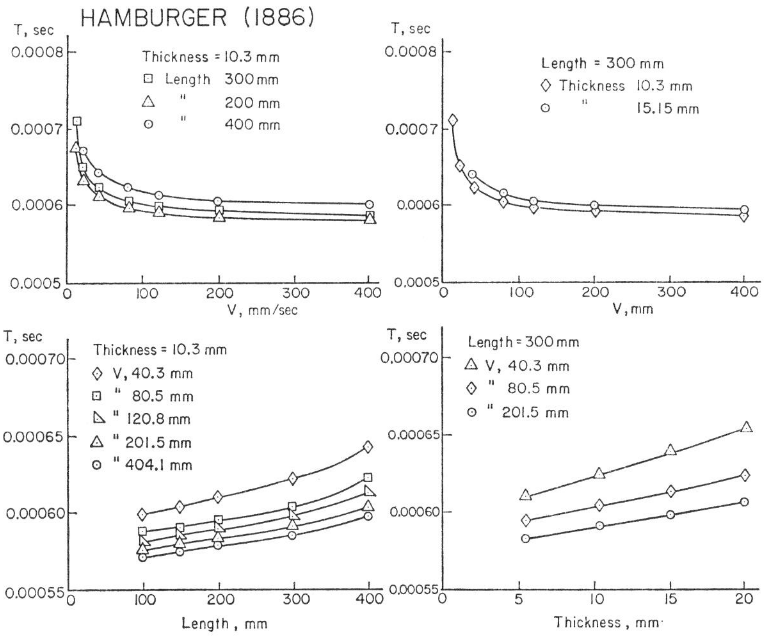

Fig. 3.51. The variation of time of contact with impact velocity for bars of different lengths. Also shown are the variations of the time of contact with length and with thickness of specimen, for specified impact velocities.

Whether or not the duration of impact was in fact a linear function of the length, and independent of the impact velocity, became a matter of discussion three decades later. All the essential features of the experiment on impact of bars were given quantitatively by HAMBURGER's data. He stressed the similarity of trends between his data and that of SCHNEEBELI while dismissing the results of HAUSMANINGER. Having found that neither SAINT-VENANT's theory nor VOIGT's elastic layer hypothesis was adequate, even in combination when various parameters were altered, he turned from cylinders, with their attendant difficulties, to spheres, with the HERTZ contact theory in mind.

On 14.75 mm radius steel spheres and on brass spheres of 13 mm radius, HAMBURGER, for five different impact velocities (given as initial heights of impacting sphere, in mm), compared the times of contact with those predicted by HERTZ's theory of impact. Faced with the 19th century dilemma on POISSON's ratio, he provided two sets of calculations for both solids, one supposing that $\nu = {}^1/_3$ and the other that $\nu = {}^1/_4$. He used the moduli $E = 20000$ kg/mm² for steel and $E = 10000$ kg/mm² for brass, with specific gravities of 7.7 and 8.39 respectively. He considered the correlation (shown in Table 96) between experiment and calculation to be very good except at small velocities, for which the data were so various that he excluded them. He noted that this should have been the region of best agreement.

Although HAMBURGER considered both the impact of cylindrical rods in relation to SAINT-VENANT's theory and the impact of spheres in relation to HERTZ's theory, he made no attempt to combine the two. This feat was left for J. E. SEARS, over two decades later.

Table 96. HAMBURGER (1886).

	Steel: Radii of the spheres, 14.75 mm				
Velocity	73.7	122.9	192.1	295.0	442.9 mm[a]
T, observed	0.000190	0.000165	0.000146	0.000134	0.000126 sec
T, calculated ($\nu = ^1/_3$)	0.000150	0.000135	0.000124	0.000114	0.000105 sec
T, calculated ($\nu = ^1/_4$)	0.000154	0.000139	0.000127	0.000119	0.000107 sec
	Brass: Radii of the spheres, 13 mm				
Velocity	73.7	122.9	192.1	295.0 mm[a]	
T, observed	0.000196	0.000173	0.000157	0.000148 sec	
T, calculated ($\nu = ^1/_3$)	0.000181	0.000164	0.000150	0.000138 sec	
T, calculated ($\nu = ^1/_4$)	0.000185	0.000167	0.000153	0.000140 sec	

[a] This is the initial distance the striking bar was displaced which is proportional to velocity.

SEARS' first series[1] of experiments on the impact of bars, in 1908, was performed for the purpose of determining an accurate dynamic value for E, which he presumed would be adiabatic for wave front data, for comparison with quasi-static isothermal moduli. Sixty-five years earlier WERTHEIM had been the first to study and compare the two moduli. He had offered a challenge which still would occupy experimentists during the first decade of the 20th century: recall, for example, the efforts of GRÜNEISEN.[2] SEARS drew upon POUILLET's experiment of 1844 but modified the technique by calibrating from the discharge of a capacitor having a known voltage. He determined times of contact for identical bars ranging in length from approximately 14 cm to 95 cm, the latter being nearly twice the length of those in any previous study. SEARS reasoned that the slope of the curves representing duration as a function of length should straighten with increasing length, approach the bar velocity of SAINT-VENANT's theory, and thus, even though not passing through the origin, should provide a dynamic value for E. Like his predecessors, SEARS rounded[3] the impact ends of his bars to avoid the difficulties occasioned by small errors in alignment of the axes. Because his bars were relatively long, what he referred to as "the end effect" was sufficiently reduced that the slopes from the data for steel, aluminum, and copper shown in Fig. 3.52 provided wave speeds of 5130 m/sec, 5070 m/sec, and 3680 m/sec, respectively. SEARS found these values in close correspondence with wave speeds calculated from quasi-static tests on the same rods, i.e. 5110 m/sec, 5050 m/sec, and 3660 m/sec, respectively, providing dynamic to static ratios of 1.0010, 1.0026, and 1.0015, on the presumption that the former corresponded to adiabatic changes, the latter to isothermal ones.[4]

The rounded ends suggested to SEARS the possibility of converting the infinite step of the SAINT-VENANT theory to a gradual increase in stress following contact, were he to assume that HERTZ's theory applied to the rounded end while SAINT-VENANT's theory applied beyond an arbitrarily chosen point in the vicinity of the impact face. The correlation with experimental results which the semi-empirical procedure provided had the unfortunate effect of creating the impression for the next three or four decades that wave initiation and propagation in cylindrical

[1] SEARS [1908 *1*].

[2] GRÜNEISEN [1906, *1*].

[3] BOLTZMANN [1882, *1*] and HAUSMANINGER [1885, *1*] equipped their impacting specimens with end caps "whose impacting faces were gently rounded."

[4] SEARS concluded in italicized type at the end of a section on elastic moduli: "The values of all the elastic constants of a metal are the same under instantaneous as under steady stress." (SEARS [1912, *1*], p. 78.)

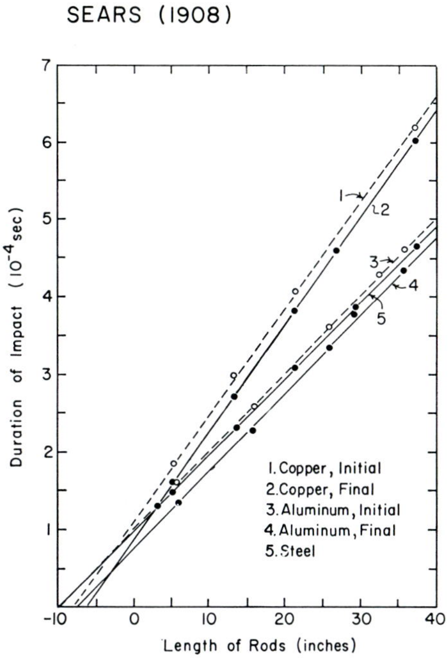

Fig. 3.52. Variation of the time of contact with the length of the specimen for rods of copper, aluminum, and steel.

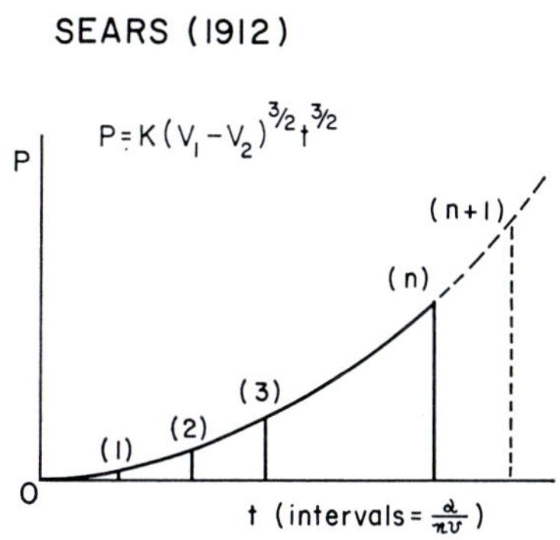

Fig. 3.53. SEARS' determination of the initial pressure vs time history in the impact of two bars with rounded ends, using the HERTZ analysis (1912).

bars following axial impact were well understood. With respect to this correlation SEARS had remembered that accuracy of the final result depended upon the assumptions made in using the static hemispherical distribution of stress in compression for the end element.[1]

Choosing a distance d at the rounded end as reference point for the HERTZ analysis, SEARS determined the initial pressure-time history shown in Fig. 3.53 where the time is plotted in sub-integral multiples n, where V is the velocity of propagation.

[1] SEARS [1912, *1*].

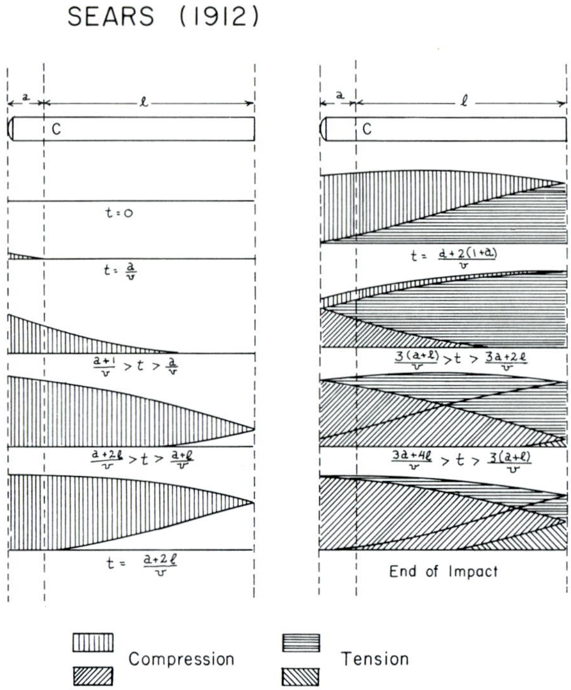

Fig. 3.54. Wave front profiles determined from the initial history of Fig. 3.53.

Wave front profiles following from this initial history are shown in Fig. 3.54 for time intervals in terms of the bar radius.

The correlation which could be achieved between the calculated (solid line) and measured (filled circles) times of contact for rods of various lengths striking a specimen of 97.4 cm is shown in Fig. 3.55. The hump occurred at approximately half the greatest length used. The dotted lines E and C were the asymptotes approached as the ratio of rod lengths increased.

The data in the three figures above were for a fixed impact velocity of 12.7 cm/sec. In Fig. 3.56 is shown the duration for these round ended rods as a function of impact velocity. The closed circles were for impact for which no permanent deformation was observed and the open circles were for impact at which some permanent deformation was observable at the point of contact. These data were compared with the calculated values of the modified combination (curve B) and with a calculation from Hertz's theory for spheres alone (curve C). The length of the rod in these tests was only ten times the diameter. (The rods were 13.97 cm long.)

Sears, like others who followed him, was concerned with how flat ended specimens which after impact exhibited the same delayed growth of the wave front with time, might be included in his explanation of the behavior of the bar in impact, based on Hertz's theory. He pondered how under these conditions Voigt's "elastic layer" hypothesis could be explored empirically.

One wonders whether Sears had come to understand the limitations of a measurement of time of contact and the necessity for a proper study of the wave profile itself, when he concluded his paper of 1912 with an acknowledgement

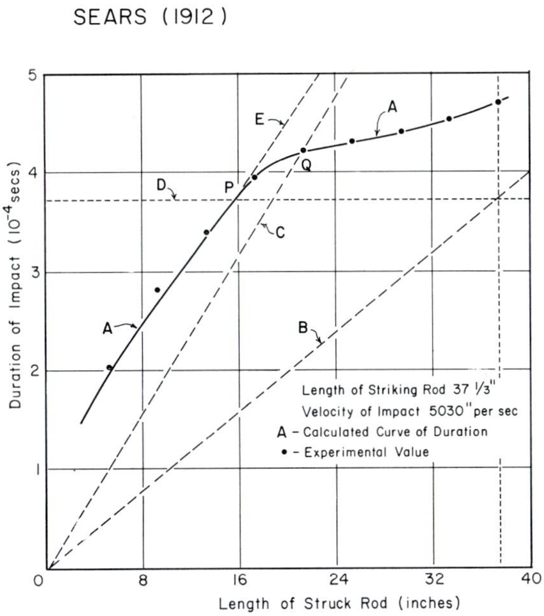

Fig. 3.55. Sears' comparison of experimental observation with calculated time of contact (solid line).

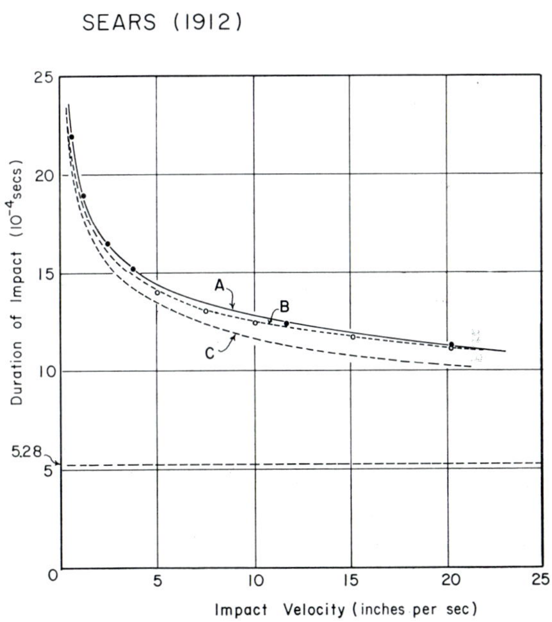

Fig. 3.56. The time of contact vs impact velocity for no permanent deformation (filled circles), and for some permanent deformation (open circles), compared with the Hertz theory for two spheres (C) and empirically modified versions of the theory, (A) and (B).

of his indebtedness to Professor BERTRAM HOPKINSON.[1] It was HOPKINSON who two years later,[2] with an ingenious experiment, was the first to explore, if indirectly, some of the detail of the wave profile of a pulse propagating along a cylindrical bar. HOPKINSON's experiment came to have very great technological importance inasmuch as it provided a method for measuring the magnitude of an explosive pulse in the atmosphere. His experiment did not delineate the actual shape of a propagating compression pulse in a bar but did provide information with respect to its height and duration.

To a one-inch diameter bar several feet in length held horizontally by four threads, HOPKINSON attached a short cylindrical piece at one end face. The ground faces of bar and piece were held together either by a fine layer of grease, or magnetically. At the opposite end, when a short stress-time pulse was applied either from the collision impact of a short specimen such as a bullet or from an adjacent explosion in the air, a compression pulse traveled along the bar and entered the attached piece. When a tensile pulse is reflected from the free end, the sum of the stresses of incoming and reflected waves eventually reaches zero; thus the piece flew off to be caught in a ballistic pendulum. By changing the length of the piece and at the same time measuring the final motion of the bar as a pendulum, he could determine the maximum stress and duration or length of the applied pulse from an analysis of the combined momentum after separation. In Fig. 3.57 are shown experimental results for the percentage momentum in the piece as a function of the length of the piece, from which he calculated the maximum pressure and duration.

In 1924, JOHN EDWARD PRETTY WAGSTAFF,[3] also of Cambridge University, used a minor modification of POUILLET's experiment of 1844, and made a type of measurement which he credited HAMBURGER with having originated in 1887.

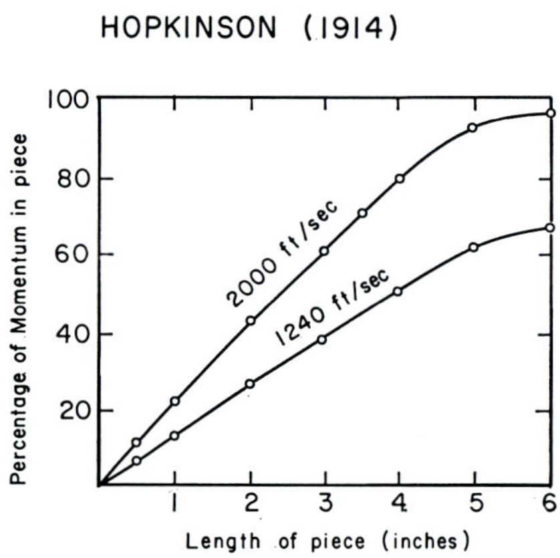

Fig. 3.57. Percentage of momentum left in piece vs the length of piece for the two velocities shown in HOPKINSON's bar experiment (1914).

[1] "In conclusion, I have to express my thanks to Prof. Hopkinson, of the Engineering Laboratory, Cambridge, where the work was carried out, for his unfailing interest and kind advice. When he first suggested that I should undertake experiments on the velocity of wave-propagation in metal rods, the developments he had in view were, I believe, of a far more practical character than those here described. I happen, however, to be interested in the abstract problem of impact, and he has always shewn himself perfectly willing that I should follow up the work on these lines." (SEARS [1912, *1*], p. 82).

[2] BERTRAM HOPKINSON [1914, *1*].

[3] WAGSTAFF [1924, *1*].

WAGSTAFF realized that the "working of the Hopkinson pressure bar seems to be directly involved in the theory, as depending on the conditions under which a sudden local pressure at one end gathers itself up at a short distance into a travelling wave uniform across the section."[1] Hence he proposed to re-examine the theoretical limits obtained by SAINT-VENANT and HERTZ and the "transition between them." He modified POUILLET's technique by substituting a charged condenser for the battery. This permitted a direct calculation of contact times through the use of a graphic plot obtained by altering the electrical circuitry.

WAGSTAFF's contribution to the problem was to vary the diameters and lengths of a series of bars. With six lengths of bars, approximately 10 to 65 cm long, and seven diameters from 1.59 to 3.175 cm, he obtained a total of 42 cases for study of the effect upon the time of contact. For the HERTZ spheres the expression relating duration and impact velocity is $t = A v^{\gamma}$ with $\gamma = -{}^1/_5$. For the SAINT-VENANT theory, $\gamma = \infty$, the duration is t, v is the velocity, and A is a constant depending on the elastic moduli, POISSON's ratio, and the radii of the sphere. If x is the distance of withdrawal of the impacting bar, T the free period of swing of the bar in its bifilar suspension, then $v = 2\pi x/T$. From the slope of straight lines, if experimentally obtained, a plot of $\log_{10} t$ vs $\log_{10} x$ reveals the value of γ. From the linearity of the data in Fig. 3.58 for 10.5, 21.3, 40.8, and 61.25 cm rods an empirical form $t = A v^{\gamma}$ is consistent. The slope varies with increasing length.

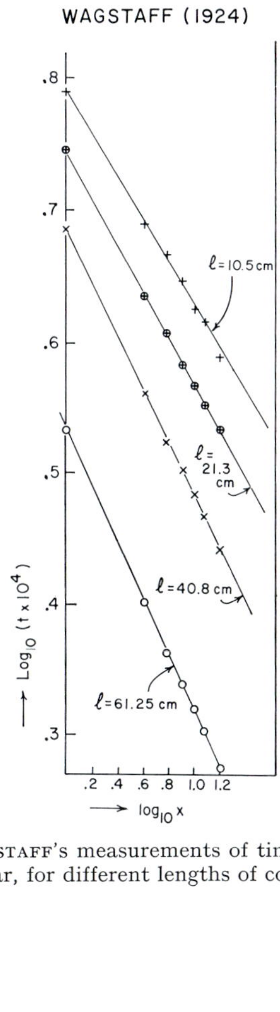

Fig. 3.58. A log-log plot of WAGSTAFF's measurements of time of contact vs displacement of impacting bar, for different lengths of colliding bars.

[1] *Ibid.*, p. 545.

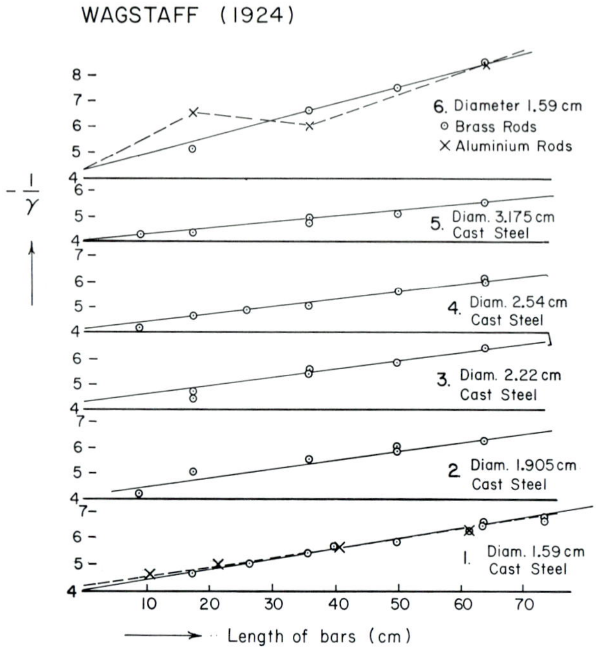

Fig. 3.59. The variation of WAGSTAFF's empirical index with the length of the colliding bars, for brass and aluminum rods, and for steel rods of different diameters.

An analysis of such data provided the values of $\gamma = -1/8$ to $-1/4$ see Fig. 3.59 for aluminum, brass, and cast steel rods of the various diameters shown. Numerical values fell below the HERTZ value of $\gamma = -1/5$ for bars 10 cm long, but the values became increasingly higher as the bar length increased.

Plots of the time of contact vs bar length for symmetrical impacts did not provide the straight lines from which SEARS had determined his dynamic values of E. Fig. 3.60 is shown to illustrate this experimental fact from the large amount of tabulated and plotted data for both increasing length and diameter in symmetrical impact at a common velocity of 10.4 cm/sec. WAGSTAFF's apparatus, Fig. 3.61, is identical with that used in the experiments of the previous 40 years. Like SEARS, WAGSTAFF withdrew the specimen and released it by electrically burning a fuse wire.

Twelve years later, in 1936, WAGSTAFF suggested to W. A. PROWSE[1] of the University of Durham that it was desirable to study the impact of bars as a function of the radius at the point of contact at the impact face. PROWSE agreed, and he proceeded to examine rods having the same diameter, 1.59 cm. Once again POUILLET's experiment on time of contact was used without reference to its origin.[2] PROWSE studied six different radii (0.159, 0.318, 0.98, 1.44, 2.58, and 4.60 cm) for five different lengths of the same diameter bar (8.5, 17.0, 35.6, 49.6, and 63.6 cm), over a range of velocities from 2.18 cm/sec to 24.00 cm/sec. To preserve similar conditions from one measurement to another, he successively cut down on the same bar the radii of the rouge polished end. The data on the duration

[1] PROWSE [1936, *1*].

[2] PROWSE referred to SEARS [1912, *1*] and WAGSTAFF [1924, *1*] as the source of the method used which in fact was almost in the same form as the experiment of POUILLET in 1844.

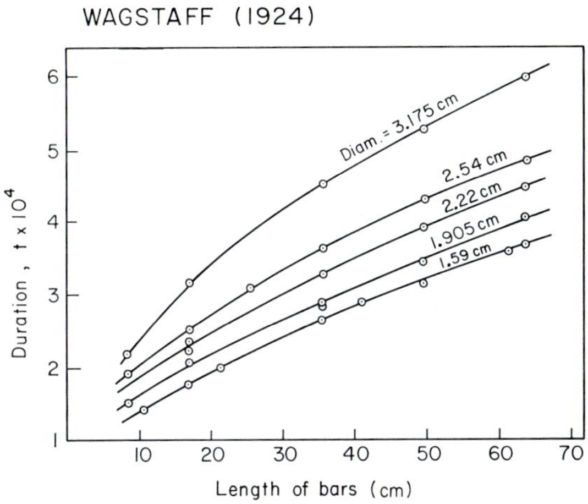

Fig. 3.60. The time of contact vs the length of the colliding bars for various rod diameters.

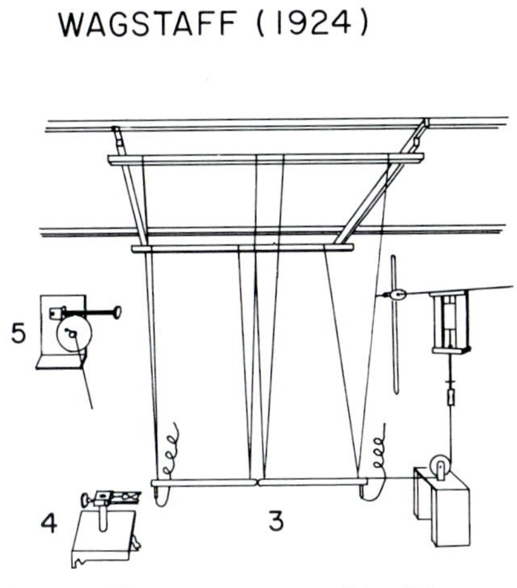

Fig. 3.61. The apparatus used by WAGSTAFF.

of the contact for the main experiments given in Table 97 demonstrate that the duration depends not only upon the length and velocity of impact as had been shown earlier, but for bars of given diameter also, and sensitively, upon the radius at the face of initial contact.

In the remainder of the paper PROWSE discussed the observed differences between hardened and soft impact surfaces. He concluded that the incremental time between the measured deviation and SAINT-VENANT's prediction of $t=2l/c_0$, $c_0=\sqrt{E/\varrho}$ being the bar velocity, and l indicating the lengths of the bar, for a double traverse of the wave front, was a function of end radii, impact velocity, and length of bar. The data had been determined from the durations given in Table 97. They did show marked increase in the time increment ΔT with decreasing end radius, decreasing impact velocity, and increasing bar length. PROWSE considered the extrapolation to flat ends for which there would be a more rapid growth of pressure. For bars more than 30 cm he expected from SAINT-VENANT's

Table 97. PROWSE (1936). *Duration of contact*, t (sec $\times$ 10^{-4})

Length of bars (cm)	63.6	49.6	35.6	17.0	8.5
Velocity (cm/sec)	t	t	t	t	t
		Radius of ball end, 4.60 cm			
2.18	4.67	4.13	3.45	2.47	1.88
3.27	4.44	3.72	3.26	2.24	—
4.36	4.23	3.69	3.11	2.12	1.59
6.45	4.03	3.46	2.89	1.97	—
8.73	3.88	3.31	2.74	1.87	1.36
13.09	3.74	3.09	2.57	1.71	1.25
17.45	3.54	2.95	2.45	1.62	1.16
24.00	3.39	2.86	2.31	1.53	1.08
		Radius of ball end, 2.58 cm			
2.18	5.05	4.03	3.60	2.51	1.73
3.27	—	3.78	3.37	2.32	1.63
4.36	4.37	3.65	3.20	2.18	1.60
8.73	3.99	3.26	2.85	1.91	1.40
13.09	3.77	3.14	2.67	1.76	1.29
17.45	3.65	2.96	2.54	1.67	1.25
24.00	3.50	2.87	2.37	—	1.16
		Radius of ball end, 1.44 cm			
2.18	5.43	4.83	4.10	2.89	2.13
4.36	4.77	4.28	3.50	2.57	1.88
8.73	4.38	3.83	3.11	2.21	1.61
13.09	4.07	3.55	—	2.05	1.47
17.45	3.90	3.42	2.80	1.93	1.37
26.17	3.70	3.15	2.62	1.78	1.31
		Radius of ball end, 0.98 cm			
2.18	5.65	5.10	4.21	3.22	2.38
3.27	5.31	4.68	3.92	2.82	2.19
4.36	5.06	4.48	3.74	2.73	2.07
6.54	4.74	4.24	3.47	2.48	1.86
8.73	4.55	4.06	3.30	2.33	1.76
13.09	4.25	3.80	3.09	2.16	1.63
17.45	4.01	3.58	2.99	2.04	1.52
24.00	—	3.38	2.79	—	—
		Radius of ball end, 0.318 cm			
2.18	7.30	5.93	4.98	3.80	2.76
3.27	6.88	5.54	4.60	3.51	2.54
4.36	6.56	5.20	4.37	3.34	2.40
8.73	5.85	4.61	3.88	2.88	2.07
13.09	5.39	4.40	3.58	2.66	1.99
17.45	—	—	3.46	2.52	1.84
24.00	4.78	3.95	3.32	2.34	1.69
		Radius of ball end, 0.159 cm			
2.18	7.45	6.91	5.73	4.12	3.02
3.27	—	6.39	5.22	3.78	2.78
4.36	—	6.03	4.96	3.56	2.64
6.54	5.99	5.55	4.62	3.27	2.43
8.73	5.68	5.27	4.37	3.08	2.29
13.09	—	4.81	4.03	2.84	2.13
17.45	4.89	4.55	3.81	2.69	2.02

theory that the linear relation between the duration and length should be reached, but this certainly was not in complete accord with all of the measurements of the previous half century.

3.36. The first use of electric resistance elements to study wave profiles in the Boltzmann experiment: Fanning and Bassett (1940).

HERBERT TOMLINSON[1] in 1883 had made extensive studies of the relation between electric resistance and tensile strain in metal wires, with no suggestion, however, that the phenomenon might be considered as the basis for the development of an extensometer. In 1931, E. C. EATON[2] used an electric resistance gage to measure strain in concrete. ROY WASHINGTON CARLSON[3] in 1935 described five years of progress in improving the "Elastic-Wire Strain Meter," and three years later DONALD S. CLARK and G. DATWYLER[4] of the California Institute of Technology described the development by EDWARD E. SIMMONS, Jr., of the first application of a bonded electric resistance gage. The bond was achieved by the use of Scotch tape and Glyptal binder. ARTHUR C. RUGE at the Massachusetts Institute of Technology at about the same time bonded the wire to a paper which in turn could be glued to the specimen, on which strain was to be measured.

FANNING and BASSETT (1940)

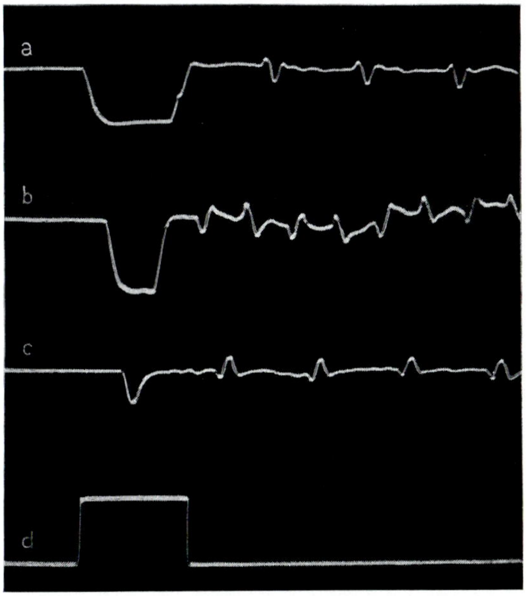

Fig. 3.62. The first measurements of the strain-time detail in the bar impact experiment. The results were obtained by means of a "carbon strip extensometer." The point of measurement was near the impact end (*a*), the center (*b*), and at the far end (*c*). (*d*) is the timing signal.

[1] TOMLINSON [1883, *1*].

[2] EATON [1931, *1*].

[3] CARLSON [1935, *1*]. Carbon pile resistance gage elements had been used 15 years earlier by BURTON MCCOLLUM and O. S. PETERS [1922, *1*].

[4] CLARK and DATWYLER [1938, *1*].

Returning to the problem of the impact of a bar and its use to determine dynamic values for E, we may note that R. Fanning and W. V. Bassett[1] in 1940 used a resistance element in the form of a "carbon strip extensometer" to examine for the first time the actual strain-time history at a point, after the symmetrical impact of bifilar suspended cylindrical bars. Measurements made at the center (b) and near the impact end (a) and the far end (c) in the struck bar are shown as oscillograms in Fig. 3.62. The bars were 152 cm long and 2.73 cm in diameter. The end of one bar was flat and the other had a known radius.

A calculation based upon Sears' analysis of 1912 was made for a comparison with measured strain (converted to stress by multiplying by E) at 14.3 cm from the impact face, which provided the result shown in Fig. 3.63 for the struck member of two 183 cm long bars. This correlation demonstrated the need to introduce a gradually rising wave front, as Sears had proposed, rather than the abrupt jump of the Saint-Venant theory.

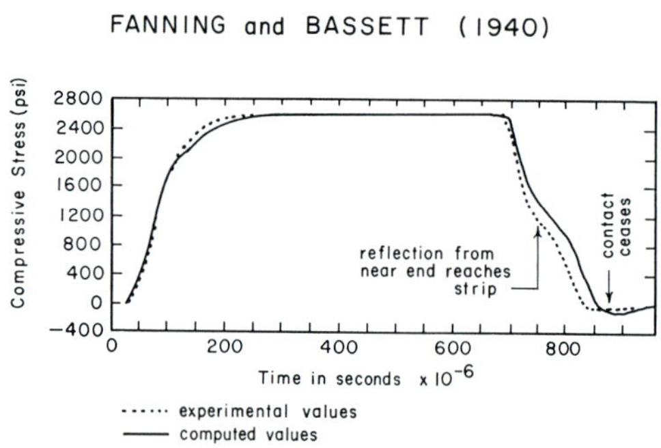

Fig. 3.63. Fanning and Bassett's comparison of a measured wave profile with values computed from Sears' analysis in 1912.

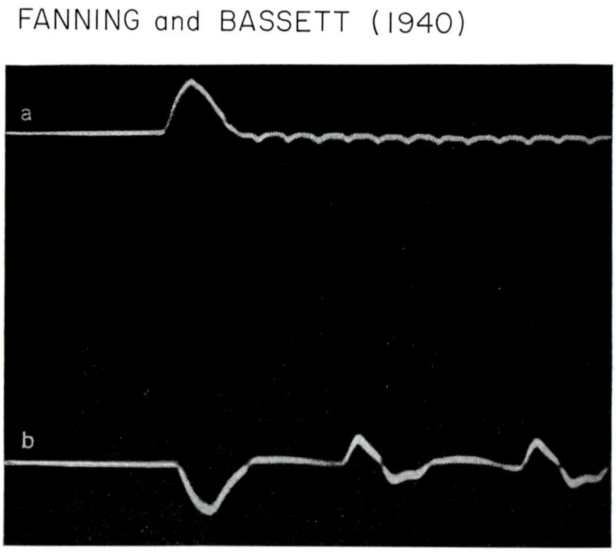

Fig. 3.64. Measurements of strain vs time for the collision of a short and long bar showing a measurement (a) near the end of the short bar, and (b) near the end of the long bar.

[1] Fanning and Bassett [1940, *1*].

Some of the complications of the earlier studies were revealed in the observations shown in Fig. 3.64 for collision of a 30.5 cm bar and a 183 cm bar; (*a*) shows a measurement near the end of the short bar and (b), near the end of the long bar.

Introducing what became a new problem in one-dimensional wave propagation, FANNING and BASSETT also made a measurement of a reflection from a "fixed" end. They expressed some concern because the reflected compression wave did not double as would be expected according to elementary theory.[1] This measurement is shown in Fig. 3.65.

FANNING and BASSETT (1940)

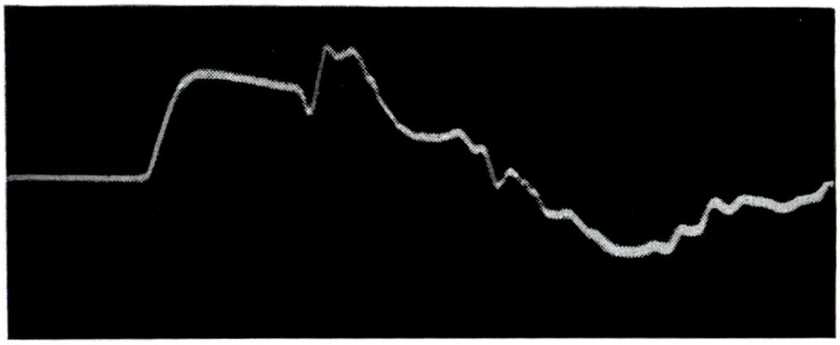

Fig. 3.65. FANNING and BASSETT's experimental demonstration that the reflection from a "fixed end" did not double as expected by elementary theory.

3.37. Davies' (1948) use of a capacitance displacement technique for the first comparison of pulse profiles with Pochhammer's (1876) three dimensional theory for cylindrical rods.

All of the experimental studies on the impact of a cylindrical bar from BOLTZMANN's in 1881 to FANNING and BASSETT's in 1940 implicitly assumed one dimensional wave propagation in attempting to explain the detail of the propagation of waves in terms of conditions at the impact face. This was despite the fact that a detailed three dimensional theory for the vibration of cylindrical rods developed by LEO POCHHAMMER[2] in 1876 and further discussed by CHARLES CHREE[3] in 1889 had been available even before BOLTZMANN in 1881 had found that SAINT-VENANT's theory did not accord with experiment.

It remained for R. M. DAVIES[4] in a review of the problem in 1948 to realize the importance of three dimensionality and its concomitant geometric dispersion in the wave propagation in a cylindrical bar. His important paper, while basically enlarging upon the studies of B. HOPKINSON, restated the problem and stimulated two decades of research. Indeed such research is still in progress, and there is an enormous array of papers on the subject.[5]

---

[1] An interesting experimental study of this problem was provided by E. A. RIPPERGER in 1952 [1952, *1*] and by E. A. RIPPERGER and H. NORMAN ABRAMSON in 1957 [1957, *1*].

[2] POCHHAMMER [1876, *1*].

[3] CHREE [1889, *1*].

[4] DAVIES [1948, *1*].

[5] Referring to elastic waves, viscoelastic waves, and plastic waves, R. M. DAVIES in 1956 stated: "At a rough estimate, some five hundred papers have been published during the past five years on these three aspects of the subject; it is clearly impossible to cover the whole field, and this review must therefore be confined to some of the more important features of current activity." (DAVIES [1956, *1*], p. 65.)

Davies' eighty-three page treatise entitled "A Critical Study of the Hopkinson Pressure Bar" is a remarkably thorough, well-written study, worth the many hours required to examine it in detail. Hopkinson had generated rapid axial stress-time histories in his long bar by firing a bullet at high velocity at the center of one end face. In the same experiment Davies introduced the displacement measurement technique of using the bar as the grounded member of a parallel plate condenser. The insulated member was a mounted metal plate. The latter was charged to high voltage by means of an electric circuit having an appropriate resistance element to produce a time constant of long duration. The change in potential produced by a very short time change in capacitance, when properly calibrated and dynamically displayed on an oscilloscope, provided a displacement vs time history at the surface of application. Such measurements were made at the free unloaded end of the bar and, by an adaptation of the device, separate determination were made of radial displacement. In Fig. 3.66 are shown oscillograms of measured profiles: (a) was a pulse from the explosion due to a one ounce charge of C. E. at the end of a 2.54 cm diameter 183 cm long bar and (b) was a 1.2 cm diameter steel ball striking the latter bar at 5 560 cm/sec.

Herbert Kolsky,[1] who further perfected this capacitance technique of Davies,[2] described in 1953 a free end response from the explosion of a No. 8 detonator (see Fig. 3.67). The timing signal for this experiment had a 5.3 microsecond period. The timing waves for the remaining data were as shown.

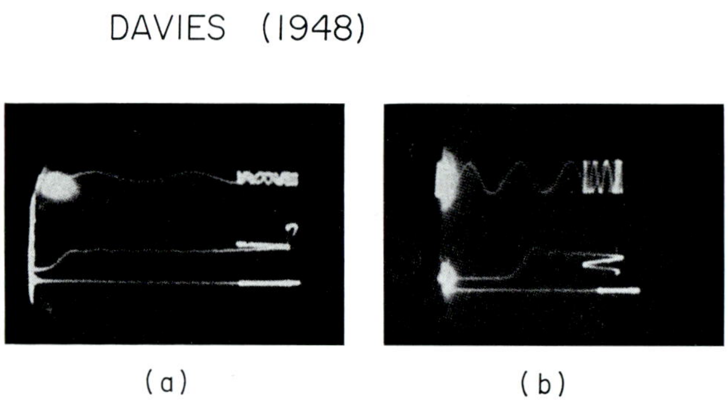

Fig. 3.66 a and b. Davies' oscillograms of displacement-time histories for a bar subjected (a) to an explosion at the opposite end, and (b) to the collision of a steel sphere.

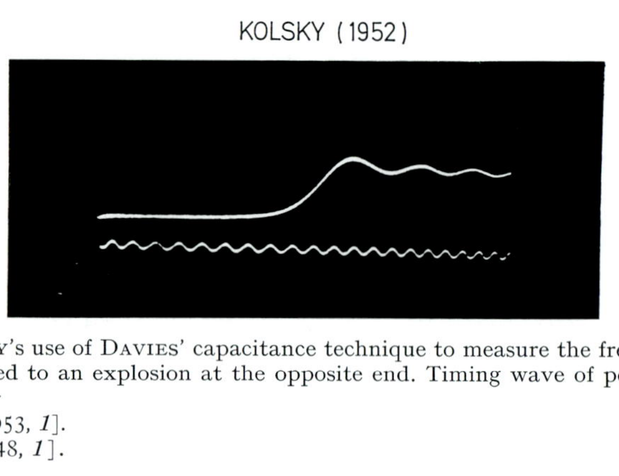

Fig. 3.67. Kolsky's use of Davies' capacitance technique to measure the free end response of a bar subjected to an explosion at the opposite end. Timing wave of period 5.3 μsec.

[1] Kolsky [1953, *1*].
[2] Davies [1948, *1*].

DAVIES had emphasized at the outset that the interpretation of his results, and also those of HOPKINSON's original experiment, depended upon the validity of the following assumptions: that linear elastic waves were involved, that the pressure pulse was propagated without distortion, and that the pressure was uniformly distributed over the cross-section of the bar. With these assumptions the displacement vs time history at the end could be differentiated to provide a stress vs time description of the pulse. POISSON's ratio ν and E for the solid were known, so the

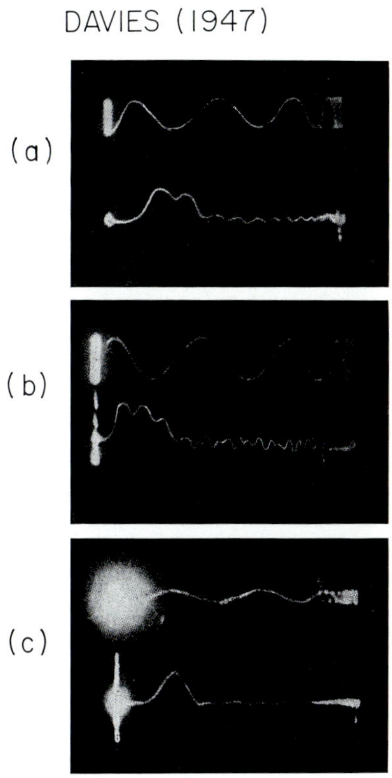

Fig. 3.68a—c. Data for radial displacement, made at three positions along a bar struck axially by a round nosed bullet.

radial displacement data also could be used directly to provide a pulse stress-time history. Thus from the radial displacement measurement made at 115 cm (a) and 35 cm (b) along a 2.54 cm diameter 183 cm long bar axially struck by a round-nosed bullet travelling at the velocity of 34595 cm/sec and 35692 cm/sec shown[1] in Fig. 3.68, the pressure-time curves of Fig. 3.69 could be calculated.[2]

A large portion of DAVIES' paper was devoted to an analysis of the second assumption, namely, that the pulse was non-dispersive. He gave the first detailed calculations of phase and group velocities for the various modes of an infinite bar in terms of the theory, then seven decades old, of POCHHAMMER and CHREE.

In a subsequent treatise in 1956 which reviewed the field of stress waves in solids, DAVIES provided the results from electric resistance gages for a pulse for

[1] DAVIES expressed appreciation to Dr. ENRICO VOLTERRA for these particular data.

[2] DAVIES [1956, *1*].

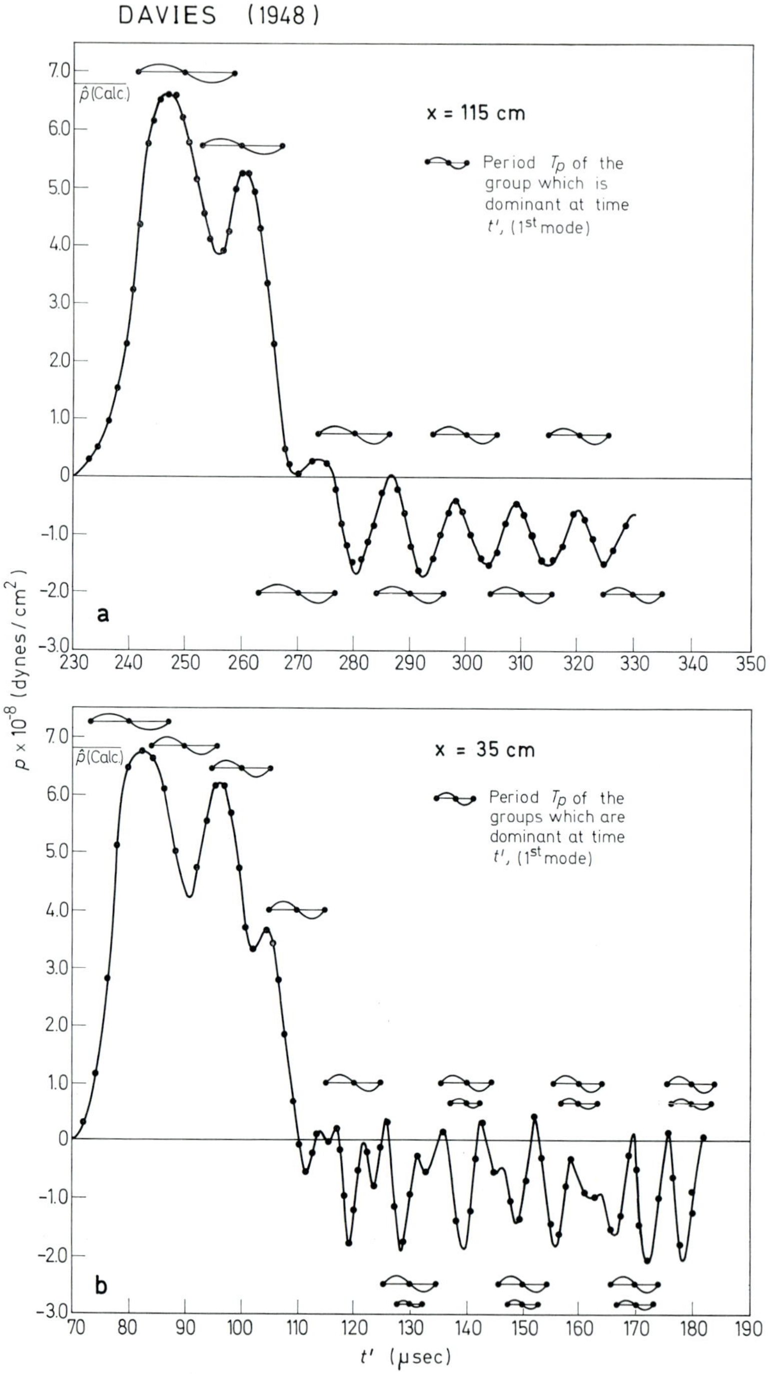

Fig. 3.69 a and b. Pressure vs time curves calculated by Davies from the measurements shown in Fig. 3.68.

DAVIES (1956)

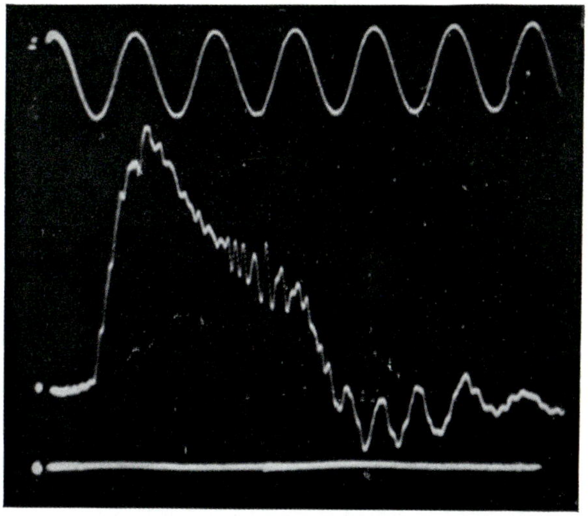

(a)

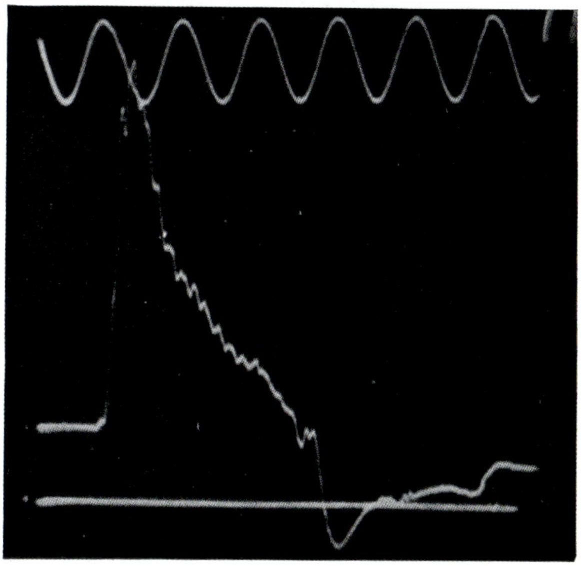

(b)

Fig. 3.70a and b. DAVIES' electric resistance measurement of pulse profile: (a) for longitudinal plus flexural components, and (b) with flexure eliminated.

longitudinal plus flexural components (a) (single gage) and for flexural elimination (b) (paired gages) shown in Fig. 3.70.

Fig. 3.70a shows a resistance measurement of pulse plus coda. DAVIES credited this experiment to Mr. N. J. REES. It involved a duralumin bar 275 cm long,

0.636 cm diameter, in which the pulse was produced by water pressure activated by firing a bullet at a piston. The experiment of Fig. 3.71 was made by the direct axial collision of a flat-nosed bullet striking a 183 cm long steel bar of 1.27 cm diameter. The gage was 79 cm from the impact end. DAVIES credited this measurement to Dr. D. E. THOMAS. These data plus those from similar measurements were compared with theoretical calculations of T_p, the mean period, for the first experiments in Fig. 3.72 and with T_p/T_a vs $t/\frac{1}{2}T_0$ for the second in Fig. 3.73. (T_a represents the time for infinitely long waves to traverse the bar radius and T_0 the distance to the point of measurement.) Fig. 3.72 gives the correlation with flexural wave prediction and Fig. 3.73 gives the correlation with extensional waves. These

DAVIES (1956)

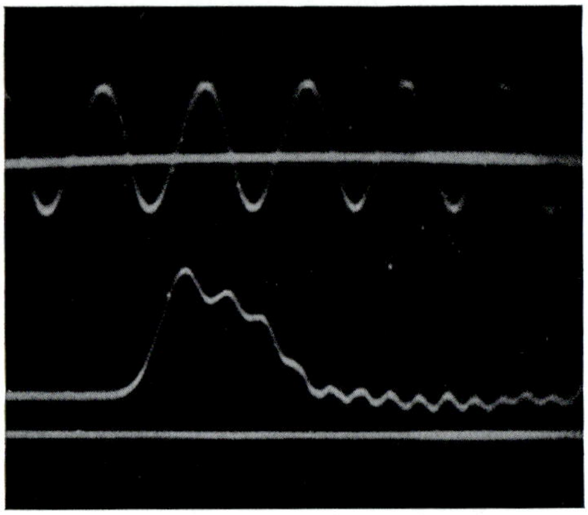

Fig. 3.71. Strain vs time profile, measured by electric resistance elements produced by a flat nosed bullet in axial collision with a long steel bar.

results demonstrate not only the presence of many modes of deformation for some situations in experiments on bars, but also the general complexity which may invalidate oversimplified interpretation of data by assuming the phenomenon to be one dimensional.

Whether the constitutive equation is linear and the exact theory for the infinite cylindrical bar is applicable, are questions difficult to answer physically. The POCHHAMMER-CHREE theory does not provide a solution for cylindrical rods with free ends, an unavoidable requirement for experiment. In 1956 RICHARD SKALAK[1] proposed a modification of the three dimensional theory, based upon an analysis of asymptotic behavior thus permitting a comparison with experiment only at relatively large distances from the impact face. However, while it simplifies the comparison between theory and experiment provided the assumed theory and constitutive equations indeed are applicable, SKALAK's development does not make it much easier to establish with certainty that the observed dispersion of the wave fronts does arise solely from the differences in wave speeds with wave lengths, as is implicit in a three dimensional theory of vibration in cylinders. If a three dimensional theory became essential to interpret experimental results, the use of data from waves propagating in a cylindrical bar to discover unknown stress-strain functions would cease to be a feasible mode of procedure.

[1] SKALAK [1957, *1*]. SKALAK reported these thoughts at a meeting in 1956; they were published in 1957.

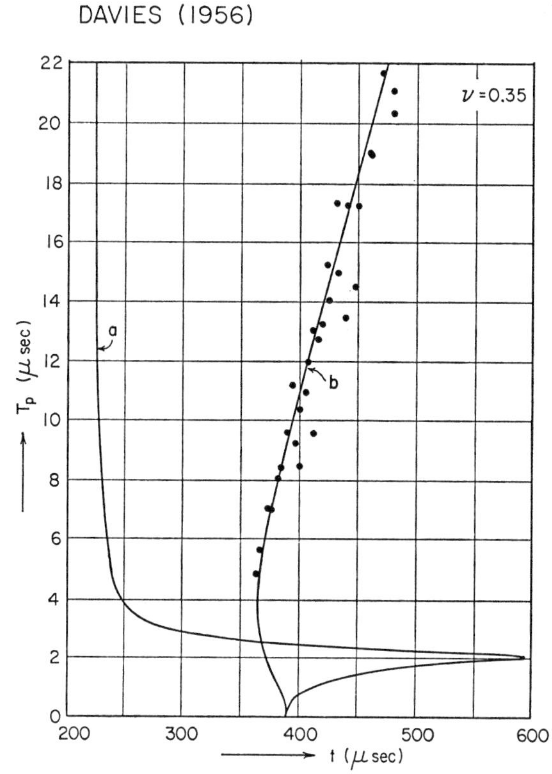

Fig. 3.72. Comparison of theory and experiment from measurements of the type shown in Fig. 3.70a.

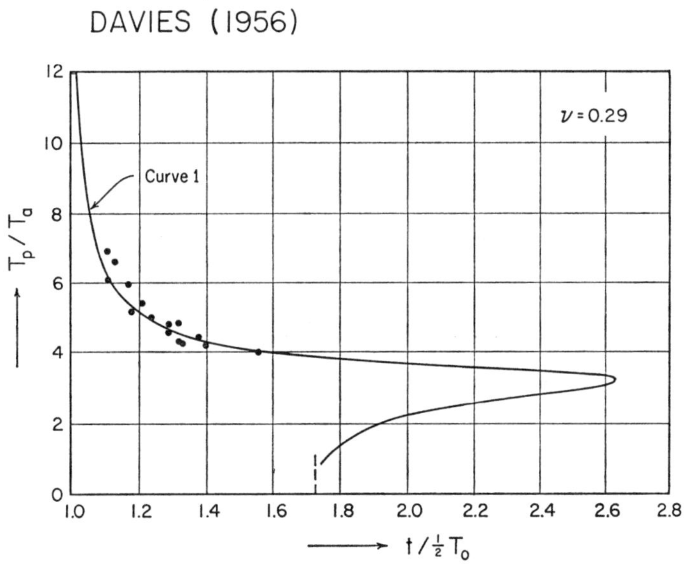

Fig. 3.73. Comparison of theory and experiment from measurements of the type shown in Fig. 3.71.

3.38. Experiments on the propagation of waves of small amplitude in metal cylinders during the past two decades: A sequence of changes in techniques and interpretation.

Eugene Ashton Ripperger,[1] in 1953, using a piezo-electric strain gage which he had developed the previous year,[2] considered the problem of axial impact of a ball against 91.4 cm long steel bars at a velocity of 130.5 cm/sec. He used steel spheres of several diameters. Triggering each strain signal from gages located at 2.54 cm, he was able to compare pulse shapes at three positions along the bar: 2.54, 15.24, and 45.72 cm from the impact face. Like Davies, Ripperger considered the problem of separating flexural and longitudinal components. Ripperger analyzed extensively the various results obtained when he altered the bar and the sphere diameters; he compared the maximum strain amplitudes with the modified Hertz theory. In Fig. 3.74 are shown the effects of dispersion for the data at three positions. Profiles were compared for the impact of a 0.318 cm diameter sphere and a 0.635 cm diameter sphere with a 2.54 cm diameter bar; in Fig. 3.75 is shown the effect of an 1.27 cm diameter sphere impacting the same bar.

RIPPERGER (1953)

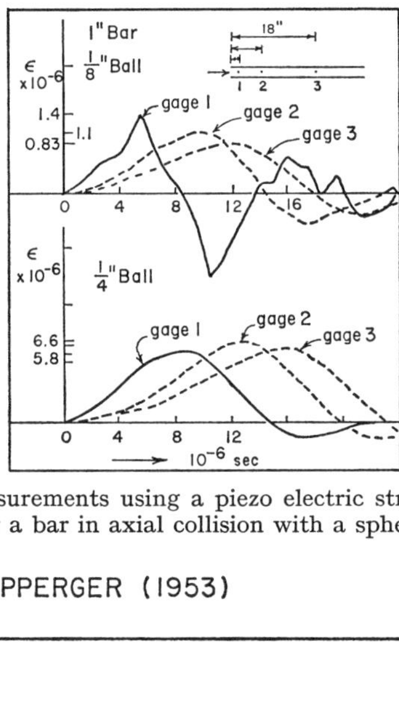

Fig. 3.74. Ripperger's measurements using a piezo electric strain gage of strain profiles at three different positions, for a bar in axial collision with a sphere of the indicated diameter.

RIPPERGER (1953)

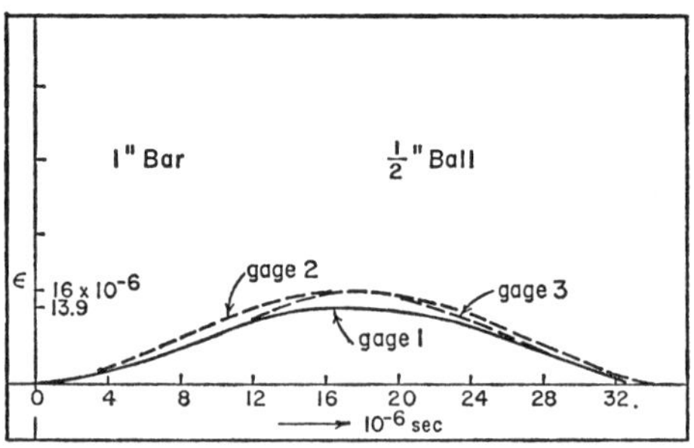

Fig. 3.75. Profiles for the bar of Fig. 3.74 struck by an 1.27 cm diameter sphere ($\frac{1}{2}$ in.).

[1] Ripperger [1953, *1*].
[2] Ripperger [1952, *1*, *2*].

In the data of Table 98, durations and pulse maxima are compared with calculation. As RIPPERGER stated, "The effects of dispersion and the decrease which takes place as the pulse duration is increased are clearly evident in Figs. 21 and 22 [here Figs. 3.74 and 3.75], as well as in the results shown in the tables."[1]

Table 98. RIPPERGER (1953).

Ball	Computed	Measured		
		Gage 1	Gage 2	Gage 3
	1" Bar—Maximum pulse amplitude			
$^1/_8$	1.84×10^{-6}	1.4	1.1	0.83×10^{-6}
$^1/_4$	7.3×10^{-6}	5.8	6.6	6.2×10^{-6}
$^1/_2$	28.9×10^{-6}	13.9	16.0	16.0×10^{-6}
	1" Bar—Pulse duration			
$^1/_8$	8.5×10^{-6} sec	8.4	14.4	17.8×10^{-6} sec
$^1/_4$	16.8×10^{-6} sec	14.6	20.0	23.4×10^{-6} sec
$^1/_2$	34.3×10^{-6} sec	33.8	33.8	33.8×10^{-6} sec

RIPPERGER concluded:

In order for a pulse to be transmitted along a bar without serious reduction in amplitude or change in form it should have a duration of approximately eight times the time required for a pulse to travel a distance equal to the bar diameter.[2]

Again, like DAVIES, RIPPERGER also expressed doubt whether the stress were uniform over the cross-section and questioned whether surface measurement alone were adequate. Finally RIPPERGER concluded that the mean velocity indeed did approximate the bar velocity $c_0=\sqrt{E/\varrho}$. From his experiments he gave a value of $c_0=16800\pm300$ ft/sec, which for steel corresponds to a dynamic value of $E=$ 21 200 kg/mm² (between the limits of 21 900 and 20 400 kg/mm²). In the propagation of large pulses, as RIPPERGER indicated, such limits were too great to allow definitive comparison of dynamic and static moduli, even 110 years after WERTHEIM had introduced the problem and nearly 50 years after GRÜNEISEN had attempted to obtain a final answer.

When experiment is contrasted with theories based upon the linear approximation, whether for presumed one dimensional waves or more generally, the fundamental difficulty is that *a priori* one is not certain how the stress and strain in the actual experiment are distributed in space and time. In addition, when one considers the dependence of the history of the deformation at a point, upon the history of the applied loading over a region on the surface, at best one can make merely an approximation regarding the effects of dimensionality and other sources of dispersion such as nonlinearity or viscosity which one might wish to delineate.

For any discussion of experiments on wave propagation in solids during the past 20 years, one must select from many hundreds of published papers.[3] Most of

[1] RIPPERGER [1953, *1*], p. 35.

[2] *Ibid.*, p. 39.

[3] In this treatise in which the main focus is upon constitutive equations, I regret that I had to omit the improvements in experimentation in photoelasticity during the century after the original discoveries of WERTHEIM. For example, pertinent to the present discussion are the work on wave propagation in photoelastic materials, such as that of M. M. FROCHT and PAUL D. FLYNN [1956, *1*] and a decage later, the studies of dynamic photoelasticity with ultra high speed photography, by FLYNN [1965, *1*].

them may be dismissed as repetitious, as trivial, or as having importance only for very specialized areas of application. There still remains, however, a large number of papers on experiment which must be studied in detail.

Certainly one of the most interesting and important of these is the study by Charles William Curtis.[1] Using the piezo-crystal technique of Ripperger, Curtis determined strain-time detail along a 5.08 cm diameter, 732 cm long magnesium bar subjected at one end to a step front from an air shock generated in a long tube following the rupture of a diaphragm. Curtis noted that, "Although the Pochhammer-Chree solutions are often referred to as exact, they were developed originally for bars extending from plus to minus infinity and only very recently have exact solutions been formulated for transient problems involving bars with an end."[2] The solution referred to was that of Richard Skalak,[3] which I noted at the end of the previous section.[4]

Curtis chose a magnesium bar 731.5 cm long and 3.81 cm diameter, because the lower E for that solid provided a higher maximum strain for the extremely low pressure pulse amplitude of 0.032 kg/mm². The assumed time for the rise of the shock step, which remained constant for milliseconds, was one microsecond. At 150 cm from the shock face, the contrast between elementary theory and observation may be seen in Fig. 3.76. In addition to a finite slope, high frequency oscillations were observed.

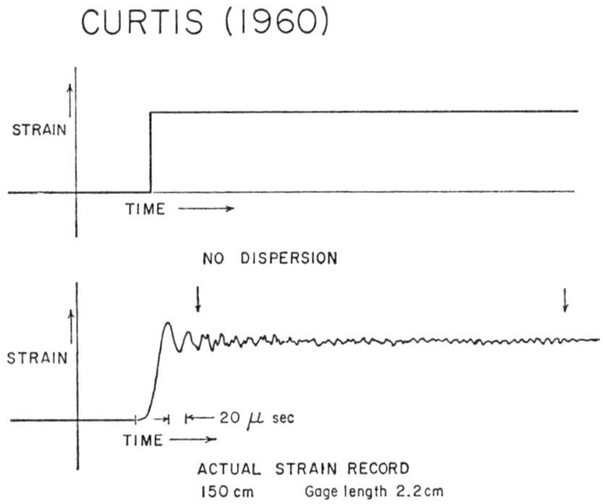

Fig. 3.76. Wave profile in a long magnesium bar subjected to an air shock front at one end. Elementary theory (above); observation (below).

From the comparison of the output of barium titanate strain gages located at the several positions along the bar, Curtis found that the wave front was dispersive. The main wave front measured at different positions was shifted to the same time origin to reveal not only the changing slope of the first major rise, but the appear-

[1] Curtis [1960, *1*].

[2] *Ibid.*, p. 17.

[3] Skalak [1957, *1*].

[4] It should be noted that Skalak's theory was for a bar of infinite length. It described the post-collision deformation of two semi-infinite bars. Since the theoretical bar was infinite, there was no problem arising from the presence of a free cross-section. The experimental bar of Curtis, on the other hand, remained semi-infinite. It was to the essentially free cross-section that he applied an air shock.

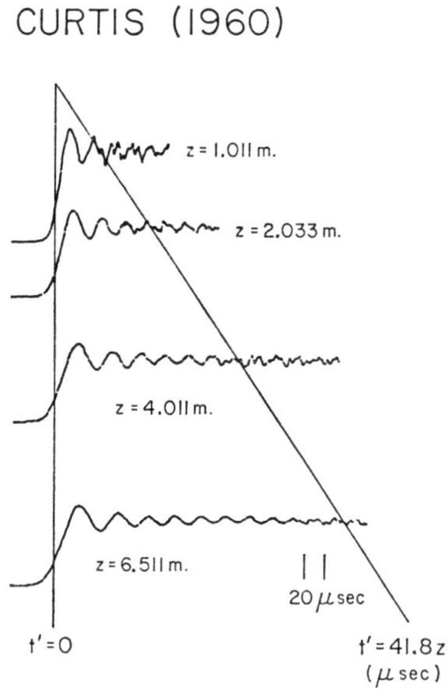

Fig. 3.77. Wave profiles at successive positions along a bar subjected to an air shock front, showing the dispersion in the main front and the onset of second mode vibrations. The main wave fronts were shifted so as to have the same origin in time.

ance of second mode vibrations at the predicted time given by the sloping solid line (see Fig. 3.77).

For those interested in unraveling by means of experiment the complex detail of the higher branches of frequency equations arising in boundary value problems of this sort, it is essential to employ experimental techniques capable of recording minute detail in microseconds. For the data of Figs. 3.76 and 3.77 above, which CURTIS credited to GEORGE FOX, a gage 2.2 cm long was used. The detail was smoothed out by the process of integrating over the gage. This is clearly seen when a comparison is made with the response of a gage 0.3 cm long. Such a comparison in shown in Fig. 3.78 at a position 150 cm from the shock face. One of the problems of ascertaining the theory sensitivity of experiment when describing a complex wave structure thus is illustrated dramatically.

CURTIS also generated shock pulses, giving rise to both longitudinal and transverse components, by masking half of the shock tube end in the manner shown

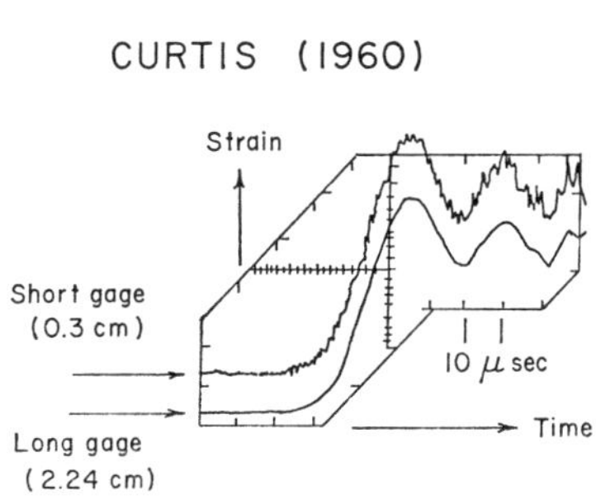

Fig. 3.78. Wave profiles for a bar subjected to an air shock front, showing the high frequency oscillations visible for a short gage length which cannot be seen with integration over a longer gage length.

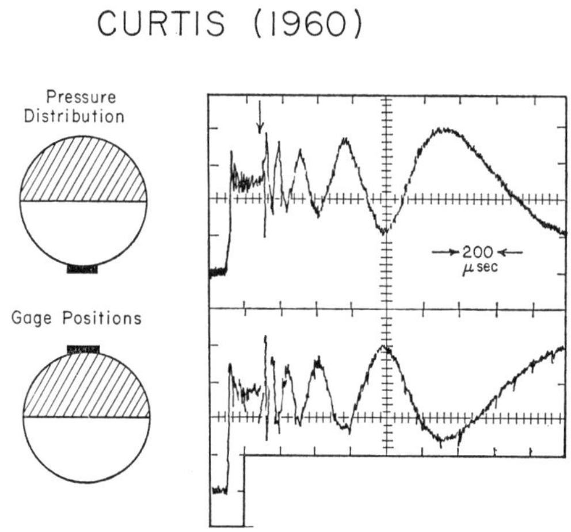

Fig. 3.79. Wave profiles for shock fronts limited to half the end face, as shown, for two different gage locations. Note the superposition of longitudinal and transverse components.

in Fig. 3.79. The strain vs time histories produced on either side of the bar are shown.

These data provided a superposition of longitudinal and transverse modes. CURTIS found that the latter were in agreement with calculations for the first branch transverse mode, and there were indications of a correlation with the second branch. The original shock wave experiments actually were performed by CURTIS in 1954, before SKALAK's asymptotic analysis using FOURIER integrals. The discussion here was taken from a summary presented[1] at a symposium in 1959.

In June 1956, six months before the presentation of SKALAK's analysis, JULIUS MIKLOWITZ and C. R. NISEWANGER[2] presented a study which repeated CURTIS's shock tube loading but used an aluminum alloy bar, 2.54 cm in diameter. They obtained the nearly seven times higher pressure maximum of 0.21 kg/mm^2. Radial displacements at many positions along the bar were made by means of a condenser microphone, and axial strains at the same positions were made by means of 20 very short electric resistance gages mounted in series. The oscillograms for both measurements at the designated positions are shown in Fig. 3.80 from which the general dispersive development of the wave is obvious. MIKLOWITZ and NISEWANGER commented upon the possible source of the differences in the fine detail of the two types of measurement. They decided the differences were due either to their instruments or to the diametral integration of radial displacement.

These data when compared with an approximate theory of R. D. MINDLIN and GEORGE HERRMANN[3] for dispersive waves in rods, provided the correlation shown in Fig. 3.81.

DAVIES had demonstrated that a three dimensional theory was essential in describing the axial collision of rods. Thus an experimentist had to be keenly alerted to the theory sensitivity of experiment in such a complex physical situation, in order to discern among many plausible theories which of them were physically applicable.

---

[1] CURTIS, published in [1960, *1*].
[2] MIKLOWITZ and NISEWANGER published in [1957, *1*].
[3] MINDLIN and HERRMANN [1952, *1*].

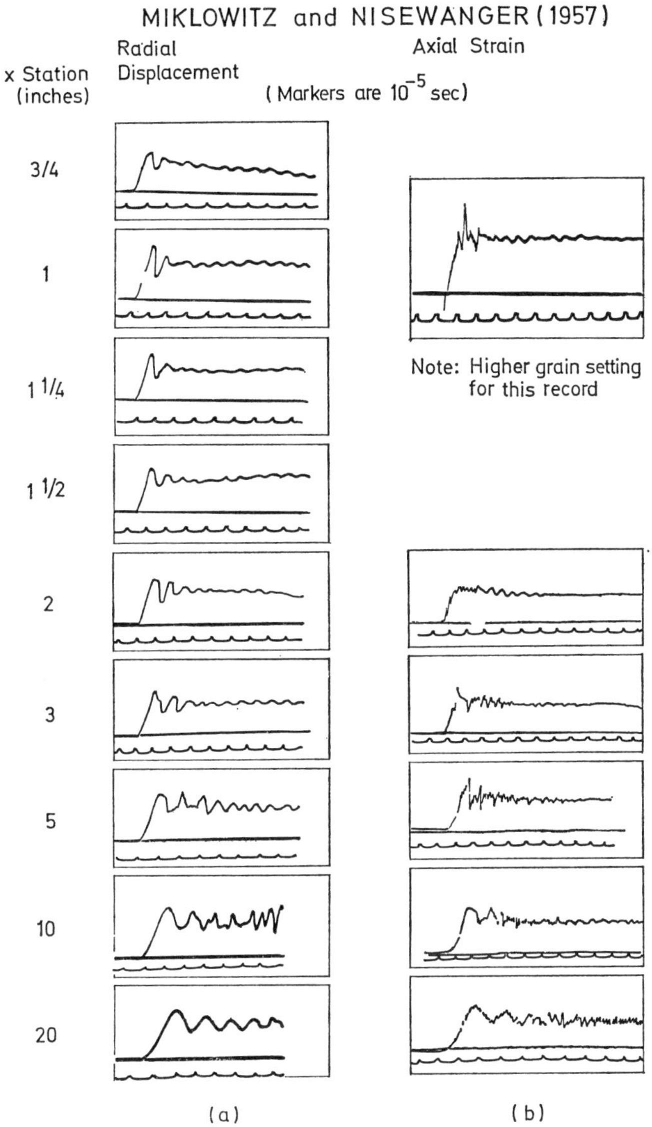

Fig. 3.80a and b. Time records for (a) radial displacement and (b) axial strain, at various distances along an 1 in. diameter aluminum alloy bar.

A different approach to the determination of dynamic elastic constants was introduced by D. S. Hughes, W. L. Pondrom, and R. L. Mims[1] in 1949. The cylindrical rod was considered as a bounded three dimensional solid in terms of microseismology. A quartz crystal was used to introduce a short pulse[2] at one end face, with a detector crystal located at the opposite end. By changing the length of the bar, Hughes, Pondrom, and Mims could trace geometrically the arrival times of the initial dilatation and the singly reflected and multiply reflected pulses

[1] Hughes, Pondrom, and Mims [1949, *1*].

[2] It actually was a 3.5 megacycle wave packet of short duration.

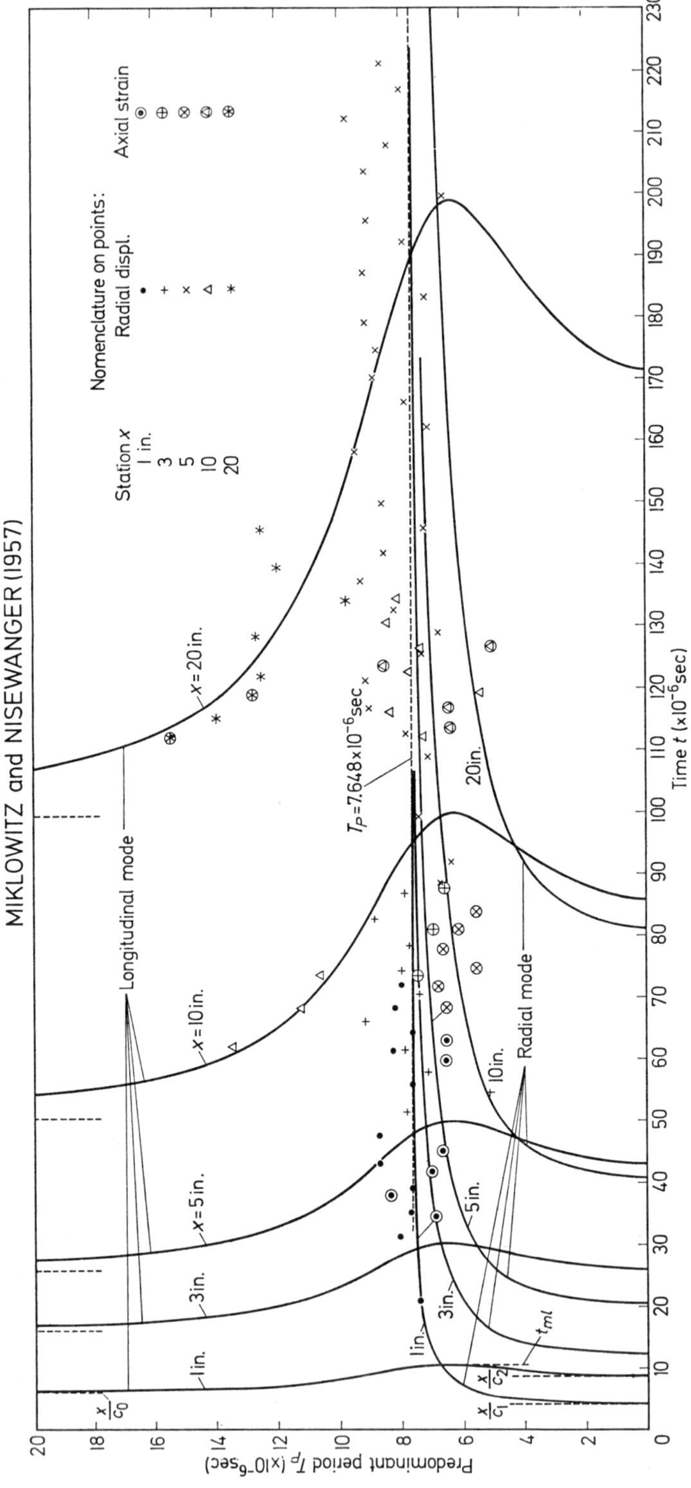

Fig. 3.81. MIKLOWITZ' and NISEWANGER'S comparison of their analysis and experiment for the results of Fig. 3.80.

when the pulse length was sufficiently short to permit separation of the various reflected components. By changing the diameter as well as the length of the specimen, they arbitrarily could check calculation and measurement before they determined elastic constants from the measured wave speeds. A representative set of their oscilloscope traces is shown in Fig. 3.82 for the bar lengths indicated; 0.00 designated the two crystals in contact. The cold rolled steel bar diameter was 3.645 cm. The traces I and VIII were 10 μsec time markers.

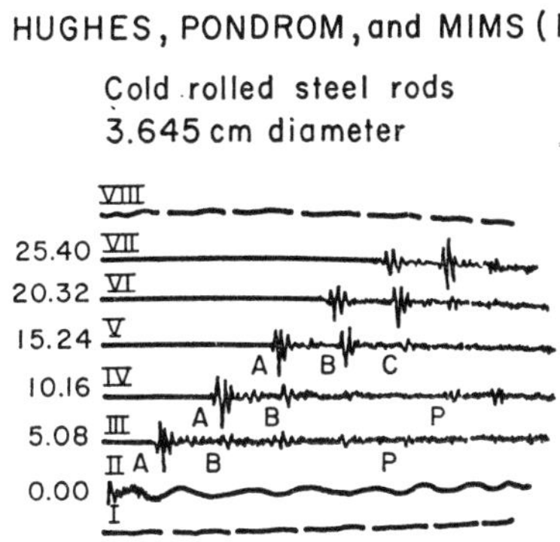

Fig. 3.82. Pulses through steel rods which were induced and detected by means of quartz crystals at either end.

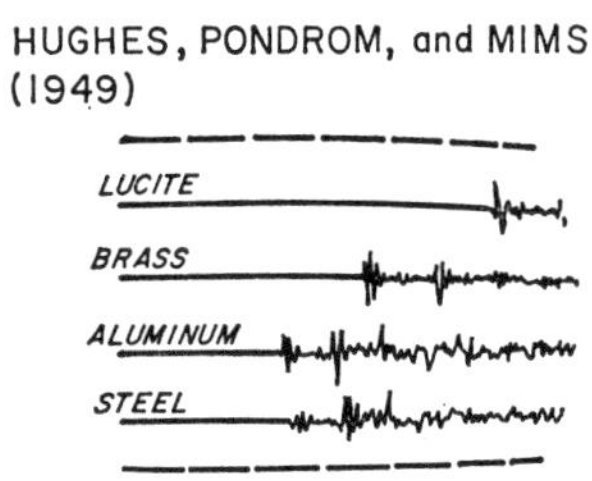

Fig. 3.83. Pulses through rods of 2.54 cm diameter and 15.240 cm length for the solids indicated. The pulses were induced and detected by quartz crystals at either end of the bar.

In Fig. 3.83 are shown comparative pulse arrival times in lucite, brass, aluminum, and steel, for rods of 2.540 cm diameter and 15.240 cm length. From these data the wave speeds and elastic constants of Table 99 were provided. The general agreement with previous values emphasized the possibility of such microseismologic studies in exploring the dynamic elasticity of solids.

Table 99. HUGHES, PONDROM, and MIMS (1949).

Material	V_D (m/sec)	V_R (m/sec)	ϱ (g/cm³)	K (kg/mm²)	μ (kg/mm²)	E (kg/mm²)	ν
C.R. Steel I	5880.0	3203.5	7.82	16685	8194	21127	0.289
C.R. Steel II	5892.2	3211.8	7.82	16736	8237	21229	0.289
Aluminum	6379.0	3100.0	2.70	7628	2655	7127	0.345
Brass	4283.0	2033.0	8.56	11253	3615	9792	0.355
Lucite	2640.0	1269.0	1.18	5820	1940	5238	0.350

In 1954 Herbert Kolsky[1] similarly considered dynamic elasticity from the viewpont of microseismology. At one end of a cylinder whose length and diameter were almost equal was a lead azide explosive pulse source of 2 to 3 μsec duration, and at the center of the other end Davies' capacitance displacement device was used in the manner shown in Fig. 3.84. The zero time was given by the response of the photocell activated by the light of the explosion.

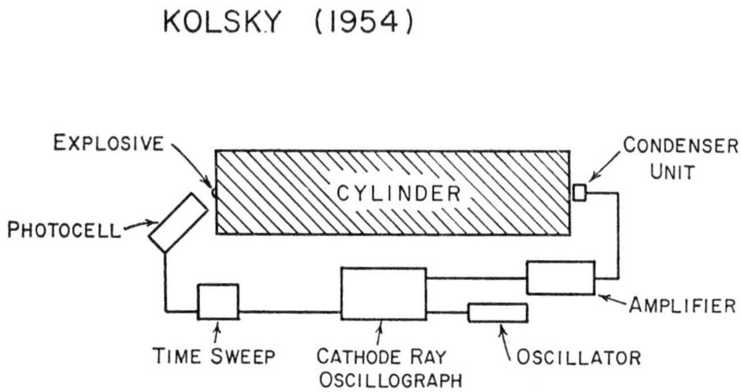

Fig. 3.84. Kolsky's apparatus for the measurement of the microseismology of pulses of very short duration in a short steel bar.

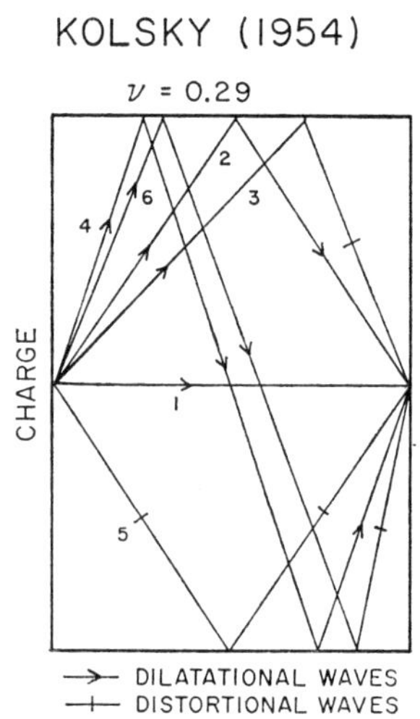

Fig. 3.85. Kolsky's determination of waves from his experiments using short pulses.

For $\nu=0.29$ in steel Kolsky calculated the reflection coefficients as a function of angle of incidence from linear elastic theory and traced out the paths shown in Fig. 3.85, for which he calculated arrival times or transit times. By changing the capacitance gap, Kolsky could change the sensitivity to record the low amplitude dilatational precursor at the dilatational velocity c, or the head of the main pulse with the bar velocity c_0. For the former, he measured 6000 m/sec which corresponded to a calculated value of $c_1=6090$ m/sec for $\nu=0.29$ from E determined from longitudinal oscillations. The bar velocity determined from the pulse was $c_0=5300$ m/sec,

[1] Kolsky [1954, *1*].

compared to the longitudinal vibration value of 5320 m/sec.[1] As KOLSKY pointed out, the POCHHAMMER-CHREE solution, which strictly applies only to infinite sinusoidal trains of waves, was inadequate when the energy was alternating between dilatational and shear waves in the multiple reflections from the free surface.

The analysis of SKALAK[2] was limited to the description of behavior at relatively large distances from the initial face of the cylinder, which is where CURTIS[3] made his comparisons of measurement and calculation. The microseismic studies of HUGHES, PONDROM and MIMS,[4] and of KOLSKY,[5] had no such limitation. Their problem was that of obtaining the required accuracy of transit time resolution of the order of 1 μ sec. Their method also required a short pulse or wave packet length to separate the arrival times.

In 1960, also from the viewpoint of microseismology, I analyzed[6] the first diameter development and subsequent growth of deformation from a sustained stress step at the center of a cylindrical aluminum bar 5.08 cm in diameter and 305 cm long. I obtained the sustained step loading by impacting the center of the large cylinder with a 0.635 cm diameter, 53.4 cm long aluminum rod; there was a 200 μsec time interval before unloading. A modified DAVIES capacitance technique was used to measure radial displacements at $^1/_4$ diameter and at every $^1/_2$ diameter to 30.8 cm from the impact surface. Electric resistance strain gages were employed to determine longitudinal strain. The apparatus is shown in Fig. 3.86 and representative radial displacement and longitudinal strain measurement are in Figs. 3.87 and 3.88, respectively.

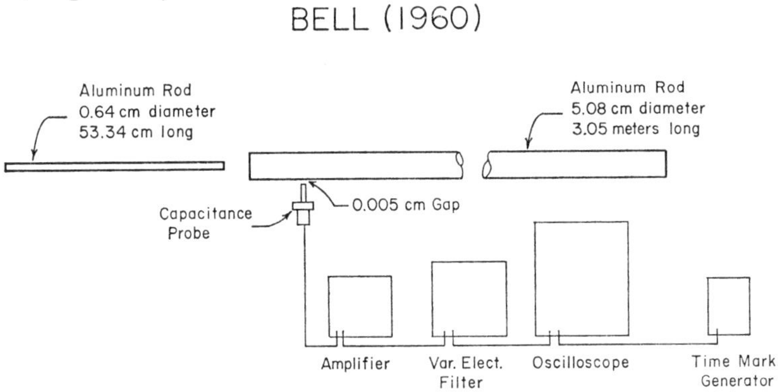

Fig. 3.86. The experimental situation used by BELL for measuring the radial displacement on a long bar subjected to a central impact at one end.

The impact velocity of 417 cm/sec in the small rod provided a maximum strain of 820×10^{-6}, well below the elastic limit. Through the area ratios, I obtained a value of 12.5×10^{-6} in the large bar. As may be seen in Fig. 3.88, this was in fair agreement with experimental results when elementary theory was compared with three-dimensional behavior. The rectangular step shown in Fig. 3.88 was that for the elementary theory with the bar velocity, $c_0=\sqrt{E/\varrho}$; (a) designated the dilatational wave speed, c_1; (b) designated c_0; and (d), the shear wave speed, $c_2=\sqrt{\mu/\varrho}$.

[1] In addition to the resonant longitudinal vibration studies, KOLSKY also determined E by sending pulses down a two meter length of the same steel.

[2] SKALAK [1957, *1*].

[3] CURTIS [1960, *1*].

[4] HUGHES, PONDROM, and MIMS [1949, *1*].

[5] KOLSKY [1954, *1*].

[6] BELL [1960, *4*].

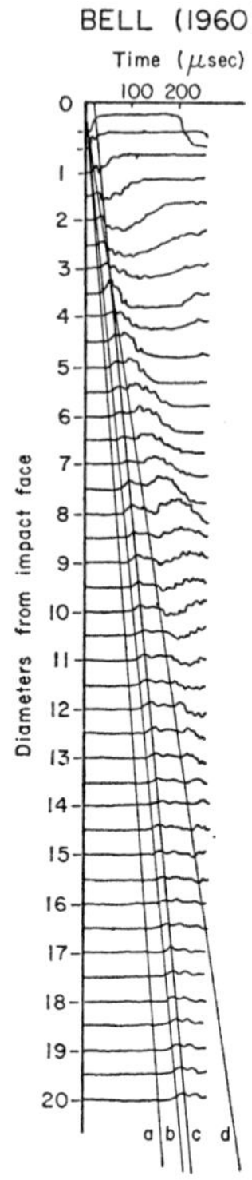

Fig. 3.87. Radial displacement measurements obtained by BELL (1960) in the experimental situation shown in Fig. 3.86. The measurements were made by means of a capacitance gage.

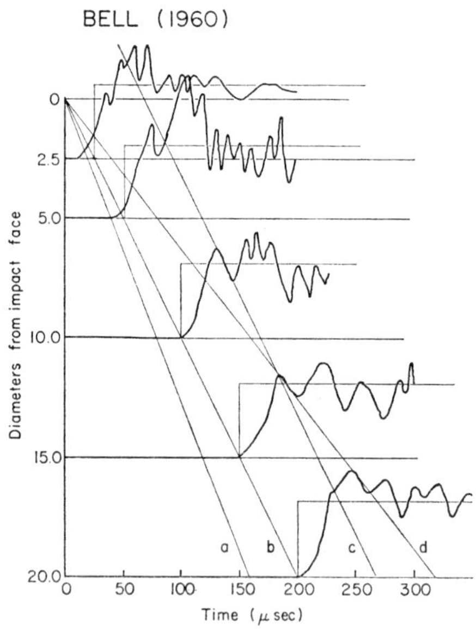

Fig. 3.88. Wave profile determined by BELL (1960) for the experimental situation shown in Fig. 3.86, at the indicated distances along the bar. The measurements were made by means of electric resistance gages.

BELL (1960)

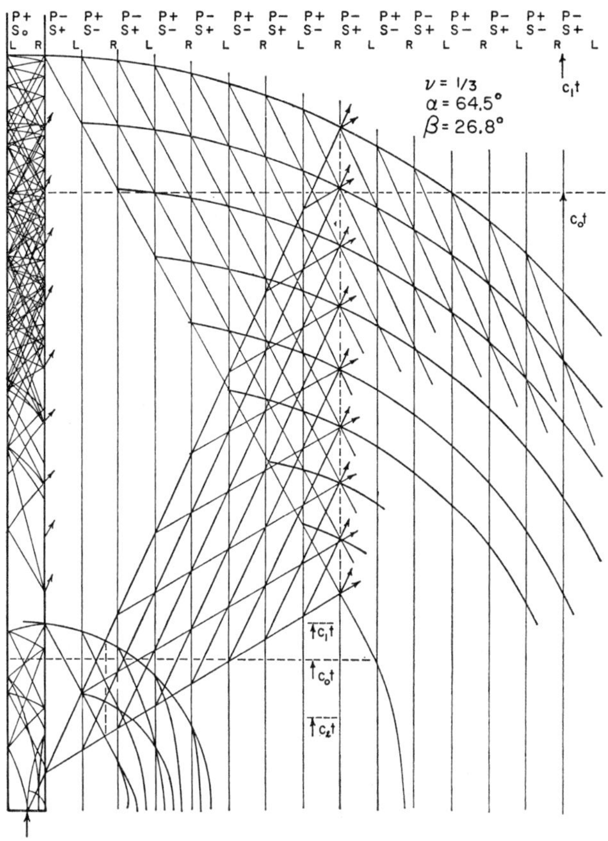

Fig. 3.89. The wave development in the microseismology of a bar subjected to an axial step front of the sort obtained in the experimental situation of Fig. 3.86. The detail is shown for one ray at an initial angle of 64.5°. The pattern was for the time when the leading dilatation wave had reached 5 diameters and 20 diameters.

I assumed[1] an initial spherical wave front, and an amplitude partition between dilatational and shear waves at the free sides of the cylinder, with spherical attenuation of the amplitude throughout the calculation. The strain-space and radial-space distributions were determined when the dilatational wave front had reached 5 diameters (25.4 cm) and 10 diameters (50.8 cm) from the impact face. The expanded (accordian-like) and projected wavelet distribution for one side of a central slice for the special situation of a ray at 64.5° initial angle is shown in Fig. 3.89. Composite double distributions for every 10° angle are shown in Fig. 3.90.

The actual calculations were performed for every five degrees. This required that careful accounting be made of the sign of the wavelet during the summation of the projections. My comparison of experimental radial displacement (circles) and calculation (solid line) when the dilatation wave had reached 5 diameters may be seen in Fig. 3.91. (I have omitted a discussion of the algebraic simplifications found in this hand calculation.)

---

[1] *Ibid.*

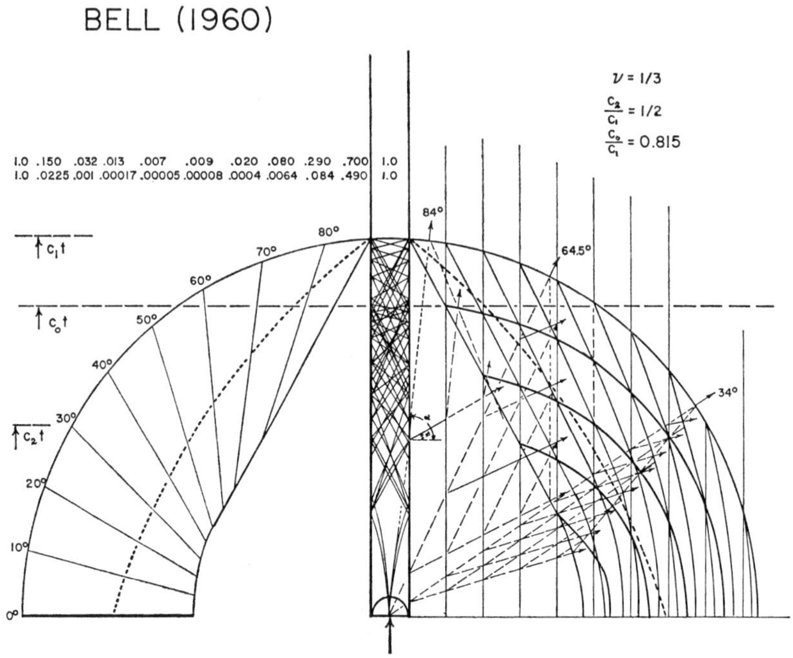

Fig. 3.90. The wave development in a bar subjected to an axial impact at the time when the leading dilatational wave front had reached 10 diameters.

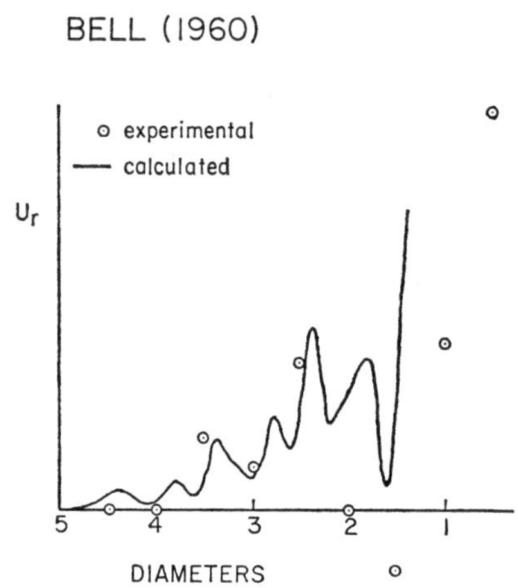

Fig. 3.91. Comparison of radial displacement measurement and calculation for the experimental situation of Fig. 3.86, when the leading dilatational wave had reached 5 diameters. The calculations were based on the analysis of the wave distribution of Figs. 3.89 and 3.90.

In Fig. 3.92 is shown the comparison of axial strain according to experiment (circles) with that calculated from wave propagation and reflection (solid line), and with that given by the elementary theory of one dimensional wave propagation in rods (dashed line) when the dilatational wave had reached 20 diameters, i.e., at a time of 200 μsec.

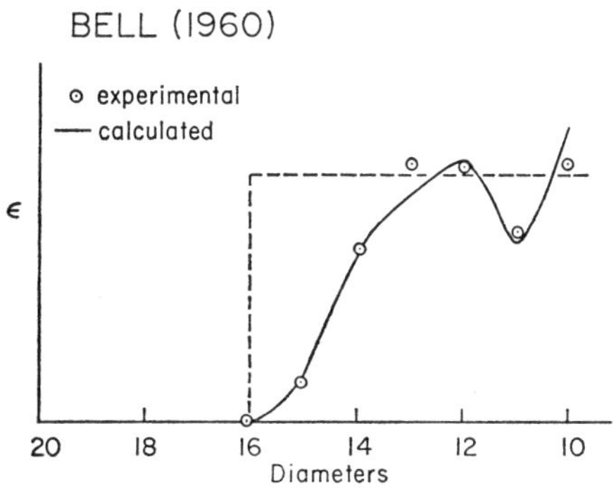

Fig. 3.92. Comparison of axial strain measurements and calculation for the experimental situation of Fig. 3.86, when the leading dilatational wave had reached 20 diameters. The calculations were based on the analysis of the wave distribution of Figs. 3.89 and 3.90 (BELL, 1960). The dashed line is the prediction of the elementary theory.

Separate experiments were performed in large blocks of aluminum to determine the amplitude distribution over the initial spherical wave front. In the somewhat forced analysis carried out because of an interest in comparisons with finite amplitude wave development in the first diameter of impacting cylinders, it was necessary to ignore the stress infinities along central axes associated with the assumed point loading. At the same velocity, impacts of small hollow cylinders of the same area as the solid rod demonstrated that beyond the first half-diameter the experimental results were insensitive to major changes in the spatial distribution of loading at the impact face; i.e., it was an experimental demonstration of a dynamic SAINT-VENANT's principle for this problem. I concluded that the correlation of experiments and calculations, shown here in Fig. 3.92, demonstrated that studies from the viewpoint of microseismology can provide the connection between the behavior near the impact face and that at the relatively large distances

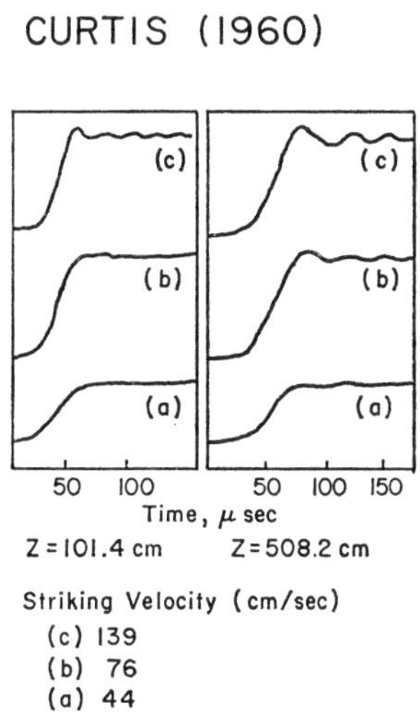

Fig. 3.93 a—c. The dispersion measured in the axial collision of almost flat bars for three different impact velocities: (a), (b), (c). The measurements were made at two different distances from the impact face to demonstrate the dispersion which was present.

from the impact face where the experimental results of CURTIS[1] and the asymptotic theory of SKALAK[2] are in fair agreement. One sees a plausible resolution of the incompatibility between experiment and theory which BOLTZMANN had found in the first experiments on the impact of bars 78 years earlier.[3] My work was a forward projection from a known pulse at the impact face. For the backward projection, CURTIS had envisaged using three dimensional bar analysis for the study of unknown applied pulse histories.

CURTIS in his paper in 1960 had shown that the impact of almost flat cylinders produced dispersive wave fronts of the type seen in Fig. 3.93. Strain measurements made at two positions for three impact velocities demonstrated the presence of changing wave forms and detail which varied with impact velocity.

3.39. Ultrasonic determination of elastic constants.

As I noted above in discussing the work of CORNU[4] (Sect. 3.21), SCHAEFER and BERGMANN[5] in 1935 introduced an experimental method for determining dynamic elastic constants for optically transparent solids. By vibrations either directly induced in piezo-electric substances or indirectly induced by means of attached driving piezo-crystals, standing waves of short wave lengths were created. With appropriate wave lengths, monochromatic light traversing such a vibrating solid produced optical interference patterns upon emission. Diffraction analysis for known crystallographic orientations of the solid permitted the calculation of elastic constants.

In a survey in 1969 of the more than three decades of development of that technique, including the impetus given by the introduction of the laser beam, JOHN MOYU LIU[6] in his Master's essay referenced 177 papers, of which nearly 90 indicated steps in the development of that ultrasonic method of measurement. More recent papers reported using that method to study third-order elastic constants and to determine temperature coefficients. R. BECHMANN and R. F. S. HEARMON[7,8] in LANDOLT-BÖRNSTEIN's *Gruppe III; Kristall- und Festkörperphysik*, Vol. 1 in 1966, and Vol. 2 in 1969, provided a massive array of numerical values for piezo-electric substances in general. The sheer bulk of BECHMANN's numbers precludes any general tabulation here.

I have included four figures from LIU's essay solely to illustrate the general nature of the optical interference studies of the type indicated. A FRAUNHOFER diffraction pattern from standing waves in a quartz crystal is shown in Fig. 3.94, and an image of acoustic wave fronts observed between crossed polaroids in fused quartz, in Fig. 3.95.

In Fig. 3.96 is a FRESNEL diffraction pattern from longitudinal standing waves in topaz; in Fig. 3.97 is an acoustical wave visualization in fused quartz using the SCHLIEREN method, which shows the spreading of acoustic energy along the path of the beam.

In general, acoustic wave speeds determined from this type of optical interference are in fair agreement with those of the more widely used ultrasonic pulse echo technique, to be discussed below.

---

[1] CURTIS [1960, *1*].
[2] SKALAK [1957, *1*].
[3] BOLTZMANN [1882, *1*].
[4] CORNU [1869, *1*].
[5] SCHAEFER and BERGMANN [1935, *1*].
[6] LIU [1969, *1*].
[7] (HEARMON) LANDOLT-BÖRNSTEIN [1966, *1*].
[8] (BECHMANN and HEARMON) LANDOLT-BÖRNSTEIN [1969, *1*].

Fig. 3.94. A FRAUNHOFER diffraction pattern due to a standing wave in a quartz crystal.

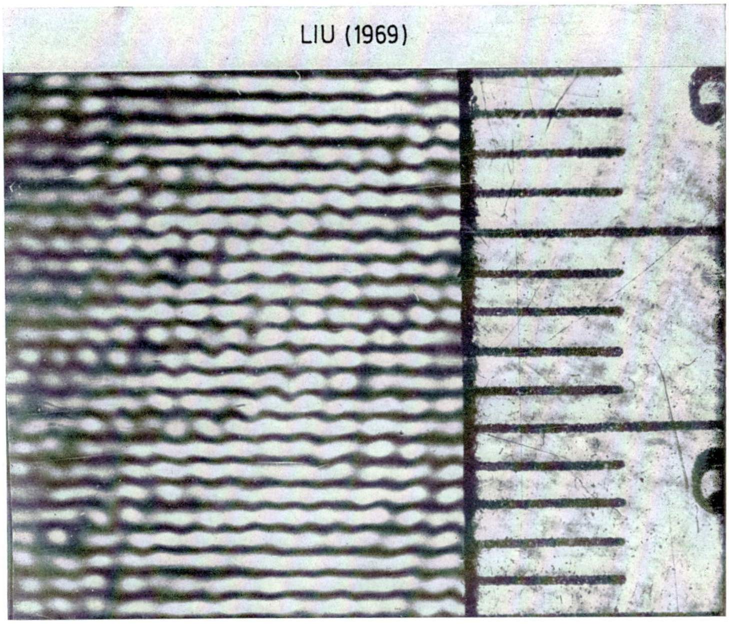

Fig. 3.95. An image of acoustic wave fronts observed between crossed polaroids in fused quartz.

In 1934 LEWIS BALAMUTH[1] presented a paper which he could entitle, "A New Method for Measuring Elastic Moduli and the Variation with Temperature of the Principal Young's Modulus of Rocksalt between 78° K and 273° K." The experiment consisted in cementing a quartz crystal of known dimensions to a crystal of rocksalt of known crystallographic orientation. Thin strips of gold leaf provided

[1] BALAMUTH [1934, *1*].

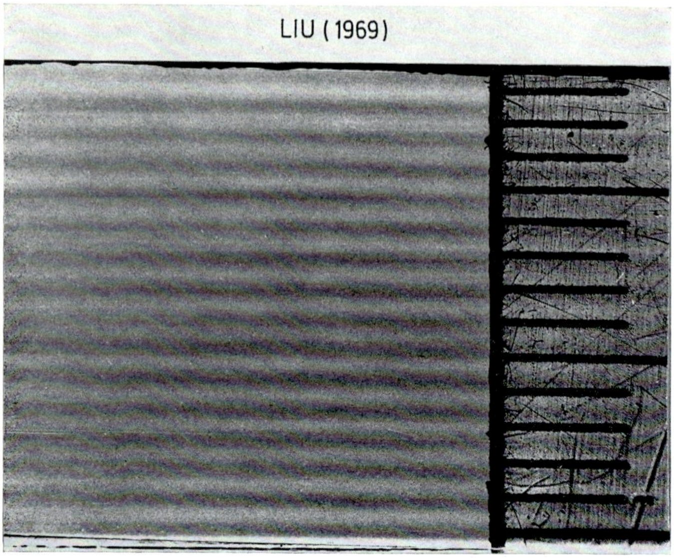

Fig. 3.96. A Fresnel diffraction pattern due to longitudinal standing waves in topaz.

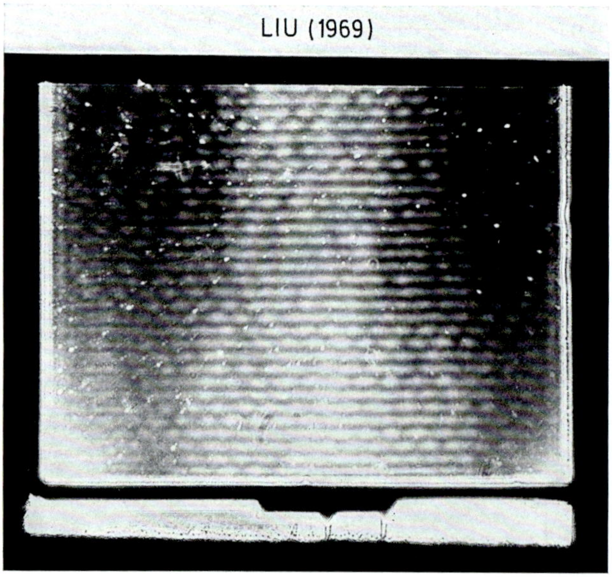

Fig. 3.97. Acoustical wave visualization in fused quartz using the Schlieren method.

contact for an electrical oscillator which, by means of the piezo-electric effect, was capable of setting the entire system in resonance. From an analysis based upon linear elasticity the resonant longitudinal vibration of the combined structure, together with estimates of the effect of the bond, provided a measurement of E along the known rocksalt specimen axis. A diagram of Balamuth's experiment is shown in Fig. 3.98.

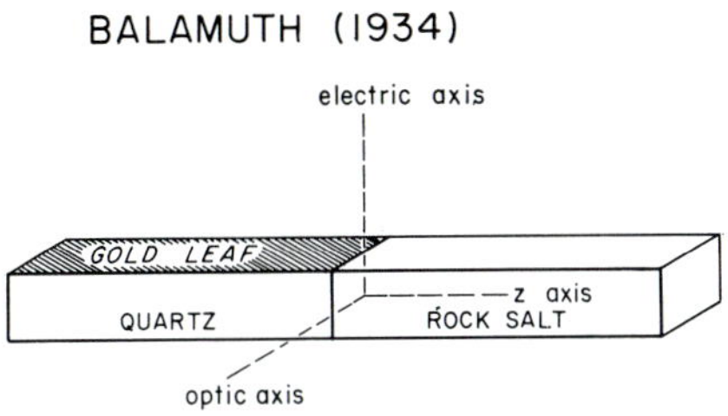

Fig. 3.98. The experimental situation for BALAMUTH's first use of a quartz crystal to measure the elastic moduli in a solid (1934).

In 1936 FRED C. ROSE[1] extended the method to include torsional oscillations by choosing an appropriately cut quartz crystal. He was able to minimize the friction effect at the bond by introducing a triple oscillator which was formed of the quartz crystal, a magnesium insert, and the rocksalt. The magnesium was chosen because of the similarity of its thermal coefficient of expansion to that of rocksalt. ROSE provided the numerical values for the three elastic constants of cubic crystal c_{11}, c_{12} and c_{44}, shown in Table 100, for a series of temperatures between 80 and 270° K, which he compared with quasi-static values for rocksalt given by WOLDEMAR VOIGT.[2]

Table 100. ROSE (1936).

T (° K)	Adiabatic constants $\times 10^{-11}$ (dyne/cm²)			Isothermal constants[a] $\times 10^{-11}$	
	c_{11}	c_{12}	c_{44}	c_{11}	c_{12}
80	5.76	1.17	1.332	5.72	1.13
90	5.73	1.18	1.330	5.69	1.14
140	5.56	1.22	1.318	5.50	1.16
150	5.52	1.23	1.315	5.45	1.17
160	5.48	1.24	1.313	5.41	1.17
170	5.44	1.24	1.310	5.37	1.17
180	5.40	1.25	1.307	5.33	1.17
190	5.37	1.26	1.304	5.29	1.17
200	5.33	1.26	1.301	5.24	1.17
210	5.29	1.27	1.298	5.20	1.17
220	5.25	1.27	1.294	5.16	1.17
230	5.22	1.28	1.291	5.12	1.17
240	5.18	1.28	1.288	5.07	1.17
250	5.14	1.29	1.285	5.03	1.17
260	5.10	1.29	1.282	4.99	1.17
270	5.06	1.30	1.278	4.95	1.17

[a] From VOIGT [1910, *1*].

ROSE, like BALAMUTH, had used the "finest grade" optical rocksalt specimen. His values of YOUNG's modulus compared with those of BALAMUTH over the same range of temperature from 80 to 270° K provided a very close agreement. It indicated that no major deviation ocurred when the double oscillator was replaced with the triple oscillator.

[1] ROSE [1936, *1*].
[2] VOIGT [1876, *1*], [1885, *1*], [1910, *1*] Chap. VII, pp. 716–763.

In 1946, FLOYD A. FIRESTONE and JULIAN R. FREDERICK[1] introduced a very high-frequency, short-pulsed, ultrasonic wave into a metal in order to detect flaws in the interior. This measurement stimulated a number of experimentists to consider the application of a short burst, ultrasonic wave packet, for the microseismological study of longitudinal and shear wave speeds in solids. The goal was the determination of elastic constants. In 1947 HILLARD BELL HUNTINGTON,[2] and in 1948 JOHN KIRTLAND GALT[3] by these means measured the elastic constants of several cubic solids at atmospheric pressure. I have described the experiments of HUGHES, PONDROM, and MIMS,[4] who in 1949 determined the isotropic elastic constants of cold rolled steel, using this ultrasonic pulse technique.

Since those beginnings slightly more than twenty years ago, a truly vast literature on the subject has systematically provided wave speed transit times for hundreds of solids. A recent bibliography on ultrasonics by ALAN B. SMITH and RICHARD W. DAMON,[5] in the *IEEE Transactions* for April 1970, gave a list 25 pages long, citing nearly 900 papers on the various aspects of the field, the great majority being drawn from the literature of the preceding ten years. HEARMON and R. BECHMANN in LANDOLT-BÖRNSTEIN[6] provided numerical values of parameters determined by piezo-crystals for well over 700 different crystalline solids. The edition of LANDOLT-BÖRNSTEIN in 1966 contained the elastic constants for nearly 300 solids, and many numbers were added three years later, in the 1969 volume. Thus, a very large proportion of this literature obviously has been concerned with the extension of the same type of experiment to an enormous number of solids. One hundred twenty-five years earlier, when such efforts at extending experimental results were confined to a single laboratory, WERTHEIM, as has been noted above, succeeded in determining quasi-static and dynamic moduli for eleven elements and 69 binary combinations. In modern practice, the number of persons who measure these elastic parameters increases at nearly the same rate as the number of solids added.

For anisotropic solids, crystals of known orientation are subjected to an ultrasonic pulse in megacycles, producing longitudinal or transverse waves, almost invariably along one of the prime crystal axes. While the orientation angles are known, and transit times are measured, the experimentists usually merely assume outright that the infinitesimal linear theory of elasticity[7] applies. Hence, the elastic constants c_{ij} or the elastic compliances s_{ij} also are assumed. Experimentists who extended earlier studies with objectives similar to those of GRÜNEISEN[8] in 1910, then concluded that the temperature dependence of these elastic constants could be determined by performing such ultrasonic measurements as far over the temperature scale as was experimentally possible.

A third field of exploration was introduced in 1949 from the experiments of DAVID LAZARUS,[9] who measured the variation of elastic constants of KCl, NaCl, CuZn, Cu, and Al at various ambient pressures, to determine the pressure coefficients of elastic constants. Recently, the ultrasonic pulse echo experiment, in which

[1] FIRESTONE and FREDERICK [1946, *1*].

[2] HUNTINGTON [1947, *1*].

[3] GALT [1948, *1*].

[4] HUGHES, PONDROM, and MIMS [1949, *1*].

[5] SMITH and DAMON [1970, *1*].

[6] (HEARMON) LANDOLT-BÖRNSTEIN [1966, *1*]; (BECHMANN and HEARMON) LANDOLT-BÖRNSTEIN [1969, *1*].

[7] These studies are unique in providing no information with respect to stress, strain, or particle velocity. Generally, the elementary theory merely is assumed *a priori* as if definitive by assertion, and all manner of phenomena are thereby amenable to such interpretation.

[8] GRÜNEISEN [1910, *1*].

[9] LAZARUS [1949, *1*].

the repeated traverse of the specimen of pulse reflections is considered for the purpose of studying the phenomenon of attenuation in solids, has come to be used widely, primarily because standardized, commercial apparatus is readily available. In such experiments, pulse frequencies of higher and higher values are sought because of an interest in the phenomenon of scatter as a function of the wave length of material parameters.

The past ten years, and particularly the past five, have witnessed much interest in using the pressure coefficients of ultrasonic elastic constants so as to determine the third-order constants of elasticity. Since the amplitudes of ultrasonic waves are in a strain range of 10^{-7} or 10^{-8}, this seems to constitute a return to the problem posed by GRÜNEISEN[1] when we recall that by 1906 he had demonstrated the essential nonlinearity of solids to near zero stress.

Ultrasonics is one of many experimental approaches available to study the deformation of solids.[2] The ramifications of expending so enormous an effort to view the subject from a single perspective obviously are still far from being understood. The pressure coefficients of elastic constants have obvious importance in the nonlinear theory of the solid continuum. The possibility of extrapolating to atomistic interpretation pressure coefficients based in the linear theory of elasticity, is a matter of current study. With all of the implications which a proper nonlinear theory I am sure will have upon the interpretation of the experimental data itself, experimentally this subject is still in its infancy, despite the vast recent literature.

The temperature dependence of elastic constants determined by ultrasonic means, the similarly determined pressure dependence of the elastic constants and the third-order constants in general, and the phenomenon of attenuation of ultrasonic waves merit a detailed review from the perspective of continuum mechanics. But, as I indicated above, the collection of elastic constants for anisotropic single crystals given in LANDOLT-BÖRNSTEIN is so extensive that I do not think it would be useful to attempt to retabulate selected results here.

While on this general subject, we should note that when the amplitude of the ultrasonic pulse increases (and hence, in relative terms, what are usually referred to as finite amplitude waves are produced), dispersion and ultraharmonics are seen. These, of course, provide further evidence of the nonlinearity of the constitutive equations at infinitesimal strain.

3.40. Short-time loading histories.

Four main types of applied loading histories to which I have referred above are shown schematically in Fig. 3.99. In most instances the time required to apply these loads to the surface of the solid is such that the reflection of waves from free surfaces on specimens of laboratory size introduces serious complications in interpreting displacement-time or strain-time histories at some point away from these surfaces. It is for this reason that in the past so much emphasis has been placed upon the determination of dynamic moduli from vibrating specimens, for which the length of pulses or wave lengths becomes the order of magnitude of the dimensions of the specimen itself. The few experimental methods by means of which it is possible to generate pulses of total length far smaller than the smallest dimension of the specimen thus are obviously of importance for the serious study

[1] GRÜNEISEN [1906, *1*].

[2] For example, WOLFGANG SACHSE in his recent Ph.D. dissertation [1970, *1*] obtained some interesting results when he used the ultrasonic technique on crystals undergoing plastic deformation.

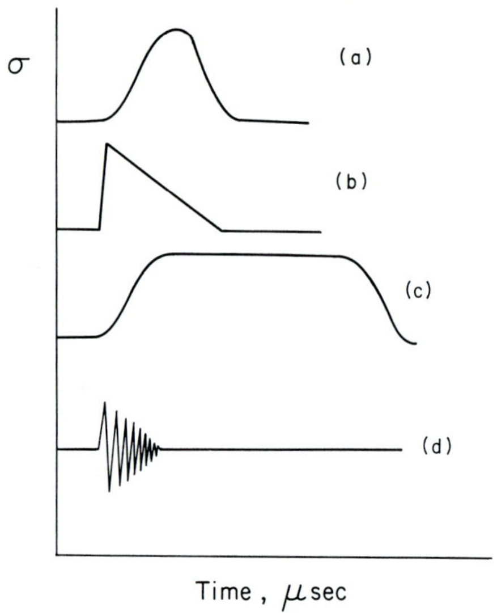

Fig. 3.99a—d. Four types of loading histories.

of wave propagation and hence for determining the form of constitutive equations.

For flat-ended bars of aluminum, Fig. 3.99c, the rise time of the wave front outside the first diameter is of the order of 20 μsec in aluminum or steel, and can vary to as much as ten times that amount when bars of rounded ends are examined at the same impact velocity. The impact of spheres on rods produces the type of pulse seen in Fig. 3.99a with rise times and durations which of course depend upon the diameter of the sphere. KOLSKY's[1] explosive pulse loading lead azide with (b), gave sharp fronted pulses whose duration was of the order of 2 μsec, and ultrasonic pulses at megacycle frequencies (d) were generated, with packet dimensions even shorter than the explosive pulse.

An interesting recent addition to the study of loading histories of the propagation of waves in solids is the thermally generated stress wave described by C. M. PERCIVAL and JAMES A. CHENEY,[2] in which an impinging laser beam produced a thermal pulse of extremely short duration at the free end of a bar. Fig. 3.100 shows their results for a series of aluminum specimens of the designated lengths.

These experiments were modified by J. A. BRAMMER and C. M. PERCIVAL[3] to permit a direct application of the laser beam on the aluminum, instead of having the intermediate glass element which had been required in the previous study. In this second series in 2024 aluminum alloy, by means of this laser pulse technique, elastic moduli were determined in the temperature range, 22 to 500° C. By use of a tourmaline crystal to ascertain arrival times, the dilatational wave speed was calculated. The problem of calculating the shear wave speed in such situations is considerable, as KOLSKY[4] had shown many years before. The assumption by BRAMMER and PERCIVAL that this wave speed might be calculated from a single

[1] KOLSKY [1953, *1*].
[2] PERCIVAL and CHENEY [1969, *1*].
[3] BRAMMER and PERCIVAL [1970, *1*].
[4] KOLSKY [1953, *1*].

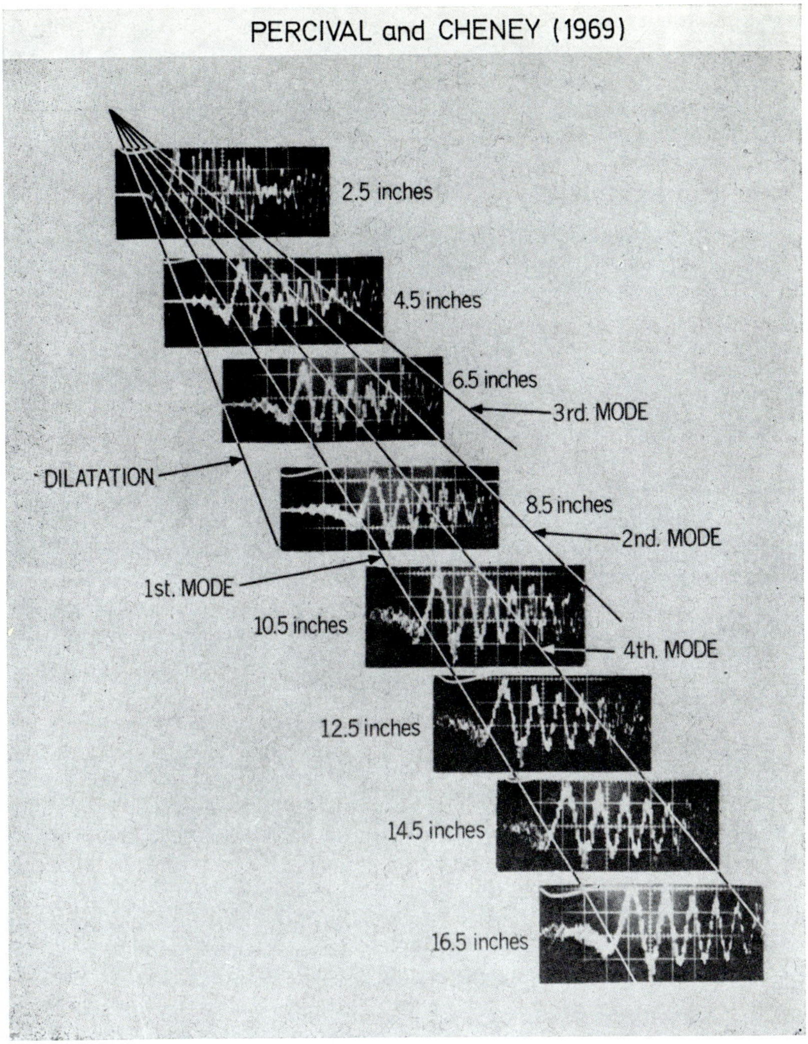

Fig. 3.100. PERCIVAL and CHENEY's measurements (1969) of wave detail at different distances along a bar for a thermally generated stress wave. The thermal pulse was produced by an impinging laser beam.

diametral traverse of the first reflected dilatationalwave front, I believe requires additional experimental investigation. For the laser technique, unlike the ultrasonic, the pulse generated cannot be varied from longitudinal to transverse. It is clearly an advance for experimental solid mechanics that one now can obtain this rapid thermal loading history along with the few short time histories of loading at boundaries previously studied.

An important feature of microseismic techniques which involve extremely short time pulse application is that experimental study is not dependent, as it was for over a century, upon specialized single stress states; instead, the generalized constitutive equations may be considered in the three-dimensional domain, albeit with a dependence upon the interpretation of surface behavior. The mid-19th century dilemma of WERTHEIM and his contemporaries on the propagation of waves in solids in a free field thus has been resolved to some extent. Legions of investigators today have very little interest in the pristine problem as they seek to extrapolate ultrasonic data from the macroscopic measurement to microscopic interpretation.

3.41. On temperature dependence of elastic constants (1843–1910).

The experimental origins of the general subjects developed thus far in this chapter may be traced back to the research of WERTHEIM. The same would be true if at this point we were to examine the development of experiment based upon photoelasticity; WERTHEIM's was the first important experimental research in this area, and it led to his discovery that the linear stress-optics law occurred in nature.[1] Similarly, the change of volume in the torsion of a hollow specimen as a function of the angle of twist later studied by POYNTING, the influence of electric and magnetic fields upon response functions, and the dependence of elastic constants upon temperature[2] were first investigated by WERTHEIM.

Concerning this last subject, we return to WERTHEIM's studies in 1843 on the temperature dependence of elastic constants.[3] From them we can trace experimental considerations of this important aspect of elastic constants during the one and a quarter century which followed. With the exception of the anomaly in iron and steel,[4] the temperature data of WERTHEIM, shown in Table 55, demonstrated not only that the modulus decreased with decreasing temperature, but also that the magnitude of the change could not be accounted for solely on the basis of the change in density due to the thermal expansion.

I already have referred briefly to the temperature studies by KUPFFER,[5] which were performed more than a decade after the measurements by WERTHEIM. The difficulties in interpreting KUPFFER's study, as I have noted above (Sect. 3.29) make meaningless any tabulation of his results.[6] However, one should note that he, too, observed, though merely qualitatively, the now familiar decrease in moduli with temperature, the decrease occurring at a faster rate than could be accounted for by the density changes due to thermal expansion.

In 1878 in his widely quoted *Encyclopaedia Britannica*[7] article on elasticity, KELVIN summarized what he considered to be the significant experimental studies to that date on the temperature dependence of elastic constants, by providing a

[1] ERNEST GEORGE COKER and LOUIS NAPOLEON GEORGE FILON, who in their *Treatise on Photo-elasticity* [1931, *1*], devoted several pages to the discussion of WERTHEIM's doctoral paper [1854, *1*] on photoelasticity, referred to the stress optics law as "Wertheim's law" (see their p. 230). See also WERTHEIM [1851, *8*], [1852, *2*].

[2] WERTHEIM [1844, *1*(a)].

[3] *Ibid.*

[4] WERTHEIM's anomaly in iron and steel concerning a maximum in the vicinity of 100° C might be dismissed on the basis of measurements from some other experiments to be described here but for the fact that such an anomaly has been observed for deformation parameters on more than one occasion under some experimental conditions. See for example the dynamic ultimate-strength measurements by MICHAEL JOSEPH MANJOINE and ARPAD LUDWIG NADAI [1941, *1*] which exhibit the same behavior, but with maxima occurring at somewhat higher temperatures, or the elastic limit data of FREDERICK CHARLES LEA [1922, *1*] in Armco iron, or the electro-thermal E-moduli data of HENRY WALKER [1907, *1*], [1908, *1*], [1911, *1*].

[5] KUPFFER [1852, *1*], [1853, *1*], [1854, *1*].

[6] These difficulties did not interfere with his being awarded a prize by the Learned Society at Göttingen in 1855 for this work, an event which probably contributed to the erroneous impressions that KUPFFER not only had been the first to study the temperature dependence but also that he had succeeded in clarifying the nature of the phenomenon.

[7] KELVIN [1878, *1*], [1880, *1*]. The publication in 1878 was a preprint of the Encyclopaedia article of 1880.

table of WERTHEIM's E-moduli data at 15, 100, and 200°C, obtained 35 years earlier. The only other reference on this subject[1] was to the torsion experiments of KOHLRAUSCH[2] (and LOOMIS) in 1870 on iron, copper, and brass. From their experiments over a range from 0 to 100°C they proposed a temperature factor of the form $\{1-\beta(t'-t)-\gamma(t'-t)^2\}$, where t' was the temperature of the test in °C, and t was the room temperature value of 20°C. Thirty years later, in 1908, CHWOLSON[3] summarized the investigations of the temperature dependence of elastic constants for this second interval, referring to many which had been derived from the research of KOHLRAUSCH and LOOMIS in 1870. As CHWOLSON pointed out, the general diminution of elastic constants with temperature was observed in all of the studies. He cited the work of ten experimentists: STREINTZ, KIEWIET, PISATI, TOMLINSON, MARY CHILTON NOYES, A. MARTENS, A. MAYER, P. A. THOMAS, C. SCHÄFER [SCHAEFER], and HORTON; "all of whom obtained similar results."

Very often, instead of attempting to find an empirical equation to fit the observed modulus-temperature relation, experimentists have reported percentage differences between two fixed temperatures, such as 0 to 100°C. For example, CHWOLSON gave the percentage diminution of E between those two temperatures for KUPFFER's data on iron, copper, and brass, as 5.5, 8.2, and 3.9%, respectively; whereas between the same two temperatures A. MAYER had provided the percentage diminutions of E as 1.16% for St. Gobain glass; 2.24 to 3.09% for various kinds of steel; 3.73% for brass; 5.5% for aluminum; in the range from 0 to 60°C, 2.47% for silver; and, from 0 to 62°C, 6.04% for zinc. CHWOLSON cited P. A. THOMAS as having established that E varied proportionally to a power x of the density, the density variation being that associated with the coefficient of expansion. For cast iron, THOMAS had obtained for x the value 31.3. Without critical comment, CHWOLSON also cited the work of DEWAR, SCHAEFER, BENTON; of DEWAR and HADFIELD;[4] and of HADFIELD, who in 1904 had made studies of E in Pt, Ni, Ag, Cu, Pd, and Fe at low temperatures, between +20°C and −186°C establishing that in general the temperature coefficient of E was a constant quantity. HADFIELD's work provided the first suggestion that the moduli of isotropic materials below room temperature varied linearly with the temperature. CHWOLSON described similar studies between +20°C and −186°C by BENTON in 1903 for copper and steel; BENTON also had obtained values of 1.180 for copper and 1.087 for steel for the lower temperature to room temperature ratio. The work of WASSMUTH, who determined experimentally the numerical value of the linear temperature coefficient β and obtained 0.000241 for iron rods, was mentioned by CHWOLSON, as were similar studies of temperature dependence by VOIGT.

[1] PEARSON (TODHUNTER and PEARSON [1893, *1*]) was critical of KELVIN (WILLIAM THOMSON) for having omitted discussion of the work of KUPFFER. (In fact PEARSON at several places in his reporting of KELVIN's studies in elasticity, attacked him for having systematically ignored KUPFFER's work, which, for reasons which I am completely unable to fathom, had enormously impressed PEARSON.) The temperature studies of KUPFFER were performed in the same manner as his studies of transverse vibration at room temperature, and thus they contained the same defects of interpretation as had been pointed out by ZÖPPRITZ [1866, *1*]. If one presumed that the error was independent of temperature, then one might note that KUPFFER's studies between 15 and 79 degrees Réaumur had provided:

$$\mu_t' = \mu_t\{1-\beta_\tau(t'-t)\}, \qquad t'>t,$$

where t' was the temperature of the measurement and t was the value at 13° R. Values of β_τ were tabulated for 22 different types of specimens for ten metallic solids.

[2] KOHLRAUSCH [1870, *1*]; KOHLRAUSCH and LOOMIS [1870, *2*].

[3] CHWOLSON [1908, *1*]. And, for example, see also LE CHATELIER [1889, *1*], [1890, *1*]; AMAGAT [1890, *1*]; KOWALSKI [1890, *1*]; MARTENS [1890, *1*]; VOIGT [1890, *1*], [1893, *3*]; WINKELMANN [1897, *1*].

[4] DEWAR and HADFIELD [1904, *1*].

My purpose in noting CHWOLSON's uncritical listing[1] of data from the second 30 year interval is to indicate the sudden, rather explosive interest in the subject which occurred toward the end of the 19th century. A detailed review of many of these papers, which I have located and examined, would be repetitive. I have thought it best to begin with the work of GILBERT ARDEN SHAKESPEAR[2] in 1899, followed by the work of EDUARD AUGUST GRÜNEISEN[3] in 1910, both of whom achieved the required accuracy for a definitive study of the temperature dependence of elastic constants through the use of the extremely accurate method of optical interference to determine the strain.

One important aspect of the effects of temperature upon the elasticity of solids is that associated with the prior thermal history, a matter which had been of interest to large numbers of persons since COULOMB[4] first investigated the subject in 1784. It had become quite well known by the mid-19th century. By 1844 WERTHEIM[5] definitively had shown that when a prior thermal history of a solid involved a cycle to a high temperature (and hence the annealing of the solid), the elastic modulus determined at room temperature in general was lower than the modulus measured in the same or similar specimens prior to the temperature cycle. Thus in an experiment to study elastic deformation at increasing ambient temperatures, one must expect differences if the measurements are made while temperatures are increasing or are decreasing after having passed through some temperature maximum. Indeed, such differences have been shown by numerous experimentists.

An important question which I considered above in the research of GAROFALO, MALENOCK, and SMITH[6] for various steels was whether or not the experimentally determined value of POISSON's ratio is a constant function of the temperature. In dealing with linear elasticity of solids and, in particular, linear thermal elasticity, it is necessary to establish experimentally that the various moduli such as the bulk modulus K, the E-modulus, and the shear modulus, μ, for the isotropic solid do indeed vary with temperature in a manner consistent with a single value of POISSON's ratio. Similarly, when one seeks to measure the elastic constants of the anisotropic single crystal on the presumption that linear elasticity applies, a discovery that the different elastic constants depend differently on the temperature as, in fact, they are known to do in many instances, will be of more than small interest for theories of thermoelasticity.

In 1870 the study by KOHLRAUSCH[7] and by KOHLRAUSCH and LOOMIS[8] on the temperature dependence of the shear modulus for iron, copper, and brass, was undertaken primarily for the purpose of establishing from more precise experiments whether or not the anomaly found by WERTHEIM 26 years before, regarding E for iron which increased with increasing temperature between 0 and 100° C, was indeed a reproducible material phenomenon. As indicated above, the investigation by KOHLRAUSCH, which had been the main reference on the subject beyond WERTHEIM's study in 1844 and had been included in the *Encyclopaedia Britannica* article

[1] CHWOLSON, more often than not, listed names without providing the date of the work or the specific reference. After presuming that the references at the end of the section contained the material referred to, I found that a third of those cited did not in fact describe temperature dependence measurements but the consequences of prior thermal annealing histories upon moduli at room temperature.

[2] SHAKESPEAR [1899, *1*].

[3] GRÜNEISEN [1910, *2*, *3*].

[4] COULOMB [1784, *1*].

[5] WERTHEIM [1844, *1*].

[6] GAROFALO, MALENOCK, and SMITH [1952, *1*]. See Sect. 3.28.

[7] KOHLRAUSCH [1870, *1*].

[8] KOHLRAUSCH and LOOMIS [1870, *2*].

by KELVIN[1] in 1880, essentially repeated, at different ambient temperatures, the experiments of COULOMB[2] with a torsion pendulum 96 years earlier.

KOHLRAUSCH and LOOMIS, like COULOMB, determined the frequency of the torsional vibration of a wire to which was attached a cylindrical weight. A mirror affixed to the weight enabled them to make their observations. They made comparisons by performing the tests first at room temperature, and then in the same chamber at 100° C obtained by heating with water steam. After the chamber had been heated, they first interrupted the supply of steam surrounding it and then observed the duration of vibration and the temperature while the specimen gradually cooled. They measured the temperature by means of three thermometers located at different positions in the chamber. Thermal lags of the thermometer with respect to the temperature of the chamber and that of the thin wire were the subject of an undiscussed, special study by a Mr. GROTRIAN. KOHLRAUSCH and LOOMIS made a few observations at 0° C using ice. To avoid permanent changes in elasticity due to annealing, they cycled the specimens a number of times between room temperature and 100° C before they actually began the tests. All three solids were initially hard wires of diameters from 0.2 to 0.3 mm. KOHLRAUSCH and LOOMIS described the copper wire as "pure, electrolytically precipitated copper," and the iron and brass as commercial wires.

The observed percentage differences in the torsion modulus obtained from the changes in the frequency of vibration were: for iron, 4.6%; for copper, 5.5%; and for brass, 5.6%, which represent the observed decrease when comparing measurements at 0 and 100° C. The measurements between those two temperatures were found to fit the following empirical equations for the designated metals:

$$\begin{aligned} &\text{Iron:} && \mu=\mu_0(1-0.000447\,T-0.00000012\,T^2).\\ &\text{Copper:} && \mu=\mu_0(1-0.000520\,T-0.00000028\,T^2).\\ &\text{Brass:} && \mu=\mu_0(1-0.000428\,T-0.00000136\,T^2).\end{aligned} \tag{3.15}$$

KOHLRAUSCH pointed out that nothing peculiar was observed in the behavior of the shear modulus in iron when compared with that of brass and copper. He could not wholly reject the idea that the differences between his own and WERTHEIM's observations on iron could be explained by the presence of volume changes which occur in uniaxial tests. As KOHLRAUSCH indicated in the beginning of his paper, it is true that WERTHEIM did not claim great accuracy in his study of the variation of E with temperature; WERTHEIM was content to provide the first demonstration that an elastic modulus indeed does vary with temperature. Finally, KOHLRAUSCH dismissed the possibility of comparing the experimental data on iron, copper, and brass with KUPFFER's[3] measurements of 18 years before, concluding that the experimental studies by KUPFFER had not been performed with depth sufficient for such a comparison to be made.[4]

---

[1] KELVIN [1880, *1*].

[2] COULOMB [1784, *1*].

[3] KUPFFER [1852, *1*].

[4] This comment on KUPFFER's work was introduced with the statement that it was the only other experimental study of the temperature dependence of the torsion modulus which existed in 1870. In the light of his subsequent critical comments, KOHLRAUSCH's statement that this was the work for which KUPFFER had been honored by a special prize from the Gesellschaft der Wissenschaften in 1855, might be interpreted as the expression of slight irony. Since KELVIN was thoroughly familiar with the work of KOHLRAUSCH and LOOMIS, having described their results in some detail in 1880 in the Encyclopaedia, this dismissal of KUPFFER's work by KOHLRAUSCH may have been a contributing factor in KELVIN's excluding KUPFFER's work from his famous article, first published in 1878. KOHLRAUSCH's comment that

The use of interference optics by CORNU[1] in 1869 had not stimulated the general use of the method during the following thirty years among experimentists interested in deformation. In 1899, CONSTANTIN RUDOLPH STRAUBEL, at the Physikalisches Institut der Universität at Jena, and G. A. SHAKESPEAR of Trinity College, Cambridge, independently reintroduced the technique of interference optics for the study of the deformation of solids.[2] As we have seen, the experiments by STRAUBEL were essentially a repetition of CORNU's experiment on a large number of glass specimens. SHAKESPEAR, on the other hand, was concerned with the temperature dependence of E and in particular with the WERTHEIM anomaly[3] in the E of iron.

SHAKESPEAR referred to a series of unsuccessful experiments on the deformation of metal rods in tension, such as that shown in Fig. 3.101 a. He was unable to maintain axiality in loading with sufficient precision to keep the interference fringes from being thrown out of the field. He achieved more success in the arrangement, shown in Fig. 3.101 b, in which the specimen consisted of three wires symmetrically attached to two circular brass plates. He had placed the optical apparatus in the center of the parallel plates with a supporting rod, as in Fig. 3.101 a, made of the same material as the rods under study. As may be seen in Fig. 3.101 c, he had provided for the initial adjustment of the optical system.

Because of the large number of interference fringes which had to be counted during the deformation, SHAKESPEAR found it was necessary to load the specimen very gradually and to minimize the small vibrations due to the application of the load. He accomplished this partly by relocating the apparatus in the Cavendish Laboratory basement, and by the use of the loading apparatus shown schematically in Fig. 3.101 d. By means of a hook, he suspended from the lower plate a cylindrical weight with a conical top and bottom. Initially, he filled the container with water at the level shown by the top dashed line. To apply the load, he removed the water by means of the indicated siphon until at the maximum load, which was the same for every test, the water level had been reduced to that illustrated by the lower dashed line. He introduced the vanes to reduce the movement of the loading cylinder. SHAKESPEAR's major difficulties included the necessity for minimizing the effects of vibration[4] and of thermal variation.

SHAKESPEAR extended his objective to study the temperature dependence of E for copper, soft iron, steel, and hard brass over the temperature range from 0 to 100° C, with one or two orders of magnitude more accuracy than had been achieved in previous studies. After trying several unsuccessful methods to heat the specimens, he finally resorted to the use of boiling water contained in a surrounding jacket. His paper contains a long description of the many problems which arose when he

the work of KUPFFER had not been performed in sufficient depth, is in decided contrast to the statement of PEARSON in 1886:

> "The experiments were apparently made with great exactitude and they confirmed Kupffer's law of reduction." [1886, *1*], Vol. 1, p. 754,

or that of TIMOSHENKO in 1953:

> "Kupffer examined the question of influence of temperature on the modulus of elasticity with great care and presented a paper on this subject to the Russian Academy of Sciences in 1852." [1953, *1*], p. 221.

[1] CORNU [1869, *1*].

[2] Whereas STRAUBEL [1899, *1*] cited the work of CORNU 30 years earlier (STRAUBEL felt he was modifying and improving upon CORNU's experiments), SHAKESPEAR [1899, *1*] did not refer to the earlier work, nor did he refer to CANTONE's [1888, *1*] use of optical interference to measure strain, eleven years before his own work.

[3] HERBERT TOMLINSON [1883, *1*], 16 years before already had shown that WERTHEIM's anomaly in iron and steel was a permanent annealing phenomenon occurring in the first heating to 100° C.

[4] For example, all measurements had to be made in the quiet of the night.

SHAKESPEAR (1899)

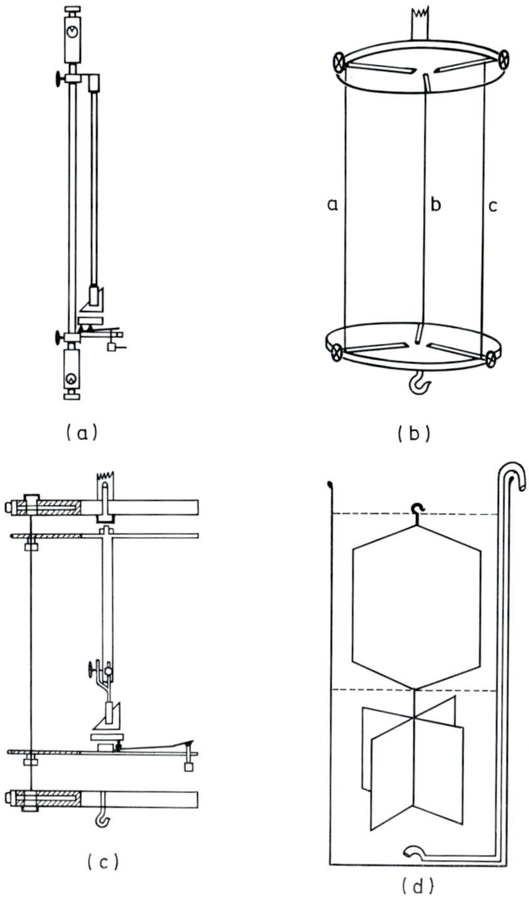

Fig. 3.101 a–d. The optical apparatus used by SHAKESPEAR (1899), showing the manner of obtaining interference rings by supporting the parallel plates with a single rod (a) and a three wire system (b). The provision for the initial adjustment of the optical system is shown in (c), and the manner of applying the load by the removal of water from container, in (d).

was attempting to make accurate interferometric measurements of displacement despite the agitation caused by the boiling water. He partly solved the problem by adding pumice.

The first series of experiments consisted in the study of the modulus of three specimens of "copper wire ($^1/_8$ in. diam.) specially drawn straight," with different initial loads, beginning with one of 2.791 g and proceeding with successive additions of 2 kg. These weights were determined very accurately "on Prof. Poynting's large balance which was kindly put at my disposal" with the water at the upper and lower levels noted in Fig. 3.101 d. He found that the modulus determined from the increments began with the value of 8.24×10^{11} dynes/cm^2 in the vicinity of the 2.791 g, and increased linearly to a maximum of 1.053×10^{12} dynes/cm^2, after which it became constant. When the load was removed, this experimental curve was retraversed to the original value. SHAKESPEAR alluded to the curious behavior as possibly explaining WERTHEIM's observations of differences in moduli when determined by different methods.

It is my opinion that the experimental situation SHAKESPEAR considered should be carefully reexamined. Because he was aware of the problems associated with the annealing of specimens during the study of the temperature dependence of moduli, SHAKESPEAR cycled his specimens a large number of times between 13 and 100°C to obtain the percentage decrease of modulus for each cycle. For these measurements he had to consider approximately 300 interference bands so that the moduli comparisons could be made for the same load at both temperature extremes. Two of SHAKESPEAR's observations are of particular significance and in fact may shed additional light on the WERTHEIM anomaly in iron. First, not only for the soft iron, but also for the copper, hard brass, and steel, the initial loading of the specimen in each instance provided a percentage *increase* of the modulus when the temperature changed from 13 to 100°C. In all successive cycles, a modulus decrease was observed, as in the data for copper shown in Table 101; after the fourth or fifth cycle, there was a constant percentage decrease of modulus. SHAKESPEAR wrote:

> In almost all cases after any change of temperature, or after a rest (of any time more than half an hour, say), the first observation of the series showed an amount of elongation different from those given by succeeding observations of the same series, the first elongation generally being greater than those succeeding.[1]

Table 101. SHAKESPEAR (1899).

	Temperature		Percentage decrease of modulus
	13° C	100° C	
No. of bands:	317.0	303.0 (1st heating)	−4.4
	291.8	298.5 (2nd heating)	+2.3
	287.5	296.4 (3rd heating)	+3.1
	283.0	293.7 (4th heating)	+3.8
	279.2	289.4 (5th heating)	+3.7
	278.0	287.5 (6th heating)	+3.4
	278.5	289.0 (7th heating)	+3.8
	279.0	289.0 (8th heating)	+3.6
	279.0	289.0 (9th heating)	+3.6
	279.0 on final cooling		

After placing the specimens under load at either the high or the low temperature for various periods of time, SHAKESPEAR made a number of measurements with respect to the memory aspects, obtaining results which exhibited significant changes in subsequent time intervals. The remainder of his experiments dealt with the consideration of another aspect of WERTHEIM's work in 1844, namely, the effect of a magnetic field on the modulus of soft iron.

That a modulus of elasticity does not necessarily decrease with temperature, as had been indicated by WERTHEIM, may be seen in the studies by FRANK HORTON in 1905,[2] of the variation of the "modulus of torsional rigidity" of quartz fibers over a temperature range from 20 to 1000°C. Repeating COULOMB's torsional pendulum experiments of 120 years earlier (1784) on 0.001 cm quartz fibers with objectives quite similar to those of COULOMB because "they are almost universally employed as suspensions in torsion instruments where accuracy is

[1] SHAKESPEAR [1899, *1*], p. 551.

[2] HORTON [1905, *1*].

required,"[1] HORTON added only two new features to the original experiments. Vibration frequencies from which the moduli were calculated, he determined with a "new method of timing by means of 'coincidences'—a method devised by Professor Poynting..." and he achieved the important determination of fiber radii to an accuracy of 0.01% by rolling a small length of it between two fine glass capillary tubes and counting the number of revolutions it made in traveling a distance of 5 mm.

HORTON noted the results of experiments on six fibers. Unlike metals, they gave a μ in very good agreement from one test to another, and from one specimen to another. The practically constant value of the shear modulus at 15° C was 3.001×10^{11} dynes/cm². He made the first series of temperature measurements at 15, 35, 55, 75, and 100° C. In every case he found that μ increased linearly with the temperature, but the temperature coefficient, unlike the initial modulus, varied from one specimen of quartz to another, with a mean value of $+0.0001235$.

The experiments were extended to the temperature range from 20 to 1000° C on rather thick fibers which were suspended inside a platinum tube which could be electrically heated, and which could be maintained at any desired temperature. HORTON made measurements at 250° C in this interval by means of a platinum and rhodo-platinum thermocouple. He found that the modulus of rigidity of the fiber increased with the temperature, at first linearly; then, as the temperature rose still higher, the rate of increase gradually diminished. There was a maximum μ at a temperature of 880° C. Above 880° C, the shear modulus decreased very rapidly with increasing temperature.

Once again in still another solid one sees the WERTHEIM anomaly of an increasing modulus, which suggests that the phenomenon of temperature dependence sometimes is more complex than WERTHEIM or his successors had realized. HORTON also had considered the internal viscosity of the fibers from the attenuation of the torsional vibrations and had found that the viscosity increased with the temperature at a rate which at first was small and constant but after about 650° C became much more rapid. "At 1060° C the internal friction of the fibres was so great that the torsional vibrations were nearly dead-beat."[2]

In 1895 MARY CHILTON NOYES became interested in WERTHEIM's anomaly in the temperature dependence of E for iron and steel and in the fact that such an anomaly had not been observed by all subsequent experimentists who had studied the phenomenon.[3] Referring to WERTHEIM, NOYES stated:

> He was such a careful experimenter, and the increase observed was in some cases so considerable, that it seems improbable that these results are altogether in error. One object of the present investigation was to see if any indication could be found of a maximum value for the modulus between 100° and 200°.
>
> Many of those who have experimented on this subject have used only a few temperatures, and few have gone to temperatures beyond 100°. There have also been considerable variations in the results. It seemed desirable to obtain the modulus for a series of temperatures extending nearly to 200°, and to secure the various temperatures by a method not previously employed in investigating the subject.[4]

NOYES's experiments were performed on wires approximately $^9/_{10}$ m long. Of particular interest in her study was the fact that she chose to apply the heat by two different methods. In the first instance, the horizontally placed wire, one of

[1] *Ibid.*, p. 401.

[2] *Ibid.*, p. 402.

[3] As I noted above in reference to the research of SHAKESPEAR, TOMLINSON [1883, *1*] had settled the matter of WERTHEIM's anomaly in iron and steel over a decade earlier, yet no reference to his then widely known studies appeared in either SHAKESPEAR's or NOYES's papers.

[4] NOYES [1895, *1*], p. 277.

whose ends passed over the circumference of a large wheel so that vertical weights could be added, was inside a glass tube around which a heating coil had been wound to produce the desired temperature. In the second type, the specimen was heated by the direct application of an electric current. Both types of experiment, of course, had been performed by WERTHEIM fifty years earlier, and hence one wonders to what "method not previously employed" NOYES was referring.

After a preliminary series of measurements to determine the coefficient of expansion, NOYES obtained E at a variety of temperatures by the application of a 0.4 kg weight. She expressed considerable surprise at the results. The E for the wire heated by the external coil decreased linearly with the temperature, and the averaged results of the many observations for each test were seen to be excellently reproducible when she compared the averaged data from each of the ten tests. However, for the specimens heated by the passage of a current through the wire, E underwent a slight increase to a maximum which was followed by a small decrease. The net result was that the modulus remained essentially constant over the same temperature range as in the first type of experiment.

Unable to accept this curious result, NOYES in the following year repeated the work in a new series of ten experiments[1] in which she revised the method of loading the specimens, and increased the maximum load from 0.4 to 1 kg. She obtained the result shown in Fig. 3.102 for a wire which had been heated rapidly by the coils before the measurements were made. In this instance, she saw no anomalous response in either type of experiment. Thus, in this second study, NOYES[2] not

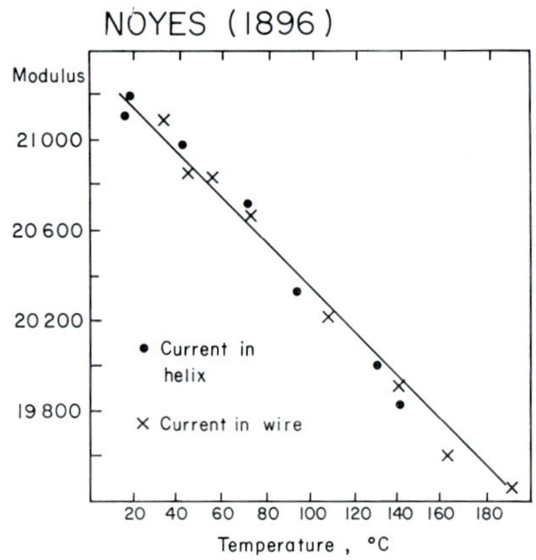

Fig. 3.102. The variation of E with temperature for two different methods of heat production.

only found that WERTHEIM's anomaly seemed not to apply to the iron wire specimens she had examined, but also she found that both manners of heating gave the same linear decrease of E; that is, the presence of an electric current produced no change in E other than that which arose from the change of temperature in the specimen caused by the current.

In 1906 HENRY WALKER, very much impressed by the observed difference in the temperature dependence of the moduli described in NOYES's first paper, and not yet having seen her second paper, repeated her experiments using a detailed

[1] NOYES [1896, *1*].

[2] This second paper was MARY CHILTON NOYES's Ph.D. dissertation (Cornell University, 1896).

duplication of her apparatus for the situation in which the wire inside the glass tube was heated by the direct application of an electric current.[1] WALKER examined steel, soft iron, copper, and platinum. He used a microscope to determine the elongations of the nearly meter long wires. In every instance, with the application of a small load (i.e., 0.5 kg for iron) E, after an initial small decrease,[2] increased by fairly large amounts, then proceeded through a pronounced maximum which was followed by an equivalent decrease. The temperature at which the maximum occurred was somewhat less than 100° C in each case. As in the experiment of NOYES, when the specimens were heated over the same temperature range from 15° C to somewhat less than 200° C in an ordinary way without the passage of an electric current through the specimen, E decreased linearly.

After having published these results, which provided a more definitive and accurate study of the behavior described in NOYES's first paper, WALKER became acquainted with the fact that in her second paper NOYES had dismissed her earlier results on experimental grounds. He saw that the chief difference between her two studies lay in the magnitude of the load, and he planned a program which soon revealed that the behavior of the increase to a maximum of the modulus, resembling the WERTHEIM anomaly in iron with furnace heating, was a function of the magnitude of the applied load. As the load increased, the temperature dependence of the moduli approached and finally became linear, as in ordinary heating. WALKER observed this dependence of the modulus both while he increased the temperature and while he decreased it. He noted some differences in the magnitude and in the location of the maximum in the two portions of the thermal cycle. He reported these first results in 1907 for iron, steel, copper, and platinum;[3] he extended and elaborated them in 1910[4] by adding studies in cobalt and nickel. This dependence of the measured modulus upon the magnitude of the applied load when in the presence of an electric current in the specimens was not confined to magnetic materials but existed in different degrees in all six metals he studied.

To consider further the load dependence of the modulus in the presence of an electrical field, WALKER performed a series of additional experiments on each of the several metals. He maintained the current constant and determined E at many loads up to that approaching the elastic limit of the solid. Again he observed that the moduli proceeded through a maximum in the low stress region, followed by a decrease as the stress increased, consistent with the temperature measurements for a constant load.

The three papers by WALKER[5] contain a considerable amount of data and include details relating to the demagnetization of the iron, steel, and nickel, and to the prestressing and preheating of all of the solids to minimize or eliminate further annealing or the producing of small permanent deformations during the actual experiments. Since essentially the same behavior was observed in iron, steel, copper, platinum, cobalt, and nickel, only the results for cobalt and copper are shown as representative. For different constant loads, the temperature was varied by means of an electric current, and the results were compared (see Fig. 3.103) with those obtained under the ordinary heating of cobalt. Fig. 3.104 shows results for the same solid, in which the modulus is determined as a function of the magnitude of the load for the fixed electrical field strengths H shown. The cobalt specimens were in the

---

[1] WALKER [1907, *1*].
[2] WALKER thus saw the same anomalous behavior as had been observed by WERTHEIM [1844, *1*].
[3] WALKER [1907, *1*], [1908, *1*].
[4] WALKER, published in [1911, *1*].
[5] WALKER [1907, *1*], [1908, *1*], [1911, *1*].

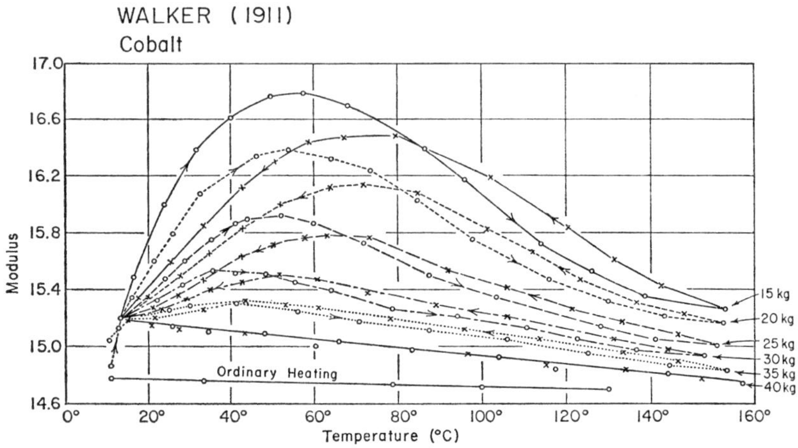

Fig. 3.103. The variation of E with temperature for cobalt when the heating is produced by electric current, compared with that of ordinary heating.

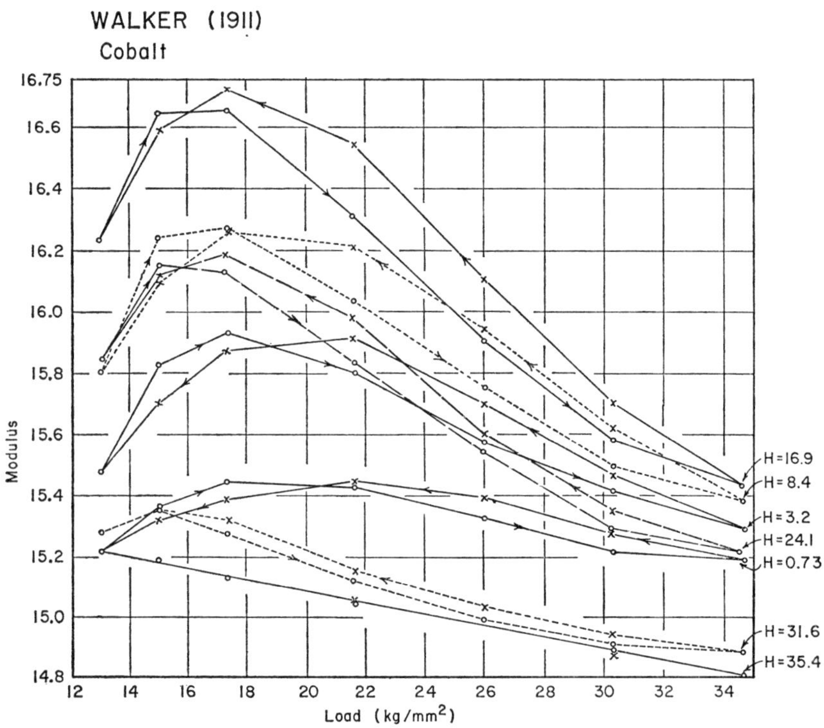

Fig. 3.104. E as a function of the load in kg/mm² for various fixed electric field strengths. The material was cobalt.

form of thin strips rather than the wires used in all his other experiments. The results for nonmagnetic copper wire, which are similar, are shown in Fig. 3.105.

It should be pointed out that TOMLINSON's explanation for the anomalous behavior in iron which WERTHEIM had observed for ordinary heating, cannot

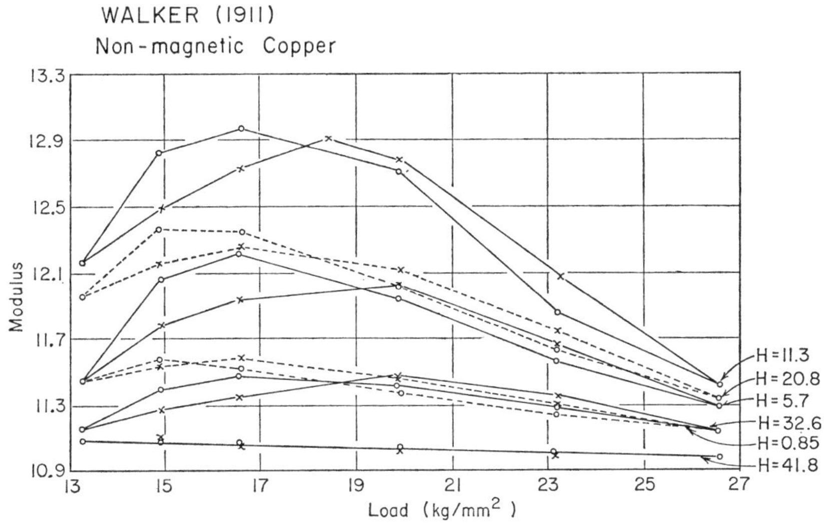

Fig. 3.105. E as a function of the load in kg/mm², for various fixed electric field strengths. The material was non-magnetic copper.

be extended to include the appearance of the behavior which WERTHEIM, NOYES, and WALKER had observed during the passage of an electric current.

WALKER, like SHAKESPEAR before him (and, as I have indicated, like TOMLINSON, 16 years before SHAKESPEAR), noted that the modulus determined during the first cycle differed from that during the second cycle, etc., with the subsequent cycles in the order of increasing magnitude. The interesting behavior observed by WALKER demonstrated that the nonlinear stress-strain function for small deformation could be varied by the presence of an electrical field of different strengths.[1] It is difficult to understand why this discovery and that of the stress dependence of the maximum in the "anomalous behavior," failed to stimulate further research during the succeeding six decades.

As we have seen in Chap. II (Sect. 2.25) with respect to the nonlinearity of the small deformation of metals in the immediate vicinity of zero stress, and in the present chapter with respect to the direct measurement of POISSON's ratio by interferometry, GRÜNEISEN's work, like WERTHEIM's was of great importance in the development of solid mechanics. GRÜNEISEN had the unusual habit of asking fundamental questions and of performing experiments which were specifically related to them. This was certainly evident in his experimental studies of the influence of temperature upon the compressibility of solids. In those experiments he had used the "MALLOCK method" described in Sect. 3.30 above. The work, published in a paper[2] in 1910, was an outgrowth of his interest in the constancy

[1] As far as I know, no serious study of this electro-elastic behavior in terms of stress magnitude has been made since 1910. Particularly during the past two decades, the emphasis has been on ultrasonics, in which the influence of imposed magnetic fields upon appropriate metals has been examined and confined to exceedingly small strain amplitudes. (The corresponding previous studies of flexural vibration similarly had been confined to small strain.) Hence, the functional dependence of the behavior upon stress has been eliminated by the nature of the experiment.

[2] GRÜNEISEN [1910, *3*].

of the ratio of the linear thermal expansion coefficient α and the specific heat c_p, i.e., α/c_p; and the volume thermal expansion coefficient β, i.e., β/c_p = constant. In particular, he had been interested in comparing the temperature dependence of the volume coefficient of expansion with an exponential formula proposed in 1908 by MAX FERDINAND THIESEN[1], and with THIESEN's exponential formula for the temperature dependence of the specific heat. These are given as

$$c_p = \varepsilon A T^{\varepsilon} \tag{3.16}$$

and

$$\beta = -\frac{dA}{dp} T^{\varepsilon}, \tag{3.17}$$

where T designates the absolute temperature and where ε is given by

$$\varepsilon = \frac{1}{B} \frac{dA}{dp}. \tag{3.18}$$

THIESEN had observed that Eqs. (3.16) and (3.17) represent the behavior of platinum for $\varepsilon = 0.18$. Later he concluded after a more critical examination of the observed data that the expansion coefficient increased more rapidly with temperature than the specific heat did with temperature, so that β/c_p was not constant with temperature, and the formulae of Eqs. (3.16) and (3.17) for the same value of ε did not apply.

GRÜNEISEN[2] in 1910 studied Al, Fe, Ni, Cu, Pd, Ag, Ir, and Pt. He made measurements of the coefficient of expansion from $-190°$ C to $+100°$ C and compared his values with those of a number of other experimentists including KARL FRANZ FRIEDRICH CHRISTIAN SCHEEL[3] (1907), SCHEEL and WILHELM HEUSE[4] (1907), FRITZ HENNING[5] (1907), and LUDWIG HOLBORN and SIEGFRIED VALENTINER[6] (1907). GRÜNEISEN also performed experiments on the temperature dependence of the coefficient of expansion of Mg, Zn, Cd, Sb, Ir, Au, Pb, and Bi, between $-183°$ C and $+100°$ C. After discussing the exceptional behavior of zinc, cadmium, and tin, which he compared with the earlier data of VOIGT,[7] GRÜNEISEN concluded that THIESEN's exponential formula did indeed fit the data for the thermal expansion coefficient as a function of temperature. He noted also that the coefficients in the exponential formulae depended upon the melting point. Plotting the elongation of a metal rod as ordinate, and as the abscissa the fractional melting point T/T_m, where T, and T_m were the temperature of the measurement and the melting point in degrees Kelvin, repectively, GRÜNEISEN provided the formula,

$$\Delta l = \mathcal{U} \left(\frac{T}{T_m}\right)^{1+\varepsilon} \tag{3.19}$$

where Δl is the elongation of the rod between 0 and T degrees absolute. Here $\mathcal{U} = V T_m^{1+\varepsilon}$ is a constant of the metal; GRÜNEISEN tabulated numerical values of this constant. These results are shown in Fig. 3.106, where the solid lines indicate the range in which the actual data were taken, and the dashed lines, the extrapolated exponential form. GRÜNEISEN noted the observed dependence on the melting point.

[1] THIESEN [1908, *1*].
[2] GRÜNEISEN [1910, *1*, *2*].
[3] SCHEEL [1907, *1*].
[4] SCHEEL and HEUSE [1907, *2*].
[5] HENNING [1907, *1*].
[6] HOLBORN and VALENTINER [1907, *1*].
[7] VOIGT [1910, *1*]. This volume refers to VOIGT's work since the 1870's.

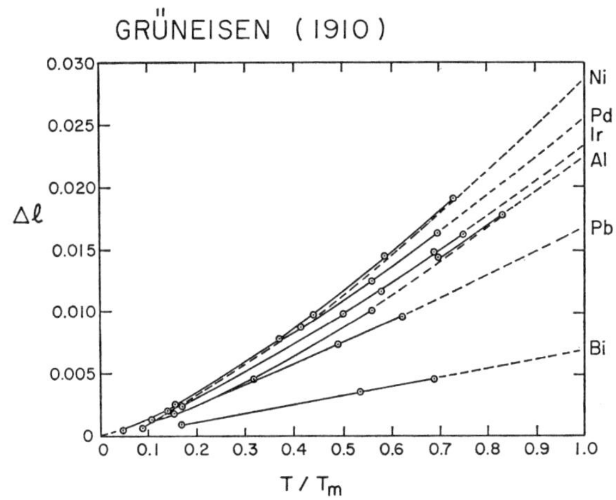

Fig. 3.106. GRÜNEISEN's (1910) plot of elongation as a function of the fractional melting point temperature for the various metals indicated.

In a second paper in 1910, GRÜNEISEN[1] considered the temperature dependence of the specific heat. He concluded that over a temperature range from $-190°$C to $+100°$C, both the specific heat and the expansion coefficient had nearly the same temperature dependence, and thus the ratio of α/c_p was constant with temperature over the range considered, and THIESEN's exponential formulae represented the facts.

In the same paper, in a section on elastic constants GRÜNEISEN raised the question which has stimulated the present interest in his work. He wished to know whether the relation α/c_p = constant was independent of pressure, and whether the constant and THIESEN'S formulae, which had been in approximate agreement at room pressure, were equally compatible at higher pressures, with or without the same constant, ε. GRÜNEISEN then proceeded to determine the compressibility of solids as a function of the temperature, using MALLOCK's method modified so as to permit measurements at different ambient temperatures. This experimental study, which is described in great detail in a third paper in 1910 by GRÜNEISEN,[2] led to a general consideration of the requirements for the linear theory of elasticity in isotropic solids, of the independently measured temperature dependence of the shear modulus μ, the E-modulus, the bulk modulus K, and POISSON's ratio ν. By 1910 several experimentists had studied the temperature dependence of E, μ, and ν, but GRÜNEISEN'S was the first investigation including a study of the compressibility which had an accuracy sufficient to provide meaningful conclusions for the isotropic solid.

The details of GRÜNEISEN's use of MALLOCK's method for determining compressibility at room temperature were given above, together with a comparison with other measured moduli in terms of the formulae for isotropic solids interrelating them. His experimental values for the compressibility using that method, over the temperature range $-191°$C to $+166°$C for aluminum, iron, copper, silver, tin, platinum, and lead, are given in Table 102, where the brackets indicate observations made without the subtraction of the elastic after-effect.

---

[1] GRÜNEISEN [1910, *2*].
[2] GRÜNEISEN [1910, *3*].

Table 102. GRÜNEISEN (1910).

Aluminum				
$t=$	$-191°$	15°	125°	
$\chi\cdot 10^6=$	1.32	1.46	[1.70 (1.90)] cm²/kg	
Iron 1				
$t=$	$-190°$	18°	128°	165°
$\chi\cdot 10^6=$	0.606	0.633	0.664	0.675
Copper				
$t=$	$-191°$	17.5°	133°	165°
$\chi\cdot 10^6=$	0.718	0.773	0.815	0.828
Silver				
$t=$	$-191°$	16°	134°	166°
$\chi\cdot 10^6=$	0.709	0.763	[0.835 (0.853)]	[0.862 (0.889)]
Tin				
$t=$	$-190°$	15.2°		
$\chi\cdot 10^6=$	2.1	[(3.1)]		
Platinum				
$t=$	$-189°$	16.8°	133°	164°
$\chi\cdot 10^6=$	0.374	0.392	0.401	0.404
Lead				
$t=$	$-191°$	14.2°		
$\chi\cdot 10^6=$	[(2.5)]	[(3.2)]		

Temperatures are in degrees Centigrade.

GRÜNEISEN enclosed in square brackets those values which he thought were too large primarily because of the permanent flow; he enclosed in parentheses those which he thought were too large primarily because of the elastic after-effect. As one of the first experimentists to recognize that the melting point T_m of the metal ought to be introduced in comparisons of the deformation parameters of solids, GRÜNEISEN plotted the product of the compressibility χ and the melting point T_m against the temperature of the test, as shown in Fig. 3.107. Obviously this product did not depend linearly upon the temperature. One notes further that for a given temperature, the product has a value inversely related to the melting points of the solids.

Having determined the temperature dependence of the compressibility, GRÜNEISEN realized that now it was possible to pose a fundamental question in linear elasticity by comparing this temperature dependence of the compressibility coefficient with the temperature dependence of E, μ, and ν as measured by previous experiments. From the relations

$$1+\nu=\frac{E}{2\mu}, \tag{3.20}$$

$$\chi=\frac{1}{K}=\frac{3(1-2\nu)}{E}, \tag{3.21}$$

for isotropic materials, elementary theory would require that the same value of POISSON's ratio ν be obtained at different ambient temperatures both in

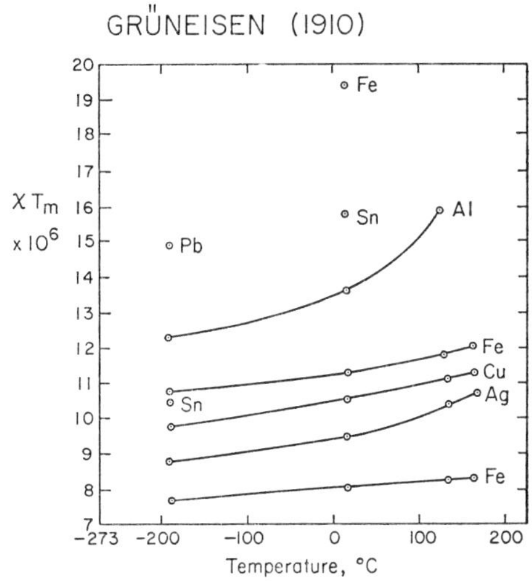

Fig. 3.107. GRÜNEISEN's (1910) results plotting the product of the compressibility and the melting point against the temperature, °C, for the various metals indicated.

comparison of the experimental values of K and E in Eq. (3.21) and in comparison of E and μ in Eq. (3.20). Introducing the four temperature coefficients k_i, namely,

$$k_\chi = \frac{1}{\chi}\frac{\partial \chi}{\partial t}, \qquad k_E = -\frac{1}{E}\frac{\partial E}{\partial t}, \qquad k_\mu = \frac{1}{\mu}\frac{\partial \mu}{\partial t}, \quad \text{and} \quad k_\nu = \frac{1}{\nu}\frac{\partial \nu}{d t}, \tag{3.22}$$

where t is the temperature, one obtains the two conditions

$$k_\chi = k_E - \frac{3K}{\mu}(k_\mu - k_E) \tag{3.23}$$

and

$$\frac{1}{\nu}\frac{\partial \nu}{\partial t} = \frac{1+\nu}{\nu}(k_\mu - k_E) = \frac{1+\nu}{\nu}\frac{\mu}{3K}(k_\mu - k_\chi). \tag{3.24}$$

From his own measurements at room temperature of all four of the elastic constants, only two of which were independent for the isotropic solid, GRÜNEISEN provided the tabulation of the coefficients of Eqs. (3.23) and (3.24) in Table 103.

From his data GRÜNEISEN noted that the compressibility χ increased with temperature in every instance considered, and that in the data of BOCK[1] (1894),

Table 103. GRÜNEISEN (1910).

	$\frac{1+\nu}{\nu}$	$\frac{3K}{\mu}$	$\frac{1+\nu}{\nu}\cdot\frac{\mu}{3K}$
Al	3.94	8.2	0.48
Fe	4.57	6.0	0.76
Cu	3.86	8.8	0.44
Ag	3.63	11.4	0.32
Sn	4.03	7.8	0.52
Pt	3.56	12.2	0.29
Pb	3.22	26.7	0.12

[1] BOCK [1894, *1*].

which I have described above in Sect. 3.25, POISSON's ratio was either independent of temperature, as for iron, or a slightly increasing function of temperature, as for some other metals which BOCK had considered. Therefore, from Eqs. (3.23) and (3.24) and the numerical values in Table 103, GRÜNEISEN stated a set of required conditions among the temperature coefficients for isotropic infinitesimal linear elasticity. These were:

(1) $k_\mu > k_E > k_\chi$ with $k_\mu = k_E = k_\chi$ as a limiting case;

(2) $k_\mu - k_E$ is only a small fraction (between $^1/_6$ and $^1/_{12}$) of $k_E - k_\chi$; it is thus a still smaller fraction of k_E;

(3) the temperature coefficient of ν is approximately four times the difference $k_\mu - k_E$, and only a fraction $\frac{1+\nu}{\nu}\,\frac{\mu}{3K}$ of the difference $k_E - k_\chi$, i.e. a larger fraction of k_E than is $k_\mu - k_E$.

The coefficients shown in Table 103 are for room temperature. GRÜNEISEN did not discuss whether they applied over the range of temperatures in which he had determined coefficients, nor did he pose the perhaps more fundamental question as to what had led him to expect the relations for isotropic materials to apply at room temperature and not at other temperatures in the range he considered.

With the data available in 1910 on elastic constants as functions of temperature, data which usually was presented in the form of a percentage decrease between two specified temperatures, in general the value of $k_\mu - k_E$ was found to be larger than permitted by statement (2), particularly for the solids with high melting points, namely, platinum and iron, since for them $k_E - k_\chi$ was small. GRÜNEISEN thought the calculation of the temperature coefficient of POISSON's ratio was rather worthless in view of the smallness of $k_\mu - k_E$. He noted that N. KATZENELSOHN's[1] and CLEMENS SCHAEFER's[2] experimental values were larger than would be permitted by statement (3). For example, for platinum and iron, for which E changed between 0 and 100° C by 2 and 3.5%, respectively, the change of ν in the same interval should have been smaller than 0.6 and 2.6%, while KATZENELSOHN had given the much higher values for POISSON's ratio of 3.07 and 3.80%. Although BOCK's[3] data on steel and nickel at 20 to 150° C were in agreement with GRÜNEISEN's requirement that POISSON's ratio be almost independent of temperature, for copper and silver the increase was such as to violate statement (3).

Because of the nature of the available data for the four elastic constants and the limitations in GRÜNEISEN's argument, the significance of this work of GRÜNEISEN is confined to the fact that he was the first to introduce a subject which, in the light of the observed contradictions, certainly should have stimulated more incisive analysis and experiment. That no such analysis and experiment did follow is largely due to the fact that during the past sixty years, experimental physicists, whether interested in anisotropy, photoelasticity, or the determination of the temperature dependence of elastic constants in crystalline solids by means of ultrasonics, etc., have not seriously questioned the applicability of the linear theory which underlay the interpretation of the data.

GRÜNEISEN must be credited also with being the first person since WERTHEIM to measure all four elastic constants of isotropic materials: E, μ, ν, and K. Lest one too casually compare these early data with the ultrasonic measurements of the past twenty years, it should be emphasized that GRÜNEISEN's measurements, like those of WERTHEIM, were made over relatively large strain amplitudes, for

[1] KATZENELSOHN, Inaug. Diss., Berlin, 1887.

[2] SCHAEFER [1901, *1*], [1902, *1*].

[3] BOCK [1894, *1*].

which GRÜNEISEN himself, among others, had demonstrated nonlinearity in small deformation. Ultrasonic measurements, having amplitudes in the range of 10^{-7}, i.e., zero stress moduli, present quite a different problem: in wave propagation, nonlinearity will appear as a change in the form of the wave profiles; in steady state vibrations, the nonlinearity gives rise to ultraharmonics. However, with respect to temperature, the questions introduced by GRÜNEISEN for quasi-static deformation also are pertinent for ultrasonic propagation of waves with amplitudes many orders of magnitude lower.

3.42. A comparison of ultrasonic and quasi-static temperature coefficients.

A correlation may or may not be accomplished when one contrasts the temperature dependence of elastic constants determined by ultrasonics at near zero stress with that for quasi-static tests at much larger strain. I have chosen to examine the experiments of CHARLES ZUCKER,[1] who in 1955 determined longitudinal and shear wave speeds in 1100 F polycrystalline aluminium (99.0% purity) in the temperature range from 20 to 400° C.

At one end of a 2 ft. long, 1 in. diameter, polycrystalline rod, ZUCKER introduced a 5 μsec long train of 5 megacycle sine waves at 5 mmsec intervals. He used either X-cut or Y-cut quartz crystals, and recorded longitudinal and transverse wave speeds. Since he encountered difficulties in cementing the crystals at higher temperatures, this end of the specimen he cooled in a water-bath with a temperature gradient to a constant temperature section at the far end of the specimen. A 3 in. longitudinal cut in the specimen at that end produced an initial reflection whose timing, compared with that from the free end, he could determine by synchronized oscillography to obtain wave speeds in the constant temperature region. The 18 in. long furnace had a 4 in. constant temperature region with a stated chrome-alumel thermocouple measured accuracy of ± 1° C. In computing density variation with temperature, he assumed an initial density of 2.70 g/cm² and a linear expansion coefficient of 25.3×10^{-6} per degree Centigrade. The stated overall accuracy in the wave speed determination was 2.0%.

Ignoring what to an experimentist in continuum mechanics is a somewhat awkward choice of simulated single crystal terminology based on a proposed averaging of the "true" elastic constants of the single crystal itself, an unnecessary procedure in view of the fact that the specimens were polycrystalline and therefore would give polycrystalline values of K and μ if the solid were statistically isotropic, we may proceed to compare ZUCKER's results at room temperature with those of GRÜNEISEN, using MALLOCK's method of 45 years earlier. We have $\varrho V_L^2 = K + \frac{4}{3}\mu$ and $\varrho V_T^2 = \mu$ for the two polycrystalline wave measurements[2] and also for the isotropic solid, since POISSON's ratio is given by

$$\nu = \frac{(V_L^2/V_T^2 - 2)}{2\,(V_L^2/V_T^2 - 1)}. \qquad (3.25)$$

Hence, measuring from the graphs since no tabulated data were given, we find at 25° C, values of K, μ, and ν of 7540 kg/mm², 2690 kg/mm², and 0.341, respectively.

Since isotropy was assumed in determining the value of POISSON's ratio, the same assumption for the calculation of E for 25° C (in ZUCKER's data) provides $E = 7220$ kg/mm². The dynamically measured compressibility $\chi = 1/K$ becomes 1.33×10^{-4} cm²/kg. Comparing these dynamically determined polycrystalline

[1] ZUCKER [1955, *1*].

[2] ZUCKER gave these as $\varrho V_L^2 = C_{11}$ and $\varrho V_T^2 = C_{44}$.

numbers at 25° C with the quasi-static values of GRÜNEISEN at 18° C which he determined by MALLOCK's method on hollow tubes of polycrystalline aluminum, we see that GRÜNEISEN observed the higher compressibility of 1.46×10^{-4} cm²/kg and the measured value of E in the tube of 7330 kg/mm². From the direct measurement of E and ν in an aluminum rod, GRÜNEISEN gave values of 7200 kg/mm² and 0.34, respectively. These *quasi-static* moduli were nearly identical with those obtained *dynamically* from ZUCKER's data on aluminum rods. ZUCKER's value for the POISSON's ratio was calculated from measured E and μ using Eq. (3.20) which holds for the linearized theory of isotropic materials, whereas GRÜNEISEN's ν was measured directly. GRÜNEISEN's calculation of POISSON's ratio from his measured E and μ, provided the slightly smaller value of $\nu = 0.32$.

KÖSTER and RAUSCHER[1] in 1948 provided a carefully determined E obtained from aluminum bars in flexural vibration; their result, $E = 7220$ kg/mm² at 20° C, is precisely the value calculated from ZUCKER's data.

During the past two decades, I often have measured E quasi-statically and inferred its value from measured wave speeds in polycrystalline aluminum. My specimens presumably were similar to ZUCKER's, since the material also was supplied as a gift by the Aluminum Corporation of America at the same time, and it had the same purity specifications and diameter. From a large number of measurements[2] at 25° C, the average value of the data for this aluminum (which will be referred to in Sect. 3.44 in more detail) was $E = 7180$ kg/mm². As a further check upon this particular set of elastic constants of ZUCKER at room temperature, for the purpose of exploring the temperature dependence, the ultrasonically measured value of the shear modulus μ provided by ZUCKER, $\mu = 2690$ kg/mm², may be compared with that predicted from the discrete distribution of zero point isotropic shear moduli of the elements which I recently discovered.[3] Measurement and prediction are in precise agreement.

As I described in considerable detail,[4] for the fractional melting point temperatures, $T/T_m < 0.4$, the shear modulus was given by the linear relation, $\mu = 1.03\,\mu(0)\,(1 - T/2T_m)$. For 25° C, at which $T/T_m = 0.32$, the calculated, stable, zero point, shear modulus for isotropic aluminum in this discrete distribution was 3110 kg/mm², which was close to that determined by introducing ZUCKER's value, i.e., $\mu = 2690$ kg/mm². It should be emphasized that my calculated value was based upon the analysis of a great many moduli measured in nearly 60 elements and binary combinations.

Thus we see that in the latter part of the 20th century, consistency of numbers in data at room temperature has been achieved despite the variety of experimental methods employed. We may note that at least in aluminum, numerical values for μ determined from quasi-static tests and from wave propagation measurements were usually very close. While agreeing that quasi-static measurement is isothermal, many experimentists and theorists tacitly assumed that wave propagation was adiabatic and thus should yield a somewhat higher value for E. However reasonable this assumption may seem, it has yet to be experimentally demonstrated.

Turning to the temperature dependence itself, we may nonetheless compare ZUCKER's[5] shear modulus as a function of temperature (circles) determined in

---

[1] KÖSTER and RAUSCHER [1948, *1*].

[2] BELL [1968, *1*]. This monograph included experimental results from my laboratory extending back to the early 1950's.

[3] *Ibid.*, Chap. V, p. 141.

[4] BELL [1968, *1*].

[5] ZUCKER [1955, *1*].

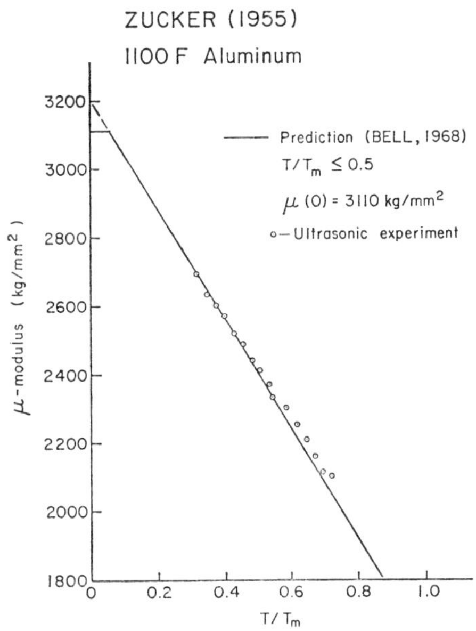

Fig. 3.108. ZUCKER's ultrasonically determined μ moduli as a function of temperature (circles), compared with prediction by BELL (solid line), Eq. (3.27), p. 396 below.

wave propagation, with my prediction[1] (solid line) shown in Fig. 3.108. Through this prediction, ZUCKER's results in aluminium thus are contrasted with the measured moduli of 57 elements and 2 binary combinations.

It is not known whether or not the small departures seen at the higher fractional melting points arose from the fact that above 200° C, i.e., $T/T_m = 0.57$, the technique had been modified in the manner described above in order that the quartz crystal could be attached to the specimen. The data below $T/T_m < 0.5$ were certainly in close agreement with my later prediction. In Fig. 3.109, E calculated from ZUCKER's determination of the bulk modulus and the shear modulus by means of the formula for isotropic materials, namely,

$$E = \frac{\varrho V_t^2 (3 V_e^2 - 4 V_t^2)}{V_e^2 - V_t^2}, \tag{3.26}$$

are compared with my later prediction, for which a constant POISSON's ratio with temperature was assumed. This comparison shows a very close agreement at room temperature, with the same small departures at higher temperatures. In the figure too, are values of E at various temperatures from the study by KÖSTER and RAUSCHER[2] for which, as was shown above, there was precise agreement between the two sets of data at room temperature.

GRÜNEISEN had determined compressibilities at -191° C, 15° C, and 125° C; the three values were 1.32×10^{-6}, 1.46×10^{-6}, and 1.70×10^{-6}, respectively,[3] as may

[1] BELL, *op. cit.*

[2] KÖSTER and RAUSCHER [1948, *1*].

[3] The value at 15° C is considerably higher than BRIDGMAN's corrected coefficient for the linear term: 1.397×10^{-6}. It is interesting that in 1915 RICHARDS, whose results were known to be in error, provided for aluminum an average compressibility value of 1.47×10^{-6}. Since this was nearly the same as the earlier value of GRÜNEISEN, it must have given RICHARDS a degree of confidence in the inaccurate number.

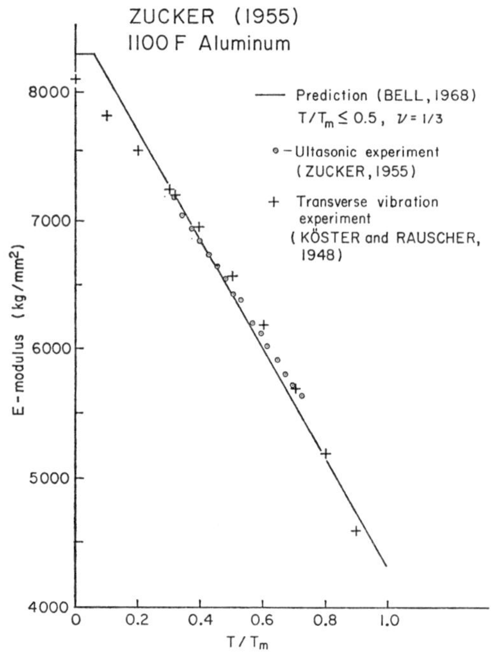

Fig. 3.109. ZUCKER's (1955) ultrasonically determined E (circles) and KÖSTER's (1948) values determined from transverse vibration experiments, plotted against the fractional melting point. These data are compared with Eq. (2.42) [p. 151] for polycrystalline aluminum.

be seen from consulting Table 102. ZUCKER's compressibility at 25° C has a considerably lower value of 1.325×10^{-6}, but this number is in far better agreement with BRIDGMAN's more accurate linear term of 1.397×10^{-6}. It is of interest to note that the dynamic value is somewhat unexpectedly below, rather than above, the quasi-static determination. ZUCKER's dynamically determined compressibility at 125° C was 1.44×10^{-6}. The lower numbers might be attributed to differences in the composition of the aluminum considered. However, as in the data of KÖSTER, the temperature ratios were not in close accord.

Thus, one may conclude from this comparison, which covers determination of the same quantities over a period of more than 45 years and by distinctly different experimental techniques, that in the 20th century greater precision in the determination of elastic constants has been achieved, and the existence of this precision now permits a close study of the fundamental questions which GRÜNEISEN first had raised.

3.43. On temperature dependence of elastic constants and damping coefficients, after 1910.

By the end of the second decade of the 20th century, major divisions of specialized interest and motivation began to be emphasized. The temperature dependence of elastic parameters offers as good an example as any of the trend toward model-oriented, compartmentalized studies which is still in progress. The development of the steam turbine, the gas turbine, the internal combustion engine, and now, space technology, for example, with their ever increasing temperatures and pressures turned one group toward the experimental study of specialized complex metal alloys with temperature coefficients and internal damping properties suitable for technological use. A second group, with but a small interest in mechanics *per se*, studied the temperature dependence of single crystal coefficients so as to compare the results with solid state models at 0° K, or to obtain numerical values of wave speeds for calculating DEBYE temperatures and examining proposed models in the field of specific heats of solids. A third group became concerned with at least semi-quantitative shear moduli data in single crystals of various structures and prior histories together with corresponding elastic limits or critically resolved shear stresses as a tool to explore thermally activated dislocation models with the hope of revealing the role of crystal imperfections in many areas. A fourth group, also technologically oriented, concentrated on dimensional complexity in the form of practical or semi-practical boundary value problems and the approximate analytical models which it was hoped would adequately conform with behavior under anticipated full scale applied loading histories. Finally, there have been those few experimentists, such as WERNER KÖSTER, who retained the earlier interest in large definitive experimental studies, in the generalized framework of which the physical applicability of any relevant theoretical formulations past or present might be considered. In rational mechanics itself, with its important emphasis upon generalized nonlinear response, the theoretical developments of the past two decades, while apparently offering a major opportunity to the experimentist, in fact have dealt only in a few instances with the bounded and controllably loaded solids upon which experiment necessarily is performed.

As representatives of the first group of technologically oriented studies, such as those of EVERETT and MIKLOWITZ,[1] or GAROFALO, MALENOCK, and SMITH,[2] already described, could be added the measurements of FREDERICK CHARLES LEA and O. H. CROWTHER[3] in 1914, who conducted quasi-static tensile experiments on mild steel from 20 to 600° C. In 1922, LEA[4] extended these investigations to Armco iron, four different percentages of carbon steel, four nickel-chrome and nickel-chrome-vanadium steels, five different commercial bronzes of stated chemical composition, and an aluminum-copper alloy used for pistons in aircraft.

The temperature dependence of E and μ moduli in dead weight loading quasi-static tests of course was only one aspect of LEA's interest in his temperature measurements by means of an electrical furnace. His mirror equipped extensometer allowed strain to be determined by telescopes outside the furnace. He also considered the temperature dependence of elastic limits, ultimate strength,[5] ductility, and final elongation. Figs. 3.110 and 3.111 show tensile tests in Armco iron, 0.5% carbon steel, and nickel-chrome-vanadium wire, while Fig. 3.112 contains dead weight torsion moduli as a function of temperature. LEA noted the major differences in behavior as a function of chemical composition including the "evidence of critical points at certain temperatures" for the torsion moduli data. From his discussion of the tabulated results for 16 solids (a 95.5% nickel alloy was one of them), he concluded that commercial use of these metals up to 250° C was permissible. In the context of the present discussion, however, the complex deformation in the presence of alloying components suggests that one should attempt

[1] EVERETT and MIKLOWITZ [1944, *1*].

[2] GAROFALO, MALENOCK, and SMITH [1952, *1*].

[3] LEA and CROWTHER [1914, *1*].

[4] LEA [1922, *1*].

[5] In view of the long discussion of WERTHEIM's temperature anomaly in the elastic constants of iron, it is perhaps worth noting that LEA in 1912, like MANJOINE and NADAI in 1940, found a similar phenomenon in a maximum of the ultimate strength at 200° C.

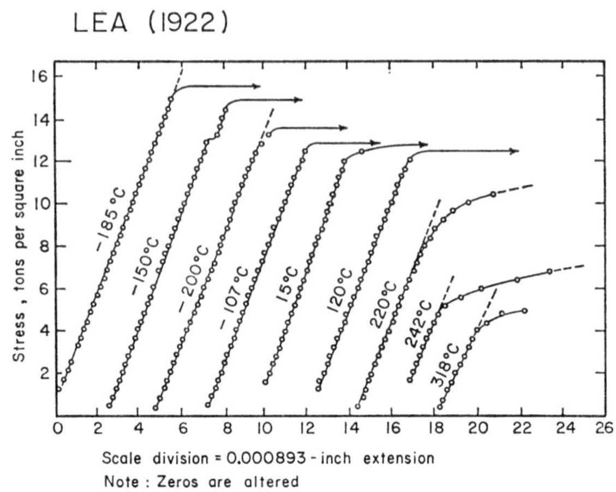

Fig. 3.110. Tensile tests on Armco iron at various temperatures.

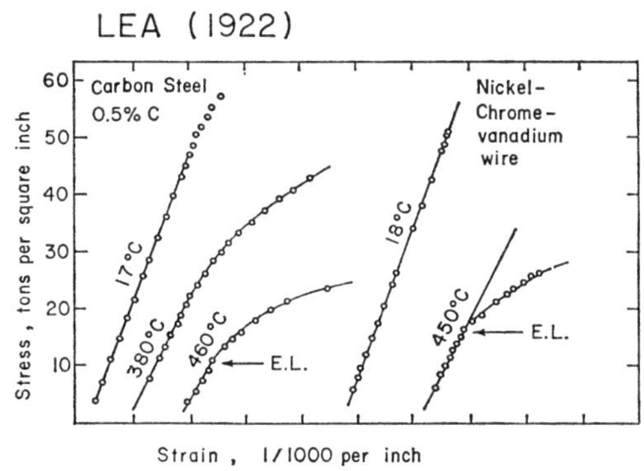

Fig. 3.111. Stress-strain curves at various temperatures for steel wires.

first to understand simpler solids before trying to correlate such heterogeneous details into fundamental theories of solids.

C. BACH and R. BAUMANN[1] in 1921 performed dead weight tests on mild steel and carbon steel from 20 to 500° C and observed a major drop at 400° C which followed the usual linear decrease of E with temperature. FRANKLIN EVERETT,[2] in his doctoral dissertation in 1931 at the University of Michigan, described results like these for the μ-modulus of steel, also obtained from dead weight testing. He observed a sudden decrease of modulus, but it occurred at the somewhat lower temperature of 300° C.

E. HONEGGER[3] in 1932 in a similarly technologically oriented study using a rotating mirror extensometer and a special Amsler testing machine performed tensile tests for E at loading rates designated as "quick" and "slow". He obtained dynamic values of E by causing a simple beam loaded by a central mass to vibrate

[1] BACH and BAUMANN [1921, *1*].

[2] EVERETT [1931, *1*]. EVERETT was a doctoral student of Professor STEPHEN PROKOFIEVITCH TIMOSHENKO.

[3] HONEGGER [1932, *1*].

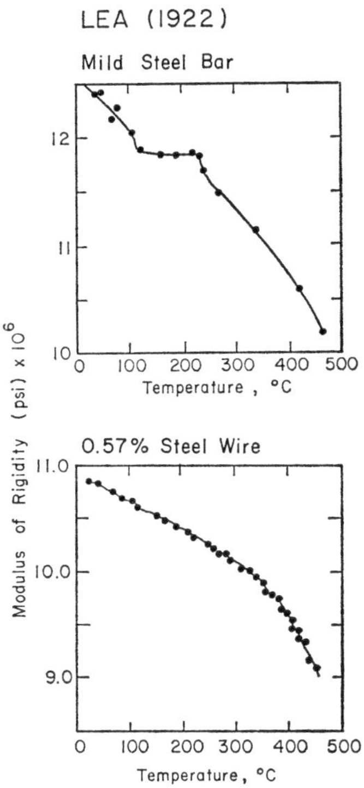

Fig. 3.112. The variation of the shear modulus with temperature for a mild steel bar and for a 0.57% steel wire, showing the differences which can be observed.

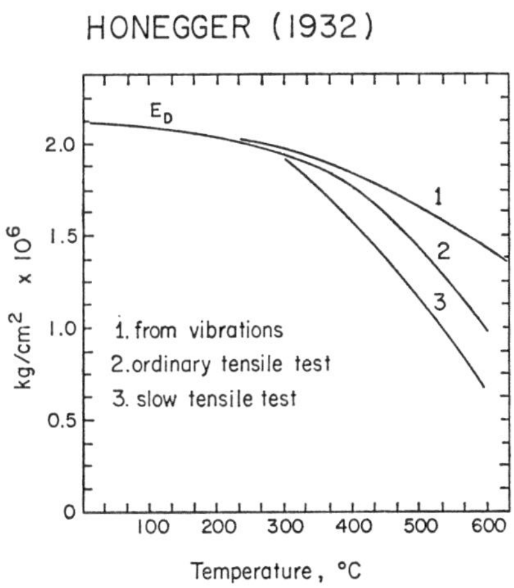

Fig. 3.113. A comparison of modulus of elasticity of stainless steel determined in the three manners indicated, showing the difference observed with increasing temperatures.

laterally. Above the 300° C critical temperature he observed that E varied with the loading rate in the manner shown in Fig. 3.113. The two quasi-static tests differed by the fact that $1^1/_2$ min instead of 30 sec elapsed before applying load increments; hence the rates of loading varied only in the sense of the average time.

A curious fact was revealed, namely that the results from the dynamic vibration tests were consistently smaller at room temperature. As usual the difference was attributed to nonhomogeneity of the material and the possibility that simple supports were not achieved for the vibrating bar.

In 1921 TIMOSHENKO[1] further modified the simple theory of the vibrating beam to include not only the rotary inertia, which had been introduced late in the 19th century, but also the deflection due to shear.

In 1931 KARL ANTON ERICH GOENS,[2] whose primary interest was to study the ramifications of atomistic models for the solid state, performed careful flexural vibration experiments on three cylindrical steel rods of different dimensions and on a prismatic rod of aluminum ($7\times10\times250$ mm). He found such close agreement between the extended theory and experiment in his analysis of frequencies to the fifth mode of vibration, for which he used an electronic tone transmitter for comparison, that he concluded he had found a method suitable for the precise determination of elastic constants. For a better evaluation he suggested the desirability of reviewing earlier measurements of the same type, including his own and those of GRÜNEISEN. Using the same method, GOENS and J. WEERTS[3] in 1936 made an extensive series of measurements on copper, gold, and lead at room temperature. The most important application of the new analysis is that described by GOENS in a paper[4] in 1940 on the temperature dependence of single crystal elastic constants of copper, gold, lead, and aluminum, between $-253°$ C and 50° C. These low temperature data still are quoted widely in the current literature, chiefly because of GOENS' interest in comparing extrapolated 0° K values with the calculations from quantum mechanics made by KLAUS FUCHS[5] in 1936.

GOENS also was interested in both the flexural and torsional vibration of single crystals so oriented as to enable him to calculate wave speeds from which to determine DEBYE temperatures at $+20°$ C and $-273°$ C. He clamped the rods centrally and cemented on the two end masses shown in Fig. 3.114. He determined frequencies in the audible range in both flexure and torsion by superposition on a standard tone transmitter.

GOENS (1940)

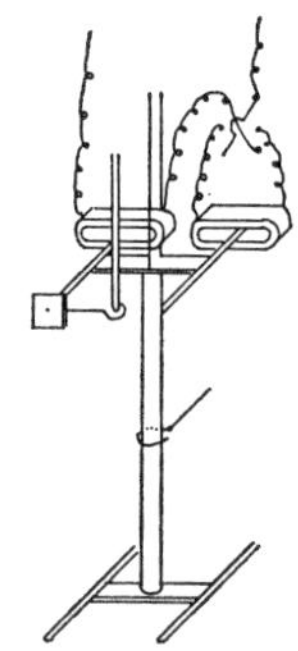

Fig. 3.114. The experimental apparatus of GOENS (1940) for the determination of elastic constants in both torsion and flexure.

[1] TIMOSHENKO [1921, *1*].
[2] GOENS [1931, *1*].
[3] GOENS and WEERTS [1936, *1*].
[4] GOENS [1940, *1*].
[5] FUCHS [1936, *1*].

The observed linearity with temperature of the elastic compliances s_{ij} suggested the feasibility of an extrapolation from $-253°$C to $-273°$C in each instance. Both these compliances and the elastic constants c_{ij} are shown in Table 104. FUCHS' calculation for copper had provided values of $c_{11}=17.5\times 10^{-11}$ dynes/cm² and $c_{12}=12.4\times 10^{-11}$ dynes/cm² at $-273°$C, which GOENS regarded as a quite satisfactory correlation.

Table 104. GOENS (1940).

Crystal	Temperature (°C)	Main elasticity constants at +20° and −273°					
		s_{ik} (10^{-13}) cm²/dyne			c_{ik} (10^{11}) dynes/cm²		
		s_{11}	s_{12}	s_{44}	c_{11}	c_{12}	c_{44}
Cu	+ 20	14.91	− 6.25	13.28	17.00	12.3	7.53
	−273	13.56	− 5.65	12.33	18.2	13.0	8.1
Au	+ 20	20.30	10.65	23.80	18.6	15.7	4.20
	−273	21.39	− 9.73	21.95	19.4	16.2	4.55
Pb	+ 20	93.0	−42.6	69.4	4.83	4.09	1.44
	−273	67.5	−31.0	52.8	6.7	5.7	1.89
Al	+ 20	15.90	− 5.80	35.16	10.8	6.22	2.84
	−273	14.20	− 5.00	31.46	11.4	6.20	3.18

Precisely a century elapsed between the pioneering investigation of elastic properties of solids by WERTHEIM in the 1840's and the culminating summaries of WERNER KÖSTER in the 1940's. KÖSTER, who depended primarily upon a precise experiment on flexural vibration, had the advantage of an extended theory for ascertaining in his studies of the fundamental mode, the magnitude of the nearly negligible contributions of rotary inertia and shear deflection. He determined E for over 30 elements, compared to WERTHEIM's 11, and for 59 binary alloys, compared to WERTHEIM's 64. The interesting difference, particularly with respect to the alloys, is the great increase in collateral information regarding crystal structure and phase phenomena, which permitted KÖSTER to sort and correlate his results from his more precisely prepared specimens and more accurately determined frequencies of vibration. WERTHEIM, with the first experimental studies of the dependence on temperature, was confined to quasi-static measurements, between $-15°$C and 100°C, and for only a few elements; KÖSTER's dynamic studies, in a wide range of solids, extended from $-185°$C to 1000°C. Both considered the continuum vs atomistic correlations or lack of them, the POISSON's ratio averages for solids, and magnetic effects upon elastic moduli where relevant. The parallel, however, lies not so much in the fact that both provided large definitive studies of the subject but that each viewed the patterns of physical behavior which emerged as a basis for objective reviewing of projected and accepted hypotheses and theories, rather than as limited studies to support some chosen explanation.

The basis of KÖSTER's experiments of interest here was the ingenious apparatus described by FRITZ FÖRSTER[1] in 1937. The objective was to suspend a specimen by means of fine wires in such a manner that energy losses to supports or coupling between supporting apparatus and specimen should become truly negligible. A variety of supporting configurations was developed, allowing flexural, torsional, or even extensional vibrations of parallelepipeds or cylinders in either driven or free vibration. One of the ends of each of the supporting wires was fixed, and the

[1] FÖRSTER [1937, *1*].

other was attached to the moving mechanical portion of an electromagnetic transducer. One system served as the driver for forced vibration, and the other as receiver. Free oscillation and damping also could be determined. The paper contained a detailed description of the several configurations considered, and an extensive study of the many problems encountered in achieving a precise measurement not only of the elastic modulus E but also of the magnitude of the resonant damping, both as functions of ambient temperature.

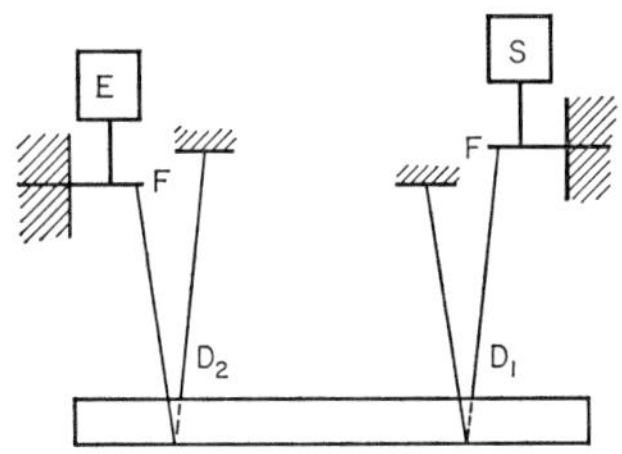

Fig. 3.115. The manner in which Förster (1937) supported his specimen in his vibration determination of elastic constants.

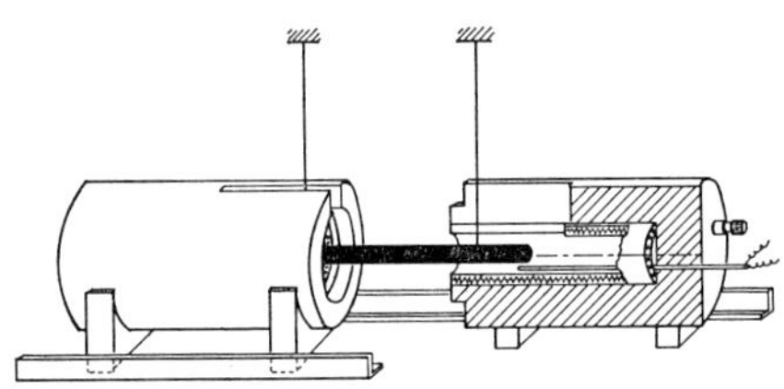

Fig. 3.116. The heating apparatus used by Förster (1937) in determining by means of vibration the variation of the E modulus with temperature.

The horizontal flexural vibration system is shown in Fig. 3.115, where E and S are the transducer elements. The furnace apparatus shown in Fig. 3.116 permitted temperature measurements over a wide range.

Leaving the interested reader to consult Förster's paper for details of the method of producing vibrations of extremely small amplitude in a sensibly support-free system, I shall point out a paper by Förster and Köster, also in 1937, which described the first series of experiments on E-moduli and on resonant damping properties of eleven pure metals and a number of binary alloys.[1]

Köster's earlier interest in the physical behavior of solids as a function of ambient temperature was evidenced[2] by his study in 1935 of the temperature dependence of magnetic and electrical parameters from room temperature to 1000° C. After the development of Förster's experiment on the vibration of beams, Köster used it to measure E for 32 of the purest metals and 59 binary combinations of elements in the temperature range from −180° C to 1000° C. In addition, at each temperature he measured the damping coefficient in the form of $\delta = \ln A_0/A_1$, i.e., the natural log of the ratio of successive amplitudes. The vast array of data from these two measured values, E and δ, was published in a

[1] Förster and Köster [1937, *2*].
[2] Köster [1935, *1*].

Table 105. FÖRSTER and KÖSTER (1937).

Metal	Elasticity modulus E (kg/mm²)	Damping δ ($\times 10^4$)	
Iron	21700	5.6	$\frac{1}{2}$ h @ 930°, aircooled
Nickel	21900	72.1	annealed $\frac{1}{2}$ h @ 700°
Copper	12820	35.5	annealed $\frac{1}{2}$ h @ 400°
Molybdenum	59100	5.1	sintered, annealed 1 h @ 900°
Aluminum	7230	0.46	99.99% Al, annealed $\frac{1}{2}$ h @ 550°
Magnesium	4530	2.1	99.99% Mg, annealed $\frac{1}{2}$ h @ 550°
Zinc	13130	7.7	99.99% Zn, test rod annealed 1 h @ 200°
Cadmium	6250	11.0	cast
Bismuth	3480	17.6	cast
Lead	1450	45.7	cast
Tin	4560	54.2	cast

series of papers between 1937 and 1948 which constitute in the 20th century the major source of polycrystalline moduli determined from the vibration of isotropic solids.

The first paper in this series was by FÖRSTER and KÖSTER.[1] It described the initial experiments, giving the values of E and δ, at room temperature, shown in Table 105 for the specified prior histories and annealing temperatures.

The temperature dependence of E and the damping coefficient presented in that initial study are shown in Fig. 3.117.

KÖSTER's interest in those parameters extended beyond the mere recording of measurements as such. He studied the discontinuities which occurred at thermally dependent polymorphic transitions, CURIE points, where appropriate, and such parameters as purity, grain size, and ordering of atomic distribution for elements

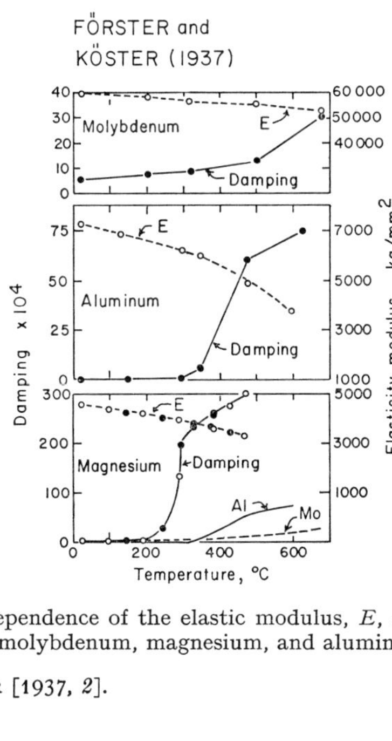

Fig. 3.117. Temperature dependence of the elastic modulus, E, and the damping coefficient of molybdenum, magnesium, and aluminum.

[1] FÖRSTER and KÖSTER [1937, 2].

Table 106. KÖSTER (1943). *Elastic characteristics of pure metals.*

1 Metal	2 χ 10^4cm²/kg	3 K (kg/mm²)	4 E (kg/mm²)	5 K/E	6 ν	7 ν (measured)	8 G (kg/mm²)
Li	7.2[b]	1390	1170[b]	1.19	0.36[b]	—	430
Na	12.0[b]	832	910[b]	0.92	0.32[b]	—	340
K	24.6[b]	406	360[b]	1.13	0.35[b]	—	130
Rb	52.0	190	240	0.8	0.3	—	—
Cs	70.0	140	175	0.8	0.3	—	—
Be	0.855	11700	29280	0.40	0.08	—	13500
Mg	2.952	3390	4515	0.75	0.28	—	1770
Ca	5.697	1750	2000	0.87	0.31	—	750
Sr	8.187	1220	1600	0.76	0.28	—	620
Ba	10.19	980	1290	0.77	0.28	—	500
Al	1.34	7460	7220	1.03	0.34	0.34[a]	2720[a]
La	3.513	2840	3820	0.74	0.28	—	1500
Ti	0.797	12500	10520	1.2	0.36	—	3870
Zr	1.097	9100	6970	1.3	0.37	—	2540
Hf	0.901	11100	8500	1.3	0.37	—	3100
Th	1.818	5500	7970	0.69	0.26	—	3160
V	0.609	16400	15000	1.1	0.35	—	5500
Nb	0.570	17500	16000	1.1	0.35	—	6000
Ta	0.479	20800	18820	1.1	0.35	—	7000
Cr	0.600	16600	19000	0.87	0.31	—	7300
Mo	0.347	28800	33630	0.85	0.31	—	12200[d]
W	0.293	34000	41520	0.82	0.30	0.28[e]	15140[e]
Mn	0.791	12600	20160	0.63	0.24	—	7800
Fe	0.587	17000	21550	0.79	0.29	0.29[a]	8280[a]
Co	0.539	18500	20380[f]	0.90	0.31	—	7630[f]
Ni	0.529	18900	19700	0.92	0.32	0.31[a]	7500[f]
Ma	0.34	29000	41500	0.70	0.26	—	16500
Re	0.27	37000	53000	0.70	0.26	—	21000
Ru	0.342	29000	44000	0.66	0.25	—	17600
Os	0.26	38000	57000	0.65	0.25	—	22800
Rh	0.372	27000	38640	0.70	0.26	—	15300
Ir	0.268	37300	53830	0.69	0.26	—	21400
Pd	0.528	19000	12360	1.54	0.39	0.39[a]	4450
Pt	0.360	27800	17320	1.59	0.39	0.39[a]	6220[a]
Cu	0.719	13900	12500	1.11	0.35	0.34[a]	4640[a]
Ag	0.987	10100	8160	1.25	0.37	0.38[a]	2940[a]
Au	0.577	17300	7900	2.19	0.42	0.42[a]	2820[a]
Zn	1.66	6000	9400	0.64	0.24	0.29[g]	3790
Cd	1.99	5000	6350	0.79	0.29	0.30[a]	2460
Hg	3.4	2900	—	—	—	—	—
Ga	~2.0	~5000	1000	5.0	0.47	—	430
In	2.5	4000	1070	3.7	0.45	—	380
Tl	3.48	2900	810	3.6	0.45	—	280
Si	0.41	24400	11500[i]	2.1	0.42	—	4050
Ge	1.41	7100	8000	0.9	0.32	—	3000
Sn	1.87	5300	5500	0.97	0.33	0.33[a]	2060
Pb	2.37	4200	1600	2.6	0.44	0.44[a]	570[h]
Sb	1.8	5550	5600	0.99	0.33	—	2000[i]
Bi	2.92	3400	3480	0.98	0.33	0.33[a]	1310

[a] E. GRÜNEISEN: Ann. Physik (4) **25**, 825 (1908).
[b] O. BENDER: Ann. Physik (5) **34**, 359 (1939); the values are for −190° and are (Na, K) calculated from single crystal data.
[c] R. L. TEMPLIN: Metals and Alloys **3**, 136 (1932).
[d] M. J. DRUYVESTEYN: Physica **8**, 439 (1941).
[e] S. J. WRIGHT: Proc. Roy. Soc. (London), Ser. A **126**, 613 (1930).
[f] K. HONDA and T. TANAKA: Sci. Rept. Tohoku Imp. Univ. **15**, 1 (1926).
[g] H. SIEGLERSCHMIDT: Z. Metall. **24**, 55 (1932).
[h] Physikalisch-Technische Reichsanstalt: Z. Metallk. **12**, 179 (1920).
[i] P. W. BRIDGMAN: Phy. Rev. (2) **9**, 138 (1917).

and binary alloys. Many of those details are of more interest for metallurgy than for continuum mechanics, but certainly here we should stress the fact that KÖSTER did provide numerical values of the two recorded parameters in terms of such various inputs.

In 1943 KÖSTER[1] gave a valuable summary of measured K, μ, E, and ν at room temperature, obtained from an analysis of the literature as well as from his own data. This list for 49 elements comprises the accumulated information on the subject of isotropic elastic constants for the elements prior to the modern overwhelming emphasis upon ultrasonics. Table 106 includes KÖSTER's list. The values for ν of column 6 were calculated from the ratio of K/E of column 5, which, as KÖSTER emphasized, was far more sensitive than E/μ for such calculation.[2] All the remaining data were directly measured. (Those for ν in column 7 are GRÜNEISEN's data given above.)

Again in 1947 KÖSTER[3] provided a summary based upon an accumulation of the measurements of the previous years. The E-moduli for the 31 elements as a function of ambient temperature are those in the E vs T/T_m plots of Figs. 3.118 and 3.119 for the low and high melting point solids, respectively.

I have omitted KÖSTER's extrapolations of these data to $T/T_m=0$ and $T/T_m=1$, i.e., zero degrees and the melting point. He considered the ratio of the two

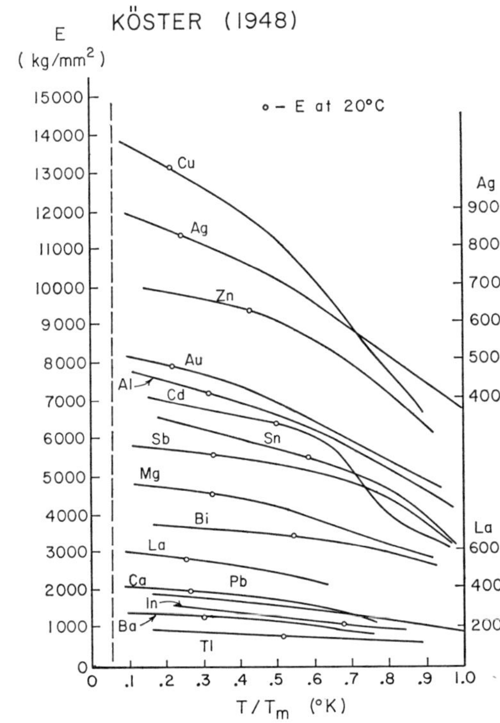

Fig. 3.118. The temperature dependence of the E moduli of low melting point elements.

---

[1] KÖSTER [1943, *1*].

[2] K/E varied from $^1/_3$ to ∞ and E/μ only from $^1/_2$ to $^1/_3$ when ν varied from 0 to $^1/_2$. More meaningfully, K/E varied from 0.66 to 3.5, while E/μ varied from 0.4 to 0.34 when ν varied from $^1/_4$ to 0.46.

[3] KÖSTER [1947, *1*].

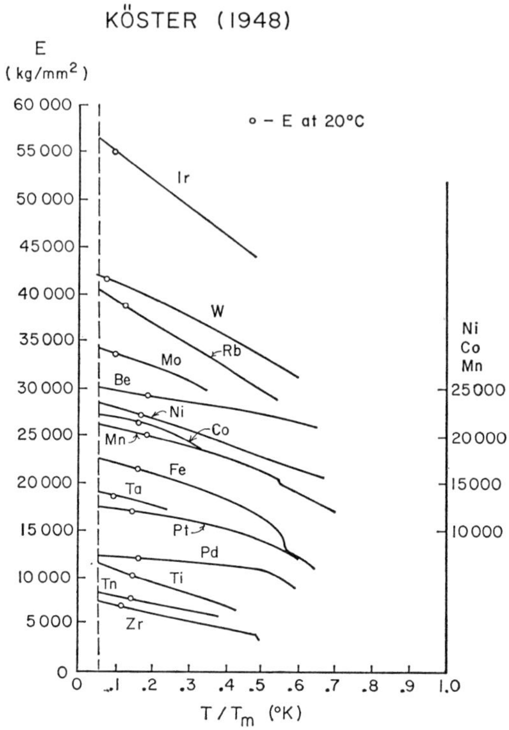

Fig. 3.119. The temperature dependence of the E moduli of elements having high melting points.

hypothetical moduli to be of considerable significance, but a simple plotting of his data without his extrapolations suggests otherwise. Because subsequent ultrasonic studies have shown the moduli to be constant below $T/T_m = 0.06$, and because modulus slopes, where measurable, were so variable at high values of T/T_m, I have omitted also KÖSTER's discussion of the behavior of extrapolated parameters at 0° K and T_m.

Many experimentists have noted that for temperatures below room temperature, the moduli depend linearly upon temperature. In Fig. 3.120 are shown KÖSTER's plots of dE/dT in kg/mm² at °C vs the ambient temperature in °C, which demonstrates this fact.

KÖSTER described the damping constants in his paper in 1948 on the behavior of pure metals.[1] He provided details of the specific origins and prior conditions of each specimen for the 32 elements considered. He also tabulated previous quasi-static and dynamic values of E for these elements, from the work of GRÜNEISEN[2] in 1907 and GUILLET[3] in 1939 for comparison with his own measurements. It is curious that nowhere in KÖSTER's work have I found any reference to his 19th century predecessor, WERTHEIM. This loss of historical contact should be noted particularly in view of the fact that in concept and objective KÖSTER essentially was repeating and extending WERTHEIM's research of a century earlier.

[1] KÖSTER [1948, *2*].
[2] GRÜNEISEN [1907, *1*].
[3] GUILLET [1939, *1*].

KÖSTER (1948)

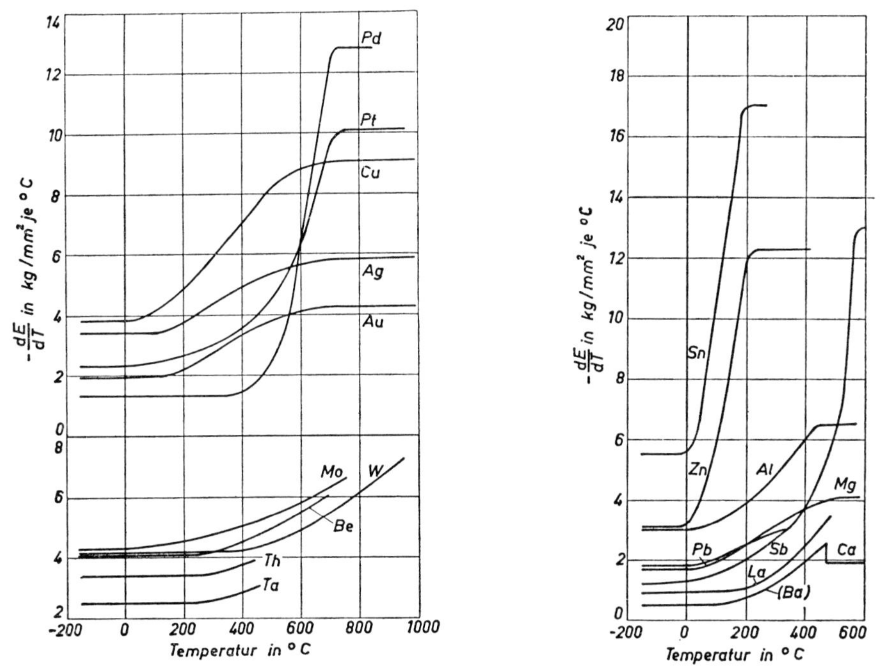

Fig. 3.120. Temperature dependence of the temperature coefficient of the elastic modulus.

All of the data in KÖSTER's early papers were presented in terms of the ambient temperature in °C rather than in terms of T/T_m which he used in his final summary. In Fig. 3.121 are shown the temperature dependence[1] of E for manganese, iron, cobalt, and nickel as well as the temperature dependence of the damping coefficient of cobalt which illustrates the interesting possibilities present in such experimental studies.

Perhaps of equal interest are his comparisons of the temperature dependence of E for different purities of aluminum and magnesium[2] shown in Fig. 3.122.

The E-moduli and damping dependence of copper and of silver,[3] with and without precaution regarding oxygen content, shown in Fig. 3.123, further illustrate the developing complexity of these parameters in terms of the variation of many factors, including, here, chemical content.

Of KÖSTER's many papers on the E-moduli, perhaps the one of greatest historical interest is that in 1948 describing the vibrationally determined values at room temperature for 59 binary alloys. In his introduction, KÖSTER,[4] with a dim view of the history of his subject, informs us that the reason for his having discussed it at such length was that his was the first general study of the dependence of the elastic moduli of binary combinations upon the chemical composition. The temperature dependence of a few binary alloys was included, but his main emphasis in that paper was upon measurements at room temperature. Although KÖSTER

[1] KÖSTER [1948, *2*].
[2] KÖSTER [1948, *3*].
[3] *Ibid.*
[4] KÖSTER and RAUSCHER [1948, *1*].

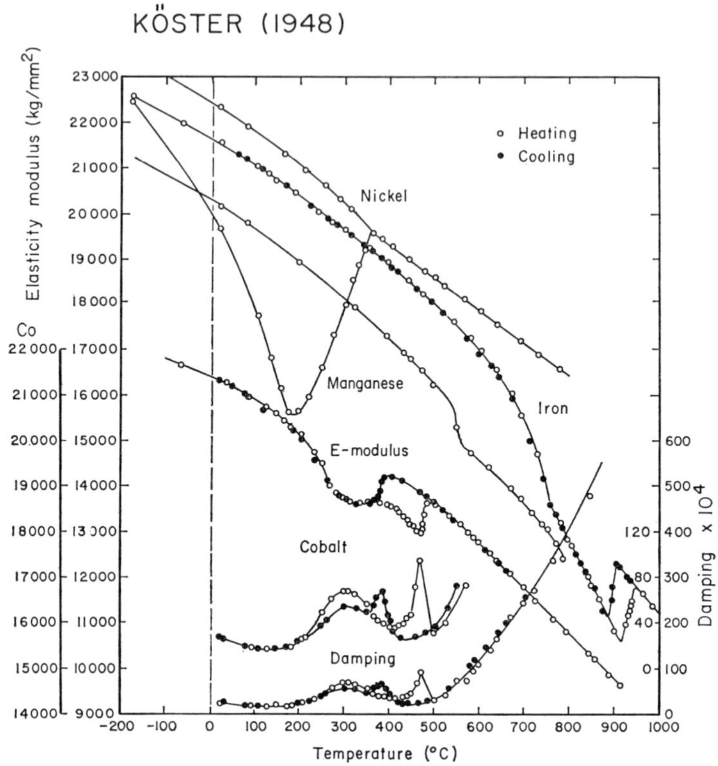

Fig. 3.121. Temperature dependence of the elastic modulus, E, of the metals manganese, iron, cobalt, and nickel, measured during heating (open circles) and during cooling (closed circles). Also shown is the temperature dependence of the damping coefficient of cobalt in two different states.

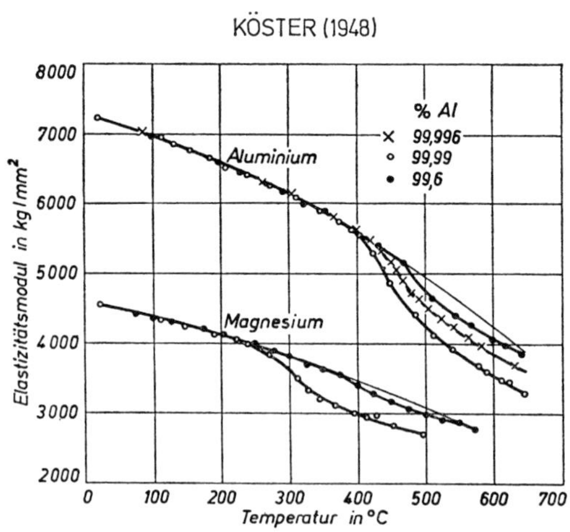

Fig. 3.122. Temperature dependence of the elastic modulus, E, of magnesium and aluminum of different purities.

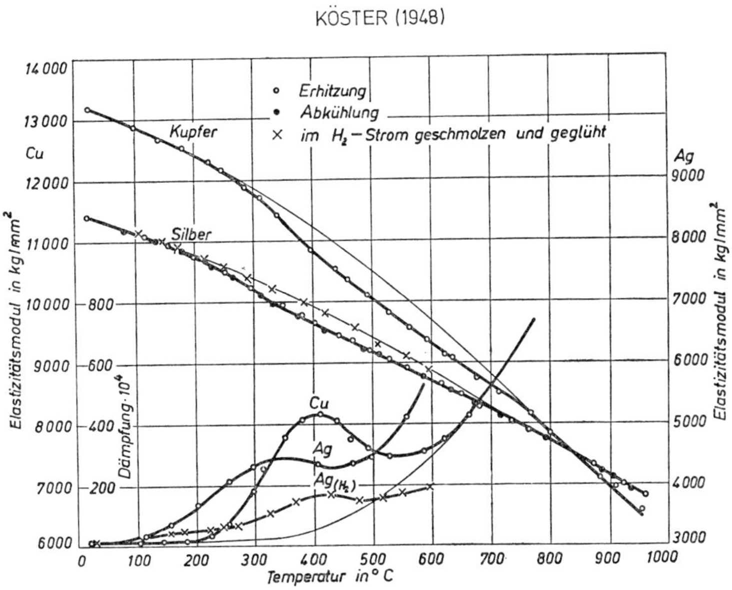

Fig. 3.123. The temperature dependence of the elastic modulus, E, of high purity copper and silver during heating (open circles) and cooling (closed circles). Of major interest was the observed maximum in the damping coefficient for these high purity metals. The crosses are for a specimen of silver which was twice re-melted and annealed in the presence of hydrogen.

considered approximately the same number of binary combination as had WERTHEIM before him, his distribution of type was far more extensive. A century of associated metallurgical study had provided a new classification and new techniques of preparation that were readily available to KÖSTER but had been unknown to WERTHEIM.

KÖSTER studied binary combinations of solids with perfect solubility, of limited solubility, eutectic alloys, and alloys with intermetallic phases. The reader interested in the deformation behavior of mixtures, is advised to consult KÖSTER'S paper of 1948 as a valuable source of detailed information.

Every experimentist from WERTHEIM'S day to ours who has accumulated moduli values ultimately has become interested in comparing the distribution of numbers with atomistic parameters. "WERTHEIM'S law", as VOIGT called it, which asserted[1] that the product of E and the seventh power of the average atomic distance was constant, was modified by VOIGT,[2] who replaced the seventh by the sixth power. THEODORE WILLIAM RICHARDS[3] in 1915 provided a subsequently widely quoted comparison of the compressibility at room temperature as a function of the atomic number, as is shown in Fig. 3.124.

KÖSTER in his summary in 1948 gave a similar plot[4] not only for the room temperature E in Fig. 3.125 but of perhaps more fundamental interest, for POISSON'S ratio ν, in terms of atomic numbers, as shown in Fig. 3.126.

The comparisons made by RICHARDS and by KÖSTER implicitly suggest that the dependence of their parameters on the fractional melting point is unimportant.

---

[1] WERTHEIM [1844, *1* (a)].
[2] VOIGT [1893, *2*].
[3] RICHARDS [1915, *1*] and see [1924, *1*].
[4] KÖSTER [1948, *4*].

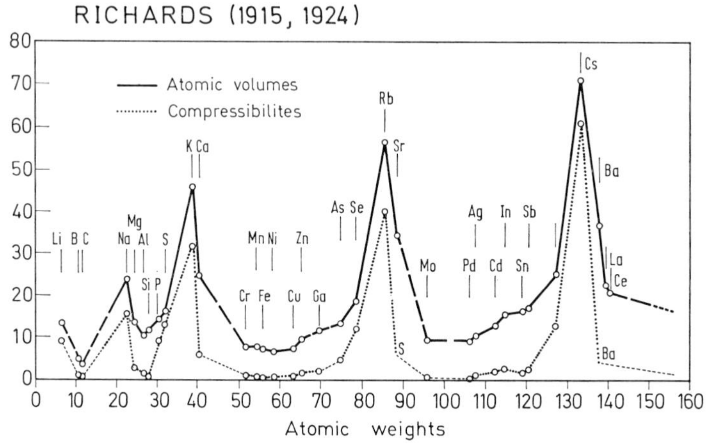

Fig. 3.124. Atomic volumes and compressibilities of solid elements, plotted in relation to atomic weights. The upper line gives the atomic volume and the lower line, the compressibilities. Atomic volumes are given as cubic centimeters per gram atom. Compressibilities are multiplied by 10^5.

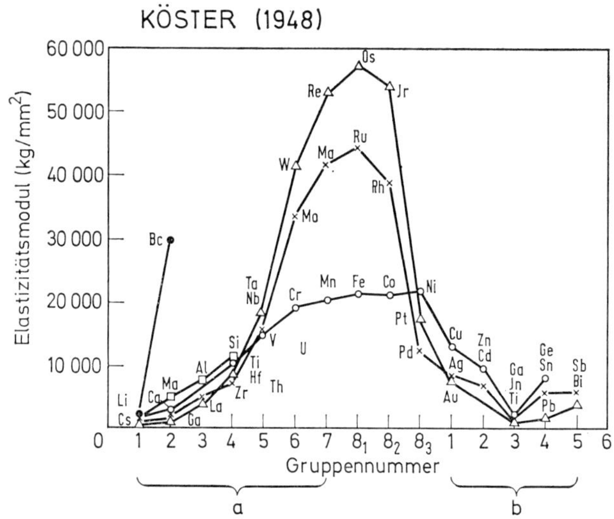

Fig. 3.125. The elastic modulus, E, in the periodic system.

In 1968 I gave[1] a plot of shear modulus $\mu(0)$ vs atomic number for zero point values for 59 elements, shown in Fig. 3.127, which removed the temperature dependence for the comparison with atomic parameters.

In the same terms I compared also the functional dependence of these zero point moduli upon atomic volume, the inverse square of the minimum measured atomic distance, and the melting point. As KÖSTER had found when he compared E with atomic volume at room temperature, I found[2] that there was no experimental basis for the analytical extension of atomistic calculation to predict the

[1] BELL [1968, *1*].
[2] *Ibid.*, pp. 147–149.

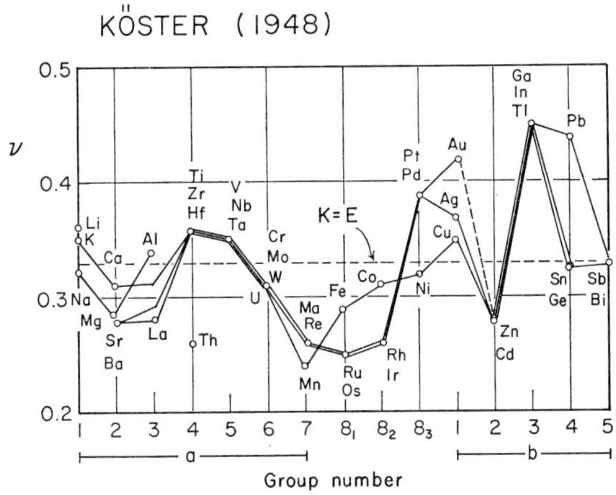

Fig. 3.126. Poisson's ratio in the periodic system.

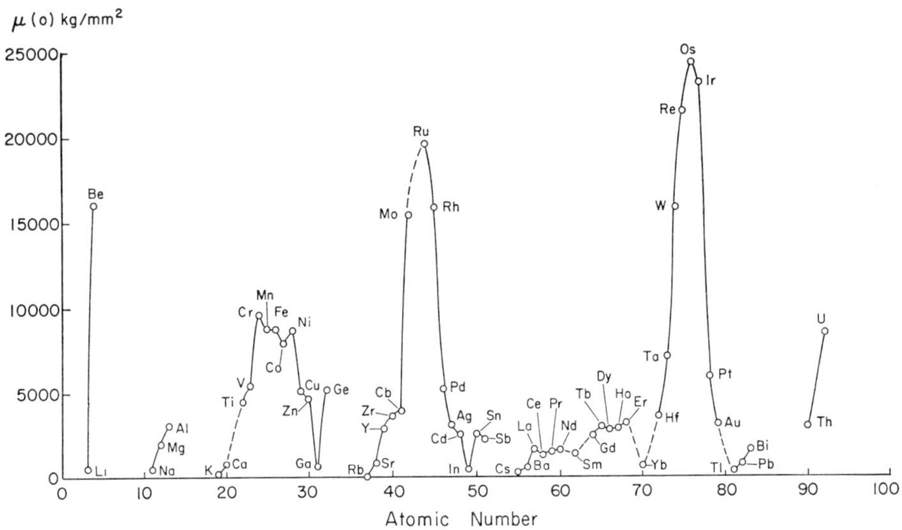

Fig. 3.127. The zero point isotropic elastic shear modulus, $\mu(0)$, of the elements, plotted against the atomic number (Bell, 1968).

macroscopic response function of the isotropic solid. It is well known that for a very small number of solids, such as copper, potassium, or rock salt, correlation with calculations has been achieved.

Below room temperature, E and μ of the isotropic solid and c_{44} one of the three constants for the cubic anisotropic single crystal which by itself designates one of the two shear moduli, all depend nearly linearly on the temperature. Ultrasonic studies have shown that for values of T/T_m below 0.06, these moduli are constant, i.e. to the left of the dashed vertical lines in Figs. 3.118 and 3.119. [T is the ambient temperature and T_m is the melting point, both in degress K.] In the mid-1960's I wished to determine zero point shear moduli of as many of the elements as possible so as to compare them with an integrally quantized distri-

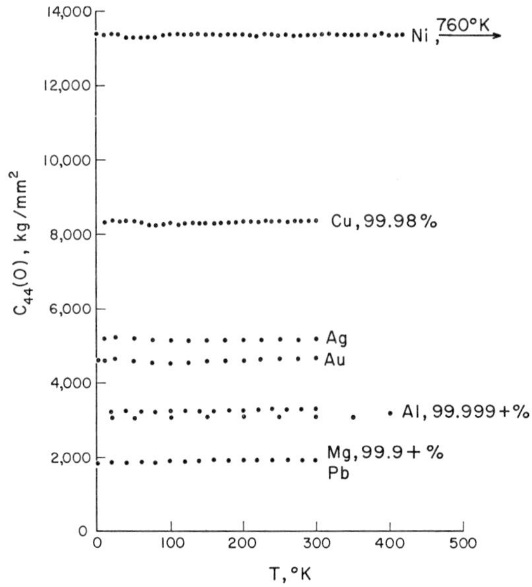

Fig. 3.128. Single crystal shear moduli $c_{44}(0)$, calculated from experimental data at various temperatures by BELL (1968) from Eq. (3.27). *Al*: G. N. KAMM and G. A. ALERS: J. Appl. Phys. **35**, No. 2, 327 (1964). — J. VALLIN, M. MONGY, K. SALAMA, and O. BECKMAN: J. Appl. Phys. **35**, No. 6, 1825 (1964); *Al (polycrystal)*. — C. ZUCKER: J. Acoust. Soc. Am. **27**, No. 2, 318 (1955). *Au*: E. VON GOENS: Ann. Physik **38**, 456 (1940). — J. R. NEIGHBOURS and G. A. ALERS: Phys. Rev. **111**, Ser. 2, 707 (1958). *Ni*: G. A. ALERS, J. R. NEIGHBOURS, and H. SATO: J. Phys. Chem. Solids **13**, 40 (1960). *Mg*: L. J. SLUTSKY and C. W. GARLAND: Phys. Rev. **107**, No. 4, 972 (1957). *Ag*: J. R. NEIGHBOURS and G. A. ALERS: Phys. Rev. **111**, Ser. 2, 707 (1958). *Cu*: E. VON GOENS: Ann. Physik **38**, 456 (1950). — W. C. OVERTON, Jr., and JOHN GAFFNEY: Phys. Rev. **98**, No. 4, 969 (1955). *Pb*: E. VON GOENS: Ann. Physik **38**, 456 (1940). — S. C. PRASAD and W. A. WOOSTER: Acta Cryst. **9**, 38 (1956).

bution of the zero point linear elastic shear modulus, which my studies of finite strain profiles in dynamic plasticity and of quasi-static stress-strain functions in the same region of large plastic strain had indicated must characterize the behavior of crystalline solids.[1] (This will be considered in Sect. 3.44). Hence I studied this linear temperature dependence of moduli for specified solids. Such linearity, as was noted above, had been observed by KÖSTER[2] and others. The initial slopes for μ and c_{44} moduli I found were as follows:

$$\begin{aligned} \mu(T/T_m) &= \mu(0), & 0 \leqq T/T_m \leqq 0.06; \\ \mu(T/T_m) &= 1.03\mu(0)\,(1 - T/2T_m), & 0.06 \leqq T/T_m \leqq \frac{300}{T_m}. \end{aligned} \tag{3.27}$$

Examples of c_{44} moduli for single crystals obtained by ultrasonic means between 4.2° K and room temperature, compiled by me[3] in 1968, are shown in Fig. 3.128. The measured values of $\mu(T/T_m)$ and T were used to calculate $\mu(0)$ at each temperature. There is obvious agreement between experiment and the temperature dependence embodied in Eq. (3.27).

[1] BELL [1968, *1*].
[2] KÖSTER [1947, *1*].
[3] BELL, *op. cit.*

3.44. The quantized distribution of elastic shear moduli at the zero point for isotropic bodies, and the multiple elasticities for a given isotropic solid: Bell (1964–1968).

Beginning with MASSON and WERTHEIM in the 1840's, a number of persons attempted to relate measured moduli of different crystalline solids in terms of atomic parameters, in particular, atomic spacing. It did not seem to occur to most of the experimentists that such comparisons should be made at the same fractional melting point temperature. (The fractional melting point temperature is the ratio of the ambient temperature of the test to the melting point of the solid, T/T_m.) KÖSTER, as we have seen, did present data for E in this form.

In 1964 and 1965, while engaged in an experimental study of stress-strain functions governing plastic wave propagation in finitely strained bodies, I found an entirely new type of distribution among the shear moduli of the elements. As will be described in detail in Chap. IV below, experiment has revealed that the loading response functions in plasticity for finite strain are parabolic for both the single crystal and the fully annealed polycrystal, with a discrete distribution of parabola coefficients expressible by an integral mode index. Searching for similarly distributed parameters among the several crystalline solids for which the large deformation response had been studied, I discovered that the ratios of parabola coefficients between one solid and another were as the elastic shear moduli at the zero point, $\mu(0)$, for the infinitesimal deformation of isotropic solids. After the first published reference to this work[1] in 1967, I presented a detailed discussion of the quantized distribution of shear moduli[2] at the zero point, in 1968. This distribution of shear moduli is expressed as follows:

$$\mu(0) = (2/3)^{s/2 + p/4} A, \tag{3.28}$$

where $s = 1, 2, 3, 4, \ldots$; p is a structure parameter, having either 0 or 1 as its value; and A is a universal constant: $A = 2.89 \times 10^4$ kg/mm².

For clarity, the comparison of experiment and prediction shown in Fig. 3.129 for 57 elements and the two binary combinations, 70—30α brass and NaCl, has been made for $(2/3)^{p/4}\mu(0)$, where $p=0$ or 1.

In Table 107 are the measured values for the solids, compared with prediction from Eq. (3.28) with the appropriate value of p in each instance. One notes that for the majority of these crystalline solids, $p=0$. I determined the universal constant, A, from numerous measurements in aluminum made at many temperatures, including 4.2° K.

In a monograph[3] in 1968, (and see Sections 3.42 and 3.43 above), I described the extensive study of the temperature dependence of elastic constants which led to the determination of reliable zero point values for these 59 isotropic solids. That such a quantized distribution characterizes the shear moduli (and, as will be seen presently, also characterizes the E-moduli) of the elements and such binary combinations as α brass and NaCl, obviously has important implications for an understanding of the structure and deformation response of crystalline solids. The fact that the distribution was discovered as a by-product of a study in finite strain, orders of magnitude larger than the maximum strain amplitudes for which these moduli were measured, strongly suggests the existence of a funda-

---

[1] BELL [1967, *2*].
[2] BELL [1968, *1*].
[3] *Ibid.*

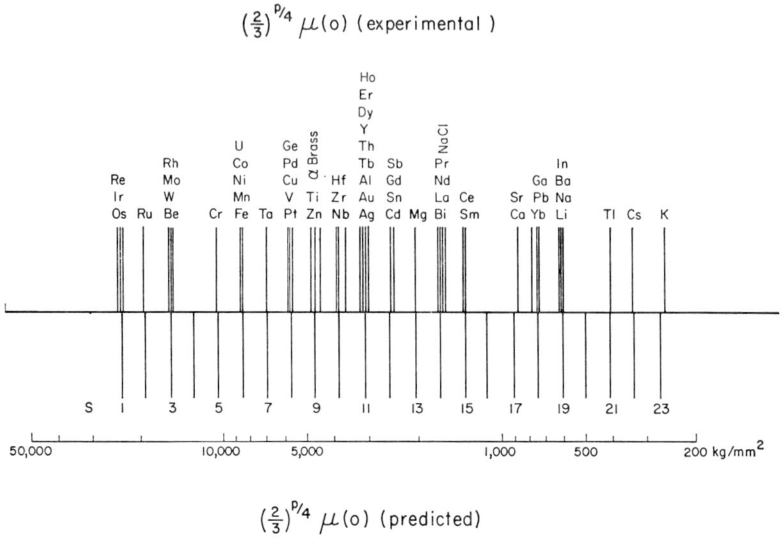

Fig. 3.129. A comparison of averaged experimental $\mu(0)$ with distribution of BELL's Eq. (3.28). Separate distributions for $p=0$ and for $p=1$ are shown as a single distribution by comparing the predicted $(2/3)^{p/4}\,\mu(0)$ with experiment.

Table 107. BELL (1968).

Crystalline solid	T_m (°K)	No. of measurements	Averaged experimental $\mu(0)$ (kg/mm²)	Predicted $\mu(0)$ (kg/mm²) Eq. (3.28)	s	p	References
Os	3273	1	24300	23600	1	0	a
Ir	2728	1	23100	23600	1	0	a
Re	3443	1	21400	21300	1	1	a
Ru	2773	1	19500	19300	2	0	a
W	3683	1	15700	15730	3	0	a
Be	1550	3	16000	15730	3	0	a, b, c
Mo	2893	1	15400	15730	3	0	e
Fe	1809	4	8700	8580	6	0	a, b, c, d
Ni	1726	2	8600	8580	6	0	a, b
U	1405	2	8440	8580	6	0	a, b
Co	1768	1	7900	7750	6	1	d
Ge	1231	1	7070	7010	7	0	h
Ta	3269	1	7040	7010	7	0	f
Pt	2042	2	5860	5720	8	0	a, b
V	2133	2	5320	5160	8	1	a, b
Cu	1356	4	5080	5160	8	1	a, b, c, d
Pd	1825	1	5200	5160	8	1	a
Zn	692.5	4	4570	4660	9	0	a, b, c, d
70−30 α-brass	1188	1	4700	4660	9	0	c
Ti	1943	2	4420	4210	9	1	a, b
Nb	2743	2	3900	3810	10	0	a, b
Zr	2125	2	3680	3810	10	0	a, b
Hf	2523	1	3600	3450	10	1	a
Ag	1233.8	3	3170	3110	11	0	a, b, c
Au	1336	4	3090	3110	11	0	a, b, c, d
Al	933	14	3110	3110	11	0	a, b, c, d, g
Th	2023	2	2960	3110	11	0	a, b

Table 107. (Continued.)

Crystalline solid	T_m (°K)	No. of measurements	Averaged experimental $\mu(0)$ (kg/mm²)	Predicted $\mu(0)$ (kg/mm²) Eq. (3.28)	s	p	References
Tb	600	1	3020	3110	11	0	f
Er	1798	1	3180	3110	11	0	f
Y	1763	1	2830	2810	11	1	f
Dy	1653	1	2770	2810	11	1	f
Ho	1461	1	2840	2810	11	1	f
Cd	593.9	2	2550	2540	12	0	c, d
Sn	504.9	4	2510	2540	12	0	a, b, c, d
Gd	1585	1	2450	2540	12	0	f
Sb	1653	1	2200	2300	12	1	d
Mg	923	4	2050	2070	13	0	a, b, c, d
La	1099	1	1700	1690	14	0	f
Bi	544	1	1600	1530	14	1	d
Nd	1113	1	1550	1530	14	1	f
Pr	1213	1	1530	1530	14	1	f
Sm	1573	1	1390	1381	15	0	f
Ce	1077	1	1380	1381	15	0	f
NaCl	1074	1	1360	1381	15	0	c
Ca	1111	1	800	830	17	1	a
Sr	1041	1	800	830	17	1	a
Yb	2073	1	754	750	18	0	f
Pb	600.4	3	750	750	18	0	a, b, c
Tl	576	1	370	370	21	1	a
Cs	301.7	1	340	340	22	0	a

[a] ALAN HOWARD COTTRELL: The Mechanical Properties of Matter. New York: John Wiley & Sons, Inc. (1964).

[b] Smithells Metals Reference Book, 3rd ed., vol. II, p. 614. Butterworth & Co. (Publ.) Ltd. 1962.

[c] R. F. S. HEARMON: Advan. Phys. 5, 370 (1955).

[d] R. F. S. HEARMON: Applied Anisotropic Elasticity. Oxford University Press 1961.

[e] Handbook of Chemistry and Physics, 43rd ed., p. 2169. Cleveland, Ohio: The Chemical Rubber Publishing Co. 1961–1962.

[f] J. F. SMITH, C. E. CARLSON, and F. H. SPEDDING: J. Metals, Trans. Met. Soc. AIME 9, 1212 (1957).

[g] C. ZUCKER: J. Acoust. Soc. Am. **27**, No. 2, 318 (1955).

[h] J. FRIEDEL: Dislocations. Reading, Massachusetts: Addison-Wesley Publishing Co. (1964).

mental unity in the mechanics of crystals. This unity comprehends a series of states in which moduli occur, apparently independent of crystal structure and of whether or not the crystalline solid is metallic, etc.; and for which comparisons of the solids must be made at the same fractional melting point temperature. The measured values of the moduli which I used in this study included the published experimental results of the metal physics literature during the past 50 years. After having presented those data, I extended the range of measured values to include a very large proportion of the reliable experimental results of the 19th and 20th centuries.

The specimens used in the pursuit of the science of the mechanics of solids are supplied from the fruits of the art of metallurgy. Generally, the objectives of the technology of metallurgy, in some contrast to those of the science of metallurgy, are to provide stable solids with high elastic limits having special properties decreed by the immediate needs of practical design. Throughout most of the past century and a half, the availability of specimens for scientific study from this category of solids has influenced, or perhaps somewhat distorted, our knowledge of this field of mechanics. For example, for metals like zinc, engineering handbooks seldom provide values of moduli, or if they do, the range of numbers is so wide that none might as well have been given. Zinc, like many other solids, is essentially unstable as far as deformation parameters are concerned, yet the more stable solids such as aluminum, copper, iron, may be brought to a condition in which they exhibit similar instabilities simply by altering their prior thermal and mechanical histories from the recipes for stable solids provided in practical metallurgy.

Nearly 20 years ago in studies of large deformation including plastic waves in finite strain, I became interested in investigating crystalline solids for which prior thermal and mechanical histories were more varied than those usually described. The results of that research, which led to my discovery of the discrete distribution of zero point isotropic moduli, significantly complemented a similar discrete distribution among the parabola coefficients I had found to characterize the loading response functions at finite strain. The results of the study also indicated the likelihood that the moduli of an individual isotropic crystalline solid could undergo transitions from the solids' stable value to other discrete values given in the quantized distribution of Eq. (3.28), for which the integer, s, changes.

Accordingly, in the mid-1960's I undertook an experimental study of moduli in polycrystalline solids with various prior deformation and prior thermal histories. I found that for all the solids considered in careful dead weight loading experiments, moduli occurred in the form of multiple elasticities described by the quantized distribution of Eq. (3.28) (see also Sect. 2.27).

I developed experiments to explore the subject of multiple elasticities, first to demonstrate that such a distribution as that of Eq. (3.28) indeed did exist for measured values in a given solid, and second, to provide the detail of such behavior in relation to thermal and mechanical histories. Polycrystalline specimens of aluminum, copper, magnesium, iron, zinc, and 70—30α brass were annealed at high temperatures, furnace-cooled, and checked for grain size and purity. Dead weight tension or compression experiments were performed on a simple loading apparatus which had been built for this purpose. The apparatus and many of the experimental results are described in the monograph[1] of 1968. Loads were applied either continuously by means of water flow, or incrementally by the addition of small quantities of sand. A few experiments also were performed on machines in which dead-weight loading was achieved by maintaining a floating beam in constant balance. During these tests many reversals of load were made both with and without small increments of permanent deformation being present following the application of the maximum load. Each new loading, therefore, had a well-defined prior thermal and mechanical history from the beginning of the first high temperature anneal.

As is illustrated in the figures which follow, the results of these experiments definitely established that the expected multiple elasticities did indeed occur and had $\sigma-\varepsilon$ slopes in close quantitative agreement with the discrete distribution I had found. In some instances, the first slope from the origin occurred with a value of the integer s different from that of the stable modulus, and after proceeding to some value of stress, underwent a second-order transition to still another value of s. In other instances, the initial slope was the stable value to some value of stress at which a series of second order discontinuities followed, each associated with a different value of the integer s.

---

[1] BELL [1968, *1*]. See Fig. 3.2 on p. 55, and Chapt. V.

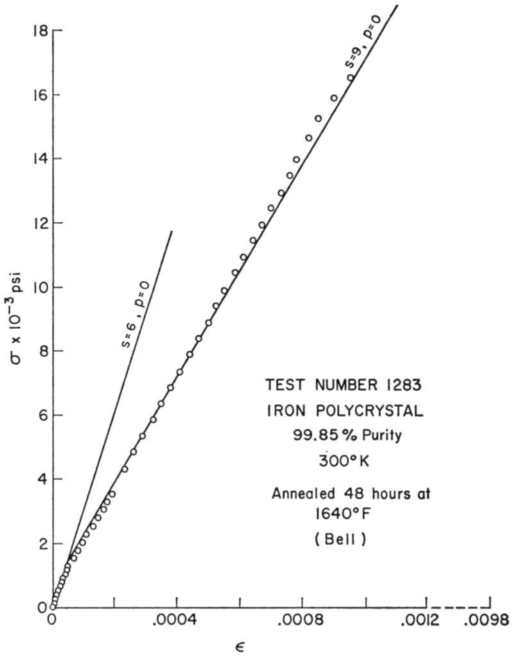

Fig. 3.130. Two moduli before discontinuous yield. The initial modulus for $s=6$, $p=0$ is $E=29.7\times10^6$ psi (20900 kg/mm²); for $s=9$, $p=0$, $E=16.2\times10^6$ psi (11400 kg/mm²) [BELL, 1968].

An example of the second situation was the experiment for polycrystalline iron, 99.85% pure, annealed 48 h in an appropriate atmosphere at 1640° F, from which I obtained the results shown in Fig. 3.130, for a test at room temperature.

The circles in Fig. 3.130 indicate the measurements, and the solid lines are the slopes given by Eq. (3.29), based upon combining Eqs. (3.27) and (3.28) in the formula relating E, μ, and ν for isotropic solids.

$$E=d\sigma/d\varepsilon=2(1+\nu)\;1.03\,(2/3)^{s/2+p/4}\,A\,(1-T/2T_m)\,. \tag{3.29}$$

The value $s=6$ is the stable room temperature value of $E=20900$ kg/mm². The change to the predicted value of $s=9$, for which $E=11370$ kg/mm² which was closely followed until yield, occurred at approximately 12 kg/mm² in this specimen; it is a dramatic example of the existence of such multiple elasticities in a well-studied, common solid. Further examples in iron, for which the initial modulus is not the stable value $s=6$, $p=0$, are shown in Fig. 3.131. These specimens were annealed for one hour at 1200° F.

I gave many examples of these multiple elasticities in the E-moduli of a number of isotropic solids.[1] For instance, magnesium, for which the stable modulus index is $s=13$, when annealed at a high temperature and furnace cooled, was shown to have a reduced value of the modulus for an integral index, $s=15$. This is not a small difference: the new modulus for $s=15$ is $^1/_3$ lower than the stable value, i.e., for $s=13$, $E=4470$ kg/mm² whereas for $s=15$, $E=2980$ kg/mm². Shown in Fig. 3.132a are the results from two experiments on magnesium, as well as from

[1] *Ibid.*, Chapt. V.

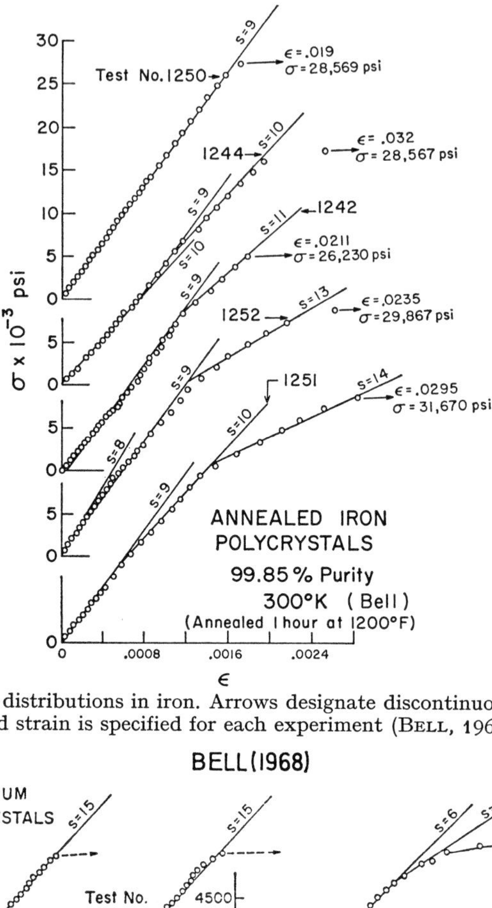

Fig. 3.131. E moduli distributions in iron. Arrows designate discontinuous yield whose stress and strain is specified for each experiment (BELL, 1968).

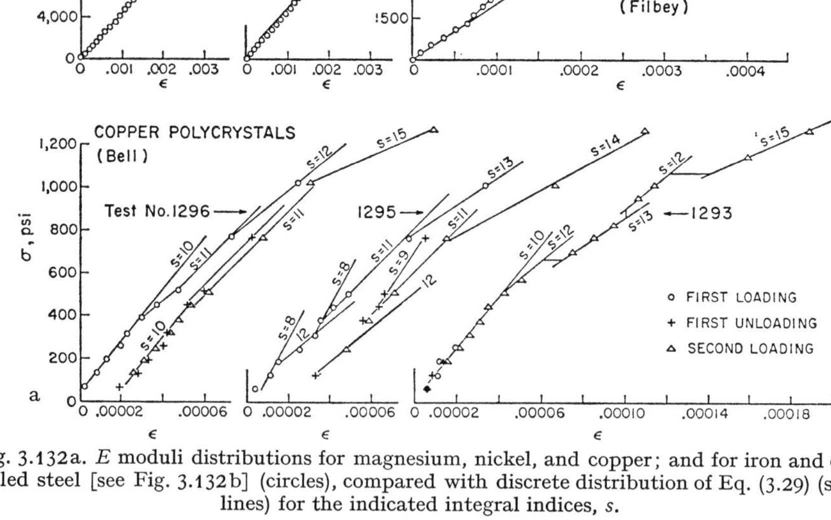

Fig. 3.132a. E moduli distributions for magnesium, nickel, and copper; and for iron and cold rolled steel [see Fig. 3.132b] (circles), compared with discrete distribution of Eq. (3.29) (solid lines) for the indicated integral indices, s.

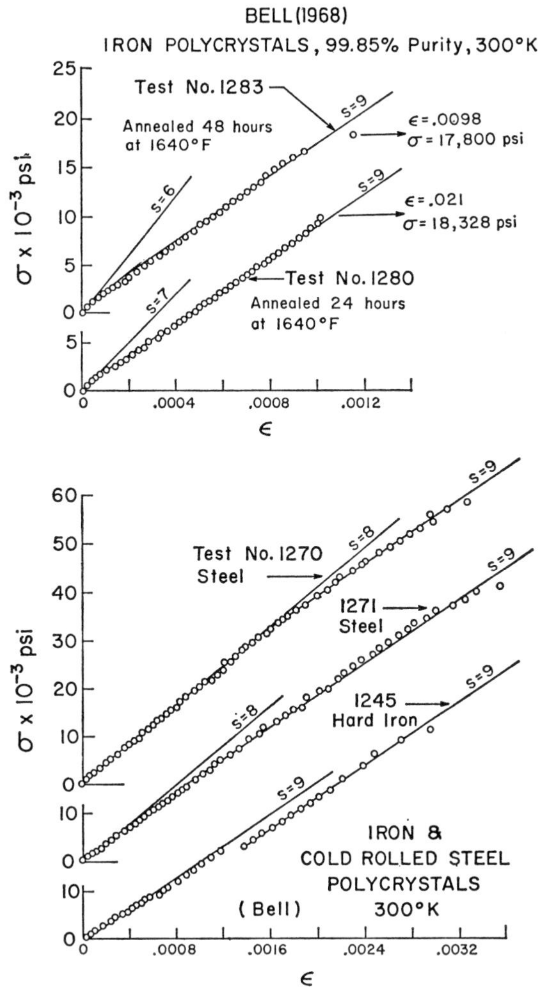

Fig. 3.132b.

experiments demonstrating multiple elasticities in nickel for which the stable modulus is $s=6, p=0$, and in copper, which has a stable modulus of $s=8, p=1$. In Fig. 3.132b are a few of the results which were obtained from experiments on cold rolled steel and hardened iron.

I also showed examples of multiple elasticities which had occurred during the loading and unloading of dead annealed aluminum polycrystals of low and high purity. One interesting correlation was that from an experiment on annealed 70—30α brass obtained a few years ago in my laboratory as part of the doctoral research of HARTMAN.[1] This was a dead weight tension test performed on the apparatus which had been used in my experiments for other solids. The results of this test were shown in Fig. 2.84 where the solid lines in each instance represent slopes predicted from the distribution of Eq. (3.29).

[1] HARTMAN [1967, *1*].

In the light of my discovery in the 1960's of this quantized isotropic moduli structure, it is now fascinating to review the experimentation of the 19th century, when handbook recipes for prior thermal and mechanical histories, standardized specimen shapes, and hard testing machines had not yet become the order of the day.[1] In reviewing the papers for this chapter of the present treatise, I found many 19th and even 20th century moduli measurements in given solids consistent with this discrete distribution of multiple elasticities. As early as 1784, for example, COULOMB[2] had obtained a shear modulus of 7580 kg/mm^2 (for an assumed weight density of 7.9) in iron from torsional vibration tests, and a value of 2730 kg/mm^2 (for a weight density of 8.6) for brass. From Eq. (3.27) these provide zero point moduli of $\mu(0)=8060$ kg/mm^2 and $\mu(0)=3040$ kg/mm^2 respectively. The former is somewhat less than the stable value of $s=6$, 8580 kg/mm^2 for iron from Eq. (3.28), but the second is far below the stable value for brass, 4660 kg/mm^2 for $s=9$. The value of 3040 kg/mm^2 is close to the value for $s=11$ of $\mu(0)=$ 3110 kg/mm^2, the stable value for aluminum, gold, and silver. Thus the low value of COULOMB's modulus for brass, which caused considerable comment early in the 19th century and which was far too large to explain away as experimental error, falls into this quantized pattern for elastic constants.

In Sect. 3.8 above, I have described in some detail the controversy in the early 19th century over the difference between CHLADNI's and BIOT's numerical values for E. CHLADNI[3] had provided the data from which E could be determined by observing the longitudinal frequency of vibrating rods. BIOT,[4] we recall, had obtained a value for E in iron from the measurement of the speed of a wave propagating in nearly a kilometer of pipe. I also noted[5] that many decades later, in the mid-19th century, the controversy was revived in the experiments of WERTHEIM and BREGUET[6] who obtained very different values of E from the measurement of wave speeds in three kilometers of iron wire, compared with those obtained from the longitudinal oscillation of a piece of the same wire. In Table 108, the measured values of E for iron, and the average value which DULEAU[7] obtained for iron, are compared with the quantized distribution of Eq. (3.28) for the indicated values of the integral index s. The measurement of multiple elasticities obviously was a common occurrence when preparation of specimens varied markedly from one laboratory to another. However, it is precisely the introducing of variations (under better defined conditions), i. e., the altering of modern technological specifications for stable materials, that provides the opportunity to observe the larger order which exists.

I gave[8] similar illustrations of multiple elasticities from the 19th century data of WERTHEIM and of KELVIN.[9] Because the discovery of the quantized distribution of Eq. (3.28) was based upon a detailed analysis of a large number of 20th century measurements of moduli of isotropic solids, it is of interest here to compare the late

---

[1] Many contemporary experimentists using standard hard testing machines over a period of years never have had the opportunity to run a dead weight test. Even for the hard machines, it is interesting to note that after I described the discovery of these multiple elasticities a few years ago, many persons informed me of similar observations which until then they had ascribed to peculiarities of the individual specimen or difficulties of the testing machine. Such data seldom were published after the first decade of the 20th century.

[2] COULOMB [1784, *1*].

[3] CHLADNI [1817, *1*], [1822, *1*].

[4] BIOT [1809, *1*].

[5] See Sect. 3.32.

[6] WERTHEIM and BREGUET [1851, *1*].

[7] DULEAU [1813, *1*], [1820, *1*].

[8] BELL [1968, *1*].

[9] KELVIN (WILLIAM THOMSON) [1880, *1*].

Table 108. *A comparison of early values of E for iron.*

s	Calculated E Eq. (3.29), BELL (1968) (kg/mm²)	WERTHEIM, and WERTHEIM and BREGUET (1851) (kg/mm²)	DULEAU (1811) (kg/mm²)	BIOT (1809) (kg/mm²)	CHLADNI (1787) (kg/mm²)
5	25 500	—	—	—	26 000 (vibration)
6	20 800	20 869 (quasi-static)	20 900 (quasi-static)	—	—
7	16 900	16 950 (quasi-static)	—	—	—
8	13 850	13 150 (vibration)	—	—	—
9	11 300[a]	—	—	10 600[b] (wave speed)	—
10	9250	9620 (wave speed)	—	—	—

[a] See Figs. 3.130 and 3.131 for modern measurements with $s=9$.
[b] BIOT gave no density for his cast iron. This calculation is based upon a density of 7.9.

Table 109.

Solid	No. of values	μ (0) calculated from GRÜNEISEN data of 1908 (kg/mm²)	μ (0) predicted by BELL in 1968 (kg/mm²)	s	p
Al	5	3113	3110	11	0
Cu	5	5117	5160	8	1
Ag	1	3222	3110	11	0
Au	4	3000	3110	11	0
Fe	2	8735	8580	6	0
Ni	1	8321	8580	6	0
Cd	1	2519	2540	12	0
Pb	1	742	750	18	0
Brass	1	3872	3810	10	0

19th, early 20th century data of VOIGT[1] and GRÜNEISEN[2] which has been described above. Such a comparison for the indicated integral index s in Eq. (3.28) is shown in Tables 109 and 110. The measured values of VOIGT and GRÜNEISEN at room temperature were reduced to absolute zero through Eq. (3.27).

In particular we should note that in GRÜNEISEN's data, brass has the value of $s=10$ instead of the more common value of $s=9$, $p=0$, or the value obtained from COULOMB's data, $s=11$, $p=0$. An equally close correlation between experimental values of shear moduli and prediction from my zero point shear moduli distribution for isotropic solids was obtained for the data of VOIGT in 1893, averaged from samples taken in several directions in a single block.

[1] VOIGT [1893, *1*].
[2] GRÜNEISEN [1908, *1*].

Table 110.

Solid	μ (0) calculated from VOIGT averaged data of 1893 (kg/mm^2)	μ (0) predicted by BELL in 1968 (kg/mm^2)	s	p
Al	2980	3110	11	0
Bronze	4520	4660	9	0
Cd	3170	3110	11	0
Fe	5540	5720	8	0
Au	3110	3110	11	0
Cu	5210	5160	8	1
Mg	1975	2070	13	0
Brass	4120	4210	9	1
Ni	8340	8580	6	0
Ag	3240	3110	11	0
Steel (Avg. 6)	8700	8580	6	0
Bi	1660	1690	14	0
Zn	4750	4660	9	0
Sn	2400	2540	12	0

HARTMAN[1] showed that the stable value of $s=9$, $p=0$ for 70—30α brass underwent a shift from $p=0$ to $p=1$ as the composition of the alloy approached 90% copper. I had shown that zinc has a stable modulus, $s=9$, $p=0$, while copper has a value, $s=8$, $p=1$.

Obviously the study of the quantized distribution of elastic constants and the related second order transitions in the tangent modulus for small deformation is fundamental to an understanding of the deformation of crystals. From the earliest measurement of elastic constants to the present, the moduli of isotrophic polycrystalline solids are seen to fall into the same quantized distribution. This discovery, however, is less than a decade[2] old, so that considerable time must elapse before all of its ramifications in both continuum and atomistic mechanics will be understood.

3.45. Anisotropy.

As we have seen, WERTHEIM was said[3] to have been planning to extend his experimental analysis to the study of anisotropic single crystals. If so, after his untimely death in 1861 the idea lay dormant for 13 years. Then, in 1874, G. BAUMGARTEN[4] measured moduli by flexure in calcspar, and in the same year VOIGT[5] in his Königsberg doctoral dissertation described quasi-static flexural and torsion experiments to determine the three elastic constants of the anisotropic cubic crystal, rock salt.

BUFFON[6] early in the 18th century had observed the importance of grain direction in the rupture of wood. WERTHEIM himself had provided[7] the first detailed study of constants for anisotropic materials by obtaining wood in the direction

[1] HARTMAN [1967, *1*], [1969, *2*].
[2] BELL [1968, *1*], [1964, *1*], [1965, *2*], [1967, *2*].
[3] VERDET [1861, *1*].
[4] BAUMGARTEN [1874, *1*].
[5] VOIGT [1874, *1*].
[6] BUFFON [1741, *1*].
[7] WERTHEIM and CHEVANDIER [1846, *1*].

of the radius, tangent, and along the trunk of the trees, from which he selected his specimens.

Certainly the success which VOIGT achieved in the study of anisotropy over a period of 36 years following his doctorate, culminating in his classic treatise[1] on the subject in 1910, makes it interesting to speculate on who else might have had the combination of analytical patience and experimental stoicism to perform the hundreds of calculations relating the thousands of measurements required to determine the numerous elastic constants for the various classes of crystals. VOIGT, particularly as a pioneer, was obliged to suffer all the false starts and repetitions familiar only to those who have inaugurated broad new fields of study. Experimental difficulties at one point required the repetition of approximately 2000 measurements on crystals of topaz and baryta, a prospect which VOIGT referred to as a weary undertaking.

VOIGT's first research which concerned rock salt[2] provided for the three elastic constants of the single crystal having cubic anisotropy, $c_{11}=8300$ kg/mm², $c_{44}=5300$ kg/mm², and $c_{12}=1292$ kg/mm². This was the first complete set thus determined. The values were entirely incorrect because the quasi-static torsion data were calculated in terms of a theory proposed by his professor, FRANZ NEUMANN, which did not apply to anisotropic materials. Eight years later[3] in 1884 a recalculation of these same measurements in terms of SAINT-VENANT's theory for the torsion of anisotropic solids provided $c_{11}=4600$ kg/mm², $c_{44}=1190$ kg/mm², and $c_{12}=1260$ kg/mm². These numbers are of particular significance because the POISSON-CAUCHY molecular theory based upon central forces predicts that $c_{12}=c_{44}$, a condition decidedly not fulfilled in the earlier erroneous results calculated from the NEUMANN analysis, but in rough agreement with calculations from the same data, based upon the correct analysis of SAINT-VENANT.

With greater care in preparing and polishing specimens and more precision in technique, VOIGT repeated[4] all the experiments on flexure and torsion of rock salt, 100 measurements of the former and 32 of the latter. The somewhat modified new values were $c_{11}=4753$ kg/mm², $c_{44}=1313$ kg/mm², and $c_{12}=1292$ kg/mm², which twenty-five years later[1] were further slightly modified to: $c_{11}=4770$ kg/mm², $c_{44}=1320$ kg/mm², and $c_{12}=1290$ kg/mm². These may be compared with summations of HEARMON in 1969 given in the LANDOLT-BÖRNSTEIN[5] tables of elastic constants: $c_{11}=4970$ kg/mm², $c_{44}=1295$ kg/mm², and $c_{12}=1285$ kg/mm².

VOIGT's student, H. KLANG,[6] in 1881 measured elastic constants in fluorspar. (In describing the work of KLANG, VOIGT referred to S. A. CORNOMILAS, whose doctoral dissertation at Tübingen in 1877 in gypsum and mica he adversely criticized on experimental grounds. CORNOMILAS, VOIGT said, was the only other person to measure elastic constants in the decade following his own studies of 1874.) In 1884 when VOIGT[4] provided the results of his recalculated and reobserved rock salt data referred to above, he gave, too, a recalculation of the fluorspar data of KLANG, which also had been interpreted erroneously in terms of NEUMANN's torsion formula. After these initial experimental and analytical ventures, VOIGT conducted an extensive program of measurement and interpretation. For example, he made the difficult determination[7] of the nine constants of the orthorhombic topaz and baryta ($BaSO_4$) in 1887. These measured values are those given for the two solids by HEAR-

[1] VOIGT [1910, *1*].
[2] VOIGT [1876, *1*].
[3] VOIGT [1884, *1*].
[4] VOIGT [1884, *1*].
[5] See BECHMANN and HEARMON: LANDOLT-BÖRNSTEIN [1969, *1*].
[6] KLANG [1881, *1*].
[7] VOIGT [1887, *1*].

MON in the 1966–1969 LANDOLT-BÖRNSTEIN tabulation of elastic constants, attesting to the well-known experimental difficulties which had been surmounted by VOIGT, 80 years earlier.[1] VOIGT also gave[2] values for cubic crystal pyrites (FeS_2) and sylvite (KCl) in 1888, and the cubic crystal,[3] sodium chlorate, in 1893. Measurements of the six elastic constants of crystals of the trigonal system, calcspar ($CaCO_3$), quartz, tourmaline, dolomite, and iron glance or haematite (Fe_2O_3), and of the orthorhombic aragonite, $CaCO_3$, were all experimentally and theoretically analyzed for inclusion in his treatise[4] in 1910.

As with all such developments primarily associated with a single laboratory over a period of years, improvements in technique were recorded at intervals. The obtaining of small crystal bars of the various necessary orientations and the measurement of their deflection on the careful application of loads, the nature of end supports, the manner of measuring the deflection and of applying the load, were all subjected to detailed scrutiny. The problems of sample condition, the obtaining of sufficiently large crystals to perform the many measurements on the same piece, the method of polishing and cutting the specimen, and the estimate of the influence of flaws, etc., presented major obstacles. As an experimentist in sympathy with the difficulties VOIGT faced, however, I find myself very critical of his methods of quite arbitrarily, it seems to me, estimating frictional and other empirical effects in terms of numerical corrections he applied to his data. I have commented upon this aspect of his experimentation in the section on impact described above, and I have similar comments to make regarding his viscosity experiments described below.

The determinations of the elastic constants for the generalized HOOKE's law, c_{ij}, whether they be for the three independent constants, c_{11}, c_{12}, or c_{44} for the cubic, or the twenty-one independent constants for the most general anisotropic material, require for quasi-static studies the measurement of an array of deflections and torsion angles. When merely tabulated by themselves as experimental results they have little meaning except to emphasize that the solid studied was not isotropic. In VOIGT's *Lehrbuch der Krystallphysik*[5] or in LOVE's *Mathematical Theory of Elasticity*[6] the analytical problems for the different crystal classes were outlined. HEARMON in his *Introduction to Applied Anisotropic Elasticity*[7] in 1961 provided an introductory view of the combined experimental-analytical problem.

One current experimental use of ultrasonics is for the measurement of longitudinal and transverse wave speeds in selected directions for the determination of what are assumed to be adiabatic values of the elastic constants for the various crystal systems. The literature describing these numbers, for the most part obtained by using standard commercial apparatus, is truly enormous. ALAN B. SMITH and RICHARD W. DAMON[8] in April 1970, as I noted above, provided a bibliography of titles on microwave ultrasonics which required 25 journal pages merely to list the papers of two decades. As was also noted above, in 1966, with a supplement in 1969, HEARMON in LANDOLT-BÖRNSTEIN[9] provided numerical values

---

[1] The values for topaz in the current tabulation are solely those from VOIGT's measurements of 1887.

[2] VOIGT [1888, *1*].

[3] VOIGT [1893, *1*].

[4] VOIGT [1910, *1*]. This work was republished in 1928. I have examined only the original volume of 1910.

[5] VOIGT [1910, *1*].

[6] LOVE [1927, *1*].

[7] HEARMON [1961, *1*].

[8] SMITH and DAMON [1970, *1*].

[9] HEARMON [1966, *1*], [1969, *1*].

from the ultrasonic literature for a large collection of crystals, including elements and compounds, together with the citation of sources from which the numbers were obtained.

In addition, HEARMON's article contained many, many diagrams of elastic constants as a function of ambient temperature. These temperature data, also drawn from the ultrasonic literature, were used to determine temperature coefficients, $T_{c_{pq}}$ defined in Eq. (3.30).

$$T_{c_{pq}} = \frac{1}{c_{pq}(300)} \frac{d c_{pq}}{d T}. \tag{3.30}$$

The tabulated values of $T_{c_{pq}}$ for the various elements and compounds of the different crystal types, since they are given in room temperature terms for various ranges of temperature, and ignore the influence of the magnitude of the melting point in comparing crystals, are very different in magnitude. When introducing Eq. (3.27) above, I showed that the relation also was applicable for the c_{44} elastic constant of cubic and hexagonal single crystals. In this instance, Eq. (3.27) becomes

$$\begin{aligned} c_{44}(T/T_m) &= c_{44}(0), & 0 \leqq T/T_m \leqq 0.06, \\ c_{44}(T/T_m) &= 1.03\, c_{44}(0)\,(1 - T/2T_m), & 0.06 \geqq T/T_m \leqq 0.4. \end{aligned} \tag{3.31}$$

As would be expected, similar results are obtained for the temperature dependence of the other shear constant in cubic metals, $c' = \frac{1}{2}(c_{11} - c_{12})$. As given in Eq. (3.30), $T_{c_{pq}}$ places undue emphasis upon room temperature. Defined in terms of the zero point value, one obtains $T_{c_{pq}} = 0$ for $T/T_m < 0.06$, and $T_{c_{pq}} = -\frac{1.03}{2T_m}$ for $T/T_m > 0.06$. A redefinition of $T_{c_{pq}}$ in terms of the fractional melting point temperature of course gives in the second instance the same value of approximately $-\frac{1}{2}$ for all the solids considered.[1]

I gave a few comparisons in Fig. 3.128 above. The major interest in this subject other than completing the study of the quantum structure of the zero point isotropic moduli of the elements was the fact that from experiments on the finite deformation of these solids (to be described in the next chapter), I found that the temperature dependence at very large strain was linear, expressible as $(1 - T/T_m)$. The shear modulus at infinitesimal strain, while also having a linear dependence on temperature, had the different numerical value of $(1 - T/2T_m)$. This difference had interesting and perhaps serious implications for atomistic theories for which parameters for finite deformation are calculated from dislocation models requiring a dependence upon the shear modulus.

VOIGT's study of anisotropic linear elasticity beginning in 1874 has led finally, by the end of the 1960's to the same conclusion concerning nonlinearity which was outlined at great length in Chap. II above, for the isotropic solid. The determination of pressure coefficients

$$P_{c_{pq}} = \frac{1}{c_{pq}(0)} \frac{d c_{pq}}{d P}$$

at constant temperature, for which a few values were given by HEARMON in 1966 in his LANDOLT-BÖRNSTEIN tables, began with the experiments of DAVID LAZARUS[2] in 1949, referred to above. Between 1966 and 1969, experimental interest in third-order elastic constants for nonlinear small deformation elasticity was reflected in a new section of the LANDOLT-BÖRNSTEIN tables compiled by BECHMANN and HEARMON.[3] The number of third-order constants, c_{pqr}, for the different crystal

[1] BELL [1968, *1*].
[2] LAZARUS [1949, *1*].
[3] BECHMANN and HEARMON [1969, *1*].

systems, let alone the additional determination of their temperature dependence, is such that the problem of interpreting the numbers is an order of magnitude beyond that posed by the enormous array of the second-order constants.

In the 20th century problem of defining third-order constants and determining how one must make the tremendous number of measurements to obtain the desired 6 to 56 of them, one may see an historically interesting, detailed parallel with the 19th century evolution of VOIGT's ideas and observations. In referring the reader to the available tabulated second and third-order constants I emphasize the experimental and theoretical dilemma of interpreting, in linear terms, wave speed data from a nonlinear domain. The interest in ultraharmonics, subharmonics, phonon interaction, energy exchange between ultrasonic wave components, and the like suggests that the importance of the linear approximation may be dwindling in one of its most important strongholds, atomistic physics. The development of nonlinear theories of wave propagation in isotropic and anisotropic solids, together with an appropriate reflection theory at free and mixed boundaries for materials in both prestressed and zero stress states characterize the 20th century just as the 19th century, we now see, was characterized largely in terms of the linear approximation.

Between 1891 and 1893 VOIGT embarked upon a program of research in polycrystalline metals which still is of considerable importance inasmuch as few persons have seen fit to continue such studies. Using specially cast blocks of 14 different metals, including six different steels, VOIGT removed specimens in the form of bars oriented in various directions in the block. He performed flexure and torsion experiments in the same manner as for the anisotropic crystals to ascertain whether or not these metals as polycrystals were homogeneous and isotropic as assumed. By determining whether or not measured E and μ values decreased in the same manner in a given direction, compared to the average, he could distinguish between inhomogeneity and anisotropy when either or both were present. To avoid the influence of the elastic after-effect and minimize permanent deformation, VOIGT performed vibratory measurements slightly reminiscent of those of KUPFFER[1] and also a bit involved with some of KUPFFER's difficulties. He attached masses to the specimen. Because of the increased inertia, vibratory measurements became possible which otherwise would have been undeterminable, but new difficulties arose in interpreting the data. Since part of VOIGT's program was a study of the damping coefficient and its possible directionality in these same specimens, numerous small empirical assumptions were needed to account for the usual matters of air friction, etc.

For each of the specimens extracted from the blocks VOIGT tabulated a number of measurements, so that he had a sizable grand total for his final averages.[2] VOIGT commented upon the observed homogeneity and isotropy of each solid and gave detailed descriptions of their thermal, mechanical, and chemical histories, and the like. Conditions varied greatly. Bronze, for example, was remarkably homogeneous and isotropic,[3] while at the other extreme, tin was neither. Aluminum and copper he found to be adequately isotropic but slightly inhomogeneous while brass was just the opposite.

Because much of the controversy in experimental solid mechanics has centered around this matter of isotropy, anisotropy, and inhomogeneity in homogeneous

[1] KUPFFER [1860, *1*].

[2] VOIGT [1893, *1*].

[3] VOIGT, therefore, used this as a sort of standard in his studies of dilatation and viscosity in solids. An inspection of his data curiously does not support his conclusion as to the superiority of bronze with respect to homogeneity and isotropy.

Table 111. VOIGT (1893). *Average values.*

Solid	E (g/mm²)	μ (g/mm²)	s_2/s	c/c_1
Al	6.56×10^6	2.578×10^6	2.55	2.66
Bronze	10.59×10^6	4.06×10^6	2.61	2.30
Cd	7.07×10^6	2.45×10^6	2.89	1.26
Fe	12.82×10^6	5.21×10^6	2.46	3.36
Au	7.58×10^6	2.85×10^6	2.66	2.02
Cu	10.85×10^6	4.78×10^6	2.27	6.43
Mg	4.261×10^6	1.710×10^6	2.49	3.06
Brass	9.217×10^6	3.695×10^6	2.50	3.04
Ni	20.34×10^6	7.82×10^6	2.60	2.33
Ag	7.79×10^6	2.96×10^6	2.63	2.16
Steel (Avg. 6 blocks)	20.87×10^6	8.217×10^6	2.54	2.78
Bi	3.19×10^6	1.24×10^6	2.58	2.46
Zn	10.30×10^6	3.88×10^6	2.66	2.04
Sn	5.407×10^6	1.726×10^6	3.13 ?	—

and isotropic polycrystalline solids, I have included in Table 111 VOIGT's averages[1] of E and μ from the large number of measurements in each of the even larger number of specimens from the various directions in the same block. The data added to those from the similar earlier studies of VOIGT in glass, make a definitive list of values in a single chunk of a solid.

VOIGT determined elastic compliances and constants whose ratios provided a quantitative measure of anisotropy. I have added those ratios, too, to Table 111, where for isotropy the ratio of compliances should be 2.50, and for the much more sensitive elastic constants, 3.00.

3.46. Thermoelasticity.

I described in Chap. II (Sect. 2.12) in some detail the remarkable experimental beginnings of this subject in the infinitesimal elasticity thermal measurements of WEBER[2] in 1830. In Chap. IV (Sect. 4.39), I shall describe the thermal behavior of rubber in finite strain, considered by GOUGH[3] in 1805, and by JOULE[4] in 1859, and of rubber and organic tissues by a series of followers to the present. Also in that chapter is a description of experimental studies of thermoplasticity in finite strain by TAYLOR[5] and by DILLON[6] in the 20th century. In the field of infinitesimal elasticity the major interest in thermoelasticity has been theoretical. JOULE in his paper of 1859 included details regarding increases of temperature, measured by a thermocouple, in the small tension and compression of wrought iron, cast iron, copper, lead, glass, a large variety of wet and dry woods, cowhide, and whalebone,[7] as well as results from experiments on thermal properties of rubber in large deformation. Among 19th century papers on experiment in solid mechanics, it was one of those most frequently cited. For all of the solids JOULE examined, except wet bay wood and rubber, the temperatures decreased as expected with the expansion of volume with axial tensile loading, and increased with compression. For wet

[1] VOIGT [1893, *1*].
[2] WEBER [1830, *1*].
[3] GOUGH [1805, *1*], [1806, *1*].
[4] JOULE [1859, *1*]; and see [1857, *1*, *2*].
[5] TAYLOR and FARREN [1925, *2*], TAYLOR and QUINNEY [1934, *2*].
[6] DILLON [1962, *1*, *2*], [1963, *2*], [1966, *2*], [1967, *2*].
[7] Whalebone, of course, had some 19th century technological importance.

bay wood and rubber, he observed an increase in temperature during tensile loading.

JOULE's main interest was to compare his values with the developing theoretical ideas of WILLIAM THOMSON (KELVIN), since the experiments were closely associated with those concepts. JOULE found sufficient general agreement for the time, i.e., differences for compression between experiment and prediction of 6% for iron, 15% for copper, 9% for lead, 40% for glass, and 30% for wood. They illustrate the maximum differences; there also were instances of relatively close correlations. JOULE's detailed discussion of method, his study of the effect of the diameter of the specimen, and his repetition of loading and unloading measurements 100 times for the same load in a given specimen, make his paper an experimental classic, commensurate with its reputation.

For the last series of measurements,[1] JOULE added a 7 lb. compression load, observing the temperature changes for a loading and unloading in compression, followed by a loading and unloading of the same specimen in tension, also with a 7 lb. weight. The results, which varied much in each test, gave an instrument deflection for the four total averages of -1.40, $+1.38$, -1.34, and $+0.88$. JOULE set this as a minimum load for study.

In 1848, C. C. PERSON[2] had attempted to use WERTHEIM's moduli data of four years earlier[3] to establish an empirical relation between the elastic constants of metals and their latent heats of fusion. His proposal was that the ratio of E among the elements was equal to the ratio of the latent heats of fusion for the same solids. Thus, for the ratios of zinc to tin he obtained 2.17 as an elastic ratio, and 1.97 for the thermal ratio; for zinc and lead these were 4.80 and 5.23; for tin and lead, 2.20 and 2.65; and for zinc and bismuth, 2.28 and 2.22.

PERSON's efforts to explain why the ratio was not exact need not be recounted here, but it is interesting that he was challenged to provide measurements of latent heats in cadmium and silver, for which he had predicted a ratio from elastic constants beforehand. The paper described his calorimetric experiments. PERSON's results agreed with his prediction sufficiently for him to state that his ratios of E and latent heats were equivalent for ratios of tin, bismuth, lead, ancet alloy, zinc, cadmium, and silver, i.e., for all the metals he tested. I have compared some 20th century ratios with PERSON's, and have found for zinc/lead 4.20 to his 5.23; tin/lead, 2.52 to his 2.42; zinc/tin 1.74 to his 1.97; and zinc/bismuth, 2.25 to his 2.22, which suggests that PERSON's proposal of 120 years ago might merit renewed attention today.

The second large study of thermal effects after JOULE's was ERIK EDLUND's,[4] which followed immediately, in 1861 and 1865. Influenced by the work of CLAUSIUS[5] twelve years before, EDLUND also used thermocouples to study thermal variations as a function of the volume changes in a deformed solid. After giving careful reference to the pioneering precedence of WEBER in thermoelasticity and to the similar measurements of JOULE, EDLUND in very careful studies located his specimens in shells of constant temperature and, with a telescope, examined them at a considerable distance, using extended thermocouple wires to avoid interference from air circulation and from the body heat of the experimentist. He demonstrated that whether the loads were applied or removed, the temperature rise and fall were in fact equivalent.

[1] JOULE [1859, *1*].
[2] PERSON [1848, *1*].
[3] WERTHEIM [1844, *1* (a)].
[4] EDLUND [1861, *1*], [1865, *1*].
[5] CLAUSIUS [1849, *1*].

EDLUND's paper in 1861 contained much detail concerning his measurements, which indicated the general difficulty of such studies of thermal elasticity in solids in small deformation. His chief contribution was in providing greater precision than in his previous observations. Obviously not having read the references he cited, EDLUND curiously believed that he was the first to find by experiment that when the volume of a metal is increased by external forces, the temperature decreases, and when the original volume is regained, the temperature rises. His demonstration of the proportionality, $x = \pm A\,P$ between the temperature change, x, and the applied load, P, was the first experimental attempt to establish a relation between temperature and deformation. By 1865, in his second paper, this relation had become $x = A\,P + B\,P^2$, and the objective of this continued study was to determine values of the constants A and B. EDLUND summarized his results for measurements in silver, steel, copper, brass, platinum, and gold by comparing observation and prediction; he obtained differences, respectively, of 0.1, 0.16, 6.4, 1.9, 9.0, and 0.86%.

Much of the attention of those interested in thermal phenomena associated with deformation was focused upon avoiding the problems of measurement arising from the thermal after-effect. I have referred to the dead weight experiments of TOMLINSON[1] in the 1880's, of J. O. THOMPSON[2] in the 1890's, and GRÜNEISEN[3] in 1906 all of whom were concerned with the problems accompanying measurements with great precision in the dead weight loading of solids. Under this general heading, of course, fall the thermal dilatation of solids under zero load and the measurement of specific heats, both of which were of interest to early 19th century experimentists. With precise values for the former, and a large recent literature on the latter, they form a part of the experimental foundations of related fields of physics.

3.47. Viscoelasticity.

It is an experimental fact, learned in the cradle, that the sound and the visible motion of the vibrating structure fade away with time. In the field of linear viscoelasticity the phenomenological observation of this behavior begins as a recorded experiment with the measurements of COULOMB[4] in 1784. His results, which we have considered above (Sect. 3.4), were given in sufficient detail to demonstrate the problems to be encountered in pursuing this subject. Some of the early experimental interest in viscoelastic damping was in controlling the difficulties arising from the decaying oscillations of galvanometer elements in the pursuit of other measurements, but some, like KELVIN's[5] in 1865, considered it an issue for its own fundamental importance. KOHLRAUSCH,[6] among others, sought to relate WEBER's[7] elastic after-effect with oscillatory damping effects. As is well known, the matter became a subject of major theoretical importance in the work of BOLTZMANN[8] and KELVIN, among others, in the late 19th century.

[1] TOMLINSON [1886, *1*].
[2] J.O. THOMPSON [1891, *1*].
[3] GRÜNEISEN [1906, *1*].
[4] COULOMB [1784, *1*]. As TRUESDELL has shown [1960, *1*], MERSENNE [1636, *1*] described the results of measurements of attenuation. Such descriptions, lacking as they were in details, as were many other very early measurements, were more suggestive than definitive from the point of view of the experimentist.
[5] KELVIN (WILLIAM THOMSON) [1865, *1*].
[6] KOHLRAUSCH [1863, *1*].
[7] WEBER [1835, *1*], [1841, *1*].
[8] BOLTZMANN [1882, *1*].

Certainly the best known 19th century experimental paper[1] was that of KELVIN in 1865. The experiments for his study were performed by DONALD MACFARLANE, the official assistant to the Professor of Natural Philosophy at the University of Glasgow. Acknowledging the stimulation of the work of COULOMB, KELVIN in the series of torsion pendulum experiments on cylindrical wires, varied the weights, moments of inertia, and initial amplitudes, all in the manner of COULOMB. In addition, however, KELVIN studied the viscous decay of specimens which had been under continuous vibration for a period of time, in contrast to similar specimens which had been at rest during the same interval. Like COULOMB, KELVIN counted the number of cycles required to reduce the amplitude from 20 units to 10 units. For aluminium, he found that it was much more rapid when the initial amplitude was 40 than when it was 20. He noted that the greater the velocity, the greater was the loss of energy through one range; he noted also the curious fact that when the loading weight was increased, the viscosity at first much increased, but then day after day it gradually decreased until it approached the value it had had with a light weight. A wire which had been in continuous oscillation all day, exhibited much greater viscous damping on later days than did a similar quiescent specimen. A reversal of roles revealed that it was indeed the prolonged vibration which produced the effect.

These few observations in the 19th century, while widely referenced, do not carry experiment in viscoelasticity much beyond COULOMB's endeavors in the 18th century. The number of 19th century experimental works on the subject listed in the *British Royal Society Catalog of Scientific Papers* is remarkably small in view of the theoretical attention viscoelasticity received in the same period. Among the former were the measurements of A. V. OBERMAYER[2] in 1877, the memoir of IGNAZ KLEMENČIČ[3] in 1879, the measurements of TOMLINSON[4] in 1886, a few notes such as those of DEWAR[5] in 1895, and of course the well known papers of VOIGT[6] in 1892. VOIGT's experimental effort must be regarded as the most significant attempt to examine the fundamentals of the subject.

OBERMAYER[7] wished to determine a coefficient of viscous friction for the flow of fresh black pitch for which he proposed a linear viscosity, the proportionality constant μ, between the applied force and the strain velocity. His work was notable experimentally for this early effort to establish linear viscosity in a solid continuum. The fresh pitch was located between two circular plates pressed together by known weights. The torques, angles, and times were recorded while the plates were turned with respect to each other. OBERMAYER compared fresh pitch with old black pitch and recorded the various results. His was of course an outgrowth of the work of TRESCA,[8] but it is of further interest partly for a presumed analogy with the inner friction of flowing glacial ice.

In his study in 1879 KLEMENČIČ was seeking to establish a relation between the logarithmic decrement of torsional vibration of an iron wire and its magnetization. This was another 19th century subject of much interest after WERTHEIM had established a relation between deformation and magnetization in the 1840's. TOMLINSON

[1] KELVIN (WILLIAM THOMSON) [1865, *1*]. These results were reprinted almost intact in KELVIN's article on "Elasticity" in the 9th edition of the *Encyclopaedia Britannica* in 1880. [1880, *1*].

[2] OBERMAYER [1877, *1*].

[3] KLEMENČIČ [1879, *1*].

[4] TOMLINSON [1886, *1*].

[5] DEWAR [1895, *1*].

[6] VOIGT [1892, *1*, *2*].

[7] OBERMAYER [1877, *1*].

[8] TRESCA [1864, *1*], [1872, *1*, *2*, *3*].

for example, included the study of the effect of magnetization upon viscosity as well as upon elasticity, among the many topics he explored experimentally in the 1880's. KLEMENČIČ noted that H. STREINTZ and P. SCHMIDT had observed earlier that for an iron wire the logarithmic decrement over a period of four weeks decreased from a value of 0.00432 immediately after annealing, to 0.00227; this was a considerable change. KLEMENČIČ stated emphatically his negative result, which in fact was directly contrary to the earlier claims of WIEDEMANN[1] that there was a relation between the logarithmic decrement and the strength of the magnetic field for iron. KLEMENČIČ found that magnetization did not influence the damping of torsional vibration in an iron wire. To have smoe positive offerings, KLEMENČIČ showed that the viscosity of the wire changed with temperature and with small amounts of permanent deformation.

DEWAR'S[2] note in 1895, which will be referred to in the next chapter, contained an attempt to divide solids into those capable of flowing from an orifice and those not capable of doing so.

As a portion of his large study of the isotropy and homogeneity condition of polycrystalline blocks of various metals, VOIGT[3] in 1892 determined logarithmic decrements in both flexural and torsional free vibrations. For the former, the rod was clamped at one end while for the latter, one end was clamped and the other was attached to a metal disk. VOIGT told of the care he exercised by proper soldering and cementing of parts to minimize losses to the apparatus. He conceded such losses were an important source of error in his results. Further difficulties were encountered in air damping, which VOIGT attempted to eliminate by correction factors.[4] He recognized that for his torsional vibration some additional energy was lost in unavoidable flexure which accompanied the vibration.

VOIGT began his study with an analysis of the linear viscoelastic solid whose properties he wished to examine in all generality. In his experiments on homogeneous isotropic bodies he sought primarily to ascertain whether a proper material constant for internal damping of a free vibration was independent of frequency, as BOLTZMANN[5] had proposed, or frequency dependent as VOIGT expected from his own linear analysis. Fom his measurements he was able to determine the logarithmic decrement or the logarithm of the ratio of two successive amplitudes. He divided his solids for separate study: those with large damping and those with minimal damping. For the latter he could ignore the dependence of the frequency upon damping. In this instance, from linear analysis he could obtain the approximate parameter $\alpha = \frac{lT}{2\pi^2}$ where l was the logarithmic decrement, and T, the period of vibration.

VOIGT chose phosphor bronze for special study because his measurements of moduli had shown it to possess nearly perfect isotropy and homogeneity. Using the subscript β for flexure and γ for torsion, VOIGT obtained definite dependence upon period for both modes of vibration. In the flexure a change in period from 0.537 to 1.023 sec in four steps gave a variation of the logarithmic decrement from 14.40×10^{-4} to 6.71×10^{-4} while the quantity, α_β, despite fluctuations, remained essentially constant with the average value, $\alpha_\beta = 37.5\times10^{-6}$.

[1] WIEDEMANN [1858, *1*], [1859, *1*].
[2] DEWAR [1895, *1*].
[3] VOIGT [1892, *1*, *2*].
[4] I already have criticized VOIGT's arbitrary, small "adjustments" of his data as he tried to account for assumed, but unobserved phenomena, before he compared his measured data with prediction.
[5] BOLTZMANN [1882, *1*].

For torsion, both l_γ and α_γ varied with the period, T, although in a more pronounced fashion for the former. Brass gave results essentially similar to those for bronze. For copper, both α_β and α_γ were nearly independent of period, although l_β and particularly l_γ varied strongly. The behavior of nickel was similar to that of copper. VOIGT thus classified these two solids as being in definite agreement with his simple laws of inner friction. For aluminum, cast steel, and cadmium, on the other hand, VOIGT's logarithmic decrements were constant with period for flexure and for torsion; for both these configurations, α was a function of frequency. Finally, in tin and silver there was such variation among the specimens that no conclusions could be drawn.[1]

In a second long and very detailed study[2] of bronze described by VOIGT in the same year, 1892, the problems present in such 19th century studies are illustrated, as is the variation of results obtainable. For torsion, he showed that the logarithmic decrement in the attentuation of vibration in bronze differed markedly with period, with a somewhat better trend toward VOIGT's prediction for α_γ. VOIGT examined a number of other aspects of the subject; he noted, for example, that he was unable to find the effect upon the logarithmic decrement of prolonged oscillation which KELVIN[3] had reported. He was much concerned with the relation between WEBER's[4] and KOHLRAUSCH's[5] elastic after-effect and vibratory damping, as the latter had been before him.

In this aspect of solid mechanics, VOIGT had delineated the experimental and analytical complexities which to the present day, if in varied form, have continued to perplex those who followed him. CLARENCE ZENER[6] in his monograph in 1948 on the *Elasticity and Anelasticity of Metals* reviewed much of the mid-20th century development. In this subject the space devoted to theory or to empirical interpretation, even in experimental papers, outstripped by far that describing experiments or experimental results.

The 20th century has seen careful suspension of the specimen in the manner of FÖRSTER[7] described above, from which KÖSTER[8] in the succeeding decade obtained the temperature dependence of the logarithmic decrement of most of the elements. In the 1950's KOLSKY[9] introduced short pulses into, among other solids, polyethylene, polymethylmethacrylate, polystyrene, and later, lead rods, from which, by a study of the variation of pulse shape with each reflection, he could determine the viscosity. For wave fronts sufficiently long that three dimensional effects could be ignored, KOLSKY was successful in analyzing his results for the polymers undergoing small deformation in linear viscoelastic terms.

Damping in crystalline solids has been used as a tool for the study of other aspects of solid state physics, usually with the assertion of linear viscosity to interpret the results obtained. TRELOAR[10] in 1958 referred to the experiments of GEHMAN, WOODFORD, and STAMBAUGH in 1941 as demonstrating that pure gum rubber had a resonance curve in forced vibration in close agreement with linear viscoelasticity. Modern ultrasonic pulse echo techniques have systematized the

[1] VOIGT [1892, *1*].
[2] VOIGT [1892, *2*].
[3] KELVIN [1865, *1*].
[4] WEBER [1835, *1*], [1841, *1*].
[5] KOHLRAUSCH [1863, *1*].
[6] ZENER [1948, *1*].
[7] FÖRSTER [1937, *1*].
[8] KÖSTER [1940, *1*], [1943, *1*], [1947, *1*], [1948, *1, 2, 3, 4*], FÖRSTER and KÖSTER [1937, *2*].
[9] KOLSKY [1953, *1*], [1954, *1*], [1956, *1*], [1959, *1*], [1960, *1*].
[10] TRELOAR [1958, *1*].

study of viscous effects in solids while at the same time introducing a number of new problems in the form of scattering and reflection.

In 1967, WILFRED E. BAKER, WILLIAM E. WOOLAM, and DANA YOUNG,[1] by performing experiments under a variety of air pressures and frequencies, provided an experimental study of what COULOMB[2] began, KUPFFER[3] attempted, and VOIGT[4] estimated, namely, the detailed division of the contributions of the solid and of the air to the observed damping of a thin cantilever beam.

In the 1950's EDWIN R. FITZGERALD[5] while studying the resonance properties of driven solids discovered an important phenomenon in the form of resonant frequencies which were a function of the solid material itself and not of the size of the specimen. In extensive investigations of this behavior FITZGERALD during the next decade removed all questions that these material resonances were associated with the driving apparatus by completely modifying and simplifying the experiment without affecting the phenomenon observed. FITZGERALD's summary in 1966 of over a decade of experimental study[6] included a proposed explanation of this discovery of resonance in terms of particle wave mechanics in the crystal lattice.

The response of solids to either free or forced vibration within the framework of the linear approximation of this section has been a subject of experimental study for nearly two centuries, and of analytical explanation, for one. It may be that KOHLRAUSCH's concern in the mid-19th century for what appeared to him to be a nonlinear phenomenon even in infinitesimal deformation, was of more importance as a prophecy than most persons in the intervening years have perceived.

3.48. Summary.

When the measurement is sufficiently gross, the initial deformation of nearly all the solids which have been studied can be represented by a linear approximation. For most of the crystalline solids for which high precision in measurement is required to detect fundamental departures from linearity, the linear approximation has assumed, theoretically and experimentally, a role of major importance during the past two centuries.

Those who have used the linear approximation and have taken advantage of its analytical simplicity have classified response to deformation into various compartments, each of which has been made the subject of special study. The characterization of solids as having perfect linear elasticity has led to a vast experimental program of the determination of elastic constants for isotropic and anisotropic, presumably homogeneous, substances. Further, it has led to the exploration of the dependence of these elastic constants, or the elastic compliances, on a variety of parameters such as the ambient temperature, applied stress rate, strain rate, the prior thermal, chemical, and mechanical history, and ambient electric and magnetic fields. For the most part, numbers were not tabulated and catalogued merely for the purpose of accumulating them (although certainly this sometimes has occurred in our time), but rather to explore and compare patterns of understanding derived from experiments, with patterns of explanation, by delineating the functional dependence of the various parameters.

[1] BAKER, WOOLAM, and YOUNG [1967, *1*].
[2] COULOMB [1784, *1*].
[3] KUPFFER [1860, *1*].
[4] VOIGT [1892, *1*, *2*].
[5] FITZGERALD (summarized 1957–1966 in [1966, *1*]).
[6] *Ibid.*

With the experiments of COULOMB, CHLADNI, and RICCATI, the 18th century initiated the study of elastic constants determined by experiments on vibrations, and in 1809 BIOT provided the first modulus determined from a measured wave speed in a solid; however, no quasi-static measurements of moduli had been made between HOOKE's observations, published in 1678, and the first such experiments of DULEAU in 1813. That DULEAU's measurements of small deformation in iron dominated the literature until the 1840's is merely another way of stating the fact that in that interval most of the experimental work related to elastic constants was of little value. Important exceptions were the experiments of VICAT in 1831. VICAT obtained a strain resolution sufficiently accurate to place his measurements in the context of the mid-20th century.

The subject became an experimental science in the modern sense with the research of the major 19th century figure in experimental solid mechanics, WERTHEIM, whose contributions in just a very few years included: the first extensive series of measurements in well defined metals and binary compounds; the first studies of elastic constants as a function of temperature and electric and magnetic fields; the first study of anisotropic elastic constants; the first experimental study of the elastic constants of various kinds of glass; the first quantitative study of photoelasticity, which led to the stress-optical law for doubly refracting solids known later as "WERTHEIM's law"; the first measurement of compressibility in solids, of longitudinal wave speeds in a wire, and of the velocity of sound in a column of water; and the disclosure of the experimental fact that the linear elasticity of isotropic solids required the determination of two constants, despite the attractive uni-constant molecular theory of elasticity almost universally accepted at the time.

Given the relative simplicity of linear elasticity, it is strange that there have been so many areas of disagreement which have persisted over long periods of time. In this treatise I have traced the details of the half century of the POISSON's ratio controversy, a subject which fascinated a large number of experimentists and generated some of the finest experiments developed in the 19th century. Another controversy had been started by the demonstration of BOLTZMANN in 1882 that SAINT-VENANT's one-dimensional theory of rod impact of 1867 simply was not in accord with experiment. This led to a debate of nearly 90 years which centered not upon the fact of the inapplicability of the theory, which was generally conceded except by the authors of elementary textbooks, but upon attempts to find experimental justification for appropriately modifying the theory. The presentation of a large proportion of the experiments after BOLTZMANN's on this subject, has enabled me to trace the growth of experiment in the general field of wave propagation.

Research based upon the linear approximation, of course, extends beyond the determination of the numerical values of elastic constants. Experimentists have been concerned with the linearity in small deformation of thermal elasticity, of viscous elasticity, and of the relation between adiabatic and isothermal behavior in solids. Of prime concern have been the fundamental questions regarding anisotropy in single crystals; whether or not isotropy indeed may be achieved in polycrystals; and the nature of the temperature dependence of elastic parameters for both single crystals and polycrystals.

With the development of a reliable experimental background of sufficient scope, it becomes possible to consider the interrelation of the various solids and classes of solids. My own experimental studies in the past decade have revealed the existence of a quantized distribution of elastic constants among the elements, with a unifying set of discrete states. These occur not only for the linear elastic shear modulus at the zero point for isotropic bodies, for which the unification was discovered, but also for the E-modulus in a given solid at ambient temperature. They have been designated as "multiple elasticities"; they exist as quantized states in a given crystalline solid, in the same form in which they occur among the 60 elements which have been studied. Hence, constitutive discontinuities, or second order transitions in infinitesimal elasticity, are of major importance; they must be included in any comprehensive explanation of the behavior of solids in small deformation.

In the past few decades, most of the measurements of linear parameters in elasticity have been obtained for secondary purposes, to provide numerical values for atomistic model analyses. The experiments in the continuum literature, while containing some emphasis on nonlinear parameters such as third order constants, continue essentially in the traditions of the past. The usefulness of the linear approximation is not eliminated by the experimental fact that it is always an approximation. The possibility of further discovery using the simplest approximation is by no means exhausted.

IV. Finite Deformation.

4.1. Paucity of experiment before 1800.

From a 20th century perspective, the 17th century experiment on large deformation of a gut string by JAMES BERNOULLI[1] closely rivals in historical importance the experiment of ROBERT HOOKE[2] on the small deformation of iron string. When one adds the curvature reversal observed in the response function obtained from the experiment on gut string by JAMES RICCATI[3] in 1721, which, as I have shown above, KARMARSCH[4] a century and a half later found was caused by the onset of permanent deformation at approximately 10% strain, it is evident that the gross features of the behavior of solids in uniaxial deformation had been delineated early. It is interesting to reflect upon the shift of emphasis which might have occurred in experimental physics in the past three centuries, had the mathematics of linearity turned out to be complex, while that of nonlinearity had been found simple.

The variety of behavior in large deformation observed in the numerous substances which have been called solid, often in loose phenomenological terms, defies simple overall generalization. Solids deform in a variety of ways. Some, such as rubber or gut, undergo finite strain along either the same or adjacent loading and unloading paths, with negligible permanent deformation on returning to the initial load; and others, such as metals and clays, invariably become permanently deformed in finite strain following stress reversal, and have markedly different loading and unloading stress-strain functions. Crystalline solids, at relatively small strain undergo a stress-strain transition, sometimes with dramatic abruptness, at a critical stress known as the yield limit, from either a recoverable "perfect" elastic or time-dependent viscoelastic state, to a thermodynamically complex plastic state. At the opposite extreme, amorphous solids may be smoothly

[1] BERNOULLI [1694, *1*].
[2] HOOKE [1678, *1*].
[3] RICCATI [1721, *1*].
[4] KARMARSCH [1841, *1*].

distorted in shape by a load of nearly any magnitude if it is applied for a sufficient length of time.

Large deformation, in which products and squares of strain components no longer may be ignored, demands more care in definition of both stress and strain. Definitions, of course, are arbitrary, but from the experimental point of view predilection and purpose in choosing them have had strong influences upon display of data and inadvertently upon interpretation and explanation.

The major interest in large deformation since the middle of the 18th century has been in recording, in addition to the all important cohesive strength, the largest deformation at which rupture occurred. COULOMB, as I have stated in Sect. 3.4, discovered experimentally the torsional elastic limit in iron and brass wires, extending his study through large deformation up to fracture. His purpose was to determine the magnitude of the unloading strain as a function of the permanent deformation, as well as to determine changes in the dynamic shear modulus around zero stress as a function of the prior permanent deformation. His discoveries that the recovery was asymptotically constant and that the shear modulus decreased as the permanent deformation increased, were the consequences of the first serious study of the plastic deformation of metals.

MUSSCHENBROEK[1] in his *Essai de Physique* (1739) described a very early example of a problem in dynamic plasticity in the experiment of RICCIOLI, who measured how far a vertical pointer subject to the impact of a wooden ball falling from various heights, h, would penetrate a barrel full of butter. MUSSCHENBROEK suggested that RICCIOLI either had a low opinion of his own experiments or had examined insufficiently what he should have concluded from them when asserting that the depth of penetration was a linear function of the velocity, i.e., proportional

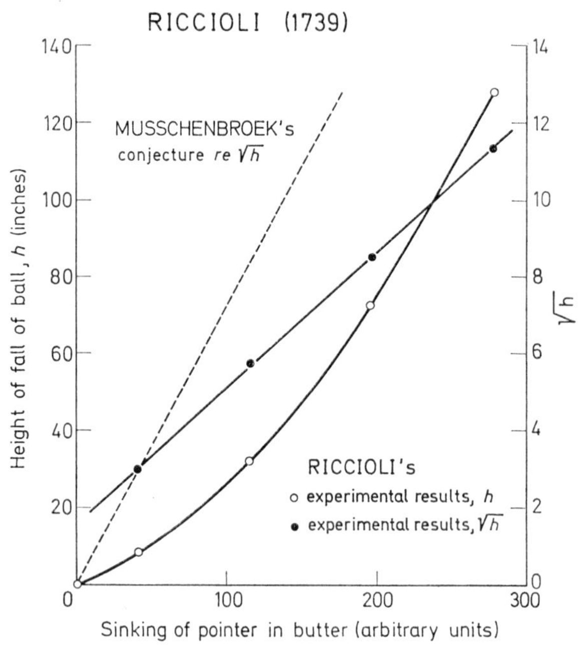

Fig. 4.1. The penetration of butter by a wooden stick subjected to the impact of a wooden sphere (1739).

[1] MUSSCHENBROEK [1739, *1*], p. 92.

to $\sqrt{h}$. MUSSCHENBROEK's statement was based upon extrapolation (dashed line, Fig. 4.1) solely from RICCIOLI's lowest height of drop, which was that producing 40 arbitrary units of penetration into the butter. In Fig. 4.1 I have given graphically RICCIOLI's tabulated results (open circles), to which I have added a plot of the distance of penetration vs $\sqrt{h}$ (closed circles), from which it may be seen that RICCIOLI indeed did find that the depth of penetration was proportional to the velocity but with an added constant, which I shall attempt to fathom no further.

4.2. 1800 to 1850: The experiments on creep of Navier and Coriolis, and the summary by Poncelet of research before 1840.

Before 1840 the description of behavior in large deformation was no more than the often casual comments of the experimentist proceeding toward the major objective of his experiment, namely, to determine the maximum load and the deformation at rupture. Typical of such comments were those of NAVIER in his memoir in 1826 on the resistance of various substances at the rupture point in longitudinal tension. Twenty-five of the twenty-seven experiments NAVIER[1] described were tensile tests on strips of iron, red copper, lead, and glass. The remaining two were on hollow iron spheres under internal pressure. NAVIER, after expressing a distrust of the use of machines in such experiments as having the effect of "invariably altering the results," made his measurements as had MARIOTTE by applying the weights directly. By suitably marking the specimen he was able to observe during the course of the test and at the time of the rupture, the changes of length and breadth. His results, which were of minor importance, described the initial shape of the specimens, their appearance at rupture, the rupture load, and in a few instances, observations with respect to the intermediate elongation. He noted that the elongation of iron before rupture was quite irregular, varying from 5 to 10% from one test to another. Before rupture copper elongated about 40%, and lead, about 10%, unless the specimen were allowed to elongate slowly and progressively under a large load which ultimately caused rupture.[2]

A decided contrast to these trivial measurements of NAVIER was the study of the compression of lead cylinders, made by CORIOLIS[3] in 1830. CORIOLIS tested lead cylinders 24 mm in diameter and 19 mm long. The scale used to measure the thickness divided the initial 19 mm length into 680 divisions, so that each division corresponded to $^1/_{36}$ mm. An undescribed lever apparatus presumably enabled him to make accurate measurements of elongations. The lead specimens were placed between two iron plates, forming a box. On the cover of the box was a steel button upon which a wheel supporting various weights could be applied by means of a jack which supported the axle. CORIOLIS determined the permanent deformation as a function of the applied load and of the length of time in which the load had been applied. Table 112 shows the measured strain.

The material was described as "poor" coin lead; no precautions concerning oxidation had been taken during the pouring. The duration of the applied load was always the same: one minute. Omitting CORIOLIS's discussion of his results related to the effects of small amounts of oxide upon the plastic resistance of lead which

---

[1] NAVIER [1826, *1*].

[2] As far as I can determine, this comment by NAVIER is the first reference to creep. Although his statement preceded by 8 years the definitive observations and discovery by VICAT (VICAT, [1834, *1*]), NAVIER should not be credited with precedence, since he made no measurements of creep and attached no special significance to his passing comment.

[3] CORIOLIS [1830, *1*].

Table 112. CORIOLIS (1830).

Load	Thickness after crushing, $^1/_{36}$ mm divisions								Average	ε
1500	464	471	462	459	462	463	—	—	463	0.32
1824	335	335	337	—	—	—	—	—	336	0.506
1950	341	336	337	339	331	—	—	—	337	0.496
3175	294	303	303	293	292	293	293	295	296	0.565

Table 113. CORIOLIS (1830).

	Duration of load placement (sec)	Thickness after crushing	ε
1500 kg	5	506	0.286
	10	503	0.291
	15	502	0.293
	20	498	0.299
	25	501	0.294
	30	501	0.294
	35	499	0.298
	40	497	0.301
	45	491	0.311
	50	487	0.317
	55	483	0.324
	60	485	0.321
	65	483	0.324
	75	483	0.324
1950 kg	30	365	0.518
	45	331	0.574
	60	322	0.589
	75	321	0.590
	90	319	0.594
	120	313	0.604
1760 kg	60	317	0.534
	3600 (1 h)	245	0.640
	24 h	223	0.663

he found to be considerable, I have selected two of his tabulations of the time dependence for loads of 1500 and 1950 kg. In Table 113, to which I have added the nominal strain, may be seen what I believe is the first measurement of short time creep at constant load.

This deformation instability in lead, shown graphically in Fig. 4.2, in what CORIOLIS himself described as incomplete experiments was regarded by him as merely complicating the measurements he wished to make. He stated that for his purposes, he had no interest in durations longer than two minutes, having maintained loads for longer times simply out of curiosity as to whether stability would be achieved in a reasonable period of time. The load of 1760 kg maintained for twenty-four hours showed that creep continued for longer than one to two hours, the duration of the previous experiments.

We already have described the experiments of GERSTNER in 1824 on elongation due to tension, which had proceeded only to a maximum strain of

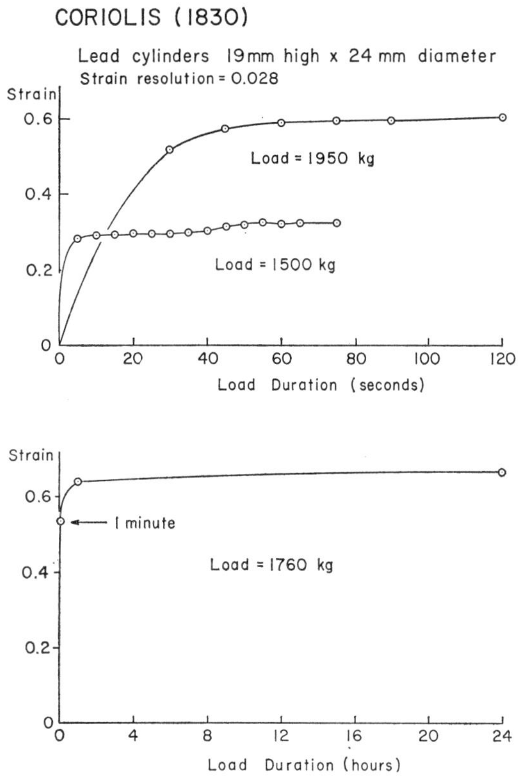

Fig. 4.2. The creep of lead under constant load.

0.007, whose envelope (*see* Fig. 2.5) was the source of GERSTNER's[1] law, $\sigma = A\varepsilon - \beta\varepsilon^2$, in that relatively small strain region. In 1823, JOHN LESLIE[2] performed a series of tensile experiments on bars of English iron 1 in. square and observed that "perfect" elasticity, in which stress was proportional to strain, held up to a load equal to $^1/_2$ the instantaneous rupture load. Beyond this limit he thought that elongations would increase according to the geometrical progression, 1, 2, 4, 8, 16, when the loads increased according to the arithmetic progression, $^4/_8$, $^5/_8$, $^6/_8$, $^7/_8$, and $^8/_8$, with the ratios referring to the fraction of the instantaneous rupture load. The region of strain he considered was from 0.001 to 0.016. LESLIE claimed that the following formula fit the data:

$$y = \frac{p\left(\frac{1}{2} + \frac{1}{8}\log 1\,000\,\varepsilon\right)}{\log 2}, \qquad \frac{p}{2} \leqq y \leqq p, \tag{4.1}$$

where y is the load; p, the rupture load; and ε, the strain.

LESLIE's proposed stress-strain relation for finite deformation interested a number of persons during the next two decades. PONCELET,[3] as in the case of elastic constants, provided in 1841 the definitive summary of early experimental results for large deformation. In his summary he gave the following numerical

---

[1] GERSTNER [1832, *1*].
[2] LESLIE [1823, *1*].
[3] PONCELET [1841, *1*].

special case of Leslie's equation for a bar of 1 mm² section whose maximum load was 50.5 kg:

$$y = 88.16 + 20.97 \text{ kg} \log \varepsilon. \tag{4.2}$$

Although Poncelet noted that Leslie's formula did not describe the data of other experimentists such as Seguin, Bornet, and Ardant, some measurements on large bars of iron by an unspecified person at St. Petersburg had indeed followed Leslie's empirical relation but linearity ceased at a load of $^2/_5$ of the rupture load rather than $^1/_2$. Equation (4.2) constituted Poncelet's empirical fit to those nebulous data. Poncelet commented upon the fact that after the initially imperceptible strain below the elastic limit, there was a rapid increase in elongation followed by a slowing down before rupture. Soft wires required a long time to achieve equilibrium, which was established only after a large number of oscillations, i.e., behavior exhibiting the Savart-Masson (Portevin-Le Chatelier) effect. He gave as an example the fact that in lead the elongation corresponding to a load less than 0.1 kg/mm², did not become established before 72 h.

Certainly the most interesting experiments on large deformation of metals during the first half of the 19th century were those by Ardant,[1] who, in 1835, had given his results for the tensile deformation of a number of metals to Poncelet for publication in the second edition (1841) of his classic treatise, *Mécanique Industrielle*. As was stated in Sect. 3.10 Poncelet chose to report these data and the data of Seguin (1826) and Bornet (1834) in quantitatively specified diagrams of stress-strain functions, the first such graphs in experimental solid mechanics. In Fig. 3.21 is shown the initial portion of 9 diagrammed stress-strain functions; in Fig. 4.3 are shown the large deformation portions of these results for iron, brass, steel, and lead wires.

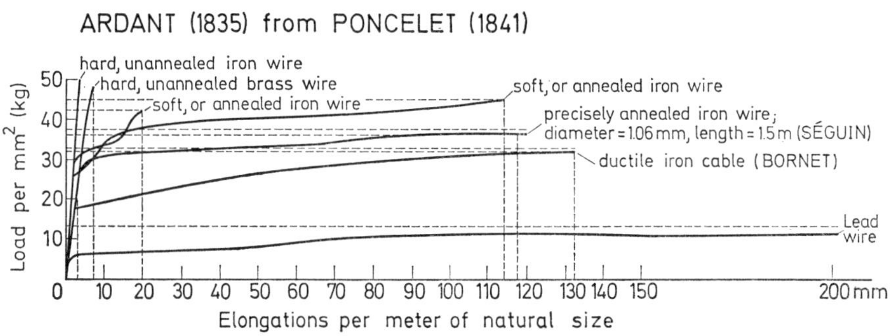

Fig. 4.3. The first graphical presentation of finite deformation in metals in which numerical values were specified. Poncelet (1841) presenting the data of Séguin (1826), Bornet (1834), and Ardant (1835).

After noting that Leslie's formula neither applied to these data nor provided for the observed difference in elastic limits for very hard and very soft wires, Poncelet commented upon the remarkable regularity obtained for large deformation experiments on hard brass wires, which was expressed by the following empirical equation, where y was the load and ε the strain:

$$\varepsilon = 0.1125\, y + 0.00039\, y\, (1.6)^{y/5}. \tag{4.3}$$

[1] Ardant (1835) *see* Poncelet [1841, *1*].

He further remarked that although with annealing there was a change in the absolute magnitude of the elastic limit, it always occured at $^1/_3$ of the rupture load, which itself varied as a function of the prior thermal history.

LAGERHJELM[1] in 1829 had proposed a law to determine the yield limit. Letting ε_Y be the strain at the elastic limit for an iron prism, and ε_R the strain or maximum elongation at the instant of rupture, LAGERHJELM claimed that for iron the two were related as follows:

$$\varepsilon_Y \sqrt{\varepsilon_R} = 0.000381\,. \tag{4.4}$$

Table 114. PONCELET (1841).

	Elongation per meter relative to natural elastic limit (mm)	Load (kg/mm²)	T' per meter length and mm² cross-section (kg)	Maximum elongation per meter before rupture (mm)	Load (rupture) (kg/mm²)	T' per meter length and mm² cross-section at rupture (kg)
Thick bar of ductile iron (BORNET)	0.55	12.0	0.00330	132.50	33.00	4.4970
Precisely annealed iron wire (SEGUIN)	—	—	—	120.00	37.84	3.9300
Iron wire, unequally annealed (ARDANT)	0.88	15.0	0.00662	20.50	42.50	0.6500
Iron wire, strong unannealed (ARDANT)	0.78	15.0	0.00585	3.10	49.00	0.0810
Steel wire coming from factory (ARDANT)	1.25	25.0	0.01560	3.15	57.50	0.0783
Ditto; tempered, blue annealed	1.20	25.0	0.01500	2.52	52.50	0.0580
Ditto; annealed, untempered and pliant	1.20	25.0	0.01500	2.40	57.50	0.0688
Ditto, strongly tempered	—	—	—	1.87	15.57	0.0125
Brass wire, soft annealed	1.35	15.0	0.01250	115.00	45.00	4.5140
Brass wire, strong unannealed	1.70	15.0	0.01275	7.19	49.00	0.2005
Lead wire, cold drawn	0.41	0.3	0.00012	324.60	1.36	0.3500

[1] LAGERHJELM [1829, *1*].

PONCELET noted that the data of ARDANT did not support any such empirical relation between the strain at the elastic limit, and the maximum elongation. On the contrary, a comparison of the behavior of hard and ductile metals could produce the apparent paradox of a reversal of LAGERHJELM's conjecture. This was evident from PONCELET's consideration of the deformation energy, T', for iron and brass of different prior thermal histories, and for lead. These data, shown in Table 114, provided measured areas under the stress-strain function at the elastic limit (third column) and at rupture (sixth column). As far as I know, they constitute the first statement regarding experimentally measured deformation energy in the plastic region.

Since PONCELET was the first person to describe a stress-strain function quantitatively in graphic form, it is no coincidence that he also was the first person

Table 115. TREDGOLD and ARDANT data from PONCELET (1841).

Designation of the metal subjected to traction experiment	Elongation relative to the natural elastic limit (m/m)	Corre-sponding load (kg/mm²)	Coeffiecient T_e' of the dynamic elastic resistance per mm² and per meter length (kg/m)	Coefficient T_r of the dynamic rupture resistance per mm² and per meter length (kg/m)	E (kg/mm²)
Iron wires or bars					
soft or annealed	0.00054	10.8	0.003000	4.00000	20000
strong or unannealed	0.00090	18.0	0.008000	0.08000	20000
Ordinary tempered and annealed steel	0.00120	25.0	0.015000	0.07000	21000
Cast English steel, 1st quality	0.002200	66.0	0.07260	0.16000	30000
Strongly tempered, very fragile steel, (ARDANT)	—	—	—	0.01250	11000
Cast iron (TREDGOLD)	0.00080	10.0	0.004000	—	12000
Annealed brass wires (ARDANT)	0.00135	15.0	0.01250	4.50000	10000
Unannealed brass wires, strong (ARDANT)	0.00170	15.0	0.012750	0.20005	—
Cast brass (TREDGOLD)	0.00075	4.8	0.001800	—	6450
Cast cannon bronze (TREDGOLD)	0.00104	7.3	0.003800	—	7000
Cast zinc (TREDGOLD)	0.00024	2.3	0.000280	—	9600
Cast English tin (TREDGOLD)	0.00063	2.0	0.000320	—	3200
Cold-drawn lead wire 4 mm diameters (ARDANT)	0.00067	0.4	0.000134	0.35000	600
Low-purity commercial lead wire, cast and cold-drawn, 6 mm diameter (ARDANT)	0.00050	0.4	0.000100	—	800
Ordinary cast lead (TREDGOLD)	0.00210	1.0	0.001050	—	500

to provide a measured value for the energy of deformation. PONCELET further explored this interesting new aspect of the plastic deformation of metals by determining what he referred to as the coefficients T_e', corresponding to the work at the elastic limit, and T_r, the work at rupture, for the extensive earlier experiments of TREDGOLD in cast iron, cast brass, cast zinc, cast tin, and cold drawn and cast lead. The coefficients of TREDGOLD were compared with the data of ARDANT, as in Table 115.

PONCELET stated that the differences in the numbers obtained still left much to be desired with respect to certitude and precision. In regard to his comment it should be noted, however, that the data of TREDGOLD actually were in flexure and were converted for comparison's sake by PONCELET in the table above.

During the 1840's, WERTHEIM's interest in finite deformation of metals was confined to the measurement of the usual ultimate or cohesion strength of each of the many elements and metal alloys studied, and to the specification of the first visible permanent deformation for a strain resolution of $\varepsilon = 10^{-5}$. For the former, his efforts to establish order among the various experimental results led him to express doubts as to the scientific value of the data. For the latter, as I have indicated in Sect. 3.12, he doubted that in general an elastic limit existed.

For the large deformation of human tissue, WERTHEIM, as has been shown in Sect. 2.14, had carried out a major experimental program which led to a quite general nonlinear, upward turning tensile stress-strain function. Neither WERTHEIM nor any of his contemporaries or immediate successors found a similar empirical simplicity for finite strain in crystalline solids; serious research on this subject lapsed for two decades.

4.3. Tresca on the flow of solids (1864–1872).

The general lack of order in experimental results for the large plastic deformation of metals was discouraging to experimentists and theorists alike. In 1864, with the strikingly original work of a single man, HENRI EDOUARD TRESCA,[1] some order was introduced. In the previous year, TRESCA had embarked upon what in the next eight years was to become an enormous number of experiments on the plastic flow of a wide variety of solids, from lead and copper to ice, paraffin, and ceramic paste. He demonstrated that there were measurable and reproducible flow coefficients which could provide the basis for a theory of large plastic deformation in solids. He made most of his measurements on lead. There were a number of different types of experiments the results of which he compared to see whether there was order among the flow coefficients. He punched cylindrical specimens axially with a cylindrical, hardened steel rod of smaller diameter; he extruded cylindrical specimens through circular, triangular, and rectangular orifices in both an enclosed and an unenclosed die; he compressed circular-cylindrical specimens between hardened plates; he examined the reverse extrusion of solid cylinders into hollow cylinders of various thicknesses, both with and without lateral constraint, etc. To observe the flow, he made the specimens in the form of a stack of individual plates.

The measurements consisted in the recording of the applied pressure, and the change in shape and location of the individual plates which it produced. Sections cut through the stack at the end of each test enabled him to describe in detail where each section of the solid had flowed during that type of experiment. For softer materials, such as clay, paraffin, etc., he used layered die markers of dif-

[1] TRESCA [1864, *1*].

ferent colors to keep track of the flow. Following his observation that the introduction of the laminated structure had affected the quantitative value of the resistance of some metals, he performed separate experiments upon solid specimens with the same geometric configuration.

The sheer bulk of the data, both described verbally and shown diagrammatically, makes the detailed study of the work of TRESCA a major undertaking. BAUSCHINGER was the only experimentist who exceeded TRESCA in the amount of data which he succeeded in getting into print. BAUSCHINGER's work during the 1880's, although different in many respects, was closely related to that of TRESCA. In fact, the studies of those two persons around one hundred years ago, still represented the major experimental basis for theories of plasticity until well beyond 1950, but certainly neither TRESCA nor BAUSCHINGER would have found acceptable the limiting constitutive assumptions of those theories: TRESCA, because of the omission of a region of plastic work hardening before the onset of constant stress plastic flow; and BAUSCHINGER, because of the oversimplification of observed yield surface phenomena.

The major flow studies of TRESCA began in 1863. Because of delays in publication and more than one presentation of the same work, it is convenient to divide the memoirs into three groups as Flow Memoir I, Flow Memoir II, and the Memoirs on Punching. Flow Memoir I was first presented as an author's summary in 1864.[1] The memoir itself was published[2] in 1868. The presentation in 1864 contained but a single drawing for the description of 40 experiments made on lead, 14 on other metals, and 40 on ceramic clays, not including several on powdery or granular materials. SAINT-VENANT[3] wrote of TRESCA's excessive reserve in the matter, complaining that while he was making his careful study of TRESCA's memoir the absence of working-drawings had been a major obstacle. He had informed TRESCA that the French Academy customarily provided facilities for making such drawings. TRESCA, accordingly, had an enormous array of superb drawings prepared, which were published in a graphic supplement[4] in 1872.

In 1867, the editors of *Comptes Rendus*[5] described TRESCA's being awarded on Monday, 11 March, the Mechanics prize of the Montyon Foundation for the year 1866. Attesting to the immediate appreciation of the importance of the study of TRESCA in 1864, the value of the prize was raised a thousand francs on that occasion.[6]

The Flow Memoir II, which first appeared as an author's summary[7] in 1867, not only extended the earlier work to new types of experiments, but also reported how TRESCA had improved the precision of the entire study. The full memoir[8] was a large paper published in 1872.

[1] TRESCA [1864, *1*].

[2] TRESCA [1868, *1*].

[3] SAINT-VENANT [1870, *1*].

[4] TRESCA [1872, *1*]. This graphic supplement also contained drawings for the punching studies.

[5] (TRESCA): Comptes Rendus [1867, *3*].

[6] That this was an unusually rapid recognition for TRESCA is seen in the fact that a translation in 1865 summarizing TRESCA's experiments on the flow of ice through orifices forming cylinders, in the *Philosophical Magazine*, cited his name as "Fresca"; the same misnomer appeared at different places in the text (TRESCA [1865, *1*]). The correct name did appear in the index of the volume. (The translation had been taken from SILLIMAN's *American Journal* for July, 1865.) The pressure required for the flow of ice was given as 126 kg/cm^2 compared to 630 kg/cm^2 for lead. The translator missed TRESCA's major point by insisting that obviously such a number must depend upon the proportions of the orifice; in addition, he misquoted TRESCA's numerical values.

[7] TRESCA [1867, *1*].

[8] TRESCA [1872, *1*].

The third collection of writings on punching initially was given in an author's presentation[1] in 1869. This was followed by the 121 page complete memoir in 1872.[2] To those two publications should be added the interesting Commissioner's Report of 1870 in which COMBES, SAINT-VENANT, and MORIN as "rapporteur" provided an important detailed study of this phase of TRESCA's work.[3] Following a suggestion of SAINT-VENANT,[4] TRESCA performed an additional series of experiments designed to provide more information on the kinematics of the interior of the flowing solid. These data are described in a supplement in 1872 to the Memoir on Punching.[5]

The main discoveries of TRESCA in those seven years of intensive effort were: 1) that solids under sufficient pressure would flow in the manner of fluids; 2) that there existed an intermediate region of plastic work-hardening beyond the elastic limit, before the constant flow began; 3) that there existed a material coefficient K in terms of the maximum shear at which, regardless of the configuration of the experiment, the solid would flow; 4) that for cylindrical punching of a cylindrical block, the length of the emitted slug L was given in terms of the radius of the punch R_1 and the radius of the specimen R by $L = R_1(1 + \log R/R_1)$; and 5) that the plastic flow of solids was isochoric.

Despite the fact that SAINT-VENANT[6] early recognized and glowingly described as remarkable the third of these, which had demonstrated the importance of a limiting shear criterion for the consideration of a plasticity theory which SAINT-VENANT proceeded to develop, TRESCA appeared to regard his calculation for slug length as his major contribution. Years later, in 1883, while examining a mechanical curiosity in the form of a 45 mm high, six sided screw nut brought to him from a Philadelphia exposition,[7] he succeeded in adapting his slug length formula to the new cross-section. He looked upon the successful adaptation as a proof of the correctness of the formula, and further stated that he regarded the discovery of this geometrical relation itself as the most significant of all his observations on the flow of solids.[8]

4.4. The punching and extrusion experiments of Tresca.

The division of papers given above refers to the titles of the memoirs rather than to the order in which TRESCA performed his experiments. The experiment of 22 April, 1864 in the first Flow Series described the punching of a rectangular block composed of 16 sheets of lead each 0.004 m thick and 0.120 m square. The stack was lightly pressed between two steel plates containing concentric central holes 0.020 m in diameter. A hydraulic press pushed a cylindrical steel punch through the lead, producing the flow patterns on the centrally cut slice shown in Fig. 4.4.

In describing this early experiment, TRESCA expressed great curiosity regarding the unexpected disappearance from the slug of such a large fraction of the metal

1 TRESCA [1869, *1*].
2 TRESCA [1872, *2*].
3 TRESCA [1870, *1*].
4 SAINT-VENANT [1870, *2*].
5 TRESCA [1872, *3*].
6 SAINT-VENANT *op. cit.*
7 TRESCA [1883, *1*].
8 One century later one is inclined to agree with TRESCA on the significance of a purely geometrical relation for the flow of various solids. In view of the ample subsequent demonstration of the physical limitations of the "TRESCA condition" in modern plasticity, there is little question that his discovery of a purely geometrical flow property, independent of material parameters, was of greater importance.

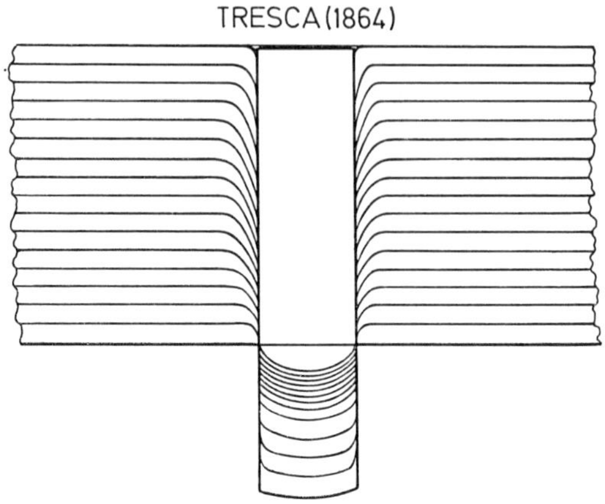

Fig. 4.4. The punching of a stack of lead plates by a cylindrical steel punch.

which had originally occupied the hole. Upon checking, he made the discovery that the density of the lead was unchanged, i.e., he had made the first observation that plastic deformation occurs isochorically.[1] The source of the discrepancy in the lateral flow into the block in the direction of least resistance he regarded as providing potential insight on the laws of the flow of liquids. The distributions of final thickness in the slugs, the members near the end having remained least deformed, he compared with the uniformity of the block. The measured thicknesses are given in Table 116.

Table 116. TRESCA (1864).

The numbers of plates in the slug	Distances of the lines of juncture from the extremity of the slug (mm)	Thicknesses of the plates measured along the axis (mm)
1	4.6	4.6
2	9.0	4.4
3	13.0	4.0
4	16.2	3.2
5	19.0	2.8
6	20.8	1.8
7	22.0	1.2
8	22.7	0.7
9	23.4	0.7
10	23.9	0.5
11	24.2	0.3
12	24.6	0.4
13	24.8	0.2
14	25.3	0.5
15	26.9	1.6
16	31.3	4.4

[1] The change in volume during plastic deformation first observed by BAUSCHINGER in 1879 (see Sect. 2.18 above), and which I described from recent experiments (see Sect. 4.35), is not observed in the type of post-deformation measurement which TRESCA made.

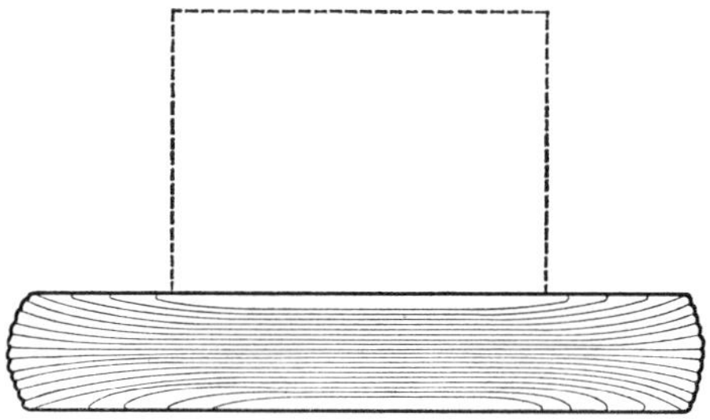

Fig. 4.5. The compression of a stack of 20 lead plates.

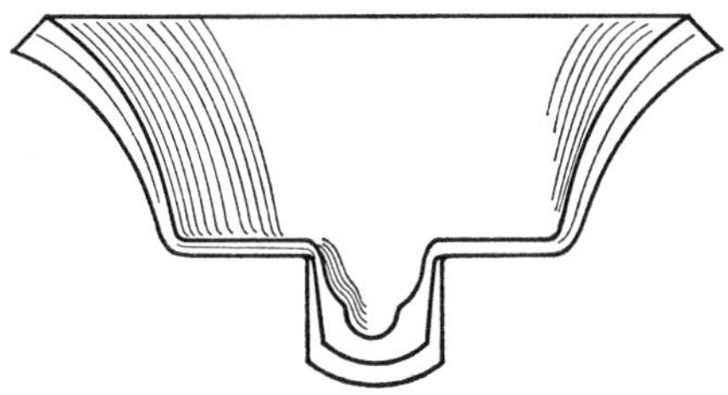

Fig. 4.6. The compression of two lead plates by a steel plate having a central orifice.

On 3 June, 1864, TRESCA compressed a block of 20 cylindrical lead sheets of overall thickness of 0.063 m to 0.018 m, as shown in Fig. 4.5.

The compression increased the central diameter from 0.060 to 0.110 m, and to 0.103 and 0.105 m at the ends. An opinion which TRESCA said he had felt bold in presenting earlier, namely that he might speak of the flow of solids, was in his judgment clearly confirmed by this experiment.

On 24 August, 1864, in the third type of preliminary experiment, he placed two lead plates upon a punched steel plate having a central orifice of 0.020 m. They were compressed by a sharp edged steel plate of 0.050 m diameter producing the "tulip shaped" flow shown in Fig. 4.6.

The two most interesting studies of the large series were those of die extrusion in which the number of plates and orifice diameters were varied, and the punching experiments on constrained and unconstrained cylindrical stacks of various numbers of plates, and on solid blocks. The die for the first series is shown in Fig. 4.7 taken from the supplemental paper[1] in 1872 on the drawings from the Flow Memoir I of 1868.

In Fig. 4.8 is shown a central slice for a 10 plate lead extrusion from the 0.020 m die[2] which may be compared with the 20 plate flow patterns for Fig. 4.9 for the same diameter orifice.

As the diameter of the orifice increased, a central hollow section appeared which TRESCA thought was analogous to vorticity in the flow of fluids. An example of this is the four plate lead extrusion with an orifice of 0.040 m shown in Fig. 4.10.

[1] TRESCA [1872, *1*].

[2] This is the lone figure referred to by SAINT-VENANT. The final memoir in 1868 contained six additional figures, including the three given above.

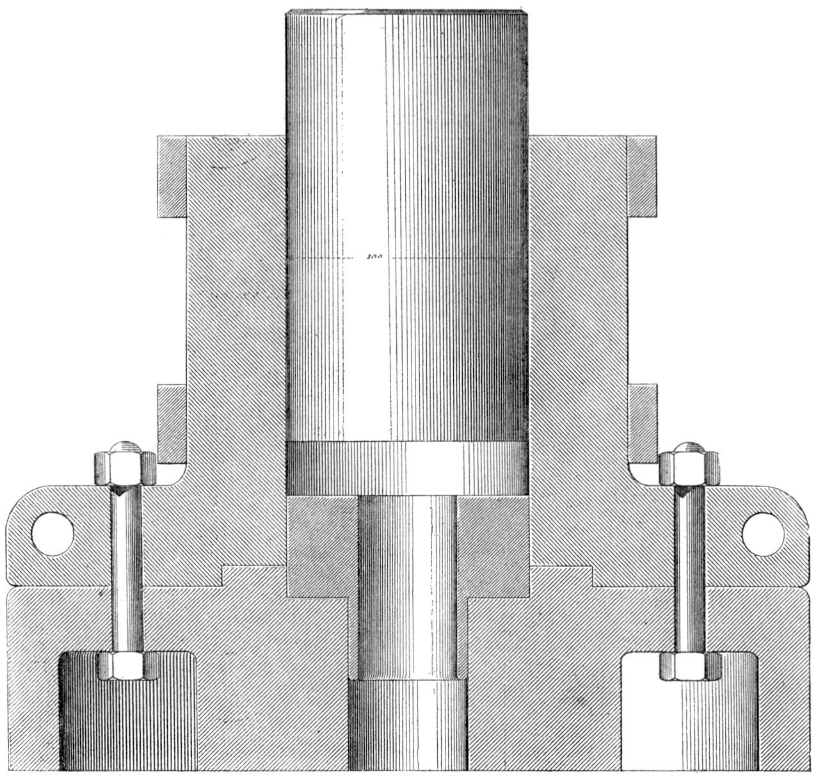

Fig. 4.7. TRESCA's drawing in 1872 of the apparatus used in his flow experiments.

This effect was readily apparent in single plate extrusion, as may be seen in the 0.030 m diameter extrusion of Fig. 4.11. Still another aspect of laminated plate extrusion was the plate separation in the five plate, 0.030 m diameter situation shown in Fig. 4.12, or the ten plate, 0.040 m diameter extrusion of Fig. 4.13, showing separation, bulging, and the presence of a central hole.

A final example of extrusion from the enormous number given by TRESCA is that for eleven plates of brick clay for a 0.040 m orifice, shown in Fig. 4.14, whose flow patterns were compared with those in lead, silver, zinc, porcelain clay, ceramic clay, kaolin, and paraffin.

TRESCA studied the influence of the cylindrical die walls by removing them and by observing the extrusion of the type shown in Fig. 4.15. He varied the radii of punch and die up to the obtaining of a thin ring of extruded material.

I find myself in complete agreement with the Commissioners of the French Academy in 1870, COMBES, SAINT-VENANT, and MORIN, that it is impossible to provide in concise form a detailed description of the enormous number of experiments and the original drawings which were on the same scale as the actual specimens "without uselessly fatiguing the attention of the Academy". The

TRESCA (1868)

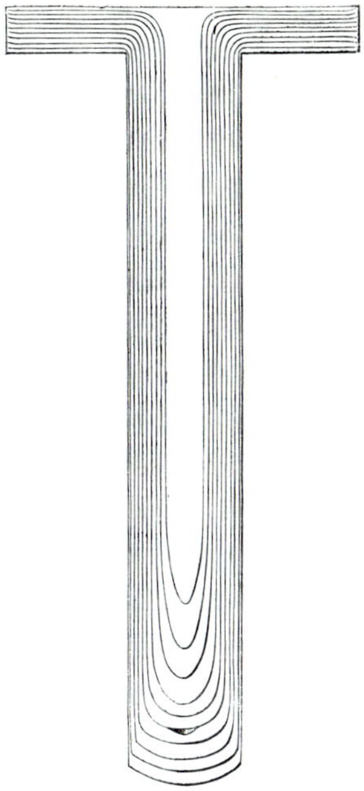

Fig. 4.8. The central slice through a stack of 10 lead plates extruded by means of the apparatus of Fig. 4.7.

chief distinction between the early memoir and the two major memoirs that followed is the increased precision of measurement and the presentation of a detailed, empirically based, generalized analysis of the behavior which the more accurate tests permitted.

Expressing the creed of the genuine experimentist, as HOOKE, COULOMB, and WERTHEIM had done before him, TRESCA stated in the introduction to his first flow memoir:[1]

> This work has not been done as a result of preconceived ideas and as an end we proposed to reach from the first. It has presented itself to us only in consequence of varied experiments realized in the first place in complicated circumstances which we have striven to reduce more and more to conditions which are simple and such as to permit the laws of the phenomena to be studied easily.

The main theoretical discussion demonstrating that TRESCA indeed had achieved this sought for experimental goal was given in the memoir[2] on punching, in 1869 published in full in 1872.[3] The long analysis considered the geometrical implications of each type of flow experiment in terms of the material constant of a fixed coefficient K prescribed as a critical shear stress.

That TRESCA had achieved his objective was demonstrated further by his discovery that subsequent flow states could be described solely in geometric terms. Thus in the punching of a cylindrical block of radius R by a punch of

[1] TRESCA [1868, *1*], pp. 733–734.
[2] TRESCA [1869, *1*].
[3] TRESCA [1872, *2*].

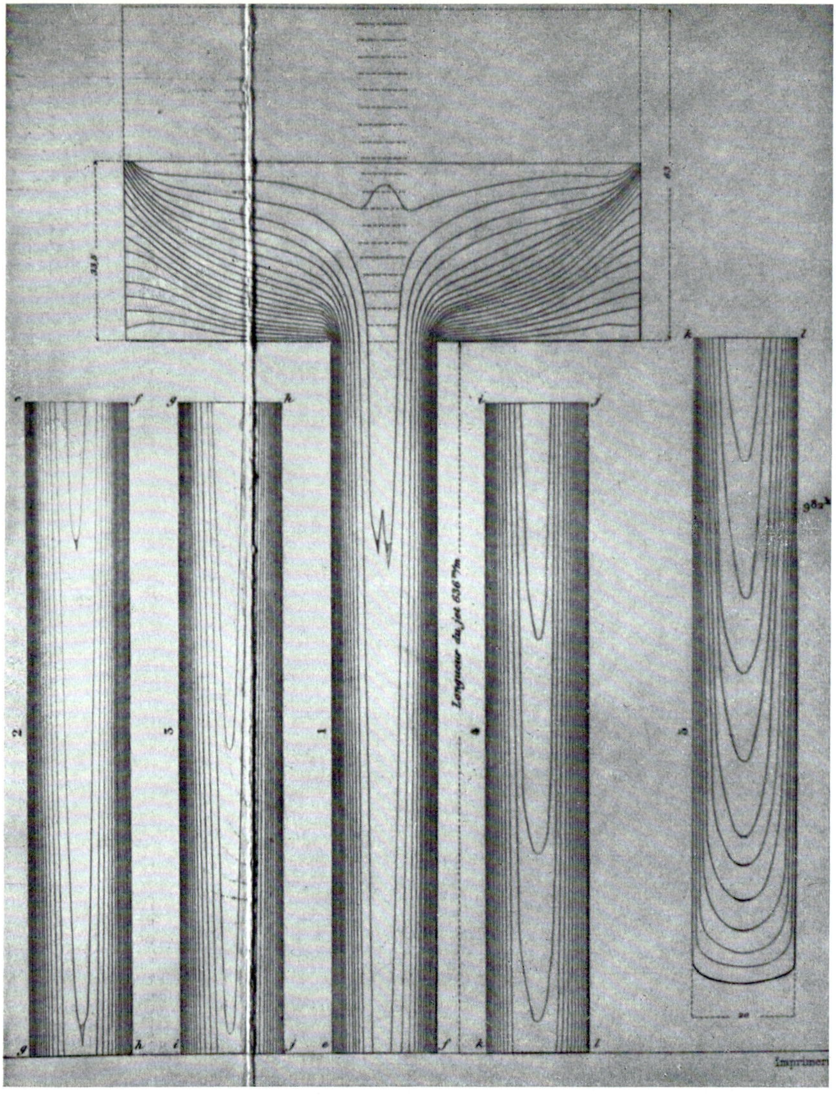

Fig. 4.9. The central slice through a stack of 20 lead plates extruded by means of the apparatus of Fig. 4.7 (TRESCA, 1872).

radius R_1, the length L of the slug emitted was given by

$$L = R_1\left(1 + \log\frac{R}{R_1}\right) \tag{4.5}$$

regardless of the initial height[1] of the block.

In the supplement to the memoir on punching in 1870, TRESCA described a new series of precise experiments performed at the request of SAINT-VENANT to provide a close correlation between formula and experiment for modelling wax,

[1] The block had to be sufficiently high that a simple cutting out of the slug did not occur.

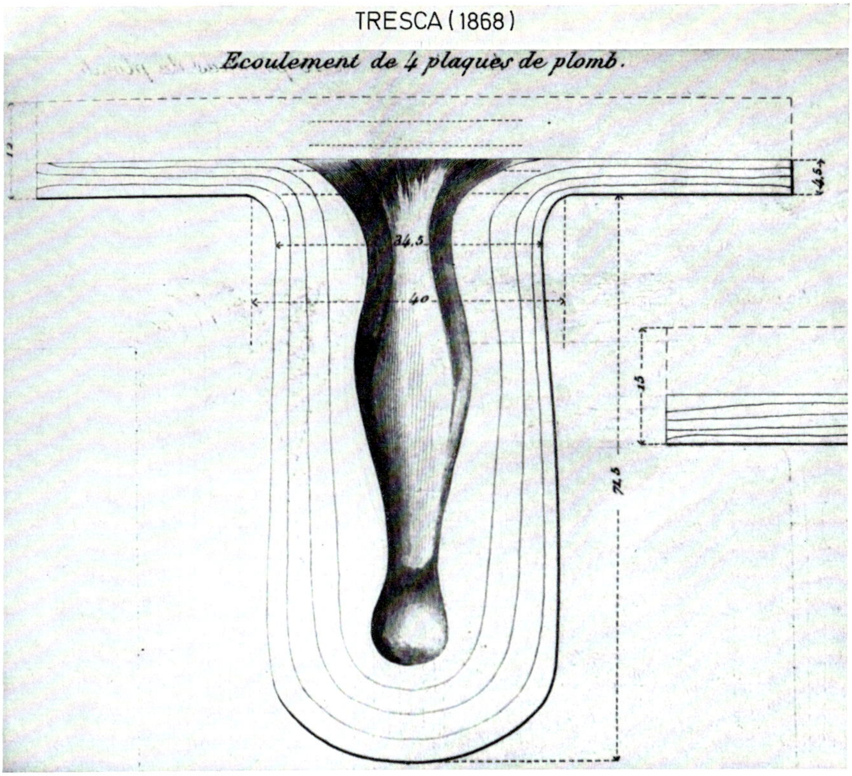

Fig. 4.10. The central slice through a stack of 4 lead plates extruded by means of the apparatus of Fig. 4.7. Note the central hollow section.

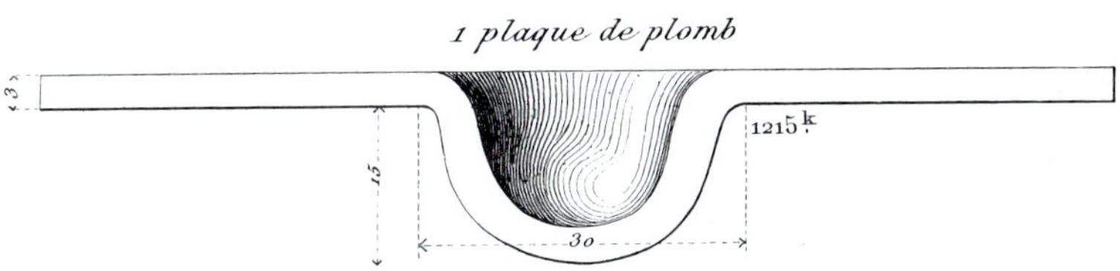

Fig. 4.11. The extrusion of a single plate of lead (TRESCA, 1872).

porcelain clay, earthenware clay, brick clay, lead, tin, copper, and iron. The values obtained for L/R_1 by these careful experiments were compared with calculated values in 41 cases as shown in Table 117.

The ratio of observed and calculated values in the final column of the table certainly justified TRESCA's enthusiasm at the result. The large variety of solids considered makes this formula, in which the length of the slug is independent of the initial height of the specimen, especially significant.

The load P required to produce the flow condition was given for each configuration. For the cylindrical punching problem TRESCA concluded that

$$P = 2K\pi R_1^2\left(1 + \log\frac{R}{R_1}\right), \tag{4.6}$$

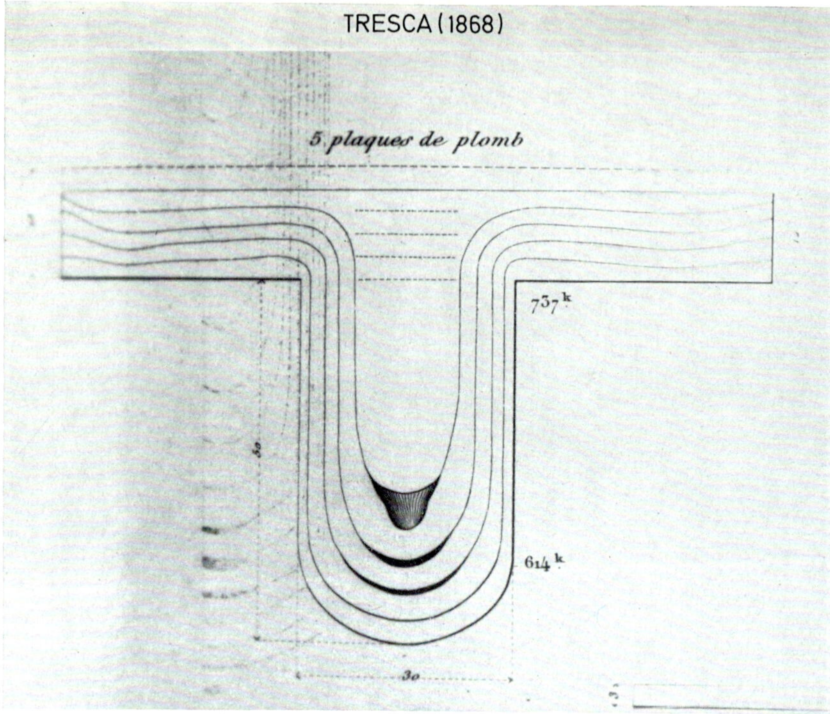

Fig. 4.12. The extrusion of 5 plates of lead. Note the separation for this orifice diameter.

from which K for a given solid could be determined since P, R, and R_1 were all measured quantities. I shall omit TRESCA's comparison of his early, less accurate data, with his later results, and refer the interested reader to the memoir of 1872 on punching for the details of the calculations[1] for the various flow situations. The values of K for the stated types of experiment are given in Table 118.

TRESCA modified Eq. (4.6) for different types of experiments. For the central punching of a cylindrical block enclosed in a cylindrical container he obtained

$$P = \pi R_1^2 K \left(3 + \frac{2R^2}{R^2 - R_1^2} \log \frac{R}{R_1}\right); \tag{4.7}$$

for the flow in a cylindrical block through a concentric orifice,

$$P = \pi (R^2 - R_1^2)\, \mathrm{K} \left(3 + \frac{2R^2}{R^2 - R_1^2} \log \frac{R}{R_1}\right). \tag{4.8}$$

The values of K were determined by means of solid specimen blocks since lower values of K invariably were obtained for the stacked plates.[2] The data for K which he obtained in 1869 for the solid block indeed demonstrated that TRESCA had discovered a material constant for the plastic flow of solids. The lower values for specimen blocks of 0.037 m and higher values for specimen blocks of 0.100 m, indicated some small dependence upon R not included in TRESCA's analysis.

[1] RODNEY HILL in his classic work on theoretical plasticity justly referred to TRESCA's analysis as "very crude". HILL [1950, *1*], p. 19.

[2] TRESCA puzzled over the differences in flow for the solid blocks and the stacked plates. He conducted experiments to compare final configurations.

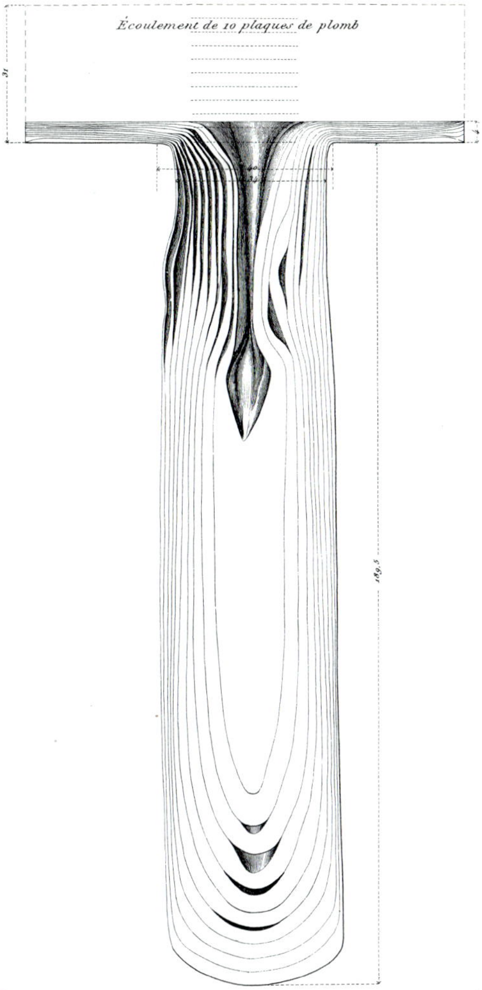

Fig. 4.13. The central section for a 10 plate extrusion. Note the separation and bulging for this orifice diameter. The solid was lead (TRESCA, 1872).

Different values of K, of course, were obtained for the different types of solids, i.e., for a 50% lead-tin alloy the average value of K from the experiments was given as 352 kg/cm². For the specimen blocks composed of plates, the corresponding average K for lead was 144 kg/cm² instead of the 200 kg/cm² for the solid block.

TRESCA compiled his data on the length of the slug for the experiments on the punching of the constrained block in the form of a plot of L/R_1 against R/R_1, which he compared in Fig. 4.16 with the theoretical calculation of Eq. (4.5). Both the Academy Commission and TRESCA considered this a close correlation, which indeed it is if one reflects upon the diversity of solids described by a simple formula in which a final length is shown to depend only on the magnitude of two initial radii.

TRESCA also provided a comparison of calculation and measurement for the position of the successive layers in the slug where Y/L referred to the distance

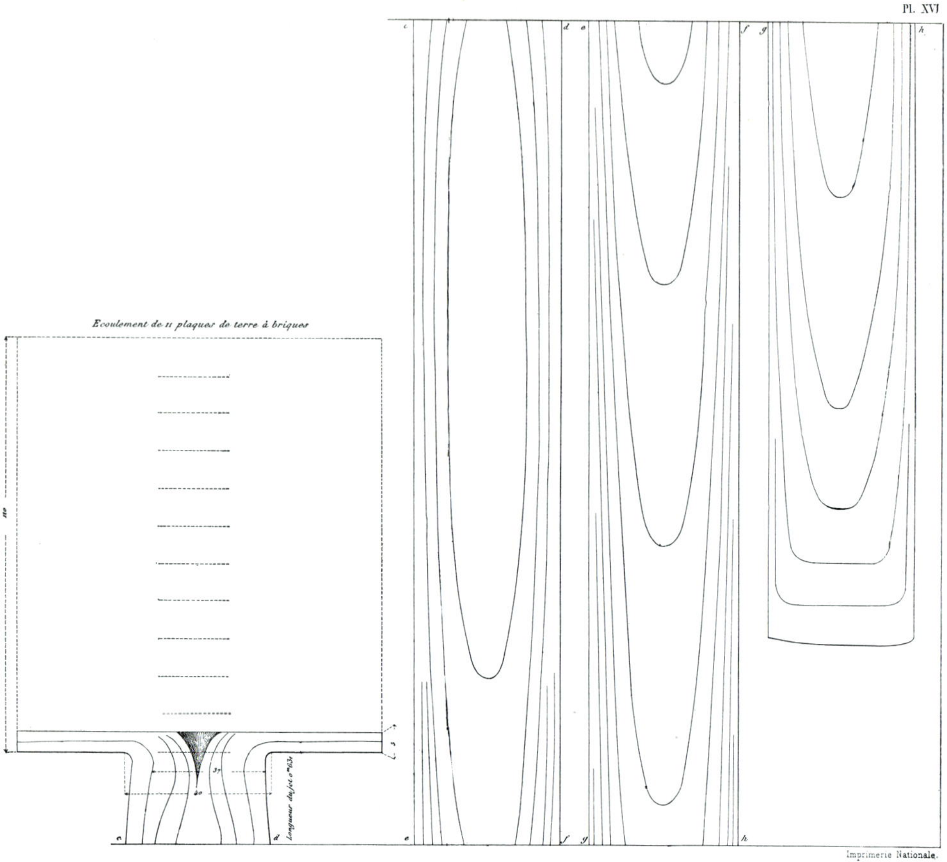

Fig. 4.14. The central section for the extrusion of 11 plates of brick clay (TRESCA, 1872).

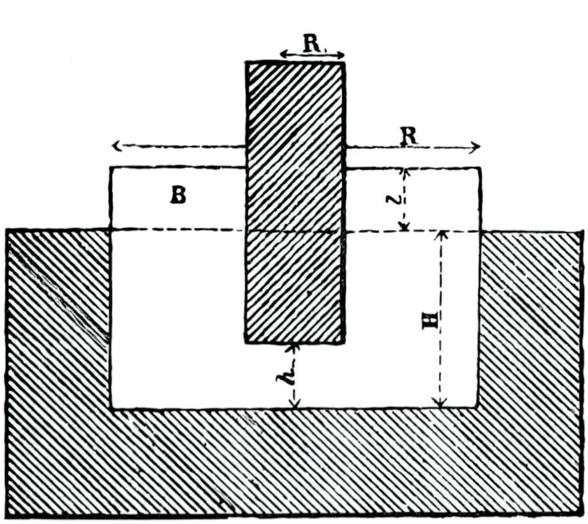

Fig. 4.15. Still another type of punch and extrusion experiment considered by TRESCA (1872).

Table 117. TRESCA (1870).

Test no.	Height of the block H	Radius of the block R	Radius of the punch R_1	Ratio $R:R_1$	Lengths of the slugs L	Observed values of L/R_1	Calculated values of L/R_1	Ratios between columns 6 and 7
				Modelling wax				
1	0.051	0.0255	0.005	5.1	0.012	2.2	2.629	0.913
2	0.049	0.0300	0.010	3.0	0.023	2.3	2.098	1.097
3	0.070	0.0505	0.020	2.525	0.0345	1.725	1.926	0.895
4	0.050	0.025	0.010	2.5	0.01525	1.525	1.916	0.795
							Average	0.925
				Ceramic clays: Porcelain clay				
5	0.049	0.029	0.010	2.9	0.023	2.3	2.064	1.114
6	0.061	0.0505	0.020	2.525	0.045	2.25	1.926	1.168
							Average	1.141
				Earthenware clay				
7	0.070	0.0505	0.010	5.05	0.029	2.9	2.619	1.107
8	0.048	0.028	0.010	2.8	0.024	2.4	2.029	1.052
9	0.072	0.0505	0.020	2.525	0.041	2.05	1.926	1.064
10	0.065	0.040	0.020	2.0	0.042	2.10	1.693	1.240
							Average	1.116
				Brick clay				
11	0.070	0.0505	0.015	3.367	0.035	2.333	2.213	1.054
12	0.070	0.050	0.015	3.333	0.032	2.204	2.204	0.968
13	0.051	0.026	0.010	2.6	0.0205	1.955	1.955	1.048
14	0.072	0.0505	0.020	2.525	0.0405	1.926	1.926	1.052
15	0.050	0.025	0.010	2.5	0.0235	1.916	1.916	1.226
16	0.052	0.025	0.010	2.5	0.0185	1.916	1.916	0.965
							Average	1.052
				Lead				
17	0.030	0.060	0.005	12.0	0.015	3.0	3.485	0.861
18	0.0295	0.055	0.005	11.0	0.015	3.0	3.398	0.883
19	0.050	0.050	0.005	10.0	0.015	3.0	3.303	0.909
20	0.073	0.060	0.010	6.0	0.030	3.0	2.792	1.074
21	0.050	0.060	0.010	6.0	0.028	2.8	2.792	1.003
22	0.0277	0.055	0.010	5.5	0.0235	2.35	2.704	0.868
23	0.065	0.050	0.010	5.0	0.026	2.6	2.609	0.997
24	0.070	0.050	0.010	5.0	0.0259	2.59	2.609	0.993
25	0.070	0.050	0.010	5.0	0.026	2.6	2.609	0.997
26	0.060	0.050	0.015	3.33	0.034	2.27	2.204	1.029
27	0.024	0.030	0.010	3.0	0.023	2.3	2.098	1.096
28	0.060	0.0505	0.020	2.525	0.039	1.95	1.926	1.012
29	0.070	0.050	0.020	2.5	0.0399	1.995	1.916	1.041
30	0.060	0.050	0.020	2.5	0.039	1.95	1.916	1.017
31	0.060	0.050	0.020	2.5	0.038	1.90	1.916	0.992
32	0.050	0.025	0.010	2.5	0.020	2.0	1.916	1.043
33	0.051	0.062	0.025	2.48	0.046	1.84	1.908	0.964
34	0.070	0.050	0.025	2.0	0.042	1.68	1.693	0.992
35	0.070	0.040	0.020	2.0	0.037	1.85	1.693	0.093
36	0.070	0.050	0.025	2.0	0.042	1.68	1.693	0.992
37	0.100	0.050	0.025	2.0	0.043	1.72	1.693	1.016
38	0.023	0.0185	0.010	1.85	0.016	1.60	1.614	0.991
							Average	0.994

Table 117. (Continued.)

Test no.	Height of the block H	Radius of the block R	Radius of the punch R_1	Ratio $R:R_1$	Lengths of the slugs L	Observed values of L/R_1	Calculated values of L/R_1	Ratios between columns 6 and 7
				Tin				
39	0.050	0.025	0.010	2.5	0.0201	2.01	1.916	1.048
				Copper				
40	0.050	0.025	0.010	2.5	0.022	2.2	1.916	1.147
				Iron				
41	0.0385	0.040	0.0175	2.285	0.0325	1.856	1.825	1.017
							General average	1.020

Table 118. TRESCA (1872). *Solid lead blocks.*

	Values of K: 1869, Blocks of 0.037 m diameter (kg)	1869, Blocks of 0.100 m diameter (kg)	Average results prior to 1869 (kg)
Concentric flow	198	201	144
Punching with cylindrical envelope	176	221	184
Punching with counter-matrix	190	211	202
Punching without counter-matrix	190	211	183
Averages	188.5	211.0	183
	200 kg (188.5 and 211.0)		

from the punch after deformation, Y_0/L to that before deformation. Both were given in ratio to the slug length. The results are shown in Fig. 4.17.

When making these measurements TRESCA used a differential manometer, cross-checked against the actual pressures in his hydraulic loading apparatus, to determine the load P in kg as a function of the depth of punch. In Fig. 4.18 is shown TRESCA's discovery, which he thought was of considerable importance, namely that his region of flow at a constant shear stress was preceded by a region of plastic work-hardening of the solid. It was because of these measurements that he insisted that the flow of solids was associated with the ultimate stress, not with the cessation of linear elasticity. In view of TRESCA's experiments, it is difficult to understand why SAINT-VENANT[1] and MAURICE LEVY[2] immediately, and others, later, assumed that the region of constant flow began at the elastic limit.

TRESCA designated five regions: 1) the HOOKEAN region up to the elastic limit; 2) a second region in which permanent deformation was observable upon removal of the load, but in which no sustained flow occurred; 3) the constant stress flow region; 4) the region of the emission of the slug; and 5) the final stage of emission

[1] SAINT-VENANT [1871, *1*].
[2] LEVY [1871, *1*].

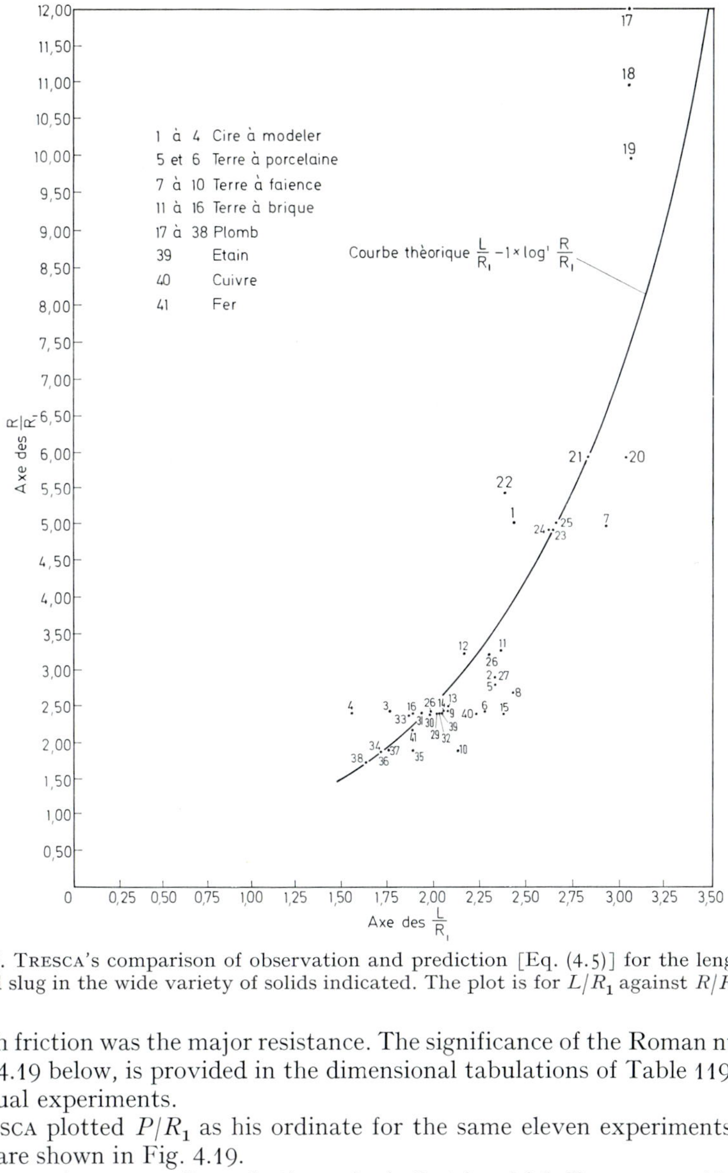

Fig. 4.16. TRESCA's comparison of observation and prediction [Eq. (4.5)] for the length of an extruded slug in the wide variety of solids indicated. The plot is for L/R_1 against R/R_1 (1872).

in which friction was the major resistance. The significance of the Roman numerals in Fig. 4.19 below, is provided in the dimensional tabulations of Table 119 for the individual experiments.

TRESCA plotted P/R_1 as his ordinate for the same eleven experiments. These results are shown in Fig. 4.19.

The most important figure in the series is that in which TRESCA considered the coefficient K as ordinate by plotting $P/2\pi R_1^2(1+\log R/R_1)$ against the penetration. These results are shown in Fig. 4.20. The values on either side of the average of $K=200$ kg/cm² provide some measure of the variation in his data on lead. At the same time they emphasize even more the significance of the generally ignored intermediate second region observed in TRESCA's research on plastic flow.

Both TRESCA and SAINT-VENANT were aware of the fact that the material parameter K was associated with an ultimate stress,[1] as may be seen from the

[1] From the actual measurements, there was no reason to infer that a maximum shear stress was the important parameter. This plausible conjecture, despite a century of its unquestioned acceptance, in fact was precisely that: a conjecture.

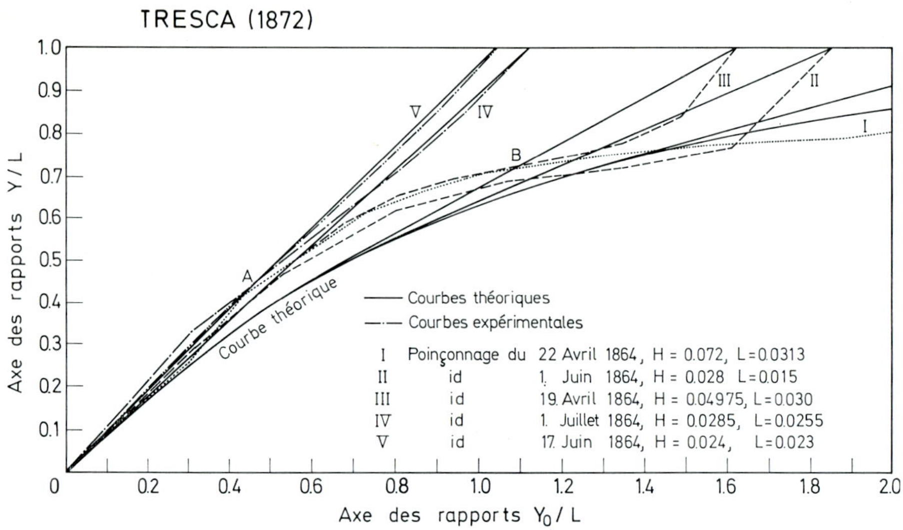

Fig. 4.17. TRESCA's comparison of calculation and measurement for the position of the different layers in the extruded slug. Y/L refers to the distance from the punched face after deformation, and Y_0/L to that before deformation (1872).

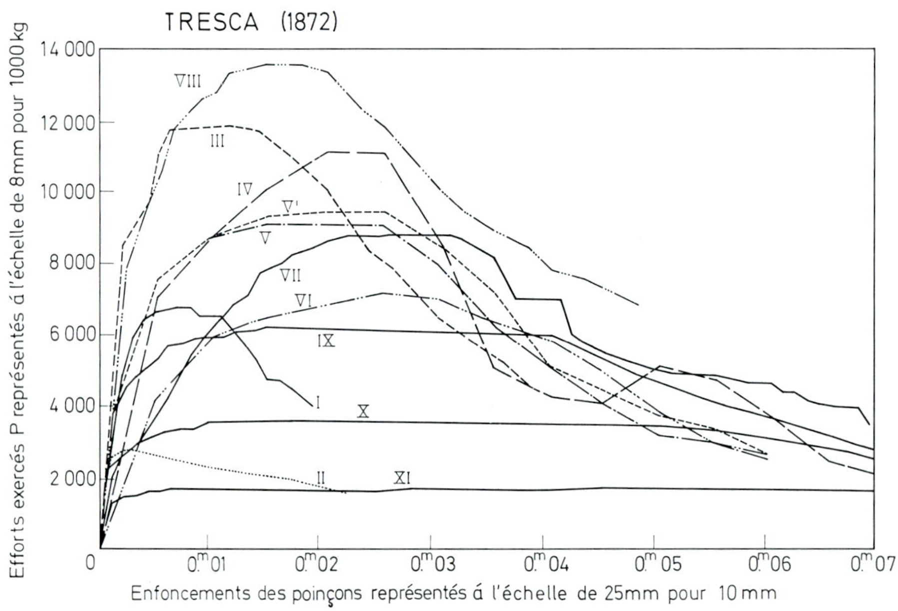

Fig. 4.18. The load, P, in kilograms vs the depth of punch (1872).

Academy Commission's comparison in 1870 of TRESCA's maximum shear stress of 37.57 kg/mm² in iron with an ultimate shear stress of 38.09 kg/mm² obtained by FAIRBAIRN in cast iron, a correlation they described as remarkable. TRESCA provided similar diagrams and analyses in several solids for which the reader is referred to the memoir in 1872 and the graphical supplements. He gave a final

Table 119. TRESCA (1872).

No. of the curves	Dimensions of the blocks (m)	Height of the blocks			Length of the slug		Minimum horizontal distance (m)	Punch diameter (m)	Maximum pressure (kg)	Maximum pressure per cm² (kg)
		Original (m)	Final exterior (m)	Final interior (m)	Exterior (m)	Along the axis (m)				
I	0.124 × 0.124	0.021	0.021	0.020	0.018	0.022	0.037	0.050	6711	342
II	$D = 0.060$	0.024	0.024	0.024	0.023	0.023	0.020	0.020	2665	849
III	0.124 × 0.124	0.038	0.038	0.0365	0.032	0.038	0.037	0.050	11887	606
IV	$D = 0.100$	0.070	0.068	0.059	0.042	0.038	0.025	0.050	11191	570
V	$D = 0.100$	0.060	0.060	0.055	0.039	0.044	0.030	0.040	9451	752
V′	$D = 0.100$	0.060	0.060	0.054	0.038	0.046	0.030	0.040	9103	724
VI	$D = 0.100$	0.060	0.060	0.0565	0.034	0.0345	0.035	0.030	7189	1017
VII	$0.085 \times x$	0.070	0.070	0.064	0.039	0.041	0.0225	0.040	8842	652
VIII	0.124 × 0.124	0.051	0.051	0.047	0.0146	0.046	0.037	0.050	13627	694
IX	$0.085 \times x$	0.070	0.070	0.070	0.036	0.036	0.0275	0.030	6232	882
X	$0.085 \times x$	0.070	0.070	0.070	0.025	0.025	0.0325	0.020	3622	1253
XI	$0.085 \times x$	0.070	0.070	0.070	0.014		0.0375	0.010	1795	2287

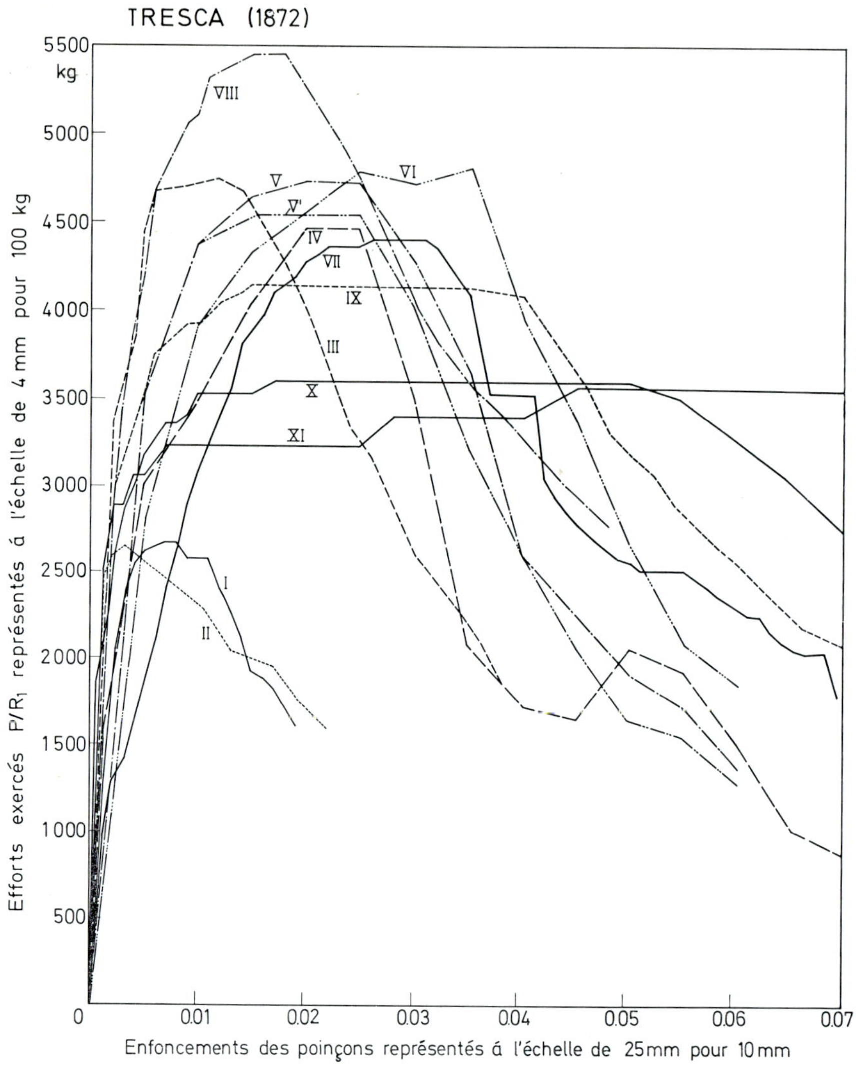

Fig. 4.19. Tresca's plot of P/R_1 vs penetration for the 11 tests of Table 119.

tabulation of the coefficient of fluidity in terms of the maximum shear stress or Tresca condition for the metals listed in Table 120.

The *Mémoires Présentés par Divers Savants of the Académie des Sciences, at Paris*, 2nd series, Tome 20 (1872)[1] contains 337 pages of text, 107 separate figures and 25 plates, with 17 of the latter having from 3 to 6 full-scale figures each. A summary of Tresca's work is contained in the 1870 Academy Commission report[2], which should be read in conjunction with the diagram supplement in the 1872 volume. Tresca also published two memoirs on applications of his flow studies.[3,4]

[1] Tresca [1872, *1*, *2*, *3*].
[2] Tresca [1870, *1*].
[3] Tresca [1867, *2*].
[4] Tresca [1868, *2*].

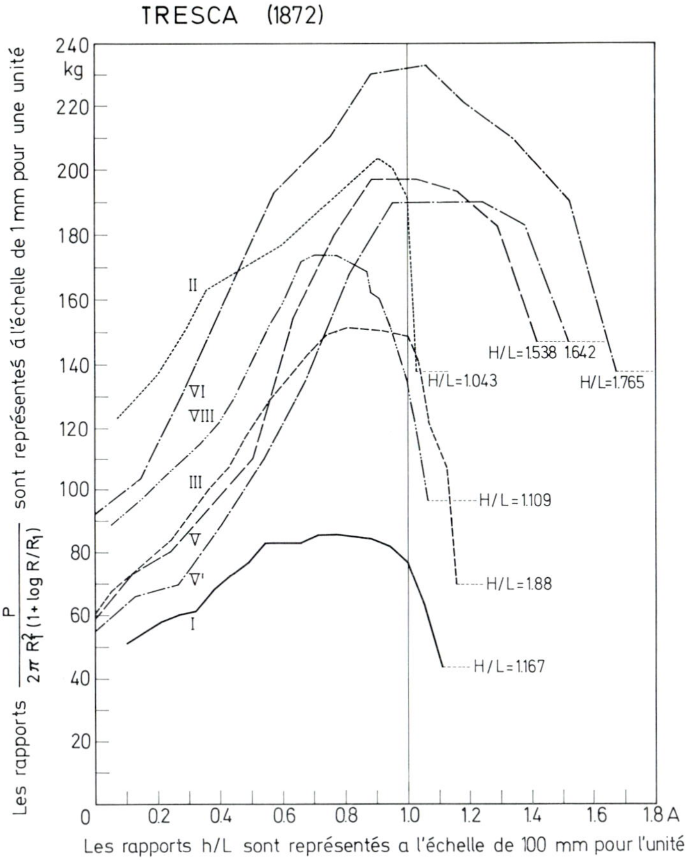

Fig. 4.20. TRESCA's plot of $P/2\pi R_1^2\left(1+\log\frac{R}{R_1}\right)$ against the penetration. This is TRESCA's plot of the constant material coefficient, K, which in the tests shown should have the value of 200 kg/cm². Note the intermediate region of work hardening before the constant plastic flow region.

Table 120. TRESCA (1872).

Substance	Flow stress (kg/mm^2)
Lead	1.82
Pure tin	2.09
Lead-tin alloy	3.39
Zinc	9.00
Copper	18.93
Iron	37.57

In June, 1878, the British Institution of Mechanical Engineers held a meeting in Paris at the Conservatoire des Arts et Métiers. An obvious high point of the meeting was the invited paper, on Wednesday 12 June, of the President of the

Société des Ingenieurs Civils of Paris, HENRI TRESCA.[1] The memoir, delivered and published in English, revealed how TRESCA's ideas on plastic flow had continued to develop after he had ceased to work in this field. TRESCA reemphasized that there were three distinctly different phases of deformation. The first two were recognized for all bodies, i.e., the perfect elastic phase in small deformation, and the imperfect elastic phase of non-linear deformation, which if carried far enough would produce for many solids at a "sufficiently great force" a third phase "designated by the author as a period of fluidity, and to which the greater part of his experiments on the flow of solids have reference."[2] After commenting on the fact that among different solid substances the magnitude of the deformation of this region of fluidity varied from "zero to extensive", TRESCA proceeded to a long review of his earlier work, with a large series of figures included for illustration.

In the controversy of the time over whether two kinds of iron existed, namely fibrous and crystalline, TRESCA noted that individual fibers of the metal were visible under acid etching, and gave examples of large deformation which could be explained only by assuming a "fibrous constitution of the metal." In this curious debate one is reminded again of the strong influence of a persistent hypothesis upon the course of experimental thought. As is noted below, ROBERT STEPHENSON[3] in 1850, in the first use of the microscope in metallurgy, had shown that "fibrous" and crystalline iron were both crystalline when viewed under the microscope, i.e., no difference existed. LÜDERS[4] in 1860 had made studies on etched surfaces, which had provided further evidence of crystallinity. Nevertheless, in 1878, a quarter of a century later, the old arguments remained undiscarded by some; as we have seen in TRESCA's statement, they still were influencing both experiment and interpretation.

Of major new interest in TRESCA's review paper, however, was his reference to a simple announcement he had made to the French Academy of Sciences on the 8th of June, 1874

> that when the bar of platinum, at the moment of forging, had just cooled down to a temperature below that of red heat, it happened several times that the blow of the steam-hammer, which at the same time made a local depression in the bar and lengthened it, also reheated it in the direction of two lines inclined to each other, forming on the sides of the piece the two diagonals of the depressed part; and so great was this reheating that the metal was along these lines fully restored to a red heat, and the form of the luminous zones could be clearly distinguished. These lines of increased heat even remained luminous for some seconds, and presented the appearance of the two limbs of the letter X. Under certain conditions as many as six of these figures of X, produced successively, could be counted simultaneously, following one another according as the piece was shifted under the hammer so as to be gradually drawn down for a certain part of its length.
>
> As to the origin of these luminous traces there could be no doubt. They were the lines of greatest sliding, and also therefore the zones of greatest development of heat—a perfectly definite manifestation of the principles of thermodynamics. That the fact had not been observed before was evidently owing to this, that the conditions necessary for its manifestation had not been combined at the same moment under such favorable circumstances. Iridised platinum requires for its deformation a large quantity of work to be expended upon it. The surface takes no scale, and is almost translucid when the metal is brought up to a red heat. The metal is but an indifferent conductor of heat, and its specific heat is low. All these are conditions favorable for rendering the phenomena visible in the forging of this metal, whilst it had remained unobserved with all others.[5]

---

[1] TRESCA [1878, *1*].
[2] TRESCA [1878, *1*], p. 302.
[3] STEPHENSON [1850, *1*].
[4] LÜDERS [1860, *1*].
[5] TRESCA [1878, *1*], pp. 314–315.

TRESCA proceeded to study this effect crudely by coating a bar of metal with wax or tallow on its two lateral faces before subjecting it to a blow of a steam hammer. He noted that the wax melted in the vicinity of the depression from the hammer and

> assumes in certain cases the form of the letter X, as was observed in the case of the platinum bar. In other cases the limbs of the cross are curved, presenting their convexities towards each other. The heat has then been more widely disseminated, and the wax melted over the whole of the interval between the curves.[1]

He calculated the volume of the metal, assuming uniform heating beneath the measured melted wax section and assuming the temperature of this volume of the bar had been raised to the melting point of the wax.

Recognizing the crudity of this first attempt at calculating experimentally the relation between work input from the hammer and the conversion of heat of the resulting plastic deformation, a feat not satisfactorily accomplished until the measurements of GEOFFREY INGRAM TAYLOR sixty years later,[2] TRESCA estimated that the amount of work converted to heat invariably exceeded 70% of the input work. As may be seen in Table 121 these approximate lower bounds of the ratio of efficiency ranged from as low as 73.1, 79.6, and 87.7% (bar blows of the hammer involved work of 579, 651, and 796 ft. lbs. respectively) to as high as 94.2% for copper subject to 434 ft. lbs. of work from the steam hammer. The temperature rises were considerable; the melting temperature of TRESCA's wax was given by him as 122° F (50° C). For its historic interest as the first effort to apply something resembling the thermodynamic principles of JOULE[3] to the flow of solids, Table 121 is included.

Table 121. TRESCA (1878).

No. of Fig. in Plate 41	32	33	34	35
Metal experimented upon	iron	iron	iron	copper
Form of impression, rectangular or curved	rect.	rect.	curv.	rect.
Area of wax melted (sq. in.)	0.22	0.23	0.34	0.27
Thickness of forging (in.)	0.98	0.98	0.98	0.78
Volume of corresponding prism (cub. in.)	0.22	0.23	0.34	0.21
Heat-units absorbed in raising prism to 122° F (units)	0.5944	0.6138	0.9003	0.5273
Equivalent work, taking 1 heat-unit = 772 ft. lbs. (ft. lbs.)	460	475	696	408
Actual work in fall of hammer (ft. lbs.)	579	651	796	434
Percentage of efficiency (%)	79.6	73.1	87.7	94.2

A second observation reported in that same memoir was one which obviously fascinated TRESCA. He had found that if he removed all of the relief from one side of a piece of money and placed this now flat face on a sheet of lead, the stamping of the opposite figured face in the press caused the detail of the relief from this side to appear on the smoothed side of the coin. Of even greater interest was the observation that this relief figure from the top of the coin was transferred to the lead, being discernible as an enlargement in the proportion of 22 to 13 when the lead sheet was $^1/_2$ in. thick. The magnitude of the enlargement depended upon the thickness of the lead sheet. Although TRESCA conceded his inability to formulate precise rules to account for this flow behavior, he recognized both its fundamental

[1] *Ibid.*, pp. 315–316.

[2] TAYLOR [1937, *1*]. During that interval of sixty years, this interesting part of TRESCA's research simply had been ignored by everyone, including TAYLOR. See Sect. 4.23.

[3] JOULE [1859, *1*].

importance and practical aspects. He provided numerous illustration of the technological implication in die or roll construction.

TRESCA's paper ended with regrets that the interests of the group and time did not permit of reference to the experiments which "call to mind, with a surprising degree of exactness, the constitution of certain rocks with their dislocations."[1] These experiments in geology, carried out in conjunction with a Mr. DAUBRÉE, were a part of his total endeavor to study all modes of deformation so as to understand the "phenomena of molecular mechanics, as well as ... the internal structure of the substances upon which the various industrial operations are performed."[2]

It is not possible to leave that memoir without commenting upon at least one aspect of the lengthy recorded audience discussion which followed. As is nearly universal in such discussions, a procession of semi-related trivia and unrelated expositions of the dicussers' own earlier pronouncements are interspersed with platitudes and minor misconceptions. With the notable exception of the French Academy panel reviews, many of which are invaluable, the only worthwhile aspect of the recording of the usual discussion miscellany is the historical insight it provides into the scholarship of the participants. Thus, JOHN HOPKINSON,[3] whose experiments in 1872 are now credited erroneously by many with initiating dynamic plasticity, asked TRESCA in 1878 to distinguish between the flow of a specimen subject to the blow either of a light hammer moving at great velocity or of a heavy hammer moving at low velocity, both doing the same work. HOPKINSON stated that he did not think that any difference could depend entirely upon the inertia of the material. From these comments, we have a better view of HOPKINSON's confusion (and his apparent ignorance of what TRESCA had accomplished) in 1872 when he had attempted to use infinitesimal linear elastic wave analysis to account for the dynamic rupture of iron wires at large deformation.[4] TRESCA replied only to HOPKINSON's reference to the importance of time effects, emphasizing that the quasi-static loading he had studied provided very different results from similar experiments which had been performed under rapid loading. A Mr. R. PRICE WILLIAMS, we also may note, was informed by TRESCA that the contention that quasi-static loading and dynamic loading should produce the same heating of the solid with reference to the temperature, was untenable because of the heat dissipation during the slow loading.

For those who subscribe to standardized educational patterns for scholarship, TRESCA offers an enigma. He was born in 1814 at Dunkirk, graduated from the École Polytechnique at the age of 19 in 1833, and sought a career in the design of civil structures. Deterred from this ambition for some time by a serious illness, he spent the next thirty years teaching, building and performing tests upon hydraulic machines, and writing on practical mechanics. From 1852 he worked at the Conservatoire des Arts et Métiers, first with the title of engineer and then as assistant director, where he organized an office for the standardization of weights and measures.[5] Thus, when in 1864 he suddenly embarked upon a career as a

---

[1] TRESCA [1878, *1*], p. 327.

[2] *Ibid.*

[3] HOPKINSON [1872, *1*, *2*].

[4] *Ibid.* HOPKINSON also spent two pages describing his own work on stress relaxation, and his interest in viscosity with reference to the difference between viscous solids and viscous fluids.

[5] MORIN, who was the director of the Conservatoire des Arts et Métiers, was a dominant figure in French science and technology at this time. Upon reviewing his published works, I found that he was singularly undistinguished as a scholar, despite his eminent positions as a member of the Academy and director of the Conservatory. Unquestionably, his major contributions to science and technology were in the opportunities and encouragement which he gave TRESCA to begin and continue his studies of plastic flow.

major experimental physicist, he was 50 years old. Those studies were interrupted by the Franco-Prussian War of 1870 when he became preoccupied with the metallurgical design of artillery and with electrical engineering. Such technological matters occupied him until his death in 1885. This brief career of but eight years in experimental physics gave him a lasting world reputation and election to the French Academy in 1872.[1]

4.5. Thurston's discovery of the dependence of the elastic limit upon the previous stress history and the elapsed time (1873).

At about the time when TRESCA was carrying on his classic studies of flow, finite deformation research began in several different directions. All were destined to have their continuing counterparts with ever changing interpretations and extensions during the century which followed. Among these were the important, now widely quoted, voluminous experiments of BAUSCHINGER in the 1870's and 1880's on the elastic limit and the yield limit; the launching of the age of automation in deformation studies between the elastic limit and fracture with ROBERT H. THURSTON's autographic testing machine in 1873; the work between the 1850's and 1880's of JOULE, EDLUND, EXNER, WINKLER, IMBERT, and others on the finite thermoelasticity of rubber; the dynamic plasticity experiments of JOHN HOPKINSON in 1872; and the increasing interest in the metallurgical science of plasticity in metals as evinced, for example, in the discovery[2] by W. LÜDERS in 1854 of his famous etched figures in deformed iron and steel, or in the suggestion of PAUL R. HODGE in 1850 that the microscope be used to study the structure of iron.[3]

Before describing THURSTON's autographic apparatus and the research it permitted on the finite deformation of wood and steel, I shall refer briefly to the work of LÜDERS although it does not fall strictly within the scope of this treatise. While testing an iron rod used for horse shoes and carriage wheels in 1854 LÜDERS had noticed the appearance on the surface of the solid of a pattern similar to the grid of a file. His subsequent study of the phenomenon included polishing and etching the surface with weak nitric acid, which revealed surface markings of the type shown in Fig. 4.21. The line $A-B$ represented the axis of the rods. He noted the similarity of the markings to the WIDMANNSTEDT figure on meteor iron. LÜDERS' speculation as to the possible crystalline origin of this phenomenon forecast a later century's concern for the general subject in the now highly diversified field of crystal plasticity.

By the middle of the second half of the 19th century in both science and technology the potentialities of college and university laboratories for the instruction of students became a widely discussed subject.[4] By the 1880's the acknowledged first and foremost German engineering materials testing laboratory was the Mechanisch-Technisches Laboratorium der Kgl. Technischen Hochschule in Munich, directorship of which had been assumed in 1871 by the 38 year old JOHANN BAU-

[1] *See* TRESCA's biography in *École Polytechnique, Livre Centenaire,* Paris, Gauthier-Villars et Fils, Paris, 1895, Vol. I, pp. 206–209.

[2] LÜDERS published in [1860, *1*].

[3] HODGE [1850, *1*]. This suggestion was acted upon immediately by ROBERT STEPHENSON [1850, *1*], President of the Institution of Mechanical Engineers, although he had not been present at the January meeting. STEPHENSON began his study in April, 1850, and was very surprised to find in this first use of the microscope in a metallurgical study that (as I noted above) under the microscope "fibrous" and "crystallized" iron were indistinguishable.

[4] See FLORIAN CAJORI [1928, *1*] for a brief review of the growth of formal physics laboratories or ALEXANDER BLACKIE WILLIAM KENNEDY [1887, *1*] for a detailed listing of the apparatus for the mechanical testing of solids in technical laboratories throughout the world.

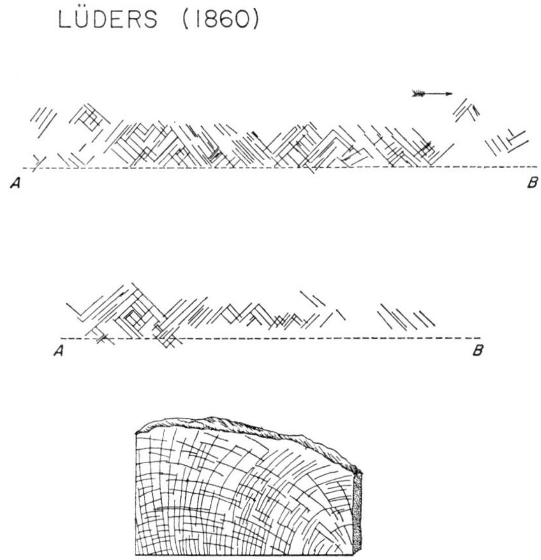

Fig. 4.21. LÜDERS' drawings in 1860 of the curious markings which appeared on etched iron when he discovered this phenomenon in 1854.

SCHINGER. BAUSCHINGER during the next two decades until his death in 1893 became the dominant continental engineering figure in the rapidly growing field of metals testing.

Somewhere near the opposite end of this spectrum was ROBERT H. THURSTON, also very technologically oriented, who in 1872 was instructing advanced students in laboratory techniques at the United States' first technical institute emphasizing mechanical engineering, Stevens Institute of Technology in Hoboken, New Jersey. THURSTON had built much of his own apparatus; indeed, he designed and developed an autographic torsion testing machine, measurements from which deeply influenced the thinking of the next forty years. His apparatus provided the prototype for the present day standardized automated materials testing devices.

THURSTON's diagram[1] of his machine is shown in Fig. 4.22. A later diagram of THURSTON's pencil recording device published by ROLLA C. CARPENTER and HERMAN DIEDERICHS[2] is reproduced in Fig. 4.23.

The cam on the loading arm gave an ordinate proportional to torque. The abscissa was the torsion angle or the relative angle between the two ends of the cylindrical specimen regardless of the mean position of the loading arm and loading weight. Typical torque-angle automatically recorded data are shown in Fig. 4.24. Despite the limited accuracy and the impossibility of ascertaining the distribution of stress in elastic and plastic zones during the finite deformation, THURSTON's invention, as was constantly pointed out during the next half-century, represented a major advance for experimentation in finite deformation.

In the fall of 1873 at the close of a meeting of the United States National Academy of Science which had taken place at Stevens Institute of Technology, THURSTON decided to see whether the flow behavior described in TRESCA's then very recently published work, could be studied in torsion by means of his machine.

[1] THURSTON [1874, *1*].

[2] CARPENTER and DIEDERICHS [1911, *1*]. This diagram was taken from an earlier work by CHURCH entitled *Mechanics of Engineering*.

THURSTON (1874)

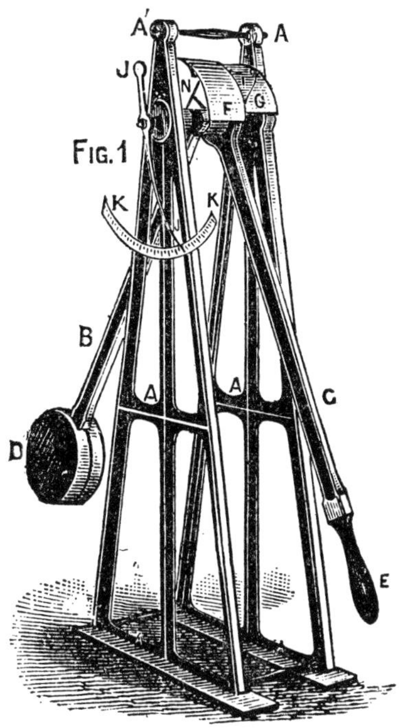

Fig. 4.22. THURSTON's diagram of his automatic recording torsion testing machine.

THURSTON'S TORSION TESTING-MACHINE

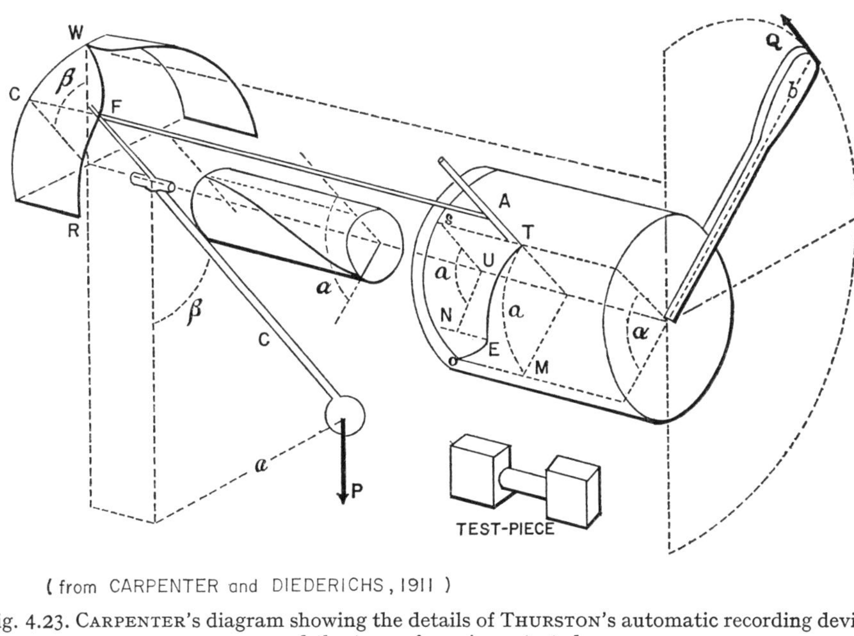

(from CARPENTER and DIEDERICHS, 1911)

Fig. 4.23. CARPENTER's diagram showing the details of THURSTON's automatic recording device and the type of specimen tested.

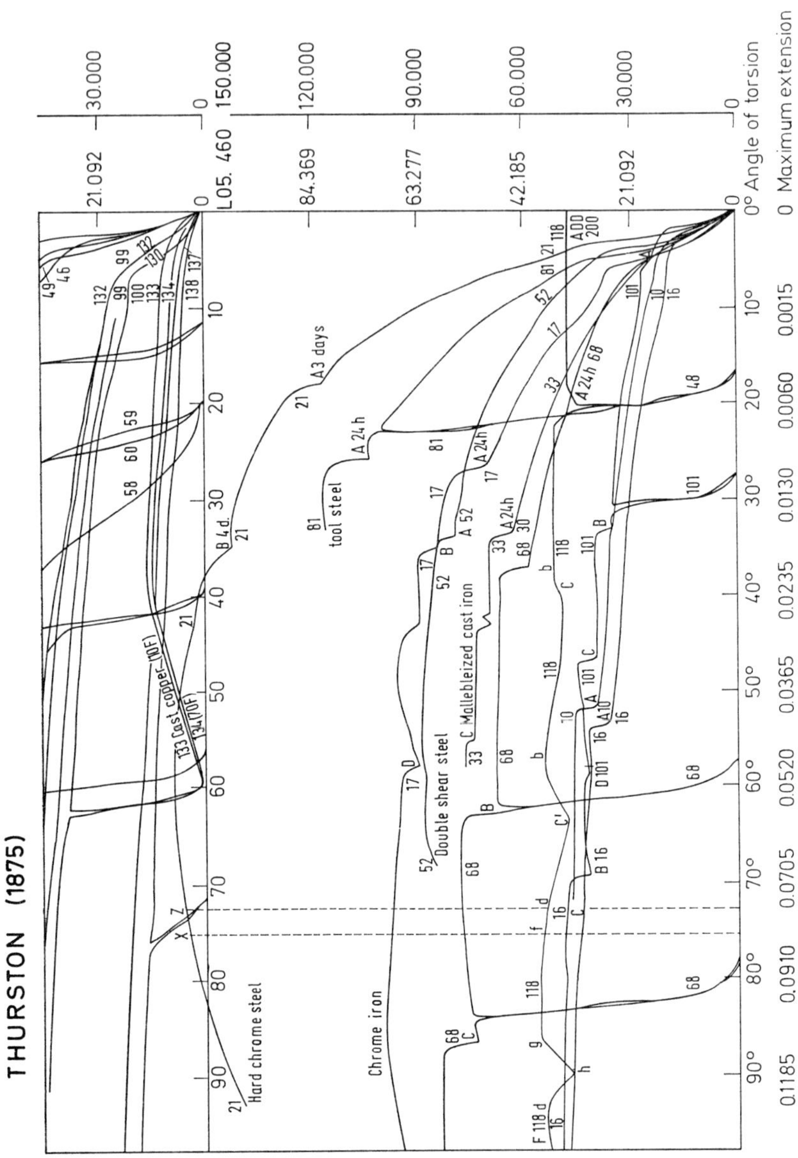

Fig. 4.24. Automatically recorded torque-angle data obtained by THURSTON from his apparatus of Fig. 4.22. Note the THURSTON effect following a pause in the loading.

Bringing a wrought-iron specimen into the plastic region, he fixed the loading arm to see whether the torque and/or the torsion angle would shift in a twenty-four hour period. On the following day, 13 November, 1873, a date which became the subject of heated later controversy, THURSTON discovered that neither the torsion angle nor the torque had changed. To his excited amazement, however, when he continued the test from this pre-stressed location a new apparently linear elastic region commenced with a new elastic limit over 25% higher than that which would have been obtained in the normal unloading and reloading of the specimen in the relatively short intervals of repeated dead load experimentation in the 19th century.

Recognizing the importance of his discovery, THURSTON six days later read a note on this work at the regular meeting of the American Society of Civil Engineers on 19 November 1873, as the published transactions of the Society duly recorded.

> The metal was left under strain for twenty-four hours and had not then yielded in the slightest degree ... after noting the result obtained as stated, it was attempted to still further distort the test piece, when the unexpected discovery was made that its resisting power was greater than when left the previous day, an increase of resistance being recorded amounting to about 25 per cent of the maximum registered the preceding day and approximating closely to the ultimate resistance of the material,[1]

and

> Releasing the piece entirely and again submitting it to the same force immediately did not produce this strengthening action.[1]

His detailed paper on this "elevation of the elastic limit" appeared in the same volume of the transactions,[2] having been presented on 4 February, 1874.

THURSTON's detailed description of the phenomenon, which referred to his original experiment as test #16, was printed in the next volume of the transactions[3] along with several of his later experiments. The first portion of the long autographic diagram from that paper also is shown in Fig. 4.24 from which the rest-time dependent phenomenon at A in test #16, and elsewhere for other tests, may be seen.

In terms of the controversy soon to be kindled by his contemporaries we must note that THURSTON did observe that if he completely unloaded and immediately reloaded a specimen which had been under sustained load for 24 h, the higher elastic limit was maintained. If THURSTON had been more the scientist than the practical engineer, he also might have included among his early experiments one in which he held the specimen in the unloaded position immediately after he had unloaded it from the plastic region. On reloading he then would have found that the rise in the elastic limit was due to the lapse of time, and that the specimen need not have remained under load during the long rest period.

In a country whose first graduate school, the Johns Hopkins University, was not to open its doors until three years later, in 1876; a nation which in the 1870's was decidedly not one from which major advances in science were to be expected, except perhaps from those savants who had studied in Europe; the immediate and wide recognition accorded THURSTON's discovery was surprising.[4] "The new method of investigation adopted and the importance of some of the conclusions deduced from the autographic records have attracted much attention and the paper has been extensively republished."[5]

A negative response also was almost instantaneous in the form of a vitriolic attack upon the experiments, the apparatus, and any and all of THURSTON's observations, by FRIEDRICH KICK, a professor of experimental mechanics at Prague. KICK's primary objection was that the inertia in the system during loading had been ignored. His claim that all results were nullified on that basis was

[1] THURSTON [1874, *1*], pp. 239–240.

[2] THURSTON [1874, *2*].

[3] THURSTON [1875, *1*].

[4] In a recent study, describing the institutional aspects of the subject, *The Mechanical Engineer in America, 1830–1910*, MONTE A. CALVERT [1967, *1*] devoted many pages to various aspects of THURSTON's professional career on the faculty of Cornell University. CALVERT wrote of THURSTON's influence upon the American Society of Mechanical Engineers and upon engineering standards in the United States but did not mention that THURSTON's work in his "moderately successful" laboratory at the Stevens Institute of Technology had made a great impact upon European experimental mechanics.

[5] THURSTON [1876, *2*], p. 9.

soundly disposed of in a reply by THURSTON,[1] and also later was dismissed as unsound by many others including BAUSCHINGER, who had different, strong objections to the device. KICK emphatically disallowed any scientific merit to the experiment while conceding the possibility that some gross technical applications might be possible. KICK was sound in his criticism of THURSTON's ignorance of such matters as the stress distribution in plastic torsion when he thought that diametral measurement allowed him to plot some of his twist data as tension and elongation, which probably contributed to the tenor of the adverse commentary on THURSTON's work during the next decade.[2] Those limitations were ignored, however, when THURSTON's invention of an autographic testing machine swept the laboratories of Europe, and particularly England, with improved versions in tension and flexure as well as torsion; its creator thereafter enjoyed a half-century of substantial recognition.[3] For example, fifteen years later, in 1887, the well-known English engineer, KENNEDY,[4] produced an autographic tension machine, and obtained the results shown in Fig. 4.25 for wrought iron and steel.

THURSTON probably was the first to comment that the work hardening curves of metals generally were parabolic and to demonstrate that the entire plastic stress-strain function increased with decreasing temperature. As far as I have been able to ascertain, THURSTON was the first experimentist to suggest that viscosity might be of significance[5] in plasticity. He noted the generally higher stress-strain function obtained upon increasing the speed of loading. These visco-plastic observations were based upon tests which were performed at relatively low strain rates in order to avoid inertia effects.

While it is undoubtedly important not only for presenting the truth but also for providing historical perspective to be as correct as possible in matters of precedence in original discovery, it is far less important, if not objectionable, to trace in detail the often bitter and, equally often, unjust tirades of the participants in precedence controversy. The current importance of the "BAUSCHINGER effect," and the experimental excellence of BAUSCHINGER in carrying out his vast, definitive study on the onset of "unstable" plasticity have prompted me to present one, perhaps extreme example of what is not an uncommon occurrence, irrespective of the calibre of the participants.

KICK in dismissing all of THURSTON's observations, at the same time wished to deny him credit for being the first to discover the increase in the elastic limit. KICK attributed precedence to Major General UCHATIUS in 1873. That UCHATIUS[6] had made an independent discovery of the effect, THURSTON acknowledged with pleasure expressed at the corroboration of his own work. However, as THURSTON pointed out, the date of the General's publication was 10 April, 1874, five months after THURSTON's note. Shortly after his own discovery THURSTON also remarked

[1] KICK [1875, *1*]. See also the translation contained in THURSTON's reply [1876, *2*], German translation [1877, *1*]. It is curious that in experiments a few years later THURSTON began striking blows on the handle to obtain "dynamic" results, thus making KICK's criticism valid whereas it had been invalid before.

[2] KICK's own research in this aspect of mechanics, some of which is described below, reveals *his* major limitations in rational mechanics. As a practical engineer he promoted rules of procedure which even then, to his consternation, and surely now, are of highly questionable value.

[3] See, for example, the research of FISCHER [1882, *1*] and MÜLLER [1882, *1*], described in Sect. 2.21; they employed an autographic device adapted by REUSCH. Or, see the extensive review of the developments on autographic testing by WILLIAM CAWTHORNE UNWIN in 1886 in the Journal of the Society of Arts [1886, *1*].

[4] KENNEDY [1887, *1*].

[5] THURSTON [1876, *1*].

[6] UCHATIUS [1874, *1*].

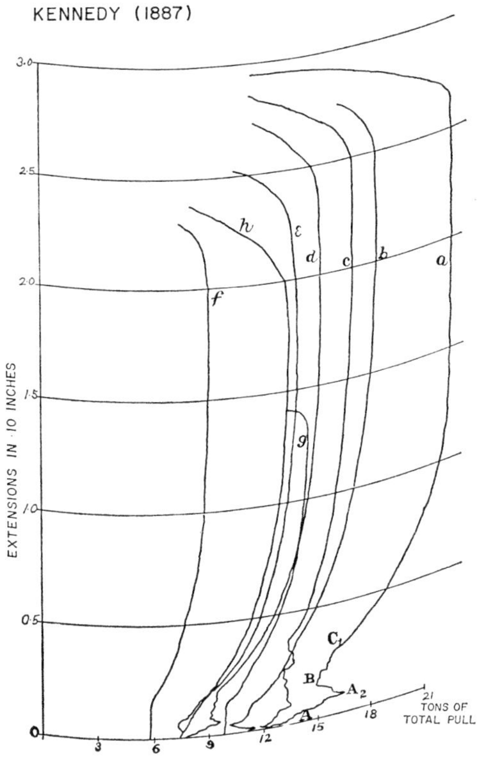

Fig. 4.25. KENNEDY's extension of THURSTON's autographic principal to a tension testing machine.

that Commodore L. W. BEARDSLEE of the Washington Navy Yard had independently discovered the phenomenon by using an entirely different method of investigation.[1] BEARDSLEE's work in tension, published late in 1874, demonstrated the important fact that the increase in the elastic limit following plastic deformation did not require that the specimen be maintained under load during the interval of time before there was a further increase in load. He observed the higher elastic limit even when he maintained the specimen under no load during this interval of time.

If it were not for the unprovoked petty vendetta which JOHANN BAUSCHINGER began in 1877, in an effort not only to deny THURSTON any credit for discovery but to warn all and sundry to ignore the man, his work, and his famous autographic testing machine, the THURSTON-BEARDSLEE-UCHATIUS affair would be just another example of the curious near simultaneity of independent discovery.

BAUSCHINGER cited General UCHATIUS,[2] who also contributed to the controversy in 1877. Realizing he had been preceded in date, UCHATIUS claimed that the phenomenon of the elastic limit increase had been known, if not published, by L. WERDER in 1854. Credit was given to WERDER on the basis of the not necessarily relevant fact that he had used prestressed rods to construct the Munich Glass Palace. Prestressed iron wires for suspension bridges, however, had been used

[1] BEARDSLEE [1874, *1*]
[2] UCHATIUS [1877, *1*].

many decades before WERDER. The aim, of course, was to raise the elastic limit by plastic deformation and thereby minimize subsequent creep.

BAUSCHINGER, too, strangely claimed some precedence for himself by referring to an earlier, entirely different type of experiment described in his famous work in 1886 on the elastic limit. On 24 November, 1874, he had submitted an iron specimen to rupture, obtaining an ultimate stress of 3200 kg/cm². He had repeated the experiment upon sections of the original specimen until it had been ruptured a total of seven times, and the ultimate stress had been raised to 4400 kg/cm². Even if this experiment had dealt with the problem of the raising of the elastic limit which was THURSTON's discovery, its being dated one year later would most certainly have precluded any claim to precedence. In successive papers BAUSCHINGER made various statements regarding the origin of one aspect of what was to be called the "BAUSCHINGER effect." In 1877[1] he credited General UCHATIUS. In 1881[2] he stated that the phenomenon had long been known and had been recently rediscovered by THURSTON and himself. Later in the same paper after spending nearly 2000 words setting aside THURSTON's contribution, he denied both THURSTON and Commodore BEARDSLEE any priority of discovery. In his classic treatise[3] in 1886 he stated that existence of the phenomenon had long been known but that UCHATIUS and he had shown that in general it occurred in many metals. In the introduction to the paper in 1886, he implied that in 1877 he had independently discovered the time influence in the increase of the elastic limit, and that only afterward had he found that in fact he had been preceded by BEARDSLEE. Now there was no mention of THURSTON. BAUSCHINGER added the comment that A. WÖHLER[4] actually had been the first, in 1858; while performing the first fatigue experiments in experimental mechanics WÖHLER had observed shifts in the endurance limit when the alternating stresses exceeded certain values. It is interesting that in the English translation[5] of BAUSCHINGER's 1886 treatise in summarized form which appeared in the Proceedings of the Institute of Civil Engineers, the translator, "W.C.U.," stated:

> On this extremely elaborate paper containing a list of experimental researches only a summary of the principal conclusions can be given. Starting from the fact that loading a bar beyond the elastic limit raises the elastic limit for a subsequent load, a fact observed first by Uchatius and himself, the Author [BAUSCHINGER] proceeds to investigate this phenomenon more closely."[6]

"W.C.U." obviously was WILLIAM CAWTHORNE UNWIN who in 1910 wrote a widely read textbook entitled, *The Testing of Materials of Construction*, which provided an entire chapter documenting and eulogizing the history of the autographic testing machine beginning with THURSTON's invention (Chap. 10, p. 239); in an earlier section he had described the various aspects of the BAUSCHINGER effect. UNWIN attributed the entire phenomenon, including precedence in discovery, to BAUSCHINGER alone.[7] It is somewhat ironic that the diagram by means of which he illustrated BAUSCHINGER's phenomenon was taken from THURSTON's autographic testing machine. UNWIN and JAMES ALFRED EWING,[8] who wrote a book on mechanical engineeering, were two of LOVE's main experimental sources for the discussion of "technical" mechanics.[9] LOVE's references to the BAUSCHIN-

[1] BAUSCHINGER [1877, *1*].
[2] BAUSCHINGER [1881, *1*].
[3] BAUSCHINGER [1886, *1*].
[4] WÖHLER [1858, *1*], [1860, *1*], [1870, *1*].
[5] UNWIN [1886, *1*].
[6] *Ibid.*, p. 463.
[7] UNWIN [1910, *1*].
[8] EWING [1899, *1*].
[9] LOVE [1927, *1*].

GER phenomenon therefore quite naturally implied that both the discovery and the definitive study of the curious properties of elastic limits and yield limits were BAUSCHINGER'S. PEARSON[1] in the second volume of the history, published in 1893, the year of BAUSCHINGER'S death, attributed the discovery of the BAUSCHINGER effect to GUSTAV HEINRICH WIEDEMANN[2] in 1859. WIEDEMANN'S torsion and flexure studies in relation to magnetism in iron certainly did contain the rudiments of the discovery. It was particularly true of his study of the permanent sets in torsion as a function of reversed loading.

The last reference to THURSTON'S discovery of the rise in the elastic limit which I have been able to find is that in an address before the American Association for the Advancement of Science by MANSFIELD MERRIMAN[3] at the 43rd meeting of the society in 1894. MERRIMAN, who was vice president of the society, provided a rather quick review of the resistance of materials under impact, from THOMAS YOUNG in 1807 through the work of NAVIER, PONCELET, HODGKINSON, WÖHLER, etc. He properly attributed the discovery of the rise of the elastic limit to THURSTON in 1873. He further referred to BAUSCHINGER'S observation in 1885 that the raising of the tensile elastic limit was accompanied by the lowering of the compressive elastic limit.

Perhaps I have spent too much time in digressing upon this curious effort upon the part of BAUSCHINGER to establish unearned precedence when he could have been well satisfied with having provided one of the most definitive studies of an important phenomenon in 19th century experimental mechanics. Every aspect of the phenomenon which he so systematically and beautifully examined had in fact been discovered by other persons. To refer to this as the "THURSTON, BEARDSLEE, UCHATIUS, and perhaps WIEDEMANN-WÖHLER effect," which of course I do not propose, would labor the commonly known fact that often several different persons discover an important phenomenon at almost the same time.

4.6. Experiments on yield limits, elastic limits, and fatigue, preceding those of Thurston and Bauschinger: Thalén (1864), Wiedemann (1859), and Wöhler (1858–1870).

BAUSCHINGER'S experiments are described in enormous detail in his various publications during the 22 years in which he was director of the Munich Laboratory. They are as fascinating to study as they are difficult to generalize briefly. I shall refer mainly to his large work of 1886, which presented his experiments between 1877 and that year.

BAUSCHINGER'S first problem was to define the elastic limit. He agreed with his predecessors, WERTHEIM, HODGKINSON *et al.*, that some small permanent deformation accompanied all deformation in metals. Like his predecessors, he emphasized that whether or not one observed this very small permanent deformation depended upon how fine a resolution could be obtained from the apparatus. When BAUSCHINGER'S mirror extensometer, invented between 1877 and 1879, finally was perfected, it allowed him to measure elongations of 1/10000 of a millimeter. This made possible a strain resolution of $\varepsilon = 7 \times 10^{-7}$, for his 15 cm long gage length in iron and steel.

As I have pointed out in Sect. 2.18, this invention also made it possible for BAUSCHINGER to perform the first definitive experiments in compression. Previous

[1] TODHUNTER and PEARSON [1893, *1*].
[2] WIEDEMANN [1859, *1*].
[3] MERRIMAN [1894, *1*].

measurements of reversed loading had had to be in torsion or in flexure because of the buckling in compression of the long specimens then being used to obtain strain resolution. BAUSCHINGER carefully distinguished between the elastic limit and the yield limit, with respect both to definition and to differences in the nature of the observed effects. Although he identified the elastic limit with the proportional limit, he did not make a purely arbitrary choice. He had noted that with a high strain resolution he could measure small permanent deformations when he loaded below the proportional limit. However, this small plastic strain was reproducible upon repeated loading of the same specimen. Above the proportional limit not only did the magnitude of this permanent deformation increase, although it still remained extremely small, but it differed from one experiment to another. This, then, was BAUSCHINGER's definition of the elastic limit, namely, the point below which microplasticity was stable, i.e., the proportional limit. He further noted that above this elastic limit there was measurable elastic after-effect over a period of time, although below the elastic limit the specimen could remain under fixed loads for a long time without any measurable increase in the deformation. He used the term "yield limit" to describe the point of onset of relatively large plastic deformation. In modern terminology the term "elastic limit" usually designates BAUSCHINGER's yield limit, a fact to be borne in mind when comparing 19th and 20th century references to the "BAUSCHINGER effect."

Before attempting to describe the many facets of BAUSCHINGER's study of the two "limits" in the plasticity in metals, we well may examine the state of experimental knowledge of elastic limits and yield limits immediately before his time. Illustrative in this regard is the memoir in 1863 of ROBERT THALÉN,[1] a Swedish experimentist, who reviewed the work of the previous three decades. THALÉN used the hydraulic testing machine with which LAGERHJELM had done his experiments of 1829 described above. THALÉN stated that he wished to resolve the controversy between HODGKINSON and MORIN[2] as to whether an increase in the elastic limit did occur after unloading from large deformation. He noted that their difference of opinion had arisen simply from MORIN's failure to realize the importance of prior mechanical history. MORIN's extruded wires already had had the elastic limit raised before he began his tests, whereas HODGKINSON's long compression bars had been essentially in an annealed state.

THALÉN performed tests upon rolled iron and steel bars of both round and square cross section, six Swedish feet long.[3] Using a 5 foot gage length and a microscope mounted on a micrometer screw which permitted direct measurement of 0.0005 mm elongation, he studied the way the elastic limit increased after plastic deformation.

THALÉN placed his maximum possible error at 0.02 mm, which for the 1484.5 mm gage length provided a strain accuracy of $\varepsilon = 10 \times 10^{-6}$ for an observed much lower resolution. After discussing the problem of defining an elastic limit as the onset of permanent deformation when the observation of the beginning of permanent deformation was a matter which depended upon the accuracy of the measurement, THALÉN followed WERTHEIM in arbitrarily defining a strain 0.00005 of the unit of length as the elastic limit.

He first noted [curve (1) of his diagram shown in Fig. 4.26] that the elastic limit could be raised by plastic deformation. This, of course, was well known from GERSTNER's experiments of forty years before. By immediately unloading and

[1] THALÉN published in [1865, *1*].

[2] The reader will recall that HODGKINSON's papers extended from the 1820's to the 1850's (see [1824, *1*] et seq.); MORIN [1862, *1*] is described in Sect. 2.16 above.

[3] 1 Swedish ft. = 29.69 cm.

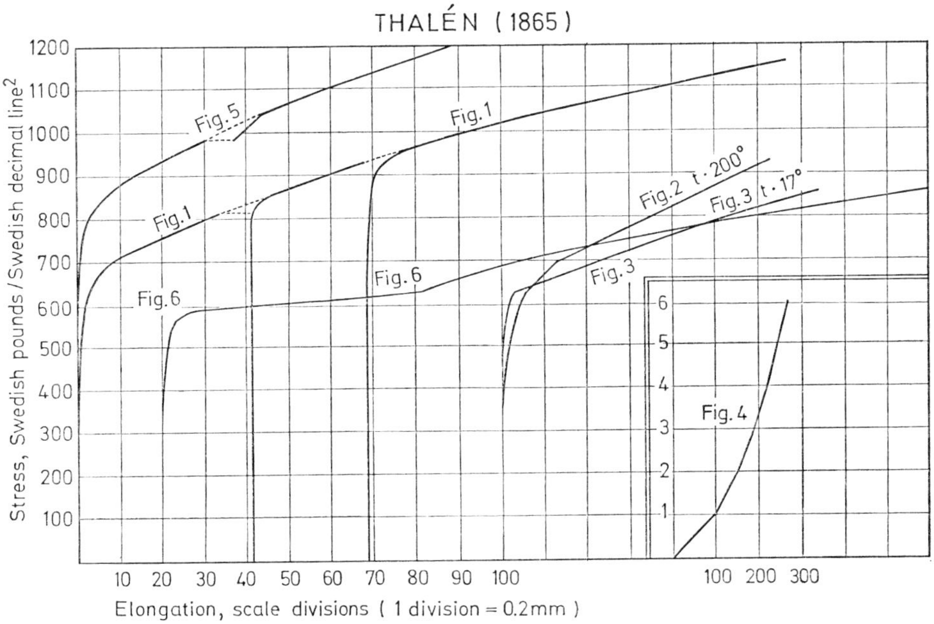

Fig. 4.26. THALÉN's observations in 1865 of the effects of repeated loading and the effects of annealing upon the plastic deformation of steel. Also shown (4) is the permanent deformation measured after each reloading. Each of the six reloadings reached a stress below that of the first loading in the plastic region.

then quickly reloading, THALÉN showed that the stress-strain function continued as though there had been no interruption.

WERTHEIM had found that when he heated the specimen, the elastic limit fell by $^1/_{11}$ to $^1/_3$ of its value in the hardened state. THALÉN's check on this lowering of the elastic limit is shown as (2) for annealing at 200° C and (3) for the unannealed specimen in Fig. 4.26. He observed that when he loaded the bar six times, each time reaching the load which was initially in the plastic region, for each new loading permanent deformation continued to be produced in successively smaller amounts. This is seen in (4) of Fig. 4.26.

Finally THALÉN reported that a "peculiar phenomenon is observed, which deserves a closer attention."[1] When he successively loaded the bar several times to a given load, before proceeding to the next higher increment of load, he observed the gradual increase in permanent deformation shown in (5) of Fig. 4.26 for 10 such reloadings to 980 Swedish pounds per Swedish decimal line square. For successive loading increments, steeper slopes occurred until the basic deformation curve was reached. This is a familiar phenomenon to those few persons today who use dead weight loading to study the SAVART-MASSON (PORTEVIN-LE CHATELIER) effect. I believe THALÉN's was the first clear demonstration of the stepped slopes, although they are contained in MASSON's[2] tabulated experimental results of 1841.

THALÉN's study in 1865, the experiments of WIEDEMANN[3] in 1858 and 1859 in reversed torsion and reversed bending, and WÖHLER's[4] remarkable experiments

[1] THALÉN [1865, *1*], p. 202.
[2] MASSON [1841, *1*].
[3] WIEDEMANN [1858, *1*], [1859, *1*].
[4] WÖHLER [1858, *1*], [1860, *1*], [1870, *1*].

from 1858 to 1870 on flexural fatigue like those of TRESCA, seriously explored phenomena only lightly investigated previously, and accurately portray what was known of elastic and yield limits before 1873, the year in which THURSTON[1] secured the handle of his autographic machine for 24 h.

WIEDEMANN's main interest was to extend the studies of magnetism vs deformation which WERTHEIM had made in iron a few years before. In attempting to show a relation between permanent deformation and magnetization, a subject extensively explored twenty-five years later by HERBERT TOMLINSON,[2] WIEDEMANN, without appreciating its significance, observed what is now known as the "BAUSCHINGER effect." Since his main interest was in the relation between magnetism and permanent deformation, he commented only briefly on the behavior of the specimens under reversed loading.

As the precursor to BAUSCHINGER's experiments of 26 years later, one may choose WIEDEMANN's flexural experiments of 1859[3] or his reversed loading torsional experiments of 1858[4], in which the same behavior was observed. Having considered in detail a large fraction of BAUSCHINGER's nearly 5000 experiments (since I confess to a penchant for rambling through masses of reliable experimental data) and having examined the writings of Major General UCHATIUS, Commodore BEARDSLEE, and Professor THURSTON and WÖHLER, it is my considered opinion that PEARSON who credited WIEDEMANN with precedence was in this instance correct. WIEDEMANN of Basel in 1858–59 in his experiments on torsion and flexure discovered, if crudely, the change which occurred in the magnitude of the elastic limit when a specimen previously loaded to plastic deformation was unloaded and reloaded by a stress of opposite sign. If precedence is the criterion, rather than definitive study, one should designate as the "WIEDEMANN effect" or perhaps the "WIEDEMANN-THURSTON effect", the phenomenon which BAUSCHINGER, perhaps with some confusion regarding its true origins, magnificently explored to the point of exhaustion in one of the most complete analyses of a single issue in the entire history of experimental solid mechanics.

WIEDEMANN characterized his objectives when he stated:

> A series of observations caused me to believe that changes of the shape of the body caused by mechanical means follow laws very similar to those which describe the changes caused by forces responsible for the magnetization of magnetic metals.[5]

In both torsion and flexure he contrasted the temporary deformation with the permanent deformation after the removal of the torque or load, comparing both with values obtained when the load was reversed. I shall leave the reader in this instance to peruse WIEDEMANN's rather sparse experimental results, but three of his conclusions in 1859 are of more than a little interest: 1) When a permanent deformation in one direction had been produced, a reversal of stress to the opposite direction required a much lower load to remove all of the visible permanent deformation. 2) When the rod had been subjected to successive loadings and *left in repose* for some time, then it tended to return somewhat, but not completely, to the behavior observed during the first loading. 3) When a rod was shaken under load, its capability of sustaining temporary deformation increased, but if it were shaken after the load was removed, its permanent deformation decreased.

A. WÖHLER was an engineer who would deserve prominence in any history of 19th century technology. In a century which learned its engineering design from

[1] THURSTON [1873, *1*]. See [1874, *1*].
[2] TOMLINSON [1883, *1*].
[3] WIEDEMANN [1859, *1*].
[4] WIEDEMANN [1858, *1*].
[5] WIEDEMANN [1859, *1*], p. 161.

disaster, he was unquestionably one of the first technologists who, from observed loading histories, attempted to develop from a sound experimental base, practical rules for analyzing the probabilities of mechanical failure. He was literally the first[1] and perhaps in some respects is still the foremost person to have studied the fatigue of metals which more than a century and unknown millions of dollars later remains a technical field of multiple mysteries.

In 1847 WÖHLER was appointed the director of the rolling stock and machine shop of the Niederschlesisch-Märkische Eisenbahn in Frankfurt an der Oder. Worried about the service failure of axles in wheels, he designed a test for determining the number of times per mile of travel the applied load reached a maximum flexural stress due to variations in roadbed and other operating track conditions. An indexing system recording by means of scratches upon a zinc plate revealed that major stresses were reached on the average of approximately once per mile. WÖHLER's work was later criticized by some for overemphasizing the occasional repetitions of large stress and ignoring the contributions of the enormous number of cycles of intervening lower stress. Be that as it may, he invented an apparatus which became known as the Wöhler Testing Machine,[2] to reproduce in the laboratory the conditions he had observed. The present interest is on the experiments in his laboratory. The apparatus, which in various versions became a standard piece of equipment in mechanical testing laboratories from the 1870's to the present day, permitted WÖHLER to explore the fatigue of the practical metals in which he was interested, including their endurance limits for various alternating stresses at prescribed mean stresses.

In 1880 it was customary when displaying data on fatigue to consider the range of alternating stress as a function of the magnitude of the lowest stress rather than

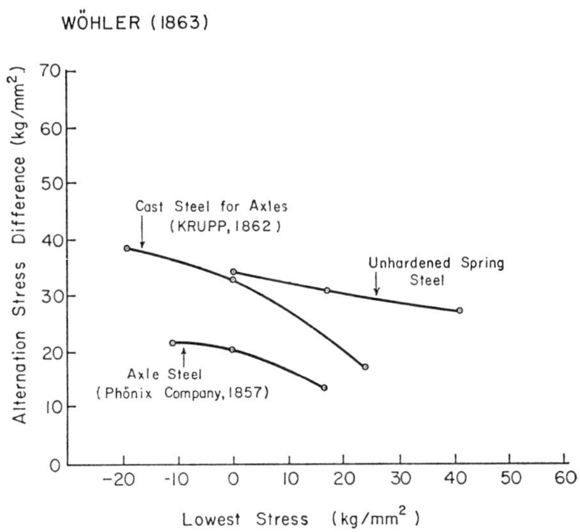

Fig. 4.27. A plot in 1863 of the alternating stress difference vs the lowest stress for the fatigue of steel. These measurements of WÖHLER were the first in the study of the phenomenon of fatigue.

[1] I am ignoring the meaningless tests of Captain HENRY JONES and Captain GALTON, who loaded and released iron bars nearly 500000 times, obtaining uninterpretable results. Their data were included in the Iron Commission Report of 1849 [1849, *1*].

[2] A fatigue testing machine also was devised by the Englishman, FAIRBAIRN, in 1864.

in terms of the mean stress. Thus WÖHLER's results in 1863 for three steels used in the axles of railway wheels were given as shown in Fig. 4.27.

BAUSCHINGER in his attempt to formalize empirically the practical design limits for fatigue, referred to such plots as "Gerber's parabolas." The number of cycles of repeated stress he obtained by using WÖHLER's machine was several million. The largest number of cycles he recorded in 1886 was sixteen million.

In 1886[1] BAUSCHINGER cited from WÖHLER's paper[2] of 1863 those aspects of the latter's work for which he accorded him precedence: that a solid in alternating stress could be ruptured without ever reaching the quasi-static rupture load; that the magnitude of the stress difference was important; and that the highest compressive stress and tensile alternating stress below which rupture would not occur was zero mean stress. BAUSCHINGER discussed WÖHLER's data at length; he credited WÖHLER with having discovered the "natural elastic limit" for tension and compression, which is lower than the original elastic limit and is reached after a number of stress reversals when the original elastic limit has been exceeded in one direction.

4.7. The experiments of Bauschinger on the yield limit and the elastic limit (1875–1886).

The documentation of laboratory apparatus, test dates, and experimental detail for BAUSCHINGER's laboratory are unusually complete. BAUSCHINGER numbered his tests consecutively. The treatise of 1886 referred to 3678 tests starting from those in 1875, numbered 938 to 1000, to test 4615, dated November 1885. In some months on a single machine he ran as many as 150 tests each involving the elaborate adjustment of his optical mirror extensometer, sometimes several times in one experiment. In 1886 BAUSCHINGER supplied KENNEDY[3] with a description of his laboratory apparatus for the latter's large study entitled "The Use and Equipment of Engineering Laboratories": 1) a 100 ton Werder testing machine equipped with BAUSCHINGER mirror extensometer strain measuring appliances (this was the main piece of equipment upon which nearly 5000 experiments were performed); 2) a WÖHLER type tension fatigue machine; 3) a WÖHLER type repeated rod flexure machine; 4) a plate bending machine; 5) a machine for testing the wear of materials; 6) appliances for testing cement in the 100 ton Werder machine; 7) machine tools for preparing specimens driven by a 2 HP Otto engine.

The list of titles of BAUSCHINGER's laboratory reports published between 1874 and 1895 indicates the growth in the interests of this late 19th century test engineer. From 1874 to 1885 the reports were concerned almost entirely with the failure properties of steel, iron, stone, wood, plaster, and cement. An exception was the experimental study in 1879 of the small elastic compression occurring during plastic deformation described in Sect. 2.18 of the present treatise. As I have noted, the important mirror extensometer, which he had developed between 1877 and 1879, had provided the opportunity for this investigation. In 1886 the great study of elastic and yield limits was described. The next nine volumes, to 1895, revealed BAUSCHINGER's interest in establishing world standards for testing engineering materials. He was instrumental in instigating international conferences to that

[1] BAUSCHINGER [1886, *1*].

[2] WÖHLER's data was given in CTR per Zoll² which have been converted to kg/mm² in Fig. 4.27. One CTR/Zoll² = 6.84 kg/cm². These are data summarized from an 1870 publication by WÖHLER [1870, *1*].

[3] KENNEDY [1887, *1*].

end. The volumes also recorded his return to studies of the failure of leather, wood, stone, and metals. Later volumes contained references to time effects in the rupture of metals and to a number of tests designed to study the influence of specimen shape on tensile testing in connection with his interest in universal standards. It is thus seen that like DAVID KIRKALDY, BAUSCHINGER had a lifelong interest in establishing practical engineering test criteria. Exceptions to this concentration were his two ventures reported in 1879 and 1886 in the field of the physics of deformation. He sometimes studied many problems simultaneously and mentioned but one of them in any particular laboratory report, as is apparent from the lack of sequence of test numbers in any given report or in his papers.

BAUSCHINGER began his study of 1886 by describing his experiments on bronze in tension and compression: he had performed them in 1875 on the 100 ton machine. In those experiments he first noted the elastic limit on the initial loading. He then increased the load to a value 25% higher than this elastic limit, thus also increasing the plastic strain. Upon immediately unloading and reloading the specimen, he found that although the elastic limit now was higher than the value obtained on the first loading, it was still slightly below the previous maximum stress. He performed his experiments wholly in tension or wholly in compression. He observed similar behavior of the elastic limit for each of the two types of simple loading.

Two years later, on 6 March, 1877, BAUSCHINGER performed the same experiment on a flat rod of Belgian zinc in tension. On the third specimen, for which the initial elastic limit was 24 kg/cm^2 and the load to which the specimen subsequently was raised was 98 kg/cm^2, he noted that with rapid unloading and reloading the elastic limit was the same 98 kg/cm^2, with a permanent strain of 0.000860. In this instance, however, he waited 22 h 48 min and then found that the permanent strain had receded to 0.000730; once again reloading the specimen, he found that the elastic limit had now risen to 108 kg/cm^2. He stated[1] that with this experiment he had discovered the influence of time in raising the elastic limit.[2]

BAUSCHINGER's crucial test was # 1765; it had been recorded on 6 March 1877. He described a series of experiments which he said were designed to study the new-found phenomenon of the effect of the lapse of time upon the elastic limit. However, the tests were #1731 and 1739; he had performed them one month earlier, on 16–17 February and 20 February 1877, and had described the same apparently then unnoticed behavior in Bessemer steel. Table 122 shows the measured loads, elongations, and times, including an eighteen hour interruption for one of these Bessemer steel specimens. There is little question in my mind that BAUSCHINGER regarded the effect of time lapse upon the elastic limit of a prestressed rod, i.e., strain aging, as the most significant observation in his experiments. His evaluation is in decided contrast to the prevailing judgment, that of prime importance was the lowering of the elastic limit in a second loading when the stress was of opposite sign from that of the first loading; this as we noted, is what has become known as the "BAUSCHINGER effect."

BAUSCHINGER then gave a long and interesting discussion of the problem of defining the elastic limit. As I noted above, he concluded that the difference in the stability of the small permanent deformation on either side of the proportional

[1] BAUSCHINGER [1886, *1*].

[2] It was at this time that BAUSCHINGER announced that he belatedly had found that BEARDSLEE and WÖHLER had made similar observations in 1874 and 1863, respectively, which indicated a curious oversight for one who was himself publishing papers in *Dingler's Journal*.

Table 122. Bauschinger (1886). *Round rod of Bessemer steel, tested in tension (Lab. # 939f; diameter 2.52 cm, gage length 15 cm).*

Load increments (tons)	Time-lapse after loading (min)	Elongation (mm)	Differences
13	1	0.195	—
	2	0.195	—
14	1	0.210	0.015
	2	0.210	0.015
15	1	0.725	0.610
	5	0.820	—
16	1	1.085	0.340
	5	1.160	—
17	1	1.420	0.350
	5	1.510	—
0	—	1.235	—
	18 h unloaded		
0	—	1.235	—
17	1	1.530	—
	2	1.530	—
18	1	1.550	0.020
	2	1.550	—
19	1	1.835	0.780
	9	2.330	—
20	1	2.650	0.520
	7	2.850	—
21	1	3.180	0.505
	5	3.355	—
22	1	3.840	0.610
	6	3.965	—
23	1	4.445	0.635
	8	4.600	—
0	1	4.165	—
	9	4.165	—
23	1	4.700	—
	4	4.730	—
24	1	5.155	0.730
	9	5.460	—
25	1	6.020	0.885
	10	6.345	—
0	1	5.870	—
	3	5.870	—
25	1	6.380	—
	4	6.420	—
0	—	5.830	—

Rupture occurred at 26.5 tons = 5300 kg/cm², which was held for several minutes. Test #1731, 16–17 February, 1877.

limit should be defined as the elastic limit, rather than the definition chosen by Wertheim, that the onset of permanent deformation occurred at an arbitrarily chosen, fixed strain. Bauschinger noted that his own definition excluded a number of solids such as cast iron and stones tested in compression, for which no such elastic limit could be found.

He next performed a series of experiments with successive loading to stresses above the elastic limit obtained in the first loading, and established that even for a large number of cycles a new, higher value remained unaltered.

BAUSCHINGER then turned his attention to the yield limit of ingot iron, wrought iron, and soft steel at which a sudden large change in strain caused his mirror extensometer to go off scale. This sudden yielding, which occurred in both tension and compression, invariably was accompanied by an elastic after-effect of considerable magnitude, sometimes lasting for several days. He noted that it introduced a certain amount of imprecision in defining the yield limit.

Two of BAUSCHINGER's observations relating to the yield limit were, first, that when unloading and reloading immediately followed the yielding, the same load value of the new yielding was observed, but if one day elapsed, a noticeably higher value was obtained, lasting for months, "perhaps years"; and second, a remarkable and still little known behavior, that during immediate reloading following yielding, the elastic limit was reduced to a value far below the original elastic limit, sometimes to zero. The elastic limit reappeared during the next several days and gradually increased until it reached the load at which yielding had occurred, which was considerably above the original pre-yield elastic limit. BAUSCHINGER expressed his belief that with sufficient time, the new elastic limit would exceed the first yield stress.

As illustrative of this behavior I have selected from BAUSCHINGER's vast array of data, his table, here Table 123, showing three tension loadings of a wrought iron specimen, the initial loading, a repetition after 7 min, and after 62 h.[1]

The single underline denoted the observed elastic limit and the double and triple underlining, the more definite yield limit. One notes that on the second 7 min immediate reloading, the elastic limit has disappeared and the yield limit has increased only very slightly. On the third loading after a time elapse of 62 h the elastic limit has reappeared, having increased to that obtained during the initial loading, and the yield limit has markedly increased beyond its initial value. BAUSCHINGER's paper then referred to tabulated experiments which demonstrated that the same type of behavior occurred in compression.

BAUSCHINGER noted that in metals in general the dramatic yield limit of iron and soft steel did not occur. He extended his studies to zinc, bronze, copper, and red brass to the point where significant yielding had taken place, to observe whether or not subsequent loading would reduce the elastic limit. In none of these metals did he observe a lowering of the elastic limit even though "significant yielding" had taken place. For copper and red brass, in fact, a raising of the elastic limit was observed which after two or three days increased beyond the previous yield limit.

In this systematic study BAUSCHINGER next raised the empirical question as to what influence, if any, intermediate mechanical or thermal histories would have upon these new elastic limits and yield limits. Violent vibrations such as occur in cold forging again lowered the elastic limit which previously had been raised due to yielding. Such a treatment also lowered the raised yield limit slightly although it remained above the original value. Annealing to 350° C affected only ingot iron. For wrought iron and steel only rapid cooling from temperatures between 350° C and 450° C had any effect. In general, rapid cooling lowered the elastic limit more than slow cooling did. Only in ingot iron below 450° C could slow cooling lower the original value. BAUSCHINGER gave tabulated results of the effect upon either virgin or prestressed elastic and yield limits of a variety of prior thermal histories. In the present context the data are interesting chiefly because they emphasize the general importance of prior thermal history on subsequent deformation.

---

[1] BAUSCHINGER [1886, *1*].

Table 123. BAUSCHINGER (1886). *Round rod of wrought iron tested in tension (Lab. # 938).*

Original condition
Round arc scale
Cross-section = 2.50 cm
Initial length = 15.00 cm

Load (tons)	Elongation ($^1/_{1000}$ cm)	Diff.	Time
0	0		
1	1.43	*143*	
2	2.89	*146*	
3	4.32	*143*	
4	5.75	*143*	
0	0.05		
4	5.73		
5	7.16	*143*	
6	8.61	*145*	
7	10.06	*145*	
8	11.48	*142*	
0	0.02		
8	11.48		
8.5	12.17	*69*	
9	12.90	*73*	
9.5	13.67	*77*	
10	14.59	*92*	
0	0.13		
10	14.72		
10.1	14.97	*25*	
10.2	15.17	*20*	
10.3	15.36	*19*	4 h 19′
10.4	15.60	*24*	4 h 22′
10.5	15.80	*20*	4 h 25′
10.6	16.15	*35*	4 h 28′
10.7	17.10	*95*	4 h 31′
10.8	Scales ran through so rapidly that no reading could bc made	9789	4 h 34′
10.8	114.99		4 h 49′
0	97.30		

Test 3202: 24 March, 1880

7 min after end of previous test
Round arc scale
Cross-section = 2.50 cm
Initial length = 15.10 cm

Load (tons)	Elongation ($^1/_{1000}$ cm)	Diff.	Time
0	0		4 h 56′
1	1.49	*149*	4 h 57′
2	2.99	*150*	4 h 58′
3	4.52	*153*	4 h 59′
4	6.07	*155*	5 h 0′
0	0.05		5 h 1′
4	6.07		5 h 2′
5	7.63	*156*	5 h 3′
6	9.20	*157*	5 h 4′
7	10.84	*164*	5 h 5
8	12.51	*167*	5 h 6′
0	0.15		5 h 7′
8	12.51		5 h 8′
8.5	13.33	*82*	5 h 9′
9	14.22	*89*	5 h 10′
9.5	15.10	*88*	5 h 11′
10	16.13	*103*	5 h 12′
0	0.45		5 h 13′
10	16.28		5 h 14′
10.5	17.50		5 h 15′
10.6	18.06	*56*	5 h 16′
10.7	18.91	*85*	5 h 19′
10.8	19.79	*88*	5 h 22′
10.9	20.86	*107*	5 h 25′
11.0	22.47	*161*	5 h 28′
11.1	Scales ran through so rapidly that no reading could be made	775	5 h 31′
11.1	30.22	824	5 h 34′
11.2	38.46	1620	5 h 40′
11.3	54.66		5 h 52′
0	35.91		5 h 54′
0	35.83		5 h 57′
0	35.67		after 15 h

Test 3202: 24 March, 1880

62 h after end of last test
Round arc scale
Cross-section = 2.495 cm
Initial length = 15.13 cm

Load (tons)	Elongation ($^1/_{1000}$ cm)	Diff.	Time
0	0		
1	1.50	*150*	
2	3.00	*150*	
3	4.51	*151*	
4	6.02	*151*	
0	0.02		
4	6.01		
5	7.55	*154*	
6	9.06	*151*	
7	10.60	*154*	
8	12.11	*151*	
0	0.04		
8	12.11		
8.5	12.88	*77*	
9	13.63	*75*	
9.5	14.41	*78*	
10	15.91	*78*	
0	0.06		
10	15.17		
10.5	15.96	*79*	
11	16.75	*79*	
11.5	17.57	*82*	
12	18.58	*101*	
0	0.32		
12	18.69		4 h 39′
12.1	19.83	*114*	4 h 42′
12.2	22.10	*227*	4 h 45′
12.3	24.10	*200*	4 h 48′
12.4	25.78	*168*	4 h 51′
12.5	27.73	*195*	4 h 54′
12.6	31.13	*340*	4 h 57′
12.7		*11145*	9 h 0′
12.7	142.58		9 h 15′
0	120.65		9 h $15^1/_4$′
0	120.53		9 h 16′
0	120.47		9 h 17′
0	120.44		9 h 18′
0	120.42		9 h 19′
0	120.41		9 h 20′
0	120.37		9 h 25′
0	120.32		9 h 30′

Test 3202: 27 March, 1880

In Sect. 5 BAUSCHINGER began to discuss what the modern metallurgist and continuum physicist refer to as the "Bauschinger effect." Since he had perfected a roller optical extensometer by 1879, he could obtain for the first time comparable strain resolution in compression and in tension. In terms of his definition of it, he now could study the elastic limit in both stress directions. When the stress went beyond the yield stress in tension or in compression, the elastic limit in either direction was greatly reduced or temporarily eliminated. When the stress exceeded the elastic limit but did not reach the yield limit in one uniaxial direction, the elastic limit in the opposite direction invariably was reduced. The more the initial stress beyond the original elastic limit exceeded this reduced value, the greater the reduction in the opposite direction. When the stress overstepped the elastic limit in the reverse direction, the new elastic limit in the original direction also was lowered. A continued overstepping by alternating between tension and compression finally resulted in an elastic limit smaller than that obtained in the first loading, which BAUSCHINGER called the "natural elastic limit."

The specimens for these opposing loads were machined as shown in Fig. 4.28. The tension was applied by clamps which fitted the trapezoidal ends. The flat end

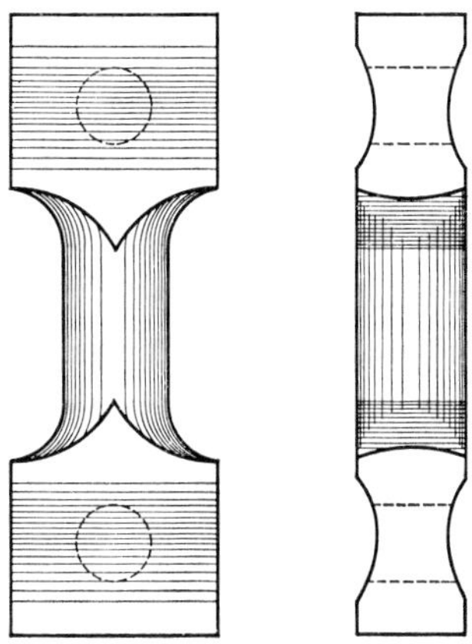

Fig. 4.28. The type of tension-compression specimens used by BAUSCHINGER in his studies of the elastic limit and the yield limit of solids subjected to reverse loading.

faces permitted the application of compression loads. BAUSCHINGER explained that contrary to his expectation, he had not had to use the drilled holes in the tension loading. The specimens were of malleable iron and Bessemer steel.

The first tension vs compression experiment in malleable iron in this series must be viewed as one of the major landmarks in the experimental mechanics of crystalline solids. WIEDEMANN undoubtedly had discovered 26 years earlier, albeit under crude circumstances, the influence of reversed loading on the elastic limit, and THURSTON 11 years earlier the time influence on the raising of the elastic limit. WIEDEMANN, as we have noted, had defined the elastic limit poorly and had made imprecise measurement of the appearance of plastic deformation during reversed flexure and reversed torsion. It was BAUSCHINGER who made the first quantitative study in his experiment of 1885, the results of which are in Table 124.

The sequence of loading cycles may be followed in Table 124. The first two series were for five reloadings of the specimen in tension at the times indicated on the top of the columns. Through repeated tension testings, after finally having established the elastic limit as 2091 kg/cm² and gone beyond it to 2214 kg/cm², BAUSCHINGER performed the compression experiment of column 3, which shows the reduction of the elastic limit to approximately $^1/_3$ the original value, i.e., 738 kg/cm². In compression, when the stress was returned to 2214 kg/cm² the elastic limit was raised to 1476 kg/cm² 5 min later, and 20 h later, to 1968 kg/cm².

Table 124. Bauschinger (1885).

Load in tons	1 Original condition Tension $d=4.55$ cm $l=6.00$ cm				2 6 days later Tension $d=4.55$ cm $l=6.00$ cm			
	Stress (kg/cm²)	Elongation ($^1/_{1000}$ cm)	Differences	Elast. Mod. (kg/cm²)	Stress (kg/cm²)	Elongation ($^1/_{1000}$ cm)	Differences	Elast. Mod. (kg/cm²)
0	0	0			0	0		
4	246	0.66	66		246	0.63	63	
8	492	1.34	68		492	1.31	68	
12	738	2.02	68		738	1.95	64	
16	984	2.70	68		984	2.62	67	
0	0	0.07			0	0.02		
16	984	2.72			984	2.63		
20	1230	3.39	67		1230	3.28	65	
24	1476	4.04	65	2,160,000	1476	3.92	64	2,260,000
28	1722	4.74	70		1722	4.58	66	
32	1968	5.42	68		1968	5.24	66	
0	0	0.16			0	0.01		
32	1968	5.46			1968	5.22		
34	2091	5.80	34		2091		93	
36					2214	6.15		
40					0	0.41		
0								
	Test 4387 6 Oct., 1885				Test 4404 12 Oct., 1885			

Load in tons	3 51 min later Compression $d=4.55$ cm $l=6.00$ cm				4 5 min later Compression $d=4.55$ cm $l=6.00$ cm			
	Stress (kg/cm²)	Contraction ($^1/_{1000}$ cm)	Differences	Elast. Mod. (kg/cm²)	Stress (kg/cm²)	Contraction ($^1/_{1000}$ cm)	Differences	Elast. Mod. (kg/cm²)
0	0	0			0	0		
4	246	0.69	69		246	0.60	60	
8	492	1.36	67		492	1.26	66	
12	738	2.04	68		738	1.90	64	
16	984	2.76	72		984	2.58	68	
0	0	0.18			0	0.02		
16	984	2.79			984	2.55		
20	1230	3.51	72		1230	3.23	68	
24	1476	4.24	73	2,170,000	1476	3.87	64	2,290,000
28	1722	5.07	83		1722	4.57	70	
32	1968	6.16	109		1968	5.27	70	
0	0	0.94			0	0.02		
32	1968	6.31			1968	5.29		
34			245				273	
36	2214	8.76			2214	8.02		
40	0	2.90			0	1.15		
0					0	1.06		
	Test 4404 12 Oct., 1885				Test 4404 12 Oct., 1885			

Load in tons	5 20 h later Compression $d=4.55$ cm $l=6.00$ cm				6 1 h later Tension $d=4.55$ cm $l=6.00$ cm			
	Stress (kg/cm²)	Contraction ($^1/_{1000}$ cm)	Differences	Elast. Mod. (kg/cm²)	Stress (kg/cm²)	Elongation ($^1/_{1000}$ cm)	Differences	Elast. Mod. (kg/cm²)
0	0	0			0	0		
4	246	0.63	63		246	0.12	72	
8	492	1.29	66		492	1.55	83	
12	738	1.93	64		738	2.41	86	
16	984	2.59	66		984	3.34	93	
0	0	0.02			0	0.76		
16	984	2.57			984	3.37		
20	1230	3.24	67		1230	4.37	100	
24	1476	3.90	66	2,260,000	1476	5.48	111	
28	1722	4.57	67		1722	6.67	119	
32	1968	5.25	68		1968	7.99	132	
0	0	0.05			0	2.88		
32	1968	5.28			1968	8.12		
34			97				147	
36	2214	6.25			2214	9.59		
40	0	0.28			0	3.71		
0								
	Test 4404 13 Oct., 1885				Test 4404 Oct., 13 1885			

Table 124. (Continued.)

Load in tons	7 46 min later Compression $d = 4.55$ cm $l = 6.00$ cm Stress (kg/cm²)	Contraction ($^1/_{1000}$ cm)	Differences	Elast. Mod. (kg/cm²)
0	0	0		
4	246	0.77	*77*	
8	492	1.65	*88*	
12	738	2.62	*97*	
16	984	3.63	*101*	
0	0	0.99		
16	984	3.69		
20	1230	4.71	*102*	
24	1476	5.82	*111*	
28	1722	7.12	*130*	
32	1968	9.03	*191*	
0	0	3.68		
32	1968	9.36		
36	2214	Scales ran through, (yield limit)		
	Test 4404 13 Oct., 1885			

Load in tons	8 $30^1/_2$ h later Compression $d = 4.55$ cm $l = 6.00$ cm Stress (kg/cm²)	Contraction ($^1/_{1000}$ cm)	Differences	Elast. Mod. (kg/cm²)	9 $15^1/_2$ h later Tension $d = 4.55$ cm $l = 6.00$ cm Stress (kg/cm²)	Elongation ($^1/_{1000}$ cm)	Differences	Elast. Mod. (kg/cm²)
0	0	0			0	0		
4	246	0.66	*66*		246	0.82	*82*	
8	492	1.31	*65*		492	1.79	*97*	
12	738	1.96	*65*	2,250,000	738	2.93	*114*	
16	984	2.62	*66*		984	4.33	*140*	
18	0	0.00			0	1.70		
0								
18								
20								
24								
28								
32								
0								
32								
36								
40								
43								
	Test 4406 14 Oct., 1885				Test 4406 15 Oct., 1885			

Load in tons	10 2 h later Tension $d = 4.55$ cm $l = 6.00$ cm Stress (kg/cm²)	Elongation ($^1/_{1000}$ cm)	Differences	Elast. Mod. (kg/cm²)	11 9 min later* Tension $d = 4.55$ cm $l = 6.00$ cm Stress (kg/cm²)	Elongation ($^1/_{1000}$ cm)	Differences	Elast. Mod. (kg/cm²)
0	0	0			0	0		
4	246	0.69	*69*		246	0.69	*69*	
8	492	1.33	*64*		492	1.32	*63*	
12	738	2.00	*67*	2,210,000	738	1.96	*64*	2,260,000
16	984	2.76	*76*		984	2.61	*65*	
18	1107	3.42	*66*		1107	2.98	*37*	
0	0	0.48			0	0.06		
18								
20								
24								
28								
32								
0								
32								
36								
40								
43								
	Test 4406 15 Oct., 1885				Test 4406 15 Oct., 1885			

Load in tons	12 27 h later Compression $d = 4.55$ cm $l = 6.00$ cm Stress (kg/cm²)	Contraction ($^1/_{1000}$ cm)	Differences	Elast. Mod. (kg/cm²)
0	0	0		
4	246	0.68	*68*	
8	492	1.39	*71*	
12	738	2.14	*75*	
16	984	2.90	*76*	
18	1107	3.30	*40*	
0	0	0.31		
18	1107	3.31		
20	1230	3.78	*47*	
24	1476	4.73	*95*	
28	1722	5.83	*110*	
32	1968	7.18	*135*	
0	0	1.79		
32	1968	7.33		
36	2214	10.40	*307*	
40	2460	25.15	*1475*	
43	2644			
	Test 4406 16 Oct., 1885			

* After the load was varied 4 times between 0 and 18 tons, and the total elongation rose to 3.65 and the permanent to 0.72/1000 cm.

Table 124. (Continued.)

Load in tons	13 30 min later Compression $d = 4.58$ cm $l = 5.93$ cm				14 3 days later Tension $d = 4.58$ cm $l = 5.93$ cm			
	Stress (kg/cm²)	Contraction ($^1/_{1000}$ cm)	Differences	Elast. Mod. (kg/cm²)	Stress (kg/cm²)	Elongation ($^1/_{1000}$ cm)	Differences	Elast. Mod. (kg/cm²)
0	0	0			0	0		
4	243	0.69	*69*		243	0.74	*74*	
8	486	1.37	*68*		486	1.82	*108*	
12	729	2.05	*68*		729	3.09	*127*	
16	971	2.70	*65*	2,100,000	971	4.74	*165*	
0	0	0.00			0	2.11		
16	971	2.70			971	5.01		
20	1214	3.40	*70*		1214	7.14	*213*	
24	1475	4.10	*70*		1457	10.07	*293*	
28	1700	4.80	*70*		1700	13.31	*324*	
32	1943	5.49	*69*		1943	20.00	*669*	
0	0	0.09			0	14.61		
32	1943	5.52			1943	20.62		
36	2186	6.38	*86*		2186	28.25	*763*	
40	2429	8.02	*164*		2429	42.25	*1400*	
42	2550	9.90	*188*					
44	2671	Scales ran through			2671	Scales ran through before 44 tons was reached		
	Test 4406 16 Oct., 1885				Test 4423 19 Oct., 1885			

Load in tons	15 2 days later Compression $d = 4.58$ cm $l = 5.94$ cm				16 2 days later Tension $d = 4.58$ cm $l = 5.94$ cm			
	Stress (kg/cm²)	Contraction ($^1/_{1000}$ cm)	Differences	Elast. Mod. (kg/cm²)	Stress (kg/cm²)	Elongation ($^1/_{1000}$ cm)	Differences	Elast. Mod. (kg/cm²)
0	0	0			0	0		
4	243	0.66	*66*		243	0.64	*64*	
8	486	1.32	*66*		486	1.29	*65*	
12	729	1.98	*66*		729	1.92	*63*	
16	971	2.81	*83*	2,190,000	971	2.57	*65*	2,250,000
0	0	0.15			0	0.02		
	Test 4444 21 Oct., 1885				Test 4457 23 Oct., 1885			

Load in tons	17 5 h later Compression $d = 4.58$ cm $l = 5.94$ cm				18 On the following day Tension $d = 4.58$ cm $l = 5.94$ cm			
	Stress (kg/cm²)	Contraction ($^1/_{1000}$ cm)	Differences	Elast. Mod. (kg/cm²)	Stress (kg/cm²)	Elongation ($^1/_{1000}$ cm)	Differences	Elast. Mod. (kg/cm²)
0	0	0			0	0		
4	243	0.62	*62*		243	0.65	*65*	
8	486	1.25	*63*		486	1.32	*67*	
12	729	1.88	*63*		729	1.96	*64*	
16	971	2.50	*62*	2,310,000	971	2.63	*67*	2,200,000
0	0	0.00			0	0.03		
16	971	2.50			971	2.65		
20	1092	2.84	*34*		1092	2.99	*34*	
24	0	0.00			0	0.13		
	Test 4462 23 Oct., 1885				Test 4468 23 Oct., 1885			

When this last value, which was still below the original elastic limit, was exceeded, Bauschinger found in a subsequent tensile test on this specimen that there was *no* elastic limit. Forty-six minutes later the same specimen again was loaded in compression. There was no change in the value of the elastic limit, 1968 kg/cm², which had been the largest value observed in the previous loading in compression, prior to the loading in tension. In none of these measurements did Bauschinger exceed the yield limit.

Table 124. (Continued.)

	19 2 days later Compression $d = 4.58$ cm $l = 5.94$ cm				20 $2^1/_2$ h later Tension $d = 4.58$ cm $l = 5.94$ cm				21 $4^1/_2$ h later Compression $d = 4.58$ cm $l = 5.94$ cm				22 1 day later Tension $d = 4.58$ cm $l = 5.94$ cm				23 9 h afterward Compression $d = 4.58$ cm $l = 5.94$ cm			
Load in tons	Stress (kg/cm²)	Contraction ($^1/_{1000}$ cm)	Differences	Elast. Mod. (kg/cm²)	Stress (kg/cm²)	Elongation ($^1/_{1000}$ cm)	Differences	Elast. Mod. (kg/cm²)	Stress (kg/cm²)	Contraction ($^1/_{1000}$ cm)	Differences	Elast. Mod. (kg/cm²)	Stress (kg/cm²)	Elongation ($^1/_{1000}$ cm)	Differences	Elast. Mod. (kg/cm²)	Stress (kg/cm²)	Contraction ($^1/_{1000}$ cm)	Differences	Elast. Mod. (kg/cm²)
0	0	0			0	0			0	0			0	0			0	0		
4	243	0.63	*63*		243	0.68	*68*		243	0.67	*67*		243	0.67	*67*		243	0.65	*65*	
8	486	1.25	*62*		486	1.36	*68*		486	1.30	*63*		486	1.35	*68*		486	1.30	*65*	
12	729	1.87	*62*	2,310,000	729	2.00	*64*	2,190,000	729	1.94	*64*	2,260,000	729	2.01	*66*	2,150,000	729	1.96	*66*	2,250,000
16	971	2.49	*62*		971	2.64	*64*		971	2.57	*63*		971	2.68	*67*		971	2.60	*64*	
0	0	0.00			0	0.06			0	0.00			0	0.06			0	0.01		
16	971	2.49			971	2.64			971	2.57			971	2.68			971	2.58		
18	1092	2.81	*32*		1092															
20	0	0.00			1214	3.32	*68*		1214	3.20	*63*		1214	3.36	*68*		1214	3.22	*64*	
22					0	0.08			1335	3.65	*45*		1335	3.69	*33*					
24									0	0.10			0	0.10			1475	4.07	*85*	
																	0	0.25		
	Test 4474 26 Oct., 1885				Test 4478 26 Oct., 1885				Test 4482 26 Oct., 1885				Test 4486 27 Oct., 1885				Test 4495 27 Oct., 1885			

BAUSCHINGER's paper contained a large number of tables similar to Table 124, in which a variety of histories were delineated. His experiments revealed that the lowered elastic limit produced by the reversal of stress remained constant for a period of several weeks. These experiments also included the presentation of many stress histories for both positive and negative values which invariably led to the same "natural elastic limit."

BAUSCHINGER included in the last part of his paper[1] nine large tables of successive experiments in which he considered the elastic limit under conditions of stress alternating to as high as 16 million cycles for many different situations. His study of fatigue was made possible because a Wöhler fatigue machine had been given to the laboratory in 1881. BAUSCHINGER had chosen a repeated tensile machine propelled by a 2 hp Otto engine which could test four, 1 cm^2 rod specimens at the same time. The testing was never continued during the night and the machine, and hence the tests, were stopped regularly for a two-hour interval during each afternoon.

Having first satisfied himself that an alternating tension stress between zero and just below the *original* elastic limit could run 16 million cycles without rupture, BAUSCHINGER observed that if one loading cycle exceeded this original elastic limit, interesting complications ensued, which he described at great length. In reverse loading, such an overstepping lowered the elastic limit in the opposite direction; with subsequent cycling, the natural elastic limit (endurance limit) became the bound for rupture. This natural elastic limit was, of course, below the original value and was determinable from quasi-static cycling of stress.

I have chosen one series of five specimens as illustrative. These data, shown in Table 125, include specimen (5) which BAUSCHINGER loaded to failure quasi-statically without alternating stress, and four other specimens (1) to (4) which he subjected to a quasi-static tensile test in the 100 ton Werder machine at the end of the number of cycles of tensile stress alternating from zero to the increasing values shown.

BAUSCHINGER determined E and the elastic limit before he applied the indicated plastic overstep, after which he again applied the alternating stress. He attached considerable significance to the increase in the elastic limit from the original value of 1043 kg/cm^2 in specimen (5) to the value of 2500 kg/cm^2 of the maximum of the alternating stress for specimen (4). He considered this rupture at 2.288 million cycles to be atypical and probably due to a flaw in the specimen. This test, of course, did not include a change of sign of stress, so that the natural elastic limit was not obtained. The gradual raising of the tensile elastic limit to values far above the original elastic limit for specimens subjected to the loading histories shown, presents obviously important implications for both science and technology.

The total behavior considered by BAUSCHINGER included the time effects in raising the elastic limit; the reduction to zero and gradual return of the elastic limit on overstepping the stress at the yield limit; the permanent lowering of the elastic limit in compression or in tension after exceeding the elastic limit in the first loading whether it was in tension or in compression; the lowering of a raised elastic limit by annealing at sufficiently high temperature; and the effect upon the elastic limit of multi-million stress cycling. All these represent material memory phenomena of utmost importance for modern continuum theory. Any experimentist of our day who does not wish to waste his time might well begin with a careful analysis of this report of BAUSCHINGER in 1886 and related papers of his in contemporary journals, in which he commented at length on the implications of

[1] BAUSCHINGER [1886, *1*].

Table 125. BAUSCHINGER (1886). *Duration experiments with 6 four-edged rods (Nos. 1–5 and 49), cut from an 11 mm thick wrought iron sheet.*

Specimen cross-section (cm^2)	Running no.	Designation no.	(No. of cycles) After loading	Between stresses kg to kg	After days of rest	Results from load change	Testing with stationary load				After rupture with changing or stationary load			
							Under test no.	Elasticity modulus (kg)	Elastic limit (kg)	Maximum load (kg)	Permanent strain (1/1000 cm)	Tensile strength (kg)	Cross-section contraction	Strain (%)
	1	5	0	—	—	—	3478	2090000	1043	—	—	3840	20	15.5
1.01 × 1.16	2	1	378971	0–1080	0	—	3503	2090000	1620	1710	0.17	—	—	—
	3	1	1043422	0–1080	0	—	3513	2010000	1620	1710	0.28	—	—	—
	4	1	1096141	0–1080	5	—	3517	2000000	1710	1880	0.13	—	—	—
	5	1	2085884	0–1080	0	—	3534	2050000	1620	1880	0.20	—	—	—
	6	1	5170523	0–1080	50	—	3571	2090000	1880	2051	0.22	3600	24	8.1
1.04 × 1.16	7	2	369800	0–1500	—	—	3502	2030000	> 1670	1670	0.09	—	—	—
	8	2	1029189	0–1500	0	—	3512	2065000	> 1670	1670	0.01	—	—	—
	9	2	1114623	0–1500	4	—	3516	2020000	> 1670	1670	0.24	—	—	—
	10	2	2104366	0–1500	0	—	3533	2025000	> 1670	1670	0.08	—	—	—
	11	2	5189005	0–1500	50	—	3570	2100000	2020	2270	0.55	3710	28	10.4
1.02 × 1.16	12	3	364456	0–2000	0	—	3501	2030000	1525	2030	0.66	—	—	—
	13	3	1025506	0–2000	0	—	3511	2070000	> 2030	2030	0.07	—	—	—
	14	3	1110940	0–2000	4	—	3515	2020000	> 2030	2030	0.11	—	—	—
	15	3	2098014	0–2000	0	—	3532	2020000	> 2030	2030	0.08	—	—	—
	16	3	5182653	0–2000	48	—	3569	2100000	2200	2460	0.49	3730	19	12.5
1.04 × 1.155	17	4	353569	0–2500	0	—	3500	2055000	2310	2480	0.42	—	—	—
	18	4	992525	0–2500	0	—	3510	1985000	2330	2500	—	—	—	—
	19	4	1086227	0–2500	4	—	3514	2050000	2290	2500	0.21	—	—	—
	20	4	2066685	0–2500	0	—	3531	2050000	2500	2500	0.25	—	—	—
	21	4	2288446	0–2500	—	Rupture	—	—	—	—	—	—	10	0.9
1.10 × 1.12	22	49	1708010	0–2000	1/4	—	4514	2060000	1790	2110	0.62	—	—	—
	23	49	3872604	0–2000	3/4	—	4612	2160000	2190	2440	0.78	—	—	—

Gage length = 10.00 cm; for final static rupture, a gage length of 15 cm was used.

these phenomena. As with the relation between the work of CORNU and STRAUBEL, that of WIEDEMANN, THURSTON *et al.* to BAUSCHINGER represents the remarkable contrast which sometimes exists between relatively crude original discovery and later definitive study. In experiment it is sometimes difficult to weigh the importance of one against the other.

4.8. On the cohesion of solids under pressure: The experiments of Spring (1880).

A third major 19th century experimental theme associated with large deformation, in addition to the study of flow by TRESCA and of yield behavior by BAUSCHINGER, concerned "regel" or "refreezing," as is seen from the remarkable experiments in 1880 performed by WALTHÈRE SPRING,[1] Correspondent to the Academy of Belgium and Professor at the University of Liège. The phenomenon of the refreezing of two pieces of ice when pressed together, which had been explained by JAMES THOMSON[2] in 1849 on the basis of the variation of the melting point with pressure, aroused further interest with the observation of MICHAEL FARADAY[3] in 1850 that the process occurred with increasing facility as the melting point at atmospheric pressure was approached.

In the thirty years of spirited controversy on the subject, which is recounted in detail by SPRING in a long introduction to his memoir, the application of JAMES THOMSON's theory to the interpretation of the formation and flow of glaciers, and the experiments of TRESCA,[4] HERMANN LUDWIG FERDINAND von HELMHOLTZ,[5] WILLIAM THOMSON[6] (KELVIN), ROBERT WILHELM BUNSEN,[7] and numerous lesser figures, generally were based upon the assumption that ice was substantially incapable of plastic flow. The distortion of the solid was thought to proceed by the process of cracking and refreezing under pressure. WILLIAM THOMSON had shown that when ice was mixed with water and then was submitted to pressures of 8.1 and 16.8 atm it melted at the temperatures, 0.059° C and 0.129° C, respectively.

SPRING referred to the experiments of IMMANUEL BURKHARD ALEXIUS FRIEDRICH PFAFF[8] in 1875, who had demonstrated that ice at 0° C did exhibit weak plasticity which decreased as the temperature dropped; to the experiments of ALBERT MOUSSON,[9] who was able to liquify ice completely at 13000 atm; and to the work of TYNDALL, who besides pressing pieces of ice in hot water succeeded in uniting two pieces in a water heated capsule under the slightest contact possible.[10] SPRING proposed numerous other explanations of the refreezing process and voiced his doubts that nature would limit such a phenomenon to a single substance, water. The discussion was a prelude to his describing his own high pressure experiments in which he succeeded in resolidifying some 83 different powdered sub-

[1] SPRING [1880, *1*], [1881, *1*]. I have used SPRING's paper published in 1881.
[2] JAMES THOMSON [1849, *1*].
[3] FARADAY [1850, *1*].
[4] TRESCA [1872, *1*].
[5] HELMHOLTZ [*1865*, *1*].
[6] WILLIAM THOMSON [1850, *1*].
[7] BUNSEN [1850, *1*].
[8] PFAFF [1875, *1*].
[9] MOUSSON [1858, *1*].
[10] TYNDALL [1871, *1*], pp. 392–393. TYNDALL's description of his experiments on ice were contained in Chap. IV of his book, and nothing was added in the German version. I could find no reference to his "uniting two pieces in a water heated capsule." He did observe pieces of ice refreezing upon contact when floating in a warm water bath.

stances, including eight different metals. On the basis of his own work, SPRING concluded that solid bodies in intimate contact enjoy the property of being able to weld themselves together.

In preliminary experiments in 1878 SPRING used a screw which pushed a piston into a cylindrical hole in a steel cylinder. He had estimated a possible pressure of as high as 20000 atm. He found that the blocks of solids packed into the container under pressure became difficult or impossible to remove.

In his main experimental series described in his paper in 1880 and in a series of six additional notes and papers between then and 1888, SPRING had a centrally split cast steel cylinder, 3.8 cm diameter, 5.0 cm long, with a central hole into which a small piston was fitted. He inserted this cylinder into the exactly matching central hole of an externally tapered cast iron cylinder. By the use of a slight taper and a massive clamp he prevented the two halves of the split cylinder from separating. He applied pressure to the piston by means of a lever on which were mounted movable cast iron weights sufficient to produce pressures in the central test section in excess of 8000 atm. SPRING estimated that the apparatus could be loaded without danger of rupture to 30000 atm.

The 83 pulverized or powdered solids were distributed as follows: 8 metals, 6 metalloids, 10 oxides and sulphurs, 32 salts, 19 carbon compounds, and 8 mixtures compressed to study the effects of pressure upon the chemical reaction of bodies.

Pressure solidification of the metals considered, from easiest to most difficult, were: lead, bismuth, tin, zinc, aluminum, copper, antimony, and platinum. Lead filings compressed in a vacuum under 2000 atm formed into a compact mass in which even under microscopic examination SPRING was unable to find the least trace of lead grains. The specific weight under pressure was 11.5013 instead of that obtained for what SPRING stated was an otherwise identical block, namely, 11.3. Increasing the pressure to 5000 atm caused the lead to flow into all the cracks of the apparatus and around the piston. The thin lead sheets which had the appearance of laminations, and the lack of resistance of the solid to 5000 atm pressure, SPRING regarded as "confirmation" of the experiments of TRESCA.

Despite the fact that bismuth is a brittle metal, SPRING found it easily amenable to pressure solidification. A fine powder compressed to 6000 atm when broken exhibited a crystalline break similar to that of the cast metal. The density of both were identical. This was the only one of the eight metals SPRING considered which exhibited the crystalline fracture when the solid had been produced under pressure.

Tin formed into a solid at 3000 atm, and like lead, flowed at 5000 atm. Unlike lead, the flow presently stopped. Increasing the pressure to 5500 atm repeated this flow behavior. It was not until 7500 atm that continuous flow was observed. This behavior has all the physical attributes of the SAVART-MASSON (PORTEVIN-LE CHATELIER) effect normally studied at very low hydrostatic pressure; SPRING was the first to observe it at high pressure.

All the metal experiments were performed at 14°C except for those on zinc which were tested not only at that temperature but also at approximately 130°C for comparison. At a pressure of 260 atm, which was the load of the unweighted lever, zinc remained in the form of the original filings. At 700 atm SPRING observed the beginning of liaison, although the block when removed easily broke into dust. Trails were visible under the microscope. Under 2000 atm the block could be filed, but broke under the hammer. By 5000 atm the block was perfect and could be pinched strongly in a vise or hammered like the usual metal.

Aluminum solidified under a pressure of 6000 atm with a density identical to the thermally formed solid. Copper behaved like aluminum. Antimony powder under a pressure of 5000 atm not only solidified but also on the surface acquired the characteristic metallic luster. When the pressure was increased further, this metallic luster appeared in central sections. Finally, platinum sponge (mousse de platine) under 5000 atm exhibited the initial stage of solidification, but increasing the pressure to some unspecified maximum, presumably between 8000 and 9000 atm, failed to produce the "perfect" solid obtained for the other metals.

SPRING noted that these results indicated that the order of ease of pressure solidification also was the order of hardness of the metals considered. Thus, as in the case of the high temperature zinc which at 130° C was more amenable to pressure solidification than at 14° C, the increase of ambient temperature which decreased the hardness should increase the facility of pressure fusion, a fact that SPRING emphasized was well known in the welding of soft iron at high temperature.

Omitting the details of the long list of substances from sealing wax, glass, rock salt, graphite, through many chlorides and bromides and sulphurs, to yellow oxide of mercury and aluminum, we come to SPRING's major conclusion, namely, that all crystalline bodies without exception in the many solids studied could be solidified under pressure. SPRING concluded that for those solids the union occurred through the mechanism of crystal growth when the interfaces were brought into sufficiently intimate contact by means of pressure, and that this growth must take place along crystal axes thus producing the observed intimate bonding of the solid.

For the amorphous solids like the waxes, pitch, acacia gum, etc., he assumed that at high pressures the basic fluidity exhibited under ordinary pressures would cease, and a coherent mass would be formed. In general, however, non-crystalline solids did not solidify under pressure as had the crystalline solids although the amorphous state did not always prevent it.

SPRING suggested that the state of a body in terms of hardness, etc., was a matter of the ambient pressure rather than an intrinsic property of the solid. At the close of his paper he discussed his observation of pressure transitions in sulphur and pointed out the general importance of his studies to the fields of geology and minerology. From his remarkable results he concluded that fusion under pressure, like rupture under tension, was a fundamental property of solids, the phenomena being in some sense opposite to one another.

The impact of SPRING's discoveries upon thought during the following thirty years was very great. His work became a classic model which stimulated additional studies as well as much controversy. BRIDGMAN as late as the mid-20th century referred to the "considerable controversy about the results of SPRING, many of which were highly spectacular, and it seems certain that many of his results were not due to pressure alone, but involve in addition the rubbing motion of one particle on another with intense shearing stress."[1] Neither SPRING nor his 19th century followers in this area foresaw the 20th century technological growth of "powdered metallurgy," the roots of which most certainly sprang from SPRING's study in 1880.

WILLIAM HALLOCK[2] in 1888 reviewed SPRING's results and claimed that neither lead nor wax actually became fluid under pressure but only appeared to flow by yielding under very high stress. JAMES DEWAR[3] in a "Note on the Viscosity of Solids," strongly influenced by the observations of SPRING, attempted to repeat the experiment to study the viscous properties of the flow of salt and organic compounds under high pressure. In an apparatus diagrammatically dupli-

[1] BRIDGMAN [1949, *1*], p. 6.
[2] HALLOCK [1888, *1*].
[3] DEWAR [1895, *1*].

cating TRESCA's in his extrusion experiment, the diameter of DEWAR's extruded specimen being $^1/_{16}$ in, DEWAR tested a variety of solids, dividing them into those which would form an extruded wire and those which would not. His estimated maximum pressure was 60 English tons per square English inch, or 8500 atm, a value similar to that obtained by SPRING thirteen years earlier. Most of the solids which formed a wire easily did so in the range of pressure between approximately 4000 atm and 5700 atm, which again were values comparable to those of SPRING.

The two groups of solids are listed in Table 126. DEWAR commented upon the variation of the rates of flow, giving a value for fused sulphocyanide of ammonia at a pressure of 8700 atm of "about one inch a minute."[1] On the other hand, some solids which failed to form a wire exploded through the narrow orifice in sudden bursts. These sudden shocks again gave evidence of the appearance of the SAVART-MASSON (PORTEVIN- LE CHATELIER) effect during the high pressure distortion of solids, with steps of sufficient magnitude so that on two occasions they caused the fracture of the strong steel cylinder. Some of the solids which did not form wires simply did not flow, or flowed too slowly under the maximum pressure of 8750 atm.

Table 126. *Extrusion results of* JAMES DEWAR (1895).

Solids which formed a wire under pressure[a]	Solids which did not form a wire up to the maximum pressure of 8500 atm
Sodium sulphate, aq	Sodium phosphate, aq
Sodium carbonate, aq	Borax, aq
Sodium thio sulphate, aq	Alums
Magnesium chloride, aq	Sodium nitrate
Ferrous sulphate, aq	Chloride of sodium
Ammonium chloride	Lithium
Potassium chloride	Zinc
Bromide	Mercury
Iodide	Sulphide of manganese, aq
Cyanide	Ammonium sulphate
Nitrate of ammonium	Sodium oxalate
Potassium and silver	Ferrous oxalate
Strontium chloride, aq	Rochelle salt
Aluminium sulphate, aq	Arsenious oxide
Caustic soda	Potassium ferocyanide
Oxalic acid, aq	Sodium acetate
Sodium acetate	Ammonium acetate, aq
Calcium chloride (slow)	Nitrates of barium
Acetamide (slow)	Strontium
Lead acetate (slow)	Caustic potash
Benzoic acid (slow)	Dry carbonate and sulphate of sodium
Graphite (easy, brittle wire)	Dry sulphide of aluminium
Iodine (easy, long continuous wires)	Sugar
Urea (easy)	Starch
Anthraquinone (very slow, brittle)	Naphthalene

[a] Hydrated salts are designated by "aq".

DEWAR concluded that the state of understanding of the subject was generally so poor as to "require prolonged investigation." He stated his belief that every substance had a limiting pressure "probably a constant" which must be reached before a "reasonable rate of flow is attained."

[1] *Ibid.*, p. 138.

4.9. Early 20th century experiments on the flow of solids under high pressure: Tammann (1902).

The deformation of solids under high stress was only a single aspect of the interest in high pressure physics which developed in the late 19th and early 20th centuries. In the historical introduction to his monograph on *The Physics of High Pressure*[1] BRIDGMAN outlined the contributions of many experimentists before the beginning of his own work in 1906. Among those described with some emphasis is GUSTAVE HEINRICH JOHANN APOLLON TAMMANN.[2] As BRIDGMAN noted, TAMMANN did not introduce new experimental methods and reached only 3000 atm pressure, thereby considerably limiting his studies. In the present context the major interest is in TAMMANN's criticism of previous flow studies including and following TRESCA's. TAMMANN emphasized that high pressure had not been maintained in the orifice during the flow. In a memoir in 1902 on the flow of ice, he described the apparatus by means of which he not only sought to maintain pressure during flow but to measure the flow velocity as a function of pressure and temperature.

In Fig. 4.29 is shown a schematic drawing of the pressure chamber. The solid under pressure flowed around the member F which in velocity studies is tapered to maintain pressure. The pressure was applied by means of a lever in contact with the piston E.

The entire apparatus, which was manufactured commercially for TAMMANN, is shown in Fig. 4.30.

The presence of an after-effect required a waiting period of two minutes before TAMMANN could begin making velocity measurements based upon the position

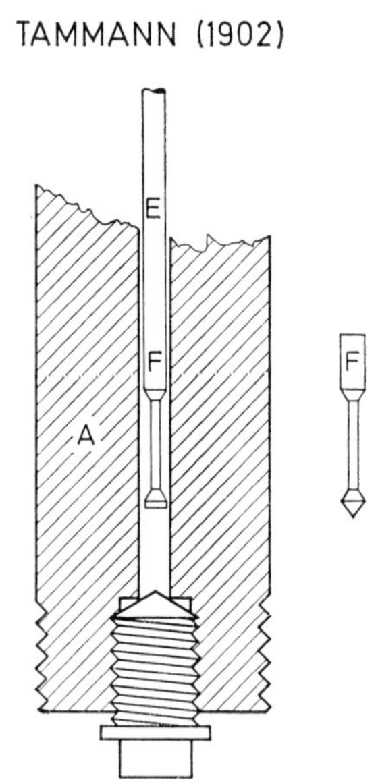

Fig. 4.29. A schematic diagram of the pressure chamber in which TAMMANN performed his flow experiments.

[1] BRIDGMAN [1931, *1*].
[2] TAMMANN [1902, *1*].

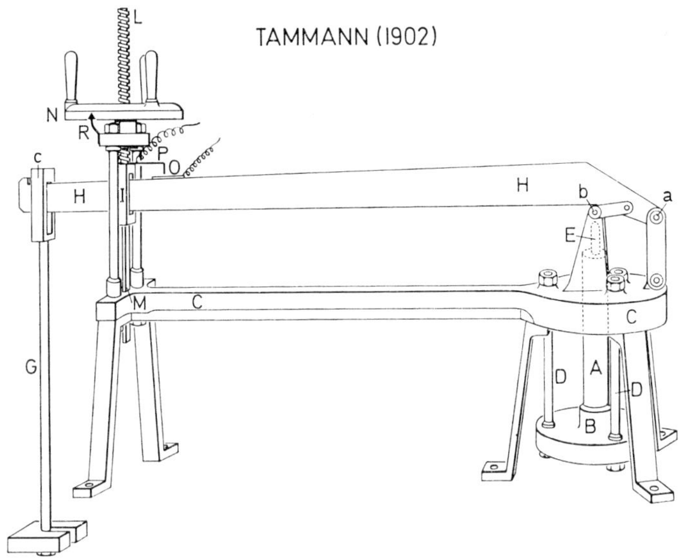

Fig. 4.30. The apparatus used by TAMMANN in his experiments in 1902.

of the lever. He measured variations in ambient temperature by enclosing the apparatus in a liquid bath at the desired values. Omitting the descriptions of the results on the flow of ice, included in TAMMANN's paper of 1902, I shall limit discussion at this point to the use of the apparatus by N. WERIGIN, J. LEWKOJEFF, and TAMMANN[1] in 1903 to consider the pressure-temperature-flow velocity relations for tin, potassium, sodium, lead, thallium, bismuth, cadmium, and zinc.

Because of the indicated fluctuations I have left the units of velocity for the abscissa in the figures which follow in the originally plotted "scale partitions per minute" although sufficient information was included, as the authors pointed out, to estimate values in cm/sec. They used two different conical configurations. For that noted as (1) in the figures, the cone angle was 95° for a cone base of 0.1444 cm^2, and that noted as (2) an angle of 100° for a base area of 0.1816 cm^2. The area of the containing steel cylinder from which the dimensions of the slug could be estimated was 0.1988 cm^2. At the beginning, the cylindrical slug was carefully fitted to the dimension of this steel cylinder.

The interested reader is referred to the original paper in which the experimental results were tabulated in detail. Figs. 4.31, 4.32, and 4.33 contain plots of representative examples for the two different cones in lead, bismuth, zinc, potassium, cadmium, and sodium.

One notes the consistent increase in flow rate with temperature in every instance, although the absolute velocity obviously is dependent upon the shape of the die. WERIGIN, LEWKOJEFF, and TAMMANN noted that a difference of 10° C could double the velocity. They were particularly interested in the abnormal behavior of tin and thallium shown in Fig. 4.34. In the former a sudden decrease in velocity occurred between 203 and 204° C, while in the latter a large increase occurred at approximately 180° C. This behavior was attributed to transitions associated with changes in crystal structure.

[1] WERIGIN, LEWKOJEFF, and TAMMANN [1903, *1*].

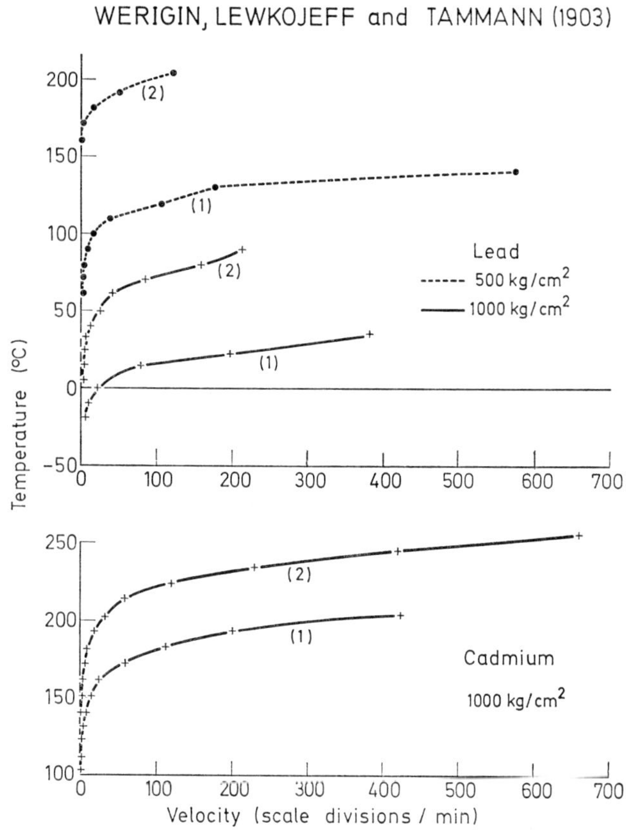

Fig. 4.31. The velocity of flow in scale partitions per minute vs the temperature for lead under two different pressures and for two cone angles, and for cadmium for two cone angles.

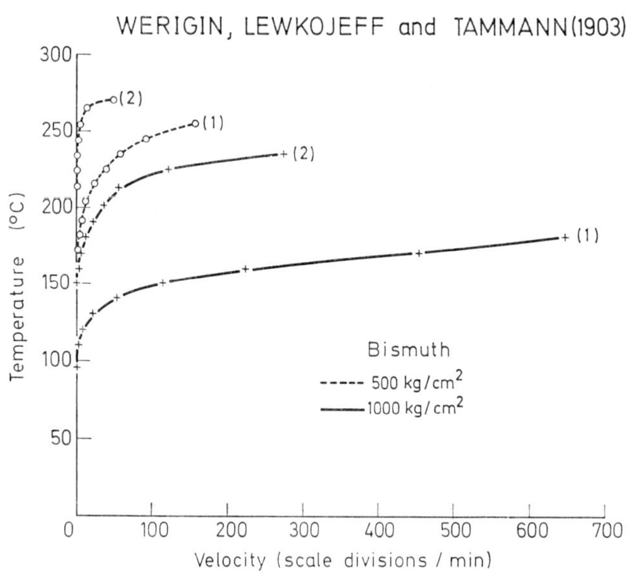

Fig. 4.32. The velocity of flow in scale partitions per minute vs the temperature for bismuth under two different pressures and two cone angles.

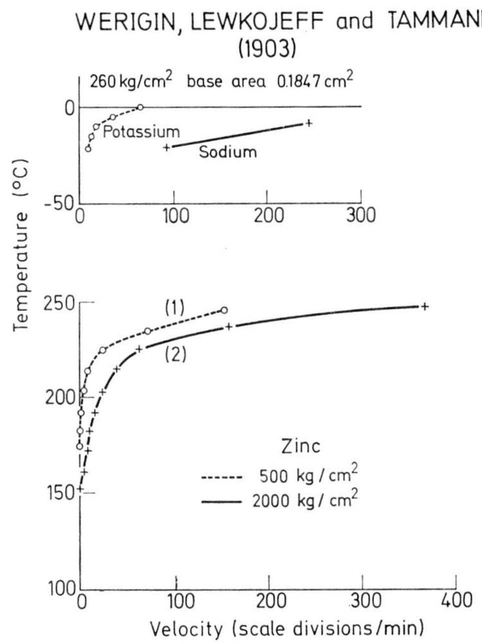

Fig. 4.33. The velocity of flow in scale partitions per minute vs the temperature for zinc, sodium, and potassium for the pressures and cone angles indicated.

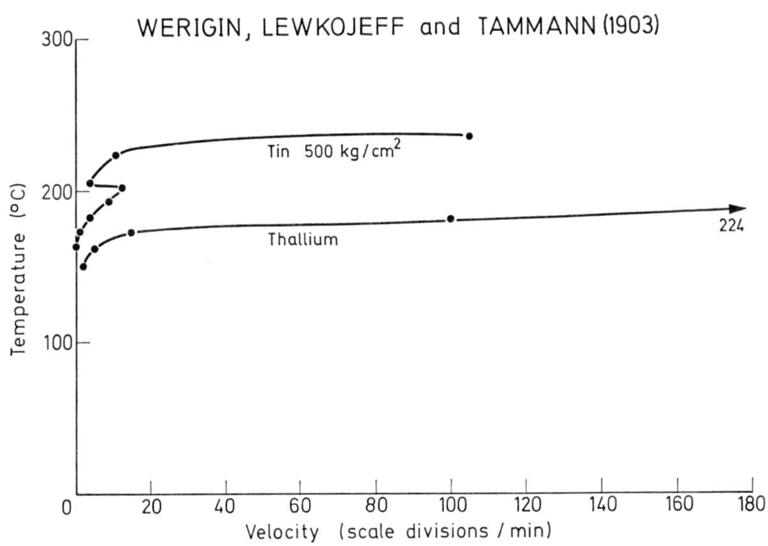

Fig. 4.34. The abnormal flow behavior of tin and thallium.

The results for tin tabulated in Table 127 demonstrate the extent to which a constant flow velocity was achieved. As may be seen, this was indeed the situation up to a temperature of approximately 200° C.

Table 127. WERIGIN, LEWKOJEFF, and TAMMANN (1903). *Tin, pressure 1000 kg/mm², cone # 1.*

Temperature (°C)	Velocity (scale partitions per minute)			Average
10.0	0.2	0.2	0.2	0.2 ± 0.0
20.0	0.4	0.4	0.5	0.4 ± 0.0
32.0	0.7	1.0	0.9	0.9 ± 0.1
41.8	1.6	1.5	1.7	1.6 ± 0.1
51.2	4.0	3.8	3.6	3.8 ± 0.1
60.5	8.3	7.7	7.7	7.9 ± 0.3
70.4	14.3	13.2	14.9	14.1 ± 0.6
83.2	32.0	33.6	31.8	32.5 ± 0.8
90.6	59.7	55.5	54.3	56.5 ± 2.1
100.3	107.6	89.2	96.0	97.6 ± 6.6
110.5	206.0	178.0	201.0	195.0 ± 11.3
120.6	—	362.0	—	362.0
131.4	—	584.0	—	584.0

In 1904, A. V. OBERMAYER,[1] while contributing nothing of his own experimentally, provided a review paper on the work of TRESCA,[2] SPRING,[3] and TAMMANN.[4] He noted that the pressure vs velocity results for ice at −21.7° C shown in Fig. 4.35, gave a fair fit to a cubed pressure vs velocity empirical relation. OBERMAYER admitted that this fit, shown in Fig. 4.36, was not characteristic of the data in general and therefore represented an approximation. OBERMAYER's

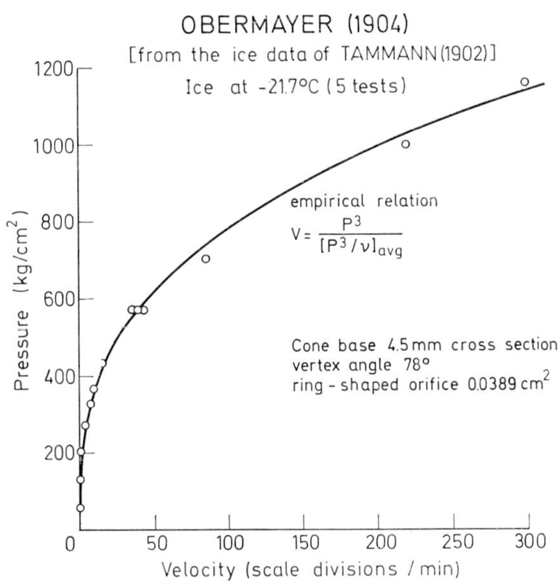

Fig. 4.35. The pressure vs flow velocity of ice at −21.7° C obtained by TAMMANN and used by OBERMAYER in his analysis in 1904.

[1] OBERMAYER [1904, *1*].
[2] TRESCA [1867, *1*], [1868, *1*], etc.
[3] SPRING [1881, *1*].
[4] TAMMANN [1902, *1*].

paper furnishes a good, brief summary of the arguments at the turn of the century on flow and on solidification under pressure.

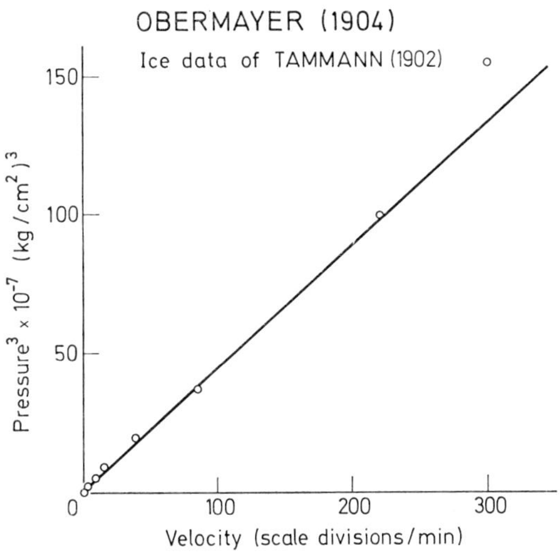

Fig. 4.36. OBERMAYER's comparison of TAMMANN's pressurevs flow velocity data with his empirical prediction.

4.10. The beginning of the experimental study of the large deformation of crystalline solids responding to loading histories with more than one non-zero stress component: Guest (1900).

COULOMB's discovery[1] in 1784 of the elastic limit of metals was definitive, as was his extensive study of torsion which showed this limit to depend upon the permanent deformation produced during cyclical loading and unloading of the specimen.

If the elastic limit were to be defined as the strain at which permanent deformation begins, then the major experimentists of the 19th century who considered the subject, HODGKINSON, WERTHEIM, and even BAUSCHINGER, had gravely doubted its existence. It was their contention that small plastic deformation, or microplasticity, would be found accompanying all strain if instrumentation were sufficient to detect it. WERTHEIM merely had chosen an arbitrary value of strain as the elastic limit, but BAUSCHINGER had chosen more carefully the proportional limit beyond which the observed permanent deformation ceased to be reproducible.

WERTHEIM's[2] observation in 1844 of the stress dependence of the initial elastic limit upon prior thermal history; WÖHLER's,[3] of its reduction on repeated stress; THURSTON's,[4] of its increase following a loading pause; and BAUSCHINGER's,[5] that if the yield limit in steel were exceeded the elastic limit could disappear upon

[1] COULOMB [1784, *1*].
[2] WERTHEIM [1844, *1* (a)].
[3] WÖHLER [1858, *1*].
[4] THURSTON [1873, *1*], et seq.
[5] BAUSCHINGER [1881, *1*], [1886, *1*].

reloading unless sufficient time elapsed, were all extensions of detail for different aspects of COULOMB's discovery.

Appropriately for the subject of crystalline plasticity, which in general terms is a 20th century development in physics, JAMES J. GUEST,[1] in the year 1900, presented the first experiments considering the initial yield limit or yield surface for loading paths involving more than one non-zero stress component. Using thin walled hollow tubes under axial tension, with added internal pressure and torsion (providing two normal stresses and a shear if one neglected the steep gradient between the inner and the outer walls when internal pressure was present), GUEST attempted to test the major yield surface hypotheses of his day. In 1900 there were three.

The first, the maximum stress theory, which GUEST stated was "adopted in the absence of experimental data, by Rankine,"[2] was exclusively English and American. Europe, on the other hand, had adopted a maximum strain theory "first advocated by St. Venant as fitting in with that molecular theory which leads to uniconstant elasticity."[3] The third theory, which GUEST gave prominence, was one which had been included among possibilities in COTTERILL's *Applied Mechanics*,[4] namely a maximum shear stress theory. GUEST noted that no formulae for a yield surface posed on maximum shear actually had been propounded, although mere acquaintance with TRESCA's observation of large deformation 35 years earlier "must have urged many towards the conclusion that this is the true criterion of elastic strength in a ductile material."[5]

TRESCA,[6] as we have seen, had established that three regions of deformation existed: the elastic region, followed by an intermediate plastic region of work hardening, and the third, a state of plastic flow under constant stress. GUEST states the case for anyone who upon reading TRESCA wonders how the concept of the ideal plastic solid, wherein plastic flow under constant stress is presumed to occur upon reaching the elastic limit, could be inferred from TRESCA's experiments.

> "St. Venant, followed by many elasticians, does not recognize the intermediate stage, and considers that Hooke's law holds up to the point at which plasticity begins; he adopts Tresca's results that in the plastic stage the shearing-force is constant, and upholds a specific maximum strain as the condition of limiting elasticity.
>
> I fail to understand how a material could have one condition for the commencement of plasticity and an entirely different one for its existence; perhaps users of these conditions tacitly admit the existence of the intermediate stage, but neglect it for the simplification of calculation and because the physical difference between the elastic and plastic states is so great; or perchance they do not admit the rigour of the deduction of the plastic law from Tresca's experiments."[7]

GUEST stressed the impossibility of obtaining results of any value from plastic torsion-tension experiments on solid rods. He performed a series of preliminary tests in small deformation for tension and for torsion on nine steel, two copper, and two brass tubes to determine E and μ for the specimens to be considered in combined stress. Since the analysis GUEST wished to perform required the knowledge of POISSON's ratio, he offered an apology for computing his values from the ratio of the two moduli, which provided the usual spread of numbers.

The experiment consisted in placing the tubes in tension, placing them in torsion, and subjecting them to internal pressure, while observing the appearance

[1] GUEST [1900, *1*].
[2] GUEST [1900, *1*], p. 77.
[3] *Ibid.*
[4] This was a late 19th century elementary textbook.
[5] GUEST, *op. cit.*, p. 79.
[6] TRESCA [1872, *2*].
[7] GUEST [1900, *1*], p. 78.

of the yield point. The procedure followed was to subject the tubes to torsion, to torsion and tension combined, to tension only, to tension and internal pressure combined, to torsion and internal pressure combined, and to internal pressure only. Measurements were made of the axial elongation, of the angle of twist, and in some cases, circumferential strain, by means of specially designed optical extensometers. I have found it impossible to condense GUEST's 64 page paper. It contains exhaustive detail on the limitations of the apparatus, the problems of friction and of calibration, the theory of the extensometers, and the method of measuring and applying pressure. In addition, GUEST gave detailed tabulations of all measurements in steel, copper, and brass tubes for which the yield point was readily apparent only for steel. He provided diagrams for the copper experiments for which, like brass, no definite yield point could be observed. I suggest that the interested reader examine these details in the original paper; I shall confine this discussion to the diagram summarizing the experimental results in steel. The results as given by GUEST are shown in Fig. 4.37. OX was the axis of principal stress most nearly coinciding with the generators of the tube. Relative values of the other principal stresses were then set off parallel to OY and OZ, those parallel to OY being the mean, and those parallel to OZ the minimum principal stress.

GUEST suggested that to visualize his construction geometrically one should conceptually fold the diagram along OX at right angles. In Fig. 4.37, the locus of points for the maximum stress then would be the line BH. For the maximum strain theory, GAH, KAL, or MAN, depend upon the value of POISSON's ratio. For

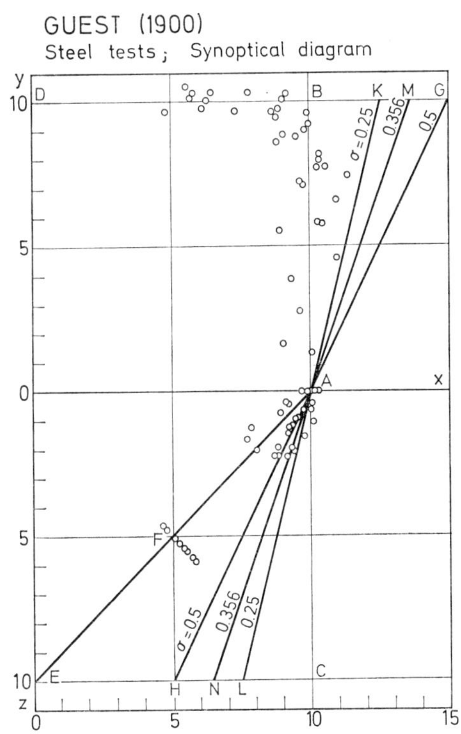

Fig. 4.37. GUEST's comparison of experimental results from tension-torsion-internal pressure tests in steel hollow tubes, compared with maximum principal stress, maximum principal strain, and maximum shear hypotheses.

the constant shearing-stress law, which must be attributed experimentally to GUEST because of those measurements, the locus was EFABD. GUEST's rejection of the maximum principal stress and maximum principal strain assumptions, together with the international engineering conflict of opinion, was in fact the preamble to a new conflict to come between this GUEST law, or TRESCA condition, for the yield surface, and the MAXWELL-von MISES maximum energy of distortion criterion. Although 75 years of subsequent experimentation has cast the die in favor of the criterion first proposed by MAXWELL but ascribed only to VON MISES since MAXWELL's paper long remained unpublished, there is historic significance in GUEST's pioneering experimental research. For copper and brass GUEST noted that despite the difficulties of locating the onset of plasticity, a comparison in terms of similar stress-strain behavior was consistent with his maximum shear hypothesis.

The hydrostatic pressure-extension experiments of VON KÁRMÁN[1] in marble in 1911, described below, represent the next step in this direction, if one conceives of finite deformation plasticity in terms of the expansion of the initial yield surface, either as a family of similar surfaces or as functionally changing with increasing deformation in some manner to be determined. GUEST[2] did not conceive of his contribution in this light, as can be ascertained from his criticism of those who attempted to apply TRESCA's large deformation results to the initiation of plasticity, and from his reference to "false yield points."

4.11. The ductility of marble and sandstone when responding to a general state of stress: von Kármán (1911).

Although both SPRING[3] and TAMMANN[4] had considered the large deformation or plastic flow of solids responding to general states of stress in which more than one stress component was non-zero, their work in this context was of limited value because they were unable to determine the magnitude of these non-zero components of stress. In fact, there had been similar limitations also in some of TRESCA's[5] results. VON KÁRMÁN[6] in 1911 was the first experimentist to subject a solid to large stresses with accompanying large deformation in which all three principal stress components were non-zero and *measured throughout the deformation.*

VON KÁRMÁN devised an experiment to permit a cylindrical specimen to be subjected to an hydrostatic stress while an independent additional axial stress was applied. Both stresses were variable and measurable. The measured deformation parameter was the axial strain. In the actual experiments the chosen hydrostatic stress was maintained constant while quasi-static axial tests were performed. The constant values for hydrostatic stress ranged from 0 to 3 260 atm, while the always larger applied axial loads went to values in excess of 5 000 atm. (The maximum obtainable for a high pressure cylinder of nickel steel having an inner diameter of 50 mm was 10000 atm.) The apparatus by means of which these independent stress states were achieved is shown in Fig. 438.

A fluid under pump pressure in (a) created in a ratio of 1 to 25 through the piston (b) the desired pressure in a fluid in chamber (c). Through an orifice (e) this pressure was transmitted to the test chamber (d) in which the cylindrical specimens whose length varied between 100 and 110 mm were located. The diameters

[1] VON KÁRMÁN [1911, *1*].
[2] GUEST [1900, *1*].
[3] SPRING [1881, *1*].
[4] TAMMANN [1902, *1*], WERIGEN, LEWKOJEFF, and TAMMANN [1903, *1*].
[5] TRESCA [1864, *1*], [1868, *1*].
[6] VON KÁRMÁN [1911, *1*].

VON KÁRMÁN (1911)

Fig. 3.

Versuchseinrichtung.

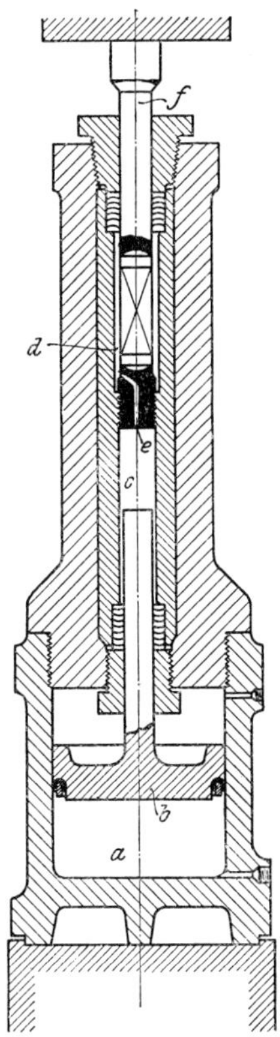

Fig. 4.38. The apparatus used by VON KÁRMÁN in his studies of axial loading of specimens in the presence of high pressure.

were 40 mm providing an L/D ratio of from 2.5 to 2.75, for which earlier experiments of LUDWIG PRANDTL and FRIEDRICH RINNE[1] were quoted to demonstrate that buckling was avoided. The length of the specimen was such that the effects of end friction were considered negligible and, as a matter of fact, despite the high pressures and the friction at the ends, the specimens in general remained uniform cylinders throughout the test. Some photographs of barreled specimens, however, were shown in the paper; their reproduction in the later publications of others led to some misconceptions.

The choice of marble and sandstone as the objects of experimental study was most fortuitous; the results were indeed dramatic. These solids at atmospheric pressure, while permitting fairly high axial stress before rupture, are brittle under compression loading. The blue-veined white Carrara marble and the red sandstone from Mutenberg which VON KÁRMÁN studied had at atmospheric pressure, ultimate stresses of 13.60 kg/mm^2 and 6.90 kg/mm^2, respectively. Fracture occurred at the order of 0.5% strain.

[1] See RINNE [1909, *1*].

Because of the porosity of these substances, to prohibit the penetration of the glycerine used as a pressure fluid the specimens were wrapped laterally in 0.1 mm brass foil with the protruding ends soldered at the pressure plates.[1]

The paper contained a detailed discussion of how the effects of friction were determined and how the influence of the brass foil was shown to be negligible. Fig. 4.39 shows the remarkable results obtained for increasing hydrostatic pressure in marble, and Fig. 4.40 shows the same sequence in red sandstone. The axial stress was σ_1, and the lateral stresses were $\sigma_2 = \sigma_3$. The axial stress always was increased above the ambient hydrostatic pressure; hence the ordinate in Figs. 4.39 and 4.40 was given as $\sigma_1 - \sigma_2$.

Every type of axial stress-strain behavior, from brittle fracture at small strain to the ductile solid proceeding to large deformation, was witnessed in the same solid. As von Kármán noted, a sufficient number of hydrostatic pressures were investigated on marble to reveal a possible relation between pressure and a maximum stress-strain function. The increase in height between 16.50 and 24.90 kg/mm² was much greater than that for the nearly equal increment between 24.90 and 32.60 kg/mm².

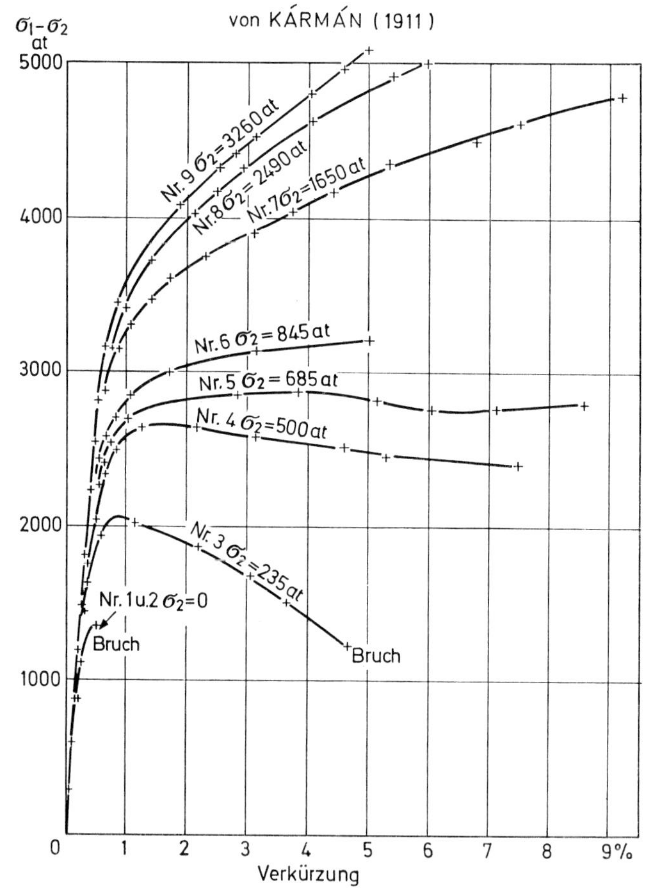

Fig. 4.39. von Kármán's axial compression measurements on marble performed at the indicated pressures. Note the remarkable increase in ductility with increasing pressure.

[1] Later experimentists who failed to observe this practice while attempting to repeat von Kármán's experiments obtained spurious results.

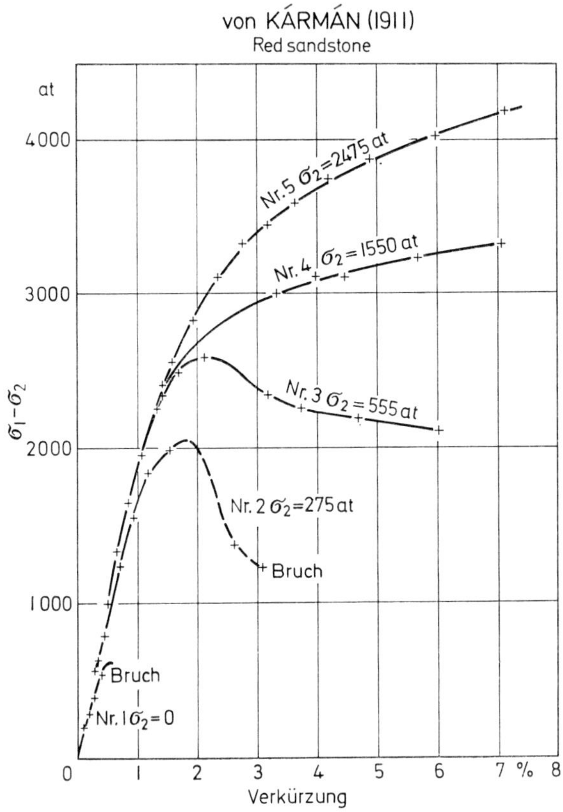

Fig. 4.40. Von Kármán's axial compression measurements on red sandstone performed at the indicated pressures.

Among the many matters pertaining to the plastic deformation of crystalline solids, yield surfaces and failure criteria early became subjects of overemphasis. This was partly because of the impetus provided by Bauschinger's studies of elastic limits in simple loading, but more because the field was dominated by those who were seeking to establish some gross criteria as an upper bound for technological design in generalized stress states. Research in the 19th century from Karl Johann Bernhard Karsten[1] in 1816 to Bauschinger[2] himself in 1886 had underscored the difficulty of defining such criteria in single stress states. Indeed most of the outstanding 19th century experimentists doubted that such a phenomenon as an elastic limit, let alone a yield surface, existed. Following the early 20th century concern for defining precise mathematical boundaries for perfect elasticity, well over a half-century of experiment, and the study of restricted plasticity theories for the "ideal solid," have not disposed of most of the original questions.

In that same paper in 1911, von Kármán,[3] after stating that in marble and sandstone it was impossible to determine an elastic limit in the sense that within

[1] Karsten [1816, *1*]. The second edition, translated from German into French by F. J. Culmann, published in Metz in 1830, is the one I used.
[2] Bauschinger [1886, *1*].
[3] von Kármán [1911, *1*].

the limit all deformation was completely reversible, further subscribed to the general 19th century attitude by stating that given sufficiently accurate instruments one could always find permanent deformation associated with each elastic deformation. Influenced by the desire of technologists to have some criterion for the elastic limit, VON KÁRMÁN described an arbitary "flow limit" which corresponded to the point at which there appeared "considerable" permanent deformation. Of course gross criteria of this sort had been introduced by HODGKINSON,[1] WERTHEIM,[2] and others for simple loading, and is used in modern technology. Today the choice of a limit usually is set at 0.2% strain instead of the equally arbitrary 0.05% strain of WERTHEIM.

KÁRMÁN compared his values of the "elastic limit" and his curves of constant strain with a representation due to OTTO MOHR.[3] KÁRMÁN then made one of the first in a long list of similar statements from then to now, independent of the particular hypothesis regarding yield surfaces, namely, that he had obtained as good an agreement between hypothesis and experiment as the imprecision of the situation would allow.

The remainder of VON KÁRMÁN's paper provided micrographs of specimens and photographs of deformed specimens which indicated the importance of compaction and binding of crystallites in producing the observed ductility at higher hydrostatic pressure.

The experimental determination of uniaxial stress-strain functions for metals in the presence of high hydrostatic pressure was studied intensively by PERCY WILLIAMS BRIDGMAN[4] three decades after VON KÁRMÁN's research for pressures under 40 kg/mm². Despite the importance of VON KÁRMÁN's experiments, they inspired little continued study. It is common now to attribute this type of research to BRIDGMAN. However, the latter's experiments in the 1940's on complicated alloys of steel were less definitive than the original work of VON KÁRMÁN.

4.12. The large deformation of solids under high hydrostatic stress: Bridgman (1909–1961).

It is obvious from what has preceded that from COULOMB's time to VON KÁRMÁN's the overwhelming emphasis in experimental solid mechanics has rested upon the study of three single stress states: uniaxial tension, compression, and torsion of cylindrical rods and tubes. Piezometer experiments from those of REGNAULT[5] to those of MALLOCK[6] and GRÜNEISEN[7] generally were designed to measure the bulk modulus in small deformation. Interpretation of results depended upon the use of linear elasticity. Therefore, such experiments could not be extended successfully to study large deformation resulting from loading paths with more than one non-zero stress component, unless a theory of finite deformation already had been well established. In the first decades of the 20th century, the experimental

[1] HODGKINSON [1843, *1*].

[2] WERTHEIM [1844, *1* (a)].

[3] For an extensive survey of both the variety of hypotheses and experiments designed to consider them, see the monograph on *Theory of Flow and Fracture of Solids*, by ARPAD LUDWIG NADAI, Vol. 1 of the revised second edition (New York: McGraw-Hill, Engineering Societies Monographs, 1950). NADAI [1950, *1*].

[4] BRIDGMAN (see the *Collected Experimental Papers of* P. W. BRIDGMAN, Vols. I to VII, Cambridge, Massachusetts [1964, *1*]).

[5] REGNAULT [1847, *1*].

[6] MALLOCK [1879, *1*].

[7] GRÜNEISEN [1906, *1*], [1907, *1*], [1908, *1*], [1910, *1*, *2*, *3*].

studies of THEODORE WILLIAM RICHARDS[1] on the atomic dependence of the hydrostatic compressibility of the elements not only were limited to pressures of 500 kg/cm² but were differential measurements. For the determination of absolute compressibility the necessity, by some means, of having knowledge of at least one solid as a reference standard was a major limiting factor in all studies, including the classic hydrostatic high pressure research of BRIDGMAN.

BRIDGMAN, like COULOMB, DULEAU, WERTHEIM, and GRÜNEISEN, is an outstanding figure in the subject of the present treatise. It is an interesting question whether fifty-six years of walking on the moon, brought about through the efforts of tens of thousands of persons, will reveal as much fundamental physics as did the efforts of a single individual, BRIDGMAN, in a similar interval, for the new world of the few cubic centimeters of his high pressure apparatus. Not to appreciate BRIDGMAN's great stature in 20th century experimental physics would be to overemphasize interpretation at the expense of experimental excellence. A sizeable fraction of the two hundred papers and the two monographs BRIDGMAN[2] contributed to high pressure physics are concerned with the deformation of solids subjected to quasi-static hydrostatic stress.

In his initial work[3] begun in 1905 and continued in the years that followed, he was to discover a method of packing[4] which would stop the leaks which had limited the earlier range of pressures. In Figs. 4.41 and 4.42 are shown the "general scheme of the packing by which pressure in the soft packing materials is automatically maintained a fixed percentage higher than in the liquid" and "the general principle of the method for giving external support to the pressure vessel in such a way that support increases automatically with the increase of the internal pressure."[5]

BRIDGMAN's early maximum pressure was the same as that of some of his immediate predecessors and his contemporaries, 6200 kg/cm². This value had been extended to 10000 kg/cm² by 1911. The earlier studies had required the disassembly of the apparatus after each individual loading, a slow, tedious process which fortunately became unnecessary. The progress of continuous improvements

---

[1] RICHARDS [1915, *1*], [1924, *1*]. RICHARDS wrote over thirty papers on compressibility from 1902 to 1930. BRIDGMAN had commented upon the lack of precision in RICHARDS' results, which RICHARDS himself suspected.

[2] BRIDGMAN [1964, *1*]. These volumes contain his collected works.

[3] BRIDGMAN [1909, *1, 2, 3*]. This was the first date of publication.

[4] BRIDGMAN [1943, *1*]. The following quotation is from pp. 3 and 4.

"When my work was started in 1905 it was my intention to study certain optical effects. I had no expectation of reaching pressures anywhere near the limits set by Amagat, since, for one thing, it was necessary to use glass for visibility. After my apparatus was constructed and some preliminary manipulations were made, there was an explosion — something very likely to happen with glass, which is most capricious. This destroyed an essential part of the apparatus, which had to be reordered from Europe; the United States had not at that time acquired its present degree of instrumental independence. In the interval of waiting for the replacement I tried to make other use of my apparatus for generating pressure. While designing a closure for a pressure vessel, so that it could be rapidly assembled or taken apart, I saw that the design hit upon did more than originally intended; the vessel automatically became tighter when pressure was increased, so that there was no reason why it should ever leak ... This at once opened an entirely new pressure field, limited only by the strength of the containing vessels and not by leak. My intended optical experiment was therefore dropped; the laboratory wrote off the expense of the replacement part and of the apparatus already constructed, and the development of the new field was begun. I have never returned to the original problem. This was a case where pertinacity of purpose would not have been good tactics."

[5] BRIDGMAN [1943, *1*]. These quotations were the captions for his Figs. 1 and 7, pp. 3 and 18, respectively. The same figures were used for his Nobel lecture in 1946, and were printed [1948, *1*] on pp. 149 and 150, as captions for Figs. 1 and 2.

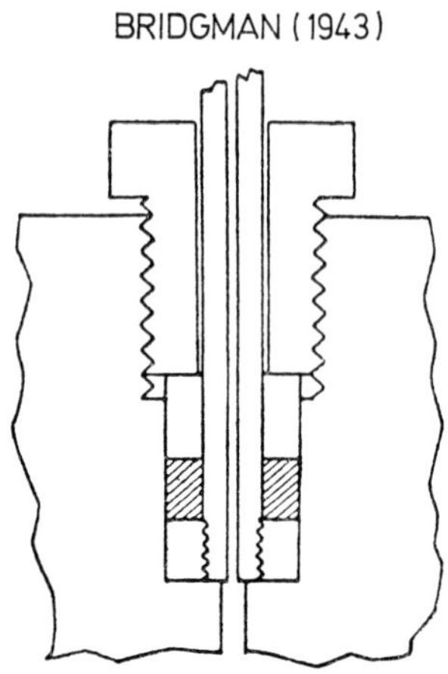

Fig. 4.41. Bridgman's diagram showing the method of packing which he used.

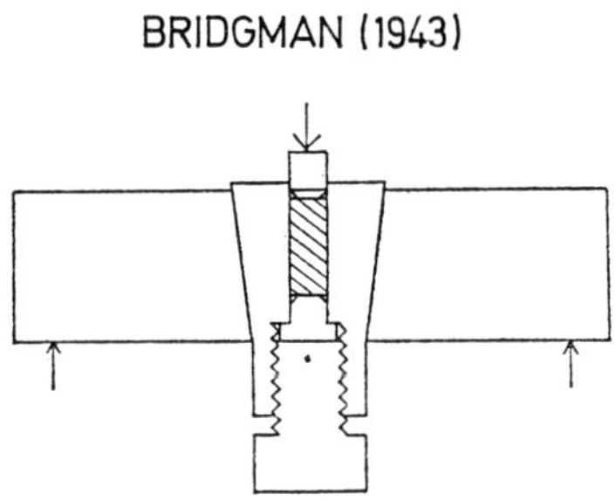

Fig. 4.42. Bridgman's diagram illustrating the general principle of giving external support to a pressure vessel so that support would increase with the increase of internal pressure.

by Bridgman is well documented in his many papers, and particularly in his monograph of 1931 *The Physics of High Pressure* which was republished with an added appendix[1] in 1949, and in a large paper in 1946 entitled "Recent Work in the Field of High Pressure"[2] which described the work of the fifteen years after the publication of his monograph.[3] Bridgman also gave an interesting short assessment of his work, "General Survey of Certain Results in the Field of High Pressure" in his lecture delivered in Stockholm in December, 1946, on his receipt of the Nobel Prize in physics in that year.[4]

Bridgman reached 12000 kg/cm². This pressure, except for a single experiment in water to 21000 kg/cm², became the limiting maximum pressure until the 1930's. In this pressure range, in an impressive, systematic experimental program resembling that of Wertheim, Bridgman studied the dependence of the volume and temperature of fluids upon the pressure, the phenomenon of melting under pressure, electric resistance under pressure, polymorphic transitions in

[1] Bridgman [1949, *1*].

[2] Bridgman [1946, *1*].

[3] Bridgman, *Ibid.* In this publication Bridgman noted that more titles had been published in the subject during the fifteen years than in all the years before his monograph of 1931. This fact was a tribute, if in some respects a dubious one, to the achievement of his pioneering effort.

[4] Bridgman [1946, *2*], see [1948, *1*].

solids under pressure, the effect of pressure on thermoelectric properties, thermal conductivities under pressure, viscosity under pressure, and the compressibility of solids.

BRIDGMAN's curious attitude toward rational mechanics is revealed in his statement as to why he waited until 1923, 18 years after his initial research on high pressure, to consider the relation between the volume of a solid and the pressure applied to it. In a review paper in 1943 for *American Scientist*,[1] after discussing liquid compressibility, polymorphic transitions, and thermoelectric properties, he arrived at the compressibility of solids, and stated:

> Not until the measurements were completed did I attempt the measurement that, from the point of view of today, one would be tempted to think the simplest of all and most immediately utilizable by theory, namely, that of the compressibility of solids, in particular, metals and simple salts. One reason why this was not done sooner is that it is only comparatively recently that there has been any stimulus from the theoretical side to do it. In the early years of the century it was thought that the order in which matter would be understood theoretically was gases, liquids, and solids. However, with the development of theories of the solid state around 1920, in the hands of Born ... and others, it became evident that solids, as well as gases, are simple, and that the study of liquids would have to be left until the last.[2]

In the first two and a half decades of his research, BRIDGMAN obtained pressures up to 12000 kg/cm². Of importance here are the measurements recording the relation between hydrostatic pressure and dilatation for 60 elements and a large number of compounds.[3] These results are described in great detail, together with an analysis of the early technique, in BRIDGMAN's monograph.[4]

By measuring the change of length of an iron rod relative to the pressure vessel and simultaneously making an independent measurement of the distortion of the pressure vessel itself, BRIDGMAN[5] achieved the absolute compressibility of pure iron by means of which the differential measurements in other solids, by comparison, could be converted to absolute. The absolute compressibility of the many solids studied was shown to be given by

$$-\frac{\Delta V}{V_0} = a p - b p^2, \tag{4.9}$$

where p is the measured hydrostatic pressure, ΔV the measured associated volume change, V_0 the initial volume at atmospheric pressure, while a and b are the two experimental constants for which BRIDGMAN tabulated values for a large number of solids.

During the 1930's, in extending the maximum pressure from 12000 kg/cm² to 30000 kg/cm² and beyond, BRIDGMAN reexamined his measurements of the absolute compressibility of iron. He found a serious descrepancy in his earlier results. He obtained a slightly different value for the first constant "a" but found that the second constant "b" differed by a factor of three. The new values for iron are given in Eq. (4.10):

$$-\frac{\Delta V}{V_0} = 5.826\times 10^{-7}\, p - 0.80\times 10^{-12}\, p^2, \tag{4.10}$$

where p is in kg/cm². BRIDGMAN had to revise all the previously published values of the two constants for Eq. (4.9), including those of the many solids given in the

---

[1] BRIDGMAN [1943, *1*].

[2] *Ibid.*, p. 12.

[3] The behavior of 30 metals as a function of temperature and pressure was published in 1923 [1923, *1*]. The paper contained the first results in the long-delayed studies of the compressibility of solids.

[4] BRIDGMAN [1931, *1*].

[5] BRIDGMAN [1940, *1*, *2*], [1946, *1*], [1949, *2*].

monograph.[1] He provided formulae for making the requisite corrections:

$$a_{\text{new}} = a_{\text{old}} - 0.033 \times 10^{-7},$$

$$b_{\text{new}} = b_{\text{old}} - 1.56 \times 10^{-12} - a_{\text{old}} \times 0.022 \times 10^{-7},$$

corrections which should be kept in mind when referring to BRIDGMAN's data on compressibility before 1940. In subsequent years BRIDGMAN merely tabulated the observed dilatation against the measured pressure. During the last 20 years of his life he continued his systematic study of physics at ever increasing high pressures and ever greater precision of measurement.

We should note particularly that BRIDGMAN measured the change in length in a single direction from which, assuming isotropy, he computed the change of volume. The assumption of isotropy could be checked in subsequent measurements by rotating the specimen. BRIDGMAN referred to this uni-directional measurement as "linear compression". To determine the change of volume of non-cubic-single crystals, he had to make successive measurement in two directions.

Using the miniature apparatus shown schematically in Fig. 4.43, which was immersed in the fluid of a larger pressure apparatus to which a pressure of 25000 kg/cm² or more was applied, BRIDGMAN by 1942 extended[2] his studies to 100000 kg/cm². The cylinder of the apparatus and the two pistons were made of carboloy. Long before the latter pressure was reached, fluids and gases had become solid; hence, in this high pressure range he no longer could assume the stress to be hydrostatic. In results obtained at such high pressures the student of continuum mechanics is likely to raise some objections to any claim to have identified and measured certain stress components. BRIDGMAN himself was the first to emphasize this limitation and the importance of the probable presence of shear stresses.

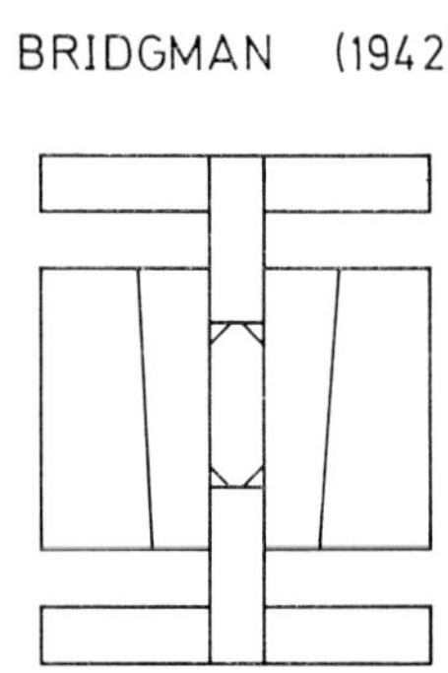

Fig. 4.43. The miniature carboloy piston device for reaching 100000 kg/cm². This apparatus was immersed in the high pressure fluid of a larger vessel.

In extending the compressibility of solids to this higher range, BRIDGMAN was interested primarily in studying transitions in the pressure-volume response functions. These occurred as first, second, and third-order transitions in the form of a discontinuity in the volume at a given pressure, a discontinuity in the slope not accompanied by a change in the volume, and a discontinuity in the second derivative. In some instances first-order transitions were found to be polymorphic. The study of changes in crystal structure at very high pressures of course presents

[1] BRIDGMAN [1931, *1*], [1949, *1*].

[2] BRIDGMAN [1942, *1*].

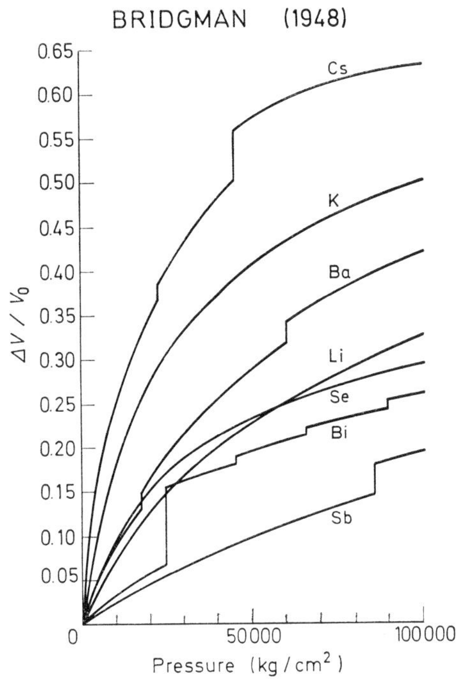

Fig. 4.44. The volume compression of several elements up to 100000 kg/cm² showing polymorphic transitions in some instances. These results are from BRIDGMAN's measurements in 1948, and differ considerably from those he showed in earlier papers.

great experimental difficulties. Second-order transitions, which, as we shall see below,[1] are very common and extremely important for more general states of stress, were not readily identifiable with phase transformations as were transitions of the first kind. In Fig. 4.44 are examples[2] of pressure-volume results, in some of which first-order transitions were present for pressures over the entire range to 100000 kg/cm².

While describing "linear compressions to 30000 kg/cm² including relatively incompressible substances," BRIDGMAN[3] in 1949 presented experimental evidence for second-order transitions or transitions of the second kind which, as just indicated, occur in the form of a discontinuity in the derivative. In that paper, he tabulated observations for 25 elements, 18 alloys and intermetallic compounds, 20 cubic and noncubic materials, and 3 organic crystals. He then discussed the data for each solid. In a section on the compression of a nickel single crystal BRIDGMAN described the discovery of a second-order transition in that solid.

Nickel. This material in single-crystal form, I also owe to the courtesy of Dr. Bozorth. Since it is cubic, measurements in a single direction suffice. The original source of the material was the International Nickel Company. The analysis was not stated, but the purity is thought to be comparable to that of a single crystal of iron, also from Dr. Bozorth, the analysis of which will be stated later. Like cobalt it had been heat treated in hydrogen near the melting point. A considerable number of measurements were made on this specimen since it was evident from the first that there was something peculiar about it. It was first pressure seasoned by two applications of 30000. Four sets of measurements to 30000 were then made in

[1] See Sect. 4.18, 4.21, 4.32. BELL [1965, *2*], [1967, *2*], [1968, *1*].

[2] BRIDGMAN [1948, *1*], p. 155; [1949, *1*], p. 421.

[3] BRIDGMAN [1949, *2*].

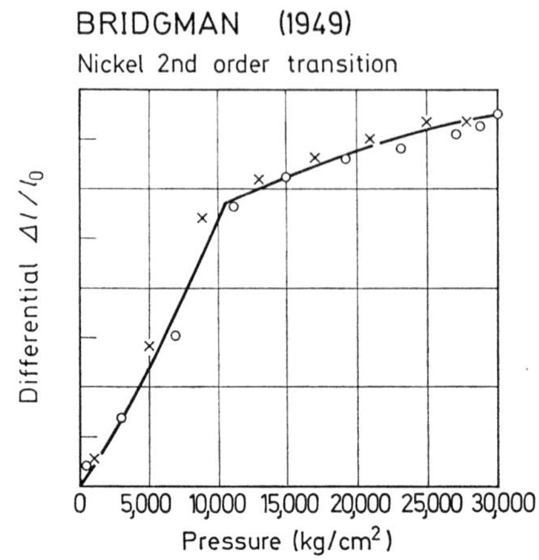

Fig. 4.45. The experimental values for the differential change of length of nickel showing a transition of the "second kind." The circles indicate increasing pressure; the crosses, decreasing pressure.

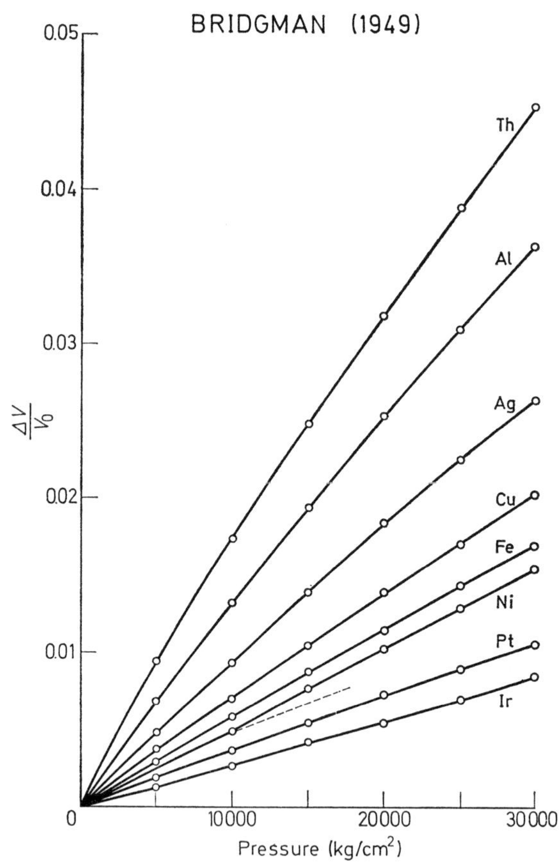

Fig. 4.46. Pressure-volume relations in 8 of 30 substances tabulated to 30000 kg/cm². Note the second order transition in nickel at 12500 kg/cm².

the former inadequately sensitive apparatus, then a measurement in the final apparatus. No consistency was obtained. The nickel was then annealed by heating to a red heat in an evacuated tube for one hour and cooled with the furnace. Two measurements to 30000 were then made with the final apparatus, then three seasoning applications were made of 30000 without measurement, and then the final measurement to 30000. All the measurements with the final apparatus indicated the same thing, a break in direction, or a transition of the second kind. The parameters of this second order transition changed with successive applications of pressure, the pressure at which the discontinuity occurred fluctuating between 10000 and 20000 kg/cm², and the amount of discontinuity also fluctuating. On the final measurements the magnitude of the discontinuity reached its largest value.[1]

He noted further:

The break at 10600 amounts to a discontinuity in $1/v_0\left(\frac{\partial v}{\partial p}\right)$ of 6.9×10^{-8} (kg/cm² units) in the direction of an *increased* compressibility at higher pressure.[2]

The experimental parts of the last run are shown here in Fig. 4.45.

I have plotted BRIDGMAN's data on pressure vs dilatation for the tabulations given for a few of the elements shown in Fig. 4.46, including the nickel single crystal with its second-order transition at $p = 10600$ kg/cm².

Similar second-order transitions in rubber (crosses) are shown in the pressure vs dilatation curves for the several solids in Fig. 4.47 for pressures to 25000 kg/cm². These results are from tabulated data from papers of BRIDGMAN[3] in 1945 and 1949.

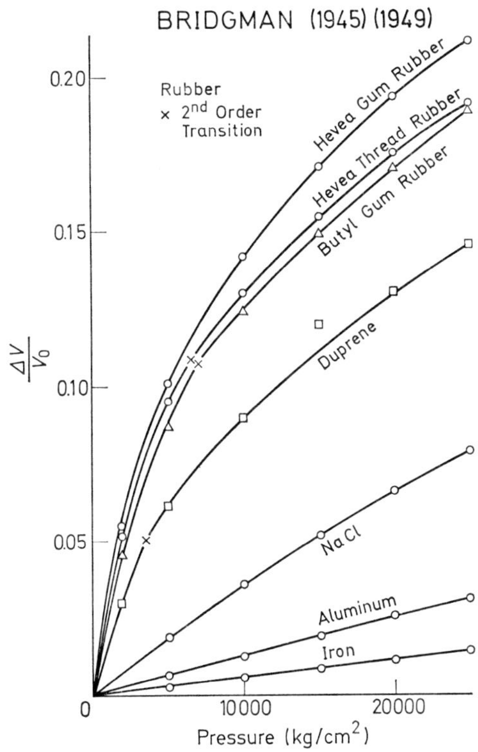

Fig. 4.47. Pressure-volume results in several solids, including rubber. Note the second order transitions at the crosses designated by BRIDGMAN.

[1] *Ibid.*, p. 203 (p. 3947 of *Collected Papers* [1964, *1*]).
[2] *Ibid.*
[3] BRIDGMAN [1945, *1*], [1949, *2*].

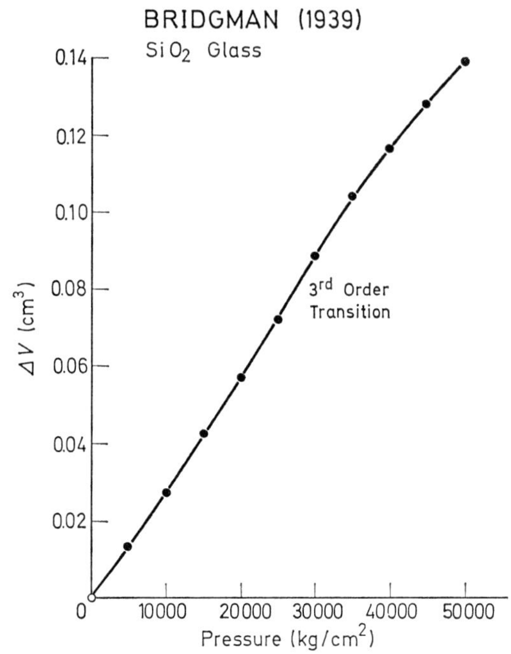

Fig. 4.48. A third order transition in the pressure-volume change results for glass.

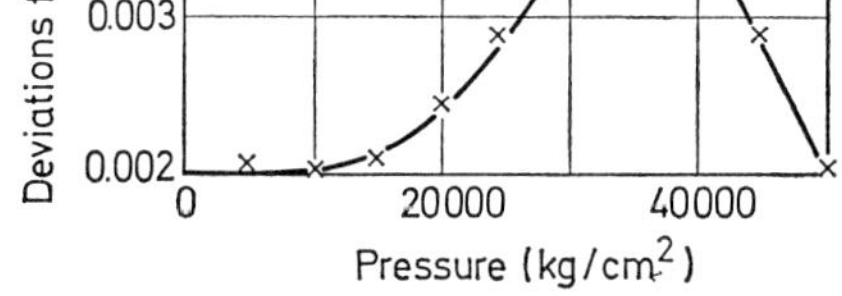

Fig. 4.49. The deviation from linearity for the incremental volume in the third order transition in glass of Fig. 4.48.

Still earlier BRIDGMAN had given an example of a third-order transition or discontinuity in the second derivative for SiO_2 glass. The pressure vs ΔV results[1] he obtained in 1939 exhibit the reversal in slope at 33 000 kg/cm² (Fig. 4.48).

BRIDGMAN[2] also gave the deviations from linearity of ΔV for this test. The data for the test plotted in this form are shown in Fig. 4.49. The sharp cusp of this transition is seen.

Third-order transitions commonly are observed in response functions for loading paths with unequal principal stress components. They were first observed for

[1] BRIDGMAN [1939, *1*].
[2] *Ibid.*

gut and other solids by JAMES RICCATI[1] in 1721. KARMARSCH[2] in 1841 showed that in gut they were associated with the onset of permanent deformation. Rubber and a number of other solids exhibit such reversals of concavity for reasons other than permanent deformation, and, of course, metals under uniaxial compression exhibit such flex points. For annealed metals in compression the stress at which the inflection occurs corresponds to that of the ultimate stress in a tensile test, provided the comparison in tension and compression is for nominal stress, referring to the original dimensions of the solid.

According to BRIDGMAN,[3] the first order or polymorphic transitions were observed first in relation to pressure, by MALLARD and LE CHATELIER[4] in 1883. The transition in ice at atmospheric pressure had been known for a considerable time. BRIDGMAN was correct in attaching great importance to this transition phenomenon under a hydrostatic state of stress. For more general states of stress, such transition phenomena both in the large and in the small deformation of crystalline solids now are known[5] to be fundamental to any complete understanding of constitutive equations for this class of materials.

4.13. Dynamic response of solids under high pressure.

A natural sequel to BRIDGMAN's quasi-static studies of solids under high pressures was the intensive interest aroused approximately 25 years ago to study the response of solids to pulses produced by explosive loading over the flat surface of a plate. Initiated by the military technology of the Second World War and supported by later industrial interest in a new manufacturing method known as explosive forming, continued fundamental study of such behavior was impossible for the lone scientist in a small laboratory. The appearance of papers by as many as seven or eight joint authors illustrates both the high cost and the "team" aspect of experimental research in which a single measurement can blow up thousands of dollars worth of apparatus and specimens.

The experiment consists in loading a solid to peak pressures four or five times those of the quasi-static range in fractions of a microsecond by means of explosions on the surface of the solid. The type of explosive, its shape, and the manner of detonation are important in the effort to produce a plane wave front traveling into a plate parallel to a flat side on which the explosive is located. The surface motion of the opposite side of the plate and at the bottom of drilled holes at designated distances from this opposite face, is measured. From the latter a shock wave speed is determined from measurements at two locations. When displacement vs time is ascertained on the free surface, one can obtain only the sum of the incident and the reflected particle velocities.

At that point, the experimental situation is similar to that of R. M. DAVIES,[6] described above in Sect. 3.37. Having made capacitance measurements of displacement vs time at the free end of his bar, DAVIES could not interpret his data without using results calculated from some particular theory alleged in advance to describe the phenomenon. Similarly, if I had been able to make *only* that measurement when I measured, optically, the displacement vs time at the free end of a cylindrical specimen undergoing finite plastic deformation, I should

[1] RICCATI [1721, *1*].
[2] KARMARSCH [1841, *1*].
[3] BRIDGMAN [1931, *1*], [1949, *1*], Chap. VIII, p. 223.
[4] MALLARD and LE CHATELIER [1883, *1*].
[5] BELL [1968, *1*].
[6] DAVIES [1948, *1*].

have had to know the theory applicable to waves of finite amplitude in order to interpret the data. In these studies, however, wave profiles I obtained dynamically by means of diffraction gratings,[1] provided *experimentally* the information necessary to evaluate such studies of surface displacement vs time for interpreting nonlinear unloading, as the POCHHAMMER-CHREE theory provided a guide for DAVIES.

Most experimentists during the two decades assumed *a priori* that when a solid is under high explosive shock, it behaves essentially like a fluid. They minimized the influence of the large shear stresses present in such a shock front. Another common assumption was that regardless of the thickness of the shock front, a steady state is reached behind it. For the given applied pressure, in order to obtain the particle velocity one must assume, or demonstrate experimentally, that unlike the reflection of plastic waves at the free face, the particle velocity of the incident wave at the surface of the specimen is doubled, as elementary linear theory for shock reflection prescribes at normal incidence. Combining the measurements of wave speed and maximum particle velocity in the framework of the assumed behavior enables one to calculate pressure vs volume data which may be compared with BRIDGMAN's quasi-static experimental results[2] in the region of overlap between the quasi-static pressures and the lower portion of the shock wave pressures.

In such comparisons it is particularly important to identify the pressures of polymorphic transitions in the HUGONIOT diagrams. Apart from the oversimplified atomistic hypotheses of such paramount interest today, understanding of the behavior of solids under high explosive shock is in a very preliminary state and probably will remain so until experiments are devised which do not depend upon introducing crude approximations for interpreting results, but will reveal the distributions of stress and of deformation which must be known before reliable constitutive equations can be ascertained.

In 1961, G. R. FOWLES,[3] for a shock wave oblique to a free surface, measured wave speeds and free surface velocities by means of streak photography. FOWLES performed the experiments, which were an interesting variation on the contact pin technique, on hardened and annealed 2024 aluminum, at pressures below 50 kilobars. He sought to examine both the elastic and the plastic behavior in shock waves. Despite the very high strain rates which were involved, he found, as had I for purely plastic waves,[4] no evidence of a significant influence of strain rate.

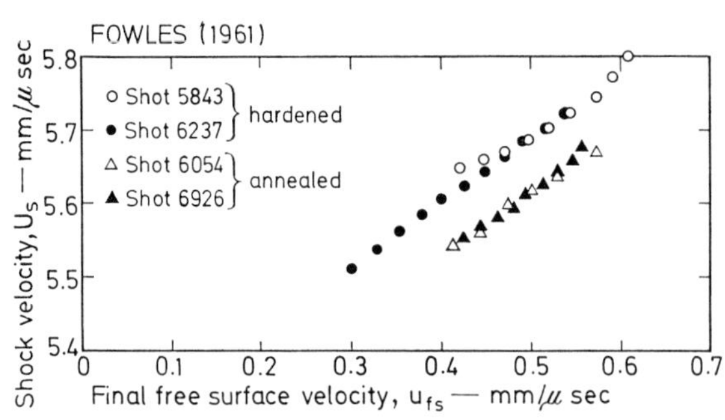

Fig. 4.50. Shock velocity (laboratory coordinates) as function of final free-surface velocity.

[1] BELL [1961, *3*, *4*] et seq. And see Sect. 4.29 below.
[2] BRIDGMAN [1949, *1*].
[3] FOWLES [1961, *1*].
[4] BELL [1956, *1*], [1960, *2*].

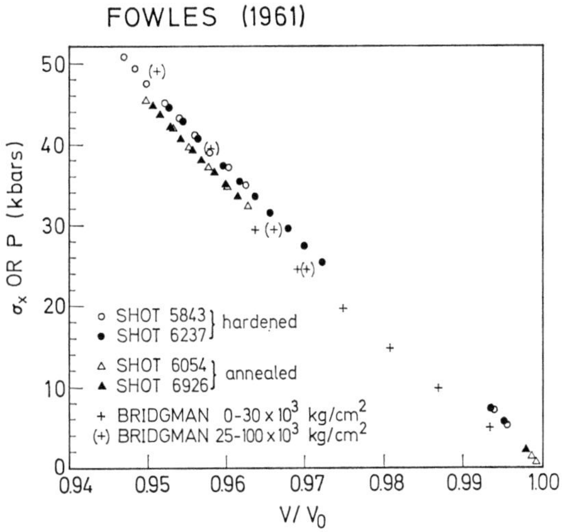

Fig. 4.51. Dynamic and quasi-static experimental HUGONIOT points.

In Fig. 4.50 are shown experimental shock velocities, U_s, in mm/μsec, vs final free surface velocity U_{fs}, also in mm/μsec. In Fig. 4.51 are shown the experimental HUGONIOT points calculated from these data and compared, as indicated above, with BRIDGMAN's quasi-static measurements for aluminum.

4.14. Further study of the Guest experiment: Lode (1926), and Taylor and Quinney (1931).

Leaving the special case of hydrostatic pressure and returning to experiments proposed to consider more general response functions arising from non-zero stress components, we examine the doctoral research of W. LODE[1] in 1926 at the University of Göttingen. LODE's experiments were a return to a more limited version of the research of GUEST,[2] after an interval of 26 years. LODE subjected steel, copper, and nickel tubes to axial tension combined with internal pressure. By varying the ratio of internal pressure to the axial load he was able to obtain a sequence of ratios of two of the principal stresses. In such experiments the third principal stress actually is not zero as is usually assumed. This stress in the radial direction varies over the thin cross-section, from the pressure on the inside of the tube to atmospheric pressure on the outside. That this steep gradient is present for one of the principal stresses places logical restrictions on the interpretation of experimental results, particularly when the loading path requires a large change in the internal pressure during the test.

For strain determinations LODE made simultaneous measurements of the accompanying change in length and diameter. He performed the experiments on a standard tensile testing machine with a commercial high pressure pump. LODE also made a small number of tension and tension-torsion tests to complete his study of the initiation of plastic yielding.

[1] LODE [1926, *1*]. (These experiments were suggested to LODE by NADAI.)
[2] GUEST [1900, *1*].

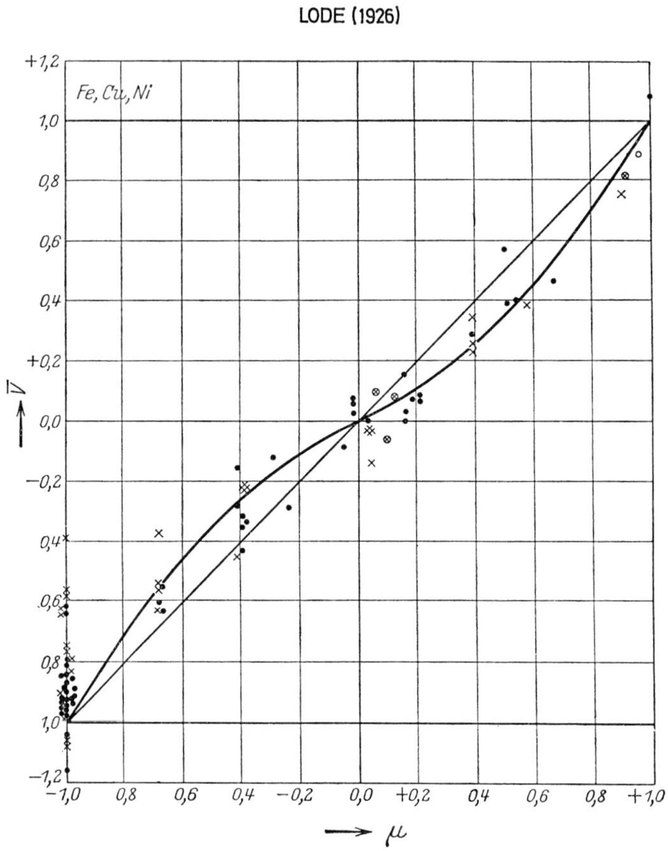

Fig. 4.52. The tensile-internal pressure experimental results of LODE in 1926.

What have become known as the LODE variables μ and ν were defined as follows:

$$\mu = 2\left(\frac{\sigma_2 - \sigma_3}{\sigma_1 - \sigma_2}\right) - 1\,, \tag{4.11}$$

where $\sigma_1 \geqq \sigma_2 \geqq \sigma_3$.

$$\nu = 2\left(\frac{\varepsilon_2 - \varepsilon_3}{\varepsilon_1 - \varepsilon_2}\right) - 1\,, \tag{4.12}$$

where σ_1, σ_2, σ_3, and ε_1, ε_2, ε_3 were the principal stresses and principal strains. LODE explored a wide selection of principal stress differences. The compiled experimental results are shown in the very familiar diagram of Fig. 4.52.

Choosing the indices 1, 2, and 3 in the axial, tangential, and radial directions respectively, and neglecting the radial stress gradient produced by the internal pressure, LODE chose $\sigma_3 = 0$; hence $\mu = \frac{2\sigma_2}{\sigma_1} - 1$. Thus for no internal pressure $\mu = -1$, and for $\sigma_2 = \sigma_1$, $\mu = 1$. As TAYLOR and QUINNEY[1] emphasized, the large scatter in experimental points for $\mu = -1$, corresponding to the simple tensile loading of the tube where symmetry requires $\mu = \nu$, suggested either experimental error or non-isotropic tubes. Therefore, in 1931 TAYLOR and QUINNEY sought to

[1] TAYLOR and QUINNEY [1931, *1*].

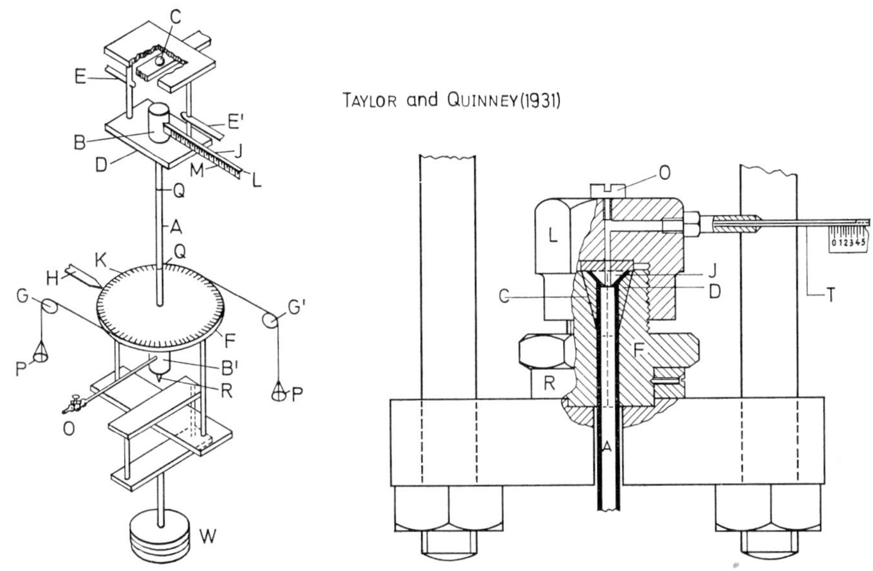

Fig. 4.53. TAYLOR and QUINNEY tension-torsion apparatus.

reexamine LODE's conclusion that $|\nu|<|\mu|$ rather than $\mu=\nu$ everywhere. Those experiments of TAYLOR and QUINNEY on aluminum, copper, lead, glass, cadmium, mild steel, and decarburized steel, for which they obtained results essentially like LODE's, except for glass and lead, have been widely quoted for four decades, sometimes even being referred to as the definitive study of the yield surface. Serious questions will be raised later in the present chapter (Sect. 4.22) with respect to the interpretation of some of the experiments in this study by TAYLOR and QUINNEY.

TAYLOR and QUINNEY saw that the chief difficulties in LODE's experiments were, first, the lack of a method for ascertaining that the material of his tubes indeed was isotropic, and, second, possible errors arising from the method he had used to measure the diameters of the strained tubes. They devised an experiment in which they hoped to overcome both difficulties simultaneously by measuring the internal volume change of the tubes during combined tension-torsion tests. Their apparatus is shown in Fig. 4.53.

For future reference, it should be noted that TAYLOR and QUINNEY's tensile loading system was not free of coupling with the torsion system, so that the weight W had to rotate during the application of the torsion moment. They annealed copper tubes for 36 h at 650° C. The aluminum tubes had been supplied in the annealed condition, having a manufacturer's purity of 99.7 to 99.8%. They annealed the mild steel tubes in a vacuum at about 920° C.

The experiment consisted in loading the specimen in tension to some total load, W, during which time they measured the internal volume by observing in a capillary tube the height of water which filled the interior of the tube. Rejecting tubes which exhibited major volume changes, they then lowered the load to some fraction, m, of the initial load and twisted the tube until they observed plastic yielding.[1]

[1] Determined in this manner, the yield surface depends upon the accuracy of the measurement. This was the reason, as we have noted, that BAUSCHINGER, 50 years earlier, had sought a more precise definition.

TAYLOR and QUINNEY compared the observed changes in volume with calculations based upon linear elasticity and a knowledge of POISSON's ratio while the solid was presumed to be responding in accordance with perfect elasticity, and they assumed that when it was responding as plastic deformation the density changes would be very small. In general, for each of the solids on which they made such measurements, they found that experimental values were not in agreement with computation. After discussing various possibilities for explaining the observed differences, and referring to the study of end conditions to interpret the changes in volume observed during pure torsion (which obviously indicated they were unfamiliar with the work of WERTHEIM on torsion in 1857), TAYLOR and QUINNEY more or less arbitrarily assumed that the source of the discrepancy was a lack of isotropy in the specimens. If one accepts this conclusion, then of course their results are of considerably less importance since they provide no measure of the details of the anisotropy they concluded was present.

As we have seen in previous chapters of this treatise, for a century and a half nearly every important experiment on isotropic solids sooner or later has been attacked on the grounds that specimens were anisotropic, when the data conflicted with popular theoretical assumptions. In some instances the criticism may have been justified. A few experimentists such as WERTHEIM, VOIGT, GRÜNEISEN, BRIDGMAN, and TAYLOR and QUINNEY by additional measurements while performing their experiments, have attempted to anticipate such post-hoc conjectures. That TAYLOR and QUINNEY did not completely succeed, is manifest in the comments in the subsequent literature. In their case, however, it is a matter of degree. How much small anisotropy was important? In Fig. 4.54 are shown the changes in volume for pure tension (A to B) for an aluminum tube in which the volume slightly increased, and a copper tube in which it slightly decreased, compared with a copper tube, C.G.C., in which relatively large changes in volume were observed. The point, B, in each instance marked the beginning of combined tension and torsion. On the basis of observations, if the change in volume of the region A to B during tension were of the same amount as that of B to C during combined tension and torsion, TAYLOR and QUINNEY rejected the tube. Thus, two of the tubes of Fig. 4.54 were accepted, and the third presumably was rejected.

Such measurements of change of volume are reminiscent of BAUSCHINGER's experiments on steel and cast iron in the 1880's. From an interest in comparing TAYLOR and QUINNEY's results[1] in 1931 with those of their 19th century predecessor[2] (see Fig. 2.36 above), I have included Fig. 4.55. It shows the internal change of volume of annealed mild steel and decarburized mild steel undergoing finite deformation. These were the very volume reversals in annealed mild steel that had excited the curiosity of BAUSCHINGER over a half century earlier.[3]

Considering values from $m=0.1$ to $m=0.9$ for aluminum, copper, lead, cadmium, glass, mild steel, and decarburized mild steel, TAYLOR and QUINNEY obtained the LODE diagram shown in Fig. 4.56, from which the behavior of μ as a function of ν was given. The departures from a straight line were consistent with LODE's[4] observations, but TAYLOR and QUINNEY's results were more limited since they considered only half the range examined by LODE, or by GUEST in the original experiment.

---

[1] TAYLOR and QUINNEY [1931, *1*].

[2] BAUSCHINGER [1879, *1*].

[3] It is common to cite the results of hydrostatic experiments as demonstrating that only minor changes of volume occur in plastic deformation. More recent research, to be discussed in Sect. 4.35, has shown that significant changes of volume associated with large deformation produced by loading paths with unequal principal stress components, indeed do occur.

[4] LODE [1926, *1*].

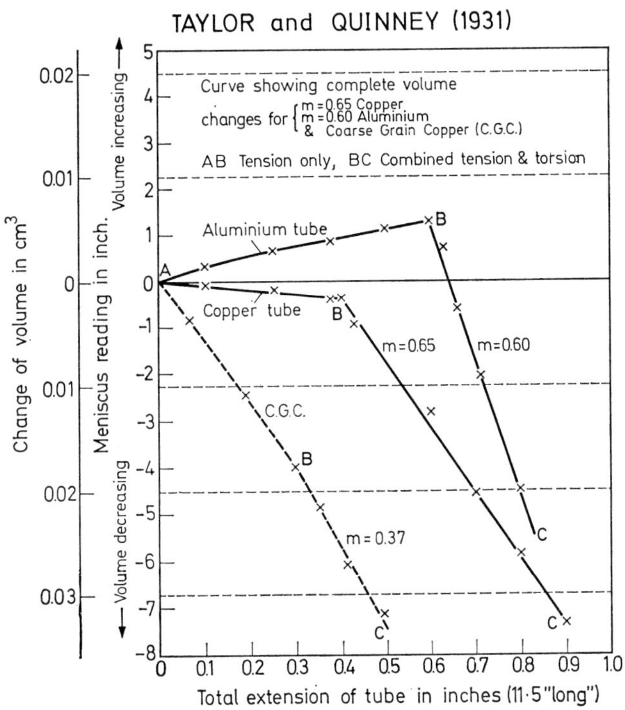

Fig. 4.54. TAYLOR and QUINNEY's internal volume measurements, testing isotropy of copper and aluminum hollow tubes.

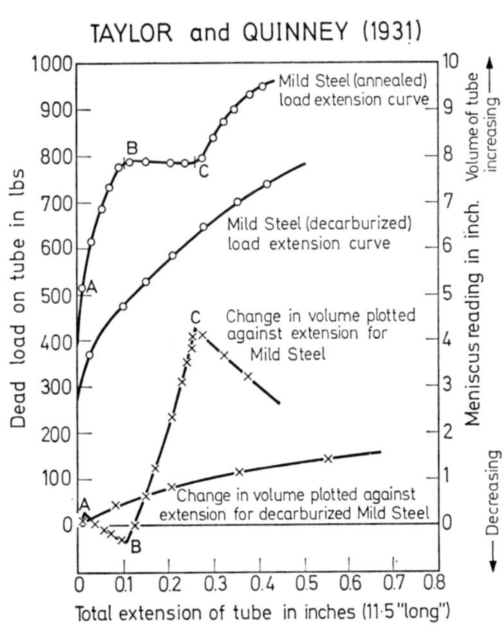

Fig. 4.55. TAYLOR and QUINNEY measurements of the internal volume change of mild steel and decarburized steel while undergoing the load-extension behavior shown.

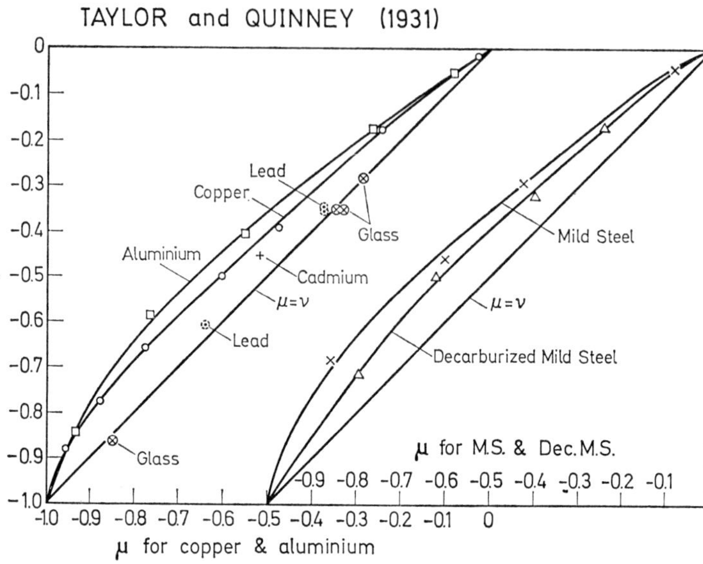

Fig. 4.56. The experimental results of TAYLOR and QUINNEY in the form of a LODE diagram, showing similar departures from linearity in μ vs ν plot which LODE had observed earlier.

However, of more consequence than TAYLOR and QUINNEY's reporting results similar to those of LODE, was their study of the yield surface in their attempt to compare GUEST's maximum shear hypothesis (designated by TAYLOR and QUINNEY as MOHR's hypothesis, after OTTO MOHR who had considered criteria for failure) and the MAXWELL-VON MISES maximum energy of distortion hypothesis. These results are shown in Figs. 4.57, 4.58, and 4.59, for copper, aluminum, and different mild steel tubes. They provide concrete evidence of the applicability of the MAXWELL-VON MISES hypothesis for copper and aluminum, a fact now well established from subsequent experiments.[1] TAYLOR and QUINNEY's data for mild steel, while not agreeing with GUEST's maximum shear hypothesis were not in complete accord

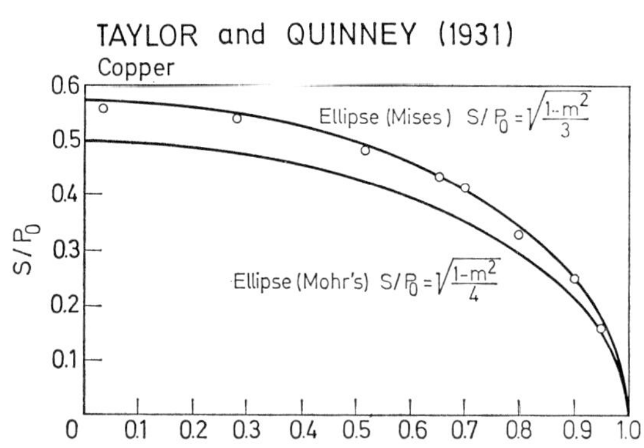

Fig. 4.57. TAYLOR and QUINNEY's comparison of experimental tension-torsion results in copper tubes (circles) compared with the MAXWELL-VON MISES hypothesis and the maximum shear hypothesis of GUEST (designated as MOHR's).

[1] BELL [1968, *1*].

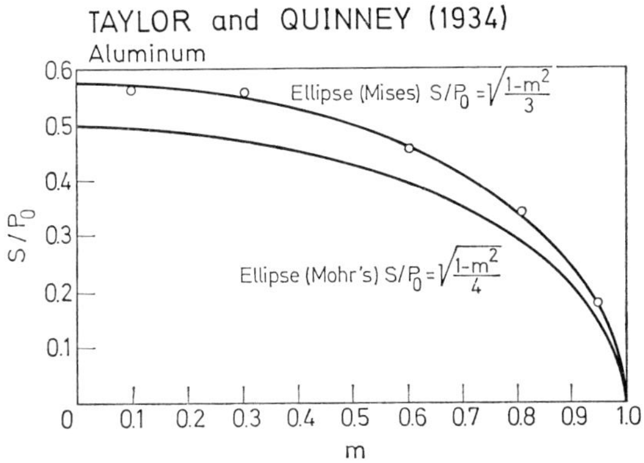

Fig. 4.58. TAYLOR and QUINNEY's comparison of experimental tension-torsion results in aluminum tubes (circles), compared with the MAXWELL-VON MISES hypothesis and the maximum shear hypothesis of GUEST (designated as MOHR's).

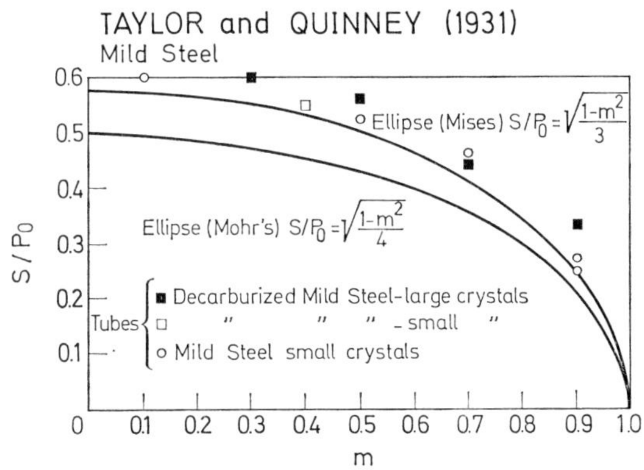

Fig. 4.59. TAYLOR and QUINNEY's comparison of experimental tension-torsion results in mild steel (circles), compared with the MAXWELL-VON MISES hypothesis and the maximum shear hypothesis of GUEST (designated as MOHR's).

with the MAXWELL-VON MISES hypothesis, as may be seen in Fig. 4.59. More recent experiments[1] have indicated that annealed mild steel does not differ from copper and aluminum in this respect. In any event, the experiments of TAYLOR and QUINNEY disposed of GUEST's maximum shear hypothesis for these solids.

The experiment of GUEST,[2] considered by TAYLOR and QUINNEY,[3] had a fundamental difficulty. The loading of a polycrystalline solid to permanent deformation along one loading path, followed by an unloading along that path and a reloading along a new path, requires that one identify the onset of permanent deformation at some point along the new loading path. All of the problems which had beset HODGKINSON,[4] WERTHEIM,[5] THALÈN,[6] THURSTON,[7] BAUSCHINGER,[8] and

[1] *Ibid.*
[2] GUEST [1900, *1*].
[3] TAYLOR and QUINNEY [1931, *1*].
[4] HODGKINSON [1843, *1*], [1844, *1*].
[5] WERTHEIM [1842, *1*], [1844, *1* (a), (b), (c)], [1857, *1*].
[6] THALÈN [1865, *1*].
[7] THURSTON [1874, *2*], [1876, *1*].
[8] BAUSCHINGER [1881, *1*], [1886, *1*].

many others in identifying the elastic limit in simple loading were magnified by the increase in the number of non-zero stress components. The greater the accuracy of the measurement, the more difficult is the decision as to when one has returned to the assumed yield surface. The detailed study of fully annealed solids demonstrates that at best there exists a yield region in which strain components are unpredictable, reminiscent of the problems which gave rise to BAUSCHINGER'S definition of the elastic limit for a simple axial loading path.

TAYLOR and QUINNEY proposed and performed an experiment which eliminated the difficulty inherent in GUEST'S experiment. They loaded specimens to very large strain along two different radial loading paths. Then, assuming given yield hypotheses, they made calculations using the measured stress-strain values from

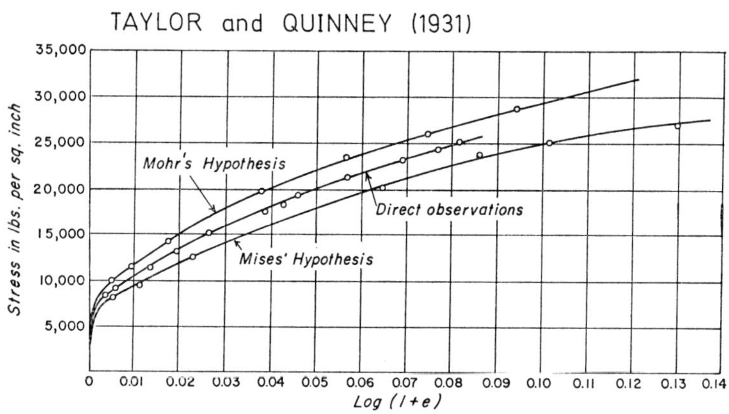

Fig. 4.60. The torsion and tension experimental results of TAYLOR and QUINNEY in 1931.

one loading path to predict that for the other path. TAYLOR and QUINNEY chose to compare simple tension and simple torsion loading paths for hollow tubes of annealed polycrystalline copper. They used stress-strain measurements in the torsion test to predict stress-strain response functions, shown in Fig. 4.60, from the MAXWELL-VON MISES hypothesis and from the GUEST-TRESCA hypothesis, which TAYLOR and QUINNEY referred to as the MOHR hypothesis. Their comparison of these two hypotheses with their direct observations from the tensile test indicated that apparently neither hypothesis was in accord with the experimental facts.

This result was in direct contradiction to that which TAYLOR and QUINNEY had obtained from the GUEST experiment. From that experiment they had concluded that the MAXWELL-VON MISES hypothesis described the yield surface of annealed copper. It should be emphasized that in the GUEST experiment the magnitude of the initial loading, and hence the yield surface considered, is arbitrary, i.e., the initial plastic deformation may be in the same range as that of the second experiment with continuous loading to large strain. However, unloading and subsequent reloading along alternate paths until the yield surface again is reached produces solely small deformation, and hence the results were given in terms of nominal stress and nominal strain. In contrast, for the second type of experiment, TAYLOR and QUINNEY described the observations in nominal stress and *logarithmic* strain. Following MOHR'S analysis, TAYLOR and QUINNEY had compared the shear strain, s, from the torsion test, with Log $(1+e)$, where e like s is referred to the original dimensions of the specimen.

In Fig. 4.104, Sect. 4.22 below,[1] I shall show that if one recalculates TAYLOR and QUINNEY's data in Fig. 4.60 so that stress and strain in tension and in torsion are defined with reference to the undeformed solid, then the MAXWELL-VON MISES hypothesis precisely agrees with experiment. TAYLOR and QUINNEY's measurements on annealed copper from the GUEST experiment and from their experiment on continuous loading to large deformation thus are in accord when all stresses and strains are stated with reference to the undeformed solid in the annealed state.[2] In the engineering literature the stress and strain so defined usually are called "nominal" stress and "nominal" strain, in contrast to the terms "true" stress and "true" strain which refer to the current configuration of the deformed solid.

4.15. On the relation between response functions for large deformation, for different radial loading paths: E. A. Davis' experiments on polycrystals (1943–1945).

The comparison of the response function of a polycrystalline solid for simple tension and for simple torsion loading paths had been considered by many persons beginning with THURSTON in the 19th century. Among those in the 20th century who provided such comparative measurements was EVAN A. DAVIS, in 1937. The experiments of DAVIS were given in the form of CAUCHY stress (or stress referring to the deformed area) and logarithmic (or natural) strain. When DAVIS' results are recalculated as nominal stress and nominal strain, one obtains the MAXWELL-VON MISES loading surface, with parabolic stress-strain functions in close quantitative accord with constitutive equations later shown to describe the large deformation of annealed crystalline solids.[3]

In 1943 DAVIS performed a series of experiments on annealed polycrystalline copper,[4] and two years later, on medium carbon steel,[5] in which he established for the first time one of the most important experimental observations since the original work of TRESCA. He carefully studied, for many loading paths, the large deformation response of the solid in TRESCA's intermediate region of plastic deformation, before fracture.

DAVIS placed hollow tubes of annealed polycrystalline solids in a standard hydraulic testing machine. To obtain more than one non-zero principal stress component he connected the tube to be studied to a high pressure pump which had a control system synchronized with the hydraulic testing machine so that the internal pressure in the tube and the axial tensile load upon the tube would produce a radial loading path of any desired ratio between the two principal stresses in the tangential and axial directions. Designating by σ_1 the principal stress in the axial direction, by σ_2 the principal stress in the tangential direction, and by σ_3 that in the radial direction, where for all but two of DAVIS' loading paths $\sigma_1 > \sigma_2 > \sigma_3$, DAVIS compared ratios of σ_2/σ_1 of 0, $^1/_4$, $^3/_8$, $^1/_2$, $^3/_4$, and 1.

---

[1] See also BELL [1968, *1*].

[2] BELL [1971, *2*], and see Sect. 4.35 below.

[3] BELL [1968, *1*]; see Sect. 4.35, below.

[4] DAVIS [1943, *1*]. The demand for rapid production of ships during World War II led to the use of an all welded construction. When a number of these ships unexpectedly developed cracks in their plating, intensive programs of applied research were begun in an attempt to find the source of the difficulty. Among these were the efforts of many persons in several laboratories to study experimentally the fracture of thin-walled tubes subjected to various loading paths having more than one non-zero stress component, i.e., specimens being subjected to what the engineers of the time referred to as "combined stress."

[5] DAVIS [1945, *1*].

For the medium carbon steel, the observations of fracture suggested the importance of the study of a number of radial loading paths in the vicinity of $\sigma_2/\sigma_1 = 0.800$. In this solid, besides obtaining the ratios of $\sigma_2/\sigma_1 = 0$, 0.50, 0.750, and 1.000, DAVIS measured large deformation for loading paths having ratios of 0.762, 0.775, 0.800, and 0.875. In addition to studying loading paths in which the maximum principal stress was in the axial direction, DAVIS observed loading paths in which the maximum principal stress, σ_1, was in the tangential direction. These were for pure internal pressure $\sigma_2/\sigma_1 = {}^1/_2$, and for the loading path for a tubular specimen in which the axial load was carried by a central rod. The pressure was maintained by U-shaped leather packing so that the only load applied to the specimen was the pure internal pressure which appeared solely as tension for the tangential principal stress component, σ_1, i.e., $\sigma_2/\sigma_1 = 0$.

In his paper in 1943, DAVIS had assumed that the radial stress, σ_3, was zero, whereas in fact, there is a steep gradient in this direction across the thin walled tube from the pressure on the inside of the tube to the near zero atmospheric pressure on the outside. The ratio of the maximum radial stress to the tangential principal stress, σ_2, varied with the thickness of the tube. For the diameters of the tubes which DAVIS used, this ratio was approximately ${}^1/_{20}$.

The error introduced by the fact that the radial stress is not zero and has a steep stress gradient through the wall of the tube, varies with the loading path from no error at $\sigma_2/\sigma_1 = 0$ to a maximum error at $\sigma_2/\sigma_1 = 1.00$.

In his paper in 1945, DAVIS[1] suggested that one introduce for this radial principal stress an empirical value of ${}^1/_2$ the internal pressure, p, i.e., $\sigma_3 = p/2$, instead of the assumption that $\sigma_3 = 0$. For the wall thickness of DAVIS' tubes, σ_3 was approximately ${}^1/_{10}$ of σ_2. The effect upon large plastic deformation of the radial stress gradient in this type of test is not known. That it is not a major contribution may be seen from a comparison of measurements for loading paths in which the gradient is present and those for which it is not, such as in pure tension and in combined tension and torsion, and for loading paths in tension and torsion to which internal pressure also is applied to maintain a constant volume, as studied by DAVIS in 1955 in medium carbon steel[2] and compared with his measurements of 1945.

In those experiments in 1943 and 1945, DAVIS measured the change in axial length and the change in the diameter of the tube as the stresses increased along the prescibed loading path. To obtain greater accuracy and to minimize the personal danger present in this type of test, he made the measurements after unloading following each increment of increased loading of stress into the plastic region. (It will be remembered that this was the type of cyclical loading which had been used to make simple tension tests, from the time of GERSTNER and HODGKINSON in the 1820's to late in the 19th century when THURSTON introduced autographic testing machines.)

DAVIS did not measure the changing thickness of the wall of the tube and thus unfortunately did not know the strain component in the third principal direction. Arbitrarily, without measurement, he *assumed* the plastic deformation was isochoric and hence, calculated his third principal strain component from his measurements in the axial and tangential directions.

To check that the tubular specimens were isotropic, DAVIS compared the response functions of specimens cut longitudinally and cut tangentially with that of a large tubular specimen. At all strains, as well as at fracture, he obtained very close agreement in the detail of the response function itself.

[1] E. A. DAVIS [1945, *1*].
[2] E. A. DAVIS [1955, *1*].

In the development of an understanding of the large deformation of crystalline solids, the importance of DAVIS' experiments was that he chose to compare the response functions of a very large number of loading paths, first, by plotting the "true" octahedral shearing stress, $\bar{\tau}_0$, against the "true" octahedral shearing strain, $\bar{\gamma}_0$; and second, by plotting the "true" maximum shearing stress $\bar{\tau}_2$ vs the "true" maximum shearing strain $\bar{\gamma}_2$ where

$$\bar{\tau}_0 = \frac{\sqrt{2}}{3}\sqrt{(\bar{\sigma}_1-\bar{\sigma}_2)^2+(\bar{\sigma}_2-\bar{\sigma}_3)^2+(\bar{\sigma}_3-\bar{\sigma}_1)^2},$$

$$\bar{\gamma}_0 = \frac{2}{3}\sqrt{(\bar{\varepsilon}_1-\bar{\varepsilon}_2)^2+(\bar{\varepsilon}_2-\bar{\varepsilon}_3)^2+(\bar{\varepsilon}_3-\bar{\varepsilon}_1)^2},$$

$$\bar{\tau}_2 = \frac{\bar{\sigma}_3-\bar{\sigma}_1}{2} \quad \text{and} \quad \bar{\gamma}_2 = \bar{\varepsilon}_3-\bar{\varepsilon}_1.$$

Shown in Figs. 4.61 and 4.62 are the results obtained in copper for the eight different loading paths for the "true" octahedral shearing and the "true" maximum shearing stress, respectively.

DAVIS used his axial and tangential strain measurements and the assumption of incompressibility to determine the octahedral shearing stress in terms of the CAUCHY stress, i.e., with reference to the area of the specimen at the time of the designated load. He gave the octahedral shearing strain as logarithmic or natural strain from the nominal strain measurements of ε_1 and ε_2. Similarly, he gave the data for the maximum shearing stress and maximum shearing strain, as CAUCHY stress and logarithmic strain. In 1943, DAVIS found that response functions thus expressed were sensibly independent of loading path. During the past decade, the discovery[1] of the general constitutive equations for the large deformation of crystalline solids has demonstrated the importance of DAVIS' observation.

To facilitate reference in Sect. 4.35 later in the present treatise, I have included in Figs. 4.61 and 4.62 along with these data of DAVIS, the response functions

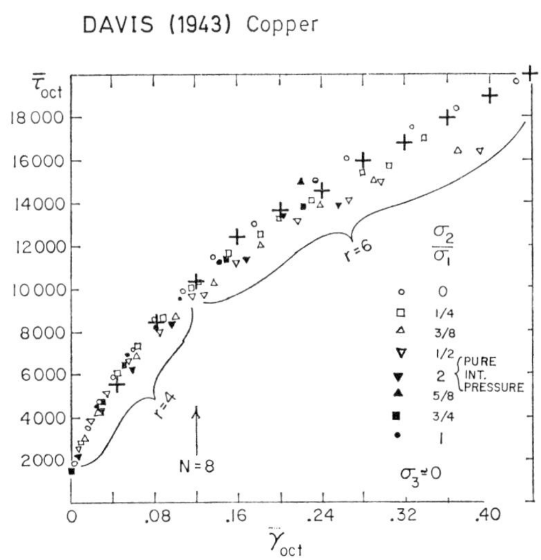

Fig. 4.61. E. A. DAVIS tension-internal fluid pressure experiments on copper tubes. The data were given as "true" octahedral shear stress vs "true" octahedral shear strain for various radial loading ratios. The crosses show the correlation with modern theory (see Sect. 4.35).

[1] BELL [1972, *2*], and see Sect. 4.35 below.

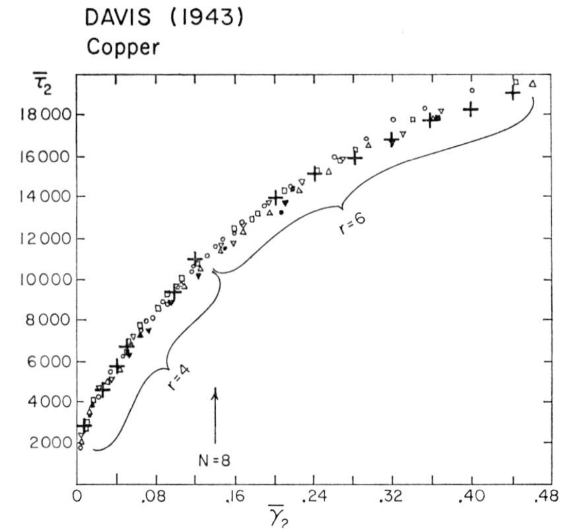

Fig. 4.62. E. A. DAVIS tension-internal fluid pressure experiments on copper tubes. The data were given as "true" maximum shear stress vs "true" maximum shear strain for various radial loading ratios. The crosses show the correlation with modern theory (see Sect. 4.35).

predicted[1] for $\bar{\tau}_0$ vs $\bar{\gamma}_0$ and $\bar{\tau}_2$ vs $\bar{\gamma}_2$, i.e., the "true" octahedral shear stress and strain and "true" maximum shear stress and strain (crosses) in the respective diagrams.

The data from DAVIS' experiments of 1945 on medium carbon steel, which had included small contributions calculated from the *assumed* radial stress $\bar{\sigma}_3 = p/2$, were presented as CAUCHY stress vs logarithmic strain for $\bar{\tau}_0$ vs $\bar{\gamma}_0$ and $\bar{\tau}_2$ vs $\bar{\gamma}_2$. These results are shown in Figs. 4.63 and 4.64, respectively, where one sees the same independence of loading path of the response functions of the large deformation

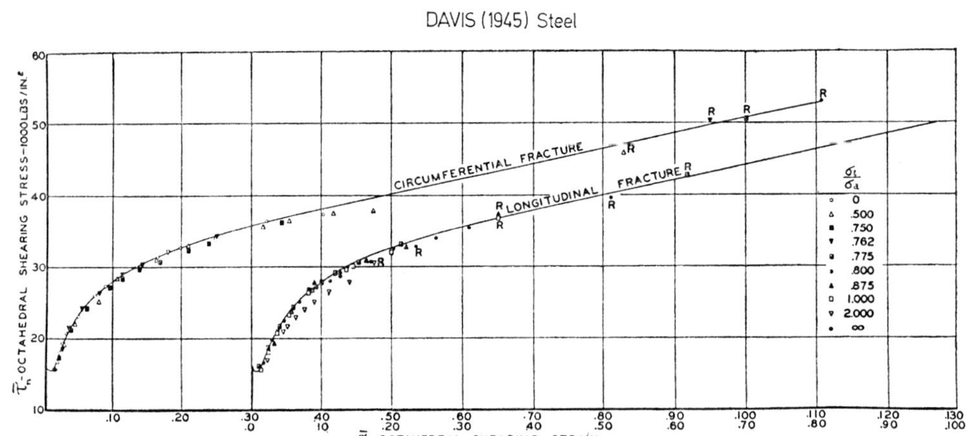

Fig. 4.63. E. A. DAVIS tension-internal fluid pressure experiments on steel tubes. The data were given as "true" octahedral shear stress vs "true" octahedral shear strain for various radial loading ratios.

[1] BELL [1968, *1*].

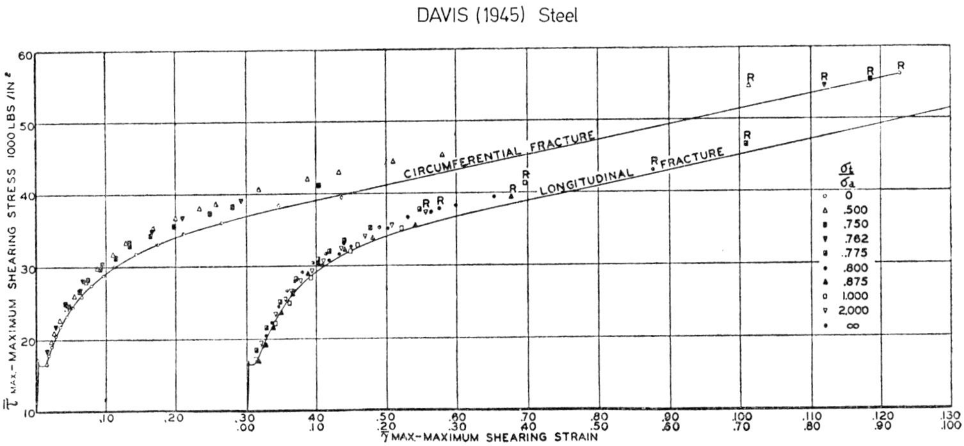

Fig. 4.64. E. A. DAVIS tension combined with internal fluid pressure experiments on steel tubes. The data were given as "true" maximum shear stress vs "true" maximum shear strain for various radial loading ratios.

for octahedral shearing stress and maximum shearing stress. DAVIS divided the results of the tests into two groups, according to the type of fracture.

I have added the integers, r, to the quantitative comparisons for copper in Figs. 4.61 and 4.62 so that reference may be made to these in Sect. 4.35, where I discuss the general constitutive equations for the large deformation of crystalline solids. (A similar correlation exists for the data of Figs. 4.63 and 4.64.) One also might add to Fig. 4.63 the data from a series of experiments DAVIS described in 1955 in a paper entitled "Combined tension and torsion tests with fixed principal directions."[1] The results of those experiments, which included radial loading paths in tension and in torsion simultaneously, further extend the generality of DAVIS' discovery.

4.16. Bridgman's experiments during the 1940's on the plastic deformation of steels.

In 1952 BRIDGMAN[2] published a monograph entitled *Studies in Large Plastic Flow and Fracture,* for the most part describing experiments related to his efforts during the Second World War to provide for technological purposes, information on the properties of steels ranging from medium carbon steel to complex steel alloys and armor plate. Of particular interest was his compilation of maximum loads, the logarithmic strain, and reduced stress at the section where necking occurs just prior to fracture.

The experiments described by BRIDGMAN in that monograph hence were performed during the same interval of time as those of DAVIS. Despite the fact that BRIDGMAN's program of research on the large deformation of polycrystalline solids under loading paths with more than one non-zero stress component was far broader in scope and in the number and variety of measurements which he made, the results he obtained have been shown to be of far less importance than those of DAVIS performed during the same years. From the perspective of nearly three decades,

[1] DAVIS [1955, *1*].
[2] BRIDGMAN [1952, *1*].

one regrets that BRIDGMAN confined his attention to the study of chemically complicated polycrystalline solids with such specialized prior thermal and mechanical histories.

BRIDGMAN's work was divided into three main sections. The first topic was that of deformation under hydrostatic pressure. Included were ultimate loading and failure for simple tension, two dimensional tension, simple compression, and punching. The pressure reached the range from 300000 to 450000 psi (212 to 317 kg/mm²). The second section, "Other tests involving large deformation," described experiments on steel, but with some reference to earlier work on soapstone, marble, copper, and duralumin, including simple compression, two dimensional compression, mixed compression, torsion combined with simple compression, and shearing combined with approximately hydrostatic pressure. In the final section BRIDGMAN described plastic flow and fracture after pre-straining, for which he had performed simple tension, compression, and torsion tests on specimens which had been subjected to the various types of loadings mentioned.

In his final chapter, "Gathering up the threads," BRIDGMAN gave a somewhat dismal portrait of the difficulties present in the interpretation of the finite deformation of complex steels subjected to the many prior mechanical and thermal histories and types of tests he had described. Since his main theme was his effort to provide technological data for the evaluation of steels subjected to the high dynamic stresses of projectile impact and penetration of wartime interest, he compared at great length various conditions of tempering and quenching for steels in the design state.

Apart from the complication of the material, many of the experiments BRIDGMAN described are of intrinsic importance. Axial tests in the presence of hydrostatic pressure, which VON KÁRMÁN[1] had been the first to accomplish in 1911, BRIDGMAN extended to a range of pressures sufficient to examine fruitfully fundamental hypotheses in simple solids. His experiments using a dilatometer to determine change of volume accompanying the finite deformation of solids in simple compression, and his unique experiments in two dimensional compression in the plastic region were particularly significant. Whatever was the technological importance of BRIDGMAN's contribution to the war, from the point of view of fundamental science his tests represented a detailed illustration of the distance to be traveled before complex solids with highly specialized prior thermal and mechanical histories could fall convincingly within the framework of plausible theory.

4.17. Experiments on single crystals: Quantitative order in response functions for the large deformation of solids.

FRIEDRICH ENGESSER[2] perhaps is best known for his having suggested in 1889 that the modulus E in EULER's buckling formula for columns be given as the tangent modulus of the axial stress-strain function when large deformation was present. By 1898, ENGESSER had provided a graphical study[3] of approximations of response functions for large deformation to be used in structural design. However, after the algebraic exercises of JOHN LESLIE[4] in 1823, there had been no other serious effort to formulate a specific functional form for measured large deforma-

[1] VON KÁRMÁN [1911, *1*].
[2] ENGESSER [1889, *1*].
[3] ENGESSER [1898, *1*].
[4] LESLIE [1823, *1*].

tion in metals[1] until the 1920's. In 1921, HAROLD C. H. CARPENTER and CONSTANCE F. ELAM[2] succeeded in growing specimen size metal crystals by the strain-anneal-recrystallization technique, and two years later BRIDGMAN[3] developed the method of producing metal single crystals from the melt by a process of slow extraction, now known as the BRIDGMAN method.

The availability of single crystal specimens and BRIDGMAN's optical orientation procedure which had been developed at the same time, and more important, the LAUE[4] method of determining by X-rays the orientation of crystal planes in single crystals, made possible the experimental study of what became known as the "resolved shear deformation" of single crystals, obtained from the uniaxial stress and strain measured in tension or in compression tests on crystals of known initial orientation. TAYLOR and ELAM,[5] and somewhat later, FRHN V. GÖLER and GEORG OSKAR SACHS,[6] assumed that plastic deformation in crystalline solids occurred from the sliding of parallel crystal planes. The specific planes and directions for a given crystal structure were determined from observation of the surface of the deformed specimen and from an analysis of X-rays of that specimen.

If only a single set of parallel planes were involved, resolved deformation was referred to as "single slip." If two sets of planes were sliding simultaneously, the term "double slip" was introduced to describe the phenomenon. Thus for face-centered cubic solids, they assumed that sliding occurred on one of the four planes known as $\{111\}$, which contained three face diagonals. The direction of the sliding would be along one of the face diagonals lying in the plane of sliding. This combination of four planes and three directions on each plane provided a total of twelve possible systems for single slip.

In a tensile or compressive test on a single crystal, the system assumed to be operative was that for which the angle between the slip direction and the specimen axis λ, and between the normal to the planes of sliding and the specimen axis ϕ, gave the largest component when the axial stress was resolved in the direction of slip. As the finite strain increases in a test in tension or compression, the values of λ and ϕ change. The stress on the planes of sliding in the direction of slip obtained from a knowledge of the angles λ and ϕ was referred to as the resolved shear stress τ. The sliding of planes in the direction of this slip was seen in macroscopic terms as a resolved shear strain γ.

When the axis of the specimen rotated to a symmetry plane, where two slip systems were equally favorable, subsequent rotation was assumed to be that for which both slip systems were equally operable. This condition was known as double slip.

For a discussion of the kinematics of crystalline slip, see, for example, the article in the *Handbuch der Physik*, Vol. VII/2, by A. SEEGER.[7]

The kinematical developments of TAYLOR and ELAM in 1923 and of GÖLER and SACHS in 1927 by means of which the resolved shear stress and resolved shear strain were determined from axial measurements, and the related studies identifying the crystallographic planes and directions for which such deformation occurred, stimulated in crystal plasticity one of the most phenomenally rapid growths of activity in the history of experimental solid mechanics. By 1935 in their superb

[1] Of course this was not the situation for tests on organic tissues and rubber, as we have seen in the second chapter of this treatise.
[2] CARPENTER and ELAM [1921, *1*].
[3] BRIDGMAN [1925, *1*], [1923, *1*].
[4] LAUE [1912, *1*]; LAUE and TANK [1913, *1*]; FRIEDRICH, KNIPPING, and LAUE [1912, *1*].
[5] TAYLOR and ELAM [1923, *1*].
[6] GÖLER and SACHS [1927, *1*].
[7] SEEGER [1958, *1*].

volume, *Plasticity of Crystals*, E. SCHMID and W. BOAS[1] revealed that in little more than a decade this aspect of metal plasticity, in technique and in scope, had enlarged beyond the experimental capacity of any individual laboratory.[2]

For single slip, with σ_s the axial stress referred to the area of the undeformed solid, ε_s the axial strain referred to the undeformed length of the specimen, τ the resolved shear stress, γ the resolved shear strain, λ the angle between the specimen axis and the shear direction, and ϕ the angle between the specimen axis and the normal to the plane upon which shear occurs, these kinematical studies provided the following equations:

$$\cos\phi = \frac{\cos\phi_0}{1+\varepsilon_s}, \qquad \sin\lambda = \frac{\sin\lambda_0}{1+\varepsilon_s}, \tag{4.13}$$

$$\gamma = \frac{\cos\lambda}{\cos\phi} - \frac{\cos\lambda_0}{\cos\phi_0}, \tag{4.14}$$

$$\tau = \sigma_s \cos\phi_0 \cos\lambda, \tag{4.15}$$

where ϕ_0 and λ_0 refer to the initial orientation before deformation. The $\cos\phi_0$ in Eq. (4.15) results from the assumption that the area of the rotating slip plane is constant, i.e., the reduction of the area of the cross-section of the specimen has been included so that the axial stress, σ_s, must be given in nominal form.[3]

If during finite deformation of cubic single crystals the specimen axis rotates to a symmetry plane, then, as indicated above, two slip systems for which resolved stress are equal, appear to become equally favorable for slip. TAYLOR and ELAM[4] in 1925 and GÖLER and SACHS[5] in 1927 made calculations for double slip when the specimen axis reached this symmetry plane. They assumed that the specimen axis of the face-centered cubic crystal upon reaching the symmetry line, followed along it to the $\langle 121 \rangle$ orientation where no further rotation occurred. Only 9 out of 49 tension tests in aluminum, silver, gold, copper, and α brass, all performed before 1932 by one of the two groups who had proposed the double slip hypothesis, gave X-ray evidence of behavior in a manner consistent with that assumption. Between 1925 and 1930, in only one of 17 tests on α brass by TAYLOR and ELAM,[6] KARNOP and SACHS,[7] MASIMA and SACHS,[8] GÖLER and SACHS,[9] and SACHS and WEERTS,[10] did X-ray measurement support the double slip assumption. Similarly, before 1930, only 4 out of 8 tests in aluminum, and 4 out of 17 in gold, copper, and silver, gave any evidence by X-ray measurement that the specimen axis rotated in double slip as postulated.

In 1967, BELL and GREEN[11] examined 168 single crystal tests performed since 1925, in all of the cubic solids for which actual X-ray measurement of crystal axes had accompanied the calculation of resolved shear deformation after the

[1] SCHMID and BOAS [1935, *1*]. I used the English translation [1950, *1*].

[2] The monograph in 1935 provided a list of 703 references, of which fewer than 40, including references to VOIGT's 19th century study of elastic constants, preceded the success in 1921 in growing large crystals.

[3] As will be seen in Sect. 4.32, this is an important point when through aggregate ratios the axial stress of the polycrystal is compared with the resolved shear stress of the single crystal. Both must be given in nominal form (referring to the undeformed state of the material) for the aggregate ratio to be independent of strain.

[4] TAYLOR and ELAM [1925, *1*].

[5] GÖLER and SACHS [1927, *1*].

[6] TAYLOR and ELAM, *op. cit.*

[7] KARNOP and SACHS [1927, *1*].

[8] MASIMA and SACHS [1928, *1*].

[9] GOLER and SACHS, *op. cit.*

[10] SACHS and WEERTS [1930, *1*].

[11] BELL and GREEN [1967, *3*].

symmetry plane had been reached. In *no* test since 1929 has the evidence supported the assumption of double slip after the symmetry plane has been reached.

Nevertheless from 1930 to the present, except for α brass, which became the object of special study, it became the custom to change automatically from a single slip to a double slip calculation when the specimen axis rotation, as calculated from single slip formula, reached the symmetry plane. Incredible as it may seem in view of the experimental facts, the double slip calculation continued to be performed throughout the 40 years, without an X-ray check during the deformation. Such is the power of an attractive hypothesis. Unfortunately, when only calculated resolved data for double slip are given, it is impossible to recover the original axial stress vs strain measurements for reexamination, and hence, hundreds of single crystal experimental results published since 1930 are fundamentally questionable.[1]

In 1964 in a study of the response functions obtained from comparing the resolved shear stress, resolved shear strain for single crystal tests in the literature on metal physics, I discovered that when tests calculated for double slip were recalculated as single slip[2] (provided sufficient information were given to do so), the response functions obtained were indistinguishable quantitatively from those obtained for tests in which the specimen axis did not rotate to the symmetry plane and only single slip had been assumed. This observation gave rise to an intensive study of the matter by Bell and Green,[3] which included not only a thorough search of the literature since 1925 for tests in which X-ray measurement had been made, but also a series of careful experiments on twenty-three, 99.47% purity aluminum single crystals, and two, 99.99% purity specimens of the same solid.

In Fig. 4.65, the X-ray measured rotations of 14 of the 23 single crystals of 99.47% purity aluminum are seen to have crossed the symmetry line apparently in continued single slip, as analysis of the measured values of parabola coefficients had indicated. The axial σ vs ε data of six of those experiments are shown in Fig. 4.66, from which the single slip resolution (heavy line) and double slip (light line) calculations of Fig. 4.67 were obtained. The measured angles λ and ϕ were introduced into the single slip relation of Eqs. (4.13), (4.14), and (4.15) above. Thus it definitely could be concluded that not only did the rotation proceed linearly across the symmetry line, but it proceeded with angles still consistent with the original single slip.

The results we[4] found from our search of the literature for completely specified tests, were perhaps even more revealing, as may be seen in Table 128. Only 9, out of 168 such experiments in the many types of single crystals listed, exhibited characteristics of double slip rotation; all of the 9 were from data obtained before 1930, the time when the double slip hypothesis was propounded. This is a fairly devastating 20th century example of the manner in which the mass acceptance of

[1] The only recourse is to consult the laboratory records of the original investigator. In some of the more recent literature, even this is not feasible, since computer calculations of resolved data are made directly from the electrical outputs of the test machine and, as I have found, the original investigator cannot supply the numerical values of axial stress and strain.

For a general study of response functions, such as that of the present treatise, the problem is not too severe since, for comparisons, one may recalculate in single slip all those test results for which measurements of axial stress and strain were recorded along with the resolved shear calculation, and one may omit all other data beyond the point where the axis reached the symmetry plane. For atomistic analysis, however, which has been the prevailing motive for performing such measurements during the past 45 years, conclusions drawn in the original papers and in papers which later cited them, may have to be reexamined.

[2] See Bell [1968, *1*].

[3] Bell and Green, *op. cit.*

[4] Bell and Green [1967, *3*].

BELL and GREEN (1967)

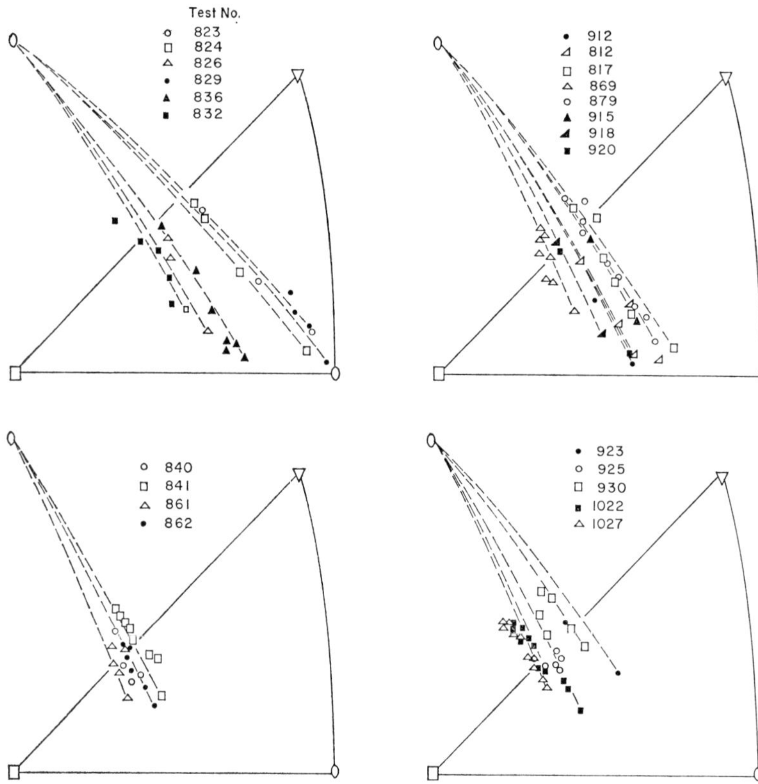

Fig. 4.65. BELL and GREEN's X-ray diffraction measurements during deformation for single crystals whose axes rotated across the symmetry line.

a plausible hypothesis can lead to four decades of potentially erroneous, unrecoverable published results.[1]

As a consequence of the above analysis and the similar, earlier studies of mine extending back to 1960, for which this research with GREEN had been the definitive sequel, the single crystal response for resolved shear in this treatise are given in single slip. For reasons given in Sect. 4.32 below, where single crystal and polycrystalline response are interrelated, it is preferable to call this behavior "macroscopic single slip."[2]

Two series of experiments dominated the early developments in the subject, and decades later still almost invariably are included in review papers. They are

[1] In some laboratories the axial data itself is fed into a computer which automatically switches from single to double slip at the appropriate angles, so, as was noted above, not even the unpublished σ vs ε data is available for a reexamination of the consequences.

[2] Since 1950, it has been realized that whereas deformation in stage I (referred to as the "easy glide" region) did occur in single slip, stage III was a region of complex, multiple slip. Hence, one wonders why, after that date, calculations continued to be made on the assumption that single slip was followed by double slip.

The term "macroscopic single slip" refers to statistical properties of the resolution of shear, for which empirical aggregate ratios relate the shear of single crystal and polycrystal.

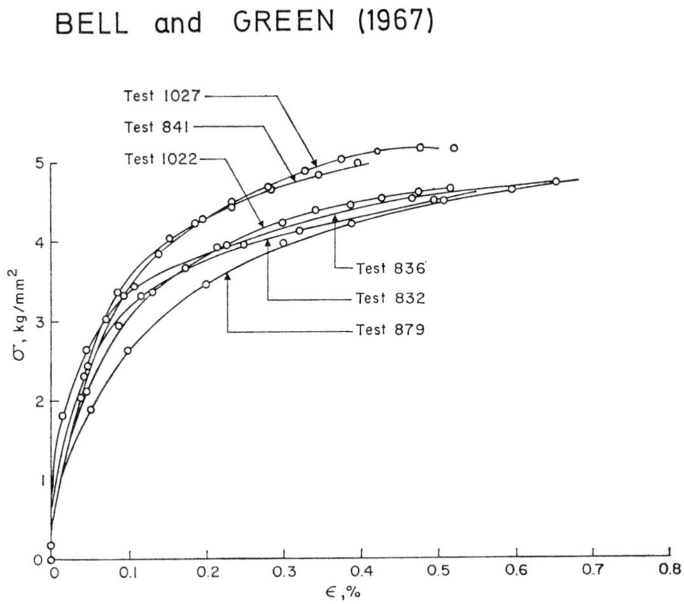

Fig. 4.66. Axial nominal tensile stress-strain data of BELL and GREEN. Circles designate X-ray diffraction measurements.

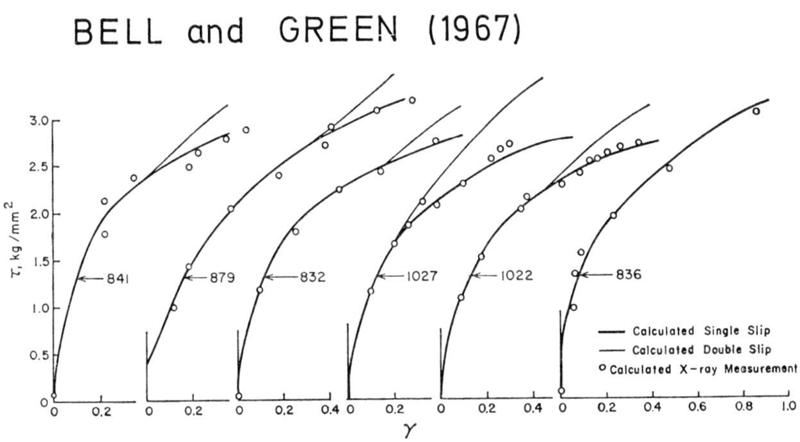

Fig. 4.67. Single slip and double slip resolved shear strain calculations, compared with τ vs γ from direct ϕ, λ measurements during deformation, showing agreement with single slip beyond the symmetry plane.

the experiments on aluminum, gold, copper, and iron single crystals by TAYLOR and ELAM[1] in 1925–1926, and the experiments on aluminum by BOAS and SCHMID[2] in 1931, over a series of temperatures ranging from $-185°$ C to 600° C. (The melting point of aluminum is 659° C.)

[1] TAYLOR and ELAM [1925, *1*], ELAM [1926, *1*], TAYLOR and ELAM [1926, *1*].
[2] BOAS and SCHMID [1931, *1*].

Table 128. Bell and Green (1967).

Year	Investigator	Material	Geometrically predicted double slip behaviour	Rotations continued linearly across symmetry line
1925	Taylor and Elam	Al	1	3
1926	Elam	Ag, Au, Cu	3	2
1927	Elam	α-brass	0	3
1927	Elam	Cu—5% Al	0	1
1927	Karnop and Sachs	Al	3	0
1928	Masima and Sachs	α-brass	0	7
1929	Goler and Sachs	Cu	1	0
		α-brass	1	6
1930	Sachs and Weerts	Ag, Au, Cu	0	8
		Ag—Au	0	7
1931	Sachs and Weerts	Cu_3Au	0	3
1933	Osswald	Cu, Ni	0	2
		Cu—Ni	0	2
	Maddin			
	Mathewson			
1949	Hibbard	α-brass	0	4
1951	Chen and Mathewson	Al	0	3
	Piercy			
	Cahn			
1955	Cottrell	α-brass	0	2
1959	Tanner and Maddin	Al	0	6
1960	Berner	Au	0	1
1962	Pfaff	Ni—Co	0	7
1963	Phillips	70% Ag—30% Zn	0	6
1963	Price and Kelly	Cu—1.8% Be	0	13
1964	Price and Kelly	Al alloys	0	57
1966	Bell and Green (present work)	Al	0	16
		Totals	9	159

It is a noteworthy index of experimental economy that G. I. Taylor, on the basis of three single crystal measurements in aluminum, four in iron, and one each in copper and gold, plus three or four polycrystalline tests in copper and aluminum, developed the kinematics of resolved shear deformation in single and double slip, proposed a physical theory of dislocations consistent with his calculated parabolic resolved shear response functions, and constructed the first plausible, if considerably restricted, theory for the plastic deformation of an aggregate based upon observations from a free single crystal component. That forty years of subsequent research have raised serious questions concerning the statistical origins of single slip and the applicability of double slip kinematics in the parabolic hardening region Taylor considered; that his dislocation theory proved too primitive to survive in the proposed form; and that the restrictive assumptions of his aggregate theory and its failure to include stress equilibrium in its formulation, precluded a complete correlation with observation; have in no way mitigated the fact that his work in a period of approximately a decade has given impetus to much of the subsequent experimental and theoretical research in crystal plasticity.

Three experiments[1] on single crystals in relatively low purity aluminum, tension test # 72 and compression tests # 61.17 and 59.9 for different initial

[1] Taylor and Elam [1925, *1*].

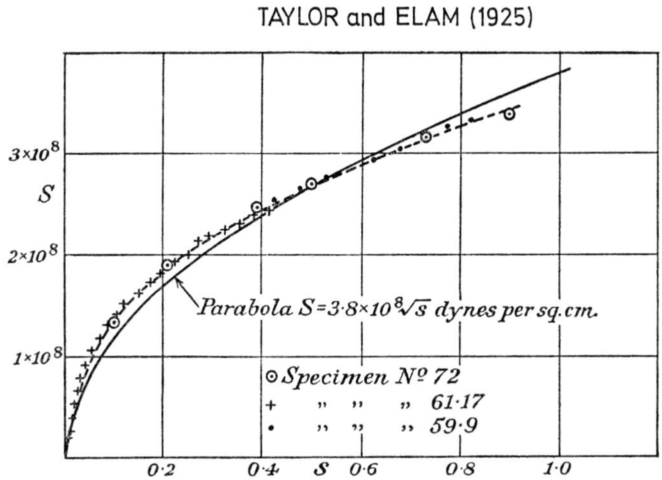

Fig. 4.68. TAYLOR and ELAM's comparison in 1925 of the resolved shear deformation in an aluminum single crystal from a tensile test (No. 72) and compression tests (Nos. 61.17 and 59.9).

orientations, demonstrated that the response function for the resolved shear deformation calculated as single slip[1] was independent of the initial orientation of the crystal planes and slip direction with respect to the specimen axis, and was independent of whether or not the axial test was in tension or in compression. For these three tests, the resolved deformation obtained by substituting measured σ_s, ε_s, ϕ_0, and λ_0 in Eqs. (4.13) to (4.15) is shown in Fig. 4.68. (In Figs. 4.68 and 4.69, TAYLOR used S to designate the resolved shear stress τ, and s for the resolved shear strain γ.)

TAYLOR included these measurements as part of his discussion of a number of different topics between 1925 and 1938.[2] In 1934, in part I of a paper entitled "The Mechanism of Plastic Deformation of Crystals,"[3] he proposed a theory of strain hardening involving dislocations, which was consistent with his observed parabolic resolved shear response function. For the purpose of obtaining numerical values for calculating a characteristic length, he fitted these data, and also the results of four tests in iron, one test in copper, and one test in gold, to their respective parabolas and obtained values of parabola coefficients of 3.8×10^8 dynes/cm^2 for the three aluminum tests; 2×10^9 dynes/cm^2 for the test in iron,[4] 8.8×10^8 dynes/cm^2 for copper; and 4.52×10^8 dynes/cm^2 for gold.

With the specification of these numerical values TAYLOR, unaware of it at the time, introduced the first quantitative precision into the study of the finite deformation of crystalline solids. One could draw a parallel with COULOMB's first obtaining a value for the shear modulus μ, 150 years earlier. Thirty years and several hundreds of tests after TAYLOR and ELAM had published their results,

[1] The region of parabolic hardening in single crystals is referred to as stage III deformation. Unlike the earlier stages of deformation, it is a region in which multiple slip systems are in operation. This matter has been considered in recent years in terms of "macroscopic single slip," and, as was noted above, will be discussed in Sect. 4.32 below.
See BELL [1964, *1*], [1965, *2*], BELL and GREEN [1967, *3*], BELL [1968, *1*].

[2] TAYLOR and ELAM [1925, *1*, *3*], TAYLOR and FARREN [1926, *2*]. TAYLOR [1927, *1*, *2*, *3*], [1934, *1*, *3*], [1938, *1*].

[3] TAYLOR [1934, *1*]. See also TAYLOR [1934, *3*, *5*].

[4] TAYLOR also provided an alternative empirical form for the iron tests shown in Fig. 4.69.

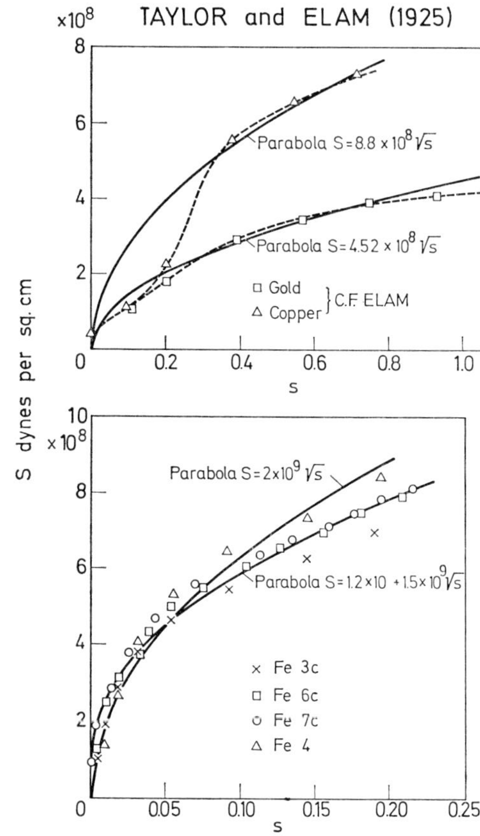

Fig. 4.69. TAYLOR and ELAM's resolved shear deformation from tensile tests on single crystals of gold and copper and four experiments on iron.

I showed through an exhaustive analysis of the literature on experiment and by an extensive series of experiments in my own laboratory that the original parabola coefficients of TAYLOR represented material parameters in close agreement with subsequent experiment, including a linear temperature dependence I found to be consistent with tests performed at many temperatures in all these solids. These correlations,[1] which in the historical development of the subject emphasize the importance of the tests of TAYLOR and ELAM, will be described in detail in Sect. 4.18.

TAYLOR and ELAM's experiments on gold, copper, and iron single crystals,[2] also repeatedly discussed between 1926 and 1934, and for which the numerical values of parabola coefficients were provided in 1934, are shown in Fig. 4.69.

The pioneering aspects of this research perhaps can be appreciated best by quoting from a paper by TAYLOR and ELAM in 1923, two years before they obtained the comparison[3] of resolved shear response functions for the loading of aluminum single crystals in tension and compression of Fig. 4.68.

Up to the present, however, the evidence of slipping is purely qualitative. It has not been shown that the deformation of a metallic crystal when the material is strained is such as

[1] BELL [1968, *1*].
[2] TAYLOR and ELAM [1926, *1*].
[3] TAYLOR and ELAM [1925, *1*].

could be produced by slipping, nor has the relationship between the crystal axes and the slip planes been determined.[1]

In 1931, BOAS and SCHMID[2] described the series of tensile experiments on aluminum single crystals at the eight ambient temperatures from −185° C to 600° C shown in Fig. 4.70a. From the observed initial orientations three calculated single slip resolved shear deformations of these same measurements are shown in Fig. 4.70b.

TAYLOR determined parabola coefficients for these single crystal tests on aluminum at the different temperatures as well as for five experiments of WALTER THIELE[3] in 1932 on rock salt crystals with ambient temperatures from 20 to 600° C. In each of these experiments of TAYLOR and ELAM and of BOAS and SCHMID, the angles of the crystal plane with respect to the specimen axis were determined by means of LAUE patterns obtained from X-ray photographs. The specimens were then extended or compressed axially. From the measured axial loads and strains, using the initial angles of orientation, resolved shear stress and resolved shear strain were computed for comparison. The data of THIELE for rock salt were given only in terms of the axial load and axial extension. The experiments of BOAS and SCHMID and of THIELE were performed in a furnace to obtain results at the ambient temperatures indicated.

As early as 1926 it had been established that the resolved shear response functions of hexagonal single crystals were not parabolic and thus not amenable to description in terms of TAYLOR's theory of work hardening.[4] In that year SCHMID's[5] experiments on zinc crystals, from which the resolved shear stress, resolved shear strain results are in Fig. 4.71, provided ample early demonstration that the response function was essentially a straight line to a resolved shear strain of 5.0 in this hexagonal solid. The angles shown in Fig. 4.71 are those between the basal plane and the specimen axis measured before the specimen was deformed. One notes that the slope of this linear response function does not change markedly with initial orientation.

Throughout this treatise, whether the subject has been the deformation properties of human tissue, metals, or the curious thermoelasticity of rubber, the emphasis has lain upon those aspects of behavior which are of significance in rational mechanics. Macroscopic continuum mechanics has its own fundamental validity. In concentrating upon constitutive relations meaningful to continuum mechanics, I have given only minor attention to the separate but related microscopic mechanics, the devising of atomistic models to interpret in an alternate manner the details which have been observed. It became obvious by the end of the 19th century, and is even more dramatically evident in the second half of the 20th century, that the construction of constitutive relations from atomistic beginnings is an endless task based upon a foundation requiring the proliferation of assumptions and many proposed mechanisms. Atomistic studies, both theoretical and experimental, have a separate validity and fascination. Progress in metals technology has been indeed closely related to atomistic analysis, as structural design technology has benefited from the growth of rational mechanics. Since the classic treatise of BOAS and SCHMID in 1935, many publications have appeared which trace the development of single crystal experimentation and the dislocation models used to interpret it. The reader is referred to such reviews for

---

[1] TAYLOR and ELAM [1923, *1*], p. 64.
[2] BOAS and SCHMID [1931, *1*].
[3] THIELE [1932, *1*].
[4] TAYLOR [1934, *1*].
[5] SCHMID [1927, *1*].

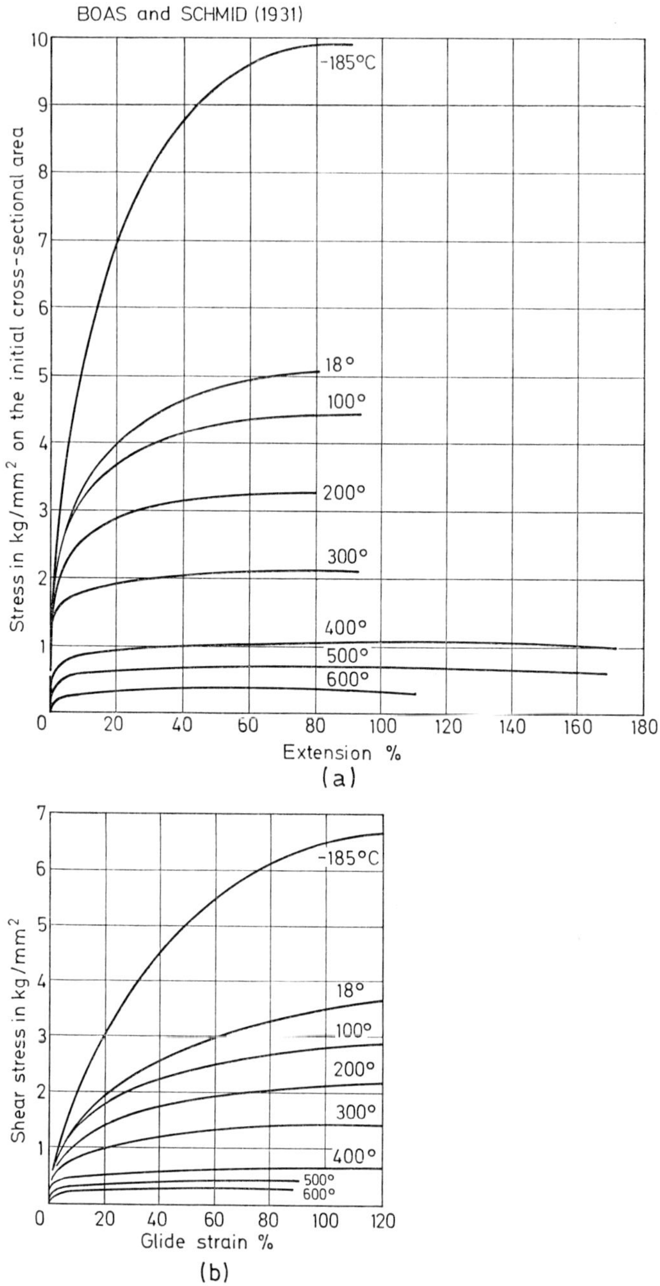

Fig. 4.70a and b. The tensile stress-strain measurements (a) of Schmid and Boas in 1931 at ambient temperatures from $-185°$ C to near the melting point, from which the resolved shear stress, resolved shear strain results (b) were calculated by single slip.

discussion and references since the emphasis here is upon the macroscopic behavior observed in such experiments, whatever were the objectives of the individual experimentists.

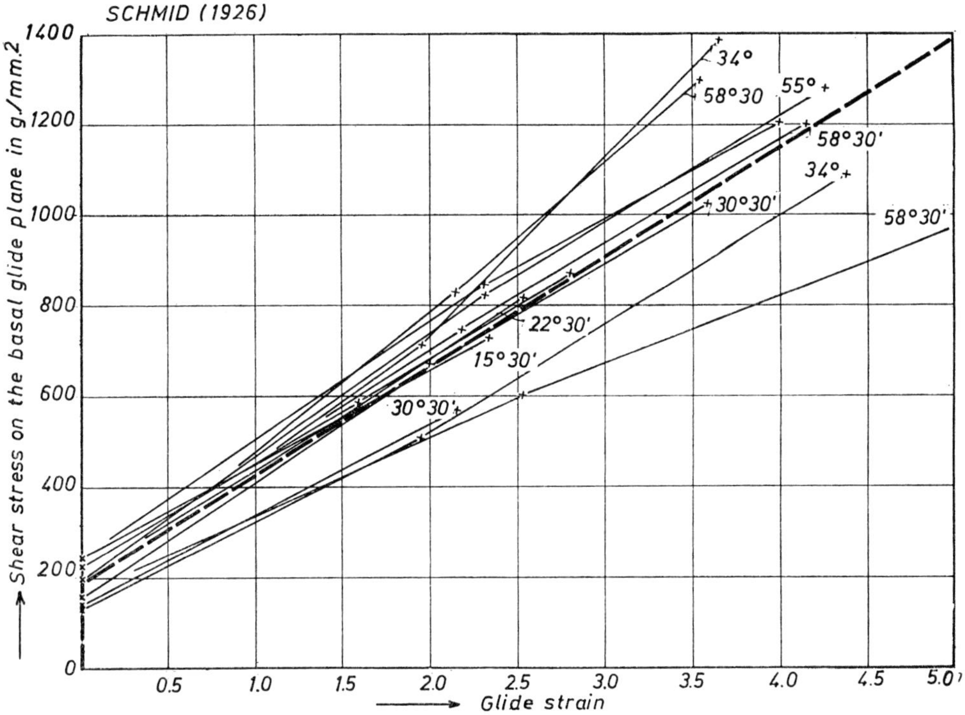

Fig. 4.71. SCHMID's resolved shear stress, resolved shear strain results in zinc single crystals.

In the beginning of the 1950's, questions were raised as to whether or not the response functions for cubic crystals were parabolic. TAYLOR's assumptions in his dislocation theory[1] also were being challenged. Of most concern was the fact that he had assumed a uniform distribution of positive and negative dislocations and that he had made no provision for the production of dislocations during deformation. A third difficulty, which was not raised at the time, is that there was no basis in either theory or experiment for the making of decisions required in the process of summation in TAYLOR's analysis. NEVILLE MOTT[2] in 1952 proposed a theory which, while still providing a parabolic response function, eliminated these questions implicit in TAYLOR's hypothesis. MOTT's theory was based upon a pile-up of dislocations at an impurity barrier which had been generated by a FRANK-READ source.

By the 1950's techniques of crystal growing had improved to the point where it became possible to produce single crystals of high purity. The calculation of resolved shear response functions in these single crystals of high purity revealed that the large deformation of cubic solids was not represented solely by a parabolic stress-strain relation.

EDWARD NEVILLE DA COSTA ANDRADE and C. HENDERSON[3] in 1951, for higher purity gold, silver, and nickel crystals, and KURT LÜCKE and HANSHEINZ LANGE[4] in 1952, with a comparison of 99.5% purity with 99.99% purity aluminum single

[1] TAYLOR [1934, *1*].
[2] MOTT [1952, *1*].
[3] ANDRADE and HENDERSON [1951, *1*].
[4] LÜCKE and LANGE [1952, *1*].

crystals, established in standard tensile tests, accompanied by the measurement, by X-rays, of the initial orientation of the crystal axes, that the region in which the response function was parabolic could be preceded by a large region in which it was not.[1]

ANDRADE and HENDERSON's measurements in high purity gold and silver, Figs. 4.72 and 4.73, show this initial region of non-parabolic deformation. In Figs. 4.72a and 4.73a are shown in expanded scale the initial region of deformation, and in Figs. 4.72b and 4.73b, the deformation to 0.8 resolved shear strain.

Experimental study of this high purity phenomenon led to the realization that there were three stages of finite deformation (designated as I, II, III) which followed the initial elastic behavior. The initial transition occurred at the subsequently much studied single crystal elastic limit or critically resolved shear stress. Stage I and stage II depended for their existence or magnitude not only upon purity but, as LÜCKE and LANGE[2] showed in their experimental study in 1952, upon the initial orientation. Even for high purity single crystals, no such behavior was found for initial orientation near the corners of the stereographic triangle. A schematic diagram of a resolved shear response function for central orientations is shown in Fig. 4.74.

Numerous experimentists showed[3] that the length of stage I deformation was dependent upon the velocity of the resolved shear strain γ, i.e., the shear strain rate $\dot{\gamma}$. SUZUKI, IKEDA, and TAKEUCHI[4] in a careful series of tensile tests using an optical technique to determine axial strain, showed experimentally in high purity copper single crystals that the slope θ_{II} of stage II deformation was orientation dependent, increasing in tension tests as the symmetry plane was approached. Generally recognizing stage I as a region of single slip, a large number of competing dislocation models were developed to account for the linear slope of stage II deformation. The experimental observation of linearity, of course, enormously simplified such model analyses.[5] Stage III designated the region having a parabolic resolved shear response function. When stage I and stage II deformation were present the origin of the parabola no longer was located at zero strain. Atomistic studies identified stage III deformation with a complex slip behavior. It is only recently that attempts have been made to consider stage III deformation at the atomistic level.

---

[1] The paper of LÜCKE and LANGE, incidentally, was particularly notable for its having been perhaps the last large-scale experimental study which provided both the measured axial results and the resolved shear calculations. The failure to continue this practice was responsible for the loss of experimental results in a fair fraction of the subsequent literature when the later experimental study and historical analysis of BELL [1964, *1*] and BELL and GREEN [1967, *3*] provided evidence which seriously questioned the arbitrarily assumed double slip behavior at symmetry planes. A continuation of the practice pursued by LÜCKE and LANGE and by many before them, of giving both the measured and computed data, would have meant that for the large deformation of single crystals, the published experimental results based upon the double slip calculations now could be recalculated as single slip. It also should be noted that the data of ANDRADE and HENDERSON in 1951 were calculated for single slip for the entire deformation, a matter which raised some subsequent criticism. In 1962 when I realized that the parabola coefficients for their data, which ignored double slip, were consistent quantitatively with a generalization I had found, based upon measurements known to be in single slip, the manner in which ANDRADE and HENDERSON chose to present their data appeared in an interesting and new light, since it provided me with additional evidence for my increasing suspicion that resolution in double slip was questionable.

[2] LÜCKE and LANGE [1952, *1*].

[3] See, for example, THORNTON, MITCHELL, and HIRSCH [1962, *1*], who made tensile tests in brass single crystals.

[4] SUZUKI, IKEDA, and TAKEUCHI [1956, *1*].

[5] I am omitting a number of similar efforts to assume simplifying linearity in stage III deformation, as not in accord with experimental fact.

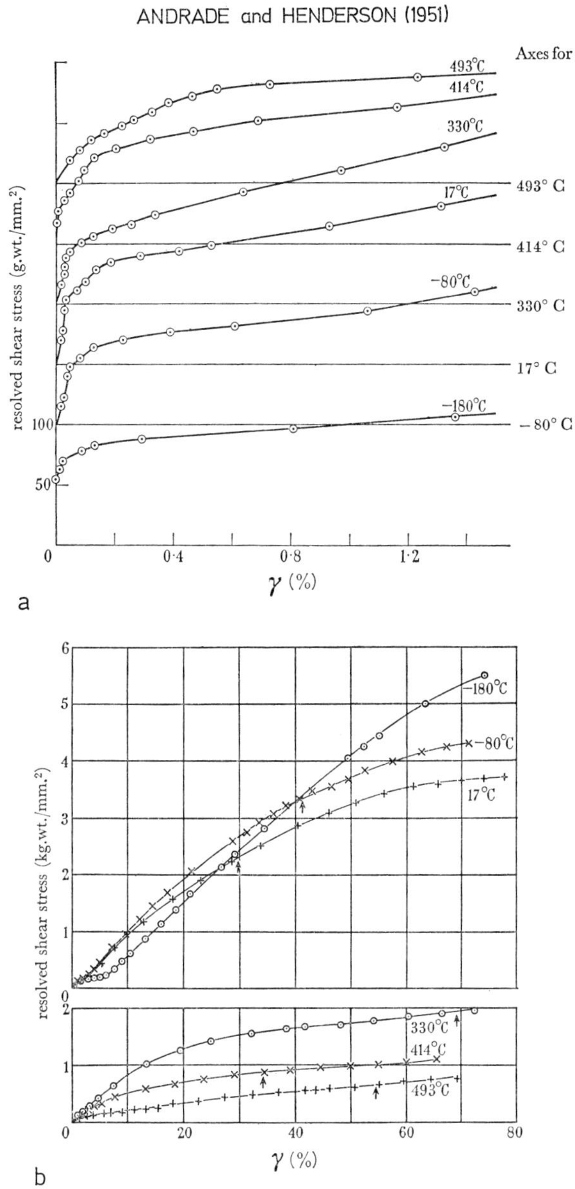

Fig. 4.72a and b. Resolved shear stress vs shear strain of ANDRADE and HENDERSON in 1951 for gold single crystals at the indicated ambient temperatures, for the initial small strain region (a) and for the large deformation region (b).

In the process of examining 430 high and low purity cubic single crystal resolved stress-strain functions of over 50 experimentists in 12 cubic solids and in the over 40 years of the metal physics and metallurgical literature, I observed[1] in 1964 that the effect of increasing the purity of the solid had been to shift the

[1] BELL [1964, *1*]. See also BELL [1965, *2*] and BELL [1968, *1*], p. 95.

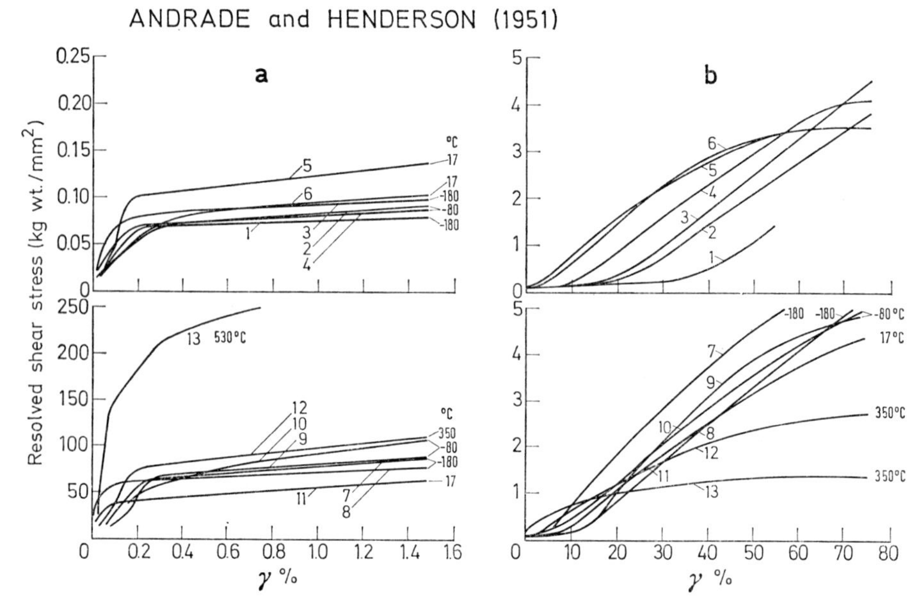

Fig. 4.73a and b. Resolved shear stress vs shear strain of ANDRADE and HENDERSON in 1951 for silver single crystals at the indicated ambient temperatures, for the initial small strain region (a) and for the large deformation region (b).

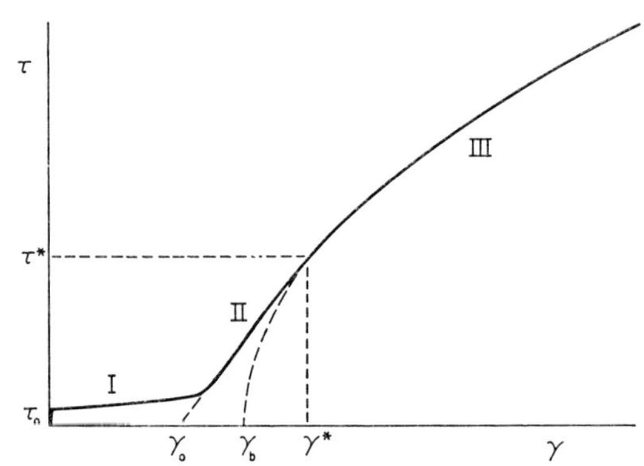

Fig. 4.74. A schematic diagram of the resolved deformation of a high purity single crystal, showing the three stages of deformation.

origin of the parabolic portion of the response function from a strain of approximately zero to a finite strain, γ_b. I further discovered at that time that the linear stage II and the parabolic stage III portions of the response function were predictably related. In the early 1960's the stage II—stage III transition stress, τ^*, and its temperature dependence at lower temperatures had assumed great importance in the dislocation models of the atomistic studies.[1] I found that this transition stress, τ^*, and the slope, θ_{II}, of the linear stage II deformation were a function of the parabola coefficient of the stage III response function. This interrelation of the response functions of stage II and stage III was independent

[1] See, for example ,the analysis of SEEGER [1958, *1*], p. 156.

of variations in purity, the loading rate, or the initial crystallographic orientation. It also was independent of the magnitude of the stage I deformation and was independent of γ_b the location of the origin of the parabola on the strain abscissa.

Of most significance in these observations was the discovery that the numerical value of the parabola coefficient of the response function for stage III deformation was independent of any of these factors, including the complete absence of stage I and stage II deformation. That the parabola coefficients of stage III deformation are independent of the presence or absence of stage I and stage II deformation in the single crystal is of prime importance for the study of the aggregate since for the polycrystalline solid, stage I and stage II deformation never are observed. Thus, from the continuum mechanics point of view, the discovery of the preliminary linear regions in the response functions of the high purity face centered cubic metals, had been merely an addendum to TAYLOR's earlier discovery of parabolic response, since the region of the parabolic response itself was in no way altered by the presence of this preliminary deformation.

For the linear stage II deformation, we have

$$\tau = \theta_{\mathrm{II}}(\gamma - \gamma_0), \tag{4.16}$$

while for the parabolic stage III deformation

$$\tau = \beta(\gamma - \gamma_b)^{\frac{1}{2}}, \tag{4.17}$$

where γ_0 is the intercept of stage II on the strain abscissa; θ_{II} the stage II slope; γ_b the intercept of stage III on the strain abscissa; and β the stage III parabola coefficient.

Equating stresses and first derivatives at the transition stress, τ^*, after noting from experiment that stage II proceeded smoothly to stage III, in 1964 I gave the following equations:

$$\theta(\gamma^* - \gamma_0) = \beta(\gamma^* - \gamma_b)^{\frac{1}{2}}, \tag{4.18}$$

$$\theta_{\mathrm{II}} = \tfrac{1}{2}\beta(\gamma^* - \gamma_b)^{\frac{1}{2}}, \tag{4.19}$$

from which follow

$$\beta = \sqrt{2\tau^* \theta_{\mathrm{II}}}, \tag{4.20}$$

$$\gamma_b = \frac{\gamma^* + \gamma_0}{2}. \tag{4.21}$$

As indicative of the experimental foundation for this predicted close interdependence of stage II and stage III, 49 studies of finite deformation at room temperature on copper single crystals by BLEWITT, COLTMAN, and REDMAN[1] in 1955, DIEHL[2] in 1956, SUZUKI, IKEDA, and TAKEUCHI[3] in 1956, BERNER[4] in 1957, and SEEGER, DIEHL, MADER, and REBSTOCK[5] in 1957 (solid lines) were compared[6] in Fig. 4.75 with prediction (circles) from my Eqs. (4.20) and (4.21).

The following year, 1965, I found[7] that this correlation for copper, extended to 318 single crystal experiments of 40 experimentists in aluminum, nickel, gold, silver, and lead. The parabolic response function for the stage III deformation was easily exhibited by plotting τ^2 vs γ, where, of course, straight lines would be anti-

---

[1] BLEWITT, COLTMAN, and REDMAN [1955, *1*].
[2] DIEHL [1956, *1*].
[3] SUZUKI, IKEDA, and TAKEUCHI [1956, *1*].
[4] BERNER [1957, *1*].
[5] SEEGER, DIEHL, MADER, and REBSTOCK [1957, *1*].
[6] BELL [1964, *1*].
[7] BELL [1965, *2*].

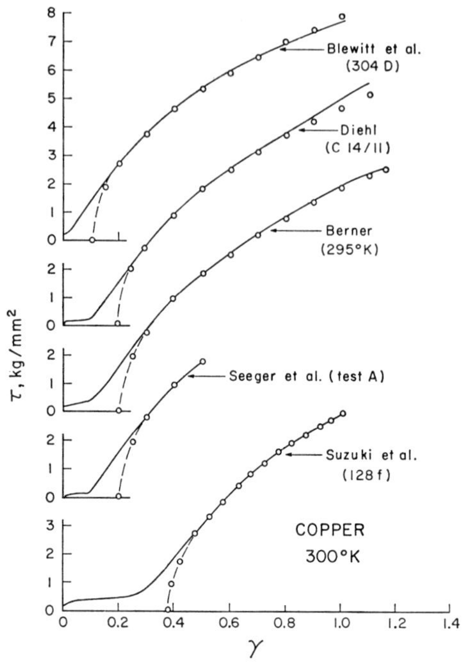

Fig. 4.75. Some of Bell's (1964) comparisons of the resolved shear high purity copper experiments of the designated experimentists (solid lines), with calculated stage III behavior (circles) from Eqs. (4.20) and (4.21). The parabola coefficient was obtained from the polycrystalline experiments by means of the aggregate ratio, $\overline{m}$.

cipated. In Fig. 4.76 are shown such plots which I provided in that study in 1964 for the experiments on copper single crystals of Fig. 4.75.

A few of the many additional such illustrations for high and low purity crystals for experiments which like those of Fig. 4.75 provided for the observed stage III response a close correlation with my prediction from θ_{II} and τ^* of stage II, were (Fig. 4.77) the high purity experiments in 1957 of T. S. Noggle and J. S. Koehler[1] in aluminum at low temperature, and the similar data of A. Kelly[2] in 1956 and Berner[3] in 1960. Plots of τ^2 vs γ for corner orientations of high purity aluminum at room temperature were given for the experiments of Priscilla W. Kingman, Robert E. Green, and Robert Barrett Pond[4] (see Fig. 4.78). Fig. 4.79 shows high and low purity copper single crystal τ^2 vs γ plots I obtained for copper single crystals, and Fig. 4.80, further such plots for the data of Lücke and Lange[5] from their study in 1952, and for the room temperature tests of Berner and Noggle and Koehler in high purity aluminum for central orientation at 153° K and at room temperature. All of these data were calculated for single slip by means of Eqs. (4.13), (4.14), and (4.15).

These few examples which I described in 1965 were chosen out of a total of 455 individual experiments on the resolved deformation of cubic single crystals performed since 1923 on aluminum, silver, gold, nickel, lead, copper, iron, tanta-

[1] Noggle and Koehler [1957, *1*].
[2] Kelly [1956, *1*].
[3] Berner [1960, *1*].
[4] Kingman, Green, and Pond [1963, *1*].
[5] Lücke and Lange [1952, *1*].

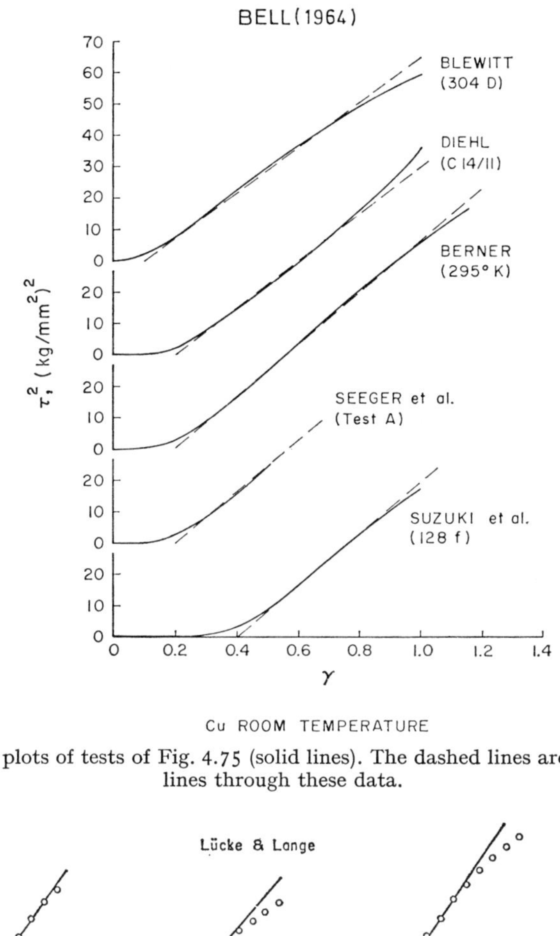

Fig. 4.76. τ^2 vs γ plots of tests of Fig. 4.75 (solid lines). The dashed lines are the best straight lines through these data.

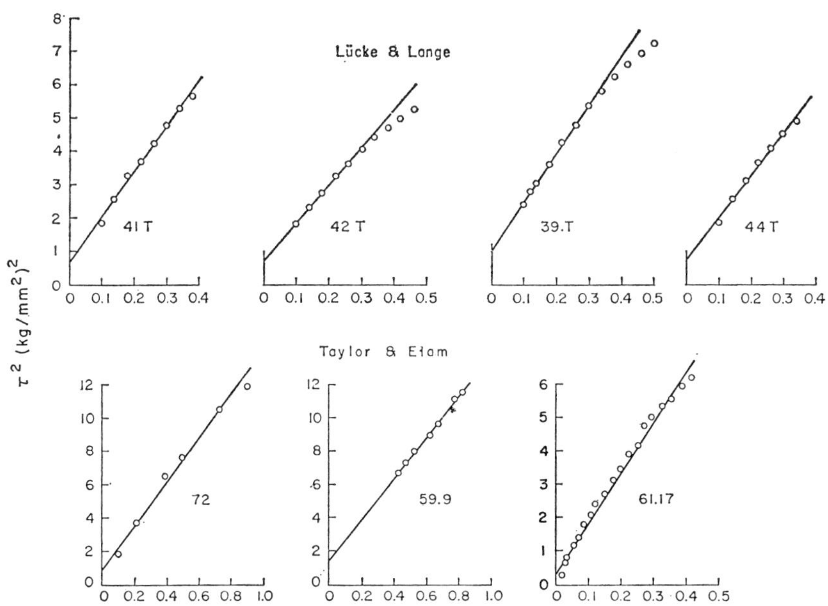

Fig. 4.77. τ^2 vs γ plots for high purity aluminum of central orientation at low temperatures (circles), compared with Bell's stage III prediction (solid lines) of Eqs. (4.20) and (4.21).

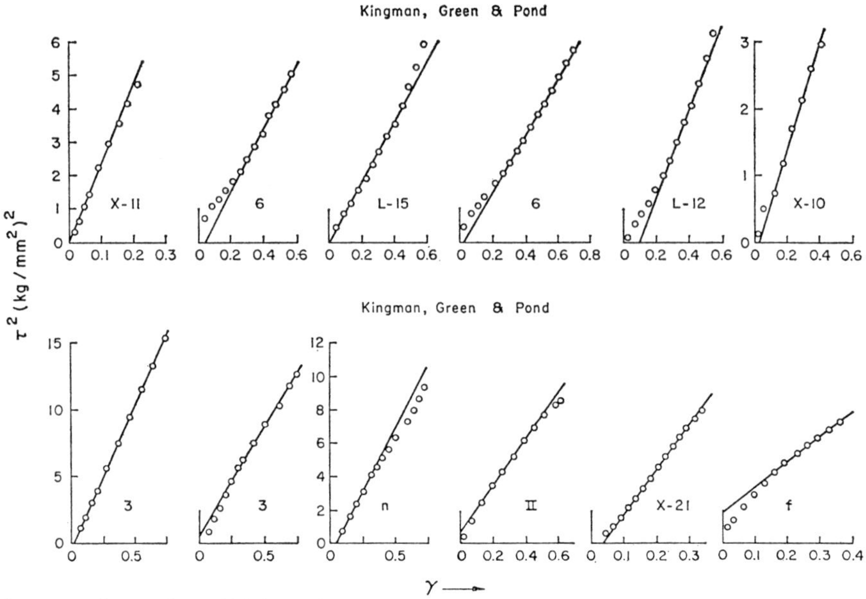

Fig. 4.78. τ^2 vs γ plots of high purity aluminum of corner orientation (circles), compared with calculation from BELL's parabolic generalization, Eqs. (4.20) and (4.21) (solid lines). These experiments were performed in compression.

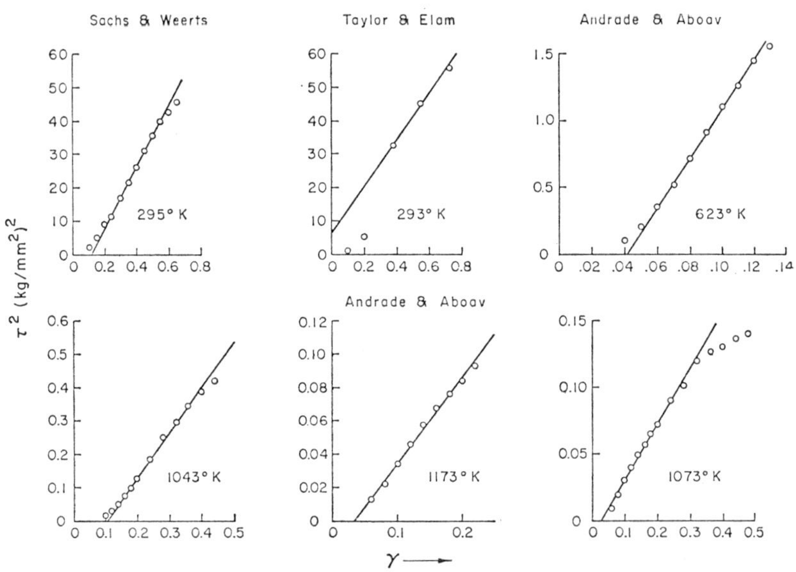

Fig. 4.79. τ^2 vs γ plots for copper of low and high purity at various temperatures (circles), compared with BELL's parabolic generalization, Eqs. (4.20) and (4.21) (solid lines).

lum, molybdenum, sodium chloride, and various percentage combinations of silver-gold and nickel-cobalt, which I summarized for all experiments, both diagrammatically in τ^2 vs γ plots and in extensive tables in 1968.[1]

[1] BELL [1968, *1*].

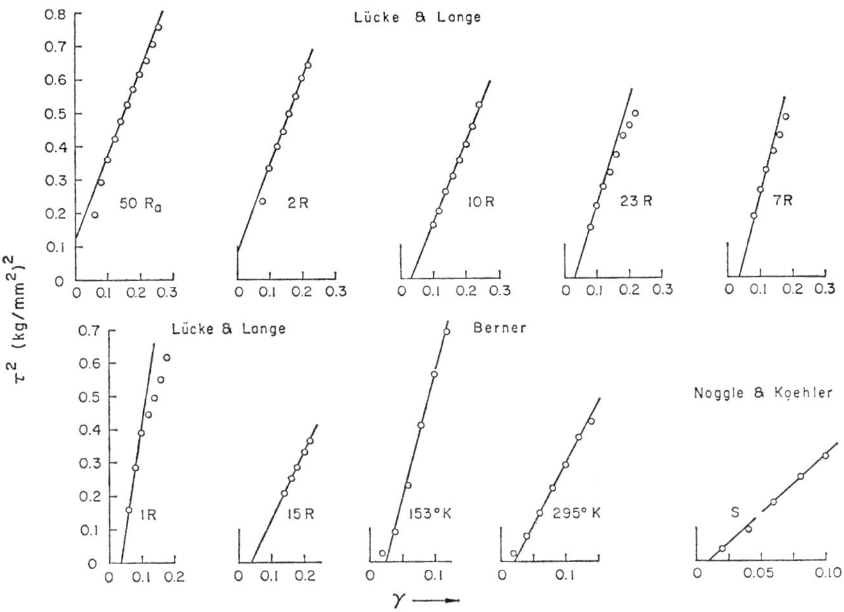

Fig. 4.80. τ^2 vs γ plots for room temperature experiments in high purity aluminum of central orientation (circles), compared with BELL's parabolic generalization, Eqs. (4.20) and (4.21) (solid lines). One experiment of BERNER was at 153° K.

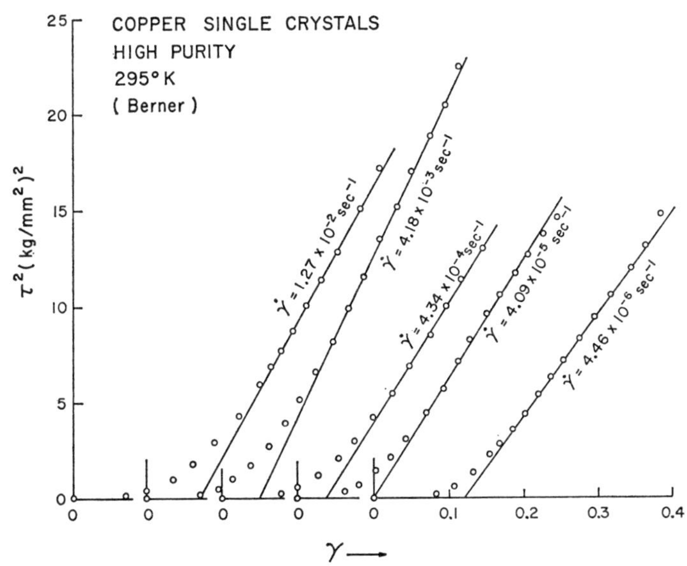

Fig. 4.81. τ^2 vs γ plots for stage III high purity copper single crystals at widely different strain rates. These comparisons of BERNER's data of 1960 (circles) and prediction from Eqs. (4.20) and (4.21) were provided by BELL in 1965. Note the lack of dependence upon strain rate for over 4 orders of magnitude.

One interesting experimental series among the high purity copper specimens were five measurements by BERNER[1] in 1960 of resolved shear deformation at

[1] BERNER [1960, *1*].

shear strain rates, varying from $\dot{\gamma} = 4.46 \times 10^{-6}$ sec^{-1} to $\dot{\gamma} = 1.27 \times 10^{-2}$ sec^{-1} over a range of strain rates of 10000. The τ^2 vs γ plots of the parabolic response in stage III, shown in Fig. 4.81 from my paper[1] in 1965, revealed that although a slight trend toward lower slopes with decreasing strain rate occurred as expected according to viscoplasticity, variations in the parabola coefficients among individual experiments exceeded the difference between the highest and lowest strain rates. This, of course, was the same strain rate range DEUTLER[2] had considered in his polycrystalline experiments 28 years earlier.

4.18. Quantized parabola coefficients and second order transitions in the resolved finite strain of single crystals: Bell (1960–1968).

In 1960, after five years of measuring the profiles of waves of finite amplitude in polycrystals,[3] I found that the response functions which governed the one-dimensional dynamic plasticity of annealed polycrystalline metals were parabolic, with coefficients which were linear functions of the ambient temperature. The results of this first successful measurement of the detail of plastic waves, made possible by my discovery of a new technique utilizing fine diffraction gratings, naturally led to a comparative study of the quasi-static response of these same solids. The uniaxial response function for the quasi-static deformation of annealed polycrystals, and for the stage III resolved shear deformation of cubic single crystals also considered in this systematic study, had the identical form as that observed in the dynamic measurements. The similarity of response functions during loading, for the finite strain of annealed crystalline solids in these three different situations, led to my undertaking a large study to compare parabola coefficients at known fractional melting points, T/T_m.

By the time I completed this study in 1968, I had described[4] over 2000 uniaxial tension or compression experiments on 27 crystalline solids. The large majority of the measurements, including all those of wave propagation, were performed in my laboratory. The others, nearly 700 quasi-static tests on polycrystals and single crystals, I obtained from published results from over 45 years of the literature of metal physics; my purpose was to ascertain that the quantitative order I had found, was indeed a generalization applicable to the sum total of experimental knowledge of the finite strain response of fully annealed crystalline solids. This was by far the largest quantitative analysis of plastic response ever undertaken. The study led to the discovery of unity and quantitative order for the parabolic response functions of fully annealed crystalline solids. At the present point in this treatise, results are introduced only for the resolved shear deformation of cubic single crystals, leaving for Sects. 4.20 and 4.21 the related discussion of the polycrystalline solids.

Each experiment for the single crystal tests consisted in the measurement of nominal axial stress in tension or in compression, nominal axial strain, and the two initial angles λ_0 and ϕ_0. Almost all of the many experimentists used hard commercial machines with standard commercial extensometers; only a few performed dead weight tests on apparatus constructed in the laboratory, and measured strain by optical means. For the details of each test of this large study,

[1] BELL [1965, *2*].
[2] DEUTLER [1932, *1*]. See Sect. 4.24 below.
[3] See Sect. 4.28 below.
[4] BELL [1968, *1*].

I refer the reader to Chap. IV and Appendix I of Vol. 14 of the Springer Tracts in Natural Philosophy.[1]

The linear temperature dependence of the parabola coefficients

$$\beta = \beta(0)(1 - T/T_m)$$

was the fact first observed. This form of the temperature dependence had been predicted for single crystals from finite amplitude wave studies in polycrystals at many temperatures, to within a few degrees of the melting point[2] (see Sect. 4.28 below). With the temperature dependence known, it became possible to determine values at the zero point from test results at any temperature so that one could compare the results from all tests, referring to a common fractional melting point, zero. This was the same procedure which I had used in the study of shear moduli of isotropic bodies at the zero point, described in Sect. 3.44. From the experiments performed in my laboratory, and from the comparison of them with the many tests published in the literature, I established that $\beta(0)$ for the resolved deformation had the form

$$\beta(0) = \left(\tfrac{2}{3}\right)^{r/2} \mu(0)\, B_0/\bar{m}^{3/2}, \tag{4.22}$$

where $r = 1, 2, 3, 4, \ldots$ is an integral mode index defining a distribution for a discrete, quantized set of parabola coefficients, paralleling for the response at large strain, the quantized distribution of shear moduli found at small strain [see Sect. 3.44]; $\mu(0)$ the linear shear modulus for isotropic solids at the zero point; B_0 a dimensionless universal constant, $B_0 = 0.0280$; and $\bar{m} = 3.06$, a dimensionless constant obtained from the ratio relating the resolved shear stress of the single crystal to the axial stress, in tension or compression, of the polycrystalline aggregate. [The original equation was written for polycrystals and thus, as will be seen below, the constant $\bar{m}$ appears in Eq. (4.22). This follows from the sequence of discovery, since I first obtained the parabola coefficients from an analysis of results from wave propagation tests on polycrystals.]

The discrete distribution of parabola coefficients among deformation modes was designated by the integral mode index r. The initial value of r in a given test for a given solid depended upon the ambient temperature of the test, the purity of the specimen, and the prior thermal and mechanical history. The initial deformation mode was found to govern the response to failure for high purity crystals at many temperatures, for most crystals at very low ambient temperatures, and, at very high strain rates, for crystals of all temperatures and purities.

Several hundred experiments might be cited to illustrate the fact that there is quantitative order in the finite strain response of cubic single crystals in stage III deformation. This order was found to exist in experimental results from the earliest measurements to those of today.

Two examples, which include the first measurements of stage III deformation ever made, suffice to demonstrate the point. In 1925 and 1926, TAYLOR and ELAM[3] performed single crystal tests on gold and aluminum. In 1934, while developing his theory of dislocations, TAYLOR[4] gave numerical values of parabola coefficients for those tests. Reexamining those results in the light of my generalization developed 35 years after those tests were performed, we note that for the two solids the

[1] *Ibid.*

[2] In historical sequence, I discovered the numerical values of parabola coefficients and second order transition phenomena from my experiments on finite amplitude waves. The correlations for the quasi-static single crystals and the polycrystal followed more or less in that order.

[3] TAYLOR and ELAM [1925, *1*], ELAM [1926, *1*].

[4] TAYLOR [1934, *1*].

shear moduli at the zero point are 3070 kg/mm² for gold, and 3110 kg/mm² for aluminum (see Sect. 3.44). The melting point of gold is $T_m = 1334°$ K, and that for aluminum is 932° K. I found that the mode index for the experiments on both solids was $r = 5$.

The parabola coefficients provided by TAYLOR from measurements at room temperature were 4.52×10^8 dynes/cm² for gold, and 3.8×10^8 dynes/cm² for aluminum. From the linear temperature dependence $\beta = \beta(0)\,(1 - T/T_m)$ and Eq. (4.22), there is sufficient information, in terms of the stated values T_m and $\mu(0)$, to predict the parabola coefficient of aluminum from a knowledge of the parabola coefficient of gold.

Introducing the measured value for gold, one calculates for aluminum a parabola coefficient of 3.94×10^8 dynes/cm², which may be compared with TAYLOR's experimental value for aluminum of 3.8×10^8 dynes/cm². This remarkable correlation, in terms of a generalization discovered in 1960, between TAYLOR and ELAM's parabola coefficients for aluminum and for gold at room temperature from tests performed in the mid-1920's, was first described in 1963 in a paper on the linear temperature dependence of response functions for finite strain.[1]

A second example, which includes TAYLOR and ELAM's tension test on gold, of 1926, are the four tests of different experimentists on that solid, shown in Fig. 4.82. The experiments were performed between 1926 and 1960. These tension tests from which the experimentists calculated the resolved deformation, were made at room temperature. The deformation mode index for all of these tests is $r = 5$, a value I had found to be the predominant stage III deformation mode for gold at room temperature. The response for gold at room temperature, predicted from the linear temperature dependence and Eq. (4.22), is the solid line in Fig. 4.82.

With the notation β_{r0}(II) for parabola coefficients determined from stage II measurements of θ_{II} and τ^* in Eq. (4.20) above, and with β_{r0}(III) for those obtained directly from the stage III slopes of τ^2 vs γ plots of the resolved deformation, the results of the examination of 375 tests of the many experimentists since 1925 are given in Table 129 for the observed initial parabolas, and in Table 130 for those

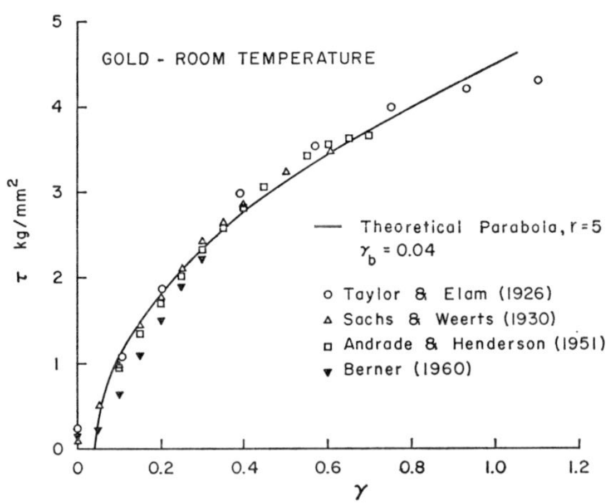

Fig. 4.82. A comparison of resolved gold single crystal data obtained during the past 40 years. The solid line is from Eq. (4.22) for the integral mode index, $r = 5$.

[1] BELL [1963, *1*].

Table 129. BELL (1968). *Initial parabolas only.*

Mode index r	No. of measurements	Theoretical $\beta_{r0}/\mu(0)$	Average experiments $\beta_{r0}(\text{II})/\mu(0)$	Average experiments $\beta_{r0}(\text{III})/\mu(0)$	Average all experiments $\beta_{r0}/\mu(0)$
1	10	0.00427	0.00426	0.00411	0.00424
2	25	0.00349	0.00353	0.00337	0.00346
3	43	0.00285	0.00287	0.00278	0.00283
4	112	0.00232	0.00228	0.00226	0.00227
5	152	0.00190	0.00193	0.00190	0.00191
6	72	0.00155	0.00158	0.00155	0.00156
7	41	0.00127	0.00129	0.00130	0.00130
8	17	0.00103	0.00108	0.00106	0.00107
9	8	0.00084	0.00082	0.00087	0.00085
10	3	0.00069	0.00074	0.00074	0.00074
11	2	0.00056	—	0.00057	0.00057

Table 130. BELL (1968). *Initial and transition parabolas.*

Mode index r	No. of measurements	Theoretical $\beta_{r0}/\mu(0)$	Average experiments $\beta_{r0}(\text{II})/\mu(0)$	Average experiments $\beta_{r0}(\text{III})/\mu(0)$	Average all experiments $\beta_{r0}/\mu(0)$
1	10	0.00427	0.00426	0.00411	0.00424
2	25	0.00349	0.00353	0.00337	0.00346
3	44	0.00285	0.00287	0.00278	0.00283
4	120	0.00232	0.00228	0.00227	0.00228
5	155	0.00190	0.00193	0.00190	0.00191
6	101	0.00155	0.00158	0.00155	0.00155
7	77	0.00127	0.00129	0.00129	0.00129
8	51	0.00103	0.00108	0.00103	0.00104
9	33	0.00084	0.00082	0.00086	0.00086
10	30	0.00069	0.00074	0.00071	0.00071
11	12	0.00056	—	0.00057	0.00057
12	2	0.00046	—	0.00045	0.00045

observed for the new deformation mode after a second order transition had occurred. Like BRIDGMAN, I defined a second order transition as a discontinuity in the derivative of the response function.

The experimental averages in these tables were compared with prediction from Eq. (4.22). Aside from BRIDGMAN's[1] study which was confined to hydrostatic pressure, the only previous attempt to present a quantitative comparison of extant experimental results for response functions at finite strain was that of PONCELET[2] in 1841. It is interesting to note how much the magnitude of the task had increased in one and a quarter centuries.

The data summarized in Tables 129 and 130, which comprise a large cross-section of the published literature on the stage III deformation of cubic crystals, did not include 55 uniaxial experiments on aluminum single crystals performed in my laboratory, which also were described in detail[3] in 1968. In addition to providing careful dead weight tests which gave precise numerical values for the quantized distribution of parabola coefficients, these experiments were performed to study another empirical aspect of response functions for finite strain, namely, whether

[1] BRIDGMAN [1923 ... 1964].
[2] PONCELET [1841, *1*].
[3] BELL [1968, *1*].

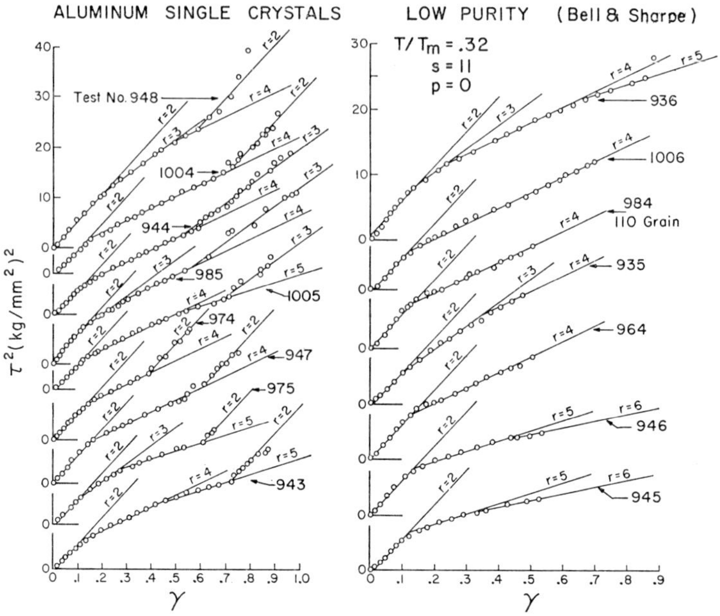

Fig. 4.83. τ^2 vs γ plots of low purity aluminum single crystals resolved in macroscopic single slip. The parabola coefficients (solid lines) are given by Eq. (4.22) for the indicated mode indices. Note second order transitions at $\gamma_N = 0.127$ where $N = 13$.

second order transitions from one discrete deformation mode to another occurred at a fixed sequence of finite resolved shear strains.

Many of the uniaxial tests in this series are shown in Fig. 4.83. The measurements were made with a dead weight loading apparatus on which a constant stream of water produced a stress rate of 0.007 kg/mm²/min. The 99.16% purity aluminum specimens were $^3/_8$ in. in diameter, 4 in. long. Plastic strain was determined by means of a calibrated clip gage, 2 in. long, located on the center of the specimen. The single crystals were grown by the strain annealing method. Initial crystallographic angles provided the specimen axis orientation[1] shown in Fig. 4.84. The resolved shear stress and resolved shear strain were calculated according to Eqs. (4.13), (4.14), and (4.15). The τ^2 vs γ plots of the experimental results (circles) were compared with predicted slopes (solid lines) for the designated mode indices. A decade ago, when I examined the first such plots of τ^2 vs γ, I found that not only did the results lie upon straight lines, in agreement with a parabolic response function, but also, second order transitions from one fixed slope to another could occur, depending upon the purity, ambient temperature, and prior thermal and mechanical histories of the specimen.[2]

The average strain for the second order transitions shown for the sixteen tests of Fig. 4.83 was $\gamma_N = 0.129$. An additional example is shown in Fig. 4.85 for seven

[1] See BARRETT [1952, *1*] for a discussion of the use of the stereographic triangle to designate the location of crystal axes with respect to λ_0 and ϕ_0.

[2] From 16 September 1958 onward, every test performed in my laboratory, including those of all of my doctoral students, has been numbered sequentially. (In accordance with the practice, standard in the past but unfortunately less universal today, the papers of my students in which they describe their doctoral research, bear their names only.) Thus, from the numbers, it is possible to determine the order of the experiments. At the present moment, test numbers have passed 1800.

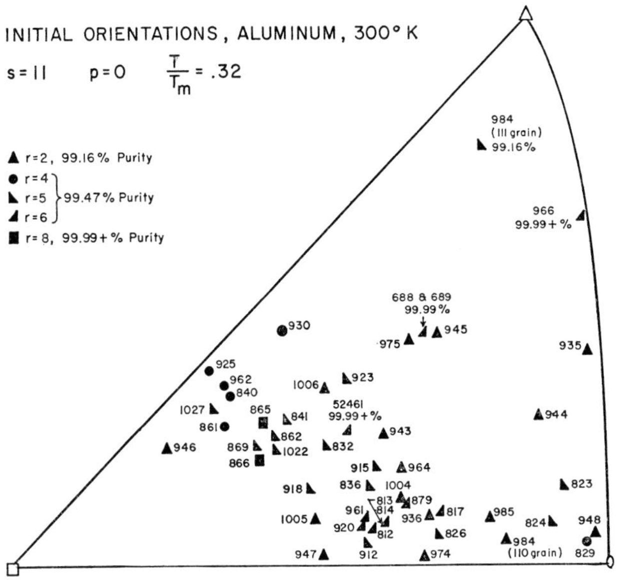

Fig. 4.84. The initial crystallographic orientations of aluminum single crystals. The numbers correspond to the tests described in Figs. 4.65, 4.66, 4.67, 4.83, and 4.85. All measurements were made in my laboratory.

axial tension tests performed on a small dead weight apparatus. The loading was incremental. These single crystals, whose initial orientation of axes also were indicated in Fig. 4.84, were grown, too, by the strain annealing technique. They had a purity of 99.47%, a specimen diameter of 1.24 mm, and lengths which ranged from 30 to 110 mm, the average length to diameter ratio being 62:1. The strain was measured by means of a traveling optical microscope. The average transition strain for the seven tests of Fig. 4.85 was $\gamma_N = 0.505$.

Another example of second order transitions is furnished by the τ^2 vs γ plots I calculated from four tension experiments on iron single crystals of Taylor and Elam[1] from the 1930's, in which the average second order transition from an initial deformation mode of $r = 2$ in Eq. (4.22) occurred at $\gamma_N = 0.046$ (Fig. 4.86).

These few tests illustrate not only the reproducibility obtained for similarly prepared specimens but also the results found in surveying the τ^2 vs γ plots of nearly 500 individual tests on 12 different cubic crystalline solids. In summarizing the results of this comparative study of resolved shear data from single crystal measurement[2] I discovered a total of eight fixed transition strains, having the values of γ_N of 0.046, 0.127, 0.232, 0.350, 0.520, 0.780, 1.176, and 1.765. The numerical values of these transition strains were *independent* of ambient temperature, of crystal structure, of purity, and of strain rate.[3]

As will be shown below, a similar study of polycrystalline tension and compression tests revealed a similar related set of eight transition strains. For the single crystal, the transition strains were found empirically to occur in the following sequence:

$$\gamma_N = \bar{n}\left(\tfrac{2}{3}\right)^{N/2}, \tag{4.23}$$

[1] Taylor and Elam [1926, *1*]; Taylor [1934, *1*].

[2] Bell [1965, *2*], [1968, *1*], [1971, *1*].

[3] Bell [1971, *1*].

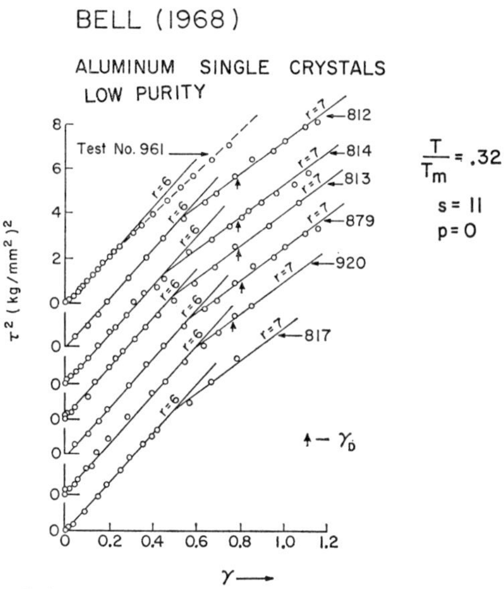

Fig. 4.85. τ^2 vs γ resolved shear plots of medium purity aluminum single crystals. The parabola coefficients (solid lines) are given by Eq. (4.22) for the indicated mode indices. Note second order transitions at $\gamma_N = 0.52$ where $N=6$. Rotations have reached the symmetry plane at γ_D.

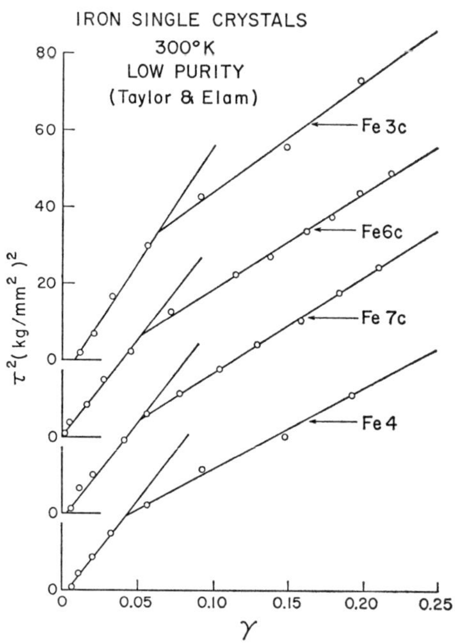

Fig. 4.86. τ^2 vs γ plots for low purity iron single crystals. Note the transition at the first single crystal transition strain of 4.6%.

where $\bar{n}=1.765$ is an aggregate ratio relating the shear stress of the polycrystal with the resolved shear stress of the single crystal, and the integer N was found to have the values: $N=0, 2, 4, 6, 8, 10, 13$, and 18.

The resolved shear response function describing the finite strain in the stage III region of cubic crystals, is based upon single slip kinematics. Even inspection by microscope reveals that stage III is a region of complex slip. The statistical analysis which gives Eq. (4.22) and Eq. (4.24) below in the form of "macroscopic single slip" remains an area for intensive study, particularly in view of the observed local inhomogeneity of the deformation.

Considerable light has been thrown on one aspect of this subject by NELSON NAI-HSING HSU,[1] who in his doctoral dissertation in 1969 described a series of experiments on high purity aluminum square specimens. To one pair of opposing sides of the specimens he applied initial stress by means of a soft rubber compressive device. Having started the deformation in one slip direction, HSU, after the removal of the initial load, proceeded to measure extension from axial loading for which the maximum shear stress lay in a different slip direction. He was seeking to determine by subsequent X-ray measurement during deformation, by inspection of the surface, and by observation of the parabola coefficient of the resolved shear response function, what changes in slip, if any, would occur.

Pertinent to interest in the behavior of the aggregate as well as the resolved shear phenoma of the single crystal, is HSU's principal conclusion:

> The shear deformation tends to continue in the active primary system even if the actual resolved shear stress on the primary plane system has been changed, during the deformation process, to a lower value than that of one of the latent systems.[2]

In final summary, HSU concluded from the results of his tensile tests on the specimens of aluminum single crystals, with or without prior lateral compression, that BELL's parabolic law [Eq. (4.24) below] could be adopted

> ... with proper r's, to describe the relation between the resolved shear stress and the resolved shear strain as calculated from the tensile data with the assumption of single slip. The actual cross section measurements of some of the specimens cannot be described as single slip, yet the parabolic law may still be applied to the resolved data. Thus it indicates that, despite the actual kinematics of the deformation, the parabolic law furnishes one with a systematic description of the relation between the strain in the axial direction and the tensile stress.[3]

The initial orientation of the specimens HSU analyzed covered the stereographic triangle. The interested reader is strongly recommended to examine the original work, which contains a very careful study of this phenomenon. That deformation mode changes were observed may be seen from one small sample of HSU's data, shown in Fig. 4.87. The capital letters refer to initial orientations.

When the number of experimentists exceeds 40 and the number of tests approaches 500 for a measurement as simple as that of extending a single crystal cylinder, recounting detail would be tedious. The manner in which crystals were prepared, the accuracy of the X-ray measurement of crystallographic axes, and the degree of precision in the measurement of axial stress and strain all too often were difficult to ascertain from the published papers. Most of those tests were performed as an aid to conjecture in the development over the years of one after another dislocation model or mechanism. Hence, often only qualitative information was sought by an experimentist who himself had no interest in continuum mechanics.

Load and large strain may be measured accurately with minimum care; errors, even of two or three degrees in measuring the crystallographic angles do not introduce large errors in calculating resolved shear stress and strain, particular-

---

[1] HSU [1969, *1*].
[2] *Ibid.*, p. 91.
[3] *Ibid.*, pp. 91–92.

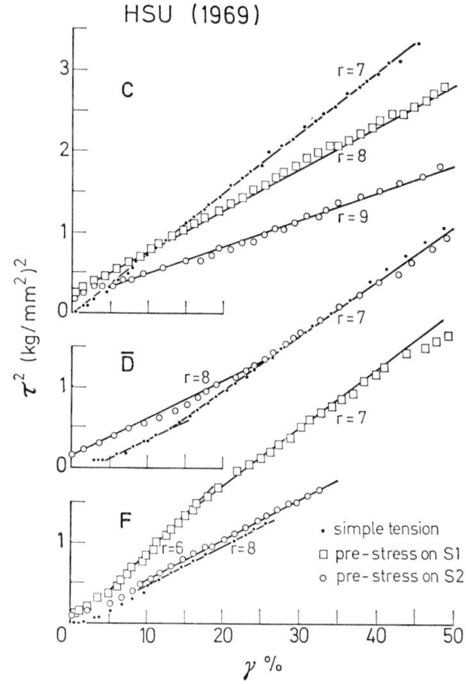

Fig. 4.87. τ^2 vs γ plots of single crystal data of Hsu showing parabolic distribution for simple tension and for lateral pre-stress. The parabola coefficients (solid lines) are given by Eq. (4.22) for the indicated mode indices.

ly when a great deal of data are averaged. A small spurious error, minimized by averaging many tests, was introduced when the data from the figures published in the literature were recalculated. Besides these factors, there was the additional fact that to obtain a parabola coefficient from the papers in the journals, I had to square the measured shear stress, replot the results as τ^2 vs γ, and measure a slope.

Hence, it is very significant that the data of the large majority of individual tests from the literature were consistent with the common quantitative pattern discovered from averaged data. Such agreement demonstrates that the experiments of the 20th century certainly are adequate for defining response functions in rational mechanics.

In sum, the results of this study of single crystal data from my own laboratory and from the general literature clearly reveal that the stage III deformation has a parabolic response quantitatively related to the linear stage II deformation, and, moreover, that the stage III deformation for specimens of high purity is not dependent upon the presence or the magnitude of the linear stage I region of "easy glide." Parabola coefficients are linearly temperature dependent and always are members of a discrete quantized set. Second order transitions may or may not occur from one discrete deformation mode to another; such transitions depend upon purity, ambient temperature, and prior history of the specimen. When second order transitions are present, they occur at one of a fixed sequence of strains, eight in total, the magnitude of which depend neither upon temperature, nor upon purity, nor upon crystal structure.

With γ_b denoting the origin of stage III deformation on the strain abscissa, these experimental discoveries[1] for cubic single crystals have been represented by the following relation:

$$\tau = \left(\tfrac{2}{3}\right)^{r/2} \mu(0)\, B_0 (\overline{m})^{-3/2} (1 - T/T_m)\, (\gamma - \gamma_b)^{\frac{1}{2}}. \qquad (4.24)$$

This generalization will be discussed further in Sect. 4.32 below, inasmuch as the response of polycrystals is derivable from measurement of single crystals.

All the cubic solids thus far studied have a resolved shear response function given by Eq. (4.24) for their single crystals. The list of solids now includes: Al, Ag, Au, Ni, Pb, Cu, Fe, Ta, Mo, Na—Cl, Ag—Au, and Ni—Co.

4.19. On stress definition and strain measure in presenting experimental results for large deformation.

In 1925, LUDWIK and SCHEU[2] compared experimental response functions for large deformation in tension, compression, torsion, and for cold rolling of polycrystalline copper. They performed their torsion experiments upon solid specimens, which required them to introduce further assumptions to interpret their results. As with their analysis of rolling experiments, the determination of elastic-plastic strain distributions demanded a far greater knowledge of plasticity theory which included work hardening, than was available.

LUDWIK and SCHEU were interested not only in the area changes from lateral contraction or expansion in unaxial testing at large strain, but also in whether displacement or rotation in an axial test were more important in comparing results for tension and compression. For tension, compression, and torsion tests on annealed copper, LUDWIK and SCHEU concluded that the rotation had the greatest influence. For alloys such as brass, the displacement seemed to be predominant. To describe their results they introduced the CAUCHY or "true" stress, which required a knowledge of the change in the area of the cross-section for each load.[3]

Confining this discussion to the results of tests for annealed copper in tension and in compression, we see in Table 131 LUDWIK and SCHEU's nominal stress vs nominal strain measurements and their calculated values of CAUCHY stress vs *nominal* strain.

LUDWIK and SCHEU concluded that a comparison of CAUCHY stress ("true" stress) with nominal strain (referred to the original area) provided common tension and compression response.[4]

TAYLOR,[5] also from simple tension and compression tests on polycrystalline copper, reached a conclusion directly opposite to that of LUDWIK and SCHEU. TAYLOR found that results for compression and tension coincided when nominal, or PIOLA-KIRCHHOFF, stress (referred to the original area) was plotted against logarithmic or "natural" strain ("true" strain). He did not give the dimensions

[1] BELL [1964, *1*], [1965, *2*], [1968, *1*].

[2] LUDWIK and SCHEU [1925, *1*].

[3] NADAI [1950, *1*], p. 253, remarked upon the fundamental error in LUDWIK and SCHEU's choice of "true shearing strains" upon which their later study of natural strains in tension and compression were based.

[4] LUDWIK and SCHEU also noted that when they changed the L/D ratio and failed to lubricate the end faces, the finite strain response markedly varied, and hence affected any comparison of response functions for tension and for compression. (If all specimens had been lubricated, the effect of changing the L/D ratio would have been reduced greatly, or eliminated.)

[5] TAYLOR [1934, *3*]. TAYLOR and QUINNEY [1934, *2*].

Table 131. LUDWIK and SCHEU (1925).

Strain $\lambda=\lambda'$ (%)	Load		Change of cross-section $f_0/f=f'/f_0'$	Effective	
	Tension P/f_0 (kg/cm²)	Compression P'/f_0' (kg/cm²)		Tensile stress $\sigma=P/f$ (kg/cm²)	Compressive stress $\sigma'=P'/f'$ (kg/cm²)
1	441	448	1.01	445	442
2	613	639	1.02	625	622
3	760	818	1.03	783	782
4	880	988	1.04	923	926
5	1005	1149	1.05	1055	1057
6	1111	1305	1.06	1177	1183
7	1204	1460	1.07	1288	1298
8	1287	1607	1.08	1390	1407
9	1363	1752	1.09	1486	1508
10	1434	1880	1.10	1578	1603
15	1715	2520	1.15	1975	2010
20	1885	3091	1.20	2262	2260
25	1995	3621	1.25	2492	2475
30	2050	4155	1.30	2667	2630
35	2075	4700	1.35	2800	2760
40	2085	5362	1.40	2915	2860
45	2085	6125	1.45	3030	2940
50	2045	7000	1.50	3130	3015
55		8100	1.55	3230	3080
60		9390	1.60	3310	3150

of the tensile specimens; the maximum nominal strain was 20%. The compression specimens had a length of 0.4770 in. and a diameter of 0.4390 in., giving an $L/D=1.087$ compared to LUDWIK and SCHEU's $L/D=3$. TAYLOR had lubricated the interfaces with grease. His comparison of compression with tension,

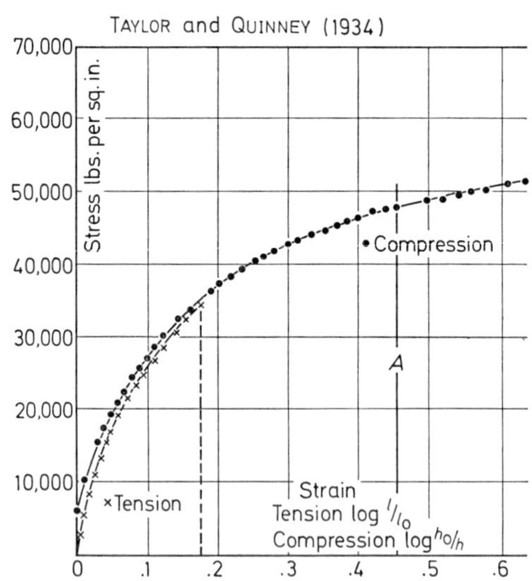

Fig. 4.88. TAYLOR and QUINNEY's comparison of tension and compression in copper in terms of nominal stress and logarithmic strain.

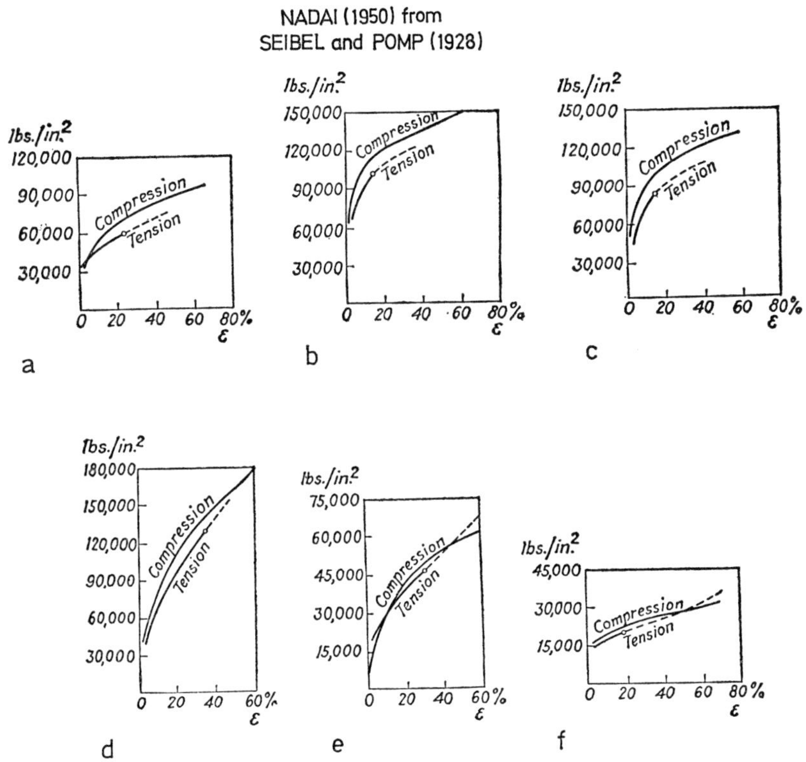

Fig. 4.89a—f. A comparison of tensile and compression experiments of SIEBEL and POMP for (a) wrought iron, (b) and (c) carbon steel, (d) nickel steel, (e) copper, and (f) aluminum. The comparison is for "true" stress vs "true" strain.

shown in Fig. 4.88, described tensile results (crosses) for nominal tension stress T vs log l/l_0 where l was the specimen length during the test, and l_0 was the initial length; and compression results (dots) for nominal compression stress P vs log h_0/h, where h was the length of specimen during the test, and h_0 was the initial length.

The correlation claimed by LUDWIK and SCHEU was for reduced stress, nominal strain, and that claimed by TAYLOR was for nominal stress, logarithmic strain. In an overwhelming proportion of the literature on experiment, the general practice since the time of those studies has been to describe results in terms of reduced stress vs natural or logarithmic strain, which technologists call "true" stress and "true" strain.

Some examples of the latter comparisons, for which close correspondence between tensile and compression data also is claimed, may be seen in Fig. 4.89, from data of SIEBEL and POMP[1] as presented by NADAI.[2]

In 1968 I showed[3] a compilation of quasi-static uniaxial tension and compression tests in several fully annealed polycrystals, which included numerous examples of correlated response functions when PIOLA-KIRCHHOFF or nominal stress was plotted against nominal strain. One example is in Fig. 4.90, in which RAMESH

[1] SIEBEL and POMP [1927, *1*].
[2] NADAI [1950, *1*].
[3] BELL [1968, *1*].

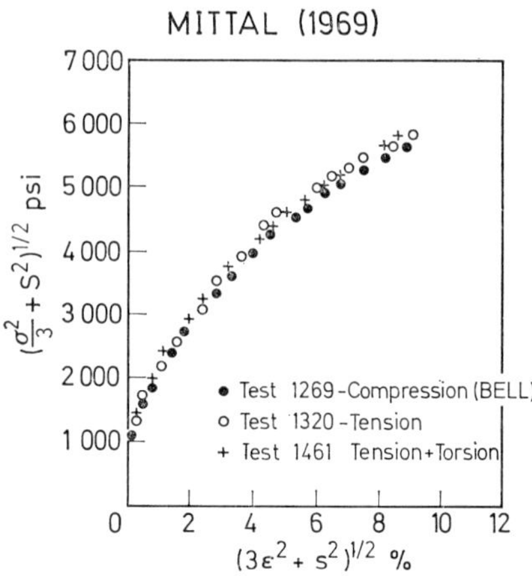

Fig. 4.90. A comparison of a simple compression experiment of BELL (No. 1269), a simple tension experiment (No. 1320) and a non-radial tension torsion experiment of MITTAL (No. 1461), showing the loading path independence for experimental response functions in polycrystalline aluminum in *nominal form*, for a Maxwell-von Mises loading surface. (See Sect. 4.32.)

MITTAL[1] as part of his doctoral research compared an uniaxial compression test of mine (# 1269) with a tension test (# 1320) and with a torsion-tension test (# 1461), all data having been given in nominal stress and nominal strain. In the compression test the specimen ends were lubricated and the L/D was 3. The tension and torsion-tension tests were performed upon thin walled tubes. All specimens were 99.16% purity aluminum; they had been annealed for two hours at 1100° F, furnace cooled, and checked for small grain size.

Of course the choice of the definition of stress and of strain measure is arbitrary. Furthermore, whatever is the choice of definition, when one is comparing the finite deformation for uniaxial stress and strain of opposite sign, i.e., for tension and for compression, there is no reason why the measured response must be an odd function. If, for a given solid, experimental results interpreted in terms of one set of definitions provide an odd response function, one does not also expect to find an odd function when stress and strain measure are defined differently. Yet, for the same solids, competent experimentists have obtained what appear to be incompatible results by finding odd response functions for uniaxial tension and compression whether stress and strain are defined as "true" stress vs nominal strain, or as nominal stress vs "true" strain, or as "true" stress vs "true" strain, or as nominal stress vs nominal strain. Hence, paralleling HARTIG's analysis of the response in tension and compression for infinitesimal deformation, described in Sect. 2.23 above, a similar study must be made for finite strain. (Efforts in this direction were described in Sect. 4.17 above, and will be considered further in Sect. 4.21 below.) Contradictory data should not be dismissed arbitrarily on the basis of results which appear logical solely from a single perspective. Among contemporary technologists it is generally believed that *only* a comparison of "true" stress and "true" strain is valid. Such a conjecture is without logical foundation, and, moreover, when support for it requires that an odd response function for tension and compression be found from experiment, the conjecture is not borne out by an inspection on a global scale of published results.

[1] MITTAL [1969, *1*].

In the measurements I am about to describe, all polycrystalline experiments from the literature (230 individual tests in 19 elements and 5 binary combinations) were "de-trued" to nominal stress and nominal strain in order to compare them with 1600 tests from my laboratory on polycrystals of the same solids which had been given in nominal form.

The choice indeed is arbitrary, but one advantage of representing data in terms of nominal stress and nominal strain is that the response functions so given occur in the same parabolic form as those for single crystals, to which they are related. The reduction of area already has been included in defining the resolved shear stress of the single crystal; thus care should be exercised in defining the response of polycrystals in terms of aggregate ratios.

4.20. Quasi-static experiments on polycrystals at finite strain: Uniaxial tests.

From 1841 when PONCELET[1] drew the diagrams of the results from quasi-static experiments on polycrystals of ARDANT, SEGUIN, and BORNET, to the present, each decade has discovered new complexity in finite deformation plasticity. The one dimensional elastic limit studies of BAUSCHINGER[2] in the 1880's evolved into the 20th century dilemma of adequately describing and understanding yield surface phenomena for six non-zero stress components. Viscoplasticity which THURSTON[3] casually had observed in the 1870's, by 1960 had become a subject comprising a vast array of sometimes opposing facts. The early 19th century discovery of creep by CORIOLIS[4] and VICAT[5] led, a century later to a field of experimental specialization occupying hundreds of persons. A variety of empirical formulae describing semi-generalized and anomalous behavior had been proposed. The SAVART-MASSON (PORTEVIN-LE CHATELIER) effect[6] which had annoyed DULEAU[7] in his experiments in 1811 and had interested SAVART and MASSON in the late 1830's, by the 1960's had come to be viewed as a fundamental feature of the finite deformation of crystalline solids. It gave rise to new and difficult problems in the form of slow waves and curious relations among the incremental step parameters.

After 1945, experimental studies were concerned with the effect of the thermomechanical history upon the finite deformation which followed. Serious questions were raised as to whether constitutive equations for the plasticity of polycrystals could be written in terms of the simultaneously measured values of stress, finite strain, temperature, and strain rate. Experimental evidence suggested that even for crystalline solids, the entire prior history of these variables to the time of measurement might be necessary for any statement of constitutive equations.

No plasticity theories have evolved which plausibly include the thermodynamic complexity of dual loading and unloading functions for crystalline solids at finite strain. The demands of technology for some manner of coping with the solid in which infinitesimal elastic regions and plastically deforming regions occur simultaneously, brought about the analytical development of what became known as "ideal plasticity." The problems of interest were limited to regions of small strain.

[1] PONCELET [1841, *1*]. The reader may recall that PONCELET had included in his paper experiments done by SEGUIN in 1826, BORNET in 1834, and ARDANT in 1835.

[2] BAUSCHINGER [1877, *1*], [1881, *1*], [1886, *1*].

[3] THURSTON [1873, *1*], [1874, *1*, *2*], [1875, *1*], [1876, *1*], [1877, *1*], [1878, *1*].

[4] CORIOLIS [1830, *1*].

[5] VICAT [1833, *1*], [1834, *1*].

[6] SAVART [1837, *1*], MASSON [1841, *1*], PORTEVIN and LE CHATELIER [1923, *1*].

[7] DULEAU [1813, *1*], see [1820, *1*].

This development, which was more "ideal" for its mathematical simplification (constant stress plasticity) than for the experimentist attempting to study the phenomenon, placed great emphasis upon an elastic-plastic transition at a definable yield surface, on which the strains were supposed small. With the restrictions of "ideal plasticity" the extremely difficult experimental study of yield surfaces, instead of the understanding of constitutive equations for finite strain, was thought to be the key to the problem. An already extremely difficult problem became almost impossible when those individuals who did undertake experiments, chose to approach the subject from the perspective of technology; their focus was upon the complex transition region for pre-deformed ferrous and non-ferrous alloys which had special properties, including that of a high yield stress. In a different context, among metallurgists the importance of the critically resolved shear stress and the subsequent "easy glide" stage I plasticity for the construction of atomistic dislocation models, also had directed a major portion of experiment on single crystals to this same small strain, elastic-plastic transition zone.

TAYLOR and QUINNEY's[1] experiment on polycrystalline copper in compression, of 1934, illustrates the contrast between this small deformation region and that for finite strain plasticity. Beginning with a specimen of annealed polycrystalline copper 0.4390 in. in diameter and 0.4770 in. long, they compressed it in 31 stages to a nominal strain of 0.37, corresponding to log h_0/h of 0.46 in Fig. 4.91.

TAYLOR and QUINNEY lubricated the interfaces of the specimen with grease. At this strain, designated as A, they unloaded the specimen. The new length of 0.3007 in. was left unchanged, but the new diameter of 0.55 in. was machined down to 0.2795 in. Further compression of the machined specimen to a length of 0.1178 in. for a nominal strain of 0.75 (referred to the original length) and log $h_0/h=1.40$ was followed by unloading at the point B. They had relubricated the specimen for the second compression. They then machined the diameter to a value of 0.1973 in. to continue to compress the specimen, which again had relubricated interfaces, to a nominal strain of 0.95 or log $h_0/h=2.91$. At the end of this third stage, designated as C, the specimen had been reduced in length from its original 0.4770 in. to 0.0260 in.

TAYLOR and QUINNEY added experimental results after C for a lubricated specimen which continued to be compressed to $^1/_{53}$ of its initial length, or a nominal strain of 0.98, with log $h_0/h=3.98$. The initial values of L/D for each compression were 1.087, 1.095, and 0.60 in that order. They noted that the maximum observed stress of 60000 lbs/in.^{2} for the originally annealed copper in compression was approximately the same as the ultimate tensile stress of hard drawn copper.

One notes that above a logarithmic strain of 1.5 the stress remained nearly constant, reminiscent of the phenomenon which had interested TRESCA in the 19th century. The elastic limit at a stress of about 10% of the maximum had a magnitude of strain $\varepsilon_Y \cong 0.000300$, a small fraction of the finite strain at either A, B, or C. Yield surfaces for small strain are reached between this strain of 0.0003 and approximately 0.0030. The more general constitutive equation for loading at finite strain must encompass for this test the far greater range of strain from 0.000300 to 0.98.

The experimentist who chooses to study yield surfaces created by a prior loading to any finite strain, must deal with problems arising from the unloading and reloading of the specimen. These problems, in addition to all of the many complications BAUSCHINGER discovered for such a test, which include among others the identification of a point on the yield surface and changes in moduli

[1] TAYLOR and QUINNEY [1934, *2*].

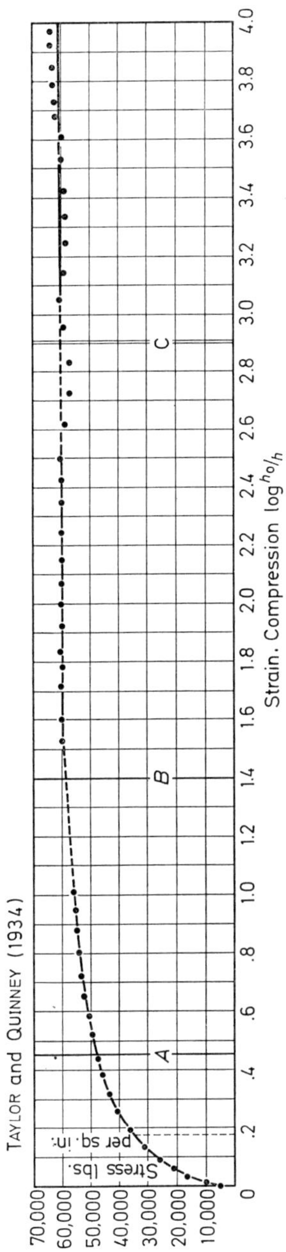

Fig. 4.91. Stress-strain curve of Taylor and Quinney for copper in compression. At maximum strain ($\log h_0/h = 3.98$) height of specimen was 1/53rd of the original height.

as a result of the prior loading and unloading, give rise to questions far beyond those propounded in theories of ideal plasticity.

The experimentist who studies loading response at finite strain may select sufficiently simple solids and prior thermal histories, as in Taylor and Quinney's experiment of Fig. 4.91 above, to mask the effects of any influence of a prior deformation history.

In this treatise, the two types of investigations, response functions and yield surfaces, are considered separately; the loading response of fully annealed polycrystals will be discussed first.

4.21. Quantized parabola coefficients and second order transitions for the finite strain of fully annealed polycrystals.

The few tensile tests on polycrystalline metals before 1841, which PONCELET examined in his review of experiments in plasticity, were for samples of metal elements with ill-defined properties and prior histories. During the years which followed, the practical metallurgist succeeded in providing precise recipes for the large scale production of metals of chemical compositions which not only had stable moduli for small strain under various prescribed, technologically important environments, but, in uniaxial tests, also had unusually high elastic limits compared to those of simple metal elements. What possible effect the prior thermal and mechanical histories which were part of these recipes, and the variously selected chemical compositions, had upon the response functions for finite strain in the plastic region beyond the yield limit, was not a matter of particular interest when such recipes were developed.

A great number of persons who wished to combine a keen interest in technology with a contribution to the science of solid mechanics, fruitlessly have performed both quasi-static and dynamic experiments at large strain on these complex, esoteric solids, as they sought to consider elementary questions. The results of such efforts filled the literature for a century. These persons tacitly assumed that memory effects, whether mechanical, thermal, or chemical, were not important for constitutive equations when plasticity and finite strain were being considered. In fact, however, the drawing, forging, rolling, or extruding, the heat treatment, the alloying, and the machining which precede the testing, provide hidden variables.

To separate the science from the technology in fundamental experimental studies, particularly in a field which has been as little understood as crystal plasticity, ultimately is fruitful for the technology and certainly is a vital avenue for advances in the science. Scientific experiments must be conducted on relatively simple crystalline solids which are in a sufficiently primordial state that prior histories of the solid are of little consequence.

In a number of the sections of this and earlier chapters of the present treatise, I have described many individual quasi-static tests in simple tension, compression, torsion, and more complex experiments with more than one non-zero stress component, for the purpose of delineating what specifically is known about large deformation. In this section it is my intention to present the results of an earlier extensive survey of quasi-static experiments, mainly on simple annealed polycrystalline metals. As in the study of the deformation of single crystals described in Sect. 4.17 and 4.18, I also compared a very large number of tests on fully annealed solids with an exhaustive collection of individual tests from other laboratories reported in the papers published during the past few decades.

Nearly all of the experiments from my laboratory were performed upon one of the various specially constructed dead weight loading apparatuses designed to produce either constant stress rate or defined incremental loading histories. Strain rates ranged from 10^{-9} sec^{-1} to 10^{-2} sec^{-1}. From the other sources, nearly all of the 177 tests in 19 metal elements and 5 binary combinations had been given in terms of "true" stress and "true" strain. To make the desired comparison, I had to recalculate, in terms of nominal stress and nominal strain, all of the measurements on annealed polycrystals which I found in the literature, excluding *none* for which the details of the original calculation were provided, so that, once again, any results obtained would be a proper global representation for constructing constitutive equations in rational mechanics.

During the interval between 1954 and 1968, when diffraction grating experiments (to be described below in Sect. 4.28) revealed that response functions for simple loading to finite strain in one fully annealed polycrystalline metal after another were parabolic, I performed quasi-static dead weight loading tests to ascertain what, if any, were the effects of increasing the strain rate by a factor of 100000 to 100000000. This was not a newly raised question. It was one which had fascinated experimentists in solid mechanics for over a century.

The experimental results found in my fourteen year survey indeed did lead to a correlation with response functions obtained from measurements of wave propagation, with those described above for single crystals. The simplest manner of illustrating this experimental fact is to state the sum of the observation, to give illustrations from published sources, and to refer the reader to the extensive detail of each of the individual tests which were described in Chap. III of an earlier monograph.[1]

In 1956, after seven years of performing experiments on the large deformation of various metals, it became apparent to me that the prior thermal and mechanical history of the specimens studied was of paramount importance. The concerted search for polycrystalline solids which either were uniform from one experiment to another or which effectively had no memory of prior history, led to the discovery that commercially pure (99.16%) hot rolled aluminium when annealed at approximately 90% of its melting point, that is, for two hours at 1100° F, and subsequently furnace cooled for 24 hours, possessed the desired properties.[2] It was necessary to check every batch of specimens to insure that the grains were small. For this fully annealed solid, and for several others for which similar recipes were established, it became possible to carry out consistent, reproducible experiments for a systematic study of finite strain plasticity at all strain rates.[3]

The comparison of experiments on wave propagation in polycrystals and resolved shear deformation in single crystals had indicated that the response functions for the axial loading in quasi-static deformation of polycrystals giving nominal stress σ as a function of nominal strain ε should be

$$\sigma = \left(\tfrac{2}{3}\right)^{r/2} \mu(0)\, B_0 (1 - T/T_m)(\varepsilon - \varepsilon_b)^{\frac{1}{2}}, \tag{4.25}$$

where $r = 1, 2, 3, 4, \ldots$ is the integral index of the deformation mode; where $\mu(0)$ is the shear modulus of isotropic solids at the zero point; and where $B_0 = 0.0280$, a dimensionless universal constant. T, the ambient temperature, and T_m, the melting point of the solid of interest, are in °K.

---

[1] Bell [1968, *1*].

[2] Bell [1956, *1*, *3*], [1960, *1*, *2*].

[3] It is of interest to note that in the past few years, commercial purity aluminum, so treated, has become the most frequently studied standard reference solid for checking proposed new experimental techniques, and variations on old ones, in dynamic plasticity. I warn the reader that the cold extruded aluminium rod now commonly manufactured in place of the former hot rolled aluminium, in general does not provide the standard reference solid when annealed in accordance with the prescribed recipe. This is not inconsequential. I myself recently wasted six months before I found that an unannounced change by the manufacturer on my standing order for the supply of specimens extending back nearly 20 years, was the source of otherwise inexplicable variations in a standard laboratory control test. The cold extrusion of one rod is sufficiently different from that of another, so that unless the influence of prior deformation can be removed, one obtains a wide scatter in experimental results.

For over a thousand wave propagation measurements in several annealed polycrystalline solids, the experimentally determined origin ε_b of the parabola on the strain abscissa was 0. However, for quasi-static measurements in polycrystalline solids where the initial linear region of deformation however small, invariably was present, in general $\varepsilon_b \neq 0$. The simplest method of illustrating the consistency of experimental results is to represent the data in σ^2 vs ε plots, so that parabolic response functions are represented by straight lines.[1]

In this treatise we are considering in the context of continuum mechanics the evaluation of results of many experimentists over many decades. For uniaxial tests in annealed polycrystals, as for finite shear strains in the stage III deformation of single crystals, a survey of over five decades of measurement already has been made.[2] For economy of space in this volume, I have selected from that survey examples illustrative of the general order which was found among test results. As when I described the results of WERTHEIM, TRESCA, BAUSCHINGER, and STRAUBEL above, I again must refer the reader to the original sources for more details of the experiments. In this case, a complete description of every one of the uniaxial tests performed in my laboratory or published in the papers of others, was provided.[3]

We begin with the σ^2 vs ε plots of 48 uniaxial tension and compression tests which I performed on dead weight loading apparatus between 1957 and 1967. These tests merit attention as being part of a large number made on annealed

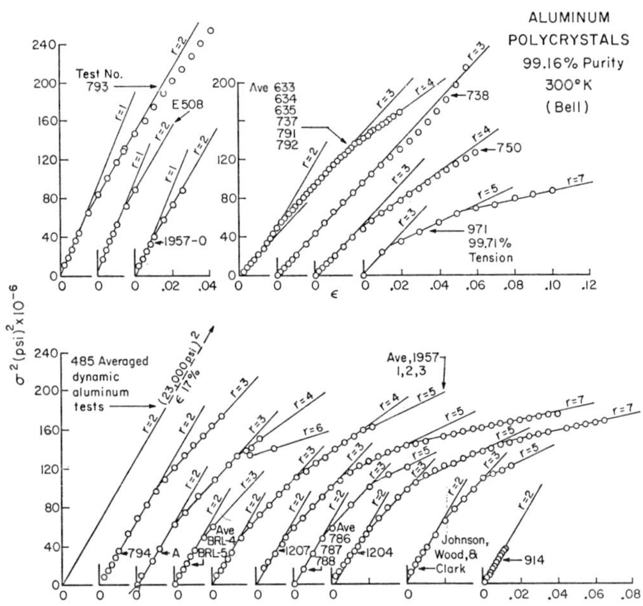

Fig. 4.92. Uniaxial stress experiments in low purity polycrystalline aluminum (circles) showing a variety of parabolic deformation modes and transitions. Experiments 786, 787, 788, 914, 971, 1204, and 1207 were in tension; all the others were in compression. The solid lines are theoretical prediction from Eq. (4.25).

[1] Since the stress is squared, correlation of experiment and prediction in this form provides an even closer check than direct plots, inasmuch as any deviation is amplified.

[2] BELL [1964, *1*], [1965, *2*], [1968, *1*].

[3] BELL [1968, *1*], Chap. III.

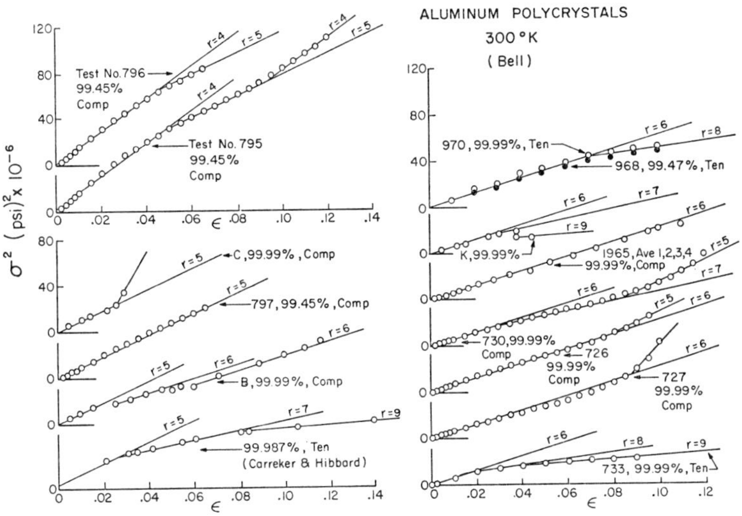

Fig. 4.93. Tension and compression experiments in polycrystalline aluminum of medium to high purity (circles) compared with prediction from Eq. (4.25) (solid lines) for the indicated mode indices, r.

aluminum between 1954 and 1968 which provided the empirical evidence that Eq. (4.25) did describe the response in finite strain. It was these results which indicated to me that it was important to make further tests in other solids, as well as to review extensively the experiments published in the literature.[1]

In Figs. 4.92, 4.93, and 4.94 are shown the data from these 48 dead weight uniaxial tests, 41 performed at room temperature, on fully annealed polycrystalline aluminum of various purities. In each instance, circles represent the measured nominal stress and nominal strain.[2]

For all designated purities and for both tension and compression tests, these data, given in σ^2 vs ε plots, have the straight lines prescribed by Eq. (4.25). Since $\mu(0)=3110$ kg/mm^2, $T_m=932$°K for aluminum, and the universal constant $B_0=0.0280$, are known, the slopes for an ambient temperature T (solid lines) predicted from Eq. (4.25) for the different mode indices r could be determined and compared with experiment.

As was found from the study of the stage III deformation of single crystals described in Sect. 4.18. above, the finite strain occurred as a series of parabolas with a quantized distribution of deformation modes, including second order transitions from one mode to another. Fig. 4.95 shows 13 tensile tests on fully annealed aluminum of 99.16% purity.[3]

[1] Unlike small deformation, for large deformation the uniaxial test is simple in concept and capable of being performed with accuracy, without demanding extreme precision. When a survey of such measurements is made, any generalization which purports to unify observation is severely tested.

[2] The uniform size of the circles in these and other figures is not related to the accuracy of measurement, the point determined being the one at the center.

[3] The aluminum is designated commercially as 1100 F − H18. These are the only tests made on that solid. All of the other hundreds of tests in my laboratory since 1954 were on polycrystalline aluminum of comparable purity, such as that of Fig. 4.92, but the material had been much less pre-hardened by the manufacturer; it was designated as 1100F − H2.

Note the shift in the initial mode index from $r=2$ for 1100F − H2 to $r=3$ for the 1100F − H18.

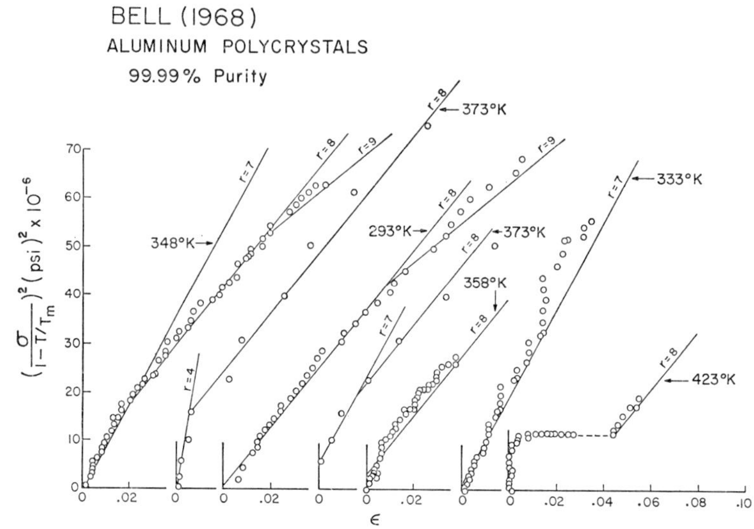

Fig. 4.94. Compression tests in high purity aluminum (circles) compared with prediction from Eq. (4.25) (solid lines) for the ambient temperatures shown.

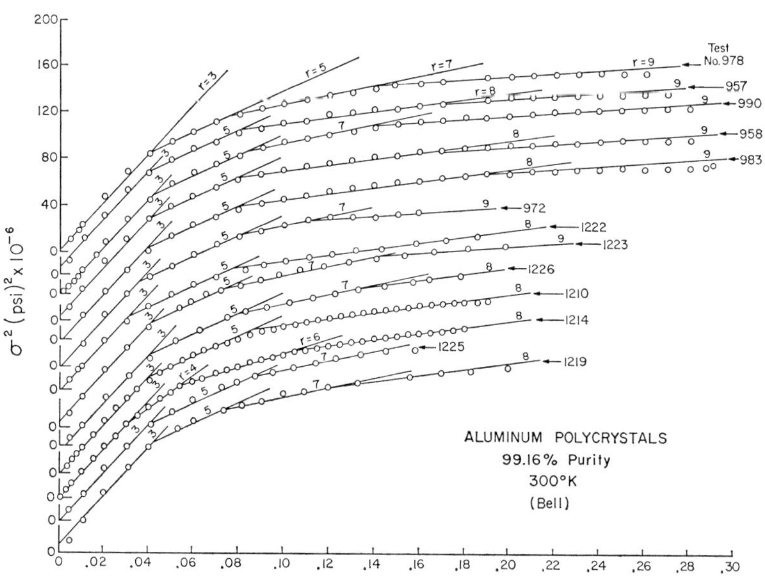

Fig. 4.95. Thirteen constant stress rate uniaxial stress tension experiments in low purity polycrystalline aluminum (circles) showing the reproducibility of deformation mode transitions at the second and third critical strains, i.e. $N=13$, $N=10$. The solid lines are theoretical predictions from Eq. (4.25).

These tests all show a second order transition from $r=3$ to $r=5$ at approximately 4% strain and an additional transition from $r=5$ to $r=7$ or $r=8$ at an average transition strain of 7.5%. Referring again to a later section on plastic wave propagation (Sect. 4.28), we see that hundreds of experiments on fully

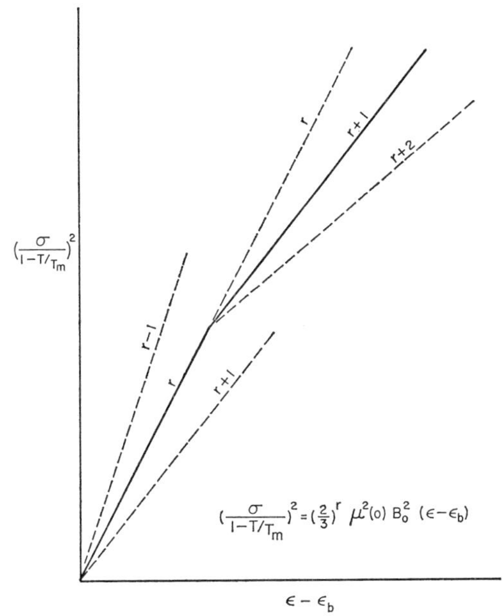

Fig. 4.96. A schematic stress[2] vs strain diagram of Eq. (4.25), showing magnitude of slope differences for adjacent values of the mode index, r.

annealed 99.16% purity aluminum polycrystals (1100 F—H2, annealed two hours at 1100° F and furnace cooled) had given at room temperature a parabola coefficient of 39.4 kg/mm² (5.60×10^4 psi) which is precisely the value for $r=2$ at that temperature.[1]

In studying this correlation, on the one hand, of finite strain during wave propagation and during the quasi-static testing of single crystals, with, on the other hand, quasi-static uniaxial measurements of annealed polycrystals, I found the schematic diagram of Fig. 4.96 to be of some value in illustrating that for σ^2 vs ε plots which exaggerate the errors in the measurement of stress, the slopes of the different deformation modes and their transitions were separated sufficiently, i.e., by $^3/_2$, to distinguish easily one slope from another as well as to determine the accuracy of individual correlations.

From Eq. (4.25) we see that if two different polycrystalline solids have the same value of $\mu(0)$ and mode index r, their response functions for simple tension or for simple compression tests at $T=0°$ K should coincide. At the zero point, shear moduli of aluminum and silver are nearly identical. For aluminum $\mu(0)=3110$ kg/mm², and for silver $\mu(0)=3170$ kg/mm². The high purity solids of both also have a common mode index $r=2$ at low temperatures. To illustrate that the coincidence of response functions near 0° K does occur, I have plotted in the same figure (Fig. 4.97) the results of a tension test on a 99.99% purity aluminum specimen of small diameter which was made at an ambient temperature of 20° K by R. P. Carreker and W. R. Hibbard[2] in 1957. Also shown is a tension test on a 99.97+% purity, 0.5 mm diameter specimen of silver which Carreker[3] made in 1957 at the

[1] Bell [1965, *2*], [1967, *2*], [1968, *1*], [1961, *1*], [1963, *1*].
[2] Carreker and Hibbard [1957, *2*].
[3] Carreker [1957, *1*].

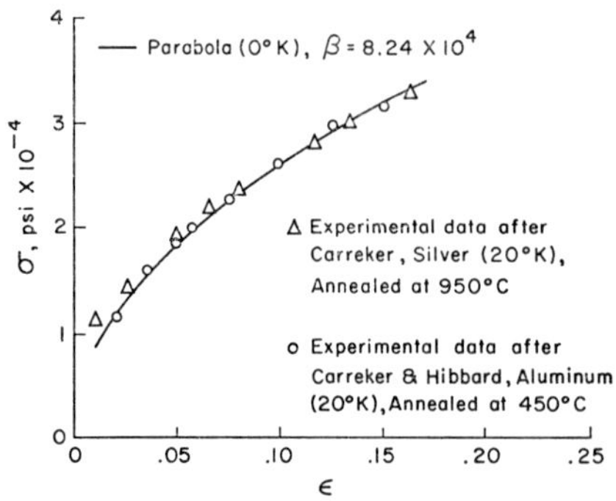

Fig. 4.97. Experimental data in silver and aluminum at 20° K compared with prediction from Eq. (4.25) for the two metals at that temperature (solid line).

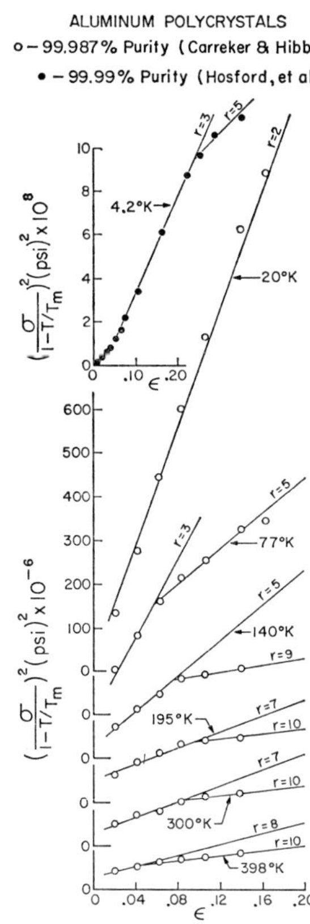

Fig. 4.98. The "de-trued" high purity polycrystalline aluminum tensile experiments of CARREKER and HIBBARD (1957) and of HOSFORD *et al.* (1960) for the indicated temperatures, reduced to absolute zero for comparison. The data are compared in these stress² vs strain plots with the prediction from Eq. (4.25).

same ambient temperature, 20° K. The responses for aluminum (circles) and for silver (triangles) not only coincide but are in close agreement with the quantitative prediction of Eq. (4.25) (solid line) for the value of $B_0 = 0.0280$ obtained from studies of wave propagation.

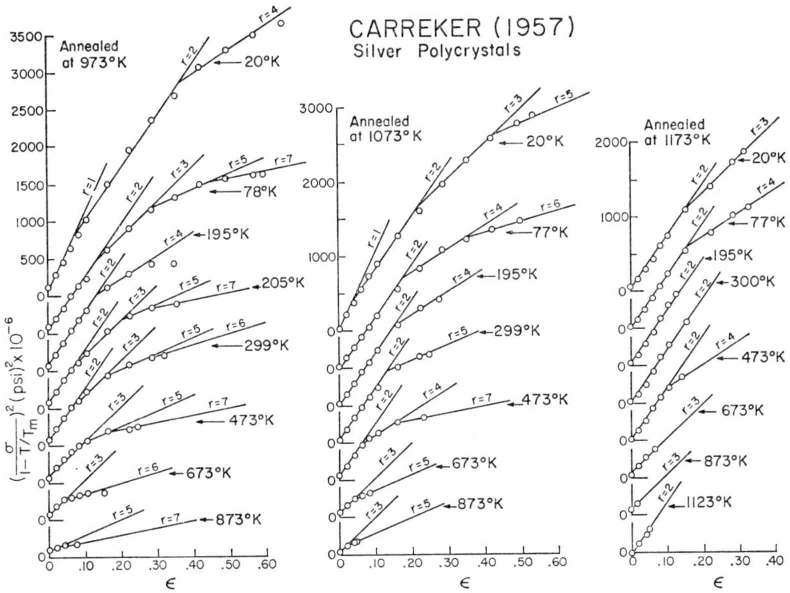

Fig. 4.99. The "de-trued" silver polycrystal data of CARREKER for the indicated temperatures, reduced to absolute zero for comparison. The data are compared in these stress² vs strain plots with the prediction (solid lines) from Eq. (4.25).

These two tests were part of a series of measurements made at many ambient temperatures. To illustrate further the applicability of the response function, Eq. (4.25) obtained from studies of wave propagation in compression, I include the rest of CARREKER's and CARREKER and HIBBARD's tensile experiments, shown in Fig. 4.98 for aluminum, and Fig. 4.99 for silver. All of these data originally were given as "true stress" and "true strain." Since the authors had included their method of computation, I was able to recalculate their results in nominal stress and nominal strain in order to compare them with the response function (straight line) of Eq. (4.25),

The number of individual tests both from my laboratory and from the literature on polycrystals for 19 elements and 5 binary combinations, was over 250. The general agreement of such data with Eq. (4.25) and with the finite deformation mode structure was amply evident.[1] Here it suffices to provide enough examples to show that the response, in general, was independent of crystal structure and of strain rate; the latter was found to have some influence on the occurrence of second order transitions and on the integral indices of changes in the deformation mode when second order transitions occurred. This of course was the source of the variations of response formerly attributed to viscoplasticity.

In Fig. 4.100 is shown a compilation of quasi-static tests at relatively high strain rate of J. E. HOCKETT[2] on orthorhombic, depleted uranium in compression. The averaged data (circles) when recalculated as nominal stress squared vs nominal strain are seen to follow closely the general pattern of Eq. (4.25).

For uranium, $\mu(0)=8450$ kg/mm² and $T_m=1405$ °K. HOCKETT's specimens with lubricated ends were 0.800 in. in diameter. They were deformed on a spe-

[1] BELL [1968, *1*], Chaps. III and VIII.
[2] HOCKETT [1959, *1*].

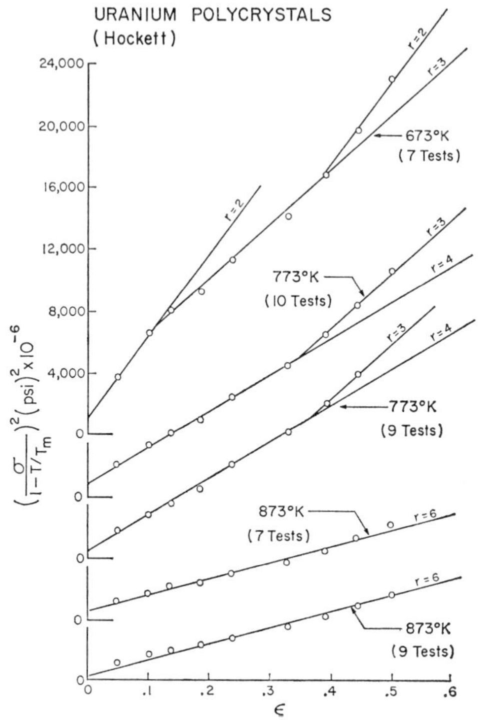

Fig. 4.100. "De-trued" results from compression experiments performed in 1959 on depleted uranium, comparing strain rates of $\dot{\varepsilon} = 1\ \text{sec}^{-1}$ (lower curve) and $\dot{\varepsilon} = 10^{-1}\ \text{sec}^{-1}$ (upper curves) at the indicated ambient temperatures. These experimental data (circles) are compared with the prediction for uranium for these temperatures from Eq. (4.25). Note the observed transitions occurring at the fourth and seventh transition strains, i.e., $N = 10$ and $N = 2$. All data have been reduced to 0° K for comparison.

cially constructed apparatus loaded by a rapidly rotating cam to obtain strain rates from $10^{-1}\ \text{sec}^{-1}$ to $1\ \text{sec}^{-1}$. The tests were performed at the designated ambient temperatures.

Eq. (4.25) contains no terms which refer to the specific crystal structure of the specimen. That the response functions do not depend upon crystal structure is illustrated by the 7 compression experiments I performed[1] in 1965 on 3 in. long, 1.in. diameter specimens of annealed zinc of two different purities are shown in Fig. 4.101. Zinc is a hexagonal metal for which elastic moduli in small deformation are known to vary widely from one specimen to the next, and for which the resolved shear deformation of the single crystal does not develop a stage III region of parabolic response. Nevertheless, for the polycrystalline uniaxial tests in quasi-static loading (and for wave propagation, as will be shown below), the observed experimental response is given by the general relation, Eq. (4.25). For zinc, $\mu(0) = 4660\ \text{kg/mm}^2$ and $T_m = 692.5°\,\text{K}$. The tests of Fig. 4.101 were performed at room temperature. The specimens were annealed at 588° K for $2^1/_2$ h and furnace cooled.

As another example of uniaxial tests which will be of some importance in later discussion of results, I have included 16 uniaxial tension measurements of CARREKER

[1] BELL, described in [1968, *1*].

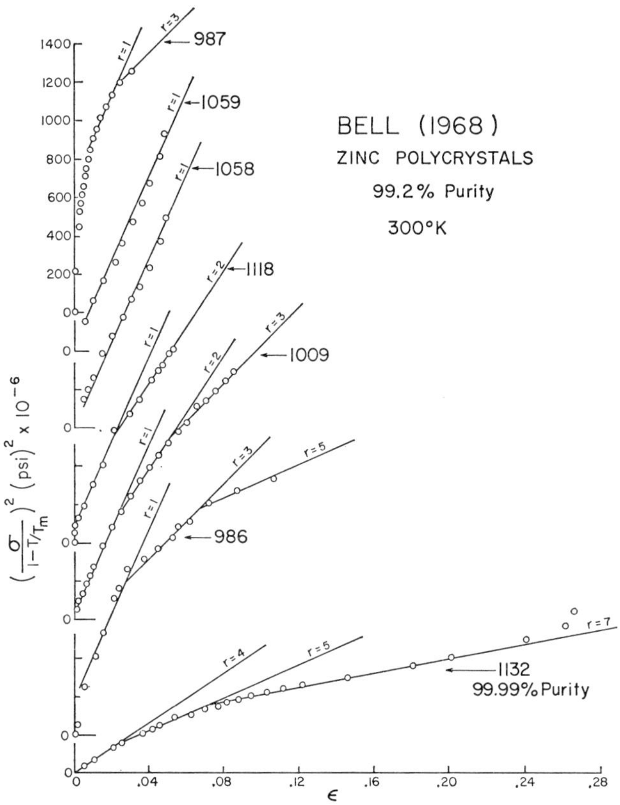

Fig. 4.101. Compression tests in low purity zinc (circles), compared at absolute zero with prediction from Eq. (4.25) (solid lines).

and HIBBARD[1] in 1953, performed upon 99.99% purity copper wires, 0.030 in. in diameter. The annealing temperatures were as shown and the ambient temperatures of the tests ranged from 20° K to 1023° K. For copper, $\mu(0) = 5080$ kg/mm² and $T_m = 1356$° K. There is general agreement with the slopes predicted from Eq. (4.25). Since it is a good conductor and easily available, high purity copppr has been studied for over a century. The mode index of $r = 4$ seen for higher anneal in the test at room temperature of Fig. 4.102, in fact has been typical of the response at finite strain ever since the experiments of SAVART in 1837.

The experiment described at the end of Sect. 4.20 was the polycrystalline compression test of TAYLOR and QUINNEY[2] in 1934 (Fig. 4.91). The results of that test, which proceeded to very large strain, were given by them in nominal stress vs *logarithmic* strain. A replotting of σ^2 vs ε for nominal strain, which I carried out for this treatise (see Fig. 4.103*a* below), revealed that the response to a *nominal* strain of approximately 60% is described by Eq. (4.25), with a single second order transition, $\varepsilon_N = 0.17$.

Starting from the annealed aluminum tests in Fig. 4.92, we have seen examples of second order transitions at $\varepsilon_N = 0.015$; for tests in Fig. 4.94, second order transitions were at $\varepsilon_N = 0.040$ and $\varepsilon_N = 0.075$; and, for TAYLOR and QUINNEY's

[1] CARREKER and HIBBARD [1953, *1*].
[2] TAYLOR and QUINNEY [1934, *2*].

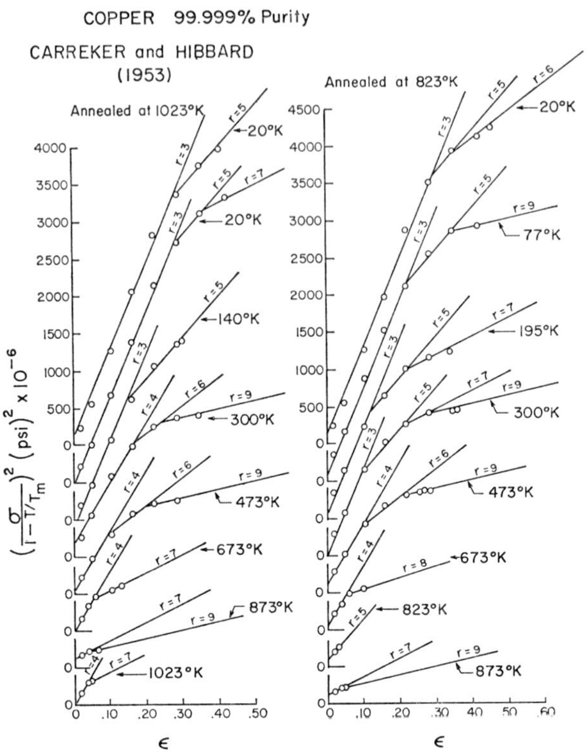

Fig. 4.102. The "de-trued" high purity copper data of CARREKER and HIBBARD for the indicated temperatures. The data (circles), reduced to absolute zero, are compared in these stress squared vs strain plots with the prediction (solid lines) from Eq. (4.25).

test on copper in Fig. 4.103 *a*, $\varepsilon_N = 0.256$. A review of the results for polycrystals has revealed 8 such transition strains which, in the manner of the stage III deformation of single crystals, may be represented empirically by

$$\varepsilon_N = \frac{1}{\sqrt{3}} \left(\tfrac{2}{3}\right)^{N/2}, \tag{4.26}$$

where N, as before, has the integral values, $N = 0$, 2, 4, 6, 8, 10, 13, and 18. These correspond to values of $\varepsilon_N = 0.577$, 0.385, 0.256, 0.171, 0.114, 0.076, 0.041, and 0.015, for all of which experimental values of second order transitions have been identified.[1] As will be seen in Sect. 4.32, the transition strains for the resolved shear deformation of the single crystal and for the polycrystal, are related. Analysis of many hundreds of tests reveal the relation as:

$$\gamma_N = \overline{m}\,\varepsilon_N \quad \text{where } \overline{m} = 3.06. \tag{4.27}$$

This experimental fact is important for the understanding of plastic deformation of aggregates as given by the summation of the response of their crystalline components.

For a comparison with TAYLOR and QUINNEY's test on annealed copper in compression, shown in Fig. 4.103 a, I have included a tensile test from my labora-

[1] BELL [1968, *1*], [1971, *1*].

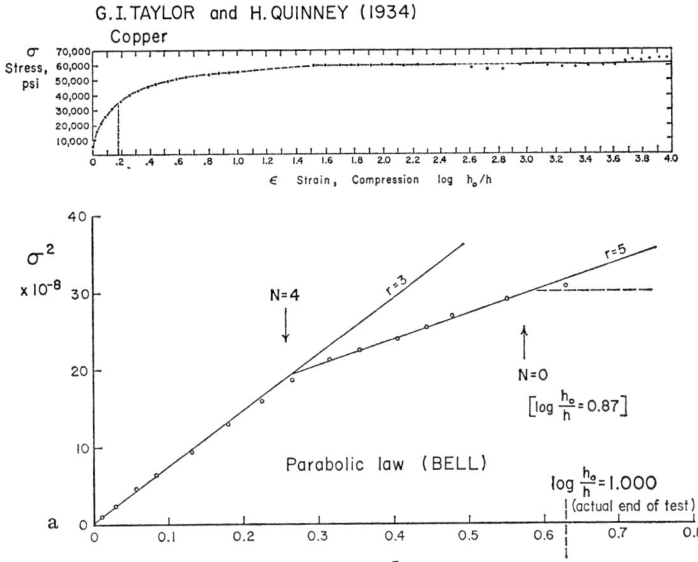

a

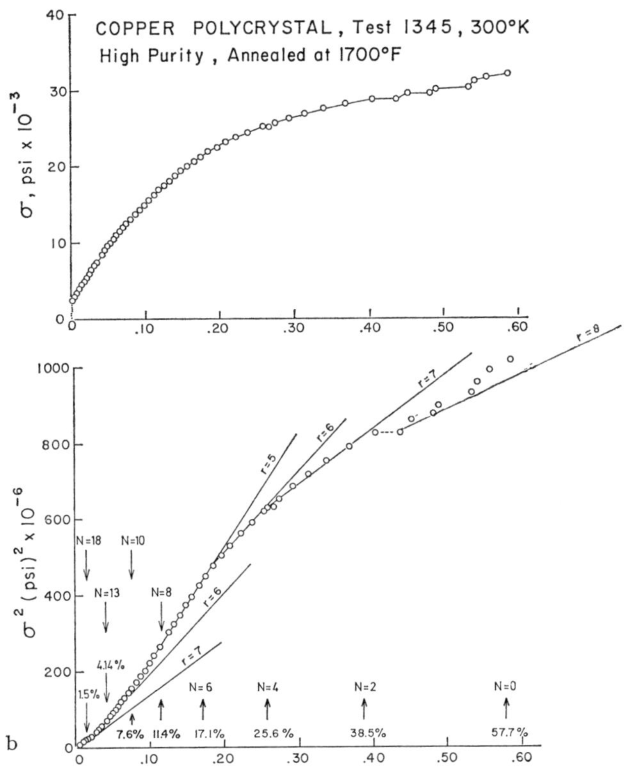

b

Fig. 4.103 a and b. (a) A compression experiment to large strain (see Fig. 4.91), replotted in nominal stress squared and nominal strain for comparison with prediction from Eq. (4.25). Note the second order transition at $N=4$. (b) σ vs ε and σ^2 vs ε plots of a dead weight tensile experiment of Bell on annealed copper, showing transitions for $N=18$, 13, 6, 4, 2, and failure at the highest experimentally known transition strain of $N=0$.

tory on the same, but now fully annealed, solid, obtained from dead weight loading on a thin walled hollow tube. The stress rate was constant; the entire test occupied approximately one hour. The results shown in Fig. 4.103b as for the TAYLOR and QUINNEY test of Fig. 4.103a are plotted as σ vs ε and as σ^2 vs ε to demonstrate the detail observable in the latter plot, which reveals not only a series of linear slopes consistent with Eq. (4.25) but also second order transitions occurring at six of the eight second order transition strains of Eq. (4.26). The final transition for $N=0$, shown in both 4.103a and 4.103b which, according to Eq. (4.26), should be at $\varepsilon_N=0.577$, is seen to be the strain at the point of ultimate stress whether in compression or in tension, when the data are plotted as nominal stress and nominal strain in terms of the generalization of Eq. (4.25).

The tests of Figs. 4.98 to 4.102, have been plotted as $(\sigma/(1-T/T_m))^2$ vs ε so that a comparison of slopes reduced to absolute zero could reveal that in each of these solids, as for many others which also were so considered, the response function varied linearly with temperature, in accordance with Eq. (4.25).

4.22. Quasi-static experiments on polycrystals at finite strain: The torsion of hollow tubes.

Compared to the many hundreds of uniaxial tests on solids which have been described during the past century, very few measurements have been made in torsion, and particularly on the torsion of hollow tubes. Tests on annealed tubes comprise but a small fraction of that number. If there is any hope that the plastic deformation of polycrystals can be described in terms of general material properties, then the order observed in results from the uniaxial testing of such solids should be extendable to data from the torsion measurement on thin walled hollow tubes. I first found that one could thus extend uniaxial results[1] when I had the opportunity to examine the experiments on the torsion of fully annealed tubes of OSCAR W. DILLON,[2] who in 1962, when he performed those tests, was a colleague at the Johns Hopkins University.

Leaving the discussion of this correlation to Sect. 4.31 on the SAVART-MASSON effect, I shall begin here by further analyzing TAYLOR and QUINNEY's[3] experiment of 40 years ago, which was described in Sect. 4.14. The experiment, shown in Fig. 4.104, consisted of comparing two tests on annealed copper tubes, one in uniaxial tension and one in simple torsion. Both tests were for monotonically increasing stress to large strain. In plotting their tension data as nominal stress vs logarithmic strain ("true strain") for comparison with nominal shear stress and shear strain for torsion, they concluded, as we have seen in Sect. 4.14, that neither the TRESCA-GUEST yield hypothesis nor the MAXWELL-VON MISES yield hypothesis applied (see Fig. 4.60). Once again we have an example in the history of experiment, of predilection in concept having influenced the presentation and interpretation of experimental results. When these same two tests were recalculated for a comparison in nominal stress and nominal strain, not only did they show an almost precise correlation with the MAXWELL-VON MISES hypothesis, but also σ^2 vs ε and S^2 vs s plots provided an equally precise agreement with the slopes predicted from Eq. (4.25) above, for the uniaxial tension test, and Eq. (4.29) below, for the torsion test, with the usual mode index of $r=4$ for annealed copper at room temperature.

We see that the torsion test on a hollow tube does indeed provide a parabolic response function for shear stress S and shear strain s, indicating the possibility

[1] BELL [1968, *1*]. See pp. 181–183.
[2] DILLON [1963, *1*].
[3] TAYLOR and QUINNEY [1931, *1*].

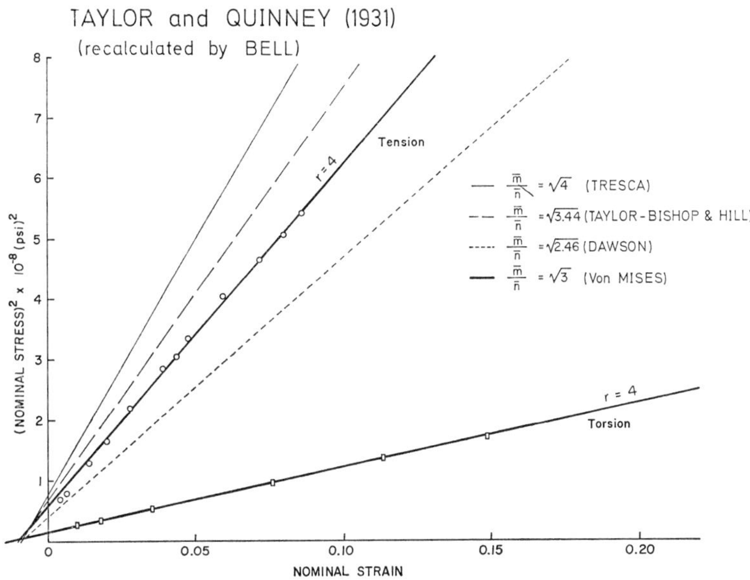

Fig. 4.104. The TAYLOR and QUINNEY experiments of Fig. 4.60 "de-trued" by BELL, showing the close agreement with the MAXWELL-VON MISES hypothesis for the nominal stress vs nominal strain data.

that Eq. (4.25) could be generalized to include any stress configuration producing plastic deformation. TAYLOR and QUINNEY's results in nominal form shown in Fig. 4.104 should be compared with their presentation of these tests shown in Fig. 4.60.

The MAXWELL-VON MISES condition requires that

$$\frac{\sigma}{S} = \sqrt{3} = \frac{s}{\varepsilon}. \tag{4.28}$$

Substituting Eq. (4.28) in Eq. (4.25) provides the following equation:

$$S = (\tfrac{2}{3})^{r/2}\, \mu(0)\, B_0 (\sqrt{3})^{-\frac{3}{2}} (1 - T/T_m)\,(s - s_b)^{\frac{1}{2}}, \qquad r = 1, 2, 3, 4, \ldots, \tag{4.29}$$

from which the prediction (solid line) for TAYLOR and QUINNEY's torsion test of Fig. 4.104 was determined.

As might be expected, torsion tests in hollow tubes exhibit the second order transitions of uniaxial measurement. In Fig. 4.105 are shown four experiments on the torsion of hollow tubes performed as part of the doctoral research on simultaneous tension-torsion of RAMESH MITTAL[1] in 1969. The specimens were fully annealed 99.16% purity aluminum polycrystals in the form of thin walled hollow tubes with an outside diameter of 0.4395 in. and with a wall thickness of 0.0322 in. The load was applied at a constant stress rate.

[1] MITTAL [1969, *1*], [1971, *1*]. The original calculation of torsion in these tests omitted a small constant torque from the buckets used in the constant torque rate dead weight tests. Since the tests had been performed in my laboratory, it was possible for me to recalculate all results from this series of measurements; these latter calculations are cited here and in Sect. 4.35 of this treatise. The omission changed none of the conclusions drawn by MITTAL in his presentation of data.

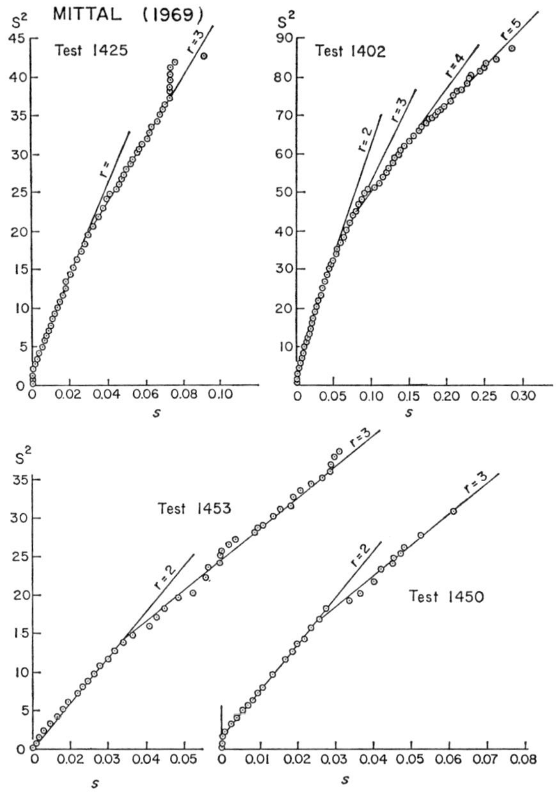

Fig. 4.105. Four dead weight torsion tests (circles) compared with Eq. (4.29) (solid lines).

As may be seen from Fig. 4.105, these experimental results (circles) are given closely in S^2 vs s plots by the slopes predicted from Eq. (4.29) with second order transitions from one deformation mode to another. This annealed aluminum in all but a very few instances provides an initial mode index of $r=2$. From Eqs. (4.25) and (4.28) one could expect that transition strains in torsion would be given by

$$s_N = \left(\tfrac{2}{3}\right)^{N/2}, \tag{4.30}$$

which corresponds to a series of 8 strains, s_N, of 0.026, 0.072, 0.132, 0.198, 0.294, 0.444, 0.667, 1.000.

Mittal averaged 38 transition strains from 19 tests on hollow tubes in simple torsion and (as will be described in Sect. 4.32) in various tension-torsion loading paths. This summary provided values of $s_N=0.025$ for 10 observations; $s_N=0.072$ for 16 observations; $s_N=0.135$ for 11 observations; and $s_N=0.198$ for 1 observation. The first 4 transition strains are seen to be very consistent with Eq. (4.30) and hence with the Maxwell-von Mises condition. The relation between transition strains for polycrystal shear and for single crystal shear becomes

$$s_N = \overline{n}\gamma_N, \qquad \text{where } \overline{n}=1.765. \tag{4.31}$$

Further discussion of the correlations of response of single crystals and of polycrystals for uniaxial and torsion tests will be given in Sect. 4.32, where the relation between the aggregate and the single crystal are considered, and in Sect. 4.35, where general constitutive equations and experiments on more complex stress configurations are described.

4.23. Experiments on thermoplasticity.

Any serious study of plastic deformation ultimately must include the thermoplastic aspects of permanent deformation, namely the associated generation of heat.[1] Whether the deformation energy lost in permanent deformation is thermally dissipated or to be accounted for otherwise, is a matter of major importance. In 1925 TAYLOR and FARREN[2] performed tests in simple tension on steel, copper, and aluminum polycrystals, and on aluminum single crystals, while at the same time observing the small rise of temperature which accompanied the deformation.[3] Because the experiment had to be performed in a time sufficiently short that thermal dissipation could be avoided, a special recording machine was built for the tests. The original paper, to which I refer the reader, contains elaborate details on the device and its calibration.

Those remarkable experiments in 1925 revealed that when the energy[4] calculated from the measured rise in temperature was compared with the deformation energy obtained from the quasi-static stress vs strain curve, there still remained an approximately 10% discrepancy, i.e., approximately 90% of the input energy appeared as measured thermal energy. Ratios of measured temperatures T_1 compared to those computed from the observed work in tension, T_2, showed that the

[1] Certainly the first experiment on heat produced by the finite deformation of metals were the measurements of CLAUDE LOUIS Le Comte BERTHOLLET in 1809 [1809, *1*].

BERTHOLLET flattened metal specimens in a coin press, and immediately following deformation threw them into a small pool of water just sufficient to cover them. With a thermometer he recorded the rise of temperature from the first, second, and third blow on the struck specimen, obtaining rises of temperature which decreased in copper from approximately 10° C for the first, to less than 1° C for the third; and for silver, from approximately 4° C for the first, to 1.25° C for the second.

BERTHOLLET ran those tests with his two friends, Messieurs PICTET and BIOT. He concluded that the heat produced by the deformation in some way had to be accounted for with a change of dimension, since the deformation produced, decreased with each blow. He emphasized the care which they had to exercise to insure that press and specimen were initially at the same temperature.

[2] TAYLOR and FARREN [1925, *2*].

[3] That except for the crude measurements of TRESCA in 1874 (see [1878, *1*] and also Sect. 4.4 of the present treatise), 95 years were to elapse before WEBER's ingenious experiments in 1830 [1830, *1*] opening the study of thermoelasticity were extended in an equally elegant manner to thermoplastic finite deformation, is a demonstration of the haphazard development of ideas. In the introduction to their paper in 1925 TAYLOR and FARREN could say:

> "It is curious that very few measurements of this type appear to have been made. The only reference which we have been able to find occurs in Dr. Rosenhain's article on 'Metals', in the *Dictionary of Physics*, where he quotes some previously unpublished observations made by Dr. Sinnat.
>
> According to these observations, only one-tenth of the work done reappears in the form of heat, the remaining 90% being presumably used up in changing the phase of the material. This result, if true, would be of very great interest, but so far no further details have been published. Dr. Rosenhain has informed us that he is carrying out further experiments on the subject." (TAYLOR and FARREN [1925, *2*], p. 85.)

TAYLOR and FARREN's discovery that 90% or more of the energy was converted to heat, i.e., the exact opposite of Dr. SINNAT's questionable results, emphasizes not only the experimental difficulties present but also the care which the theorist must exercise in establishing his contacts with Mother Nature. TAYLOR and FARREN were unaware that TRESCA, albeit with crude experiments, in the 1870's actually had been the first to discover that for polycrystalline copper, over 90 % of the energy of plastic deformation is converted to heat.

[4] Since these were quasi-static tests, small corrections were made for the loss of heat.

ratio essentially was independent of the magnitude of the strain. The averages of these ratios for three steel specimens were 0.865, 0.865, and 0.865; for three copper specimens, 0.92, 0.905, and 0.905; for aluminum, 0.93, 0.92, and 0.935. For two aluminum single crystals, the ratios were 0.95 and 0.945. In the single crystals also, the magnitude of the ratios was independent of strain. The maximum strains, ε, considered were: 0.131, 0.1183, and 0.1622 in steel; 0.1745, 0.1990, and 0.2020 in copper; 0.2306, 0.2195, and 0.2188 in aluminum; and 0.5272 and 0.5572 in the aluminum single crystals.[1]

In 1934 TAYLOR and QUINNEY[2] performed torsion and compression experiments to determine the thermal behavior by means of both thermocouples and calorimeters. They obtained comparable results from both. The deformed specimens were removed rapidly from the torsion lathe and dropped in to the calorimeter. For annealed pure copper and mild steel the order of magnitude of the latent heat was the same, but instead of remaining constant with deformation it underwent a percentage decrease at very large strain. From comparing the calorimeter results for pure copper with the stress maximum reached at log $h_0/h=1.45$ of the test of Fig. 4.91, they showed that the cold work necessary to saturate copper with latent energy at room temperature was roughly the same as that necessary to raise the metal to its maximum strength.

In 1962, OSCAR W. DILLON[3] described the first of a series of experiments extending this work of the 1930's beyond single loading. In repeated torsion tests in dead annealed commercial purity aluminum polycrystals, he studied different maximum twist angles and specimen lengths. DILLON could state with accuracy in 1962 that "Except for Taylor's work there appears to be no experimental data on the coupling effect in real materials."[4] The coupling referred to was that between the temperature field and the deviatoric components of strain when the material is deformed beyond the elastic region.[5] DILLON's first experiments were primarily on solid rods. While exhibiting the main features of the thermoplastic behavior, the rods he reported upon in that paper gave evidence of a heat sink for the cooler elastic central section. This he ascertained by comparing the results with preliminary experiments on thin walled tubes.

That same year, 1962, saw the publication by DILLON[6] of what must be regarded as one of the most revealing papers on experiment in this subject to date. He used low purity aluminum polycrystalline tubes, annealed for two hours at 1100° F and furnace cooled.[7] The strain was defined as $\varepsilon=\phi\frac{a}{L}$ where ϕ was the angle of twist, a was the mean radius, and L was the length of the tube. For those experiments, a was 0.218 in. and the wall thickness was 0.064 in. Observing

[1] Such experiments are an interesting contrast to the thermal speculations concerning permanent deformation which followed the 19th century studies of TRESCA; or the equally vague references to the subject by CORIOLIS after he had made his compression tests on lead in the 1830's.

[2] TAYLOR and QUINNEY [1934, *2*].

[3] DILLON [1962, *2*].

[4] *Ibid.*, p. 3100.

[5] In 1942, DART, ANTHONY, and GUTH [1942, *1*], and in 1943, JAMES and GUTH [1943, *1*] performed some experiments reversibly relating elongation and temperature in natural and synthetic rubber, some of which were essentially a repetition of JOULE's experiments in 1859 [1859, *1*] examining JOHN GOUGH's discovery in 1802 [1805, *1*], but some of their other measurements proceeded to far greater strain and thus were of fresh importance. The interest of these experimentists in the 1940's, however, was not in the deviatoric components of strain.

[6] DILLON [1962, *1*].

[7] He stated that he chose this solid and prior thermal history so that a correlation of data could be made with the dynamic plastic wave studies of BELL. DILLON concluded: "As described below, this has been a fruitful choice."

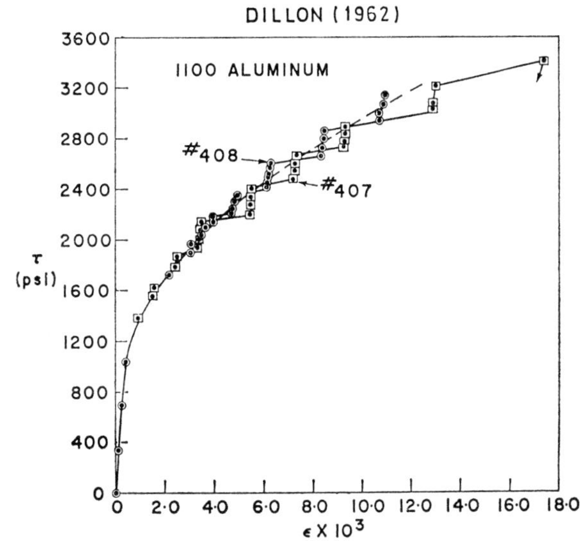

Fig. 4.106. DILLON's torsion data for annealed material.

that this virgin solid did not creep, but that for previously worked materials creep occurred in the region of interest, DILLON noted that below a stress of 4500 lbs./in² (3.17 kg/mm²) this creep ceased in a matter of minutes.

In the virgin annealed solid, two torsion stress-strain tests containing SAVART-MASSON (PORTEVIN-LE CHATELIER) steps, in Fig. 4.106, demonstrate the initial reproducibility. DILLON showed that for strain above 0.0025 the relation between the torsional shearing stress, τ, and the torsional shearing strain, ε, was of the form $\tau=\beta\varepsilon^{\frac{1}{2}}$, "where the value $\beta=2.96\times10^4$ psi [21 kg/mm²] can be related to single crystal tests of TAYLOR and FARREN (1925) and to dynamic tests of BELL (1961.)"[1] Of major importance was the fact that after a few cycles DILLON obtained the new, reproducible stress-strain relation which included the temporary creep, seen in Fig. 4.107.

In Fig. 4.107 are shown DILLON's quasi-static comparisons of τ vs ε after dynamic oscillations with amplitudes of $\varepsilon_{\mathrm{dyn}}=0.0025$ and $\varepsilon_{\mathrm{dyn}}=0.0060$. The comparison of quasi-static and dynamic cycling by means of the initial tangent and unloading slope was given by $\tau=\mu\varepsilon$, with μ the linear elastic shear modulus given[2] as 3.0×10^6 psi (2110 kg/mm²).

DILLON's apparatus and test specimen are diagrammed in Fig. 4.108. The ends of the specimen were wrapped in a loose fiberglass insulation. The variable speed motor permitted frequency variations from 400 to 2200 rev/min, with rotation amplitudes from 5.5° to 30°.

[1] DILLON [1962, *1*], p. 238. DILLON's parabola coefficient of 2.96×10^4 psi was in remarkably close agreement with that which I had predicted at the time, for the torsion of low purity aluminum: 2.93×10^4 psi. In the mode index terms, this corresponds to $r=1$. See Sect. 4.22, Eq. (4.29).

[2] In part, this is the reduction of modulus following permanent deformation, consistent with the multiple elasticities described in Chap. III, Sect. 3.44 above. The stable initial shear modulus for aluminum is $\mu=2690$ kg/mm²; i.e., I found that for isotropic solids the quantized shear moduli were $s=11$, $p=0$. After permanent deformation, DILLON's measurement of 2110 kg/mm² may be seen as a multiple elastic shift in torsion from $s=11$ to $s=12$. The shear modulus predicted for $s=12$, $p=0$, is 2190 kg/mm² (BELL [1968, *1*]). See above, Sect. 3.43, Eq. (3.27) and Sect. 3.44, Eq. (3.28).

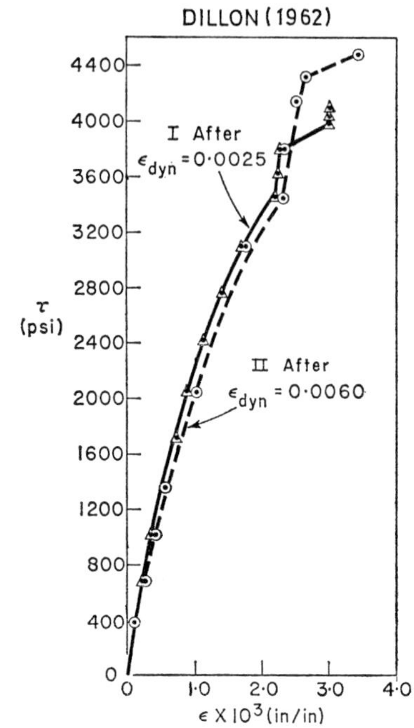

Fig. 4.107. **Static** stress-strain tests following oscillations, with the indicated shear strain amplitudes showing the reproducibility obtainable.

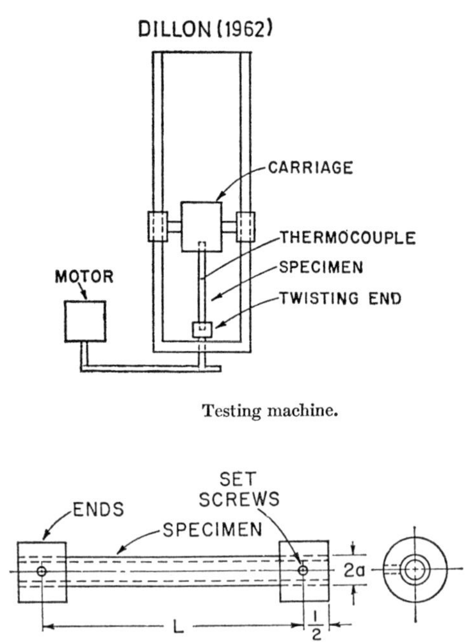

Fig. 4.108. Schematic diagrams of Dillon's torsion testing machine and test specimen for his thermal studies.

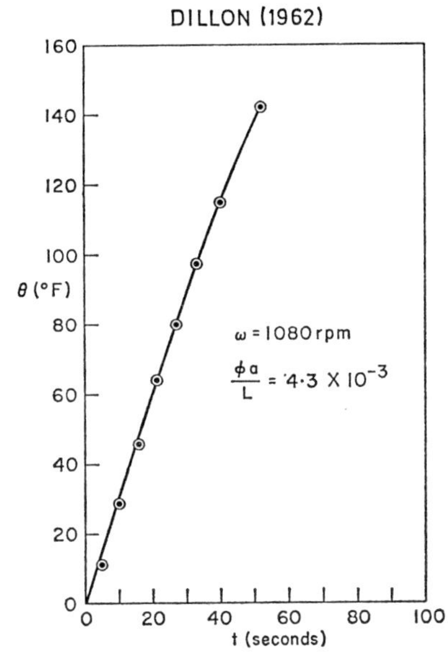

Fig. 4.109. Typical experimental temperature history.

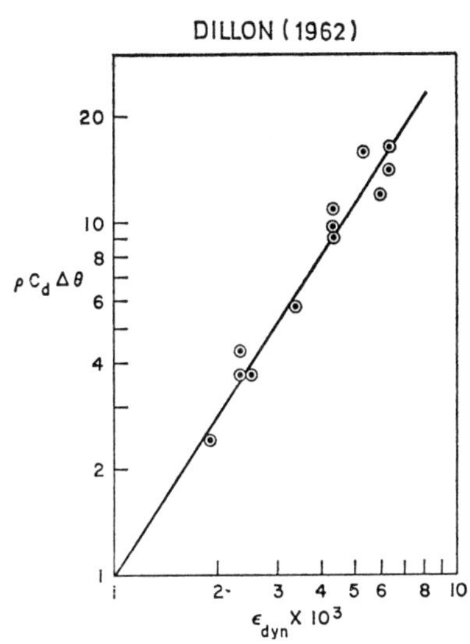

Fig. 4.110. Heat generated per quarter cycle in aluminum.

In Fig. 4.109 is shown an history of temperatures obtained by DILLON for an angular velocity $\omega = 1080$ rpm and a strain amplitude of 0.0043, with the temperature rise θ being given in °F. From such experimental results he found the temperature increment per quarter cycle $\Delta\theta$ which, multiplied by the mass density ϱ and specific heat at constant deformation, C_d, he plotted against the amplitude of the dynamic shear strain, ε_{dyn}, as in Fig. 4.110.

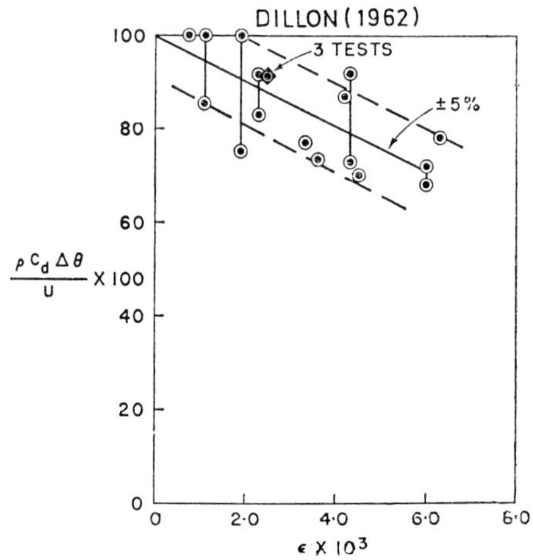

Fig. 4.111. Ratio of heat generated per quarter cycle to mechanical energy, as a function of shear strain ε.

From the logarithmic plot, choosing $\varrho C_d = 205$ in.lb./in.[3] per °F, Dillon obtained the empirical relation

$$\varrho C_d \Delta\theta = 34000\, \varepsilon^{\frac{3}{2}}_{\text{dyn}}. \tag{4.32}$$

Calculating the mechanical energy per unit volume, U, Dillon obtained the percentage ratio as a function of shear strain, ε, in Fig. 4.111. These results showed that even with large numbers of reversed cycles, for a single loading the conclusion of Taylor and Farren[1] in 1925 applied. Most of the residual energy was transformed into heat.

In 1966 Dillon[2] extended this study to annealed polycrystalline copper, and in 1967 he[3] analyzed in detail the rise of temperature in very slow torsional cycling for which he described the cooling of the unloading interval. He found that after the first two slow cycles the temperature rise was related to the present amplitude rather than the previous one.

> Except for the first two cycles from the fully annealed state, we could find no significant changes in the percent of the mechanical work that appeared as heat. It is always between 95 and 100%. The magnitude of the experimental error is estimated to be five percent.[4]

4.24. Viscoplasticity in metals: Experiments before 1940.

In the 1870's, Thurston[5] when studying the large deformation of metals had noted that if the speed of quasi-static testing were increased from one relatively low loading rate to another, a generally higher stress-strain function resulted. No attempt at a serious experimental study of this phenomenon was made until the experiments in 1909 of Ludwik.[6] Today, Ludwik's probably is the most frequently

[1] Taylor and Farren [1925, *2*].
[2] Dillon [1966, *2*].
[3] Dillon [1967, *2*].
[4] *Ibid.*, p. 56.
[5] Thurston [1874, *1*].
[6] Ludwik [1909, *1*].

cited, if rarely read, paper on experimental solid mechanics of the early 20th century.

LUDWIK was interested in the relation between deformation velocity, permanent deformation, and after-effect. Referring to preliminary investigations in an effort to find a suitable material for the study of such behavior, since he had found that for most metals viscous effects appeared to be negligible, LUDWIK finally selected tin. He did two kinds of experiments. In the first, tin wires 3 mm in diameter, 3 m long were loaded in tension with weights ranging from 2 kg to 15 kg, so he could observe the velocities at which the specimens elongated.

In the second experimental series[1] LUDWIK, with the assistance of Dr. ALFONS LEON, on an Amster-Laffon testing machine in the mechanical-technical laboratory of the K. K. Technical University of Vienna, at various velocities pulled in tension, to failure, tin wires 6 mm in diameter, 20 cm long. In these experiments he varied the velocity of the movable grip by changing the gear ratios and drive velocity. He attempted to provide constant velocity tests through the use of a metronome and a brake. He increased the velocity of the movable grip in 7 steps, from 0.00875 mm/sec to 9.7 mm/sec, corresponding to strain rates for his 200 mm long specimens of from 4.375×10^{-5}, to 4.85×10^{-2}, a ratio of 1000. LUDWIK's strain rate ratio of ten million fold came from a comparison of the extremes of the two types of test. Strain rates, which of course were not constant for the constant load experiments, were observed from values of $\dot{\varepsilon}=5\times10^{-9}\text{sec}^{-1}$ to $1.6\times10^{-5}\text{sec}^{-1}$, when the creep strain under each of 6 weights had reached a value of 0.15. Four specimens were considered for each situation, providing a total of 52 tests in the combined series. LUDWIK's example of an experiment for which the velocity of elongation was constant, is shown in Fig. 4.112.

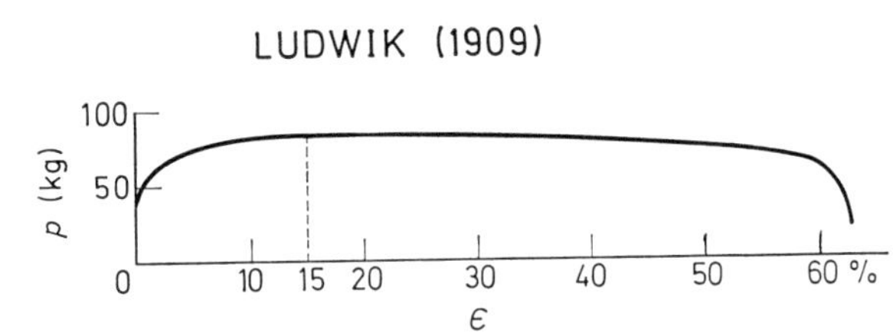

Fig. 4.112. Quasi-static stress-strain curve in a tin wire at constant strain rate, showing maximum stress at 15% elongation.

LUDWIK noted that in tin at all 6 loading velocities, the long flat constant region of ultimate stress had been reached by a strain of 0.15. He decided to compare strain rate with stress at 0.15 strain for creep tests at constant load, i.e., for tests of the first type, in which the strain rates were variable, and for tests of the second type at constant strain rate, for which the stress at a strain of 0.15 was the ultimate stress. The load rate was variable in the experiments of the second type. These results, including the scatter among the four repetitions of each type of measurement, are given in Table 132 for the constant load tests, and Table 133 for the constant strain rate tests.[2] It was solely from these observations in tin that LUDWIK proposed, for that solid, a visco-plastic uniaxial response function.

Because the ultimate strength for each velocity was referred to the same finite specific strain of 0.15, LUDWIK called it the specific deformation resistance or inter-

---

[1] *Ibid.*

[2] *Ibid.*

Table 132. LUDWIK (1909). *Tension experiments with constant load.*

Load in kg/cm² of the original cross-section	Elongation per sec at 15% strain referring to original measurement length × 10⁹		
	Observation	From Eq. (1)	From Eq. (2)
28.3	5–8	0.0000000002	5.7
42.5	13–19	0.00000098	16
56.5	25–35	0.00012	33
70.75	60–75	0.0037	63
141.5	1000–1200	53.1	1030
212.0	14000–16000	9332.6	15060

Table 133. LUDWIK (1909). *Tension experiments with constant elongation velocity.*

Elongation per sec referring to the original meas. length × 10⁹	Tensile strength in kg/cm² at 15% strain referred to the original cross-section		
	Observation	From Eq. (1)	From Eq. (2)
43750	238–242	240	240
175000	269–276	268	277
700000	304–315	299.5	313
2100000	336–346	327.5	342
6060000	360–371	358	370
24250000	396–410	400.5	407
48500000	422–428	425	425

nal friction, which he designated with the symbol R. He obtained two empirical relations "after some unsuccessful trials with other curves,"[1] which equally well fit the constant velocity data: a parabolic curve, Eq. (4.33), and a logarithmic curve, Eq. (4.34). Only one of them, however, could be extended to the much lower strain rates of the experiments for creep at constant load. These empirical relations are

$$R = R_0 + k\sqrt[n]{\dot{\varepsilon}}, \tag{4.33}$$

$$\dot{\varepsilon} = k\left(a^{R-R_0} - 1\right), \tag{4.34}$$

where R_0 was the unmeasurable ultimate tensile stress for a test with $\dot{\varepsilon} = 0$; and k, a, and n were material constants. LUDWIK's comparison of calculations based upon each of these empirical relations also are seen in Tables 132 and 133.

LUDWIK's logarithmic relation was not a viscous stress-strain function to describe the general response of solids, as is so often stated, but rather, was a comparison for a single solid, tin, of creep rates at constant stress at a specified strain, with the strain rate for a measured ultimate stress at the same specified strain in a constant strain rate test. That the ultimate stress of tin varies with the rate of the applied strain, unfortunately provides no information on the dynamic response function for TRESCA's intermediate region which preceded his third region of constant ultimate stress where LUDWIK's measurement was made. LUDWIK had chosen the specific strain of 0.15 because for all strain rates the ultimate stress had been reached. He observed no "necking" in any specimen at that strain.

[1] *Ibid.*

In fact, LUDWIK noted that uniform strain along the flat plateau of the ultimate stress proceeded to far larger strains than this, before localized deformation commenced. He stated in his introduction that since he was certain that a marked boundary did not exist between liquid and solid bodies, there should be no fundamental difference in the relation between internal friction and velocity in the two material groups. He speculated that perhaps the linear viscosity of the fluid, with $R=kv$, in fact might be interpreted more validly as a very flat logarithmic curve with $a\sim 1$ within certain velocity limits. He suggested that temporarily this had to remain uncertain.

LUDWIK regarded his inability to obtain a measured value for R_0, his neglect of the influence of the specific weight of the wire, the month-long interval required for the experiments during which there occurred temperature fluctuations, vibrations, etc., and especially the inhomogeneity of the matrial, as sources of serious error for the small strain rates of the experiments at constant load. Such errors made it difficult to compare results from the two types of experiments.

LUDWIK did not measure stress-strain curves in tension at different constant strain rates; instead, he measured the variation of ultimate stress with strain rate. He did provide a diagram, Fig. 4.113, based upon his measurements of ultimate stress, which he proceeded to analyze in terms of the raising of the elastic limit. He also considered in these terms the THURSTON[1] effect, which he erroneously attributed to BAUSCHINGER,[2] i.e., the increase in elastic strain following a pause in plastic deformation. He gave as examples of materials which he thought might exhibit viscous behavior: fluids, many resins, lead, tin, and zinc; he emphasized that such behavior was negligible in copper, bronze, brass, tombac, pinchbeck (an alloy of zinc and copper), iron, and steel. Presumably, the basis of these judgments was in his preliminary survey.

LUDWIK pointed out that even in tin, viscous behavior would be negligible in the higher strain rates of impact (i.e., in dynamic plasticity), and he cited the experiments of KICK,[3] who in the late 19th century had been a leading advocate of the point of view that impact energy could be calculated from the quasi-static stress-strain curve. LUDWIK also referred to an experiment in tin at a higher velocity. By shortening the specimen to 10 cm and increasing the stroke of the machine,

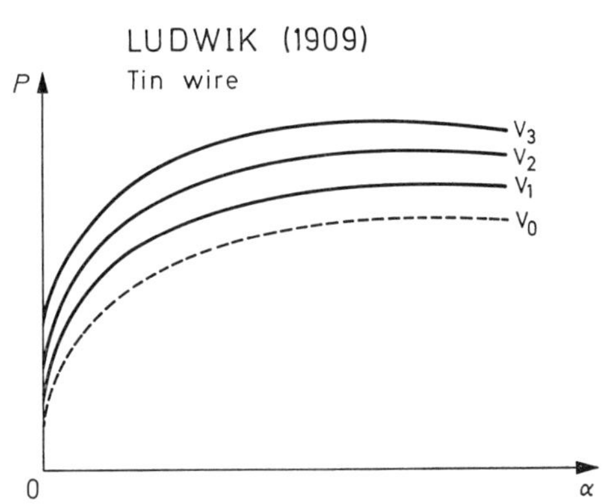

Fig. 4.113. LUDWIK's representation of load vs elongation diagrams for the tension of tin wire, based on his observations at 15% strain, for different loading velocities.

[1] THURSTON [1873, *1*].
[2] BAUSCHINGER [1881, *1*]. See Sect. 4.5 above.
[3] KICK [1875, *1*].

an increase in strain rate resulted in less than a 5% increase in stress, a trend which from the rest of his data also led him to conclude that viscosity was unimportant at higher strain rates.

Although LUDWIK's influence was considerable in the realm of attitudes, it was nearly two decades before anyone again investigated seriously plastic deformation as a function of strain rate. E. SIEBEL and A. POMP[1] did so in 1927. In the interim, however, there had been a great deal of interest in the increase of the elastic limit with strain rate, observed in some metals. This of course was an infinitesimal elastic strain rate phenomenon rather than one in finite strain viscoplasticity. Much discussion and some experimentation centered around the "double yield" in iron and steel as a function of the velocity of loading. On this particular topic ELAM[2] gave nearly twenty references for the decade before 1937. These studies had given evidence that the elasticity of the testing machines was at least as important in providing the "double yield" as the variation of any material parameter of the solid tested.

SIEBEL and POMP reconsidered LUDWIK's problem. LUDWIK's studies,[3] because of the low melting point of tin, were for a solid with a fractional melting point of $T/T_m=0.59$ at room temperature. For lead and zinc, which he conjectured were also viscoplastic, the fractional melting points were $T/T_m=0.50$, and $T/T_m=0.43$, respectively. On the other hand, for steel, copper, and brass, for which he declared the velocity effects were negligible, $T/T_m=0.17$, 0.22, and 0.25, respectively. LUDWIK's conclusions thus had been influenced by his having chosen a particular value of T, i.e., room temperature, for all of his comparisons.

SIEBEL and POMP, who also did experiments only at room temperature, tested soft steel, copper, and lead. Their experiments in tension and compression at different strain rates revealed that not only the lead but also the soft steel and copper exhibited a measurable increase in stress with velocity. They proposed from their data that the parameter k in Eq. (4.34) also was velocity dependent, with $k_{\text{dyn}}=k_{\text{st}}+c\,(v^*)^n$, where v^* corresponded to an often interrupted test at $v^*\simeq 2.5\times 10^{-4}\,\text{sec}^{-1}$. This was the beginning[4] of a series of descriptions of viscoplastic response which have appeared frequently ever since.

H. DEUTLER[5] in 1932, while a doctoral student of LUDWIG PRANDTL, at PRANDTL's suggestion undertook an investigation of LUDWIK's conjecture regarding the unimportance of viscosity at strain rates above $10^{-2}\,\text{sec}^{-1}$. Using a testing machine at the Institute of Applied Mechanics at the University of Göttingen of the same type as LUDWIK's, he obtained strain rates between $1\times 10^{-5}\,\text{sec}^{-1}$ and $5\times 10^{-3}\,\text{sec}^{-1}$. For strain rates above this range, DEUTLER, accepting the quasistatic hypothesis of uniform stress and strain during impact, calculated average strain rates as usual from a pendulum impact test based upon the energy absorbed. The apparatus was a modified Charpy notched bar impact machine, from which DEUTLER estimated his strain rates went to $10\,\text{sec}^{-1}$, which provided a total range of a million fold between 10^{-5} and $10\,\text{sec}^{-1}$.

DEUTLER lamented the wide scatter of his data from impact tests, which he partly attributed to difficulties in adjusting the zero line of the recording device. Plotting the percentage difference in ultimate stress vs log v/v_1 where v_1 was the velocity of elongation for the slowest test performed at $\dot{\varepsilon}=1\times 10^{-5}\,\text{sec}^{-1}$, DEUTLER

[1] SIEBEL and POMP [1928, *1*].
[2] ELAM [1935, *1*].
[3] LUDWIK [1909, *1*].
[4] SEIBEL and POMP [1928, *1*].
[5] DEUTLER [1932, *1*].

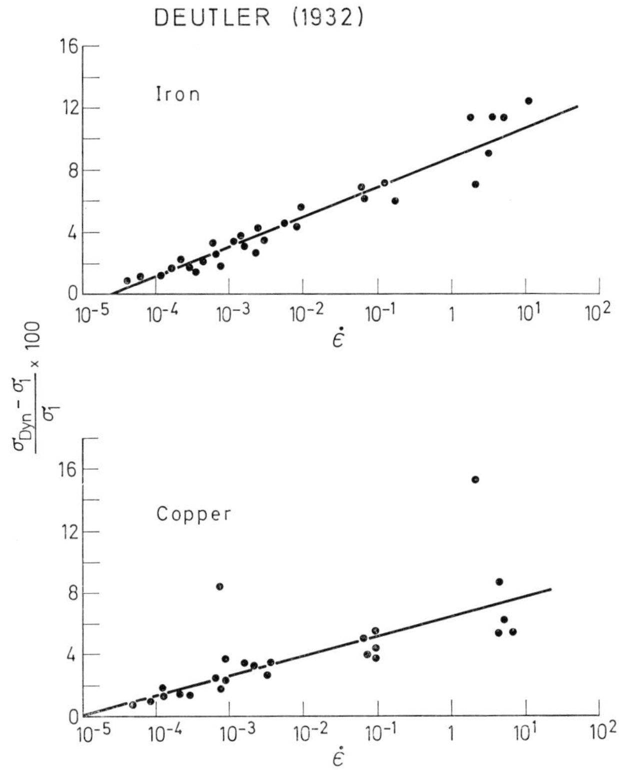

Fig. 4.114. The percentage increase in maximum stress based upon a quasi-static stress-strain curve at a strain rate $\dot{\varepsilon} = 10^{-5}\,\mathrm{sec}^{-1}$ for iron and copper, plotted against the estimated strain rate of the test.

concluded that PRANDTL's[1] formula $\sigma - \sigma_1 = \text{constant} \times \log v/v_1$, was satisfactory in the region of higher strain rate. The semi-log plot, for his experiments on iron and copper are given in Fig. 4.114.

One notes that the abscissa also is given as v/v_1 where $v_1 = 1 \times 10^{-5}\,\mathrm{sec}^{-1}$; thus the Amsler machine data ended just above $\log v/v_1 = 2$. DEUTLER initially had attempted to use a base of $v_1 = 1 \times 10^{-4}\,\mathrm{sec}^{-1}$ but found the range of the results at low speed insufficient to draw any conclusions.

DEUTLER was very concerned about the inaccuracies observed in the impact results, which in addition to his inability to establish a zero line, he attributed to time effects and other kinds of influences on the impact test. The general trend of an increase of ultimate stress with strain rate was evident even though, by depending solely on the motor drive of his apparatus, he could not be assured of constant strain rates. The paper[2] contained a thorough discussion of experimental details on every aspect of the problem *except* the conceptual limitation of the pendulum impact test. In Fig. 4.115 are shown DEUTLER's measurements of elongation velocity, $v = f(\varepsilon_0)$, during a test which produced the $\sigma = f(\varepsilon_0)$ stress-strain results shown. He worried that merely assuming constant strain rate in such tests was insufficient.

[1] PRANDTL [1928, *1*].
[2] DEUTLER *op. cit.*

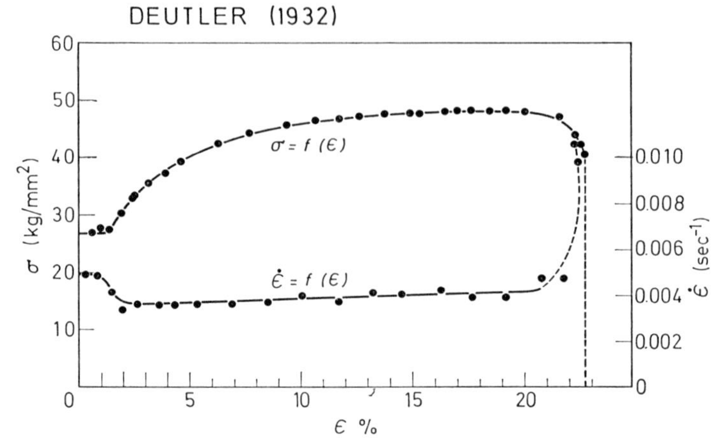

Fig. 4.115. Measured strain rate as a function of strain during a tensile test in iron.

In sum, DEUTLER'S experiments were inadequate to effect a reliable comparison between response functions for quasi-static loading and response functions for high strain rates. However, they were sufficient to raise doubts concerning LUDWIK'S conjecture that dynamic plasticity was non-viscous for most metals.

In a paper in 1938, whose primary objective was to reexamine the effect of rate of deformation on the yield point of iron and steel, C. F. ELAM[1] provided what was also a sequel to the experiments of LUDWIK[2] and DEUTLER.[3] After a discussion of the problems associated with her use of the 50 ton Buchton apparatus and the two autographic testing machines of the Cambridge University Engineering Laboratories, ELAM concluded that although errors precluded the comparison of absolute values, a comparison of relative values might be valid since all measurements had been made in the same way.

Rejecting "streamlined" specimen shapes, ELAM used cylindrical specimens from 0.30 to 0.40 inch in diameter and 6 inches long, with ends 0.75 inch in diameter and 0.5 inch long. In addition, she did some experiments on the 50 ton machine with bars 18 inches in length and with a diameter of 0.75 inch. She used ball shackles in all tests to ensure axial loading. In every instance, for Armco iron and steel the elastic limit and initial portion of the finite deformation increased, as may be seen in Fig. 4.116. Beyond the initial deformation, the stress-strain curve for the higher strain rate lay below that for the lower rate for Armco iron.

For steel, the stress-strain curve for the fast strain rate lay above the stress-strain curve for the very slow strain rate, as in Fig. 4.116. However, for slow strain rates the opposite situation was observed for steel, as is shown in Fig. 4.117.

A comparison of fast and slow rates in Armco iron and copper for finite deformation revealed, as is seen in Fig. 4.118, that although in copper stress increased with strain rate, in Armco iron the plastic response for the higher strain rate lay below that of the lower strain rate. This was an observation not only contrary to that of DEUTLER but also one indicating that standard viscoplasticity hypotheses were inadequate to account for differences in plastic response as a function of loading rate.

[1] ELAM [1938, *1*].
[2] LUDWIK [1909, *1*].
[3] DEUTLER [1932, *1*].

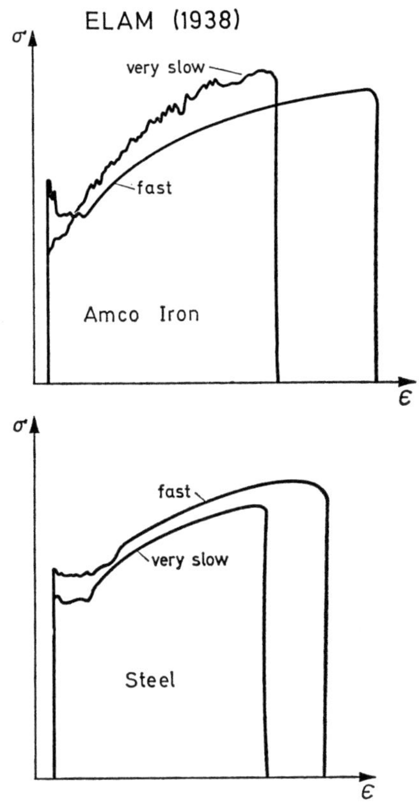

Fig. 4.116. ELAM's quasi-static tensile experiments on Armco iron and steel for relatively fast and relatively slow strain rates. Note that for the large plastic strain in iron, the higher curve is the lower strain rate.

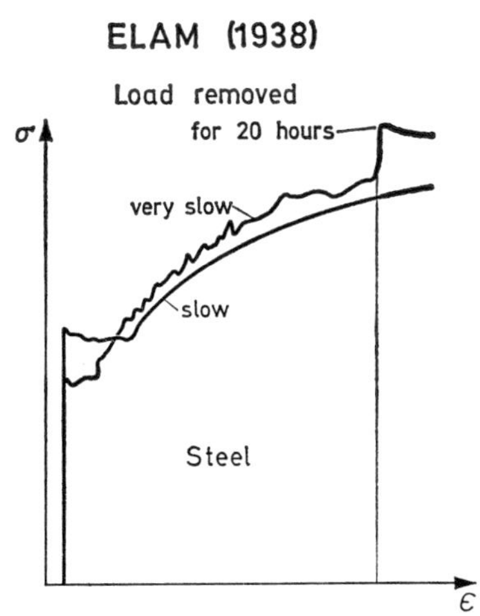

Fig. 4.117. A comparison of tensile tests on steel for slow and very slow strain rates; at large strain the higher stress occurs for the lower strain rate. Note the THURSTON effect after a 20 hour period of unloading.

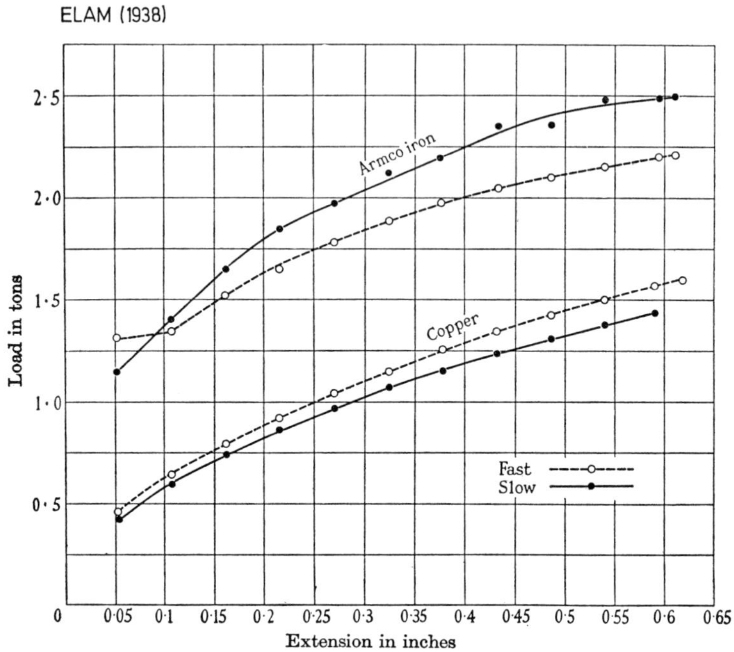

Fig. 4.118. ELAM's comparison of fast and slow quasi-static load extension measurements on Armco iron and copper.

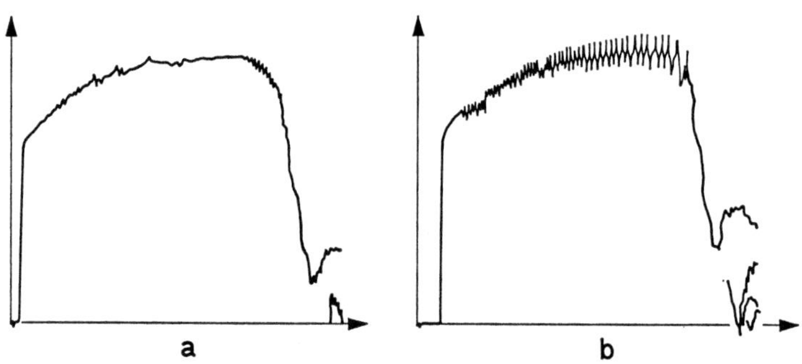

Fig. 4.119a and b. Fast (a) and very slow (b) tensile tests of ELAM in an aluminum alloy, showing the serrations of the discontinuous deformation phenomenon for the hard testing machine.

Following the description of a series of experiments in which she considered the "THURSTON effect" for pauses of 0.5 to, for the final step, 22 h in the loading throughout the finite deformation, ELAM presented a comparison of very slow and fast tests in an aluminum alloy designated as "high duty alloy," which exhibited the type of SAVART-MASSON (PORTEVIN-LE CHATELIER) effect seen when tests are performed upon hard machines. These results, shown in Fig. 4.119, provided the same stress-strain response for the very slow test when compared to the fast test. Thus the anomaly in viscoplastic terms was not limited to Armco iron. Hence we find by 1938, in the face of contradictory experimental evidence, that the role of viscosity in the plastic deformation of crystalline solids had not yet been defined.

4.25. The impact experiments of J. Hopkinson (1872) and Dunn (1897).

For linear elasticity, it was recognized early in the 19th century that elastic moduli for high strain rates could not be determined from an extension of a quasi-static test. Such parameters defining linear elastic response functions at high strain rates were found in experiments either on vibration of solids or on wave propagation. For dynamic plasticity, studies of vibration were not possible, of course, because of the large difference between the response functions for loading and unloading, and because of the nonlinearity and permanent deformation associated with the loading. It was not until 1956, after the diffraction grating experiment[1] had been conceived and developed, that wave profiles for highly dispersive plastic waves could be determined accurately. Hence, for a century, between the 1850's and 1956, experimentists attempted with very small success somehow to extrapolate assumptions and analyses from quasi-static tests in uniaxial tension and compression or in torsion, to the region of high strain rates of plastic waves of finite amplitude.

Shortly after the mid-19th century, metal specimens were subjected to rapid tension by being blown apart by gunpowder while in gun tubes. In the latter part of the century, falling weight and pendulum devices came into engineering use. The amount of permanent deformation produced by a specified blow was used as a design criterion. Near the end of the century in an effort to establish some sort of reproducibility in these tests and, in particular, to study the serious engineering problem of brittle fracture at lower temperature, the notched specimen test was introduced which, in varied form, still is part of the gross test procedures for industrial materials. Many writers, I being one,[2] have emphasized the scientific and possibly technological limitations of those tests and the caution which must be exercised in presuming they correspond with engineering field conditions.

In 1897, however, a new phase of experiment in this domain was introduced by Lieutenant B. W. Dunn[3] of the United States Army. He succeeded by optical-photographic means in measuring the displacement-time history of the ends of a short cylindrical copper specimen during impact from a falling weight. The analysis of this measurement in quasi-static terms was the prototype for a continuous succession of basically similar tests during the seven decades which followed.

Before briefly evaluating those "extended quasi-static impact tests" and the interpretations of them which evoked so much controversy in the mid-1960's despite the fact that for ten years it had been possible to study directly the propagation of finite amplitude wave fronts, it is interesting to consider the first efforts of John Hopkinson[4] in 1872 to treat dynamic plasticity as a wave phenomenon. The experiments of Hopkinson were unique at that time, and since then they have been esteemed highly by persons who neither have tried to repeat the tests nor have subjected Hopkinson's analysis of them to close scrutiny. Otherwise, not too much would be lost if discussion of Hopkinson's experiments were omitted. Except for John Hopkinson, however, no one attempted[5] to consider dynamic plasticity as a problem in wave propagation until the introduction of the

[1] Bell [1956, *1*, *2*, *3*].

[2] Bell [1962, *2*, *3*].

[3] Dunn [1897, *1*].

[4] John Hopkinson [1872, *1*].

[5] We must note however, that Bertram Hopkinson in the early 20th century reexamined his father's measurements and carried out some experiments related to dynamic elastic limits [1901, *1*], [1905, *1*].

the bilinear analysis of L. H. DONNELL[1] in 1930 and the more plausible finite amplitude wave theory of Sir GEOFFREY INGRAM TAYLOR[2] and of THEODORE VON KÁRMÁN[3] in 1942.

JOHN HOPKINSON, ignoring TRESCA'S intermediate region, assumed that the linear theory of elasticity applied until the specimen ruptured. Thus, for a long wire "fixed" at one end and subjected to a blow at the other, the first failure, as the height of a falling weight was increased, should occur near the upper clamp or point of fixture, since in elementary terms the stress would double upon reflection. The further increase of the height of fall to four times the value then would cause the wire to break immediately at the end where the blow was applied; i.e., doubling the initial velocity would provide the same stress as that from the reflection at the fixed end. In the first of his two papers on the subject[4] HOPKINSON also was interested in determining whether the rupture for different falling weights followed a rule of *vis viva*, mv^2, momentum, mv, or was independent of the mass of the falling weight, depending only on the amplitude of the velocity in the wire as elementary wave theory suggested.[5]

In the initial experiment the wires were 27 feet (8.23 m) long and had a diameter described as # 13 gauge, or 0.092 inches (0.234 cm). The wire was clamped at the top and held taut by a 56 lb. (25.4 kg) weight. A clamp was attached to the wire at some undesignated distance from the weight on which a metal sphere, with the wire through its central hole, fell in impact from a predetermined height. The measured quasi-static load for rupture was 350 lbs. (159.kg), corresponding to an ultimate tensile stress of 52700 psi (37 kg/mm²).

In that paper HOPKINSON reported a total of eleven measurements with four weights. First a weight of $7^1/_4$ lbs. (3.29 kg) falling from 6 ft. and 6 ft. 6 in. (1.83 and 1.98 m) failed to break the wire, but falls from both 7 ft. 0 in. and 7 ft. 2 in. (2.13 and 2.18 m), contrary to expectation, broke the wire at the bottom. He was unable to find a lower velocity at which failure would occur from the reflection at the fixed end. The second weight of 16 lbs. (7.26 kg) falling from 5 ft. 6 in. (1.67 m) broke the wire at the upper clamp, and although no further measurements were reported for this weight, this number was erroneously included in his comparison of the height of fall required to produce failure at the lower clamp. A 28 lbs. (12.70 kg) weight broke the wire at the upper clamp with falls of 2 ft., 3 ft., and 5 ft., while 4 ft. 6 in. (1.37 m) broke it 3 ft. up the wire in a "wounded place",[6] but a 6 ft. fall broke the wire at the lower clamp. Lastly 41 lbs. broke the wire at the upper clamp for 4 ft. 6 in. and at the lower clamp for 5 ft. 6 in.

> In problems of this kind it has been assumed by some that two blows were equivalent when their vis vivas were equal, by others when the momenta were equal; my result is that they are equal when the velocities or heights of fall are equal.[7]

Arbitrarily rounding off these curious results,[8] HOPKINSON claimed that rupture was independent of mv^2 and of mv.

---

[1] DONNELL [1930, *1*].
[2] TAYLOR [1942, *1*].
[3] VON KÁRMÁN [1942, *1*].
[4] JOHN HOPKINSON [1872, *1*].
[5] See TAYLOR'S collected papers [1958, *1*], pp. 516–519 (from TAYLOR [1946, *1*]), which contain a caustic comment concerning JOHN HOPKINSON'S two papers. Said TAYLOR: "I must confess that if my knowledge of the subject was confined to these experiments I should not have felt that they justified the statement that 'blows are equivalent when their velocities are equal.'" *Ibid.*, *p.* 517.
[6] JOHN HOPKINSON, *op. cit.*, p. 319.
[7] *Ibid.*
[8] *Ibid*; and see footnote no. 5 on p. 581.

HOPKINSON's results were inadequate for any firm conclusion. His velocities to break the wire at the lower clamp for the $7^1/_4$ lb., 28 lb., and 41 lb. weights were, if the fall actually were free, 250 in/sec, 236 in./sec, and 226 in./sec, respectively. He was considerably puzzled by the fact that the specimens did not fail upon the reflection of the wave at the upper clamp under the $7^1/_4$ lb. weight when a 2 ft. drop was sufficient for failure with the 41 lb. weight.[1] He stated that he had unsuccessfully attempted to ascribe this discrepancy to friction in the wire itself, comparable to such a phenomenon for sound in air. Finally he did ascribe it to momentary overstresses[2] and want of rigidity in the supports of the upper clamp which would favor the heavier weight.

Having judged that the results for the 28 lb. weight were sufficient to demonstrate the approximate doubling of the velocity, HOPKINSON concluded[3] that the linear theory he provided was adequate.

In his second paper on the subject,[4] after commenting on the "confirmation" of theory and experiment in the first, HOPKINSON conceded that the weight hanging below the clamp and the mass and elasticity of the clamp itself "which have a material effect on the results" had been wholly neglected. After stating, "these I have taken into consideration," he proceeded to present a new series of measurements in wires "from 9 ft. to 12 ft. long," with a 61 lb. suspended prestress weight, the results of which are given in Table 134.

The experiments are identical to those of the first paper, except that he supplied a column of corrected fall heights given as a multiplying factor $\left(\frac{M}{M+M'}\right)^2$ where M was the weight of the falling mass and M' was that of the 26 oz. clamp, and a series of measurements (noted by an asterisk) in which the lower clamp and adjacent wire had been submerged in ether, prior to the test, to lower the temperature. The all-important prestress weight still was ignored. The introduction of a spring at the upper clamp and a series of dry and soaked leather washers were not constructive except to indicate HOPKINSON's concern over the inadequacy of the experiment.

An examination of those additional data reveals that the overall average of the height of drop, whether measured or calculated as a reduced height, differed by only a small amount between that required for failure at the upper clamp and the bottom. For the actual heights, these averages were 72 in. for failure at the top, and 73.9 in. for failure at the bottom clamp, with a trend for the latter for the 7 lb., 16 lb., and 28 lb. weights of 84 in., 78 in., and 69.9 in., respectively. This decrease with increasing mass was not referred to in HOPKINSON's paper, nor was the fact that a simple calculation from the theory revealed that the dynamic stress plus the prestress fell considerably below the quasi-static rupture stress. The height of fall would have had to have been in excess of 100 in. to have been equal to that rupture stress, let alone higher, as his argument with respect to momentary overstresses would imply.[5]

---

1 This, of course, disagreed with his stated measurement requiring 4′6″ for failure at the upper clamp for the 41 lbs. weight.

2 This was a comment which undoubtedly influenced the experimental studies of his son, BERTRAM HOPKINSON, to consider such overstress in 1905. See BERTRAM HOPKINSON [1905, *1*].

3 JOHN HOPKINSON [1872, *1*].

4 JOHN HOPKINSON [1872, *2*].

5 Several years ago, in repeating HOPKINSON's experiments under what was hoped were similar conditions, inasmuch as he had not given the distance from the lower clamp to the prestressed weight, I found a similar lack of reproducibility. However, most of the failures occurred at the clamps themselves, which probably accounts for HOPKINSON's not reporting most of the measured distances from the clamp.

Table 134. JOHN HOPKINSON (1872).

Inches	Inches	Point of Rupture
	First series 16 lbs. weight	
72	60	18″ from top
78	65	12″ from bottom
78	65	24″ from top
81	$67^1/_2$	at top and bottom
82	$68^1/_2$	21″ from top
84	70	at bottom
84	70	at bottom
*48	40	did not break
*54	45	at bottom
*60	50	at bottom
*72	60	at bottom
	28 lbs. weight	
72	65	20″ from top
78	70	close to top
$79^1/_2$	$71^1/_2$	at bottom
81	73	at bottom
	7 lbs. weight	
81	54	at top
84	56	at bottom
*72	48	at bottom
*75	50	at bottom
	Second series 28 lbs. weight	
54	48	broke at top
60	$53^1/_2$	bottom and half-way up
60	$53^1/_2$	at top
63	56	at bottom
66	59	at bottom
69	$61^1/_2$	at bottom
72	$64^1/_2$	at bottom
*36	32	at top
*48	43	at bottom
	16 lbs. weight	
60	50	half-way up
66	55	at bottom
	With one dry leather washer	
72	60	4″ from bottom
66	55	near top
	Two dry washers	
72	60	6″ from bottom
	Three soaked washers	
78	65	broke in middle
83	69	at top

In recording my rather severe adverse criticism of the experiments of JOHN HOPKINSON I have had the objective of pointing up the dangers implicit in developing experiments intimately dependent upon the unchecked applicability of a theory, in his case linear elasticity, adopted solely for the sake of imposing analytical simplicity upon a situation known to be complex. The fact that during the past 30 years the obviously unread study of HOPKINSON has been cited as definitive in the introductions to literally hundreds of papers on dynamic plasticity, serves to emphasize a major tenet in the present treatise, namely, that it is important to distinguish between what may be merely the legendary roots and what are in fact the sound experimental bases upon which physically plausible theories may be built.

The paper of DUNN in 1897 characterized the experimental dilemma of his time in the second sentence of his introduction: "As ordinarily understood, an impulsive force is one whose intensity changes too rapidly to permit successive measures of it."[1] DUNN's ingenious experiment not only revised the definition, but certainly provided a major landmark in experimental solid mechanics by the introduction of high speed optical-photographic techniques which permitted the resolution of continuous displacement-time behavior in microsecond intervals.

DUNN's experimental objective was to provide a finite deformation dynamic stress-strain curve from a copper cylinder subjected to impact from a falling weight. The fundamental quasi-static assumptions made in interpreting his results underlie every extended quasi-static test from 1897 until the most recent "split HOPKINSON bar" tests of the 1960's. Until the 1940's, when electric resistance and capacitance techniques appeared, there was no essential change from DUNN's fundamental experiment, despite the large number in each decade who either claimed to discover the experiment or ascribed its origin to someone else.

By 1890 the method of POUILLET[2] from the 1840's was available to measure single events of duration of the order of $^1/_{10\,000}$ sec. The best resolution of a vibrating tuning fork recording on the lampblack or pine dust of a rotating metal cylinder, was about 500 vibrations/sec, with a recording surface velocity of 10 ft./sec. This limit in milliseconds was insufficient for the study of phenomena occurring in microseconds. DUNN referred to unsuccessful efforts to improve this technique, and in the summer of 1891 he tried a new approach.[3]

Using the sun for a light beam source and a hole with perfectly smooth edges bored in the end of the tuning fork as a reflector, DUNN obtained a trace of the vibrating fork on photographic paper wound on a rapidly rotating glass cylinder. A proper combination of lenses was used to get a sufficient concentration of the light beam on the photographic film. DUNN's diagram of this method as applied to the problem of determining for the copper specimen the displacement-time history of the falling weight throughout impact is shown in Fig. 4.120.

The new optical system allowed a tuning fork calibration for $^1/_{4000}$ sec on a recording surface traveling about 100 ft./sec. DUNN pointed out that these were by no means upper limits, but they were adequate values for his objective.

The obtaining of a freely rotating glass cylinder carefully ground to accurate diameter, the development of a dual shutter system to permit of the simultaneous observation of mass displacement and calibrated tuning fork, the optical calibration of the tuning fork system, and the electromagnetic triggering scheme which placed the entire system in automatic operation in a darkened room, were all described in careful detail by DUNN. He observed that this was most certainly an experiment for the research laboratory and not for the ordinary engineer. He then devoted a considerable portion of the remainder of an otherwise remarkable

---

[1] DUNN [1897, *1*], p. 321.

[2] POUILLET [1844, *1*].

[3] That DUNN stood on the watershed of a new age in more than experimental mechanics may be noted from the description on the page before his article in the Journal of the Franklin Institute, of the new invention of the industrial time clock which "renders unnecessary the services of a clerk timekeeper"; it was recommended that the John Scott Legacy Premium and Medal be given the inventor.

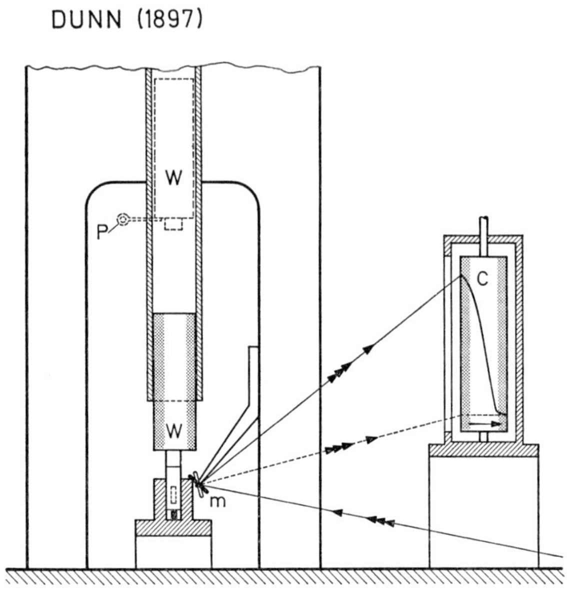

Fig. 4.120. The apparatus for obtaining the first "dynamic" stress-strain curve from an extended quasi-static impact experiment.

paper to suggestions for simplified guessing procedures for the practical technologist.

The idea of the experiment was simple and is basic to all such extended quasi-static tests to the present day. If the displacement vs time history of the two ends of a specimen are determined experimentally, the difference provides an overall average strain-time history if the whole specimen is subject to one dimensional uniform strain during the entire history of the impact. In DUNN's experiment he had assumed zero displacement for the anvil upon which the specimen rested, and the displacement of the opposing end in contact with the falling mass was given by the measured displacement-time history of the mass. The double differentiation of this measured curve, presuming a rigid mass devoid of linear elastic wave propagation and reflection and not undergoing multiple impacts from the combined wave behavior in specimen and mass, provided an acceleration-time curve from which the stress-time history for the specimen was given. Again, as for strain, a uniform distribution of stress was assumed during the entire impact history. Plotting this calculated stress vs the calculated average strain gave what was hoped would be a plausible dynamic stress-strain curve.

In Fig. 4.121a is shown a measured mass displacement-time curve given by DUNN from which, by graphically differentiating, he obtained the velocity-time curve of Fig. 4.121b. Again graphically differentiating the velocity-time curve he obtained the acceleration-time plot of Fig. 4.121c which he referred to as the "retardation curve" from which he obtained the dynamic stress-strain curve of Fig. 4.121d. In both differentiations DUNN smoothed the results as shown. The dynamic stress-strain curve compared with a quasi-static test in a similar copper specimen is the first such plot in a literature which by now contains literally thousands of such comparisons from data obtained by similar or nearly similar means and assumptions.

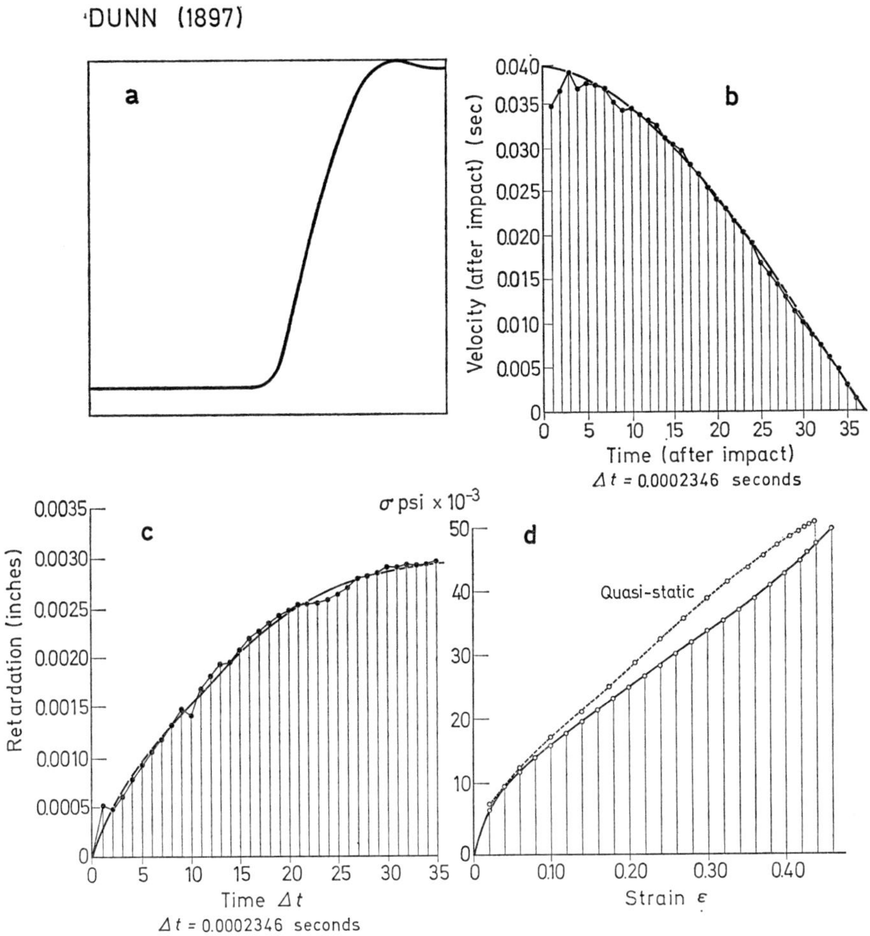

Fig. 4.121 a–d. DUNN's double graphical differentiation. (a) The experimental displacement-time data; (b) The smoothed initial slopes; (c) The smoothed second slopes; (d) The calculated "dynamic" stress-strain curve compared with a measured quasi-static test in the same solid, copper.

One of the marked characteristics of such dynamic stress-strain functions from DUNN's to those today is the lack of reproducibility from one experiment to another. Nearly every paper presenting such results, from RUDOLPH PLANK's[1] in 1912 in Berlin (PLANK, incidentally, attributed the origin of DUNN's experiment[2] to the Berlin Doctoral dissertation of a student named HOENINGER in 1910) to recent publications, comment upon the inadequacy of previous experimental results and then suggest some modification of the extended quasi-static test so as to get more reliable data.

That DUNN's stress-strain curve from impact testing lies below the quasi-static curve could have very important technological implications if the simple quasi-

[1] PLANK [1912, *1*].

[2] The only reference to DUNN's work I have run across in later years was that in the paper of KÖRBER and STORP [1925, *1*]. The DUNN experiment is most often erroneously attributed to HATT who in fact merely repeated DUNN's test seven years later. HATT [1904, *1*].

static assumptions of uniformly distributed, one dimensional stress and strain were indeed correct.[1]

Although as noted above, each successive experimentist emphasized the limitations of such experiments, especially for the initial plastic strain region, almost none when considering their results gave more than passing comment to their neglect of the influence of finite amplitude wave propagation, wave interaction, and complex plastic-elastic and elastic-plastic boundary phenomena.

4.26. On the effort for seventy-five years to extend Dunn's experiment, based upon his assumption that quasi-static and impact tests on short specimens are identical.

In every decade for over sixty years, extensive efforts were made to improve or extend DUNN's experiment. Because technologists continue to seek a trustworthy version of DUNN's test, and because the existence of the large early literature has been forgotten, in an earlier draft of this treatise I had traced in detail most of the variations and extensions of DUNN's experiment to the present time. I now judge such details to be superfluous. It seems more appropriate to limit the description of the sequel to DUNN's experiment, and, with one or two exceptions, to refer only briefly to the few experiments which modified the original, since the seven decades of variations on DUNN's experiment have witnessed the raising of the same questions, assertions, and doubts as had been raised over and over throughout the entire interval; indeed, they are still under constant discussion.

In 1912, RUDOLPH PLANK[2] repeated DUNN's experiment. PLANK's substitution of mechanical recording for photographic actually was a step backward. His discussion of sources of error, about which he did nothing, was referred to many times by later experimentists who did attempt to consider those problems.

In 1914, H. SEEHASE,[3] attributing the idea to the inspiration of his professor, Dr. EUGEN MEYER, repeated the DUNN experiment but used a projection lamp instead of the sun as a source of light. Although he believed he had designed an original experiment SEEHASE in fact incorporated all of the improvements in photographic recording and calibration which DUNN himself had suggested.

After some earlier difficulties,[4] FRIEDRICH KÖRBER[5] in 1925 succeeded in developing the DUNN experiment in its original form, but with the grave limitation that KÖRBER used the falling mass to calibrate his apparatus. Like SEEHASE, he included the deflection of the anvil in his analysis.

Throughout these years, experimentists[6,7] also were performing pre-DUNN measurements in which the kinetic energy of the falling mass at the beginning of impact was compared with the permanent deformation of the specimen after impact was over.

---

[1] CONSIDÈRE in 1885 had tested thin wires dynamically by a falling weight, estimating the work and force at rupture, He suggested that the dynamic values were 42 % higher than the quasi-static for the work, and 37 % higher for the force (CONSIDÈRE [1885, *1*]). According to KÖRBER and SACK [1922, *1*] a German engineer, A. MARTENS, in 1891 and a French engineer, LEBASTEUR in 1892, also from consideration of strain energy in axial impact tests, concluded just the opposite, namely, that the quasi-static test governed all loading rates.

[2] PLANK [1912, *1*].

[3] SEEHASE [1914, *1*].

[4] KÖRBER and SACK [1922, *1*].

[5] KÖRBER and STORP [1925, *1*].

[6] KÖRBER and SACK [1922, *1*].

[7] TAFEL and VIEHWEGER [1931, *1*].

Between 1931 and 1936, MITITOSI ITIHARA[1] in a series of five reports issued by the Tohoku Imperial University described many comparisons of quasi-static and impact torsion tests on very short specimens of mild steel and copper. His was an adaptation of DUNN's experiment to torsion, using the same photographic technique. ITIHARA performed his tests at ambient temperatures between $-170°$ C and 1 000° C.

PLANK, SEEHASE, KÖRBER, and ITIHARA all attempted to compare dynamic and quasi-static response at finite strain, using the same quasi-static assumptions, experimental technique, and method of graphical analysis which had been introduced by DUNN.

In 1937, D. W. GINNS[2] modified DUNN's experiment by using a spring loading system to produce the plastic deformation and a carbon pile resistance stack to measure elongation. In a written discussion to GINNS' paper, TAYLOR described an alternate modification of DUNN's experiment. The details of his ballistic pendulum, his method of analysis, and the results he obtained in these tests performed with H. QUINNEY, TAYLOR gave much later[3] in a James Forrest lecture to the Institute of Civil Engineers in London in 1946. TAYLOR said that by these means he had achieved strain rates in tension as high as $\dot{\varepsilon} = 3\,000\ \text{sec}^{-1}$ for annealed copper and mild steel. His results will be shown below in Fig. 4.126, where such data are compared.

In the same paper TAYLOR described experiments performed with R. M. DAVIES,[4] which it seems to me provided the most convincing evidence in the literature of the time that for some steels the yield stress, for whatever reason, could be increased in impact loading. TAYLOR and DAVIES dropped steel balls on a steel plate resting on rubber supports, with ever increasing heights of drop until permanent deformation indentation appeared. For the highest height of elastic behavior, the maximum normal pressure, P_d, could be calculated from the HERTZ[5] contact theory and compared with the pressure required from a quasi-static loading, P_s, to produce an indentation of the ball on the plate. The HERTZ theory for the two situations provided:

$$\frac{P_d}{P_s} = 40.4\, \mathrm{h}^{\frac{1}{5}}\, R^{\frac{2}{3}}\, M^{-\frac{1}{2}} \tag{4.35}$$

where h was the height of drop; R was the ball radius; and M was the mass.

Table 135 gives TAYLOR and DAVIES' results for one particular steel, where the ratio of P_d/P_s was nearly unity in two tests, and 1.18 in two others, indicative of the often observed variation in this matter in all types of testing.

Such experiments give indications that when the infinitesimal elastic strain rate is increased there is an increase in the elastic limit. However, TAYLOR and DAVIES did not provide any information with respect to the importance or unimportance of viscosity in finite strain plasticity at high strain rates.

In 1938 DONALD S. CLARK and G. DÄTWYLER[6] performed an impact pendulum experiment. For the first time in a version of DUNN's experiment, electric resistance gages and oscillography were used on a dynamometer bar to infer a stress-time history at one end of a plastically deforming specimen. The measurement, of

[1] ITIHARA [1933, *1*], [1935, *1*], [1936, *1*].
[2] GINNS [1937, *1*].
[3] TAYLOR [1946, *1*].
[4] TAYLOR and DAVIES, see TAYLOR [1946, *1*]; in TAYLOR's collected works [1958, *1*], see p. 521 et seq.
[5] HERTZ [1882, *1*].
[6] CLARK and DÄTWYLER [1938, *1*].

Table 135. TAYLOR and DAVIES (1946).

Steel	Static				Dynamic					
	Brinell no.	Least mass for indentation (g)	Greatest mass for no indentation	M (g)	Least height for no indentation	Greatest height for no indentation	h (cm)	R (cm)	P_d/P_s [from Eq. (11)]	Yield stress in tensile test (tons/sq. in.)
WTM	351	4510	3610	4060	1.1	0.9	1.0	0.137	1.18	74
WTN	321	8020	6330	7180	0.4	0.3	0.35	0.476	1.03	69
WTN	321	2400	1910	2150	0.4	0.3	0.35	0.317	1.18	69
WTN	321	1100	890	1000	0.5	0.4	0.45	0.159	1.01	69

course, was made on an adjacent bar in which elementary linear elasticity could be inferred and extrapolated to the specimen interface. Instead of a measured overall elongation like DUNN's, CLARK and DÄTWYLER made a double integration of their force vs time measurement to obtain what was assumed to be an average elongation vs time history. In contrast with the data TAYLOR and QUINNEY had obtained that same year, the yield stress of copper now presumably increased markedly. Mild steel and stainless steel exhibited the usual increases in yield stress, while the SAE 6140 steel showed a dynamic curve considerably below the quasi-static, a state of affairs observed more often than one might be led to believe from most review papers. Aluminum exhibited no increase in dynamic response over quasi-static until strains above 0.04, although many experimentists using extended quasi-static impact tests claim to have seen higher stresses below this strain.

In 1935, H. C. MANN[1] who did not attempt to determine dynamic response functions because he was interested solely in the thermodynamics of energy criteria, introduced the rapidly rotating flywheel to apply a nearly constant velocity to one end of a specimen in tension. MANN hoped to use the experiment to study the thermodynamics of rapid deformation. For this reason he concentrated on measuring the loss of energy. Although he contributed little to thermoplasticity, he did succeed in finding a "transition velocity" (an experimental discovery in 1935 of VON KÁRMÁN's critical velocity of 1942)[3] above which the absorbed deformation energy decreased. Although MANN did not attempt to determine response functions in impact loading, it was not long before other experimentists tried to do so, such as a group at the California Institute of Technology in 1938 and MANJOINE and NADAI[2] in 1940.

Probably the most widely discussed experimental results of the past 30 years which were based upon DUNN's experiment and hypotheses, were those of MANJOINE and NADAI in 1940–1941. They were published in three parts, the first in the *Proceedings of the American Society for Testing Materials*, and Parts II and III in a single paper in the *Journal of Applied Mechanics* in June, 1941. By 1940 DUNN's unique contributions in 1897 to experimental mechanics, both in technique

[1] MANN [1935, *1*].
[2] MANJOINE and NADAI [1940, *1*], NADAI and MANJOINE [1941, *1*].
[3] VON KÁRMÁN [1942, *1*].

and in this particular experiment, had been completely forgotten. MANJOINE and NADAI referred to PLANK as the originator of the experiment[1] in 1912.

After a brief resume of the results of a few of their predecessors who had used this type of test, MANJOINE and NADAI discussed in some detail the torsion studies of ITIHARA,[2] as the main precursor for their own studies since he had performed tests at ambient temperatures from −170° C to 1000° C. They were correctly critical of the extreme shortness of ITIHARA's torsion specimens with a ratio of length to diameter of 1.25 which prohibited a meaningful description of a stress-strain response for torsion but, strangely, they obviously considered it of importance that the results of ITIHARA "had a great resemblance to the tension test diagrams described below."[3]

MANJOINE and NADAI performed the experiments with the high speed flywheel which had been introduced by MANN.[4] The massive spinning wheel contained a device which could be triggered to engage one end of a specimen. They attached the other end of the specimen to a metal load bar, the linear elastic deformation of which was recorded optically by means of a slit. The upper edge of the slit followed the motion of the bar, and the elongation was recorded by a beam of light which passed through the 0.002 in. slit and fell on a phototube. At the anvil end of the specimen, which was engaged by the tups from the wheel, a second beam of light was cut by the moving end of the specimen. With the voltage output of the two tubes presented as rectangular motion on the oscilloscope, a dynamic stress-strain curve was provided which required no auxiliary analysis except the introduction of the DUNN hypotheses and the calibration of the two displacement signals as stress and strain. In their paper[5] MANJOINE and NADAI gave sample oscillograms from 35 mm photographs. The ambient temperature had been varied by means of an induction furnace.

In Fig. 4.122 are shown in (a), (b), and (c), temperature variations for three different strain rates, 135 sec^{-1}, 450 sec^{-1}, and 900 sec^{-1}; and in (d), all three strain rates and a quasi-static test at room temperature. The material tested was pure copper. The strain rates were inferred from the wheel speed which remained nearly constant during the test.

MANJOINE and NADAI's main interest was in ultimate strength, which they noted tended to increase with increasing strain rate. In 1966 in plotting their data at room temperature, I noted[6] however, that for stresses below their maxima the stress-strain curves to large strain were directly the reverse, i.e., 135 sec^{-1} was the highest, and 900 sec^{-1} was the lowest. None of those who have widely quoted the tabulated ultimate stress have commented upon this fact. In TRESCA's intermediate region, MANJOINE and NADAI's data do not support visco-plasticity hypotheses. More detail on this point will be given below when I summarize the extended quasi-static experiments.

Part II of this series by MANJOINE and NADAI saw their extension of these experiments to the study of steel and aluminum for which considerably more difficulty was encountered from major oscillations.

---

[1] As I have noted, since PLANK [1912, *1*] ignored DUNN's work, only 15 years after the original experiment, once again we see, through errors of precedence, how records of experimental excellence, fade into oblivion until extracted by hoary encyclopedists.

[2] ITIHARA [1933, *1*], [1935, *1*], [1936, *1*].

[3] MANJOINE and NADAI [1940, *1*], p. 826.

[4] MANN [1935, *1*].

[5] *op. cit.*

[6] BELL [1966, *1*].

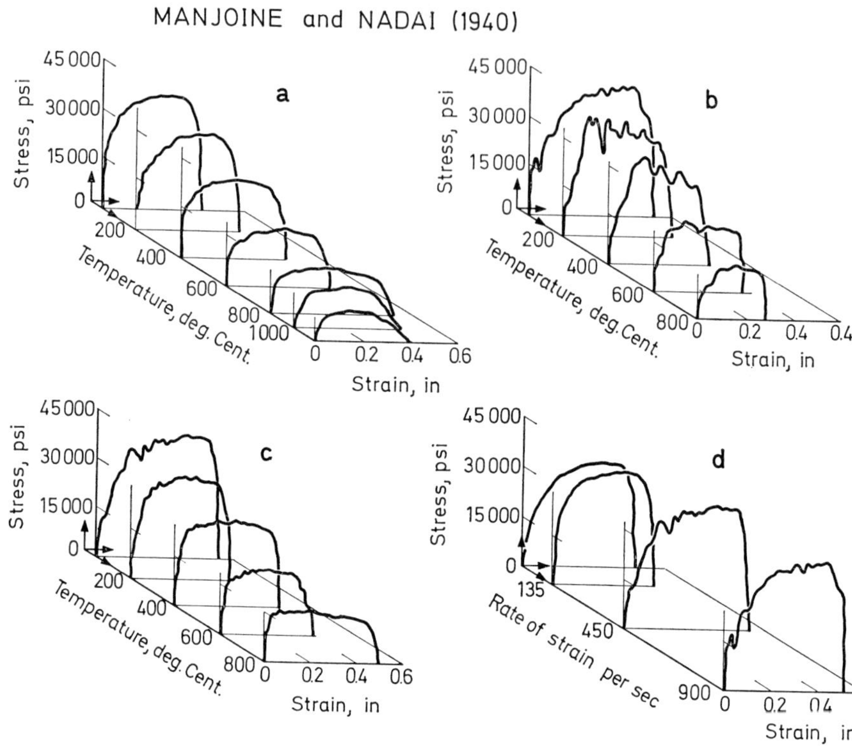

Fig. 4.122a–d. The high speed fly wheel version of the Dunn experiment for pure copper from room temperature to the high temperatures indicated for estimated strain rates of (a) 135 $\sec^{-1}$, (b) 450 $\sec^{-1}$, (c) 900 $\sec^{-1}$. In (d) is shown a comparison of the static stress-strain curve with dynamic results at the three strain rates, all tests being at room temperature.

A. F. C. Brown and N. D. G. Vincent[1] in 1941, just before dynamic plasticity began to be considered in the context of theories of wave propagation,[2–6] in their brief historical introduction described in disparaging terms a few of the many results from extended quasi-static impact tests preceding their own work. Brown and Vincent's version of the Dunn impact experiment compared the stress-time history at one end of the specimen, obtained by means of a pair of piezo crystals, with the average strain, obtained by means of a slit system and photoelectric cell. Although they no longer had to perform graphical integration, they included all of Dunn's other assumptions as to the similarity between quasi-static and impact tests. Once again the impact machine was a pendulum apparatus capable of producing strain rates which they estimated were as high as 800 $\sec^{-1}$. Their

[1] Brown and Vincent [1941, *1*].
[2] Taylor [1942, *1*].
[3] von Kármán [1942, *1*].
[4] White and Griffis [1942, *1*], [1947, *1*], [1948, *1*].
[5] Rakhmatulin [1945, *1*].
[6] The waste of war not only led to the uncorrelated, separate development of an issue by so many, but delayed for several years, through the secrecy of publication, the general knowledge that such a theory, with related experiments by Duwez, actually existed. (See Duwez, Wood, and Clark [1942, *1*]; von Kármán and Duwez [1946, *1*]; Duwez and Clark [1947, *1*]; von Kármán and Duwez [1950, *1*].)

results were not much more reliable than those which had preceded them, due in part to the severe oscillations recorded.

That the state of the DUNN experiment had not advanced significantly by 1948 may be seen in the experiments of E. T. HABIB[1] who, unaware of his many predecessors, made an historically interesting return to the 19th century pre-DUNN studies.[2] The permanent deformation of copper cylinders again was being used as a standard for dynamic pressure.

Steel pistons of different masses were accelerated by an air gun to strike small cylinders of oxygen free, high conductivity copper which had been annealed in an atmosphere of hydrogen at 950° F. The temperature was maintained for $2^1/_2$ to 3 h, following which the specimens were furnace cooled with the flow of hydrogen being maintained until the temperature dropped below 250° F. By streak photography, HABIB determined the velocity of the steel piston before and after impact. In the manner of earlier studies he used these differences to calculate an energy loss to the specimen; he plotted this loss against the measured permanent deformation in the specimen. With the usual assumptions of one dimensionality, etc., HABIB converted the results of many tests to a proposed response function for high strain rates.

In 1949, HERBERT KOLSKY[3] introduced another version of the DUNN experiment, still based upon DUNN's hypotheses. In an elegant adaptation of the original BERTRAM HOPKINSON[4] pressure bar study, KOLSKY provided an experiment which to some technologists over a decade later seemed likely to provide an answer to the century-old search for an impact test to evaluate materials. If many persons came to view the results of such tests as of limited or very questionable value in fundamental physics, it was largely because of the contemporaneous experiments[5] which finally permitted the detailed and accurate study of non-linear wave propagation at finite strain. Particularly during the 1960's, KOLSKY's version of DUNN's experiment, while subjected to much criticism, continued to be developed as an impact test for materials technology, primarily because its simplicity permitted its being used widely, whereas the obviously far more fundamental experiments on finite amplitude waves seemed to demand so large an investment of time and of expertness from the experimentist as to limit their use to the university research laboratory. (The reader will recall that in 1897 DUNN had had the opinion that his experiment, too, was too complicated to leave the research laboratory!)

KOLSKY's experiment,[6] which soon came to be known as the "split HOPKINSON bar test," was based upon the behavior in axial impact of long hard bars in the presence of an extremely short soft specimen, referred to as a "wafer," mounted between the hard bars. A detonator, fired electrically, was located at the far end of one hard bar, as shown in Fig. 4.123, while a parallel plate condenser microphone to provide displacement-time histories through the variation of capacitance, was located on the face at the far end of the second bar. A cylindrical condenser microphone[7] was located at some distance from the detonator to provide informa-

[1] HABIB [1948, *1*].

[2] Again, in this return to an earlier type test for which there is a very large 19th and 20th century literature, no reference extended back more than eight years, and there was no allusion to the earlier versions of this test.

[3] KOLSKY [1949, *1*].

[4] B. HOPKINSON [1905, *1*].

[5] BELL [1956, *1*, *2*], [1958, *1*], [1960, *1*, *2*, *3*], [1966, *1*].

[6] KOLSKY [1949, *1*].

[7] This device was similar to that described in the linear elastic wave studies of R. M. DAVIES the previous year [1948, *1*].

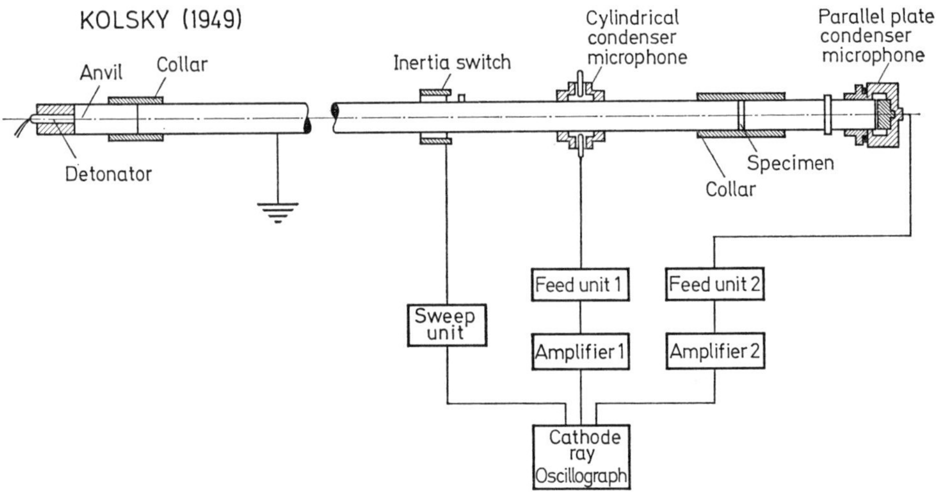

Fig. 4.123. The apparatus for the Kolsky wafer experiment.

tion on the radial displacement of the initial and reflected waves from the wafer, as was an inertia switch to trigger the oscilloscope upon which the displacement-time histories were recorded.

As is typical of Kolsky's writing, there is a thorough, meaningful discussion of experimental details, difficulties, and limitations, which unfortunately is not typical of most of the papers of those who subsequently described variations on this experiment. Kolsky compared displacement-time histories at the far end of the second bar, with and without the short wafer specimen between the bars; by these means he calculated the displacement-time histories which were present on either side of the wafer. This calculation, of course, was based upon linear elastic wave theory assuming no viscous or geometric dispersion of pulses traveling along the hard bars.

Differentiation of these same displacement-time histories from the parallel plate condenser provided initial and transmitted stress-time histories by means of the stress velocity function of the linear theory. To interpret the data in terms of stress as a function of strain at high strain rate, it was necessary to introduce in the usual manner Dunn's hypothesis that quasi-static conditions hold in dynamic plastic flow. As usual, too, nonlinear wave propagation, reflection, and interaction (including that of small amplitude and finite amplitude components) were ignored.

Kolsky's stress-strain calculations from the tests for polythene (a), natural rubber (b), Perspex (c), copper (d), and lead (e), are shown in Fig. 4.124. The experimental points were at 2 μsec intervals for specimens of copper 0.05 cm thick, and lead 0.05 cm and 0.10 cm thick, which gave the different results shown.

Kolsky was aware of the importance of interface friction. As he stated in his conclusions, however, he found that the introduction of lubricants made no essential difference. This fact, plus the facts that the specimens remained cylindrical and that the measured average permanent deformation in copper did not differ markedly in final value from the average found by subtracting the two parallel plate condenser displacement-time histories, were deemed sufficient in 1949 (as they were so regarded among test engineers who used variations on Kolsky's

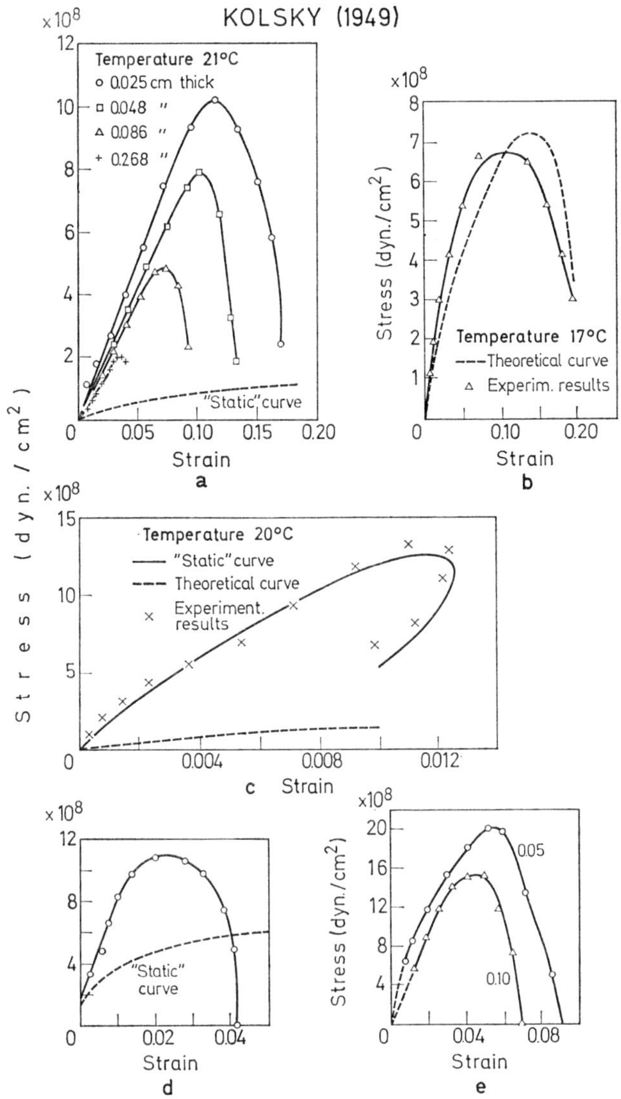

Fig. 4.124a—e. Dynamic results for KOLSKY's version of the DUNN experiment. The data was for thin wafers of (a) polythene, (b) natural rubber, (c) perspex, (d) copper, and (e) lead.

experiment in the 1960's) to justify the one dimensional extended quasi-static interpretation of the data, i.e. the DUNN hypothesis.

KOLSKY had emphasized the necessity of using an extremely short specimen to minimize the effects of wave interaction. This point, as well as some of the three dimensional aspects of this problem, were explored by E. D. H. DAVIES and STEPHEN C. HUNTER[1] several years later in their experiments in 1963 with KOLSKY's split HOPKINSON bar technique. They showed that the length of the

[1] E. D. H. DAVIES and HUNTER [1963, *1*].

wafer could not be chosen arbitrarily as some others had done with the hope of changing the strain rate.

During the 1950's most experimentists turned in one form or another to the effort to study direct or indirect aspects of wave propagation in the context of the nonlinear wave theory which had been developed in the 1940's. Unwittingly reversing this step forward however, at a Symposium in 1960 on the Response of Metals to High Velocity Deformation, HAUSER, SIMMONS and DORN[1] in a paper curiously titled, "Strain Rate Effects in Plastic Wave Propagation" (since the quasi-static hypotheses of DUNN were used to interpret the experimental results) presented a modification of KOLSKY's experiment. The specimen no longer was required to be a short wafer but was allowed to range from 0.1 to 1.6 in. to provide, by a change in the gage length of the specimen, different strain rates in the quasi-static sense, for different applied maximum stresses. Instead of using KOLSKY's short time explosive pulse, they introduced a third hard bar whose collision with the initial bar produced a trapezoidal shaped loading pulse whose duration could be made as long as desired by changing the length of the impacting bar. This pulse traveling along the struck hard bar eventually reached the interface between the hard bar and the relatively short plastically deforming specimen, where reflected and transmitted waves occurred. The former were reflected back in the struck bar, and the latter deformed the specimen to finite strain and emerged to travel along the second hard bar.

The gist of the experiment was to measure the incident and reflected waves in the first hard bar and the final transmitted wave in the second hard bar by means of flexure eliminating electric resistance strain gages. Interpreting such data by extrapolation as stress and displacement at each interface, and introducing all of DUNN's hypotheses on the identity of quasi-static and impact tests, HAUSER, SIMMONS and DORN plotted averaged end loads and averaged strain from end displacements so as to get a dynamic stress-strain function.

In HAUSER, SIMMONS, and DORN's paper, for the first time in over sixty years an effort was made to examine the DUNN experiment in terms of reflecting plastic waves. Since they used the DONNELL[2] bilinear hypothesis of 1930, and, curiously, one of the "dynamic" response functions based upon DUNN's hypotheses, this first effort was unsuccessful. They concluded that the effects of wave propagation and reflection were unimportant. In a doctoral dissertation in 1964, ANDREW CONN,[3] using the same "dynamic" stress-strain function and averaged data, showed that a better approximation in terms of the more general KÁRMÁN-TAYLOR theory of waves of finite amplitude, proposed in 1942, led to results on wave propagation and reflection much closer to experimental data and thus revealed that the main source of error in the DUNN experiment lay in the neglect of nonlinear wave propagation and reflection. As CONN further emphasized, any wave analysis employing the complex interaction of the short specimens in experiments in dynamic plasticity, demands that the desired stress-strain function be known in advance in order that the long computer calculation can be performed.

The KOLSKY experiment in the modified form described above was so simple to devise and interpret in quasi-static terms that it rapidly became a technological standard for the evaluation of materials in the 1960's, but not without considerable controversy as to the validity of the results obtained.

---

[1] HAUSER, SIMMONS, and DORN [1961, *1*].
[2] DONNELL [1930, *1*].
[3] CONN [1964, *1*], [1965, *1*].

For example, J. L. CHIDDISTER and LAWRENCE E. MALVERN[1] in 1963 performed split bar experiments on aluminum wafers over a wide range of temperatures. E. D. H. DAVIES and STEPHEN C. HUNTER[2] in addition to providing an interesting evaluation of the experiment, gave results for aluminum, copper, and lead. ULRIC S. LINDHOLM[3] in a series of papers described split HOPKINSON bar results for aluminum, copper, lead, and iron. LINDHOLM and L. M. YEAKLEY[4] obtained such data in 1965 in single crystals, and in 1967 they modified the wafer to measure split bar results in tension. H. P. TARDIF and H. MARQUIS[5] in 1963 added to KOLSKY's earlier results in polythene by performing split HOPKINSON bar tests on eight different plastics, including lucite, nylon, and teflon. WILLIAM BAKER and C. H. YEW[6] in 1966, and E. CONVERY and H. L. PUGH[7] in 1968 performed torsion versions of the test by releasing an elastic prestress to produce the initial wave.

Experimental support for the physical applicability of viscoplastic analyses of dynamic plasticity and, to a lesser extent, model production in the field of dislocations, in the main rests upon the validity of DUNN's hypotheses. In 1966, in the initial portion of a paper describing a series of split HOPKINSON bar experiments in which, for the first time, a direct diffraction grating measurement of local strain was made on the wafer *during* the test,[8] I compiled for the purpose of comparing them, the results of quasi-static impact tests of the DUNN type by eight experimentists between 1937 and 1964, on the same solid, annealed copper. In Fig. 4.125 are shown the quasi-static stress-strain curves for annealed copper of six of those experimentists. Except for KOLSKY's[9] test, which was performed upon a wafer 0.05 in. thick, there was sufficient general agreement among the quasi-static tests of all the experimentists to ensure that they indeed were studying the large deformation of the same solid.

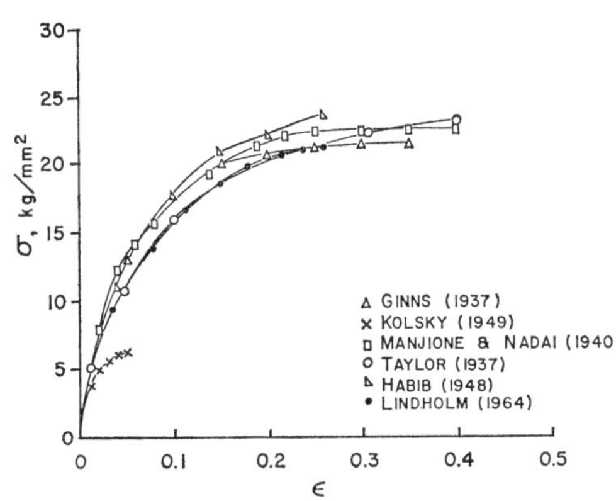

Fig. 4.125. Quasi-static tests in annealed copper provided by the experimentists listed for comparison with their dynamic data obtained from the assumption of the DUNN hypotheses shown in Fig. 4.126. See BELL [1966, *1*].

[1] CHIDDISTER and MALVERN [1963, *1*].
[2] E. D. H. DAVIES and HUNTER [1963, *1*].
[3] LINDHOLM [1964, *1*].
[4] LINDHOLM and YEAKLEY [1965, *1*], [1967, *1*].
[5] TARDIF and MARQUIS [1963, *1*].
[6] BAKER and YEW [1966, *1*].
[7] CONVERY and PUGH [1968, *1*].
[8] BELL [1966, *1*].
[9] KOLSKY [1949, *1*].

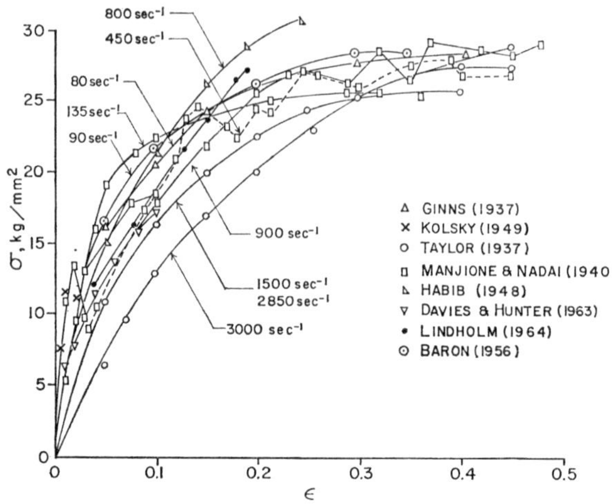

Fig. 4.126. Stress-strain data from tension and compression "extended quasi-static" impact tests in annealed polycrystalline copper. The strain rates are the individual estimates provided by the experimentists whose data are included. See BELL [1966, *1*].

In Fig. 4.126 is my compilation from that paper in 1966, of the dynamic stress-strain curves of those same experimentists obtained from their version of the DUNN experiment.

A detailed examination of the wide scatter in these published[1] "dynamic" stress-strain curves for the same solid as for the quasi-static tests of Fig. 4.125, reveals the inadequacy both of the DUNN experiment and of the conjecture that quasi-static and impact tests are identical. If these data were taken seriously, one would have to reach the inescapable conclusion that the height of the dynamic stress-strain curve *decreases* throughout all but the final maximum region with *increasing* strain rate. The highest strain rate of TAYLOR,[2] 3000 sec^{-1}, was for the lowest curve reported, with 1500 sec^{-1} and 2850 sec^{-1} coinciding above it. The experiments of MANJOINE and NADAI[3] at 135 sec^{-1}, 450 sec^{-1}, and 900 sec^{-1}, lay precisely in the reverse order of stress magnitude until strains of well over 20%. The lowest strain rate data of H. B. BARON,[4] $\dot{\varepsilon} = 90$ sec^{-1}, and GINNS'[5] $\dot{\varepsilon} = 80$ sec^{-1}, were among the highest stressed curves to finite strains of 20%.

The above is fairly typical of comparisons of extended quasi-static tests in other solids at intermediate strain for tension, as well as for compression. Yet, almost inconceivably, it is from such experimental data that many persons accept the

[1] Many of them had to be recalculated back from reduced stress, logarithmic strain, to their original nominal form for the comparison. See BELL [1966, *1*].

[2] TAYLOR [1937, *2*].

[3] MANJOINE and NADAI [1940, *1*].

[4] BARON [1956, *1*]. BARON performed his experiments using the pendulum of a Charpy notched bar machine. The specimens were in a small, tensile testing attachment. Electric resistance gages on a hollow steel tube were in line with the specimen. From these measurements, interpreted as a force-time diagram, the usual double integration provided what was assumed to be the average strain-time history. The chief feature of the tests was that they were performed at −196° C, −78° C, as well as at the usual 20° C. The materials considered were Armco iron, mild steel, three more complex steels, annealed aluminum, annealed copper, and annealed brass (39% zinc). These extended quasi-static data for the most part exhibited higher dynamic values. The data for copper, shown in Fig. 4.126, oddly enough differed far more than those for brass which only slightly increased at all temperatures.

[5] GINNS [1937, *1*].

importance of viscosity in dynamic plasticity. The main experimental point for the theorist interested in these matters is that from such comparisons one must conclude that 70 years of experimental compromise and approximation based upon DUNN's hypotheses have been inadequate to justify any physical assumptions either in viscous plasticity or nonviscous plasticity, let alone theories of dislocation and of finite amplitude waves.[1]

As with the notched bar impact test, I leave the judgment of the technological usefulness of the DUNN experiment, whatever its form, to those who need a simplified evaluation of materials to face the demanding problems of engineering design. Seventy-five years of experimentation in this direction have contributed little toward understanding the physics of solids.

4.27. Finite amplitude wave propagation in annealed polycrystals: Experiments from 1942 to 1956.

Unquestionably, the scant achievement in experiment on impact before 1940 can be attributed to the failure to conceive the subject in terms of nonlinear wave propagation, as well as to the lack of an experimental technique to study such waves. Although a method for the direct experimental study of profiles of waves of finite amplitude in solids did not appear until the first measurement of finite strain by means of diffraction gratings[2] in 1956, guidance from a plausible nonlinear theory became possible after 1942. It was a theory which extended to isochoric deformation some of the ideas contained in a nonlinear theory of dilatational waves in a solid discussed in the 1880's by HUGONIOT. It was proposed by TAYLOR[3] in 1942, by VON KÁRMÁN[4] in 1942, and in more restricted forms by RAKHMATULIN[5] in 1945 and by WHITE and GRIFFIS[6] in 1942.

[1] Objections to the use of DUNN's hypotheses have been common. JAMES J. GUEST was an outspoken critic in the 1930's (see [1930, *1*]). An illustration in this vein, after the development of a proper wave theory in 1942, was the statement of DONALD SHERMAN CLARK in 1950.

> "Some investigators have expressed the results of tensile impact tests in terms of strain rate. It can be shown readily with the aid of the theory of plastic strain propagation that such a practice is untenable. In such a test, the strain rate varies in all parts of the specimen nonuniformly from a relatively low value to almost infinity. By this same reasoning it is not permissible to convert a force-time diagram obtained at either end of the test specimen in an impact test into a stress-strain diagram. This difficulty can be surmounted only by the elimination of the propagation effect from the test. Such a requirement dictates a test of entirely different character." (CLARK and WOOD [1950, *1*], p. 48.)

Of equal interest is the further comment of CLARK when he was giving a CAMPBELL Memorial Lecture to the American Society of Metals in 1953.

> "The most fundamental indication of the resistance of a metal to deformation is a stress-strain relation. But how does one obtain a stress-strain relation during impact? Attempts have been made by many investigators to secure data that would permit the establishment of this relation. Until recently, all of these attempts have produced questionable results. Such results are disqualified because consideration was not given to the propagation phenomenon which prevails during impact. The tests have been made by measuring the force acting at one end of a specimen and the corresponding strain over some specified length of the specimen. In view of what has been said here about the propagation of strain, it should now be clear that the results of impact tests in which propagation effects are not taken into account will provide only average values and will not allow the establishment of the true relationship between stress and strain during the impact test." (CLARK, published [1954, *1*], pp. 38–39.)

[2] BELL [1956, *1*].

[3] TAYLOR [1942, *1*].

[4] VON KÁRMÁN [1942, *1*].

[5] RAKHMATULIN [1945, *1*].

[6] WHITE and GRIFFIS [1942, *1*], [1947, *1*], [1948, *1*].

The one dimensional theory of TAYLOR and VON KÁRMÁN for loading waves in a solid is, of course, a special case of the classical theory of finite elasticity. It is not until unloading, with the accompanying permanent deformation, occurs that plasticity *per se* must be taken into account. With σ a single valued function of the strain ε, x a LAGRANGIAN coordinate along the specimen axis, the particle velocity $v=\partial u/\partial t$ and the strain $\varepsilon=\partial u/\partial x$ being given in terms of the corresponding displacement u, t the time, and ϱ the mass density, the problem considered is given by the following equations:

$$\begin{aligned} \varrho\frac{\partial v}{\partial t} &= \frac{\partial \sigma}{\partial x} \\ \frac{\partial v}{\partial x} &= \frac{\partial \varepsilon}{\partial t} \\ \sigma &= \sigma(\varepsilon). \end{aligned} \tag{4.36}$$

L. H. DONNELL[1] in 1930 had suggested that the lower wave speeds of plastic deformation might be introduced by adopting a stress-strain function consisting of two linear slopes, thereby providing a two wave structure. In 1942, the proposals of TAYLOR, in EULERIAN coordinates, and of VON KÁRMÁN in LAGRANGIAN coordinates, removed this arbitrary restriction upon the relation between stress and strain. The response function was not assumed to be given *a priori*. Stress was assumed to be a single valued function of strain and the otherwise arbitrary response function was assumed to be concave toward the axis of strain. TAYLOR and VON KÁRMÁN did not refer to the fact that the plastic wave motion is isochoric; therefore, they did not take into account the fact that isochoric motion for a one dimensional wave in a cylinder implies lateral motion, and thus, changes in the linear density and the presence of radial acceleration. Proceeding beyond the earlier statement of HUGONIOT for dilatational waves, they treated the problem of an impact of constant velocity at the finite end of a bar. In terms of the constant wave speeds, they gave the integrals which related plastic strain and particle velocity and which related stress and strain. Introducing $c_p(\varepsilon)$ for these constant wave speeds, the magnitude of which depended upon the amplitude of the strain or of the particle velocity, they gave:

$$c_p(\varepsilon)=\sqrt{\frac{d\sigma(\varepsilon)}{d\varepsilon}\Big/\varrho} \tag{4.37}$$

and for the particle velocity:

$$v=\int_0^{\varepsilon} c_p(\varepsilon)\,d\varepsilon \tag{4.38}$$

and the stress

$$\sigma=\int_0^{\varepsilon}\varrho\, c_p^2\,d\varepsilon. \tag{4.39}$$

The specific form of the governing stress-strain function could be determined only by experiment. Such experiments would have to measure strain-time profiles and velocity-time profiles at many positions.

In dynamic linear elasticity, when the one dimensional theory for waves propagating along a cylinder is implied, the constancy of wave shape must be established experimentally before the numerical value of E can be determined. Similarly, in dynamic plasticity a theory ought not be assumed in advance but must be established as applicable before constitutive equations are determinable. The simpler theories of materials, particularly those that presume some kind of

[1] DONNELL [1930, *1*].

material symmetry such as isotropy, contain certain universal relations independent of choice of constants and, more generally, functions. If these are not satisfied, the theory cannot apply, so there is no use even trying to fit constants or functions. The one dimensional theory of plastic waves in rods requires that particle velocity be a function of strain and that wave speeds for each amplitude of strain be constant. When experiment demonstrates that these conditions do apply in a given solid, the response function governing the dynamic deformation follows directly. Thus, *measured* constant wave speeds introduced into Eq. (4.39) provide the form of such a governing response function.

Since for nearly a decade and a half after the theory was proposed it was not possible to determine either $\varepsilon(x, t)$ or $v(x, t)$ during wave propagation, a much weaker approach was adopted before 1956, namely, to *assume* some particular governing response function and to compare the calculations from the guess with measurable secondary effects. Initially, with little questioning of its import, the response function assumed was the quasi-static stress-strain function of the material. VON KÁRMÁN had noted[1] that since this stress-strain function when given in nominal stress and nominal strain underwent a maximum at the ultimate stress, where the horizontal slope would give a zero wave speed, there must exist, in terms of the integral of Eq. (4.38), a limiting velocity v_1. This is now known as the "VON KÁRMÁN critical velocity," above which failure would ensue.

Since an unloading wave was assumed to travel at the much faster wave speed of the linear theory of elasticity, two experimental possibilities based upon the quasi-static stress-strain curve were conceived. In the first, by POL DUWEZ[2] in 1942, an impacting mass produced a rapid increase in particle velocity at one end of a very long copper wire. For the response function assumed, wave speeds decreased rapidly as the strain increased. As a result of the large difference in wave speeds, the finite strain varied with position along the wire. If after a certain time the mass broke a grooved tup, producing a fast traveling unloading wave along the specimen, a frozen pattern of permanent deformation would result, the measurement of which could be compared with calculation if the finite strain response function were known. Otherwise, of course, no conclusion could be drawn.

One feature of the experiment, however, was independent of the choice of $\sigma(\varepsilon)$. If the experiment did consist of a constant velocity impact, as assumed, the maximum strain would appear as a flat plateau from the impact end to the point along the wire which the wave speed of this particular strain and particle velocity had reached when the unloading wave, traveling in the same direction, had arrived. This became the main point of subsequent discussions of the data of DUWEZ.

A second significant experiment designed to consider the theory of plastic waves before it was possible to measure the shape of wave profiles, was that performed by WILLIAM H. HOPPMANN[3] in 1947. As far as I know, his was the first and only use of the DUNN experiment to seek information regarding a prediction from a plausible theory of waves of finite amplitude. Guided down an 80 ft. long "guillotine" impact testing machine, a falling mass to which was attached a specimen with an additional weight at its opposite end was interrupted by an anvil containing a hole sufficiently large for the specimen and its added weight to fall through. Between the hammer and the weight, in series with the specimen, a weigh bar instrumented with electric resistance gages provided,

[1] VON KÁRMÁN [1942, *1*].
[2] DUWEZ, WOOD, and CLARK [1942, *1*].
[3] HOPPMANN [1947, *1*].

by what already had become a standard technique, a force-time history of the event. In addition, a synchronous spark recorder generating signals on wax paper attached to a rotating disk gave a displacement-time history of the falling hammer.

As in earlier experiments described above, plots of impact energy lost vs impact velocity could be obtained. In HOPPMANN's experiments, however, these plots were used to determine whether above v_1 (the VON KÁRMÁN critical velocity), the energy rapidly decreased, as would be expected from the theory. From the slopes of the quasi-static stress-strain curve of the hard drawn polycrystal he investigated, v_1 had the value of 47.3 ft./sec (15.5 m/sec).

These experiments of DUWEZ and of HOPPMANN merit examination in greater detail. For DUWEZ' experiment, the device used to stop the impact after a given deformation is shown in Fig. 4.127. The v-shaped groove designated as N ruptured when the face of the tup A holding the wire struck the bar B. By choosing the distance D for a known velocity of the mass H, the duration of the impact or the time at which the unloading wave traveled into the wire to produce the "frozen" distribution could be determined. A number of years ago, in repeating the DUWEZ experiment in my laboratory as part of a program of investigating earlier experiments in dynamic plasticity, I observed by means of an optical displacement apparatus that the time of the breaking of the tup changed from one test to another and that there was a measurable deceleration prior to rupture.

For the 100 in. long annealed copper, 0.071 in. diameter wires DUWEZ studied, interacting loading waves from the fixed end made difficult the interpretation of results for an impact longer than $1^1/_2$ to 2 msec.[1] In Fig. 4.128 are shown the permanent deformation distributions measured after impact by observing the changes in 1 in. spaced markings on the wire. Also shown is the impact velocity in ft./sec

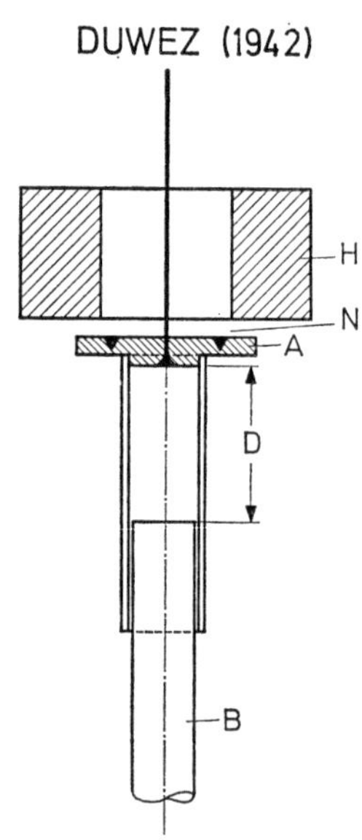

Fig. 4.127. Experimental device used to stop the impact after a given deformation of the specimen is reached.

[1] In DUWEZ and CLARK's paper of 1947 [1947, 1], they gave experimental results for cold rolled steel. I found that the data for this pre-stressed solid were not representable in the simple form of the parabolic stress-strain function from a zero stress state.

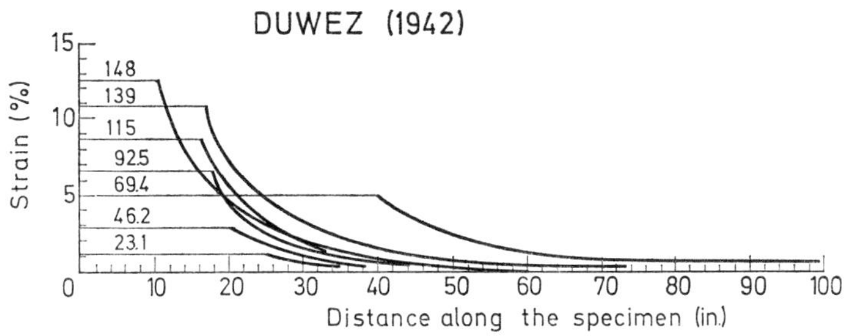

Fig. 4.128. Strain distribution curves for annealed copper specimens subjected to different impact velocities, indicated on each curve in ft/sec.

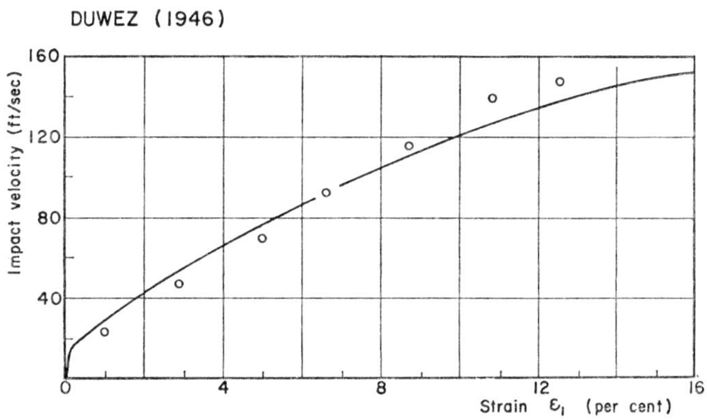

Fig. 4.129. Variation of the maximum permanent strain ε_1 measured after impact, with the impact velocity, v_1 (circles), compared with prediction from the slopes of the quasi-static stress-strain curve.

which produced the observed strain. In each instance the observation of the flat plateau was in agreement with the behavior expected from the theory of waves of finite amplitude; this behavior was independent of the particular form of the stress-strain function,[1] as was indicated above. Whether or not the guess of the quasi-static stress-strain function was good is irrelevant as far as the theory is concerned.

In Fig. 4.129 are shown VON KÁRMÁN and DUWEZ's plot[2] of experimental impact velocity vs maximum strain results (circles) compared with the prediction from the quasi-static stress-strain curve (solid line); they viewed the correlation as satisfactory in view of the nature of the experiment. Since the theory was

[1] Seventeen years later, at a Symposium on the propagation of stress waves in 1959, ERASTUS H. LEE described a reinterpretation of these measurements, showing that in the first 25 % of the plateau of DUWEZ some small increase of permanent deformation might be present. (LEE [1960, *1*], pp. 221–223 and Fig. 15.) L. EFRON in 1964 [1964, *1*] and MALVERN [1965, *1*], and EFRON and MALVERN [1969, *1*] have shown that for the strain rate dependent response function proposed by MALVERN, in which LEE was interested, if computer calculations were carried on for sufficient lengths of time, which up to then they had not been, then MALVERN's conjecture for strain rate dependence also predicted the same flat plateau as the TAYLOR-VON KÁRMÁN theory which assumed an independence of strain rate. The existence of a maximum strain plateau in experiment thus was not necessarily a theory sensitive issue in the arbitrary choice of stress-strain function.

[2] VON KÁRMÁN and DUWEZ [1946, *1*].

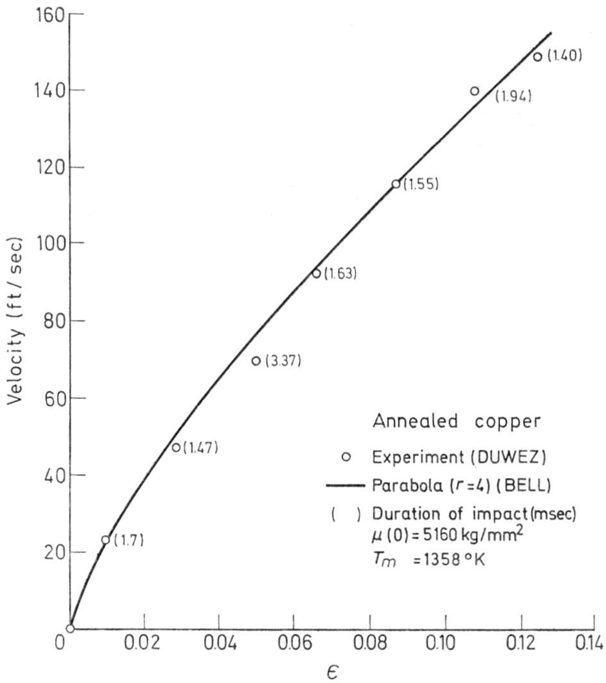

Fig. 4.130. Experimental velocity-strain results of DUWEZ in annealed copper (circles) compared with prediction from BELL's parabolic generalization (solid line).

for an infinite bar, and the experiments required a finite bar elastically restrained at one end, the problem of reflected loading waves for the longer duration impacts obviously was of concern. Efforts to account for this by extending the theory to assume a fixed end were inadequate to describe definitively what was observed. A quarter of a century elapsed before interface studies unraveled some of the complexities of this behavior. Such a development had to be preceded by a precise knowledge of the governing stress-strain function for annealed copper.

I have compared those early results with the impact velocity vs maximum stress prediction (see Sect. 4.28 below) obtained from the parabolic response function, Eq. (4.25), which I found fifteen years later from direct studies of waves of finite amplitude.[1] The predicted velocity vs strain (solid lines) are compared in Fig. 4.130 with DUWEZ's results (circles) for annealed copper, and in Fig. 4.131, for annealed mild steel, which DUWEZ also examined. For the data on copper, the stated durations of impact in milliseconds given by DUWEZ are included in parentheses. It should be noted that the experimental value at 5% strain had the longest duration, 3.37 msec. That the strain is larger than predicted is entirely consistent with the greater influence of reflected loading waves from the "fixed" end.

The importance of determining the governing response function from the detailed study of profiles of waves of finite amplitude was emphasized when VON KÁRMÁN and DUWEZ's experimental distribution of permanent deformation (solid line in Fig. 4.132) for an impact velocity of 92.5 ft./sec and duration of

[1] BELL [1961, *1*] et seq.

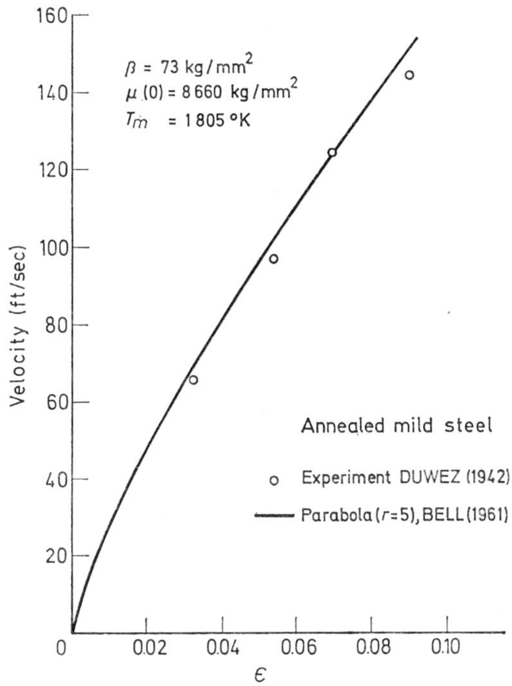

Fig. 4.131. Experimental velocity-strain results of DUWEZ in annealed mild steel (circles) compared with prediction from BELL's parabolic generalization (solid line).

0.83 msec in annealed copper, were compared with predictions from the measured slopes of their quasi-static stress-strain curve (dashed line). From the parabolic response function, Eq. (4.25) for copper, the final strain distribution which now is derived from wave profiles measured later in experiments on waves of finite amplitude, provided the much closer correlation (circles in Fig. 4.132). In retrospect, thus made clearly evident, are the limitations of an *a priori* guessing of stress-strain functions in dynamic plasticity.

HOPPMANN's experiments in 1947 were performed on hard copper,[1] for which the calculated VON KÁRMÁN critical velocity lay well below the maximum speed of 125 ft./sec for those tests. HOPPMANN's plots of energy vs impact velocity and measured permanent elongation vs impact velocity, shown in Fig. 4.133 a and b, demonstrated dramatically the existence of a maximum near the VON KÁRMÁN critical velocity of 47.3 ft./sec calculated from the slopes of the quasi-static stress-strain curve of the material.

By 1947, the experimental results of DUWEZ[2] and of HOPPMANN,[3] although testing secondary phenomena and not the detail of the wave profiles, created the general belief that plausible theory and plausible experiment had combined to establish a basis for further study of a nonviscous dynamic plasticity probably governed by the quasi-static response function.

[1] HOPPMANN [1947, *1*]. HOPPMANN has since informed me that he did perform such experiments on annealed copper, but the higher critical velocity, between 150 ft/sec and 171 ft/sec according to VON KÁRMÁN and DUWEZ, discouraged such a study of that solid at that time.

[2] DUWEZ, WOOD, and CLARK [1942, *1*].

[3] HOPPMANN *op. cit.*

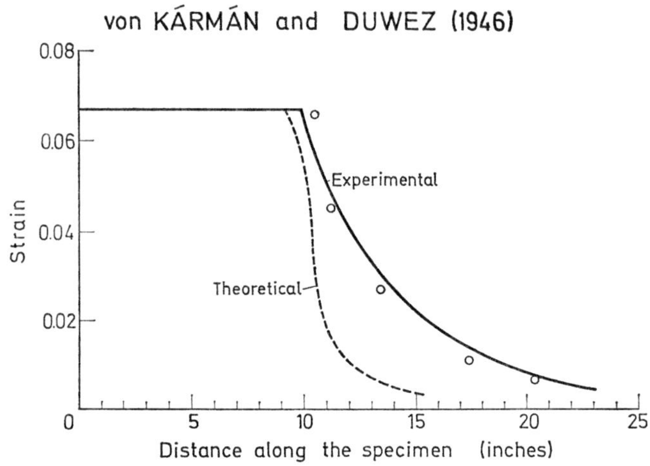

Fig. 4.132. DUWEZ' experimental permanent strain distribution after impact (solid line) compared with his prediction from slopes of his quasi-static stress-strain curve (dashed line) and with prediction from BELL's parabolic generalization (circles).

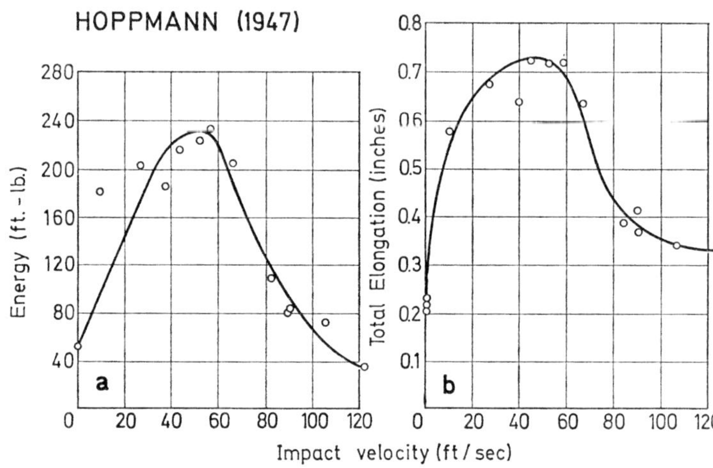

Fig. 4.133a and b. (a) Variation of energy with impact velocity for hard copper. (b) Variation of elongation with impact velocity for hard copper.

VON KÁRMÁN and DUWEZ's[1] observation from experiment, that plastic deformation in iron did not exhibit permanent deformation until velocities considerably above those calculated from the quasi-static elastic limit, provided some link with previous experiment and were important in giving impetus to "delay time" studies which followed. The oft-quoted statement of VON KÁRMÁN that the discrepancies between experiment and the prediction of plastic strain distribution from the quasi-static response function (Fig. 4.132) might suggest a small strain rate effect, was a *non sequitur* in view of the fact that the quasi-static curve had been chosen arbitrarily as the governing stress-strain function.

[1] VON KÁRMÁN and DUWEZ [1946, *1*].

The immediate reaction of experimentists to these wave propagation developments was a series of papers reporting experiments performed upon specimens L/D ratios of which were increased by an order of magnitude. The experiments were intended primarily to explore the possible increase in the elastic limit in impact loading of steel, following VON KÁRMÁN and DUWEZ's observation in that solid.

A further pursuit of a method of studying experimentally the consequences of the finite amplitude wave theory before 1956 when wave profiles could first be examined, was the experiment independently introduced by J. D. CAMPBELL[1] in England and J. E. JOHNSON, D. S. WOOD, and DONALD S. CLARK[2] in the United States. In essence the experiment was the B. HOPKINSON[3] pressure bar test of 1913, analyzed in the light of the then existing theory of waves of finite amplitude and refined by the introduction of electric resistance gages to replace HOPKINSON's pellets, but still involving an indirect measurement of wave propagation.

A relatively long specimen to be plastically deformed was made to collide axially with a much longer hard bar. Since the elastic limit of the hard bar was not exceeded, electric resistance gage instrumentation provided, through application of the elementary linear theory, a stress-time history in the anvil bar. A POISSON's ratio of $^1/_2$ was assumed in the soft bar and the measurement of diameters of specimens before and after impact constituted the opposing variable in the experiment. A stress-strain function was calculated by using many specimens, each struck at a different impact velocity. The maximum velocity in the hard bar after impact for this nonsymmetrical impact was subtracted from the measured velocity of the specimen prior to impact, which provided a value for the maximum particle velocity in the plastically deforming specimen.

The solid was commercial purity polycrystalline aluminum annealed for two hours at 670° F after machining. The anvil bar was phosphor bronze of the same diameter, having a length of 88 in. in contrast to the specimens which were either 23 or 43 in. long. In these tests, as in the almost precise repetition of the experiment nine years later by KOLSKY and L. S. DOUCH,[4] the general conclusion reached from comparing the maximum stress in the anvil bar with the maximum permanent deformation in the specimen after impact, for different impact velocities, was that the propagation of plastic waves in annealed solids could be approximated by a single stress-strain curve. From the tests the curve was thought to lie above the quasi-static response for the same solid. The measured stress vs velocity and stress vs maximum permanent deformation results of JOHNSON, WOOD, and CLARK[5] are shown in Fig. 4.134a and b.

Measuring and squaring the slopes of the stress vs velocity curve in Fig. 4.134a for introduction into

$$\varepsilon_1 = \int_0^{\sigma_1} \frac{\varrho\, d\varepsilon}{\left(\dfrac{d\sigma}{dv}\right)^2}$$

[1] CAMPBELL [1953, *1*].

[2] JOHNSON, WOOD, and CLARK [1953, *1*].

[3] B. HOPKINSON, published [1914, *1*].

[4] KOLSKY and DOUCH [1962, *1*] used a hardened steel anvil bar instead of phosphor bronze, and considered a series of specimens with lengths ranging from 6 to 12 inches. In repeating JOHNSON, WOOD, and CLARK's experiments of 1953, KOLSKY and DOUCH in 1962, besides examining annealed aluminium (the solid studied by JOHNSON, WOOD, and CLARK), performed tests on annealed copper.

[5] JOHNSON, WOOD, and CLARK [1953, *1*].

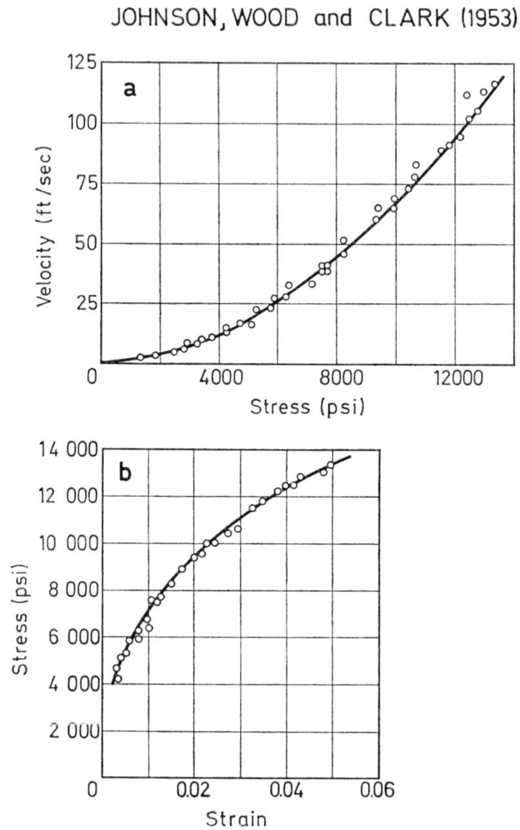

Fig. 4.134a and b. JOHNSON, WOOD, and CLARK's experimental results. (a) Impact stress vs particle velocity, and (b) impact stress vs maximum permanent deformation after impact.

provided the stress-strain curve, A, of Fig. 4.135 which may be compared with that from the measurements of permanent deformation, B, and the quasi-static curves, C and D.

No particular significance could be attached to the higher stress-strain function in view of the graphical calculation for A. Also adversely affecting the results was the auxiliary assumption that the permanent strain, $\varepsilon(x, \infty)$, after unloading waves traversed the specimen, was identical with that of the unmeasured actual maximum strain of the wave front; the assumption that at the impact face in the early stage of impact, measured $\sigma(0, t)$ was uninfluenced by the effects of three dimensional wave initiation; the assumptions that radial measurements could be interpreted as longitudinal strain in the presence of the radial constraint of the hard bar at the impact face and that the frictional effects also present at that interface were unimportant.

Realizing that they were dealing with a finite specimen, JOHNSON, WOOD, and CLARK carried out a graphical solution as proposed earlier by VON KÁRMÁN, H. F. BOHNENBLUST, and D. H. HYERS[1] and by BOHNENBLUST, J. V. CHARYK, and

[1] VON KÁRMÁN, BOHNENBLUST, and HYERS [1942, 2].

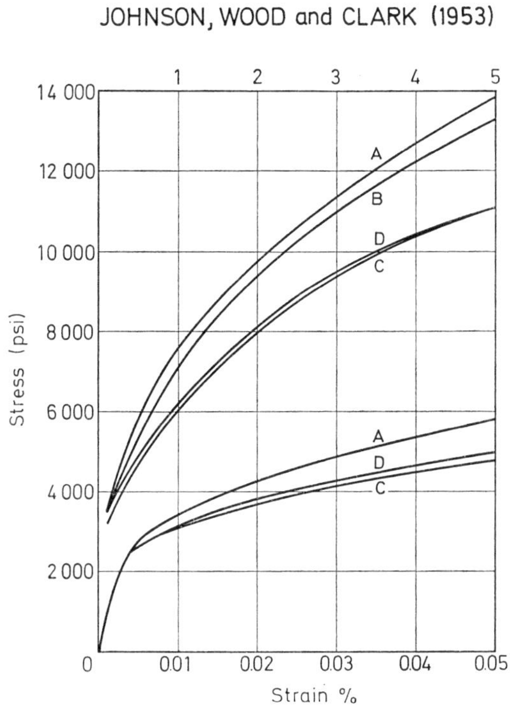

Fig. 4.135. Calculated dynamic stress-strain relations for situations designated, compared with quasi-static experiments at two strain rates. *A* Stress-velocity measurements. *B* Stress-plastic strain measurements. *C* Static. *D* Strain rate 0.040 per min.

D. H. HYERS.[1] They obtained the LAGRANGIAN plot of Fig. 4.136 from which the calculated stress-time history at the impact face (dashed line, Fig. 4.137) was compared with the extrapolated measured history in the anvil bar.[2]

Load bar results were obtained by KOLSKY and DOUCH in 1962, as shown in Fig. 4.138. Here the stress of the anvil bar was plotted against the permanent deformation.[3] The solid line accords with my load bar test results of 1960–61.

Once the governing response function became known from my direct study of the profiles of waves of finite amplitude during their propagation, this load bar experiment and all its attendant assumptions became interesting as an example of nonsymmetrical impact.[4]

The first experimentist to attempt to determine a stress-strain function from the measurement of finite amplitude wave profiles was WILLIAM R. CAMPBELL.[5] It has been my practice in this treatise to omit "feasibility studies" which are, in

---

[1] BOHNENBLUST, CHARYK, and HYERS [1942, *1*].

[2] JOHNSON, WOOD, and CLARK [1953, *1*].

[3] KOLSKY and DOUCH [1962, *1*] predicted the final strain distributions on the basis of the assumption of simple linear elastic unloading; thus, they ignored the absorption of the unloading wave predicted by LEE's theory of unloading [1953, *1*] and which I had demonstrated experimentally in 1961 (BELL [1961, *3*, *4*]). The analysis in terms of nonlinear theory gives for the maximum strain a much larger penetration into the specimen than the simple linear theory permits. (See Sects. 4.29 and 4.34 below.)

[4] BELL [1960, *3*], !1961, *1*, *4*], [1963, *2*], [1968, *2*], [1969, *1*].

[5] W. R. CAMPBELL [1951, *1*], [1952, *1*].

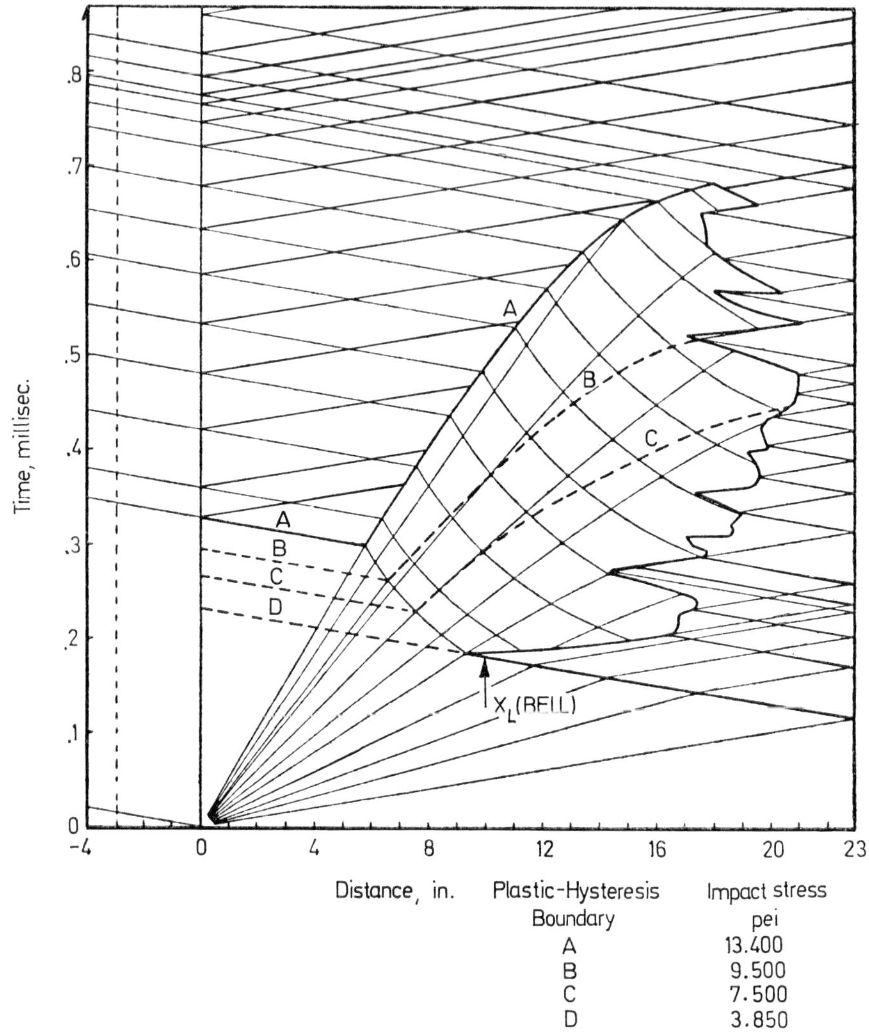

Fig. 4.136. The Lagrangian diagram of Johnson, Wood, and Clark (1953) based upon their dynamic stress-strain curve. x_L denotes the position of complete absorption of the first reflected wave from the experiments of Bell in 1961 and the theory of Lee in 1953.

fact, the premature publication of results. One notable exception is the experiment of W. R. Campbell in the late 1940's, presented to a meeting of the Society for Experimental Stress Analysis in the spring of 1951.

In Campbell's experiment a vertically supported bar, $^1/_2$ in. in diameter, was subjected to a tensile impact at its lower end by means of a falling weight, as shown in Fig. 4.139. Flexure eliminating electric resistance gages at positions 1–2 and 3–4 were 10 in. apart.

The strains in the electrolytically annealed and hard copper solids which he studied, Fig. 4.140, were limited to the very low values of 0.005 and 0.003 strain, respectively.

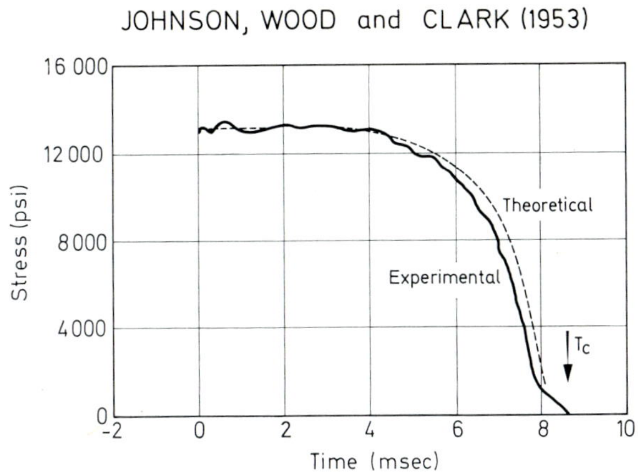

Fig. 4.137. The experimental stress-time history of JOHNSON, WOOD, and CLARK at the impact face, compared with their prediction from the calculations of Fig. 4.136 (dashed line). The time of contact, T_c, has been added, which is the value predicted by BELL in 1961 from the parabolic generalization.

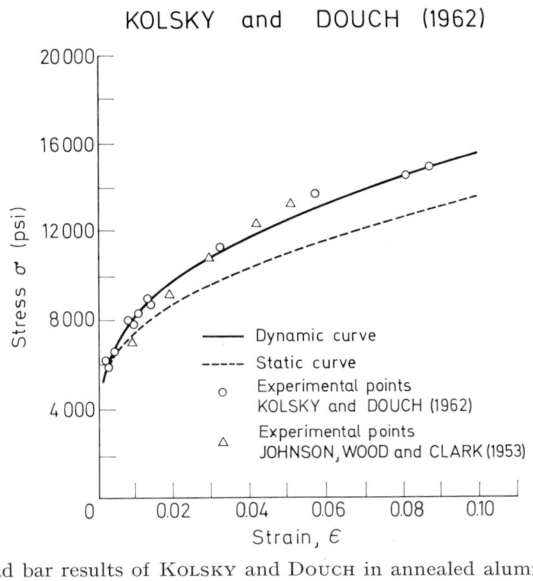

Fig. 4.138. The load bar results of KOLSKY and DOUCH in annealed aluminum, which I have compared with those of JOHNSON, WOOD, and CLARK.

CAMPBELL failed to appreciate, or at least failed to state, that it was necessary to establish that wave speeds were constant, and that measured finite strain was a single valued function of measured particle velocity, before one could determine a functional relation between finite strain and stress. The wave speeds given by strain profiles at two positions are of little value unless the experimentist knows from other measurements the answer he is seeking.

Asserting the applicability of the TAYLOR-VON KÁRMÁN theory, CAMPBELL used the elapsed time from two electric resistance measurements to provide wave speeds for Eq. (4.39). The solid lines in Fig. 4.141 are the quasi-static results in his hard and soft copper rods.

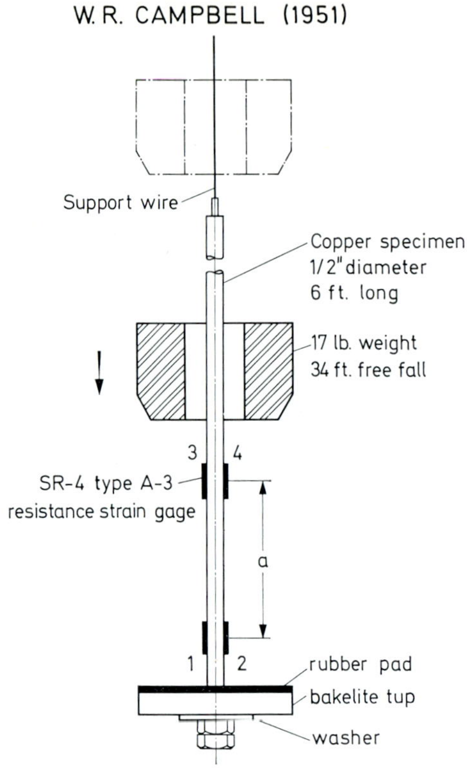

Fig. 4.139. The apparatus of W. R. Campbell.

Campbell was perplexed by the enigma of the dynamic stress-strain curves of annealed copper lying so far below the quasi-static curve, an observation he quite correctly considered as highly dubious. "It is believed that the calculated stress in these tests are uncertain by as much as 25% ", he wrote.[1]

Although the results Campbell obtained have no value other than historical importance as a first attempt to explore finite strain profiles in relation to the Taylor-von Kármán theory, there is an additional reason for including them here. Every experimentist since Campbell, including myself in the early 1950's, and Ripperger, Malvern, and others at various times from then to now, has concluded that electric resistance gages, even those which perform satisfactorily to large strain quasi-statically, are unreliable for the study of dynamic plasticity. It is not merely because of the problem of integrating an unknown, highly dispersive function over a relatively large gage length, but also because such measurements invariably are late and contain errors of magnitudes ranging from 5 to 30%, depending upon the situation and their station along the bar. This point was demonstrated decisively by William J. Gillich[2] in 1960 in a Master's essay describing over 30 measurements with electric resistance gages, which he compared with strains determined[3] optically over very short gage lengths at the same

[1] W. R. Campbell [1951, *1*], p. 8.

[2] Gillich [1960, *1*].

[3] These were diffraction grating tests from a series which I had performed before Gillich made his measurements by means of electric resistance gages. See Bell [1960, *2*].

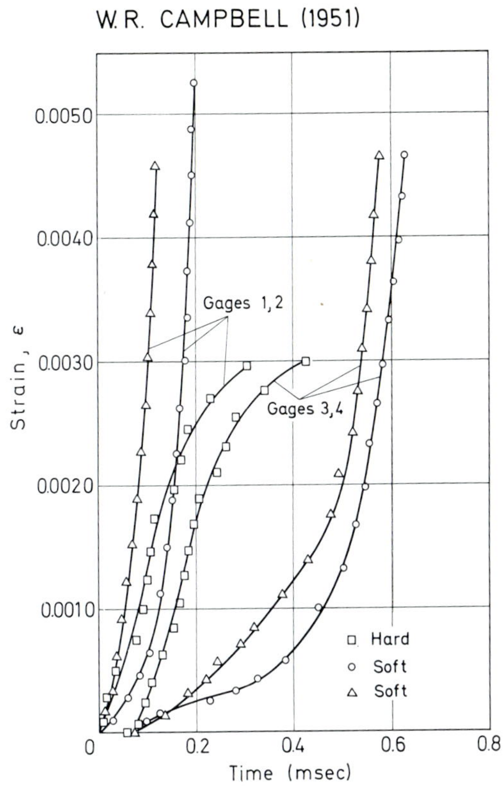

Fig. 4.140. W. R. CAMPBELL's wave profiles for strains measured by means of electric resistance gages.

position on the same solid. MALVERN[1] when comparing data from electric resistance gages with results for particle velocity obtained by magnetic induction in a study in 1965 with EFRON, reached the same conclusion. My rejection of such measurements in the early 1950's was based upon inconsistencies observed while analyzing the results of a large number of impact experiments I performed at that time.

Because of these difficulties, which in 1952 were instrumental in my continuing efforts to find some alternate method of making finite strain measurements during wave propagation, I have not included in the present treatise the numerous studies of many experimentists whose investigations were based upon the use of electric resistance measurements at the higher strain rates of dynamic plasticity.

In the fall of 1948, I conceived an experiment[2] which seemed at the time to offer a direct experimental check upon the use of the quasi-static response function if

[1] MALVERN [1965, *1*].

[2] It may be of some interest that this idea occurred to me while listening to a lecture WILLIAM R. CAMPBELL gave at The Johns Hopkins University in the fall of 1948, in which he related some of the discouraging preliminaries of the experiments described above, and particularly of other experiments in which long aluminum specimens were projected along a trough by 50 stretched door springs. Being then a young professor concerned solely with theory in other areas, I was unable to interest any experimentist in my idea for an incremental wave experiment on a long prestressed bar; this resulted in my undertaking the problem myself, thus inaugurating a series of experiments which are still in progress nearly a quarter of a century later.

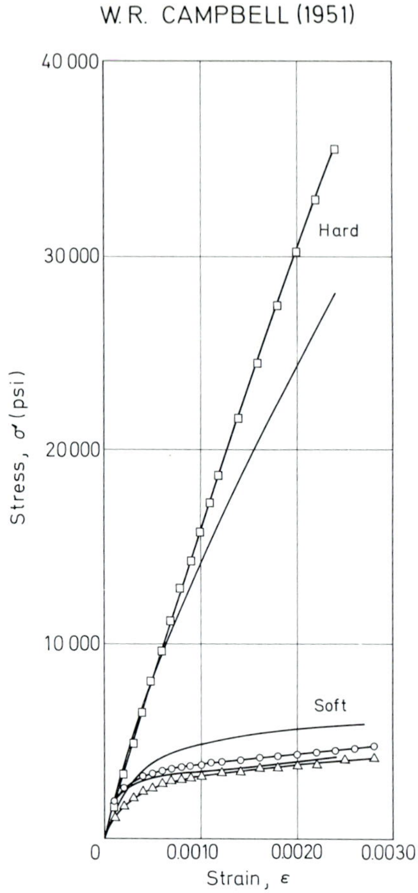

Fig. 4.141. W. R. Campbell's stress-strain curves in hard and soft copper calculated from the wave speeds of Fig. 4.140 compared with quasi-static results.

the nonlinear wave theory indeed did apply. The idea was extremely simple. The theory of waves of finite amplitude stated that the constant wave speeds for a given large strain were specified by the tangent modulus of an unknown stress-strain function. The prestressing of a long specimen quasi-statically to the desired strain and the introduction of an incremental loading wave at that point, should, by measurement of the incremental wave speed, provide the desired tangent modulus experimentally. Since only waves of small amplitude were to be measured, it was possible that the errors of electric resistance elements for measuring large strain might not be present. By creating an elastic island on which the electric resistance gages were located, the arrival times of incremental waves which had traversed a large plastically deformed zone could be measured without subjecting the electric elements to large strain.[1]

The apparatus used to prestress 6 ft. long, mild steel bars to any desired value up to rupture is shown in Fig. 4.142.

A steel load bar in series with the specimen, and a clip gage, both instrumented with electric resistance gages undergoing small elastic deformation, provided

[1] Bell [1951, *1*].

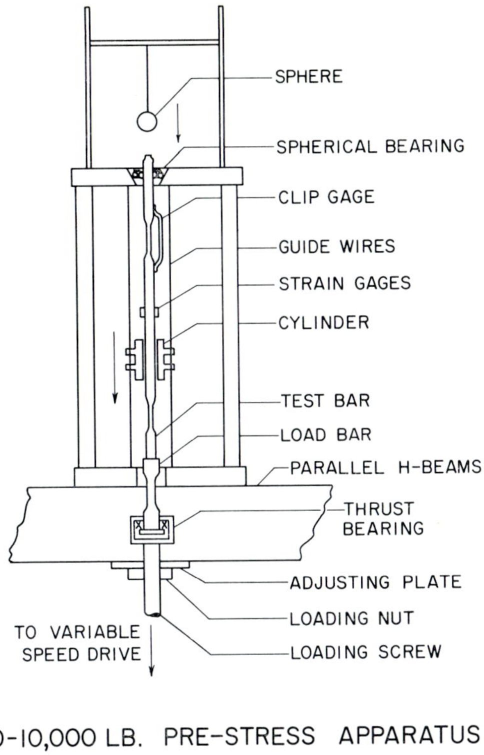

Fig. 4.142. BELL's (1951) apparatus for incremental loading wave and incremental unloading wave in pre-stressed bars. The cylinder was for loading increments, and the sphere was for unloading increments. The pre-stress was in tension.

measurements of the quasi-static stress and strain of the prestress. The 6 in. long clip gage was calibrated by a micrometer device. The extension of the bar was accomplished by a variable speed motor accompanied by a gear reduction system, with the specimen suspended in a spherical bearing at the top and a special thrust bearing at the bottom. Since the stress and strain of the quasi-static prestress were recorded simultaneously on separate charts, it was possible to obtain a variety of loading histories by manual control of the speed of the drive motor.

The incremental loading wave was produced by the hollow cylinder shown; it slid along taut guide wires to impact the hardened steel projection of the specimen holder at the bottom. By extending the specimen bar through the spherical bearing at the top, it also was possible to introduce an incremental unloading pulse by the impact of the falling steel sphere shown. The prestress was in tension.

The first experiments performed during the summer of 1949 revealed that for both loading and unloading waves at all values of prestress nearly up to rupture of the bar, the incremental wave speed was that of the elastic bar, $c_0=\sqrt{E/\varrho}$, rather than the expected wave speed of the tangent modulus of the governing stress-strain curve for the theory of waves of finite amplitude.

To make certain that the unexpected experimental discovery of the high incremental wave speed was not due to unseen, small unloading at the quasi-

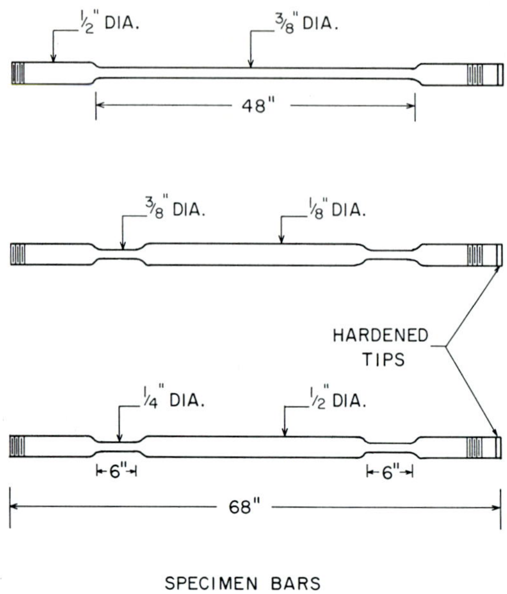

Fig. 4.143. Three of the several shapes of specimens BELL studied in 1951.

static prestress, I performed experiments upon specimens which were undergoing low strain rate loading as the incremental impact occurred. In every instance the same elastic bar velocity was observed. A variety of specimen types were considered including those shown in Fig. 4.143, using the apparatus of Fig. 4.142.

The fact that the specimens had two plastic zones on either side of the linear elastic island and that wave speeds were measured by means of flexure eliminating electric resistance gages at the center of the island, removed the possibility that the observed arrival times were in any way related to the steel structure of the test apparatus. The results indicated that beyond question, the incremental loading wave in a plastically prestressed mild steel bar propagated at the linear elastic bar wave speed of the solid, no matter how high had been the prestress.

The electric resistance gages which recorded the arrival times were never less than a meter from the point of impact, so that accuracy in the determination of arrival times and hence incremental wave speed, was achieved. Although in the 1960's electronic signal triggering devices had become standard oscilloscope equipment, in 1949 the triggering of the sweeps at the moment of impact presented considerable difficulty. Other than the preparation of 6 ft. specimens, it was the most difficult problem encountered in that experiment. In Fig. 4.144 are shown a few of the many oscillograms for the incremental loading waves, which demonstrate the independence of the arrival time from the magnitude of the prestress.

One interesting phenomenon observed at high stress, and shown at a prestress of 88000 psi in Fig. 4.144 was the suppression of the complex wave reflection and propagation which followed the initial wave front at that value of prestress. When the stress was increased further, the amplitude of the wave train again increased, as may be seen in Fig. 4.144. Of interest also was the shift in the base of the oscillations, with increasing strain. Fortunately, that complex behavior occurred after the desired wave speed information had been recorded.

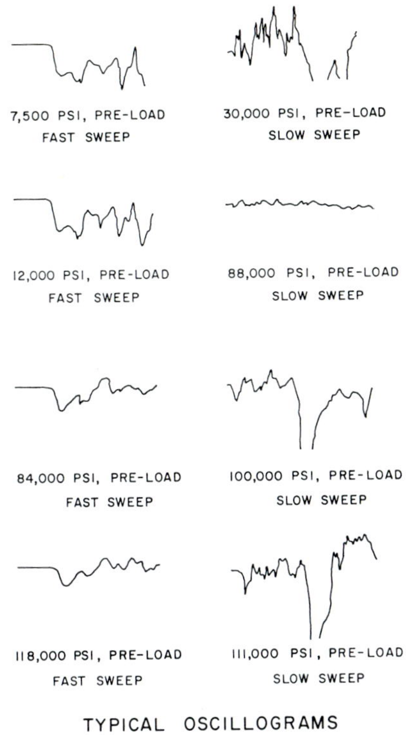

Fig. 4.144. Typical oscillograms obtained by BELL between 1949 and 1951 for different sweep speeds. The initial straight portion gave the arrival time of the incremental wave from which its wave speed was determined.

The results of this discovery,[1] which I had described in 1951, at that time appeared to be in conflict with the accepted one-dimensional theory of plastic waves. In 1953, R. J. RUBIN[2] suggested that the incremental wave behavior which I had described was consistent with a rate-dependent uniaxial stress-strain function proposed by MALVERN,[3] a conjecture which experiment in the 1960's amply demonstrated was untenable.[4] (The results of experiments on waves of finite amplitude in the 1950's and 1960's indicated that uniaxial response functions of the MALVERN type did not apply. The large linear wave front behind which plastic deformation was assumed to grow, was not observed in any crystalline solid.) In 1951, CHANG S. HAHN,[5] then a student at Cornell University, performed similar experiments on incremental waves in copper and extended the original discovery by demonstrating that the high speed of the incremental wave was not confined to mild steel where elastic limit anomalies had been observed at lower strain rates.

---

[1] BELL [1951, *1*].
[2] RUBIN [1953, *1*], [1954, *1*].
[3] MALVERN [1951, *1*].
[4] BELL and STEIN [1962, *7*].
[5] Knowledge of the unpublished experimental work of CHANG S. HAHN came to me from reading the introduction to GIOVANNI BIANCHI'S Cornell University Master's eassy [1953, *1*]; from a footnote in the paper of STERNGLASS and STUART [1953, *1*], p. 429; and from a reference in a paper by CARLOS RIPARBELLI [1953, *1*]. I was unable to obtain from Cornell University any further information about this work.

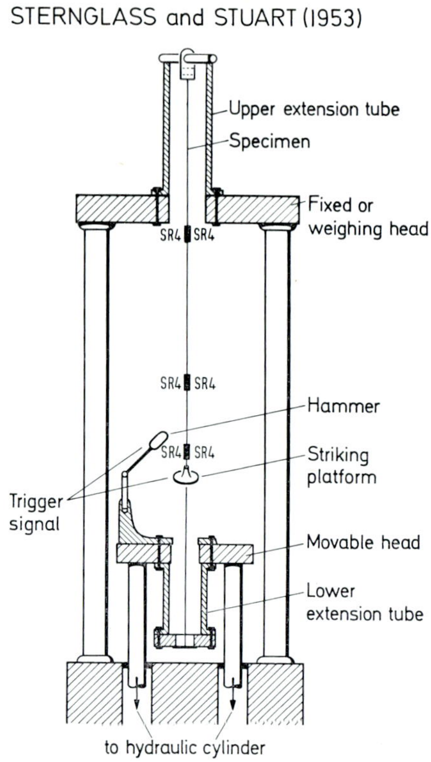

Fig. 4.145. The incremental wave apparatus of Sternglass and Stuart.

Carlos Riparbelli[1] appreciated the importance of the incremental wave experiment for developing constitutive relations in dynamic plasticity. He urged E. J. Sternglass and D. A. Stuart[2] of the Department of Engineering Materials at Cornell University, and also Giovanni Bianchi,[3] then a graduate student studying for a Master's degree in mechanical engineering at Cornell, to continue the study of the incremental wave. Bianchi in 1953 undertook a study of incremental waves in 30 ft. long prestressed, annealed copper ribbons. Sternglass and Stuart in 1953 tested flat strips, $^1/_2$ in. wide by $^1/_8$ in. thick, in a standard testing machine to which had been added the extension tubes shown in Fig. 4.145, to permit the examination of 120 in. long specimens.

A 4.7 lb. hammer dropped 8 in. on to a striking platform attached to the specimen; with electric resistance gages Sternglass and Stuart made measurements on the plastically deformed specimen at three different distances from the impact platform. They initially studied the effect of varying the weight of the platform upon tension loading pulses above the point of impact, and compressive unloading pulses below that point. Since they also were interested in the dispersion indicated by a change of the incremental pulse shape, reproducibility was of importance. A series of measurements are shown in Fig. 4.146a–c.

---

[1] Riparbelli [1953, *1*].
[2] Sternglass and Stuart [1953, *1*].
[3] Bianchi [1953, *1*].

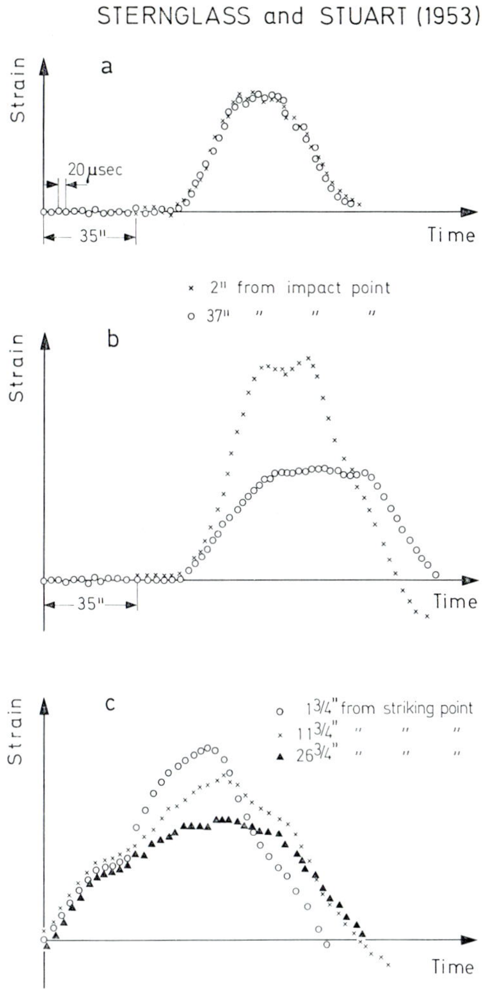

Fig. 4.146a—c. Incremental wave results of STERNGLASS and STUART in 1953: (a) linear elastic pre-stress; (b) plastic pre-stress; (c) three successive positions for plastic pre-loading. Note that all electric resistance traces were shifted to coincide at the beginning of the pulse.

For (a) and (b) the distance between points of measurement was 35 in., the first electric resistance element being located 2 in. from the point of impact. In (a), for 9250 psi prestress, the pulse of about 200×10^{-6} strain propagated in accordance with one dimensional linear theory. The second pulse was shifted in time to demonstrate the lack of dispersion observed. For (b), where the quasi-static prestress of 36300 psi was well into the plastic region with strains over 0.005, a similar incremental impact resulted in the dispersion shown, for measurement between the same two positions. Once again, they shifted the second signal in time so as to make the pulses coincide for comparison. In a third set of results, also shifted in time, for an unspecified amount of plastic prestress they compared the pulse dispersion at $1^3/_4$ in., $11^3/_4$ in., and $26^3/_4$ in. from the point of impact.

What effect dynamic pulse loading had upon electric resistance gages already loaded into the plastic region is uncertain, so not too much attention should be paid to Sternglass and Stuart's quantitative results. On the matter of arrival times, for which the actual amplitudes are of less importance, their results were consistent with my earlier observation[1] that incremental waves in prestressed solids propagated at high wave speeds.

Bianchi's experiments were described at a Symposium on "La Plasticità nella Scienza delle Costruzioni," at Varenna in 1956, and were published the following year.[2] Except for the fact that the specimens were in the form of annealed copper ribbons about 8 m long, loaded by a weight over a wheel, as shown in Fig. 4.147, the experiment did not differ significantly from that of Sternglass and Stuart.[3] The impact occurred at an attachment at a point on the ribbon, electric resistance gages being employed to measure loading and unloading pulses on either side of the impact point, as shown in Fig. 4.147.

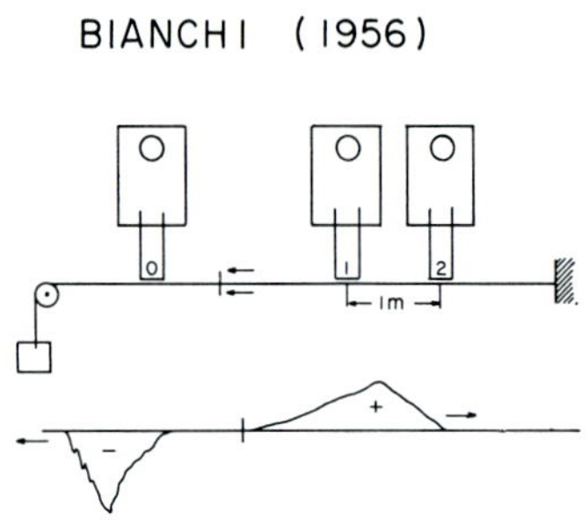

Fig. 4.147. The incremental wave apparatus of Bianchi and representative electric resistance determined strain-time. See Bianchi [1953, *1*], [1957, *1*].

The conclusions reached by Bianchi were similar to those of Sternglass and Stuart: there was a general attentuation and dispersion of the pulse. With the threefold increase in the length of the specimen, greater changes of wave shape were visible. The large ribbon permitted the study of the long tail following the incremental pulse under prestress, since reflections from the ends of the specimen were delayed for a much longer time.

Influenced by the serious questions W. R. Campbell had raised[4] regarding the use of electric resistance gages in the study of dynamic plasticity, I had gone to considerable pains to make long specimens which had an elastic island at the center so that the maximum strain in that portion of the specimen on which the gages were placed would not exceed the elastic limit. Later experimentists failed to explore fully this aspect of the experiment in their studies, and hence a period of 10 to 15 years elapsed before reliable measurements of incremental loading wave detail became available.

The first important advance beyond the original experiments was the experiment of B. E. K. Alter and C. W. Curtis,[5] using a dynamic prestress in lead, in 1956. Their lead specimens, 13 in. long, $^1/_2$ in. in diameter, unfortunately were instrumented with electric resistance gages at a series of four or five positions, from 0.25 in. from impact to 3.5 in. from impact. A stepped steel bar was in

[1] Bell [1951, *1*].

[2] Bianchi [1957, *1*].

[3] Sternglass and Stuart [1953, *1*].

[4] W. R. Campbell [1951, *1*], [1952, *1*]. Campbell's maximum strains were of the same order of magnitude as those of Sternglass and Stuart and Bianchi.

[5] Alter and Curtis [1956, *1*].

axial collision with the specimen. The elastic wave in the bar upon being reflected from the step, returned as an incremental loading wave following the dynamic prestress of the specimen from the initial impact. ALTER and CURTIS' diagram of this situation is shown in Fig. 4.148.

The impact bar and also an anvil bar at the end of the lead specimen were each 10 ft. long to eliminate unloading effects from the free ends of the hard bars. In Fig. 4.149a may be seen an experiment with measurements at the five positions shown, for an impact of the hard bar without the stepped section. This, then, was the dynamic prestress propagating as a dispersive nonlinear finite amplitude wave front. The dashed line marked E was the slope of the elastic bar velocity for lead where the very small initial strains were recorded. For a

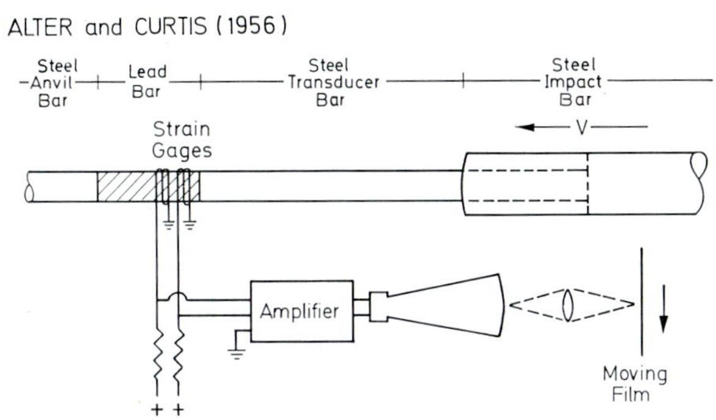

Fig. 4.148. ALTER and CURTIS' diagram for their incremental plastic wave studies in lead bars.

velocity of about 15% of the linear elastic wave speed, the dashed line P was drawn through what appears to be approximately the mean plastic strain.

In Fig. 4.149b is shown the results of a similar impact experiment in which ALTER and CURTIS had introduced the stepped section of the impact bar The expected double wave is readily visible. The TAYLOR and VON KÁRMÁN theory would predict that the flat between the maximum of the first wave and the initiation of the incremental wave should remain unchanged as the finite amplitude waves propagate down the bar. The fact that the two wave fronts joined demonstrated the more rapid velocity of the initial portion of the increment. The dashed lines, E_1 and E_2, represented the arrival times of initial strains for each wave, although the second must have been difficult to separate from the small variation in the maximum of the first.

It is interesting that the dashed lines through the mean plastic strain, P_1 and P_2, definitely represented slower plastic wave speeds, indicating that to the limits of such strain gage measurements the unloading portion of the much smaller maximum amplitude pulses of STERNGLASS and STUART and of BIANCHI was absent. The incremental wave beyond the leading portion of the initial wave had wave speeds consistent with the magnitudes of tangent moduli. There was no attenuation in the magnitude of the incremental wave front that was not unloaded immediately. That fact strongly suggests that unloading from the back of the pulse was present in the measurements of STERNGLASS and STUART and of BIANCHI.

In Fig. 4.150 are shown ALTER and CURTIS' comparison of their mean strain slopes with and without the incremental wave. The high wave speed is limited

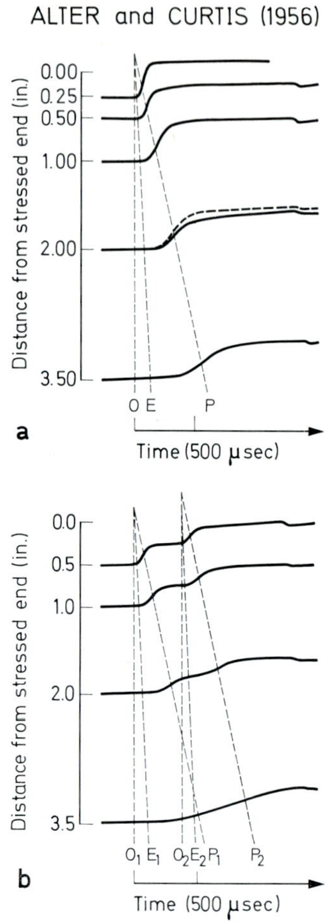

Fig. 4.149a and b. Strain-time records for (a) pulse produced by single pressure step, and (b) pulse produced by double pressure step. Measurements were at the indicated positions. Arrival times for elastic and plastic strain are shown as dashed lines.

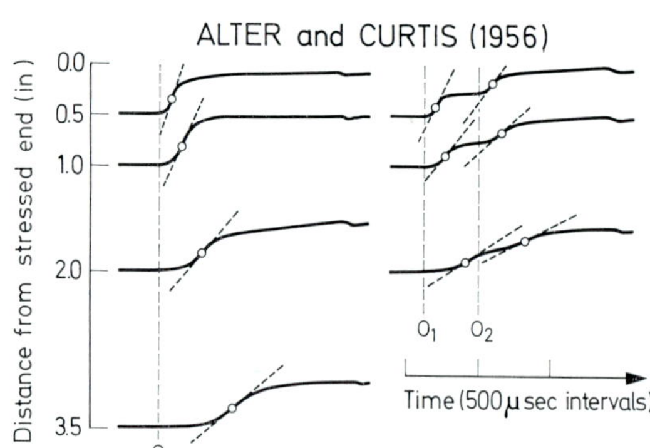

Fig. 4.150. Alter and Curtis' comparison of mean strain-time slopes at the indicated distance from the impact face, and with and without the incremental wave. O and O_1 represent zero time at the impact face; O_2, the time of arrival of the increment at the impact face.

to the small initial strain; the lower plastic wave speeds govern strain above that value. This observation[1] became the main focus for incremental wave experiments in the 1960's.

After an hundred years of study of impact, from the mid-19th century tests in gun barrels and calculations of energy based upon the crushing of cylindrical samples, through the seven decades of the 20th century variations on DUNN's extended quasi-static experiment, to the measurement of secondary effects in the post-deformation tests of DUWEZ and of HOPPMANN, to the load bar tests of J. D. CAMPBELL and of JOHNSON, WOOD, and CLARK, and to my incremental wave experiments and those of ALTER and CURTIS, no successful direct study of the finite amplitude wave profiles had been made as required for a proper comparison between experiment and any nonlinear theory. That important milestone in the study of dynamic plasticity[2] was achieved in 1956.

4.28. On the direct measurement of strain profiles during finite amplitude wave propagation: Bell (1956–1972).

The use of NEWTON's rings by CORNU in 1869 and the interferometer by GRÜNEISEN in 1906 had demonstrated amply the possibilities for the use of interference optics in the study of the deformation of solids. In 1956, by an entirely different approach, a new experiment, but one also based upon optical interference, successfully resolved the century old dilemma in impact studies by allowing a direct and accurate measurement of finite strain wave profiles in microsecond intervals of time.[3]

In principle, the new experiment was extremely simple. The angle of diffraction of the various orders of a reflection grating depends upon the spacing of the rulings. If, as during strain, this spacing were altered, the diffraction angles also would be changed in a determinable fashion. Thus, observation of the angular changes of diffraction images from incident monochromatic light from gratings ruled on the specimen could provide a direct measure of dynamic finite strain as a large amplitude wave propagated through the solid.

This experiment was conceived by me in the spring of 1950 near the completion of the incremental wave studies. It was the direct result of a deliberate search for an experiment to eliminate the objections to previous studies by examining the phenomena of dynamic plasticity in the same terms as the nonlinear theory, i.e., as a propagating dispersive wave front. That five years of continuous labor were required to obtain usable experimental results, and four more years to obtain a high degree of precision in the observation, was due primarily to the necessity of designing and building a cylindrical diffraction grating ruling engine capable of ruling 30000 lines per inch, and to the need to conceive and develop a recording apparatus capable of measuring small angle changes of diffraction images occurring in microsecond time.

To obtain high density gratings on a cylinder, a 50 year old precision Rivett lathe from the Johns Hopkins University machine shop, originally designed to produce a maximum of 241 threads per inch, was modified, in a series of stages, to

---

[1] As will be noted below, Sect. 4.34, recent experimental and theoretical studies have shown that the SAVART-MASSON (PORTEVIN-LE CHATELIER) effect cannot be ignored when interpreting incremental wave results in annealed polycrystals.

[2] BELL [1956, *1*, *2*].

[3] BELL [1956, *1*, *2*, *3*].

produce several thousand threads per inch.[1] By 1954 the number of lines per inch had reached 8400, which was sufficient to measure finite strain profiles on one inch diameter annealed aluminum and annealed copper cylinders several inches long. I described the results of the first series of measurements on annealed aluminum in 1956. By 1958 the cylindrical ruling engine had been modified further so that excellent gratings of from 30000 lines per inch to 35000 lines per inch could be ruled.[2] Since that time several thousand rulings at 30720 lines per inch have been made on several metals.[3] With this number of lines per inch and with a refinement of the method of observing moving images, and new calibration procedures also developed in 1958, a precision of measurement and flexibility of application of the method was reached so that the experimental technique has remained essentially unaltered for fourteen years.

In this experiment, a 30720 lines per inch grating from 0.0025 to 0.0125 cm long, depending upon the gage length desired, was ruled on the cylindrical specimen at a specified distance from the impact face. Normally incident 5461 λ monochromatic light produced two first order images at 41°18′ on either side of the normal. (For this number of lines per inch, only the two first orders are present.) Each image fell through an appropriate displacement-eliminating cylindrical lens to a focus upon a V shaped slit behind which was a ten stage photomultiplier tube, 5 in. in diameter. During finite strain the images moved in the direction of the apex of the slits, providing voltage changes related to the changing angles of diffraction, β_1 and β_2. The measurement of two images was required, since in addition to the strain, the normal to the surface of the specimen underwent a small change of surface angle α due to the lateral motion of the surface as the wave propagated through the grating.

By a simple calculation, the exact expression for this surface angle and for the strain ε was obtained:

$$\tan\alpha = \operatorname{Sin}\left(\frac{\beta_1+\beta_2}{2}\right)\operatorname{Cos}\left(\frac{\beta_1-\beta_2}{2}+\theta_0\right), \tag{4.40}$$

$$\varepsilon = 1 - \frac{\operatorname{Sin}\theta_0}{\operatorname{Sin}\left(\frac{\beta_1-\beta_2}{2}+\theta_0\right)\operatorname{Cos}\left(\frac{\beta_1+\beta_2}{2}-\alpha\right)}, \tag{4.41}$$

where θ_0 is the known angle of diffraction before strain.

Since surface angles were small, for strains below 5%, Eqs. (4.40) and (4.41) could be approximated sufficiently by

$$\alpha = \frac{\beta_1+\beta_2}{2}\cdot\frac{\operatorname{Cos}\theta_0}{1+\operatorname{Cos}\theta_0}, \tag{4.42}$$

$$\varepsilon = \frac{\beta_1-\beta_2}{2}\operatorname{Cot}\theta_0. \tag{4.43}$$

---

[1] The exasperating labor required for the building of a high density diffraction grating ruling engine has not been exaggerated by any of the few persons who have undertaken such a task. Although there are some advantages for a cylindrical engine over the reciprocating engine, there are a sufficient number of disadvantages that the traditional number of "seven demons" is not diminished. Unsuccessful rulings numbering in the hundreds dictated what seemed like an endless series of modifications until successful rulings were obtained consistently.

[2] BELL [1956, *1*], [1958, *1*].

[3] Each new metal required a period of experimentation to determine proper diamond pressure and speed of ruling, and to develop an adequate means of polishing the surface of the specimen. This surface polish, obtained by applying a series of rouges, was made in the axial direction, perpendicular to the rulings

Thus, only the two angles β_1 and β_2 needed to be measured in order to obtain both[1] α and ε. It remained only to relate the voltage changes with angle, a calibration procedure performed in place, before impact, by rotating the light source by a high precision screw to produce known angular incidence from which, by means of Eq. (4.44) the desired calibration was obtained. The light was chopped by means of a rotating disk so that the absolute change from zero to full light could be determined:

$$\operatorname{Sin}\theta - \operatorname{Sin}\alpha = n\lambda m_0, \tag{4.44}$$

where m_0 was the initial number of lines per inch, and $n = \pm 1$ designated the diffraction order.

In Fig. 4.151 is shown a drawing of the apparatus for a symmetrical free flight impact of two identical bars, and in Fig. 4.152, the oscillograph recording of the two photomultiplier outputs from a wave front on a fully annealed aluminum specimen having a maximum strain of 2.6%. Each small division in Fig. 4.152 represents one microsecond.

A recent monograph[2] and several earlier papers,[3] to which the interested reader is referred, contain much additional detail on this experimental technique.[4]

In the original experiments in 1956, the impact was produced by a colliding steel mass. Although those tests were adequate to demonstrate at the time that the nonlinear wave theory of TAYLOR and VON KÁRMÁN described the observed propagation of finite amplitude waves, I immediately began another experiment which gave a much greater opportunity to explore other ramifications of the theory. In this experiment, two identical specimens underwent a free flight axial collision. The diffraction grating was ruled on the struck specimen. Since identical waves were initiated and propagated along the bars, the impact face acted as an infinite mechanical, electrical, and thermal barrier, assuring no transfer of energy from one specimen to the other until unloading from the free ends occurred.

The results of over a thousand such experiments on free flight impact in many solids, both polycrystals and single crystals, have been described in numerous

---

[1] To my knowledge, this was the first determination of the small angle α during wave propagation; it subsequently was found to have an importance approaching that of the measured strain itself.

[2] BELL [1968, *1*].

[3] BELL [1956, *1*, *2*, *3*], [1958, *1*], [1960, *1*, *2*], [1962, *4*], [1967, *1*].

[4] Besides permitting the first measurement of the changing finite strain and of the angle of the normal to the surface (surface angle) in time intervals as short as fractions of a microsecond, this technique has several other advantages which are worth noting here (see BELL [1960, *1*]):

(1) Either static or dynamic strains, from a few microinches per inch to 10%, may be measured on a single gage;

(2) Accurate measurements of strain may be made at temperatures well in excess of 1000° F;

(3) In addition to the measurement of strain, the method may be used to determine simultaneously surface angle, the strain gradient and surface angle gradient in both LAGRANGIAN and EULERIAN frames of reference, and the absolute displacement;

(4) Gage lengths as short as 0.001 inch are feasible, although most of the experiments involved gage lengths from 0.005 to 0.030 inch;

(5) Accuracies of better than 1% may be achieved;

(6) The gage may be calibrated in place, and, since it is integral with the surface, possesses a non-drifting zero over indefinite intervals of time;

(7) Measurement of strain in moving projectiles or machine elements may be made without the necessity for electrical contact with the moving member.

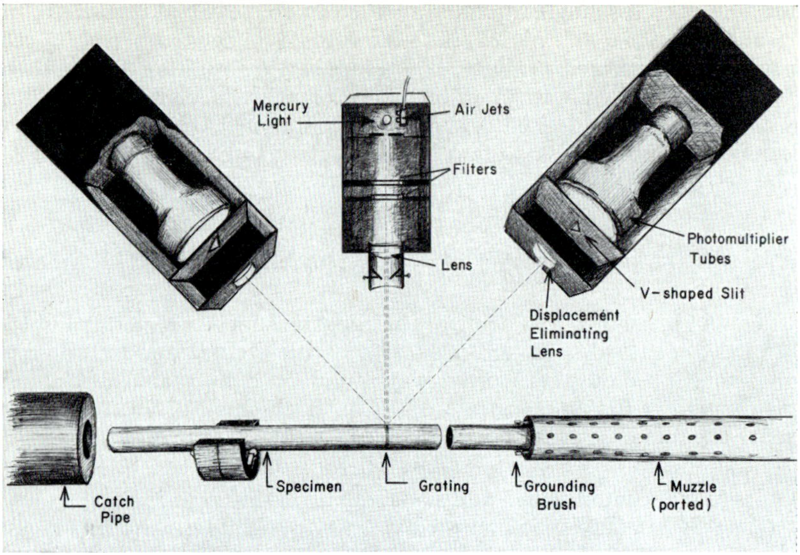

Fig. 4.151. Bell's apparatus for diffraction grating measurement of finite strain and surface angle in free-flight symmetrical impact experiments.

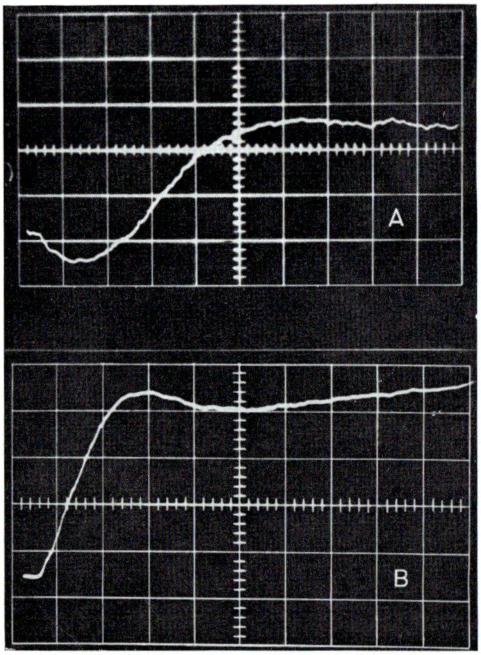

Fig. 4.152 A and B. The photomultiplier voltage outputs for one of Bell's symmetrical free-flight impact experiments. The measurement was made at 1.26 cm from the impact face on a 25.4 cm long, 2.5 cm diameter aluminum specimen. Each small division is one microsecond.

papers since 1960. The discussion here is limited to a few illustrations and the summaries of results which have been published. In Fig. 4.153 may be seen an

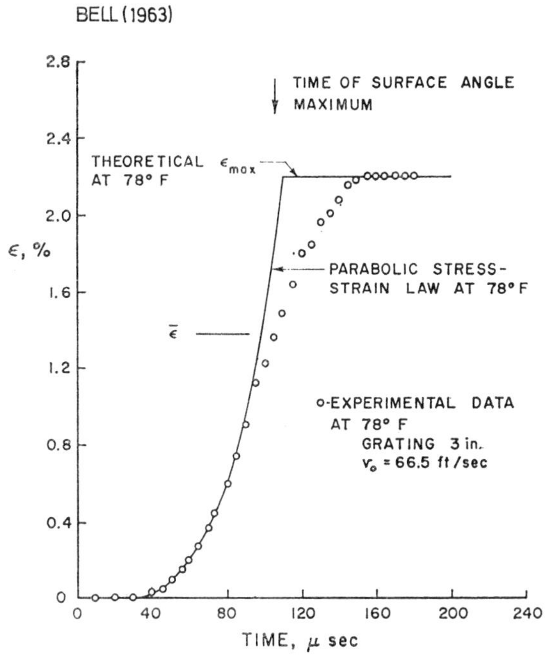

Fig. 4.153. Strain-time data (circles) for a free-flight symmetrical impact at 78° F, compared with predicted curve (solid line) from parabolic stress-strain law.

experimental strain-time profile[1] for a symmetrical free flight impact in fully annealed aluminum[2] at room temperature (circles).

An increase in temperature to 811° K for a similar symmetrical free flight impact provided the result shown in Fig. 4.154a (circles), where the dashed line near the origin indicated what would have been seen at room temperature for the same impact velocity; in Fig. 4.154b for the same solid, is a diffraction grating strain profile at 695° K.

Altering another variable, I obtained the results, also in annealed aluminum, for symmetrical impacts in free flight with velocities which produced the maximum strains of 1.3, 3.3, and 7.5%. The corresponding maximum particle velocities were 1337, 2743, and 5000 cm/sec, respectively.[3] Altering still other variables, i.e., the solid itself, diffraction grating measurements provided the representative results shown in Fig. 4.155 for a symmetrical free flight test in annealed copper; those in Fig. 4.156 for two tests in the alloy, α brass; and, after altering the crystal structure as well as the element, the results from four tests in zinc, shown in Fig. 4.157.

In Figs. 4.156 and 4.157, unlike the preceding figures, the results are plotted as solid lines for the sake of reference to other aspects of the problem discussed below. For the copper specimen of Fig. 4.155, the simultaneous measurement of surface angle obtained in all such experiments was included (dashed line).

[1] Bell [1963, *1*].

[2] This was 99.16% purity, hot rolled aluminum, annealed for two hours at 1100° F, furnace cooled, and checked for small grain size.

[3] Since the maximum particle velocity in a symmetrical collision is one half the impact velocity, the initial velocity of the impacting specimen was twice these values.

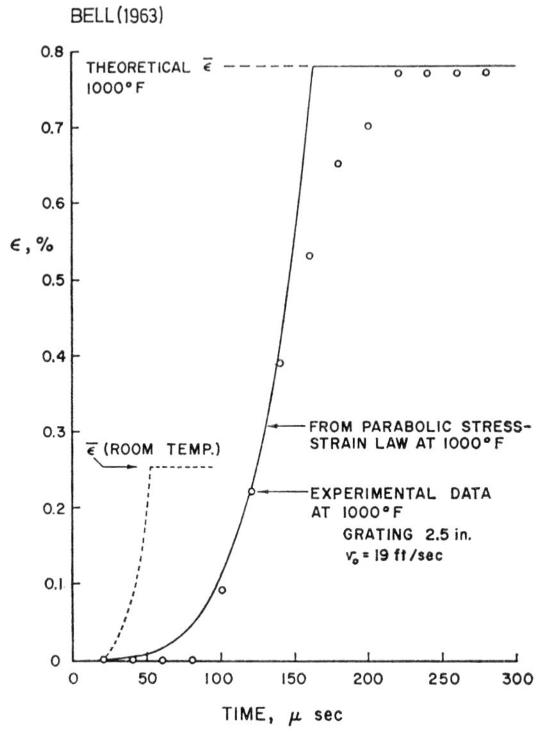

Fig. 4.154a. Strain-time data (circles) for a free-flight symmetrical impact at 1000° F, compared with predicted curve (solid line) from parabolic stress-strain law.

The measurement of finite strain in an impact experiment at constant velocity provides a measurement of maximum strain which can be compared with the known maximum particle velocity obtained from halving the measured impact velocity. Thus, a velocity vs strain relation can be determined from changing the impact velocity and measuring the corresponding maximum strain. (The opposite, unfortunately, is not true; the measurement of velocity vs time profiles alone, whether optically or by magnetic induction, does not provide a value of maximum strain to establish whether the velocity vs strain relation is that expected from measured wave speeds.)

As an additional check upon the velocity vs strain relation, I also measured by means of an optical displacement technique I developed in the late 1950's, the actual velocity vs time profiles for a direct comparison with finite strain profiles at the same location. Later independent checks were made from measurements of particle velocity using an alternate technique by MALVERN[1] and by WILLIAM J. GILLICH and WILLIAM O. EWING,[2] as well as by me.[3]

[1] MALVERN [1965, *1*].

[2] GILLICH and EWING [1968, *1*].

[3] BELL [1965, *1*]. MALVERN's paper and mine were both presented at the same Symposium. That the latter paper and not the former contained the comparison between experimental results and the parabolic stress-strain function I had found, arose from the fact that MALVERN had described his experiments in a report issued earlier, and I used that report in presenting that particular independent check. It was at this time that MALVERN concluded, as I had earlier, that electric resistance gage measurements were worthless in providing meaningful finite strain profiles.

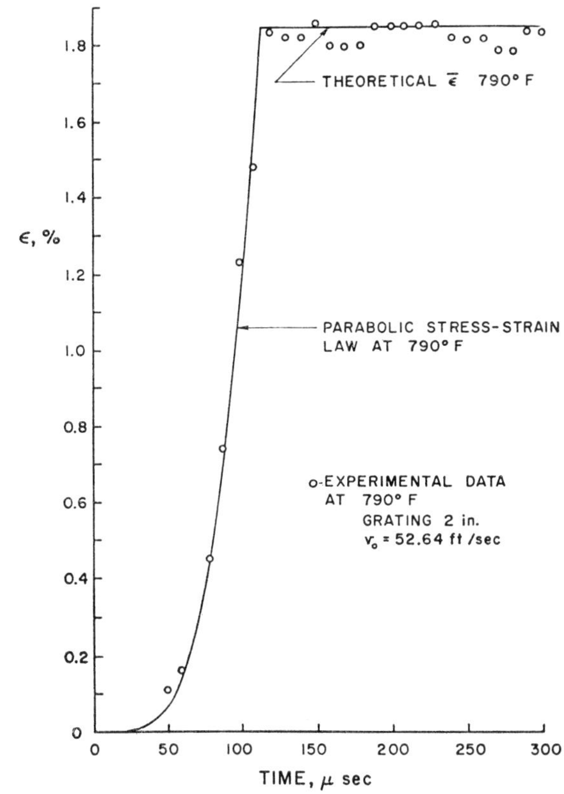

Fig. 4.154b. Strain-time data (circles) for a free-flight symmetrical impact at 790° F, compared with predicted curve (solid line) from parabolic stress-strain law.

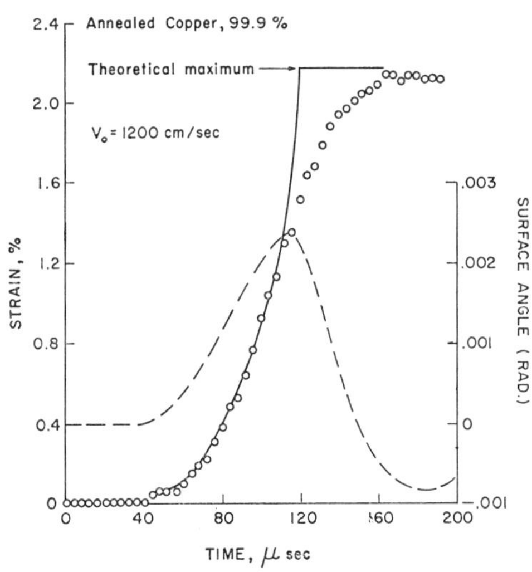

Fig. 4.155. A diffraction grating strain-time (circles) and surface angle-time (dashed line) measurement by BELL at 5.1 cm from the impact face, from the symmetrical free-flight impact of 25.4 cm long, 2.50 cm diameter cylinders, compared with prediction for $r = 4$. The impact velocity was 1 200 cm/sec. The dashed line is the simultaneously measured surface angle vs time.

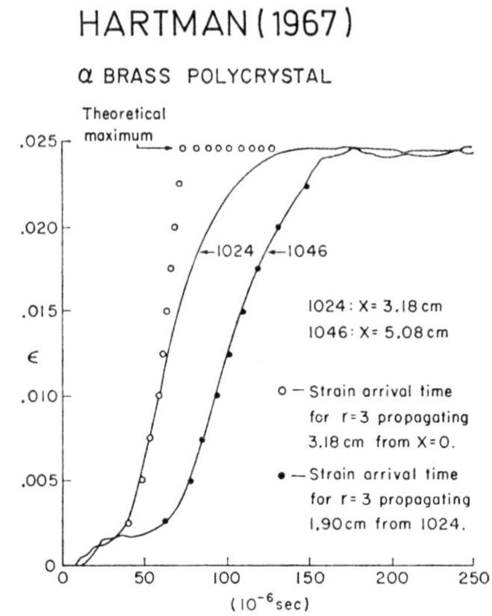

Fig. 4.156. Two diffraction grating experiments in 70–30α brass, compared with prediction from the parabolic generalization of Bell, assuming an infinite step at the impact face (open circles) and with predicted traverse times from one position to the other (closed circles).

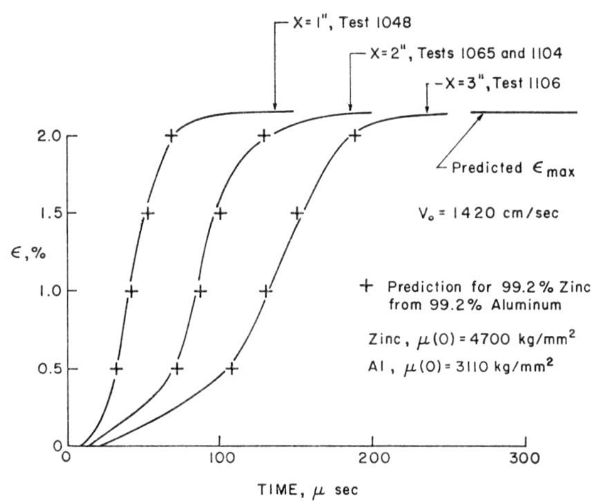

Fig. 4.157. Diffraction grating strain-time data of Bell for four experiments in annealed polycrystalline zinc (solid line) compared with prediction for $r = 2$ from aluminum (crosses).

The optical displacement technique which I introduced in 1960 for the examination of particle velocity profiles during wave propagation is shown diagrammatically in Fig. 4.158.

One half the specimen was painted with a non-reflecting black paint and the other, with white. As the specimen was displaced, the percentage of diffuse reflected light changed as seen through the rectangular window shown, and thus gave a measure of displacement vs time. The slopes of these measured displacement vs time curves gave the desired particle velocity vs time, wave profiles at a point.

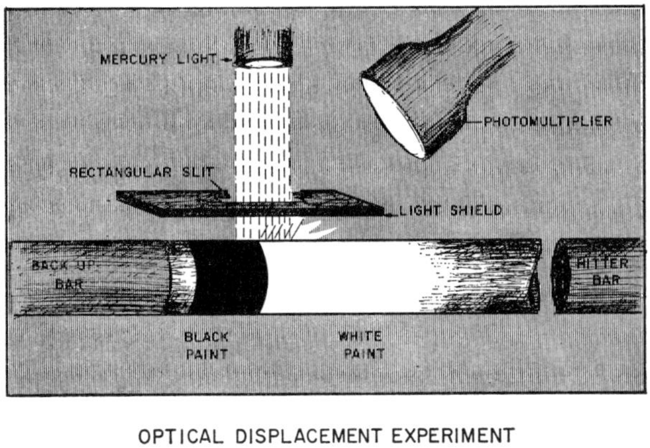

Fig. 4.158. Drawing of BELL's apparatus of 1961 for the optical determination of particle displacement during wave propagation.

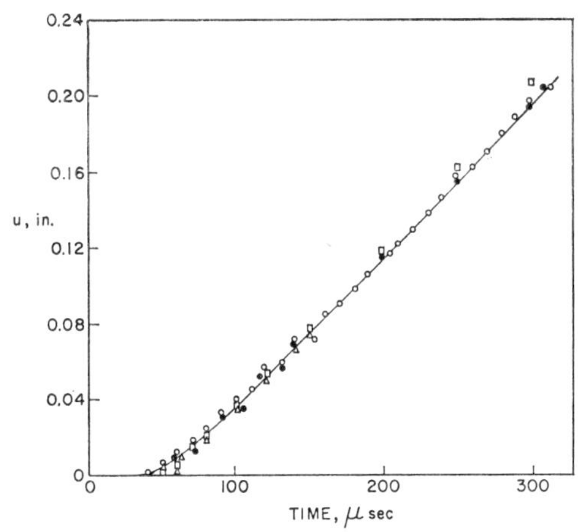

Fig. 4.159. Four displacement-time measurements of BELL (1961) in aluminum for an impact velocity of $v_0 = 2030$ cm/sec at 2 diameters from the impact face, compared with prediction from finite amplitude wave theory (solid line).

Light intensity changes produced by the displacement of the specimen were calibrated to provide displacement vs time histories such as the representative four shown in Fig. 4.159 (circles). The solid line through these data was the displacement vs time, wave profile predicted from the measured constant wave speeds of finite strain vs time profiles.

Thus, all parameters for the finite amplitude wave had been measured directly, establishing without specific *a priori* reference to a nonlinear wave theory the conditions predicted in the solution to the theory of TAYLOR and VON KÁRMÁN. Both finite strain vs time profiles and particle velocity profiles had been measured at the same point, including the measurement of maximum values for each. That particle velocity was a single valued function of finite strain, $v(\varepsilon)$, and wave speeds $c_p(\varepsilon)$ were constant for each value of strain during wave propagation in annealed polycrystals, had been ascertained by the measurement of *both* variables at the same position during nonlinear wave propagation. The two conditions of the theory were given above in Sect. 4.27, by Eqs. (4.38) and (4.37). After it had been shown, without having assumed it in advance, that the theory applied, the integral of Eq. (4.39), without further assumption, led to the governing response function. In every annealed solid studied, this was found to be the parabolic relation, Eq. (4.25) with $\varepsilon_b=0$. [See Sect. 4.21 above.][1]

In 1957, R. WALTER RAMBERG and L. K. IRWIN[2] had introduced a method for determining velocity vs time measurements by means of an electromagnetic transducer. The technique was used in 1960 by BIANCHI[3] for a new series of experiments on the incremental wave (Sect. 4.34) and by RIPPERGER and L. M. YEAKLEY[4] in 1963 for experiments at small strain for linear elastic wave propagation. In 1964, L. EFRON,[5] a doctoral student of MALVERN at Michigan State University, and in 1965, MALVERN,[6] and in 1968 GILLICH and EWING[7] used this technique to provide an independent check upon my earlier optical observations of velocity vs time profiles during wave propagation at finite strain in annealed polycrystals.

In MALVERN and EFRON's experiment, insulated wires looped around the specimen at four different positions provided induced voltages when during the passage of a wave front the particle velocity of the specimen caused them to move in calibrated magnetic fields. MALVERN and EFRON did not have symmetrical free flight impact tests, since the impact bar was made of steel. The specimens of commercial purity aluminum upon which measurements were made were bars $^1/_2$ in. in diameter, 58 in. long. Unlike the fully annealed aluminum of my experiments (annealed for two hours at 1100° F and furnace cooled) for which the elastic limit was $Y=1100$ psi, their specimens were annealed for one hour at the much lower value of 650° F and furnace cooled. The resulting elastic limit was much higher, i.e., $Y\cong 3800$ psi. Their first transducer was located 3 in. (6 diameters) from the impact face. Subsequent locations were approximately at 3.25 in. intervals. MALVERN and EFRON provided the four velocity vs time profiles for a single typical test, as shown in Fig. 4.160.

From the observed constancy of wave speeds in this region far beyond the first diameter (the final reading was at 25 diameters from the impact face), they determined from the averages of many tests the wave speed vs particle velocity (solid line) shown in Fig. 4.161.

As was pointed out above, from particle velocity wave profiles alone, one can check only the constancy of wave speeds provided by the finite amplitude wave theory. Without the simultaneous measurement of strain, the second condition of the theory namely, that particle velocity is a function of the strain, cannot be demonstrated, let alone the fact that the form of such a function cannot be determined. In this instance, however, both particle velocity and finite strain profiles in annealed aluminum had been obtained earlier by me, and thus the new data could be considered in terms of the nonlinear theory. Although MALVERN and EFRON did not compare their results with my measurements, beyond noting that

[1] Eq. (4.25), of Sect. 4.21 was given from the analysis of quasi-static response functions in the same annealed metals. In historical sequence, however, as was noted above, I discovered this parabolic response function from the wave propagation experiments, and in later research I showed that it also described the quasi-static reponse at strain rates from 10000 to over 100000000 times lower.

[2] RAMBERG and IRWIN [1957, *1*].

[3] BIANCHI [1960, *1*].

[4] RIPPERGER and YEAKLEY [1963, *1*].

[5] EFRON [1964, *1*].

[6] MALVERN [1965, *1*], EFRON and MALVERN [1969, *1*].

[7] GILLICH and EWING [1968, *1*].

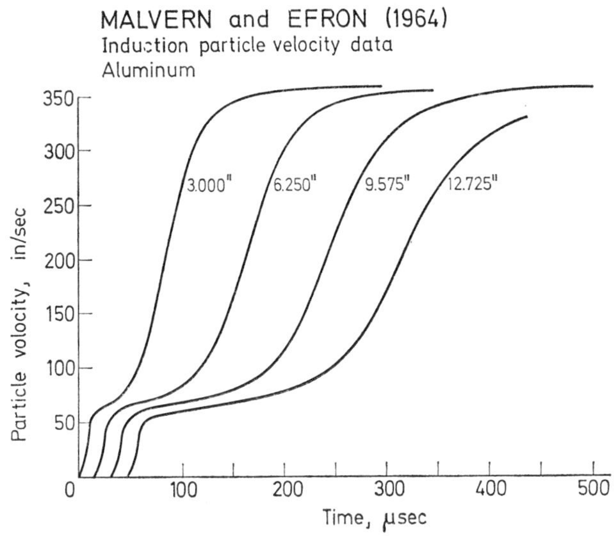

Fig. 4.160. Typical record of particle velocity vs time at different positions, obtained by MALVERN and EFRON.

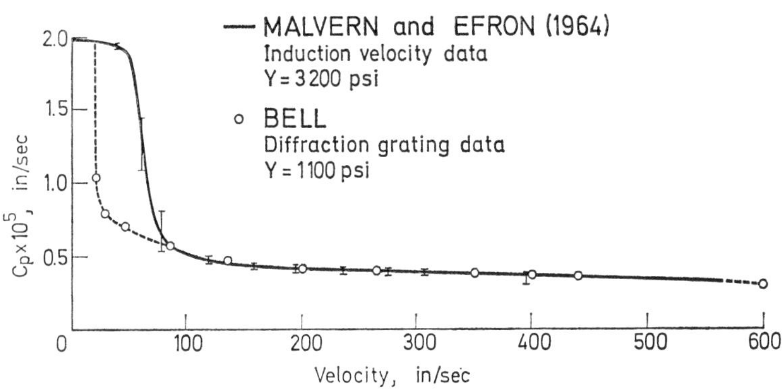

Fig. 4.161. Wave speed vs velocity data of EFRON and MALVERN, compared with prediction from BELL's diffraction grating strain measurements.

the wave speeds in annealed aluminum indeed were constant as I had found in 1956, I made such a comparison[1] in 1965. The solid circles in Fig. 4.161 were the predicted values of wave speeds at the various particle velocities shown, based upon my previous diffraction grating and optical displacement measurements, and consistent with the governing parabolic response function, which appears in Eq. (4.25), for annealed aluminum at room temperature.

This independent check upon my earlier work was of additional interest in view of the fact that there had been an increase of the elastic limit by a factor of nearly 4, showing that for the annealed solid the plastic wave propagation was not altered by a significant change in the amplitude of the linear elastic precursor. Similar electromagnetic transducer studies of particle velocity by GILLICH and EWING[2] of the Ballistic Research Laboratory, Aberdeen Proving Ground, Mary-

[1] BELL [1965, *1*]
[2] GILLICH and EWING [1968, *1*].

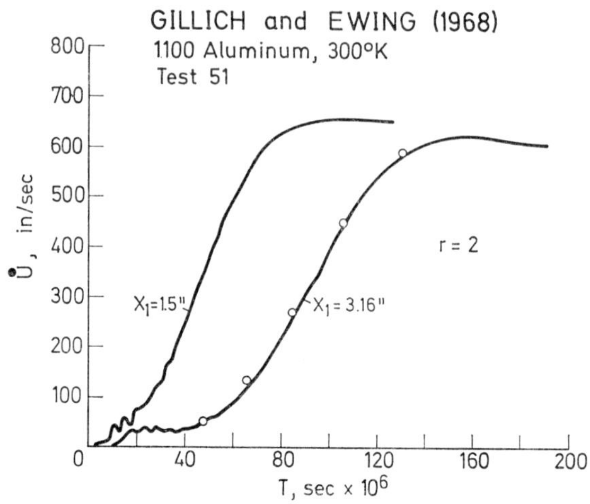

Fig. 4.162. GILLICH and EWING's velocity-time profiles at the two positions indicated (solid lines). The circles on the second profile correspond to arrival times provided by BELL from his parabolic generalization applied to annealed aluminum, for $r=2$.

land, on dead annealed aluminum and chemical lead in 1968, provided a further independent check; because of their expressed desire to explore some of the results I had obtained a decade earlier, their experiments were carried out on dead annealed hot rolled aluminum and lead, for symmetrical impact collisions in free flight. In Fig. 4.162 are shown two velocity vs time measurements in annealed aluminum (solid lines). The circles I have added on the second profile indicate the arrival time for a traverse of the wave between the first and the second positions, based upon my study of wave profiles in the same solid in terms of Eq. (4.25) for the parabolic response function at room temperature.

Similarly, in Fig. 4.163 may be seen the same close correlation of measured velocity vs time at room temperature with the parabolic stress-strain function for lead. The experiments in this instance were SPERRAZZA's[1] diffraction grating studies on lead for his doctoral dissertation in 1961. Again the solid lines were experimental results, and the circles I have added represent the arrival times at the second position predicted from the measured profile at the first, by means of Eq. (4.25). [$r=4$ for lead].

These results of MALVERN and EFRON and of GILLICH and EWING, of course, were only a corroboration of earlier measurements of particle velocity, but the magnetic induction technique was preferable to the optical displacement method because it determined the particle velocity directly, and thus eliminated the errors caused by taking the slopes from curves of displacement vs time. All measurements of particle velocity along the specimen before those of GILLICH and EWING had been made on the surface of the solid. Those authors provided an important extension of earlier work by comparing velocity profiles from wires located on the surface with those integrated along the diameter of the specimen. They accomplished this by placing the wires in a small hole drilled through the specimen.

My first experiments which were on 99.16% purity polycrystalline aluminum annealed for two hours at 1100°F, furnace cooled, and checked for grain size,

[1] SPERRAZZA [1961, *1*], and see SPERRAZZA [1962, *1*] and BELL [1963, *1*].

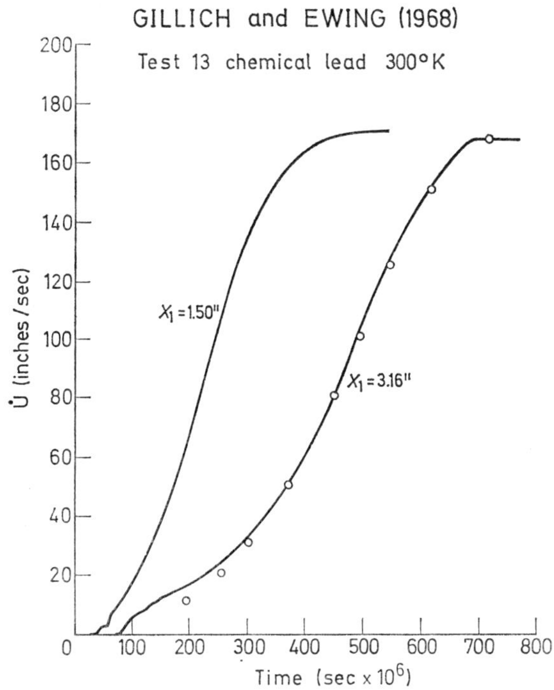

Fig. 4.163. GILLICH and EWING's experimental particle velocity-time profiles (solid lines) at the two positions indicated, compared with prediction of traverse times by BELL (circles) from his parabolic generalization for $r=4$, applied to lead.

revealed that wave speeds were constant for each strain.[1] Since a finite strain profile for a large number of positions along cylindrical specimens had to be determined for each of many impact velocities, and for specimens of different diameters, hundreds of tests were performed to establish this most important point, namely, that within a fraction of a per cent, at each magnitude of strain considered, wave speeds in fact were constant.

Early in my studies measuring both finite strain and particle velocity, I observed that in the region of the first diameter, wave propagation was not one-dimensional. Thus, measurement for a minimum of three positions beyond the first diameter was essential for any study of wave speeds since the entire wave front did not, as others had assumed, proceed from a one dimensional infinite step at the impact face; i.e., the wave front at the first diameter was distorted by the initiation process. In order to determine the governing constitutive equation to three or more significant figures, beyond the first diameter a far larger number of positions than the minimum three were explored. After my experiments in fully annealed aluminum, similar results were obtained by me or by my graduate students in annealed copper, annealed zinc, lead, magnesium, and α brass polycrystals, and in aluminum and copper single crystals of known orientations.

[1] BELL [1960, *2*]. In 1956 (BELL [1956, *1*]) experiments with steel mass impacts also had been conducted on annealed aluminum, with 8300 lines/inch diffraction gratings. The experiments reported in 1960 were symmetrical free flight impacts, using 30720 lines/inch diffraction gratings.

In my diffraction grating experiments the observed wave profiles at given values of strain went through three orders of magnitude of strain rate without any alteration in this functional relation, hence it was obvious that the influence of viscosity indeed was negligible. In reaching this conclusion, an important experimental fact sometimes overlooked by those who think only in terms of wave speeds, is that the agreement of measured maximum strain with measured maximum particle velocity, both compared calculations from the measured wave speeds, provides a sensitive measure of the absence of viscosity.

From the constancy of the wave speeds, and the knowledge of their magnitude, and from the separately measured stress and particle velocity, every parameter in the integral of Eq. (4.38) could be checked by

$$v = \int_0^{\varepsilon} c_p(\varepsilon)\, d\varepsilon . \qquad (4.38)\ \text{(repeated)}$$

As already noted, the velocity vs strain relation together with the constant wave speeds of Eq. (4.45):

$$c_p(\varepsilon) = \sqrt{\frac{d\sigma}{d\varepsilon}\Big/\varrho} = \text{constant}, \qquad (4.45)$$

established that the nonlinear wave theory could be used to describe the dynamic plasticity of a given solid without the necessity for having presumed the form of the response function in advance.

Since these two conditions had been shown to apply to the annealed metals considered, it became possible, by squaring and integrating Eq. (4.45), to obtain a direct experimental determination of the governing response function for finite strain:

$$\sigma = \int_0^{\varepsilon} \varrho\, c_p^2\, d\varepsilon . \qquad (4.39)\ \text{(repeated)}$$

In every metal studied, at all ambient temperatures considered, I found, as is now well known, that the experimentally determined stress-strain function was parabolic, having the form $\sigma = \beta\, \varepsilon^{\frac{1}{2}}$. Perhaps not so well known is the fact that also by experiment I was able to examine the final remaining question, that is, whether the phenomenon was indeed one dimensional and isochoric. This I accomplished by utilizing the surface angle vs time histories simultaneously obtained in every diffraction grating measurement. A series of 59 impact experiments from the early 1960's at the single maximum particle velocity of 2030 cm/sec with gratings located at four different positions from the impact face are shown in Fig. 4.164. The number of tests averaged at each position was indicated. I had noted that these were a representative sample of experiments on annealed aluminum in symmetrical free flight impact all of which demonstrated, from the first diameter on, the constancy of wave speeds and common maximum strain at each position.

My analysis was as follows: assuming isochoric motion, one has before and after strain

$$\pi R_0^2 = \pi (R_0 + u_r)^2 (1 - \varepsilon), \qquad (4.46)$$

from which follows

$$u_r = \frac{R_0}{(1-\varepsilon)^{\frac{1}{2}}} - R_0 . \qquad (4.47)$$

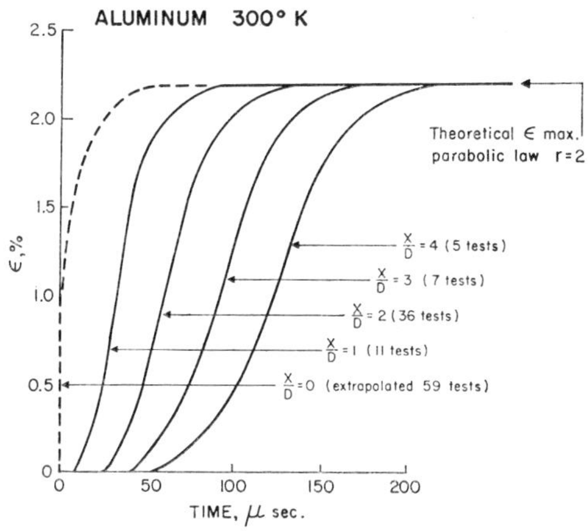

Fig. 4.164. Averaged strain-time data of BELL for 59 tests performed in 1960 on fully annealed aluminum, for an impact velocity, $v_0 = 2030$ cm/sec. (For this compilation see BELL [1967, *2*] or [1968, *1*]. See also [1960, *2*.].) The specimens were 25.4 cm long, with a diameter, $D = 2.50$ cm. X denotes the distance from the impact face.

The surface angle $\alpha = \frac{1}{1-\varepsilon}\frac{\partial u_r}{\partial x}$ where u_r is the radial displacement at the surface. Differentiating u_r in Eq. (4.47) with respect to x yields

$$\alpha = \frac{R_0 \frac{\partial \varepsilon}{\partial x}}{2(1-\varepsilon)^{\frac{5}{2}}}, \tag{4.48}$$

where, from the compatibility relation $\partial v/\partial x = \partial \varepsilon/\partial t$ and for $v = v(\varepsilon)$

$$\frac{\partial \varepsilon}{\partial x} = \frac{\partial \varepsilon}{\partial t} \bigg/ \frac{\partial v}{\partial \varepsilon}, \tag{4.49}$$

whence finally

$$\alpha = \frac{R_0 \dot{\varepsilon}}{2(1-\varepsilon)^{\frac{5}{2}}} \frac{1}{dv/d\varepsilon}. \tag{4.50}$$

Since $dv/d\varepsilon$ was established experimentally, and every other quantity in this relation also was measured, including $\alpha(t)$, an experimental test was provided for one dimensionality and for isochoric motion at each and every point of the many considered. From Eq. (4.45), Eq. (4.50) also could be written in terms of the known wave speeds:

$$\alpha = \frac{R_0 \dot{\varepsilon}}{2(1-\varepsilon)^{\frac{5}{2}} c_p(\varepsilon)}. \tag{4.51}$$

The surface angle was that change from the horizontal produced by the wave front in a purely axial impact, as shown in Fig. 4.165.

To eliminate errors which could arise from small non-axiality in some of the impact tests the data were averaged for many tests.[1] The results of 50 averaged measurements in annealed aluminum at the positions shown are given in Fig. 4.166. Results from individual tests of known axiality are shown in Fig. 4.167 for elastic-plastic boundary experiments where the axiality is closely controlled. In Fig. 4.168 are shown comparisons between predicted surface angle (circles) and measured surface angles, for symmetrical free flight impact tests in 99.2% purity zinc. These

[1] BELL [1972, *1*].

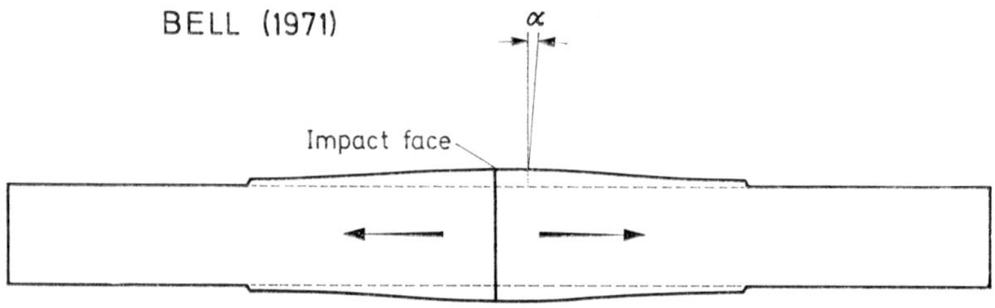

Fig. 4.165. A schematic diagram showing the variation of the surface angle, α during wave propagation.

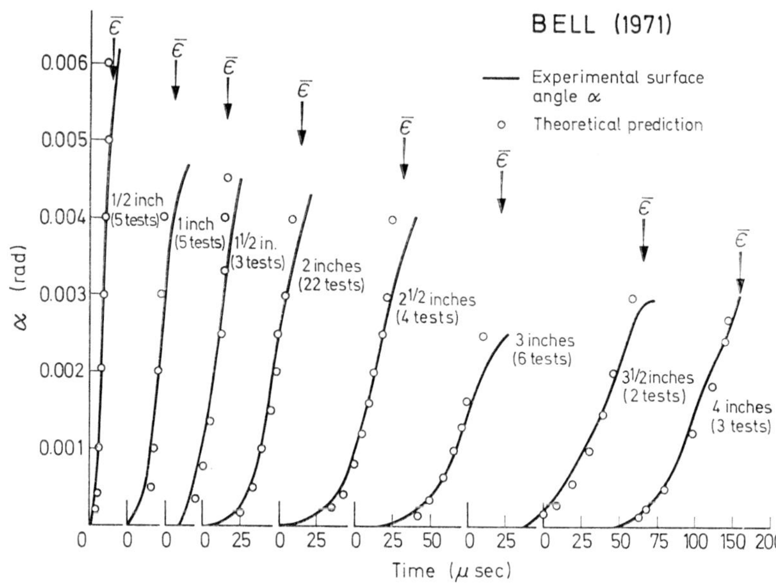

Fig. 4.166. Averaged experimental surface angle diffraction grating data for annealed aluminum, compared with prediction from Eq. (4.51).

data thus provided a demonstration, independent of the theory of wave propagation, of one-dimensionality and isochoric motion.[1]

In 1959 I had constructed an impact apparatus which included a furnace for both the striking specimen and the initially stationary specimen, thus permitting diffraction grating measurements of finite dynamic strain and surface angle to be made through quartz windows at all ambient temperatures to within a few degrees of the melting point. The details of this development, which considerably increased the problems of conducting an impact test with the diffraction grating technique, were described in a publication[2] in 1962. The results of the high temperature tests revealed that at every ambient temperature explored, to within a few degrees of the melting point in fully annealed aluminum, the theory of waves of

[1] I extended this independent demonstration of one dimensionality and isochoric motion, from the study of surface angles, to the dynamic plasticity of polycrystals at high temperatures, single crystals, copper, α brass, etc.

[2] Bell [1962, 4].

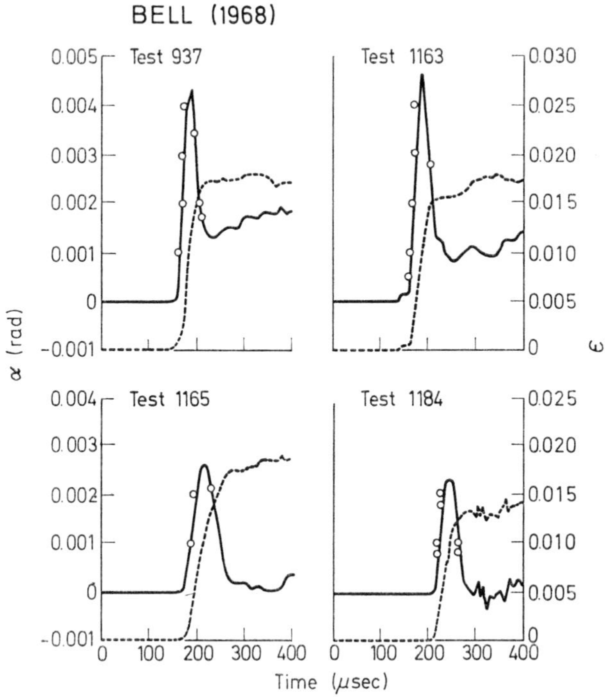

Fig. 4.167. Diffraction grating surface angle measurements in annealed aluminum (solid lines) for known axial elastic-plastic interface experiments, compared with prediction (circles) from Eq. (4.51). The simultaneously measured strain-time profile is shown by the dashed lines.

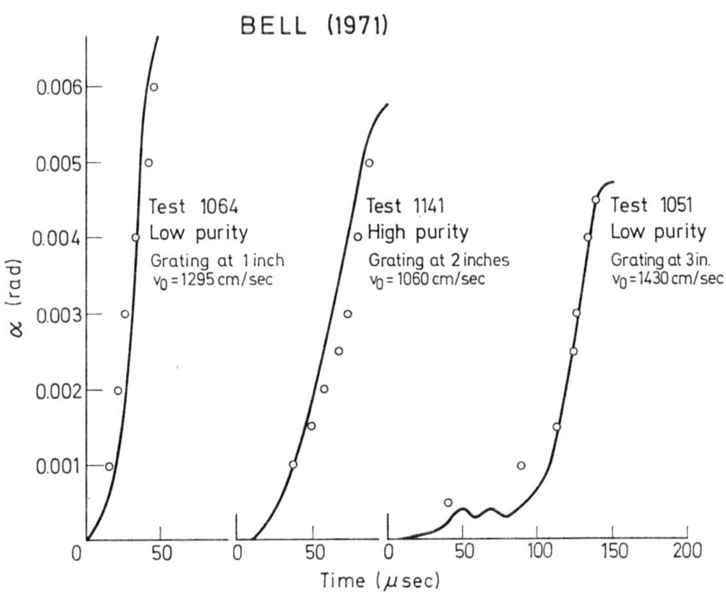

Fig. 4.168. Three diffraction grating surface angle measurements (solid lines) at 1, 2, and 3 diameters from the impact face in the hexagonal metal, zinc, compared with prediction (circles) from Eq. (4.51).

finite amplitudes, as stated by TAYLOR[1] and VON KÁRMÁN,[2] applied, and the response function which followed from the observation was parabolic. The first 63 measurements in annealed aluminum at eight different temperatures above room temperature had measured parabola coefficients which were a linear function of the fractional melting point temperature, T/T_m. Thus, as I showed[3] in 1963, the governing uniaxial response function for propagation at finite strain in that solid was that of Eq. (4.52).

$$\sigma = \beta(0)\,(1 - T/T_m)\,\varepsilon^{\frac{1}{2}} \tag{4.52}$$

where σ was the stress referred to the undeformed area; ε, the strain referred to the undeformed length; and where $\beta(0)$ represented a predicted value at absolute zero; T was the ambient temperature and T_m was the melting point, both in degrees

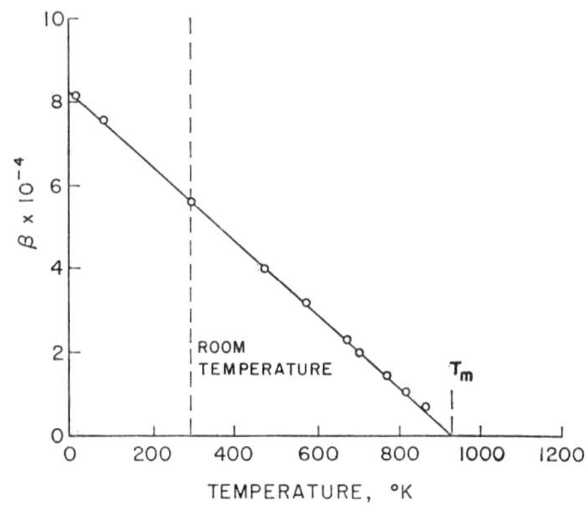

Fig. 4.169. Experimental parabola coefficients obtained by BELL in 1963 from 63 commercial purity polycrystalline aluminum free-flight impact experiments above room temperature.

Kelvin. The results of these experiments of 1959 to 1963 are in Fig. 4.169, where the measured parabolic coefficients from experiments on waves of finite amplitude were plotted against the ambient temperature (circles). These were averaged experimental data.

By 1962 symmetrical free flight impact tests in which finite strain and surface angle were recorded had been performed on aluminum, copper, lead, and magnesium. Subsequent experiments provided results in zinc and, much later, α brass. The early experiments on copper, the experiments on lead and magnesium, and the much later experiments on α brass were on the same laboratory apparatus, using the same diffraction grating ruling engine, and were part of the research of a number of my graduate students at that time. Those on copper were measurements by W. MEADE WERNER,[4] those on lead by JOSEPH SPERRAZZA,[5] those on magnesium by ANDREW CONN,[6] and those on α brass by WILLIAM FRANCIS HARTMAN.[7] The examination of finite strain profiles of 25 to 35 tests on symmetrical impact in free flight in each of these annealed metals revealed that in every instance, particle velocity was a single valued function of strain and wave speeds at

[1] TAYLOR [1942, *1*].
[2] VON KÁRMÁN [1942, *1*].
[3] BELL [1963, *1*].
[4] WERNER [1960, *1*] (Master's essay); and see BELL and WERNER [1962, *5*].
[5] SPERRAZZA [1961, *1*], [1962, *1*].
[6] CONN [1959, *1*].
[7] HARTMAN [1967, *1*], [1969, *1*, *2*].

each strain were constant. The nonlinear wave theory of TAYLOR and VON KÁRMÁN hence was applicable and, as a consequence, response functions could be determined.

In comparing the large body of data from these solids with the data of aluminum, I discovered[1] in 1961 that in every instance the measured response function was parabolic and expressible in terms of Eq. (4.52). Since the values of $\beta(0)$ were different from one solid to another and, in annealed aluminum differed for low and high purity specimens, an intensive and systematic search for parameters which would relate empirically the different values of $\beta(0)$ led to my discovery[2] from 1962 to 1965 that all known data on wave propagation were accounted for by the following form for $\beta(0)$:

$$\beta(0) = (\tfrac{2}{3})^{r/2}\,\mu(0)\,B_0, \tag{4.53}$$

where $r=1, 2, 3, \ldots$; $\mu(0)$ is the shear modulus at the zero point for the isotropic solid of interest; and $B_0=0.0280$, a dimensionless universal constant. Thus the response function for finite strain during wave propagation had the experimentally established form

$$\sigma = (\tfrac{2}{3})^{r/2}\,\mu(0)\,B_0\,(1-T/T_m)\,\varepsilon^{\frac{1}{2}}. \tag{4.54}$$

Quasi-static uniaxial tension and compression tests performed in my laboratory at the same time on fully annealed specimens of the same solids which had been examined for wave detail, revealed, as was shown above in Sect. 4.21, that measured response functions for slow loading in dead weight tests also were expressible in the form of Eq. (4.54). Given the ambient temperature T one merely had to have experimental values for $\mu(0)$, T_m, and the integral mode index r, to describe the response of a given solid at any strain rate.

The possibility that a universal response function described the finite deformation of the entire class of annealed crystalline solids was directly responsible for my undertaking the large programs of slow loading experimentation and the extensive literature analysis described in Sect. 3.44 for zero point shear moduli; in Sect. 4.18 for the stage III deformation of cubic single crystals, and in Sect. 4.21 for the dead weight uniaxial loading of annealed polycrystals. I presented numerous papers describing different details during the six years, 1962 to 1968, while this study was in progress. In a monograph on *The Physics of Large Deformation of Crystalline Solids* in 1968, I gave a complete summary in the form of a compilation and comparison of the details of over 2000 individual tests, mainly from my own laboratory, but including over 700 tests described in the previous half century of the metal physics literature. The discovery that the finite response of 19 elements and 8 binary combinations, 27 crystalline solids of 5 different crystal structures, was always parabolic, is paralleled in experiment by the exhaustive analysis of WERTHEIM in 1844 who discovered, also for the dynamic and quasi-static loading, that all members of this same class of elements and binary combinations had linear response functions in infinitesimal deformation.

The present study, however, revealed an order for this class of solids far beyond the observation that for both infinitesimal deformation and finite deformation the analytical form of response functions was in every case common to the entire class. This was the discovery that the elastic constants for the linear infinitesimal deformation and the parabola coefficients for the finite deformation of all studied members of this class of solids were universally related by a quantized distribution of coefficients. For both elastic constants and parabola coefficients

[1] BELL [1961, *1*].
[2] BELL [1962, *4*], [1963, *1*], [1965, *2*].

these specific distributions interrelated the different solids and, in terms of second order transitions, different states of the same solid.

The relation of elastic constants among the elements was given by Eq. (3.28) in Sect. 3.44, and that for their second order transitions in a given solid, which I have called "multiple elasticities," by Eq. (3.29) in Sect. 3.44. The similarly quantized distribution of parabola coefficients for the resolved finite strain and second order transition of single crystals was described in Sect. 4.18; that for the finite deformation of polycrystals in uniaxial loading in Eq. (4.25) in Sect. 4.21; and for polycrystalline solids in simple torsion, by Eq. (4.29) in Sect. 4.22.

From the experimental point of view, the deformation, both small and large, of annealed crystals thus are qualitatively and quantitatively unified. As with all such advances in physics, related additional discoveries, such as, in this instance, quantized distribution of coefficients and the second order transition, offer a stimulus for experimental and theoretical research in hitherto unsuspected directions.

A final point should be noted here. HOOKE's law, of course, is expressible as a completely generalized set of constitutive equations relating all stresses and strains. In Sect. 4.35 of this treatise, I shall describe recent experiments which provided the basis for my development of similarly generalized constitutive equations for the parabolic response of finite deformation.

4.29. The experimental study of unloading waves in dynamic plasticity: Bell (1961).

The nonlinear theory for plastic loading waves which experiment has shown is applicable for annealed crystalline solids, is indistinguishable from a statement in nonlinear elasticity. Unless the heat generated during the passage of the wave front can be observed, the first indication of plastic waves is given when the unloading waves arrive. In slow loading, according to ample experimental evidence given by TAYLOR[1] and by DILLON,[2] approximately 95% of the energy of deformation was dissipated as heat. Many experimentists in the 19th century had shown that the unloading response function for quasi-static tests was approximately linear, with slopes much lower than before permanent deformation, but still in the same order of magnitude.

What form an unloading response function might have when loading had occurred a few microseconds before, is solely a matter for experiment to reveal. Despite the interest generated by the theoretical effort to describe unloading waves, by BOHNENBLUST[3] in 1942, RAKHMATULIN[4] in 1945, and particularly by LEE[5] in 1953, no direct experimental study of unloading wave fronts was made until my experiments[6] in 1961. BOHNENBLUST[7] and BOHNENBLUST, CHARYK, and HYERS[8] in 1942 were interested in providing a correction for the analysis of the "frozen" strain distribution of DUWEZ's[9] experiment. It was not until 1953, however, that plausible description of unloading, based upon the same assumed

---

[1] TAYLOR and FARREN [1925, *2*]; TAYLOR and QUINNEY [1934, *2*].
[2] DILLON [1962, *1*, *2*], [1966, *2*].
[3] BOHNENBLUST [1942, *1*, *2*].
[4] RAKHMATULIN [1945, *1*].
[5] LEE [1953, *1*].
[6] BELL [1961, *3*, *4*].
[7] BOHNENBLUST [1942, *2*].
[8] BOHNENBLUST, CHARYK, and HYERS [1942, *1*].
[9] DUWEZ, WOOD, and CLARK [1942, *1*].

linear unloading response function which by then had become known as the "BOHNENBLUST equation," was presented in a form appropriate for experimental study. As a plastic wave formed by a constant velocity impact propagates toward the free end of a rod, it ultimately is met by unloading waves propagating in the opposite direction, generated by the reflection from the free end of faster moving portions of the original wave. The fastest moving component of the original wave front, which first interacts with the loading front, is the linear elastic precursor, having an amplitude given by the dynamic elastic limit Y.

The theory of LEE, utilizing the method of characteristics to examine the momentum of interacting wave fronts, required not only a knowledge of Y but also the precise form of the response function at finite strain, governing the original loading wave. Using a quasi-static response function for a one dimensional problem LEE[1] found that the reflected loading wave was absorbed as it propagated into the dispersive plastic wave. Letting $\Delta\sigma$ be the magnitude of this unloading increment, and σ the stress of the loading wave before interaction, and v the corresponding particle velocity, LEE found that

$$\Delta\sigma = Y - \tfrac{1}{2}(\varrho\, c_0\, v - \sigma), \tag{4.55}$$

where $c_0 = \sqrt{E/\varrho}$ and where ϱ is the mass density.

Since v increased more rapidly than σ, if Y were sufficiently small compared to the maximum stress of the loading wave, this unloading wave front would be completely absorbed.

Having by 1961 determined by diffraction gratings the governing loading response function for the dynamic plasticity of annealed aluminum and annealed copper,[2] I took advantage of the opportunity thus provided to examine this prediction of LEE's theory.[3] First, it was necessary to establish experimentally the magnitude of the dynamic elastic limit Y for use in the analysis. I obtained this by studying the initial detail of waves in the free flight impact experiment. One such measurement is shown in Fig. 4.170. These studies revealed that in low purity, completely annealed aluminum the quasi-static and dynamic elastic limits were the same.[4] They both had the value of $Y = 0.774$ kg/mm^2 (1100 psi). Since $E = 7180$ kg/mm^2 (10.2×10^6 psi), we have for the strain at Y, $\varepsilon_y = 108 \times 10^{-6}$.

By introducing this measured value of Y for the elastic limit, and my parabolic response function for the loading wave, I calculated from Eq. (4.55) that complete absorption of the unloading wave would occur at a distance of 0.43 L from the impact face, where L is the length of the specimen. In Fig. 4.171 a–c are shown electric resistance measurements of the small strain of the reflected unloading wave at 0.6 L, 0.5 L, and 0.4 L. For aluminum an elastic wave travels one inch in 5 µsec. The unloading wave for $L = 10$ in. arrived at 0.6 L, as predicted, at 70 µsec, Fig. 4.171 a; at 0.5 L the then reduced unloading wave arrived at the predicted 75 µsec, Fig. 4.171 b; and at 0.4 L where, if absorption had *not* taken place at 0.43 L an unloading wave would be seen at 80 µsec, Fig. 4.171 c. That no evidence of any unloading wave was seen at a time of 80 µsec at a distance of

[1] LEE [1953, *1*].

[2] BELL [1961, *1*].

[3] BELL [1961, *3*, *4*].

[4] These measurements had provided additional experimental evidence as early as 1960 that viscoplasticity stress-strain functions of the MALVERN [1951, *1*] type, containing the bar modulus E, could not govern the dynamic plasticity of this solid since a high elastic wave front attenuating into plastic deformation simply was not observed. BELL [1960, *2*], [1961, *3*].

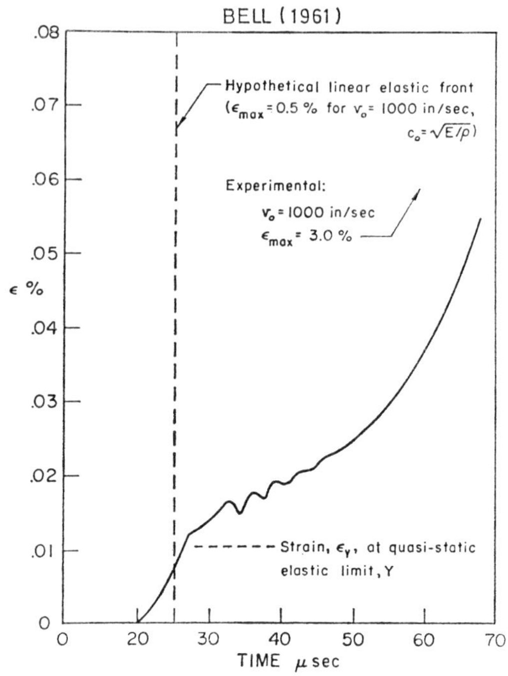

Fig. 4.170. Experimental detail of initial elastic and initial plastic portion of 3% finite amplitude wave, showing that the maximum of the former agrees with the quasi-static value rather than with the larger amplitude predicted by the SOKOLOVSKY-MALVERN viscous hypothesis. The material is completely annealed aluminum.

0.4 L in these experiments lent credence both to LEE's theory[1] and the assumption that the response function for unloading waves was linear, having the slope E.

Experiment and theory were finally in accord for what was now indeed dynamic plasticity, since loading waves and interacting unloading waves had been accounted for. In addition, the total time of contact T_c that the specimens remained together in a symmetrical impact in free flight and the final velocity of each specimen after impact had been determined. Since the initial velocities also were known, the measurement of final velocity provided an experimental coefficient of restitution, e, for plastically deforming specimens in collision. The problem first posed by HODGKINSON in the 1830's and intermittently studied with minor success for 130 years, finally could be described in experiment in complete detail.

Placing the impact faces between an uniform light field on one side and a photomultiplier tube on the opposite side, after a proper calibration, permitted me to determine the coefficient of restitution and time of contact of the impact, from the relative slopes of the changes in light intensity before and after impact, and the time of the extinction of light.[2] In addition, by placing the light source

[1] The experiments of KOLSKY and DOUCH [1962, *1*], which were a repetition of the hard load bar tests of JOHNSON, WOOD, and CLARK [1953, *1*] and of J. D. CAMPBELL [1953, *1*], ignored both the unloading wave absorption of the LEE theory [1953, *1*] and my experiments of the previous year (BELL [1961, *3*, *4*]) in assuming that the maximum distance of penetration of strain was given in linear elasticity by the arrival of the undiminished elastic unloading wave front.

[2] BELL [1961, *3*, *4*].

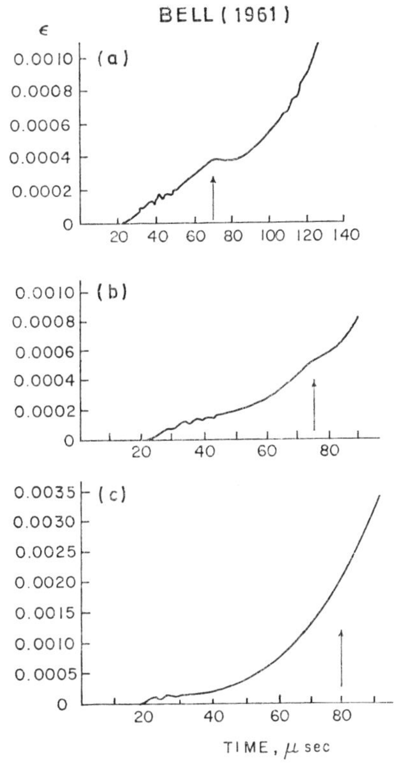

Fig. 4.171a—c. Experimental strain-time data from electric resistance gage measurements. The vertical arrows indicate the predicted time of arrival of the reflected unloading wave for $\sigma_Y = 1\,100$ psi. Note that by $4D$ the wave has been absorbed, in agreement with LEE's analysis.

and the photomultiplier tube at the free end of the struck specimen, I obtained a displacement vs time trace, the slopes of which provided a description of the velocity vs time history of the free end throughout the impact and after the specimens had separated. This final velocity could be checked against measurements of final velocity given by the time for the end of the specimen to traverse the carefully measured distance between two fine wires it successively contacted. Such a measurement of velocity by means of an electronic chronograph had been made for nearly every one of over a thousand symmetrical impact tests in my laboratory on all of the solids studied.

In Fig. 4.172a—c are shown a few of my displacement measurements obtained by the use of the optical technique at the interface and at the free end of the specimen. One sees the results of measurements of time of contact (a), and of coefficient of restitution (b), for different impact velocities; and the displacement vs time history (c) at the free end of a specimen 25 cm long, at the particle velocity of 2032 cm/sec. In each situation, the solid was 99.16% purity dead annealed polycrystalline aluminum for which the mode index was $r=2$ in Eq. (4.54), which provided, for the governing uniaxial response function, the parabola coefficient of $\beta=39.4$ kg/mm^2.

The experimental results (circles) in Fig. 4.172 were compared with prediction (solid lines) based upon an empirical equation, Eq. (4.56), which I found described

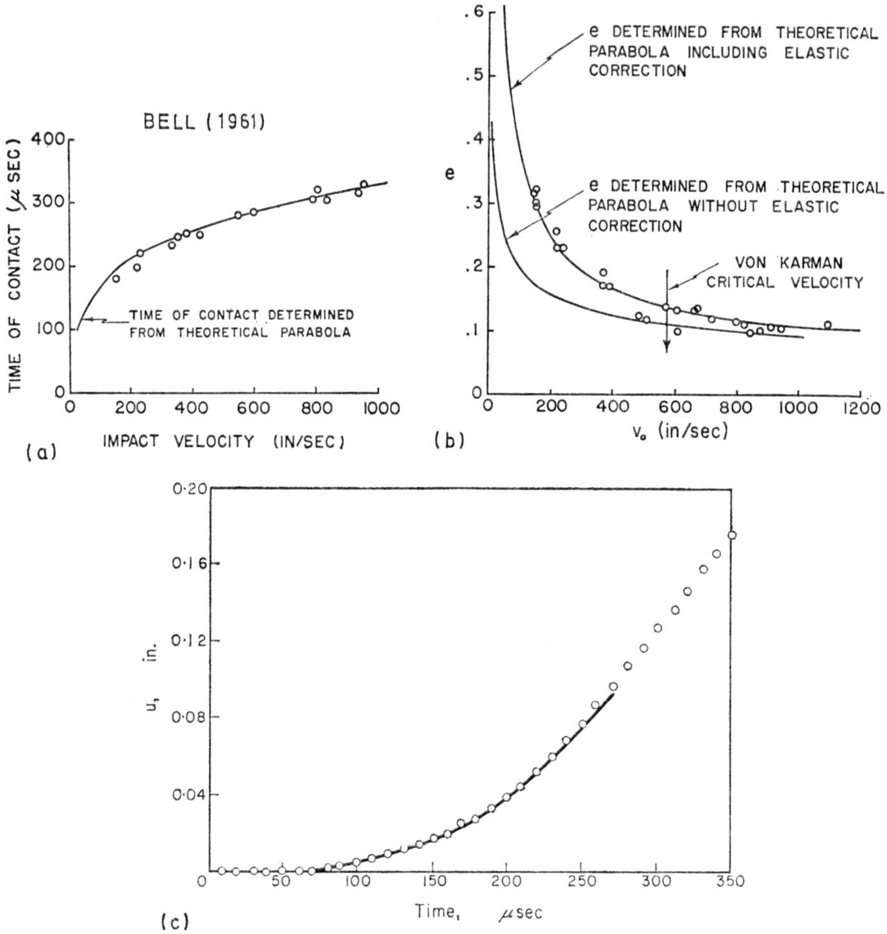

Fig. 4.172a—c. BELL's measurement by optical displacement technique of (a) time of contact vs impact velocity, (b) coefficient of restitution vs impact velocity, and (c) displacement-time behavior at free end of specimen, compared with computation from linear equivalent approximation. (solid lines). The elastic correction for the coefficient of restitution referred to the consideration of the small particle velocity, v_Y, of the elastic precursor. The solid was completely annealed aluminium.

all such measurements. The time of contact T_c in fully annealed aluminum and copper for a specimen of any length L at all impact velocities to 15000 cm/sec is given by the simple formula

$$T_c = \frac{L}{c_p^*} + \frac{L}{c_0}, \tag{4.56}$$

where $c_p^* = c_p\left(\frac{\sigma\max}{2}\right)$, the wave speed for one half the maximum stress according to the parabolic law for the annealed solid considered; and $c_0 = \sqrt{E/\varrho}$, the linear elastic bar velocity of the solid.

The empirical equation for the coefficient of restitution e was Eq. (4.57).

$$e = \frac{\sigma c_0}{2E v_0} + \frac{v_e - v_p}{v}. \tag{4.57}$$

In this instance it was found necessary to include in the calculation a correction for the small particle velocity of the initial elastic wave front (see BELL [1961, *1*]). The correction appeared in the difference between the elastic velocity v_e and the velocity v_p from the parabolic response at the point for very small strain, where the linear response function and the parabolic response function intersect. For fully annealed aluminum at room temperature, $v_e - v_p = 17$ in./sec, regardless of the magnitude of the maximum particle velocity of the finite amplitude wave.

In an historical context, probably the most important observation from this series of measurements was the following: given measured particle velocity and plastic strain distributions throughout the specimen at all times both during and after impact, and knowing the amount of kinetic energy at the first instant of impact, one could achieve an energy balance at any time after impact. In 1961 I described[1] two such energy balances for fully annealed aluminum polycrystalline cylinders 2.5 cm in diameter and 25.4 cm long. One was at a time of 50 μsec when the linear elastic precursor had just reached the free end of the specimens, and the second was at 310 μsec at the instant the two unloaded specimens separated. Both calculations, the detail of which was described in one of my two papers in 1961 on unloading response,[2] gave an energy balance of closer than 2%. The speculations of KICK[3] in the 1870's and of GUEST[4] in the 1930's, and a few others in both the 19th and 20th centuries, had been justified. They had been unable to proceed beyond pure speculation, however, because the impact problem simply could not be interpreted in the quasi-static terms to which their experimentation was limited.

In a recent reexamination of all the aspects of unloading for which I had provided experimental data in 1961, CRISTESCU and I in 1970 made[5] a detailed computer calculation for the symmetrical impact of two bars. After having inserted the parabolic response function I had found, both into TAYLOR and VON KÁRMÁN's nonlinear theory of waves and into LEE's theory of unloading, we calculated strain and particle velocity distribution throughout the bar (including the free end), time of contact T_c, coefficient of restitution e, and maximum stress as a function of impact velocity, during the entire period of specimen contact until complete unloading had been effected. We studied three different initial conditions in an attempt to simulate the wave growth in the immediate vicinity of the impact face where experiment has shown one dimensionality may not be assumed.[6]

That all the major features of the loading and unloading waves for the dynamic plasticity of annealed polycrystals were understood sufficiently to find accord between experiment and nonlinear theory, was amply demonstrated. Some discrepancies, such as the fact that experiment reveals that the maximum strain propagates further into the specimen than such calculations have shown, and that maximum strains in the immediate vicinity of the free end of the specimen were slightly less than estimated, are due partly to the limitations of such computer analyses and partly to the need to study further the dynamics of the small strain transition region just beyond the elastic limit.

Referring the reader to the many details described in that paper in 1970 and to my earlier papers in 1961 on the experiments on which it was based, I have in-

[1] BELL [1961, *3*, *4*].
[2] *Ibid.*, *3*.
[3] KICK [1875, *1*].
[4] GUEST [1930, *1*].
[5] CRISTESCU and BELL [1970, *1*].
[6] BELL [1960, *3*], [1960, *4*], [1961, *1*], [1962, *6*], [1963, *2*], [1967, *2*].

Table 136. CRISTESCU and BELL (1970).

	Y	σ_{max}	ε_{max}	V	e	v_f	T_c
Calculation	77.34	594.12	0.022491	40.74	0.116 to 0.1262	22.74 to 22.94	307.5 to 320
Experiment	77.34	583.57	0.0219	40.64	0.1235	22.63	310.8

where Y is the measured elastic limit Kg/cm²;
σ_{max} is the maximum loading stress Kg/cm²;
ε_{max} is the maximum loading strain;
V is the initial velocity of the impacting specimen m/sec;
e is the coefficient of restitution;
v_f is the final velocity of the struck specimen after impact, m/sec; and
T_c is the time of specimen contact μsec.

cluded in Table 136 the range of calculations for the three assumed initial conditions, compared with the experimental results. The experimental data are the averages of several hundred individual tests in completely annealed aluminum, for which the velocity of the impacting specimens in all instances was 40.64 m/sec.

4.30. The dynamic elastic limit.

Perhaps the most important aspect of dynamic plasticity for the technologist is whether or not in the practical metals in which he is interested, he may assume in his design calculations a higher elastic limit in the impact loading of his structures than in quasi-static loading; and, if so, what differences will be found in the plastic deformation which follows if this value of stress is exceeded.

A number of experiments, such as those of R. M. DAVIES and TAYLOR,[1] or those of BERTRAM HOPKINSON[2] in 1905, briefly described above, have provided evidence that in at least some solids some such increase, for whatever reason, is to be expected. Other experiments in solids like annealed aluminum where initial wave fronts were examined, included no such increase at high strain rates. It is an entirely separate question in this context whether or not such an increase in the elastic limit, if present, would have any measurable effect upon the dynamic plasticity which follows when the upper limit is exceeded.

I considered two methods of studying this subject. The first was to examine the propagation of waves of finite amplitude in a given solid in which prior annealing histories and mechanical histories had provided different quasi-static elastic limits. The second, and perhaps more important, was to study finite amplitude wave propagation in a solid whose dynamic elastic limit as measured from strain amplitudes of the initial wave front by diffraction gratings, was significantly higher than the quasi-static value in the same solid.

The experimentally established parabolic response function of Eq. (4.54) makes no specific allowance for the presence or absence of a small linear elastic region. The experimental evidence from my studies of waves of finite amplitude has revealed, for the solids examined thus far, that whatever the magnitude of the dynamic elastic limit, the finite loading wave which followed, once the elastic limit was exceeded, behaved as if no initial linear region had been present. The parabolic stress-strain function provided in terms of the integrals of the theory

[1] See TAYLOR [1946, *1*].
[2] B. HOPKINSON [1905, *1*].

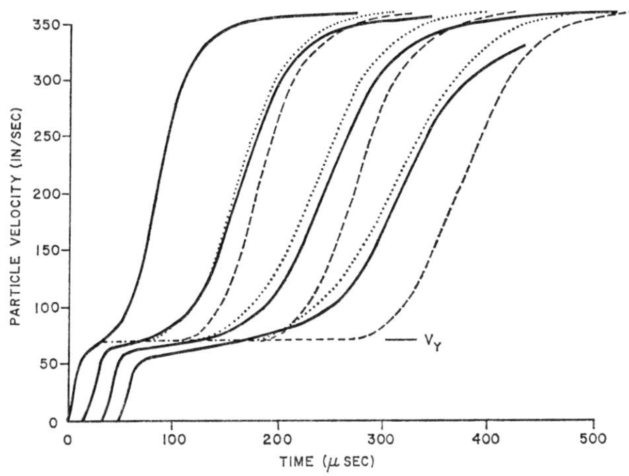

Fig. 4.173. The particle velocity data of MALVERN (solid lines) which I have compared with transit times from the first position to the other three positions based upon calculations assuming $Y=0$ (dotted lines), and $Y=3500$ psi (dashed lines), consistent with the particle velocity, v_Y, of the elastic precursor. The solid was partially annealed, low purity aluminum.

of waves of finite amplitude the following relations for wave speeds c_p and particle velocity vs finite strain:

$$c_p = \frac{\beta}{(2\varrho\sigma)^{\frac{1}{2}}}, \tag{4.58}$$

$$v = \left(\frac{8}{9}\frac{\sigma^3}{\varrho\beta^2}\right)^{\frac{1}{2}} \quad \text{and} \quad v = \left(\frac{8}{9}\frac{\beta}{\varrho}\right)^{\frac{1}{2}} \varepsilon^{\frac{3}{4}}. \tag{4.59}$$

If there were an increase in the elastic limit Y with a corresponding increase in plastic stress at a given plastic strain and reduction in the particle velocity available for the plastic portion of the wave, then, from Eqs. (4.58) and (4.59) one would expect slower wave speeds for a given finite strain, and a change in the maximum plastic strains for a given impact velocity. Two illustrations[1] of the fact that these changes were *not* found experimentally in annealed polycrystals, are given in Figs. 4.173 and 4.174.

In Fig. 4.173 the particle velocity profiles obtained by EFRON and MALVERN[2] (Sect. 4.28) in aluminum for which lower annealing temperatures provided an elastic limit of approximately 3800 psi instead of the 1100 psi for my fully annealed solid, were compared with calculations which included the supposed influence on the wave speeds of the higher quasi-static elastic limit. That no such slowing of wave speeds was observed is immediately obvious.

Closer to the physical problem considered by R. M. DAVIES and by B. HOPKINSON are the symmetrical free flight impact tests shown in Fig. 4.174. As part of his doctoral research in 1967, HARTMAN[3] performed a series of diffraction grating measurements of strain in polycrystals of annealed α brass. The measured quasi-static elastic limit of his annealed brass was $Y=14500$ psi (10.2 kg/mm²). The dynamic value determined from the initial wave fronts of the two diffraction grating measurements of strain profiles shown in Fig. 4.174 was $Y=27700$ psi

[1] BELL [1972, *6*].
[2] EFRON and MALVERN [1969, *1*]. See also EFRON [1964, *1*] and MALVERN [1965, *1*].
[3] HARTMAN [1967, *1*], [1969, *1*].

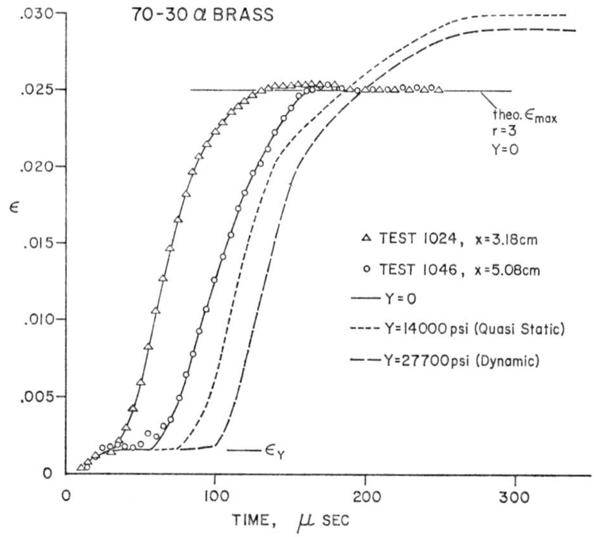

Fig. 4.174. Two diffraction grating experiments of Hartman in α brass. I have added calculated transit times and theoretical maximum strains for $Y=0$ (solid line), for $Y=14000$ psi (the measured quasi-static elastic limit), and for $Y=27700$ psi (the dynamic elastic limit measured from the observed ε_Y).

(19.5 kg/mm²), an increase by a factor of almost two. By comparing the experimental data (solid lines) with calculations based upon the lowering of wave speeds and maximum strain in terms of the elastic limit Y, I found that neither the quasi-static value of 10.2 kg/mm² nor the increased dynamic value of 19.5 kg/mm² described the behavior. The wave speeds and maximum strains observed experi-

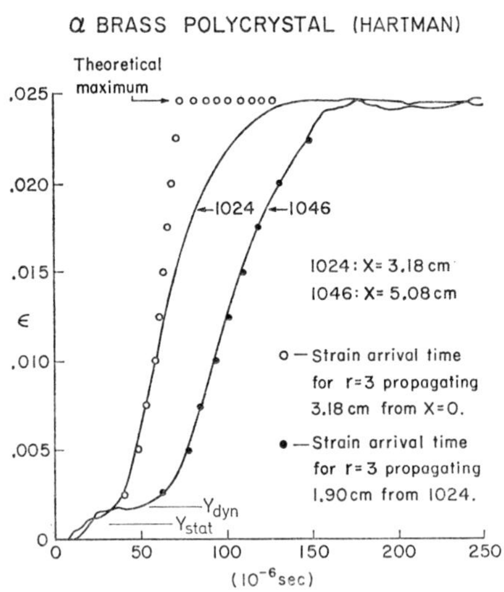

Fig. 4.175. Diffraction grating measurement of finite strain showing that the plastic wave front is unchanged by the increase of the dynamic elastic limit, Y_{dyn}, over the quasi-static value, Y_{stat}. Open circles are values predicted for an infinite step on impact; closed circles are traverse times predicted from first to second position.

Fig. 4.176. Schematic diagram of the dynamic stress-strain behavior based on the observed correlation of wave speeds and maximum strains, with calculations based on $Y=0$.

mentally, as in every solid for which finite strain profiles have been examined, were for an elastic limit, $Y=0$. In Fig. 4.175 the traverse times (filled circles) from one position to another, and the maximum strain at both positions, are shown to agree with prediction from my parabolic generalization for $r=3$. Thus, one is led to the type of behavior diagrammatically shown in Fig. 4.176. Experiments on polycrystalline magnesium, for which major changes in Y are easily produced, gave results[1] identical with those described for aluminum and α brass.

For annealed crystalline solids, the experimental evidence to date indicates that the magnitude of the elastic limit, even when increased by the high strain rate of the linear elastic precursor, does not influence the loading wave which follows. We already have seen in the previous section, however, that the magnitude of Y does have an important effect upon the process of unloading. Whether such a conclusion is applicable to the hardened, esoteric metals of practical interest in technology simply is not known; experimental diffraction grating studies of finite amplitude waves in such solids have been undertaken only recently.

Although the properties of yield surfaces for both radial and nonradial quasi-static loading have been studied extensively from the experiments of BAUSCHINGER and of GUEST to the present, similar studies of the dynamics of yield phenomena are still in their infancy.

4.31. Discontinuous finite deformation: The Savart-Masson (Portevin-Le Chatelier) effect.

After the classic experiments of DULEAU[2] in 1813, who first remarked upon irregularities observed during the measurement of strain which occurred in dead weight testing, nearly every decade has seen commentary on the phenomenon

[1] BELL [1968, *1*].
[2] DULEAU [1813, *1*].

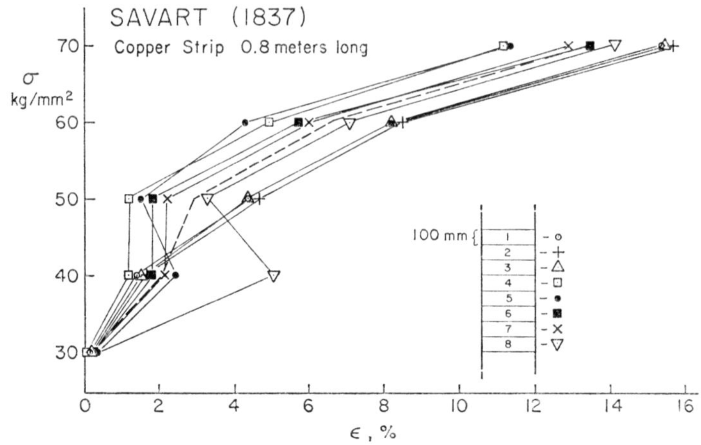

Fig. 4.177. Early tensile test showing both inhomogeneous deformation and the discontinuous step of the SAVART-MASSON effect.

of discontinuous deformation. When the applied force is an increasing dead weight load, these discontinuities appear as steps in the response, which has given rise to the occasional use of the term "staircase phenomenon" in describing the behavior. Comments have varied from that of MASSON[1], who referred to the fact that in 1837 SAVART[2] had considered the phenomenon as a fundamental material stability property, to that of WERTHEIM[3] in 1844, who suggested that a proper choice of experiment could eliminate the effect. Because the "hard" testing machines which came into existence in the late 19th century could mask the phenomenon and, except in exceptional circumstances, provide the smooth curves preferred in technology, it generally was believed by then that this was a machine-produced effect of small, or of no fundamental, importance.

In citing as the origin of the experimental study of this phenomenon, the work of SAVART in 1837 and MASSON in 1841, I refer to the tension test of the former on a copper strip, 0.8 m long, shown in Fig. 4.177 (the original data were tabulated), and to the comments of the latter when he wrote of SAVART's tests, which, as I showed in Sect. 2.11, MASSON compared with later measurements of his own on the same specimens. For the data of the illustrative test in Fig. 4.177, SAVART had measured the strain over eight different sections of the specimen, obtaining the evidence of inhomogeneity and the discontinuity shown. Because I was interested in the form of the response of the overall specimen, I plotted the averaged results of this test in the σ^2 vs ε plot shown in Fig. 4.178. The parabolic response of this test performed 135 years ago provides the $r=4$ mode index in Eq. (4.25), which I have found describes the data of nearly every experimentist who has studied the finite strain of polycrystalline annealed copper at room temperature. In the inset of Fig. 4.178, I have included τ vs γ resolved shear results in copper from a single crystal test of ELAM[4] in 1926 which, for a parabolic deformation mode index of $r=4$, also exhibits the same discontinuity as SAVART found 89 years earlier for polycrystalline copper.

---

[1] MASSON [1841, *1*].
[2] SAVART [1837, *1*].
[3] WERTHEIM [1844, *1* (a).]
[4] ELAM [1926, *1*]. See also TAYLOR [1934, *3*].

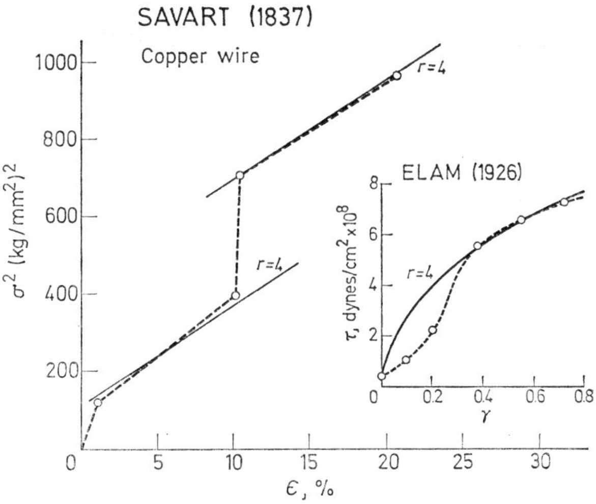

Fig. 4.178. A σ^2 vs ε plot of the averaged data of a test on copper wire, showing the correspondence with Eq. (4.25) and the first observation of a SAVART-MASSON discontinuity in dead weight loading. The insert shows a similar discontinuity for a single crystal test in copper of ELAM in 1926.

SAVART was interested in explaining from quasi-static tests the shift in nodal patterns which he observed during the vibration of such solids. This was an extension of the work of CHLADNI,[1] with which SAVART was engrossed during a large part of his life. It was with much regret that I found it necessary to omit from this treatise an account of both his and CHLADNI's extensive and fascinating experimental study of the vibration of solids, as well as the later work of Lord RAYLEIGH in that area. Among the enormous number of papers which SAVART published, I have not been able to find the observation to which MASSON referred when he attributed to SAVART the statement that the observed discontinuities were a fundamental phenomenon in the deformation and that further study of it was essential for an understanding of the physics of solids.

In 1912 W. ROSENHAIN and S. L. ARCHBUTT,[2] in opposition to prevailing opinion of that time, described the fluctuations in stress, which in fact are the way discontinuities appear on "hard" testing machines, as a material property of the finite deformation of some non-ferrous alloys. PORTEVIN and LE CHATELIER[3] presented in 1923 the first effort to study the discontinuous behavior as a prime objective, eighty-six years after SAVART's insistence that it was a matter of fundamental importance. PORTEVIN and LE CHATELIER made no reference to SAVART or to MASSON. In their tests on aluminum—4.5% copper, aluminum—4.5% copper—0.5% magnesium alloys, using a "hard" testing machine at a strain rate of 8% strain per minute, they noted the appearance of specimen marking and occasionally an audible sound associated with the occurrence of each discontinuity.

A "hard" testing machine for which the *strain* history is prescribed and the load is the measured variable provides discontinuities of the form shown by ELAM[4] for normal and slow tensile loading of an aluminum alloy (see Fig. 4.119,

[1] CHLADNI [1787, *1*].
[2] ROSENHAIN and ARCHBUTT [1912, *1*].
[3] PORTEVIN and LE CHATELIER [1923, *1*].
[4] ELAM [1938, *1*].

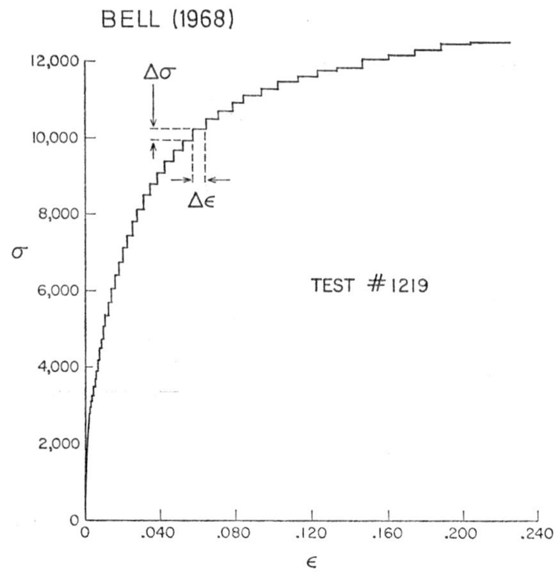

Fig. 4.179. Typical SAVART-MASSON effect for a constant stress rate tension experiment in annealed polycrystalline aluminum.

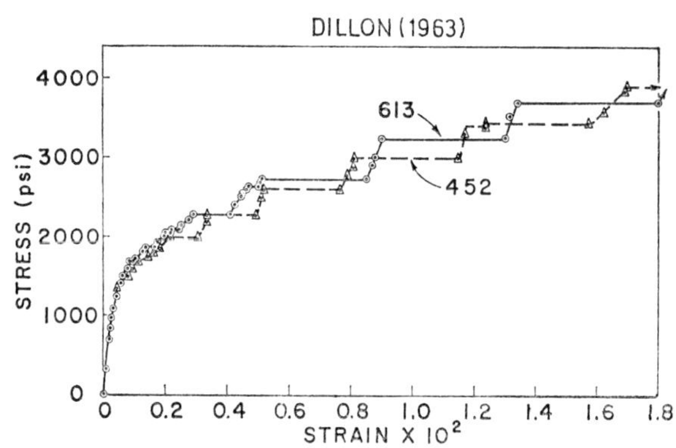

Fig. 4.180. Typical SAVART-MASSON effect for intermittent loading of annealed aluminum polycrystalline drilled tubes in torsion.

Sect. 4.24). For dead weight loading where the loading history is prescribed and strain is the measured variable, the discontinuities appear in the form which I illustrate from one of my own experiments on low purity polycrystalline aluminum for a tensile test at constant load rate (Fig. 4.179), or from the slow loading torsion test of DILLON[1] for successive very small increment loading of the same solid (Fig. 4.180).

The steps seen when the dead weight loading apparatus is used rather than serrations from the hard testing machine, offer far greater opportunities for studying discontinuous behavior. Since one knows the accelerations which are present in the $\Delta\varepsilon$ portion of the step, the precise influence of the machines can be

[1] DILLON [1963, *1*]; and see [1964, *1*], [1966, *1*], [1967, *2*], and KENIG and DILLON [1966, *1*].

calculated and shown to be negligible. The relatively long time required to traverse the vertical portion of the step, $\Delta\sigma$, in either constant load rate or small incremental loading, allows for a precise determination of the magnitude of the small elastic loading increment required to trigger a step at a given total finite strain.

The first 20th century experimental study of discontinuous steps in dead weight loading, and still one of the most important, was that of HANSON and WHEELER[1] in 1931. Over a period of approximately one and a half years (573 days), they added at intervals very small weights to a tensile specimen of polycrystalline aluminum, obtaining the results shown in Fig. 2.16, in Chap. II, Sect. 2.11, above. The integers in Fig. 2.16 refer to the number of days involved for each interval. In one instance, for a period of 153 days, the specimen which previously had undergone creep at constant load, remained at a fixed strain.

TOMIYA SUTOKI[2] in 1941 performed a series of tests on carbon steel of various compositions; duralumin, copper, brass, nickel, aluminum, zinc, and single crystals of iron. Since he used a "hard" testing machine it is not possible to separate the influence of the apparatus from the influence of the material, so his results, like those of PORTEVIN and LE CHATELIER and of ELAM, were only qualitative. The ambient temperatures varied from that of liquid nitrogen to red heat. SUTOKI performed the experiments on a standard 5 ton testing machine at elongation rates ranging from 0.01% per minute to 0.1% per minute. He found that prior thermal histories influenced serration amplitudes, as did changes of the ambient temperature. In this latter context, the serrations sometimes were visible only in specific temperature ranges, although this range also was influenced by prior thermal history.

Rather than describe SUTOKI's more or less random observations of the serrations, it is more fruitful to examine the results of the first major scientific study of this SAVART-MASSON effect, that by ANDREW W. MCREYNOLDS[3] in 1949. MCREYNOLDS' experiments using a dead weight loading apparatus included measurements at different ambient temperatures. His experiment became a model for most of the extensive study of the step phenomenon which was pursued in the 1960's. Most of the main features of discontinuous finite deformation first were observed by MCREYNOLDS, including his fundamental discovery of the slow wave which accompanies discontinuous deformation and now bears his name.

MCREYNOLDS in his first experiments examined the effect of purity upon the magnitude and upon the visibility of discontinuous steps at room temperature. Two tests in low purity (2 S) aluminum and high purity aluminum are shown in Fig. 4.181. For high purity aluminum, steps were visible only when the load had been held constant for a period of 15 min. This, of course, is the effect discovered by THURSTON[4] in steel in 1873. Both experiments of MCREYNOLDS were for simple tension in dead weight loading, performed at the same constant stress rate of 120 psi/min.

In a second series of measurements MCREYNOLDS examined the influence of ambient temperature on the SAVART-MASSON effect in low purity aluminum polycrystals, also loaded in tension at the same constant stress rate. He found in aluminum that the visibility of steps in dead weight loading had the same depend-

[1] HANSON and WHEELER [1931, 1].

[2] SUTOKI [1941, 1].

[3] MCREYNOLDS [1949, 1]. Certainly if one has to name the effect after a 20th century scientist and ignore SAVART and MASSON, one should select either MCREYNOLDS for his definitive study, or HANSON and WHEELER, for their precedence in dead weight loading.

[4] THURSTON [1873, 1].

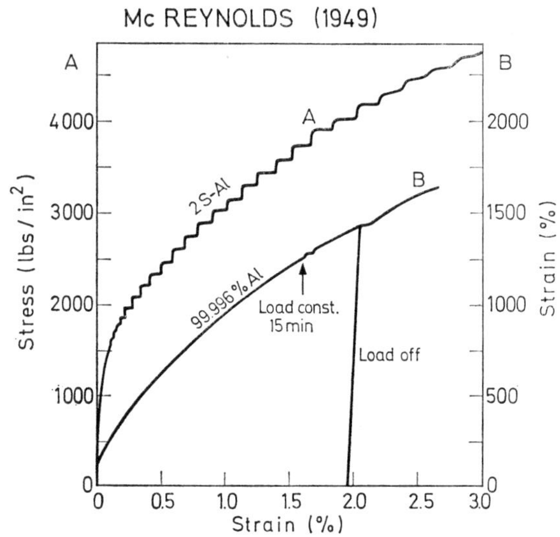

Fig. 4.181. McREYNOLDS' comparison of discontinuous deformation in low and high purity aluminum. The constant loading rate was 120 psi/min.

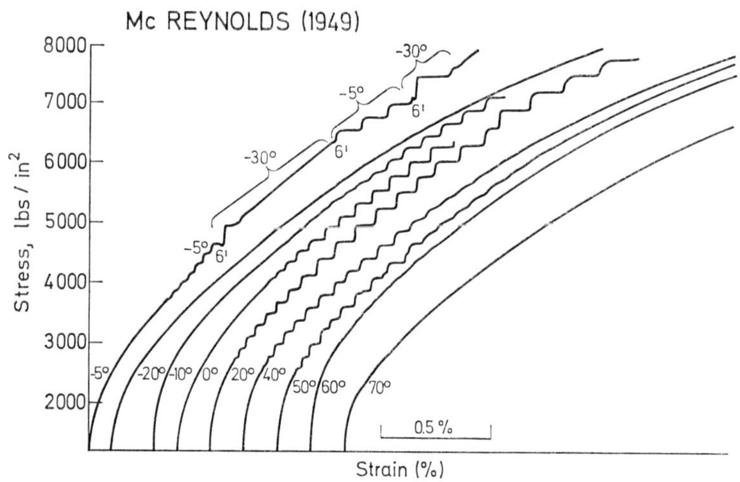

Fig. 4.182. McREYNOLDS' demonstration in low purity aluminum of the dependence of discontinuous deformation upon the ambient temperature.

ence upon ambient temperature that SUTOKI, in several other metals, had observed earlier for the serrations from the "hard" testing machine. McREYNOLDS' results are shown in Fig. 4.182.

McREYNOLDS also studied the effect of abruptly changing the constant loading rate in four ranges from 44 lbs./min to 257 lbs./min, and the effect of ageing for 20 h at 1000° F. In either case he obtained negligible differences in the step behavior in low purity aluminum.

The principal discovery in McREYNOLDS' work was that obtained from the simultaneous observation of four electric resistance gages equally spaced along

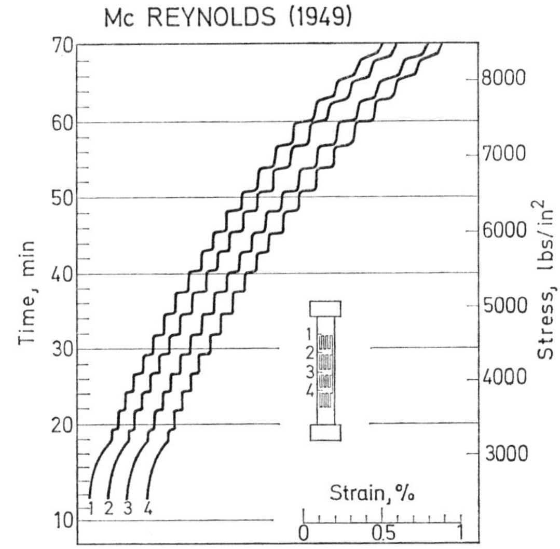

Fig. 4.183. McReynolds' measurement of the occurrence of discontinuous deformation at the indicated positions along the specimen. From these data McReynolds' slow waves were discovered.

the specimen. From the simultaneous recording of the output of these gages during a constant load rate, dead weight loading experiment, he observed the behavior shown in Fig. 4.183. These data demonstrated that the $\Delta\varepsilon$ portion of the step began at one point and propagated through the specimen at velocities which from one step to another ranged from 0.5 cm/sec to 80 cm/sec.

Many explanations of discontinuous behavior have been proposed, including the solute atom diffusion-vacancy hypothesis of A. H. Cottrell[1] in 1953; the analysis of B. M. Lempriere[2] who in 1962 concluded, in a return to previous decades, that: "The Portevin-le Chatelier effect is seen to be due to an unstable strain-rate phenomenon characteristic of certain materials which permits oscillation of the testing machine";[3] the Coulomb friction analog of Saul R. Bodner and A. Rosen[4] in 1967, for which the "stopped motion" depended upon the presence of a negative strain rate; and the particle wave hypothesis of Edwin R. Fitzgerald[5] in 1966. For one reason or another, none of these proposed explanations is consistent with more than a small fraction of the known physical facts. In Fig. 4.184 are shown increments of stress and increments of strain plotted against the stress at which the step occurred. In comparing McReynolds'[6] experimental results with those of later studies it should be remembered that because of the limitations of his electric resistance elements, his maximum strain at room temperature was only about 3%, while at 170° C "the useful strain range was only of the order of 1 to 1.5 pct."[7] McReynolds, like others before him, had noted that specimen markings appeared in conjunction with the discontinuities.

[1] Cottrell [1953, *1*].
[2] Lempriere [1962, *1*].
[3] *Ibid.*, p. 183.
[4] Bodner and Rosen [1967, *1*].
[5] Fitzgerald [1966, *1*].
[6] McReynolds [1949, *1*].
[7] *Ibid.*, p. 35.

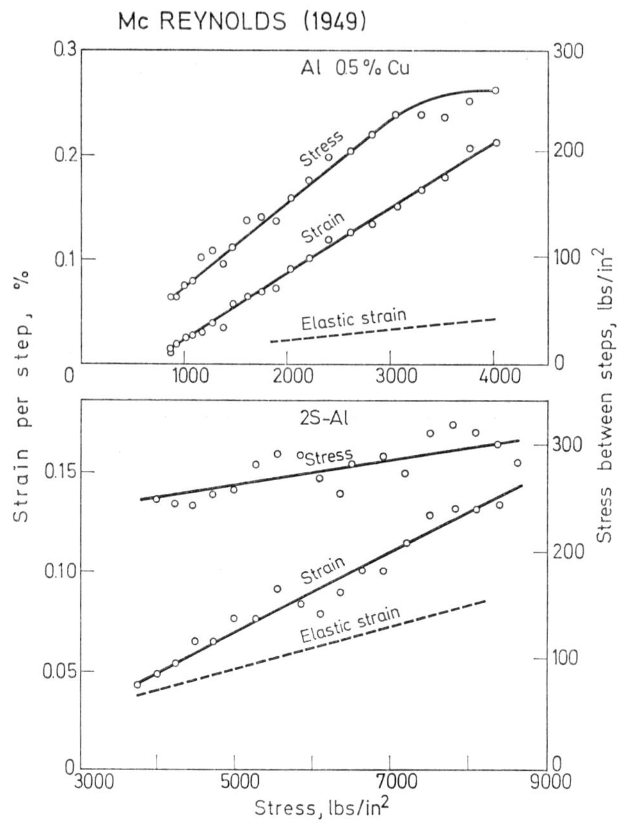

Fig. 4.184. McReynolds' plot of $\Delta\sigma$ vs σ and $\Delta\varepsilon$ vs σ for an aluminum copper alloy and for low purity aluminum.

McReynolds measurements were for integrated strain over several small areas of the specimen. In contrast, William N. Sharpe[1] and I,[2] in experiments performed seventeen years later, measured finite strain over a much larger region by means of a "clip" gage, and obtained reproducibility from one test to another.

Sharpe in his doctoral dissertation in 1966 gave a comparison of the clip gage results for three tests in annealed low purity polycrystalline aluminum designated 1100 F-H 18, which had been annealed for two hours at 1100° F and furnace cooled. Three such tensile tests using clip gages for a constant stress rate ($\sigma = 5$ psi/sec) are shown in Fig. 4.185 for specimens with sections 4 in. long, $^1/_2$ in., $^3/_8$ in., and $^1/_4$ in. in diameter. Not only were the tests demonstrably reproducible, but also, when the diameters of the specimens were changed, the results provided another demonstration that the discontinuity did not originate in an interaction between the apparatus and the specimen.

In two papers submitted to the *Journal of the Institute of Metals* within eight days of each other in 1953, N. Krupnik and Hugh Ford[3] in the one, and V. A. Phillips, A. J. Swain, and R. Eborall[4] in the other, performed McReynolds'

[1] Sharpe [1966, *1*].
[2] Bell [1968, *1*].
[3] Krupnik and Ford [1953, *1*].
[4] Phillips, Swain, and Eborall [1953, *1*].

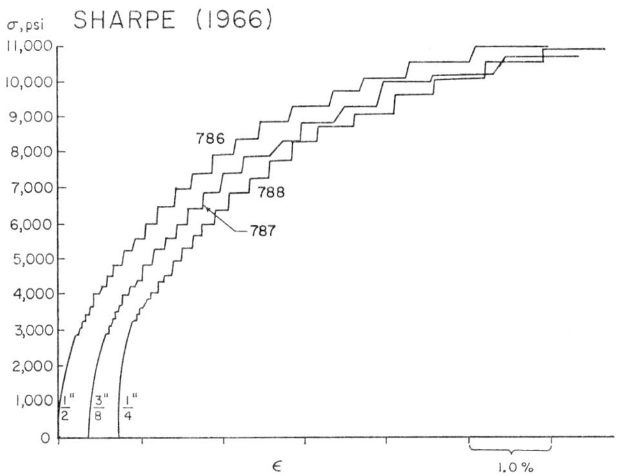

Fig. 4.185. Three constant dead weight tensile tests on low purity aluminum for specimens of 3 different diameters, indicating from reproducibility, lack of dependence of discontinuous deformation upon specimen cross-section and testing machine.

dead weight experiment in aluminum alloys and in aluminum and magnesium alloys, respectively. KRUPNIK and FORD considered polycrystals of duralumin, aluminum—1.35% copper alloy, aluminum—3.83% copper alloy, and aluminum—35% magnesium alloy. A dial indicator over a $1^5/_{32}$ in. gage length measured specimen extension to a strain of 10%, which was considerably in excess of McREYNOLDS' maximum strain of 1.5 to 3%, although the reading of extension of a dial gage "every 5 or 10 sec depending on the rate of loading"[1] also was a factor which limited accuracy in the new experiments.

The apparatus was a lever system like that of McREYNOLDS, and KRUPNIK and FORD also obtained a constant load rate by the flow of water into a bucket. For higher rates of loading the dial gage and a stop watch were on the same movie film and could be correlated with the constant water flux to provide simultaneous stress vs strain information.

In this repetition of McREYNOLDS' experiment for additional solids, the only new observation was that in varying constant loading rates (and hence strain rates) from 0.027 tons/in.²/sec to 1.235 tons/in.²/sec, an increase of 1:46, KRUPNIK and FORD found for the 1.5% copper-aluminum alloy no essential difference in either the steps or the underlying response function upon which the steps were superposed, as is seen in Fig. 4.186. For some of the other solids, the amplitude of the steps decreased to a smooth response as the loading rate increased.

Of equal interest is an observation I made when looking back over these results. Plotting σ^2 vs ε for the envelope of the response at the *bottom* of the steps, as shown in Fig. 4.187 (circles), demonstrates the now firmly established fact that the steps of the SAVART-MASSON effect in dead weight loading consist in departures from and returns to my parabolic response function (solid line) of Eq. (4.25). For this alloy a second order transition occurs at all strain rates from the mode index $r=3$ to $r=4$, near the second transition strain, $N=13$, of Eq. (4.26).

[1] KRUPNIK and FORD, *op. cit.*, p. 602.

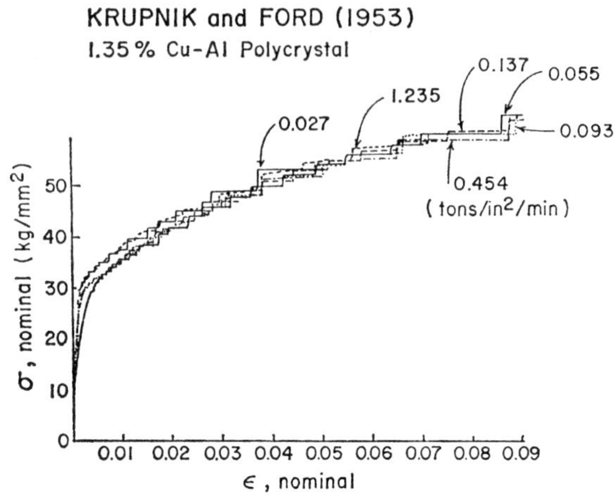

Fig. 4.186. KRUPNIK and FORD's dead weight loading tensile tests in a copper aluminum alloy, showing the reproducibility obtained from the widely differing constant loading rates.

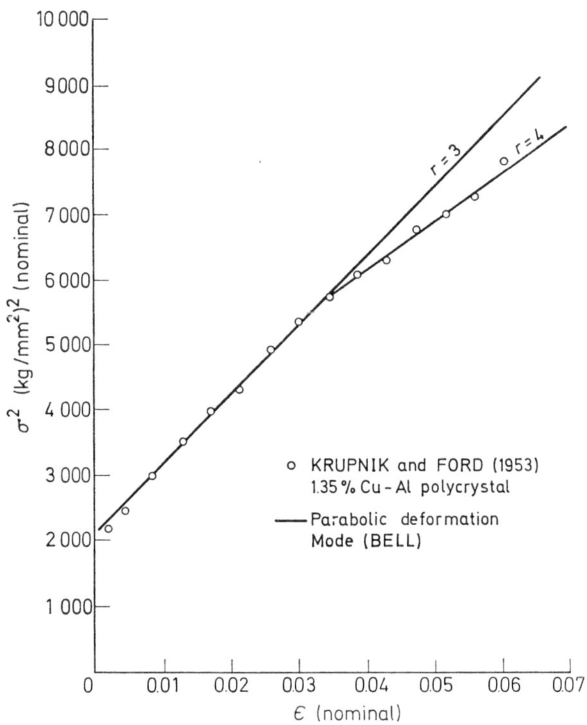

Fig. 4.187. The common envelope for the KRUPNIK and FORD measurements of Fig. 4.186 (circles) in a σ^2 vs ε plot, compared with prediction from BELL's parabolic generalization for a response function, Eq. (4.25) in nominal stress and strain.

The paper of PHILLIPS, SWAIN, and EBORALL[1] was concerned with a different aspect of MCREYNOLDS'[2] experiment, namely the appearance of surface markings

[1] PHILLIPS, SWAIN, and EBORALL [1953, *1*].
[2] MCREYNOLDS [1949, *1*].

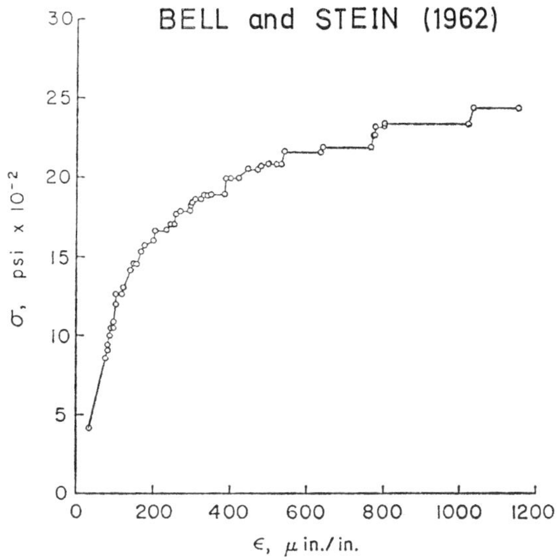

Fig. 4.188. A dead weight static compression test in annealed polycrystalline aluminum. The time for the total test, of which only the first 1200 μin./in. are shown, was five days. The increments added at long time intervals were 9 psi. The several increments added to produce the vertical portion of a step are not shown.

which accompanied the step production in the aluminum alloys. They identified two types of surface markings on solids which were in the form of strips and were aluminum-magnesium alloys with a magnesium content of $3^1/_2\%$.

In dead weight loading and small incremental loading at regular or irregular time intervals, the largest portion of the time in the "staircase" stress-strain behavior is consumed with the rise of the step, $\Delta\sigma$. When a horizontal increment $\Delta\varepsilon$ is triggered, there is a gradual increase in strain at constant stress, followed by a rapid increase of strain and a slow final approach to the point on the parabolic response function where for periods of observation extending to many hours or days, no further strain is seen.

The first study of the strain vs time behavior of $\Delta\varepsilon$ was that given by BELL and STEIN[1] in 1962. These experiments were the first study of the SAVART-MASSON effect in compression. They were performed on a dead weight lever apparatus to which 0.006 kg/mm^2 was added at intervals of the order of 30 min over a period of 5 days. On one of the steps of the test shown in Fig. 4.188, the strain vs time detail observed in Fig. 4.189 was described. Whatever may be the origin of the mechanism which produced the vertical departure from the parabolic stress-strain relation, the return to this basic curve in the increment $\Delta\varepsilon$ was shown to have a time dependent detail of other than a simple form. In further studies, to be referred to briefly below, the maximum slope of this incremental strain detail, i.e., the maximum incremental straining rate as contrasted with the usually studied average strain rate, was shown by me in 1968 to be stress dependent.

The incremental wave studies of BELL and STEIN, which also were part of our study in 1962, and further experiments of mine[2] in 1971 revealed the connection

[1] BELL and STEIN [1962, 7].
[2] BELL [1971, 7].

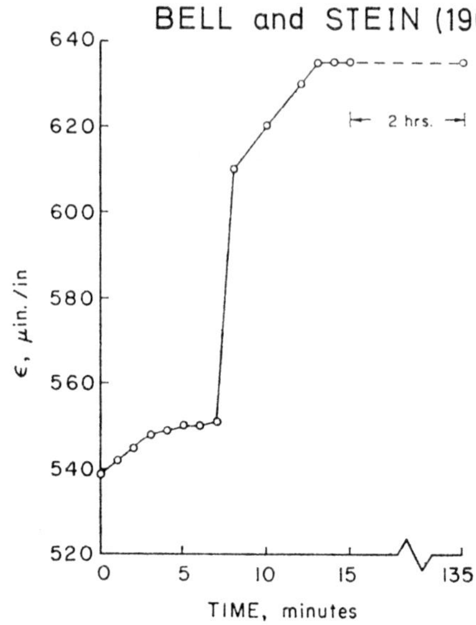

Fig. 4.189. The strain-time behavior at constant stress for the horizontal portion of a typical step of the stress-strain curve of Fig. 4.188.

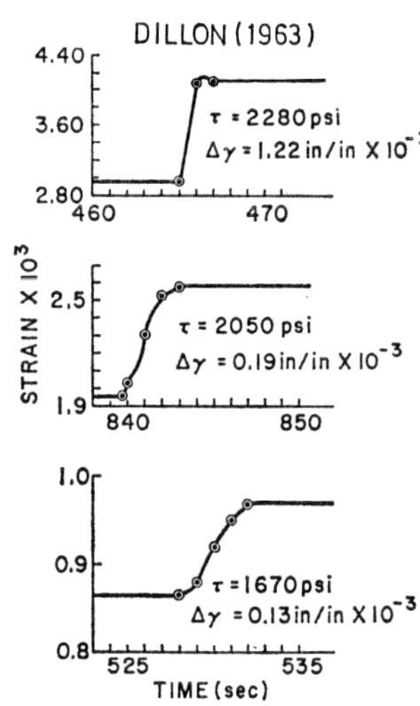

Fig. 4.190. Typical time histories of jumps in strain. Time was measured from the last load increment.

between the higher wave speed of the incremental wave (Sect. 4.34) and the SAVART-MASSON effect, even when dynamic pre-stress remained constant for an interval as short as 10 μsec. In quasi-static tests the triggering time for a step was considerably longer, as the $\Delta\varepsilon$ vs time measurements in torsion obtained by DILLON[1] in 1963 demonstrated. DILLON's results were in dead annealed low purity aluminum, shown in Fig. 4.190, for increments such as those shown in

[1] DILLON [1963, *1*].

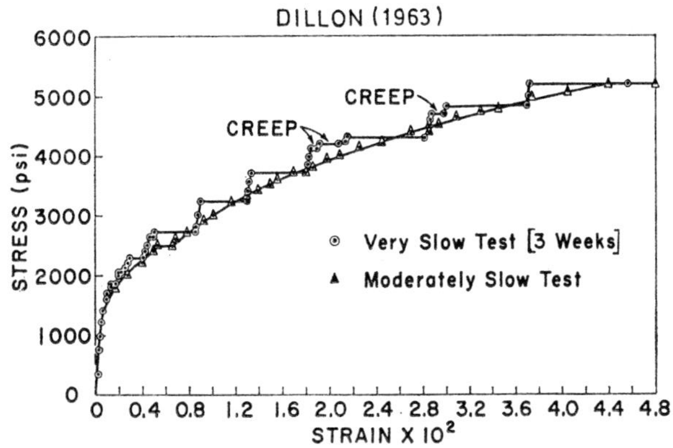

Fig. 4.191. Very slow and moderately slow torsion tests in drilled hollow tubes of low purity aluminum from which the detail of Fig. 4.190 was obtained.

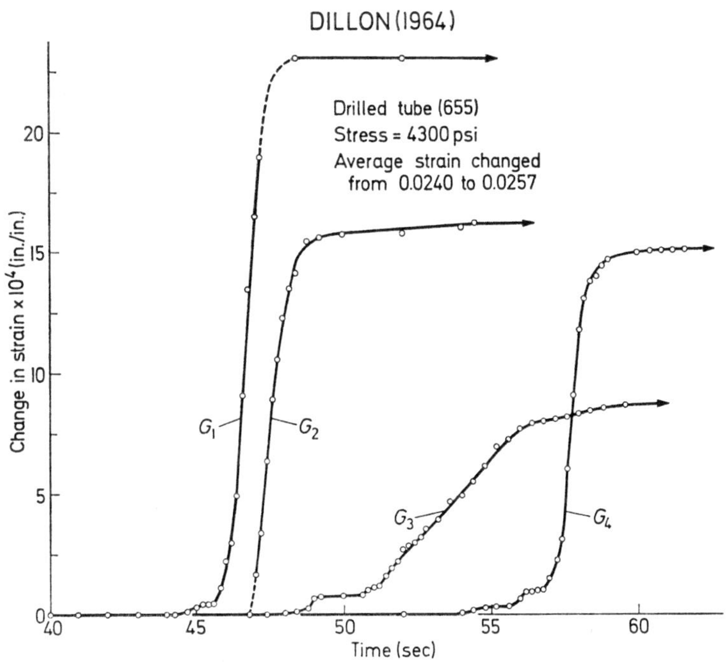

Fig. 4.192. Strain-time histories at four positions, exhibiting the detail of the McREYNOLDS' slow wave during a discontinuous step.

Fig. 4.191 for a very slow test (3 weeks) and a moderately slow test. The load which he applied at intervals consisted in pouring lead shot into a can.

In 1964 DILLON[1] described a detailed study of the "slow wave" discovered by McREYNOLDS fifteen years before. DILLON used 5 electric resistance gages located 3.81 cm apart along a tube of 1.27 cm in diameter with a wall thickness of 0.149 cm. The successive gage locations from one end of the dead annealed

[1] DILLON [1964, *1*].

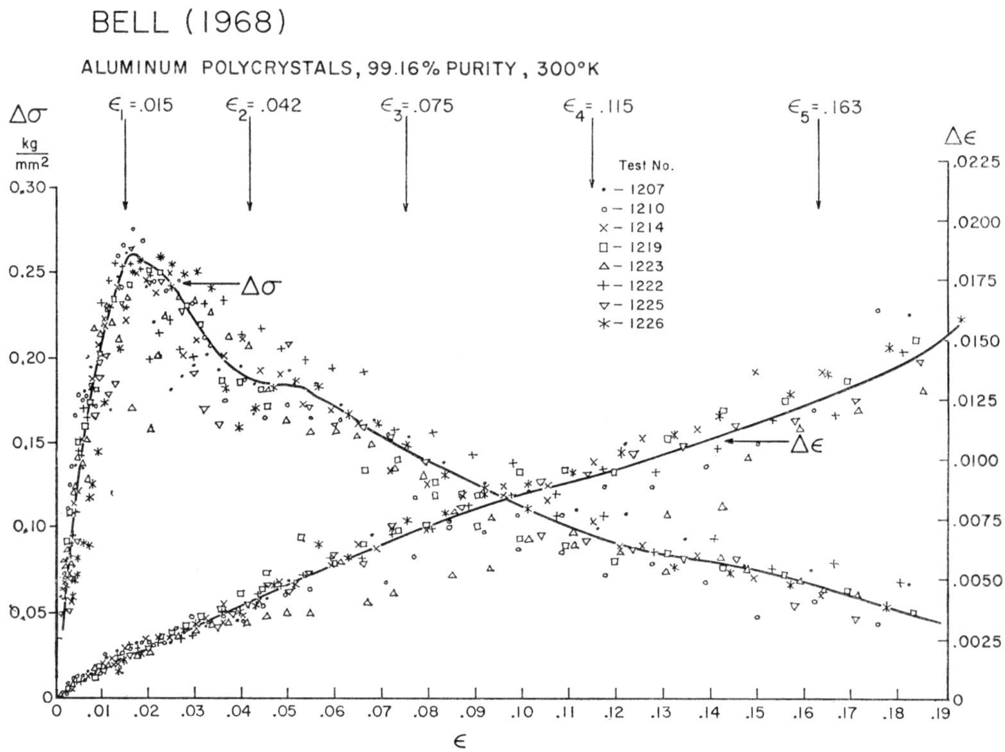

Fig. 4.193. Plots of $\varDelta\sigma$ and $\varDelta\varepsilon$ vs ε for discontinuous jumps in annealed aluminum. Note the peak at the first transition strain of 0.015 ($N=18$) even though a second order transition did not occur until the second transition strain of 0.0414 (see Fig. 4.95).

aluminum tube to the other were G_1 to G_5 consecutively. One example of the many slow wave experiments performed by Dillon in that study is shown in Fig. 4.192 for a pre-stress of 4300 psi.

Additional experimental studies of these important phenomena were reported by Dillon[1] in 1966 and by Kenig and Dillon[2] that same year. They demonstrated the range of small velocities of a few cm/sec encountered and the fact that such waves could be generated at one point and propagate in either direction, in a few instances being absorbed between measuring positions.

The staircase phenomenon, or the Savart-Masson (Portevin-le Chatelier) effect in dead weight loading for either single crystal or polycrystal consists in a departure from the governing parabolic response function in the vertical step, $\varDelta\sigma$, and a return to it in the horizontal position $\varDelta\varepsilon$. In Fig. 4.95 above, in Sect. 4.21, for annealed aluminum, all tests proceeded from the origin to the second transition strain of 0.0414 on a single parabolic deformation mode, $r=3$. Nevertheless for these same tests, as shown in Fig. 4.193, in both my data[3] and those of Sharpe,[4] a major peak at the first transition strain $\varepsilon_N=0.015$ occurs, although no second order transition had occurred for the parabolic response function. Obviously

[1] Dillon [1966, *1*].
[2] Kenig and Dillon [1966, *1*].
[3] Bell [1968, *1*].
[4] Sharpe [1966, *1*, *2*].

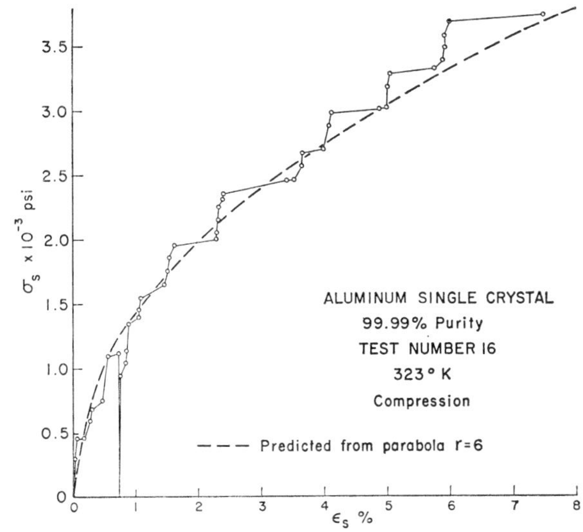

a

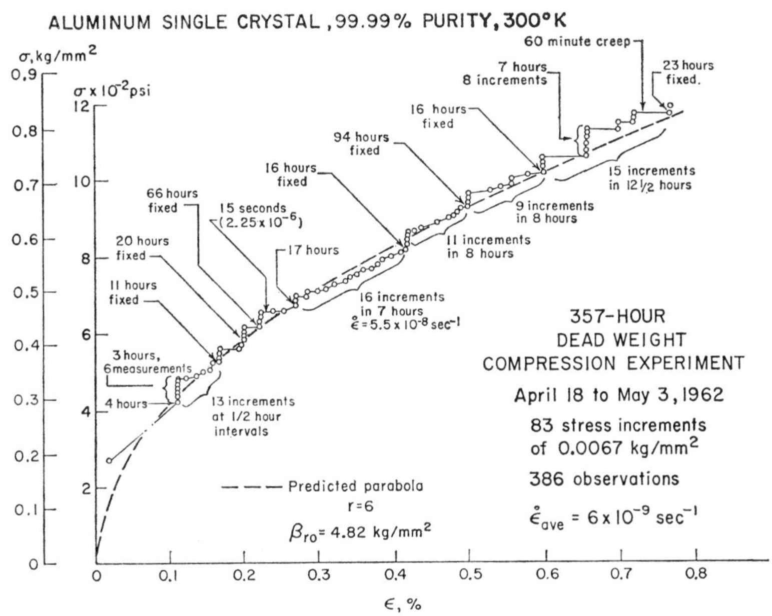

b

Fig. 4.194 a and b. (a) The unresolved stress vs strain for the incremental dead weight loading of an aluminum single crystal, compared with prediction from the anticipated $r = 6$ parabola. (b) A continuously monitored incremental compression experiment exhibiting the Savart-Masson effect at $\dot{\varepsilon} = 6 \times 10^{-9}\ \text{sec}^{-1}$ in a high purity aluminum single crystal (Bell, 1962).

this dramatic peak at a transition strain where no corresponding second order transition actually occurred in the response function, is of importance for any understanding of second order transitions.

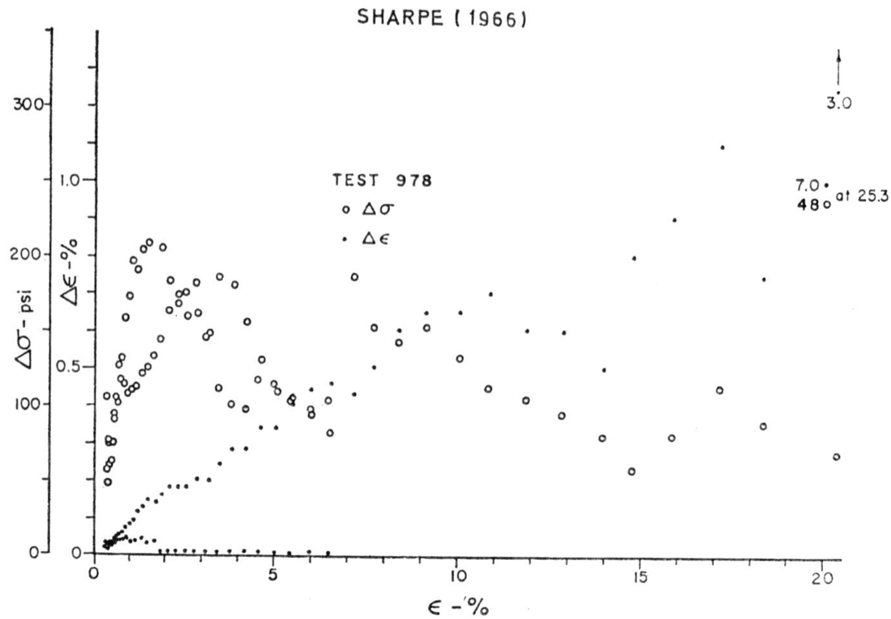

Fig. 4.195. A double step discontinuous dead weight loading test in low purity aluminum.

Furthermore, the incremental strain, $\Delta\varepsilon$, varied approximately linearly with strain, undergoing a change of slope at this same first transition strain, $\varepsilon_N = 0.015$ (where $N = 18$). A linear relation, $\Delta\varepsilon = K\varepsilon + C$, noted by SHARPE in aluminum, also has been observed for copper and for α brass. $\Delta\sigma$ as a function of total strain varies markedly from one solid to another.

That the SAVART-MASSON steps are superposed on the stage III parabolic response function [Eq. (4.25), Sect. 4.21] is shown in an experiment of mine in Fig. 4.194a which was for a 99.99% purity aluminum single crystal with an $r = 6$ mode index consistent with that for the dead weight compression of high purity aluminum at room temperature and at the 323° K ambient temperature of that test. In Fig. 4.194b is the result of a 16 day experiment in compression at room temperature which I performed in 1962, also on a 99.99% purity aluminum single crystal, where for an average strain rate of $\dot{\varepsilon} = 6\times10^{-9}\ \text{sec}^{-1}$ the staircase is seen to be superposed upon the parabolic response for the usual $r = 6$ deformation mode for that temperature. (It is worth emphasizing that the parabola coefficient for this response function at an average strain rate of $\dot{\varepsilon} = 6\times10^{-9}\ \text{sec}^{-1}$ was obtained originally from finite strain profiles of propagating wave fronts at strain rates of $10^3\ \text{sec}^{-1}$ and $10^4\ \text{sec}^{-1}$, giving a range of applicability for this parabolic response function of 13 orders of magnitude.)

SHARPE showed in his study that in general the density of steps was proliferated by the presence of grain boundaries. The SAVART-MASSON effect also could occur as two or three distinctly different series of steps. For double steps in polycrystalline aluminum, SHARPE described the manner in which the grouped step could be compared from one specimen to another. Fig. 4.195 is an example of a double step in a tensile test of SHARPE. My inspection of the peak of the first step system disclosed that it occurred as the first transition strain of $\varepsilon_N = 0.015$ ($N = 18$) and the second step series at the second transition strain of $\varepsilon_N = 0.0414$ ($N = 13$) where a second order transition in parabolic deformation mode did occur.

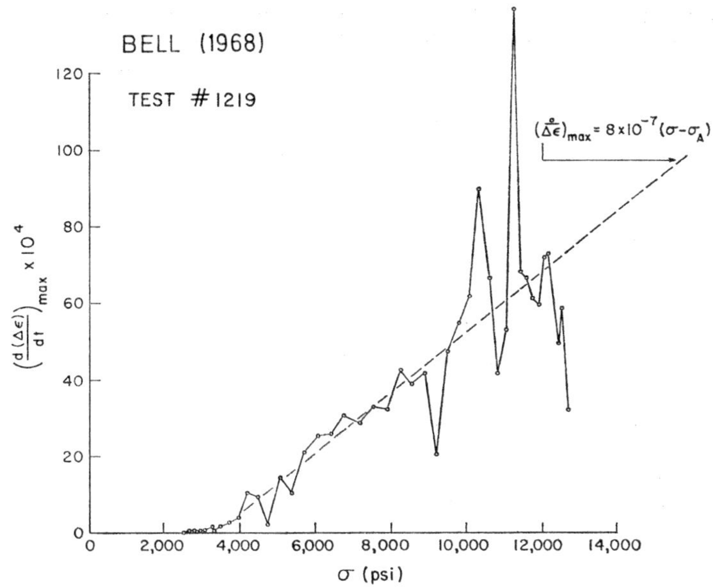

Fig. 4.196. Maximum straining rate vs stress for a constant stress rate tension experiment on annealed aluminum.

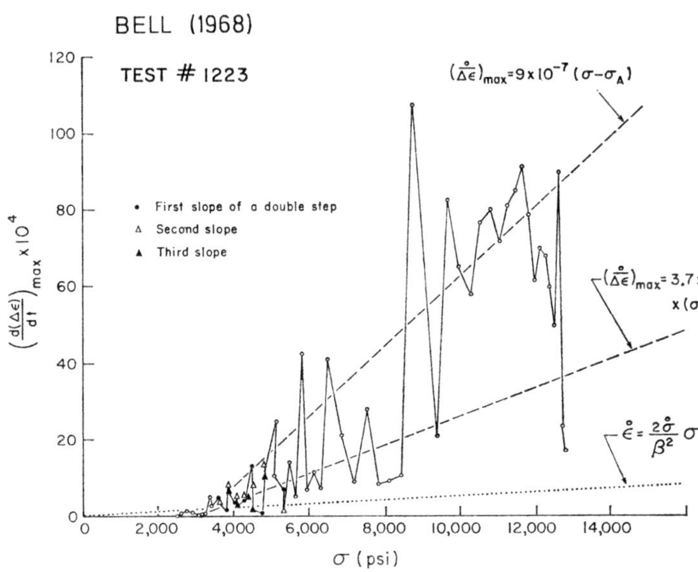

Fig. 4.197. Maximum straining rate vs stress for a double-stepped Savart-Masson effect experiment on annealed aluminum.

One final experimental study of discontinuous deformation was that comparing the measured maximum straining rate during the $\Delta\varepsilon$ portion of the step, $\dot{\overline{\Delta\varepsilon}}$, with the total stress at which the step occurred. In Fig. 4.196 is shown such a comparison which I reported in 1968 for an experiment in which a single step series was present; and in Fig. 4.197 for a double step system. Further reference to the appearance and decay of double, single, and even triple step systems during the finite deformation are given in SHARPE's paper and in mine.

At the time of this writing there is no satisfactory explanation of the SAVART-MASSON effect, which very recently has been and is still being studied intensively in many laboratories. In dynamic plasticity the underlying smooth parabolic response function governs wave propagation at finite strain. Independently of the magnitude of $\Delta\sigma$, the source of the constraint during which sufficient energy is present for $\Delta\varepsilon$ to return precisely to the underlying parabolic response function and remain there for hours or days if there is no additional load, is still a matter of conjecture.

4.32. On the prediction of the response functions and second order transitions of the aggregate, from a knowledge of the deformation of the free crystal.

Once the response function for the resolved shear of the single crystal, $\tau=f(\gamma)$ was known, as it was in 1923, the interesting possibility arose that the response function of the polycrystalline aggregate might be deduced if a proper theory could be found. Except for the primitive suggestion of SACHS[1] in 1928, who ignored continuity of both stress and strain at grain boundaries and merely summed the SCHMID factors or orientation coefficients after assuming a random orientation for the grains, fifteen years elapsed before an aggregate theory was proposed by TAYLOR[2] in 1938.

To examine the final objectives of an aggregate theory, it is simplest to refer to what has been found from experiment.[3] Experiment gave for the resolved shear function[4] of $\tau=\beta_s\gamma^{\frac{1}{2}}$, with $\int_0^\gamma \tau\,d\gamma=\frac{2}{3}\tau\gamma$, the stress and strain ratios

$$\frac{\sigma}{\tau}=\overline{m}-\frac{\gamma}{\varepsilon}, \tag{4.60}$$

where σ and ε were the uniaxial nominal stress and nominal strain for tension or compression tests in the polycrystal. Similarly, for a polycrystal in torsion, for the shear stress and strain S and s, Eq. (4.61) followed.

$$\frac{S}{\tau}=\overline{n}=\frac{\gamma}{s} \tag{4.61}$$

where the aggregate ratios $\overline{m}$ and $\overline{n}$ were constants.

Thus, since from $\tau=f(\gamma)$ it followed that $\sigma=\overline{m}f(\overline{m}\varepsilon)$ and $S=\overline{n}f(\overline{n}s)$ for the resolved shear response of Eq. (4.62), the response functions for polycrystalline aggregates, [Eqs. (4.63) and (4.64)], provided:

$$\tau=\beta_s\gamma^{\frac{1}{2}}, \tag{4.62}$$

$$\sigma=(\overline{m})^{\frac{3}{2}}\beta_s\varepsilon^{\frac{1}{2}}, \tag{4.63}$$

$$S=(\overline{n})^{\frac{3}{2}}\beta_s s^{\frac{1}{2}}. \tag{4.64}$$

[1] SACHS [1928, *1*].
[2] TAYLOR [1938, *1*, *2*].
[3] BELL [1961, *1*], [1962, *5*], [1964, *1*], [1965, *2*], [1968, *1*].
[4] As noted in Sect. 4.18, Eq. (4.24) the parabola coefficient is

$$\beta_s=(\tfrac{2}{3})^{r/2}\mu(0)\frac{B_0}{(\overline{m})^{\frac{3}{2}}}(1-T/T_m).$$

For the response to increased loading only, where $\int_0^\varepsilon \sigma\, d\varepsilon = \frac{2}{3}\sigma\varepsilon$ and $\int_0^s S\, ds = \frac{2}{3} S s$, it followed that

$$\tau\gamma = \sigma\varepsilon = S s. \tag{4.65}$$

SACHS, who in his calculation ignored the fact that his grains would separate and overlap through rotation, obtained the lower bound $\overline{m} = 2.238$. TAYLOR in 1938, introducing the 12 slip systems of the face centered cubic solid for which only 5 were independent, and assuming homogeneous strain, uniform deformation of grain, and continuity of displacement at grain boundaries, made a summation based upon a minimum energy principle which provided $\overline{m} = 3.06$. J. F. W. BISHOP and RODNEY HILL[1] in 1951 reexamined and extended TAYLOR's theory, viewing the subject in terms of the components of shear stress, and summed on a principle of maximum virtual work. They also obtained TAYLOR's value of $\overline{m} = 3.06$ but provided additional calculations leading to a ratio of $\overline{n} = 1.65$ for the torsion of polycrystals.

An aggregate theory should consider the conditions of both continuity and equilibrium. The theories of TAYLOR and of BISHOP and HILL explicitly included only the first of these. In 1968, THOMAS H. DAWSON[2] described in his doctoral research an analysis which explicitly included both conditions; otherwise, DAWSON made the same assumptions and performed his summations in the same manner as had TAYLOR. The results of DAWSON's analysis, obviously more acceptable as a basis for theory, were the experimentally unacceptable values of $\overline{m} = 3.27$, and $\overline{n} = 2.08$.

As I showed in Sects. 4.18, 4.20, 4.21, and 4.22, the comparison of experimental results for hundreds of tests on single crystals and polycrystals provided values for the aggregate ratios of $\overline{m} = 3.06$, and $\overline{n} = 1.765$. For the MAXWELL-VON MISES condition to hold over the entire plastic region of deformation for the polycrystalline solid $\overline{m}/\overline{n}$ must equal $\sqrt{3}$. This value is precisely that obtained from experiments in fully annealed polycrystals. In TAYLOR's and in BISHOP and HILL's analysis, $\overline{m}/\overline{n} = \sqrt{3.44}$; and in DAWSON's, $\overline{m}/\overline{n} = \sqrt{2.47}$.

The discrepancy between experiment and prediction is so large[3] that obviously a different theory is essential. For DAWSON's analysis, the discrepancy is even more serious in view of the fact that the equilibrium condition forbids any rotation of grains for uniaxial tests, a result most certainly not in accord with observation during the deformation of polycrystals.

In Sect. 4.22 above I showed from the examination of many tests that the MAXWELL-VON MISES condition, for which $\overline{m}/\overline{n} = \sqrt{3}$, was applicable only if both the shear and axial stresses and strains were described with reference to the undeformed state of the solid. TAYLOR and QUINNEY's[4] effort to effect a comparison in "true" strain was inconclusive (see Fig. 4.60, Sect. 4.14) until I recalculated the data, as shown in Fig. 4.104, Sect. 4.22, after which a close agreement was obtained not only with the MAXWELL-VON MISES condition but also with the quantitative prediction of the response function,[5] Eqs. (4.25) [(4.63)] and (4.29)

[1] BISHOP and HILL [1951, *1*, *2*].

[2] DAWSON [1968, *1*], [1970, *1*].

[3] My comparison of observed response functions from tension, compression, and torsion tests was for the ratio of the square of parabola coefficients in (stress)^2 vs strain plots. From Eq. (4.63), for example, a comparison of prediction and observation involves $(\overline{m})^3$. From experiment $(\overline{m})^3 = 28.8$; for SACHS, the much lower value of $(\overline{m})^3 = 11.25$; for TAYLOR, the same number as the experimental value, $(\overline{m})^3 = 28.8$; and for DAWSON, the high value of $(\overline{m})^3 = 35.0$. From experiments in torsion, $(\overline{n})^3 = 5.5$; from DAWSON, $(\overline{n})^3 = 9.0$ and from BISHOP and HILL $(\overline{n})^3 = 4.49$.

[4] TAYLOR and QUINNEY [1931, *1*].

[5] See Sect. 4.21 and 4.22.

[(4.64)]. In his aggregate theory TAYLOR presumed that the uniaxial stress and strain had to be given in "true" stress and "true" strain. Possibly the reason that such an assumption is in complete disagreement with observation is that in the calculation of the resolved deformation of the single crystal, Eq. (4.24) [(4.62)], the change of dimension during deformation already has been included.

Experiment amply demonstrated that an aggregate theory with ratios $\bar{m}$ and $\bar{n}$ constant to very large strain, indeed was a plausible possibility. A proper theory obviously must be based upon assumptions different from those considered up to now.

Looking to further experiment for suggestions, we see that one obvious source of difficulty in earlier hypotheses was the assumption of homogeneous strain in individual grains. A number of experimental investigations of the behavior of polycrystalline grains, such as those of BARRETT and LEVENSON[1] in 1940, GREENOUGH[2] in 1952, BATEMAN[3] in 1954, BOAS and OGILVIE[4] in 1954, and ROSENTHAL and GRUPEN[5] in 1962, have demonstrated that aggregate grains do not rotate as predicted and that they deform inhomogeneously. The results of such investigators also showed that shear deformation was not confined to the crystallographic planes assumed for the face centered cubic crystal, and that surface effects were significant. Furthermore, the stress as determined from X-ray measurement after uniaxial deformation, was not necessarily uniform among the members of the aggregate.

Possibly of greatest importance, however, in raising serious questions with respect to theories depending upon the operation of prescribed crystallographic slip systems, is the experimental discovery I made when comparing the resolved deformation of the body centered cubic single crystal with polycrystals of the same solid.[6] Here, too, the aggregate ratio for uniaxial loading was $\bar{m}=3.06$, but the number of possible slip systems was far greater than for the face centered cubic solid.

In DAWSON's paper[7] of 1970, in which the details of his analysis may be found, he omitted the description of an experiment included in his dissertation[8] of 1968 which, in view of the unsatisfactory state of the theory, is of greater importance for further study. In that experiment he divided a rectangular cross-section of a large grained, 99.99% purity fully annealed aluminum polycrystal into two pieces. He inscribed rectangular grids on the orthogonal faces of one corner grain and made X-ray measurements of crystallographic orientations of this grain. The reassembled specimen with lubricated end faces, shown in Fig. 4.198, he loaded in axial compression to 6% total strain, obtaining a parabolic response function for the expected mode index $r=6$ for that solid for an aggregate ratio of $\bar{m}=3.06$. The measurement of crystallographic angles before and after deformation revealed that angle changes had occurred which were a combination of those expected for the deforming free crystal and a rigid body rotation. This, of course, was not in accord with DAWSON's own theory in which the equilibrium conditions demanded that there be no rotations for an uniaxial test. From the parallel rulings shown in Fig. 4.199, DAWSON measured inhomogeneous deformation for

[1] BARRETT and LEVENSON [1940, *1*].
[2] GREENOUGH [1952, *1*].
[3] BATEMAN [1954, *1*].
[4] BOAS and OGILVIE [1954, *1*].
[5] ROSENTHAL and GRUPEN [1962, *1*].
[6] BELL [1961, *1*], [1963, *1*], [1964, *1*], [1965, *2*], [1968, *1*].
[7] DAWSON [1970, *1*].
[8] DAWSON [1968, *1*].

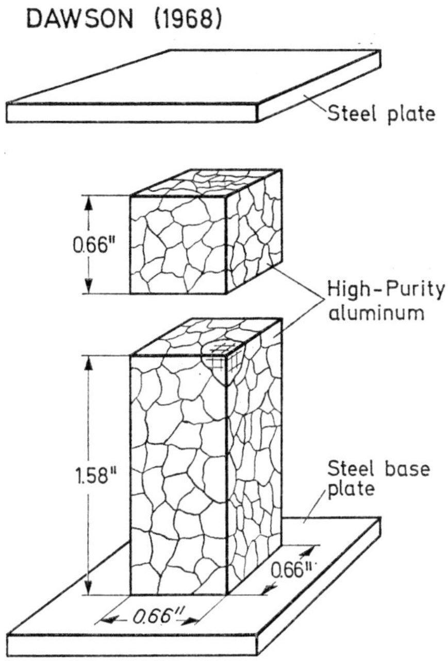

Fig. 4.198. The experimental configuration used to study aggregate grain deformation.

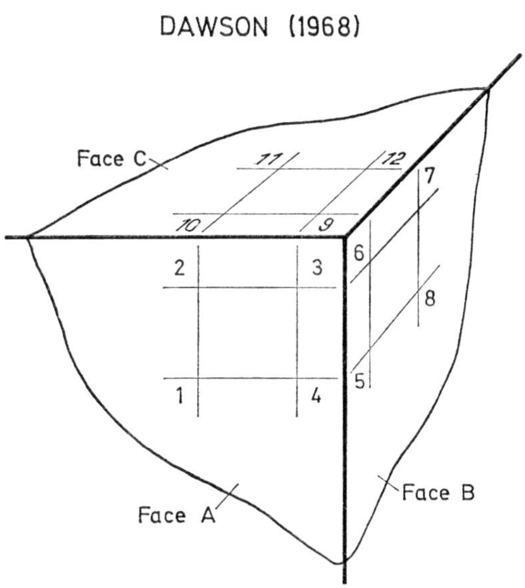

Fig. 4.199. The system of grain markings prior to deformation from which the evidence of inhomogeneous deformation was observed for the individual grain in an aggregate having the parabolic behavior predicted from Eq. (4.25).

the portion of the grain studied, but the total deformation of the aggregate averaged in both axial and lateral directions gave no evidence of local irregularity.

In 1968 I found that the resolved deformation of a large grain averaged over the whole grain in relation to the overall deformation of the polycrystal of which

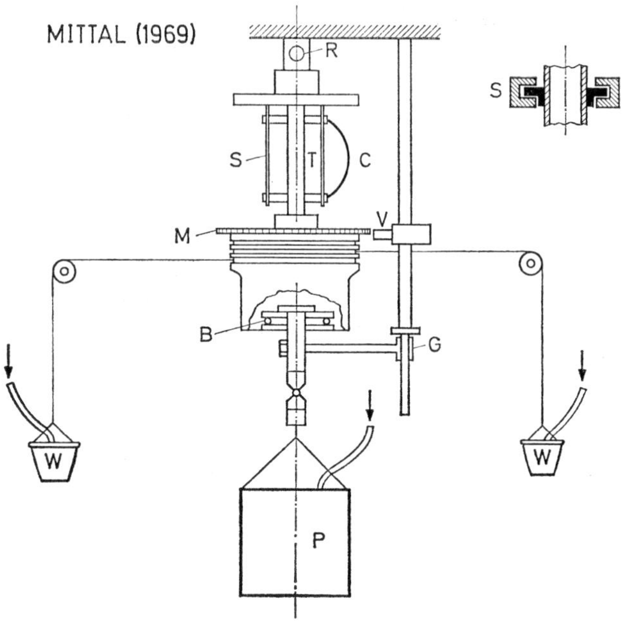

Fig. 4.200. Tension, torsion experimental apparatus in which the tension and torsion components were de-coupled.

it was a member, gave an aggregate ratio $\bar{m} = 3.06$ obtained when the free crystal response was compared with that of the polycrystal.[1] From further experimental study of the embedded single crystal, which has not been completed at this writing, local inhomogeneity in strain and rotation over the single crystal was found to be important in maintaining continuity at grain boundaries. It seems fairly obvious that a successful theory must postulate a plausible distribution of inhomogeneous deformation in the single crystal.

Despite the limitations of theory for the study of the aggregate in terms of the deformation of the single crystal, experiment has succeeded in establishing a number of important further generalizations in this area. In his doctoral dissertation in 1969, RAMESH MITTAL[2] described the results of an experimental study on tubes of fully annealed low purity polycrystalline aluminum, undertaken to ascertain whether the parabolic response functions[3] Eqs. (4.25) and (4.29) which I had found for simple tension and simple torsion could be extended to radial and non-radial loading paths in combined tension, torsion tests. MITTAL's apparatus, diagrammed in Fig. 4.200, was unique in that for the first time in this type of experiment he decoupled the tension and torsion loading systems, a factor of major importance for the study of a solid exhibiting the inertial influences of the SAVART-MASSON effect.

The tubular specimen T was supported at R so that rotation was prevented while vertical alignment was assured. The tensile load P and the torsional load W were controlled by water flow to provide the desired $\dot{\sigma}$ and $\dot{S}$ constant loading rates. The bearing at B allowed the torsion to occur without the rotation of the

[1] BELL [1968, *1*], Fig. 4.20, p. 119.
[2] MITTAL [1969, *1*], [1971, *1*].
[3] See Sects. 4.21 and 4.22.

tensile load. The element S containing a calibrated clip gage C for the accurate determination of nominal strain by means of a semiconductor gage bonded to it, also allowed for the rotation of the specimen so that the total angle of the torsional twist could be determined optically. This latter reading provided the desired shear strain s. The fully annealed polycrystalline aluminum tubes had an outside diameter of 0.4395 in., with a tube thickness of 0.0322 in. They were of 99.16% purity, of small grain; they had been annealed for two hours at 1100° F and furnace cooled, that is, the solid was fully annealed so that it would be possible to make a comparison with response from my finite amplitude wave experiments. MITTAL performed simple tension and simple torsion tests to check the apparatus by showing that the ratio $\bar{m}/\bar{n}$ obtained in these tests agreed with earlier observations that $\bar{m}/\bar{n}=\sqrt{3}$.

Combining my Eqs. (4.25) and (4.29)[1] for simple tension and simple torsion, MITTAL introduced a general stress $T=\sqrt{\frac{\sigma^2}{3}+S^2}$ and $E=\sqrt{3\varepsilon^2+s^2}$, which gave parabola coefficients as follows when the loading was simple:

$$T=(\bar{n}/\bar{m})^{\frac{3}{2}}\,\mu(0)\,B_0(1-T/T_m)(E-E_b)^{\frac{1}{2}}, \tag{4.66}$$

or, in terms of the single crystal coefficient and aggregate ratios,

$$T=(\bar{n})^{\frac{3}{2}}\,\beta_s\,E^{\frac{1}{2}}, \tag{4.67}$$

and

$$\frac{T}{\tau}=\bar{n}=\frac{\gamma}{E}. \tag{4.68}$$

MITTAL performed 20 tests in this series: 4 in simple torsion, 4 in simple tension, 7 for radial loading with various constant loading ratios $\dot{\sigma}/\dot{S}$, and 5 non-radial loading tests in which simple loading in either S or σ to some specified stress was followed by simple loading of the opposite type.

E. A. DAVIS[2] in 1943 had discovered that response in the form of octahedral shear or maximum shear at finite strain for *radial* loading paths at the various ratios of principal stresses obtainable from the combined tension and internal pressure loading of hollow tubes was independent of the principal stress ratio.[3] MITTAL extended DAVIS' discovery by finding a general response function independent of path not only for radial loading but also for *non-radial* loading.

For DAVIS' data, it was necessary for me to recalculate the results given in CAUCHY stress and logarithmic strain to PIOLA-KIRCHHOFF stress and strain referred to the undeformed state of the material for the experiments on annealed copper of Fig. 4.61 and mild steel, Fig. 4.63. I showed the results of such a calculation, using DAVIS' definitions of octahedral stress and strain and maximum shear stress and strain; the unspecified response functions of DAVIS were not only independent of radial loading path but in retrospect also were closely given by my parabolic response function for finite strain, with the usual initial mode index of $r=4$ for copper, and a second order transition to $r=6$ at the transition strain of $N=8$. To illustrate further, I have extracted one test for each solid. In Fig. 4.201 is a σ^2 vs ε plot which I have recalculated in nominal stress and nominal strain, compared with prediction of Eq. (4.25) (solid line). In contrast, in Fig. 4.202 for simple tension in mild steel in a tube under internal pressure, I have calculated from Eq. (4.25) in nominal form the predicted values (crosses) for the representation in "true" stress and "true" strain.

[1] See Sects. 4.21, 4.22, and 4.28 above.
[2] E. A. DAVIS [1943, *1*].
[3] See Sect. 4.15, Figs. 4.61, 4.62, 4.63, and 4.64.

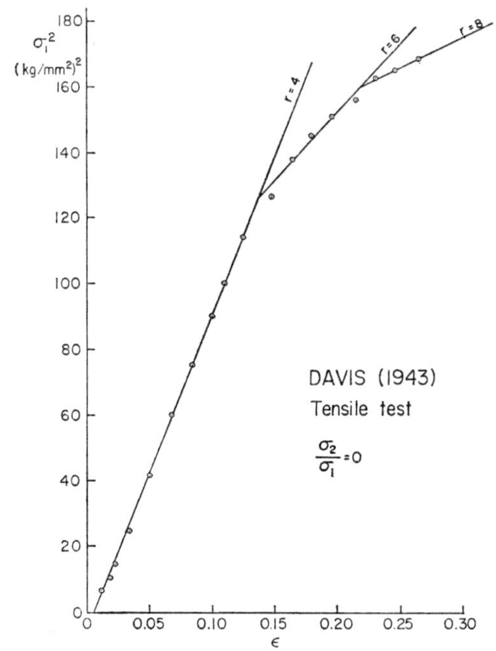

Fig. 4.201. A σ_1^2 vs ε plot for copper tubes in uniaxial tension in nominal stress and strain from the tests of Davis shown in Fig. 4.61 (circles) compared with prediction of Eq. (4.25) (solid lines).

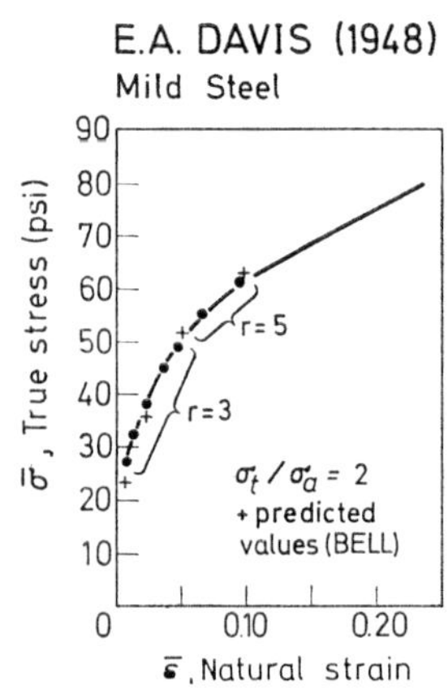

Fig. 4.202. A $\bar{\sigma}$ vs $\bar{\varepsilon}$ plot in "true" stress and "true" strain for a mild steel tube (solid line) in pure tension under internal pressure, compared with prediction calculated from Eq. (4.25) (crosses).

In Mittal's experiments on hot rolled annealed low purity aluminum, the initial mode index was the usual $r=2$ for the solid for 19 of the 20 tests he performed, with a second order transition to the mode index $r=3$ in 17 of the tests, and $r=4$ in 2 tests. The 20th test had an initial mode index $r=1$, with a second order transition to $r=3$ precisely at $N=18$.

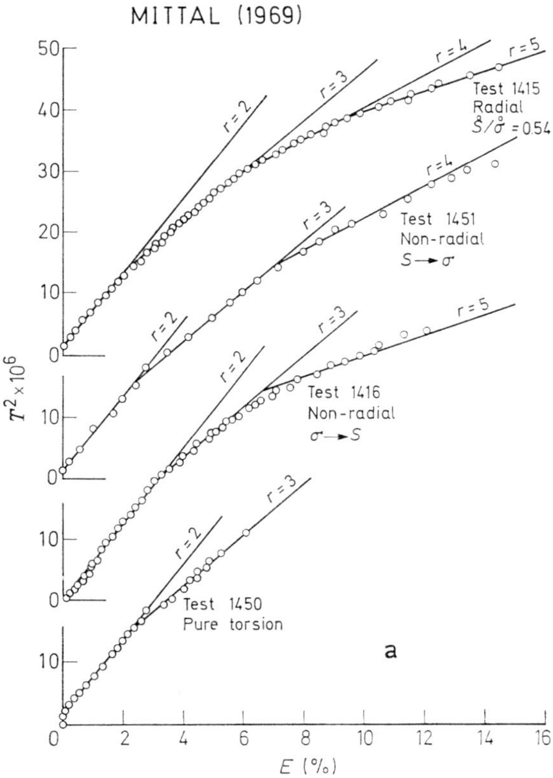

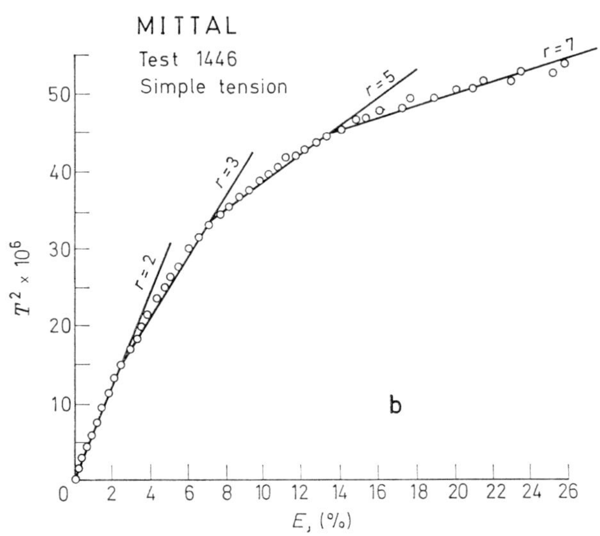

Fig. 4.203a and b. T^2 vs E plots for tests on hollow tubes of fully annealed aluminum for radial and non-radial loading paths with torsion component (a) and simple tension (b), compared with prediction of Eq. (4.66) (solid line).

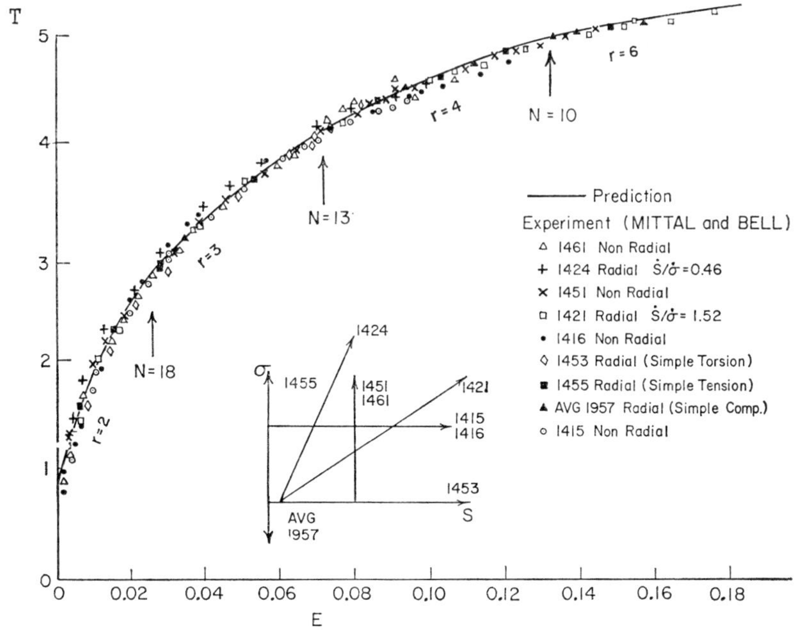

Fig. 4.204. T vs E compared with Eq. (4.66), showing independence of loading path.

Illustrative radial and non-radial measurements[1] of Mittal are shown in Figs. 4.203 and 4.204, which demonstrate not only the complete independence of the response function $T = \phi(E)$ from loading path and its agreement with the extension of the original parabolic response function in the form of Eq. (4.66) and with the aggregate ratio expansion, Eq. (4.68), but also that the second order transition structure is independent of loading path, whether radial or non-radial.

In Fig. 4.203 are compared T^2 vs E plots of a representative result for simple tension, simple torsion; a radial load path for combined tension and torsion with $\dot{\sigma}/\dot{S} = 0.57$; a non-radial loading path, first of tension, then of torsion; and a non-radial loading path, first of torsion, then of tension. In Fig. 4.204 are shown the T vs E plots of many tests for the loading paths shown, to which I have added averaged experiments of mine in compression to illustrate further that when stress and strain are referred to the undeformed solid the response function for finite strain, as determined from the single crystal through the aggregate, is completely general for loading in any combination of two stress components determinable from tension-torsion tests. The data of Davis given above, demonstrated that a similar experimental result prevailed for any ratio of two principal stresses when tension and internal pressure tests were considered. In Mittal's two sets of tests, a radial loading path was intersected at a point in $\sigma - S$ space by non-radial loading paths $\sigma \to S$ and $S \to \sigma$. The values of T and E coincided at the point of intersection, providing a general response function.

Of equal importance in Mittal's experimental study is that the transition strains at which I had found mode changes occurred in tension, compression, and

[1] In Mittal's publication of these results, a small contribution to the torsion load was omitted inadvertently. The omission changed none of his conclusions. In presenting his results here, I have recalculated the data to include this small correction for the torsion component.

Table 137. MITTAL (1969). *Transition strain, tension-torsion experiments (aluminum).*

Transition no.	N	Tension average	Torsion average	Radial tension-torsion average	Non-radial tension-torsion average	Total no. of meas.	Total average	Predicted (BELL) $E_N=(\frac{2}{3})^{N/2}$
I	18	0.026	0.026	0.025	0.022	10	0.025	0.026
II	13	0.070	0.072	0.073	0.072	16	0.072	0.072
III	10	0.136	0.138	0.136	—	11	0.135	0.132
IV	8	—	—	0.198	—	1	0.198	0.198

torsion for simple loading of polycrystals and in the resolved shear strain of single crystals, also were shown to be completely independent of loading path. Table 137 compiles MITTAL's measurements of these transition strains for each type of loading, and their total average. Also shown is the close agreement of the experimental average with my transition strain relation, $E_N=(\frac{2}{3})^{N/2}$.

It is of further interest that MITTAL found that in every instance torsional buckling either in simple torsion or in radial loading occurred at or near these transition strains. The observed values in different measurements were: 0.026, 0.076, 0.078, 0.132, 0.198. This corresponds with my earlier observation that critical strains were associated with ultimate stress in tension and compression.

Transition strains for second order transitions in the polycrystal for all types of simple and combined loading are determinable from the empirical aggregate ratios from the transition strains of the resolved deformation of the single crystal.[1] The experimental results described in Sect. 4.21 for the uniaxial loading of polycrystals, those in Sect. 4.22 for the torsion of polycrystals, the data of MITTAL and of DAVIS for radial and non-radial loading paths in Sect. 4.15, all reveal the presence of transition strains which are related to those of the single crystal by

$$\begin{aligned} \varepsilon_N &= \overline{m}\,\gamma_N, \\ s_N &= \overline{n}\,\gamma_N, \\ E_N &= \overline{n}\,\gamma_N. \end{aligned} \tag{4.69}$$

Since, as I have shown, $E_N=(\frac{2}{3})^{N/2}$ where $N=0, 2, 4, 6, 8, 10, 13, 18$, all eight known transition strains for the plastic deformation of crystals are designated

Table 138. BELL (1971). $E_N=(\frac{2}{3})^{N/2}$.

Transition no.	N	$\varepsilon_N=\frac{n}{m}E_N$	$\gamma_N=\overline{n}\,E_N$	$E_N=s_N$	Measured dynamic ε_N
I	18	0.015	0.046	0.026	0.015
II	13	0.041	0.127	0.072	0.041
III	10	0.076	0.233	0.132	0.076
IV	8	0.114	0.350	0.198	—
V	6	0.171	0.524	(0.296)	—
VI	4	0.256	0.784	(0.444)	—
VII	2	0.385	1.178	(0.667)	—
VIII	0	0.577	1.765	(1.000)	—

[1] BELL [1968, *1*], [1971, *2*].

numerically. These values of the transition strains are summarized in Table 138. All these strains are given with reference to the undeformed configuration. All values in Table 138, except those enclosed in parentheses for torsion at large strain, have been documented experimentally.

4.33. On the study of the yield surface after 1948, based upon extensions of the experiment of Guest and the measurements of Bauschinger.

Since the 1950's there has been a resurgence of interest in the study of yield surfaces, motivated by concurrent developments in the theory of the "ideal" plastic solid. All but one of the experiments referred to here involved a prescription of the loading path for tension vs torsion either of strains for a hard testing device or for stresses for a dead weight apparatus. The experiments of BAUSCHINGER[1] in the 1880's for simple radial loading in tension and compression provided a measure of only two points on the yield surface. They are of importance in the present context because they delineated the difficulties inherent in recognizing a point on the yield surface upon reloading. Like other 19th century experimentists such as KELVIN and WERTHEIM, and a few in the 20th century, BAUSCHINGER, in addition to having the difficulty of defining the point of re-arrival at a yield surface, observed that the moduli of the reloading response was generally lower and, in some instances, markedly so. In this context I refer to Sect. 3.44, in which we saw from more recent studies that not only do moduli vary on reloading following plastic deformation, but also that second order transitions referred to as "multiple elasticities" may occur, thus even further complicating the interpretation of cyclical loading where plastic deformation is present.

One other phenomenon influencing the results has been observed in all types of tests in some solids, and particularly with respect to yield surfaces, namely the SAVART-MASSON (PORTEVIN-LE CHATELIER) effect, which until very recently largely has been ignored. (This will be discussed in Sect. 4.35 below.)

In 1951, BERNARD BUDIANSKY, NORRIS F. DOW, ROGER W. PETERS, and ROLAND P. SHEPARD[2] tested thin walled cylinders of 14 S-T 4 aluminum alloy by loading specimens in compression to a strain of approximately 0.005, after which they introduced a prescribed ratio of compression to shear loading. Their results, which gave rise to considerable discussion as to whether or not they could assume linearity for response functions, were found to be in accord neither with their version of a deformation theory, nor with flow theory, nor with their proposed slip theory of plastic deformation. Anisotropy in the large cylinders machined from forgings, the peculiarities of the alloy studied, and the "hard" testing apparatus for which the strains were prescribed, must have been factors affecting their results.

In 1954, PAUL M. NAGHDI and J. C. ROWLES[3] described results from 10 thin walled tubular specimens of 24 S-T 4 aluminum alloy which they noted definitely was not an isotropic solid. They studied various loading paths consisting of tension alone to a specified stress followed by torsion with different amounts of accompanying tension. Again the apparatus was a "hard" testing machine for which the strain was prescribed. NAGHDI and ROWLES gave their results in complete detail,

[1] BAUSCHINGER [1881, *1*, *2*], [1886, *1*].
[2] BUDIANSKY, DOW, PETERS, and SHEPARD [1951, *1*].
[3] NAGHDI and ROWLES [1954, *1*].

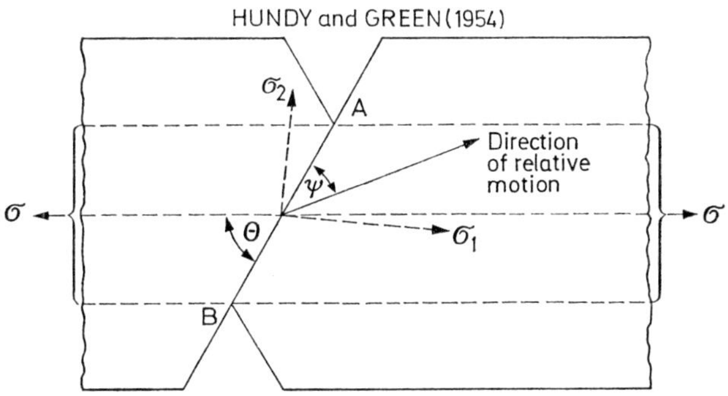

Fig. 4.205. Diagram of the specimen.

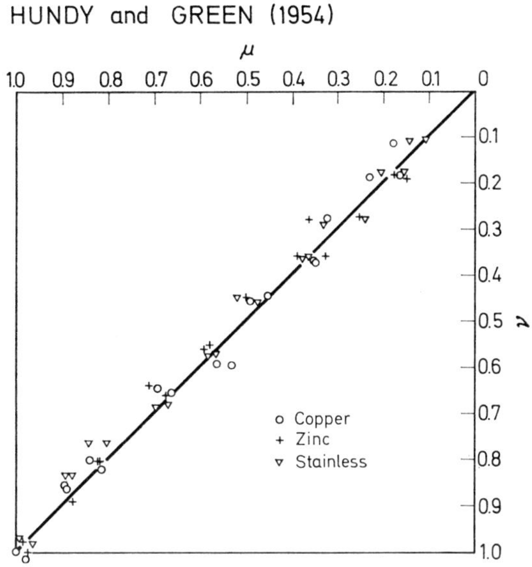

Fig. 4.206. The LODE diagram obtained from tests on specimens such as that shown in Fig. 4.205.

the most interesting aspect being the large reduction (a factor of 2) in the initial shear modulus they observed when twist began, compared to that for initial torsion alone. (By 1954 the reduction in modulus with permanent deformation had been forgotten.) Once again, in addition to the unknown effects of anisotropy, the results from the GUEST experiment were unable to provide conclusive evidence for the direction which the developing theory might adopt.

An interesting new experiment related to the earlier investigations of LODE[1] and of TAYLOR and QUINNEY[2] (Sect. 4.14) involved measurements of the pulling of notched strips of copper, zinc, and stainless steel past their yield points, by B. B. HUNDY and A. P. GREEN[3] in 1954. On specimens of the type shown in Fig. 4.205, pulled in tension, HUNDY and GREEN obtained different combinations of the LODE variables μ and ν by altering the angle θ. Referring the reader to

[1] LODE [1926, *1*].
[2] TAYLOR and QUINNEY [1931, *1*].
[3] HUNDY and A. P. GREEN [1954, *1*].

their paper for details, I include in Fig. 4.206 the results in the three solids for which $\mu = \nu$ is approximated closely.

V. S. Lensky[1] in 1960 described a series of tests upon copper and mild steel which also were performed upon "hard" testing machines, in this case, semi-automated to provide prescribed strain histories in combined tension and torsion of relatively small thin walled tubes. Lensky's strain paths which included unloading as well as loading, were shown along with the slope of the increments of shear to tension stresses at various points in strain space. I have included in Fig. 4.207 results for two tests in copper having strain trajectories with corners, and in Fig. 4.208 results for two copper specimens having curvilinear strain trajectories which are self-explanatory as illustrative of what is observed when the common engineering test on "hard" machines is extended. The subscript 3 refers to the torsion component, and the subscript 1, to the tension.

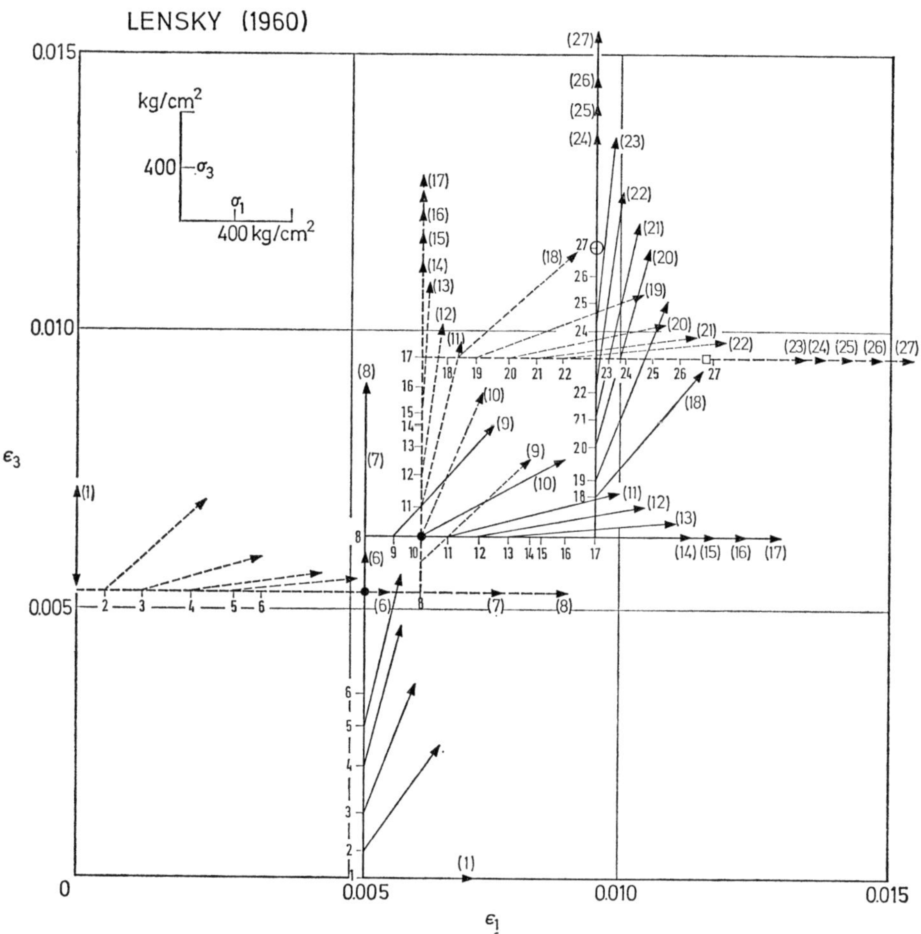

Fig. 4.207. Incremental stress ratios for the rectangular tension-torsion non-radial strain paths prescribed by a "hard" testing machine.

[1] Lensky [1960, *1*].

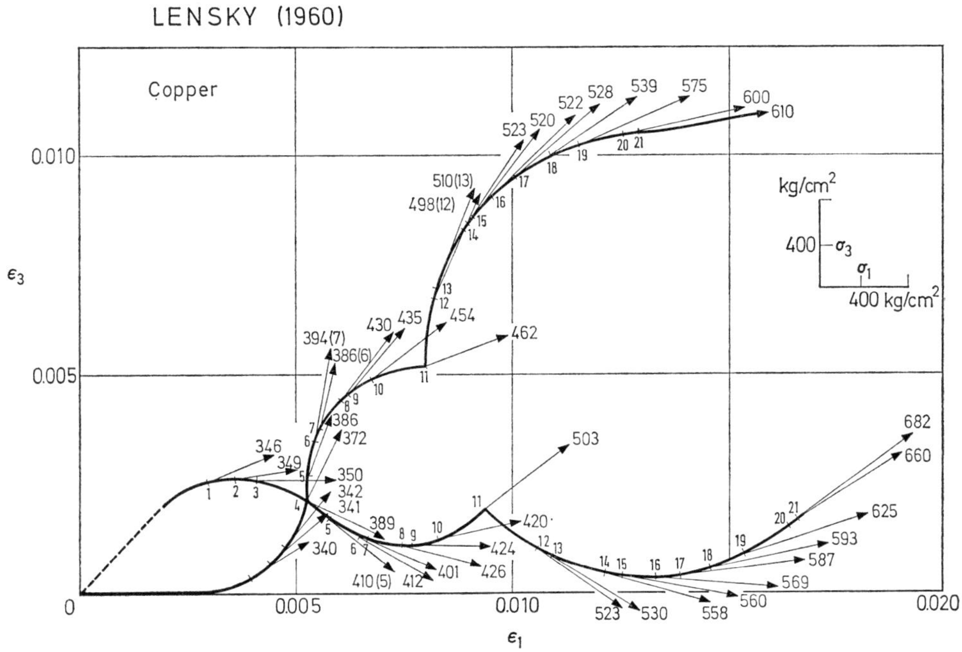

Fig. 4.208. Incremental stress ratios for the curvilinear tension-torsion non-radial **strain** paths prescribed by a "hard" testing machine.

Throughout this treatise whenever experiments requiring slow loading were needed to consider a fundamental physical question, the superiority of the dead weight test has been evident. Once again this point is emphasized for dead weight tests obtained by ARIS PHILLIPS[1] in a series of tension vs torsion measurements from 1957 to the present. PHILLIPS provided a thorough description of his dead weight apparatus[2] shown in Fig. 4.209, the method of applying the loads, and the precise manner in which the yield surface was identified upon reentry. The solid he studied in dead weight tests for fifteen years fortunately was annealed commercial purity aluminum for which, as we have seen above, the influence of unknown prior thermal and mechanical histories is negligible.

In contrast to LENSKY's measurements, PHILLIPS prescribed the loading path for his two tension vs torsion stress components σ and τ, measuring the resulting axial strain and shear strain on tubes with very thin walls where the wall thickness to diameter ratio was approximately 1:16. In 1957, 1960, and 1961 PHILLIPS[3] described the results obtained for different loading paths which he compared with the slopes or ratio of strain increments at various points in stress space. In Figs. 4.210 and 4.211 are shown some of the loading paths chosen and the strain ratios (solid lines) compared with prediction (dashed lines) based upon the assumption of incremental theories of plasticity that this ratio is given by the ratio of the total stresses and not by the ratio of increments.

[1] PHILLIPS [1957, *1*], [1960, *1*], [1972, *1*]. PHILLIPS and GRAY [1961, *1*]. SIERAKOWSKI and PHILLIPS [1965, *1*], [1968 *1*].

[2] In this apparatus the torsion loading system was not decoupled from the tension, which becomes a serious limitation when the steps of the SAVART-MASSON effect are present, as is possible in annealed aluminum.

[3] PHILLIPS [1957, *1*], [1960, *1*]; PHILLIPS and GRAY [1961, *1*].

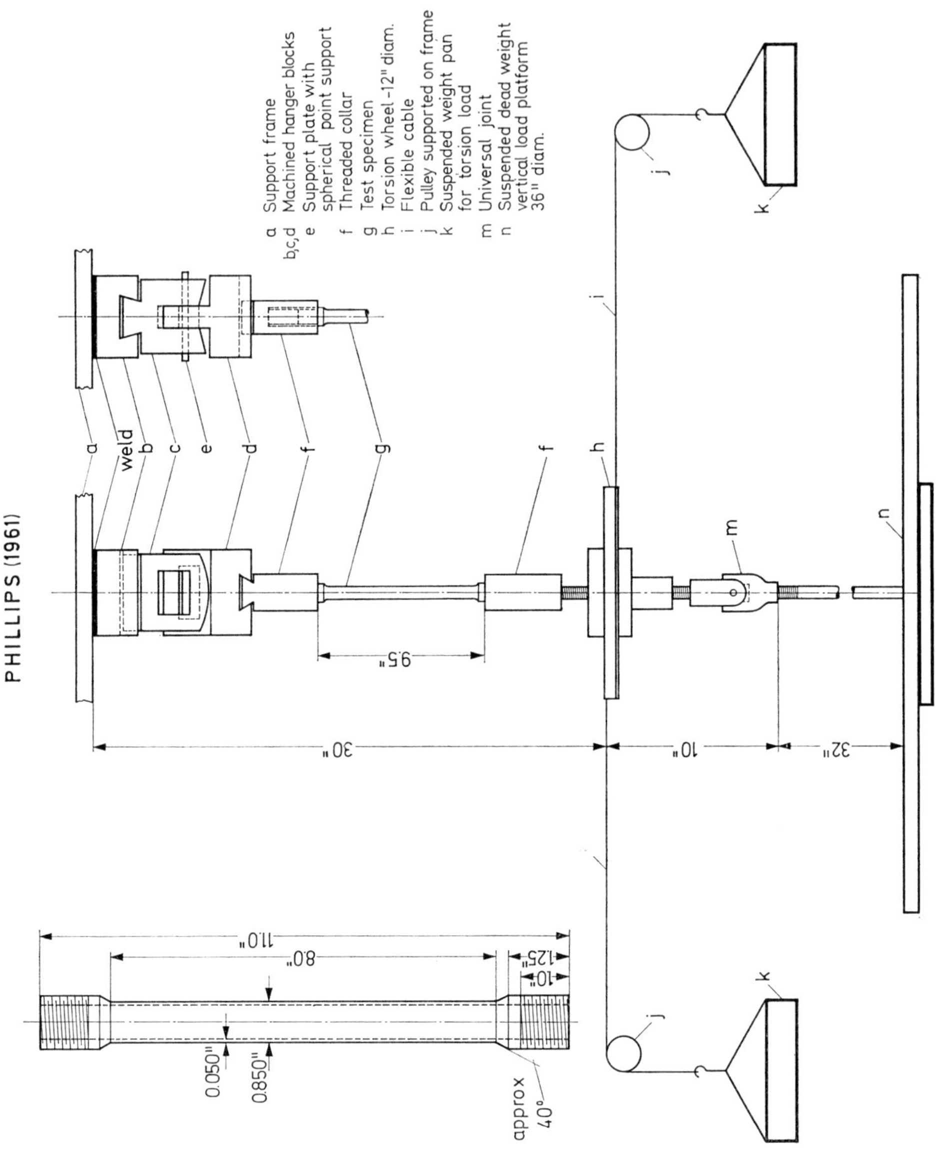

Fig. 4.209. Diagram of the dead weight tension-torsion apparatus used by Phillips. The tension-torsion components were not de-coupled.

In general, Phillips' results, apart from their intrinsic importance, do not settle the claims of competing theories, but for some non-radial loading paths there would appear to be closer agreement than for others. Since these results do not include the tension and torsion ratios of the components of the Savart-Masson steps, which are not always visible but which certainly are observed in most experiments on the dead weight radial loading of this solid, it is difficult to place too much emphasis on the detail of these comparisons. However, it is clear from Phillips' experiments that ratios for stress and for incremental strain do not coincide.

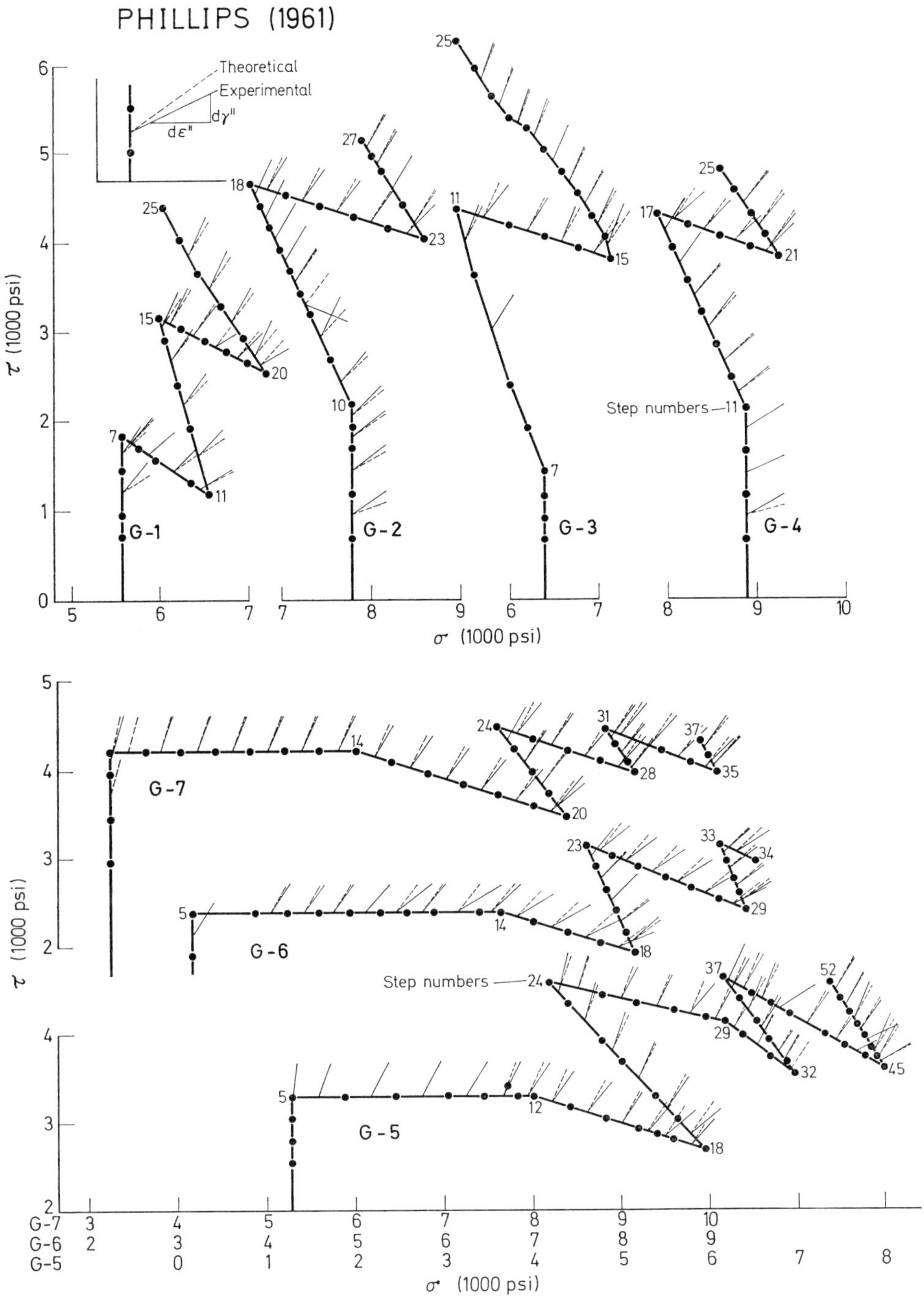

Fig. 4.210. Incremental strain ratios for the non-radial stress paths prescribed by dead weight loading.

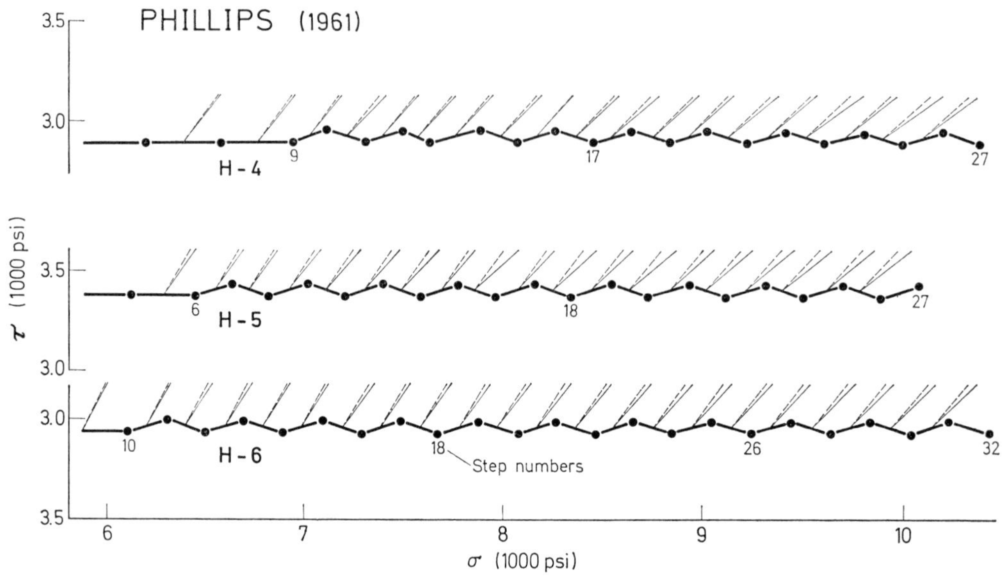

Fig. 4.211. Incremental strain ratios for increasing tension stress at constant torsion stress.

In 1968, R. L. SIERAKOWSKI and PHILLIPS[1] studied the effect of repeated loading on the yield surface, and PHILLIPS[2] studied the variation of the yield surface of annealed aluminum with ambient temperature. These experiments and the temperature studies which followed[3] in 1972, were a return to the original experiment of GUEST,[4] in that, following unloading from an initial pre-stress the specimen was reloaded along various loading paths until the yield surface again was reached. With the use of the GUEST experiment, once again the thorny question arises as to how the yield surface is to be identified on reentry. From the 1840's to the 1880's proposals included the adoption of a fixed strain (0.2% in the 20th century in contrast to the 0.05% of the mid-19th), sometimes with the added condition that the specimen be unloaded repeatedly to ascertain the permanent deformation. PHILLIPS retraced[5] in detail the logical arguments made by BAUSCHINGER ninety years before, and reached the same conclusion: that the only acceptable definition of the yield surface, the BAUSCHINGER elastic limit, was the proportional limit at which there was an observable erratic small change in the plastic strain component in contrast to the more regular microplastic strain below this limit.

I have included two examples of these measurements of PHILLIPS. One, shown in Fig. 4.212 indicates how the yield surface in annealed aluminum, obtained through the repeated use of the GUEST experiment at different temperatures, varied as a function of temperature. The second is a single example[6] from many described in two of PHILLIPS' papers in 1972, in which both the temperature and

[1] SIERAKOWSKI and PHILLIPS [1968, *1*].
[2] PHILLIPS [1968, *1*].
[3] PHILLIPS [1972, *1*].
[4] GUEST [1900, *1*].
[5] See Sect. 4.7 above. In PHILLIPS' discussion, there is no reference to BAUSCHINGER'S analysis.
[6] PHILLIPS and TANG [1972, *1*].

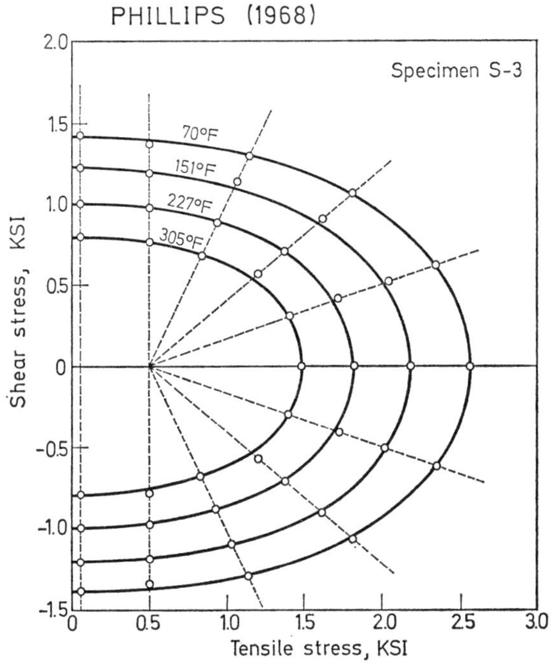

Fig. 4.212. Yield surfaces determined from the GUEST experiment at the temperatures shown.

the loading path, in terms of a pre-stress, were considered by means of the GUEST experiment. The results shown in Fig. 4.213 are for pre-stresses in both tension and torsion, with yield surfaces determined for four ambient temperatures around each value of pre-stress.

This important research of PHILLIPS during the past decade and a half is a continuation and long-awaited extension, after a lapse of nearly ninety years, of BAUSCHINGER's definitive study of but two points on the yield surface.

The dead weight tension vs torsion tests of MITTAL[1] at room temperature performed on fully annealed hollow tubes, of the same solid as that studied by PHILLIPS, described above in Sect. 4.32, Fig. 4.204, and the experiments of E. A. DAVIS[2] on annealed copper, Sect. 4.15, Fig. 4.61, have shown that for monotonically increasing loading along many radial and non-radial loading paths, response functions in generalized stress and generalized strain are independent of path. Experiments such as those, together with studies of the type made by PHILLIPS, must be considered simultaneously if general constitutive equations for plasticity which include both loading and unloading response are to be established experimentally. This point will be elaborated further in Sect. 4.35, with some very recent experiments which, in addition, include the SAVART-MASSON effect.

Also pertinent to studies of the yield surface is the fact that elastic moduli during reloading are changed by the permanent deformation of the previous prestress. There is evidence that the multiple elasticities for the fully annealed solid, described above in Sect. 3.44 for simple loading, are of importance in more general circumstances. I shall include only two illustrations, drawn from the

[1] MITTAL [1969, *1*], [1971, *1*].
[2] E. A. DAVIS [1943, *1*], [1945, *1*].

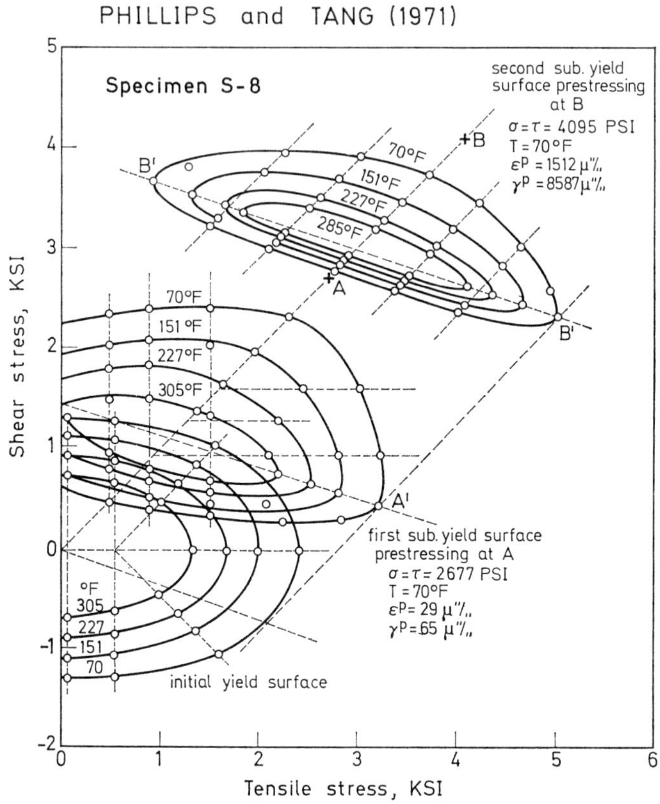

Fig. 4.213. The effect of prestress, loading path, and temperature on the yield surface measured in the Guest experiment.

current doctoral research of Hahngue Moon in my laboratory at the Johns Hopkins University. Moon is examining the general properties of second order transitions in the unloading and reloading of fully annealed aluminum.

In Fig. 4.214a is shown the S^2 vs s plot of a dead weight torsion test of Moon on a fully annealed aluminum tube in which the usual initial mode index of $r=2$ is followed by a second order transition to $r=3$ and $r=4$. At three plastic strains, the specimen was unloaded and reloaded. The detail of the reduction in shear moduli following plastic deformation, first discovered by Coulomb in 1784 and delineated by Kelvin in 1865, is shown in Fig. 4.214b in the context of the multiple elasticities [Sect. 3.44, Eq. (3.28)]. In this test, the stable index for aluminum observed at the initial loading was $s=11$. [Note the unfortunate conflict of symbols in Figs. 4.214 and 4.215, where s designates the indices for elastic moduli from Eq. (3.28) and shear strain in torsion from Eq. (4.29).] In Fig. 4.215a are the results of a test Moon performed in reversed torsion, where the S^2 vs s plot revealed that the parabolic response for stress, Eq. (4.29), still continues to govern plastic deformation after reversal. An enlarged diagram of Moon's careful measurement of elastic slopes during unloading and reloading, shown in Fig. 4.215b emphasizes that in understanding the physics of solids, the important role Kelvin predicted for such reductions in moduli following plastic deformation indeed has a basis in fact.

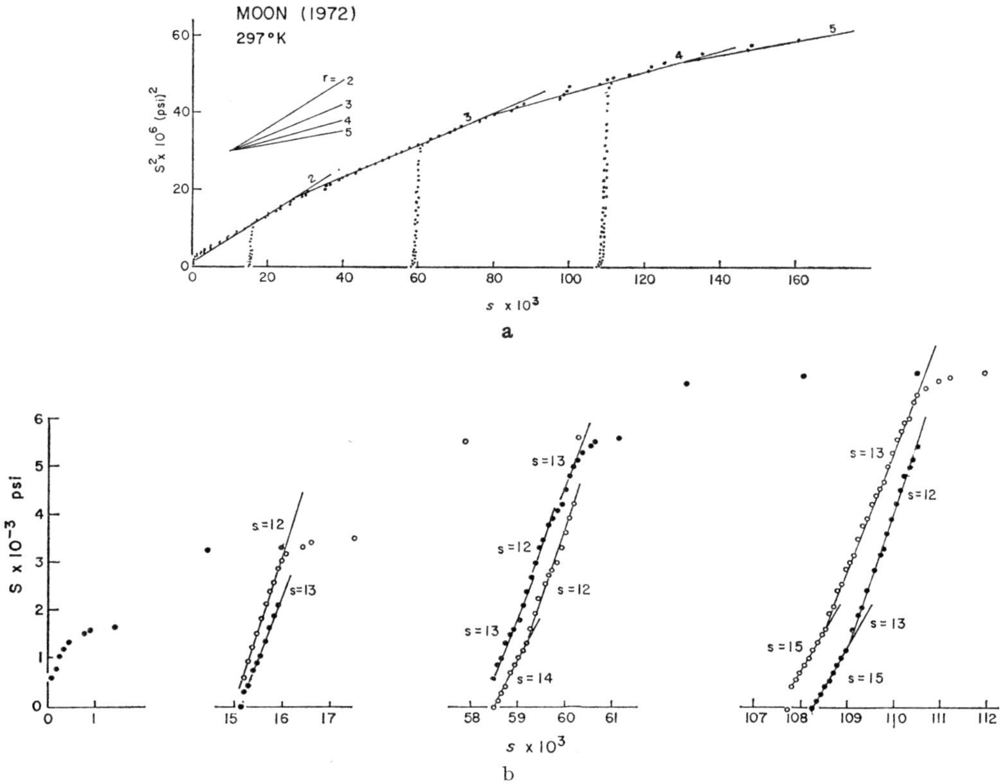

Fig. 4.214. (a) A torsion test in a fully annealed aluminum tube, compared with prediction, Eq. (4.29) (solid line). There are 3 unloading cycles. (b) The measurement of multiple elasticities from Eq. (3.29) during the unloading cycles.

In 1848 JAMES THOMSON[1] warned his contemporary experimentists that the unknown prior thermal and mechanical histories of the materials they were studying was "sufficient to account for many very discordant and perplexing results which have been arrived at by different experimenters on the strength of materials."[2]

HERBERT TOMLINSON[3] in 1887 regarded the study of the influence of one deformation event involving permanent deformation upon subsequent response

---

[1] JAMES THOMSON [1848, *1*].

[2] *Ibid.*, p. 255.

[3] TOMLINSON [1887, *1*, *2*]. While I was writing the present treatise, TOMLINSON's experimental research on "The Influence of Stress and Strains on the Action of Physical Forces," presented me with a problem. In five huge papers in the *Philosophical Transactions of the Royal Society of London* between 1883 and 1891, TOMLINSON presented a 313 page account of the manner in which his £ 4000 grant had been spent. He considered a large number of experimental situations or topics, 61 in all, with a total of 136 separate measurements or experiments.

The range of subjects included moduli measurements; the effects of permanent deformation on subsequent elastic rigidity; volume elasticity; the influence of electrical fields, magnetic fields, and thermal histories upon the deformation; the effect of various types of deformation (including prior loading histories) upon electrical conductivity; magnetic induction; and, finally, the combined effect of all of these upon the viscoelastic properties of the solid. A conclusion concerning his possible superficiality is inevitable when one adds up all this and reckons that he conducted an average of $2\,^1/_4$ experiments or measurements per topic. My

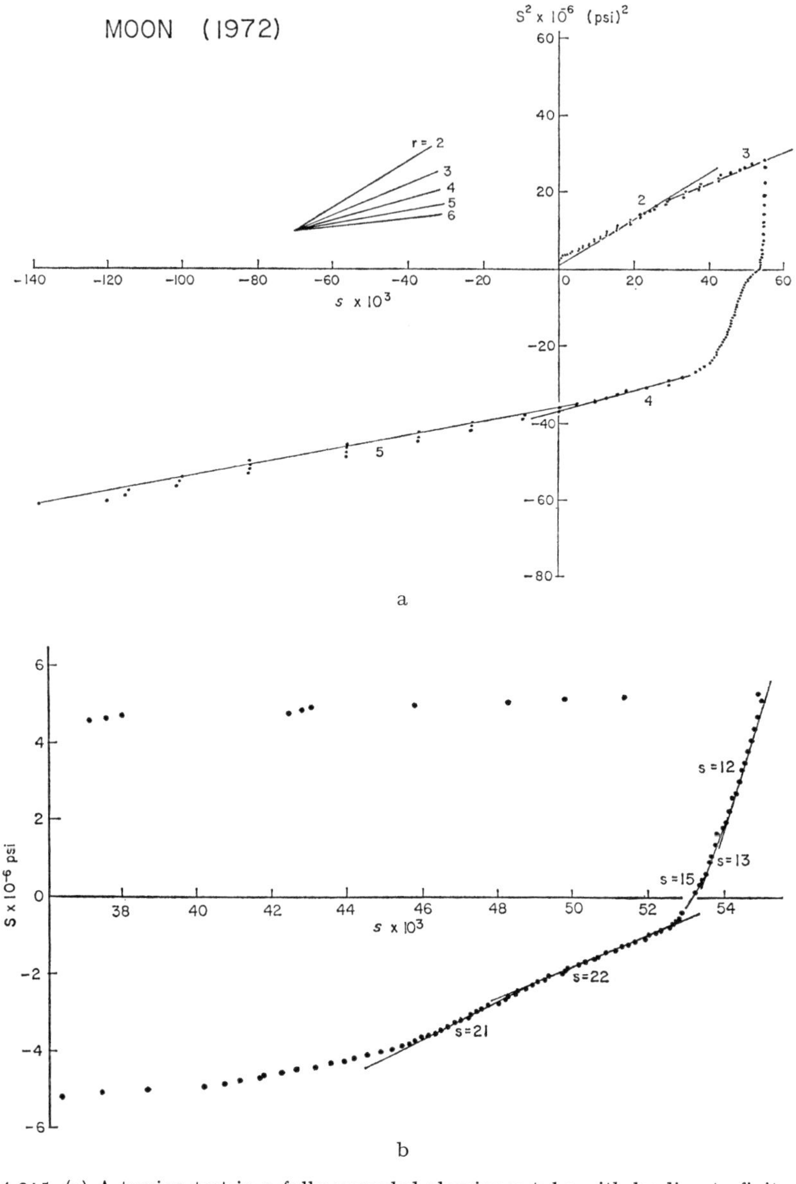

Fig. 4.215. (a) A torsion test in a fully annealed aluminum tube with loading to finite strain in both directions. The solid lines show correspondence with prediction from Eq. (4.29). (b) The "multiple elasticities" observed during the reversal of the torsion, compared with prediction from Eq. (3.29).

problem has been that the 61 topics he selected were of much interest, but his experiments more often than not were merely suggestive of possible behavior. When they were more than that, they usually were repetitions, with or without greater insight, of the work of earlier experimentists. This is my evaluation after many hours of detailed study of his test results.

I have omitted his list of topics and tabulated conclusions only because of their length, and I strongly recommend that the reader, whether a novice or a scientist of proven erudition, peruse them as a fruitful source for research in this field. Tomlinson: [1883, *1*], [1886, *1*], [1887, *1, 2*], [1889, *1*].

as of major importance, particularly when a change of temperature was included. He vertically suspended copper, iron, aluminum, silver, and platinum wires, 1 mm in diameter, of lengths ranging from 60 to 90 cm, clamping the upper end and inserting the lower end into a small brass block to which he had attached a hook at the bottom, thereby permitting the addition of arbitrarily chosen large weights.

In the first experiments,[1] permanent torsional deformation was produced by an initial twisting of the axially unloaded wire. By means of a mirror attached to the brass block and a lamp and scale appropriately located, TOMLINSON could determine the subsequent temporary twisting or untwisting which accompanied the application of an axial load. These experiments were followed by measurements in which he first twisted the wire to a permanent torsional deformation and then heated it in steam. During a period of $^1/_2$ to 2 h he measured a *permanent* untwisting of the wire. The chamber around the specimen he then filled with cold water, which produced a temperature drop of 85° C, invariably accompanied by a *temporary* twist or untwist of the solid. Repeated heating and cooling caused little or no variation in the amount of temporary twist or untwist. The increase in temperature from 15 to 100° C produced in permanently twisted iron, aluminum, and silver, a temporary untwist, whereas in copper and platinum temporary twist resulted. As TOMLINSON emphasized, the magnitude of these effects was so small that relatively long wires were needed to disclose them. The length of the 1 mm diameter wires in the tests was 120 cm.

TOMLINSON followed that paper in the *Philosophical Magazine* for September 1887 with a second paper[2] entitled, "Remarkable Effect on Raising Iron when under Temporary Stress or Permanent Strain to a Bright-red Heat." The "remarkable" aspect of the six tests, which also belong in the category of studies on the influence of prior history, was the observation of the sudden change in torsional, longitudinal, or flexural strain at either of two critical points whose actual temperature TOMLINSON hoped in the future to discover. He described them as "dull red" and "bright red" critical temperatures.

In 1949, JOHN E. DORN, A. GOLDBERG, and T. E. TIETZ,[3] just 101 years after JAMES THOMSON[4] had issued his call for caution, did find indeed a "perplexing result" in what they described from experiments with imposed thermal histories as a "failure of the mechanical equation of state."[5] To measure the load, they placed flat tensile specimens of 3 in. long reduced section, in series with an instrumented "proving ring." To measure strain, they used a "rack and pinion"[6] strain gage with a 2 in. gage length. Like so many experimental papers in the second half of the 20th century, theirs contained no description of the tensile apparatus except the statement that "All tests were conducted at a true strain rate of approximately 0.0011 per sec."[7] They performed the main series of their experiments on low purity aluminum polycrystals annealed at 1 000° F for 20 min. They also made measurements upon 99.98% purity aluminum, brass, copper, and stainless steel specimens.

The experiment consisted in loading a metal in tension to a specified finite strain at one temperature, unloading to a zero stress state, changing the ambient temperature either up or down to a new ambient temperature, then reloading the specimen to large finite strain. It was expected that the comparison of

[1] TOMLINSON [1887, *1*].
[2] TOMLINSON [1887, *2*].
[3] DORN, GOLDBERG, and TIETZ [1949, *1*].
[4] THOMSON [1848, *1*].
[5] DORN, GOLDBERG, and TIETZ, *op. cit.*, p. 212.
[6] *Ibid.*, p. 208.
[7] *Ibid.*

the response function at the second ambient temperature with that for an experiment which had proceeded from zero stress at that ambient temperature would provide a measure of the dependence upon thermal history. They considered four ambient temperatures: 78, 194, 260, and 292° K. Only the comparison of the two extremes will be discussed here. In Fig. 4.216, in terms of "true" stress vs "true" strain, are shown a series of experiments initially loaded at 292° K to the indicated finite strain, and, after unloading, reloaded to finite strain at 78° K. DORN, GOLDBERG, and TIETZ noted that finite deformation for the second loading at the lower temperature did not follow the measured response function (dashed line) for a test performed solely at that temperature. That the two response functions did not coincide led them to the conclusion that a finite strain response function $\sigma=\sigma(\varepsilon, \dot{\varepsilon}, T)$, to which they referred as an "equation of state," was insufficient to describe the plastic deformation of crystalline solids which must be very sensitive to the thermal-mechanical history. Such a conclusion, of course, was based upon the assumption that the elastic unloading, subsequent change of ambient temperature, and then elastic reloading to the new yield surface, were of no importance in producing the observed shift of response function for finite strain shown in Fig. 4.216.

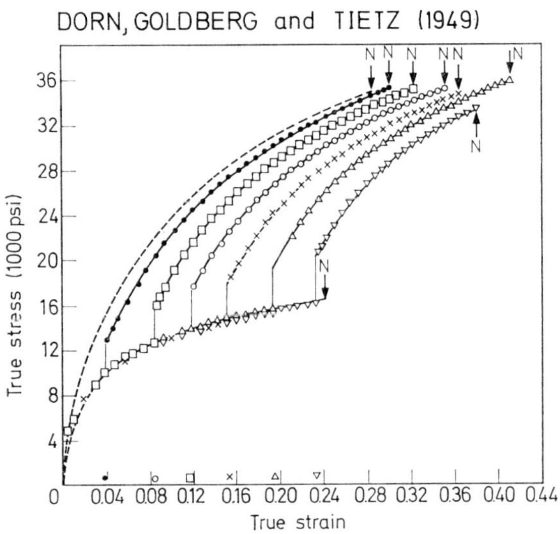

Fig. 4.216. "True" stress vs "true" strain tension tests with pre-straining at 292° K, unloading and reloading at 78° K at the positions shown. The material was annealed low purity polycrystalline aluminum.

To ascertain whether or not the major influence was in the unloading and reloading cycle, as research of the previous century would indicate, I recalculated the "true" stress vs "true" strain data of DORN, GOLDBERG, and TIETZ, in nominal stress and nominal strain referred to the undeformed specimen, for the purpose of comparing with the general parabolic response function, Eq. (4.25), Sect. 4.21. The results (circles) for the σ^2 vs ε plots of Fig. 4.217 show that not only does a single response function (solid lines) describe finite strain when imposed thermal changes are present, but also that after plastic deformation, unloading and reloading accompanied by changing temperatures result solely in a shift of the origin ε_b in Eq. (4.25).

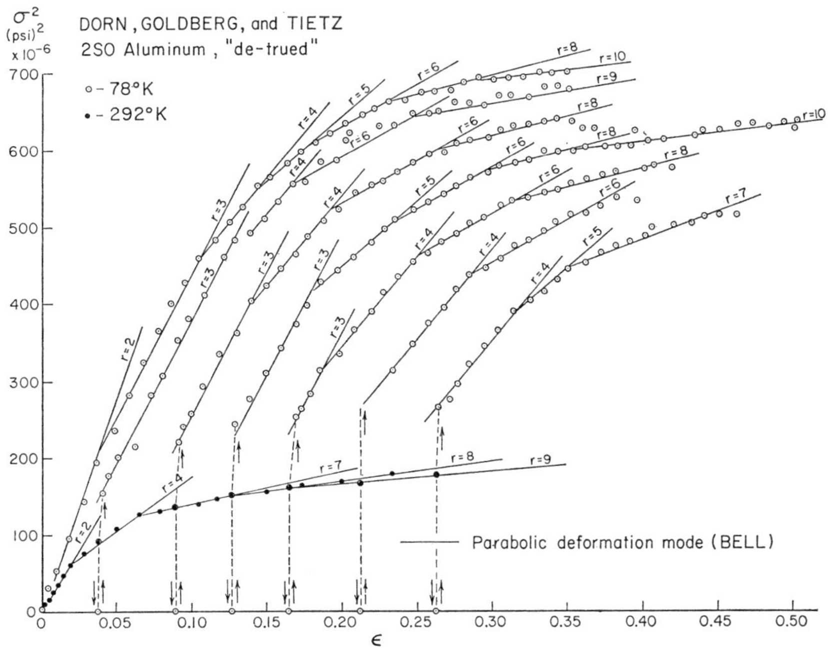

Fig. 4.217. σ^2 vs ε plot of "de-trued" data of Fig. 4.216, showing the agreement in all instances with the parabolic generalization, Eq. (4.25). The pre-stress was at 292° K.

The lowest and highest responses in Fig. 4.217 are for tests performed without intermediate unloading, at 292 and 78° K respectively. A comparison of Figs. 4.216

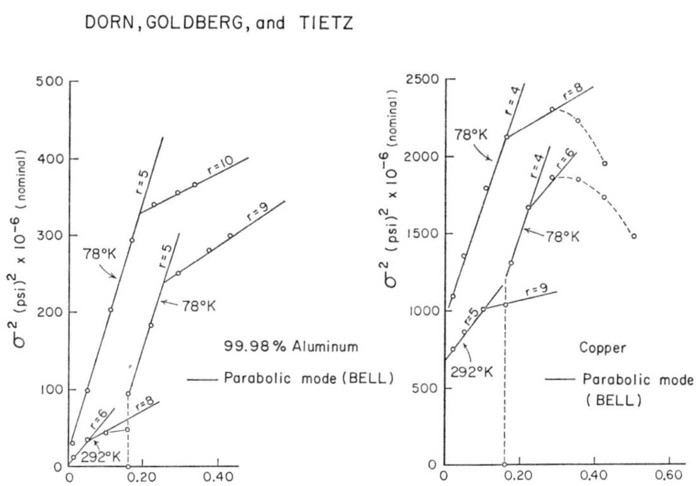

Fig. 4.218. σ^2 vs ε plots of tensile tests in high purity aluminum and in annealed copper, which I have recalculated to nominal stress and strain. The initial loading was at 292° K, followed by reloading at 78° K. These data are compared with results obtained from Eq. (4.25) [solid lines] for the indicated mode indices.

and 4.217 once again illustrates how the manner of presenting experimental results influences the interpretation.

The specimen of Fig. 4.217 first was loaded at 292° K and then reloaded at 78° K. In Fig. 4.218 are shown σ^2 vs ε plots from the data of DORN, GOLDBERG, and TIETZ, which I have recalculated in nominal stress and nominal strain. In these tests for high purity aluminum and for copper, the initial loading also was at 292° K and the reloading was at 78° K. A comparison with the parabolic response function, Eq. (4.25), reveals that similar conclusions may be drawn for all three solids.

I have omitted the results of DORN, GOLDBERG, and TIETZ for intermediate temperatures. Also omitted are their similar results from tests on brass and on stainless steel. Calculation in nominal stress and strain, which I performed for all low purity aluminum tests and for those on high purity aluminum and copper, gave results in close agreement with Eq. (4.25) whatever had been the imposed thermal and mechanical history.

4.34. A brief summary of experiments after 1960 describing additional aspects of propagation of waves of finite amplitude in crystalline solids.

Experimental research on waves of finite amplitude during the past decade has extended into many more aspects of this subject than it is possible to include in the present chapter. I shall give a few examples of measurements drawn from that research to illustrate certain features, such as the maximum stress during wave propagation, the nature of tension waves at finite strain, some of the evidence related to the transition structure at high strain rates, more recent work on incremental waves, and the exploration of the complex development of the finite strain wave close to impact surfaces.

In his doctoral dissertation in 1961 GORDON LUTHER FILBEY[1] studied the formation of plastic waves for very high impact velocities. I shall omit his description of diffraction grating measurements of strain rates as high as 7×10^4 sec^{-1}, but I have included in Fig. 4.219 the results of a measurement of stress in a symmetrical free flight test on bars of fully annealed aluminum, which he accomplished by determining the voltage change from a 0.005 in. thick piezo crystal attached to the struck specimen.

The maximum stress for a velocity of the impact specimen of 14200 cm/sec (maximum particle velocity after impact: 7100 cm/sec) predicted from the parabolic response function of Eq. (4.54) was 19200 psi, which is seen to be in close agreement with measurement. The calculated maximum strain of the wave front for this stress was nearly 12%. FILBEY discovered the initial high peak stress shown, which became the subject of much study later.[2]

A somewhat similar investigation at much lower impact velocities was described by me[3] in 1968. A hard aluminum bar of length L_2, in which was propagating an elastic wave of known magnitude produced by the impact of a second hard bar of length L_1, had a lubricated interface with a fully annealed aluminum bar. Electric resistance measurements of the incident wave σ_I and of the maximum stress σ_T after the passage of the reflected wave in the hard bar of length L_3,

[1] FILBEY [1961, *1*].
[2] See, for example, BELL [1962, *1*], BELL and SUCKLING [1962, *6*]; BELL [1968, *1*].
[3] BELL [1968, *2*].

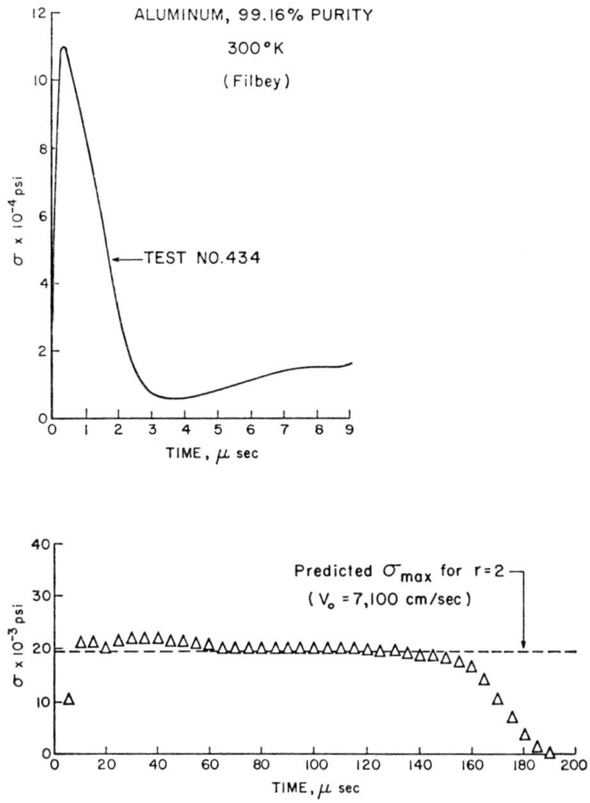

Fig. 4.219. FILBEY's piezo crystal determination of stress in high velocity impact, showing the initial peak stress from the first microsecond and the subsequent maximum stress of BELL's parabolic stress-strain function.

Fig. 4.220, were compared with diffraction grating measurements in the soft bar in terms of a one dimensional analysis based upon the use of HOOKE's law in the hard bar and the parabolic law of Eq. (4.54) in the soft bar. The comparison of measurement and calculation revealed that both the maximum stress and the maximum strain indeed were in accord with prediction based upon the general response function, Eq. (4.54).

Still another type of experiment related to the measurement of maximum stress was the repetition of the load bar test of JOHNSON, WOOD, and CLARK[1] and of J. D. CAMPBELL[2] of 1953 (see Sect. 4.27), but now accompanied by an actual measurement of strain in the soft bar, as is shown for example in the experiment of Fig. 4.221 from the doctoral research of HARTMAN[3] in 1967 for a diffraction grating measurement of strain on an annealed brass specimen struck by a hard brass bar. Again, an analysis of the boundary condition based upon the parabolic response function, Eq. (4.54) revealed a close agreement between experiment and prediction in this nonsymmetrical impact.

[1] JOHNSON, WOOD, and CLARK [1953, *1*].
[2] J. D. CAMPBELL [1953, *1*].
[3] HARTMAN [1967, *1*].

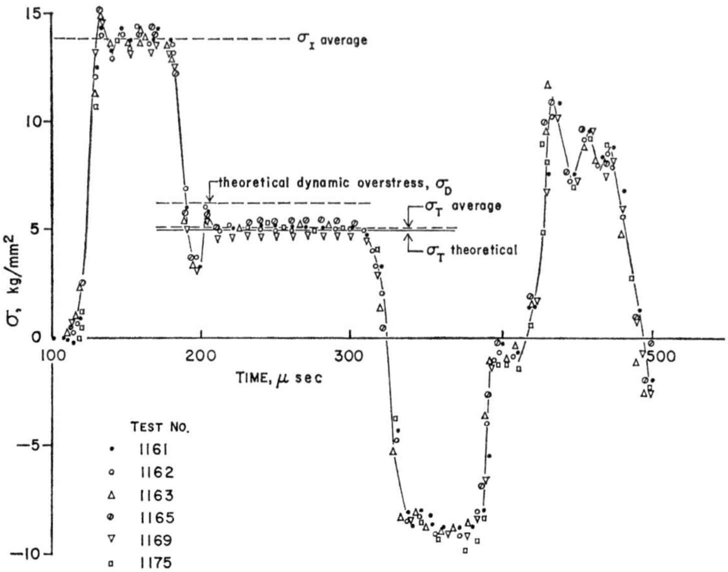

Fig. 4.220. The hard bar stress-time histories for 6 experiments by Bell in 1968. $L_1 = 20$ in., $L_2 = 30$ in., and $L_3 = 10$ in.

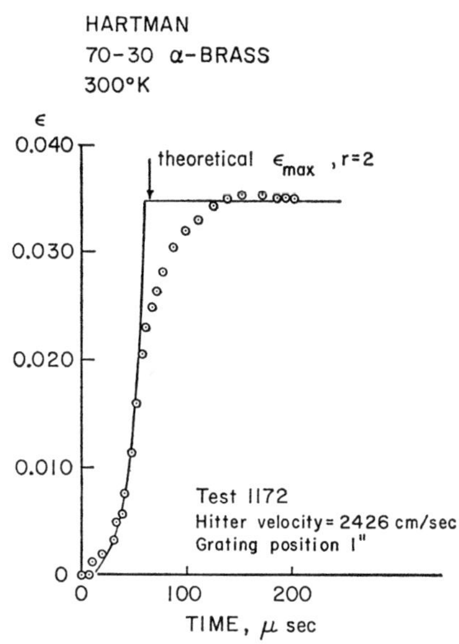

Fig. 4.221. Diffraction grating strain vs time profile (circles) compared with prediction from Eq. (4.54) [solid line] for a 30.5 cm long, hard, α-brass projectile specimen striking an annealed 25.4 cm long α-brass polycrystalline specimen.

Since 1959, I have performed a large number of such tests on different solids.[1] From these I have selected for illustration two experiments on high purity

[1] Bell [1960, *3*], [1961, *3, 4*], [1963, *2*], [1967, *2*] [1968, *1*], [1969, *1*].

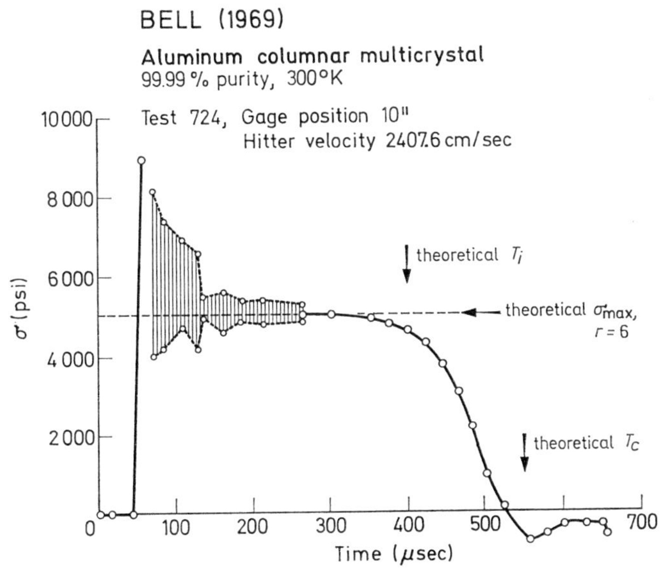

Fig. 4.222. Stress vs time data in the hard bar (circles) compared with prediction from Eq. (4.54) [dashed line], for an experiment in which a high-purity, annealed aluminum polycrystal strikes an aluminum hard bar.

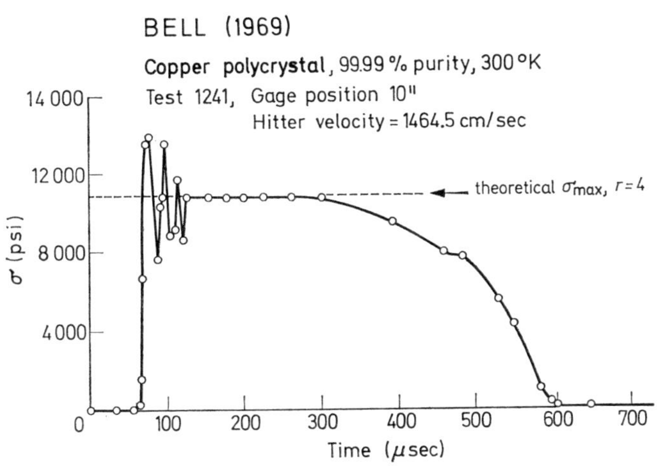

Fig. 4.223. Stress vs time data in the hard bar (circles) compared with prediction from Eq. (4.54) [dashed line], for an experiment in which a high purity, annealed copper polycrystal strikes a hard copper bar.

aluminum and copper, shown in Figs. 4.222 and 4.223, in which fully annealed specimens struck hard bars of the same solids on which electric resistance measurements revealed that the stress for both were in close agreement with the parabolic response function of Eq. (4.54).

One very important study of this type was that of WILLIAM J. GILLICH[1] in his doctoral research in 1964; by means of the diffraction grating experiment for which he measured strain on large single crystals of known orientations during the propagation of waves of finite amplitude, he found that the wave profiles revealed that the parabolic response function for resolved shear, Eq. (4.24), could be extended to strain rates over a million times higher than the slow loading tests from which I had discovered it.

One of GILLICH's other measurements, in which an aluminum single crystal of known orientation struck a hard polycrystalline bar, is shown in Fig. 4.224, for which a calculation of stress based upon the parabolic response function for the single crystal, Eq. (4.24), provided the close correlation between the predicted and measured stress.

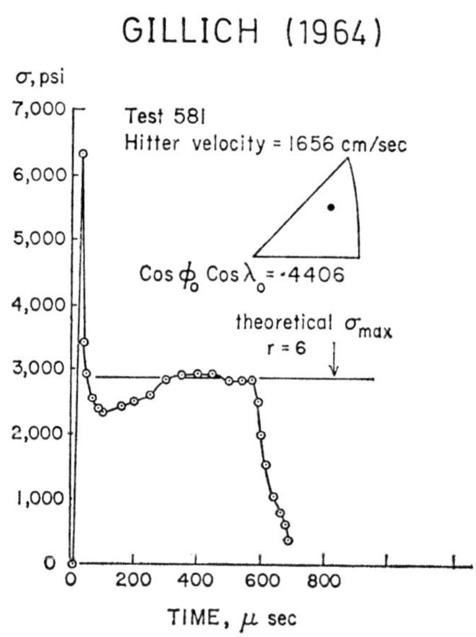

Fig. 4.224. Measured stress vs time history in a 6 ft. long hard polycrystalline struck specimen. The projectile specimen was a 10 in. long, high purity single crystal with the designated initial orientation and projectile velocity.

Most of the fully annealed polycrystalline solids for which wave propagation at finite strain has been studied, required a fairly radical change in their prior thermal and mechanical histories to alter the initial mode index r in Eq. (4.54) for the response function governing nonlinear wave propagation. An interesting exception was α brass, carefully studied by HARTMAN[2] in 1967. In every instance, diffraction grating wave profiles were in accord with the TAYLOR-VON KÁRMÁN theory, but the mode index r of the parabolic response function determined after that fact had been established, occurred in the distribution shown on the left of Fig. 4.225. The average of these experimental parabola coefficients for each group was compared with prediction from my Eq. (4.54). This sensitivity of the mode index r for α brass to small changes in prior history makes it an interesting solid for continued study.

[1] GILLICH [1964, *1*], [1967, *1*].
[2] HARTMAN [1967, *1*], [1969, *1*, *2*].

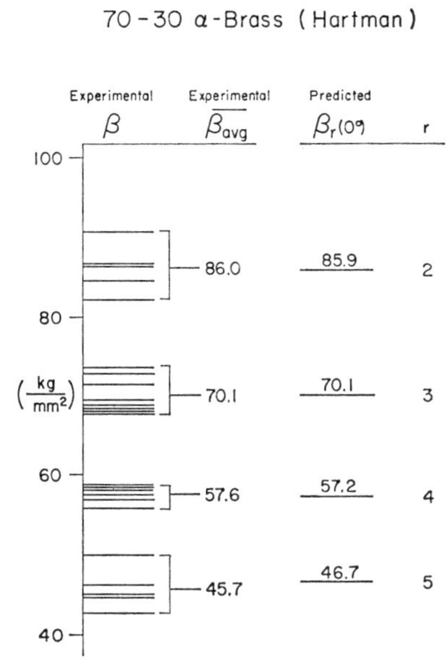

Fig. 4.225. Experimental parabola coefficients from diffraction grating wave speed determinations of HARTMAN in 70–30 α brass. The averaged values of each group are compared with prediction from BELL's parabolic generalization, for the mode index, r, shown.

The difficulty of introducing a wave front with finite strain in tension has been such that in over a thousand measurements for eighteen years after the diffraction grating technique was developed, all measurements were made in compression. In his doctoral research in 1972, AKHTAR SALAMAT KHAN[1] succeeded in developing an experiment which not only allowed for the diffraction grating measurement of finite tensile strain during wave propagation, but also permitted a simultaneous measurement of the maximum stress of the wave obtained in the same test.

The experiment, involving a hard load bar in which an elastic wave initially propagated to the interface of a soft bar, was the same in conception as the compression experiment on the elastic-plastic boundary, described above, Fig. 4.220. A compression wave was transmitted in the soft bar and a reflected wave in the hard bar. By machining the soft specimen so that one half was a hollow tube, KHAN obtained both a tension wave and a compression wave propagating in opposite directions from the interface with the hard bar. In Fig. 4.226 are shown a diagram of the experimental arrangement of KHAN and his measurement, by electric resistance gages, of incident and reflected waves in the hard bar, i.e., σ_I and σ_R respectively. In Fig. 4.227 are shown wave profiles of finite strain he measured by diffraction gratings at two positions in the tension section of the bar and at two positions in the compression section.

From his study of wave profiles for finite strains at known particle velocities, KHAN first established that the nonlinear theory of TAYLOR and VON KÁRMÁN applied to tensile waves. KHAN then was able to determine the governing response function. He found that this response function was the very one that I had found

[1] KHAN [1972, *1*].

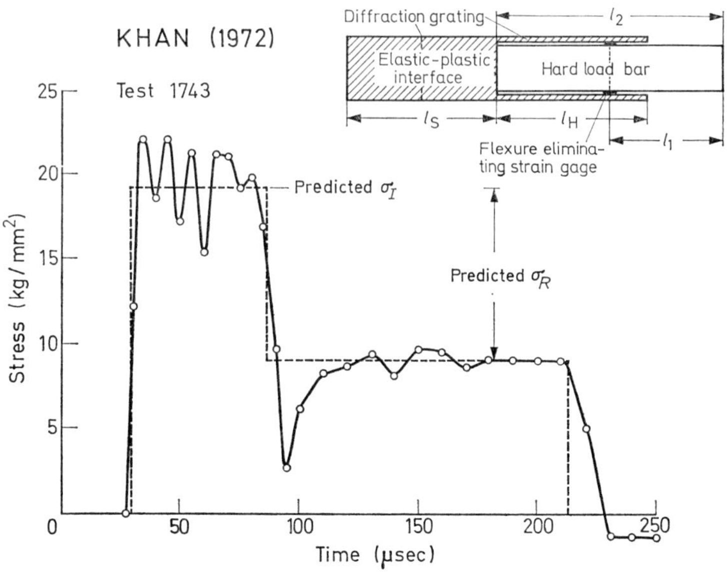

Fig. 4.226. Diagram of experiment and comparison of stress-time in hard bar with prediction from Eq. (4.54).

for compression waves, i.e. Eq. (4.54), Sect. 4.28. The measured and predicted arrival times of the wave fronts in tension as well as in compression were closely given by this same response function, as were the measured maximum strains in each instance and the maximum stress of the reflected wave in the hard bar shown in Fig. 4.226.

Thus, as for quasi-static measurement, when the stress and strain are chosen with reference to the undeformed solid (i.e., nominal stress, nominal strain), the detail of tensile wave fronts is given by the same response function as that of the compressive wave front.

All of the incremental wave experiments which I described in Sect. 4.27 above, from the time of my original experiments in 1951 to those of ALTER and CURTIS in 1956, were performed before the obtaining of wave profiles by the diffraction grating method had made possible the definitive experimental study of the propagation of waves of finite amplitude. BIANCHI[1] in 1963 and BELL and STEIN[2] in 1962 again performed experiments on incremental waves, but now with the background of new experimental techniques and theory not available during the earlier years.

BIANCHI modified the experiment on long copper ribbons which he had reported in his Master's essay[3] in 1953. This time he used the magnetic induction technique to measure particle velocity at different positions along the ribbon. Despite the distortions in the initial wave profile from BIANCHI's method of producing finite strain through impact on two connected ribbons, except for the small initial front which propagated at the same higher velocity whatever had

[1] BIANCHI [1963, *1*]. See also [1960, *1*, *2*].
[2] BELL and STEIN [1962, *7*].
[3] BIANCHI [1953, *1*].

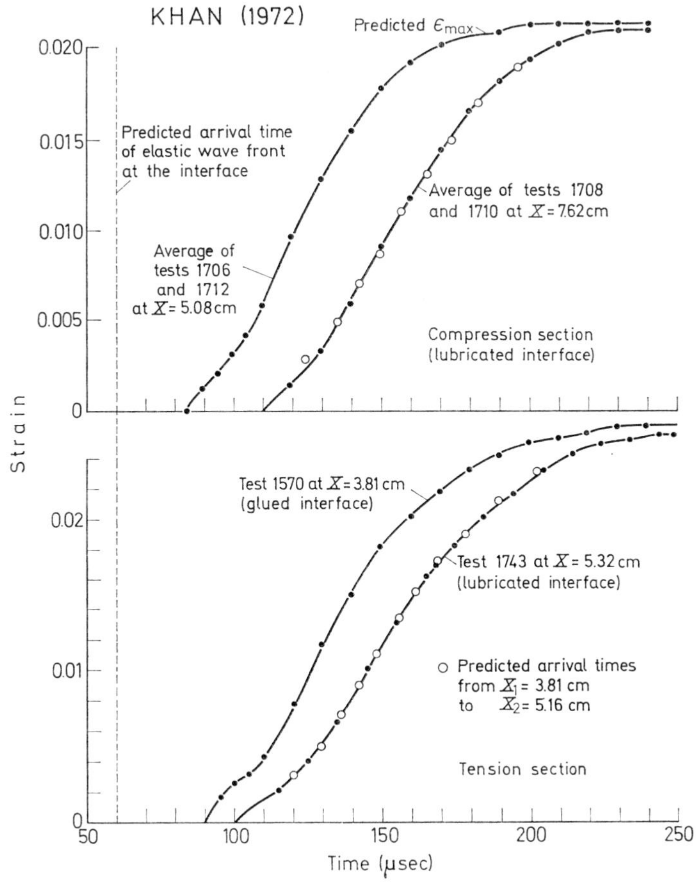

Fig. 4.227. Finite strain profiles measured by diffraction gratings in the tension and compression section of the specimen of Fig. 4.226, compared with prediction from Eq. (4.54).

been the quasi-static pre-stress, once the waves were generated in what he referred to as the asymptotic region, he found that they propagated in agreement with the nonlinear theory of TAYLOR[1] and VON KÁRMÁN.[2]

The experiments of BELL and STEIN in 1962 were based upon the incremental wave experiment in 1956 of ALTER and CURTIS[3] in which a dynamic pre-stress from the impact of a hard bar with the soft specimen was followed, after a few microseconds, by an incremental increase in load produced by an abrupt change of cross-sectional area at some point on the hard bar. However, now it was possible to use diffraction gratings to study the profile of the incremental wave. Fully annealed hot rolled aluminum was the solid studied. In Fig. 4.228 is shown a diagram of the specimens used. Measurements by means of electric resistance gages in the hard bar provided the correlation between experimental results and

[1] TAYLOR [1942, *1*].

[2] VON KÁRMÁN [1942, *1*].

[3] ALTER and CURTIS [1956, *1*].

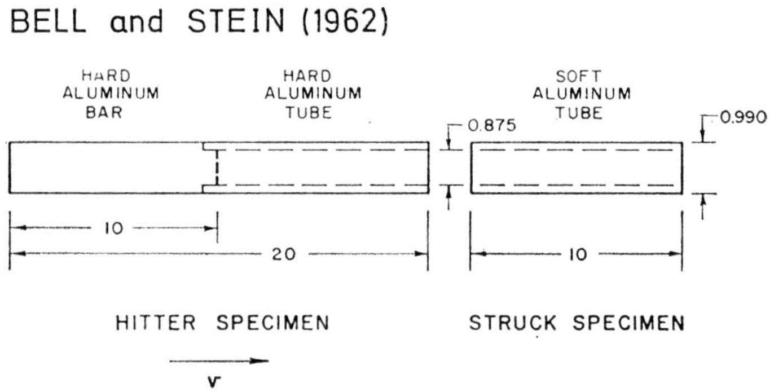

Fig. 4.228. The composite hard aluminum hitter specimen and the soft aluminum tubular struck specimen. The diffraction grating is located on the soft aluminum struck specimen.

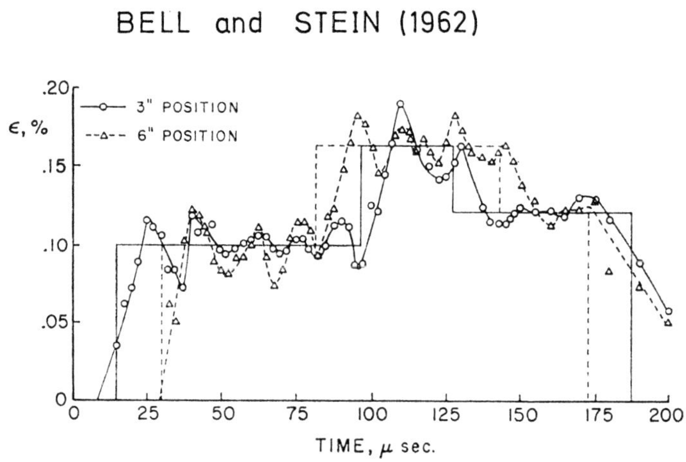

Fig. 4.229. Strain-time experimental data in the composite elastic aluminum hitter specimen compared with the computed results at 3 and 6 in. from the impact face. The impact was with a soft aluminum tubular struck specimen.

prediction at a position 3 in. (solid line) from the impact face and at 6 in. (dashed line). For a hard bar, in which the wave is governed by linear elasticity, the stress is proportional to strain. One sees that the magnitude of the applied increment of stress was only slightly less for the measurement of the test shown in Fig. 4.229 than that of the original dynamic pre-stress.

An illustrative measurement[1] of the incremental portion of the strain profiles, shown in Fig. 4.230, obtained by the use of diffraction gratings at two positions in the fully annealed aluminum bar, revealed that only the very small initial portion of the wave front traveled at velocities of the magnitude of those in linear elasticity, whereas large strains of the incremental wave had wave speeds commensurate with prediction[2] for the slower moving plastic wave.

[1] BELL and STEIN [1962, 7].
[2] Eq. (4.54), Sect. 4.28.

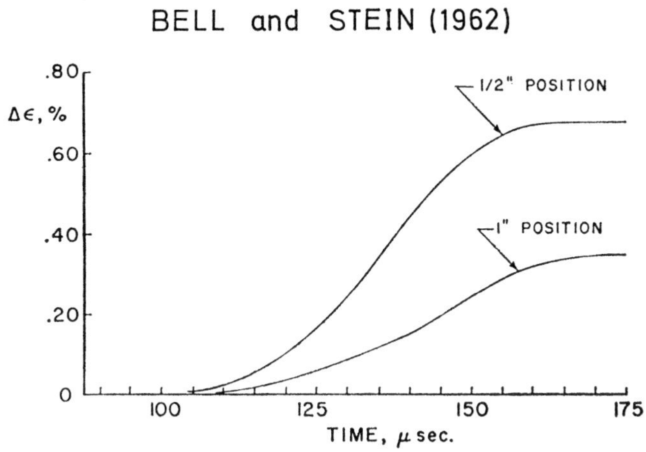

Fig. 4.230. A comparison of typical diffraction grating measurements of incremental waves at indicated positions for a dynamic plastic pre-stress of 4.50 kg/mm².

From a large number of measurements, many of them still unpublished, I found that all of the details beyond the initial linear elastic precursor were given from the parabolic response function, Eq. (4.54). My conjecture that the high speed of the incremental waves was associated with the SAVART-MASSON (PORTEVIN-LE CHATELIER) effect[1] even though the pre-stress had been constant for times as short as 10 μsec, has been demonstrated in great detail. The initial wave front of high speed was limited to the known vertical height of a SAVART-MASSON step and was followed by an incremental strain step, after which the plastic wave speeds were the same as those obtained when no incremental wave was present. In Fig. 4.231 is shown one of many tests from this later study, also in fully annealed aluminum, in which the surface angle vs time and finite strain vs time profiles of the dynamic pre-stress and the incremental wave obtained by the diffraction grating technique were compared with calculation based upon the parabolic response function, Eq. (4.54). I have omitted the measurement of stress in the hard bar which also provided a detailed correlation between experimental results and prediction from the nonlinear wave theory as stated by TAYLOR and VON KÁRMÁN.

In dynamic plasticity, the study of the growth of wave fronts near impacting surfaces has received almost as much attention in experiment as has the study of the loading and unloading waves. Hundreds of experiments attacking this difficult problem from many very different perspectives have shown that to a distance from one half to one diameter beyond the impact face, the problem can be considered only in three dimensional terms.

In a number of papers during the past twelve years I have included[2] the results of various experiments which led to a fairly complete description of the complex detail of wave growth and reflection from the side walls of the cylinder: there was a two wave structure involving an equipartition of energy between the first wave, a nearly infinite step, and a second wave front which developed more slowly from the reflection at the free surface. Instead of describing the many

[1] See Sect. 4.31 above.

[2] BELL [1960, *4*], [1961, *1*], [1962, *1*, *6*], [1963, *2*], [1965, *1*], [1967, *2*].

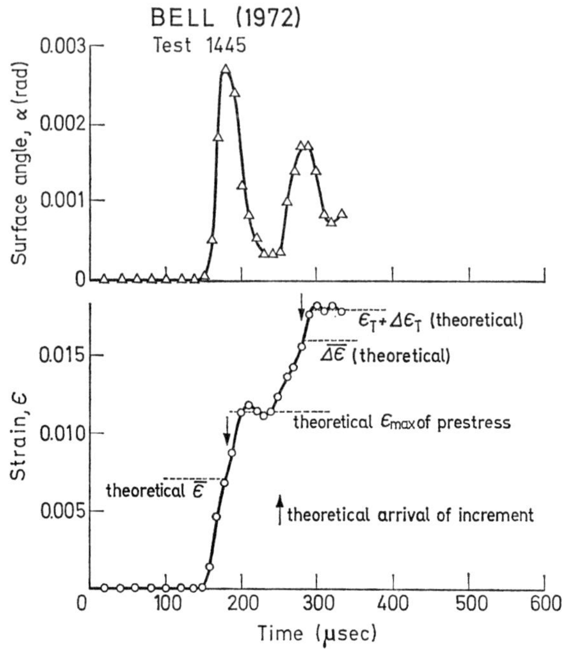

Fig. 4.231. The strain and surface angle measured by diffraction gratings on a solid fully annealed aluminum rod, compared with prediction for an incremental wave using Eq. (4.54) for both the dynamic prestress and the increment.

types of experiments, I shall limit this discussion to a single series. With the exception of this one, the remaining experimental studies have not as yet led to a successful method of establishing mixed response functions capable of accounting for all the details observed.

The series of experiments I include here, consisted in the determination of traverse times for initial wave fronts, obtained from the average of many strain profiles from diffraction gratings located at a distance of $^1/_4$ diameter beyond the impact face in a symmetrical collision of two identical, fully annealed aluminum cylinders, and a similar average of measurements at $^1/_2$ diameter. These results[1] were compared with wave speeds given by TRUESDELL[2] in 1961 in his "General and Exact Theory of Waves in Finite Elastic Strain."

JERALD LAVERNE ERICKSEN pointed out that TRUESDELL's wave speeds for principal waves in an isotropic solid "assume a simpler form if expressed in terms of the relation giving the principal stresses ... as functions of the principal stretches."[3] Experiment and theory could be compared for longitudinal and transverse wave speeds, $U_{\|}$ and $U_{\perp}$, respectively, for waves propagating along a principal axis. For hydrostatic stress, TRUESDELL's results reduced to a particularly simple form. From a knowledge of the wave speeds as given by my measurement from the traverse times, shown in Fig. 4.232, between $^1/_4$ and $^1/_2$ diameters from the impact face, a knowledge of the strain ε and mass density ϱ was sufficient to calculate

[1] BELL [1962, *1*].
[2] TRUESDELL [1961, *1*].
[3] See TRUESDELL, *ibid.*, p. 275, footnote 7.

a stress vs time history from TRUESDELL's equations. The results of such a calculation (circles in Fig. 4.233) based upon the data in Fig. 4.232 show that the development of the peak stress and its subsequent collapse described in terms of these theoretical wave speeds, were in close agreement with a piezo crystal measurement (triangles and squares) of the stress vs time history at the impact face itself for the same type of experiment.

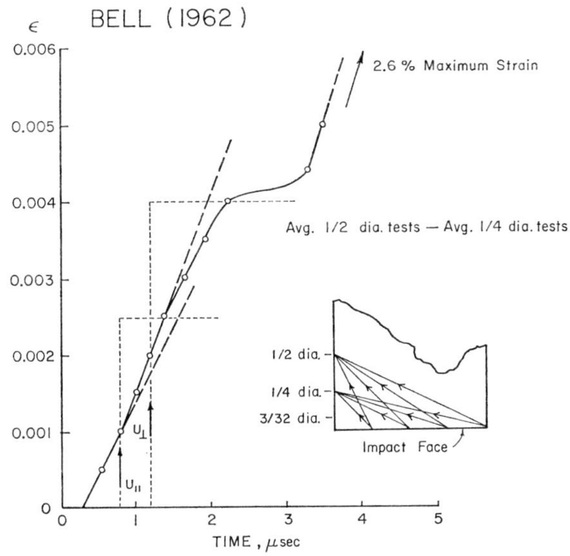

Fig. 4.232. Initial part of the strain-time curve obtained from subtracting arrival times for averaged $^1/_4$ diameter data from averaged $^1/_2$ diameter data. The lower dashed step is the hypothetical shock front. The upper dashed step is that of the elementary bar theory with a wave speed $c_0 = \sqrt{E/\varrho_0}$.

The delay of approximately 5 μsec between the calculated stress vs time histories from the diffraction grating measurement, and the direct measurement of stress vs time histories at the impact face by piezo crystals, arises from the difference in the location of these two types of measurement. Even though the stress levels do not exceed 35 kg/mm², the measured dilatational wave speed of 8128 cm/sec during this first few microseconds of wave initiation exceeded, as TRUESDELL had predicted, the value of 6350 cm/sec obtained from elementary theory. Beyond the immediate vicinity of impact, except for the very smallest strain, wave speeds higher than those given in elementary linear elasticity from ultrasonic measurement or quasi-static tests were not found.

Since TRUESDELL in his theory considered weak waves propagating in an elastic solid arbitrarily deformed, whereas my experiments were for wave fronts of large amplitude propagating in an initially stress-free polycrystalline aluminum, the correlations I have made are purely empirical. In referring the interested reader to my papers describing experiments on wave initiation and growth following the collision of solids, I emphasize that in my judgment this area of study will prove to be one of the most fruitful for continued experimental and theoretical research.

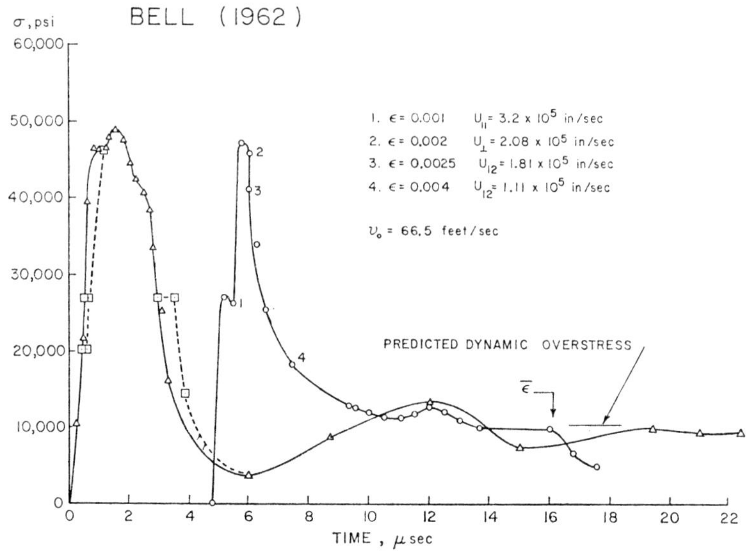

Fig. 4.233. The calculated stress-time behavior at $^3/_8$ diameter (circles) compared with two piezo crystal measurements at the impact face (triangles and squares). $\bar{\varepsilon}$ is the strain of the mean energy from the parabolic stress-strain law, also given experimentally from surface angle data.

I chose the few illustrations in this section to indicate, once again, that numerous diverse studies may quickly follow the discovery of new techniques of measurement. Research on these various aspects of propagation of waves of finite amplitude currently is in a stage of such rapid development that a critical appraisal at this point soon would require revision.

4.35. On experiments leading toward a general theory of plasticity for the loading response of annealed crystalline solids.

In the spring of 1971 in Bologna, while I was summarizing for an earlier draft of this treatise the details of experimental research on the finite deformation of annealed crystalline solids, a review and comparison of all the known experimental facts led me to formulate a general theory of plasticity, which since that time has been fully developed as an internally consistent statement in rational mechanics.[1] The experimental data include the measurement of response at finite strain for principal stress ratios from -1 to $+1$, but do not yet include measurements for finite strain at large rotation nor measurements for which all three principal stress components differ from zero. Therefore, it is convenient here to present these experimental results in terms of approximations within the general

[1] Bell [1972, 2]. This previously unpublished research was part of a program of study carried out under the sponsorship of the United States Air Force Office of Scientific Research, as were portions of the earlier research in my laboratory, described in Sects. 4.17 to 4.22 and 4.27 to 4.34, which led to the discovery of this generalization. Other parts of the research described in those sections, including the very early work of the 1950's and the initial use of my diffraction grating technique, were early and long sponsored by the United States Army Ballistics Research Laboratories, Aberdeen Proving Ground, and by the U.S. Army Research Office, Durham.

theory, which apply for finite strain with relatively small rotations, and to note that at least one principal stress component is zero.

The theory which is intended to describe the more general *loading* response of fully annealed crystalline solids is formally a statement in nonlinear elasticity. In terms of the principal stretches $\lambda_i = 1 + E_i$, the strain energy W is given by

$$W = \tfrac{2}{3}\beta\left[(E_1 - E_2)^2 + (E_2 - E_3)^2 + (E_3 - E_1)^2\right]^{\frac{3}{4}} = \tfrac{2}{3}\tau\gamma, \tag{4.70}$$

subject to the following constraint: $\sum E_i = 0$ during deformation. In the present context, with large strain at relatively small rotation, one may introduce the approximation

$$T \cong \frac{1}{\sqrt{3}}\left[\left(\frac{\partial W}{\partial E_1} - \frac{\partial W}{\partial E_2}\right)^2 + \left(\frac{\partial W}{\partial E_2} - \frac{\partial W}{\partial E_1}\right)^2 + \left(\frac{\partial W}{\partial E_3} - \frac{\partial W}{\partial E_1}\right)^2\right]^{\frac{1}{2}}. \tag{4.71}$$

For simple tension, for simple torsion, and for combined tension and torsion experiments on hollow tubes, Eq. (4.71) is given in terms of the axial stress σ and the torsion shear stress S, both referred to the undeformed configuration, as

$$T \cong \sqrt{2}\sqrt{\frac{\sigma^2}{3} + S^2}\,. \tag{4.72}$$

For experiments combining internal pressure with axial tension in hollow tubes, and for experiments in double compression, Eq. (4.71) is given in terms of the two non-zero principal stresses, σ_1 and σ_2, also referred to the undeformed configuration, as

$$T \cong \sqrt{\frac{2}{3}}\sqrt{\sigma_1^2 + \sigma_2^2 - \sigma_1\sigma_2}\,. \tag{4.73}$$

In accord with the precept that the best plausible hypotheses follow, rather than precede, experimental observation, again it must be noted that irrespective of whether one or more non-zero stress components are considered, the constant ratio between the stresses and strains of the aggregate and those of the single crystal are obtained only when all these stresses and strains are referred to the undeformed reference state of the fully annealed solid. The importance of this reference point had been established earlier, when I found that unorthodox annealing recipes produced a new genesis in which the memory of prior thermal and deformation history was minimized.

First, we may note the pertinent form of the response function[1] for the single crystal discussed in Sects. 4.17 and 4.18 of this treatise, namely that the resolved shear deformation is given by

$$\tau = \beta_s \gamma^{\frac{1}{2}}, \qquad \text{(4.62) repeated}$$

where

$$\beta_s = \left(\tfrac{2}{3}\right)^{r/2} \mu(0)\,\frac{B_0}{m^{\frac{3}{2}}}\,(1 - T/T_m)\,.$$

Note that in this equation, T is the ambient temperature and T_m is the melting point of the solid.

From the experiments of E. A. DAVIS[2] in 1943, Sect. 4.15 above, for radial loading paths with two non-zero principal stress components having ratios from zero to one in annealed polycrystalline copper, and from the experiments of MITTAL[3] in 1969, Sect. 4.22, for many radial and non-radial loading paths in

[1] BELL [1961, *1*], [1963, *1*], [1964, *1*], [1965, *2*], [1968, *1*].
[2] E. A. DAVIS [1943, *1*].
[3] MITTAL [1969, *1*], [1971, *1*].

which tension and torsion loading of fully annealed aluminum were applied simultaneously, we see that when the stress components of Eqs. (4.73) and (4.72) are referred to the undeformed solid, and the measured strain components likewise are defined with respect to the undeformed solid, then the general response function is parabolic, independent of loading path and with parabola coefficients and linear temperature dependence directly given by experimentally determined aggregate ratios (Sect. 4.32) which I had established a decade ago for the simple loading components. [Mittal's tests were uncoupled, in that the dead weight tension load did not have to rotate when the dead weight torsion load was applied.]

It is of equal importance, as will be seen below, that for isotropic solids, including the various combinations of stress components, the Savart-Masson steps may be superposed on this general response function. The vertical departure in the rising incremental stress portion of the step at nearly constant strain is followed by the incremental strain portion at nearly constant stress, which returns precisely to the general parabolic response function. This is invariably the case for the combination of incremental strain components. That it is not necessarily so for the ratios of the individual components is an important new discovery affecting any consideration of constitutive equations governing finite strain in crystals.

For simple loading in tension, compression, or torsion, it was shown in Sect. 4.32 that from experiment, for nominal stress and strain Eq. (4.60) for tension and compression, and Eq. (4.61) for torsion, were applicable.

$$\frac{\sigma}{\tau}=\bar{m}=\frac{\gamma}{c} \quad \text{where } \bar{m}=3.06, \qquad \text{(4.60) repeated}$$

$$\frac{S}{\tau}=\bar{n}=\frac{\gamma}{s} \quad \text{where } \bar{n}=1.765, \qquad \text{(4.61) repeated}$$

Combining the experimentally observed response functions from previous simple loading studies with the new results for more complex situations, and introducing Γ as

$$\Gamma=\frac{1}{\sqrt{3}}\left[(E_1-E_2)^2+(E_2-E_3)^2+(E_3-E_1)^2\right]^{\frac{1}{2}}$$

which for combined tension and torsion has the form $\Gamma=\frac{1}{\sqrt{2}}\sqrt{3\,\varepsilon^2+s^2}$, furnishes for the aggregate ratios for single crystal and polycrystal obtained from all types of experiment the following expression:

$$\frac{T}{\tau}=\bar{k}=\frac{\gamma}{\Gamma}, \quad \text{where } \bar{k}=2.50. \tag{4.74}$$

We note that

$$\bar{m}=\sqrt{\tfrac{3}{2}}\,\bar{k} \quad \text{and} \quad \bar{n}=\frac{\bar{k}}{\sqrt{2}}.$$

From Eqs. (4.62) and (4.74) we have

$$T=\bar{k}^{\frac{3}{2}}\beta_s\,\Gamma^{\frac{1}{2}}. \tag{4.75}$$

In Fig. 4.234 I have replotted the tension, compression, and torsion tests of Fig. 4.204 in Sect. 4.32 in terms of T and Γ to illustrate the correspondence be-

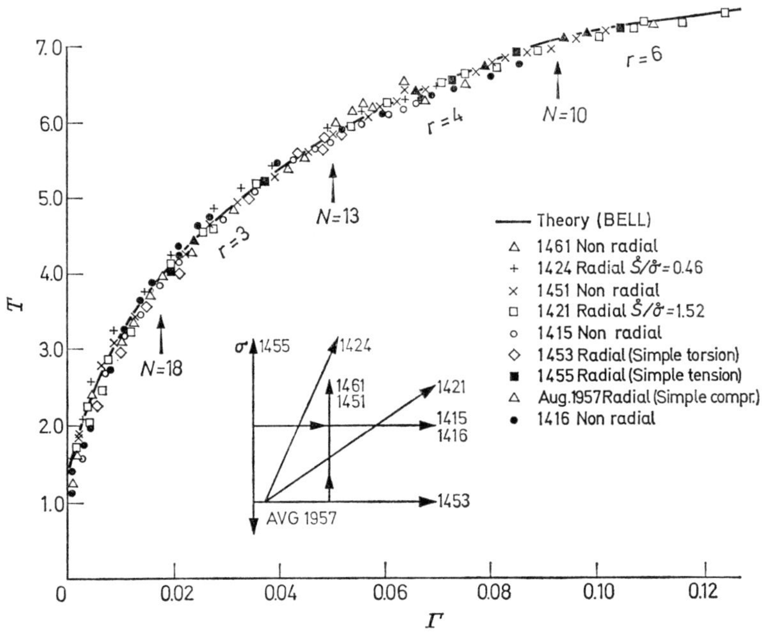

Fig. 4.234. Radial and non-radial tension-torsion tests in fully annealed aluminum tube, compared with predicted T vs Γ from Eq. (4.75). T and Γ are defined by Eqs. (4.70) and (4.71). The initial mode index is $r=2$.

tween experiment and prediction.[1] The experiments of DAVIS in 1943 for various ratios of principal stress were in annealed copper, and those of MITTAL for simultaneous tension and torsion loading were in fully annealed aluminum.

In my analysis of single crystal and polycrystalline response functions, summarized in 1968, the fact that second order transitions occur at a fixed sequence of finite strains was demonstrated amply. The values of γ_N, ε_N, and s_N (see Sect. 4.32 above) were found to be related through the aggregate ratios of Eq. (4.69). More detail from subsequent study was given in a recent paper.[2] In terms of Γ, the observed series of fixed transition strains were given by

$$\Gamma_N = \frac{1}{\sqrt{2}}\left(\frac{2}{3}\right)^{N/2} \tag{4.76}$$

where $N=0, 2, 4, 6, 8, 10, 13, 18$.

From the strain ratio of Eq. (4.74) we have, for simple loading,

$$\gamma_N = \bar{n}\left(\frac{2}{3}\right)^{N/2}, \quad \varepsilon_N = \frac{\bar{n}}{\bar{m}}\left(\frac{2}{3}\right)^{N/2}, \quad s_N = \left(\frac{2}{3}\right)^{N/2}. \tag{4.76a}$$

[1] In describing his experimental results, MITTAL [1969, *1*], [1971, *1*], as I noted above, omitted a small contribution to the torsion component from the weight of the empty buckets into which water flowed at constant rates. While this small correction affected none of MITTAL's conclusions, I have included this correction in recalculating from those tests the Γ vs T plots of Fig. 4.234.

[2] BELL [1971, *1*].

It was from the observation of resolved shear in the single crystal and axial loading and simple torsion response functions in the polycrystal, all given by Eqs. (4.76a), that the general result of Eq. (4.76) was obtained.

MEIR FLORENZ[1] in a Master's essay in 1969 described a series of over 40 tests on cubes of fully annealed aluminum using a modified version of BRIDGMAN's[2] double compression experiment (Sect. 4.16 above). After establishing that careful measurement with lubricated interfaces in uniaxial compression of equal sided cubes also were described by the parabolic response function, Eq. (4.25), for slow loading, both for constant stress rate and for incremental dead weight loading, FLORENZ obtained a parabolic response function for the tests having two principal stress components. In examining the parabola coefficient FLORENZ obtained from the average of his many tests, I found that the response function was in close accord with Eq. (4.75) and with Eqs. (4.78), below.[3]

While compiling and analyzing more complex loading experiments from my laboratory and from the literature of the past few decades for this treatise on experimental foundations, I came to see that when plastic response functions are expressed in nominal form, as above, the ratios of individual stress and strain components for the radial loading of annealed aluminum, copper, and mild steel satisfy the relation

$$\frac{E_i}{P_i} = \frac{\Gamma}{T}, \quad \text{where } P_i = \sigma_1 - \frac{\sigma_1 + \sigma_2}{3}$$

for tension with internal pressure and for double compression;

and (4.77)

$$\frac{\varepsilon}{\frac{2}{3}\sigma} = \frac{s}{2S} = \frac{\Gamma}{T},$$

for tension with torsion, where the constraint $\sum E_i = 0$ is imposed.

For large strain with relatively small rotations, since Eq. (4.75) applied in all instances, Eqs. (4.77) led to

$$E_i = \frac{T}{\bar{k}^3 \beta_s^2} P_i,$$ for tension with internal pressure and for double compression;

and (4.78)

$$\varepsilon = \frac{T}{\bar{k}^3 \beta_s^2} \frac{2}{3} \sigma$$

$$s = \frac{T}{\bar{k}^3 \beta_s^2} 2S,$$ for tension with torsion.

Eqs. (4.78) are compatible with over 2000 measurements of stress and strain for simple loading in 28 different annealed solids. As will be shown below, Eqs. (4.78) also describe the experimental results obtained in fully annealed aluminum for combined tension and torsion along some non-radial loading paths in which simple tension is followed by torsion at constant tension. A very recent[4] series of tension and torsion measurements on fully annealed copper and aluminium along non-radial loading paths chosen to provide a more severe test of

[1] FLORENZ [1969, *1*].

[2] BRIDGMAN [1949, *1*].

[3] BELL and FLORENZ [1972, *5*].

[4] This series of 50 tests were performed when this treatise was in page proof, and hence no detailed report of them can be included here.

the applicability of Eqs. (4.78), has shown that these equations are one of the integrated forms of modified incremental constitutive equations. The aggregate ratios and the loading surfaces still are given by Eqs. (4.74) and (4.75). Of course, for all *radial* loading paths, Eqs. (4.77) and (4.78) describe the response of fully annealed copper and aluminum.

Of as much interest as the result (4.78) for annealed metals, which has been developed from a theory based upon experiments which require the compressibility constraint $\sum E_i = 0$ and the obvious loading constraint $\Delta T \geqq 0$, are two new constraints themselves.

The "staircase" phenomenon for dead weight loading, or the SAVART-MASSON effect, was found to occur in the T vs Γ plots of loading surfaces, which provided the experimental basis for Eq. (4.75). This is true for simple loading or for radial and non-radial loading paths with more than one non-zero stress component. For non-simple loading, the $\Delta\Gamma$ incremental components corresponding to the ΔT of the SAVART-MASSON step are given by Eq. (4.79), since $\sum E_i = 0$.

$$\Delta\Gamma = \frac{E_1\,\Delta E_1 + E_2\,\Delta E_2 + E_3\,\Delta E_3}{\Gamma} + \frac{(\Delta E_1)^2 + (\Delta E_2)^2 + (\Delta E_3)^2 - (\Delta\Gamma)^2}{2\Gamma}. \tag{4.79}$$

My analysis of experiment for the response of annealed aluminum in finite strain led to the discovery not only that the constitutive Eqs. (4.78) apply, but also that their applicability to the physical situation was subject to the following constraint:

$$\frac{\Delta E_i}{E_i} = \frac{\Delta\Gamma}{\Gamma}, \quad \text{for tension with internal pressure and for double compression;}$$

and (4.80)

$$\frac{\Delta s}{s} = \frac{\Delta\varepsilon}{\varepsilon} = \frac{\Delta\Gamma}{\Gamma}, \quad \text{for tension with torsion.}$$

In fact, whether this constraint holds or not, the loading surfaces given by Eq. (4.75) still are applicable, and $\Delta\Gamma$ is given by a rearrangement of values in Eq. (4.79).

A second new constraint with respect to unloading was found by HAHNGUE MOON in his current doctoral research in my laboratory, namely that for the constitutive Eqs. (4.78) to hold during combined tension and torsion, one must exclude non-radial loading paths, for which principal deviatoric stress components undergo a reversal of sign, i.e.

$$\frac{\Delta P_i}{P_i} \geqq 0, \quad \text{where} \quad P_{1,2} = \frac{\sigma}{6} \pm \sqrt{\frac{\sigma^2}{4} + S^2}, \quad \text{and} \quad P_3 = -\frac{\sigma}{3}. \tag{4.81}$$

Here, as for the constraint of Eq. (4.80) from the SAVART-MASSON material instability, the response for the loading surface still is described by the parabola, Eq. (4.75), and the aggregate ratios still are given by Eq. (4.74), i.e., the loading surface for $\Delta T \geqq 0$ still is unaltered.

For simple radial loading with only one non-zero stress component, the conditions of Eq. (4.80) and (4.81) automatically are satisfied, and Eq. (4.75) and the only non-zero member of the constitutive Eqs. (4.78) coalesce in every instance, to provide an appropriate one dimensional response function.

The data of Fig. 4.234, and Figs. 4.61 and 4.62 of Sect. 4.15, demonstrate that for annealed aluminum and copper, the loading surfaces described by Eq. (4.75) indeed are independent of loading path, and have the predicted value of the coefficient β_s, which I earlier had discovered for simple loading.

Since $\mu(0)$ and the melting point T_m were known for annealed copper, it was possible to convert Eq. (4.75) in nominal stress and strain to "true" stress and "true" strain for comparison with the results E. A. DAVIS had obtained in 1943 for that solid. These comparisons for the various principal stress ratios were shown in Fig. 4.61 (crosses).

The parabola coefficient for copper was obtained from the same Eq. (4.75) as for the aluminum of Fig. 4.234. Thus annealed solids subject to various loading paths when there is more than one non-zero stress component conform with the experimental generalization found earlier for simple loading.[1]

In Fig. 4.235 is shown the T^2 vs Γ plot (circles) of the non-radial test 1451 in fully annealed aluminum which violated the loading constraint of Eq. (4.81) and thus, as will be seen in the ε vs s plot of Fig. 4.237, did not satisfy the constitutive Eqs. (4.78). Nevertheless, as in all other instances, the loading surface given by Eq. (4.75) (solid lines) still are in accordance with prediction. One notes further that the critical strains at which the second order transitions occur for $N=18$ and for $N=13$ still are where predicted from Eq. (4.76) (arrows).

In Fig. 4.236 for fully annealed aluminum is shown the radial loading test 1457 in which both the constraints of Eqs. (4.80) and (4.81) are satisfied so that in the ε vs s plot the experimental results (circles) were in close accord with prediction from the constitutive Eqs. (4.78) (solid line).

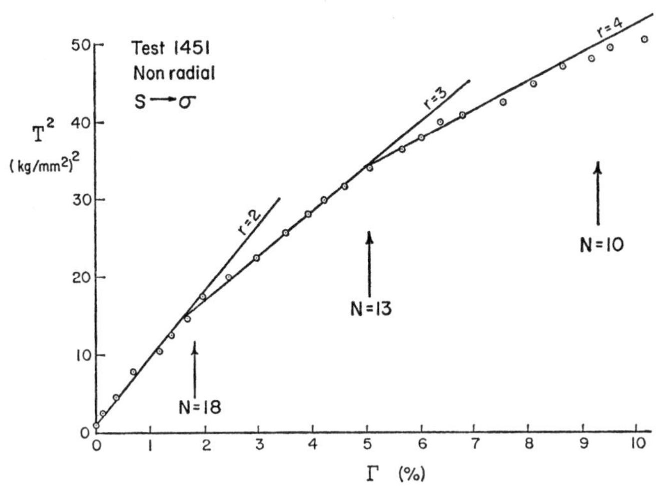

Fig. 4.235. A T^2 vs Γ plot of a non-radial loading test in fully annealed aluminum tube, for which the loading constraint of Eq. (4.81) is not satisfied. The experimental results [circles] are compared with the loading surface given by Eq. (4.75) [solid lines].

[1] BELL [1968, *1*].

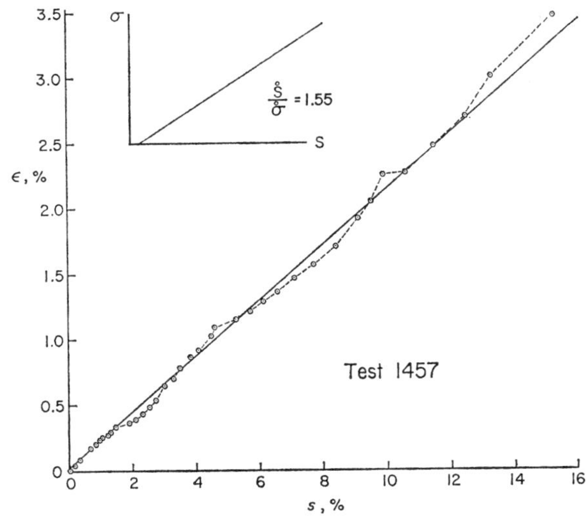

Fig. 4.236. The ε vs s plot for a radial loading test in fully annealed aluminum, compared with prediction from Eq. (4.78) (solid line).

For the non-radial loading paths of test 1416 and test 1451, also for annealed aluminum, shown in Fig. 4.234, both constraints of Eqs. (4.80) and (4.81) were satisfied for the former, while, as indicated above, the constraint of Eq. (4.81) was not satisfied for the latter. [For this loading path, Eq. (4.81) does not

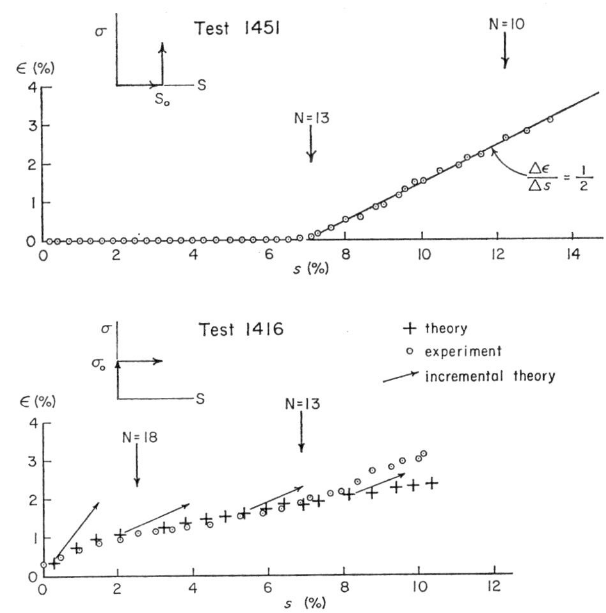

Fig. 4.237a and b. Non-radial loading tests. For test 1416, Eq. (4.81) is satisfied; and for test 1451, Eq. (4.81) is not satisfied.

apply until $\sigma \geqq \frac{3}{2} S$.] The ε vs s plots of Fig. 4.237 demonstrate for this type of non-radial loading path where loading in only one component is followed by loading in only the other component, that the prediction from the constitutive Eqs. (4.78) (crosses) agrees with the experimental data (circles) when both constraints, Eqs. (4.80) and (4.81), are satisfied (as is shown in test 1416), and does not agree when one or the other of these conditions is violated. For strains above 8% in test 1416 the constraint on the SAVART-MASSON effect, Eq. (4.80), does not hold, although Eq. (4.81) still is satisfied. For this non-radial loading test, I have added the arrows to indicate the slopes which would be expected if an incremental theory were applicable rather than the constitutive Eqs. (4.78).

The consequences of violating Eq. (4.81) currently are under intensive study in my laboratory. [This work has provided modified incremental constitutive equations which retain condition (4.80) but remove condition (4.81).] One notes from Fig. 4.237 for test 1451, in which Eq. (4.81) does not hold for this non-radial loading path, that one obtains a linear slope having the value of $^1/_2$. Experiments just completed have established that the components of strain do occur in definable terms when Eq. (4.81) does not hold.

In Fig. 4.238 is shown the non-radial loading path for a recent experiment in the decoupled tension-torsion of a thin-walled tube of fully annealed aluminum, the T^2 vs Γ plot of which is shown in Fig. 4.239.

In Fig. 4.240, measured σ vs ε and S vs s are compared with prediction from the appropriate constitutive equations (4.78). In such a calculation, of course, one must take due cognizance of the fact that second order transitions have occurred. Both conditions of Eqs. (4.80) and (4.81) were satisfied for this test in a fully annealed aluminum tube.

The constraint $\sum E_i = 0$ inferred from experimental observation requires at finite strain that the infinitesimal strains be isochoric. For finite strain, this constraint prescribes the change of volume. That changes of volume do occur

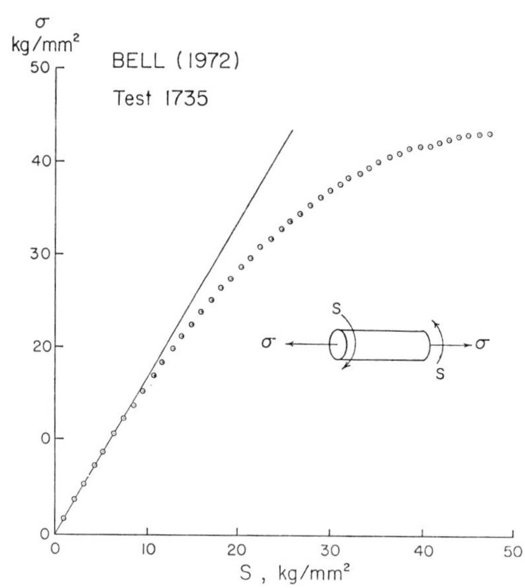

Fig. 4.238. The tension-torsion non-radial loading path for the test of Figs. 4.239 and 4.240.

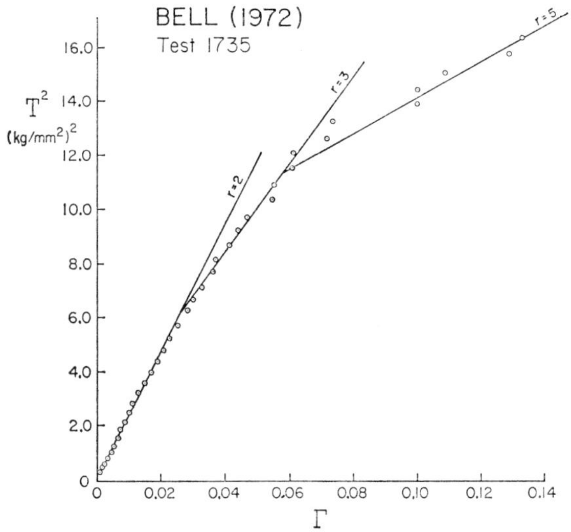

Fig. 4.239. A T^2 vs Γ plot for a non-radial loading path from a tension-torsion test in a fully annealed aluminum tube, compared with prediction (solid line) from Eq. (4.75).

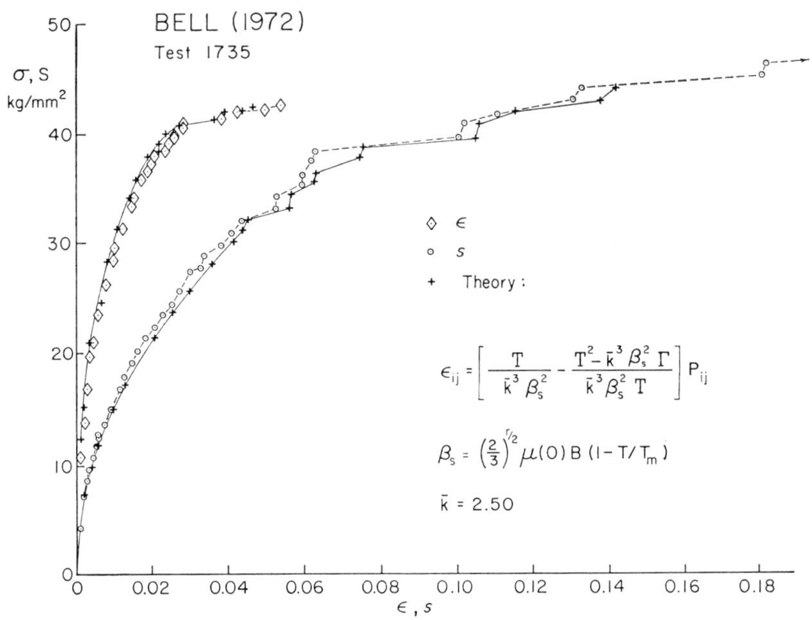

Fig. 4.240. σ vs ε and S vs s from the test of Fig. 4.237, compared with prediction (crosses) from Eq. (4.78).

during plastic deformation has been known since 1879 when BAUSCHINGER did his research (see Sect. 2.18). An interesting cross-check upon the observation from the measurement of response functions at finite strain, that $\sum E_i = 0$, is given by comparing with prediction a direct measurement of change of volume

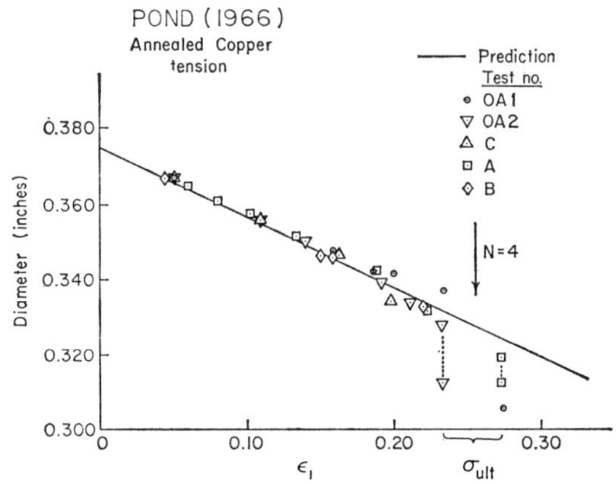

Fig. 4.241. A comparison of measured diameters for tensile tests in copper, compared with prediction from the constraint of Eq. (4.70) (solid line).

during plastic strain. Such a measurement requires extreme accuracy. Professor ROBERT POND of the Johns Hopkins University has made available[1] the experimental results in tension for annealed copper shown in Fig. 4.241. His diametral measurements during deformation were sufficiently accurate for my comparison of them with the predicted diameter (solid line) from tr $E=0$. In these tests the ultimate stress was reached at the sixth transition strain, $N=4$.

In Fig. 4.242 are shown the T^2 vs Γ and the ε vs s plots for a near radial loading experiment with $\dot{S}/\dot{\sigma}=0.46$. One notes that the experimental strains are in agreement with the theoretical prediction of Eqs. (4.78) to the second transition strain, $N=13$. Below this transition strain the condition on the components of the SAVART-MASSON effect is satisfied. Above this strain, this condition is not satisfied inasmuch as for an interval of deformation, $\Delta\varepsilon$ is zero, followed by an interval in which Δs is approximately zero until there is a return to the predicted point, after which there is a second departure. Although the condition of Eq. (4.80) obviously does not hold above $N=13$, the value of $\Delta\Gamma$ in Eq. (4.79) is unaltered, i.e., when the condition of Eq. (4.80) is not satisfied the two incremental components of strain are altered so as to give the same value of $\Delta\Gamma$ required by the loading surface of Eq. (4.75). The loading condition of Eq. (4.81) is satisfied throughout the test.

In Fig. 4.243 is shown the ε vs s plot for the experimental results of another nearly radial experiment, # 1421, which had the ratio $\dot{S}/\dot{\sigma}=1.52$, nearly three times that of the experiment of Fig. 4.242 (see Fig. 4.234). In every other respect, the two tests were identical. In Fig. 4.243 we see that the experimental values are in agreement with prediction to the third transition strain, $N=10$, satisfying both the conditions of Eqs. (4.80) and (4.81). Near this transition strain, the condition of Eq. (4.80) ceases to apply. At a shear strain s, slightly above 12%, the slope predicted from Eqs. (4.78) again is obtained, although the absolute values of the strain now are altered.

Of perhaps as much interest as the fact that for two dimensional loading paths, experimental data exists correlating with the constitutive Eqs. (4.78)

[1] Private communication.

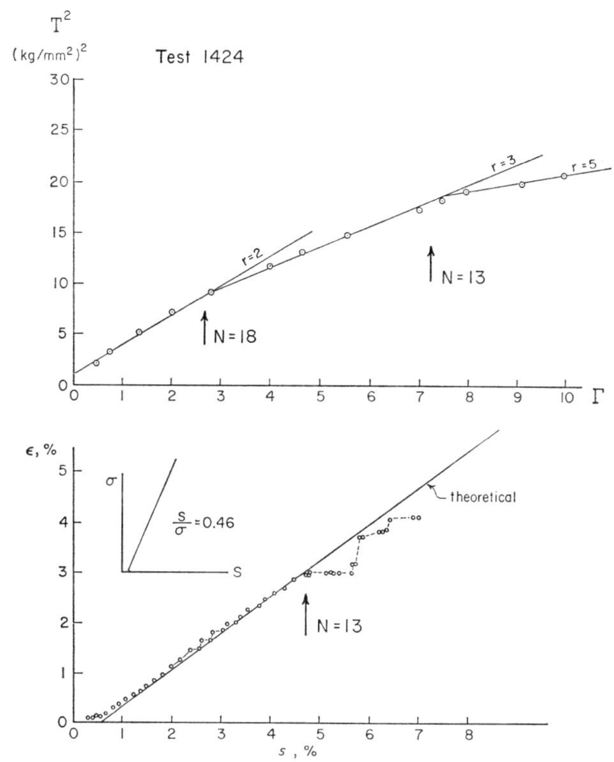

Fig. 4.242. The T^2 vs Γ plot for a nearly radial tension-torsion test in a fully annealed aluminum tube, compared with prediction from Eq. (4.75). Also shown is the ε vs s plot compared with prediction from Eq. (4.78). Note the decomposition of strain components in the SAVART-MASSON effect after the transition strain $N = 13$.

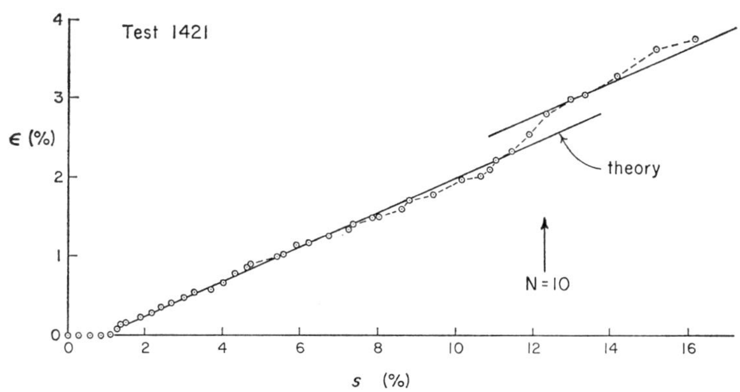

Fig. 4.243. An ε vs s plot for a nearly radial loading test (circles) showing agreement with prediction from Eq. (4.78) until the third transition strain, $N = 10$. The loading path for this test is shown in Fig. 4.234.

and the loading surface of Eq. (4.75) in terms of the predicted parabola coefficient from simple loading and the second order transition structure, is the fact that even for radial loading paths for which incremental and deformation laws are indistinguishable there can occur in the vicinity of a second order transition strain material instability phenomena such that the condition of Eq. (4.80) on the SAVART-MASSON effect no longer holds, although the loading surface response function of Eq. (4.75) still applies. The importance of this fact, which can be seen only if there are at least two non-zero stress components, is of obvious interest if more general response functions are to be evolved which include the constitutive Eqs. (4.78) as a special case when the condition of Eq. (4.80) on the material instability phenomenon holds.

In Fig. 4.244a–c are shown the results from radial loading tests for which the condition of Eq. (4.80) does not apply. For test 1415 of Fig. 4.244a in the ε vs s plot, instead of the predicted slope of $\Delta s/\Delta\varepsilon = 5.5$, the observed slope was 3.0. As may be seen from the dashed lines at successive intervals during the deformation, the slope for a series of brief intervals approximates prediction. Throughout the test, however, the loading surface given by Eq. (4.75) applied, as is seen from the T^2 vs Γ plot, also shown in Fig. 4.244a.

In the radial loading tests shown in Fig. 4.244b and c, interesting departures from, and return to the theoretical slope from Eq. (4.78) dramatically illustrate the importance of material instability in describing the response at finite strain for two non-zero stress components in fully annealed aluminum. An inspection of Fig. 4.234 reveals no evidence of the anomalous behavior of Fig. 4.244a–c in either the aggregate ratios, Eq. (4.74) or the loading surface given by Eq. (4.75),

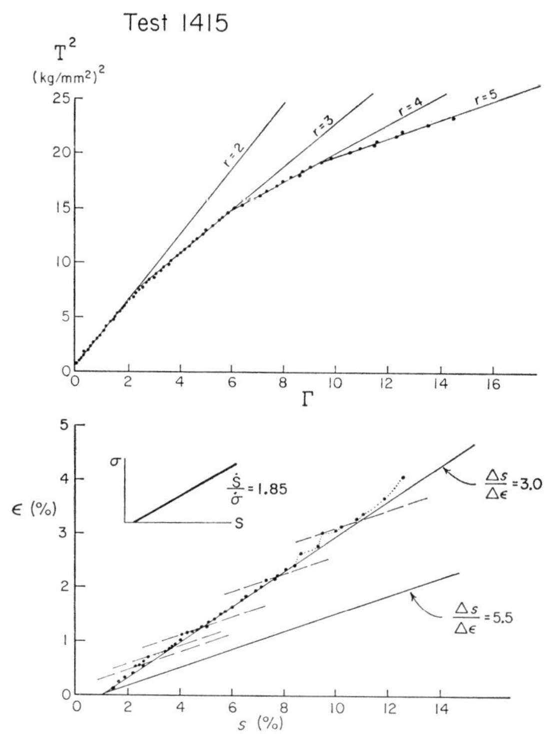

Fig. 4.244a. A T^2 vs Γ plot for a radial loading test showing correspondence with prediction from Eq. (4.75); and an ε vs s plot showing departures from theory when the condition of Eq. (4.80) is not satisfied.

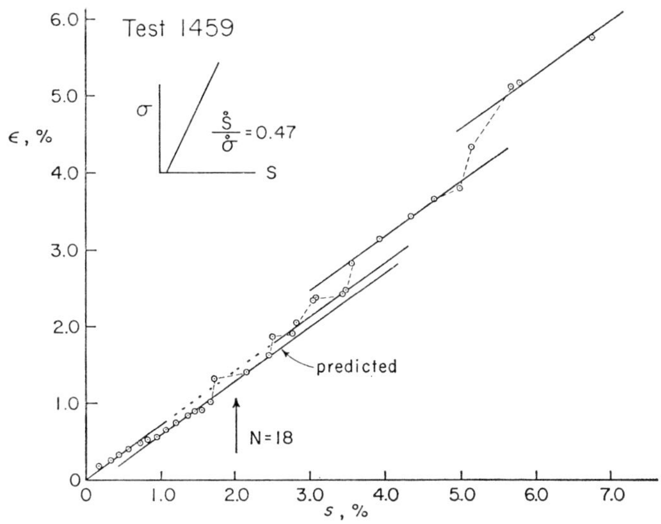

Fig. 4.244b The decomposition of strain components in the SAVART-MASSON effect for a nearly radial test. In this test Δs was less than predicted, although $\Delta \Gamma$ was unaffected.

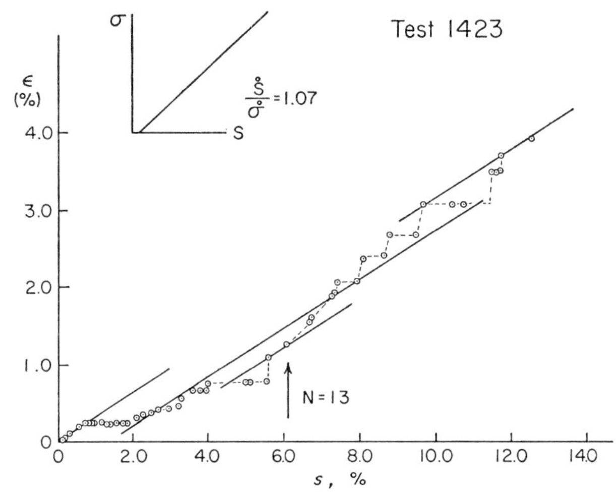

Fig. 4.244c. The decomposition of strain components in the SAVART-MASSON effect for a nearly radial test. In this test $\Delta\varepsilon$ was less than predicted, although $\Delta \Gamma$ was unaffected.

both of which are in close agreement with experiment whether or not the conditions of Eqs. (4.80) and (4.81), and thus the constitutive Eqs. (4.78), apply.

Material instability in the form of changing ratios of incremental components of the SAVART-MASSON steps and second order transitions have been observed both along non-radial loading paths in fully annealed aluminum for which the response is given by Eqs. (4.78) from a deformation theory, and along non-radial loading paths in fully annealed copper and aluminum for which experimental results are in close agreement with prediction from modified incremental constitutive equations, also based upon Eqs. (4.74) and (4.75). The appearance of similar instability in radial loading, and, in addition, the observations of ARIS PHILLIPS[1] on aluminum, described in Sect. 4.33, emphasize that it is necessary to learn more about material instability in solids.[2] Thus, I view this independence of strain components sometimes appearing in the SAVART-MASSON steps, as offering an important clue to the understanding of the nature of the local inhomogeneous strain which experiment has shown accompanies large plastic deformation in annealed crystals (see Sect. 4.32).

The constitutive equations,[3] Eqs. (4.78), apply for the results obtained in fully annealed aluminum whether SAVART-MASSON steps are absent, or whether they are present, provided in the second instance the ratios for the components of incremental strain are consistent with the conditions of Eq. (4.80), which they sometimes are not. The second order transitions and the fixed sequence of strains at which the transitions occur, while representing the observed physical state of annealed crystalline solids, have yet to be represented satisfactorily in terms of the undeformed reference state. The constitutive equations for annealed solids, generated from Eq. (4.77) can be viewed as the norm, representing the stable solid against which the effects of material instability can be measured and understood. This situation is not too different from that for infinitesimal deformation in annealed crystals for which second order transitions among moduli, referred to above in Sect. 3.44 as "multiple elasticities," also represent departures from the stable moduli of the generalized HOOKE's law.

This discussion of current study emphasizes what must be concluded from experiment regarding the plasticity of annealed crystalline solids, and what in my judgment remain open questions for the experimentist to answer if general constitutive equations evolve which include this discovery with respect to material instability. A representation of the finite deformation of fully annealed crystalline solids must include the origin of these instabilities in the local, inhomogeneous strain and rotation; it must encompass these new experimental facts, which seem discordant only when viewed from the perspective of pure homogeneous deformation.

4.36. On the discovery of shock waves in the tensile deformation of rubber strings: Kolsky (1969).

The laboratory measurement of actual wave speeds in solids other than metals began with the work of EXNER[4] on stretched rubber strings in 1874. Insofar as they were unloading waves from a deformed tensile state, they were compression waves. It is interesting to contrast EXNER's results, which were described above in Sect. 3.33, with the study of incremental waves, also in stretched rubber strings, made nearly a century later by KOLSKY,[5] in 1969. The latter's direct study of wave profile provided evidence of the development of a tensile shock front in a solid with a known quasi-static stress-strain function concave to the stress axis. KOLSKY's discovery of these tensile pulse shock fronts is a landmark in experi

[1] PHILLIPS [1957, *1*], [1960, *1*]; PHILLIPS and GRAY [1961, *1*].

[2] When the SAVART-MASSON effect is present, the importance of decoupling the tension and torsion components in the experiment is evident. As far as I know, the only experiments in which this has been accomplished in dead weight testing, were those of the past few years in my laboratory. (A decoupled experiment is one in which the dead-weight tension load does not rotate when the dead-weight torsion load is applied.)

[3] BELL [1972, *2*].

[4] EXNER [1874, *1*].

[5] KOLSKY [1969, *1*].

mental solid mechanics, easily recognizable as such even in this day of mass measurement of wave speeds in all manner of solids.

As we have seen, EXNER had shown that the unloading wave speed increased as the pre-stressed tension increased. He measured only transit times. KOLSKY, with 20th century oscillography, was able to determine the shape of particle velocity vs time profiles by observing the electrical outputs of light wires attached to the rubber as they cut the lines of force of magnetic fields of constant strength.

KOLSKY stretched an 1.27×1.27 cm square section of "vulcanized natural rubber gumstock" to 5 times its original length. By means of a length of piano wire, he subjected a 25.4 cm long pre-stressed end section of the 396 cm pre-stressed rubber string to an additional tensile elongation. Volatilizing the steel wire by applying an electric current, he introduced a tensile pulse from the short section into the long stretched string section where he could analyze wave profiles at several positions. He could vary the magnitude of the tensile pulse by altering the magnitude of the additional pre-loading in the short section. KOLSKY observed that small amplitude pulses, i.e., less than 10%, propagated along the rubber without measurable change of shape. Larger pulses, however, sharpened in profile to form a shock wave after they had traveled about 274 cm.

KOLSKY provided a representative example, shown in Fig. 4.245, in which the 25.4 cm loading section was subjected to a 2.54 cm additional elongation, producing a 40% additional strain. (KOLSKY used English feet and inches in describing his experiment.)

The steepening of the particle velocity profile as the wave front proceeded from 30.5 to 274 cm from the point of initiation is readily seen. EXNER's unloading wave in vulcanized rubber string which had stretched 5 times the initial length, was 65.9 m/sec (see Sect. 3.33) compared to KOLSKY's approximately 122 m/sec, demonstrating once again, as MALLOCK[1] had emphasized in 1904, the variation

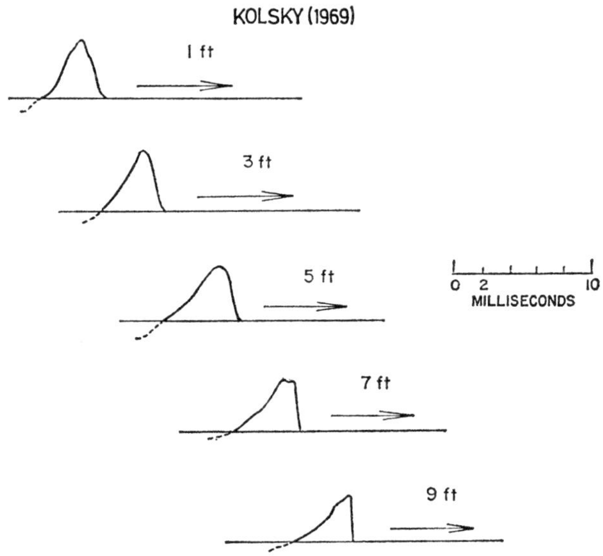

Fig. 4.245. KOLSKY's incremental tensile wave profiles in pre-stressed rubber at the designated positions. Note the formation of the tensile shock wave.

[1] MALLOCK [1904, *1*].

of rubber from lot to lot, as well as the possible difference between tensile loading waves and compressive unloading waves in pre-stressed rubber. Shock waves in solids often have been postulated, but these measurements provided the first direct evidence of the steepening of wave fronts during wave propagation.

4.37. Experiments on the finite deformation of strings subjected to transverse impact.

Before leaving the subject of waves for finite strain, it is interesting to refer to two studies of large deformation which were different in conception from all of the above. They involved a long wire or strip of a solid, i.e., aluminum and rubber, subjected to a transverse high velocity impact from a projectile.

The experiments of A. B. Schultz[1] were in pre-stressed wires of partially annealed low purity aluminum. (The elastic limit, $Y = 4700$ psi was higher than that of the fully annealed aluminum described above.) The experimental situation, which is that of the incremental wave in the pre-stressed solid, is shown in Fig. 4.246.

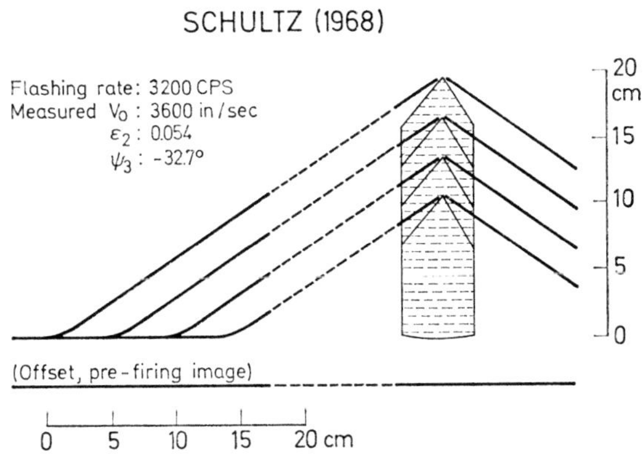

Fig. 4.246. Schematic diagram of Schultz' transverse wire impact experiment with a prescribed tensile pre-stress.

The experimental details were described in a paper in 1967 by Schultz, P. A. Tuschak, and A. A. Vicario.[2] Microflash photography of bands of black ink marked on the specimen gave gage lengths of "approximately 2 or 4.5 in.," for the measurement of strain. Interpretation of measurements required the assumption of a perfect string. Schultz[3] provided an analysis of the angle of the string vs impact velocity by means of which strain as a function of impact velocity was estimated for comparison with quasi-static observation for strain to 20%. The values of pre-stress were from 740 to 11000 psi. He obtained the results shown in Fig. 4.247.

The study of this subject in the terms suggested by Schultz is severely hampered by the lack of an adequate means of interpreting the observed results.

Jack C. Smith and Carl A. Fenstermaker[4] in 1967 described a series of experiments on strings consisting of strips of natural rubber subjected to high

[1] Schultz [1968, *1*].
[2] Schultz, Tuschak, and Vicario [1967, *1*].
[3] Schultz, *op. cit.*
[4] J. C. Smith and Fenstermaker [1967, *1*].

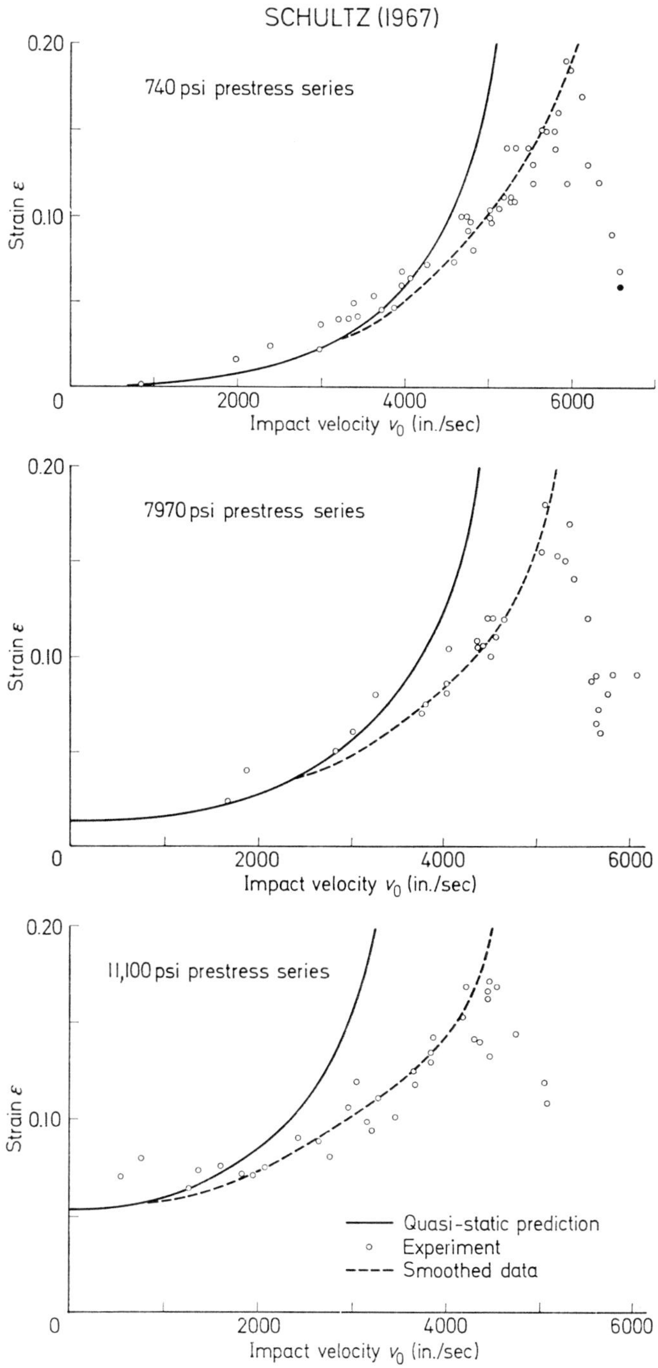

Fig. 4.247. SCHULTZ' strain vs impact velocity for three different pre-stresses, compared with prediction from the slopes of the quasi-static stress-strain curve.

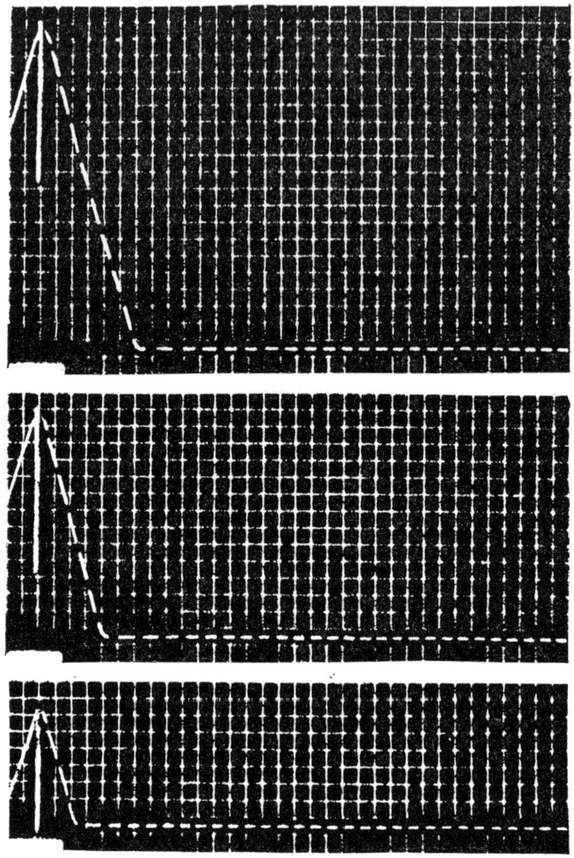

Fig. 4.248. Configurations of a natural rubber filament after transverse impact at 42.7 m/sec velocity. Grid spacing was 1 cm. Times after impact were: bottom frame, 1.430 msec; middle frame, 2.886 msec; top frame, 4.340 msec.

velocity transverse impact. They presented their experimental results in a form which allows the interested reader to examine them as measured. In Fig. 4.248 are shown photographs of successive positions of an impacted rubber string superposed upon an underlying grid.

The strain vs distance distributions in Lagrangian coordinates for the nearly equal millisecond intervals are shown in Fig. 4.249.

SMITH and FENSTERMAKER's detailed analysis of wave speeds vs strain at nearly equally spaced millisecond time intervals were given in three large tables, for impact velocities of 20.6, 42.7, and 66.3 m/sec. The strain vs velocity distribution compared with quasi-static values shown in Fig. 4.250 for the three different impact velocities, interestingly enough, were lower in each instance.

The direct study of finite amplitude wave propagation in rubber, as these studies[1] and those of KOLSKY[2] suggest, obviously should lead to further successful

[1] *Ibid.*
[2] KOLSKY [1969, *1*].

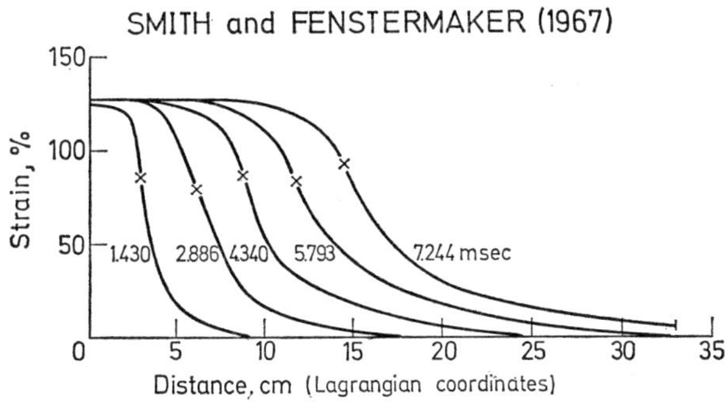

Fig. 4.249. Strain-distance distributions at various times after impact for a natural rubber filament impacted at 42.7 m/sec velocity. Crosses indicate positions of the transverse wave front.

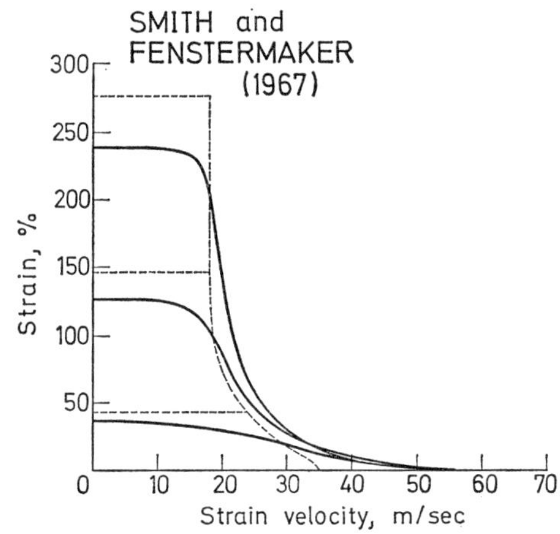

Fig. 4.250. Strain-velocity distributions resulting from transverse impact tests of natural rubber. Solid line curves give distributions obtained from impact tests. Dashed line curves give distributions calculated from the quasi-static stress-strain curve. Impact velocities were: bottom curve, 20.6 m/sec; middle curve, 42.7 m/sec; top curve, 66.3 m/sec.

research. I have examined the wave speeds of SMITH and FENSTERMAKER for one impact velocity. The results of my calculations suggest that it would be worth the effort to attempt to extend what has been learned from nonlinear wave experiments in crystals to this member of a different class of solids.

4.38. Poynting's experiments (1909–1912).

COULOMB[1] in 1784 noted the decrease in the period of torsional oscillations for a wire under large tensile load. As TRUESDELL[2] has emphasized, COULOMB's attributing this to diametral decrease was an insufficient explanation of the phenomenon. "COULOMB has observed an effect of non-linear elasticity related to the POYNTING effect."[3] As was noted in Sect. 2.19, WERTHEIM[4] measured the changes

[1] COULOMB [1784, *1*].

[2] TRUESDELL [1960, *1*]. The term "POYNTING effect" was introduced by TRUESDELL in 1952. [1952, *1*].

[3] TRUESDELL [1960, *1*], p. 407.

[4] WERTHEIM [1857, *1*, *2*].

of volume unexpectedly occurring in the small deformation torsion of hollow specimens together with the associated nonlinearity of the torsional elasticity. KELVIN[1] in 1865 and TOMLINSON[2] in 1883 noted the changes of length in long wires undergoing finite torsional strain. In the non-radial loading path experiments on finite strain of MITTAL[3] in 1969 when the initial tension stress was held constant and the subsequent torsion applied, the finite axial strain, of course, continued to increase. This also was the case for the torsional strain following initial torsion and subsequent tension.

WERTHEIM indicated the importance he attached to his measurements of volume for hollow tubes in torsion in his own unsuccessful efforts to provide an explanation and his plea to other geometers to attempt to do so. Fifty years were to pass, however, before a serious experimental study of the phenomenon itself was undertaken. In a paper of 1909 entitled, "On Pressure Perpendicular to the Shear Planes in Finite Pure Shears, and on the Lengthening of Loaded Wires

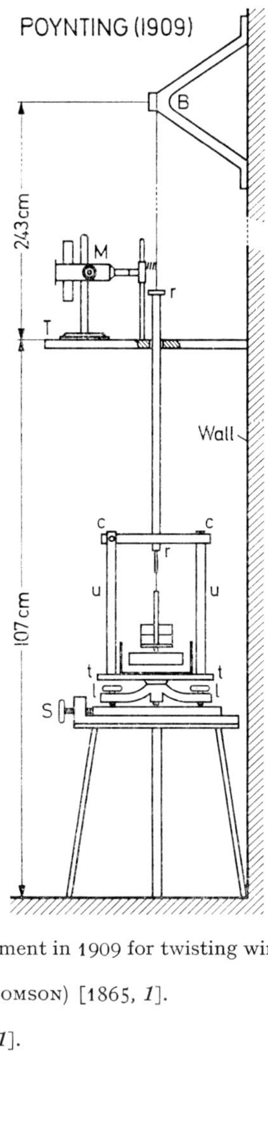

Fig. 4.251. POYNTING's arrangement in 1909 for twisting wires and measuring their elongation.

[1] KELVIN (Sir WILLIAM THOMSON) [1865, *1*].
[2] TOMLINSON [1883, *1*].
[3] MITTAL [1969, *1*], [1971, *1*].

when Twisted,"[1] and in 1912, "On the Changes in the Dimensions of a Steel Wire when Twisted, and on the Pressure of Distortional Waves in Steel,"[2] POYNTING took up the problem. He accompanied his report of experiments with a special and far from sufficient theoretical explanation. Nearly another half century was to elapse before satisfactory explanations of COULOMB's, WERTHEIM's, and POYNTING's experimental results were discovered within the classical theory of finite elastic strain.

Confining the discussion to POYNTING's experiments, we see in Fig. 4.251 the apparatus he used in 1909.

The wire was 231 cm long. Its lower end was clamped in jaws in the upper end of a steel rod, $r\,r$, 51 cm long, which passed through a hole in the table, T, in which was an observing microscope and a parallel plate micrometer. The system was enclosed in a wooden tube to avoid temperature changes during the test. POYNTING's expected results for elongation occurring between a starting twist of $^1/_4$ turn and a maximum twist of $4^1/_4$ turns, are shown in Table 139.

Table 139. POYNTING (1909).

Micrometer readings at lower end			Micrometer readings half-way up the wire		
$^1/_4$ turn	$4^1/_4$ turns	Lowering	$^1/_4$ turn	$4^1/_4$ turns	Lowering
22.3	18.6	3.7	30.4	28.3	2.1
22.5	19.0	3.5	30.5	28.4	2.1
23.0	19.2	3.8	30.2	28.1	2.1
22.6	19.4	3.2	30.5	28.5	2.0
22.9	19.6	3.3	30.5	28.0	2.5
22.6	19.5	3.1	30.4	28.7	1.7
23.0	19.5	3.5	31.0	28.4	2.6
23.0	18.9	4.1	30.6	28.5	2.1
22.7	19.2	3.5	31.8[a]	29.9	1.9
22.9	19.6	3.3	31.6[a]	29.9	2.5
Mean lowering, 3.50 divisions One divisions of micrometer = 0.00974 mm The lowering was 0.0341 mm			Mean lowering, 2.16 divisions One division of micrometer = 0.00751 mm The lowering was 0.0162 mm		

[a] Another point on the needle sighted.

As may be seen, measurements were made at the lower end and half-way up the wire. The mean lowering at the end was 0.0341 mm, indicating a value half-way up, of 0.0171 mm. The mean experimental value at that position was 0.0162 mm.

In Table 140, I have compiled POYNTING's observations for various axial loads and angles of twist in steel, copper, and brass. His major conclusions are expressed by the relation

$$d\,l = s\,\frac{a^2\,\theta^2}{2l}\,, \tag{4.82}$$

where $d\,l$ was the observed small elongation; a, the radius of the wire; l, the length; θ, the angle of twist; and s, a constant having the value $s \cong 1$ for steel, and $s \cong 1.5$ for copper and brass.

This result, that the incremental elongation was proportional to the *square* of the angle of twist for any given load, parallels the observation of WERTHEIM[3]

[1] POYNTING [1909, *1*]. POYNTING referred to his "faulty analysis" presented in an earlier paper on the subject in 1905.

[2] POYNTING [1912, *1*].

[3] WERTHEIM [1857, *1*, *2*].

Table 140. POYNTING (1909).

Steel:	0.720 mm; total load 7081 g
	Clockwise twist, 0–4 turns; lowering 0.0181 mm, mean of 10 observations
	Clockwise twist, 0–8 turns; lowering 0.0732 mm, mean of 10 observations
	The ratio of these is 4.04:1
Steel:	0.970 mm; total load 19504 g
	Clockwise twist, $^1/_4$–$2^1/_4$ turns; lowering 0.0087 mm, mean of 10 observations
	Clockwise twist, $^1/_4$–$4^1/_4$ turns; lowering 0.0339 mm, mean of 10 observations
	Counterclockwise twist, $^1/_4$–$2^1/_4$ turns; lowering 0.0089 mm, mean of 10 observations
	Counterclockwise twist, $^1/_4$–$4^1/_4$ turns; lowering 0.0340 mm, mean of 10 observations
	Mean lowering, $^1/_4$–$2^1/_4$, 0.0088 mm
	Mean lowering, $^1/_4$–$4^1/_4$, 0.0340 mm
Copper:	0.655 mm; total load 7081 g
	Clockwise twist, $^1/_4$–$2^1/_4$ turns; lowering 0.0066 mm, mean of 10 observations
	Counterclockwise twist, $^1/_4$–$2^1/_4$ turns; lowering 0.0083 mm, mean of 10 observations.
	("It was not safe to give a greater twist owing to the largeness of the permanent set. With $2^1/_4$ turns the set was still small.")
Brass:	0.928 mm; total load 19504 g
	Clockwise twist, $^1/_4$–$2^1/_4$ turns; lowering 0.0169 mm, mean of 10 observations
	Clockwise twist, $^1/_4$–$4^1/_4$ turns; lowering 0.0540 mm, mean of 10 observations
	Counterclockwise twist, $^1/_4$–$2^1/_4$ turns; lowering 0.0135 mm, mean of 10 observations
	Counterclockwise twist, $^1/_4$–$4^1/_4$ turns; lowering 0.0479 mm, mean of 10 observations

half a century earlier that the incremental change of volume was proportional to the square of the angle of twist per unit length as well as to the length of the specimen.

POYNTING remarked upon the small magnitude of the quantity to be measured, and the subsequent possible error involved. This was not so large, however, that the distinct differences between the magnitude of the incremental elongation in clockwise and counterclockwise twist in copper and brass could be attributed to measurement error, nor could the *decrease* in length for steel under small load instead of the *increase* for larger loads be so dismissed. POYNTING subjected the steel wires to an electric current which produced "red heat", after which he rubbed off the surface oxidation. He repeated the experiment to study a possible source of such deviations. The results also are included in Table 140.

Finally, POYNTING noted that the apparent shear modulus as determined by torsional vibrations would be decreased about 0.1% for an axial load, producing an axial strain of 2×10^{-4}. He commented further upon the longitudinal motion proportional to ε^2, to be expected in distortional waves in a solid if the pressure in the direction of propagation were neglected.

There is no mention in POYNTING's paper[1] that COULOMB,[2] WERTHEIM,[3] KELVIN,[4] TOMLINSON,[5] and BAUSCHINGER,[6] to name a few, had observed a small decrease with increasing axial load, for the shear modulus determined from torsional vibration, nor that WERTHEIM had, in addition, considered volume as a function of the angle of twist.

[1] POYNTING [1909, *1*].
[2] COULOMB [1784, *1*].
[3] WERTHEIM [1857, *1*, *2*].
[4] KELVIN (Sir WILLIAM THOMSON) [1865, *1*].
[5] TOMLINSON [1883, *1*].
[6] BAUSCHINGER [1881, *2*].

In his paper in 1912, POYNTING[1] succeeded in demonstrating that the source of differences he had found earlier between large and small loads was simply a matter of making certain the wire was sufficiently straight before conducting the experiment. He described experiments in which he surrounded the 106.5 cm long wires by a long narrow tube filled with water.[2] When he twisted the wire, the change in the level of water in an associated capillary tube provided a measure of the change in diameter produced by the twist. This contraction of diameter also was found to be proportional to the square of the angle of twist.

Under the heading "Subsidiary Experiment," POYNTING determined E-moduli and POISSON's ratio, ν, for different twist angles. He further demonstrated that his dimensional changes did not arise from thermal sources.

In 1951 RONALD RIVLIN[3] by appeal to the theory of finite strain of incompressible elastic materials provided a theoretical foundation for the phenomenon observed by POYNTING and several earlier experiments. In 1953 RIVLIN calculated the magnitude of both the KELVIN and the POYNTING effects. RIVLIN's related experiments in the finite elasticity of rubber will be described below. H.W. SWIFT[4] in 1947 in a paper entitled "Length Changes in Metals under Torsional Overstress," reconsidered the type of plastic strain which, as we have seen, had been the subject of much study in the 19th century. AMNON FOUX[5] at an International Symposium on Second-Order Effects in 1962 presented a paper entitled "An Experimental Investigation of the Poynting Effect." FOUX's experiments remind one that the quantities POYNTING was obliged to measure in the determination of nonlinear effects were extremely small. Despite the passage of 53 years, all of the difficulty and fluctuation of data of POYNTING's work were still present.

In 1969 ERWIN J. SAXL and MILDRED ALLEN[6] with a "precision torsion pendulum" restudied the dynamic aspect of the normal stress effect in their paper en-

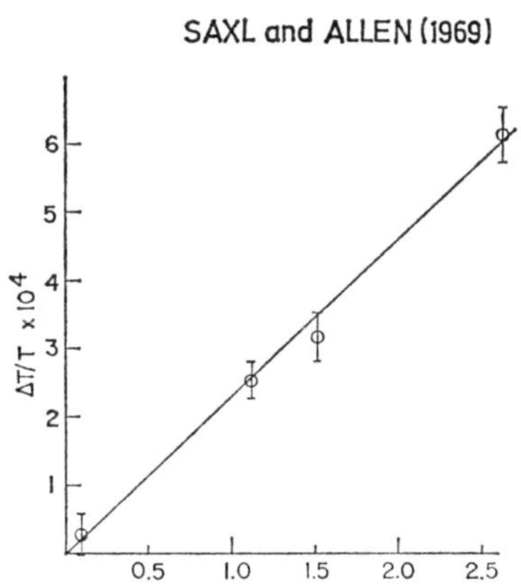

Fig. 4.252. Relative changes in period vs added weights, corrected for the increases in period caused by the increased moments of inertia.

[1] POYNTING [1912, *1*].

[2] It is interesting again to compare this experiment and its details with that of CAGNIARD DE LATOUR in 1828 [1828, *1*], who first attempted without success to measure the changes of volume of a wire in a tube.

[3] RIVLIN and SAUNDERS [1951, *1*], [1953, *1*]. See also TRUESDELL and NOLL [1965, *1*], Sects. 57, 66, and 70.

[4] SWIFT [1947, *1*].

[5] FOUX [1962, *1*].

[6] SAXL and ALLEN [1969, *1*].

titled "Period of a Torsion Pendulum as Affected by Adding Weights." In Fig. 4.252 are shown their comparisons of prediction and four experimental measurements after the former had been corrected for the increases in period caused by the increased moment of inertia of the added weights.

In the same issue of the *Journal of Applied Physics*, ALLEN and SAXL[1] restudied the quasi-static version of the POYNTING experiment, demonstrating that for five wires of different alloys and diameter, an initial residual twist always caused asymmetry which in general could be considered consistent with POYNTING's observations of 60 years earlier.

POYNTING's experiments and those few repetitions of them in metal referred to here, illustrated nonlinearity in relatively small deformation. They were included in this section on finite deformation to emphasize the universality of what is a fundamental nonlinear phenomenon in finite strain. This is amply demonstrated in the finite elasticity of rubber described immediately below.

4.39. Experiments on the finite elasticity of rubber: From Joule to Rivlin (1850's to 1950's).

Generations of experimentists interested in the deformation of the solid continuum have focused primarily upon the metallic elements and their binary combinations. This has not been solely because of their associated technological importance, which certainly has been a factor, but because it is here that variations from one specimen to another have not precluded observations of patterns of behavior among them.

At one point or another in this treatise we have considered the response to deformation of glass, gut, rubber, wood, silk, human tissue, paint, enamel, lacquers, ice, leather, cork, marble, sandstone, brick, ceramic paste, clay, frog muscles, and concrete. The literature on experiments related to solid mechanics includes a far larger list of substances. R. HOUWINK[2] in his interesting portrayal of elastic and plastic properties of solid substances in his monograph in 1953, *Elasticity, Plasticity and Structure of Matter*, extended the list to baker's dough, resin, asphalt, gutta percha, balata, cellulose, gelatin, glue, casein, wool, urea-formaldehyde, and sulphur. Industrial interest in the deformation characteristics of synthetic yarns, meat, plywood, and many other products, either for their further development or as a means of controlling desired specifications, extended the list of substances for which stress and deformation properties have been described.

It is characteristic of most of these solids that they are so sensitive to their preparation and that many of them are so variable in their properties over a period of time that study of them has been mainly qualitative. Certainly one of the most interesting of such substances is rubber. The curious thermal behavior discovered by JOHN GOUGH[3] in 1805 fascinated the second half of the 19th century after WILLIAM THOMSON[4] (the later Lord KELVIN) and JAMES PRESCOTT JOULE[5] in the 1850's pointed out the importance of GOUGH's observations.

Whether or not THOMSON and JOULE, as seems likely, were influenced in turning their attention to rubber by the developing concern for the properties of vulcanized rubber railway bumpers which had come into use in the 1840's, P.

[1] ALLEN and SAXL [1969, *1*].

[2] HOUWINK [1953, *1*].

[3] GOUGH [1805, *1*], [1806, *1*]; and see *supra*, Sects. 2.12 and 3.33.

[4] WILLIAM THOMSON (KELVIN) [1855, *1*].

[5] JOULE [1857, *1*, *2*], [1859, *1*].

BOILEAU[1] in his experiments in 1853 certainly was. In his introduction to a paper in 1856 describing a compression experiment BOILEAU stated that undoubtedly earlier experiments had been performed on rubber since railway use had extended back a decade, but he was unable to obtain any evidence of their existence.[2]

BOILEAU tested a stack of rubber disks, 93 mm in diameter, 23 mm thick, with a central hole of 39 mm diameter. Iron disks 5 mm thick were interposed between each pair of rubber disks. Such stacks were being used at that time as compression springs. The experiments were performed at a testing facility of the Railroad of the East at Montigny. In Fig. 4.253 I have converted BOILEAU's tabulated load vs elongation data to stress vs strain. I have presented his data not only for their precedence but because of his emphasis that his data were indicative of only general behavior. He realized that in order to compare

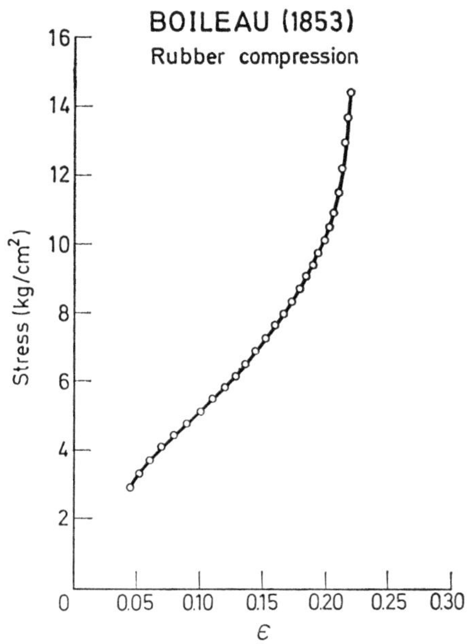

Fig. 4.253. The compression behavior of a mid-19th century rubber railway spring.

---

[1] BOILEAU: published in [1856, *1*].

[2] BOILEAU thus missed not only the discovery of GOUGH of 1805, but the experimental rediscovery of part of GOUGH's observations by C. C. PAGE in 1847, the torsion experiments of SAINT-VENANT in 1847, also in rubber, and the compressibility experiments on rubber bars by WERTHEIM in 1848.

Perhaps of more specific significance was the fact that BOILEAU also missed three experimental papers on railway springs: one by W. C. CRAIG in 1853, one by HOVINE, and one by DEBONNEFOY in 1854 within three years of his own study. In examining the Royal Society Catalogue of Scientific Papers between 1800 and 1900 on the subject of rubber or caoutchouc, from which, incidentally, BOILEAU's paper was omitted, I found among the 34 studies listed in the 19th century, reference only to the work of GOUGH, CRAIG, and DEBONNEFOY preceding 1856, the date of BOILEAU's paper. When to these one adds the paper of WILLIAM THOMSON in 1855, the two brief papers of JOULE in 1857, JOULE's classic experimental paper in 1859, the experiments of DIETZEL in 1857, who also was interested in railway springs, and the experiments of CLAPEYRON on rubber in 1858, it seems rather obvious that rubber railway bumpers played more than a small role in the sudden increase in experimentists' interest in the substance.

PAGE [1847, *1*], SAINT-VENANT [1847, *1*], WERTHEIM [1848, *1*], CRAIG [1853, *1*], HOVINE [1854, *1*], DEBONNEFOY [1854, *1*], DIETZEL [1857, *1*], CLAPEYRON [1858, *1*].

experimental results it would be necessary to have samples obtained from identical fabrication procedures, particularly with respect to the sulphurization.

BOILEAU noted that when he successively removed the loading weights the elongations were not identical. He observed that a relaxation or elastic after-effect was present, but finally, under zero stress, there was a return to the original dimensions without permanent deformation. The crude experiments of BOILEAU were adversly criticized by C. F. DIETZEL[1] in a paper in 1857, as indicating that BOILEAU had insufficiently examined WEBER's[2] after-effect which DIETZEL had studied on a vulcanized rubber thread.

JOHN GOUGH's[3] observation in 1805 certainly should be referred to solely as the "GOUGH effect" in view of the fact that JOULE[4] was not the first person to observe anew some aspects of GOUGH's discoveries. GOUGH not only had observed the curious heating of rubber when stretched while in contact with his lips and the fact that a rubber specimen under tensile load contracted when heated, but also that if a length of rubber were stretched and then immediately quickly placed in cold water it would lose some of its elasticity. He had noted that even when removed from the water it failed to return to its original shape, but upon being heated it regained its original shape and elasticity. The last of those unexpected effects was independently reobserved by C. C. PAGE[5] in 1847, who did not know GOUGH had preceded him. PAGE's paper in SILLIMAN's *American Journal of Science* contained a description of the peculiarities of such an inelastic rubber strip when compressed at moderate temperatures.

In a classic paper of 1859 JOULE[6] described experiments on vulcanized rubber from which he found that that material, while sharing the other peculiar properties of ordinary rubber, differed from it by maintaining its elasticity unimpaired at low temperatures. JOULE stated:

> I observed the principal physical character of this substance before I was acquainted with Mr. Gough's discovery of the properties of simple india-rubber, above noticed.[7]

JOULE's experiments had begun somewhat earlier, coinciding in time with KELVIN's interest[8] in thermoelasticity theory, beginning in 1855. Two notes[9] of JOULE's appeared in the *Proceedings of the Royal Society* for 1857. In the first of these he briefly described some aspects of the thermoelectric power of solids relevant to his use of thermocouples to measure temperature changes in deformed metals and the cooling observed when stretching iron wire, cast iron, hard steel, copper, lead, and even gutta percha. Then he stated:

> With gutta percha also a cooling effect on extension was observed; but a reverse action was discovered in the case of vulcanized india-rubber, which became *heated* when the weight is laid on, and *cooled* when the weight was removed. On learning of this curious result, Professor Thomson, who had already intimated the probability of a reverse action being observed under certain circumstances with india-rubber, suggested to the author experiments to ascertain whether vulcanized india-rubber stretched by a weight is shortened by increase of temperature. Accordingly, on trial, it was found that this material, when stretched by a weight capable of doubling its length, has that length diminished by one-tenth when its temperature is raised 50° Centigrade. This shortening effect was found to increase rapidly with the stretching

[1] DIETZEL [1857, *1*].
[2] WEBER [1835, *1*].
[3] GOUGH [1805, *1*], [1806, *1*].
[4] JOULE [1857, *1*, *2*], [1859, *1*].
[5] PAGE [1847, *1*].
[6] JOULE [1859, *1*].
[7] *Ibid.*, p. 104.
[8] KELVIN (WILLIAM THOMSON) [1855, *1*].
[9] JOULE [1857, *1*, *2*].

weight employed; and, exactly according with the heating effects observed with different stretching weights, entirely to confirm the theory of Professor Thomson.[1]

In his second short report[2] in 1857 JOULE referred to the measurement of the rise and fall of temperature accompanying the compression loading and unloading of every substance studied, metals and India rubber. The increases invariably were slightly in excess of values predicted by WILLIAM THOMSON. JOULE therefore turned to what he erroneously believed were the first measurements of the temperature dependence of elastic constants, in order to account for the observed thermal discrepancy.

He wound steel and copper wires into a helical spring, finding the change in length with temperature from which he computed a shift in the shear modulus of 0.00041 for each degree Centigrade for steel, and 0.00047 per degree Centigrade for copper. In an appended remark on August 1 (the paper had been submitted on June 18) JOULE noted that he had been unaware of the prior work of KUPFFER[3] in 1856 who had obtained the similar numbers of 0.000471 and 0.000478 for the thermal dependence per degree of steel and copper. Both JOULE's and KUPFFER's numbers were low for these two metals, as an examination of the first such experiments of WERTHEIM[4] in 1844 and modern studies demonstrate. JOULE was unaware of WERTHEIM's work in the area.

By 1859 JOULE[5] in the most significant of his papers on thermoelasticity indicated he had finally become familiar with the contribution of GOUGH[6] 54 years earlier.

In addition to the finite thermoelastic vulcanized India rubber experiments of interest in the present context, the paper described a series of small deformation thermoelasticity measurements[7] in steel, iron, copper, lead, gutta percha, various woods (including wet bay wood which alone exhibited the curious temperature reversal of rubber), wheat-straw, cane, cardboard, vine in the green state, cowskin leather, and whalebone. In Fig. 4.254 are seen: (a) JOULE's load-strain results for an experiment in which (b) by means of a copper-iron junction thermocouple, he simultaneously measured the temperature change associated with the deformation. This figure is often reproduced; it shows the initial cooling for a slightly over 20% nominal strain followed by heating during further tensile loading.[8]

In a second experiment, in which JOULE successively loaded a double rubber band to rupture at nearly six times its original length, the reversal of slope in his tensile diagram shown in Fig. 4.255a is seen to be accompanied by a maximum in the incremental temperature increase (Fig. 4.255b) for the 2 lb. weights which had been added in all but the first and last instance. (A 4 lb. weight also was added at 14.5 lbs., giving a temperature change of only 0.001° C.)

Finally, in examining the second GOUGH effect of a decrease in elongation being associated with an increase in temperature, for the same specimen JOULE obtained

---

[1] JOULE [1857, *1*], pp. 355–356. At the time of this quotation JOULE, as he stated, was unfamiliar with the observations GOUGH had made. However, it would seem likely that KELVIN was not unaware of GOUGH's papers.

[2] JOULE [1857, *2*].

[3] KUPFFER [1856, *1*]. It is far more significant that JOULE was ignorant of WERTHEIM's measurements 12 years before those of KUPFFER, since KUPFFER's results were in fact erroneous.

[4] WERTHEIM [1844, *1* (*a*)].

[5] JOULE [1859, *1*].

[6] GOUGH [1805, *1*], [1806, *1*].

[7] As noted above, JOULE obviously considered his measurements to be the pioneering effort in experimental thermoelasticity. He apparently was completely unaware of the experiments of WEBER in 1830, nearly 30 years earlier. See Sect. 2.12 above.

[8] JOULE rendered GOUGH's experiment quantitative, and used a thermocouple instead of lips. GOUGH had referred with disdain to this type of development (see Sect. 2.12).

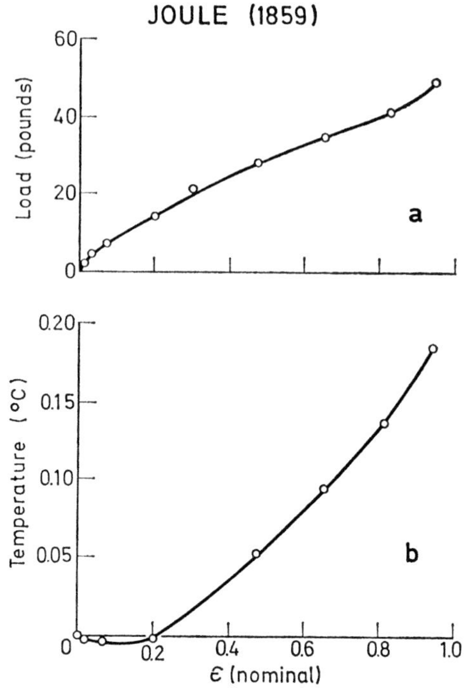

Fig. 4.254a and b. A load vs strain experiment in vulcanized rubber (a) in which Joule measured the temperature variation shown in (b).

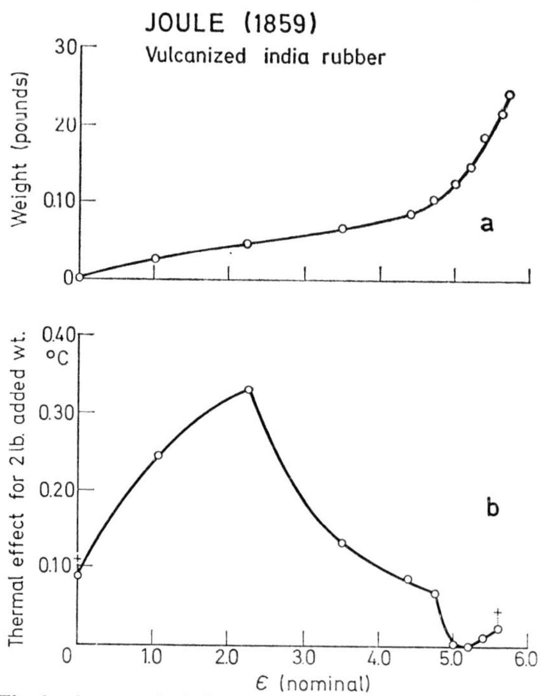

Fig. 4.255a and b. The load vs strain behavior in vulcanized India rubber (a) for which the corresponding thermal effect was measured for each two pound increment (b).

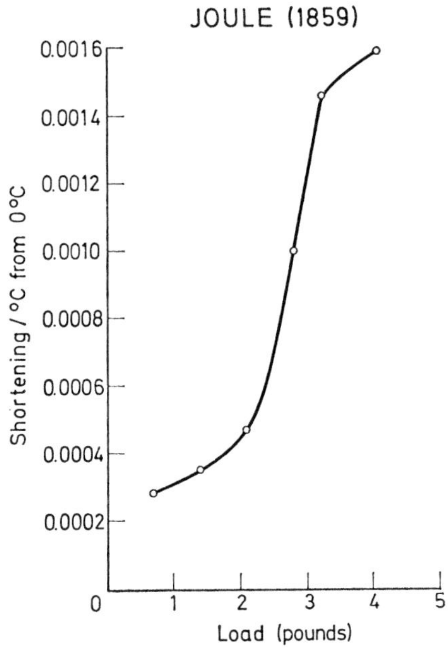

Fig. 4.256. The second GOUGH effect in rubber, showing the observed shortening per degrees C, vs the load in pounds.

the results shown in Fig. 4.256. The amount of shortening per °C from 0° C is seen to increase markedly with increased tensile load.

These experiments of JOULE, which had been suggested by KELVIN for a comparison with his analysis of thermoelasticity, stimulated a succession of experiments and proposed explanations during the next few decades.

I have referred above to SCHMULEWITSCH's[1] experimental efforts to simulate in vulcanized rubber, behavior like that of muscle tissue. The basis of his analogue was the finding of the thermal anomaly of rubber in the leg muscle of the frog. His explanation for the GOUGH effect, as we saw in Chap. III, was disposed of by the dynamic wave propagation experiment of EXNER,[2] whose main interest was in studying incremental stress-strain behavior on either side of the "neutral point" of the thermal effect. SCHMULEWITSCH had performed his experiments upon the apparatus of A. FICK of Zurich, who with W. DYBKOWSKY[3] in the same year turned his attention back to the muscular tissue itself in an effort to draw upon the GOUGH effect to explain some aspects of the behavior of living tissue.

In 1873, BERTIN[4] provided the first review of prior research on the finite thermoelasticity of rubber. It is interesting partly for its having noted the fact that in a paper in 1855 W. THOMSON (KELVIN) had cited rubber as being like other bodies, dilating and cooling with tension.[5] GOUGH's article thus had been rediscovered between 1855 and 1857.

A series of 19th century papers on the finite elasticity of rubber explored most of the single stress states. As we have seen in Chap. II, Sect. 2.23, WINKLER[6] in

[1] SCHMULEWITSCH [1866, *1*]; see Sect. 3.33.
[2] EXNER [1874, *1*]; see Sect. 3.33 above.
[3] DYBKOWSKY and FICK [1866, *1*].
[4] BERTIN [1873, *1*].
[5] WILLIAM THOMSON (KELVIN) [1855, *1*].
[6] WINKLER [1878, *1*].

1878 compared results for tension and compression and showed that the origin at zero stress was of no particular significance. IMBERT[1] in 1880, whose tensile stress-strain formula for rubber was among the stress-strain functions listed by MEHMKE[2] in 1897, also was referred to above (Sect. 2.23). IMBERT had made a series of experiments determining tangent moduli. He found for weak loads an increase in the tangent moduli with temperature, and for heavy loads, a decrease with increasing temperature. He was interested chiefly in those experiments which showed that there existed a load such that the tangent moduli were independent of temperature variations. The most original experimental contribution of IMBERT was his study of rubber membranes inflated inside of spherical glass bottles. By means of capillary tubes connected with either side of such a water filled system, he determined tensile forces on the membrane. This work was performed primarily because of his interest not only in the behavior of rubber balloons and fish bladders under pressure, but also in the behavior of the human foetal sack at birth.

In 1886 we find C. PULFRICH[3] repeating WERTHEIM's[4] experiments of 1848 related to POISSON's ratio or the compressibility of rubber. PULFRICH, however, was cognizant of the importance of WEBER's[5] elastic after-effect in rubber based upon the experiments of KOHLRAUSCH[6] in this substance in 1876. PULFRICH's purpose was the ambitious one of attempting to determine experimentally whether POISSON's ratio was affected by the elastic after-effect.

PULFRICH performed the experiments on a rubber tube. He measured the change of volume of the water filled interior by its rise or fall in an attached glass tube; by means of a cathetometer he determined elongation of the specimen under tensile load. He loaded and unloaded the specimen and observed the deformation over a period of many hours. The striking feature of the results was that invariably the contractions were larger than the elongations, which had been observed earlier by KOHLRAUSCH. PULFRICH found no significant change of volume during the elastic after-effect.

Referring to the correction by W. C. RÖNTGEN[7] in 1876 of WERTHEIM's[8] erroneous calculation of compressibility, and the subsequent claim by others that rubber was indeed a compressible solid, PULFRICH[9] showed that when properly calculated (see Chap. III, Sect. 3.16 above) his data, like WERTHEIM's, demonstrated the incompressibility of rubber when under tensile load.

The "batch" sensitivity of rubber was made evident in the experiments of MALLOCK[10] in 1889, who determined dynamic and quasi-static values of the E, μ, and K moduli in soft grey, red and hard grey rubber. The E and μ were determined in the usual way. He obtained the volume elasticity by immersing the India rubber under a pressure of 550 lbs./in.^{2} in a water filled glass tube. He determined moduli for small stress and the tangent modulus, E, for large extension. The results for the former are listed in Table 141 where the units are the inch, the pound, and the second. One notes the large variation with the type of rubber, the difference between dynamic and static values depending upon the

[1] IMBERT [1880, *1*].
[2] MEHMKE [1897, *1*].
[3] PULFRICH [1886, *1*].
[4] WERTHEIM [1848, *1*].
[5] WEBER [1835, *1*].
[6] KOHLRAUSCH [1876, *1*].
[7] RÖNTGEN [1876, *1*].
[8] WERTHEIM [1848, *1*].
[9] PULFRICH [1886, *1*].
[10] MALLOCK [1889, *1*].

Table 141. MALLOCK (1889).

Description of india-rubber	Density	Young's modulus (lbs./sq. in.)		Simple rigidity (lbs./sq. in.)	
		Statical	Dynamical	Statical	Dynamical
Soft grey	1.289	124	195	65	80 to 127
Red	1.407	166	217	50	57
Hard grey	2.340	495	500	158	156 to 202
	Volume elasticity lbs./ sq. in.	Viscosity lbs./sq. in.	Limit of stretching of unit length	Breaking strain lbs./sq. in.	Breaking strain for a square inch of unstrained material (lbs.)
				about	about
Soft grey	198000	13.74	9.9	8100	820
Red	115000	2.578	7.3	6400	820
Hard grey	940000	7.725	4.4	4400	820

density. The viscosity measurements demonstrated at least one of BOLTZMANN's[1] difficulties in 1882 when he had chosen rubber as the solid to test by experiment SAINT-VENANT's theory of the impact of rods (Chapt. III, Sect. 3.34).

In 1897, THURSTON[2] commented upon the "Singular Stress-Strain Relations of India Rubber," by returning to the study of the inflection point in the large tensile deformation of rubber. This aspect had been largely ignored since JOULE's experiments in 1859. In Fig. 4.257 is shown THURSTON's plot of stress in lbs./in.^{2} vs nominal strain, which emphasizes the flex point in the tensile diagram whose technological possibilities it was the purpose of his paper to make evident. In addition to seeing the practical engineering uses of such behavior, THURSTON also referred to the fact that he had been unable to induce in this solid his elastic incremental increase, which he had discovered in metals in 1873, i.e., the "THURSTON effect."[3]

In 1907, LUDWIG SCHILLER[4] returned to PULFRICH's problem of two decades earlier. Referring to RÖNTGEN's calculation in 1876, and apparently unaware of

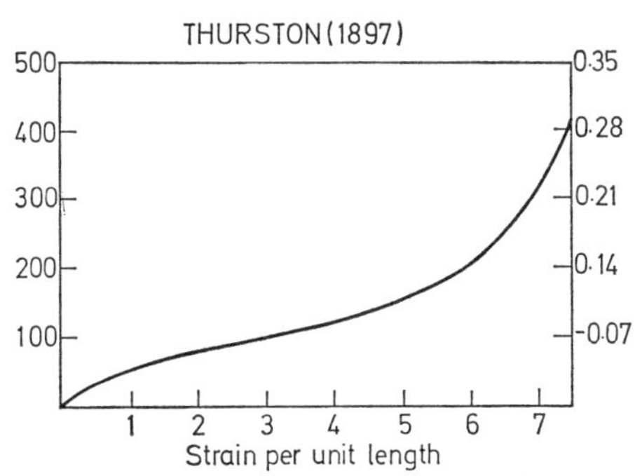

Fig. 4.257. THURSTON's (1897) tensile stress-strain diagram of India rubber.

[1] BOLTZMANN [1882, *1*].
[2] THURSTON [1897, *1*].
[3] THURSTON [1873, *1*].
[4] SCHILLER [1907, *1*].

PULFRICH's contribution, SCHILLER performed experiments on a rubber plate to determine the thickness at a given elongation. Calculating an equivalent value for ν for 20 elongations ranging from $\varepsilon = 0.023$ to $\varepsilon = 1.021$, SCHILLER obtained an average value of $\nu = 0.480$ by RÖNTGEN's[1] incompressibility relation for finite strain. He referred to a similar series of experiments performed the year before, in 1906, by O. FRANK which had provided a value of $\nu = 0.460$. SCHILLER attributed part of the difference in values to FRANK's strip length to width ratio of 40:10 compared to his own ratio of 35:1.6. He considered significant the small difference between RÖNTGEN's incompressibility value of $\nu = 0.5$ for all strain and SCHILLER's $\nu = 0.480$. (RÖNTGEN's calculation provided an apparent value of ν at all strain to be constant with the value $\nu = 0.5$ for incompressibility for infinitesimal strain.)

The monograph on *The Physics of Rubber Elasticity* by L. R. G. TRELOAR[2] in 1958 makes somewhat superfluous a detailed discussion of experimental research in finite rubber elasticity since about 1930.[3] A review of this work reveals major advances in understanding some aspects of the physics of rubber, including the development of a crystalline structure at large deformation; however, there was a surprising amount of mid-20th century repetition of 19th century measurements. Volume changes upon extension were shown to be determinable. WINKLER's[4] original observation that the tangent modulus continued to increase when a decrease in tension was followed by an increase in compression after the stress had passed through zero, was rediscovered. The GOUGH[5] effect at small and very large extension was studied, including both the extension and retraction features which JOULE[6] had examined.

Certainly the most important 20th century experimental development in the finite elasticity of rubber was the experiments of RONALD S. RIVLIN and D. W. SAUNDERS[7] in 1951, which are detailed in TRELOAR's book, but of course I shall refer here to the original paper.

In what was given as Part VII of a series of papers in the *Philosophical Transactions of the Royal Society of London* in 1948 and 1949, RIVLIN and SAUNDERS described experiments on finite deformation performed for comparison with the theory of large elastic deformation of incompressible isotropic materials developed by RIVLIN in the previous sections of the total work. The mathematical theory[8]

> ... describes the deformation, under the action of applied forces, of bodies of ideal highly elastic materials which are incompressible and isotropic in their undeformed state. The relevant physical properties of the material are specified in terms of a stored-energy function W which must be a function of two strain invariants I_1 and I_2.
>
> These invariants are expressible in terms of the principal extension ratios λ_1, λ_2 and λ_3 at the point of the deformed body considered, by the formulae
>
> $$I_1 = \lambda_1^2 + \lambda_2^2 + \lambda_3^2 \quad \text{and} \quad I_2 = \frac{1}{\lambda_1^2} + \frac{1}{\lambda_2^2} + \frac{1}{\lambda_3^2},$$
>
> in which, since the material of the body is assumed incompressible, $\lambda_1 \lambda_2 \lambda_3 = 1$.

[1] RÖNTGEN [1876, *1*].

[2] TRELOAR [1958, *1*].

[3] Out of a total of 201 references, the only papers of the 19th century TRELOAR cited were those of GOUGH, JOULE, and KELVIN, already noted above. Those three plus six papers from the 1920's constitute all the references TRELOAR cited prior to 1930; all but 31 references were within twenty years of the date of publication of the book.

[4] WINKLER [1878, *1*].

[5] GOUGH [1805, *1*].

[6] JOULE [1859, *1*].

[7] RIVLIN and SAUNDERS [1951, *1*].

[8] *Ibid.*, p. 252.

The experiments considered by RIVLIN and SAUNDERS were: (1) the pure homogeneous deformation of a thin sheet of rubber in which the deformation was varied in such a manner that one of the invariants of the strain, I_1 or I_2 was maintained constant; (2) pure shear of a thin piece of rubber; (3) simultaneous simple extension and pure shear of a thin sheet; (4) simple extension of a strip; (5) simple compression; (6) simple torsion of a right-circular cylinder; and, (7) superposed axial extension and torsion of a right-circular cylindrical rod. Since the experiments had as a major purpose the comparison of different cases in terms of expected conformity to the theory, attention had to be paid to the control and variation of the proper quantities.

RIVLIN realized that within the accuracy of experiment, an analysis of load and deformation for the various simple types of deformation would provide the response function for this solid. Assuming a stored energy,[1] W, RIVLIN examined experimentally the quantities $\partial W/\partial I_1$ and $\partial W/\partial I_2$ as functions of I_1 and I_2. These data led him to approximate W as follows:

$$W = C(I_1 - 3) + f(I_2 - 3) \tag{4.83}$$

where $C =$ constant for the undeformed solid, $\lambda_1 = \lambda_2 = \lambda_3 = 1$ so that $I_1 = I_2 = 3$.

RIVLIN had shown that the principal components of stress t_i associated with the deformation of the material for which extension ratios were λ_i, was given by

$$t_i = 2\left[\lambda_i^2 \frac{\partial W}{\partial I_1} - \frac{1}{\lambda_i^2}\frac{\partial W}{\partial I_2}\right] + P, \quad i = 1, 2, 3, \tag{4.84}$$

where $-P$ is an arbitrary hydrostatic pressure.

In the first experiment, which is shown diagrammatically in Fig. 4.258c, the springs not only permitted of a determination of the forces applied, but allowed an easy adjustment of λ_1 and λ_2 to maintain either I_1 or I_2 constant while the other was varied.

The results obtained in this experiment are shown in Fig. 4.259. $\partial W/\partial I_1$ is here approximately a constant function of both I_1 and I_2, whereas $\partial W/\partial I_2$ decreases approximately linearly with I_1 and I_2. RIVLIN and SAUNDERS also gave their results in tabulated form. The decrease in $\partial W/\partial I_2$ with I_1 while $\partial W/\partial I_1$ remains constant with I_2, requires further experimental study of the assumption of the existence of a stored energy since the relation $\frac{\partial}{\partial I_2}\left(\frac{\partial W}{\partial I_1}\right) = \frac{\partial}{\partial I_1}\left(\frac{\partial W}{\partial I_2}\right)$ is not satisfied by the experimental results.

The experiments on pure shear and on combined shear and simple extension were performed on the apparatus shown in Fig. 4.260 where $AA'B'B$ was the specimen loaded in shear by the horizontal motion of bar R, and in tension by the load on the bar, H. Plots were given of the tabulated values of $\left(\frac{\partial W}{\partial I_1} + \lambda_2 \frac{\partial W}{\partial I_2}\right)$ vs I_2 for pure shear and pure shear with simple tension. These results were found to be in accord with the experiments on simple torsion and simple torsion superposed upon an elongation.

Experiments on simple tension provided the measured values of $\frac{\partial W}{\partial I_1} + \frac{1}{\lambda}\frac{\partial W}{\partial I_2}$ given in Table 142 which are compared with calculated values from the pure shear experiments above.

[1] TRUESDELL and NOLL in 1965 [1965, *1*] presented RIVLIN and SAUNDERS' results in a form which did not require the assumption of a stored energy.

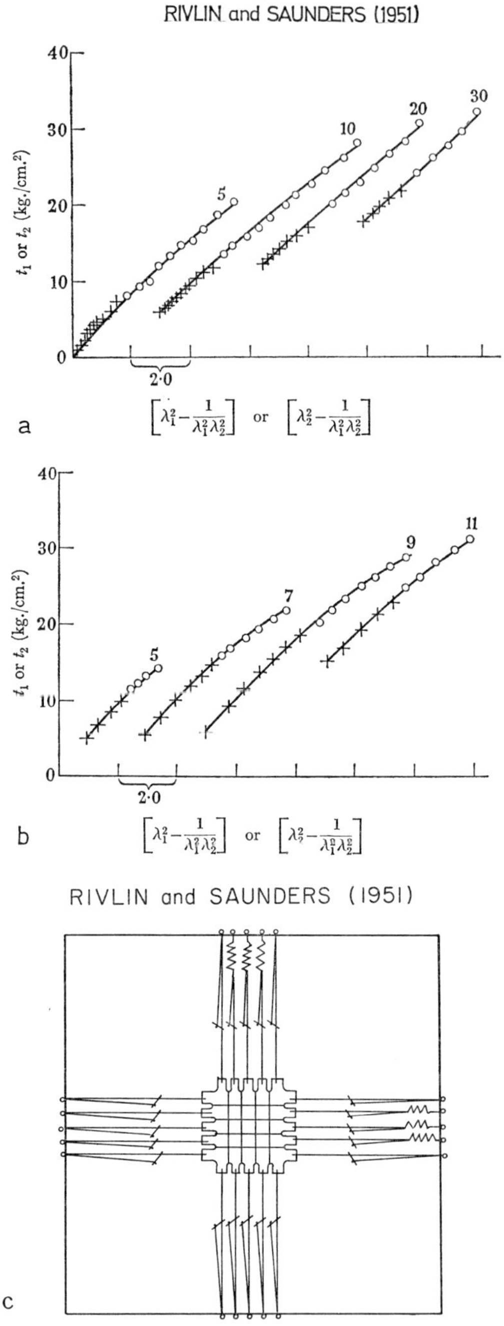

Fig. 4.258. (a) Plot of t_1 and t_2 against $(\lambda_1^2 - 1/\lambda_1^2\,\lambda_2^2)$ and $(\lambda_2^2 - 1/\lambda_1^2\,\lambda_2^2)$ respectively at constant I_2. The crosses denote t_2 and the circles t_1; the origin is shifted by two units parallel to the abscissa for each increment in the value of I_2. (b) Plot of t_1 and t_2 against $(\lambda_1^2 - 1/\lambda_1^2\,\lambda_2^2)$ and $(\lambda_2^2 - 1/\lambda_1^2\,\lambda_2^2)$ respectively at constant I_1. The crosses denote t_2 and the circles t_1; the origin is shifted by two units parallel to the abscissa for each increment in the value of I_1. (c) The experiment from which these data were obtained.

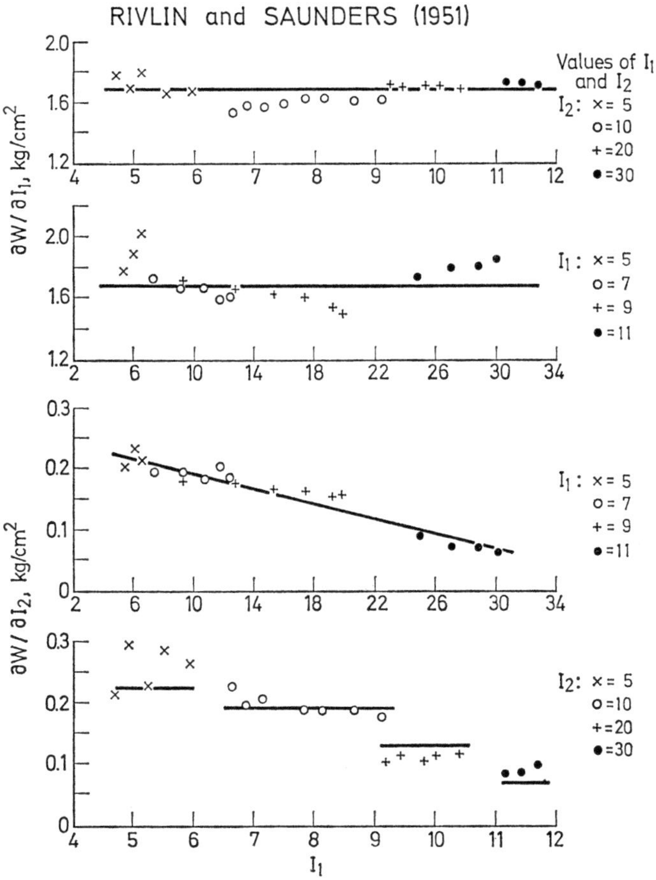

Fig. 4.259. Plot of $\partial W/\partial I_1$ and $\partial W/\partial I_2$ against I_1 and I_2.

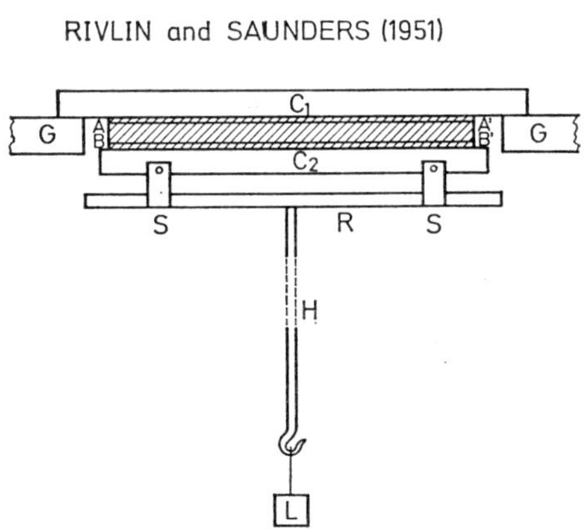

Fig. 4.260. Experimental arrangement for experiment on pure shear and combined shear and simple extension.

Table 142. Rivlin and Saunders (1951).

I_2		3.2	3.6	4.0	4.4	4.8	5.2
$\left(\frac{\partial W}{\partial I_1}+\frac{1}{\lambda}\frac{\partial W}{\partial I_2}\right)$	calculated (in kg/cm^2)	2.18	2.08	2.02	1.98	1.94	1.92
$\left(\frac{\partial W}{\partial I_1}+\frac{1}{\lambda}\frac{\partial W}{\partial I_2}\right)$	measured (in kg/cm^2)	2.24	2.08	1.96	1.92	1.88	1.85

As Rivlin and Saunders pointed out, the agreement is within experimental error.

For brevity I shall omit the experiments comparing simple compression and simple tension. Those on torsion of a cylinder were achieved by means of the apparatus of Fig. 4.261.

After checking that the hysteresis was small by comparing the loading and unloading behavior, Rivlin and Saunders provided the important experimental results of Fig. 4.262, in which the normal force, N, was plotted against the angle

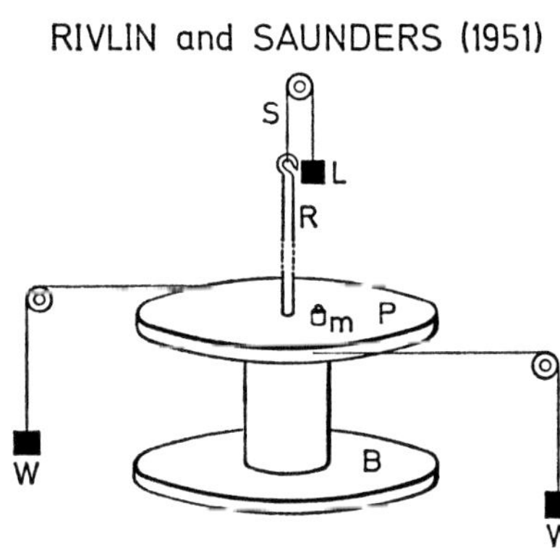

Fig. 4.261. Experimental arrangement for experiment on torsion of a cylinder.

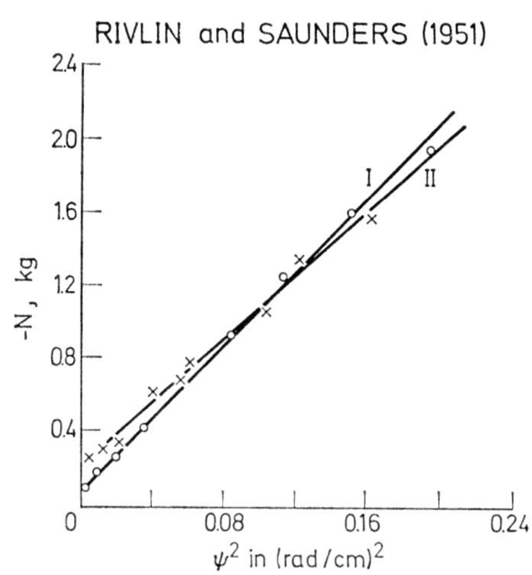

Fig. 4.262. Plot of normal force $-N$ against ψ^2. I, load increasing; II, load decreasing.

of twist squared, ψ^2, for loading and unloading. This correlation of the normal force with the square of the angle of twist provided experimentally for finite strain the phenomenon which POYNTING[1] had endeavored to determine at small strain. In addition it provided a proper theoretical foundation for the effect, since this behavior was predicted by the theory under consideration.

Finally, in the apparatus shown in Fig. 4.263, RIVLIN and SAUNDERS[2] performed simultaneous simple extension and torsion experiments.

The torque, M, depended upon the angle of twist, ψ, for various values of the extension ratio, λ, as is seen in Fig. 4.264, from which the plot of $\partial W/\partial I_2$ vs I_2-3 of Fig. 4.265 was given.

All of the above experiments were performed at deformations below that at which crystallization occurs.

These experiments of RIVLIN and SAUNDERS are a landmark in the history of experimental mechanics, as RIVLIN's theoretical counterpart is in rational mechanics. Few of the subjects considered in this treatise have seen anything

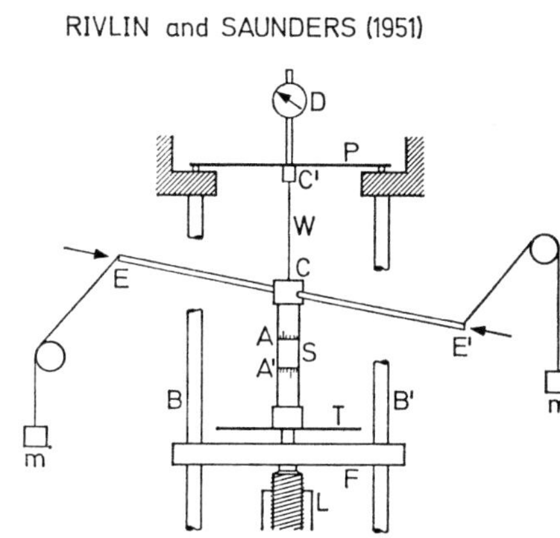

Fig. 4.263. Experimental arrangement for experiment on torsion superposed on simple extension.

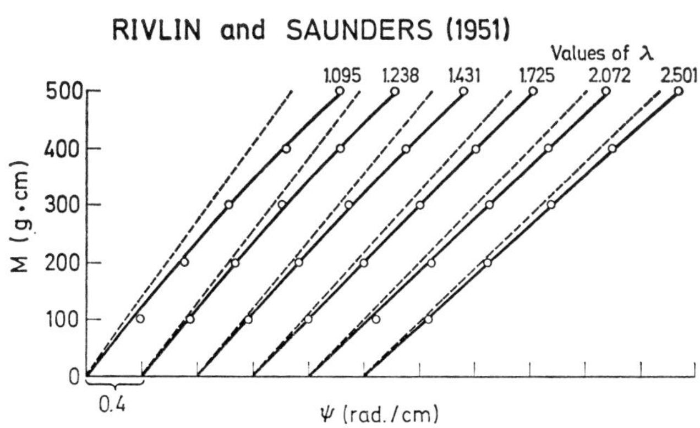

Fig. 4.264. Plot of torsional couple M against amount of torsion ψ for various values of the extension ratio λ. Large range of variation of ψ; the origin is shifted by 0.4 unit parallel to the abscissa for each increment in the value of λ.

[1] POYNTING [1909, *1*].
[2] RIVLIN and SAUNDERS [1951, *1*].

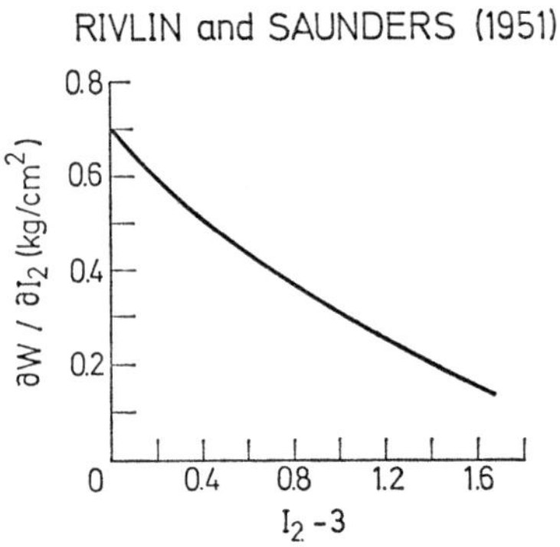

Fig. 4.265. Plot of $\partial W/\partial I_2$ against (I_2-3).

approximating the successful confluence of experimental observation and theoretical explanation which this mid-20th century study achieved. RIVLIN and SAUNDERS described with thoroughness and clarity the details of specimen preparation, the apparatus, the method of performing the experiments, and commentary on the important aspects of the correlations and the experimental limitations of certain of the measurements and calculations.

It is impossible to leave this subject without referring to the continued experiments on the finite elasticity of rubber by A. N. GENT and RIVLIN[1] the following year, 1952. These included the experiment of everting a hollow rubber tube 6 in. long, 1 in. in diameter, having external and internal radii of a_1 and a_2, respectively, before eversion. After first comparing a tube in simple extension with RIVLIN and SAUNDERS' results, inasmuch as the tubes were produced by a different vulcanization process, they everted the tubes and then determined their diameters. The diameters were given as $\mu_1 a_1$ and $\mu_2 a_2$, all measurements being in inches. After eversion the tube also underwent a change in length. GENT and RIVLIN found that the strain energy function given by Eq. (4.85), proposed by M. MOONEY[2] some years earlier, fit the data fairly well. Their comparison of experimental results and calculation shown in Table 143, where λ designates the extension ratio, with $\varepsilon_1 = -\lambda\mu_1^2$ and $\varepsilon_2 = -\lambda\mu_2^2$, and C_1 and C_2 are the constants in Eq. (4.85).

$$W = C_1(I_1-3) + C_2(I_2-3). \tag{4.85}$$

Table 143. GENT and RIVLIN (1952).

a_1 (cm)	a_2 (cm)	$\mu_1 a_1$ (cm)	$\mu_2 a_2$ (cm)	$-\lambda$	Measured ε_1	Measured ε_2	Calculated ε_1	Calculated ε_2	C_2/C_1
1.27	0.78	0.86	1.30	1.0275	0.47	2.84	0.4583	2.866	0.29
1.27	0.85	0.91	1.31	1.0175	0.52	2.42	0.5110	2.373	0.32
1.27	0.945	0.99_3	1.30	1.008	0.62	1.91	0.5926	1.876	0.42

The paper on the eversion experiments was "I" of a series. During the same year, GENT and RIVLIN published two additional papers,[3] designated as II and III. Their second paper described the torsion, inflation, and extension of rubber

[1] GENT and RIVLIN [1952, *1*].
[2] MOONEY [1940, *1*].
[3] GENT and RIVLIN [1952, *2*, *3*].

tubes for various degrees of vulcanization. Again, under these added conditions close correlation was provided between theory and prediction for the various cases. Their final paper in this area, III, described experiments on the small torsion of stretched rubber prisms. Four rubber rods, each 7.46 cm in length, had various rectangular cross-sectional ratios ranging from 1.001 to 3.980. Measurements of the torque M, the twist angle ψ, and the extension ratio λ, provided a further correlation with observation, and extended theoretical details.

Rubber, despite its sensitivity to specimen preparation and its peculiar thermal properties which stimulated 19th century interest in the substance, has become one of the solids whose response to large deformation is now the best defined by theory.

4.40. Summary

Experiment on the finite deformation of solids is a development mainly of the 20th century. Its relative newness and, in important areas such as plastic deformation and flow, the paucity of plausible theories to parallel the mass of reliable experimental knowledge, have made this a field ripe for speculation and construction of models. Hence, it is particularly advantageous to present the experimental foundations clearly, apart from the conflicting tenets which came from the perspectives both of applied mechanics and of atomistic mechanics.

The flow studies of Tresca of a century ago still are not satisfactorily explained for the experimentist who thinks in quantitative terms. More recently, new dimensions have been added to our knowledge of the physics of large deformation because of experiments in areas such as thermoplasticity, dynamic plasticity, and the plasticity of single crystals. Among the many fundamental physical facts disclosed is that the plastic deformation of crystals is inhomogeneous. And, for fully annealed crystalline solids, experiments have established that constitutive equations must comprehend second order transitions at fixed shear angles, a quantized distribution of deformation modes, and the Savart-Masson effect. Sooner or later, for this class of solids, parallel developments in continuum theory must include these phenomena. For the finite elasticity of rubber, on the other hand, success has been achieved in relating nonlinear elastic theory and experiment, but the substance is so dependent upon its prior thermal, mechanical, and chemical history that it often is difficult to compare the numerical results of one experimentist with those of another.

Sixty years of oversimplified impact experiments on finite deformation in crystalline solids led to a long, fruitless controversy on the role of viscosity in plasticity. In no way did those years of indirect measurement lay a foundation for the startling discovery in the 1950's that for fully annealed crystalline solids the dynamic plasticity at high strain rates was sensibly non-viscous over the entire temperature scale, nearly to the melting point of the solid. That discovery became possible only after a method had been devised to study finite amplitude waves directly, and accurately, independently of quasi-static hypotheses.

The pressure of technology for rapid answers generated highly restrictive approaches to the study of finite deformation. Before the response to loading in uniform plastic deformation was understood, efforts were made to analyze simultaneous non-uniform distributions of stress and of strain in both elastic and plastic regions. For the "ideal" plastic solid posited, such efforts overemphasized the importance of the initial yield surface and hence overemphasized the small deformation region of transition between the approximately linear infinitesimal deformation and the finite plastic strain. It is obvious to any serious experimentist that the physical situation is very different, much more complex, and far more interesting than can be seen in terms of such imposed analytical constraints.

There were, moreover, contradictions between such oversimplified theory and the results of the overly restricted experiments. The problem of adequately describing finite deformation was resolved only when experiments on large deformation for monotonic increasing, uniformly distributed stress were performed for both radial and non-radial loading paths and for strain rates from the lowest recorded value of 10^{-9} sec^{-1} to the highest for which it is possible to define the wave front, 10^4 sec^{-1}. Many of the earlier conjectures simply do not have a counterpart in nature. For fully annealed crystalline solids, the finite deformation, whether reversible elastic or irreversible plastic, now can be expressed by experimentally based constitutive equations derivable from known strain energy relations. At the same time, the need for further study of material stability for both small and large deformation has been made quite evident. For plasticity in which 95% of the energy at finite strain is experimentally known to be converted to heat, the vastly different loading and unloading response functions for crystalline solids offers a decidedly complex challenge for future theorists. A major objective of the present treatise has been to present and develop the experimental foundations of this relatively recent area in the physics of solids for which detailed, reliable, experimental knowledge at this point has far outstripped plausible explanation.

We are living today with the distortions of an overemphasis upon technique and apparatus, the excessive publication of shallowly conceived experimentation, and the detailed description of trivia; nevertheless, at no time in the history of experimental solid mechanics have the opportunities for scholarship been greater. In this treatise it has been demonstrated amply that experimentists in physics, by an independent study of nature, can critically evaluate heretofore accepted explanations, provide the basis for distinguishing between the physically trivial and the physically significant, and, at the same time, be the major source of new understanding and pivotal physical discovery.

References.

The sources used to compare the various units of measure were: ADAMS [1821, *1*]; BLIND [1906, *1*]; KENNELLY [1928, *1*]; LEMALE [1875, *1*]; NELKENBRECHER [1820, *1*]; TATE [1842, *1*].

1636 *1.* MERSENNE, MARIN: Harmonie Universelle, contenant la théorie et la pratique de la musique ... Paris: Sebastien Cramoisy. Facsimile reprint. Paris: Centre Nat. Rech. Sci. (1965).

1676 *1.* HOOKE, ROBERT: A Description of Helioscopes, and Some Other Instruments. London: John Martyn. Reprinted in: Early Science in Oxford (ed. R. T. GUNTHER), Vol. VIII. Oxford: R. T. Gunther 1931.

1678 *1.* HOOKE, ROBERT: Lectures De Potentia Restitutiva, or of Spring, Explaining the Power of Springing Bodies. London: John Martyn. Reprinted in: Early Science in Oxford (ed. R. T. GUNTHER), Vol. VIII, pp. 331–356. Oxford: R. T. Gunther 1931.

1687 *1.* BERNOULLI, JAMES: Letter (in 1687) to G. W. LEIBNIZ, appearing in: G. W. LEIBNIZ, Mathematische Schriften (ed. G. E. GERHARDT), Vol. III, Part 1, pp. 10–13. Heidelsheim: Georg Olms Verlagsbuchhandlung 1855. [Reprinted in 1962.]

1690 *1.* LEIBNIZ, GOTTFRIED WILHELM: Letter to JAMES BERNOULLI, appearing in G. W. LEIBNIZ, Mathematische Schriften (ed. G. E. GERHARDT), Vol. III, Part 1, pp. 13–20. Heidelsheim: Georg Olms Verlagsbuchhandlung 1855. [Reprinted in 1962.]

1694 *1*. BERNOULLI, JACOBUS (JAMES): Curvatura laminae elasticae ... Acta Eruditorum Lipsiae. June, 262–276. [Opera JACOBI BERNOULLI, 576–600.] See also TRUESDELL [1960, *1*], pp. 88–96.

1700 *1*. MARIOTTE, EDMÉ: Traité du Mouvement des Eaux et des Autres Corps Fluides. Nouvelle édition, Paris. MARIOTTE had died in 1684. See Eulogy by CONDORCET [1799, *1*].

1705 *1*. BERNOULLI, JACQUES (JAMES): Veritable hypothèse de la resistance des solides, avec la démonstration de la courbure des corps qui font ressort ..., Mémoires Acad. Sci. Paris 1705, 4to ed., Paris, 176–186 (1706). [Second 4to ed., Paris, 176–186 (1730). 12mo ed., Amsterdam, 230–244 (1707). Opera JACOBI BERNOULLI, 976–989.] See also TRUESDELL [1960, *1*], pp. 105–109.

1720 *1*. s'GRAVESANDE, WILHELM JACOB STORM: Physices Elementa Mathematica, Experimentis Confirmata. Sive Introductio ad Philosophiam Newtonianam. Lugduni Batavorum, Vander Aa, Vol. 1. [A second edition was published in 1725.]

1721 *1*. RICCATI, JAMES: Excerpt of letter reproduced in PAUL HEINRICH FUSS: Correspondence Mathématique et Physique du XVIIIème Siècle, Vol. 2. St. Petersburg 1843. Reprinted in: The Sources of Science, Vol. 2, No. 35, pp. 101–102. New York and London: Johnson Reprint Corporation 1968. See also TRUESDELL [1960, *1*], p. 115.

1725 *1*. s'GRAVESANDE, WILHELM JACOB STORM: Physices Elementa Mathematica ... Second edition. [See 1720, first edition.]

1729 *1*. BÜLFFINGER, GEORG BERNHARD: De solidorum resistentia specimen. Commentarii Academiae Scientiarum, Petrop. **4**, 140–155.

1. MUSSCHENBROEK, PIETER VAN: Physicae Experimentales et Geometricae de Magnete, Tuborum Capillarium Vitreorumque Speculorum Attractione, Magnitudine Terrae, Cohaerentia Corporum Firmorum Dissertationes. Apud Samuelem Luchtmans, Lugduni Batavorum.

1739 *1*. MUSSCHENBROEK, PIETER VAN: Essai de Physique, Vol. I (translated by P. MASSUET). Leyden.

1741 *1*. BUFFON, GEORGES LOUIS LECLERC, Comte de: Expériences sur la force du bois. Sécond Mémoire. Mémoires de l'Académie Royale des Sciences, Paris, pp. 292–334.

1742 *1*. DUHAMEL DU MONCEAU, HENRI LOUIS: Reflexions et expériences sur la force des bois. Mémoires de l'Académie Royale des Sciences, Paris, pp. 335–370.

1751 *1*. BERNOULLI, DANIEL: De sonis multifariis quos laminae elasticae diversimode edunt disquisitiones mechanico-geometricae experimentis acusticis illustratae et confirmatae. Commentarii Academiae Scientiarum, Petrop. **13** (1741/1743), 167–196. See also [1770, *1*].

1765 *1*. D'ARCY, PATRICK: Mémoire sur la durée de la sensation de la vue. Mémoires de l'Académie Royale des Sciences, Paris, **37**, 700–718. Reprinted, Amsterdam: J. Schreuder (1773).

1770 *1*. BERNOULLI, DANIEL: Examen physico-mechanicum de motu mixto qui laminis elasticis a percussione simul imprimitur. Novi Commentarii Academiae Petropol. **15**, 361–380.

1773 *1*. COULOMB, CHARLES AUGUSTIN: Sur une application des règles de maximis et minimis à quelques problèmes de statique relatifs à l'architecture. Mémoires de Mathématique et de Physique, Acad. Sci. Paris **7**, 343–382.

1774 *1*. GAUTHEY, EMILAND MARIE: Mémoire sur la charge que peuvent porter les pierres. Observations sur la Physique, sur l'Histoire Naturelle et sur les Arts, Vol. 4, Paris.

1775 *1*. BELIDOR, BERNARD FOREST DE: La Science des Ingénieurs dans la Conduite des Travaux de Fortification et d'Architecture Civile. Nouvelle édition à la Haye. P. F. Gosse.

1780 *1*. COULOMB, CHARLES AUGUSTIN: Recherches sur la meilleure manière de fabriquer les aiguilles aimantées. Mémoires de Mathématique et de Physique, Acad. Sci. Paris **9**, 165–264.

1782 *1*. RICCATI, JORDAN (GIORDANO): Delle vibrazione sonore dei cilindri. Memorie Matematica e Fisica Società Italiana **1**, 444–525.

1784 *1*. COULOMB, CHARLES AUGUSTIN: Recherches théoriques et expérimentales sur la force de torsion et sur l'élasticité des fils de métal. Histoire de l'Académie Royale des Sciences, Paris, pp. 229–269. [Reprinted in Mémoires de Coulomb, Vol. 1, Collections de Mémoires Relatifs à la Physique, publiés par la Société Française de Physique, Paris: Gauthier-Villars (1884), pp. 63–103.]

1787 *1*. CHLADNI, ERNST FLORENS FRIEDRICH: Entdeckungen über die Theorie des Klanges. Leipzig: Weidemanns, Erben und Reich.

1799 *1*. CONDORCET, MARIE JEAN ANTOINE NICOLAS CARITAT: Éloge de Mariotte. Éloges des Académiciens de l'Académie Royale des Sciences, Paris **1**, 74–91.

1802 *1*. CHLADNI, ERNST FLORENS FRIEDRICH: Die Akustik. Leipzig.

1805 *1.* GOUGH, JOHN: A description of a property of Caoutchouc, or India rubber; with some reflections on the cause of the elasticity of this substance. Memoirs of the Literary and Philosophical Society of Manchester, Second series **1**, 288–295. [Also in Phil. Mag. **24**, 39–43 (1806).]

1806 *1.* GOUGH, JOHN: See Phil. Mag. **24**, 39–43. This is substantially the paper published in [1805, *1*].

1807 *1.* YOUNG, THOMAS: A Course of Lectures on Natural Philosophy and the Mechanical Arts, Vols. I and II. London.

1809 *1.* BERTHOLLET, CLAUDE LOUIS: Notes sur la chaleur produit par le choc et la compression. Mémoires de Physique et de Chimie de la Société d'Arcueil **2**, 441–448.

1. BIOT, JEAN BAPTISTE: Expériences sur la propagation du son à travers les corps solides et à travers l'air, dans des tuyaux très-allongés. Mémoires de Physique et de Chimie de la Société d'Arcueil **2**, 405–423.

1811 *1.* BENZENBERG, D.: Auszug aus einem Schreiben des D. Benzenberg an den Herausgeber über seine und Biot's Schallversuche. Annalen der Physik und Chemie, first series **37**, 221–225.

1. LÉHOT, DÉSORMES, and CLÉMENT: Versuche über den Widerstand, welchen Luft in langen Röhren in ihrer Bewegung leiden soll. Annalen der Physik und Chemie, first series **39**, 142–149. Also in: J. de Physique **73**, 36–40.

1813 *1.* DULEAU, ALPHONSE JEAN CLAUDE BOURGUIGNON: Essai théorique et expérimental sur la résistance du fer forgé. See DULEAU [1820, *1*].

1814 *1.* MUNKE, GEORG WILHELM: Über die Fortpflanzung des Schalles durch Wasser. Annalen der Physik und Chemie (Gilbert's) **48**, 66–104.

1815 *1.* DUPIN, PIERRE CHARLES FRANÇOIS: Expériences sur la flexibilité, la force et l'élasticité des bois. Journal de l'École Royale Polytechnique **10**, 137–211.

1816 *1.* BIOT, JEAN BAPTISTE: Dédication à M. Berthollet. Traité de Physique Expérimentale et Mathématique **1**, i–xxiv, Paris.

1. BREWSTER, Sir DAVID: On the communication of the structure of doubly-refracting crystals to glass, muriate of soda, fluor spar, and other substances by mechanical compression and dilatation. Phil. Trans. Roy. Soc. London, 156–178.

1. KARSTEN, KARL JOHANN BERNHARD: Manuel de la Métallurgie de Fer. [See second edition (1830).]

1. LAPLACE, PIERRE SIMON (Marquis) DE: Sur la vitesse du son dans l'air et dans l'eau. Annales de Chimie et de Physique **3**, 238–241. Also in: Phil. Mag. **49**, 260–262 (1817).

1817 *1.* BARLOW, PETER: Essay on the Strength and Stress of Timber. London. See also RENNIE [1818, *1*].

1. CHLADNI, ERNST FLORENS FRIEDRICH: Neue Beiträge zur Akustik. Leipzig.

1818 *1.* RENNIE, GEORGE: Account of experiments made on the strength of materials. (Letter to THOMAS YOUNG, M.D.; read 12 February, 1818.) Phil. Trans. Roy. Soc. London, 118–136.

1819 *1.* [DULEAU] CAUCHY, AUGUSTIN LOUIS (rapporteur, with POISSON and GIRARD): Analyse d'un mémoire intitulé: Essai théorique et expérimental sur la résistance du fer forgé. Annales de Chimie et de Physique, second series **12**, 133–148.

1820 *1.* DULEAU, ALPHONSE JEAN CLAUDE BOURGUIGNON: Essai Théorique et Expérimental sur la Résistance du Fer Forgé. Paris. [DULEAU presented the memoir to the French Academy 28 June, 1819; CAUCHY, on 9 August 1819, read the analysis written by CAUCHY, POISSON, and GIRARD.]

1. NELKENBRECHER, JOHANN CHRISTIAN: J. C. Nelkenbrechers allgemeines Taschenbuch der Munz-, Maass-, und Gewichtskunde, 13th edition, J. P. Schellenberg, editor. Berlin: Sanderschen Buchhandlung.

1. SAVART, FÉLIX: Mémoire sur la communication des mouvements entre les corps solides. Annales de Chimie et de Physique **14**, 113–172.

1821 *1.* ADAMS, JOHN QUINCY: Report upon Weights and Measures. Philadelphia: Abraham Small.

1822 *1.* CHLADNI, ERNST FLORENS FRIEDRICH: Remarques conçernant le Mémoire de M. Savart sur les communications des mouvements vibratoires entre les corps solides, imprimé dans les Annales de Chimie et de Physique, tome XIV, Juin. Annales de Chimie et de Physique, second series **20**, 74–78.

1. POUILLET, CLAUDE SERVAIS MATHAIS MARIE ROLAND: Extrait des séances de l'Académie Royale des Sciences: Séance du lundi 26 Août. Annales de Chimie et de Physique, second series **21**, 77–79.

1. ROBISON, JOHN: Strength of materials. A System of Mechanical Philosophy, Vol. 1, pp. 369–495. Edinburgh.

1. SAVART, FÉLIX: Mémoire sur les vibrations ... [See: 1824, *1*].

1. Tredgold, Thomas: A Practical Essay on the Strength of Cast Iron. London. [See: 1842, *1*.]

1823 *1.* Leslie, Sir John: Elements of Natural Philosophy. Edimburg (sic).

1824 *1.* Hodgkinson, Eaton: On the transverse strain, and strength of materials (presented in 1822). Memoirs of the Literary and Philosophical Society of Manchester, second series **4**, 225–289.

1. Savart, Félix: Mémoire sur les vibrations des corps solides, considérées en général. Annales de Chimie et de Physique, second series **25**, Part I, 12–50; Part II, 138–178; Part III, 225–269. [Presented to the Academy 22 April 1822].

1. Tredgold, Thomas: Experiments on the elasticity and strength of hard and soft steel. Phil. Trans. Roy. Soc. London, Part I, 354–359. See also [1822, *1*].

1826 *1.* Babbage, Charles: On electrical and magnetic rotations. Phil. Trans. Roy. Soc. London, Parts II and III, 494–528.

1. Bevan, Benjamin: Account of an experiment on the elasticity of ice. (In a letter to Dr. Thomas Young read before the Royal Society on 27 April, 1826.) Phil. Trans. Roy. Soc. London, Part III, 304–306.

1. Cauchy, Louis Augustin: Mémoire sur le choc des corps élastiques. Bulletin de la Société Philomatique, Paris 180–182 (December).

1. Dulong, Pierre Louis: Untersuchungen über das Brechungsvermögen der elastischen Flüssigkeiten. Annalen der Physik und Chemie, second series **6**, 393–424.

1. Navier, Claude Louis Marie Henri: Expériences sur la résistance de divers substances à la rupture causée par une tension longitudinale. Annales de Chimie et de Physique **33**, 225–240.

1. Young, Thomas: Note by Dr. Young. Phil. Trans. Roy. Soc. London, Part III, 306.

1827 *1.* Poisson, Siméon Denis: Note sur l'extension des fils et des plaques élastiques. Annales de Chimie et de Physique, second series **36**, 384–387.

1828 *1.* Cagniard de Latour, Charles: Note sur l'élasticité des cordes métalliques. La Globe **6**, No. 19, 107–108, Paris. [Translated and published in English in the Edinburgh Journal of Science **8**, 201–203 (1828).]

1829 *1.* Girard, P. S.: Mémoire sur la posé des conduites d'eau ... [See: 1831, *1*].

1. Lagerhjelm, Pehr: Schreiben des Hrn. Lagerhjelm an den Herausgeber, in Betreff seiner Untersuchungen über die Cohaesion u.s.w. des Eisens. Annalen der Physik und Chemie (Poggendorff), second series **17**, 348–351.

1. Poncelet, Jean Victor: Cours de Mécanique Industrielle, Metz. [See: 1841, *1*, for title of second edition.]

1. Savart, Félix: Mémoire sur la réaction de torsion des lames et des verges rigides. Annales de Chimie et de Physique **41**, 373–397.

1830 *1.* Cauchy, Augustin Louis: Mémoire sur la torsion et les vibrations tournantes d'une verge rectangulaire. Mémoires de l'Académie des Sciences, Paris, **9**, 119–124. [Presented at the Academy, 9 February, 1829.]

1. Coriolis, Gustave Gaspard: Expériences sur la résistance du plomb à l'écrasement, et sur l'influence qu'a sur sa dureté une quantité inappréciable d'oxide. Annales de Chimie et de Physique **44**, 103–111.

1. Karsten, Carl Johann Bernhard: Manuel de la Métallurgie du Fer **1**, Metz. Second edition, F. J. Culmann translator. (First edition: 1816.)

1. Savart Félix: Recherches sur l'élasticité des corps qui crystallisent régulièrement. Mémoires de l'Académie Royale des Sciences **9**, 405–453.

1. Weber, Wilhelm: Über die specifische Wärme fester Körper, insbesondere der Metalle. Annalen der Physik und Chemie, second series **20**, 177–213.

1831 *1.* Gerstner, Franz Joseph: Handbuch der Mechanik, Vol. 1. Leipzig: Herbig.

1. Girard, P. S.: Mémoire sur la posé des conduites d'eau dans la ville de Paris, tableaux et discussions d'expériences entreprises à ce sujet sur la dilatabilité de la fonte de fer. Mémoires de l'Académie Royale des Sciences de l'Institute de France **10**, 405–456.

1. Hodgkinson, Eaton: Theoretical and experimental researches to ascertain the strength and best forms of iron beams. Memoirs of the Literary and Philosophical Society of Manchester, second series **5**, 407–544.

2. Hodgkinson, Eaton: On the chain bridge at Broughton. Memoirs of the Literary and Philosophical Society of Manchester, second series **5**, 384–397; 545–553, and see Appendix.

3. Hodgkinson, Eaton: On the forms of the catenary in suspension bridges. Memoirs of the Literary and Philosophical Society of Manchester, second series **5**, 354–383.

1. Vicat, Louis Joseph: Ponts suspendus en fil de fer sur le Rhône, Rapport. Annales des Ponts et Chaussées, Mémoires; first series, first semester **1**, 93–144.

1832 *1.* Gerstner, Franz Anton (Ritter) von: Über die Festigkeit der Körper. Annalen der Physik und Chemie (Poggendorff), second series **26**, 269–279.

1833 *1.* Gerstner, Franz Joseph: Handbuch der Mechanik. Vol. I, Mechanik fester Körper.

1. Poisson, Siméon Denis: Chapter entitled: Choc longitudinal des verges élastiques. Traité de Mécanique **2**, 331–343, Paris.

1. Navier, Claude Louis Marie Henri: Resumé des Leçons données à l'École des Ponts et Chaussées, sur l'application de la mécanique à l'établissement des constructions et des machines. Part I, contenant les leçons sur la résistance des matériaux et sur l'établissement des constructions en terre, en maçonnerie et en charpente; second edition, Paris. [See: 1851, *1*; 1864, *1*.]

1. Vicat, Louis Joseph: Note sur l'allongement progréssif du fil de fer soumis à divers tensions. Annales de Chimie et de Physique, second series **54**, 35–40.

2. Vicat, Louis Joseph: Recherches expérimentales sur les phénomènes physiques qui précèdent et accompagnent la rupture ou l'affaissement d'une certaine classe de solides. Annales des Ponts et Chaussées, Mémoires, first series **6**, second semester, 201–286.

1834 *1.* Vicat, Louis Joseph: Note sur l'allongement progressif du fil de fer soumis à divers tensions. Annales des Ponts et Chaussées, Mémoires, first series, first semester 40–44. And see [1833, *1*, *2*].

1835 *1.* Weber, Wilhelm: Ueber die Elasticität der Seidenfäden. Annalen der Physik und Chemie, second series **34**, 247–257.

1837 *1.* Barlow, Peter: A Treatise on the Strength of Timber, Cast Iron, Malleable Iron, and Other Materials; with rules for application in architecture, construction of suspension bridges, railways, etc. London: J. Weale.

1. Savart, Félix: Recherches sur les vibrations longitudinales. Annales de Chimie et de Physique, second series **65**, 337–402.

1838 *1.* Colladon, Jean Daniel, and Jacques Charles François Sturm: Mémoire sur la compression des liquides. Académie des Sciences, Mémoires présentés par divers savants, second series **5**, 267–347. [This paper had been presented to the French Academy in 1827.]

1839 *1.* Hodkginson, Eaton: On the relative strength and other mechanical properties of cast iron obtained by hot and cold blast. J. Franklin Inst. **24**, 184–196, 238–257.

1. Leblanc, Félix: Expériences sur la résistance du fil de fer et la fabrication des cables. Annales des Ponts et Chaussées, Mémoires, first series, second semester, 300–334.

1841 *1.* Karmarsch, Karl: Ueber die Festigkeit und Elastizität der Darmsaiten. Dingler's Polytechnisches Journal **81**, 427–432.

1. Masson, Antoine Philibert: Sur l'élasticité des corps solides. Annales de Chimie et de Physique, third series **3**, 451–462.

1. Poncelet, Jean Victor: Introduction à la Mécanique Industrielle, Physique ou Expérimentale. Metz (second edition). [The first edition of this book appeared in 1829 with the title: Cours de Mécanique Industrielle. See: 1829, *1*.]

1. Weber, Wilhelm: Ueber die Elastizität fester Körper. Annalen der Physik und Chemie, second series **54**, 1–18.

1. Wertheim, Guillaume: Paquet cacheté addressé par M. Wertheim en 1841 ... [See: 1842, *2*.]

1842 *1.* Colladon, Jean Daniel: Nouvelles expériences sur la propagation du son dans l'eau. L'Institut, No. 401.

1. Giulio, Carlo Ignazio: Sur la torsion des fils métalliques et sur l'élasticité des ressorts en hélices. Mémoires de l'Académie de Turin **4**, 329–384.

1. Regnault, Henri Victor: Note sur la dilatation du verre. Annales de Chimie et de Physique, third series **4**, 64–67.

1. Tate, William: The Modern Cambist, 4th edition. London: Effington Wilson.

1. Tredgold, Thomas: Practical Essay on the Strength of Cast Iron and Other Metals (fourth edition). London 1842–1846.

1. Wertheim, Guillaume: Recherches sur l'élasticité et la ténacité des métaux. Extrait. Compt. Rend. Acad. Sci. Paris **15**, 110–115.

2. Wertheim, Guillaume: Paquet cacheté addressé par M. Wertheim en 1841, et dont le dépôt a été accepté par l'Académie dans sa séance du 19 juillet. Compt. Rend. Acad. Sci. Paris **15**, 115–117.

1843 *1.* Hodgkinson, Eaton: Experiments to prove that all bodies are in some degree inelastic, and a proposed law for estimating the deficiency. Report of the 13th Meeting of the British Association for the Advancement of Science, Cork, 23–25. See also, [1846, *1*]. [Published in London in 1844.]

1844 *1.* Babinet, *et al.*: Rapport sur deux Mémoires de M. Wertheim, intitulés: Recherches sur l'élasticité. Compt. Rend. Acad. Sci. Paris **18**, 921–932 .

1. BAUDRIMONT, A.: Sur des procédés mécaniques destinés à donner la mesure d'intervalles de temps très courts. Compt. Rend. Acad. Sci. Paris **19**, 1454–1455. [Supplement to POUILLET's Note sur un moyen ...]

1. HODGKINSON, EATON: Experimental inquiries into the falling-off from perfect elasticity in solid bodies. Report of the Fourteenth Meeting of the British Association for the Advancement of Science, York, Section 2, Transactions, 25–27.

1. POUILLET, CLAUDE SERVAIS MATHAIS ROLAND: Note sur un moyen de mesurer des intervalles de temps extrêmement courts, comme la durée du choc des corps élastiques, celle du débandement des ressorts, de l'inflammation de la poudre, etc., et sur un moyen nouveau de comparer les intensités des courants électriques, soit permanents, soit instantanés. Compt. Rend. Acad Sci. Paris **19**, 1384–1389. [Also published in German: Über ein Mittel zur Messung äußerst kurzer Zeiträume, wie die Dauer des Stoßes elastischer Körper, der Auslösung von Springfedern, der Entzündung von Schießpulver usw., und über ein neues Mittel, die Intensität elektrischer Ströme, permanenter wie instantaner, zu messen. Annalen der Physik und Chemie, 3rd series **4**, 452–459 (1845).] See also, D'ARCY [1765, *1*] and BAUDRIMONT [1844, *1*].

1. WERTHEIM, GUILLAUME: Recherches sur l'élasticité. Annales de Chimie et de Physique, third series **12**.
(a) Première Mémoire: De l'élasticité et de la cohésion des métaux, 385–454. (Presented 18 July, 1842.)
(b) Deuxième Mémoire: De l'élasticité et de la cohésion des alliages, 581–610. (Presented 8 May, 1843.)
(c) Troisième Mémoire: De l'influence du courant galvanique et de l'électromagnetisme sur l'élasticité des métaux, 610–624. (Presented 22 July, 1844.)
[Also published in German: Untersuchungen über die Elasticität. Annalen der Physik und Chemie, Leipzig, Ergänzungsband **11**/II, 1–114 (1848).]

2. WERTHEIM, GUILLAUME: Recherches sur l'élasticité. Troisième Mémoire. Extrait. Compt. Rend. Acad. Sci. Paris **19**, 229–231.

3. WERTHEIM, GUILLAUME: Note sur l'influence des basses températures sur l'élasticité des métaux. (Extrait par l'auteur.) Compt. Rend. Acad. Sci. Paris **19**, 231–233.

1845 *1.* POUILLET. See [1844, *1*].

1. SULLIVAN, WILLIAM: On currents of electricity produced by the vibration of wires and metallic rods. Phil. Mag., third series **27**, 261–264.

1. WERTHEIM, GUILLAUME: Note sur l'influence des basses températures sur l'élasticité des métaux. Annales de Chimie et de Physique **15**, 114–120. [Presented in 1844.]

2. WERTHEIM, GUILLAUME, and JEAN PIERRE EUGÈNE NAPOLÉON CHEVANDIER: Note sur l'élasticité et la cohésion des différentes espèces de verre. [See: 1847, *3*].

1846 *1.* HODGKINSON, EATON: Experimental Researches. Appended as the second part to the 4th edition of TREDGOLD, TH. Practical Essay on the Strength of Cast Iron. London.

1. WEBER, EDUARD: Art. Muskelbewegung. WAGNER's Handwörterbuch der Physiologie, Vol. 3, Part 2. (Reference was given in WUNDT [1858, *1*.])

1. WERTHEIM, GUILLAUME, and JEAN PIERRE EUGÈNE NAPOLÉON CHEVANDIER: Sur les propriétés mécaniques du bois. Compt. Rend. Acad. Sci. Paris **23**, 663–674. (An Abstract, presented to the Academy in 1845; and see WERTHEIM, 1847, *2*; and CHEVANDIER and WERTHEIM, 1848, *1*.) [Also published in German: Über die mechanischen Eigenschaften des Holzes. Annalen der Physik und Chemie (POGGENDORFF), Ergänzungsband II, 481–496 (1848).]

1847 *1.* HOPKINS, WILLIAM: Report on the geological theories of elevation and earthquakes. Report of the British Association for the Advancement of Science 33–92.

1. PAGE, C. G.: Singular property of Caoutchouc, illustrating the value of latent heat in giving elasticity to solid bodies, and the distinct functions in this respect of latent and free or sensible heat. Silliman's Journal of Science **4**, 341–342.

1. REGNAULT, HENRI VICTOR: Relations des expériences ... pour déterminer les principales lois et les données numériques qui entrent dans les calculs des machines à vapeur. Septième Mémoire: De la compressibilité des liquides. Mémoires de l'Académie des Sciences, Paris **21**, 429–464. And see [1842, *1*].

1. SAINT-VENANT, ADHÉMARD JEAN CLAUDE BARRÉ DE: Mémoire sur l'équilibre des corps solides, dans les limites de leur élasticité, et sur les conditions de leur résistance, quand les déplacements éprouvés par leur points ne sont pas très petits. Compt. Rend. Acad. Sci. Paris **24**, 260–263.

1. WERTHEIM, GUILLAUME: Mémoire sur l'élasticité et la cohésion des principaux tissues du corps humain. Annales de Chimie et de Physique, third series **21**, 385–414. [Presented in 1846.]

2. (WERTHEIM, GUILLAUME): Rapport sur un Mémoire de MM. EUG. CHEVANDIER et WERTHEIM ayant pour objet la recherche expérimentale des propriétés mécaniques du bois. Compt. Rend. Acad. Sci. Paris **24**, 537–541.

3. WERTHEIM, GUILLAUME, and JEAN PIERRE EUGÈNE NAPOLÉON CHEVANDIER: Note sur l'élasticité et la cohésion des différentes espèces de verre. Annales de Chimie et de Physique, third series **19**, 129–138. (Presented to the French Academy on 2 June, 1845.) [Also published in German: Über die Elasticität und Cohäsion verschiedener Glassorten. Annalen der Physik und Chemie (POGGENDORFF), Ergänzungsband II, 115–118 (1848).]

1848 *1.* CHEVANDIER, JEAN PIERRE EUGÈNE NAPOLÉON, and GUILLAUME WERTHEIM: Mémoire sur les Propriétés Mécaniques du Bois. (Pamphlet), Paris. (Presented to the French Academy in 1846; reported upon by an Academy Commission in 1847; see WERTHEIM, 1846, *1*; 1847, *2*.)

1. PERSON, C. C.: Relation entre le coéfficient d'élasticité des métaux et leur chaleur latente de fusion; chaleur latente de cadmium et de l'argent. Annales de Chimie et de Physique, third series **23**, 265–277.

1. THOMSON, JAMES: On the strength of materials, as influenced by the existence or nonexistence of certain mutual strains among the particles composing them. Cambridge and Dublin Mathematical Journal **3**, 252–266.

1. WERTHEIM, GUILLAUME: Mémoire sur l'équilibre des corps solides homogènes. Annales de Chimie et de Physique, third series **23**, 52–95.

2. WERTHEIM, GUILLAUME: Mémoire sur la vitesse du son dans les liquides. Annales de Chimie et de Physique, third series **23**, 434–475. See also [1851, *7*].

3. WERTHEIM, GUILLAUME: Note sur la torsion des verges homogènes. (Author's extract) Compt. Rend. Acad. Sci. Paris **27**, 649–650.

4. WERTHEIM, GUILLAUME: Mémoire sur les sons produits par le courant électrique. Annales de Chimie et de Physique, third series **23**, 302–327.

1849 *1.* CLAUSIUS, RUDOLF JULIUS EMMANUEL: Ueber die Veränderungen, welche in den bischer gebräuchlichen Formeln über das Gleichgewicht und die Bewegung elastischer fester Körper durch neuere Beobachtungen nothwendig geworden sind. Annalen der Physik und Chemie, second series **76**, 46–67.

1. Iron Commission Report: *Report* of a Commission appointed by the English government to conduct a study on the use of iron and cast iron in the construction of railroads and bridges. Members of the Commission: JOHN WROTTESLEY, ROBERT WILLIS, HENRI JAMES, GEORGE RENNIE, WILLIAM CUBITT, and EATON HODGKINSON. Whitehall, 26 July, 1849. [See translation into French for reprinting of the Report in 1851.]

1. THOMSON, JAMES: Theoretical considerations regarding the effect of pressure in lowering the freezing point of water. Trans. Edinburgh Roy. Soc. **16**, 575–580.

1. WERTHEIM, GUILLAUME: Note sur la torsion des verges homogènes. Annales de Chimie et de Physique, third series **25**, 209–215.

1850 *1.* BAUDRIMONT, A.: Expériences sur la ténacité des métaux malléables, faites aux températures, 0, 100 et 200 dégrés. Annales de Chimie et de Physique, third series **30**, 304–311.

1. BUNSEN, ROBERT WILHELM: Ueber den Einfluss des Drucks auf die chemische Natur der plutonischen Gesteine. Annalen der Physik und Chemie, second series **81**, 562–567.

1. FAIRBAIRN, WILLIAM: An experimental inquiry into the strength of wrought-iron plates and their riveted joints as applied to ship-building and vessels exposed to severe strains. Phil. Trans. Roy. Soc. London **140**, 677–725.

1. FARADAY, MICHAEL: On certain conditions of freezing water. The Athenaeum, No. 1181, 640–641 (June 15).

1. HODGE, PAUL R.: See: Proceedings of the Third Annual General Meeting of the Institution of Mechanical Engineers, Birmingham, 23 January, 1850; p. 16, (1850–1851).

1. KIRCHHOFF, GUSTAV ROBERT: Ueber das Gleichgewicht und die Bewegung einer elastischen Scheibe. Journal für die reine und angewandte Mathematik (CRELLE) **40**, 51–88.

1. STEPHENSON, ROBERT: See: Proceedings of the Third Annual General Meeting of the Institution of Mechanical Engineers, Birmingham, 24 April, 1850; pp. 6–9, (1850–1851).

1. THOMSON, WILLIAM: The effect of pressure in lowering the freezing-point of water experimentally demonstrated. Phil. Mag., third series **37**, 123–127.

1. WERTHEIM, GUILLAUME: Ueber die Hauptresultate seiner Untersuchungen der allgemeinen Gesetze des Gleichgewichtes und Bewegung der festen und flüssigen Körper. Sitz.ber., Math.-Naturw. Cl. Kaiserl. Akad. Wiss., Wien **5**, Hefte 6–10, pp. 19–31.

2. WERTHEIM, GUILLAUME: Remarques à l'occasion du Mémoire de M. BAUDRIMONT sur la ténacité des métaux. Annales de Chimie et de Physique, third series **30**, 507–508.

1851 *1.* [Iron Commision Report] Rapport d'une commission nommée par le gouvernement anglais, pour faire une enquête sur l'emploi du fer et de la fonte dans les constructions dépendant des chemins de fer. (Translated by BUSCHE.) Annales des Ponts et Chaussées, Mémoires, third series, first semester **1**, 193–220. [See: 1849, *1*.]

1. NAVIER, CLAUDE LOUIS MARIE HENRI: Resumé des Leçons données à l'École des Ponts et Chaussées, sur l'application de la mécanique à l'établissement des constructions et des machines (Third edition, with the Appendix by SAINT-VENANT). [See NAVIER, 1833, *1*; 1864, *1*.]

1. WERTHEIM, GUILLAUME, and LOUIS FRANÇOIS CLÉMENT BREGUET: Expériences sur la vitesse du son dans le fer. Compt. Rend. Acad. Sci. Paris **32**, 293–294. See also: WERTHEIM [1851, *10*].

2. WERTHEIM, GUILLAUME: Mémoire sur les vibrations sonores de l'air. Annales de Chimie et de Physique, third series **31**, 385–432.

3. WERTHEIM, GUILLAUME: Mémoire sur les vibrations des plaques circulaires. Annales de Chimie et de Physique, third series **31**, 5–19. [Presented to the French Academy, 1 October, 1849.]

4. WERTHEIM, GUILLAUME: Mémoire sur la propagation du mouvement dans les corps solides et dans les liquides. Annales de Chimie et de Physique, third series **31**, 19–36. [Presented to the French Academy, 10 December, 1849.]

5. WERTHEIM, GUILLAUME: Sur les effets optiques de la compression du verre. Compt. Rend. Acad. Sci. Paris **32**, 144–145. [Sealed packet deposited on 23 December, 1850 and opened on 27 January, 1851.]

6. WERTHEIM, GUILLAUME: Mémoire sur les vibrations sonores de l'air. (Extrait.) Compt. Rend. Acad. Sci. Paris **32**, 14–16 [Presented to the French Academy, 6 January, 1851]. See also [1856, *1*].

7. WERTHEIM, GUILLAUME: Description d'un appareil pour la détermination de la vitesse du son dans les gaz. (Extrait.) Compt. Rend. Acad Sci. Paris **32**, 16. [This note was read to the French Academy on 6 January.]

8. WERTHEIM, GUILLAUME: Mémoire sur la polarisation chromatique produite par le verre comprimé. Compt. Rend. Acad. Sci. Paris **32**, 289–292. [Extract read to the French Academy, 25 February].

9. (WERTHEIM, GUILLAUME): CAUCHY, AUGUSTIN LOUIS; HENRI VICTOR REGNAULT; and JEAN MARIE CONSTANT DUHAMEL: Rapport sur divers Mémoires de WERTHEIM. Compt. Rend. Acad. Sci. Paris **32**, 326–330.

10. WERTHEIM, GUILLAUME: Note sur la vitesse du son dans les verges. Annales de Chimie et de Physique, third series **31**, 36–39.

1852 *1.* KUPFFER, ADOLF THEODORE: Bemerkungen über das mechanische Aequivalent der Wärme. Annalen der Physik und Chemie, second series **86**, 310–314. [Also in: Bulletin de la Classe Physico-math. de l'Académie, St. Petersburg **10**, 193.]

1. WERTHEIM, GUILLAUME: Note sur des courants d'induction produits par la torsion du fer. Compt. Rend. Acad. Sci. Paris **35**, 702–704.

2. WERTHEIM, GUILLAUME: Deuxième note sur la double réfraction artificiellement produite dans les cristaux du système regulier. Compt. Rend. Acad. Sci. Paris **35**, 276–278.

3. WERTHEIM, GUILLAUME: Remarques à l'occasion d'une note récente de M. GARNIER sur les chaleurs specifiques des corps composés. Compt. Rend. Acad. Sci. Paris **35**, 300–301.

1853 *1.* CRAIG, W. C.: Ueber verbesserte Kautschuk-Federn für Locomotiven und Eisenbahnwagen. (Translation.) Dingler's Polytechnisches Journal **129**, 264–272.

1. HODGKINSON, EATON: On the elasticity of stone and crystalline bodies. Report of the British Association for the Advancement of Science, 23rd Meeting, held at Hull, in September, 1853, 36–37.

1. KUPFFER, ADOLF THEODORE: Recherches expérimentales sur l'élasticité des métaux. Mémoires de l'Académie de St. Petersburg, Sciences Mathématiques, Physiques et Naturelles, sixth series **7**, 232–305.

1. SAINT-VENANT, ADHÉMARD JEAN CLAUDE BARRÉ DE: Mémoire sur la torsion des prismes. Compt. Rend. Acad. Sci. Paris **36**, 1028–1031.

1854 *1.* DEBONNEFOY, E.: Note sur le caoutchouc vulcanisé et sur son emploi comparé à celui de l'acier fondu pour les ressorts des voitures et wagons dans les chemins de fer. Mémoires et Comptes Rendus des Travaux de la Société des Ingénieurs Civils, Paris, 86–95.

1. Hovine, M.: Note sur l'emploi des rondelles en caoutchouc vulcanisé comme ressort de choc et de traction. Mémoires et Comptes Rendus des Travaux de la Société des Ingénieurs Civils, Paris, 80–85.

1. Kupffer, Adolf Theodore: Untersuchungen über die Flexion elastischer Metallstäbe. Bulletin de la Classe Physico-Mathématique de l'Académie Impériale des Sciences de St. Pétersbourg **12**, 161–167.

1. Soret, J. Louis: Recherches sur l'élasticité et la cohésion des corps solides. Extrait des Mémoires de M. Wertheim et de M. Kupffer. Archives des Sciences Physiques et Naturelles, Genève **25**, 40–58.

1. Wertheim, Guillaume: Mémoire sur la double réfraction temporairement produite dans les corps isotropes, et sur la relation entre l'élasticité mécanique et entre l'élasticité optique. Annales de Chimie et de Physique, third series **40**, 156–221. [Also published in English: On the double refraction temporarily produced in isotropic bodies. Phil. Mag. **8**, fourth series, 241–263.] And see [1851, *5*].

2. Wertheim, Guillaume: Mémoire sur la capillarité. Presented to the French Academy in 1854. [See: 1861, *1*.]

1855 *1.* Thomson, Sir William (Lord Kelvin): On the thermo-elastic and thermo-magnetic properties of matter. Quarterly Jnl. Math., **1**, 55–77.

1. Wertheim, Guillaume: Mémoire sur la torsion. Presented to the French Academy, 19 February. Compt. Rend. Acad. Sci. Paris **40**, 411–414. [See: 1857, *1*].

1856 *1.* Boileau, Pierre Prosper: Note sur l'élasticité du caoutchouc vulcanisé. Compt. Rend. Acad. Sci. Paris **42**, 933–937.

1. Cox, Homersham: The deflection of imperfectly elastic beams and the hyperbolic law of elasticity. Trans. Cambridge Phil. Soc. **9**, Part II, 177–190.

1. Kupffer, Adolf Theodore: Travaux exécutés à l'Observatoire Physique Central. Comptes Rendus Annuelles, St. Pétersbourg 57–66.

1. Saint-Venant, Adhémard Jean Claude Barré de: Mémoire sur la flexion des prismes, sur les glissements transversaux et longitudinaux qui l'accompagnent lorsqu'elle ne s'opère pas uniformément ou en arc de cercle, et sur la forme courbe affectée alors par leurs sections transversales primitivement planes. Journal de Mathématiques de Liouville, second series **1**, 89–189.

2. Saint-Venant, Adhémard Jean Claude Barré de: Mémoire sur la torsion des prismes, avec des considérations sur leur flexion, ainsi que sur l'équilibre intérieur des solides élastiques en général. Académie des Sciences, Paris, Mémoires Presentés par Divers Savants, second series **14**, 233–560.

1. Wertheim, Guillaume: Remarques à l'occasion d'une Note de M. Zamminer sur le mouvement vibratoire de l'air dans les tuyaux. Compt. Rend. Acad. Sci. Paris **42**, 493–494.

1857 *1.* Dietzel, Carl Franz: Ueber die Elasticität des vulkanisierten Kautschuks und Bemerkungen über die Elasticität fester Körper überhaupt. Polytechnisches Centralblatt, Cols. 689–694.

1. Joule, James Prescott: On the thermo-electricity of ferruginous metals, and on the thermal effects of stretching solid bodies. Proc. Roy. Soc. (London) **8**, 355–356 (June, 1857).

2. Joule, James Prescott: On the thermal effects of longitudinal compression of solids. Proc. Roy. Soc. (London) **8**, 564–565 (June, 1857).

1. Thomson, Sir William (Lord Kelvin): On the thermo-elastic and thermo-magnetic properties of matter. Quarterly Journal of Mathematics, Cambridge **1**, 57–77.

1. Wertheim, Guillaume: Mémoire sur la torsion, Première Partie. (Presented to the French Academy of Science, February, 1855.) Annales de Chimie et de Physique, third series **50**, 195–321.

2. Wertheim, Guillaume: Mémoire sur la torsion, Deuxième Partie. Annales de Chimie et de Physique, third series **50**, 385–431. And see [1852, *1*].

1858 *1.* Biot, Jean Baptiste: Sur le charlatanisme. Mélanges Scientifiques et Litteraires, Paris **2**, 69–86. [This article appeared in 1808 in Mercure de France, The Mélanges ... was published in 1858 by Biot himself.]

2. Biot, Jean Baptiste: Discours sur l'ésprit d'invention et de recherche dans les sciences. Mélanges Scientifiques et Littéraires, Paris **2**, 87–95.

1. Clapeyron, E. (Benoit Paul Émile): Mémoire sur le travail des forces élastiques dans un corps solide élastique déformé par l'action de forces extérieures. Compt. Rend. Acad. Sci. Paris **46**, 208–212.

1. Mousson, Albert: Einige Thatsachen betreffend das Schmelzen und Gefrieren des Wassers. Annalen der Physik und Chemie (Poggendorff) **105**, 161–174.

1. Rankine, W. J. M.: A Manual of Applied Mechanics. London, 8 vols.

1. WIEDEMANN, GUSTAV HEINRICH: Über die Torsion. Annalen der Physik und Chemie (POGGENDORFF) **103**, 563–577.

1. WÖHLER, A.: Berichte über die Versuche, welche auf der Königl. Niederschlesisch-Märkischen Eisenbahn mit Apparaten zum Messen der Biegung und Verdrehung von Eisenbahnwagen-Achsen während der Fahrt angestellt wurden. Zeitschrift für Bauwesen **8**, 642–652.

1. WUNDT, WILHELM MAX: Die Lehre von der Muskelbewegung. Braunschweig.

1859 *1.* FARADAY, MICHAEL: On certain conditions of freezing water, etc. Researches in Chemistry and Physics, London: Taylor and Francis, pp. 372–374.

1. JOULE, JAMES PRESCOTT: On some thermo-dynamic properties of solids. Phil. Trans. Roy. Soc. London **149**, 91–131.

1. KARMARSCH, KARL: Ueber die absolute Festigkeit der Metalldrähte. Polytechnisches Centralblatt, Leipzig 1272–1276.

1. KIRCHHOFF, GUSTAVE ROBERT: Ueber das Verhältnis der Quercontraction zur Längendilatation bei Stäben von federhartem Stahl. Annalen der Physik und Chemie (POGGENDORFF), second series **108**, 369–392.

1. VOLKMANN, ALFRED WILHELM: Ueber die Elasticität der organischen Gewebe. Archiv für Anatomie, Physiologie und Wissenschaftliche Medizin, 293–313.

1. WIEDEMANN, GUSTAV HEINRICH: Ueber die Biegung. Annalen der Physik und Chemie (POGGENDORFF), second series **107**, 439–448.

1860 *1.* KUPFFER, ADOLF THEODORE: Recherches Expérimentales sur l'Élasticité des Métaux, Vol. 1. St. Pétersbourg. See also [1854, *1*].

2. (KUPFFER, ADOLF THEODORE): État du Personnel. Bulletin de l'Académie Impériale des Sciences de St. Pétersbourg **11**, supplément 1, 15–16 (1 July, 1860).

1. LÜDERS, W.: Ueber die Äusserung der Elasticität an stahlartigen Eisenstäben und Stahlstäben, und über eine beim Biegen solcher Stäbe beobachtete Molecularbewegung. Dingler's Polytechnisches Journal, Stuttgart, fourth series **5**, 18–22.

1. MORIN, ARTHUR JULES, and HENRI ÉDOUARD TRESCA: Détermination du coefficient d'élasticité de l'aluminium. (Extrait des Annales du Conservatoire des Arts et Métiers, No. 2.) Ann. Mines, fifth series, **18**, 63–66.

1. WERTHEIM, GUILLAUME: Mémoire sur la compresssibilité cubique de quelques corps solides et homogènes. Compt. Rend. **51**, 969–974. [Also published in English: On the cubical compressibility of certain solid homogeneous bodies. Phil. Mag., fourth series **21**, 447–451 (1861).] See also [1852, *3*].

2. WERTHEIM, GUILLAUME: Remarques au sujet d'une communication de M. CAVAILLÉ-COLL sur les tuyaux d'orgues. Compt. Rend. Acad. Sci. Paris **50**, 309–311.

1. WÖHLER, A.: Versuche der Ermittlung der auf die Eisenbahnwagen-Achsen einwirkenden Kräfte und der Widerstandsfähigkeit der Wagen-Achsen. Zeitschrift für Bauwesen **10**, 583–616.

1861 *1.* EDLUND, ERIK: Untersuchung über die bei Volumveränderung fester Körper entstehenden Wärme-Phänomene, sowie deren Verhältniss zu der dabei geleisteten mechanischen Arbeit. Annalen der Physik und Chemie (POGGENDORFF), second series **114**, 1–40.

1. VERDET, MARCEL ÉMILE: Notices sur les travaux scientifiques de M. GUILLAUME WERTHEIM, membre de la Société Philomathique de Paris, redigée sur la demande de la Société. L'Institut **29**, Nos. 1432, 1433, 1434, Footnotes on pp. 198–201; 205–209; and 213–216.

1. WERTHEIM, GUILLAUME: Mémoire sur la capillarité. Annales de Chimie et de Physique **63**, 129–201. [Presented to the French Academy in 1854.]

1862 *1.* MORIN, ARTHUR JULES: Résistance des Matériaux, 3ème edition; Extrait. Compt. Rend. Acad. Sci. Paris **54**, 235–239.

1863 *1.* KOHLRAUSCH, FRIEDRICH: Ueber die elastische Nachwirkung bei der Torsion. Annalen der Physik und Chemie (POGGENDORFF), second series **119**, 337–368.

1. OKATOW, MICHAIL: Ueber das Verhältniss der Quercontraction zur Längendilatation bei Stahlstäben. Annalen der Physik und Chemie (POGGENDORFF), second series **119**, 11–42.

1. POGGENDORFF, JOHANN CHRISTIAN: Biographisch-Literarisches Handwörterbuch zur Geschichte der Exacten Wissenschaften, Vol. 2, pp. 1302–1303, Leipzig. [On GUILLAUME WERTHEIM.]

1. WINKLER, EMIL: Die Elasticitäts- und Festigkeitscoefficienten. Civilingenieur 405–436.

1. WÖHLER, A.: Über die Versuche zur Ermittlung der Festigkeit von Achsen, welche in den Werkstätten der Niederschlesisch-Märkischen Eisenbahn zu Frankfurt a.d.O. angestellt sind. Zeitschrift für Bauwesen **13**, 234–258.

1864 *1.* FIZEAU, HIPPOLYTE LOUIS: Recherches sur la dilatation et la double refraction du crystal de roche échauffé. Compt. Rend. Acad. Sci. Paris **58**, 923–932. [See also: Untersuchungen über die Ausdehnung und Doppelbrechung des erhitzten Bergkrystalls. Annalen der Physik und Chemie (POGGENDORFF) **123**, 515–526.]

1. NAVIER, CLAUDE LOUIS MARIE HENRI: De la résistance des corps solids; Première section. Resumé des Leçons données à l'École des Ponts et Chaussées, sur l'application de la mécanique á l'établissement des constructions et des machines. Third edition, with Notes and Appendices by SAINT-VENANT. Paris: Dunod. [See 1833, *1*; 1851, *1*.]

1. TRESCA, HENRI ÉDOUARD: Mémoire sur l'écoulement des corps solides soumis à de fortes pressions. (Extrait par l'auteur.) Compt. Rend. Acad. Sci. Paris **59**, 754–758.

1865 *1.* EDLUND, ERIK: Quantitative Bestimmung der bei Volumveränderung der Metalle entstehenden Wärmephänomene und des mechanischen Wärme-Aequivalents, unabhängig von der inneren Arbeit des Metalls. Annalen der Physik und Chemie (POGGENDORFF), second series **126**, 539–579.

1. HELMHOLTZ, HERMANN LUDWIG FERDINAND VON: Populäre Wissenschaftliche Vorträge, Braunschweig: F. Vieweg und Sohn. [See Eis und Gletscher, pp. 93–134.]

1. THALÉN, ROBERT: On the determination of the limit of elasticity in metals. Phil. Mag., fourth series **30**, 194–207.

1. THOMSON, Sir WILLIAM (Lord KELVIN): On the elasticity and viscosity of metals. Proc. Roy. Soc. (London) **14**, 289–297. (See also: [1878, *1*], [1880, *1*].

1. TRESCA, HENRI ÉDOUARD: Experiments on the production of cylinders of ice by pressure through orifices. Phil. Mag., fourth series **30**, 239–240.

1866 *1.* DYBKOWKSY, W., and A. FICK: Ueber die Wärmeentwicklung beim Starrwerden des Muskels. Vierteljahresschr. Naturforsch. Ges., Zürich **11–12**, 317–348.

1. EVERETT, JOSEPH DAVID: Account of experiments on the flexural and torsional rigidity of a glass rod, leading to the determination of the rigidity of glass. Phil. Trans. Roy. Soc. London **156**, 185–191.

1. SCHMULEWITSCH, JACOB: Ueber das Verhalten des Kautschuks zur Wärme und zur Belastung. Vierteljahresschr. Naturforsch. Ges. Zürich **11–12**, 201–224.

1. ZÖPPRITZ, KARL: Berechnung von Kupffer's Beobachtungen über die Elasticität schwerer Metallstäbe. Annalen der Physik und Chemie (POGGENDORFF), second series **129**, 219–237. See also SORET [1854, *1*].

1867 *1.* EVERETT, JOSEPH DAVID: Account of experiments on torsion and flexure for the determination of rigidities. Phil. Trans. Roy. Soc. London **157**, 139–153. See also [1868, *1*].

1. SAINT-VENANT, ADHÉMARD JEAN CLAUDE BARRÉ DE: Mémoire sur le choc longitudinal de deux barres élastiques de grosseurs et de matières semblables ou différentes, et sur la proportion de leur force vive qui est perdue pour la translation ultérieure; et généralement sur le mouvement longitudinal d'un système de deux ou plusieurs prismes élastiques. J. Math. Pures Appl., second series **12**, 237–376.

1. TRESCA, HENRI ÉDOUARD: Sur l'écoulement des corps solides soumis à de fortes pressions. Compt. Rend. Acad. Sci. Paris **64**, 809–812.

2. TRESCA, HENRI ÉDOUARD: Applications de l'écoulement des corps solides au laminage et au forgeage. Extrait. Compt. Rend. Acad. Sci. Paris **64**, 1132–1136.

3. [TRESCA, HENRI ÉDOUARD]: Compt. Rend. Acad. Sci. Paris **64**, 442.

1868 *1.* EVERETT, JOSEPH DAVID: Account of experiments on torsion and flexure for the determination of rigidities. Phil. Trans. Roy. Soc. London **158**, 363–370. [Also in: Proc. Roy. Soc. (London) **15**, 356 (1867); **16**, 248 (1868).]

1. TRESCA, HENRI ÉDOUARD: Mémoire sur l'écoulement des corps solides. Mémoires Presentés par Divers Savants, Académie des Sciences, Paris **18**, 733–799.

2. TRESCA, HENRI ÉDOUARD: Sur l'application des formules générales du mouvement permanent des liquides à l'écoulement des corps solides. Compt. Rend. Acad. Sci. Paris **66**, 1027–1032; 1244–1246.

1869 *1.* CORNU, A. [MARIE ALFRED]: Méthode optique pour l'étude de la déformation de la surface extérieure des solides élastiques. Compt. Rend. Acad. Sci. Paris **69**, 333–337.

1. TRESCA, HENRI ÉDOUARD: Mémoire sur le poinçonnage et la théorie mécanique de la déformation des métaux. (Extrait par l'auteur.) Compt. Rend. Acad. Sci. Paris **68**, 1197–1201.

1870 *1.* KOHLRAUSCH, FRIEDRICH: Über den Einfluß der Temperatur auf die Elasticitätscoefficienten einiger Metalle. Nachrichten von der königlichen Gesellschaft der Wissenschaften und der Georg-Augustus-Universität, Göttingen 257–262.

2. KOHLRAUSCH, FRIEDRICH, and F. E. LOOMIS: Ueber die Elasticität des Eisens, Kupfers, und Messings, insbesondere ihre Abhängigkeit von der Temperatur. Annalen der Physik und Chemie (POGGENDORFF) **141**, 481–503.

1. SAINT-VENANT, ADHÉMARD JEAN CLAUDE BARRÉ DE: Rapport sur un complément, présenté par M. TRESCA le 7 fevrier 1870 à son Mémoire du 27 novembre 1864 relatif à l'écoulement des corps solides malléables poussés hors d'un vase cylindrique par un orifice circulaire. Compt. Rend. Acad. Sci. Paris **70**, 368–369.

2. SAINT-VENANT, ADHÉMARD JEAN CLAUDE BARRÉ DE: Preuve théorique de l'égalité des deux coefficients de résistance au cisaillement et à l'extension ou à la compression dans le mouvement continu de déformation des solides ductiles au delà des limites de leur élasticité. Compt. Rend. Acad. Sci. Paris **70**, 309–311.

1. [TRESCA, HENRI EDOUARD]: Recherches sur le poinçonnage et sur la théorie mécanique de la déformation des corps solides. (Rapport sur l'ensemble de ces travaux; Rapporteur, M. MORIN: M. M. COMBES; DE SAINT-VENANT.) Compt. Rend. Acad. Sci. Paris **70**, 288–308.

2. TRESCA, HENRI ÉDOUARD: Mémoire sur le poinçonnage des métaux et des matières plastiques. Compt. Rend. Acad. Sci. Paris **70**, 27–31.

1. WÖHLER, A.: Ueber die Festigkeits-Versuche mit Eisen und Stahl. Zeitschrift für Bauwesen **20**, 73–106.

1871 *1.* LÉVY, MAURICE: Extrait du mémoire sur les équations générales des mouvements intérieurs des corps solides ductiles au delà des limites où l'élasticité pourrait les ramener à leur premier état. J. Math. Pures Appl. **16**, 369–372.

1. SAINT-VENANT, ADHÉMARD JEAN CLAUDE BARRÉ DE: Mémoire sur l'établissement des équations différentielles des mouvements intérieurs opérés dans les corps solides ductiles au delà des limites où l'élasticité pourrait les ramener à leur premier état. J. Math. Pures Appl. **16**, 308–316.

1. SCHNEEBELI, HEINRICH: Ueber den Stoss elastischer Körper und eine numerische Bestimmung der Stosszeit. Annalen der Physik und Chemie (POGGENDORFF) **143**, 239–250.

1. TRESCA, HENRI ÉDOUARD: Étude sur la torsion prolongée au delà de la limite d'élasticité. (Extrait par l'auteur.) Compt. Rend. Acad. Sci. Paris **73**, 1104–1105.

2. TRESCA, HENRI ÉDOUARD: Résultats des expériences de flexions faites sur des rails en fer et en acier au delà de la limite d'élasticité. Compt. Rend. Acad. Sci. Paris **73**, 1153–1155.

1. TYNDALL, JOHN: Hours of Exercise in the Alps. London: Longmans, Green, & Co.

1872 *1.* HOPKINSON, JOHN: On the rupture of iron wire by a blow. Proceedings of the Manchester Literary and Philosophical Society **11**, 40–45. [Also in: Original Papers of John Hopkinson, Vol. II, Scientific Papers, ed. by B. HOPKINSON; pp. 316–320, Cambridge, England (1901).]

2. HOPKINSON, JOHN: Further experiments on the rupture of iron wire. Proceedings of the Manchester Literary and Philosophical Society **11**, 119–121. [Also in: Original Papers ... (see supra), pp. 321–324.]

1. KOHLRAUSCH, FRIEDRICH WILHELM GEORG: Leitfaden der praktischen Physik, second edition. Leipzig: B. G. Teubner Verlag. [First edition, 1870; editions 1 to 9 were entitled as above; editions 10 to 19, were entitled Lehrbuch der praktischen Physik. (I have made particular use of the edition in 1905.) A condensed version appeared in 1900, entitled: Kleiner Leitfaden der praktischen Physik.] For English translation, see [1886, *1*].

1. SCHNEEBELI, HEINRICH: Stossversuche mit Kugeln aus verschiedenem Metall. Annalen der Physik und Chemie (POGGENDORFF) **145**, 328–331.

1. TRESCA, HENRI ÉDOUARD: Mémoire sur l'écoulement des corps solides. Mémoires présentés par Divers Savants à l'Académie des Sciences, Paris, second series **20**, 75–135.

2. TRESCA, HENRI ÉDOUARD: Mémoire sur le poinçonnage des métaux et des matières plastiques. Mémoires présentés par Divers Savants, l'Académie des Sciences, Paris, second series **20**, 619–828. [Presented to the French Academy, March, 1869.]

3. TRESCA, HENRI ÉDOUARD: Mémoire complémentaire sur le poinçonnage des métaux et des matières plastiques. Mémoires présentés par Divers Savants à l'Académie des Sciences, Paris, second series **20**, 829–838.

1873 *1.* BERTIN, A.: Révue des travaux publiés à l'étranger. See Chapter I: Chaleur; Section 3: Sur les propriétés thermiques du caoutchouc. Annales de Chimie et de Physique, fourth series **28**, 398–415.

1. THURSTON, ROBERT HENRY: Torsional resistance of materials determined by a new apparatus with automatic registry. J. Franklin Inst. **95** (January–June 1873), 254–260.

1874 *1.* BAUMGARTEN, G.: Die Elasticität von Kalkspathstäbchen. Annalen der Physik und Chemie (POGGENDORFF) **152**, 369–397.

1. BEARDSLEE, (Commander) L. W.: Increase of resisting power of metals under stress. J. Franklin Inst., third series **67**, 150–151, 302–304.

1. EXNER, FRANZ: Über die Abhängigkeit der Elasticität des Kautschuks von der Temperatur. Sitz.ber. Math.-Naturwiss. Cl. Kaiserl. Akad. Wiss., Wien **69**, Part II, 102–114.

1. THURSTON, ROBERT HENRY: A note on the resistance of materials. Trans. Am. Soc. Civil Engrs. **2**, Supplement, 239–240.

2. THURSTON, ROBERT HENRY: On the strength, elasticity, ductility and resilience of materials of machine construction. Trans. Am. Soc. Civil Engrs **2**, Supplement, 349–378.

1. UCHATIUS, (General-Major) FRANZ (Ritter) VON: Die Stahlbronce. Oesterreich, Zeitschrift Bergwesen **22**, 445–447, 458–460, 466–468, 479–481. [Also in: Dingler's Polytechnisches Journal **217**, 122–133.]

1. VOIGT, WOLDEMAR: Doctoral dissertation, Königsberg.

1875 *1.* EVERETT, JOSEPH DAVID: Illustrations of the C.G.S. System of Units. [See EVERETT (1891, *1*), 4th edition.]

1. KICK, FRIEDRICH: Kritik über R. H. Thurston's Untersuchungen über Festigkeit und Elasticität der Constructions-Materialien. Dingler's Polytechnisches Journal **218**, No. 3, 185–191.

1. LEMALE, ALEX GUISLAIN: Monnaies, Poids, Mesures et Usages Commerciaux de tous les Etats du Monde, 2nd edition. Paris: Hachette Cie; Havre: A. Lemale, Ainé.

1. PFAFF, IMMANUEL BURKHARD ALEXIUS FRIEDRICH: Versuch über die Plasticität des Eises. Annalen der Physik und Chemie (POGGENDORFF), second series **155**, 169–174.

1. THURSTON, ROBERT HENRY: On the mechanical properties of materials of construction. Trans. Am. Soc. Civil Engrs. **3**, 1–30. [Also in: J. Franklin Inst. **67**, 273–292, 344–356, 419–430; **68**, 47–66. (1874).]

1876 *1.* KOHLRAUSCH, FRIEDRICH WILHELM GEORG: Experimental-Untersuchungen über die elastische Nachwirkung bei der Torsion, Ausdehnung, und Biegung. Dritte Mittheilung. Annalen der Physik und Chemie (POGGENDORFF) **158**, 337–375.

1. POCHHAMMER, LEO: Ueber die Fortpflanzungsgeschwindigkeiten kleiner Schwingungen in einem unbegrenzten isotropen Kreiscylinder. Journal für reine und angewandte Mathematik (CRELLE) **81**, 327–336.

1. RÖNTGEN, WILHELM CONRAD: Ueber das Verhältniss der Quercontraction zur Längendilatation bei Kautschuk. Annalen der Physik und Chemie (POGGENDORFF) **159**, 601–616.

1. THURSTON, ROBERT HENRY: Note on the resistance of materials, as affected by flow and by rapidity of distortion. Trans. Am. Soc. Civil Engrs. **5**, 199–214.

2. THURSTON, ROBERT HENRY: The strength and other properties of materials of construction, as deduced from strain diagrams automatically produced by the autographic recording testing machine. [Answer to Professor KICK.] Trans. Am. Soc. Civil Engrs. **5**, 9–18. [See also German translation: Über die Festigkeit der Constructionsmaterialien bestimmt durch die selbstätig registrirende Festigkeitsmaschine. Dingler's Polytechnisches Journal **223**, 16–18.]

1. VOIGT, WOLDEMAR: Bestimmung der Elasticitätsconstanten des Steinsalzes. Annalen der Physik und Chemie (POGGENDORFF) **7**, 1–53.

1877 *1.* BAUSCHINGER, JOHANN: Ueber die Erhöhung der Elasticitätsgrenze der Metalle. Dingler's Polytechnisches Journal **224**, 1–13, 129–134.

1. OBERMAYER, A. VON: Ein Beitrag zur Kenntniss der zähflüssigen Körper. Sitz.ber. Math.-Naturw. Cl. Kaiserl. Akad. Wiss., Wien **75**, 665–678.

1. TYNDALL, JOHN: Six Lectures on Light. Second edition. New York.

1. UCHATIUS (General-Major), FRANZ (Ritter) VON: Ueber die Erhöhung der Elasticitätsgrenze der Metalle durch dauernde Spannung. Dingler's Polytechnisches Journal **223**, 242–246.

1878 *1.* CORNU, A. [MARIE ALFRED], and J. B. BAILLE: Étude de la résistance de l'air dans la balance de torsion. Compt. Rend. Acad. Sci. Paris **86**, 571–574.

1. SAINT-VENANT, ADHÉMARD JEAN CLAUDE BARRÉ DE: Des paramètres d'élasticité des solides, et de leur détermination expérimentale. Compt. Rend. Acad. Sci. Paris **86**, 781–785.

1. THOMSON, Sir WILLIAM (Lord KELVIN): On the thermoelastic, thermomagnetic, and pyroelectric properties of matter. Phil. Mag., fifth series **5**, 4–27.

2. THOMSON, Sir WILLIAM (Lord KELVIN): Elasticity. Edinburgh: Adam and Charles Black. (Pre-print of [1880, *1*].)

1. THURSTON, ROBERT HENRY: Salisbury Iron. Its Composition, Qualities and Uses. Salisbury, Connecticut, U.S.A.

1. TRESCA, HENRI ÉDOUARD: On further application of the flow of solids. Proc. Inst. Mech. Engrs (London) 301–345 (June, 1878).

1. WINKLER, EMIL: Deformationsversuche mit Kautschuk-Modellen. Civilingenieur **24**, 81–100. For an earlier summary of work on elasticity and strength of many solids, see WINKLER [1863, *1*].

1879 *1.* BAUSCHINGER, JOHANN: Ueber die Quercontraction und -Dilatation bei der Längenausdehnung und -Zusammendrückung prismatischer Körper. Civilingenieur, Leipzig **25**, 81–124.

1. BERTRAND, JOSEPH LOUIS FRANÇOIS: Éloge de JEAN-VICTOR PONCELET. Mémoires de l'Académie des Sciences, Paris **41**, i–xxv.

1. KLEMENČIČ, IGNAZ: Beitrag zur Kenntniss der inneren Reibung in Eisen. Sitz.ber. Math.-Naturw. Cl. Kaiserl. Akad. Wiss., Wien **78**, 11. Abth. 935–942.

1. MALLOCK, ARNULPH: The measurement of the ratio of lateral contraction to longitudinal extension in a body under strain. Proc. Roy. Soc. (London) **29**, 157–161.

1880 *1.* AMAGAT, ÉMILE HILAIRE: Sur la déformation des tubes de verre sous de fortes pressions. Compt. Rend. Acad. Sci. Paris **90**, 863–864 (April, 1880).

1. BUCHANAN, JOHN YOUNG: Preliminary note on the compressibility of glass. Trans. Roy. Soc. Edinburgh **29**, 589–598.

1. IMBERT, A.: Recherches Théoriques et Expérimentales sur l'Élasticité du Caoutchouc. Lyon.

1. SPRING, WALTHÈRE: Recherches sur la propriété que possèdent les corps solides de se souder par l'action de la pression. Bulletin de l'Académie Royale des Sciences, des Lettres, et des Beaux-Arts de Belgique, Bruxelles **49**, 323–379. [Also in Annales de Chimie et de Physique; fifth series **22**, 170–217 (1881).]

1. THOMSON, Sir WILLIAM (Lord KELVIN): Elasticity. Encyclopaedia Britannica; ninth edition, page 796. [Reprinted in: THOMSON, Sir WILLIAM, Elasticity and Heat. Edinburgh: Adam and Charles Black (1880).] (See also [1865, *1*], [1878, *1*].)

1881 *1.* BAUSCHINGER, JOHANN: Ueber die Veränderung der Elasticitätsgrenze und des Elasticitätsmoduls verschiedener Metalle. Civilingenieur **27**, 289–348.

2. BAUSCHINGER, JOHANN: Experimentelle Prüfung der neueren Formeln für die Torsion prismatischer Körper. Civilingenieur **27**, 115–130.

1. KLANG, HERRMANN: Die Elasticitätsconstanten des Flußspathes. Annalen der Physik und Chemie (WIEDEMANN) **12**, 321–335.

1. SPRING, WALTHÈRE: Recherches sur la propriété que possèdent les corps de se souder sous l'action de la pression. Annales de Chimie et de Physique, fifth series **22**, 170–217. This is substantially the same paper as [1880, *1*].

1882 *1.* BOLTZMANN, LUDWIG: Einige Experimente über den Stoss von Cylindern. Sitz.ber. Math.-Naturw. Cl. Kaiserl. Akad. Wiss., Wien **84**, Abt. II, 1225–1229.

1. FISCHER, (FRIEDRICH) HUGO ROBERT: Untersuchungen über das Verhalten des Phosphorbronzedrahtes bei der Beanspruchung durch Zugkräfte. Dingler's Polytechnisches Journal **245**, 64–75.

1. HERTZ, HEINRICH RUDOLPH: Ueber die Berührung fester elastischer Körper. Journal für die reine und angewandte Mathematik (CRELLE) **92**, 156–171.

1. MILLER, ANDREAS: Untersuchung über den Einfluß der Temperatur auf Aeußerungen von Molekularkräften. Sitz.ber. Math.-Physik. Cl. Akad. Wiss., München **12**, 377–462.

1. MÜLLER, ERNST: Ueber bleibende und elastische Formänderungen der Rohseide. Civilingenieur **28**, 631–640.

1. VOIGT, WOLDEMAR: Ueber das Verhältnis der Quercontraction zur Längendilatation bei Stäben von isotropem Glas. Annalen der Physik und Chemie (WIEDEMANN), Neue Folge **15**, 497–513.

1883 *1.* MALLARD, E., and H. LE CHATELIER: Sur le dimorphisme de l'iodure d'argent. Compt. Rend. Acad. Sci. Paris **97**, 102–105.

1. TOMLINSON, HERBERT: The influence of stress and strain on the action of physical forces. Phil. Trans. Roy. Soc. London **174**, Part I, 1–172.

1. TRESCA, HENRI ÉDOUARD: Contribution à l'étude du poinçonnage et des proues dont il detérmine la formation. Compt. Rend. Acad. Sci. Paris **96**, 816–822.

1. VOIGT, WOLDEMAR: Die Theorie des longitudinalen Stosses cylindrischer Stäbe. Annalen der Physik und Chemie (WIEDEMANN), Neue Folge **19**, 44–65.

1884 *1.* AMAGAT, ÉMILE HILAIRE: Sur la valeur du coefficient de Poisson relative au caoutchouc. Compt. Rend. Acad. Sci. Paris **99**, 130–133. See also [1880, *1*].

1. BACH, CARL: Festigkeit und Dehnung von Treibriemenleder. Zeitschrift des Vereins deutscher Ingenieure **28**, 740–742.

1. COULOMB. [See 1784, *1*.]

1. FISCHER, HUGO: Experimentelle Untersuchungen über die Zugfestigkeit und Zugelasticität von Metalldrähten. Civilingenieur **30**, 392–410.

1. HAUSMANINGER, VICTOR: Versuch über den Stoss von Kautschukcylindern. Sitz.ber. Math.-Naturw. Cl. Kaiserl. Akad. Wiss., Wien **88**, Part II, Nos. 1–5, 768–781.

1. VOIGT, WOLDEMAR: Neue Bestimmungen der Elasticitätsconstanten von Steinsalz und Flusspath. Sitz.ber. Akad. Wiss., Berlin 989–1004 .

1885 *1.* CONSIDÈRE, ARMAND GABRIEL: Mémoire sur l'emploi du fer et de l'acier dans les constructions. Annales des Ponts et Chaussées, Mémoire No. 34, sixth series, Vol. 9, first semester, Paris: Vve Ch. Dunod; pp. 574–775. (See [1888, *1*] for German Translation.)

1. HAUSMANINGER, VICTOR: Zur Theorie des longitudinalen Stosses cylindrischer Körper. Annalen der Physik und Chemie, Neue Folge **25**, 189–202. [Also in: Sitz.ber. Akad. Wiss., Berlin 49–62 (1885).]

1886 *1.* BAUSCHINGER, JOHANN: Über die Veränderung der Elasticitätsgrenze und der Festigkeit des Eisens und Stahls durch Strecken und Quetschen, durch Erwärmen und Abkühlen und durch oftmal wiederholte Beanspruchung. Mitteilungen aus dem mechanisch-technischen Laboratorium der k. polytechnischen Schule, München, Hefte 7–14 (1877–1886); Heft 13, pp. 1–115.

1. HAMBURGER, MAX: Untersuchungen über die Zeitdauer des Stosses von Cylindern und Kugeln. Annalen der Physik und Chemie. Neue Folge **28**, 653–665.

1. KOHLRAUSCH, FRIEDRICH: An Introduction to Physical Measurements. Translated from the second German edition by THOMAS HUTCHINSON WALLER and HENRY RICHARDSON PROCTOR; New York: D. Appleton and Company.

1. PULFRICH, CARL: Ueber die elastische Nachwirkung eines Kautschukschlauches und deren Einfluß auf die Constante μ. Annalen der Physik und Chemie, Neue Folge **28**, 87–107.

1. TODHUNTER, ISAAC, and KARL PEARSON: A History of the Theory of Elasticity and of the Strength of Materials, Vol. I. Cambridge: University Press. (Vol. II, Parts I and II were first published in 1893.) Volumes I and II were reprinted in Dover Publications, New York (1960).

1. TOMLINSON, HERBERT: Internal Friction of Metals. Phil. Trans. Roy. Soc. London **177**, Part II, 802–807.

1. UNWIN, WILLIAM CAWTHORNE: Translation summary of "On the change of the elastic limit and strength of iron and steel, by drawing out, by heating and cooling, and by repetition of loading" (by J. BAUSCHINGER, Mitteilung aus dem Mechanisch-technischen Laboratorium der k. Hochschule in München, 1886, p. 1–115). Proc. Inst. Civil Engrs **87**, 463–465.

2. UNWIN, WILLIAM CAWTHORNE: The employment of autographic records in testing materials. J. Soc. Arts, 1885–1886 **34**, 336–347.

1887 *1.* BACH, CARL VON: Elasticität von Treibriemen und Treibseilen. Zeitschrift des Vereins deutscher Ingenieure **31**, 221–225.

1. KENNEDY, ALEXANDER BLACKIE WILLIAM: The use and equipment of engineering laboratories. Proc. Inst. Civil Engrs, session 1886–1887 **88**, Part II, 1–80.

1. TOMLINSON, HERBERT: The effect of change of temperature in twisting or untwisting wires which have suffered permanent torsion. Phil. Mag. **24**, No. 148, 253–256 (September, 1887).

2. TOMLINSON, HERBERT: Remarkable effect on raising iron under temporary stress or permanent strain to a bright-red heat. Phil. Mag. **24**, No. 148, 257–267 (September, 1887).

1. VOIGT, WOLDEMAR: Bestimmung der Elasticitätsconstanten von Beryll und Bergkrystall. Annalen der Physik und Chemie, Neue Folge **31**, 474–501.

1888 *1.* AMAGAT, ÉMILE HILAIRE: Sur la vérification expérimentale des formules de LAMÉ et la valeur du coefficient de POISSON. Compt. Rend. Acad. Sci. Paris **106**, 479–482 (February, 1888).

2. AMAGAT, ÉMILE HILAIRE: Recherches sur l'élasticité du cristal. Compt. Rend. Acad. Sci. Paris **107**, 618–620 (October, 1888).

1. CANTONE, MICHELE: Nuovo metodo per la determinazione delle due costanti di elasticità. Rendiconti della Reale Accademia dei Lincei, Roma. Scienze fisiche, matematiche e naturali, fourth series 4, Note I, 220–227; Note II, 292–297. [See also in the same volume, the papers by CANTONE: Ricerche intorno alle deformazioni dei condensatori. Note I, 344–353; Note II, 471–477.]

1. CONSIDÈRE, ARMAND GABRIEL: Anwendung von Eisen und Stahl bei Konstruktionen. German translation by E. HAUFF; Vienna. (See [1885, *1*].)

1. HALLOCK, WILLIAM: The flow of solids: a Note. Am. J. Sci. **36**, 59–60.

1. MERCADIER, ERNEST JULES: Sur la détermination des constants et du coefficient d'élasticité dynamique de l'acier. Compt. Rend. Acad. Sci. Paris **107**, 27–29. And see [1889, *1*].

1. VOIGT, WOLDEMAR: Bestimmung der Elasticitätsconstanten von Topaz und Baryt. Annalen der Physik und Chemie, Neue Folge **34**, 981–1028.

1. WEYRAUCH, JACOB JOHANN VON: Die Festigkeitseigenschaften und Methoden der Dimensionsberechnung von Eisen- und Stahlconstructionen. Leipzig: Teubner Verlag.

1889 *1.* AMAGAT, ÉMILE HILAIRE: Compressibilité du mercure et élasticité du verre. Compt. Rend. Acad. Sci. Paris **108**, 228–231 (February, 1889).

2. AMAGAT, ÉMILE HILAIRE: Recherches sur l'élasticité des solides. Compt. Rend. Acad. Sci. Paris **108**, 1199–1202 (June, 1889).

3. AMAGAT, ÉMILE HILAIRE: Détermination direct (c'est-à-dire sans faire usage d'aucune formule) de la compressibilité du verre du cristal et des métaux jusqu'à 2000 atm. Compt. Rend. Acad. Sci. Paris **108**, 727–730 (April, 1889).

1. CHREE, CHARLES: The equations of an isotropic elastic solid in polar and cylindrical coordinates, their solution and application. Trans. Cambridge Phil. Soc. **14**, 250–369.

1. ENGESSER, FRIEDRICH: Ueber die Knickfestigkeit gerader Stäbe. Zeitschrift des Architekten- und Ingenieur-Vereins, Hannover **35**, 456–462.

1. LE CHATELIER, ANDRÉ: Influence de la température sur les propriétés mécaniques du fer et de l'acier. Compt. Rend. Acad. Sci. Paris **109**, 58–61.

1. MALLOCK, ARNULPH: The physical properties of vulcanized India-rubber. Proc. Roy. Soc. London **46**, 233–249.

1. MERCADIER, ERNEST JULES PIERRE: Études expérimentales sur l'élasticité dynamique et statique des fils métalliques. Compt. Rend. Acad. Sci. Paris **108**, 344–346.

1. TACKE, EDUARD: Über den Wert der s'Gravesande'schen Methode zur Bestimmung der Elasticitätscoefficienten dünner Drähte. Greifswald: J. Abel.

1. TOMLINSON, HERBERT: The influence of stress and strain on the physical properties of matter. Part I. Elasticity (continued). The effect of magnetisation on the elasticity and the internal friction of metals. Phil. Trans. Roy. Soc. London **179** A, 1–26. See also [1892, *1*].

1890 *1.* AMAGAT, ÉMILE HILAIRE: Variation de l'élasticité du verre et du cristal avec la température. Compt. Rend. Acad. Sci. Paris **110**, 1246–1249 (June, 1890).

1. BERTRAND, JOSEPH LOUIS FRANÇOIS: Éloge historique de PIERRE-CHARLES-FRANÇOIS DUPIN. Éloges Académiques, Paris pp. 221–246.

1. KOWALSKI, JOSEPH VON: Elasticität und Festigkeit des Glases bei höheren Temperaturen. Annalen der Physik und Chemie, Neue Folge **39**, 155–158.

1. LE CHATELIER, ANDRÉ: Les lois du recuit et leur conséquences au point de vue des propriétés mécaniques des métaux. Compt. Rend. Acad. Sci. Paris **110**, 705–708. See also [1889, *1*].

1. STRADLING, GEORG: Ueber s'Gravesande's Verfahren zur Bestimmung des Elasticitätsmoduls. Annalen der Physik und Chemie, Neue Folge **41**, 330–333.

1891 *1.* EVERETT, JOSEPH DAVID: Illustrations of the C.G.S. System of Units, fourth edition. London and New York: Macmillan & Co. [First edition, 1875.]

1. KIRKALDY, WILLIAM G.: Strength and Properties of Materials with Description of the System of Testing. London.

1. MARTENS, ADOLF: Investigations on the influence of heat on the strength of iron. Translated and abstracted by GEORGE RUDOLPH BODMER. Proc. Inst. Civil Engrs, London **104**, 209–224 (1890–1891). Original German in: Mittheilungen aus den Koeniglichen technischen Versuchsanstalten zu Berlin, p. 159 (1890).

1. THOMPSON, JOSEPH OSGOOD: Ueber das Gesetz der elastischen Dehnung. Annalen der Physik und Chemie, Neue Folge **44**, 555–576.

1. VOIGT, WOLDEMAR: Ueber einen einfachen Apparat zur Bestimmung der thermischen Dilatation fester Körper, speciell der Krystalle. Annalen der Physik und Chemie, Neue Folge **43**, 831–834.

1892 *1.* BEARE, THOMAS HUDSON: Notes from an engineering laboratory; elastic tests of iron and steel. Engineering **53**, 276–278, 310–311.

1. LOVE, AUGUSTUS EDWARD HOUGH: A Treatise on the Mathematical Theory of Elasticity. [Fourth edition was published in 1927 and reprinted in 1944, New York: Dover.]

1. TOMLINSON, HERBERT: The influence of stress and strain on the physical properties of matter. Part III. Magnetic induction (continued). The internal friction of iron, nickel, and cobalt, studied by means of magnetic circles of very minute range. Phil. Trans. Roy. Soc., London **182** A, 341–369.

1. VOIGT, WOLDEMAR: Bestimmung der Constanten der Elasticität und Untersuchung der inneren Reibung für einige Metalle. Abhandlungen der Gesellschaft der Wissenschaften zu Göttingen **38**, 1–87.

2. VOIGT, WOLDEMAR: Über die innere Reibung der festen Körper, insbesondere der Krystalle. Annalen der Physik und Chemie, Neue Folge **47**, 671–693.

1893 *1.* AMAGAT, ÉMILE HILAIRE: Mémoire sur l'élasticité et la dilatabilité des fluides jusqu'aux très hautes pressions. Annales de Chimie et de Physique **29**, 505–574.

1. HARTIG, ERNST KARL: Der Elasticitätsmodul des gerades Stabes als Funktion der spezifischen Beanspruchung. Civilingenieur **39**, 113–138.

1. TODHUNTER, ISAAC, and KARL PEARSON: A History of the Theory of Elasticity and of the Strength of Materials, Vol. II, Parts I and II (1893). Cambridge: University Press. [Vol. I was first published in 1886.] Fourth edition of both vols. reprinted: New York: Dover (1960).

1. VOIGT, WOLDEMAR: Bestimmung der Elasticitätsconstanten einiger quasi-isotroper Metalle durch langsame Schwingungen von Stäben. Annalen der Physik und Chemie, Neue Folge **48**, 674–707.

2. VOIGT, WOLDEMAR: Ueber ein von WERTHEIM aufgestelltes Gesetz für die Elasticitätsconstanten fester Körper. Annalen der Physik und Chemie, Neue Folge **49**, 396–400.

3. VOIGT, WOLDEMAR: Bestimmung der Constanten der thermischen Dilatation und des thermischen Druckes für einige quasi-isotrope Metalle. Annalen der Physik und Chemie, Neue Folge **49**, 697–708.

1894 *1.* AUERBACH, FELIX: Über die Härte- und Elasticitätsverhältnisse des Glases. Annalen der Physik und Chemie, Neue Folge **53**, 1000–1038.

1. BACH, CARL: Elasticität und Festigkeit. Berlin: J. Springer. [There were 6 editions of this book, between 1894 and 1911.]

1. BOCK, ADALBERT MICHAEL: Ueber das Verhältnis der Quercontraction zur Längendilatation bei Stäben von verschiedenen Metallen als Funktion der Temperatur. Annalen der Physik und Chemie, Neue Folge **52**, 607–620.

1. MERRIMAN, MANSFIELD: The resistance of materials under impact. Proceedings of the American Association for the Advancement of Science 175–189 (August, 1894).

1. RAYLEIGH, Lord (JOHN WILLIAM STRUTT, 3rd): Theory of Sound. London: Macmillan.

1. STROMEYER, C. E.: Experimental determinations of Poisson's ratio. Proc. Roy. Soc. (London) **55**, 373–383.

1. WINKELMANN, ADOLPH AUGUST, and FRIEDRICH OTTO SCHOTT: Über die Elasticität und über die Zug- und Druckfestigkeit verschiedener neuer Gläser in ihrer Abhängigkeit von der chemischen Zusammensetzung. Annalen der Physik und Chemie, Neue Folge **51**, 697–746.

1895 *1.* DEWAR, Sir JAMES: Note on the viscosity of solids. Proc. Chem. Soc. **10**, 136–138.

1. NOYES, MARY CHILTON: The influence of heat and the electric current upon Young's modulus for a piano wire. Phys. Rev. **2**, 277–297.

1. Le Premier Siècle de l'Institut de France, 1795–1895, 2 Vols. (by le COMTE DE FRANQUEVILLE) Paris.

1. [Re: TRESCA] École Polytechnique, Livre Centenaire, Vol. 1: Histoire de l'Enseignement de l'École Polytechnique, pp. 206–209. Paris: Gauthier-Villars et Fils.

1896 *1.* NOYES, MARY CHILTON: The influence of heat, of the electric current, and of magnetization upon Young's Modulus. Phys. Rev. **3**, 432–447.

1897 *1.* BACH, CARL: Allgemeines Gesetz der elastischen Dehnungen. Zeitschrift des Vereins deutscher Ingenieure **41**, especially pp. 248–252.

1. DUNN, B.: A photographic impact testing machine for measuring the varying intensity of an impulsive force. J. Franklin Inst. **144**, No. 5, 321–348 (November, 1897).

1. MEHMKE, RUDOLF: Zum Gesetz der elastischen Dehnungen. Zeitschrift für Mathematik und Physik, Leipzig **42**, 327–338.

1. THURSTON, ROBERT HENRY: Singular stress-strain relations of India-rubber. Science **6**, 758–760 (July–December, 1897).

1. WINKELMANN, ADOLPH AUGUST: Ueber die Änderung des Elasticitätscoefficienten des Platins mit wachsender Temperatur. Annalen der Physik und Chemie, Neue Folge **63**, 117–123.

1898 *1.* BRILLOUIN, MARCEL: Les écarts apparents de la loi de Hooke. Poinçons et couteaux de pendules, chaines, vis calantes. Annales de Chimie et de Physique, 7th series **13**, 231–264.

2. BRILLOUIN, MARCEL: Théorie des déformations permanentes des métaux industriels. Annales de Chimie et de Physique, 7th series **13**, 377–404.

1. Engesser, Friedrich: Widerstandsmomente und Kernfiguren bei beliebigem Formänderungsgesetz (Spannungsgesetz). Zeitschrift des Vereins deutscher Ingenieure **42**, 903–907 and 927–931.

1. Haweis, Hugh Reginald: Old Violins and Violin Lore. London and New York. I am not absolutely certain of the first date of publication of this volume. On page 13 of the prelude, written in 1898, Haweis said: "I may perhaps be pardoned if I close this prelude with some words which I used before The Royal Institution in 1872." Apparently the book was reprinted in 1922.

1. Pulfrich, Carl: Über einen Interferenzmeßapparat. Zeitschrift für Instrumentenkunde **18**, 261–267.

1899 *1.* Ewing, James Alfred: The Strength of Materials. Cambridge University Press, England.

1. Shakespear, Gilbert Arden: The application of an interference-method to the investigation of Young's modulus for wires, and its relation to changes of temperature and magnetization; and a further application of the same method to the study of the change in dimensions of iron and steel wires by magnetization. Phil. Mag., London, fifth series **47**, 539–556.

1. Straubel, Constantin Rudolph: Ueber die Elasticitätszahlen und Elasticitätsmoduln des Glases. Annalen der Physik und Chemie, Neue Folge **68**, 369–413.

1900 *1.* Benton, John Robert: Abhängigkeit des spezifischen Torsionswiderstandes einiger Metalldrähte von der Spannung. Annalen der Physik und Chemie, fourth series **3**, 471–491.

1. Guest, James J.: On the strength of ductile materials under combined stress. Phil. Mag., fifth series **50**, 69–132.

1. Searle, George Frederick Charles: Apparatus for measuring the extension of a wire. Proc. Cambridge Phil. Soc. **10**, 318–323.

1901 *1.* Benton, John Robert: Determination of Poisson's ratio by means of an interference apparatus. Phys. Rev. **12**, 36–42.

1. Hopkinson, John: Original Papers of John Hopkinson, Vol. II: Scientific Papers, edited by B. Hopkinson, Cambridge, England.

1. Kohlrausch, Friedrich Wilhelm Georg, and Eduard August Grüneisen: Über die durch sehr kleine elastische Verschiebungen entwickelten Kräfte. Sitz.ber. Königl.-preuss. Akad. Wiss., Berlin, 1086–1091.

1. Schaefer, Clemens: Ueber den Einfluss der Temperatur auf die Elasticität der Metalle. Annalen der Physik und Chemie, fourth series **5**, 220–233.

1. Voigt, Woldemar: Zur Festigkeitslehre. Annalen der Physik und Chemie, fourth series **4**, 567–591.

1902 *1.* Bach, Carl von: Elasticität und Festigkeit, 4th edition. Berlin: J. Springer. [First edition was published in 1889.]

1. Schaefer, Clemens: Ueber den Einfluß der Temperatur auf die Elasticität der Elemente. Annalen der Physik und Chemie, fourth series **9**, 665–676.

2. Schaefer, Clemens: Ueber die Elasticitätszahlen einiger Stoffe mit niedrigem Schmelzpunkt. Annalen der Physik und Chemie, fourth series **9**, 1124–1127.

1. Schüle, Wilhelm: Die Biegungslehre gerader Stäbe mit veränderlichem Dehnungskoeffizienten. Dingler's Polytechnisches Journal **317**, 149–154.

1. Tammann, Gustave Heinrich Johann Apollon: Ueber die Ausflußgeschwindigkeit krystallisierter Stoffe. Annalen der Physik und Chemie, fourth series **7**, 198–224.

1903 *1.* Duhem, Pierre Maurice Marie: L'Évolution de la Mécanique. Paris.

1. Morrow, J.: On an instrument for measuring the lateral contraction of tie-bars, and on the determination of Poisson's ratio. Phil. Mag. **6**, 417–424.

1. Perot, A.: Sur les efforts développés dans le choc d'éprouvettes entaillées. Compt. Rend. Acad. Sci. Paris **137**, 1044–1046 (July–December, 1903).

1. Werigin, N., J. Lewkojeff, and Gustave Tammann: Ueber die Ausflussgeschwindigkeit einiger Metalle. Annalen der Physik und Chemie, fourth series **10**, 647–654.

1904 *1.* Buchanan, John Young: On the compressibility of solids. Proc. Roy. Soc. Edinburgh **73**, 296–310.

1. Dewar, Sir James, and Robert Asholt Hadfield: Action de la température de l'air liquide sur les propriétés mécaniques du fer de ses alliages. Annales de Chimie et de Physique, 8th series **4**, 556–574.

1. Hatt, Kendrick: Tensile impact tests of metals. Proc. Am. Soc. Testing Materials **4**, 282–315.

1. Mallock, Arnulph: On a direct method of measuring the coefficient of volume elasticity of metals. Proc. Roy. Soc. (London) **74**, 50–52.

1. OBERMAYER, A. VON: Versuche über den Ausfluß fester Körper, insbesondere des Eisens unter hohem Druck. Sitz.ber. Math.-Naturw. Kl. Kaiserl. Akad. Wiss., Wien **113**, Part IIa, 511–566.

1. PEROT, A., and HENRI MICHEL LÉVY: Sur la fragilité des métaux. Compt. Rend. Acad. Sci. Paris **138**, 474–476.

2. PEROT, A., and HENRI MICHEL LÉVY: Sur la fragilité de certains aciers. Compt. Rend. Acad. Sci. Paris **139**, 1198–1200.

1905 *1.* CHREE, CHARLES: Note on the determination of the volume elasticity of elastic solids. Proc. Roy. Soc. (London) **74**, 518–523.

1. HOPKINSON, BERTRAM: The effects of momentary stresses in metals. Proc. Roy. Soc. (London) **74**, 498–506.

1. HORTON, FRANK: On the modulus of torsional rigidity of quartz fibers and its temperature coefficient. Proc. Roy. Soc. (London) **74**, 401–402.

1. KOHLRAUSCH, FRIEDRICH WILHELM GEORG: Lehrbuch der praktischen Physik, Leipzig. [See: KOHLRAUSCH, 1872, *1.*]

1906 *1.* BLIND, AUGUST: Maß-, Münz-, und Gewichtswesen. Leipzig: G. J. Göschen. (This is Vol. 288 in a series of books entitled Sammlung Göschen.)

1. BORN, MAX: Untersuchungen über die Stabilität der elastischen Linie in Ebene und Raum, unter verschiedenen Grenzbedingungen. Doctoral dissertation, 13 June, Göttingen.

1. GRÜNEISEN, EDUARD AUGUST: Über das Verhalten des Gusseisens bei kleiner elastischer Dehnung. Deutsche Physikalische Gesellschaft **8**, 469–477.

1907 *1.* GRÜNEISEN, EDUARD AUGUST: Die elastischen Konstanten der Metalle bei kleinen Deformationen. I. Der dynamisch und statisch gemessene Elastizitätsmodul. Annalen der Physik und Chemie, fourth series **22**, 801–851.

1. HENNING, FRITZ: Über die Ausdehnung fester Körper bei tiefer Temperatur. Annalen der Physik und Chemie, fourth series **22**, 631–639.

1. HOLBORN, LUDWIG, and SIEGFRIED VALENTINER: Eine Vergleichung der optischen Temperaturskale mit dem Stickstoffthermometer bis 1600°. Annalen der Physik und Chemie, fourth series **22**, 1–48.

1. LEBER, E.: Ueber den gegenwärtigen Stand der Schlagbiegeprobe mit eingekerbten Stäben. Stahl und Eisen **27**, especially pp. 1121–1125, 1160–1164.

1. RICHARDS, THEODORE WILLIAM [in collaboration with W. N. STULL, F. N. BRINK, and F. BONNET, Jr.]: The Compressibilities of the Elements and Their Periodic Relations. Carnegie Institution of Washington, Publication No. 76.

1. SCHEEL, KARL FRANZ FRIEDRICH CHRISTIAN: Versuche über die Ausdehnung fester Körper, insbesondere von Quartz in Richtung der Hauptachse, Platin, Palladium und Quartz bei der Temperatur der flüssigen Luft. Verhandlungen der Deutschen Physikalischen Gesellschaft **9**, 3–23.

2. SCHEEL, KARL FRANZ FRIEDRICH CHRISTIAN, and WILHELM HEUSE: Bestimmung der Ausdehnung des Platins zwischen −183° und Zimmertemperatur mit dem Komparator und dem Fizeauschen Apparat. Verhandlungen der Deutschen Physikalischen Gesellschaft **9**, 449–459.

1. SCHILLER, LUDWIG: Ueber die Poissonsche Konstante des Kautschuks. Annalen der Physik und Chemie, fourth series **22**, 204–208.

1. WALKER, HENRY: The variation of Young's modulus under an electric current. Proc. Roy. Soc. Edinburgh **27**, 343–356 (November 1906 to July 1907).

1908 *1.* CHWOLSON, OREST DANULOVICH: Traité de Physique **1**; translated from the Russian and German editions by E. DAVAUX. Paris.

1. GRÜNEISEN, EDUARD AUGUST: Torsionmodul, Verhältnis von Querkontraktion zu Längsdilatation und kubische Kompressibilität. Annalen der Physik und Chemie, fourth series **25**, 825–851.

1. SEARLE, GEORGE FREDERICK CHARLES: Experimental Elasticity. Cambridge University Press.

1. SEARS, J. E.: On the longitudinal impact of metal rods with rounded ends. Part I. Proc. Cambridge Phil. Soc. **14**, 257–286.

1. THIESEN, MAX FERDINAND: Die Zustandsgleichung der Metalle. Verh. Deut. Phys. Ges. **10**, 410–417.

1. WALKER, HENRY: The variation of Young's modulus under an electric current (Part II). Proc. Roy. Soc. Edinburgh **28**, 652–675 (November 1907–July 1908).

1909 *1.* BRIDGMAN, PERCY WILLIAMS: The measurement of high hydrostatic pressure, I. A simple primary gauge. Proc. Am. Acad. Arts Sci. **44**, 201–217. [Also in BRIDGMAN (1964, *1*), Vol. I, pp. 1–17.]

2. Bridgman, Percy Williams: The measurement of high hydrostatic pressure, II. A secondary mercury resistance gauge. Proc. Am. Acad. Arts Sci. **44**, 221–251. [Also in Bridgman (1964, *1*), Vol. I, pp. 19–49.]

3. Bridgman, Percy Williams: An experimental determination of certain compressibilities. Proc. Am. Acad. Arts Sci. **44**, 255–279. (Also in Bridgman [1964, *1*], Vol. I, pp. 51–75.)

1. Ludwik, Paul: Über den Einfluß der Deformationsgeschwindigkeit bei bleibenden Deformationen mit besonderer Berücksichtigung der Nachwirkungserscheinungen. Physikalische Zeitschrift **10**, 411–417.

1. Poynting, J. Henry: On pressure perpendicular to the shear planes in finite pure shears, and on the lengthening of loaded wires when twisted. Proc. Roy. Soc. (London), Ser. A **82**, 546–559.

1. Rinne, Friedrich: Vergleichende Untersuchungen über die Methoden zur Bestimmung der Druckfestigkeit von Gesteinen. (Zweiter Bericht über von L. Prandtl, und F. Rinne durchgeführte Versuche.) Neues Jahrb. Mineral., Geol. Paläonthol. **2**, 121–128.

1. Royal Society of London, Catalogue of Scientific Papers (1800–1900) Subject Index, Vol. II, Mechanics. Cambridge.

1910 *1.* Andrade, Edward Neville da Costa: The viscous flow in metals and allied phenomena. Proc. Roy. Soc. (London), Ser. A **84**, 1–12.

1. Grüneisen, Eduard August: Ueber die thermische Ausdehnung der Metalle. Annalen der Physik und Chemie, fourth series **33**, 33–64.

2. Grüneisen, Eduard August: Ueber den Einfluß von Temperatur und Druck auf Ausdehnungskoeffizient und spezifische Wärme der Metalle. Annalen der Physik und Chemie, fourth series **33**, 65–78.

3. Grüneisen, Eduard August: Einfluß der Temperatur auf die Kompressibilität der Metalle. Annalen der Physik und Chemie, fourth series **33**, 1239–1274.

1. Unwin, William Cawthorne: The Testing of Materials of Construction, 3rd edition, London: Longmans, Green & Co. See also [1886, *2*].

1. Voigt, Woldemar: Lehrbuch der Krystallphysik, especially Chapter VII, pp. 716–763. Leipzig und Berlin: B. G. Teubner. [Reprinted in 1928.]

1911 *1.* Carpenter, Rolla C., and Herman Diederichs: Experimental Engineering and Manual for Testing. New York.

1. Kármán, Theodore von: Festigkeitsversuche unter allseitigem Druck. Zeitschrift des Vereins deutscher Ingenieure **55**, 1749–1757.

1. Walker, Henry: The variation of Young's modulus under an electric current. Part III. Proc. Roy. Soc. Edinburgh **31**, 186–250.

1912 *1.* Friedrich, W., P. Knipping and M. von Laue: Interferenz-Erscheinungen bei Röntgenstrahlen. ("Theoretischer Teil" by M. von Laue, pp. 303–311; "Experimenteller Teil" by Friedrich and Knipping, pp. 311–322). Sitz.ber. math.-phys. Klasse der Bayer. Akad. Wiss. München. Substantially the same paper also in: Annalen der Physik und Chemie, fourth series **41**, 971–1002 (1913).

1. Laue, M. von: Eine quantitative Prüfung der Theorie für die Interferenz-Erscheinungen bei Röntgenstrahlen. Sitz.ber. math.-phys. Klasse der Bayerischen Akademie der Wissenschaften, München, pp. 363–373. [Also in: Annalen der Physik und Chemie, fourth series, **41**, 989–1002 (1913).]

1. Plank, Rudolph: Betrachtungen über dynamische Zugbeanspruchung. Zeitschrift des Vereins deutscher Ingenieure **56**, 17–24, 46–51.

1. Poynting, J. Henry: On the changes in the dimensions of a steel wire when twisted, and on the pressure of distorsional waves in steel. Proc. Roy. Soc. (London), Ser. A **86**, 534–561.

1. Rosenhain, Walter, and S. L. Archbutt: Tenth Report to the Alloys Research Committee: On the alloys of aluminum and zinc. Proc. Inst. Mech. Engrs 313–515.

1. Sears, J. E.: On the longitudinal impact of metal rods with rounded ends. Part II. Trans. Cambridge Phil. Soc. **21**, 49–105.

1913 *1.* Laue, M. von, and F. Tank: Die Gestalt der Interferenzpunkte bei den Röntgenstrahlinterferenzen. Annalen der Physik und Chemie, fourth series **41**, 1003–1011.

1914 *1.* Colonnetti, Gustavo: Esperienze sulla elasticità a trazione del rame. Atti della Reale Accademia dei Lincei, Rendiconti Classe di scienze fisiche, matematiche e naturali **23**. See: pp. 165–171, 225–231, 421–427.

1. Desch, Cecil H.: Abstract of paper by Colonnetti [1914, *1*] J. Inst. Met. No. 2, 272–273.

1. Hopkinson, Bertram: A method of measuring the pressure produced in the detonation of explosives or by the impact of bullets. Phil. Trans. Roy. Soc. London, Ser. A **213**, 437–456.

1. LEA, FREDERICK CHARLES, and O. H. CROWTHER: The change of the modulus of elasticity and of other properties of metals with temperature. Engineering **98**, 487–489.

1. SEEHASE, H.: Die experimentelle Ermittlung des Verlaufes der Stoßkraft und die Bestimmung der Deformationsarbeit beim Stauchversuch. Zeitschrift des Vereins deutscher Ingenieure, Berlin **58**, 1345–1348.

1915 *1.* RICHARDS, THEODORE WILLIAM: Concerning the compressibilities of the elements, and their relations to other properties. J. Am. Chem. Soc. **37**, No. 7, 1643–1657 (July–December, 1915).

1921 *1.* BACH, CARL, and RICHARD WILHELM BAUMANN: Festigkeitseigenschaften und Gefügebilder der Konstruktionsmaterialien. Berlin: J. Springer.

1. CARPENTER, H. C. H., and CONSTANCE F. ELAM: The production of single crystals of aluminium and their tensile properties. Proc. Roy. Soc. (London) **100**A, 329–353.

1. JESSOP, H. T.: On Cornu's method of determining the elastic constants of glass. Phil. Mag., sixth series **42**, 551–568.

1. NELSON, HARLEY A.: Stress-strain measurements on films of drying oils, paints and varnishes. Proceedings of the 24th Annual Meeting of the American Society for Testing and Materials **21**, 1111–1138. See also [1923, *1*, *2*].

1. TIMOSHENKO, STEPHEN PROKOFIEVITCH: On the correction for shear of the differential equation for transverse vibrations of prismatic bars. Phil. Mag., sixth series **41**, 744–746.

1922 *1.* KÖRBER, FRIEDRICH, and RUDOLF H. SACK: Vergleichende statische und dynamische Zugversuche. Mitteilungen des Kaiser-Wilhelm-Instituts für Eisenforschung, **4**, 11–29.

1. LEA, FREDERICK CHARLES: The effect of temperature on some of the properties of metals. Engineering **113**, 829–832.

1. MCCOLLUM, BURTON, and O. S. PETERS: A new electrical telemeter. U.S. Bur. Standards Technological Papers, No. 221, **17**, 737–777 (1922–1924).

1. WHITTEMORE, H. L.: Resume of impact testing of materials, with bibliography. Symposium on Impact Testing of Materials, American Society of Testing Materials 6–36.

1923 *1.* BAUMANN, RICHARD WILHELM: see LANDOLT, HANS HEINRICH: LANDOLT-BÖRNSTEIN'S Physikalisch-Chemische Tabellen, Vol. 1, RICHARD BÖRNSTEIN, editor, pp. 79–93. Berlin: J. Springer.

1. BRIDGMAN, PERCY WILLIAMS: The compressibility of thirty metals as a function of pressure and temperature. Proc. Am. Acad. Arts Sci. **58**, 165–242. (Also in BRIDGMAN [1964, *1*], Vol. III, paper No. 45, pp. 1583–1660.)

1. NELSON, HARLEY A.: Physical properties of varnish films indicated by stress-strain measurements. Proc. of the 26th Annual Meeting of the American Society for Testing and Materials **23**, Part I, Committee Report 290–299.

2. NELSON, HARLEY A., and GEORGE W. RUNDLE: Further studies of the physical properties of drying-oil, paint and varnish films. Proc. Am. Soc. Testing Materials **23**, Part II: Technical Papers: pp. 356–368.

1. PORTEVIN, ALBERT, and FRANÇOIS LE CHATELIER: Sur un phénomène observé lors de l'essai de traction d'alliages en cours de transformation. Compt. Rend. Acad. Sci. Paris **176**, 507–510.

1. TAYLOR, Sir GEOFFREY INGRAM, and CONSTANCE F. ELAM: The distortion of an aluminium crystal during a tensile test. Proc. Roy. Soc. (London), Ser. A **102**, 643–647. (Also in: TAYLOR [1958, *1*], Paper No. 5, pp. 63–84.)

1924 *1.* RICHARDS, THEODORE WILLIAM: Compressibility, internal pressure and change of atomic volume. J. Franklin Inst. **198**, 1–27 (July, 1924).

1. WAGSTAFF, JOHN EDWARD PRETTY: Experiments on the duration of impacts, mainly of bars with rounded ends, in elucidation of the elastic theory. Proc. Roy. Soc. (London), Ser. A **105**, 544–570.

1925 *1.* BRIDGMAN, PERCY WILLIAMS: Certain physical properties of single crystals of tungsten, antimony, bismuth, tellurium, cadmium, zinc, and tin. Proc. Am. Acad. Arts Sci. **60**, 305–383. (Also in BRIDGMAN [1964, *1*], Vol. III, paper No. 58, pp. 1851–1929.)

1. KÖRBER, FRIEDRICH, and HANS ARNOLD STORP: Ueber den Kraftverlauf bei der Schlagpruefung. Mitt. Kaiser-Wilhelm-Institut für Eisenforschung **7**, 81–97.

1. LUDWIG, PAUL, and R. SCHEU: Vergleichende Zug-, Druck-, Dreh- und Walzversuche. Stahl und Eisen **45**, 373–381.

1. TAYLOR, Sir GEOFFREY INGRAM, and CONSTANCE F. ELAM: The plastic extension and fracture of aluminium crystals. Proc. Roy. Soc. (London), Ser. A **108**, 28–51. (Also in TAYLOR [1958, *1*], paper No. 7, pp. 109–129.)

2. TAYLOR, Sir GEOFFREY INGRAM, and W. S. FARREN: The heat developed during plastic extension of metals. Proc. Roy. Soc. (London), Ser. A **107**, 422–451. (Also in TAYLOR [1958, *1*], paper No. 6, pp. 85–108.)

3. TAYLOR, Sir GEOFFREY INGRAM, and CONSTANCE F. ELAM: Notes on the "Navier effect". Paper written for the Aeronautical Research Committee. (Also in TAYLOR [1958, *1*], Vol. I, paper No. 8, pp. 130–132.)

1926 *1.* ELAM, CONSTANCE F.: Tensile tests of large gold, silver and copper crystals. Proc. Roy. Soc. (London), Ser. A **112**, 289–296.

1. LODE, W.: Versuche über den Einfluß der mittleren Hauptspannung auf das Fließen der Metalle Eisen, Kupfer und Nickel. Z. Physik **36**, 913–939.

1. RICHARDS, THEODORE WILLIAM: A brief history of the investigations of internal pressures. Chem. Rev. **2**, 315–348.

1. TAYLOR, Sir GEOFFREY INGRAM, and CONSTANCE F. ELAM: The distortion of iron crystals. Proc. Roy. Soc. (London), Ser. A **112**, 337–361. (Also in TAYLOR [1958, *1*], paper No. 10, pp. 153–173.) See also DOHI [1960, *1*].

2. TAYLOR, Sir GEOFFREY INGRAM, and W. S. FARREN: The distortion of crystals of aluminium under compression. Part I. Proc. Roy. Soc. (London), Ser. A **111**, 529–551. (Also in TAYLOR [1958, *1*], paper No. 9, pp. 133–152.)

1927 *1.* GÖLER, V., and GEORG OSKAR SACHS: Das Verhalten von Aluminiumkrystallen bei Zugversuchen. I. Geometrische Grundlagen. Z. Physik **41**, 103–115.

1. KARNOP, R. VON, and GEORG OSKAR SACHS: Das Verhalten von Aluminiumkrystallen bei Zugversuchen. II. Experimenteller Teil. Z. Physik **41**, 116–134.

1. LOVE, AUGUSTUS EDWARD HOUGH: A Treatise on the Mathematical Theory of Elasticity. 4th edition. [First published in 1892. See also Dover, New York, reprint of 4th edition (1944).]

1. SACHS, GEORG OSKAR: See LANDOLT-BÖRNSTEIN, Physikalisch-chemische Tabellen, Ergänzungsband I, pp. 13–54.

1. SCHMID, E.: Über die Schubverfestigung von Einkristallen bei plastischer Deformation. Z. Physik **40**, 54–74.

1. SIEBEL, E., and A. POMP: Die Ermittlung der Formänderungsfestigkeit von Metallen durch den Stauchversuch. Mittheilungen, Kaiser-Wilhelm-Inst. Eisenforsch. (Max-Planck-Inst.) **9**, 157.

1. TAYLOR, Sir GEOFFREY INGRAM: The distortion of single crystals of metals. Proceedings of the 2nd International Congress for Applied Mechanics Zürich: Orell Füssli. (Also in TAYLOR [1958, *1*], Vol. I, paper No. 11, pp. 174–184.)

2. TAYLOR, Sir GEOFFREY INGRAM: The distortion of crystals of aluminium under compression. Part II. Distortion of double slipping and changes in orientation of crystal axes during compression. Proc. Roy. Soc. (London), Ser. A **116**, 16–38. (Also in TAYLOR [1958, *1*], Vol. I, paper No. 12, pp. 185–204.)

3. TAYLOR, Sir GEOFFREY INGRAM: The distortion of crystals of aluminium under compression. Part III. Measurements of stress. Proc. Roy. Soc. (London), Ser. A **116**, 39–60. (Also in TAYLOR [1958, *1*], Vol. I, paper No. 13, pp. 205–224.)

1928 *1.* CAJORI, FLORIAN: A History of Physics. New York: Dover Publications.

1. KENNELLY, ARTHUR EDWIN: Vestiges of Pre-Metric Weights and Measures Persisting in Metric-System Europe. New York: Macmillan Co.

1. PRANDTL, LUDWIG: Ein Gedankenmodell zur kinetischen Theorie der festen Körper. Z. Angew. Math. Mech. **8**, 85–106.

1. MASIMA, M., and GEORG OSKAR SACHS: Mechanische Eigenschaften von Messingkristallen. Z. Physik **50**, 161–186.

1. RIEKERT, P. VON: Tafeln der Elastizitätskonstanten und Festigkeitszahlen. Mechanik der elastischen Körper, pp. 623–627. In: Handbuch der Physik, H. GEIGER and KARL SCHEEL, eds., Vol. 6. Berlin: J. Springer.

1930 *1.* DONNELL, L. H.: Longitudinal wave transmission and impact. J. Appl. Mech., Transactions, American Society of Mechanical Engineers **52**, No. 1, 153–167.

1. GUEST, JAMES J.: Effects of rapidly acting stress. Proc. Inst. Mech. Engrs 1273–1296.

1. SACHS, G., and J. WEERTS: Zugversuche an Gold-Silberkristallen. Z. Physik **62**, 473–493.

1. SAYRE, MORTIMER FREEMAN: Elastic behavior of spring materials. Proceedings of the Thirty-third Annual Meeting of the American Society for Testing and Materials **30**, Part II, Technical Papers, pp. 546–558.

1931 *1.* BOAS, W., and E. SCHMID: Über die Temperaturabhängigkeit der Kristallplastizität: III. Aluminium. Z. Physik **71**, 703–712.

1. BRIDGMAN, PERCY WILLIAMS: The Physics of High Pressure. New York: The Macmillan Co.; also: London: G. Bell & Sons, Ltd. (1949).

1. COKER, ERNEST GEORGE, and LOUIS NAPOLEON GEORGE FILON: A Treatise on Photo-Elasticity. Cambridge University Press.
1. EATON, ERIK C.: Resistance strain gage measures stresses in concrete. Engineering News-Record **107**, 615–616 (Oct. 15, 1931).
1. EVERETT, FRANKLIN L.: Strength of materials subjected to shear at high temperatures. Trans. Am. Soc. Mech. Engrs **53**, APM-53-10, 117–134. [From a dissertation for the degree of Doctor of Philosophy in the University of Michigan, with the supervision of Prof. S. TIMOSHENKO.]
1. GOENS, KARL ANTON ERICH: Über die Bestimmung des Elasticitätsmoduls von Stäben mit Hilfe von Biegeschwingungen. Annalen der Physik und Chemie, 5th Series **11**, 649–678.
1. HANSON, DANIEL, and M. A. WHEELER: The deformation of metals under prolonged loading. Part I: The flow and fracture of aluminium. J. Inst. Metals, London **45**, 229–264.
1. TAFEL, WILHELM, and E. VIEHWEGER: Einfluß der Verformungsgeschwindigkeit auf den Formänderungswiderstand. Zeitschrift des Vereins deutscher Ingenieure **75**, No. 49, 1479–1483 (December, 1931).
1. TAYLOR, Sir GEOFFREY INGRAM, and H. QUINNEY: The plastic distortion of metals. Phil. Trans. Roy. Soc. London, Ser. A **230**, 323–362. (Also in: TAYLOR [1958, *1*], paper No. 16, pp. 252–290.)
1932 *1.* DEUTLER, H.: Experimentelle Untersuchungen über die Abhängigkeit der Zugspannungen von der Verformungsgeschwindigkeit. Physikalische Zeitschrift **33**, 247–259.
1. HONEGGER, E.: The modulus of elasticity of steel at high temperatures. Brown Boveri Review **19**, 143–147.
1. THIELE, WALTER: Temperaturabhängigkeit der Plastizität und Zugfestigkeit von Steinsalzkristallen. Z. Physik **75**, 763–776.
1933 *1.* ITIHARA, MITITOSI: Impact torsion test. The Technology Reports of the Tôhoku Imperial University, Sendai, Japan **11**, No. 1, 16–50.
1934 *1.* BALAMUTH, LEWIS: A new method for measuring elastic moduli and the variation with temperature of the principal Young's modulus of rocksalt between 78° K and 273° K. Phys. Rev., second series **45**, 715–720. (January–June, 1934.)
1. SAYRE, MORTIMER FREEMAN: Plastic behavior in light of creep and elastic recovery phenomenon. Trans. Am. Soc. Mech. Engrs **56**, 559–561.
1. SCHAEFER, CLEMENS, and LUDWIG BERGMANN: Laue-Diagramme mit optischen Wellen. Sitz.ber. Akad. Wiss., Berlin 152–153.
2. SCHAEFER, CLEMENS, and LUDWIG BERGMANN: Neue Interferenzerscheinungen an schwingenden Piezoquarzen. Sitz.ber. Akad. Wiss., Berlin 192–193.
1. TAYLOR, Sir GEOFFREY INGRAM: The mechanism of plastic deformation of crystals. Part I: Theoretical. Proc. Roy. Soc. (London), Ser. A **145**, 362–387. (Also in TAYLOR [1958, *1*], paper No. 21, pp. 344–366.)
2. TAYLOR, Sir GEOFFREY INGRAM, and H. QUINNEY: The latent energy remaining in a metal after cold working. Proc. Roy. Soc. (London), Ser. A **143**, 307–326. (Also in TAYLOR [1958, *1*], paper No. 19, pp. 310–328.)
3. TAYLOR, Sir GEOFFREY INGRAM: The mechanism of plastic deformation of crystals. Part II. Comparison with observations. Proc. Roy. Soc. (London), Ser. A **145**, 388–404. (Also in TAYLOR [1958, *1*], Vol. I, paper No. 22, pp. 367–380.)
4. TAYLOR, Sir GEOFFREY INGRAM: The strength of rock salt. Proc. Roy. Soc. (London), Ser. A **145**, 405–415. (Also in TAYLOR [1958, *1*], Vol. I, paper No. 23, pp. 381–389.)
5. TAYLOR, Sir GEOFFREY INGRAM: A theory of the plasticity of crystals. Z. Kristallographie. Ser. A **89**, 375–385. (Also in TAYLOR [1958, *1*], Vol. I, paper No. 24, pp. 390–398.)
1935 *1.* CARLSON, ROY WASHINGTON: Five years' improvement of the elastic-wire strain meter. Engineering News-Record **114**, 696–697 (May 16, 1935).
1. CHALMERS, BRUCE: An interference extensometer and some observations on the elasticity of lead. Proc. Phys. Soc. **47**, 352–370. See also SAYRE [1934, *1*].
1. ELAM, CONSTANCE F.: Distortion of Metal Crystals. Oxford, England: Clarendon Press.
1. FUES, ERWIN, and HANFRIED, LUDLOFF: Weitere Untersuchungen über die Beugungserscheinungen an schwingenden Kristallen. (II. Theoretischer Teil.) Sitz.ber. Akad. Wiss., Berlin, Phys.-Math. Kl. 225–239.
1. ITIHARA, MITITOSI: Impact torsion test, No. 4. The Technology Reports of the Tôhoku Imperial University, Sendai, Japan **11**.

1. KÖSTER, WERNER: Eigenschaftsveränderungen irreversibler Ternärer Eisenlegierungen durch Wärmebehandlung. Arch. Eisenhüttenw. **8**, No. 11, Gruppe E, No. 439, 491–498 (May, 1935).

1. MANN, H. C.: The relation between the tension static and dynamic tests. Proc., Am. Soc. Testing Materials **35**, Part II, 323–335.

1. SCHAEFER, CLEMENS, and LUDWIG BERGMANN: Weitere Untersuchungen über die Beugungserscheinungen an schwingenden Kristallen. (I. Experimenteller Teil.) Sitz.ber. Akad. Wiss., Berlin, Phys.-Math. Kl. 222–225. See also [1934, *2*].

1. SCHMID, ERICH, and W. BOAS: Krystallplastizität. Berlin: J. Springer. See [1950, *1*].

1. SICHEL, FERDINAND J. M.: The elasticity of isolated resting skeletal muscle fibers. J. Cellular and Comparative Physiology **5**, 21–41.

1936 *1.* CHALMERS, BRUCE: Micro-plasticity in crystals of tin. Proc. Roy. Soc. (London), Ser. A **156**, 427–443.

1. FUCHS, KLAUS: The elastic constants and specific heats of the alkali metals. Proc. Roy. Soc. (London), Ser. A **157**, 444–450.

1. GOENS, E. (KARL ANTON ERICH), and J. WEERTS: Hauptelastizitätskonstanten des Einkristalls von Kupfer, Gold und Blei. Physikalische Zeitschrift **37**, 321–326.

1. ITIHARA, MITITOSI: Impact torsion test, No. 5. The Technology Reports of the Tôhoku Imperial University, Sendai, Japan **12**, No. 1.

1. JONES, PAUL G., and FRANK EDWIN RICHART: The effect of testing speed on strength and elastic properties of concrete. Proceedings of the Thirty-ninth Meeting of the American Society for Testing and Materials **36**, Part II, Technical Papers, 380–391.

1. PHILLIPS, ALBERT J., and ALBERT ALONZO SMITH, JR.: Effect of time on tensile properties of hard-drawn copper wire. Proceedings of the Thirty-ninth Meeting of the American Society for Testing and Materials **36**, Part II, Technical Papers, 263–273.

1. PROWSE, W. A.: The development of pressure waves during the longitudinal impact of bars. Phil. Mag., seventh series **22**, 209–239.

1. ROSE, FRED C.: The variation of the adiabatic elastic moduli of rocksalt with temperature between 80° K and 270° K. Phys. Rev., second series **49**, 50–54 (January–June, 1936).

1937 *1.* CHALMERS, BRUCE: Precision extensometer measurements on tin. J. Inst. Metals, London **61**, 103–118.

1. FÖRSTER, FRITZ: Ein neues Meßverfahren zur Bestimmung des Elastizitätsmoduls und der Dämpfung. Z. Metallk. **29**, 109–115.

2. FÖRSTER, FRITZ, and WERNER KÖSTER: Elastizitätsmodul und Dämpfung in Abhängigkeit vom Werkstoffzustand. Z. Metallk. **29**, 116–123.

1. GINNS, D. W.: The mechanical properties of some metals and alloys broken at ultra high speeds. J. Inst. Metals **61**, 61–71.

1. MURNAGHAN, FRANCIS D.: Finite deformation of an elastic solid. Am. J. Math. **59**, 235–260.

1. TAYLOR, Sir GEOFFREY INGRAM: The emission of the latent energy due to previous cold working when a metal is heated. Proc. Roy. Soc. (London), Ser. A **163**, 157–181. (Also in TAYLOR [1958, *1*], paper No. 26, pp. 402–423.)

2. TAYLOR, Sir GEOFFREY INGRAM: See GINNS [1937, *1*], and TAYLOR [1946, *1*].

1938 *1.* CLARK, DONALD SHERMAN, and G. DATWYLER: Stress-strain relations under tension impact loadings. American Society of Metals, Proceedings of the 41st Annual Meeting **38**, Part II, 98–111.

1. ELAM, CONSTANCE F.: The influence of rate of deformation on the tensile test with special reference to the yield point in iron and steel. Proc. Roy. Soc. (London) **165**, 568–592.

1. POWERS, TREVAL CLIFFORD: Measuring Young's modulus of elasticity by means of sonic vibrations. Proceedings of the Thirty-eighth Annual Meeting of the American Society for Testing and Materials **38**, 460–467. See also JONES and RICHART [1936, *1*].

1. PRAGER, W.: On isotropic materials with continuous transition from elastic to plastic state. Proc., 5th International Congress of Applied Mechanics (Cambridge), pp. 234–237.

1. TAYLOR, Sir GEOFFREY INGRAM: Plastic strain in metals. J. Inst. Metals **62**, 307–324. (Also in TAYLOR [1958, *1*], paper No. 27, pp. 424–438.)

2. TAYLOR, Sir GEOFFREY INGRAM: Analysis of plastic strain in a cubic crystal. Stephen Timoshenko 60th Anniversary Volume, pp. 218–224. New York: Macmillan. (Also in TAYLOR [1958, *1*], paper No. 28, pp. 439–445.)

1939 *1.* BRIDGMAN, PERCY WILLIAMS: The high pressure behavior of miscellaneous minerals. Am. J. Sci. **237**, 7–18. (Also in BRIDGMAN [1964, *1*], Vol. VI, paper No. 127, pp. 3363–3374.)

1. BENDER, OTTMAR: Elastizitätsmessungen an Alkalimetall-Einkristallen in tiefer Temperatur. Ann. Physik und Chemie, fifth ser. **34**, 359–376.

1. GUILLET, LÉON: Contribution à l'étude du module d'élasticité des alliages métallurgiques **36**, 497–521.

1. MEISSER, OTTO FRANZ: Die Ausbiegung schlanker, gerader Stäbe bei Beanspruchung auf Knickung und ihre meßtechnische Verwendung für statische Elastizitätsmoduluntersuchungen. Physikalische Zeitschrift **40**, 551–556.

1940 *1.* BARRETT, CHARLES S., and L. H. LEVENSON: The structure of aluminum after compression. Transactions of the American Institute of Mining and Metallurgical Engineers **137**, 112–126.

1. BRIDGMAN, PERCY WILLIAMS: The linear compression of iron to 30000 kg/cm². Proc. Am. Acad. Arts Sci. **74**, 11–20. (Also in BRIDGMAN [1964, *1*], Vol. VI, paper No. 133, pp. 3401–3410.)

2. BRIDGMAN, PERCY WILLIAMS: The compression of 46 substances to 50000 kg/cm². Proc. Am. Acad. Arts Sci. **74**, 21–51. (Also in BRIDGMAN [1964, *1*], Vol. VI, paper No. 134, pp. 3411–3441.)

1. FANNING, R., and W. V. BASSETT: Measurement of impact strains by a carbon-strip extensometer. Trans. Am. Soc. Mech. Engrs **62**, A-24–A-28.

1. GOENS, KARL ANTON ERICH: Über die Temperaturabhängigkeit der Hauptelastizitätskonstanten des Einkristalls von Kupfer, Gold, Blei, Aluminium bei tiefen Temperaturen. Annalen der Physik und Chemie, fifth ser. **38**, 456–465.

1. KÖSTER, WERNER: Elasticitätsmodul und Dämpfung von Eisen und Eisenlegierungen. Arch. Eisenhüttenwes. **14**, No. 6, Group E, No. 836, 271–278 (December, 1940).

1. MANJOINE, MICHAEL JOSEPH, and ARPAD LUDWIG NADAI: High-speed tension tests at elevated temperatures. Proc. Am. Soc. Testing Materials **40**, 822–839.

1. MOONEY, M.: A theory of large elastic deformation. J. Appl. Phys. **11**, 582–592.

1. SMITH, CYRIL STANLEY: Proportional limit tests on copper alloys. Proceedings of the Forty-third Annual Meeting of the American Society for Testing Materials **40**, 864–872. (Discussions by A. R. ANDERSON, M. F. SAYRE, L. B. TUCKERMAN *et al.*, pp. 873–874.)

1941 *1.* BRIDGMAN, PERCY WILLIAMS: Compressions and polymorphic transitions of seventeen elements to 100000 kg/cm². Phys. Rev. **60**, 351–354. (Also in BRIDGMAN [1964, *1*], Vol. VI, paper No. 136, pp. 3453–3457.)

1. BROWN, A. F. C., and N. D. G. VINCENT: The relationship between stress and strain in the tensile impact test. Proc. Inst. Mech. Engrs (London) **145**, 126–134.

1. NADAI, ARPAD LUDWIG, and MICHAEL JOSEPH MANJOINE: High speed tension tests at elevated temperatures, Parts II and III. Trans. Am. Soc. Mech. Engrs **63**, A-77 to A-91.

1. PRAGER, W.: A new mathematical theory of plasticity. Rev. Fac. Sci. Univ. Istanbul, A 5, 215–226.

1. SMITH, CYRIL STANLEY, and R. W. VAN WAGNER: The tensile properties of some copper alloys. Proc. Am. Soc. Testing Materials **41**, 825–845.

1. SUTOKI, TOMIYA: On the serrated elongation in different metals. The Science Reports of the Tôhoku Imperial University, first ser. **29**, 673–688.

1942 *1.* BOHNENBLUST, H. F., J. U. CHARYK, and D. H. HYERS: Graphical solutions for problems of strain propagation in tension. U.S.A. National Defense Research Council Report A-131, OSRD No. 1204.

2. BOHNENBLUST, H. F.: Propagation of plastic waves. A comparison of Reports NRRC A-29 and R.C. 329. U.S.A. National Research Committee, Memorandum No. A-53, pages 1–4.

1. BRIDGMAN, PERCY WILLIAMS: Pressure-volume relations for seventeen elements to 100000 kg/cm². Proc. Am. Acad. Arts and Sci. **74**, 425–440. (Also in BRIDGMAN [1964, *1*], Vol. VI, paper No. 139, pp. 3489–3504.) See also [1941, *1*], [1948, *2*].

1. DART, S. L., R. L. ANTHONY, and EUGENE GUTH: Rise of temperature on fast stretching of synthetics and natural rubbers. Industrial Engineering Chemistry, Industrial Edition **34**, No. 11, 1340–1342.

1. DUWEZ, POL E., D. S. WOOD, and D. S. CLARK: The propagation of plastic strain in tension. U.S.A. National Defense Research Council, Progress Report, No. A-99, OSRD No. 931 (October, 1942).

1. KÁRMÁN, THEODORE VON: On the propagation of plastic deformation in solids. U.S.A. National Defense Research Council, Report A-29 (February, 1942).

2. KÁRMÁN, THEODORE VON, H. F. BOHNENBLUST, and D. H. HYERS: The propagation of plastic waves in tension specimens of finite length. Theory and methods of integration. U.S.A. National Defense Research Council, Report A-103, OSRD No. 946.

1. PRAGER, W.: Fundamental theorems of a new mathematical theory of plasticity. Duke Math. J. **9**, 228–233.

1. TAYLOR, Sir GEOFFREY INGRAM: The plastic wave in a wire extended by an impact load. British Ministry of Home Security, Civil Defense Research Committee Report, R.C. 329. (Also in TAYLOR [1958, *1*], paper No. 32, pp. 467–479.)

1. WHITE, M. P., and LE VAN GRIFFIS: U.S.A. National Defense Research Council Progress Report No. A-7, OSRD No. 742. See: WHITE and GRIFFIS [1947, *1*] and [1948, *1*].

1943 *1*. BRIDGMAN, PERCY WILLIAMS: Recent work in the field of high pressures. Am. Sci. **31**, 1–35. (Also in BRIDGMAN [1964, *1*], Vol. VI, paper No. 140, pp. 3505–3539.)

1. DAVIS, EVAN A.: Increase of stress with permanent strain and stress-strain relations in the plastic state for copper under combined stresses. J. Appl. Mech. **10**, A-187 to A-196 (December, 1943).

1. JAMES, HUBERT M., and EUGENE GUTH: Theory of the elastic properties of rubber. J. Chem. Phys. **11**, No. 10, 455–481 (October, 1943).

1. KÖSTER (KOESTER), WERNER: Die Querkontraktionszahl im periodischen System. Z. Elektrochem. **49**, 233–237.

1944 *1*. EVERETT, FRANKLIN L., and JULIUS MIKLOWITZ: Poisson's ratio at high temperatures. J. Appl. Phys. **15**, 592–598.

1. SZYMANOWSKI, W. T.: Rapid determination of elastic constants of glass and other transparent substances. J. Appl. Phys. **15**, 627.

1945 *1*. BRIDGMAN, PERCY WILLIAMS: The compression of sixty-one solid substances to 25000 kg/cm^2, determined by a new rapid method. Proc. Am. Acad. Arts and Sci. **76**, 9–24. (Also in BRIDGMAN [1964, *1*], Vol. VI, paper No. 148, pp. 3609–3624.)

1. DAVIS, EVAN A.: Yielding and fracture of medium-carbon steel under combined stress. J. Appl. Mech. (Transactions of the American Society of Mechanical Engineers) **12**, No. 1, A-13 to A-24 (March, 1945).

1. MACGREGOR, C. W., and J. C. FISHER: Tension tests at constant true strain rates. J. Appl. Mech. (Transactions of the American Society of Mechanical Engineers) **12**, No. 4, A-217 to A-227 (December, 1945).

1. RAKHMATULIN, K. A.: Propagation of a wave of unloading. Soviet J. Appl. Math. Mech. (Prikl. Mat. Mekh.) **9**, No. 1, 91–100.

1946 *1*. BRIDGMAN, PERCY WILLIAMS: Recent work in the field of high pressures. Rev. Mod. Phys. **18**, 1–93. (Also in BRIDGMAN [1964, *1*], Vol. VI, paper No. 151, pp. 3647–3739.)

2. BRIDGMAN, PERCY WILLIAMS: Nobel Prize lecture, see [1948, *1*]. (Also in BRIDGMAN [1964, *1*], Vol. VI, paper No. 164.)

1. FIRESTONE, FLOYD AUBURN: The supersonic reflectoscope, an instrument for inspecting the interior of solid parts by means of sound waves. J. Acoust. Soc. Am. **17**, 287–299.

2. FIRESTONE, FLOYD AUBURN, and JULIAN R. FREDERICK: Refinements in supersonic reflectroscopy. Polarized sound. J. Acoust. Soc. Am. **18**, 200–211.

1. KÁRMÁN, THEODORE VON, and POL E. DUWEZ: The propagation of plastic deformation in solids. Paper presented at the Sixth International Congress for Applied Mechanics; Paris, France; September. See: [1950, *1*].

1. TAYLOR, Sir GEOFFREY INGRAM: The testing of materials at high rates of loading. J. Inst. Civil Engrs **26**, 486–518. (Also in TAYLOR [1958, *1*], paper No. 36, pp. 516–545.)

1947 *1*. DUWEZ, POL E., and DONALD S. CLARK: An experimental study of the propagation of plastic deformation under conditions of longitudinal impact. Proc. Am. Soc. Testing and Materials **47**, 502–532.

1. HOPPMANN, WILLIAM H., II: The velocity aspect of tension-impact testing. Proc. Am. Soc. Testing and Materials **47**, 533–545.

1. HUNTINGTON, HILLARD BELL: Ultrasonic measurements on single crystals. Phys. Rev. **72**, 321–331.

1. KÖSTER (KOESTER), WERNER: Elastizität fester Körper. FIAT Rev. Ger. Sci., Part I, pp. 119–125, Wiesbaden.

1. SWIFT, H. W.: Length changes in metals under torsional overstrain. Engineering **163**, 253.

1. WHITE, M. P., and LE VAN GRIFFIS: The permanent strain in a uniform bar due to longitudinal impact. J. Appl. Mech. **14**, A-337 to A-342 (December, 1947). See 1942 Report.

1948 *1*. BRIDGMAN, PERCY WILLIAMS: General survey of certain results in the field of high pressure physics. Nobel Lecture, delivered at Stockholm, December 11, 1946. Reprinted in the J. Wash. Acad. Sci. **38**, 149–166 (May, 1948). (Also in BRIDGMAN [1964, *1*], Vol. VI, paper No. 164, pp. 3873–3890.)

2. Bridgman, Percy Williams: The compression of 39 substances to 100000 kg/cm². Proc. Am. Acad. Arts Sci. **76**, 55–70. (Also in Bridgman [1964, *1*], Vol. VI, paper No. 160, pp. 3819–3834.)

1. Brown, A. F. C., and R. Edmonds: The dynamic yield strength of steel at an intermediate rate of loading. Inst. Mech. Engrs, Appl. Mech. Proc. **159** (War Emergency Issue No. 37), 11–16.

1. Clark, Donald S., and Pol E. Duwez: Discussion of the forces acting in tension impact tests of materials. J. Appl. Mech. **15**, 243.

1. Davies, R. M.: A critical study of the Hopkinson Pressure Bar. Phil. Trans. Roy. Soc. London, Ser. A **240**, 375–457.

1. Davis, Evan A.: The effect of size and stored energy on fracture of tubular specimens. J. Appl. Mech. **15**, No. 3, 216–221 (September, 1948).

1. Dudzinski, N., J. R. Murray, B. W. Mott, and Bruce Chalmers: The Young's modulus of some aluminum alloys. J. Inst. Metals **74**, 291–310.

1. Galt, John Kirtland: Mechanical properties of NaCl, KBr, KCl. Phys. Rev. **73**, 1460–1462.

1. Grover, S. F., W. Munro, and Bruce Chalmers: The moduli of aluminum alloys in tension and compression. J. Inst. Metals **74**, 310–314. Appendix to paper by Dudzinski *et al.* [1948, *1*].

1. Habib, E. T.: A method of making high-speed compression tests on small copper cylinders. J. Appl. Mech. **15**, No. 3, 248–255.

1. Köster (Koester), Werner, and Walter Rauscher: Beziehungen zwischen dem Elasticitätsmodul von Zweistofflegierungen und ihrem Aufbau. Z. Metallk. **39**, No. 4, 111–120.

2. Köster, Werner: Die Temperaturabhängigkeit des Elasticitätsmoduls reiner Metalle. Z. Metallk. **39**, No. 1, 1–9.

3. Köster, Werner: Über eine Sondererscheinung im Temperaturgang von Elasticitätsmodul und Dämpfung der Metalle Kupfer, Silber, Aluminium und Magnesium. Z. Metallk. **39**, No. 1, 9–12.

4. Köster, Werner: Betrachtungen über den Elasticitätsmodul der Metalle und Legierungen. Z. Metallk. **39**, No. 5, 145–158.

1. Warnock, F. V., and J. B. Brennan: The tensile yield strength of certain steels under suddenly applied loads. The Institution of Mechanical Engineers, Applied Mechanics Proceedings **159** (War Emergency Issue No. 37), 1–10.

1. White, M. P., and Le van Griffis: The propagation of plasticity in uniaxial compression. Trans. Am. Soc. Mech. Engrs **70**, 256–260.

1. Zener, Clarence: Elasticity and Anelasticity of Metals. The University of Chicago Press.

1949 *1.* Bridgman, Percy Williams: The Physics of High Pressure. London, G. Bell & Sons Ltd. [This and later editions contain a supplement which did not appear in the first edition (1931).] (Also reprinted in 1952, 1958.)

2. Bridgman, Percy Williams: Linear compressions to 30000 kg/cm², including relatively incompressible substances. Proc. Am. Acad. Arts Sci. **77**, 189–234. (Also in Bridgman [1964, *1*], Vol. VI, paper No. 168, pp. 3933–3978.)

1. Dorn, John E., A. Goldberg, and T. E. Tietz: The effect of thermal-mechanical history on the strain hardening of metals. Transactions of the American Institute of Mining and Metallurgical Engineers **180**, Institute of Metals Division, 205–224.

1. Hughes, Darrell Stephen, Walter Lewis Pondrom, and R. L. Mims: Transmission of elastic pulses in metal rods. Phys. Rev. **75**, 1552–1556.

1. Kolsky, Herbert: An investigation of the mechanical properties of materials at very high rates of loading. Proc. Phys. Soc. (London) B **62**, 676–700.

1. Lazarus, David: The variation of the adiabatic elastic constants of KCl, NaCl, CuZn, Cu, and Al with pressure to 10000 bars. Phys. Rev. **76**, second series, 545–553.

1. McReynolds, Andrew Wetherbee: Plastic deformation waves in aluminum. Transactions of the American Institute of Mining and Metallurgical Engineers **185**, 32–45 (January, 1949).

1. Muhlenbruch, Carl W.: Elastic and fracture toughness studies of a stainless steel. Proc. Am. Soc. Testing and Materials **49**, 738–753.

1. Rivlin, Ronald S.: Large elastic deformations of isotropic materials. Phil. Trans. Roy. Soc. London, Ser. A **242**, 173–195.

1. Warnock, F. V., and D. B. C. Taylor: The yield phenomena of a medium carbon steel under dynamic loading. Applied Mechanics, Proceedings of the Institution of Mechanical Engineers **161**, 165–175.

1950 *1.* Clark, Donald S., and D. S. Wood: The tensile impact properties of some metals and alloys. Trans. Am. Soc. Metals **62**, 45–74. See also Clark and Duwez [1948, *1*].

1. Hill, Rodney: The Mathematical Theory of Plasticity. Oxford, England; Clarendon Press, Oxford Engineering Series.

1. Kármán, Theodore von, and Pol E. Duwez: The propagation of plastic deformation in solids. J. Appl. Phys. **21**, 987–994 (October, 1950). (Substantially the same content as [1946, *1*].)

1. Nadai, Arpad Ludwig: Theory of Flow and Fracture of Solids, Vol. 1, revised edition. New York: McGraw Hill Engineering Societies Monographs.

1. Schmid, Erich, and W. Boas: Plasticity of Crystals. [English translation (London, F. A. Hughes and Co., Ltd.) of the volume which was published in German in 1935.]

1951 *1.* Andrade, Edward Neville da Costa, and C. Henderson: The mechanical behaviour of single crystals of certain face-centred cubic metals. Phil. Trans. Roy. Soc., London, **244**, A 880, 177–202 (December, 1951).

1. Bell, James Frederick: Propagation of plastic waves in pre-stressed bars. Technical Report No. 5, U.S. Naval Contract. The Johns Hopkins University (June, 1951).

1. Bishop, J. F. W., and Rodney Hill: A theoretical derivation of the plastic properties of polycrystalline face-centred metals. Phil. Mag. **42**, 1298–1307.

2. Bishop, J. F. W., and Rodney Hill: A theory of plastic distortion of a polycrystalline aggregate under combined stresses. Phil. Mag. **42**, 414–427.

1. Budiansky, Bernard, Norris F. Dow, Roger W. Peters, and Roland P. Shepherd: Experimental studies of polyaxial stress-strain laws of plasticity. Proceedings of the First U.S. National Congress of Applied Mechanics, held at Illinois Institute of Technology, Chicago, Illinois, June 11–16, 1951; pp. 503–512.

1. Campbell, William R.: Determination of dynamic stress-strain curves from strain waves in long bars. U.S. Nat. Bur. Std. Rep. No. 1017 (NBS Project 0604–31–0614), pp. 1–12, plus 14 figures (May, 1951).

1. Malvern, Lawrence E.: The propagation of longitudinal waves of plastic deformation in a bar of material exhibiting a strain rate effect. J. Appl. Mech. **18**, 203–208.

1. Rivlin, Ronald S., and D. W. Saunders: Large elastic deformations of isotropic materials. VII. Experiments on the deformation of rubber. Phil. Trans. Roy. Soc. London, Ser. A, No. 865, **243**, 251–288 (April, 1951). See also Rivlin [1949, *1*].

1952 *1.* Barrett, Charles S.: Structure of Metals: Crystallographic Methods, Principles, and Data, second edition. New York-Toronto-London: McGraw-Hill Book Co., Inc.

1. Bridgman, Percy Williams: Studies in Large Plastic Flow and Fracture. (Metallurgy and Metallurgical Engineering Series.) New York-Toronto-London: McGraw-Hill Book Co.

1. Campbell, William R.: Determination of dynamic stress-strain curves from strain waves in long bars. Proc. Soc. Exp. Stress Anal. **10**, No. 1, 113–124.

1. Garofalo, Frank, P. R. Malenock, and G. V. Smith: The influence of temperature on the elastic constants of some commercial steels. American Society for Testing Materials, Symposium on Determination of Elastic Constants, Special Technical Publication No. 129.

1. Gent, A. N., and Ronald S. Rivlin: Experiments on the mechanics of rubber. I: Eversion of a tube. Proc. Phys. Soc. (London), Ser. B **65**, 118–121.

2. Gent, A. N., and Ronald S. Rivlin: Experiments on the mechanics of rubber. II: The torsion, inflation and extension of a tube. Proc. Phys. Soc. (London), Ser. B **65**, 487–501.

3. Gent, A. N., and Ronald S. Rivlin: Experiments on the mechanics of rubber. III: Small torsions of stretched prisms. Proc. Phys. Soc. (London), Ser. B **65**, 645–648.

1. Greenough, G. B.: Quantitative X-ray diffraction observations on strained metal aggregates. Progr. Metal Phys. **3**, 176–219.

1. Lücke, Kurt, and Hansheinz Lang: Über die Form der Verfestigungskurve von Reinstaluminiumkristallen und die Bildung von Deformationsbändern. Z. Metallk. **43**, 55–66.

1. Mindlin, Raymond David, and George Herrmann: A one-dimensional theory of compressional waves in an elastic rod. Proceedings of the First U.S. National Congress of Applied Mechanics, Ann Arbor, Michigan, pp. 187–191.

1. Mott, Sir Neville F.: A theory of work-hardening of metal crystals. Phil. Mag., seventh series **43**, 1151–1178.

1. Partington, James Reddick: An Advanced Treatise on Physical Chemistry, Vol. 3. London-New York-Toronto: Longmans, Green & Co.

1. RICHARDS, JOHN T.: An evaluation of several static and dynamic methods for determining elastic moduli. American Society for Testing Materials, Symposium on the Determination of Elastic Constants, Special Technical Publication No. 129.

1. RIPPERGER, EUGENE ASHTON: Longitudinal impact of cylindrical bars. Proc. Soc. Exp. Stress Anal. **10**, No. 1, 209–226.

2. RIPPERGER, EUGENE ASHTON: A technique for elastic wave measurements. Technical Report No. 12, N6-ONR-251 Task Order 12, Stanford University.

1. TRUESDELL, CLIFFORD AMBROSE: The mechanical foundations of elasticity and fluid dynamics. J. Rational Mech. Anal. **1**, 125–300. Corrections and Additions, *ibid.* **2**, 595–616 (1953) and **3**, 801 (1954). Corrected reprint, with preface and notes, Continuum Mechanics, Part 1, Internat. Sci. Review Ser. New York-London-Paris: Gordon & Breach (1966).

1953 *1.* BIANCHI, GIOVANNI: On the propagation of longitudinal strain pulses in a bar prestressed into plastic region. Master's essay, Cornell University, Ithaca, New York (September, 1953).

1. CAMPBELL, J. D.: An investigation of the plastic behavior of metal rods subjected to longitudinal impact. J. Mech. Phys. Solids **1**, 113–123.

1. CARREKER, R. P., JR., and W. R. HIBBARD: Tensile deformation of high purity copper as a function of temperature, strain rate and grain size. Acta Met. **1**, 654–663 (November, 1953).

1. COTTRELL, A. H.: A note on the Portevin-le Chatelier effect. Phil. Mag., seventh series **44**, No. 335, 829–832 (August, 1953).

1. HOUWINK, R.: Elasticity, Plasticity and Structure of Matter, second edition. Washington, D. C.: Harren Press.

1. JOHNSON, J. E., D. S. WOOD, and D. S. CLARK: Dynamic stress-strain relations for annealed 2S aluminum under compression impact. J. Appl. Mech. **20**, No. 4, 523–529.

1. KOLSKY, HERBERT: Stress Waves in Solids. Oxford, England: Clarendon Press. See also [1963, *1*] for Dover republication.

1. KRUPNIK, N., and HUGH FORD: The stepped stress/strain curve of some aluminum alloys. J. Inst. Metals **81**, 601–615.

1. LEE, E. H.: A boundary value problem in the theory of plastic wave propagation. Quart. Appl. Math. **10**, No. 4, 335–346.

1. PHILLIPS, V. A., A. J. SWAIN, and R. EBORALL: Yield-point phenomena and stretcher-strain markings in aluminum-magnesium alloys. J. Inst. Metals **81**, 625–647.

1. RIPARBELLI, CARLOS: On the time lag of plastic deformation. Proceedings, First Midwestern Conference on Solid Mechanics, University of Illinois, pp. 148–157.

1. RIPPERGER, EUGENE ASHTON: The propagation of pulses in cylindrical bars—an experimental study. Proceedings, First Midwestern Conference on Solid Mechanics, University of Illinois, pp. 29–39 (April, 1953).

1. RIVLIN, RONALD S.: The solution of problems in second order elasticity theory. J. Rational Mech. Anal. **2**, 53–81. Reprinted in Problems of Non-Linear Elasticity. Int'l Sci. Rev. Ser. New York: Gordon & Breach (1965).

1. RUBIN, R. J.: Some problems connected with propagation of stresses above the yield stress in a material exhibiting strain-rate effect. Proceedings, First Midwestern Conference on Solid Mechanics, University of Illinois, pp. 133–135.

1. STERNGLASS, E. J., and D. A. STUART: An experimental study of the propagation of transient longitudinal deformations in elastoplastic media. J. Appl. Mech. **20**, 427–434.

1. TIMOSHENKO, STEPHEN PROKOFIEVITCH: History of Strength of Materials. New York: McGraw-Hill.

1954 *1.* BATEMAN, CATHERINE M.: Residual lattice strains in plastically deformed aluminum. Acta Met. **2**, 451–455 (May, 1954).

1. BOAS, W., and G. J. OGILVIE: The plastic deformation of a crystal in a polycrystalline aggregate. Acta Met. **2**, No. 5, 655–659.

1. CLARK, DONALD S.: The behavior of metals under dynamic loading. [1953 Edward de Mille Campbell Memorial Lecture.] Trans. Am. Soc. Metals **46**, 34–62.

1. CURTIS, CHARLES WILLIAM: Second mode vibrations of the Pochhammer-Chree frequency equation. J. Appl. Phys. **25**, No. 7, page 928.

1. HUNDY, B. B., and A. P. GREEN: A determination of plastic stress-strain relations. J. Mech. Phys. Solids, **3**, 16–21 (October, 1954).

1. KOLSKY, HERBERT: The propagation of longitudinal elastic waves along cylindrical bars. Phil. Mag., seventh series **45**, 712–726.

1. NAGHDI, PAUL M., and J. C. ROWLEY: An experimental study of biaxial stress-strain relations in plasticity. J. Mech. Phys. Solids **3**, 63–80.

1. Rubin, R. J.: Propagation of longitudinal deformation waves in a pre-stressed rod of material exhibiting a strain rate effect. J. Appl. Phys. **25**, No. 4, 528–536.

1. Zwikker, Cornelis: Physical Properties of Solid Materials. London and New York: Pergamon Press. [Second edition, 1955.]

1955 *1.* Blewitt, T. H., R. R. Coltman, and J. K. Redman: Defects in Crystalline Solids. London: The Physical Society (1955).

1. Davis, Evan A.: Combined tension-torsion tests with fixed principal directions. J. Appl. Mech. **22**, No. 3, 411–415 (September, 1955). See also Davis and Connelly [1959, *1*].

1. Truesdell, Clifford Ambrose: Hypo-elasticity. J. Rational Mech. Anal. **4**, No. 1, 83–133. Reprinted in Foundations of Elasticity Theory. Int'l Sci. Rev. Ser. New York: Gordon & Breach (1965).

1. Zucker, Charles: Elastic constants of aluminum from 20° C to 400° C. J. Acoust. Soc. Am. **27**, No. 2, 318–320.

1956 *1.* American Society for Testing Materials, 58th Annual Meeting, Atlantic City, New Jersey, U.S.A., June 27, 1955: Symposium on Impact Testing, ASTM Special Technical Publication No. 176.

1. Alter, B. E. K., and C. W. Curtis: Effect of strain rate on the propagation of a plastic strain pulse along a lead bar. J. Appl. Phys. **27**, No. 9, 1079–1085.

1. Baron, H. G.: Stress/strain curves of some metals and alloys at low temperatures and high rates of strain. J. Iron Steel Inst. (London) **182**, 354–365.

1. Bell, James Frederick: Determination of dynamic plastic strain through the use of diffraction gratings. J. Appl. Phys. **27**, No. 10, 1109–1113.

2. Bell, James Frederick: 10000 threads to the inch. American Machinist **100**, No. 16, 112–113.

3. Bell, James Frederick: Plastic wave propagation in rods subject to longitudinal impact. U.S. Army Ballistics Research Laboratory Technical Report No. 4. The Johns Hopkins University.

1. Davies, R. M.: Stress waves in solids. Surveys in Mechanics, The G. I. Taylor 70th Anniversary Volume, pp. 64–138. Cambridge Monographs on Mechanics and Applied Mathematics, G. K. Batchelor and H. Bondi, editors. Cambridge University Press.

1. Diehl, Jörg: Zugverformung von Kupfer-Einkristallen I. Verfestigungskurven und Oberflächenerscheinungen. Z. Metallk. **47**, 331–343.

1. Driscoll, David E.: Reproducibility of Charpy Impact Test. American Society for Testing Materials, Special Technical Publication No. 176, 70–75.

1. Frocht, M. M., and Paul D. Flynn: Studies in dynamic photoelasticity. J. Appl. Mech. **23**, No. 1, 116–122. [Also in Frocht, M. M.: Photoelasticity: The Selected Scientific Papers of M. M. Frocht (Pergamon Press, 1969), paper No. 17, pp. 301–317.]

1. Kelly, Anthony: The mechanism of work softening in aluminium. Phil. Mag., 8th series, **1**, 835–845.

1. Kolsky, Herbert: The propagation of stress pulses in viscoelastic solids. Phil. Mag., eighth series **1**, No. 7, 693–710.

1. Lauriente, Michael, and Robert B. Pond: Effect of growth imperfections on the strength of aluminum single crystals. J. Appl. Phys. **27**, No. 8, 950–954 (August, 1956).

1. Suzuki, Hideji, Susmu Ikeda, and Sakae Takeuchi: Deformation of thin copper crystals. J. Phys. Soc. Japan, **11**, No. 4, 382–393.

1. Taylor, Sir Geoffrey Ingram: Strains in crystalline aggregates. Proceedings of the Colloquium on Deformation and Flow of Solids (Madrid, 1955), pp. 3–12. Berlin-Göttingen-Heidelberg: Springer. (Also in Taylor [1958, *1*], Vol. I, paper No. 41, pp. 586–593.)

1957 *1.* Bianchi, Giovanni: La propagazione di onde d'urto in regime plastico. Presented at Symposium su la Plasticità nella Scienza della Construzioni, Varenna, Italia. Issued by Politecnico di Milano, 22 pages.

1. Berner, Rolf: See Seeger [1958, *1*], p. 53.

1. Carreker, R. P., Jr.: Tensile deformation of silver as a function of temperature, strain rate, and grain size. J. Metals **9**, 112–115 (January, 1957).

2. Carreker, R. P., Jr., and W. R. Hibbard, Jr.: Tensile deformation of aluminum as a function of temperature, strain rate, and grain size. Trans. AIME, J. Metals 1157–1163 (October, 1957). See also Hosford, *et al.* [1960, *1*].

1. Miklowitz, Julius, and C. R. Nisewanger: The propagation of compressional waves in a dispersive elastic rod. Part II: Experimental results and comparison with theory. J. Appl. Mech. **24**, 240–244.

1. NOGGLE, T. S., and J. S. KOEHLER: Electron microscopy of aluminum crystals deformed at various temperatures. J. Appl. Phys. **28**, No. 1, 53–55 (January, 1957).
1. PHILLIPS, ARIS: An experimental investigation on plastic stress-strain relations. IXe Congrès International de Mécanique Appliquée **8**, 23–33. Université de Bruxelles.
1. RAMBERG, R. WALTER, and L. K. IRWIN: A pulse method for determining dynamic stress-strain relations. Proceedings, Ninth International Congress of Applied Mechanics, Brussels; pp. 480–489.
1. RIPPERGER, EUGENE ASHTON, and H. NORMAN ABRAMSON: Reflection and transmission of elastic pulses in a bar at a discontinuity in cross section. Proceedings of the Third Midwestern Conference on Solid Mechanics (Ann Arbor, Michigan), pp. 135–145. University of Michigan Press.
1. SEEGER, A., JÖRG DIEHL, S. MADER, and H. REBSTOCK: Work-hardening and work-softening of face-centred cubic metal crystals. Phil. Mag., **2**, 323–350.
1. SKALAK, RICHARD: Longitudinal impact of a semi-infinite circular elastic bar. Trans. A.S.M.E., J. Appl. Mech. **24**, 59–64.
1958 *1.* BELL, JAMES FREDERICK: Normal incidence in the determination of large straïn through the use of diffraction gratings. Proceedings of the Third U.S. National Congress of Applied Mechanics, Brown University, Providence, Rhode Island, pp. 489–493.
1. FOX, GEORGE, and CHARLES WILLIAM CURTIS: Elastic strain produced by sudden application of pressure to one end of a cylindrical bar. II. Experimental observations. J. Acoust. Soc. Am. **30**, 559–563.
1. SEEGER, ALFRED VON: Kristallplastizität. In: Handbuch der Physik, Vol. VII/2, Kristallphysik II, pp. 1–210. Berlin-Göttingen-Heidelberg: Springer.
1. TAYLOR, Sir GEOFFREY INGRAM: The Scientific Papers of Sir GEOFFREY INGRAM TAYLOR, Vol. I: Mechanics of Solids, edited by GEORGE K. BATCHELOR, at the University Press, Cambridge, England.
1. TRELOAR, L. R. G.: The Physics of Rubber Elasticity, second edition. Oxford: Clarendon Press.
1959 *1.* BOWSHER, JOHN M.: On "plastic" waves in solids. Can. J. Phys. **37**, 1017–1035.
1. CONN, ANDREW F.: Plastic waves in cylindrical magnesium rods. Master's essay. The Johns Hopkins University, Baltimore, Maryland.
1. DAVIS, EVAN A., and F. M. CONNELLY: Stress distribution and plastic deformation in rotating cylinders of strain-hardening material. Trans. A.S.M.E., J. Appl. Mech. **26**, No. 1, 25–30 (March, 1959).
1. HOCKETT, J. E.: Compression testing at constant true strain rates. Proc. Am. Soc. Testing Materials **59**, 1309–1317.
1. KOLSKY, HERBERT: The mechanical testing of high polymers. Progress in Non-Destructive Testing, **2**, pp. 29–59. London: Heywood & Co. Ltd.
1. THOMAS, D. A., and BENJAMIN LEWIS AVERBACH: The early stages of plastic deformation in copper. Acta Met. **7**, 69–75.
1960 *1.* BELL, JAMES FREDERICK: Diffraction grating strain gauge. Proc. Soc. Exp. Stress Anal. **17**, No. 2, 51–64.
2. BELL, JAMES FREDERICK: Propagation of large amplitude waves in annealed aluminum. J. Appl. Phys. **31**, No. 2, 277–282.
3. BELL, JAMES FREDERICK: Study of initial conditions in constant velocity impact. J. Appl. Phys. **31**, No. 12, 2188–2195.
4. BELL, JAMES FREDERICK: The initial development of an elastic strain pulse propagating in a semi-infinite bar. U.S. Army Ballistics Research Laboratory Technical Report No. 6, The Johns Hopkins University (November, 1960).
5. BELL, JAMES FREDERICK: An experimental development of the applicability of the strain rate independent theory for plastic wave propagation in annealed aluminum, copper, magnesium, and lead. Technical Report No. 5, U.S. Army, OOR Contract No. D-36-034-ORD 2366, The Johns Hopkins University.
6. BELL, JAMES FREDERICK: Discussion—Proceedings of the Second Symposium on Naval Structural Mechanics. New York: Pergamon Press, p. 485.
1. BERNER, ROLF: Die Temperatur und Geschwindigkeitsabhängigkeit der Verfestigung kubisch-flächenzentrierter Metalleinkristalle. Z. Naturforsch. A **15**, 689–706.
1. BIANCHI, GIOVANNI: Il progetto di una attrezzatura per prove d'urto. Ingegneria Meccanica **9**, No. 11, 3–11 (November, 1960).
2. BIANCHI, GIOVANNI: Il comportamento elasto-viscoso dei metalli sollecitati dinamicamente al di sopra del limite di elasticità. Rend. Ist. Lombardo, Accad. Sci. Lettere, A **94**, 511–526.

1. Curtis, Charles William: Propagation of an elastic strain pulse in a semi-infinite bar. International Symposium on Stress Wave Propagation in Materials, pp. 15–44. New York: Interscience Publishers, Inc. See also [1954, *1*], and Fox and Curtis [1958, *1*].

1. Dohi, Shoso: Tensile straining and deformation bands of single crystal of pure iron. J. Sci., Hiroshima University, Ser. A **24**, No. 1, 97–106 (July, 1960).

1. Gillich, William J.: The response of bonded wire resistance strain gauges to large amplitude waves in annealed aluminum. Master's essay. The Johns Hopkins University, Baltimore, Maryland. See also Bell [1960, *6*].

1. Hosford, W. F., R. L. Fleischer, and W. A. Backofen: Tensile deformation of aluminum single crystals at low temperatures. Acta Met. **8**, 187–199.

1. Kolsky, Herbert: Viscoelastic waves. International Symposium on Stress Wave Propagation in Materials, Proceedings, pp. 59–90. New York: Interscience Publishers, Inc.

1. Lee, E. H.: The theory of wave propagation in anelastic materials. International Symposium on Stress Wave Propagation in Materials, edited by Norman Davids, pp. 199–228. London.

1. Lensky, V. S.: Analysis of plastic behavior of metals under complex loading. Proceedings of the Second Symposium on Naval Structural Mechanics, held at Brown University, April 5–7, 1960; pp. 259–278.

1. Papirno, Ralph, and George Gerard: Dynamic stress-strain phenomena and plastic wave propagation in metals. American Society for Metals, 42nd Annual Convention, Philadelphia, Pennsylvania, Proceedings, pp. 381–406.

1. Phillips, Aris: Pointed vertices in plasticity. Proceedings of the Second Symposium on Naval Structural Mechanics, held at Brown University, April 5–7, 1960; pp. 202–214.

1. Rivlin, Ronald S.: Some topics in finite elasticity. Proceedings, First Symposium, Naval Structural Mechanics, Stanford University; Oxford: Pergamon, pp. 169–198.

1. Roberts, John Melville, and Norman Brown: Microstrain in zinc single crystals. Trans. Met. Soc. Am. Inst. Mech. Engrs **218**, 454–463 (June, 1960).

1. Todhunter, Isaac, and Karl Pearson: A History of the Theory of Elasticity and the Strength of Materials, Dover edition. (Vols. were first published in 1886–1893.)

1. Truesdell, Clifford Ambrose: The Rational Mechanics of Flexible or Elastic Bodies, 1638–1788: Introduction to Leonhardi Euleri Opera Omnia, Vol. X et XI, Seriei Secundae. Zürich: Orell Füssli.

1. Werner, W. Meade: The applicability of the strain rate independent theory of plastic wave propagation in annealed copper. Master's essay. The Johns Hopkins University, Baltimore, Maryland.

1961 *1.* Bell, James Frederick: Experimental study of the interrelation between the theory of dislocations in polycrystalline media and finite amplitude wave propagation in solids. J. Appl. Phys. **32**, No. 10, 1982–1993.

2. Bell, James Frederick: Discussion. Response of Metals to High Velocity Deformation, Proceedings, 1960 Conference of Metallurgical Societies, Vol. 9, pp. 112–113.

3. Bell, James Frederick: An experimental study of the unloading phenomenon in constant velocity impact. J. Mech. Phys. Solids **9**, 1–15.

4. Bell, James Frederick: Further experimental study of the unloading phenomenon in constant velocity impact. J. Mech. Phys. Solids **9**, 261–278.

1. Filbey, Gordon Luther: Intense plastic waves. Ph.D. dissertation, The Johns Hopkins University, Baltimore, Maryland.

1. Fowles, G. R.: Shock wave compression of hardened and annealed 2024 aluminum. J. Appl. Phys. **32**, No. 8, 1475–1487 (August, 1961).

1. Handbook of Chemistry and Physics, 43rd edition, Cleveland, Ohio: The Chemical Rubber Publishing Co. (1961–1962).

1. Hauser, F. E., J. A. Simmons, and J. E. Dorn: Strain rate effects in plastic wave propagation. Response of Metals to High Velocity Deformation, Proceedings, of July 1960, Conference of Metallurgical Societies, Vol. 9, pp. 93–114. See Bell [1961, *2*].

1. Hearmon, R. F. S.: An Introduction to Applied Anisotropic Elasticity. Oxford, England: Oxford University Press.

1. Heinrichs, Joseph A.: An experimental study of 0.483 inch diameter aluminum bars under constant velocity impact. Master's essay. The Johns Hopkins University, Baltimore, Maryland.

1. Phillips, Aris, and G. A. Gray: Experimental investigation of corners of the yield surface. Trans. A.S.M.E., Ser. D, J. Basic Engineering **83**, 275–288 (June, 1961).

1. SPERRAZZA, JOSEPH: Propagation of large amplitude waves in pure lead. Dr. Eng. dissertation. The Johns Hopkins University, Baltimore, Maryland.

1. TRUESDELL, CLIFFORD AMBROSE: General and exact theory of waves in finite elastic strain. Arch. Rational Mech. Anal. **8**, No. 3, 263–352. Corrected reprint in Continuum Mechanics, Part IV, Problems of Non-linear Elasticity; Internat. Sci. Review Ser. New York-London-Paris: Gordon & Breach 1965. Also in Wave Propagation in Dissipative Materials. New York: Springer 1965.

1962 *1.* BELL, JAMES FREDERICK: Experiments on large amplitude waves in finite elastic strain. Proceedings, IUTAM Symposium on Second-Order Effects in Elasticity, Plasticity, and Fluid Dynamics, Haifa, Israel, pp. 173–186. Oxford-Paris-New York: Pergamon Press [Proceedings published in 1964.]

2. BELL, JAMES FREDERICK: Impact strength. Encyclopedic Dictionary of Physics, **3**, p. 799. Pergamon Press.

3. BELL, JAMES FREDERICK: Impact testing. Encyclopedic Dictionary of Physics, **3**, p. 780. Pergamon Press.

4. BELL, JAMES FREDERICK: Experimental study of dynamic plasticity at elevated temperatures. Exp. Mech. **2**, No. 1, 1–6 (June, 1962).

5. BELL, JAMES FREDERICK, and W. MEADE WERNER: Applicability of the Taylor theory of the polycrystalline aggregate to finite amplitude wave propagation in annealed copper. J. Appl. Phys. **33**, No. 8, 2416–2425.

6. BELL, JAMES FREDERICK, and JOHN H. SUCKLING: The dynamic overstress and the hydrodynamic transition velocity in the symmetrical free-flight plastic impact of annealed aluminum. Proceedings, 4th U.S. National Congress of Applied Mechanics, pp. 877–883.

7. BELL, JAMES FREDERICK, and ALBERT STEIN: The incremental loading wave in the pre-stressed plastic field. J. de Mécanique **1**, No. 4, 395–412.

1. BROWN, NORMAN, and R. A. EKVALL: Temperature dependence of the yield points in iron. Acta Met. **10**, 1101–1107.

1. DILLON, OSCAR W., JR.: An experimental study of the heat generated during torsional oscillations. J. Mech. Phys. Solids **10**, 235–244.

2. DILLON, OSCAR W., JR.: Temperature generated in aluminum rods undergoing torsional oscillations. J. Appl. Phys. **33**, No. 10, 3100–3105 (October, 1962).

1. FOUX, AMNON: An experimental investigation of the Poynting effect, 1. Determination of changes in length. Proceedings, International Symposium on Second-Order Effects in Elasticity, Plasticity, and Fluid Dynamics, Haifa, Israel, pp. 228–251 (April, 1962). [Oxford-Paris-New York: Pergamon Press 1964.]

1. KOLSKY, HERBERT, and L. S. DOUCH: Experimental studies in plastic wave propagation. J. Mech. Phys. Solids **10**, 195–223 .

1. LARSEN, EGON: The Cavendish Laboratory: A Nursery of Genius. London: E. Ward; New York: F. Watts.

1. LEMPRIERE, B. M.: Oscillations in tensile tests. Intern. J. Mech. Sci. **4**, 171–184.

1. ROSENTHAL, D., and W. B. GRUPEN: Second order effect in crystal plasticity: Deformation of surface layers in face-centred cubic aggregates. I.U.T.A.M. International Symposium on Second-Order Effects in Elasticity, Plasticity, and Fluid Dynamics, Haifa, Israel, pp. 391–415 (April, 1962). [Oxford-Paris-New York: Pergamon Press 1964.]

1. SPERRAZZA, JOSEPH: Propagation of large amplitude waves in pure lead. Proceedings, 4th U.S. National Congress of Applied Mechanics **2**, 1123–1129.

1. STEIN, ALBERT: An experimental study of incremental plastic wave propagation. Master's essay. The Johns Hopkins University, Baltimore, Maryland.

1. THORNTON, P. R., T. E. MITCHELL, and P. B. HIRSCH: The strain-rate dependence of the flow stress of copper single crystals. Phil. Mag., **7**, 337–358. See also MITCHELL and THORNTON [1963, *1*].

1963 *1.* BELL, JAMES FREDERICK: Single temperature-dependent stress-strain law for the dynamic plastic deformation of annealed face-centered cubic metals. J. Appl. Phys. **34**, No. 1, 134–141 (January, 1963).

2. BELL, JAMES FREDERICK: The initiation of finite amplitude waves in annealed metals. Proceedings, I.U.T.A.M. Symposium on Stress Waves in Anelastic Solids, Brown University, Providence, Rhode Island, pp. 166–182. [Berlin-Göttingen-Heidelberg: Springer 1964.]

1. BIANCHI, GIOVANNI: Some experimental and theoretical studies on the propagation of longitudinal plastic waves in a strain-rate-dependent material. Stress Waves in Anelastic Solids; IUTAM Symposium held at Brown University, Providence, Rhode Island, April 3–5, 1963, pp. 101–117. Berlin-Göttingen-Heidelberg: Springer 1964.

1. BRIDGMAN, PERCY WILLIAMS: General outlook in the field of high-pressure research. Solids under Pressure, edited by WILLIAM PAUL and DOUGLAS M. WARSCHAUER, pp. 1–13. New York: McGraw-Hill.

1. CHIDDISTER, J. L., and LAWRENCE E. MALVERN: Compression-impact testing of aluminum at elevated temperatures. Exptl. Mech. **3**, 81–90.

1. DAVIES, E. D. H., and STEPHEN C. HUNTER: The dynamic compression testing of solids by the method of the Split Hopkinson Pressure Bar. J. Mech. Phys. Solids **11**, 155–179.

1. DILLON, OSCAR W.: Experimental data on aluminum as a mechanically unstable solid. J. Mech. Phys. Solids **11**, 289–304.

2. DILLON, OSCAR W.: Coupled thermoplasticity. J. Mech. Phys. Solids **11**, 21–33.

1. KINGMAN, PRISCILLA W., ROBERT E. GREEN, and ROBERT BARRETT POND: Reaction of fine metal wires to imposed loads. U.S. Army Terminal Report, Ballistics Research Laboratories, Aberdeen Proving Ground, Maryland. The Johns Hopkins University, Baltimore, Maryland.

1. KOLSKY, HERBERT: Stress Waves in Solids. New York: Dover Publications, Inc. See [1953, *1*].

1. MITCHELL, T. E., and P. R. THORNTON: The work-hardening characteristics of Cu and α-brass single crystals between 4.2 and 500° K. Phil. Mag. **8**, 1127–1159.

1. PAUL, WILLIAM, and DOUGLAS M. WARSCHAUER, editors: Solids under Pressure. New York-San Francisco-Toronto-London: McGraw-Hill Book Co. Inc.

1. RIPPERGER, EUGENE ASHTON, and L. M. YEAKLEY: Measurement of particle velocities associated with waves propagating in bars. Exptl. Mech. **3** 47–56.

1. TARDIF, H. P., and H. MARQUIS: Some dynamic properties of plastics. Can. Aeronautics and Space J. 205–213 (September 1963).

1964 *1.* BELL, JAMES FREDERICK: A generalized large deformation behaviour for face-centred cubic solids—high purity copper. Phil. Mag. **10**, No. 103, 107–126.

2. BELL, JAMES FREDERICK: Theory vs. experiment for finite amplitude stress waves. Symposium, Society for Engineering Science (November 1964). See BELL [1967, *4*].

1. BRIDGMAN, PERCY WILLIAMS: Collected Experimental Papers of P. W. BRIDGMAN. Vols. I to VII. Cambridge, Massachusetts: Harvard University Press.

1. CONN, ANDREW: On impact testing for dynamic properties of metals. Ph.D. dissertation, The Johns Hopkins University, Baltimore, Maryland.

1. DILLON, OSCAR W.: The response of pre-stressed aluminum. Intern. J. Engr. Sci. **2**, 327–339.

1. EFRON, L.: Longitudinal plastic wave propagation in annealed aluminum bars. Ph.D. dissertation, Michigan State University, East Lansing, Michigan.

1. GILLICH, WILLIAM J.: Plastic wave propagation in high purity single crystals of aluminum. Ph.D. dissertation. The Johns Hopkins University, Baltimore, Maryland.

1. LINDHOLM, ULRIC S.: Some experiments with the split Hopkinson pressure bar. J. Mech. Phys. Solids **12**, 317–335.

1965 *1.* BELL, JAMES FREDERICK: The dynamic plasticity of metals at high strain rates: An experimental generalization. Behavior of Materials under Dynamic Loading. American Society of Mechanical Engineers, Colloquium, pp. 19–41. See also BELL [1964, *1*], [1967, *4*].

2. BELL, JAMES FREDERICK: Generalized large deformation behaviour for face-centred cubic solids: Nickel, aluminum, gold, silver, and lead. Phil. Mag. **11**, No. 114, 1135–1156.

1. CONN, ANDREW F.: On the use of thin wafers to study dynamic properties of metals. J. Mech. Phys. Solids **13**, 311–327.

1. FILBEY, GORDON LUTHER: Longitudinal plastic deformation waves in bars. Proceedings, Princeton University Conference on Solid Mechanics, pp. 111–126.

1. FLYNN, PAUL D.: Dynamic Photoelasticity using a dual-beam polarscope and ultra-high-speed photography. Kurzzeitpolographie, Proceedings of the 7th International Congress on High-Speed Photography, pp. 351–357, Zurich, September 12–18.

1. LINDHOLM, ULRIC S., and L. M. YEAKLEY: Dynamic deformation of single and polycrystalline aluminium. J. Mech. Phys. Solids **13**, 41–53.

1. MALVERN, LAWRENCE E.: Experimental studies of strain-rate effects and plastic wave propagation in annealed aluminum. Behavior of Materials under Dynamic Loading. American Society of Mechanical Engineers, Colloquium, pp. 81–92.

1. SIERAKOWSKI, R. L., and ARIS PHILLIPS: On the concept of the yield surface. Acta Mech. **1**, No. 1, 29–35.

1. TRUESDELL, CLIFFORD AMBROSE, and WALTER NOLL: The non-linear field theories of mechanics. In: Handbuch der Physik, Vol. III/3, pp. 1–602. Berlin-Heidelberg-New York: Springer.

1966 *1.* BAKER, W. E., and C. H. YEW: Strain-rate effects in the propagation of torsional plastic waves. J. Appl. Mech. Trans. Am. Soc. Mech. Engrs **33**, No. 4, 917–923 (December, 1966).

1. BELL, JAMES FREDERICK: An experimental diffraction grating study of the quasi-static hypothesis of the Split Hopkinson Bar experiment. J. Mech. Phys. Solids **14**, 309–327.

2. BELL, JAMES FREDERICK: The relevance of dynamic finite distortion research to high energy rate forming processes. Proceedings, International Symposium on High Energy Rate Forming, Prague, pp. 1–12.

1. DILLON, OSCAR W., JR.: Waves in bars of mechanically unstable materials. J. Appl. Mech. **33**, 267–274. See also [1967, *1*].

2. DILLON, OSCAR W., JR.: The heat generated during torsional oscillations of copper tubes. Intern. J. Solids and Structures **2**, 181–204.

1. FITZGERALD, EDWIN R.: Particle Waves and Deformation in Crystalline Solids. New York-London-Sydney: Interscience Publishers.

1. HEARMON, R. F. S.: Die elastischen Konstanten nicht-piezoelektrischer Kristalle. LANDOLT-BÖRNSTEIN, Zahlenwerte und Funktionen aus Naturwissenschaften und Technik. Neue Serie. Gruppe III: Kristall- und Festkörperphysik. Band 1. Elastische, piezoelektrische, piezooptische und elektrooptische Konstanten von Kristallen, pp. 1–37. [English title: The elastic constants of non-piezoelectric crystals. LANDOLT-BÖRNSTEIN, Numerical Data and Functional Relationships in Science and Technology. New Series. Group III: Crystal and Solid State Physics. Vol. 1. Elastic, Piezo-electric, Piezooptic and Electrooptic Constants of Crystals]. Berlin-Heidelberg-New York: Springer.

1. KENIG, M. J., and OSCAR W. DILLON, JR.: Shock waves produced by small stress increments in annealed aluminum. J. Appl. Mech. **33**, 907–916.

1. SHARPE, WILLIAM N.: The Portevin-le Chatelier effect in aluminum single crystals and polycrystals. Ph.D. dissertation. The Johns Hopkins University, Baltimore, Maryland.

2. SHARPE, WILLIAM N.: The Portevin-le Chatelier effect in aluminum single crystals and polycrystals. J. Mech. Phys. Solids **14**, 187–202.

1967 *1.* BAKER, WILFRED, WILLIAM E. WOOLAM, and DANA YOUNG: Air and internal damping of thin cantilever beams. Intern. J. Sci. **9**, 743–766.

1. BASINSKI, Z. S., and S. SAIMOTO: Resistivity of deformed crystals. Can. J. Phys. **45**, 1161–1176.

1. BELL, JAMES FREDERICK: On the direct measurement of very large strain at high strain rates. Exptl. Mech. **7**, No. 1, 1–8 (January, 1967). [Paper was presented at the meeting of the Society for Experimental Stress Analysis in 1965.]

2. BELL, JAMES FREDERICK: An experimental study of instability phenomena in the initiation of plastic waves in long rods. Proceedings, Symposium on the Mechanical Behavior of Materials under Dynamic Loads, San Antonio, Texas, September, 1967, pp. 10–20. New York: Springer. [Proceedings published in 1969.]

3. BELL, JAMES FREDERICK, and ROBERT E. GREEN, JR.: An experimental study of the double slip deformation hypothesis for face-centred cubic single crystals. Phil. Mag. **15**, 469–476.

4. BELL, JAMES FREDERICK: Theory vs. experiment for finite amplitude stress waves. Presented at the Society for Engineering Science Symposium (1964), Recent Advances in Engineering Science, pp. 565–592. New York: Gordon & Breach Science Publishers, Inc.

1. BODNER, S. R., and A. ROSEN: Discontinuous yielding of commercially-pure aluminum. J. Mech. Phys. Solids **15**, 63–77.

1. CALVERT, MONTE A.: The Mechanical Engineer in America, 1830–1910. The Johns Hopkins Press.

1. DILLON, OSCAR W., JR.: The dynamic elastic-plastic interface and related topics. J. Mech. Phys. Solids **15**, 341–358.

2. DILLON, OSCAR W., JR.: Plastic deformation waves and heat generated near the yield point of annealed aluminum. Proceedings, Symposium on the Mechanical Behavior of Materials under Dynamic Loads, San Antonio, Texas, September, 1967, pp. 21–60. New York: Springer.

1. GILLICH, WILLIAM J.: Propagation of finite amplitude waves in single crystals of high-purity aluminium. Phil. Mag. **15**, No. 136, 659–671.

1. HARTMAN, WILLIAM FRANCIS: The applicability of the generalized parabolic deformation law to a binary alloy. Ph.D. dissertation. The Johns Hopkins University, Baltimore, Maryland.

1. Lindholm, Ulric S., and L. M. Yeakley: A dynamic biaxial testing machine. Exptl. Mech. **7**, No. 1, 1–7.

1. Schultz, A. B., P. A. Tuschak, and A. A. Vicario: Experimental evaluation of material behavior in a wire under transverse impact. J. Appl. Mech.; Transactions, A.S.M.E. **34**, No. 2, Series E, 392–396 (June, 1967).

1. Smith, Jack C., and Carl A. Fenstermaker: Strain-wave propagation in strips of natural rubber subjected to high-velocity transverse impact. J. Appl. Phys. **38**, No. 11, 4218–4224 (October, 1967).

1. Troisième Centenaire de l'Académie des Sciences. 1666–1966. 2 Vols. Paris: Gauthier-Villars.

1968 *1.* Bell, James Frederick: The Physics of Large Deformation of Crystalline Solids. Springer Tracts in Natural Philosophy, Vol. 14. Berlin-Heidelberg-New York: Springer.

2. Bell, James Frederick: Large deformation dynamic plasticity at an elastic-plastic interface. J. Mech. Phys. Solids **16**, 295–313.

1. Convery, E., and H. Ll. D. Pugh: Velocity of torsional waves in metals stressed statically into the plastic range. J. Mech. Engr. Sci. **10**, No. 2, 153–164.

1. Dawson, Thomas H.: The plastic deformation of crystalline aggregates. Ph. D. dissertation. The Johns Hopkins University, Baltimore, Maryland.

1. Gillich, William J., and William O. Ewing: Measurement of particle velocities associated with plastic waves propagating in bars. U.S. Army Aberdeen Research and Development Center, Ballistics Research Laboratory, Aberdeen Proving Ground, Maryland; Report No. BRL-R-1405, pp. 1–46 (June, 1968).

1. Hardie, D., and R. N. Parkins: A study of the errors due to shear and rotary inertia in the determination of Young's modulus by flexural vibrations. Brit. J. Appl. Phys., second series **1**, 77–85.

1. Hart, S.: Comments on the paper: A study of the errors due to shear and rotary inertia in the determination of Young's modulus by flexural vibrations. Brit. J. Appl. Phys., second series **1**, 1763–1766.

1. Phillips, Aris: Yield surfaces of pure aluminum at elevated temperatures. Proceedings of the IUTAM Symposium, East Kilbride, June 25–28, 1968; pp. 241–258.

1. Schultz, A. B.: Material behavior in wires of 1100 aluminum subjected to transverse impact. J. Appl. Mech. Transactions of the A.S.M.E. **35**, No. 2, 342–348 (June, 1968). See also [1969, *1*].

1. Sierakowski, R. L., and Aris Phillips: The effects of repeated loading on the yield surface. Acta Mech. **6**, 217–231.

1. Strutt, Robert John (4th Baron Rayleigh): Life of John William Strutt, Third Baron Rayleigh. Republication by the University of Wisconsin Press, Madison, Wisconsin. [First published by Edward Arnold Co. (1924).]

1. Truesdell, Clifford Ambrose: Essays in the History of Mechanics. Berlin-Heidelberg-New York: Springer.

1969 *1.* Allen, Mildred, and Erwin J. Saxl: Elastic torsion in wires under tension. J. Appl. Phys. **40**, No. 6, 2505–2509 (May, 1969).

1. (Bechmann and Hearmon) Landolt-Börnstein: The third-order elastic constants. (See Hearmon [1966, *1*] for full title.) Group III, Vol. 2 (Vol. 2 is a supplement and extension of Vol. 1, and includes the additional topics: Electropic Constants, and, Nonlinear Dielectric Susceptibilities of Crystals), pp. 102–123.

1. Bell, James Frederick: The dynamic plasticity of non-symmetrical free-flight collision impact. International J. Mech. Sci. **11**, 633–657.

1. Bilello, John, and Marvin Metzger: Microyielding in polycrystalline copper. Trans. Met. Soc. A.I.M.E. **245**, 2279–2284 (October, 1969).

1. Efron, L., and Lawrence E. Malvern: Electromagnetic velocity-transducer studies of plastic waves in aluminum bars. Exptl. Mech. **9**, No. 6, 1–8 (June, 1969).

1. Florenz, Meir: Two dimensional plastic compression of polycrystalline aluminum. Master's essay. The Johns Hopkins University, Baltimore, Maryland.

1. Hartman, William Francis: Propagation of large amplitude waves in annealed brass. Intern. J. Solids Structures **5**, 303–317. Pergamon Press.

2. Hartman, William Francis: Generalized parabolic work hardening during tensile deformation of brass. J. Mat. Am. Soc. Testing Materials **4**, No. 1, 104–116.

1. (Hearmon) Landolt-Börnstein: The elastic constants of non-piezoelectric crystals. Group III, Vol. 2, pp. 1–37. (See Hearmon [1966, *1*] for full title.)

1. Hsu, Nelson Nai-Hsing: Experimental studies of latent work hardening of aluminum single crystals. Ph. D. dissertation. The Johns Hopkins University, Baltimore, Maryland.

1. Kolsky, Herbert: Production of tensile shock waves in stretched natural rubber. Nature **224**, No. 5226, 1301 (December, 1969).
1. Liu, John Moyu: Some optical experiments in the study of acoustic waves in solids. Master's essay. The Johns Hopkins University, Baltimore, Maryland (June, 1969).
1. Maguire, John R.: On the dynamic plasticity of aluminum polycrystals at very high temperatures. Master's essay. The Johns Hopkins University, Baltimore, Maryland.
1. Mittal, Ramesh Kumar: Biaxial loading of dead annealed commercial purity aluminum and the generalization of the parabolic law. Ph.D. dissertation. The Johns Hopkins University, Baltimore, Maryland.
1. Palmer, C. Harvey: Differential optical strain gauge for torsion measurements. Appl. Opt. **8**, 1015–1019.
2. Palmer, C. Harvey: Differential angle measurements with moire fringes. Opt. Technol. **1**, 150–152.
1. Percival, C. M., and J. A. Cheney: Thermally generated stress waves in a dispersive elastic rod. Exptl. Mech. **9**, No. 2, 49–57 (February, 1969).
1. Saxl, Erwin J., and Mildred Allen: Period of a torsion pendulum as affected by adding weights. J. Appl. Phys. **40**, No. 6, 2499–2503 (May, 1969).
1. Schultz, A. B.: Dynamic behavior of metals under tensile impact. Part III: Annealed and cold worked materials. U.S. Air Force Materials Laboratory, Ohio; Technical Report AFML-TR-69-76, Part II, 39 pages (June, 1969).
1970 *1.* Brammer, J. A., and C. M. Percival: Elevated-temperature elastic moduli of 2024 aluminum obtained by a laser-pulse technique. Exptl. Mech. **10**, No. 6, 245–250 (June, 1970).
1. Cristescu, Nicolae, and James Frederick Bell: On unloading in the symmetrical impact of two aluminum bars. Battelle Memorial Institute. Symposium Proceedings, Columbus, Ohio, October, 1969. pp. 397–421.
1. Dawson, Thomas H.: The mechanics of crystalline aggregate deformations. Intern. J. Mech. Sci. **12**, 197–204.
1. Sachse, Wolfgang: The effects of deformation rate on the attenuation behavior during loading, unloading and microplastic reloading of single crystals of aluminum. Ph.D. dissertation. The Johns Hopkins University, Baltimore, Maryland.
1. Smith, Alan B., and Richard W. Damon: A bibliography of microwave ultrasonics. IEEE Transactions on Sonics and Ultrasonics SU-17, No. 2, 86–111 (April, 1970).
1971 *1.* Bell, James Frederick: On experiments revealing the distribution of critical strains in the large deformation of solids. Rendiconti del Seminario Matematico dell'Università e del Politecnico di Torino; Italia, **30**, 49–61.
1. Mittal, Ramesh: Biaxial loading of aluminum and a generalization of the parabolic law. J. Materials **6**, No. 1. 67–81.
1972 *1.* Bell, James Frederick: The plane wave and isochoric deformation hypotheses in one-dimensional dynamic plasticity. J. Mech. Phys. Solids. In press.
2. Bell, James Frederick: On experiments revealing the importance of material instability, for modern theories of plasticity. U.S. Air Force Office of Scientific Research Scientific Report AFOSR-TR-72-1194; The Johns Hopkins University (May, 1972).
3. Bell, James Frederick: Material stability and second order transitions for crystals under finite strain. Manuscript.
4. Bell, James Frederick: An experimental study of the reflection of finite waves at a plastic-elastic boundary. Manuscript.
5. Bell, James Frederick, and Meir Florenz: A modification of the double compression experiment for the plastic deformation of crystalline solids. Manuscript.
6. Bell, James Frederick: On the dynamic elastic limit in plastic wave propagation. Manuscript.
7. Bell, James Frederick: New experiments on the incremental waves in the prestressed crystalline solid. Manuscript.
1. Khan, Akhtar S.: The theoretical and experimental study of tensile plastic waves at finite strain. Ph.D. dissertation. The Johns Hopkins University, Baltimore, Maryland.
1. Palmer, C. Harvey, and Bruce Z. Hollmann: Transmission of incoherent light between Ronchi grids. Applied Optics **11**, 780–785.
1. Phillips, Aris, and Juh-Ling Tang: The effect of loading path on the yield surface at elevated temperatures. Intern. J. Solids and Structures, **8**, No. 4, 463–474.
2. Phillips, Aris, C. S. Liu, and J. W. Justusson: An experimental investigation of yield surfaces at elevated temperatures. Acta Mech. In press.

Namenverzeichnis. — Author Index.

Sachverzeichnis.

(Deutsch-Englisch.)

Bei gleicher Schreibweise in beiden Sprachen sind die Stichwörter nur einmal aufgeführt.

Subject Index.

(English-German.)

Where English and German spellings of a word are identical, the German version is omitted.